ENCYCLOPEDIA OF MATHEMATICS AND ITS APPLICATIONS

FOUNDING EDITOR G. C. ROTA

All the titles listed below can be obtained from good booksellers or from Cambridge University Press. For a complete series listing visit http://publishing.cambridge.org/stm/mathematics/eom/.

88. Teo Mora *Solving Polynomial Equation Systems, I*
89. Klaus Bichteler *Stochastic Integration with Jumps*
90. M. Lothaire *Algebraic Combinatorics on Words*
91. A. A. Ivanov & S. V. Shpectorov *Geometry of Sporadic Groups, 2*
92. Peter McMullen & Egon Schulte *Abstract Regular Polytopes*
93. G. Gierz *et al. Continuous Lattices and Domains*
94. Steven R. Finch *Mathematical Constants*
95. Youssef Jabri *The Mountain Pass Theorem*
96. George Gasper & Mizan Rahman *Basic Hypergeometric Series, 2nd ed.*
97. Maria Cristina Pedicchio & Walter Tholen *Categorical Foundations*
100. Enzo Olivieri & Maria Eulalia Vares *Large Deviations and Metastability*
102. R. J. Wilson & L. Beineke *Topics in Algebraic Graph Theory*

Well-Posed Linear Systems

OLOF STAFFANS

Department of Mathematics

Åbo Akademi University, Finland

CAMBRIDGE
UNIVERSITY PRESS

CAMBRIDGE UNIVERSITY PRESS
Cambridge, New York, Melbourne, Madrid, Cape Town, Singapore, São Paulo

Cambridge University Press
The Edinburgh Building, Cambridge CB2 2RU, UK

www.cambridge.org
Information on this title: www.cambridge.org/9780521825849

First published 2005

Printed in the United Kingdom at the University Press, Cambridge

A catalog record for this book is available from the British Library

Library of Congress Cataloging in Publication data

Staffans, Olof J., 1947–
Well-posed linear systems / Olof Staffans.
p. cm. – (Encyclopedia of mathematics and its applications)
Includes bibliographical references and index.
ISBN 0-521-82584-9
1. Linear systems. 2. System theory. I. Staffans, Olof J., 1947–
II. Title. III. Series.
QA372.S784 2004
003′.74–dc22 2004051868

ISBN-13 978-0-521-82584-9 hardback
ISBN-10 0-521-82584-9 hardback

Contents

Figures

Preface

This main purpose of this book is to present the basic theory of well-posed linear systems in a form which makes it available to a larger audience, thereby opening up the possibility of applying it to a wider range of problems. Up to now the theory has existed in a distributed form, scattered between different papers with different (and often noncompatible) notation. For many years this has forced authors in the field (myself included) to start each paper with a long background section to first bring the reader up to date with the existing theory. Hopefully, the existence of this monograph will make it possible to dispense with this in future.

My personal history in the field of abstract systems theory is rather short but intensive. It started in about 1995 when I wanted to understand the true nature of the solution of the quadratic cost minimization problem for a linear Volterra integral equation. It soon became apparent that the most appropriate setting was not the one familiar to me which has classically been used in the field of Volterra integral equations (as presented in, e.g., Gripenberg *et al.* [1990]). It also became clear that the solution was not tied to the class of Volterra integral equations, but that it could be formulated in a much more general framework. From this simple observation I gradually plunged deeper and deeper into the theory of well-posed (and even non-well-posed) linear systems.

One of the first major decisions that I had to make when I began to write this monograph was how much of the existing theory to include. Because of the nonhomogeneous background of the existing theory (several strains have been developing in parallel independently of each other), it is clear that it is impossible to write a monograph which will be fully accepted by every worker in the field. I have therefore largely allowed my personal taste to influence the final result, meaning that results which lie closer to my own research interests are included to a greater extent than others. It is also true that results which blend more easily into the general theory have had a greater chance of being included than those which are of a more specialist nature. Generally speaking, instead of borrowing results directly from various sources I have reinterpreted and reformulated many existing results into a coherent setting and, above all, using a coherent notation.

The original motivation for writing this book was to develop the background which is needed for an appropriate understanding of the quadratic cost minimization problem (and its indefinite minimax version). However, due to page and time limitations, I have not yet been able to include any optimal control in this volume (only the background needed to attack optimal control problems). The book on optimal control still remains to be written.

Not only was it difficult to decide exactly what parts of the existing theory to include, but also in which form it should be included. One such decision was whether to work in a *Hilbert space* or in a *Banach space* setting. Optimal control is typically done in Hilbert spaces. On the other hand, in the basic theory it does not matter if we are working in a Hilbert space or a Banach space (the technical differences are minimal, compared to the general level of difficulty of the theory). Moreover, there are several interesting applications which require the use of Banach spaces. For example, the natural norm in population dynamics is often the L^1-norm (representing the total mass), parabolic equations have a well-developed L^p-theory with $p \neq 2$, and in nonlinear equations it is often more convenient to use L^∞-norms than L^2-norms. The natural decision was to present the basic theory in an arbitrary Banach space, but to specialize to Hilbert spaces whenever this additional structure was important. As a consequence of this decision, the present monograph contains the first comprehensive treatment of a well-posed linear system in a setting where the input and output signals are continuous (as opposed to belonging to some L^p-space) but do not have any further differentiability properties (such as belonging to some Sobolev spaces). (More precisely, they are continuous apart from possible jump discontinuities.)

The first version of the manuscript was devoted exclusively to *well-posed* problems, and the main part of the book still deals with problems that are well posed. However, especially in H^∞-optimal control, one naturally runs into non-well-posed problems, and this is also true in circuit theory in the impedance and transmission settings. The final incident that convinced me that I also had to include some classes of non-well-posed systems in this monograph was my discovery in 2002 that every passive impedance system which satisfies a certain algebraic condition can be represented by a (possibly non-well-posed) *system node*. System nodes are a central part of the theory of well-posed systems, and the well-posedness property is not always essential. My decision not to stay strictly within the class of well-posed systems had the consequence that this monograph is also the the first comprehensive treatment of (possibly non-well-posed) systems generated by arbitrary system nodes.

The last three chapters of this book have a slightly different flavor from the earlier chapters. There the general Banach space setting is replaced by a standard Hilbert space setting, and connections are explored between well-posed linear systems, Fourier analysis, and operator theory. In particular, the admissibility of scalar control and observation operators for contraction semigroups is characterized by means of the Carleson measure theorem, and systems theory interpretations are given of the basic dilation and model theory for contractions and continuous-time contraction semigroups in Hilbert spaces.

It took me approximately six years to write this monograph. The work has primarily been carried out at the Mathematics Institute of Åbo Akademi, which has offered me excellent working conditions and facilities. The Academy of Finland has supported me by relieving me of teaching duties for a total of two years, and without this support I would not have been able to complete the manuscript in this amount of time.

I am grateful to several students and colleagues for helping me find errors and misprints in the manuscript, most particularly Mikael Kurula, Jarmo Malinen and Kalle Mikkola.

Above all I am grateful to my wife Marjatta for her understanding and patience while I wrote this book.

Notation

Basic sets and symbols

$\mathbb{C}$	The complex plane.
$\mathbb{C}^+_\omega$, $\overline{\mathbb{C}}^+_\omega$	$\mathbb{C}^+_\omega := \{z \in \mathbb{C} \mid \Re z > \omega\}$ and $\overline{\mathbb{C}}^+_\omega := \{z \in \mathbb{C} \mid \Re z \geq \omega\}$.
$\mathbb{C}^-_\omega$, $\overline{\mathbb{C}}^-_\omega$	$\mathbb{C}^-_\omega := \{z \in \mathbb{C} \mid \Re z < \omega\}$ and $\overline{\mathbb{C}}^-_\omega := \{z \in \mathbb{C} \mid \Re z \leq \omega\}$.
$\mathbb{C}^+$, $\overline{\mathbb{C}}^+$	$\mathbb{C}^+ := \mathbb{C}^+_0$ and $\overline{\mathbb{C}}^+ := \overline{\mathbb{C}}^+_0$.
$\mathbb{C}^-$, $\overline{\mathbb{C}}^-$	$\mathbb{C}^- := \mathbb{C}^-_0$ and $\overline{\mathbb{C}}^- := \overline{\mathbb{C}}^-_0$.
$\mathbb{D}^+_r$, $\overline{\mathbb{D}}^+_r$	$\mathbb{D}^+_r := \{z \in \mathbb{C} \mid \Re z > r\}$ and $\overline{\mathbb{D}}^+_r := \{z \in \mathbb{C} \mid \lvert z\rvert \geq r\}$.
$\mathbb{D}^-_r$, $\overline{\mathbb{D}}^-_r$	$\mathbb{D}^-_r := \{z \in \mathbb{C} \mid \Re z < r\}$ and $\overline{\mathbb{D}}^-_r := \{z \in \mathbb{C} \mid \lvert z\rvert \leq r\}$.
$\mathbb{D}^+$, $\overline{\mathbb{D}}^+$	$\mathbb{D}^+ := \mathbb{D}^+_1$ and $\overline{\mathbb{D}}^+ := \overline{\mathbb{D}}^+_1$.
$\mathbb{D}^-$, $\overline{\mathbb{D}}^-$	$\mathbb{D}^- := \mathbb{D}^-_1$ and $\overline{\mathbb{D}}^- := \overline{\mathbb{D}}^-_1$.
$\mathbb{R}$	$\mathbb{R} := (-\infty, \infty)$.
$\mathbb{R}^+$, $\overline{\mathbb{R}}^+$	$\mathbb{R}^+ := (0, \infty)$ and $\overline{\mathbb{R}}^+ := [0, \infty)$.
$\mathbb{R}^-$, $\overline{\mathbb{R}}^-$	$\mathbb{R}^- := (-\infty, 0)$ and $\overline{\mathbb{R}}^- := (-\infty, 0]$.
$\mathbb{T}$	The unit circle in the complex plane.
$\mathbb{T}_T$	The real line $\mathbb{R}$ where the points $t + mT, m = 0, \pm 1, \pm 2, \ldots$ are identified.
$\mathbb{Z}$	The set of all integers.
$\mathbb{Z}^+$, $\mathbb{Z}^-$	$\mathbb{Z}^+ := \{0, 1, 2, \ldots\}$ and $\mathbb{Z}^- := \{-1, -2, -3, \ldots\}$.
j	$j := \sqrt{-1}$.
0	The number zero, or the zero vector in a vector space, or the zero operator, or the zero-dimensional vector space $\{0\}$.
1	The number one and also the identity operator on any set.

Operators and related symbols

A, B, C, D	In connection with an $L^p\lvert Reg$-well-posed linear system or an operator node, A is usually the main operator, B the control

	operator, C the observation operator and D a feedthrough operator. See Chapters 3 and 4.
$C\&D$	The observation/feedthrough operator of an $L^p\|Reg$-well-posed linear system or an operator node. See Definition 4.7.2.
$\mathfrak{A}$, $\mathfrak{B}$, $\mathfrak{C}$, $\mathfrak{D}$	The semigroup, input map, output map, and input/output map of an $L^p\|Reg$-well-posed linear system, respectively. See Definitions 2.2.1 and 2.2.3.
$\widehat{\mathfrak{D}}$	The transfer function of an $L^p\|Reg$-well-posed linear system or an operator node. See Definitions 4.6.1 and 4.7.4.
$\mathcal{B}(U;Y)$, $\mathcal{B}(U)$	The set of bounded linear operators from U into Y or from U into itself, respectively.
$\mathcal{C}$, $\mathcal{L}$	The Cayley and Laguerre transforms. See Definition 12.3.2.
τ^t	The bilateral time shift operator $\tau^t u(s) := u(t+s)$ (this is a left-shift when $t > 0$ and a right-shift when $t < 0$). See Example 2.5.3 for some additional shift operators.
γ_λ	The time compression or dilation operator $(\gamma_\lambda u)(s) := u(\lambda s)$. Here $\lambda > 0$.
π_J	$(\pi_J u)(s) := u(s)$ if $s \in J$ and $(\pi_J u)(s) := 0$ if $s \notin J$. Here $J \subset \mathbb{R}$.
π_+, π_-	$\pi_+ := \pi_{[0,\infty)}$ and $\pi_- := \pi_{(-\infty,0)}$.
Я	The time reflection operator about zero: $(\mathbf{Я}u)(s) := u(-s)$ (in the L^p-case) or $(\mathbf{Я}u)(s) := \lim_{t\downarrow -s} u(t)$ (in the *Reg*-case). See Definition 3.5.12.
$\mathbf{Я}_h$	The time reflection operator about the point h. See Lemma 6.1.8.
$\boldsymbol{\sigma}$	The discrete-time bilateral left-shift operator $(\boldsymbol{\sigma}\mathbf{u})_k := u_{k+1}$, where $\mathbf{u} = \{u_k\}_{k\in\mathbb{Z}}$. See Section 12.1 for the definitions of $\boldsymbol{\sigma}_+$ and $\boldsymbol{\sigma}_-$.
$\boldsymbol{\pi}_J$	$(\boldsymbol{\pi}_J\mathbf{u})_k := u_k$ if $k \in J$ and $(\boldsymbol{\pi}_J\mathbf{u})_k := 0$ if $k \notin J$. Here $J \subset \mathbb{Z}$ and $\mathbf{u} = \{u_k\}_{k\in\mathbb{Z}}$.
$\boldsymbol{\pi}_+$, $\boldsymbol{\pi}_-$	$\boldsymbol{\pi}_+ := \boldsymbol{\pi}_{\mathbb{Z}^+}$ and $\boldsymbol{\pi}_- := \boldsymbol{\pi}_{\mathbb{Z}^-}$.
w-lim	The weak limit in a Banach space. Thus $\text{w-lim}_{n\to\infty} x_n = x$ in X iff $\lim_{n\to\infty} x^* x_n = x^* x$ for all $x^* \in X^*$. See Section 3.5.
$\langle x, x^*\rangle$	In a Banach space setting $x^*x := \langle x, x^*\rangle$ is the continuous linear functional x^* evaluated at x. In a Hilbert space setting this is the inner product of x and x^*. See Section 3.5.
$E^\perp$	$E^\perp := \{x^* \in X^* \mid \langle x, x^*\rangle = 0 \text{ for all } x \in E\}$. This is the annihilator of $E \subset X$. See Lemma 9.6.4.
${}^\perp F$	${}^\perp F := \{x \in X \mid \langle x, x^*\rangle = 0 \text{ for all } x^* \in F\}$. This is the pre-annihilator of $F \subset X^*$. See Lemma 9.6.4. In the reflexive case ${}^\perp F = F^\perp$, and in the nonreflexive case ${}^\perp F = F^\perp \cap X$.

A^*	The (anti-linear) dual of the operator A. See Section 3.5.
$A \geq 0$	A is (self-adjoint and) positive definite.
$A \gg 0$	$A \geq \epsilon$ for some $\epsilon > 0$, hence A is invertible.
$\mathcal{D}(A)$	The domain of the (unbounded) operator A.
$\mathcal{R}(A)$	The range of the operator A.
$\mathcal{N}(A)$	The null space (kernel) of the operator A.
$\operatorname{rank}(A)$	The rank of the operator A.
$\dim(X)$	The dimension of the space X.
$\rho(A)$	The resolvent set of A (see Definition 3.2.7). The resolvent set is always open.
$\sigma(A)$	The spectrum of A (see Definition 3.2.7). The spectrum is always closed.
$\sigma_p(A)$	The point spectrum of A, or equivalently, the set of eigenvalues of A (see Definition 3.2.7).
$\sigma_r(A)$	The residual spectrum of A (see Definition 3.2.7).
$\sigma_c(A)$	The continuous spectrum of A (see Definition 3.2.7).
$\omega_{\mathfrak{A}}$	The growth bound of the semigroup $\mathfrak{A}$. See Definition 2.5.6.
TI, TIC	TI stands for the set of all time-invariant, and TIC stands for the set of all time-invariant and causal operators. See Definition 2.6.2 for details.
$A\&B$, $C\&D$	$A\&B$ stands for an operator (typically unbounded) whose domain $\mathcal{D}(A\&B)$ is a subspace of the cross-product $\left[\begin{smallmatrix} X \\ U \end{smallmatrix}\right]$ of two Banach spaces X and U, and whose values lie in a third Banach space Z. If $\mathcal{D}(A\&B)$ splits into $\mathcal{D}(A\&B) = X_1 \dotplus U_1$ where $X_1 \subset X$ and $U_1 \subset U$, then $A\&B$ can be written in block matrix form as $A\&B = [A \ \ B]$, where $A = A\&B_{\vert X_1}$ and $B = A\&B_{\vert U_1}$. We alternatively write these identities in the form $Ax = A\&B\left[\begin{smallmatrix} x \\ 0 \end{smallmatrix}\right]$ and $Bu = A\&B\left[\begin{smallmatrix} 0 \\ u \end{smallmatrix}\right]$, interpreting $\mathcal{D}(A\&B)$ as the cross-product of X_1 and U_1.

Special Banach spaces

U	Frequently the input space of the system.
X	Frequently the state space of the system.
Y	Frequently the output space of the system.
X_n	Spaces constructed from the state space X with the help of the generator of a semigroup $\mathfrak{A}$. In particular, X_1 is the domain of the semigroup generator. See Section 3.6.
X_n^*	$X_n^* := (X^*)_n = (X_{-n})^*$. See Remark 3.6.1.
$\dotplus$	$X = X_1 \dotplus X_2$ means that the Banach space X is the direct sum of X_1 and X_2, i.e., both X_1 and X_2 are closed subspaces

	of X, and every $x \in X$ has a unique representation of the form $x = x_1 + x_2$ where $x_1 \in X_1$ and $x_2 \in X_2$.
$\oplus$	$X = X_1 \oplus X_2$ means that the Hilbert space X is the orthogonal direct sum of the Hilbert spaces X_1 and X_2, i.e., $X = X_1 \dot{+} X_2$ and $X_1 \perp X_2$.
$\left[\begin{smallmatrix} X \\ Y \end{smallmatrix}\right]$	The cross-product of the two Banach spaces X and Y. Thus, $\left[\begin{smallmatrix} X \\ Y \end{smallmatrix}\right] = \left[\begin{smallmatrix} X \\ 0 \end{smallmatrix}\right] \dot{+} \left[\begin{smallmatrix} 0 \\ Y \end{smallmatrix}\right]$.

Special functions

χ_I	The characteristic function of the set I.
1_+	The Heaviside function: $1_+ = \chi_{\mathbb{R}^+}$. Thus $(1_+)(t) = 1$ for $t \geq 0$ and $(1_+)(t) = 0$ for $t < 0$.
B	The Beta function (see (5.3.1)).
Γ	The Gamma function (see (3.9.7)).
e_ω	$\mathrm{e}_\omega(t) = \mathrm{e}^{\omega t}$ for $\omega, t \in \mathbb{R}$.
log	The natural logarithm.

Function spaces

$V(J;U)$	Functions of type V (= L^p, BC, etc.) on the interval $J \subset \mathbb{R}$ with range in U.
$V_{\mathrm{loc}}(J;U)$	Functions which are locally of type V, i.e., they are defined on $J \subset \mathbb{R}$ with range in U and they belong to $V(K;U)$ for every bounded subinterval $K \subset J$.
$V_c(J;U)$	Functions in $V(J;U)$ with bounded support.
$V_{c,\mathrm{loc}}(J;U)$	Functions in $V_{\mathrm{loc}}(J;U)$ whose support is bounded to the left.
$V_{\mathrm{loc},c}(J;U)$	Functions in $V_{\mathrm{loc}}(J;U)$ whose support is bounded to the right.
$V_0(J;U)$	Functions in $V(J;U)$ vanishing at $\pm\infty$. See also the special cases listed below.
$V_\omega(J;U)$	The set of functions u for which $(t \mapsto \mathrm{e}^{-\omega t}u(t)) \in V(J;U)$. See also the special cases listed below.
$V_{\omega,\mathrm{loc}}(\mathbb{R};U)$	The set of functions $u \in V_{\mathrm{loc}}(\mathbb{R};U)$ which satisfy $\pi_- u \in V_\omega(\mathbb{R}^-;U)$.
$V(\mathbb{T}_T;U)$	The set of T-periodic functions of type V on $\mathbb{R}$. The norm in this space is the V-norm over one arbitrary interval of length T.
BC	Bounded continuous functions; sup-norm.
BC_0	Functions in BC that tend to zero at $\pm\infty$.
BC_ω	Functions u for which $(t \mapsto \mathrm{e}^{-\omega t}u(t)) \in BC$.

$BC_{\omega,\text{loc}}(\mathbb{R};U)$	Functions $u \in C(\mathbb{R};U)$ which satisfy $\pi_{-}u \in BC_{\omega}(\mathbb{R}^{-};U)$.
$BC_{0,\omega}$	Functions u for which $(t \mapsto e^{-\omega t}u(t)) \in BC_0$.
$BC_{0,\omega,\text{loc}}(\mathbb{R};U)$	Functions $u \in C(\mathbb{R};U)$ which satisfy $\pi_{-}u \in BC_{0,\omega}(\mathbb{R}^{-};U)$.
BUC	Bounded uniformly continuous functions; sup-norm.
BUC^n	Functions which together with their n first derivatives belong to BUC. See Definition 3.2.2.
C	Continuous functions. The same space as BC_{loc}.
C^n	n times continuously differentiable functions. The same space as BC^n_{loc}.
C^∞	Infinitely many times differentiable functions. The same space as BC^∞_{loc}.
L^p, $1 \le p < \infty$	Strongly measurable functions with norm $\left\{\int |u(t)|^p\,dt\right\}^{1/p}$.
L^∞	Strongly measurable functions with norm $\operatorname{ess\,sup}|u(t)|$.
L^p_0	$L^p_0 = L^p$ if $1 \le p < \infty$, and L^∞_0 consists of those $u \in L^\infty$ which vanish at $\pm\infty$, i.e., $\lim_{t\to\infty} \operatorname{ess\,sup}_{\lvert s\rvert \ge t} |u(s)| = 0$.
L^p_ω	Functions u for which $(t \mapsto e^{-\omega t}u(t)) \in L^p$.
$L^p_{\omega,\text{loc}}(\mathbb{R};U)$	Functions $u \in L^p_{\text{loc}}(\mathbb{R};U)$ which satisfy $\pi_{-}u \in L^p_\omega(\mathbb{R}^{-};U)$.
$L^p_{0,\omega}$	Functions u for which $(t \mapsto e^{-\omega t}u(t)) \in L^p_0$.
$L^p_{0,\omega,\text{loc}}(\mathbb{R};U)$	Functions $u \in L^p_{\text{loc}}(\mathbb{R};U)$ which satisfy $\pi_{-}u \in L^p_{0,\omega}(\mathbb{R}^{-};U)$.
$W^{n,p}$	Functions which together with their n first (distribution) derivatives belong to L^p. See Definition 3.2.2.
Reg	Bounded right-continuous functions which have a left hand limit at each finite point.
Reg_0	Functions in Reg which tend to zero at $\pm\infty$.
Reg_ω	The set of functions u for which $(t \mapsto e^{-\omega t}u(t)) \in Reg$.
$Reg_{\omega,\text{loc}}(\mathbb{R};U)$	The set of functions $u \in Reg_{\text{loc}}(\mathbb{R};U)$ which satisfy $\pi_{-}u \in Reg_\omega(\mathbb{R}^{-};U)$.
$Reg_{0,\omega}$	The set of functions u for which $(t \mapsto e^{-\omega t}u(t)) \in Reg_0$.
$Reg_{0,\omega,\text{loc}}(\mathbb{R};U)$	Functions $u \in Reg_{\text{loc}}(\mathbb{R};U)$ which satisfy $\pi_{-}u \in Reg_{0,\omega}(\mathbb{R}^{-};U)$.
Reg^n	Functions which together with their n first derivatives belong to Reg. See Definition 3.2.2.
$L^p\lvert Reg$	This stands for either L^p or Reg, whichever is appropriate.

1
Introduction and overview

We first introduce the reader to the notions of a system node and an L^p-well-posed linear system with $1 \le p \le \infty$, and continue with an overview of the rest of the book.

1.1 Introduction

There are three common ways to describe a finite-dimensional linear time-invariant system in continuous time:

(i) the system can be described in the *time domain* as an *input/output map* $\mathfrak{D}$ from an input signal u into an output signal y;
(ii) the system can be described in the *frequency domain* by means of a *transfer function* $\widehat{\mathfrak{D}}$, i.e., if $\hat{u}$ and $\hat{y}$ are the Laplace transforms of the input u respectively the output y, then $\hat{y} = \widehat{\mathfrak{D}}\hat{u}$ in some right half-plane;
(iii) the system can be described in *state space form* in terms of a set of first order linear differential equations (involving matrices A, B, C, and D of appropriate sizes)

$$\begin{aligned} \dot{x}(t) &= Ax(t) + Bu(t), \\ y(t) &= Cx(t) + Du(t), \qquad t \ge 0, \\ x(0) &= x_0. \end{aligned} \tag{1.1.1}$$

In (i)–(iii) the input signal u takes its values in the *input space* U and the output signal y takes its values in the *output space* Y, both of which are finite-dimensional real or complex vector spaces (i.e., $\mathbb{R}^k$ or $\mathbb{C}^k$ for some $k = 1, 2, 3, \ldots$), and the state $x(t)$ in (iii) takes its values in the *state space* X (another finite-dimensional vector space).

All of the three descriptions mentioned above are important, but we shall regard the third one, the state space description, as the most fundamental one. From a state space description it is fairly easy to get both an input/output description and a transfer function description. The converse statement is more difficult (but equally important): to what extent is it true that an input/output description or a transfer function description can be converted into a state space description? (Various answers to this question will be given below.)

The same three types of descriptions are used for infinite-dimensional linear time-invariant systems in continuous time. The main difference is that we encounter certain technical difficulties which complicate the formulation. As a result, there is *not just one general infinite-dimensional theory*, but a *collection of competing theories that partially overlap each other* (and which become more or less equivalent when specialized to the finite-dimensional case). In this book we shall concentrate on two quite general settings: the case of a system which is either *well-posed in an L^p-setting* (for some $p \in [1, \infty]$) or (more generally), it has a differential description resembling (1.1.1), i.e., it is induced by a *system node*.

In order to give a definition of a *system node* we begin by combining the four matrices A, B, C, and D into one single block matrix $S = \left[\begin{smallmatrix} A & B \\ C & D \end{smallmatrix}\right]$, which we call the *node* of the system, and rewrite (1.1.1) in the form

$$\begin{bmatrix} \dot{x}(t) \\ y(t) \end{bmatrix} = S \begin{bmatrix} x(t) \\ u(t) \end{bmatrix}, \quad t \geq 0, \quad x(0) = x_0. \tag{1.1.2}$$

For a moment, let us ignore the original matrices A, B, C, and D, and simply regard S as a linear operator mapping $\left[\begin{smallmatrix} X \\ U \end{smallmatrix}\right]$ into $\left[\begin{smallmatrix} X \\ Y \end{smallmatrix}\right]$ (recall that we denoted the input space by U, the state space by X, and the output space by Y). If U, X and Y are all finite-dimensional, then S is necessarily bounded, but this need not be true if U, X, or Y is infinite-dimensional. The natural infinite-dimensional extension of (1.1.1) is to replace (1.1.1) by (1.1.2) and to allow S to be an *unbounded linear operator* with some additional properties. These properties are chosen in such a way that (1.1.2) generates some reasonable family of *trajectories*, i.e., for some appropriate class of initial states $x_0 \in X$ and input functions u the equation (1.1.2) should have a well-defined state trajectory $x(t)$ (defined for all $t \geq 0$) and a well-defined output function y. The set of additional properties that we shall use in this work is the following.

Definition 1.1.1 We take U, X, and Y to be Banach spaces (sometimes Hilbert spaces), and call S a *system node* if it satisfies the following four conditions:[1]

(i) S is a closed (possibly unbounded) operator mapping $\mathcal{D}(S) \subset \left[\begin{smallmatrix} X \\ U \end{smallmatrix}\right]$ into $\left[\begin{smallmatrix} X \\ Y \end{smallmatrix}\right]$;

[1] It follows from Lemma 4.7.7 that this definition is equivalent to the definition of a system node given in 4.7.2.

(ii) if we split S into $S = \left[\begin{smallmatrix} S_X \\ S_Y \end{smallmatrix}\right]$ in accordance with the splitting of the range space $\left[\begin{smallmatrix} X \\ Y \end{smallmatrix}\right]$ (S_X is the 'top row' of S and S_Y is the 'bottom row'), then S_X is closed (with $\mathcal{D}(S_X) = \mathcal{D}(S)$);
(iii) the operator A defined by $Ax = S_X \left[\begin{smallmatrix} x \\ 0 \end{smallmatrix}\right]$ with domain $\mathcal{D}(A) = \{x \in X \mid \left[\begin{smallmatrix} x \\ 0 \end{smallmatrix}\right] \in \mathcal{D}(S)\}$ is the generator of a strongly continuous semigroup on X;
(iv) for every $u \in U$ there is some $x \in X$ such that $\left[\begin{smallmatrix} x \\ u \end{smallmatrix}\right] \in \mathcal{D}(S)$.

It turns out that when these additional conditions hold, then (1.1.2) has trajectories of the following type. We use the operators S_X and S_Y defined in (ii) to split (1.1.2) into

$$
\begin{aligned}
\dot{x}(t) &= S_X \begin{bmatrix} x(t) \\ u(t) \end{bmatrix}, \quad t \geq 0, \quad x(0) = x_0, \\
y(t) &= S_Y \begin{bmatrix} x(t) \\ u(t) \end{bmatrix}, \quad t \geq 0.
\end{aligned}
\tag{1.1.3}
$$

If (i)–(iv) hold, then for each $x_0 \in X$ and $u \in C^2([0, \infty); U)$ such that $\left[\begin{smallmatrix} x_0 \\ u(0) \end{smallmatrix}\right] \in \mathcal{D}(S)$, there is a unique function $x \in C^1([0, \infty); X)$ (called a *state trajectory*) satisfying $x(0) = x_0$, $\left[\begin{smallmatrix} x(t) \\ u(t) \end{smallmatrix}\right] \in \mathcal{D}(S)$, $t \geq 0$, and $\dot{x}(t) = S_X \left[\begin{smallmatrix} x(t) \\ u(t) \end{smallmatrix}\right]$, $t \geq 0$. If we define the output $y \in C([0, \infty); Y)$ by $y(t) = S_Y \left[\begin{smallmatrix} x(t) \\ u(t) \end{smallmatrix}\right]$, $t \geq 0$, then the three functions u, x, and y satisfy (1.1.2) (this result is a slightly simplified version of Lemma 4.7.8).

Another consequences of conditions (i)–(iv) above is that it is almost (but not quite) possible to split a system node S into $S = \left[\begin{smallmatrix} A & B \\ C & D \end{smallmatrix}\right]$ as in the finite-dimensional case. If X is finite-dimensional, then the operator A in (iii) will be bounded, and this forces the full system node S to be bounded, with $\mathcal{D}(S) = \left[\begin{smallmatrix} X \\ U \end{smallmatrix}\right]$. Trivially, in this case S can be decomposed into four bounded operators $S = \left[\begin{smallmatrix} A & B \\ C & D \end{smallmatrix}\right]$. If X is infinite-dimensional, then a partial decomposition still exists. The operator A in this partial decomposition corresponds to an extension $A_{|X}$ of the *semigroup generator* A in (iii).[2]This extension is defined on all of X, and it maps X into a larger 'extrapolation space' X_{-1} which contains X as a dense subspace. There is also a *control operator* B which maps U into X_{-1}, and the operator S_X defined in (ii) (the 'top row' of X) is the restriction to $\mathcal{D}(S)$ of the operator $\left[A_{|X} \ B\right]$ which maps $\left[\begin{smallmatrix} X \\ U \end{smallmatrix}\right]$ into X_{-1}. (Furthermore, $\mathcal{D}(S) = \{\left[\begin{smallmatrix} x \\ u \end{smallmatrix}\right] \in \left[\begin{smallmatrix} X \\ U \end{smallmatrix}\right] \mid \left[A_{|X} \ B\right]\left[\begin{smallmatrix} x \\ u \end{smallmatrix}\right] \in X\}$.) Thus, S_X always has a decomposition (after an appropriate extension of its domain and also an extension of the range space). The 'bottom row' S_Y is more problematic, due to the fact that it is not always possible to embed Y as a dense subspace in some larger space Y_{-1} (for example, Y may be finite-dimensional). It is still true,

[2] We shall also refer to A as the *main operator* of the system node.

however, that it is possible to define an *observation operator* C with domain $\mathcal{D}(C) = \mathcal{D}(A)$ by $Cx = S_Y \left[\begin{smallmatrix} x \\ 0 \end{smallmatrix}\right]$, $x \in \mathcal{D}(A)$. The *feedthrough operator* D in the finite-dimensional decomposition $A = \left[\begin{smallmatrix} A & B \\ C & D \end{smallmatrix}\right]$ need not always exist, and it need not be unique. However, this lack of a unique well-defined feedthrough operator is largely compensated by the fact that every system node has a *transfer function*, defined on the resolvent set of the operator A in (iii). See Section 4.7 for details.[3]

The other main setting that we shall use (and after which this book has been named) is the *L^p-well-posed setting* with $1 \le p \le \infty$. This setting can be introduced in two different ways. One way is to first introduce a system node of the type described above, and then add the requirement that for all $t > 0$, *the final state $x(t)$ and the restriction of y to the interval* $[0, t)$ *depend continuously on x_0 and the restriction of u to* $[0, t)$. This added requirement will give us an L^p-well-posed linear system if we use the X-norm for x_0 and $x(t)$, the norm in $L^p([0, t); U)$ for u, and the norm in $L^p([0, t); Y)$ for y.[4] (See Theorem 4.7.13 for details.)

However, it is also possible to proceed in a different way (as we do in Chapter 2) and to introduce the notion of an L^p-well-posed linear system *without any reference to a system node*. In this approach we look directly at the mapping from the initial state x_0 and the input function (restricted to the interval $[0, t)$) to the final state $x(t)$ and the output function y (also restricted to the interval $[0, t)$). Assuming the same type of continuous dependence as we did above, the relationship between these four objects can be written in the form (we denote the restrictions of u and y to some interval $[s, t)$ by $\pi_{[s,t)}u$, respectively $\pi_{[s,t)}y$)

$$\begin{bmatrix} x(t) \\ \pi_{[0,t)}y \end{bmatrix} = \left[\begin{array}{c|c} \mathfrak{A}_0^t & \mathfrak{B}_0^t \\ \hline \mathfrak{C}_0^t & \mathfrak{D}_0^t \end{array}\right] \begin{bmatrix} x_0 \\ \pi_{[0,t)}u \end{bmatrix}, \qquad t \ge 0,$$

for some families of bounded linear operator $\mathfrak{A}_0^t\colon X \to X$, $\mathfrak{B}_0^t\colon L^p([0, t); U) \to X$, $\mathfrak{C}_0^t\colon X \to L^p([0, t); Y)$, and $\mathfrak{D}_0^t\colon L^p([0, t); U) \to L^p([0, t); Y)$. If these families correspond to the trajectories of some system node (as described earlier), then they necessarily *satisfy some algebraic conditions*, with can be stated without any reference to the system node itself. Maybe the simplest way to list these algebraic conditions is to look at a slightly extended version of (1.1.2)

[3] Another common way of constructing a system node is the following. Take any semigroup generator A in X, and extend it to an operator $A_{|X} \in \mathcal{B}(X; X_{-1})$. Let $B \in \mathcal{B}(U; X_{-1})$ and $C \in \mathcal{B}(X_1; U)$ be arbitrary, where X_1 is $\mathcal{D}(A)$ with the graph norm. Finally, fix the value of the transfer function to be a given operator in $\mathcal{B}(U; Y)$ at some arbitrary point in $\rho(A)$, and use Lemma 4.7.6 to construct the corresponding system node.

[4] Here we could just as well have replaced the interval $[0, t)$ by $(0, t)$ or $[0, t]$. However, we shall later consider functions which are defined pointwise everywhere (as opposed to almost everywhere), and then it is most convenient to use half-open intervals of the type $[s, t)$, $s < t$.

where the initial time zero has been replaced by a general initial time s, namely

$$\begin{bmatrix} \dot{x}(t) \\ y(t) \end{bmatrix} = S \begin{bmatrix} x(t) \\ u(t) \end{bmatrix}, \quad t \geq s, \quad x(s) = x_s, \tag{1.1.4}$$

and to also look at the corresponding maps from x_s and $\pi_{[s,t)}u$ to $x(t)$ and $\pi_{[s,t)}y$ which we denote by

$$\begin{bmatrix} x(t) \\ \pi_{[s,t)}y \end{bmatrix} = \left[\begin{array}{c|c} \mathfrak{A}_s^t & \mathfrak{B}_s^t \\ \hline \mathfrak{C}_s^t & \mathfrak{D}_s^t \end{array}\right] \begin{bmatrix} x_s \\ \pi_{[s,t)}u \end{bmatrix}, \quad s \leq t.$$

These two-parameter families of bounded linear operators $\mathfrak{A}_s^t$, $\mathfrak{B}_s^t$, $\mathfrak{C}_s^t$, and $\mathfrak{D}_s^t$ have the properties listed below. In this list of properties we denote the left-shift operator by

$$(\tau^t u)(s) = u(t+s), \qquad -\infty < s, t < \infty,$$

and the identity operator by 1.

Algebraic conditions 1.1.2 *The operator families $\mathfrak{A}_s^t$, $\mathfrak{B}_s^t$, $\mathfrak{C}_s^t$, and $\mathfrak{D}_s^t$ satisfy the following conditions:*[5]

(i) *For all $t \in \mathbb{R}$,*

$$\left[\begin{array}{c|c} \mathfrak{A}_t^t & \mathfrak{B}_t^t \\ \hline \mathfrak{C}_t^t & \mathfrak{D}_t^t \end{array}\right] = \left[\begin{array}{c|c} 1 & 0 \\ \hline 0 & 0 \end{array}\right].$$

(ii) *For all $s \leq t$,*

$$\left[\begin{array}{c|c} \mathfrak{A}_s^t & \mathfrak{B}_s^t \\ \hline \mathfrak{C}_s^t & \mathfrak{D}_s^t \end{array}\right] = \left[\begin{array}{c|c} \mathfrak{A}_s^t & \mathfrak{B}_s^t \pi_{[s,t)} \\ \hline \pi_{[s,t)}\mathfrak{C}_s^t & \pi_{[s,t)}\mathfrak{D}_s^t \pi_{[s,t)} \end{array}\right].$$

(iii) *For all $s \leq t$ and $h \in \mathbb{R}$,*

$$\left[\begin{array}{c|c} \mathfrak{A}_{s+h}^{t+h} & \mathfrak{B}_{s+h}^{t+h} \\ \hline \mathfrak{C}_{s+h}^{t+h} & \mathfrak{D}_{s+h}^{t+h} \end{array}\right] = \left[\begin{array}{c|c} \mathfrak{A}_s^t & \mathfrak{B}_s^t \tau^h \\ \hline \tau^{-h}\mathfrak{C}_s^t & \tau^{-h}\mathfrak{D}_s^t \tau^h \end{array}\right].$$

(iv) *For all $s \leq r \leq t$,*

$$\left[\begin{array}{c|c} \mathfrak{A}_s^t & \mathfrak{B}_s^t \\ \hline \mathfrak{C}_s^t & \mathfrak{D}_s^t \end{array}\right] = \left[\begin{array}{c|c} \mathfrak{A}_r^t \mathfrak{A}_s^r & \mathfrak{B}_r^t + \mathfrak{A}_r^t \mathfrak{B}_s^r \\ \hline \mathfrak{C}_r^t \mathfrak{A}_s^r + \mathfrak{C}_s^r & \mathfrak{D}_r^t + \mathfrak{C}_r^t \mathfrak{B}_s^r + \mathfrak{D}_s^r \end{array}\right].$$

All of these conditions have natural interpretations (see Sections 2.1 and 2.2 for details): (i) is an initial condition, (ii) says that the system is causal, (iii)

[5] By Theorem 2.2.14, these algebraic conditions are equivalent to those listed in Definition 2.2.1.

says that the system is time-invariant, and (iv) gives a formula for how to patch two solutions together, the first of which is defined on $[s, r]$ and the second on $[r, t]$, and with the initial state of the second solution equal to the final state of the first solution at the 'switching time' r. For example, if we take a closer look at the family $\mathfrak{A}_s^t$, then (iii) says that $\mathfrak{A}_s^t = \mathfrak{A}_0^{t-s}$ for all $s \le t$, (i) says that $\mathfrak{A}_0^0 = 1$, and (iv) says that $\mathfrak{A}_0^t = \mathfrak{A}_0^r \mathfrak{A}_0^{t-r}$ for all $0 \le r \le t$. This means that the family $\mathfrak{A}_0^t$ is simply a semigroup (it is the semigroup generated by the operator A of the corresponding system node).

Not only are the conditions (i)–(iv) above *necessary* for the family $\begin{bmatrix} \mathfrak{A}_s^t & \mathfrak{B}_s^t \\ \mathfrak{C}_s^t & \mathfrak{D}_s^t \end{bmatrix}$ to be generated by a system node S through the equation (1.1.4), but they are *sufficient* as well (when combined with the appropriate continuity assumptions). This will be shown in Chapters 3 and 4 (out of which the former deals exclusively with semigroups). However, it is possible to develop a fairly rich theory by simply appealing to the algebraic conditions (i)–(iv) above (and appropriate continuity conditions), without any reference to the corresponding system node. Among other things, every L^p-well-posed linear system has a finite growth bound, identical to the growth bound of its semigroup $\mathfrak{A}_0^t$. See Chapter 2 for details.

Most of the remainder of the book deals with extensions of various notions known from the theory of finite-dimensional systems to the setting of L^p-well-posed linear systems, and even to systems generated by arbitrary system nodes. Some of the extensions are straightforward, others are more complicated, and some finite-dimensional results are simply not true in an infinite-dimensional setting. Conversely, many of the infinite-dimensional results that we present do not have any finite-dimensional counterparts, in the sense that these statements become trivial if the state space is finite-dimensional. In many places the case $p = \infty$ is treated in a slightly different way from the case $p < \infty$, and the class of L^∞-well-posed linear systems is often replaced by another class of systems, the *Reg*-well-posed class, which allows functions to be evaluated everywhere (recall that functions in L^∞ are defined only almost everywhere), and which restricts the set of permitted discontinuities to jump discontinuities.

The last three chapters have a slightly different flavor from the others. We replace the general Banach space setting which has been used up to now by a standard Hilbert space setting, and explore some connections between well-posed linear systems, Fourier analysis, and operator theory. In particular, in Section 10.3 we establish the standard connection between the class of bounded time-invariant causal operators on L^2 and the set of bounded analytic functions on the right half-plane, and in Sections 10.5–10.7 the admissibility and boundedness of scalar control and observation operators for contraction semigroups are characterized by means of the Carleson measure theorem. Chapter 11 has a distinct operator theory flavor. It contains among others a systems theory interpretation of the basic dilation and model theory for continuous-time contraction semigroups on Hilbert spaces.

Chapter 12 contains a short introduction to discrete-time systems (and it also contains a section on continuous-time systems). Some auxiliary results have been collected in the appendix.

After this rough description of what this book is all about, let us also tell the reader what this book is *not* about, and give some indications of where to look for these missing results.

There are a number of examples of L^p-well-posed linear system given in this book, but these are primarily of a mathematical nature, and they are not the true physical examples given in terms of partial differential equations which are found in books on mathematical physics. There are two reasons for this lack of physical examples. One of them is the lack of space and time. The present book is quite large, and any addition of such examples would require a significant amount of additional space. It would also require another year or two or three to complete the manuscript. The other reason is that the two recent volumes Lasiecka and Triggiani (2000a, b) contain an excellent collection of examples of partial differential equations modeling various physical systems. By Theorem 5.7.3(iii), most of the examples in the first volume dealing with parabolic problems are *Reg*-well-posed. Many of the examples in the second volume dealing with hyperbolic problems are L^2-well-posed. Almost all the examples in Lasiecka and Triggiani (2000a, b) are generated by system nodes. (The emphasis of these two volumes is quite different from the emphasis of this book. They deal with *optimal control*, whereas we take a more general approach, focusing more on input/output properties, transfer functions, coprime fractions, realizations, passive and conservative systems, discrete time systems, model theory, etc.)

Our original main motivation for introducing the class of systems generated by arbitrary system nodes was that this class is a very natural setting for a study of *impedance passive* systems. Such systems need not be well-posed, but under rather weak assumptions they are generated by system nodes. The decision *not* to include a formal discussion of impedance passive systems in this book was not easy. Once more this decision was dictated partly by the lack of space and time, and partly by the fact that there is another recently discovered setting which may be even more suitable for this class of systems, namely the continuous time analogue of the state/signal systems introduced in Arov and Staffans (2004, see also Ball and Staffans 2003). Impedance passive systems are discussed in the spirit of this book in Staffans (2002a, b, c).

Another obvious omission (already mentioned above) is the lack of results concerning quadratic optimal control. This omission may seem even more strange in light of the fact that the original motivation for writing this book was to present a general theory that could be used in the study of optimal control problems (of definite and indefinite type). However, also this omission has two valid reasons. The first one is the same as we mentioned above, i.e., lack of space and time. The other reason is even more fundamental: the theory of optimal control is at this very moment subject to very active research, and it has not yet reached the needed maturity to be written down in the form of a monograph. We are here thinking about a general theory in the spirit of this book. There do, of course, exist quite mature theories for various subclasses of systems. One such class is the one which assumes that the system is of the 'classical' form (1.1.1), where A is the generator of a strongly continuous semigroup and the operators B, C, and D are bounded. This class is thoroughly investigated in Curtain and Zwart (1995). Systems of this type are easy to deal with (hence, they have a significant pedagogical value), but they are too limited to cover many of the interesting boundary control systems encountered in mathematical physics. (For example, the models developed in Sections 11.6 and 11.7 have bounded B, C, and D only in very special cases.) Other more general (hence less complete) theories are found in, e.g., Lions (1971), Curtain and Pritchard (1978), Bensoussan *et al.* (1992), Fattorini (1999), and Lasiecka and Triggiani (2000a, b). Quadratic optimal control results in the setting of L^2-well-posed linear systems are found in Mikkola (2002), Staffans (1997, 1998a, b, c, d), Weiss (2003), and Weiss and Weiss (1997).

There is a significant overlap between some parts of this book and certain books which deal with 'abstract system theory', such as Fuhrmann (1981) and Feintuch and Saeks (1982), or with operator theory, such as Lax and Phillips (1967), Sz.-Nagy and Foiaş (1970), Brodskiĭ (1971), Livšic (1973), and Nikol'skiĭ (1986). In particular, Chapter 11 can be regarded as a natural continuous-time analogue of one of the central parts of Sz.-Nagy and Foiaş (1970, rewritten in the language of L^2-well-posed linear systems).

1.2 Overview of chapters 2–13

Chapter 2 In this chapter we develop the basic theory of L^p-well-posed linear systems starting from a set of algebraic conditions which is equivalent to 1.1.2. We first simplify the algebraic conditions 1.1.2 by using a part of those conditions to replace the original two-parameter families $\mathfrak{A}_s^t$, $\mathfrak{B}_s^t$, $\mathfrak{C}_s^t$, and $\mathfrak{D}_s^t$ introduced in Section 1.1 by a semigroup $\mathfrak{A}^t$, $t \geq 0$, and three other operators, the *input map* $\mathfrak{B} = \mathfrak{B}_{-\infty}^0$, the *output map* $\mathfrak{C} = \mathfrak{C}_0^\infty$, and the *input/output map* $\mathfrak{D} = \mathfrak{D}_{-\infty}^\infty$. The resulting algebraic conditions that $\mathfrak{A}$, $\mathfrak{B}$, $\mathfrak{C}$, and $\mathfrak{D}$ have to satisfy are listed in 2.1.3 and again in Definition 2.2.1. The connection between the quadruple $\mathfrak{A}$, $\mathfrak{B}$, $\mathfrak{C}$, and $\mathfrak{D}$ and the original four operator families $\mathfrak{A}_s^t$, $\mathfrak{B}_s^t$, $\mathfrak{C}_s^t$, and $\mathfrak{D}_s^t$ is explained informally in Section 2.1 and more formally in Definition 2.2.6 and Theorem 2.2.14. Thus, we may either interpret an L^p-well-posed linear system as a quadruple $\Sigma = \left[\begin{array}{c|c} \mathfrak{A} & \mathfrak{B} \\ \hline \mathfrak{C} & \mathfrak{D} \end{array}\right]$, or as a two-parameter family of operators $\Sigma_s^t = \left[\begin{array}{c|c} \mathfrak{A}_s^t & \mathfrak{B}_s^t \\ \hline \mathfrak{C}_s^t & \mathfrak{D}_s^t \end{array}\right]$, where s represents the initial time and t the final time.

In the case where $p = \infty$ we often require the system to be *Reg*-well-posed instead of L^∞-well-posed. Here *Reg* stands for the class of *regulated functions* (which is described in more detail in Section A.1). By a regulated function we mean a function which is locally bounded, right-continuous, and which has a left-hand limit at each finite point. The natural norm in this space is the L^∞-norm (i.e., the sup-norm). In this connection we introduce the following terminology (see Definition 2.2.4). By an *$L^p|Reg$-well-posed linear system* we mean a system which is either *Reg*-well-posed or L^p-well-posed for some p, $1 \le p \le \infty$, and by a *well-posed linear system* we mean a system which is either *Reg*-well-posed or L^p-well-posed for some p, $1 \le p < \infty$. Thus, the L^p-case with $p = \infty$ is included in the former class but not in the latter. The reason for this distinction is that not all results that we present are true for L^∞-well-posed systems. Whenever we write $L^p|Reg$ we mean either L^p or *Reg*, whichever is appropriate at the moment.

In our original definition of the operators $\mathfrak{B}$ and $\mathfrak{D}$ we restrict their domains to consist of those input functions which are locally in $L^p|Reg$ with values in U, and whose supports are bounded to the left. The original range spaces of $\mathfrak{C}$ and $\mathfrak{D}$ consist of output functions which are locally in $L^p|Reg$ with values in Y. However, as we show in Theorem 2.5.4, every $L^p|Reg$-well-posed linear system has a *finite exponential growth bound* (equal to the growth bound of its semigroup). This fact enables us to extend the operators $\mathfrak{B}$ and $\mathfrak{D}$ to a larger domain, and to confine the ranges of $\mathfrak{C}$ and $\mathfrak{D}$ to a smaller space. More precisely, we are able to relax the original requirement that the support of the input function should be bounded to the left, replacing it by the requirement that the input function should belong to some exponentially weighted $L^p|Reg$-space. We are also able to show that the ranges of $\mathfrak{C}$ and $\mathfrak{D}$ lie in an exponentially weighted $L^p|Reg$-space (the exponential weight is the same in both cases, and it is related to the growth bound of the system). In later discussions we most of the time use these extended/confined versions of $\mathfrak{B}$, $\mathfrak{C}$, and $\mathfrak{D}$.

As part of the proof of the fact that every $L^p|Reg$-well-posed linear system has a finite growth bound we show in Section 2.4 that every such system can be interpreted as a discrete-time system $\mathbf{\Sigma} = \left[\begin{array}{c|c} \mathbb{A} & \mathbb{B} \\ \hline \mathbb{C} & \mathbb{D} \end{array}\right]$ with infinite-dimensional input and output spaces, and with bounded operators $\mathbb{A}$, $\mathbb{B}$, $\mathbb{C}$, and $\mathbb{D}$. More precisely, $\left[\begin{array}{c|c} \mathbb{A} & \mathbb{B} \\ \hline \mathbb{C} & \mathbb{D} \end{array}\right] = \left[\begin{array}{c|c} \mathfrak{A}_0^T & \mathfrak{B}_0^T \\ \hline \mathfrak{C}_0^T & \mathfrak{D}_0^T \end{array}\right]$, the discrete-time input space is $L^p|Reg([0, T); U)$, and the output space is $L^p|Reg([0, T); Y)$, for some $T > 0$. We achieve this by regarding $L^p|Reg([0,\infty);U)$ as an infinite product of the spaces $L^p|Reg([kT,(k+1)T);U)$, $k=0,1,2,\ldots$, and treating $L^p|Reg([0,\infty);Y)$ in a similar manner.

In Section 2.6 we show that a linear time-invariant causal operator which maps $L^p|Reg_{\mathrm{loc}}([0,\infty);U)$ into $L^p|Reg_{\mathrm{loc}}([0,\infty);U)$ can be interpreted as the input/output map of some $L^p|Reg$-well-posed linear system if and only if it is exponentially bounded. In Section 2.7 we show how to re-interpret an L^p-well-posed linear system with $p<\infty$ as a strongly continuous semigroup in a suitable (infinite-dimensional) state space. This construction explains the connection between a well-posed linear system and the semigroups occurring in scattering theory studied in, e.g., Lax and Phillips (1967).

Chapter 3 Here we develop the basic theory of C_0 (i.e., strongly continuous) *semigroups and groups*. The treatment resembles the one found in most textbooks on semigroup theory (such as Pazy (1983)), but we put more emphasis on certain aspects of the theory than what is usually done. The generator of a C_0 semigroup and its resolvent are introduced in Section 3.2, and the celebrated Hille–Yosida generating theorem is stated and proved in Section 3.4, together with theorems characterizing generators of contraction semigroups. The primary examples are shift semigroups in (exponentially weighted) L^p-spaces. Dual semigroups are studied in Section 3.5, both in the reflexive case and the nonreflexive case (in the latter case the dual semigroup is defined on a closed subspace of the dual of the original state space). Here we also explain the duality concept which we use throughout the whole book: in spite of the fact that most of the time we work in a Banach space instead of a Hilbert space setting, we still use the conjugate-linear dual rather than the standard linear dual (to make the passage from the Banach space to the Hilbert space setting as smooth as possible).

The first slightly nonstandard result in Chapter 3 is the introduction in Section 3.6 of "Sobolev spaces" with positive and negative index induced by a semigroup generator A, or more generally, by an unbounded densely defined operator A with a nonempty resolvent set.[6] If we denote the original state space by $X=X_0$, then this is a family of spaces

$$\cdots\subset X_2\subset X_1\subset X\subset X_{-1}\subset X_{-2}\subset\cdots,$$

where each embedding is continuous and dense, and $(\alpha-A)$ maps X_{j+1} one-to-one onto X_j for all α in the resolvent set of A and all $j\ge 0$. A similar statement is true for $j<0$: the only difference is that we first have to *extend* A

[6] In the Russian tradition these spaces are known as spaces with a 'positive norm' respectively 'negative norm'. Spaces with positive index are sometimes referred to as 'interpolation spaces,' and those with negative index as 'extrapolation spaces'.

to an operator $A_{|X_{j+1}}$ mapping X_{j+1} into X_j (such an extension always exists and it is unique). We shall refer to this family as the family of *rigged spaces* induced by A. The most important of these spaces with *positive index* is X_1, which is the domain of A equipped with (for example) the graph norm. The most important of these spaces with *negative index* is X_{-1}, which will contain the range of the control operator induced by a system node whose semigroup generator is the operator A above.

Standard resolvent and multiplicative approximations of the semigroup are presented in Section 3.7. We then turn to a study of the *nonhomogeneous Cauchy problem*, i.e., the question of the existence of solutions of the nonhomogeneous differential equation

$$\begin{aligned} \dot{x}(t) &= Ax(t) + f(t), \qquad t \geq s, \\ x(s) &= x_s. \end{aligned} \tag{1.2.1}$$

More generally, we often replace A by the extended operator $A_{|X_{-1}}$ in the equation above, or by $A_{|X_j}$ for some other $j \leq -1$. We show that under fairly mild assumptions on the *forcing function* f in (1.2.1) the solution produced by the variation of constant formula

$$x(t) = \mathfrak{A}^{t-s} x_s + \int_s^t \mathfrak{A}^{t-v} f(v)\, dv, \tag{1.2.2}$$

is indeed a more or less classical solution of (1.2.1), provided we work in a rigged space X_j with a sufficiently negative value of j (most of the time it will suffice to take $j = -1$).

In Section 3.9 we develop a *symbolic calculus* for semigroup generators. This calculus enables us to introduce rigged spaces X_α of fractional order $\alpha \in \mathbb{R}$. The same calculus is also needed in Section 3.10, where we develop the theory of *analytic semigroups* (whose generators are sectorial operators). The *spectrum determined growth property*, i.e., the question of to what extent the growth bound of a semigroup can be determined from the spectrum of its generator, is studied in some detail in Section 3.11. We then take a closer look at the *Laplace transform*, and present some additional symbolic calculus for Laplace transforms. This leads eventually to frequency domain descriptions of the shift semigroups that we originally introduced in the time domain. Finally, we study *invariant* and *reducing subspaces* of semigroups and their generators, together with two different kinds of *spectral projections*.

Chapter 4 In Chapter 2 we developed the theory of $L^p|Reg$-well-posed linear systems starting from a set of algebraic conditions equivalent to 1.1.2 combined with appropriate continuity conditions. Here we replace these algebraic conditions by a set of differential/algebraic conditions, i.e., we try to recover as much as possible of the system (1.1.1) that we used to motivate the algebraic conditions (1.1.2) in the first place. We begin by proving in Section 4.2 the existence of a *control operator* B mapping the input space U into the extrapolation space X_{-1}. This operator is called *bounded* if $\mathcal{R}(B) \subset X$. In the next section we give conditions under which the state trajectory $x(t)$ of a $L^p|Reg$-well-posed linear system is a solution of the non homogeneous Cauchy problem

$$\begin{aligned} \dot{x}(t) &= A_{|X}x(t) + Bu(t), \qquad t \geq s, \\ x(s) &= x_s. \end{aligned} \tag{1.2.3}$$

Here the values in the first of these equations (including $\dot{x}(t)$) lie in X_{-1}, and $A_{|X}$ is the extension of the semigroup generator A to an operator which maps the original state space X into X_{-1}. Under suitable additional smoothness assumptions x will be continuously differentiable in X (rather than differentiable almost everywhere in X_{-1}), but it will not, in general, be possible to replace $A_{|X}$ by A in (1.2.3) (i.e., it need not be true that $x(t) \in \mathcal{D}(A) = X_1$). The results of this section depend heavily on the corresponding results for the non homogeneous Cauchy problem proved in Chapter 3.

The existence of an *observation operator* C mapping the interpolation space X_1 into the output space Y is established in Section 4.4. This operator is called *bounded* if it can be extended to a bounded linear operator from X into Y.

The question of how to define a *feedthrough operator*, i.e., how to find an operator corresponding to the operator D in (1.1.1), is more complicated. (This question is the main theme of Chapter 5.) Two cases where this question has a simple solution are discussed in Section 4.5: one is the case where the control operator is bounded, and the other is the case where the observation operator is bounded.

In Section 4.6 we prove that every $L^p|Reg$-well-posed linear system has an analytic *transfer function*. It is operator-valued, with values in $\mathcal{B}(U;Y)$ (where U is the input space and Y is the output space). Originally it is defined on a right half-plane whose left boundary is determined by the growth bound of the system, but it is later extended to the whole resolvent set of the main operator. In this section we also prove the existence of a *system node* of the type described in Definition 1.1.1. Here we introduce a slightly different notation compared to the one in (1.1.3): we denote the 'top row' of S by $A\&B$ instead of S_X, and the 'bottom row' of S by $C\&D$ instead of S_Y. The reason for this notation is that intuitively $A\&B$ can be regarded as a combination of two operators A and B which cannot be completely separated from each other, and analogously, $C\&D$ can intuitively be regarded as a combination of two other operators C and D which cannot either be completely separated from each other. We call $C\&D$ the *combined observation/feedthrough operator*. Actually, the splitting of $A\&B$

into two independent operators is always possible in the sense that $A\&B$ is the restriction of the operator $\begin{bmatrix} A_{|X} & B \end{bmatrix}$ (which maps $\left[\begin{smallmatrix} X \\ U \end{smallmatrix}\right]$ continuously into X_{-1}) to its domain $\mathcal{D}(A\&B) = \mathcal{D}(S) = \left\{\left[\begin{smallmatrix} x \\ u \end{smallmatrix}\right] \in \left[\begin{smallmatrix} X \\ U \end{smallmatrix}\right] \mid A_{|X}x + Bu \in X\right\}$. Thus, this separation is based on the fact that the domain of $A\&B$ can be extended to all of $\left[\begin{smallmatrix} X \\ U \end{smallmatrix}\right]$ at the expense of also extending the range space from X to X_{-1}. The question to what extent $C\&D$ can be split into two operators C and D is more difficult, and it is discussed in Chapter 5.

Motivated by the preceding result we proceed in Section 4.7 to study linear systems which are not necessarily $L^p|Reg$-well-posed, but which still have a dynamics which is determined by a system node. In passing we introduce the even more general class of *operator nodes*, which differs from the class of system nodes in the sense that the operator A in Definition 1.1.1 must still be densely defined and have a non-empty resolvent set, but it need not generate a semigroup. It is still true that every operator node has a *main operator* $A \in \mathcal{B}(X_1; X)$ (i.e., the operator A in Definition 1.1.1), a control operator $B \in \mathcal{B}(U; X_{-1})$, an observation operator $C \in \mathcal{B}(X_1; Y)$, and an analytic transfer function defined on the resolvent set of A.

The system nodes of some of our earlier examples of $L^p|Reg$-well-posed linear systems are computed in Section 4.8, including the system nodes of the delay line and of the Lax–Phillips semigroup presented in Section 2.7. *Diagonal* and *normal systems* are studied in Section 4.9.

Finally, in Section 4.10 it is shown how one can 'peel off' the inessential parts of the input and output spaces, namely the null space of the control operator and a direct complement to the range of the observation operator. These subspaces are of less interest in the sense that with respect to these subspaces the system acts like a *static* system rather than a more general dynamic system (a system is static if the output is simply the input multiplied by a fixed bounded linear operator; thus, it has no memory, and it does not need a state space). The same section also contains a different type of additive decomposition: to any pair of reducing subspaces of the semigroup generator, one of which is contained in its domain, it is possible to construct two independent subsystems in such a way that the original system is the *parallel connection* of two separate subsystems.

Chapter 5 In this chapter we take a closer look at the existence of a *feedthrough operator*, i.e., an operator $D \in \mathcal{B}(U; Y)$ corresponding to the operator D in (1.1.1). We begin by defining a *compatible system*. This is a system whose combined observation/feedthrough operator $C\&D$ (this is the same operator which was denoted by S_Y in Definition 1.1.1) can be split into two independent operators $C_{|W}$ and D in the following sense. There exists a Banach space W, $X_1 \subset W \subset X$, and two operators $C_{|W} \in \mathcal{B}(W; Y)$ and $D \in \mathcal{B}(U; Y)$ such that

$C\&D$ is the restriction of $\begin{bmatrix} C_{|W} & D \end{bmatrix}$ to its domain $\mathcal{D}(C\&D) = \mathcal{D}(S)$. We warn the reader that *neither is the space W unique, nor are the operators $C_{|W}$ and D corresponding to a particular space W unique* (except in the case where X_1 is dense in W).[7] Note that this splitting of $C\&D$ differs from the corresponding splitting of $A\&B$ described earlier in the sense that *the operators $C_{|W}$ and D have the same range space Y as the original observation/feedthrough operator.*[8] Also note that $C_{|W}$ is an extension of the original observation operator C, whose domain is $X_1 \subset W$. There is a minimal space W, which we denote by $(X + BU)_1$. This is the sum of X_1 and the range of $(\alpha - A_{|X})^{-1}B$, where α is an arbitrary number in $\rho(A)$. Often it is enough to work in this smallest possible space W, but sometimes it may be more convenient to use a larger space W (for example, in the case where X_1 is not dense in $(X + BU)_1$, or in the regular case which will be introduced shortly). One of the most interesting results in Section 5.1 (only recently discovered) says that *most $L^p|Reg$-well-posed linear systems are compatible*. In particular, this is true whenever the input space U and the state space X are Hilbert spaces.

Section 5.2 deals with *boundary control systems*. These are systems (not necessarily well-posed) whose control operator B is strictly unbounded in the sense that $\mathcal{R}(B) \cap X = 0$. It turns out that every boundary control system is compatible, and that it is possible to choose the operator D in a compatible splitting of $C\&D$ in an arbitrary way. (The most common choice is to take $D = 0$.)

As a preparation for the next major subject treated in Chapter 5 we study various approximations of the identity operator acting on the state space in Section 5.3. By using these approximations and *summability methods* we *extend the observation operator* of an $L^p|Reg$-well-posed linear system Σ to a larger domain in Section 5.4. Apart from using different summability methods we also distinguish between limits in the weak, the strong, or the uniform operator topology. The system Σ is called *regular* if this extension of the observation operator is a compatible extension of the type described above, i.e., together with some operator $D \in \mathcal{B}(U; Y)$ it provides us with a compatible splitting of the combined observation/feedthrough operator $C\&D$. In this case it is possible to develop some explicit formulas for the operator D. Maybe the simplest of these formulas is the one which says that if we denote the transfer function of Σ by $\widehat{\mathfrak{D}}$, then $D = \lim_{\alpha\to+\infty} \widehat{\mathfrak{D}}(\alpha)$ (here α is real, and the limit is taken in the weak, strong, or uniform sense). It turns out that all L^1-well-posed systems are weakly regular, and they are even strongly regular whenever their state space

[7] However, D is determined uniquely by $C_{|W}$, and $C_{|W}$ is determined uniquely by D.

[8] This is important, e.g., in the case where X is infinite-dimensional but Y is finite-dimensional, in which case Y does not have any nontrivial extension in which Y is dense.

is reflexive (see Theorem 5.6.6 and Lemma 5.7.1(ii)). All L^∞-well-posed and *Reg*-well-posed systems are strongly regular (see Lemma 5.7.1(i)). The standard delay line is uniformly regular (with $D = 0$), and so are all typical L^p-well-posed systems whose semigroup is analytic. Roughly speaking, in order for an $L^p|Reg$-well-posed linear system *not* to be regular both the control operator B and the observation operator C must be 'maximally unbounded' [see Weiss and Curtain (1999, Proposition 4.2) or Mikkola (2002) for details].

Chapter 6 Here we introduce various transformations that can be applied to an $L^p|Reg$-well-posed linear system or to a system or operator node. Some of these transformations produce systems which evolve in the *backward* time direction. We call these systems *anti-causal*, and describe their basic properties in Section 6.1. A closely related notion is the *time-inversion* discussed in Section 6.4. By this we mean the reversal of the direction of time. The time-inverse of a (causal) $L^p|Reg$-well-posed linear system or system node is always an *anti-causal* $L^p|Reg$-well-posed linear system or system node. However, it is sometimes possible to alternatively interpret the new system as a *causal* system, of the same type as the original one. This is equivalent to saying that the original causal system has an alternative interpretation as an anti-causal system. This will be the case if and only if the *system semigroup can be extended to a group*, and (only) in this case we shall call the original system *time-invertible*. Compatibility is always preserved under time-inversion, but none of the different types of regularity (weak, strong, or uniform) need be preserved.

In Section 6.2 we present the *dual* of an L^p-well-posed linear system with $p < \infty$ in the case where the input space U, the output space Y, and the state space X are reflexive. This dual can be defined in two different ways which are time-inversions of each other: the *causal dual* evolves in the *forward time direction*, and the *anti-causal dual* evolves in the backward time direction. Both of these are L^q-well-posed with $1/p + 1/q = 1$ ($q = \infty$ if $p = 1$). We also present the dual of a system or operator node S. Here the *causal dual* is simply the (unbounded) adjoint of S, whereas the *anti-causal dual* is the adjoint of S with an additional change of sign (due to the change of the direction of time).

In the rest of this chapter we discuss three different types of *inversions* which can be carried out under suitable additional assumptions on the system, namely *flow-inversion*, *time-inversion*, and *time-flow-inversion*. We have already described time-inversion above. Flow-inversion is introduced in Section 6.3. It amounts to interchanging the input with the output, so that the old input becomes the new output, and the old output becomes the new input. For this to be possible the original system must satisfy some additional requirements. A well-posed linear system (recall that we by this mean an L^p-well-posed linear system with $p < \infty$ or a *Reg*-well-posed linear system) has a well-posed flow-inverse if and only if the input/output map has a locally bounded inverse. In this case we call the system *flow-invertible* (in the well-posed sense). Also system and operator nodes can be flow-inverted under suitable algebraic assumptions described in Theorems 6.3.10 and 6.3.13. Under some mild conditions, compatibility and strong regularity are preserved in flow-inversion.[9] Weak regularity is not always preserved, but uniform regularity is.

Time-flow-inversion is studied in Section 6.5. It amounts to performing *both the preceding inversions at the same time*. If the original system is flow-invertible and the flow-inverted system is time-invertible, then we get the time-flow-inverted system by carrying out these two inversions in sequence. A similar statement is true if the original system is time-invertible and the time-inverted system is flow-invertible. However, a system *may be time-flow-invertible even if it is neither flow-invertible nor time-invertible*. The exact condition for time-flow-invertibility in the well-posed case is that the block operator matrix $\left[\begin{array}{c|c} \mathfrak{A}_0^t & \mathfrak{B}_0^t \\ \hline \mathfrak{C}_0^t & \mathfrak{D}_0^t \end{array}\right]$ introduced in Section 1.1 should have a bounded inverse for some (hence, for all) $t > 0$. For example, all conservative scattering systems (defined in Chapter 11) are time-flow-invertible. It is an interesting fact that the conditions for flow-invertibility, time-invertibility, and time-flow-invertibility are all independent of each other in the sense that any one of these conditions may hold for a given system but not the other two, or any two may hold but not the third (and there are systems where none of these or all of these hold).

Finally, in Section 6.6 we study *partial flow-inversion*. In ordinary flow-inversion we exchange the roles of the full input and the full output, but in partial flow-inversion we only interchange a part of the input with a part of the output, and keep the remaining parts of the input and output intact. This transformation is known under different names in different fields: people in H^∞ control theory call this a *chain scattering transformation*, and in the Russian tradition a particular case is known under the name *Potapov–Ginzburg transformation*. The technical difference between this transformation and the original flow-inversion is not very big, and it can be applied to a wider range of problems. In particular, the *output feedback* which we shall discuss in the next chapter can be regarded as a special case of partial-flow-inversion (and the converse is true, also).

Chapter 7 This chapter deals with *feedback*, which is one of the most central notions in control theory. The most basic version is *output feedback* discussed in Section 7.1. In output feedback the behavior of the system is modified by adding a term Ky to the input, where y is the output and K is a bounded linear

[9] At the moment there are no counter-examples known where strong regularity would not be preserved.

operator from the output space Y to the input space U. As we mentioned above, output feedback can be regarded as a special case of partial flow-inversion, which was discussed in Section 6.6, and it would be possible to prove all the results in Section 7.1 by appealing to the corresponding results in Section 6.6. However, since feedback is of such great importance in its own right, we give independent proofs of most of the central results (the proofs are slightly modified versions of those given in Section 6.6). In particular, an operator $K \in \mathcal{B}(Y; U)$ is called an *admissible feedback operator* for a well-posed linear system with input space U, output space Y, and input/output map $\mathfrak{D}$ if the operator $1 - K\mathfrak{D}$ has a locally bounded inverse (or, equivalently, $1 - \mathfrak{D}K$ has a locally bounded inverse); in this case the addition of K times the output to the input leads to another well-posed liner system, which we refer to as the *closed-loop system*. Some alternative feedback configurations which are essentially equivalent to the basic output feedback are presented in Section 7.2.

From this simple notion of output feedback it is possible to derive some more advanced versions by first adding an input or an output to the system, and then using the new input or output as a part of a feedback loop. The case where we add another output which we feed back into the original input is called *state feedback*, and the case where we add another input to which we feed back the original output is called *output injection*. Both of these schemes are discussed in Section 7.3.

Up to now we have in this chapter only dealt with the well-posed case. In Section 7.4 we first investigate how the different types of feedback described above affect the corresponding system nodes, and then we use the resulting formulas to define generalized feedback which can be applied also to non-well-posed systems induced by system nodes. This type of feedback is defined in terms of operations involving only the original system node, feedback operators, and extensions of the original system node corresponding to the addition of new inputs and outputs. To save some space we do not give independent proofs of most of the results of this section, but instead reduce the statements to the corresponding ones in Section 6.6.

In Section 7.5 we investigate to what extent compatibility and regularity are preserved under feedback (the results are analogous to those for flow-inversion). As shown in Section 7.6, output feedback commutes with the duality transformation (but state feedback becomes output injection under the duality transform, since the duality transform turns inputs into outputs and conversely). Some specific feedback examples are given in Section 7.7, with a special emphasis on the preservation of compatibility.

Chapter 8 So far we have not said much about the stability of a system (only well-posedness, which amounts to local boundedness). Chapter 8 is devoted to *stability* and various versions of *stabilizability*. In our interpretation, stability implies well-posedness, so here we only discuss well-posed systems.[10]

By the *stability* of a system we mean that *the maps from the initial state and the input function to the final state and the output* are not just locally bounded (which amounts to well-posedness), but that they are *globally bounded*. In other words, in the L^p-case, an arbitrary initial state x_0 and an arbitrary input function in $L^p([0,\infty);U)$ should result in a bounded trajectory $x(t)$, $t \geq 0$, and an output in $L^p([0,\infty);Y)$. The system is *weakly* or *strongly stable* if, in addition, the state $x(t)$ tends weakly or strongly to zero as $t \to \infty$.[11] As shown in Section 8.1, to some extent the stability of the system is reflected in its frequency domain behavior. In particular, the transfer function is defined in the full open right-half plane. *Exponential stability* means that the system has a negative growth rate.

A (possibly unstable) system is *stabilizable* if it is possible to make it stable through the use of some *state feedback*. It is *detectable* if it is possible to make it stable through the use of some *output injection*. (Thus, every stable system is both stabilizable and detectable.) When we add adjectives such as 'exponentially,' 'weakly,' or 'strongly' we mean that the resulting system has the indicated additional stability property. A particularly important case is the one where the system is both stabilizable and detectable, and each type of feedback stabilizes not only the original system, but the extended system which we get by adding the new input and the new output (thus, it is required that the state feedback also stabilizes the new input used for the output injection, and conversely). We refer to this situation by saying that the system is *jointly stabilizable and detectable*.

A very important fact is that the transfer function of every jointly stabilizable and detectable system has a *doubly coprime factorization*, and that this factorization can be computed directly from a jointly stabilizing and detecting state feedback and output injection pair. This is explained in Section 8.3, together with the basic definitions of coprimeness and coprime fractions. Both time domain and frequency domain versions are included. We interpret coprimeness throughout in the strongest possible sense, i.e., in order for two operators to be coprime we require that the corresponding Bezout identity has a solution.

In applications it can be very important that a particular input/output map (or its transfer function) has a doubly coprime factorization, but it is often irrelevant

[10] We regret the fact that we have not been able to include a treatment of the important case where the original system is non-well-posed, but can be made well-posed by appropriate feedback. The reason for this omission is simply the lack of space and time. Most of the necessary tools are found in Chapters 6 and 7.

[11] In the *Reg*-well-posed case we add the requirements that the input function and output function should also tend to zero at infinity. The same condition with the standard limit replaced by an essential limit is used in the L^∞ case as well.

how one arrives at this factorization. The existence of a doubly coprime factorization is a pure input/output property which can be stated without any reference to an underlying system. Moreover, it is easy to construct examples of systems which are not jointly stabilizable and detectable, but whose input/output map still has doubly coprime factorizations. We address this question in Section 8.4, where we introduce the notions of *coprime stabilizability and detectability*. We call a state feedback *right coprime stabilizing* if the closed-loop system corresponding to this feedback is stable and produces a right coprime factorization of the input/output map. Analogously, an output injection is *left coprime detecting* if the closed-loop system corresponding to this feedback is stable and produces a left coprime factorization of the input/output map.

The last theme in this chapter is the *dynamic stabilization* presented in Section 8.5. Here we show that every well-posed jointly stabilizable and detectable system can be stabilized by means of a *dynamic controller*, i.e., we show that there is another well-posed linear system (called the *controller*) such that the interconnection of these two systems produces a stable system. We also present the standard *Youla parametrization* of all stabilizing controllers.

Chapter 9 By a *realization* of a given time-invariant causal map $\mathfrak{D}$ we mean a (often well-posed) linear system whose input/output map is $\mathfrak{D}$. In this chapter we study the basic properties of these realizations. For simplicity we stick to the $L^p|Reg$-well-posed case. We begin by defining what we mean by a *minimal realization*: this is a realization which is both *controllable* and *observable*. Controllability means that the range of the input map (the map denoted by $\mathfrak{B}$ above) is dense in the state space, and observability means that the output map (the map denoted by $\mathfrak{C}$ above) is injective. As shown in Section 9.2, any two $L^p|Reg$-well-posed realizations of the same input/output map are *pseudo-similar* to each other. This means roughly that there is a closed linear operator whose domain is a dense subspace of one of the two state spaces, its range is a dense subspace of the other state space, it is injective, and it intertwines the corresponding operators of the two systems. Such a pseudo-similarity is not unique, but there is one which is maximal and another which is minimal (in the sense of graph inclusions). There are many properties which are *not* preserved by a pseudo-similarity, such as the spectrum of the main operator, but pseudo-similarities are still quite useful in certain situations (for example, in Section 9.5 and Chapter 11).

In Section 9.3 we show how to construct a realization of a given input/output map from a factorization of its Hankel operator.

The notions of controllability and observability that we have defined above are often referred to as *approximate* controllability or observability. Some other notions of controllability and observability (such as *exact*, or *null in finite time*, or *exact in infinite time*, or *final state observable*) are presented in Section 9.4, and the relationships between these different notions are explained. In particular, it is shown that every controllable L^p-well-posed linear system with $p < \infty$ whose input map $\mathfrak{B}$ and output map $\mathfrak{C}$ are (globally) bounded can be turned into a system which is exactly controllable in infinite time by replacing the original state space by a subspace with a stronger norm. If it is instead observable, then it can be turned into a system which is exactly observable in infinite time by completing the original state space with respect to a norm which is weaker than the original one. Of course, if it is minimal, then both of these statements apply.

Input normalized, output normalized, and balanced realizations are presented in Section 9.5. A minimal realization is *input normalized* if the input map $\mathfrak{B}$ becomes an isometry after its null space has been factored out. It is *output normalized* if the output map $\mathfrak{C}$ is an isometry. These definitions apply to the general L^p-well-posed case in a Banach space setting. In the Hilbert space setting with $p = 2$ a minimal system is input normalized if its *controllability gramian* $\mathfrak{B}\mathfrak{B}^*$ is the identity operator, and it is output normalized if its *observability gramian* $\mathfrak{C}^*\mathfrak{C}$ is the identity operator. We construct a (Hankel) *balanced realization* by interpolating half-way between these two extreme cases (in the Hilbert space case with $p = 2$ and a bounded input/output map). This realization is characterized by the fact that its controllability and observability gramians coincide. All of these realizations (input normalized, output normalized, or balanced) are unique up to a unitary similarity transformation in the state space. The balanced realization is always strongly stable together with its dual.

A number of methods to test the controllability or observability of a system in frequency domain terms are given in Section 9.6, and some further time domain test are given in Section 9.10. In Section 9.7 we discuss *modal controllability and observability*, i.e., we investigate to what extent it is possible to control or observe different parts of the spectrum of the main operator (the semigroup generator). We say a few words about *spectral minimality* in Section 9.8. This is the question about to what extent it is possible to construct a realization with a main operator whose spectrum essentially coincides with the points of singularities of the transfer function. A complete answer to this question is not known at this moment (and it may never be).

Some comments on to what extent controllability and observability are preserved under various transformations of the system (including feedback and duality) are given in Sections 9.9 and 9.10.

Chapter 10 In Chapter 4 we saw that every $L^p|Reg$-well-posed linear system has a control operator B mapping the input space U into the extrapolation space X_{-1}, and also an observation operator C mapping the domain X_1 of the semigroup generator into the output space. Here we shall study the converse question: given a semigroup generator A and an operator B or C of the type described above, when can B or C be interpreted as the control operator, respectively, observation operator of an $L^p|Reg$-well-posed linear system whose main operator is A? We call B or C *admissible* whenever the answer to this question is positive. It is called *stable* if, in addition, the corresponding input or output map is bounded. The input map $\mathfrak{B}$ of the system is determined uniquely by A and B, and the output map $\mathfrak{C}$ is determined uniquely by A and C. Note that in this formulation there is no coupling between B and C, i.e., they need not be the control and observation operators of the *same* $L^p|Reg$-well-posed linear system. If they are, then we call them *jointly admissible*. In this case they do not only determine (together with A) the input map $\mathfrak{B}$ and the output map $\mathfrak{C}$ of the system uniquely, but also the input/output map $\mathfrak{D}$, up to an arbitrary static constant (i.e., an undetermined feedthrough term in $\mathcal{B}(U;Y)$).

After some preliminaries presented in Section 10.1 we proceed to show in Section 10.2 that the two questions about the admissibility of a control, respectively, observation operator are dual to each other.

In Sections 10.3–10.7 we restrict our focus to the L^2-well-posed Hilbert space case. We begin by showing in Section 10.3 that there is a one-to-one correspondence between the space $TIC(U;Y)$ of all time-invariant causal continuous operators mapping $L^2(\mathbb{R}^+;U)$ into $L^2(\mathbb{R}^+;Y)$ and the space of frequency domain multiplication operators with a symbol in $H^\infty(U;Y)$ (the space of all $\mathcal{B}(U;Y)$-valued bounded analytic function on the open right-half plane $\mathbb{C}^+$). The correspondence between the time and frequency domain operators is the same as before, i.e, the frequency domain multiplier is the transfer function of the time-domain operator. However, in the new setting we can from the boundedness of the transfer function conclude that the corresponding time-domain operator is bounded as well (which is not true in the general $L^p|Reg$-well-posed Banach space case). Here we also state and prove the well-known fact that, in the L^2-well-posed Hilbert space case, the Laplace transform maps $L^2(\mathbb{R}^+;U)$ one-to-one onto $H^2(\mathbb{C}^+;U)$. A related result is that an output map $\mathfrak{C}$ is bounded from X to $L^2(\mathbb{R}^+;Y)$ if and only if the function $\lambda \mapsto C(\lambda - A)^{-1}x$ belongs to $H^2(\mathbb{C}^+;Y)$ for each $x \in X$. The analogous dual result for input maps is also valid.

An input map $\mathfrak{B}$ is bounded on $L^2(\mathbb{R}^-, U)$ if and only if the *controllability gramian* $\mathfrak{B}\mathfrak{B}^*$ is a bounded operator on X, and an output map $\mathfrak{C}$ is bounded from X into $L^2(\mathbb{R}^+, Y)$ if and only if the *observability gramian* $\mathfrak{C}^*\mathfrak{C}$ is a bounded operator on X (here we require U, X, and Y to be Hilbert spaces, and we identify them with their own duals). These two operators on the state space X can be characterized in several different ways, as shown in Section 10.4. Among others, they are the solutions of the *controllability*, respectively, *observability*

Lyapunov equations. This gives us an alternative way of testing the admissibility and stability of control and observation operators.

There are some special admissibility and stability tests that can be applied in the Hilbert space case with one-dimensional input and output cases. These tests require the semigroup of the system to be a contraction semigroup (after a possible exponential rescaling). To present these results we need the notion of a *Carleson measure* over the closed right half-plane $\overline{\mathbb{C}}^+$. The most basic results about such measures are presented in Section 10.5. By using these results we are able to give necessary and sufficient conditions for the admissibility and stability of control and observation operators with scalar input, respectively, output spaces in the case where the system semigroup is diagonal or normal (see Section 10.6; by a diagonal semigroup we mean a semigroup whose generator is normal and has a complete set of eigenvectors). The same result can be extended to the more general case of a contraction semigroup, and this is done in Section 10.7.

Finally, in Section 10.8 we return to the general Banach space L^p-well-posed case with $p < \infty$ and give some necessary and sufficient conditions for the admissibility and stability of a control or observation operator in terms of conditions on the corresponding Lax–Phillips semigroup introduced in Section 2.7.

Chapter 11 In this chapter we study passive and conservative systems in a scattering setting, and we assume throughout that the input space U, the state space X, and the output space Y are Hilbert spaces. Intuitively, passivity means that the system has no internal energy sources, and conservativity means that neither the system itself nor the dual system has any energy sources or sinks.

We begin by presenting the basic results for passive systems in Section 11.1. A system node S is (scattering) *passive* if the trajectories of this system node satisfy the energy inequality (here u is the input, x the state, and y the output)

$$|x(t)|_X^2 + \int_0^t |y(s)|_Y^2\,ds \le |x_0|_X^2 + \int_0^t |u(s)|_U^2\,ds, \quad t \ge 0. \tag{1.2.4}$$

This inequality is *stronger than the corresponding well-posedness inequality* (where the right-hand side is multiplied by a finite positive constant $M(t)$), so *every (scattering) passive system is L^2-well-posed.* The same inequality implies that the system is *stable* (in the sense that we attached to this word in Chapter 4). It is even true that the semigroup $\mathfrak{A}$ is a contraction semigroup, and that the other system operators $\mathfrak{B}$, $\mathfrak{C}$, and $\mathfrak{D}$ are contractions. A system is passive if and only if the dual system is passive, and this is true if and only if the L^2-version of the corresponding Lax–Phillips semigroup is a contraction semigroup. It is easy to formulate conditions directly on a system node in order for the corresponding system to be passive; see Theorem 11.1.5. To each passive system there corresponds a number of *deficiency operators*, which measure how much $\mathfrak{A}$, $\mathfrak{B}$, $\mathfrak{C}$, and $\mathfrak{D}$ differ from isometries or co-isometries. Many properties of the system can be expressed in terms of these deficiency operators.

We end Section 11.1 with a decomposition of an arbitrary passive system into three independent subsystems. The semigroup of the first subsystem is unitary, and there is no interaction between the state space and the surrounding world (the control and observation operators are zero, and this part of the state space is both uncontrollable and unobservable). We shall refer to this subsystem as the *invisible unitary part*. The second subsystem is *static* and *unitary*, i.e., it has no state space (meaning that the dimension of its state space is zero), and it is represented by a plain unitary feedthrough operator from a part of the input space to a part of the output space. All the nontrivial interaction between the state space and the surrounding world takes place in the remaining third subsystem. This part is *completely nonunitary* in the sense that its semigroup does not have any reducing subspace on which it is unitary, and its transfer function is *purely contractive* (it has no eigenvalues with absolute value one). More generally, using the terminology introduced above, a system is called *purely passive* if it has no static unitary part, and it is called completely nonunitary if it has no invisible unitary part.[12]

Energy preserving and *conservative systems* are presented in Section 11.2. A system is energy preserving if (1.2.4) holds in the form of an equality instead of an inequality, and it is conservative if both the system itself and the dual system are energy preserving. Equivalently, a system is energy preserving if and only if the L^2-version of the corresponding Lax–Phillips semigroup is isometric, and it is conservative if and only if the Lax–Phillips semigroup is unitary. Various direct conditions on a given system node to generate an energy preserving system are also presented.

In an energy preserving system no energy is lost, but it may be first transferred from the input to the state, and then 'trapped' in the state space forever, so that it can no longer be retrieved from the outside. Thus, from the point of view of an external observer, a conservative system may be 'lossy.' To specifically exclude this case we introduce the notion of losslessness in Section 11.3. A system is *semi-lossless* if its input/output map is an isometry (as a map from $L^2(\mathbb{R}^+; U)$ into $L^2(\mathbb{R}^+; Y)$), and it is *lossless* if the input/output map is unitary (thus, both the original system and its dual are semi-lossless). Equivalently, a system is semi-lossless if its transfer function is inner (from the left), and it is lossless if the transfer function is bi-inner (inner from both sides). We show that an energy preserving system is lossless if and only if the restriction of its semigroup to the reachable subspace is strongly stable, and that a completely nonunitary

[12] In the conservative case a completely nonunitary system is often called *simple*.

conservative system is lossless if and only if its semigroup is strongly stable together with its adjoint.

In Section 11.4 we first define what we mean by an orthogonal *dilation* or *compression* of a semigroup: the semigroup $\widetilde{\mathfrak{A}}$ acting on $\widetilde{X}$ is an orthogonal dilation of $\mathfrak{A}$ acting on X, or equivalently, $\mathfrak{A}$ is an orthogonal compression of $\widetilde{\mathfrak{A}}$, if X is a closed subspace of $\widetilde{X}$ and $\mathfrak{A}^t = \pi_X \widetilde{\mathfrak{A}}^t_{|X}$, $t \geq 0$, where π_X is the orthogonal projection of $\widetilde{X}$ onto X. We then prove that every contraction semigroup can be dilated into an isometric semigroup. This dilation is unique up to a unitary similarity transformation in the state space if we require it to be orthogonal and *minimal* in the sense that the orbits of (forward) trajectories starting in X are dense in the larger state space $\widetilde{X}$. This isometric semigroup can in turn be dilated into a unitary semigroup, which is also unique up to a unitary similarity transformation in the state space if we require it to be minimal (here we define minimality by using both forward and backward trajectories). Combining the two transformations we get an orthogonal unitary dilation of the original contraction semigroup, which is unique up to unitary similarity. All of this is well known, but it is less well known that these dilations have natural interpretations as well-posed linear systems: to get an isometric dilation we simply add a suitable output to the original semigroup and pass to the corresponding Lax–Phillips semigroup (with no input). To get the unitary dilation we further add a suitable input. This dilation theorem can be re-interpreted in the following way. Every contraction semigroup gives rise to a conservative system (whose semigroup is the given one). Without loss of generality, we may take this system to be purely passive, and with this extra requirement the resulting system is unique up to unitary similarity transformations in the input and output spaces. The (purely contractive) transfer function of this conservative system is usually called the *characteristic function* of its main operator. Some of the results of this section are expanded to dilations and compressions of *systems* in Section 11.5.

In Section 11.6 we proceed to develop a *universal model* for a contraction semigroup (on a Hilbert space), i.e., we show that every contraction semigroup is unitarily similar to a compression of a bilateral shift defined on a suitable L^2-space. At the same time we get a universal model for an arbitrary completely nonunitary conservative system (it is unitarily similar to a particular system whose semigroup is the compression of a bilateral shift). It is actually possible to get two different unitary similarities by starting either from the original system or from its dual. These models become especially simple in the case where the system is semi-lossless (or lossless).

The transfer function of every passive system (hence of every conservative system) is a contraction on the open right half-plane $\mathbb{C}^+$. In Section 11.7 we prove the converse: every contractive analytic operator-valued function (between two Hilbert spaces) on $\mathbb{C}^+$ is the transfer function of a conservative system, which we, without loss of generality, may take to be completely nonunitary. This completely nonunitary conservative system is unique up to a unitary similarity transformation in the state space.

Controllable energy preserving and *minimal passive realizations* of a given contractive analytic operator-valued function are studied in Section 11.8. A controllable energy preserving realization is unique up to unitary similarity. Among all minimal passive realizations there are two extreme ones, one whose norm in the state space is the weakest possible one (the *optimal realization* whose norm is called the *available storage*), and another whose norm in the state space is the strongest possible one (the *$*$-optimal realization* whose norm is called the *required supply*). Both of these are determined uniquely by the transfer function up to unitary similarity. By interpolating half-way between these two extreme cases we get a *balanced passive realization* (sometimes also called the *balanced bounded real realization*), which is also unique up to unitary similarity.

Finally, in Section 11.9 we say a few words about the spectrum of a conservative system, relating it to the invertibility of the transfer function.

Chapter 12 The main part of this chapter is a short overview of the theory of discrete time systems of the type

$$\begin{aligned} x_{k+1} &= \mathbf{A}x_k + \mathbf{B}u_k, \\ y_k &= \mathbf{C}x_k + \mathbf{D}u_k, \qquad k \in \mathbb{Z}^+ = \{0, 1, 2, \ldots\}, \end{aligned} \tag{1.2.5}$$

where $\mathbf{A} \in \mathcal{B}(X)$, $\mathbf{B} \in \mathcal{B}(U; X)$, $\mathbf{C} \in \mathcal{B}(X; Y)$, $\mathbf{D} \in \mathcal{B}(U; Y)$ and U, X, and Y are Banach spaces. Here $\mathbf{A}$ is the *main operator*, $\mathbf{B}$ is the *control operator*, $\mathbf{C}$ is the *observation operator*, and $\mathbf{D}$ is the *feedthrough operator*. The discrete time semigroup $\mathbb{A}$, input map $\mathbb{B}$, output map $\mathbb{C}$, and input/output map $\mathbb{D}$ are given by

$$\begin{aligned} (\mathbb{A}x)_k &= \mathbf{A}^k x, && k \in \mathbb{Z}^+, \\ \mathbb{B}\mathbf{u} &= \sum_{k=0}^{\infty} \mathbf{A}^k \mathbf{B} u_{-k-1}, && \\ (\mathbb{C}x)_k &= \mathbf{C}\mathbf{A}^k x, && k \in \mathbb{Z}^+, \\ (\mathbb{D}\mathbf{u})_k &= \sum_{i=0}^{\infty} \mathbf{C}\mathbf{A}^i \mathbf{B} u_{k-i-1} + \mathbf{D}u_k, && k \in \mathbb{Z} = \{0, \pm 1, \pm 2, \ldots\}, \end{aligned}$$

where $\mathbf{u} = \{u_k\}_{k\in\mathbb{Z}}$ represents a U-valued sequence with finite support and $x \in X$.

The *local* discrete time theory is much simpler than the corresponding continuous time theory due to the boundedness of the generators $\mathbf{A}$, $\mathbf{B}$, $\mathbf{C}$, and $\mathbf{D}$. However, this simplicity disappears when we look at the *global* behavior of solutions (over the full half-axis $\mathbb{Z}^+$). There is actually a very close analogy between the global discrete time and continuous time theories.

Section 12.1 contains a presentation of the basic discrete time setting, including the definition of the Z-transform (which is the discrete time analogue of the Laplace transform). We use the 'engineering' version of this transform, meaning that the Z-transform of the sequence $\{u_k\}_{k\in\mathbb{Z}}$ is $\sum_{k\in Z} z^{-k}u_k$.[13] The advantage of this version of the Z-transform is that the formula for the transfer function becomes the same as in continuous time, namely

$$\widehat{\mathbb{D}}(z) = \mathbf{C}(z - \mathbf{A})^{-1}\mathbf{B} + \mathbf{D}, \qquad z \in \rho(\mathbf{A}).$$

Clearly, this transfer function is analytic in a neighborhood of infinity, and the feedthrough operator $\mathbf{D}$ is the value of $\widehat{\mathbb{D}}$ at infinity. The section ends with a short description of how our earlier continuous time results can be translated into discrete time (with essentially the same proofs or simplified versions of the earlier proofs). This includes notions such as duality, flow-inversion, time-inversion, time-flow-inversion, feedback, stabilization and detection, ℓ^p-stability, controllability, observability, admissibility, passivity, and conservation of energy.

In Section 12.2 we study the frequency domain *linear fractional transformations* of a continuous time system, interpreting the transformed system as a discrete time system. Here the discrete time main operator $\mathbf{A}$ is a linear fractional transformation of the continuous time main operator A, and the discrete time transfer function $\widehat{\mathbb{D}}$ can be obtained from the continuous time transfer function $\widehat{\mathfrak{D}}$ by applying the same linear transformation to the argument of $\widehat{\mathfrak{D}}$. This linear fractional transformation preserves the reachable and observable subspaces whenever the image of infinity lies in the unbounded component of the resolvent set of the discrete time generator (but it need not preserve these subspaces if the above condition is violated). We also point out that if we interpret a discrete time system (with bounded generating operators) as a continuous time system (with the same generating operators), then the reachable and observable subspaces are the same in both interpretations.

In Section 12.3 we specialize to the ℓ^2-bounded Hilbert space case, where the main operator is a contraction which does not have -1 as an eigenvalue. We show that this class of systems can be mapped one-to-one onto the class of all L^2-stable continuous time (well-posed) linear systems whose semigroup is a contraction semigroup. The mapping between these two classes even preserves the norms of most of the involved (integral level) operators. We shall refer to the time-domain version of this map as the *Laguerre transform*, and to the frequency domain version as the *Cayley transform*. The discrete time main

[13] In the 'mathematical version' of this transform one replaces z^{-k} by z^k.

operator is called the *co-generator* of the continuous time semigroup. The Laguerre transform does not preserve *local* properties, i.e., if we know a finite part of a continuous time trajectory, then this does not tell us much about a finite part of the corresponding discrete time trajectory. However, most *global* properties are preserved, such as stability and strong stability, invariant subspaces of the main operator, observability and controllability (and more generally, the reachable and unobservable subspaces), passivity, energy preservation, conservation of energy, deficiency operators, and so on.

Section 12.4 is again devoted to continuous time systems. It uses the theory of Section 12.3 to develop the continuous time *reciprocal transform* (the same transform has a discrete time interpretation too: there it stands for *time inversion*). This transformation corresponds to the linear fractional transformation $z \mapsto 1/z$ in the complex plane, and it requires the main operator A to be injective (so that A^{-1} exists). Like the Laguerre transform the reciprocal transform does not preserve local properties, but it preserves most global properties, at least if the original semigroup is a contraction semigroup (e.g., all the properties listed at the end of the preceding paragraph are preserved).

Appendix Section A.1 describes the most important properties of the class of regulated functions, which we use as a substitute for L^∞ in many places. In Section A.2 we develop the polar decomposition of a closed linear operator between two Hilbert spaces, and show that every positive (possibly unbounded) operator on a Hilbert space has a unique positive square root. Section A.3 lists a number of basic results about convolutions, and in Section A.4 we study the inverses of block matrices (where each entry is a bounded operator).

2

Basic properties of well-posed linear systems

In this chapter we describe the basic properties of well-posed linear systems. We work directly with the operators that map the initial state and the input function into the final state and the output function, and describe their algebraic properties. Different continuity assumptions give different types of well-posed systems. In particular, we show that a well-posed linear system may be interpreted as a strongly continuous semigroup in a suitable state space. The alternative description of a well-posed linear system by means of a differential system will be given later in Chapter 4.

2.1 Motivation

To motivate the notion of a well-posed linear system that will be introduced in the next section we first take a closer look at the traditional state space system

$$\begin{aligned} \dot{x}(t) &= Ax(t) + Bu(t), \\ y(t) &= Cx(t) + Du(t), \qquad t \geq s, \\ x(s) &= x_s. \end{aligned} \tag{2.1.1}$$

Here s is a specified initial time, often taken to be zero. We shall occasionally use diagrams of the type drawn in Figure 2.1 to describe these systems. The operator A is supposed to generate a strongly continuous semigroup $\mathfrak{A}^t$ on a Banach space X, *the state space*. For the moment we assume that *the control operator* B and *the observation operator* C are bounded linear operators, i.e., $B \in \mathcal{B}(U; X)$ and $C \in \mathcal{B}(X; Y)$, where U and Y are two more Banach spaces, *the input and output spaces*. *The feedthrough operator* D is also bounded, i.e., $D \in \mathcal{B}(U; Y)$. We call u *the input function* (or control), x *the state trajectory*, and y *the output function* (or observation) of this system. The state trajectory

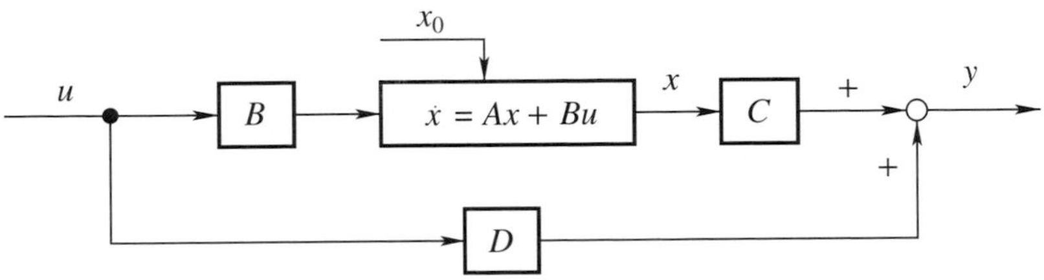

Figure 2.1 Regular well-posed linear system

x is required to be a strong solution of (2.1.1), i.e., the state x and output y are given by

$$\begin{aligned} x(t) &= \mathfrak{A}^{t-s}x_s + \int_s^t \mathfrak{A}^{t-v}Bu(v)\,dv, && t \geq s, \\ y(t) &= C\mathfrak{A}^{t-s}x_s + C\int_s^t \mathfrak{A}^{t-v}Bu(v)\,dv + Du(t), && t \geq s. \end{aligned} \tag{2.1.2}$$

The preceding formulas define the state trajectory $x(t)$ and the output function $y(t)$ at time $t \geq s$ in terms of the given initial state x_s and the input function $u(v)$, $v \geq s$. By separating the contributions of x_s and u to $x(t)$ and y from each other we get a total of four different maps,

$$\begin{aligned} x(t) &= \mathfrak{A}^{t-s}x_s + \mathfrak{B}_s^t u, && t \geq s, \\ y &= \mathfrak{C}_s x_s + \mathfrak{D}_s u, \end{aligned} \tag{2.1.3}$$

where $\mathfrak{B}_s^t$ (the mapping from the input to the final state), $\mathfrak{C}_s$ (the mapping from the initial state to the output function), and $\mathfrak{D}_s$ (the mapping from the input to the output) are given by

$$\begin{aligned} \mathfrak{B}_s^t u &:= \int_s^t \mathfrak{A}^{t-v}Bu(v)\,dv, && t \geq s, \\ (\mathfrak{C}_s x_s)(t) &:= C\mathfrak{A}^{t-s}x_s, && t \geq s, \\ (\mathfrak{D}_s u)(t) &:= C\int_s^t \mathfrak{A}^{t-v}Bu(v)\,dv + Du(t), && t \geq s. \end{aligned} \tag{2.1.4}$$

So far we have made only marginal use of *the time-invariance* of the system (2.1.1), i.e., of the fact that none of the operators A, B, C and D in (2.1.1) depend on t. The important variable in (2.1.3) and (2.1.4) is *the time difference* $t - s$, not s or t separately, and by a simple time shift we can make either the starting time s or the final time t equal to zero. By using these facts we can express the operators $\mathfrak{B}_s^t$, $\mathfrak{C}_s$ and $\mathfrak{D}_s$ in terms of three 'master' operators

$\mathfrak{B}$, $\mathfrak{C}$, and $\mathfrak{D}$ which do not depend on t or s. These master operators are defined by[1]

$$\begin{aligned}
\mathfrak{B}u &:= \int_{-\infty}^{0} \mathfrak{A}^{-s} Bu(s)\, ds, \\
\mathfrak{C}x &:= \big(t \mapsto C\mathfrak{A}^{t}x, \qquad t \ge 0\big), \\
\mathfrak{D}u &:= \left(t \mapsto \int_{-\infty}^{t} C\mathfrak{A}^{t-s} Bu(s)\, ds + Du(t), \qquad t \in \mathbb{R}\right).
\end{aligned} \tag{2.1.5}$$

Thus, $\mathfrak{B}$ is the mapping from an input u defined on $\mathbb{R}^- = (-\infty, 0)$ to the final state $x(0) \in X$ at time zero (take $s = -\infty$, $x_s = 0$, and $t = 0$, and let u have a finite support so that the integral converges), $\mathfrak{C}$ is the mapping from the initial state $x_0 \in X$ at time zero to the output y defined on $\overline{\mathbb{R}}^+ = [0, \infty)$ (take $s = 0$ and $u = 0$), and $\mathfrak{D}$ is the mapping from the input u defined on $\mathbb{R}$ to the output y, also defined on $\mathbb{R}$ (suppose that the support of u is bounded to the left so that the integral converges). For obvious reasons we call $\mathfrak{B}$ *the input map* (with final time zero), $\mathfrak{C}$ *the output map* (with initial time zero), and $\mathfrak{D}$ *the input/output map* of (2.1.1).

In order to rewrite the operators $\mathfrak{B}_s^t$, $\mathfrak{C}_s$ and $\mathfrak{D}_s$ in terms of $\mathfrak{B}$, $\mathfrak{C}$, and $\mathfrak{D}$ we need two more auxiliary operators. For each $J \subset \mathbb{R}$ we define the projection operator π_J by

$$(\pi_J u)(s) := \begin{cases} u(s), & s \in J, \\ 0, & s \notin J, \end{cases} \tag{2.1.6}$$

and for each $t \in \mathbb{R}$ we define the *time shift operator* τ^t by

$$(\tau^t u)(s) := u(t+s), \qquad s \in \mathbb{R}. \tag{2.1.7}$$

Then

$$\begin{aligned}
\mathfrak{B}_s^t &= \int_s^t \mathfrak{A}^{t-v} Bu(v)\, dv = \int_{-\infty}^{t} \mathfrak{A}^{t-v} B\pi_{[s,t)}u(v)\, dv \\
&= \int_{-\infty}^{0} \mathfrak{A}^{-v} B\pi_{[s,t)}u(v+t)\, dv = \int_{-\infty}^{0} \mathfrak{A}^{-v} B\tau^t\pi_{[s,t)}u(v)\, dv \\
&= \mathfrak{B}\tau^t\pi_{[s,t)}u,
\end{aligned}$$

and

$$(\mathfrak{C}_s x_s)(t) = C\mathfrak{A}^{t-s}x_s = (\mathfrak{C}x_s)(t-s) = (\tau^{-s}\mathfrak{C}x_s)(t).$$

[1] For the moment these definitions are only formal. We shall describe the exact domains and ranges of these operators later.

A similar computation can be carried out for $\mathfrak{D}_s$, and we find that, for all $t \geq s$,

$$\begin{aligned} \mathfrak{B}_s^t &= \mathfrak{B}\tau^t \pi_{[s,t)} u, \\ (\mathfrak{C}_s x_s)(t) &= (\tau^{-s} \mathfrak{C} x_s)(t), \\ (\mathfrak{D}_s u)(t) &= (\mathfrak{D}\pi_{[s,\infty)} u)(t). \end{aligned} \tag{2.1.8}$$

Thus x and y can be written in the form

$$\begin{aligned} x(t) &= \mathfrak{A}^{t-s} x_s + \mathfrak{B}\tau^t \pi_{[s,t)} u, \qquad t \geq s, \\ y &= \tau^{-s} \mathfrak{C} x_s + \mathfrak{D}\pi_{[s,\infty)} u, \end{aligned} \tag{2.1.9}$$

or alternatively (use the fact that $\mathfrak{B} = \mathfrak{B}\pi_{(-\infty,0)}$),

$$\begin{aligned} x(t) &= \mathfrak{A}^{t-s} x_s + \mathfrak{B}\tau^t \pi_{[s,\infty)} u, \qquad t \geq s, \\ y &= \tau^{-s} \mathfrak{C} x_s + \mathfrak{D}\pi_{[s,\infty)} u. \end{aligned} \tag{2.1.10}$$

Let us emphasize the following fact that we have just established:

Statement 2.1.1 *In order to know the state trajectory x and the output function y of the system* (2.1.1) *with arbitrary initial time s, arbitrary initial state x_s, and arbitrary input function u, it suffices to know the semigroup $\mathfrak{A}$, the input map $\mathfrak{B}$, the output map $\mathfrak{C}$, and the input/output map $\mathfrak{D}$. When these are known the state and the output at an arbitrary time $t \geq s$ can be recovered from* (2.1.9) *or* (2.1.10).

If the operators B, C, and D in (2.1.1) are bounded, then it is possible to avoid the operators $\mathfrak{B}$, $\mathfrak{C}$, and $\mathfrak{D}$; we can simply work with the variation of constants formula (2.1.2) all the time. However, we are primarily interested in the case where not only A, but also B and C are allowed to be unbounded, and D is not necessarily well-defined. In this case formulas (2.1.9) have one great advantage over (2.1.2): under appropriate assumptions all the operators in (2.1.9) will be *bounded*[2]linear operators.

Of course, the word 'bounded' always refers to some topology on the spaces of input and output signals and on the state space, and it is not at all obvious which are the 'best' topologies: too weak assumptions lead to mathematical difficulties, and too strong assumptions lead to a limited applicability. One choice that works well in many cases is to impose the following requirements:

Well-posedness requirement 2.1.2

(i) *The input space U, the state space X, and the output space Y are Banach spaces.*
(ii) *The input u belongs locally to $L^p([s,\infty;U)$, for some p, $1 \leq p \leq \infty$.*

[2] Locally in time.

(iii) *The state $x(t) \in X$ is well-defined at each time instance $t \geq s$ (where $s \in \mathbb{R}$ is the initial time).*
(iv) *The output y belongs locally to $L^p([s,\infty);Y)$ for the same value of p as in (ii).*
(v) *The state $x(t) \in X$ and the output $y \in L^p_{loc}([s,\infty);Y)$ depend continuously on the initial state $x_0 \in X$, on the input $u \in L^p_{loc}([s,\infty);U)$, and (in the case of $x(t)$) on the time parameter $t \geq 0$.*

This is the choice that we make throughout most of this book, and it leads to a well-posedness notion of L^p-type. The most important case is $p = 2$, and the second most important cases are $p = 1$ and $p = \infty$. The main reason for taking the same value of p in (ii) and (iv) is that we want to be able to study (static) *feedback* connections where a part of the output y is fed back into the input u.

Up to now the discussion has focused on the following two questions: how to construct the solution operators $\mathfrak{B}$, $\mathfrak{C}$, and $\mathfrak{D}$ in Statement 2.1.1 from a given set of generators (A, B, C, D) (in the case where A generates a semigroup and B, C, and D are bounded), and what type of continuity requirements are appropriate. There is a third question which is at least as interesting: when does a given quadruple $(\mathfrak{A}, \mathfrak{B}, \mathfrak{C}, \mathfrak{D})$ arise from a system of the type (2.1.1), and how can we compute the characterizing operators (A, B, C, D) of this system from $(\mathfrak{A}, \mathfrak{B}, \mathfrak{C}, \mathfrak{D})$?

There are certain algebraic conditions that are necessary in order for the operators $(\mathfrak{A}, \mathfrak{B}, \mathfrak{C}, \mathfrak{D})$ to have representations of the type given in (2.1.5) and in order for (2.1.9) to be valid for all permitted inputs u. To formulate these algebraic conditions we need the projection operators π_+ and π_- and the time-shift operators τ^t_+ and τ^t_- on the positive, respectively, negative half-line, defined by

$$\pi_+ := \pi_{[0,\infty)}, \qquad \pi_- := \pi_{(-\infty,0)}, \qquad \tau^t_+ := \pi_+\tau^t, \qquad \tau^t_- := \tau^t\pi_-. \tag{2.1.11}$$

Algebraic conditions 2.1.3 *The operators $\mathfrak{A}$, $\mathfrak{B}$, $\mathfrak{C}$, and $\mathfrak{D}$ satisfy the following conditions:*[3]

(i) *$\mathfrak{A}^0 = 1$ and $\mathfrak{A}^{s+t} = \mathfrak{A}^s\mathfrak{A}^t$ for all $s, t \geq 0$ (i.e, $\mathfrak{A}$ is a semigroup),*
(ii) *$\mathfrak{A}^t\mathfrak{B}u = \mathfrak{B}\tau^t_-u$ for all $t \geq 0$,*
(iii) *$\mathfrak{C}\mathfrak{A}^tx = \tau^t_+\mathfrak{C}x$ for all $t \geq 0$,*
(iv) *$\tau^t\mathfrak{D}u = \mathfrak{D}\tau^tu$, $\pi_-\mathfrak{D}\pi_+u = 0$, and $\pi_+\mathfrak{D}\pi_-u = \mathfrak{C}\mathfrak{B}u$ for all $t \in \mathbb{R}$.*

[3] Technically, condition (ii) says that the input map $\mathfrak{B}$ 'intertwines' the semigroup $\mathfrak{A}^t$ on X and the left-shift semigroup τ^t_- on $L^p_c(\mathbb{R}^-;U)$, and condition (iii) says that the output map $\mathfrak{B}$ intertwines the semigroup $\mathfrak{A}^t$ on X and the left-shift semigroup τ^t_+ on $L^p_{loc}(\mathbb{R}^+;Y)$.

The first of the conditions above says that, in the absence of an input, the state trajectory x behaves like the state of a time-invariant linear dynamical system.

To derive the second condition (ii) we compute

$$\begin{aligned}\mathfrak{B}\tau^t\pi_-u &= \int_{-\infty}^{-t} \mathfrak{A}^{-s}Bu(s+t)\,ds = \int_{-\infty}^{0} \mathfrak{A}^{t-v}Bu(v)\,dv \\ &= \mathfrak{A}^t \int_{-\infty}^{0} \mathfrak{A}^{-v}Bu(v)\,dv = \mathfrak{A}^t\mathfrak{B}u.\end{aligned}$$

This formula, too, can be regarded a consequence of the time-invariance of the system. Suppose that the input u vanishes on $\overline{\mathbb{R}}^+$. Then the state at time zero is given by $x(0) = \mathfrak{B}u$, and, for each $t > 0$, the state at time t is given by $x(t) = \mathfrak{A}^t x(0) = \mathfrak{A}^t\mathfrak{B}u$. Formula (ii) says that this is equal to the state at time zero corresponding to an input that otherwise looks identical to the original one, but it has been advanced by t time units (in particular, it vanishes on $(-t, \infty)$).

Formula (iii) is derived in a similar way, and it has a similar interpretation: if we restrict the output y corresponding to a given initial state x_0 to the time interval $[t, \infty)$ and then shift it back to $[0, \infty)$, then the resulting output is identical to the one that we get by first letting the system develop freely for t time units to get the state $x(t) = \mathfrak{A}^t x_0$, and then observing the output $\mathfrak{C}x(t)$ of the system with this initial state and initial time zero.

This leaves condition (iv) to be accounted for. The first condition $\tau^t\mathfrak{D}u = \mathfrak{D}\tau^t u$ in (iv) is another consequence of time-invariance: if the input u is delayed or advanced by t time units, then the output is also delayed or advanced by the same amount, but it does not change in any other way.

The condition $\pi_-\mathfrak{D}\pi_+u = 0$ in (iv) is a *causality requirement*: future inputs are not allowed to have any effect on past outputs.

Maybe the least obvious part of (iv) is the condition $\pi_+\mathfrak{D}\pi_-u = \mathfrak{C}\mathfrak{B}u$, but it is easy to verify: for all $t > 0$ we have

$$(\mathfrak{D}\pi_-u)(t) = C\int_{-\infty}^{0} \mathfrak{A}^{t-s}Bu(s)\,ds = C\mathfrak{A}^t \int_{-\infty}^{0} \mathfrak{A}^{-s}Bu(s)\,ds = (\mathfrak{C}\mathfrak{B}u)(t).$$

It says that if the input u vanishes on $\overline{\mathbb{R}}^+$, then the output y on $\overline{\mathbb{R}}^+$ is the product of the input map (which maps the past input u into the state $x(0) = \mathfrak{B}u$ at time zero) and the output map (which maps $x(0)$ into the future output $y = \mathfrak{C}x(0) = \mathfrak{C}\mathfrak{B}u$).

It is a very interesting and important fact that the algebraic conditions 2.1.3 are not only necessary, but they are in fact sufficient (together with the well-posedness requirement 2.1.2) for the quadruple $(\mathfrak{A}, \mathfrak{B}, \mathfrak{C}, \mathfrak{D})$ to be interpreted as the semigroup, input map, output map, and input/output map of a well-posed linear system. Every such system has a 'differential' representation similar to the one given in (2.1.1) but slightly more complicated. The picture is roughly the following. We combine the four operators A, B, C, and D into one single block matrix operator $S = \left[\begin{smallmatrix} A & B \\ C & D \end{smallmatrix}\right]$, which we call the *node* of the system, and rewrite (2.1.1) in the form

$$\begin{bmatrix} \dot{x}(t) \\ y(t) \end{bmatrix} = S \begin{bmatrix} x(t) \\ u(t) \end{bmatrix}, \quad t \geq s, \quad x(s) = x_s. \tag{2.1.12}$$

As we shall see in Chapter 4, every well-posed linear system has a representation of this type, valid for a restricted set of data: the input function should be, for example, two times continuously differentiable, and the initial state x_s should satisfy a certain compatibility condition involving the value $u(s)$. The operator S is closed and (typically) unbounded from $\left[\begin{smallmatrix} X \\ U \end{smallmatrix}\right]$ to $\left[\begin{smallmatrix} X \\ Y \end{smallmatrix}\right]$, and it has a natural splitting into an operator matrix $S = \left[\begin{smallmatrix} A & B \\ \multicolumn{2}{c}{C\&D} \end{smallmatrix}\right]$ (where the combined observation/feedthrough operator $C\&D$ stands for a certain operator from the domain of S to Y). Under weak additional conditions even the operator $C\&D$ can be split into $C\&D = \left[C \; D\right]$, so that S can be written in the familiar form $S = \left[\begin{smallmatrix} A & B \\ C & D \end{smallmatrix}\right]$. The splitting of $C\&D$ into $\left[C \; D\right]$ is, unfortunately, not always unique, and different applications may require different splittings.

2.2 Definitions and basic properties

Without further ado, let us give a formal definition of *an L^p-well-posed linear system*. In the terms of the discussion in the preceding section, such a system consists of a semigroup $\mathfrak{A}$ and three maps $\mathfrak{B}$, $\mathfrak{C}$, and $\mathfrak{D}$ satisfying the algebraic conditions 2.1.3, and having enough continuity in order for the well-posedness requirement 2.1.2 to hold.

Definition 2.2.1 Let U, X, and Y be Banach spaces, and let $1 \leq p \leq \infty$. A (causal, time-invariant) *L^p-well-posed linear system* Σ on (Y, X, U) consists of a quadruple $\Sigma = \left[\begin{smallmatrix} \mathfrak{A} & \mathfrak{B} \\ \mathfrak{C} & \mathfrak{D} \end{smallmatrix}\right]$ satisfying the following conditions:[4]

(i) $t \mapsto \mathfrak{A}^t$ is a strongly continuous *semigroup* on X (see Definition 2.2.2);
(ii) $\mathfrak{B}\colon L^p_c(\mathbb{R}^-; U) \to X$ satisfies $\mathfrak{A}^t\mathfrak{B}u = \mathfrak{B}\tau^t_- u$ for all $u \in L^p_c(\mathbb{R}^-; U)$ and all $t \geq 0$;
(iii) $\mathfrak{C}\colon X \to L^p_{\mathrm{loc}}(\mathbb{R}^+; Y)$ satisfies $\mathfrak{C}\mathfrak{A}^t x = \tau^t_+ \mathfrak{C}x$ for all $x \in X$ and all $t \geq 0$;
(iv) $\mathfrak{D}\colon L^p_{c,\mathrm{loc}}(\mathbb{R}; U) \to L^p_{c,\mathrm{loc}}(\mathbb{R}; Y)$ satisfies $\tau^t\mathfrak{D}u = \mathfrak{D}\tau^t u$, $\pi_-\mathfrak{D}\pi_+ u = 0$, and $\pi_+\mathfrak{D}\pi_- u = \mathfrak{C}\mathfrak{B}u$ for all $u \in L^p_{c,\mathrm{loc}}(\mathbb{R}; U)$ and all $t \in \mathbb{R}$.

[4] The notation used here is explained immediately after the definition. We shall see in Theorem 2.5.4 that these operators have some additional continuity properties, which are related to the fact that the system has a finite exponential growth bound.

We use the following names for the different components of Σ: U is the *input space*, X is the *state space*, Y is the *output space*, $\mathfrak{A}$ is the *semigroup*, $\mathfrak{B}$ is the input (or controllability, or reachability) map (with final time zero), $\mathfrak{C}$ is the output (or observability) map (with initial time zero), and $\mathfrak{D}$ is the *input/output map*.

In the preceding definition we used the following notation. We let $\mathbb{R} := (-\infty, \infty)$, $\mathbb{R}^- := (-\infty, 0)$, $\mathbb{R}^+ := (0, \infty)$,

$$(\pi_J u)(s) := \begin{cases} u(s), & s \in J, \\ 0, & s \notin J, \end{cases} \qquad \text{for all } J \subset \mathbb{R},$$

$$\begin{aligned} \pi_+ u &:= \pi_{[0,\infty)}, & \pi_- u &:= \pi_{(-\infty,0)}, \\ (\tau^t u)(s) &:= u(t+s), & -\infty &< t, s < \infty, \\ \tau_+^t &:= \pi_+ \tau^t, & \tau_-^t &:= \tau^t \pi_-, \quad t \geq 0. \end{aligned}$$

The space $L^p_{c,\mathrm{loc}}(\mathbb{R}; U)$ consists of functions $u : \mathbb{R} \to U$ that are locally in L^p and have a support that is bounded to the left. A sequence of functions u_n converges in $L^p_{c,\mathrm{loc}}(\mathbb{R}; U)$ to a function u if the common support of all the functions u_n is bounded to the left and u_n converges to u locally in L^p with values in U. The space $L^p_c(\mathbb{R}^-; U)$ contains those $u \in L^p_{c,\mathrm{loc}}(\mathbb{R}; U)$ which vanish on $\mathbb{R}^+$, and the space $L^p_{\mathrm{loc}}(\mathbb{R}^+; Y)$ contains those $u \in L^p_{c,\mathrm{loc}}(\mathbb{R}; Y)$ which vanish on $\mathbb{R}^-$. The continuity of $\mathfrak{B}$, $\mathfrak{C}$ and $\mathfrak{D}$ is with respect to this convergence.[5]

The preceding definition refers to the notion of a strongly continuous semigroup. This notion (and the notion of a strongly continuous group) is defined as follows:

Definition 2.2.2 Let X be a Banach space.

(i) A family $\mathfrak{A}^t$, $t \geq 0$, of bounded linear operators $X \to X$ is a *semigroup* on X if $\mathfrak{A}^0 = 1$ and $\mathfrak{A}^s \mathfrak{A}^t = \mathfrak{A}^{s+t}$ for all $s, t \geq 0$.
(ii) A family $\mathfrak{A}^t$, $t \in \mathbb{R}$, of bounded linear operators $X \to X$ is a *group* on X if $\mathfrak{A}^0 = 1$ and $\mathfrak{A}^s \mathfrak{A}^t = \mathfrak{A}^{s+t}$ for all $s, t \in \mathbb{R}$.
(iii) The semigroup in (i) is *locally bounded* if $\sup_{0 \leq s \leq t} \|\mathfrak{A}^s\|$ is bounded for each finite $t > 0$. The group in (ii) is *locally bounded* if $\sup_{-t \leq s \leq t} \|\mathfrak{A}^s\|$ is bounded for each finite $t > 0$.
(iv) The semigroup in (i) is strongly continuous (at zero) if $\lim_{t \downarrow 0} \mathfrak{A}^t x = x$ for all $x \in X$. The group in (ii) is strongly continuous (at zero) if $\lim_{t \to 0} \mathfrak{A}^t x = x$ for all $x \in X$.
(v) We abbreviate 'strongly continuous semigroup' to 'C_0 *semigroup*' and 'strongly continuous group' to 'C_0 *group*'.

[5] In the terminology of Köthe (1969), these spaces are strict (LF)-spaces, i.e., they are the strict inductive limits of the Fréchet spaces $L^p_{\mathrm{loc}}([T, \infty); U)$ which we identify with the subspace of functions in $L^p_{\mathrm{loc}}(\mathbb{R}; U)$ which vanish on $(-\infty, T)$.

In some parts of the theory the value of p plays an important role. Especially the case $p = \infty$ differs significantly from the other cases. The main difficulty in L^∞ is that the set of continuous functions is not dense as it is in L^p with $p < \infty$. This complicates some of the proofs: it is not enough to first prove a result for the class of continuous (or even continuously differentiable) inputs, and to then extend the result to arbitrary inputs in L^∞ by using a density argument. For this reason we shall introduce yet another class of well-posed linear systems that in many cases replaces the class of L^∞-well-posed systems. The simplest choice would be to restrict all the inputs to be continuous, but this leads to a difficulty with the algebraic conditions in Definition 2.2.1: even if u is continuous the functions $\pi_- u$ and $\pi_+ u$ are not continuous (unless $u(0) = 0$). Therefore we need a slightly larger class of functions to work in. The most natural class is then the set of regulated functions, i.e., functions that are right-continuous and have a left hand limit at each finite point.[6] We denote this class of functions by Reg_{loc}. The space $Reg_{c,\text{loc}}$ consists of functions locally in Reg whose support is bounded to the left. Convergence in this space means that the common support of all the functions should be bounded to the left, and the convergence is uniform on each bounded interval. Observe, in particular, that functions in Reg_{loc} have a well-defined value at *every* point and not just almost everywhere.

Definition 2.2.3 Let U, X, and Y be Banach spaces. A (causal, time-invariant) *Reg-well-posed linear system* Σ on (Y, X, U) consists of a quadruple $\Sigma = \left[\begin{smallmatrix} \mathfrak{A} & \mathfrak{B} \\ \mathfrak{C} & \mathfrak{D} \end{smallmatrix}\right]$ satisfying the same conditions as in Definition 2.2.1 but with $L^p_{c,\text{loc}}$ replaced by $Reg_{c,\text{loc}}$. More precisely, $L^p_{c,\text{loc}}(\mathbb{R})$ is replaced by $Reg_{c,\text{loc}}(\mathbb{R})$, $L^p_c(\mathbb{R}^-)$ is replaced by $Reg_c(\mathbb{R}^-)$, and $L^p_{\text{loc}}(\mathbb{R}^+)$ is replaced by $Reg_{\text{loc}}(\overline{\mathbb{R}}^+)$. We call the different components of Σ by the same names as in Definition 2.2.1.

Here $\overline{\mathbb{R}}^+ := [0, \infty)$ is the *closed* positive real half-line. (Analogously, we denote the closed negative real half-line by $\overline{\mathbb{R}}^-$.) To make the formulas in the L^p-well-posed case look more similar to the formulas in the *Reg*-well-posed case we sometimes write $L^p_{\text{loc}}(\overline{\mathbb{R}}^+)$ instead of $L^p_{\text{loc}}(\mathbb{R}^+)$, etc.

Definition 2.2.4

(i) By $L^p|Reg$ we mean either L^p or Reg and by $L^p|Reg_0$ we mean either L^p_0 or Reg_0, depending on the context.[7]

[6] See Section A.1 for a short introduction to this class of function. By *Reg* we denote the set of all bounded regulated functions. A function in *Reg* is not required to have a limit at $\pm\infty$.

[7] The space L^p_0 is the same as L^p if $p < \infty$, and in the case $p = \infty$ it consists of those $u \in L^\infty$ which vanish at $\pm\infty$, i.e., $\lim_{t\to\infty} \operatorname{ess\,sup}_{|s|\ge t} |u(s)| = 0$. The space *Reg* consists of all bounded regulated functions, and Reg_0 consists of those functions in *Reg* which vanish at $\pm\infty$.

(ii) By an *$L^p|Reg$-well-posed linear system* we mean a system which is either *Reg*-well-posed or L^p-well-posed for some p, $1 \le p \le \infty$. (Thus, the case $p = \infty$ is included, except when explicitly excluded.)

(iii) By a *well-posed linear system* we mean a system which is either *Reg*-well-posed or L^p-well-posed for some p, $1 \le p < \infty$. (Thus, the case $p = \infty$ is excluded.)

Remark 2.2.5 To begin with we shall work with $L^p|Reg$-well-posed linear systems, but later on we shall exclude the L^∞-well-posed case, due to the fact that some of our main results are not necessarily valid for inputs $u \in L^\infty_{c,\mathrm{loc}}(\mathbb{R}; U)$ in the L^∞-well-posed case. See, in particular, Theorems 4.2.1, 4.5.2, 4.5.4, 4.6.9, and Corollaries 4.5.5, 4.5.6.

See Section 1.1 for an intuitive explanation of the algebraic conditions (i)–(iv) in Definition 2.2.1.

The primary reason for the introduction of the operators $\mathfrak{B}$, $\mathfrak{C}$, and $\mathfrak{D}$ in Section 2.1 was that we wanted to make the algebraic conditions in Definition 2.2.1 as simple as possible. However, when we deal with *the state trajectory and the output function* of the system $\Sigma = \left[\begin{smallmatrix} \mathfrak{A} & \mathfrak{B} \\ \mathfrak{C} & \mathfrak{D} \end{smallmatrix}\right]$ it is more convenient to use the operators $\mathfrak{B}^t_s$, $\mathfrak{C}_s$, and $\mathfrak{D}_s$ in (2.1.3) with initial time s and final time t. Let us formally reintroduce these operators and some related operators as follows:

Definition 2.2.6 Let $\Sigma = \left[\begin{smallmatrix} \mathfrak{A} & \mathfrak{B} \\ \mathfrak{C} & \mathfrak{D} \end{smallmatrix}\right]$ be a $L^p|Reg$-well-posed linear system on (Y, X, U).

(i) We interpret $\mathfrak{B}$ as an operator $L^p|Reg_{c,\mathrm{loc}}(\mathbb{R}; U) \to X$ by defining $\mathfrak{B}u = \mathfrak{B}\pi_- u$ for all $u \in L^p|Reg_{c,\mathrm{loc}}(\mathbb{R}; U)$.

(ii) We interpret $\mathfrak{C}$ as an operator $X \to L^p|Reg_{c,\mathrm{loc}}(\mathbb{R}; Y)$ by defining $\mathfrak{C}x = \pi_+\mathfrak{C}x$ for all $x \in X$.

(iii) The *state transition map* $\mathfrak{A}^t_s$, the *input map* $\mathfrak{B}^t_s$, the *output map* $\mathfrak{C}^t_s$, and the *input/output map* $\mathfrak{D}^t_s$ with initial time $s \in \mathbb{R}$ and final time $t \ge s$ are defined by

$$\left[\begin{array}{c|c} \mathfrak{A}^t_s & \mathfrak{B}^t_s \\ \hline \mathfrak{C}^t_s & \mathfrak{D}^t_s \end{array}\right] := \left[\begin{array}{c|c} \mathfrak{A}^{t-s} & \mathfrak{B}\tau^t\pi_{[s,t)} \\ \hline \pi_{[s,t)}\tau^{-s}\mathfrak{C} & \pi_{[s,t)}\mathfrak{D}\pi_{[s,t)} \end{array}\right], \qquad t \ge s.$$

(iv) The *input map* $\mathfrak{B}^t$ and the *input/output map* $\mathfrak{D}^t$ with final time $t \in \mathbb{R}$ (and initial time $-\infty$) are defined by

$$\mathfrak{B}^t = \mathfrak{B}^t_{-\infty} := \mathfrak{B}\tau^t, \qquad \mathfrak{D}^t = \mathfrak{D}^t_{-\infty} := \pi_{(-\infty,t)}\mathfrak{D}, \qquad t \in \mathbb{R}.$$

(v) The *output map* $\mathfrak{C}_s$ and the *input/output map* $\mathfrak{D}_s$ with initial time $s \in \mathbb{R}$ (and final time $+\infty$) are defined by

$$\mathfrak{C}_s = \mathfrak{C}^\infty_s := \tau^{-s}\mathfrak{C}, \qquad \mathfrak{D}_s = \mathfrak{D}^\infty_s := \mathfrak{D}\pi_{[s,\infty)}, \qquad s \in \mathbb{R}.$$

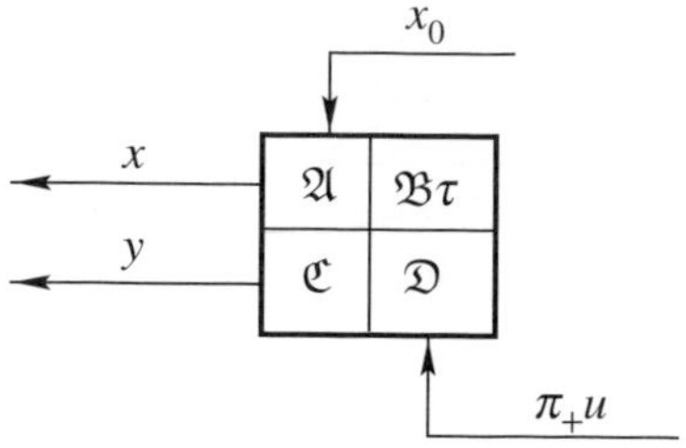

Figure 2.2 Well-posed linear system

The conventions (i) and (ii) are implicitly contained in Definition 2.2.1(ii)–(iii) (take $t = 0$).

Definition 2.2.7 Let U, X, and Y be Banach spaces, and let $\Sigma = \left[\begin{smallmatrix} \mathfrak{A} & \mathfrak{B} \\ \mathfrak{C} & \mathfrak{D} \end{smallmatrix}\right]$ be a $L^p|Reg$-well-posed linear system on (Y, X, U). For each $s \in \mathbb{R}$, $x_s \in X$, $t \geq s$, and $u \in L^p|Reg_{\text{loc}}([s, \infty); U)$ we define the *state* $x(t)$ *at time* t and the *output function* y of Σ with initial time s, initial state x_s, and *input function* u by (cf. Definition 2.2.6)

$$\begin{aligned} x(t) &= \mathfrak{A}_s^t x_s + \mathfrak{B}_s^t u, \qquad t \geq s, \\ y &= \mathfrak{C}_s x_s + \mathfrak{D}_s u. \end{aligned} \tag{2.2.1}$$

In particular, if the initial time s is zero and the initial state x_0, then

$$\begin{aligned} x(t) &= \mathfrak{A}_0^t x_0 + \mathfrak{B}_0^t u = \mathfrak{A}^t x_0 + \mathfrak{B}^t \pi_+ u, \qquad t \geq 0, \\ y &= \mathfrak{C}_0 x_0 + \mathfrak{D}_0 u = \mathfrak{C} x_0 + \mathfrak{D} \pi_+ u. \end{aligned} \tag{2.2.2}$$

We use diagrams of the type drawn in Figure 2.2 to represent the relation between the state x, the output y, the initial value x_0, and the input u of Σ with initial time zero. In our diagrams we use the following conventions throughout:

(i) Initial states and inputs enter at the top or bottom, and they are acted on by all the operators located in the column to which they are attached. In particular, note that x_0 is attached to the first column and u to the second.
(ii) Final states and outputs leave to the left or right, and they are the sums of all the elements in the row to which they are attached. In particular, note that x is attached to the top row, and y to the bottom row.

We can reformulate the algebraic conditions in Definition 2.2.1 in terms of the operators introduced in Definition 2.2.6 as follows.

Lemma 2.2.8 *Let* $\Sigma = \left[\begin{smallmatrix} \mathfrak{A} & \mathfrak{B} \\ \mathfrak{C} & \mathfrak{D} \end{smallmatrix}\right]$ *be a* $L^p|Reg$*-well-posed linear system on* (Y, X, U). *Then the operators* $\mathfrak{A}_s^t$, $\mathfrak{B}_s^t$, $\mathfrak{C}_s^t$, *and* $\mathfrak{D}_s^t$ *introduced in Definition 2.2.6 have the following properties.*

(i) *For all* $t \in \mathbb{R}$,

$$\left[\begin{array}{c|c} \mathfrak{A}_t^t & \mathfrak{B}_t^t \\ \hline \mathfrak{C}_t^t & \mathfrak{D}_t^t \end{array}\right] = \left[\begin{array}{c|c} 1 & 0 \\ \hline 0 & 0 \end{array}\right]. \tag{2.2.3}$$

(ii) *For all* $s \le t$,

$$\begin{aligned} \left[\begin{array}{c|c} \mathfrak{A}_s^t & \mathfrak{B}_s^t \\ \hline \mathfrak{C}_s^t & \mathfrak{D}_s^t \end{array}\right] &= \begin{bmatrix} 1 & 0 \\ 0 & \pi_{[s,t)} \end{bmatrix} \left[\begin{array}{c|c} \mathfrak{A}_s^t & \mathfrak{B}_s^t \\ \hline \mathfrak{C}_s^t & \mathfrak{D}_s^t \end{array}\right] \begin{bmatrix} 1 & 0 \\ 0 & \pi_{[s,t)} \end{bmatrix} \\ &= \left[\begin{array}{c|c} \mathfrak{A}_s^t & \mathfrak{B}_s^t \pi_{[s,t)} \\ \hline \pi_{[s,t)} \mathfrak{C}_s^t & \pi_{[s,t)} \mathfrak{D}_s^t \pi_{[s,t)} \end{array}\right]. \end{aligned} \tag{2.2.4}$$

(iii) *For all* $s \le t$ *and* $h \in \mathbb{R}$,

$$\begin{aligned} \left[\begin{array}{c|c} \mathfrak{A}_{s+h}^{t+h} & \mathfrak{B}_{s+h}^{t+h} \\ \hline \mathfrak{C}_{s+h}^{t+h} & \mathfrak{D}_{s+h}^{t+h} \end{array}\right] &= \begin{bmatrix} 1 & 0 \\ 0 & \tau^{-h} \end{bmatrix} \left[\begin{array}{c|c} \mathfrak{A}_s^t & \mathfrak{B}_s^t \\ \hline \mathfrak{C}_s^t & \mathfrak{D}_s^t \end{array}\right] \begin{bmatrix} 1 & 0 \\ 0 & \tau^{h} \end{bmatrix} \\ &= \left[\begin{array}{c|c} \mathfrak{A}_s^t & \mathfrak{B}_s^t \tau^h \\ \hline \tau^{-h} \mathfrak{C}_s^t & \tau^{-h} \mathfrak{D}_s^t \tau^h \end{array}\right]. \end{aligned} \tag{2.2.5}$$

(iv) *For all* $s \le r \le t$,

$$\begin{aligned} \left[\begin{array}{c|c} \mathfrak{A}_s^t & \mathfrak{B}_s^t \\ \hline \mathfrak{C}_s^t & \mathfrak{D}_s^t \end{array}\right] &= \left[\begin{array}{c|cc} \mathfrak{A}_r^t & 0 & \mathfrak{B}_r^t \\ \hline \mathfrak{C}_r^t & 1 & \mathfrak{D}_r^t \end{array}\right] \left[\begin{array}{c|c} \mathfrak{A}_s^r & \mathfrak{B}_s^r \\ \hline \mathfrak{C}_s^r & \mathfrak{D}_s^r \\ 0 & 1 \end{array}\right] \\ &= \left[\begin{array}{c|c} \mathfrak{A}_r^t \mathfrak{A}_s^r & \mathfrak{B}_r^t + \mathfrak{A}_r^t \mathfrak{B}_s^r \\ \hline \mathfrak{C}_r^t \mathfrak{A}_s^r + \mathfrak{C}_s^r & \mathfrak{D}_r^t + \mathfrak{C}_r^t \mathfrak{B}_s^r + \mathfrak{D}_s^r \end{array}\right]. \end{aligned} \tag{2.2.6}$$

We shall refer to (ii) as the *causality property*, to (iii) as the *time-invariance property* and to (iv) as the *composition property*. See Theorem 2.2.11 for an interpretation of these properties, and see also Theorem 2.2.14 for a slightly modified set of algebraic conditions where (ii) and (iv) are combined into one single condition.

Both in the proof of Lemma 2.2.8 and later we shall need to manipulate expressions involving the operators τ^t and π_J and their combinations. For this the following lemma is useful:

Lemma 2.2.9 *Let* $a, h, s, t \in \mathbb{R}$, *and* $-\infty < b \le \infty$. *Then*

$$\begin{aligned} \tau^{s+t} &= \tau^s \tau^t, \\ \pi_{[a,b)} &= \pi_{[a,\infty)} \pi_{(-\infty,b)}, & \pi_{[a,b)} &= \pi_{(-\infty,b)} \pi_{[a,\infty)}, \\ \pi_{[a,b)} \tau^h &= \tau^h \pi_{[a+h,b+h)}, & \tau^h \pi_{[a,b)} &= \pi_{[a-h,b-h)} \tau^h, \\ \pi_{(-\infty,b)} \tau^h &= \tau^h \pi_{(-\infty,b+h)}, & \tau^h \pi_{(-\infty,b)} &= \pi_{(-\infty,b-h)} \tau^h \end{aligned}$$

(we define $\pi_{[a,b)} = 0$ *when* $b \le a$*).*

The easy proof of this lemma is left to the reader.

Proof of Lemma 2.2.8 We leave the easy proofs of (i)–(iii) to the reader, and only prove (iv). Thus, in the sequel we take $s \le r \le t$.

The identity $\mathfrak{A}_s^t = \mathfrak{A}_r^t \mathfrak{A}_s^r$ follows from the semigroup property $\mathfrak{A}^{t-s} = \mathfrak{A}^{t-r}\mathfrak{A}^{r-s}$ and the definition of $\mathfrak{A}_s^t$.

By Definitions 2.2.1 and 2.2.6 and Lemma 2.2.9

$$\begin{aligned}
\mathfrak{B}_s^t &= \mathfrak{B}\tau^t \pi_{[s,t)} \\
&= \mathfrak{B}\tau^t(\pi_{[s,r)} + \pi_{[r,t)}) \\
&= \mathfrak{B}\tau^{t-r}\pi_{[s-r,0)}\tau^r + \mathfrak{B}\tau^t\pi_{[r,t)} \\
&= \mathfrak{A}^{t-r}\mathfrak{B}\pi_{[s-r,0)}\tau^r + \mathfrak{B}\tau^t\pi_{[r,t)} \\
&= \mathfrak{A}^{t-r}\mathfrak{B}\tau^r\pi_{[s,r)} + \mathfrak{B}\tau^t\pi_{[r,t)} \\
&= \mathfrak{A}_r^t\mathfrak{B}_s^r + \mathfrak{B}_r^t.
\end{aligned}$$

Likewise,

$$\begin{aligned}
\mathfrak{C}_s^t &= \pi_{[s,t)}\tau^{-s}\mathfrak{C} \\
&= (\pi_{[s,r)} + \pi_{[r,t)})\tau^{-s}\mathfrak{C} \\
&= \pi_{[s,r)}\tau^{-s}\mathfrak{C} + \tau^{-r}\pi_{[0,t-r)}\tau^{r-s}\mathfrak{C} \\
&= \pi_{[s,r)}\tau^{-s}\mathfrak{C} + \tau^{-r}\pi_{[0,t-r)}\mathfrak{C}\mathfrak{A}^{r-s} \\
&= \pi_{[s,r)}\tau^{-s}\mathfrak{C} + \pi_{[r,t)}\tau^{-r}\mathfrak{C}\mathfrak{A}^{r-s} \\
&= \mathfrak{C}_s^r + \mathfrak{C}_r^t\mathfrak{A}_s^r,
\end{aligned}$$

and

$$\begin{aligned}
\mathfrak{D}_s^t &= \pi_{[s,t)}\mathfrak{D}\pi_{[s,t)} \\
&= (\pi_{[s,r)} + \pi_{[r,t)})\mathfrak{D}(\pi_{[s,r)} + \pi_{[r,t)}) \\
&= \mathfrak{D}_s^r + \pi_{[r,t)}\tau^{-r}\mathfrak{D}\tau^r\pi_{[s,r)} + \pi_{[s,r)}\tau^{-r}\mathfrak{D}\tau^r\pi_{[r,t)} + \mathfrak{D}_r^t \\
&= \mathfrak{D}_s^r + \tau^{-r}\pi_{[0,t-r)}\mathfrak{D}\pi_{[s-r,0)}\tau^r + \tau^{-r}\pi_{[s-r,0)}\mathfrak{D}\pi_{[0,t-r)}\tau^r + \mathfrak{D}_r^t \\
&= \mathfrak{D}_s^r + \tau^{-r}\pi_{[0,t-r)}\mathfrak{C}\mathfrak{B}\pi_{[s-r,0)}\tau^r + \mathfrak{D}_r^t \\
&= \mathfrak{D}_s^r + \pi_{[r,t)}\tau^{-r}\mathfrak{C}\mathfrak{B}\tau^r\pi_{[s,r)} + \mathfrak{D}_r^t \\
&= \mathfrak{D}_s^r + \mathfrak{C}_r^t\mathfrak{B}_s^r + \mathfrak{D}_r^t.
\end{aligned}$$

□

Occasionally we shall also need the following version of Lemma 2.2.8 where $\mathfrak{B}_s^t$ is replaced by $\mathfrak{B}^t$, $\mathfrak{C}_s^t$ is replaced by $\mathfrak{C}_s$, and $\mathfrak{D}_s^t$ is replaced by $\mathfrak{D}_s$ or $\mathfrak{D}^t$.

Lemma 2.2.10 *Let* $\Sigma = \left[\begin{smallmatrix}\mathfrak{A} & \mathfrak{B} \\ \mathfrak{C} & \mathfrak{D}\end{smallmatrix}\right]$ *be a* $L^p|Reg$*-well-posed linear system on* (Y, X, U). *With the notation of Definition 2.2.6, the following claims are true:*

(i) *For all* $s, t \in \mathbb{R}$, $x \in X$, *and* $u \in L^p|Reg_{c,\mathrm{loc}}(\mathbb{R}; U)$,

$$\mathfrak{B}^t u = \lim_{r\to-\infty} \mathfrak{B}^t_r u, \qquad \mathfrak{C}_s u = \lim_{r\to\infty} \mathfrak{C}^r_s x,$$
$$\mathfrak{D}^t u = \lim_{r\to-\infty} \mathfrak{D}^t_r u, \qquad \mathfrak{D}_s u = \lim_{r\to\infty} \mathfrak{D}^r_s u,$$

Moreover, the original system $\Sigma = \left[\begin{smallmatrix}\mathfrak{A} & \mathfrak{B} \\ \mathfrak{C} & \mathfrak{D}\end{smallmatrix}\right]$ *can be recovered as follows:* $\mathfrak{A}^t = \mathfrak{A}^t_0$ *for* $t \ge 0$, *and*

$$\mathfrak{B} = \mathfrak{B}^0, \qquad \mathfrak{C} = \mathfrak{C}_0, \qquad \mathfrak{D} = \lim_{\substack{s\to-\infty \\ t\to\infty}} \mathfrak{D}^t_s.$$

(ii) *For all* $s, t \in \mathbb{R}$

$$\left[\begin{array}{c}\mathfrak{B}^t \\ \hline \mathfrak{D}^t\end{array}\right] = \left[\begin{array}{c}\mathfrak{B}^t \pi_{(-\infty,t)} \\ \hline \pi_{(-\infty,t)}\mathfrak{D}^t \pi_{(-\infty,t)}\end{array}\right],$$
$$\left[\begin{array}{c|c}\mathfrak{C}_s & \mathfrak{D}_s\end{array}\right] = \left[\begin{array}{c|c}\pi_{[s,\infty)}\mathfrak{C}_s & \pi_{[s,\infty)}\mathfrak{D}_s \pi_{[s,\infty)}\end{array}\right].$$

(iii) *For all* $s, t \in \mathbb{R}$ *and* $h \in \mathbb{R}$,

$$\left[\begin{array}{c}\mathfrak{B}^{t+h} \\ \hline \mathfrak{D}^{t+h}\end{array}\right] = \left[\begin{array}{c}\mathfrak{B}^t \tau^h \\ \hline \tau^{-h}\mathfrak{D}^t \tau^h\end{array}\right],$$
$$\left[\begin{array}{c|c}\mathfrak{C}_{s+h} & \mathfrak{D}_{s+h}\end{array}\right] = \left[\begin{array}{c|c}\tau^{-h}\mathfrak{C}_s & \tau^{-h}\mathfrak{D}_s \tau^h\end{array}\right].$$

(iv) *For all* $r \le t$,

$$\left[\begin{array}{c}\mathfrak{B}^t \\ \hline \mathfrak{D}^t\end{array}\right] = \left[\begin{array}{c|c}\mathfrak{A}^t_r & 0 \ \mathfrak{B}^t_r \\ \hline \mathfrak{C}^t_r & 1 \ \mathfrak{D}^t_r\end{array}\right] \left[\begin{array}{c}\mathfrak{B}^r \\ \hline \mathfrak{D}^r \\ 1\end{array}\right]$$
$$= \left[\begin{array}{c}\mathfrak{B}^t_r + \mathfrak{A}^t_r \mathfrak{B}^r \\ \hline \mathfrak{D}^t_r + \mathfrak{C}^t_r \mathfrak{B}^r + \mathfrak{D}^r\end{array}\right].$$

and for all $s \le r$,

$$\left[\begin{array}{c|c}\mathfrak{C}_s & \mathfrak{D}_s\end{array}\right] = \left[\begin{array}{c|c}\mathfrak{C}_r & 1 \ \mathfrak{D}_r\end{array}\right] \left[\begin{array}{c|c}\mathfrak{A}^r_s & \mathfrak{B}^r_s \\ \hline \mathfrak{C}^r_s & \mathfrak{D}^r_s \\ 0 & 1\end{array}\right]$$
$$= \left[\begin{array}{c|c}\mathfrak{C}^r_s + \mathfrak{C}_r \mathfrak{A}^r_s & \mathfrak{D}_r + \mathfrak{C}_r \mathfrak{B}^r_s + \mathfrak{D}^r_s\end{array}\right].$$

Proof Part (i) follows trivially from Definition 2.2.6 (since $\lim_{r\to-\infty}\pi_{[r,t)} = \pi_{(-\infty,t)}$, $\lim_{r\to\infty}\pi_{[s,r)} = \pi_{[s,\infty)}$, and $\lim_{s\to-\infty,t\to\infty}\pi_{[s,t)} = 1$ in $L^p|Reg_{c,\text{loc}}$). We get (ii), (iii), and (iv) by letting $s \to -\infty$ or $t \to \infty$ in the appropriate parts of Lemma 2.2.8(ii)–(iv). □

We still refer to (ii) as the *causality property*, to (iii) as the *time-invariance property* and to (iv) as the *composition property* for the operators in question.

The essential content of Lemmas 2.2.8 and 2.2.10 is that it is possible to separate, at each time instant, the past and the future behavior of the system Σ from each other in a clean way:

Theorem 2.2.11 *Let* $\Sigma = \left[\begin{smallmatrix}\mathfrak{A} & \mathfrak{B}\\ \mathfrak{C} & \mathfrak{D}\end{smallmatrix}\right]$ *be a* $L^p|Reg$*-well-posed linear system on* (Y, X, U)*, and let* $s \in \mathbb{R}$, $x_s \in X$, *and* $u \in L^p|Reg_{\text{loc}}([s,\infty); U)$. *Let x and y be the state trajectory x and the output function y of Σ with initial time s, initial value x_s and input function u. Then, for each $r > s$,*

$$\begin{aligned} x(r) &= \mathfrak{A}_s^r x_s + \mathfrak{B}_s^r u, \\ \pi_{[s,r)} y &= \mathfrak{C}_s^r x_s + \mathfrak{D}_s^r u \end{aligned} \tag{2.2.7}$$

do not depend on $u(t)$ for $t \geq r$ (i.e., 'the future has no influence on the past'), and

$$\begin{aligned} x(t) &= \mathfrak{A}_r^t x(r) + \mathfrak{B}_r^t u, \qquad t \geq r, \\ \pi_{[r,\infty)} y &= \mathfrak{C}_r x(r) + \mathfrak{D}_r u, \end{aligned} \tag{2.2.8}$$

can be interpreted as the state trajectory and the output function of Σ with initial time r, initial value $x(r)$, and input function u (i.e., 'the past influences the future only through the present state $x(r)$').

Proof of Theorem 2.2.11 That $x(r)$ and $\pi_{[s,r)}y$ do not depend on $u(t)$ for $t \geq r$ follows from the causality property in Lemma 2.2.8(ii).

By the composition properties in Lemmas 2.2.8(iv) and 2.2.10(iv),

$$\begin{aligned} x(t) &= \mathfrak{A}_s^t x_s + \mathfrak{B}_s^t u \\ &= \mathfrak{A}_r^t \mathfrak{A}_s^r x_s + (\mathfrak{A}_r^t \mathfrak{B}_s^r + \mathfrak{B}_s^r) u \\ &= \mathfrak{A}_r^t (\mathfrak{A}_s^r x_s + \mathfrak{B}_s^r) + \mathfrak{B}_s^r u \\ &= \mathfrak{A}_r^t x(r) + \mathfrak{B}_s^r u, \\ y &= \mathfrak{C}_s x_s + \mathfrak{D}_s u \\ &= (\mathfrak{C}_s^r + \mathfrak{C}_r \mathfrak{A}_s^r) x_s + (\mathfrak{D}_s^r + \mathfrak{C}_r \mathfrak{B}_s^r + \mathfrak{D}_r) u \\ &= (\mathfrak{C}_s^r x_s + \mathfrak{D}_s^r u) + \big(\mathfrak{C}_r (\mathfrak{A}_s^r x_s + \mathfrak{B}_s^r u) + \mathfrak{D}_r u\big) \\ &= \pi_{[s,r)} y + \pi_{[r,\infty)} (\mathfrak{C}_r x(r) + \mathfrak{D}_r u). \end{aligned}$$

□

Another important property of the solution is the following:

Theorem 2.2.12 *Let $\Sigma = \left[\begin{smallmatrix} \mathfrak{A} & \mathfrak{B} \\ \mathfrak{C} & \mathfrak{D} \end{smallmatrix}\right]$ be an L^p-well-posed linear system where $p < \infty$, and let $s \in \mathbb{R}$, $x_s \in X$, and $u \in L^p_{\mathrm{loc}}([s,\infty);U)$. Then the state $x(t)$ of Σ with initial time s, initial value x_s and input u depends continuously on t for $t \geq s$, and $x(s) = x_s$.*

Proof The state consists of two terms,

$$\begin{aligned} x(t) &= \mathfrak{A}_s^t x_s + \mathfrak{B}_s^t u = \mathfrak{A}^{t-s} x_s + \mathfrak{B}\tau^t \pi_{[s,t)} u \\ &= \mathfrak{A}^{t-s} x_s + \mathfrak{B}\tau^t \pi_{[s,\infty)} u. \end{aligned}$$

That $x(s) = x_s$ follows from Lemma 2.2.8(ii). The continuity of the first term follows from Lemma 2.2.13 below, and the continuity of the second term from Lemma 2.3.3 below. □

A similar result for *Reg*-well-posed and L^∞-well-posed linear systems is given in Theorem 4.2.7.

In the preceding proof we used a part of the following lemma:

Lemma 2.2.13 *Let $\mathfrak{A}^t$ be a C_0-semigroup on X.*

(i) *$\|\mathfrak{A}^t\| \leq M e^{\omega t}$ for some $M > 0$, some $\omega \in \mathbb{R}$ and all $t \geq 0$. In particular, $\mathfrak{A}$ is locally bounded.*
(ii) *For each $x \in X$, $t \mapsto \mathfrak{A}^t x$ is continuous on $\overline{\mathbb{R}}^+$.*

A sharper version of part (i) is given in Theorem 2.5.4(i) below.

Proof (i) We begin by showing that there is some $T > 0$ and $M > 0$ such that $\|\mathfrak{A}^t\| \leq M$ on $[0, T]$. If not, then there is some sequence $t_n \to 0$ such that $\|\mathfrak{A}^{t_n}\| \to \infty$. But this contradicts the uniform boundedness principle since $\mathfrak{A}^{t_n} x \to x$ for each $x \in X$. Thus indeed, there exist $T > 0$ and $M \geq 1$ such that $\|\mathfrak{A}^t\| < M$ for $t \in [0, T]$. If $t \geq 0$ is arbitrary, then we can choose some $n = 0, 1, 2, \ldots$ such that $nT \leq t \leq (n+1)T$. By the semigroup property

$$\|\mathfrak{A}^t\| = \|(\mathfrak{A}^T)^n \mathfrak{A}^{t-nT}\| \leq \|\mathfrak{A}^T\|^n \|\mathfrak{A}^{t-nT}\| \leq M^{n+1} = M e^{\omega n T} \leq M e^{\omega t},$$

where $\omega = 1/T \log M$.

(ii) Let $t, h > 0$. The right continuity of $\mathfrak{A}^t x$ follows from

$$\begin{aligned} |\mathfrak{A}^{t+h} x - \mathfrak{A}^t x| &= |\mathfrak{A}^t(\mathfrak{A}^h x - x)| \\ &\leq \|\mathfrak{A}^t\| |\mathfrak{A}^h x - x| \leq M e^{\omega t} |\mathfrak{A}^h x - x|, \end{aligned}$$

and the left continuity from (take $0 \leq h \leq t$)

$$\begin{aligned} |\mathfrak{A}^{t-h} x - \mathfrak{A}^t x| &= |\mathfrak{A}^{t-h}(x - \mathfrak{A}^h x)| \\ &\leq \|\mathfrak{A}^{t-h}\| |\mathfrak{A}^h x - x| \leq M e^{\omega t} |x - \mathfrak{A}^h x|. \end{aligned}$$

□

Above we first defined what we mean by a well-posed linear system $\Sigma = \left[\begin{smallmatrix} \mathfrak{A} & \mathfrak{B} \\ \mathfrak{C} & \mathfrak{D} \end{smallmatrix}\right]$, then defined the operator families $\mathfrak{A}_s^t$, $\mathfrak{B}_s^t$, $\mathfrak{C}_s^t$, and $\mathfrak{D}_s^t$ (and a few more) in Definition 2.2.6, and finally used these operator families to define what we mean by the state trajectory and the output function of Σ corresponding to a given initial time s, a given initial state x_s, and a given input function u. Conversely, suppose that we have some independent method of constructing the state $x(t)$ and output y corresponding to any given initial time $s \le t$, any given initial state x_s, and any given input function u. If the mapping from x_s and u to $x(t)$ and $\pi_{[s,t)}y$ is linear, then it can be written in the form (compare this to (2.2.7))

$$\begin{aligned} x(t) &= \mathfrak{A}_s^t x_s + \mathfrak{B}_s^t u, \\ \pi_{[s,t)} y &= \mathfrak{C}_s^t x_s + \mathfrak{D}_s^t u. \end{aligned} \qquad t \ge s, \tag{2.2.9}$$

This equation then serves as a *definition* of the operator families $\mathfrak{A}_s^t$, $\mathfrak{B}_s^t$, $\mathfrak{C}_s^t$, and $\mathfrak{D}_s^t$. If these families satisfy the crucial parts of the identities listed in Lemma 2.2.8, then, as the following theorem shows, there is an underlying system $\Sigma = \left[\begin{smallmatrix} \mathfrak{A} & \mathfrak{B} \\ \mathfrak{C} & \mathfrak{D} \end{smallmatrix}\right]$.

Theorem 2.2.14 *Let U, X, and Y be Banach spaces, and let $\mathfrak{A}_s^t \colon X \to X$, $\mathfrak{B}_s^t \colon L^p|Reg_{c,\mathrm{loc}}(\mathbb{R};U) \to X$, $\mathfrak{C}_s^t \colon X \to L^p|Reg_{c,\mathrm{loc}}(\mathbb{R};Y)$, and $\mathfrak{D}_s^t \colon L^p|Reg_{c,\mathrm{loc}}(\mathbb{R};U) \to L^p|Reg_{c,\mathrm{loc}}(\mathbb{R};Y)$ be four families of bounded linear operators indexed by $-\infty < s \le t < \infty$. Suppose that $\mathfrak{A}_0^0 x = \lim_{t\downarrow 0} \mathfrak{A}_0^t x = x$ for all $x \in X$, that for all $s \le t$ and $h \in \mathbb{R}$ the time invariance condition* (2.2.5) *holds, and that for all $s \le r \le t$,*

$$\begin{aligned} \left[\begin{array}{c|c} \mathfrak{A}_s^t & \mathfrak{B}_s^t \\ \hline \mathfrak{C}_s^t & \mathfrak{D}_s^t \end{array}\right] &= \begin{bmatrix} 1 & 0 & 0 \\ 0 & \pi_{[s,r)} & \pi_{[r,t)} \end{bmatrix} \left[\begin{array}{c|c|c} \mathfrak{A}_r^t & 0 & \mathfrak{B}_r^t \\ \hline 0 & 1 & 0 \\ \hline \mathfrak{C}_r^t & 0 & \mathfrak{D}_r^t \end{array}\right] \\ &\quad \times \left[\begin{array}{c|c|c} \mathfrak{A}_s^r & \mathfrak{B}_s^r & 0 \\ \hline \mathfrak{C}_s^r & \mathfrak{D}_s^r & 0 \\ 0 & 0 & 1 \end{array}\right] \begin{bmatrix} 1 & 0 \\ 0 & \pi_{[s,r)} \\ 0 & \pi_{[r,t)} \end{bmatrix}. \end{aligned} \tag{2.2.10}$$

Then we get a $L^p|Reg$-well-posed linear system $\Sigma = \left[\begin{smallmatrix} \mathfrak{A} & \mathfrak{B} \\ \mathfrak{C} & \mathfrak{D} \end{smallmatrix}\right]$ by defining $\mathfrak{A}^t = \mathfrak{A}_0^t$ for $t \ge 0$ and, for all $x \in X$ and $u \in L^p|Reg_{c,\mathrm{loc}}(\mathbb{R};U)$,

$$\mathfrak{B}u = \lim_{s\to-\infty} \mathfrak{B}_s^0 u, \qquad \mathfrak{C}x = \lim_{t\to\infty} \mathfrak{C}_0^t x, \qquad \mathfrak{D}u = \lim_{\substack{s\to-\infty \\ t\to\infty}} \mathfrak{D}_s^t u \tag{2.2.11}$$

(in particular, these limits exist in X, $L^p|Reg_{\mathrm{loc}}(\mathbb{R}^+;Y)$, respectively $L^p|Reg_{c,\mathrm{loc}}(\mathbb{R};Y)$). Moreover, the given operator families $\mathfrak{A}_s^t$, $\mathfrak{B}_s^t$, $\mathfrak{C}_s^t$, and $\mathfrak{D}_s^t$ are identical to those derived from Σ as described in Definition 2.2.6.

Proof We begin by expanding (2.2.10) into

$$\begin{aligned}\mathfrak{A}_s^t &= \mathfrak{A}_r^t\mathfrak{A}_s^r,\\ \mathfrak{B}_s^t &= \mathfrak{B}_r^t\pi_{[r,t)} + \mathfrak{A}_r^t\mathfrak{B}_s^r\pi_{[s,r)},\\ \mathfrak{C}_s^t &= \pi_{[r,t)}\mathfrak{C}_r^t\mathfrak{A}_s^r + \pi_{[s,r)}\mathfrak{C}_s^r,\\ \mathfrak{D}_s^t &= \pi_{[r,t)}\mathfrak{D}_r^t\pi_{[r,t)} + \pi_{[r,t)}\mathfrak{C}_r^t\mathfrak{B}_s^r\pi_{[s,r)} + \pi_{[s,r)}\mathfrak{D}_s^r\pi_{[s,r)}.\end{aligned} \tag{2.2.12}$$

From here we observe that (2.2.4) holds for all $s \le t$. Thus, (2.2.10) is equivalent to the combination of the causality property (2.2.4) and the composition property (2.2.6).

Next we show that $t \mapsto \mathfrak{A}^t$ is a C_0 semigroup. The strong continuity requirement and the identity $\mathfrak{A}^0 = 1$ were assumed separately. The top left corners of (2.2.5) and (2.2.6) give, for all $s, t \ge 0$,

$$\mathfrak{A}^{s+t} = \mathfrak{A}_0^{s+t} = \mathfrak{A}_s^{s+t}\mathfrak{A}_0^s = \mathfrak{A}_0^t\mathfrak{A}_0^s = \mathfrak{A}^t\mathfrak{A}^s.$$

Thus, $\mathfrak{A}$ is a C_0 semigroup.

From (2.2.12) we observe that for all $s \le r \le t$,

$$\mathfrak{B}_s^t\pi_{[r,t)} = \mathfrak{B}_r^t,\ \ \pi_{[s,r)}\mathfrak{C}_s^t = \mathfrak{C}_s^r,\ \ \mathfrak{D}_s^t\pi_{[r,t)} = \mathfrak{D}_r^t,\ \ \pi_{[s,r)}\mathfrak{D}_s^t = \mathfrak{D}_s^r.$$

Thus, trivially, for all $x \in X$ and $u \in L^p|Reg_{c,\text{loc}}(\mathbb{R}; U)$, the limits in (2.2.11) exist. Moreover, still by (2.2.12), for all $s \le r \le t$,

$$\mathfrak{B}_s^t\pi_{[s,r)} = \mathfrak{A}_r^t\mathfrak{B}_s^r, \qquad \pi_{[r,t)}\mathfrak{C}_s^t = \mathfrak{C}_r^t\mathfrak{A}_s^r.$$

We rewrite these equations, using (2.2.4) and Lemma 2.2.9, into

$$\mathfrak{B}_s^t\tau^{-r}\pi_{[s-r,0)} = \mathfrak{A}^{t-r}\mathfrak{B}_{s-r}^0, \qquad \pi_{[0,t-r)}\tau^r\mathfrak{C}_s^t = \mathfrak{C}_0^{t-r}\mathfrak{A}^{r-s}.$$

In the first equation we take $t = 0$ and let $s \to -\infty$ to get $\mathfrak{B}\tau^{-r}\pi_- = \mathfrak{A}^{-r}\mathfrak{B}$ for all $r \le 0$. In the second equation we take $s = 0$ and let $t \to \infty$ to get $\pi_+\tau^r\mathfrak{C} = \mathfrak{C}\mathfrak{A}^r$ for all $r \ge 0$.

We have now verified conditions (i)–(iii) in Definitions 2.2.1 and 2.2.3, and this only leaves the final condition (iv) concerning $\mathfrak{D}$. By (2.2.12),

$$\pi_{[r,t)}\mathfrak{D}_s^t\pi_{[s,r)} = \mathfrak{C}_r^t\mathfrak{B}_s^r, \qquad \pi_{[s,r)}\mathfrak{D}_s^t\pi_{[r,t)} = 0.$$

Taking $r = 0$ and letting $s \to -\infty$ and $t \to \infty$ we get $\pi_+\mathfrak{D}\pi_- = \mathfrak{C}\mathfrak{B}$ and $\pi_-\mathfrak{D}\pi_+ = 0$. Finally, letting $s \to -\infty$ and $t \to \infty$ in the identity $\tau^h\mathfrak{D}_{s+h}^{t+h} = \mathfrak{D}_s^t\tau^h$ (which is part of (2.2.5)) we get $\tau^h\mathfrak{D} = \mathfrak{D}\tau^h$ for all $h \in \mathbb{R}$. □

2.3 Basic examples of well-posed linear systems

Our first example of a well-posed linear system is the one discussed in Section 2.1:

Proposition 2.3.1 *Let U, X, and Y be Banach spaces. Let $\mathfrak{A}$ be a C_0 semigroup on X, and let $B \in \mathcal{B}(U; X)$, $C \in \mathcal{B}(X; Y)$, and $D \in \mathcal{B}(U; Y)$. Define $\mathfrak{B}$, $\mathfrak{C}$, and $\mathfrak{D}$ by* (2.1.5). *Then $\left[\begin{smallmatrix} \mathfrak{A} & \mathfrak{B} \\ \mathfrak{C} & \mathfrak{D} \end{smallmatrix}\right]$ is both a Reg-well-posed and an L^p-well-posed linear system on $(Y, X; U)$ for every p, $1 \le p \le \infty$. Moreover, the state trajectory x and the output function y of this system, as defined in Definition 2.2.7, coincide with the state trajectory and the output function of the system* (2.1.1) *defined in* (2.1.2).

Proof The algebraic conditions are easy to verify; see the calculations after the algebraic conditions 2.1.3. The definitions of the state trajectory and the output function have been chosen in such a way that (2.1.3) coincides with (2.2.1). It is assumed explicitly that $\mathfrak{A}$ is a C_0 semigroup. Thus, only the continuity requirements on $\mathfrak{B}$, $\mathfrak{C}$, and $\mathfrak{D}$ in Definitions 2.2.1 and 2.2.3 need to be checked.

To prove the continuity of $\mathfrak{B}\colon L^p|Reg_c(\mathbb{R}^-; U) \to X$ it suffices to prove continuity when Σ is L^1-well-posed, since $L^p|Reg([-T, 0); U) \subset L^1((-T, 0); U)$. In this case, for each $T > 0$,

$$\begin{aligned}\left\| \int_{-T}^{0} \mathfrak{A}^{-s} Bu(s)\, ds \right\|_X &\le \int_{-T}^{0} \|\mathfrak{A}^{-s}\| \|B\| |u(s)|\, ds \\ &\le \left(\sup_{0 \le t < T} \|\mathfrak{A}^{-s}\| \right) \|B\| \|u\|_{L^1((-T,0);U)}.\end{aligned}$$

Thus, $\mathfrak{B}$ is continuous $L^p|Reg_c(\mathbb{R}^-; U) \to X$. The continuity of $\mathfrak{C}$ follows from the strong continuity of $\mathfrak{A}$ and the fact that $C(\overline{\mathbb{R}}^+; U) \subset L^p|Reg_{\text{loc}}(\mathbb{R}^+; Y)$. The continuity of $\mathfrak{D}$ follows from the continuity of $\mathfrak{B}$ and the fact that $(\mathfrak{D}u)(t) = C\mathfrak{B}\tau^t u + Du(t)$ for all $t \in \mathbb{R}$. □

The book by Curtain and Zwart (1995) is devoted to systems of the type described in Proposition 2.3.1, and it contains several examples of processes with distributed actuators and sensors which can be modeled in this way. (Boundary control systems require B to be strictly unbounded and boundary observation systems require C to be strictly unbounded.)

Our next example is a *delay line*. The model for this line is very simple: a signal entering the right end of the line reappears T time units later at the left end. No signals pass in the opposite direction. The simplest way to model this as a well-posed linear system is to build it around a shift (semi)group on a suitable L^p-space. We shall need these shift semigroups later, too, so let us first take a closer look at them.

Example 2.3.2 *Let U be a Banach space, let $1 \le p \le \infty$, and let $T > 0$.*

(i) *The family τ^t, $t \in \mathbb{R}$, defined by*

$$(\tau^t u)(s) := u(s+t), \qquad s \in \mathbb{R},$$

is a group on $L^p(\mathbb{R};U)$ and on $Reg(\mathbb{R};U)$. It is strongly continuous on $L^p(\mathbb{R};U)$ for $1 \le p < \infty$ and on $BUC(\mathbb{R};U)$, but not on $L^\infty(\mathbb{R};U)$ and on $Reg(\mathbb{R};U)$. More precisely, $\tau^t u \to u$ in $L^\infty(\mathbb{R};U)$ or $Reg(\mathbb{R};U)$ as $t \to 0$ if and only if $\tau^t u \to u$ in $L^\infty(\mathbb{R};U)$ or $Reg(\mathbb{R};U)$ as $t \downarrow 0$, and this happens if and only if $u \in BUC(\mathbb{R};U)$. We call τ^t the bilateral left shift *on $L^p|Reg(\mathbb{R};U)$.*

(ii) *The family τ_+^t, $t \ge 0$, defined by*

$$(\tau_+^t u)(s) := (\pi_+ \tau^t u)(s) = \begin{cases} u(s+t), & s \ge 0, \\ 0, & \text{otherwise}, \end{cases}$$

is a semigroup on $L^p(\mathbb{R}^+;U)$ and on $Reg(\overline{\mathbb{R}}^+;U)$. It is strongly continuous on $L^p(\mathbb{R}^+;U)$ for $1 \le p < \infty$ and on $BUC(\overline{\mathbb{R}}^+;U)$, but not on $L^\infty(\mathbb{R}^+;U)$ and on $Reg(\overline{\mathbb{R}}^+;U)$. More precisely, $\tau_+^t u \to u$ in $L^\infty(\mathbb{R}^+;U)$ or $Reg(\overline{\mathbb{R}}^+;U)$ as $t \downarrow 0$ if and only if $u \in BUC(\overline{\mathbb{R}}^+;U)$. We call τ_+^t the incoming left shift *on $L^p|Reg(\overline{\mathbb{R}}^+;U)$.*

(iii) *The family τ_-^t, $t \ge 0$, defined by*

$$(\tau_-^t u)(t) := (\tau^t \pi_- u)(s) = \begin{cases} u(s+t), & s < -t, \\ 0, & \text{otherwise}, \end{cases}$$

is a semigroup on $L^p(\mathbb{R}^-;U)$ and on $Reg(\overline{\mathbb{R}}^-;U)$. It is strongly continuous on $L^p(\mathbb{R}^-;U)$ for $1 \le p < \infty$ and on $\{u \in BUC(\overline{\mathbb{R}}^-;U) \mid u(0) = 0\}$, but not on $L^\infty(\mathbb{R}^-;U)$ and on $Reg(\mathbb{R}^-;U)$. More precisely, $\tau_-^t u \to u$ in $L^\infty(\mathbb{R}^-;U)$ or $Reg(\mathbb{R}^-;U)$ as $t \downarrow 0$ if and only if $u \in BUC(\overline{\mathbb{R}}^-;U)$ and $u(0) = 0$. We call τ_-^t the outgoing left shift *on $L^p|Reg(\mathbb{R}^-;U)$.*

(iv) *The family $\tau_{[0,T)}^t$, $t \ge 0$, defined by*

$$\begin{aligned}(\tau_{[0,T)}^t u)(s) &:= (\pi_{[0,T)} \tau^t \pi_{[0,T)} u)(s) \\ &= \begin{cases} u(s+t), & 0 \le s < T-t, \\ 0, & \text{otherwise}, \end{cases}\end{aligned}$$

is a semigroup on $L^p((0,T);U)$ and on $Reg([0,T);U)$. It is strongly continuous on $L^p((0,T);U)$ for $1 \le p < \infty$ and on $\{u \in C([0,T];U) \mid u(T) = 0\}$, but not on $L^\infty((0,T);U)$ and on $Reg([0,T);U)$. More precisely, $\tau_{[0,T)}^t u \to u$ in $L^\infty((0,T);U)$ or $Reg([0,T);U)$ as $t \downarrow 0$ if and *only if* $u \in C([0,T];U)$ *and* $u(T)=0$. *We call* $\tau^t_{[0,T)}$ *the* finite left shift *on* $L^p|Reg((0,T);U)$.

(v) *The family* $\tau^t_{\mathbb{T}_T}$, $t \in \mathbb{R}$, *defined by*

$$(\tau^t_{\mathbb{T}_T} u)(s) := u(s+t), \qquad s \in \mathbb{R},$$

is a group on $L^p(\mathbb{T}_T;U)$ *and on* $Reg(\mathbb{T}_T;U)$. *It is strongly continuous on* $L^p(\mathbb{T}_T;U)$ *for* $1 \le p < \infty$ *and on* $C(\mathbb{T}_T;U)$, *but not on* $L^\infty(\mathbb{T}_T;U)$ *and on* $Reg(\mathbb{T}_T;U)$. *More precisely,* $\tau^t_{\mathbb{T}_T} u \to u$ *in* $L^\infty(\mathbb{T}_T;U)$ *or* $Reg(\mathbb{T}_T;U)$ *as* $t \to 0$ *if and only if* $\tau^t_{\mathbb{T}_T} u \to u$ *in* $L^\infty(\mathbb{T}_T;U)$ *or* $Reg(\mathbb{T}_T;U)$ *as* $t \downarrow 0$, *and this happens if and only if* $u \in C(\mathbb{T}_T;U)$. *We call* τ^t_{-} *the* circular left shift *on* $L^p|Reg(\mathbb{T}_T;U)$.

Thus, in each case, the (semi)group shifts the function on which it operates t time units to the left, replaces missing values by zero, and restricts the result to the appropriate interval. The notation $\mathbb{T}_T$ stands for the real line $\mathbb{R}$ where the points $t+mT$, $m=0,\pm1,\pm2,\ldots$, are identified, and $L^p(\mathbb{T}_T)$ and $Reg(\mathbb{T}_T)$ represent the spaces of T-periodic functions of type L^p or Reg. We remark that $L^p(\mathbb{T}_T)$, $Reg(\mathbb{T}_T)$ and $BC(\mathbb{T}_T)$ can be identified with $L^p((0,T))$, $Reg([0,T))$, and $\{u \in C([0,T]) \mid u(T)=u(0)\}$, respectively (that it, we restrict the periodic function u to some interval of length T).

Proof of Example 2.3.2. All the proofs are very similar, so we treat only case (i), and leave the others to the reader.

It is obvious that $\tau^0 = 1$, and it is trivial to verify the semigroup property $\tau^{s+t} = \tau^s\tau^t$.

To prove the strong continuity for $1 \le p < \infty$ we let $u \in L^p(\mathbb{R};U)$ be arbitrary and let $\epsilon > 0$. Choose some $v \in C(\mathbb{R};U)$ supported in some finite interval $[-T,T]$ such that $\|u-v\| \le \epsilon$ (this is possible since C_c is dense in L^p when $p<\infty$). Then $\tau^t v$ tends to v uniformly as $t \to 0$, and $\tau^t v - v$ vanishes outside of $[-T-1, T+1]$ for $|t| \le 1$. This implies that $\tau^t v \to v$ in $L^p(\mathbb{R};U)$, so we can make $\|\tau^t v - v\| \le \epsilon$ by choosing t small enough. Then

$$\begin{aligned}\|\tau^t u - u\| &\le \|\tau^t u - \tau^t v\| + \|\tau^t v - v\| + \|v-u\| \\ &= \|u-v\| + \|\tau^t v - v\| + \|v-u\| \le 3\epsilon.\end{aligned}$$

Thus, τ^t is strongly continuous on $L^p(\mathbb{R};U)$.

By definition, a *continuous* function u is uniformly continuous iff $\tau^t u(s) - u(s) \to 0$ uniformly in s as $t \to \infty$. Thus, τ^t is strongly continuous on $BUC(\mathbb{R};U)$. To show that τ^t is not strongly continuous on $L^\infty(\mathbb{R};U)$ and on $Reg(\mathbb{R};U)$ it suffices to consider the counter-example $u(t)=0$ for $t<0$ and $u(t)=u \ne 0$ for $t \ge 0$.

The more specific claims about the existence of the limits $\lim_{\tau\to 0}\tau^t u$ and $\lim_{\tau\downarrow 0}\tau^t u$ in $L^\infty(\mathbb{R};U)$ or $Reg(\mathbb{R};U)$ remain to be proved. The L^∞-case is more difficult, so let us concentrate on this case (to get the *Reg*-case we simply replace ess sup by sup everywhere).

The equivalence of the existence of the two different limits follows from the fact that (by a change of the parameter s),

$$\lim_{t\downarrow 0}\operatorname*{ess\,sup}_{s\in\mathbb{R}}|u(s+t)-u(s)| = \lim_{t\downarrow 0}\operatorname*{ess\,sup}_{s\in\mathbb{R}}|u(s)-u(s-t)|.$$

Thus, to prove the final claim in (i) it suffices to show that u is (a.e. equal to) a continuous function if $\tau^t u \to u$ in $L^\infty(\mathbb{R};U)$ as $t\downarrow 0$.

Suppose $\tau^t u\to u$ in $L^\infty(\mathbb{R};U)$ as $t\downarrow 0$. For $k=0,1,2,\dots$ we define

$$u_k(t) = k\int_t^{t+1/k} u(s)\,ds = k\int_0^{1/k} u(s+t)\,ds.$$

Then, $\sup_{t\in\mathbb{R}}|u_k(t)| \le \operatorname{ess\,sup}_{t\in\mathbb{R}}|u(t)|$ and, for all h,

$$\begin{aligned}|u_k(t+h)-u_k(t)| &\le k\int_0^{1/k}|u(s+t+h)-u(s+t)|\,ds\\ &\le k\int_0^{1/k}|\tau^h u(s+t)-u(s+t)|\,ds,\end{aligned}$$

which tends to zero as $h\downarrow 0$, uniformly in t. Thus, the sequence u_k is uniformly bounded and equicontinuous, so it converges uniformly on bounded intervals to a continuous limit. This limit is equal to u a.e. since $u_k(t)\to u(t)$ at every Lebesgue point of u as $k\to\infty$ (see, e.g., Gripenberg *et al.* 1990, Lemma 7.4, p. 67). We conclude that we can make u continuous by redefining it on a set of measure zero. □

We shall also need the following modification of this result:

Lemma 2.3.3 *Let $u\in L^p_{\mathrm{loc}}(\mathbb{R};U)$, where $1\le p\le\infty$.*

(i) *$\tau^t u\to u$ in $L^p_{\mathrm{loc}}(\mathbb{R};U)$ as $t\to 0$ if $p<\infty$.*
(ii) *$\tau^t u\to u$ in $L^\infty_{\mathrm{loc}}(\mathbb{R};U)$ as $t\to 0$ if and only if $\tau^t u\to u$ in $L^\infty_{\mathrm{loc}}(\mathbb{R};U)$ as $t\downarrow 0$, and this is true if and only if $u\in C(\mathbb{R};U)$.*
(iii) *$\tau^t_+ u\to u$ in $L^\infty_{\mathrm{loc}}(\mathbb{R}^+;U)$ as $t\downarrow 0$ if and only if $u\in C(\overline{\mathbb{R}}^+;U)$.*
(iv) *$\tau^t_- u\to u$ in $L^\infty_{\mathrm{loc}}(\mathbb{R}^-;U)$ as $t\downarrow 0$ if and only if $u\in C(\overline{\mathbb{R}}^-;U)$ and $u(0)=0$.*

Proof This follows from Example 2.3.2, because as long as we are only interested in the values of u on an interval $[-T,T]$ we can multiply u by a continuous function η satisfying $\eta(t)=1$ for $|t|\le T+1$ and $\eta(t)=0$ for $|t|\ge T+2$. The resulting function belongs to $L^p(\mathbb{R},U)$, or in the continuous case to $BUC(\mathbb{R};U)$.

(In (iii) and (iv) we identify $L^\infty_{\mathrm{loc}}(\mathbb{R}^+;U)$ with the subspace of functions in $L^\infty_{\mathrm{loc}}(\mathbb{R};U)$ which vanish on $\mathbb{R}^-$, and vice versa.) □

We could use any one of the four semigroups τ^t, τ^t_+, τ^t_-, or $\tau^t_{[0,T)}$ in Example 2.3.2(i)–(iv) as the central piece of a well-posed linear system which realizes the delay line example. For simplicity, let us use the one that 'stores the minimal amount of information', namely $\tau^t_{[0,T)}$.

Example 2.3.4 *Let U be a Banach space, and let $T > 0$ and $1 \le p \le \infty$. Define $Y = U$, $X = L^p((0,T);U)$, and*

$$(\mathfrak{A}^t x)(s) := (\tau^t_{[0,T)}x)(s) = \begin{cases} x(s+t), & 0 \le s < T - t, \\ 0, & \textit{otherwise}, \end{cases}$$

$$(\mathfrak{B}u)(s) := (\pi_{[0,T)}\tau^{-T}u)(s) = \begin{cases} u(s-T), & 0 \le s < T, \\ 0, & \textit{otherwise}, \end{cases}$$

$$\mathfrak{C}x := \pi_{[0,T)}x = \begin{cases} x(s), & 0 \le s < T, \\ 0, & \textit{otherwise}, \end{cases}$$

$$(\mathfrak{D}u)(s) := (\tau^{-T}u)(s) = u(s-T), \qquad s \in \mathbb{R}.$$

If $p < \infty$, then this is an L^p-well-posed linear system. If $p = \infty$, then it satisfies all the requirements of an L^∞-well-posed linear system except for the strong continuity of its semigroup.

The continuity of $\mathfrak{B}$, $\mathfrak{C}$, and $\mathfrak{D}$ is obvious, and it is not difficult to check that the algebraic conditions in Definition 2.2.1 hold. Instead of giving a formal proof, let us therefore give an informal interpretation of how this system works. For example, let us look at the initial value problem with initial time zero. The initial state consists of an old input to the system that has entered during the time interval $[-T,0)$. It is traveling to the left in the delay line, and shows up in the output during the time interval $[0,T)$. If the input u is zero on $\overline{\mathbb{R}}^+$, then the output will vanish on $[T,\infty)$. A nonzero input u enters the delay line at its right end, so that at time T the state consists of the restriction of u to $[0,T)$. This part will show up in the output during the time interval $[T,\infty)$. It will be an exact copy of the input function u, apart from the fact that it has been delayed by T time units.

We shall see later in Example 4.5.7 that it is impossible to realize the delay line as an L^∞-well-posed or *Reg*-well-posed linear system. It is also impossible to realize it as an L^1-well-posed linear system with a reflexive state space; this follows from Example 4.5.13.

Another method to construct a well-posed linear system is to use one or several well-posed linear systems to construct a new one. The following examples are of this type. We leave the easy verifications that these examples are well-posed linear systems to the reader.

Example 2.3.5 *Let* $\Sigma = \left[\begin{smallmatrix} \mathfrak{A} & \mathfrak{B} \\ \mathfrak{C} & \mathfrak{D} \end{smallmatrix}\right]$ *be a* $L^p|Reg$*-well-posed linear system on* (Y, X, U)*. For each* $\alpha \in \mathbb{C}$*, let* e_α *be the scalar function* $\mathrm{e}_\alpha(t) := \mathrm{e}^{\alpha t}$, $t \in \mathbb{R}$. *Then*

$$\Sigma_\alpha = \left[\begin{array}{c|c} \mathfrak{A}_\alpha & \mathfrak{B}_\alpha \\ \hline \mathfrak{C}_\alpha & \mathfrak{D}_\alpha \end{array}\right] := \left[\begin{array}{c|c} \mathrm{e}_\alpha \mathfrak{A} & \mathfrak{B}\mathrm{e}_{-\alpha} \\ \hline \mathrm{e}_\alpha \mathfrak{C} & \mathrm{e}_\alpha \mathfrak{D}\mathrm{e}_{-\alpha} \end{array}\right]$$

is a linear system which is well-posed in the same sense. We call Σ_α *the* exponential shift *of* Σ *by the amount* α.

Example 2.3.6 *Let* $\Sigma = \left[\begin{smallmatrix} \mathfrak{A} & \mathfrak{B} \\ \mathfrak{C} & \mathfrak{D} \end{smallmatrix}\right]$ *be a* $L^p|Reg$*-well-posed linear system on* (Y, X, U)*. For each* $\lambda > 0$*, let* γ_λ *be the* time compression operator

$$(\gamma_\lambda u)(s) := u(\lambda s), \qquad s \in \mathbb{R}.$$

Let $\mathfrak{A}_\lambda^t := \mathfrak{A}^{\lambda t}$ *for* $t \geq 0$*, and*

$$\Sigma_\lambda = \left[\begin{array}{c|c} \mathfrak{A}_\lambda & \mathfrak{B}_\lambda \\ \hline \mathfrak{C}_\lambda & \mathfrak{D}_\lambda \end{array}\right] := \left[\begin{array}{c|c} \mathfrak{A}_\lambda & \mathfrak{B}\gamma_{1/\lambda} \\ \hline \gamma_\lambda \mathfrak{C} & \gamma_\lambda \mathfrak{D}\gamma_{1/\lambda} \end{array}\right].$$

Then Σ_λ *is a linear system which is well-posed in the same sense. We call* Σ_λ *the* time compression *of* Σ *by the amount* λ.

Example 2.3.7 *Let* $\Sigma = \left[\begin{smallmatrix} \mathfrak{A} & \mathfrak{B} \\ \mathfrak{C} & \mathfrak{D} \end{smallmatrix}\right]$ *be a* $L^p|Reg$*-well-posed linear system on* (Y, X, U)*, and let* $E \in \mathcal{B}(X_1; X)$ *have an inverse in* $\mathcal{B}(X; X_1)$*. Define*

$$\Sigma_E := \left[\begin{array}{c|c} E^{-1}\mathfrak{A}E & E^{-1}\mathfrak{B} \\ \hline \mathfrak{C}E & \mathfrak{D} \end{array}\right].$$

Then Σ_E *is well-posed on* (Y, X_1, U) *in the same sense, and it has the same input/output map as* Σ*. We call* Σ_E *the* similarity transform *of* Σ *with similarity operator* E.

Example 2.3.8 *Let* $\Sigma = \left[\begin{smallmatrix} \mathfrak{A} & \mathfrak{B} \\ \mathfrak{C} & \mathfrak{D} \end{smallmatrix}\right]$ *be a* $L^p|Reg$*-well-posed linear system on* (Y, X, U)*, and let* $E \in \mathcal{B}(U_1; U)$*. Define*

$$\Sigma_E := \left[\begin{array}{c|c} \mathfrak{A} & \mathfrak{B}E \\ \hline \mathfrak{C} & \mathfrak{D}E \end{array}\right]$$

Then Σ_E *is well-posed on* (Y, X, U_1) *in the same sense.*

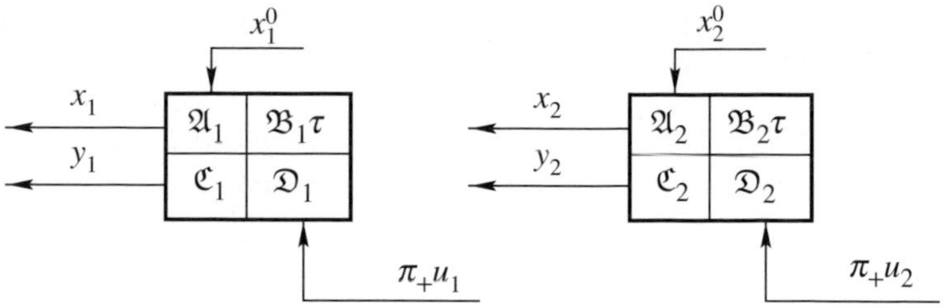

Figure 2.3 Cross-product (the union of two independent systems)

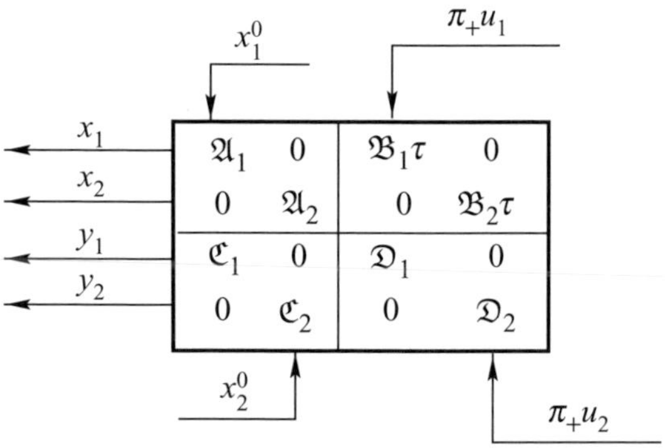

Figure 2.4 Cross-product in block form

Example 2.3.9 *Let* $\Sigma = \left[\begin{smallmatrix} \mathfrak{A} & \mathfrak{B} \\ \mathfrak{C} & \mathfrak{D} \end{smallmatrix}\right]$ *be a* $L^p|Reg$*-well-posed linear system on* (Y, X, U)*, and let* $E \in \mathcal{B}(Y; Y_1)$*. Define*

$$\Sigma_E := \left[\begin{array}{c|c} \mathfrak{A} & \mathfrak{B} \\ \hline E\mathfrak{C} & E\mathfrak{D} \end{array}\right].$$

Then Σ_E *is well-posed on* (Y_1, X, U) *in the same sense.*

Example 2.3.10 *Let* $\Sigma_1 = \left[\begin{smallmatrix} \mathfrak{A}_1 & \mathfrak{B}_1 \\ \mathfrak{C}_1 & \mathfrak{D}_1 \end{smallmatrix}\right]$ *be a* $L^p|Reg$*-well-posed linear system on* (Y_1, X_1, U_1)*, and let* $\Sigma_2 = \left[\begin{smallmatrix} \mathfrak{A}_2 & \mathfrak{B}_2 \\ \mathfrak{C}_2 & \mathfrak{D}_2 \end{smallmatrix}\right]$ *be another linear system on* (Y_2, X_2, U_2) *which is well-posed in the same sense (i.e., both are* L^p*-well-posed with the same value of* p*, or both are Reg-well-posed). Define*

$$U := \left[\begin{smallmatrix} U_1 \\ U_2 \end{smallmatrix}\right], \qquad X := \left[\begin{smallmatrix} X_1 \\ X_2 \end{smallmatrix}\right], \qquad Y := \left[\begin{smallmatrix} Y_1 \\ Y_2 \end{smallmatrix}\right],$$
$$\mathfrak{A} := \left[\begin{smallmatrix} \mathfrak{A}_1 & 0 \\ 0 & \mathfrak{A}_2 \end{smallmatrix}\right], \qquad \mathfrak{B} := \left[\begin{smallmatrix} \mathfrak{B}_1 & 0 \\ 0 & \mathfrak{B}_2 \end{smallmatrix}\right], \qquad \mathfrak{C} := \left[\begin{smallmatrix} \mathfrak{C}_1 & 0 \\ 0 & \mathfrak{C}_2 \end{smallmatrix}\right], \qquad \mathfrak{D} := \left[\begin{smallmatrix} \mathfrak{D}_1 & 0 \\ 0 & \mathfrak{D}_2 \end{smallmatrix}\right].$$

Then Σ *is a linear system on* (Y, X, U) *which is well-posed in the same sense as* Σ_1 *and* Σ_2*. See the equivalent Figures 2.3 and 2.4. We call* Σ *the* cross-product *of* Σ_1 *and* Σ_2*.*

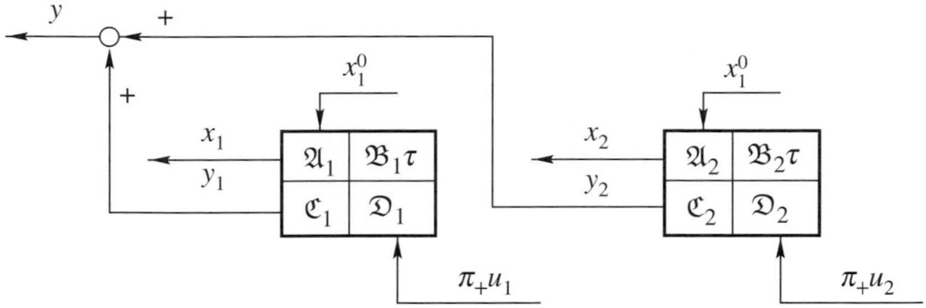

Figure 2.5 Sum junction

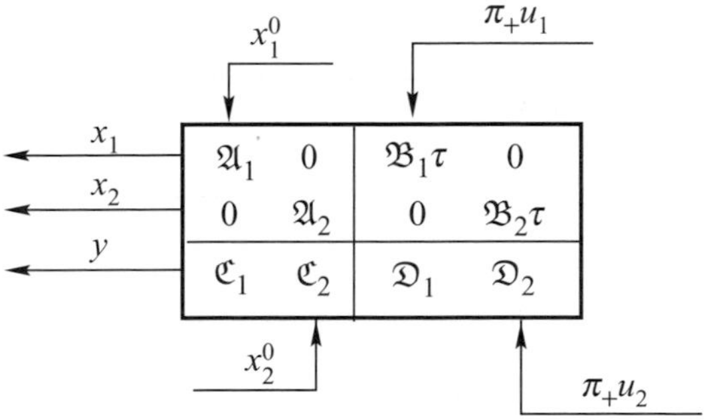

Figure 2.6 Sum junction in block form

Example 2.3.11 *Let* $\Sigma_1 = \left[\begin{array}{c|c} \mathfrak{A}_1 & \mathfrak{B}_1 \\ \hline \mathfrak{C}_1 & \mathfrak{D}_1 \end{array}\right]$ *be a* $L^p|Reg$*-well-posed linear system on* (Y_1, X_1, U)*, and let* $\Sigma_2 = \left[\begin{array}{c|c} \mathfrak{A}_2 & \mathfrak{B}_2 \\ \hline \mathfrak{C}_2 & \mathfrak{D}_2 \end{array}\right]$ *be another linear system on* (Y_2, X_2, U) *which is well-posed in the same sense. Define*

$$U := \begin{bmatrix} U_1 \\ U_2 \end{bmatrix}, \qquad X := \begin{bmatrix} X_1 \\ X_2 \end{bmatrix},$$
$$\mathfrak{A} := \begin{bmatrix} \mathfrak{A}_1 & 0 \\ 0 & \mathfrak{A}_2 \end{bmatrix}, \qquad \mathfrak{B} := \begin{bmatrix} \mathfrak{B}_1 & 0 \\ 0 & \mathfrak{B}_2 \end{bmatrix}, \qquad \mathfrak{C} := \begin{bmatrix} \mathfrak{C}_1 & \mathfrak{C}_2 \end{bmatrix}, \qquad \mathfrak{D} := \begin{bmatrix} \mathfrak{D}_1 & \mathfrak{D}_2 \end{bmatrix}.$$

Then Σ *is a linear system on* (Y, X, U) *which is well-posed in the same sense as* Σ_1 *and* Σ_2*. See the equivalent Figures 2.5 and 2.6. We call* Σ *the* sum junction *of* Σ_1 *and* Σ_2*.*

Example 2.3.12 *Let* $\Sigma_1 = \left[\begin{array}{c|c} \mathfrak{A}_1 & \mathfrak{B}_1 \\ \hline \mathfrak{C}_1 & \mathfrak{D}_1 \end{array}\right]$ *be a* $L^p|Reg$*-well-posed linear system on* (Y, X_1, U_1)*, and let* $\Sigma_2 = \left[\begin{array}{c|c} \mathfrak{A}_2 & \mathfrak{B}_2 \\ \hline \mathfrak{C}_2 & \mathfrak{D}_2 \end{array}\right]$ *be another linear system on* (Y, X_2, U_2) *which is well-posed in the same sense. Define*

$$X := \begin{bmatrix} X_1 \\ X_2 \end{bmatrix}, \qquad Y := \begin{bmatrix} Y_1 \\ Y_2 \end{bmatrix},$$
$$\mathfrak{A} := \begin{bmatrix} \mathfrak{A}_1 & 0 \\ 0 & \mathfrak{A}_2 \end{bmatrix}, \qquad \mathfrak{B} := \begin{bmatrix} \mathfrak{B}_1 \\ \mathfrak{B}_2 \end{bmatrix}, \qquad \mathfrak{C} := \begin{bmatrix} \mathfrak{C}_1 & 0 \\ 0 & \mathfrak{C}_2 \end{bmatrix}, \qquad \mathfrak{D} := \begin{bmatrix} \mathfrak{D}_1 \\ \mathfrak{D}_2 \end{bmatrix}.$$

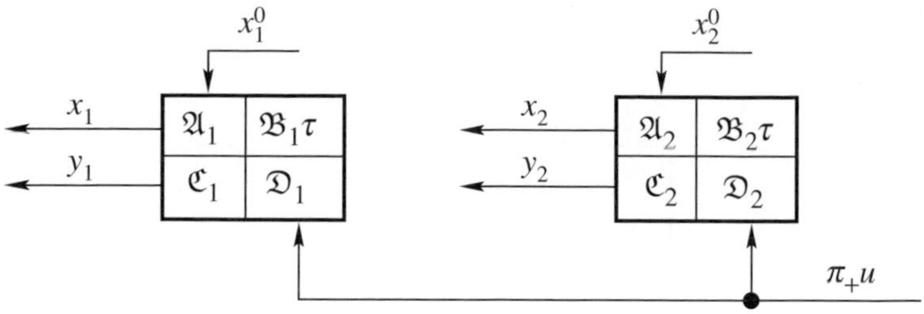

Figure 2.7 T-junction

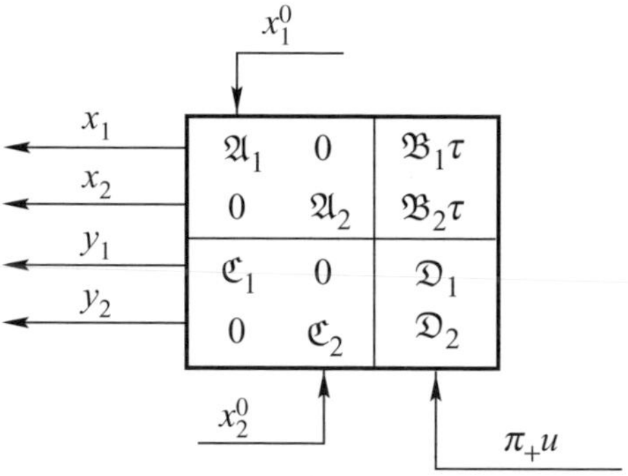

Figure 2.8 T-junction in block form

Then Σ is a linear system on (Y, X, U) which is well-posed in the same sense as Σ_1 and Σ_2. See the equivalent Figures 2.7 and 2.8. We call Σ the T-junction *of Σ_1 and Σ_2.*

Example 2.3.13 *Let $\Sigma_1 = \left[\begin{array}{c|c} \mathfrak{A}_1 & \mathfrak{B}_1 \\ \hline \mathfrak{C}_1 & \mathfrak{D}_1 \end{array}\right]$ be a $L^p|Reg$-well-posed linear system on (Y, X_1, U), and let $\Sigma_2 = \left[\begin{array}{c|c} \mathfrak{A}_2 & \mathfrak{B}_2 \\ \hline \mathfrak{C}_2 & \mathfrak{D}_2 \end{array}\right]$ be another linear system on (Y, X_2, U) which is well-posed in the same sense. Define*

$$X := \begin{bmatrix} X_1 \\ X_2 \end{bmatrix},$$
$$\mathfrak{A} := \begin{bmatrix} \mathfrak{A}_1 & 0 \\ 0 & \mathfrak{A}_2 \end{bmatrix}, \qquad \mathfrak{B} := \begin{bmatrix} \mathfrak{B}_1 \\ \mathfrak{B}_2 \end{bmatrix}, \qquad \mathfrak{C} := \begin{bmatrix} \mathfrak{C}_1 & \mathfrak{C}_2 \end{bmatrix}, \qquad \mathfrak{D} := \mathfrak{D}_1 + \mathfrak{D}_2.$$

Then Σ is a linear system on (Y, X, U) which is well-posed in the same sense as Σ_1 and Σ_2. See the equivalent Figures 2.9 and 2.10. We call Σ the parallel connection *of Σ_1 and Σ_2.*

We postpone the presentation of the more complicated cascade and feedback connections to Section 7.2.

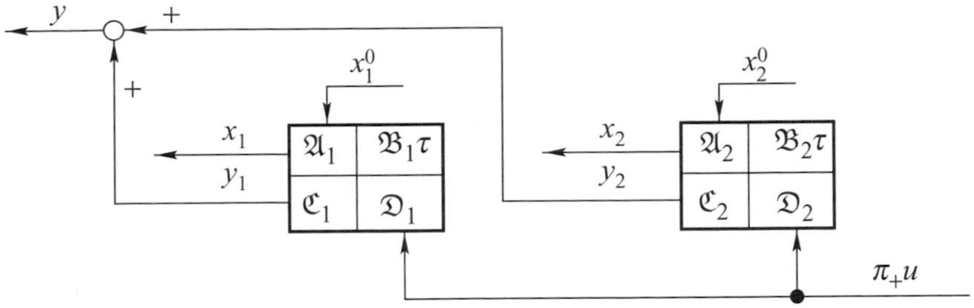

Figure 2.9 Parallel connection

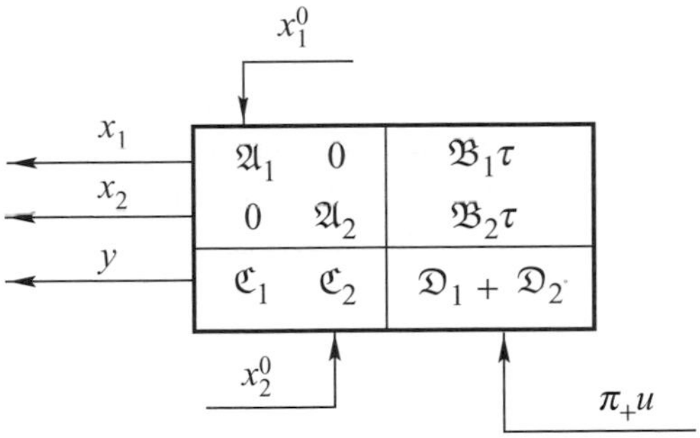

Figure 2.10 Parallel connection in block form

2.4 Time discretization

Our next theorem shows that it is possible to turn a well-posed linear system into a *discrete time system* by discretizing the time.

Theorem 2.4.1 *Let U, X, and Y be Banach spaces, let $\Sigma = \left[\begin{smallmatrix} \mathfrak{A} & \mathfrak{B} \\ \mathfrak{C} & \mathfrak{D} \end{smallmatrix}\right]$ be a $L^p|Reg$-well-posed linear system on (Y, X, U), and let $T > 0$. For each $x_0 \in X$, $u \in L^p|Reg_{\mathrm{loc}}(\overline{\mathbb{R}}^+; U)$ and $n = 0, 1, 2, \ldots$, we define*

$$\begin{aligned} u_n &:= \pi_{[0,T)}\tau^{nT} u = \tau^{nT}\pi_{[nT,(n+1)T)}u, \\ x_n &:= x(nT), \\ y_n &:= \pi_{[0,T)}\tau^{nT} y = \tau^{nT}\pi_{[nT,(n+1)T)}y, \end{aligned}$$

where x and y are the state trajectory and the output function of Σ with initial time zero, initial value x_0 and input function u. Then, for each $n \geq 0$,

$$\begin{aligned} x_{n+1} &= \mathbf{A}x_n + \mathbf{B}u_n, \\ y_n &= \mathbf{C}x_n + \mathbf{D}u_n, \end{aligned} \tag{2.4.1}$$

where the operators (the notation is explained in Definition 2.2.6)

$$\mathbf{A} := \mathfrak{A}_0^T = \mathfrak{A}^T, \qquad \mathbf{B} := \mathfrak{B}_0^T = \mathfrak{B}\tau^T\pi_{[0,T)},$$
$$\mathbf{C} := \mathfrak{C}_0^T = \pi_{[0,T)}\mathfrak{C}, \qquad \mathbf{D} := \mathfrak{D}_0^T = \pi_{[0,T)}\mathfrak{D}\pi_{[0,T)},$$

are bounded linear operators between the following spaces:

$$\mathbf{A}\colon X \to X, \qquad \mathbf{B}\colon L^p((0,T);U) \to X,$$
$$\mathbf{C}\colon X \to L^p((0,T);Y), \qquad \mathbf{D}\colon L^p((0,T);U) \to L^p((0,T);Y)$$

(in the Reg-well-posed case we replace $L^p((0,T))$ by $Reg([0,T))$ throughout).

Proof This follows immediately from Lemma 2.2.8 and Theorem 2.2.11. □

Thus, we get a standard discrete time system with bounded operators **A**, **B**, **C**, and **D**, and with state space X, input space $L^p|Reg([0,T);U)$, and output space $L^p|Reg([0,T);Y)$. Observe that the new input and output spaces are always infinite-dimensional, even if U and Y are finite-dimensional.

It is also possible to go in the opposite direction: If we know the solution sequence x_n, y_n of the discrete time system (2.4.1), then it is possible to recreate the state trajectory x and the output function y of the original system Σ using only the operators **A** and **B**:

Theorem 2.4.2 *Let $\Sigma = \left[\begin{smallmatrix}\mathfrak{A} & \mathfrak{B}\\ \mathfrak{C} & \mathfrak{D}\end{smallmatrix}\right]$ be a $L^p|Reg$-well-posed linear system on (Y,X,U), and let $T > 0$. Let $x_0 \in X$, let u_n be a sequence in $L^p|Reg([0,T);U)$, let x_n, $n \ge 1$, and y_n, $n \ge 0$, be the solution of* (2.4.1) *with initial value x_0, and define $u \in L^p|Reg_{\mathrm{loc}}(\overline{\mathbb{R}}^+;U)$ by*

$$u = \sum_{n=0}^{\infty} \tau^{-nT}\pi_{[0,T)}u_n.$$

Let x and y be the state trajectory and the output function of Σ with initial time zero, initial state x_0, and input function u. Then,

$$\begin{aligned} x(t) &= \mathfrak{A}^{t-nT}x_n + \mathfrak{B}\tau^{t-nT}\pi_{[0,T)}u_n \\ &= \mathfrak{A}^{t-nT}x_n + \mathbf{B}\tau^{t-(n+1)T}\pi_{[0,T)}u_n, \qquad nT \le t < (n+1)T, \\ y &= \sum_{n=0}^{\infty} \tau^{-nT}\pi_{[0,T)}y_n. \end{aligned}$$

Thus, in order to recreate the state trajectory and the output function of the original system it suffices to know $\mathfrak{A}^t$ for $0 \le t \le T$ and $\mathfrak{B}\pi_{[-T,0)}$ (in addition to the discrete time solution).

The straightforward proof of this theorem is left to the reader.

The one-to-one correspondence between the original system and its discrete time counterpart established in Theorems 2.4.1 and 2.4.2 is important for two reasons: it is fundamental in the study of time discretizations of the original system Σ, and it is also useful in the general study of the original system Σ. At the moment the second property is the one that interests us the most.

If we know the solution and output of Σ for arbitrary initial states x_0 and inputs u, then we actually know all the four basic operators $\mathfrak{A}$, $\mathfrak{B}$, $\mathfrak{C}$, and $\mathfrak{D}$. It follows from Theorems 2.4.1 and 2.4.2 that these operators can be rewritten in terms of the operators **A**, **B**, **C**, and **D**, or equivalently, in terms of the 'local' operators $\mathfrak{A}^t$, $0 \le t \le T$, $\mathfrak{B}^0_{-T} = \mathfrak{B}\pi_{[-T,0)}$, $\mathfrak{C}_0^T = \pi_{[0,T)}\mathfrak{C}$, and $\mathfrak{D}_0^T = \pi_{[0,T)}\mathfrak{D}\pi_{[0,T)}$. The exact correspondence is the following:

Lemma 2.4.3 *Let* $\Sigma = \left[\begin{smallmatrix}\mathfrak{A} & \mathfrak{B}\\ \mathfrak{C} & \mathfrak{D}\end{smallmatrix}\right]$ *be a* $L^p|Reg$*-well-posed linear system on* (Y, X, U)*, and let* $T > 0$*. Then the operators* $\mathfrak{A}$, $\mathfrak{B}$, $\mathfrak{C}$, *and* $\mathfrak{D}$ *satisfy*

$$\mathfrak{A}^t = \mathfrak{A}^{nT}\mathfrak{A}^{t-nT}, \quad n = 0, 1, 2, \ldots, \quad t \ge nT,$$

$$\mathfrak{B} = \sum_{n=0}^{\infty} \mathfrak{A}^{nT}\mathfrak{B}\pi_{[-T,0)}\tau^{-nT},$$

$$\mathfrak{C} = \sum_{n=0}^{\infty} \tau^{-nT}\pi_{[0,T)}\mathfrak{C}\mathfrak{A}^{nT},$$

$$\begin{aligned}
\mathfrak{D} &= \sum_{n=-\infty}^{\infty}\sum_{k=0}^{\infty} \tau^{-nT}\pi_{[0,T)}\Big(\mathfrak{C}\mathfrak{A}^{kT}\mathfrak{B}\pi_{[-T,0)}\tau^{-kT} + \mathfrak{D}\pi_{[0,T)}\Big)\tau^{nT}\\
&= \sum_{n=-\infty}^{\infty} \tau^{-nT}\pi_{[0,T)}\left(\mathfrak{C}\mathfrak{B} + \mathfrak{D}\pi_{[0,T)}\right)\tau^{nT}\\
&= \sum_{n=-\infty}^{\infty} \tau^{-nT}\left(\mathfrak{C}\mathfrak{B}\pi_{[-T,0)} + \pi_{[0,T)}\mathfrak{D}\pi_{[0,T)}\right)\tau^{nT}.
\end{aligned}$$

Thus, the operators $\mathfrak{A}$, $\mathfrak{B}$, $\mathfrak{C}$, *and* $\mathfrak{D}$ *can be reconstructed from the preceding formulas if we know* $\mathfrak{A}^t$ *for* $0 \le t \le T$, $\mathfrak{B}\pi_{[-T,0)}$, $\pi_{[0,T)}\mathfrak{C}$, *and* $\pi_{[0,T)}\mathfrak{D}\pi_{[0,T)}$.

Before proving Lemma 2.4.3, let us comment on the convergence of the infinite sums. This convergence is actually trivial in the sense that in each case of interest there are only finitely many nonzero terms in each sum, due to the fact that we are working in $L^p|Reg_{c,\mathrm{loc}}$. This means that every u to which we apply $\mathfrak{B}$ and $\mathfrak{D}$ vanishes on some interval $(-\infty, -M)$, and that we are only interested in the values of $\mathfrak{C}x_0$ and $\mathfrak{D}u$ on the finite interval $[-M, M)$. In other words, we may multiply $\mathfrak{B}$ and $\mathfrak{D}$ by $\pi_{[-M,\infty)}$ to the right and multiply $\mathfrak{C}$ and $\mathfrak{D}$ by $\pi_{[-M,M)}$ to the left. Then only finitely many terms in each sum will be nonzero because

$$\pi_{[-T,0)}\tau^{-nT}\pi_{[-M,\infty)} = \pi_{[-T,0)}\pi_{[nT-M,\infty)}\tau^{-nT}$$

vanishes for all $n \geq M/T$ and

$$\pi_{[-M,M)}\tau^{-nT}\pi_{[0,T)} = \tau^{-nT}\pi_{[-M-nT,M-nT)}\pi_{[0,T)}$$

vanishes for all $n \geq M/T$ and for all $n \leq -M/T - 1$.

Proof of Lemma 2.4.3 The formula for $\mathfrak{A}$ is an immediate consequence of the semigroup property $\mathfrak{A}^{s+t} = \mathfrak{A}^s\mathfrak{A}^t$ for $s, t \geq 0$.

To prove the formula for $\mathfrak{B}$ we solve (2.4.1) recursively to get for all $n = 1, 2, 3 \ldots,$

$$x_n = \mathbf{A}^n x_0 + \sum_{k=0}^{n-1} \mathbf{A}^k \mathbf{B} u_{n-k-1},$$

or, if we rewrite this in terms of the original state and the original operators $\mathfrak{A}$ and $\mathfrak{B}$,

$$x_n = x(nT) = \mathfrak{A}^{nT}x_0 + \sum_{k=0}^{n-1} \mathfrak{A}^{kT}\mathfrak{B}\pi_{[-T,0)}\tau^{(n-k)T}u.$$

On the other hand, by Theorem 2.2.11,

$$x(nT) = \mathfrak{A}^{nT}x_0 + \mathfrak{B}\tau^{nT}\pi_{[0,nT)}u.$$

This being true for all x_0 and u, we find that $(\mathfrak{A}^T)^n = \mathfrak{A}^{nT}$ and that

$$\mathfrak{B}\pi_{[-nT,0)} = \sum_{k=0}^{n-1} \mathfrak{A}^{kT}\mathfrak{B}\pi_{[-T,0)}\tau^{-kT},$$

from which we get the desired formula (with n replaced by k) by letting $n \to \infty$.

The formulas for $\mathfrak{C}$ and $\mathfrak{D}$ follow in a similar way from the fact that (cf. Theorem 2.4.1)

$$y_n = \mathbf{C}x_n + \mathbf{D}u_n = \mathbf{C}\left(\mathbf{A}^n x_0 + \sum_{k=0}^{n-1} \mathbf{A}^k \mathbf{B} u_{n-k-1}\right) + \mathbf{D}u_n,$$

hence (cf. Theorems 2.4.1 and 2.4.2)

$$\begin{aligned}
y &= \sum_{n=0}^{\infty} \tau^{-nT}\pi_{[0,T)}y_n \\
&= \sum_{n=0}^{\infty} \tau^{-nT}\pi_{[0,T)}\Bigg(\mathfrak{C}\mathfrak{A}^{nT}x_0 \\
&\quad + \mathfrak{C}\sum_{k=0}^{n-1} \mathfrak{A}^{kT}\mathfrak{B}\pi_{[-T,0)}\tau^{(n-k)T}u + \mathfrak{D}\pi_{[0,T)}\tau^{nT}u\Bigg).
\end{aligned}$$

On the other hand, the output y is given by

$$y = \mathfrak{C}x_0 + \mathfrak{D}\pi_+ u.$$

Equating the coefficients for x_0 in these formulas we get the desired formula for $\mathfrak{C}$, whereas the coefficients for u give

$$\mathfrak{D}\pi_+ = \sum_{n=0}^{\infty}\sum_{k=0}^{n-1} \tau^{-nT}\pi_{[0,T)}\Big(\mathfrak{C}\mathfrak{A}^{kT}\mathfrak{B}\pi_{[-T,0)}\tau^{-kT} + \mathfrak{D}\pi_{[0,T)}\Big)\tau^{nT}.$$

Multiply this by τ^{-NT} to the right and by τ^{NT} to the left, use the shift-invariance of $\mathfrak{D}$, and make a change of summation variable to get

$$\begin{aligned} \mathfrak{D}\pi_{[-NT,\infty)} = & \sum_{n=-N}^{\infty}\sum_{k=0}^{N+n-1} \tau^{-nT}\pi_{[0,T)} \\ & \times \Big(\mathfrak{C}\mathfrak{A}^{kT}\mathfrak{B}\pi_{[-T,0)}\tau^{-kT} + \mathfrak{D}\pi_{[0,T)}\Big)\tau^{nT}. \end{aligned} \tag{2.4.2}$$

We get the first of the three given formulas for $\mathfrak{D}$ by letting $N \to \infty$. The other two formulas follow from this one and the formulas for $\mathfrak{B}$ and $\mathfrak{C}$. □

The preceding lemma has several important consequences, the first of which is the following:

Theorem 2.4.4 *Let $\Sigma = \left[\begin{smallmatrix} \mathfrak{A} & \mathfrak{B} \\ \mathfrak{C} & \mathfrak{D} \end{smallmatrix}\right]$ be an L^p-well-posed linear system on (Y, X, U) for some p, $1 \le p \le \infty$, and let $T > 0$.*

(i) *Let $q \ge p$. If $\pi_{[0,T)}\mathfrak{C}$ maps X into $L^q([0,T);Y)$, then $\mathfrak{C}$ maps X continuously into $L^q_{\text{loc}}(\mathbb{R}^+;U)$, and if, in addition, $\pi_{[0,T)}\mathfrak{D}\pi_{[0,T)}$ maps $L^q((0,T);U)$ into $L^q((0,T);Y)$, then $\mathfrak{D}$ maps $L^q_{c,\text{loc}}(\mathbb{R};U)$ into $L^q_{c,\text{loc}}(\mathbb{R};Y)$. Thus, in this case Σ is also an L^q-well-posed linear system on (Y, X, U).*

(ii) *Let $1 \le q \le p$. If $\mathfrak{B}\pi_{[-T,0)}$ can be extended to a continuous map from $L^q([-T,0);U)$ into X, then $\mathfrak{B}$ can be extended to a continuous map from $L^q_c(\mathbb{R}^-;U)$ into X, and if, in addition, $\pi_{[0,T)}\mathfrak{D}\pi_{[0,T)}$ can be extended to a continuous map from $L^q((0,T);U)$ into $L^q((0,T);Y)$ then $\mathfrak{D}$ can be extended to a continuous map from $L^q_{c,\text{loc}}(\mathbb{R};U)$ into $L^q_{c,\text{loc}}(\mathbb{R};U)$. Thus, in this case Σ can be extended to an L^q-well-posed linear system on (Y, X, U).*

This follows immediately from Lemma 2.4.3 and the fact that

$$L^q_{c,\text{loc}}(\mathbb{R};U) \subset L^p_{c,\text{loc}}(\mathbb{R};U) \text{ for } q \ge p.$$

2.5 The growth bound

Our next task will be to employ Lemma 2.4.3 to develop a *global growth estimate* on the operators $\mathfrak{A}$, $\mathfrak{B}$, $\mathfrak{C}$, and $\mathfrak{D}$. The growth estimates on $\mathfrak{B}$, $\mathfrak{C}$, and $\mathfrak{D}$ are given in terms of a weighted L^p-space:

Definition 2.5.1 Let $1 \le p \le \infty$, $J \subset \mathbb{R}$, $\omega \in \mathbb{R}$, and let U be a Banach space.

(i) The space $L^p_\omega(J;U)$ consists of all functions $u\colon J \to U$ for which the function $e_{-\omega}u$ belongs to $L^p(J;U)$ where $e_{-\omega}(t) := e^{-\omega t}$, $t \in \mathbb{R}$. The norm of u in $L^p_\omega(J;U)$ (which we often denote by $\|u\|_\omega$) is equal to the norm of $e_{-\omega}u$ in $L^p(J;U)$.

(ii) The space $L^p_{\omega,\mathrm{loc}}(\mathbb{R};U)$ consists of all functions $u \in L^p_{\mathrm{loc}}(\mathbb{R};U)$ which satisfy $\pi_- u \in L^p_\omega(\mathbb{R}^-;U)$.

(iii) The spaces $L^p_{0,\omega}(J;U)$, $L^p_{0,\omega,\mathrm{loc}}(\mathbb{R};U)$, $BC_\omega(J;U)$, $BC_{\omega,\mathrm{loc}}(\mathbb{R};U)$, $BC_{0,\omega}(J;U)$, $BC_{0,\omega,\mathrm{loc}}(\mathbb{R};U)$, $BUC_\omega(J;U)$, $BUC_{\omega,\mathrm{loc}}(\mathbb{R};U)$, $Reg_\omega(J;U)$, $Reg_{\omega,\mathrm{loc}}(\mathbb{R};U)$, $Reg_{0,\omega}(J;U)$, and $Reg_{0,\omega,\mathrm{loc}}(\mathbb{R};U)$, are defined in an analogous way, with L^p replaced by L^p_0, BC, BC_0, BUC, Reg, or Reg_0, respectively.[8]

The operators π_J and τ^t act on these ω-weighted spaces as follows:

Lemma 2.5.2 *Let* $1 \le p \le \infty$, $\omega \in \mathbb{R}$, *and* $u \in L^p_\omega(\mathbb{R};U)$. *Define* $e_{-\omega}(t) := e^{-\omega t}$ *for* $t \in \mathbb{R}$.

(i) *For each* $J \subset \mathbb{R}$ *(of positive measure) the operator* π_J *is a projection operator in* $L^p_\omega(J;U)$ *and in* $Reg_\omega(J;U)$ *(i.e.,* $\pi_J = \pi_J^2$*) with norm* $\|\pi_J\|_\omega = 1$.

(ii) $e_{-\omega}\tau^t u = e^{\omega t}\tau^t(e_{-\omega}u)$ *for* $t \in \mathbb{R}$. *In particular,* $\tau^t u \to u$ *in* $L^p_\omega(\mathbb{R};U)$ *as* $t \to 0$ *if and only if* $\tau^t(e_{-\omega}u) \to e_{-\omega}u$ *in* $L^p(\mathbb{R};U)$ *as* $t \to 0$. *The same claim is true if we replace* L^p *by* BC, BUC, BC_0, Reg, *or* Reg_0.

Proof (i) This is obvious.

(ii) For all $s \in \mathbb{R}$,

$$\big(e^{\omega t}\tau^t(e_{-\omega}u)\big)(s) = e^{\omega t}e^{-\omega(s+t)}u(s+t) = e^{-\omega s}u(s+t),$$

hence $e_{-\omega}\tau^t u = e^{\omega t}\tau^t(e_{-\omega}u)$. To prove the second claim it suffices to observe that $\tau^t u \to u$ in $L^p_\omega(\mathbb{R};U)$ iff $e_{-\omega}(\tau^t u - u) \to 0$ in $L^p(\mathbb{R};U)$, and that

$$e_{-\omega}(\tau^t u - u) = e^{\omega t}\big(\tau^t(e_{-\omega}u) - e_{-\omega}u\big) + \big(e^{\omega t} - 1\big)e_{-\omega}u,$$

where $e^{\omega t} \to 1$ as $t \to 0$. The same argument remains valid if we replace L^p by BC, BUC, BC_0, Reg, or Reg_0. □

[8] The space L^p_0 is the same as L^p if $p < \infty$, and in the case $p = \infty$ it consists of those $u \in L^\infty$ which vanish at $\pm\infty$, i.e., $\lim_{t\to\infty} \operatorname{ess\,sup}_{|s|\ge t} |u(s)| = 0$. The space *Reg* consists of all bounded regulated functions, and Reg_0 consists of those functions in *Reg* which vanish at $\pm\infty$.

Example 2.5.3 *All the claims in Example 2.3.2(i)–(iv) remain true if we replace the spaces L^p, Reg, and BUC by the corresponding weighted spaces L^p_ω, Reg_ω, and BUC_ω, respectively, where $\omega \in \mathbb{R}$ is arbitrary. Moreover, if we denote the ω-weighted L^p-norm or sup-norm by $\|\cdot\|_\omega$, then with the notation of Example 2.3.2,*

(i) $\|\tau^t u\|_\omega = e^{\omega t}\|u\|_\omega$ *for* $t \in \mathbb{R}$,
(ii) $\|\tau^t_+ u\|_\omega \le e^{\omega t}\|u\|_\omega$ *for* $t \ge 0$,
(iii) $\|\tau^t_- u\|_\omega = e^{\omega t}\|u\|_\omega$, $t \ge 0$,
(iv) $\|\tau^t_{[0,T)} u\|_\omega \le e^{\omega t}\|u\|_\omega$ *for* $0 \le t < T$ *and* $\|\tau^t_{[0,T)} u\|_\omega = 0$ *for* $t \ge T$.

We leave the easy proof of Example 2.5.3 to the reader (the additional claims about the norms of the shift operators are obvious, and the rest of Example 2.5.3 can be reduced to Example 2.3.2 by use of Lemma 2.5.2(ii)).

The following theorem gives us global growth estimates on $\mathfrak{A}$, $\mathfrak{B}$, $\mathfrak{C}$, and $\mathfrak{D}$.

Theorem 2.5.4 *Let* $\Sigma = \left[\begin{smallmatrix}\mathfrak{A} & \mathfrak{B}\\ \mathfrak{C} & \mathfrak{D}\end{smallmatrix}\right]$ *be a* $L^p|Reg$*-well-posed linear system on* (Y, X, U).

(i) *The limit* $\lim_{t\to\infty} \frac{1}{t}\log(\|\mathfrak{A}^t\|)$ *exists, and*

$$\omega_{\mathfrak{A}} := \lim_{t\to\infty} \frac{\log(\|\mathfrak{A}^t\|)}{t} = \inf_{t>0} \frac{\log(\|\mathfrak{A}^t\|)}{t} < \infty$$

(but possibly $\omega_{\mathfrak{A}} = -\infty$*). In particular, for all* $t \ge 0$ *it is true that* $\|\mathfrak{A}^t\| \ge e^{\omega_{\mathfrak{A}} t}$ *and that the spectral radius of* $\mathfrak{A}^t$ *is equal to* $e^{\omega_{\mathfrak{A}} t}$. *Moreover, for each* $\omega > \omega_{\mathfrak{A}}$ *there is a constant* $M \ge 1$ *such that*

$$\|\mathfrak{A}^t\| \le M e^{\omega t} \text{ for } t \ge 0,$$

and $e^{-\omega t}\|\mathfrak{A}^t\| \to 0$ *as* $t \to \infty$.

(ii) *If* Σ *is* L^p*-well-posed then, for each* $\omega > \omega_{\mathfrak{A}}$, $\mathfrak{C}$ *is a continuous linear operator* $X \to L^p_\omega(\mathbb{R}^+, Y)$, *and* $\mathfrak{B}$ *and* $\mathfrak{D}$ *have unique extensions (that we still denote by the same letters) to continuous linear operators* $\mathfrak{B}\colon L^p_\omega(\mathbb{R}^-, U) \to X$ *and* $\mathfrak{D}\colon L^p_{\omega,\mathrm{loc}}(\mathbb{R}, U) \to L^p_{\omega,\mathrm{loc}}(\mathbb{R}, Y)$. *The latter operator maps* $L^p_\omega(\mathbb{R}, U)$ *continuously into* $L^p_\omega(\mathbb{R}, Y)$. *In the case* $p = \infty$, $\mathfrak{C}$ *maps* X *into* $L^\infty_{0,\omega}(\mathbb{R}^+, Y)$ *and the extended operator* $\mathfrak{D}$ *also maps* $L^\infty_{0,\omega,\mathrm{loc}}(\mathbb{R}, U)$ *into* $L^p_{0,\omega,\mathrm{loc}}(\mathbb{R}, Y)$ *and* $L^\infty_{0,\omega}(\mathbb{R}, U)$ *into* $L^p_{0,\omega}(\mathbb{R}, Y)$.

(iii) *If* Σ *is* L^p*-well-posed,* $\omega > \omega_{\mathfrak{A}}$, $s \in \mathbb{R}$, $x_s \in X$, *and* $u \in L^p_\omega([s, \infty); U)$, *then the output* y *of* Σ *satisfies* $y \in L^p_\omega([s, \infty); Y)$. *If, in addition,* $p < \infty$, *then the state trajectory* x *of* Σ *satisfies* $x \in BC_{0,\omega}([s, \infty); X)$.

(iv) *If* Σ *is Reg-well-posed, then (ii) and (iii) remain true if we replace* L^p *by* Reg_0 *throughout, including the statement in (iii) that* $x \in BC_{0,\omega}([s, \infty); X)$.

Note that the semigroup $\tau_{[0,T)}$ in Example 2.3.2(iv) has a growth bound equal to minus infinity, because $\|\tau^t_{[0,T)}\| = 0$ for $t \geq T$.

Proof (i) Recall that, by Lemma 2.2.13, $\|\mathfrak{A}^t\|$ is locally bounded. Define

$$\omega_{\mathfrak{A}} := \inf_{t>0} \frac{\log(\|\mathfrak{A}^t\|)}{t}.$$

Trivially, $\omega_{\mathfrak{A}} < \infty$. Let $\omega > \omega_{\mathfrak{A}}$, and choose some $T > 0$ such that $\frac{1}{T}\log(\|\mathfrak{A}^T\|) < \omega$. For each $t \geq 0$ we can choose an $n = 0, 1, 2, \ldots$ so that $nT \leq t \leq (n+1)T$. Then

$$\begin{aligned}\omega_{\mathfrak{A}} &\leq \frac{\log(\|\mathfrak{A}^t\|)}{t} \leq \frac{nT}{t}\frac{\log(\|\mathfrak{A}^T\|)}{T} + \frac{\log(\|\mathfrak{A}^{t-nT}\|)}{t} \\ &< \frac{nT\omega}{t} + \frac{\log(\|\mathfrak{A}^{t-nT}\|)}{t}.\end{aligned}$$

Letting $t \to \infty$ we get

$$\limsup_{t\to\infty} \frac{\log(\|\mathfrak{A}^t\|)}{t} \leq \omega,$$

and this shows that $\lim_{t\to\infty} \frac{1}{t}\log(\|\mathfrak{A}^t\|) = \omega_{\mathfrak{A}}$. The claim that the spectral radius of $\mathfrak{A}^t$ is equal to $e^{\omega_{\mathfrak{A}} t}$ is trivial if $t = 0$, and for $t > 0$ the logarithm of the spectral radius is given by

$$\lim_{n\to\infty} \log(\|(\mathfrak{A}^t)^n\|^{1/n}) = t \lim_{n\to\infty} \frac{\log(\|\mathfrak{A}^{nt}\|)}{nt} = \omega_{\mathfrak{A}} t.$$

The remaining claims in (i) follow from the facts that $\mathfrak{A}^t$ is locally bounded and that $\lim_{t\to\infty} \frac{1}{t}\log(\|\mathfrak{A}^t\|) = \omega_{\mathfrak{A}}$.

(ii) We claim that the expression for $\mathfrak{B}$ in Lemma 2.4.3 is an absolutely converging series in $\mathcal{B}(L^p_\omega(\mathbb{R}, U); X)$; hence it defines $\mathfrak{B}$ as an operator in $\mathcal{B}(L^p_\omega(\mathbb{R}, U); X)$. We prove this claim as follows. Choose some α, $\omega_A < \alpha < \omega$ and some $M \geq 1$ so that $\|\mathfrak{A}^t\| \leq M e^{\alpha t}$ for all $t \geq 0$ (cf. Lemma 2.5.2). Then

$$\sum_{n=0}^{\infty} \left\| \mathfrak{A}^{nT}\mathfrak{B}\pi_{[-T,0)}\tau^{-nT} \right\|_\omega \leq \sum_{n=0}^{\infty} M e^{\alpha T n} \left\| \mathfrak{B}\pi_{[-T,0)} \right\|_\omega e^{-\omega T n} < \infty.$$

Thus, $\mathfrak{B}$ can be extended to an operator in $\mathcal{B}(L^p_\omega(\mathbb{R}^-, Y); X)$. If $p < \infty$, then this extension is unique since $L^p_c(\mathbb{R}^-; U)$ is dense in $L^p_\omega(\mathbb{R}^-; U)$. In the case $p = \infty$ we choose some ω' satisfying $\omega_{\mathfrak{A}} < \omega' < \omega$ and observe that $\mathfrak{B}$ has a unique extension to $L^\infty_{0,\omega'}(\mathbb{R}^-; U)$ (since $L^\infty_c(\mathbb{R}^-; U)$ is dense in this space). But $L^\infty_\omega(\mathbb{R}^-; U) \subset L^\infty_{0,\omega'}(\mathbb{R}^-; U)$, hence the extension to $L^\infty_\omega(\mathbb{R}^-; U)$ is also unique.

The proof of the fact that $\mathfrak{C} \in \mathcal{B}(X; L^p_\omega(\mathbb{R}^+, Y))$ (and that $\mathfrak{C} \in \mathcal{B}(X; L^\infty_{0,\omega}(\mathbb{R}^+, Y))$ when $p = \infty$) is very similar to the one give above for $\mathfrak{B}$, and it is left to the reader.

The claims about the extension of $\mathfrak{D}$ remain to be proved. The proof of the uniqueness of the extension is essentially the same as the proof of the uniqueness of the extension of $\mathfrak{B}$, since (by causality) we know $\mathfrak{D}$ uniquely once we know $\pi_{(-\infty,t)}\mathfrak{D} = \pi_{(-\infty,t)}\mathfrak{D}\pi_{(-\infty,t)}$ for all $t \in \mathbb{R}$. If we can extend $\mathfrak{D}$ to an operator in $\mathcal{B}(L^p_\omega(\mathbb{R}, U); L^p_\omega(\mathbb{R}, Y))$, then we can use the identity $\pi_{(-\infty,t)}\mathfrak{D} = \pi_{(-\infty,t)}\mathfrak{D}\pi_{(-\infty,t)}$ to extend $\mathfrak{D}$ to an operator in $\mathcal{B}(L^p_{\omega,\mathrm{loc}}(\mathbb{R}, U); L^p_{\omega,\mathrm{loc}}(\mathbb{R}, Y))$. Thus, it suffices to show that we can extend $\mathfrak{D}$ to an operator in $\mathcal{B}(L^p_\omega(\mathbb{R}, U); L^p_\omega(\mathbb{R}, Y))$. Below we treat only the case $p < \infty$, and leave the analogous case $p = \infty$ to the reader. (The reader should also check that the extended operator maps $L^\infty_{0,\omega}(\mathbb{R}, U)$ into $L^\infty_{0,\omega}(\mathbb{R}, Y)$ in the L^∞-case.) All the L^p-type norms below are interpreted as norms in the weighted space L^p_ω, including the L^p-norms over the interval $[0, T)$, and to stress this fact we denote these norms by $\|\cdot\|_\omega$.

Take $u \in L^p_\omega(\mathbb{R}^+; U)$. We fix some $T > 0$ and define u_n, y_n, $\mathbf{A}$, $\mathbf{B}$, $\mathbf{C}$, and $\mathbf{D}$ as in Theorems 2.4.1 and 2.4.2, but this time we take $x_0 = 0$. Then

$$\begin{aligned}
\|u\|^p_\omega &= \int_0^\infty |\mathrm{e}^{-\omega s}u(s)|^p\,ds = \sum_{n=0}^\infty \int_{nT}^{(n+1)T} |\mathrm{e}^{-\omega s}u(s)|^p\,ds \\
&= \sum_{n=0}^\infty \int_0^T |\mathrm{e}^{-\omega(s+nT)}u(s+nT)|^p\,ds = \sum_{n=0}^\infty \|\mathrm{e}^{-\omega nT}u_n\|^p_\omega.
\end{aligned}$$

A similar formula is valid for $\|y\|_\omega$. As we saw in the proof of Lemma 2.4.3 (with $x_0 = 0$),

$$\mathrm{e}^{-\omega nT}y_n = \mathrm{e}^{-\omega nT}\left(\mathbf{C}\sum_{k=0}^{n-1}\mathbf{A}^k\mathbf{B}u_{n-k-1} + \mathbf{D}u_n\right).$$

Choose some $\omega_{\mathfrak{A}} < \alpha < \omega$ and some M such that $\|\mathfrak{A}^t\| \le M\mathrm{e}^{\alpha t}$ for all $t \ge 0$. Then $\|\mathbf{A}^k\| \le M\mathrm{e}^{\alpha kT}$, and

$$\begin{aligned}
\|\mathrm{e}^{-\omega nT}y_n\|_\omega &\le \|\mathbf{D}\|_\omega\|\mathrm{e}^{-\omega nT}u_n\|_\omega \\
&\quad + \mathrm{e}^{-\omega nT}\|\mathbf{C}\|_\omega\sum_{k=0}^{n-1}M\mathrm{e}^{\alpha kT}\|\mathbf{B}\|_\omega\|u_{n-k-1}\|_\omega \\
&= \|\mathbf{D}\|_\omega\|\mathrm{e}^{-\omega nT}u_n\|_\omega \\
&\quad + \mathrm{e}^{-\omega nT}\|\mathbf{C}\|_\omega\sum_{l=0}^{n-1}M\mathrm{e}^{\alpha(n-l-1)T}\|\mathbf{B}\|_\omega\|u_l\|_\omega \\
&= \|\mathbf{D}\|_\omega\|\mathrm{e}^{-\omega nT}u_n\|_\omega \\
&\quad + \|\mathbf{C}\|_\omega\|\mathbf{B}\|_\omega M\mathrm{e}^{-\alpha T}\sum_{l=0}^{n-1}\mathrm{e}^{(\alpha-\omega)(n-l)T}\|\mathrm{e}^{-\omega lT}u_l\|_\omega.
\end{aligned}$$

This can be interpreted as a convolution of the l^1-sequence $\{a_n\}_{n=0}^\infty$ and the l^p-sequence $\{\|e^{-\omega nT}u_n\|_\omega\}_{n=0}^\infty$, where

$$a_n = \begin{cases} \|\mathbf{D}\|_\omega, & n = 0, \\ \|\mathbf{C}\|_\omega \|\mathbf{B}\|_\omega M e^{-\alpha T} e^{(\alpha-\omega)nT}, & n = 1, 2, 3, \ldots \end{cases}$$

The convolution $\{(a * b)_n\}_{n=0}^\infty$ of an l^1-sequence $\{a_n\}_{n=0}^\infty$ and a l^p-sequence $\{b_n\}_{n=0}^\infty$ belongs to l^p, and $\|a * b\|_p \le \|a\|_1 \|b\|_p$ (the continuous time version of this result is found in, e.g., Gripenberg *et al.* (1990, Theorem 2.2, p. 39), and the proof of the discrete time version is essentially the same). Thus, the sequence $\{\|e^{-\omega nT}u_n\|_\omega\}_{n=0}^\infty$ belongs to l^p, and

$$\begin{aligned} \|\mathfrak{D}\pi_+ u\|_\omega^p &= \|y\|_\omega^p = \sum_{n=0}^\infty \|e^{-\omega nT} y_n\|_\omega^p \\ &\le \left(\sum_{n=0}^\infty a_n\right)^p \|e^{-\omega nT} u_n\|_\omega^p = \left(\sum_{n=0}^\infty a_n\right)^p \|\pi_+ u\|_\omega^p. \end{aligned}$$

In other words, we have found a constant $K = \sum_{n=0}^\infty a_n$ such that $\|\mathfrak{D}u\|_\omega \le K\|u\|_\omega$ for all $u \in L^p_\omega(\mathbb{R}; U)$ vanishing on $(-\infty, 0)$. If $u \in L^p_{c,\text{loc}}(\mathbb{R}; U) \cap L^p_\omega(\mathbb{R}; U)$, then we can choose some $t \in \mathbb{R}$ such that $\tau^t u$ is supported on $\overline{\mathbb{R}}^+$ and use this result (and the time-invariance of $\mathfrak{D}$ and Lemma 2.5.2) to get

$$\|\mathfrak{D}u\|_\omega = \|\tau^{-t}\mathfrak{D}\tau^t u\|_\omega = e^{-\omega t}\|\mathfrak{D}\tau^t u\|_\omega \le e^{-\omega t} K \|\tau^t u\|_\omega = K\|u\|_\omega.$$

This proves that $\mathfrak{D}$ can be extended to a bounded linear operator from $L^p_\omega(\mathbb{R}; U)$ to $L^p_\omega(\mathbb{R}; Y)$.

(iii) The claim that $y \in L^p_\omega([s, \infty); Y)$ follows from (ii), Definition 2.2.7, and Theorem 2.2.12.

By Definition 2.2.7, $e^{-\omega t}x(t) = e^{-\omega t}\left(\mathfrak{A}^{t-s}x_s + \mathfrak{B}\tau^t \pi_{[s,\infty)}u\right)$. According to Theorem 2.2.12, $t \mapsto x(t)$ is continuous. By the estimate in (i), $e^{-\omega t}\mathfrak{A}^{t-s}x_s \to 0$ as $t \to \infty$. Suppose for the moment that $u \in L^p_\omega(\mathbb{R}; U) \cap L^p_{\omega'}(\mathbb{R}; U)$ for some $\omega_\mathfrak{A} < \omega' < \omega$. Then by Example 2.5.3, $\|\tau^t \pi_{[s,\infty)}u\|_{\omega'} \le e^{\omega' t}\|u\|_{\omega'}$, and this combined with part (ii) gives $\mathfrak{B}\tau^t \pi_{[s,\infty)}u = O(e^{\omega' t})$ as $t \to \infty$. In particular, $e^{-\omega t}\mathfrak{B}\tau^t \pi_{[s,\infty)}u \to 0$ as $t \to \infty$. The same claim must then be true for all $u \in L^p_\omega(\mathbb{R}; U)$ since $L^p_\omega(\mathbb{R}; U) \cap L^p_{\omega'}(\mathbb{R}; U)$ is dense in $L^p_\omega(\mathbb{R}; U)$. Thus $x \in BC_{0,\omega}([s, \infty); X)$.

(iv) The proofs are the same as in the L^p-case, except for the fact that we have to replace the reference to Theorem 2.2.12 by a forward reference to Theorem 4.3.1 to get the continuity of x. □

Remark 2.5.5 The conclusion of Theorem 2.5.4 remains valid if we replace the strong continuity assumption on $\mathfrak{A}$ by a local boundedness assumption (cf. Lemma 2.2.13), except for the continuous dependence of $x(t)$ on t. The proof remains the same.

Definition 2.5.6 Let $\Sigma = \left[\begin{smallmatrix} \mathfrak{A} & \mathfrak{B} \\ \mathfrak{C} & \mathfrak{D} \end{smallmatrix}\right]$ be a $L^p|Reg$-well-posed linear system on (Y, X, U).

(i) The *growth bound* of Σ is the same as the growth bound of its semigroup $\mathfrak{A}$, and it is given by the number $\omega_{\mathfrak{A}}$ defined in Theorem 2.5.4(i). We denote the growth bound of the semigroup $\mathfrak{A}$ and the growth bound of a system Σ with semigroup $\mathfrak{A}$ by $\omega_{\mathfrak{A}}$ throughout.

(ii) By *ω-boundedness* of Σ or one of its components we mean the following:
   (a) $\mathfrak{A}$ is *ω-bounded* if $\mathfrak{A}$ satisfies $\sup_{t\geq 0} \left\| e^{-\omega t}\mathfrak{A}^t \right\| < \infty$;
   (b) $\mathfrak{B}$ is *ω-bounded* if $\mathfrak{B}$ can be extended to a continuous linear operator $L^p|Reg_{0,\omega}(\mathbb{R}^-, U) \to X$;
   (c) $\mathfrak{C}$ is *ω-bounded* if $\mathfrak{C}$ is a continuous linear operator $X \to L^p|Reg_\omega(\overline{\mathbb{R}}^+, Y)$;
   (d) $\mathfrak{D}$ is *ω-bounded* if $\mathfrak{D}$ can be extended to a continuous linear operator $L^p|Reg_{0,\omega}(\mathbb{R}^-, U) \dotplus L^p|Reg_\omega(\overline{\mathbb{R}}^+, U) \to L^p|Reg_{0,\omega}(\mathbb{R}^-, Y) \dotplus L^p|Reg_\omega(\overline{\mathbb{R}}^+, Y)$;
   (e) Σ is *ω-bounded* if (a)–(d) above hold.

With the help of Theorem 2.5.4 we can prove the following analogue of Theorem 2.2.11, which corresponds to the case where the initial time is $-\infty$ and the initial state is zero.

Theorem 2.5.7 *Let $\Sigma = \left[\begin{smallmatrix} \mathfrak{A} & \mathfrak{B} \\ \mathfrak{C} & \mathfrak{D} \end{smallmatrix}\right]$ be a $L^p|Reg$-well-posed linear system on (Y, X, U) with growth bound $\omega_{\mathfrak{A}}$, let $\omega > \omega_{\mathfrak{A}}$, and let $u \in L^p|Reg_{0,\omega,\mathrm{loc}}(\mathbb{R}; U)$. Define $x(t) = \mathfrak{B}^t u = \mathfrak{B}\tau^t u$, $t \in \mathbb{R}$, and $y = \mathfrak{D}u$. Then $x \in BC_{0,\omega,\mathrm{loc}}(\mathbb{R}; X)$, $y \in L^p|Reg_{0,\omega,\mathrm{loc}}(\mathbb{R}; Y)$, and for all $s \in \mathbb{R}$ (cf. Definition 2.2.6)*

$$
\begin{aligned}
x(t) &= \mathfrak{A}_s^t x(s) + \mathfrak{B}_s^t u, \qquad t \geq s, \\
\pi_{[s,\infty)} y &= \mathfrak{C}_s x(s) + \mathfrak{D}_s u.
\end{aligned} \tag{2.5.1}
$$

Thus, $x(t)$ for $t \geq s$ and $\pi_{[s,\infty)} y$ can be interpreted as the state trajectory and the output function of Σ with initial time s, initial value $x(s)$, and input u. If $u \in L^p|Reg_{0,\omega}(\mathbb{R}; U)$ then $x \in BC_{0,\omega}(\mathbb{R}; X)$ and $y \in L^p|Reg_{0,\omega}(\mathbb{R}; Y)$.

Proof By Theorem 2.2.11, (2.5.1) is true if $u \in L^p|Reg_{c,\mathrm{loc}}(\mathbb{R}; U)$. This set of functions u is dense in $L^p|Reg_{0,\omega,\mathrm{loc}}(\mathbb{R}; U)$, so the general case then follows from the continuity of $\mathfrak{B}$ and $\mathfrak{D}$; cf. Theorem 2.5.4(ii),(iv). The additional claims about the growth bounds of x and y follow from Theorem 2.5.4(iii),(iv) (observe that $\mathfrak{B}\tau^t u \to 0$ in $L^p|Reg_{0,\omega}(\mathbb{R}; U)$ as $t \to -\infty$). □

Definition 2.5.8 We call the functions x and y in Theorem 2.5.7 the *state trajectory* and the *output function* of Σ with initial time $-\infty$ and input u.

Example 2.5.9

(i) *The growth bound of the delay line in Example 2.3.4 is* $-\infty$.
(ii) *The growth bound of the exponentially shifted system* Σ_α *in Example 2.3.5 is* $\omega_{\mathfrak{A}_\alpha} = \omega_{\mathfrak{A}} + \Re\alpha$. *Thus, it is possible to make* Σ_α *exponentially stable by choosing* $\Re\alpha < -\omega_{\mathfrak{A}}$.
(iii) *The growth bound of the time compressed system* Σ_λ *in Example 2.3.6 is* $\omega_{\mathfrak{A}_\lambda} = \lambda\omega_{\mathfrak{A}}$.
(iv) *The growth bound of the similarity transformed system* Σ_E *in Example 2.3.7 is* $\omega_{\mathfrak{A}_E} = \omega_{\mathfrak{A}}$.
(v) *The growth bound of the systems in Examples 2.3.10– 2.3.13 is* $\omega_{\mathfrak{A}} = \max\{\omega_{\mathfrak{A}_1}, \omega_{\mathfrak{A}_2}\}$.
(vi) *The growth bound of the Lax–Phillips model in Definition 2.7.2 is* $\omega_{\mathfrak{A}} = \max\{\omega_{\mathfrak{A}}, \omega\}$.

We leave the easy proofs of these claims to the reader.

Example 2.5.10 *In the case of the system presented in Proposition 2.3.1, the extended input map* $\mathfrak{B}$ *and the extended input/output map* $\mathfrak{D}$ *constructed in Theorem 2.5.4(ii) and (iv) are given by (for all* $t \in \mathbb{R}$*)*

$$\begin{aligned} \mathfrak{B}u &= \int_{-\infty}^{0} \mathfrak{A}^{-s} B u(s)\, ds, && u \in L^p|Reg_{0,\omega}(\mathbb{R}^-; U), \\ (\mathfrak{D}u)(t) &= C \int_{-\infty}^{t} \mathfrak{A}^{t-s} B u(s)\, ds + D u(t), && u \in L^p|Reg_{0,\omega,\mathrm{loc}}(\mathbb{R}; U). \end{aligned} \tag{2.5.2}$$

Proof Fix some $\omega' \in (\omega_{\mathfrak{A}}, \omega)$. Then $\|\mathfrak{A}^t\| \le M e^{\omega' t}$ for some $M < \infty$. This implies that the integral $\int_{-\infty}^0 \mathfrak{A}^{-s} B u(s)\, ds$ converges absolutely, because (by Hölder's inequality, with $1/p + 1/q = 1$),

$$\begin{aligned} \int_{-\infty}^{0} \left|\mathfrak{A}^{-s} B u(s)\right| ds &\le M\|B\| \int_{-\infty}^{0} e^{(\omega-\omega')s} |e^{-\omega s} u(s)|\, ds \\ &\le \left(\frac{1}{(\omega-\omega')q}\right)^{1/q} \|u\|_{L^p|Reg_\omega(\mathbb{R}^-)}. \end{aligned}$$

The same computation shows that if we define the operator $\widetilde{\mathfrak{B}}$ by $\widetilde{\mathfrak{B}} = \int_{-\infty}^0 \mathfrak{A}^{-s} B u(s)\, ds$, then $\widetilde{\mathfrak{B}} \in \mathcal{B}(L^p|Reg_{0,\omega}(\mathbb{R}^-; U); X)$. Since this operator coincides with the extended operator $\mathfrak{B}$, which also belongs to $\mathcal{B}(L^p|Reg_{0,\omega}(\mathbb{R}^-; U); X)$, on the dense subset of functions u vanishing outside of some finite interval, we must have $\widetilde{\mathfrak{B}} = \mathfrak{B}$.

Essentially the same proof can be used to prove the second half of (2.5.2). By the preceding argument,

$$(\widetilde{\mathfrak{D}}u)(t) := C \int_{-\infty}^{t} \mathfrak{A}^{t-s} Bu(s)\, ds + Du(t)$$

is well-defined for all $u \in L^p|Reg_{0,\omega,\mathrm{loc}}(\mathbb{R}; U)$ and all $t \in \mathbb{R}$, and by Theorem A.3.4, $\widetilde{\mathfrak{D}}$ maps $L^p|Reg_{0,\omega,\mathrm{loc}}(\mathbb{R}; U)$ continuously into $L^p|Reg_{0,\omega,\mathrm{loc}}(\mathbb{R}; U)$. In addition it coincides with $\mathfrak{D}$ on a dense subset. □

2.6 Shift realizations

We have seen in Theorem 2.5.4 that every input/output map $\mathfrak{D}$ of an L^p-well-posed linear system Σ can be extended to a time-invariant linear operator $L^p_\omega(\mathbb{R}; U) \to L^p_\omega(\mathbb{R}; Y)$ for some $\omega \in \mathbb{R}$. Here we shall study the converse question: given such a time-invariant operator, is it possible to construct a well-posed linear system with this particular input/output map? As we shall see in a moment, the answer is yes if $1 \le p < \infty$. An analogous result is also true if we replace L^p by *Reg*.

Definition 2.6.1 Let $1 \le p \le \infty$, let $\omega \in \mathbb{R}$, let U and Y be Banach spaces, and let $\mathfrak{D}$ be a linear operator $L^p_{\mathrm{loc}}(\mathbb{R}; U) \supset \mathcal{D}(\mathfrak{D}) \to L^p_{\mathrm{loc}}(\mathbb{R}; Y)$ or $Reg_{\mathrm{loc}}(\mathbb{R}; U) \supset \mathcal{D}(\mathfrak{D}) \to Reg_{\mathrm{loc}}(\mathbb{R}; Y)$.

(i) $\mathfrak{D}$ is *time-invariant* if $\tau^t \mathfrak{D} u = \mathfrak{D} \tau^t u$ for all $u \in \mathcal{D}(\mathfrak{D})$ and all $t \in \mathbb{R}$ (in particular, $\tau^t \mathcal{D}(\mathfrak{D}) = \mathcal{D}(\mathfrak{D})$).
(ii) A time-invariant operator $\mathfrak{D}$ is *causal* if $\pi_- \mathfrak{D} \pi_+ = 0$ and it is anti-causal if $\pi_+ \mathfrak{D} \pi_- = 0$.
(iii) A time-invariant operator $\mathfrak{D}$ is *static* if it is both causal and anti-causal.
(iv) The *Hankel operator* induced by a time-invariant operator $\mathfrak{D}$ is the operator $\pi_+ \mathfrak{D} \pi_-$, and the *anti-Hankel operator* induced by $\mathfrak{D}$ is the operator $\pi_- \mathfrak{D} \pi_+$.
(v) The *Toeplitz operator* induced by a time-invariant operator $\mathfrak{D}$ is the operator $\pi_+ \mathfrak{D} \pi_+$, and the *anti-Toeplitz operator* induced by $\mathfrak{D}$ is the operator $\pi_- \mathfrak{D} \pi_-$.

Thus, a time-invariant operator is causal iff its anti-Hankel operator vanishes, it is anti-causal iff its Hankel operator vanishes, and it is static if both the Hankel operator and the anti-Hankel operator vanish. The condition imposed on the input/output map $\mathfrak{D}$ in Definition 2.2.1(iv) requires $\mathfrak{D}$ to be time-invariant and causal with a Hankel operator equal to $\mathfrak{C}\mathfrak{B}$.

Definition 2.6.2

(i) For all $1 \le p < \infty$, the notation $TI^p_\omega(U;Y)$ stands for the space of all bounded time-invariant operators $\mathfrak{D}\colon L^p_\omega(\mathbb{R};U) \to L^p_\omega(\mathbb{R};Y)$, with the operator norm. The notation $TI^\infty_\omega(U;Y)$ stands for the space of all bounded time-invariant operators $\mathfrak{D}\colon L^\infty_{0,\omega}(\mathbb{R}^-;U) \dotplus L^\infty_\omega(\mathbb{R}^+;U) \to L^\infty_{0,\omega}(\mathbb{R}^-;Y) \dotplus L^\infty_\omega(\mathbb{R}^+;Y)$, with the operator norm. In both cases we abbreviate $TI^p_\omega(U;U)$ to $TI^p_\omega(U)$.

(ii) We denote the space of all causal operators in $TI^p_\omega(U;Y)$ by $TIC^p_\omega(U;Y)$, and abbreviate $TIC^p_\omega(U;U)$ to $TIC^p_\omega(U)$.

(iii) By TI^p and TIC^p we mean TI^p_ω and TIC^p_ω with $\omega = 0$.

(iv) By TIC^p_∞ we mean $\bigcup_{\omega\in\mathbb{R}} TIC^p_\omega$.

(v) We denote the space of all continuous causal time-invariant operators $\mathfrak{D}\colon L^p_{c,\mathrm{loc}}(\mathbb{R};U) \to L^p_{c,\mathrm{loc}}(\mathbb{R};Y)$ by $TIC^p_{\mathrm{loc}}(U;Y)$, and abbreviate $TIC^p_{\mathrm{loc}}(U;U)$ to $TIC^p_{\mathrm{loc}}(U)$.

(vi) The spaces TI^{Reg}_ω, TIC^{Reg}_ω, TI^{Reg}, TIC^{Reg}, TIC^{Reg}_∞ TI^{Reg}_{loc}, and TIC^{Reg}_{loc} are defined in the same way, with L^p_ω replaced by $Reg_{0,\omega}(\mathbb{R}^-) \dotplus Reg_\omega(\overline{\mathbb{R}}^+)$ and $L^p_{c,\mathrm{loc}}$ replaced by $Reg_{c,\mathrm{loc}}$.

(vii) We use TI (with different subindices) to represent either TI^p or TI^{Reg} and TIC (with different subindices) to represent either TIC^p or TIC^{Reg}, depending on the context.

Definition 2.6.3 Let $1 \le p \le \infty$, let U and Y be Banach spaces, and let $\mathfrak{D} \in TIC^p_{\mathrm{loc}}(U;Y)$. By a *$L^p$-realization* of $\mathfrak{D}$ we mean an L^p-well-posed linear system on (Y, X, U) (for some Banach space X) with input/output map $\mathfrak{D}$. A *Reg-realization* of an operator $\mathfrak{D} \in TIC^{Reg}_{\mathrm{loc}}(U;Y)$ is defined in the same way, with L^p replaced by Reg.

We shall also apply this definition in the case where $\mathfrak{D}$ belongs to $TIC_\omega(U;Y)$ instead of $TIC_{\mathrm{loc}}(U;Y)$. To do this we need the following fact:

Lemma 2.6.4 *If $\mathfrak{D} \in TIC_\omega(U;Y)$ for some $\omega \in \mathbb{R}$, then $\mathfrak{D}$ has a unique (restriction followed by an) extension to an operator in $TIC_{\mathrm{loc}}(U;Y)$, and this extension determines $\mathfrak{D}$ uniquely. Moreover, the extension does not depend on ω, in the sense that if $\mathfrak{D} \in TIC_\alpha \cap TIC_\omega(U;Y)$, then we get the same extension if we interpret $\mathfrak{D}$ as an operator in $TIC_\alpha(U;Y)$ or as an operator in $TIC_\omega(U;Y)$.*

Proof By the time-invariance and causality of $\mathfrak{D}$, the restriction of $\mathfrak{D}$ to $L^p|Reg_{c,\omega}(\mathbb{R};U)$ maps this space into $L^p|Reg_{c,\omega}(\mathbb{R};Y)$ (these are the spaces of functions in $L^p|Reg_\omega$ whose support is bounded to the left). Moreover, $\mathfrak{D}$ is determined uniquely by this restriction since $L^p|Reg_c((-\infty,t);U)$ is dense in $L^p|Reg_\omega((-\infty,t);U)$ (or in $L^\infty_{0,\omega}((-\infty,t);U)$ in the L^∞-case) for every $t \in \mathbb{R}$, and since we know $\mathfrak{D}$ uniquely as soon as we know

$\pi_{(-\infty,t)}\mathfrak{D} = \pi_{(-\infty,t)}\mathfrak{D}\pi_{(-\infty,t)}$ for all $t \in \mathbb{R}$. Next we use the identity $\pi_{(-\infty,t)}\mathfrak{D} = \pi_{(-\infty,t)}\mathfrak{D}\pi_{(-\infty,t)}$ to extend $\mathfrak{D}$ to an operator in TIC_{loc}. This extension is unique since $L^p|Reg_{c,\omega}(\mathbb{R};U)$ is dense in $L^p|Reg_{c,\text{loc}}(\mathbb{R};U)$. □

Example 2.6.5 *Let $\mathfrak{D} \in TIC^p_\omega(U;Y)$ where $1 \le p < \infty$, $\omega \in \mathbb{R}$, and U and Y are Banach spaces.*

(i) *The system Σ defined by (cf. Example 2.3.2)*

$$\Sigma = \left[\begin{array}{c|c} \mathfrak{A} & \mathfrak{B} \\ \hline \mathfrak{C} & \mathfrak{D} \end{array}\right] = \left[\begin{array}{c|c} \tau_- & 1 \\ \hline \pi_+\mathfrak{D}\pi_- & \mathfrak{D} \end{array}\right]$$

is a ω-bounded L^p-well-posed linear system on $(Y, L^p_\omega(\mathbb{R}^-;U), U)$. This is the exactly controllable shift realization of $\mathfrak{D}$.

(ii) *The system Σ defined by*

$$\Sigma = \left[\begin{array}{c|c} \mathfrak{A} & \mathfrak{B} \\ \hline \mathfrak{C} & \mathfrak{D} \end{array}\right] = \left[\begin{array}{c|c} \tau_+ & \pi_+\mathfrak{D}\pi_- \\ \hline 1 & \mathfrak{D} \end{array}\right]$$

is an ω-bounded L^p-well-posed linear system on $(Y, L^p_\omega(\mathbb{R}^+;Y), U)$. It is strongly stable when $\omega = 0$ (see Definition 8.1.1). This is the exactly observable shift realization of $\mathfrak{D}$.

(iii) *The system Σ defined by (cf. Example 2.3.2)*

$$\Sigma = \left[\begin{array}{c|c} \mathfrak{A} & \mathfrak{B} \\ \hline \mathfrak{C} & \mathfrak{D} \end{array}\right] = \left[\begin{array}{c|c} \tau & \pi_- \\ \hline \pi_+\mathfrak{D} & \mathfrak{D} \end{array}\right]$$

is a ω-bounded L^p-well-posed linear system on $(Y, L^p_\omega(\mathbb{R};U), U)$. This is the bilateral input shift realization of $\mathfrak{D}$.

(iv) *The system Σ defined by*

$$\Sigma = \left[\begin{array}{c|c} \mathfrak{A} & \mathfrak{B} \\ \hline \mathfrak{C} & \mathfrak{D} \end{array}\right] = \left[\begin{array}{c|c} \tau & \mathfrak{D}\pi_- \\ \hline \pi_+ & \mathfrak{D} \end{array}\right]$$

is an ω-bounded L^p-well-posed linear system on $(Y, L^p_\omega(\mathbb{R};Y), U)$. This is the bilateral output shift realization of $\mathfrak{D}$.

We leave the easy verifications of these claims to the reader. Strictly speaking, we should replace $\mathfrak{D}$ in (i)–(iv) by the operator in TIC^p_{loc} induced by $\mathfrak{D}$; see Lemma 2.6.4. Controllability and observability will be studied in Chapter 9. See, in particular, Example 9.4.12. The semigroups used in these realizations have quite different spectral properties; see Example 3.3.1.

This example provides us with a simple proof of the following theorem:

Theorem 2.6.6 *Let $1 \le p < \infty$, let U and Y be Banach spaces, and let $\mathfrak{D}$ be an operator $L^p_{c,\mathrm{loc}}(\mathbb{R}; U) \to L^p_{c,\mathrm{loc}}(\mathbb{R}; Y)$. Then the following conditions are equivalent:*

(i) *$\mathfrak{D}$ is the* input/output map *of an L^p-well-posed linear system;*
(ii) *$\mathfrak{D}$ is linear, continuous, time-invariant, and causal, and $\mathfrak{D}$ can be extended to a bounded operator $L^p_\omega(\mathbb{R}; U) \to L^p_\omega(\mathbb{R}; Y)$ for some $\omega \in \mathbb{R}$;*
(iii) *$\mathfrak{D}$ is linear, continuous, time-invariant, and causal, and the Hankel operator $\pi_+\mathfrak{D}\pi_-$ of $\mathfrak{D}$ can be extended to a bounded operator $L^p_\omega(\mathbb{R}^-; U) \to L^p_{\mathrm{loc}}(\mathbb{R}^+; Y)$ for some $\omega \in \mathbb{R}$;*
(iv) *$\mathfrak{D}$ is linear, continuous, time-invariant, and causal, and the Hankel operator $\pi_+\mathfrak{D}\pi_-$ of $\mathfrak{D}$ is bounded from $L^p_c(\mathbb{R}^-; U)$ into $L^p_\omega(\mathbb{R}^+; Y)$ for some $\omega \in \mathbb{R}$.*

Thus, Theorem 2.6.6 says that $\mathfrak{D} \in TIC_{\mathrm{loc}}(U; Y)$ has an L^p-realization iff $\mathfrak{D}$ can be extended to a bounded operator $L^p_\omega(\mathbb{R}; U) \to L^p_\omega(\mathbb{R}; Y)$, or equivalently, $\pi_+\mathfrak{D}\pi_-$ can be extended to a bounded operator $L^p_\omega(\mathbb{R}^-; U) \to L^p_\omega(\mathbb{R}^+; Y)$.

Proof By Theorem 2.5.4(ii), (i) $\Rightarrow$ (ii), and obviously (ii) $\Rightarrow$ (iii) and (ii) $\Rightarrow$ (iv). If (iii) holds, then the system in Example 2.6.5(i) is an L^p-well-posed linear system with the input/output map $\mathfrak{D}$, and if (iv) holds, then the system in Example 2.6.5(ii) is an L^p-well-posed linear systems with the input/output map $\mathfrak{D}$. Thus (iii) $\Rightarrow$ (i) and (iv) $\Rightarrow$ (i). □

There is also a *Reg*-well-posed version of Theorem 2.6.6:

Theorem 2.6.7 *Let U and Y be Banach spaces, and let $\mathfrak{D}$ be an operator $Reg_{c,\mathrm{loc}}(\mathbb{R}; U) \to Reg_{c,\mathrm{loc}}(\mathbb{R}; Y)$. Then the following conditions are equivalent:*

(i) *$\mathfrak{D}$ is the* input/output map *of an Reg-well-posed linear system;*
(ii) *$\mathfrak{D}$ is linear, continuous, time-invariant, and causal, and there is an operator $D \in \mathcal{B}(U; Y)$ and a constant $\omega \in \mathbb{R}$ such that the mapping $u \mapsto \big(t \mapsto (\mathfrak{D}u)(t) - Du(t)\big)$ maps $Reg_{c,\mathrm{loc}}(\mathbb{R}; U)$ into $C_c(\mathbb{R}; Y)$ and can be extended to a continuous linear mapping $Reg_{0,\omega}(\mathbb{R}; U) \to BC_{0,\omega}(\mathbb{R}; Y)$;*
(iii) *$\mathfrak{D}$ is linear, continuous, time-invariant, and causal, and the Hankel operator $\pi_+\mathfrak{D}\pi_-$ maps $Reg_c(\mathbb{R}^-; U)$ continuously into $BC_{0,\omega}(\overline{\mathbb{R}}^+; Y)$ for some $\omega \in \mathbb{R}$.*

Proof By Definition 2.2.3, Theorem 2.5.4 and Corollary 4.5.6 below, (i) $\Rightarrow$ (ii), and obviously (ii) $\Rightarrow$ (iii).

To prove that (iii) $\Rightarrow$ (i) it suffices to construct a *Reg*-well-posed realization of $\mathfrak{D}$. For this we can use the same exactly observable shift realization as in

Example 2.6.5(ii), i.e.,

$$\left[\begin{array}{c|c}\mathfrak{A} & \mathfrak{B} \\ \hline \mathfrak{C} & \mathfrak{D}\end{array}\right] = \left[\begin{array}{c|c}\tau_+ & \pi_+\mathfrak{D}\pi_- \\ \hline 1 & \mathfrak{D}\end{array}\right]$$

but this time with state space $BC_{0,\omega}(\overline{\mathbb{R}}^+; Y)$, since the left-shift semigroup τ_+ is strongly continuous on this space (see Example 2.5.3). □

Example 2.6.5 also provides us with a simple proof of the following lemma:

Lemma 2.6.8 *For each $\alpha > \omega$, $TIC_\omega(U; Y)$ is continuously embedded in $TIC_\alpha(U; Y)$, i.e., every operator $\mathfrak{D} \in TIC_\omega(U; Y)$ has a continuous (restriction followed by an) extension to an operator in $\mathfrak{D} \in TIC_\alpha(U; Y)$, and there is a one-to-one correspondence between the original operator $\mathfrak{D} \in TIC_\omega(U; Y)$ and its extended version $\mathfrak{D} \in TIC_\alpha(U; Y)$.*

Proof In the L^p-case with $1 \le p < \infty$ the existence of a continuous (restriction followed by an) extension of $\mathfrak{D}$ follows from Example 2.6.5 and Theorem 2.5.4(ii), and, according to Lemma 2.6.4, there is a one-to-one correspondence between the original operator and its extended version. In the L^∞-case and *Reg*-case we can use exactly the same proof: the realization in Example 2.6.5(ii) is still valid apart from the fact that the semigroup is not strongly continuous, and the strong continuity was not used in the proof of Theorem 2.5.4(ii) (it was used in the proof of Theorem 2.5.4(i) in the form of Lemma 2.2.13, but this time we know in advance that the semigroup has growth bound ω). □

2.7 The Lax–Phillips scattering model

Instead of using a $L^p|Reg$-well-posed linear system to formalize the idea of having an output and state at time $t > 0$ which depend continuously on the input and the initial state we can proceed in a different way which leads to a generalized *Lax–Phillips scattering model*. This is a particular semigroup $\mathfrak{T}$ defined on $\left[\begin{smallmatrix}\mathcal{Y} \\ X \\ \mathcal{U}\end{smallmatrix}\right]$, where $\mathcal{Y} = L^p|Reg_\omega(\mathbb{R}^-; Y)$ and $\mathcal{U} = L^p|Reg_\omega(\overline{\mathbb{R}}^+; U)$. We call $\mathcal{U}$ the *incoming subspace*, X the *inner state space*, and $\mathcal{Y}$ the *outgoing subspace*. In the classical cases treated in Lax and Phillips (1967, 1973) ω is taken to be zero and $\mathfrak{T}$ is required to be unitary (the conservative case) or a contraction semigroup (the nonconservative case).

Theorem 2.7.1 *Let $\Sigma = \left[\begin{smallmatrix}\mathfrak{A} & \mathfrak{B} \\ \mathfrak{C} & \mathfrak{D}\end{smallmatrix}\right]$ be a $L^p|Reg$-well-posed linear system on (Y, X, U). Let $\omega \in \mathbb{R}$, $\mathcal{Y} = L^p|Reg_\omega(\mathbb{R}^-; Y)$ (the outgoing subspace) and $\mathcal{U} = L^p|Reg_\omega(\overline{\mathbb{R}}^+; U)$ (the incoming subspace). For each $\left[\begin{smallmatrix}y_0 \\ x_0 \\ u_0\end{smallmatrix}\right] \in \left[\begin{smallmatrix}\mathcal{Y} \\ X \\ \mathcal{U}\end{smallmatrix}\right]$*

and $t \geq 0$, *define (the notation is explained in Definition 2.2.6 and Example 2.3.2)*

$$\mathfrak{T}^t := \begin{bmatrix} \tau^t & 0 & 0 \\ 0 & 1 & 0 \\ 0 & 0 & \tau_+^t \end{bmatrix} \begin{bmatrix} 1 & \mathfrak{C}_0^t & \mathfrak{D}_0^t \\ 0 & \mathfrak{A}^t & \mathfrak{B}_0^t \\ 0 & 0 & 1 \end{bmatrix} = \begin{bmatrix} \tau_-^t & \tau^t\mathfrak{C}_0^t & \tau^t\mathfrak{D}_0^t \\ 0 & \mathfrak{A}^t & \mathfrak{B}_0^t \\ 0 & 0 & \tau_+^t \end{bmatrix}.$$

Then $\mathfrak{T}$ *is a semigroup on* $\left[\begin{smallmatrix} \mathcal{Y} \\ X \\ \mathcal{U} \end{smallmatrix}\right]$. *It is strongly continuous on* $\left[\begin{smallmatrix} \mathcal{Y} \\ X \\ \mathcal{U} \end{smallmatrix}\right]$ *iff* Σ *is* L^p-*well-posed with* $1 \leq p < \infty$. *If* x *and* y *are the state trajectory and the output function of* Σ *corresponding to the initial state* $x_0 \in X$ *and the input function* $u_0 \in \mathcal{U}$, *and if we define* $y(t) = y_0(t)$ *for* $t < 0$, *then for all* $t \geq 0$,

$$\begin{bmatrix} \pi_{(-\infty,t]}y \\ x(t) \\ \pi_{[t,\infty)}u_0 \end{bmatrix} = \begin{bmatrix} \tau^{-t} & 0 & 0 \\ 0 & 1 & 0 \\ 0 & 0 & \tau^{-t} \end{bmatrix} \mathfrak{T}^t \begin{bmatrix} y_0 \\ x_0 \\ u_0 \end{bmatrix}. \tag{2.7.1}$$

Formula (2.7.1) shows that at any time $t \geq 0$, the *first component* of $\mathfrak{T}_t \left[\begin{smallmatrix} y_0 \\ x_0 \\ u_0 \end{smallmatrix}\right]$ represents the *past output*, the *second component* represents the *present state* and the *third component* represents the *future input*.

Proof That $\mathfrak{T}(0)$ is the identity operator follows from Lemma 2.2.8(i). The claim about the strong continuity can be reduced to the strong continuity of the two shift operators, discussed in Example 2.3.2. Thus, only the semigroup property $\mathfrak{T}^{s+t} = \mathfrak{T}^s\mathfrak{T}^t$ for $s, t \geq 0$ remains to be shown. For this we use Definition 2.2.6, the composition property in Lemma 2.2.8(iv), and Example 2.3.2, which give

$$\begin{aligned}
\mathfrak{T}^s\mathfrak{T}^t &= \begin{bmatrix} \tau_-^s & \tau^s\mathfrak{C}_0^s & \tau^s\mathfrak{D}_0^s \\ 0 & \mathfrak{A}^s & \mathfrak{B}_0^s \\ 0 & 0 & \tau_+^s \end{bmatrix} \begin{bmatrix} \tau_-^t & \tau^t\mathfrak{C}_0^t & \tau^t\mathfrak{D}_0^t \\ 0 & \mathfrak{A}^t & \mathfrak{B}_0^t \\ 0 & 0 & \tau_+^t \end{bmatrix} \\
&= \begin{bmatrix} \tau_-^s\tau_-^t & \tau_-^s\tau^t\mathfrak{C}_0^t + \tau^s\mathfrak{C}_0^s\mathfrak{A}^t & \tau_-^s\tau^t\mathfrak{D}_0^t + \tau^s\mathfrak{C}_0^s\mathfrak{B}_0^t + \tau^s\mathfrak{D}_0^s\tau_+^t \\ 0 & \mathfrak{A}^s\mathfrak{A}^t & \mathfrak{A}^s\mathfrak{B}_0^t + \mathfrak{B}_0^s\tau_+^t \\ 0 & 0 & \tau_+^s\tau_+^t \end{bmatrix} \\
&= \begin{bmatrix} \tau_-^{s+t} & \tau^{s+t}(\mathfrak{C}_0^t + \mathfrak{C}_t^{s+t}\mathfrak{A}_0^t) & \tau^{s+t}(\mathfrak{D}_0^t + \mathfrak{C}_t^{s+t}\mathfrak{B}_0^t + \mathfrak{D}_t^{s+t}) \\ 0 & \mathfrak{A}^{s+t} & \mathfrak{A}_t^{s+t}\mathfrak{B}_0^t + \mathfrak{B}_t^{s+t} \\ 0 & 0 & \tau_+^{s+t} \end{bmatrix} \\
&= \begin{bmatrix} \tau_-^{s+t} & \tau^{s+t}\mathfrak{C}_0^{s+t} & \tau^{s+t}\mathfrak{D}_0^{s+t} \\ 0 & \mathfrak{A}^{s+t} & \mathfrak{B}_0^{s+t} \\ 0 & 0 & \tau_+^{s+t} \end{bmatrix} \\
&= \mathfrak{T}^{s+t}.
\end{aligned}$$

□

The semigroup $\mathfrak{T}$ in Theorem 2.7.1 has an additional 'causality' property, which in the Hilbert space case where $p = 2$ and U, X, and Y are Hilbert spaces can be described as follows: for all $t \geq 0$, the images of the inner and incoming states under $\mathfrak{T}^t$ are orthogonal to the image of the outgoing state, and the null space of $\mathfrak{T}^t$ projected onto the inner and outgoing spaces is orthogonal to the null space of $\mathfrak{T}^t$ projected onto the incoming space. In the general case these properties can most easily be characterized in the following way.

Definition 2.7.2 A *Lax–Phillips model* of type $L^p|Reg_\omega$ is a semigroup on $\left[\begin{smallmatrix} \mathcal{Y} \\ X \\ \mathcal{U} \end{smallmatrix}\right]$, where $\mathcal{Y} = L^p|Reg_\omega(\mathbb{R}^-; Y)$ and $\mathcal{U} = L^p|Reg_\omega(\overline{\mathbb{R}}^+; U)$, with the structure

$$\mathfrak{T}^t = \begin{bmatrix} \tau_-^t & \mathfrak{C}^t & \mathfrak{D}^t \\ 0 & \mathfrak{A}^t & \mathfrak{B}^t \\ 0 & 0 & \tau_+^t \end{bmatrix},$$

where $\mathfrak{A}$ is strongly continuous and $\mathfrak{B}^t$, $\mathfrak{C}^t$, and $\mathfrak{D}^t$ satisfy the *causality conditions*

$$\mathfrak{B}^t = \mathfrak{B}^t \pi_{[0,t)}, \qquad \mathfrak{C}^t = \pi_{[-t,0)} \mathfrak{C}^t, \qquad \mathfrak{D}^t = \pi_{[-t,0)} \mathfrak{D}^t \pi_{[0,t)}. \tag{2.7.2}$$

Corollary 2.7.3 *The semigroup $\mathfrak{T}$ constructed in Theorem 2.7.1 is a Lax–Phillips model of type $L^p|Reg_\omega$.*

This is immediate from Theorem 2.7.1 and Definition 2.7.2.

Definition 2.7.4 We call the semigroup $\mathfrak{T}$ in Theorem 2.7.1 the Lax–Phillips model (of type $L^p|Reg_\omega$) induced by Σ.

Remark 2.7.5 Above we have absorbed both the input and the output of an $L^p|Reg$-well-posed linear system $\Sigma = \left[\begin{smallmatrix} \mathfrak{A} & \mathfrak{B} \\ \mathfrak{C} & \mathfrak{D} \end{smallmatrix}\right]$ on (Y, X, U) into the Lax–Phillips model. It is also possible to absorb only the input or the output. If Σ is L^p-well-posed with $p < \infty$ and $\omega \in \mathbb{R}$, then

$$\left[\begin{array}{cc} \mathfrak{A}^t & \mathfrak{B}_0^t \\ 0 & \tau_+^t \\ \hline \mathfrak{C} & \mathfrak{D}\pi_+ \end{array}\right] \tag{2.7.3}$$

is an L^p-well-posed linear system on $\left(Y, \left[\begin{smallmatrix} X \\ \mathcal{U} \end{smallmatrix}\right], 0\right)$ (with no input) where $\mathcal{U} = L^p_\omega(\mathbb{R}^+; U)$, and

$$\left[\begin{array}{cc|c} \tau_-^t & \tau^t \mathfrak{C}_0^t & \mathfrak{B} \\ 0 & \mathfrak{A}^t & \pi_- \mathfrak{D} \end{array}\right] \tag{2.7.4}$$

is an L^p-well-posed linear system on $\left(0, \left[\begin{smallmatrix} \mathcal{Y} \\ X \end{smallmatrix}\right], U\right)$ (with no output) where $\mathcal{Y} = L^p_\omega(\mathbb{R}^-; Y)$. We leave the easy proof to the reader.

It is only slightly more difficult to prove the converse: to every Lax–Phillips model there corresponds a well-posed linear system which induces this Lax–Phillips model:

Theorem 2.7.6 *Let $\mathfrak{T}$ be a Lax–Phillips model of type $L^p|Reg_\omega$. With the notation of Definition 2.7.2, define*

$$\mathfrak{B} = \lim_{s\to-\infty} \boldsymbol{\mathfrak{B}}^{-s}\tau^s, \quad \mathfrak{C} = \lim_{t\to\infty} \tau^{-t}\boldsymbol{\mathfrak{C}}^t, \quad \mathfrak{D} = \lim_{\substack{t\to\infty \\ s\to-\infty}} \tau^{-t}\boldsymbol{\mathfrak{D}}^{t-s}\tau^s. \tag{2.7.5}$$

Then $\Sigma = \left[\begin{array}{c|c} \mathfrak{A} & \mathfrak{B} \\ \hline \mathfrak{C} & \mathfrak{D} \end{array}\right]$ is a $L^p|Reg$-well-posed linear system on (Y, X, U), and $\mathfrak{T}$ is the Lax–Phillips model induced by this system.

Proof of Theorem 2.7.6. This proof is based on Theorem 2.2.14. We begin by defining the needed operator families (indexed by $-\infty < s \le t < \infty$) as

$$\left[\begin{array}{c|c} \mathfrak{A}_s^t & \mathfrak{B}_s^t \\ \hline \mathfrak{C}_s^t & \mathfrak{D}_s^t \end{array}\right] = \left[\begin{array}{c|c} \mathfrak{A}^{t-s} & \boldsymbol{\mathfrak{B}}^{t-s}\tau^s \\ \hline \tau^{-t}\boldsymbol{\mathfrak{C}}^{t-s} & \tau^{-t}\boldsymbol{\mathfrak{D}}^{t-s}\tau^s \end{array}\right].$$

This family has the time-invariance property (2.2.5), and it follows from (2.7.2) that it also has the causality property (2.2.4). The semigroup property $\mathfrak{T}^{s+t} = \mathfrak{T}^s\mathfrak{T}^t$ with $s, t \ge 0$ gives us four nontrivial identities, namely

$$\begin{aligned} \mathfrak{A}^{s+t} &= \mathfrak{A}^s\mathfrak{A}^t, & \boldsymbol{\mathfrak{B}}^{s+t} &= \mathfrak{A}^s\boldsymbol{\mathfrak{B}}^t + \boldsymbol{\mathfrak{B}}^s\tau^t, \\ \boldsymbol{\mathfrak{C}}^{s+t} &= \tau^s\boldsymbol{\mathfrak{C}}^t + \boldsymbol{\mathfrak{C}}^s\mathfrak{A}^t, & \boldsymbol{\mathfrak{D}}^{s+t} &= \tau^s\boldsymbol{\mathfrak{D}}^t + \boldsymbol{\mathfrak{C}}^s\boldsymbol{\mathfrak{B}}^t + \boldsymbol{\mathfrak{D}}^s\tau^t \end{aligned}$$

(where we have omitted some redundant projections) and this implies that the operators $\mathfrak{A}_s^t$, $\mathfrak{B}_s^t$, $\mathfrak{C}_s^t$, and $\mathfrak{D}_s^t$ have the composition property (2.2.6). As we noticed in the proof of Theorem 2.2.14, the two conditions (2.2.4) and (2.2.6) together imply (2.2.10). By Theorem 2.2.14, $\Sigma = \left[\begin{array}{c|c} \mathfrak{A} & \mathfrak{B} \\ \hline \mathfrak{C} & \mathfrak{D} \end{array}\right]$ is an $L^p|Reg$-well-posed linear system, and the corresponding Lax–Phillips model is the given one. □

Corollary 2.7.7 *For each $\omega \in \mathbb{R}$, there is a one-to-one correspondence between the class of all $L^p|Reg$-well-posed linear systems and all Lax–Phillips models of type $L^p|Reg_\omega$: every $L^p|Reg$-well-posed linear system Σ induces a unique Lax–Phillips model $\mathfrak{T}$ of type $L^p|Reg_\omega$, and conversely, every Lax–Phillips model $\mathfrak{T}$ of type $L^p|Reg_\omega$ induces a unique $L^p|Reg$-well-posed linear system Σ.*

Proof See Corollary 2.7.3 and Theorem 2.7.6. □

There are a number of important ingredients in the Lax–Phillips *scattering theory*, such as the backward and forward wave operators, the scattering operator, and the scattering matrix. All of these have natural analogs in the context of well-posed linear systems. In the discussion below we suppose that Σ is

ω-bounded (see Definition 2.5.6), or at least that $\mathfrak{B}$, $\mathfrak{C}$, and $\mathfrak{D}$ are ω-bounded. This is true, for example, if $\omega > \omega_{\mathfrak{A}}$, where $\omega_{\mathfrak{A}}$ is the growth bound of $\mathfrak{A}$.

The *backward wave operator* W_- (denoted by W_2 in Lax and Phillips 1973, Theorem 1.2) is the limit of the last column of $\mathfrak{T}\tau^{-t}$ as $t \to \infty$. It maps $L^p|Reg_\omega(\mathbb{R}; U)$ into $\left[\begin{smallmatrix} L^p|Reg_\omega(\mathbb{R}^-;Y) \\ X \\ L^p|Reg(\overline{\mathbb{R}}^+;U) \end{smallmatrix}\right]$, and it is given by (cf. Theorems 2.5.4 and 2.7.6)

$$W_- u = \begin{bmatrix} \pi_- \mathfrak{D} \\ \mathfrak{B} \\ \pi_+ \end{bmatrix} u. \tag{2.7.6}$$

Thus, it keeps the future input $\pi_+ u$ intact, and maps the past input $\pi_- u$ into the past output $\pi_- \mathfrak{D} u$ and the present inner state $\mathfrak{B} u$.

The *forward wave operator* W_+ (denoted by W_1 in Lax and Phillips 1973, Theorem 1.2) is the limit of the first row of $\tau^{-t}\mathfrak{T}$ as $t \to \infty$. It maps $\left[\begin{smallmatrix} L^p|Reg_\omega(\mathbb{R}^-;Y) \\ X \\ L^p|Reg(\overline{\mathbb{R}}^+;U) \end{smallmatrix}\right]$ into $L^p|Reg_\omega(\mathbb{R}; Y)$, and it is given by (cf. Theorems 2.5.4 and 2.7.6)

$$W_+ \begin{bmatrix} y \\ x_0 \\ u \end{bmatrix} = \begin{bmatrix} \pi_- & \mathfrak{C} & \mathfrak{D}\pi_+ \end{bmatrix} \begin{bmatrix} y \\ x_0 \\ u \end{bmatrix}. \tag{2.7.7}$$

Thus, it keeps the past output $\pi_- y$ intact, and maps the present inner state x_0 and the future input $\pi_+ u$ into the future output $\mathfrak{C} x_0 + \mathfrak{D}\pi_+ u$.

The two wave operators W_- and W_+ play very important roles in scattering theory and also in the theory of passive and conservative systems (see Chapter 11). Their most important property is that they intertwine the Lax–Phillips semigroup with the *bilateral left-shift* τ^t on $L^p|Reg_\omega(\mathbb{R}; U)$, respectively, $L^p|Reg_\omega(\mathbb{R}; Y)$ (just as the input map $\mathfrak{B}$ intertwines the semigroup $\mathfrak{A}^t$ on X with the outgoing left-shift τ_-^t on $L^p|Reg_\omega(\mathbb{R}^-; U)$ and the output map $\mathfrak{C}$ intertwines $\mathfrak{A}^t$ with the incoming left-shift τ_+^t on $L^p|Reg_\omega(\mathbb{R}^+; Y)$). This is the content of the following lemma.

Lemma 2.7.8 *Let $\omega \in \mathbb{R}$, let $\Sigma = \left[\begin{smallmatrix} \mathfrak{A} & \mathfrak{B} \\ \mathfrak{C} & \mathfrak{D} \end{smallmatrix}\right]$ be an ω-bounded $L^p|Reg$-well-posed linear system on (Y, X, U), and let $\mathfrak{T}$ be the corresponding Lax–Phillips model of type $L^p|Reg_\omega$. Then the two wave operators W_+ and W_- intertwine $\mathfrak{T}$ with the bilateral left-shift τ as follows: for all $t \ge 0$,*

$$\mathfrak{T}^t W_- u = W_- \tau^t u, \qquad y \in L^p|Reg_\omega(\mathbb{R}; U),$$

$$W_+ \mathfrak{T}^t \left[\begin{smallmatrix} y \\ x \\ u \end{smallmatrix}\right] = \tau^t W_+ \left[\begin{smallmatrix} y \\ x \\ u \end{smallmatrix}\right], \qquad \left[\begin{smallmatrix} y \\ x \\ u \end{smallmatrix}\right] \in \left[\begin{smallmatrix} L^p|Reg_\omega(\mathbb{R}^-;Y) \\ X \\ L^p|Reg_\omega(\mathbb{R}^+;U) \end{smallmatrix}\right].$$

In particular, $\mathcal{R}\left(\mathfrak{T}^t W_-\right) = \mathcal{R}(W_-)$ and $\mathcal{N}\left(W_+ \mathfrak{T}^t\right) = \mathcal{N}(W_-)$ for all $t \ge 0$.

Proof We leave the straightforward proof of the intertwining properties to the reader (see Definition 2.7.4 and Lemma 2.2.10(iii)–(iv)). The invariance of $\mathcal{R}\left(\mathfrak{T}^t W_-\right)$ and $\mathcal{N}\left(W_+\mathfrak{T}^t\right)$ follow from the identities $\mathfrak{T}^t W_-\tau^{-t} = W_-$ and $\tau^{-t} W_+\mathfrak{T}^t = W_+$ and the fact that τ^{-t} is injective and onto. $\square$

In this lemma the ω-boundedness of $\mathfrak{A}$ is irrelevant, but we need $\mathfrak{B}$, $\mathfrak{C}$, and $\mathfrak{D}$ to be ω-bounded in order for the two wave operators $W_-\colon L^p|Reg_\omega(\mathbb{R};U) \to \begin{bmatrix} L^p|Reg_\omega(\mathbb{R}^-;Y) \\ X \\ L^p|Reg_\omega(\mathbb{R}^+;U) \end{bmatrix}$ and $W_+\colon \begin{bmatrix} L^p|Reg_\omega(\mathbb{R}^-;Y) \\ X \\ L^p|Reg_\omega(\mathbb{R}^+;U) \end{bmatrix} \to L^p|Reg_\omega(\mathbb{R};Y)$ to be bounded.

The *scattering operator* in Lax–Phillips theory is the product W_+W_-, and it is given by

$$W_+W_- = \begin{bmatrix}\pi_- & \mathfrak{C} & \mathfrak{D}\pi_+\end{bmatrix}\begin{bmatrix}\pi_-\mathfrak{D} \\ \mathfrak{B} \\ \pi_+\end{bmatrix} = \pi_-\mathfrak{D} + \mathfrak{C}\mathfrak{B} + \pi_+\mathfrak{D} = \mathfrak{D}. \qquad (2.7.8)$$

Thus, the scattering operator is nothing but the (bilaterally shift-invariant) input/output map $\mathfrak{D}$ of the corresponding well-posed linear system.

To get the *scattering matrix* of the Lax–Phillips system we apply the scattering operator $\mathfrak{D}$ to an input of the form $u(t) = \mathrm{e}^{zt}u_0$, where $z \in \mathbb{C}$ has a sufficiently large real part and $u_0 \in U$ is fixed; see Lax and Phillips (1973, pp. 187–188). Because of the shift-invariance of $\mathfrak{D}$, the resulting output is of the type $y(t) = \mathrm{e}^{zt}y_0$ for some $y_0 \in Y$. The scattering matrix (evaluated at z) is defined to be the operator that maps $u_0 \in U$ into $y_0 \in Y$. By Definition 4.6.1, the scattering matrix of a Lax–Phillips system is equal to the transfer function $\widehat{\mathfrak{D}}$ of the corresponding well-posed linear system.

The generator of the Lax–Phillips semigroup $\mathfrak{T}$ and its resolvent are described in Theorem 4.8.3.

2.8 The Weiss notations

In the L^2-case the notion of a well-posed linear system that we have introduced in Definition 2.2.1 goes back to Salamon (1989) and Staffans (1997, 1998a). There is another commonly used notion which was introduced by G. Weiss (1989a, b, 1994a). Here the starting point is the input/state/output relations in Definition 2.2.7 with initial time zero, which are written in the form, for every $t \ge 0$,

$$\begin{aligned} x(t) &= \mathbb{T}_t x_0 + \Phi_t u = \mathfrak{A}^t x_0 + \mathfrak{B}_0^t u, \\ \pi_{[0,t)}y &= \Psi_t x_0 + \mathbb{F}_t u = \mathfrak{C}_0^t x_0 + \mathfrak{D}_0^t u, \end{aligned} \qquad (2.8.1)$$

where

$$\begin{bmatrix} \mathbb{T}_t & \Phi_t \\ \Psi_t & \mathbb{F}_t \end{bmatrix} := \begin{bmatrix} \mathfrak{A}^t & \mathfrak{B}_0^t \\ \mathfrak{C}_0^t & \mathfrak{D}_0^t \end{bmatrix} \tag{2.8.2}$$

is the notation used by Weiss, and the operators on the right hand side are defined in Definition 2.2.6. Comparing these operators to those appearing in the Lax–Phillips model we find that

$$\begin{aligned} \mathbb{T}_t &= \mathfrak{A}^t, & \Phi_t &= \mathfrak{B}_0^t = \boldsymbol{\mathfrak{B}}^t, \\ \Psi_t &= \mathfrak{C}_0^t = \tau^{-t}\boldsymbol{\mathfrak{C}}^t, & \mathbb{F}_t &= \mathfrak{D}_0^t = \tau^{-t}\boldsymbol{\mathfrak{D}}^t. \end{aligned} \tag{2.8.3}$$

Substituting this into the composition property in Lemma 2.2.8(iv) we get, for all $s, t \geq 0$,

$$\begin{aligned} \mathbb{T}_{s+t} &= \mathbb{T}_s \mathbb{T}_t, \\ \Phi_{s+t} &= \mathbb{T}_s \Phi_t + \Phi_s \tau^t, \\ \Psi_{s+t} &= \Psi_t + \tau^{-t} \Psi_s \mathbb{T}_t, \\ \mathbb{F}_{s+t} &= \mathbb{F}_t + \tau^{-t} \Psi_s \Phi_t + \tau^{-t} \mathbb{F}_s \tau^t. \end{aligned} \tag{2.8.4}$$

(Weiss uses a special concatenation operator to rewrite (2.8.4) in a more compact form, and he further usually denotes $\tau^{-t}\pi_+$ by $\mathbf{S}_t$, $\pi_{[0,t)}$ by $\mathbf{P}_t$, and τ^{-t} by $\mathcal{S}_t$.)

Theorem 2.8.1 *Let Σ be a $L^p|Reg$-well-posed linear system on (Y, X, U), and, for each $t \geq 0$, define $\left[\begin{smallmatrix} \mathbb{T}_t & \Phi_t \\ \Psi_t & \mathbb{F}_t \end{smallmatrix}\right]$ by* (2.8.2). *Then $\mathbb{T}$ is a C_0 semigroup on X,*

$$\begin{aligned} &\Phi_t \in \mathcal{B}(L^p|Reg(\mathbb{R}; U); X), \\ &\Psi_t \in \mathcal{B}(X; L^p|Reg(\mathbb{R}; Y)), \\ &\mathbb{F}_t \in \mathcal{B}(L^p|Reg(\mathbb{R}; U); L^p|Reg(\mathbb{R}; Y)), \end{aligned} \tag{2.8.5}$$

$$\begin{aligned} \Phi_t &= \Phi_t \pi_{[0,t)}, & \mathbb{F}_t &= \mathbb{F}_t \pi_{[0,t)}, \\ \mathbb{F}_t &= \pi_{[0,t)} \mathbb{F}_t, & \Psi_t &= \pi_{[0,t)} \Psi_t, \end{aligned} \tag{2.8.6}$$

and (2.8.4) *hold for all $s, t \geq 0$. Conversely, if $\left[\begin{smallmatrix} \mathbb{T}_t & \Phi_t \\ \Psi_t & \mathbb{F}_t \end{smallmatrix}\right]$ is a family of operators defined for $t \geq 0$ such that $\mathbb{T}$ is a C_0 semigroup on X, and* (2.8.4)–(2.8.6) *hold for all $s, t \geq 0$, then $\left[\begin{smallmatrix} \mathfrak{A} & \mathfrak{B} \\ \mathfrak{C} & \mathfrak{D} \end{smallmatrix}\right]$ is a $L^p|Reg$-well-posed linear system on (Y, X, U), where*

$$\mathfrak{B} = \lim_{s \to -\infty} \Phi_{-s}\tau^s, \quad \mathfrak{C} = \lim_{t \to \infty} \Psi_t, \quad \mathfrak{D} = \lim_{\substack{t \to \infty \\ s \to -\infty}} \tau^{-s}\mathbb{F}_{t-s}\tau^s. \tag{2.8.7}$$

Proof Use (2.8.3) and Theorems 2.7.1 and 2.7.6. □

We remark that the composition property (2.8.4) and the causality property (2.8.6) are equivalent to the block matrix identity (with $s, t \geq 0$)

$$\begin{bmatrix} \mathbb{T}_{s+t} & \Phi_{s+t} \\ \Psi_{s+t} & \mathbb{F}_{s+t} \end{bmatrix} = \begin{bmatrix} 1 & 0 & 0 \\ 0 & \pi_{[0,t)} & \tau^{-t}\pi_{[0,s)} \end{bmatrix} \begin{bmatrix} \mathbb{T}_s & 0 & \Phi_s \\ 0 & 1 & 0 \\ \Psi_s & 0 & \mathbb{F}_s \end{bmatrix} \times \begin{bmatrix} \mathbb{T}_t & \Phi_t & 0 \\ \Psi_t & \mathbb{F}_t & 0 \\ 0 & 0 & 1 \end{bmatrix} \begin{bmatrix} 1 & 0 \\ 0 & \pi_{[0,t)} \\ 0 & \pi_{[0,s)}\tau^t \end{bmatrix}. \tag{2.8.8}$$

2.9 Comments

Sections 2.1–2.2 The class of $L^p|Reg$-well-posed linear systems which we present here is by no means the only possible setting for an infinite-dimensional systems theory. Over time several related theories have been developed within different fields, often independently of each other.

One classical approach is to start as we did in Section 2.1 with the formal system

$$\begin{aligned} \dot{x}(t) &= Ax(t) + Bu(t), \\ y(t) &= Cx(t) + Du(t), \qquad t \geq s, \\ x(s) &= x_s, \end{aligned} \tag{2.1.1}$$

and to then impose more or less stringent conditions on A, B, C, and D. It is quite natural to require A to be the generator of a strongly continuous semigroup. The notion of a C_0 semigroup has its background in *parabolic* and *hyperbolic* partial differential equations, and we refer the reader to Davies (1980), Dunford and Schwartz (1958, 1963, 1971), Goldstein (1985), Lunardi (1995), Nagel (1986), Hille and Phillips (1957), Pazy (1983), and Yosida (1974) for the history and theory of C_0 semigroups beyond what we present in this book.

We get the mathematically simplest version of an infinite-dimensional systems theory by taking A in (2.1.1) to be the generator of a C_0 semigroup, and to take B, C, and D to be bounded linear operators, i.e., $B \in \mathcal{B}(U; X)$, $C \in \mathcal{B}(X; Y)$, and $D = \mathcal{B}(U; Y)$. Here U, X and Y are usually taken to be Hilbert spaces (instead of Banach spaces). By Proposition 2.3.1, this leads to a system which is both Reg-well-posed and L^p-well-posed for every p, $1 \leq p \leq \infty$. The big drawback with this class of systems is that the impulse response of such a system is always a continuous function, and this severely limits their applicability to boundary control and point observation processes. Systems of this type have been studied and used by, e.g., Baras and Brockett

(1975), Brockett and Fuhrmann (1976), Baras *et al.* (1974), Baras and Dewilde (1976), Bucci and Pandolfi (1998), Callier and Dumortier (1998), Callier *et al.* (1995), Callier and Winkin (1990, 1992), Curtain (1993), Curtain and Glover (1986b), Curtain and Oostveen (1998), Curtain and Rodman (1990), Curtain and Zwart (1994, 1995), Fuhrmann (1972, 1974, 1981), Lukes and Russell (1969), Oostveen (1999), and Pandolfi (1992), to mention only a few works.

To partly overcome the the limitations of the class described above, the notion of an L^p-well-posed input map $\mathfrak{B}$ (induced by A and B) and an L^p-well-posed output map $\mathfrak{C}$ (induced by A and C) gradually evolved in Chang and Lasiecka (1986), Curtain (1984), Curtain and Glover (1986a), Curtain and Salamon (1986), Da Prato *et al.* (1986), Desch *et al.* (1985), Dolecki and Russell (1977), Fattorini (1968), Flandoli (1984), Glover *et al.* (1988), Ho and Russell (1983), Lasiecka (1980), Lasiecka and Triggiani (1981, 1983a, b, c, 1986, 1991a), Lions (1971, 1988), Pritchard and Wirth (1978), Russell (1975), Washburn (1979), etc. These input and output maps were combined into the so-called *Pritchard–Salamon class* of infinite-dimensional systems by Salamon (1984) and Pritchard and Salamon (1985, 1987). This class achieved a certain popularity for a number of years. The characteristic feature of this class is that two different norms are used in the state space. With respect to the weaker norm, the control operator B is bounded and the output map $\mathfrak{C}$ is L^2-well-posed, and with respect to the stronger norm the input map $\mathfrak{B}$ is L^2-well-posed and the observation operator C is bounded. It turns out that these conditions are sufficiently strong that virtually all of the finite-dimensional systems theory can be extended to this class. However, the Pritchard–Salamon class is still not general enough for the most interesting boundary control and point observation processes. The impulse response is locally strong in L^2, and so is the dual impulse response (see Theorems 4.3.4 and 4.4.8). In particular, the delay line in Example 2.3.4 does not have a Pritchard–Salamon realization. The Pritchard–Salamon class of systems has been studied and used by, e.g., Curtain (1985, 1988, 1990, 1992, 1996), Curtain *et al.* (1994), Curtain and Pritchard (1994), Curtain and Ran (1989), Curtain, Weiss and Zhou (1996), Curtain and Zwart (1994), Kaashoek *et al.* (1997), van Keulen (1993), and M. Weiss (1994, 1997).

The first 'modern' version of an L^2-well-posed linear system on three Hilbert spaces (Y, X, U) was presented by Salamon (1987, 1989), G. Weiss (1989c), and Curtain and Weiss (1989) (also found in Helton (1976) is a more implicit way). As we mentioned above, at that time it was well-known how to create L^2-well-posed input and output maps from the operators A, B, and C, but the general construction of an input/output map from A, B, and C, was still incomplete. Such a construction was given by Salamon (1987), Šmuljan (1986), G. Weiss (1989c), and Curtain and Weiss (1989) (see also Helton (1976), Arov and Nudelman (1996), Ober and Montgomery-Smith (1990), and Ober and Wu (1996)). The converse question of how to construct B and C from the input and output maps $\mathfrak{B}$ and $\mathfrak{C}$ (see Chapter 4) was addressed in (Salamon, 1989), (Weiss, 1989a, b). Since then L^2-well-posed linear systems in the Hilbert space setting (sometimes regular; cf. Chapter 5) have been studied and used by a large number of authors, such as Avalos *et al.* (1999), Curtain (1989a, b, 1997), Curtain, Weiss, and Weiss (1996), Hinrichsen and Pritchard (1994), Logemann *et al.* (1998), Jacob and Zwart (2001a, b, 2002), Logemann *et al.* (1996), Logemann and Ryan (2000), Logemann *et al.* (1998, 2000), Logemann and Townley (1997a, b), Morris (1994, 1999), Rebarber (1993, 1995), Staffans (1996, 1997, 1998a, b, c, d, 1999a), G. Weiss (1994a, b), Weiss and Curtain (1997), Weiss and Häfele (1999), Weiss and Rebarber (1998), Staffans and Weiss (2002, 2004), M. Weiss (1994, 1997), and Weiss and Weiss (1997).

In parallel with the work described above there was intensive research going on in optimal control of partial differential equations, which uses much the same technique, but which does not fit into the general framework described above. There the authors usually work directly with the partial differential equation, often rewritten in the form (2.1.1), instead of introducing the operators $\mathfrak{B}$, $\mathfrak{C}$, and $\mathfrak{D}$. Sometimes this approach is replaced by a 'direct' approach based on the study of an integral or integro-differential equation. It is true that most examples of parabolic type are L^2-well-posed if we choose the state space appropriately (see Theorem 5.7.3), but it is usually more important that these systems are *Reg*-well-posed in a smaller state space (where the observation operator C is bounded). Most hyperbolic examples have an L^2-well-posed input map $\mathfrak{B}$ and a bounded control operator C (thus, they are both L^2-well-posed and *Reg*-well-posed). However, there do exist exceptions where the system is not $L^p|Reg$-well-posed in any sense. There are a number of books on optimal control of partial differential equations, such as Bensoussan *et al.* (1992), Curtain and Pritchard (1978), Lasiecka and Triggiani (1991a, 2000a, b), and Lions (1971). Systems of parabolic type are studied by, e.g., Curtain and Ichikawa (1996), Da Prato and Ianelli (1985), Da Prato and Ichikawa (1985, 1993), Da Prato and Lunardi (1988, 1990), Flandoli (1987), Lasiecka (1980), Lasiecka *et al.* (1995), Lasiecka *et al.* (1997), Lasiecka and Triggiani (1983a, b, 1987a, b, 1992b), and McMillan and Triggiani (1994a, b). Systems of hyperbolic type are studied by, e.g., Chang and Lasiecka (1986), Da Prato *et al.* (1986), Flandoli *et al.* (1988), Hendrickson and Lasiecka (1993, 1995), Lasiecka and Triggiani (1981, 1983c, 1986, 1987c, 1988, 1989a, b, 1990a, b, 1991b, c, d, 1992a, c, 1993a, b), and Lions (1988).

Above we have described how the theory of well-posed linear systems has developed out of the theory for the state space system (2.1.1) with bounded control operator B and observation operator C. In parallel to this development an algebraic theory evolved at an early stage, based on the interplay between the input map, the output map, and the input/output map (see the algebraic conditions in Definition 2.2.1). In particular the input/output map $\mathfrak{D}$ and its Hankel operator $\pi_+\mathfrak{D}\pi_-$ play important roles, and the main objective is often to construct a (spectrally) *minimal realization* of a given input/output map or its transfer function. A good early representative for this class of work is Kalman *et al.* (1969, Part 4). The approach in Kalman *et al.* (1969, Part 4) is algebraic, somewhat similar in spirit to our definition of a $L^p|Reg$-well-posed linear system, but it is mainly discrete time, more abstract, and it puts less emphasis on the exact continuity requirements of the different parts of the system. A somewhat different approach was taken by Balakrishnan (1966). He starts with a system with well-defined initial state, input and output, and constructs the corresponding semigroup $\mathfrak{A}$ in roughly the same way as we do in Theorem 9.3.1(iv). Theorem 3.1 in Balakrishnan (1966) describes a *Reg*-well-posed linear system with finite-dimensional input and output spaces. Further works in the same direction have been done by Baras and Brockett (1975) and Baras *et al.* (1974) (bounded B and C), Baras and Dewilde (1976) (bounded B and C, and frequency domain), Dewilde (1971) (frequency domain approach), Fuhrmann (1974) (discrete and continuous time with bounded B and C), Fuhrmann (1981) (many different settings), Feintuch (1998) (input/output approach), Feintuch and Saeks (1982) (a general theory based on Hilbert resolution spaces), Kamen (1975) (an algebraic approach), and Yamamoto (1981) (very weak continuity assumptions). In addition, a significant amount of corresponding results have been obtained for discrete time systems, and these can be turned into continuous time results by use of the Cayley transform (see Theorem 12.3.5). See, e.g., Arov (1974a, b, 1979a, b, c), Fuhrmann (1981), Helton (1974), Ober and Montgomery-Smith (1990), and Ober and Wu (1993, 1996).

So far we have primarily discussed publications with a dominating control theory background (although some of the work in optimal control is quite mathematically oriented). In the early 1960s a complementary theory evolved in the field of pure mathematics and mathematical physics. This theory was infinite-dimensional at the outset, and it uses a very different language. The book by Sz.-Nagy and Foiaş (1970) can be viewed after a translation of terms as a treatise on infinite-dimensional discrete-time systems (Chapter 11 can be regarded as a natural continuous-time analogue of one of the central parts of Sz.-Nagy and Foiaş (1970), rewritten in the language of L^2-well-posed linear systems). In Section 2.7 we describe the close connection which exists between the theory of L^2-well-posed linear systems and Lax–Phillips scattering theory, as presented in Lax and Phillips (1967, 1973). Adamajan and Arov (1970) proved the Sz.Nagy–Foiaş and Lax–Phillips theories to be equivalent. The strong connection between the Lax–Phillips theory and the theory of well-posed linear systems was brought forward by Helton (1974, 1976), and their work was continued by Ober and Montgomery-Smith (1990) and Ober and Wu (1996). This connection was also noticed in the Soviet Union, where the theory of well-posed linear systems evolved independently of the the work by Salamon and Weiss in the West (see, e.g., Arov (1974a, b, 1979a, b, c, 1999), Brodskiĭ (1971), Livšic (1973), Livšic and Yantsevich (1977), Šmuljan (1986); further references are found in Arov and Nudelman (1996). For a more detailed description of the early history of this subject we refer the reader to Arov (1995) and Helton (1976).

In this book we have chosen to let the input space U, the state space X, and the output space Y be Banach spaces, and we use either a local L^p-norm or a local sup-norm on the input and output functions. This is less general than the Fréchet spaces used by, for example, Yamamoto (1981), but more general than the usual Hilbert space setting where U, X, and Y are Hilbert spaces and $p = 2$. There are two main reasons for our choice of setting. We do not want to leave the Banach space context, because then we lose the finite growth bound in Theorem 2.5.4 (see Yamamoto, 1981), and this finite growth bound is a very important technical tool. Among others, it implies the existence of a transfer function, defined on some half-plane (see Section 4.6). On the other hand, in the present book we would gain very little by restricting ourselves to the Hilbert space setting (the vast majority of the formulas and the proofs are the same in the Banach and Hilbert space cases), and there are applications where it is quite useful to be able to work in L^p with $p \neq 2$. This is, in particular, true for systems with an analytic semigroup (see Theorem 5.7.3), and in perturbation theory involving certain types of nonlinearities. However, some of the results that we present are valid only in the Hilbert space setting.

Section 2.3 Shift groups and semigroups are found in one form or another in many books on semigroup theory, such as Hille and Phillips (1957). Books on harmonic analysis often exploit the properties of the shift operator; see, e.g., Foiaş and Frazho (1990), Nikol'skiĭ (1986), Rosenblum and Rovnyak (1985), and Sz.-Nagy and Foiaş (1970). The delay line example is classical, and it is found in, e.g., Helton (1976), Salamon (1987), G. Weiss (1989c), G. Weiss (1994a), and Weiss and Zwart (1998). The systems in Examples 2.3.5–2.3.13 have been modeled after the corresponding classic finite-dimensional examples (and several of them are also found in Weiss and Curtain (1997)).

Section 2.4–2.5 That every C_0-semigroup has a finite growth bound is well-known; see, e.g., Hille and Phillips (1957) or Pazy (1983). The rest of Theorem 2.5.4 is due to G. Weiss (1989a, b, 1994a) (it is also stated in Salamon (1989) with a partial proof). Our proof is a rewritten version of G. Weiss's proof, where we emphasize the possibility to discretize the time.

Section 2.6 Different versions of the exactly observable and exactly controllable shift realizations in Example 2.6.5 are found in, e.g., Baras and Brockett (1975), Baras and Dewilde (1976), Fuhrmann (1974, Theorem 2.6), Fuhrmann (1981, Section 3.2), Helton (1974, p. 31), Jacob and Zwart (2002, Theorem A.1), Ober and Montgomery-Smith (1990), and Ober and Wu (1996, Sections 5.2–5.3). The first time that they appear in (almost) this generality is Salamon (1989, Theorem 4.3). The bilateral input and output shift realizations are related to the incoming and outgoing translations representations used by Lax and Phillips (1967, 1973).

The exactly controllable and the exactly observable shift realizations have also been known in the integral and functional equations communities in a somewhat different setting. The 'initial function semigroup' in (Gripenberg *et al.* 1990, Section 8.2) can be interpreted as a flow-inverted version of the exactly controllable shift realization, and the 'forcing function semigroup' described there is a flow-inverted version of the exactly observable shift realization. (Flow-inversion is described in Section 6.3.) Analogously, the extended semigroups in Gripenberg *et al.* (1990, Section 8.3) are flow-inverted versions of the bilateral shift semigroups in Example 2.6.5. See Gripenberg *et al.* (1990) for the history of these semigroups.

Section 2.7 Our presentation of the Lax–Phillips scattering model has been modeled after Helton (1976). (The corresponding discrete time version is given in Helton (1974).) To make the connection more transparent we have replaced the right shift used in the shift representations of the incoming and outgoing subspaces in Lax and Phillips (1967, 1973) and Helton (1976) by a left shift.

As Corollary 2.7.7 shows, there is a very strong connection between the Lax–Phillips scattering theory and the theory of $L^p|Reg$-well-posed linear systems. This connection does not seem to have had a significant influence on the work by Curtain and Weiss (1989), Salamon (1987, 1989) and Weiss (1989a, b, c, 1994a, b), but it has clearly influenced the work by Arov (1979b) and Arov and Nudelman (1996). We shall use very little of the actual theory from Lax and Phillips (1967, 1973), although we shall rederive many of the results given there from the present theory of well-posed linear systems. The main motivation behind Lax and Phillips (1967, 1973) is quite different from the motivation behind this book. There the main object is to create a theory which can be applied to scattering. The incoming and outgoing subspaces are defined in an abstract way, and they are intrinsically infinite-dimensional. There the typical incoming and outgoing subspaces consist of initial data for solutions of the wave equation in free space which vanish in a truncated backward or forward light cone (the amount of truncation is proportional to the size of the scatterer). In Lax and Phillips (1967) the incoming and outgoing subspaces are defined in such a way that the corresponding well-posed linear system is always controllable and observable (cf. Chapter 9).

Parts of Corollary 2.7.7 (where either the input operator or output operator vanishes) were proved by Grabowski and Callier (1996) and Engel (1998). It is also implicitly contained in Arov and Nudelman (1996) and in Helton (1976).

Section 2.8 The Weiss notation appears for the first time in Weiss (1989a) (the input operator), Weiss (1989b) (the output operator), and Curtain and Weiss (1989) (the input/output operator), and has since then been used by many workers in the field. The notation used in Definition 2.2.1 was introduced in Staffans (1997), and resembles the notation used by Salamon (1989) (and by Fuhrmann (1981) in the discrete time case). The Weiss notation is convenient as long we confine ourselves to the initial value problem with initial time zero and positive final time, but the resulting algebraic conditions (2.8.4) are more complicated than those used in Definition 2.2.1, and they are less flexible if one wants to work with some other notions of state and output. These alternative notions of state are important, for example, in optimal control theory and for some of the transformations described in Chapter 6. In this work we have throughout replaced the Weiss notation by that in Definition 2.2.6.

3

Strongly continuous semigroups

The most central part of a well-posed linear system is its semigroup. This chapter is devoted to a study of the properties of C_0 semigroups, both in the time domain and in the frequency domain. Typical time domain issues are the generator of a semigroup, the dual semigroup, and the nonhomogeneous initial value problem. The resolvent of the generator lives in the frequency domain.

3.1 Norm continuous semigroups

We begin by introducing the notion of the *generator* of a C_0 (semi)group (cf. Definition 2.2.2).

Definition 3.1.1

(i) The *generator* A of a C_0 semigroup $\mathfrak{A}$ is the operator

$$Ax := \lim_{h \downarrow 0} \frac{1}{h}(\mathfrak{A}^h - 1)x,$$

defined for all those $x \in X$ for which this limit exists.

(ii) The *generator* A of a C_0 group $\mathfrak{A}$ is the operator

$$Ax := \lim_{h \to 0} \frac{1}{h}(\mathfrak{A}^h - 1)x,$$

defined for all those $x \in X$ for which this limit exists.

Before we continue our study of C_0 semigroups and their generators, let us first study the smaller class of uniformly continuous semigroups, i.e., semigroups $\mathfrak{A}$ which satisfy (cf. Definition 2.2.2)

$$\lim_{t \downarrow 0} \|\mathfrak{A}^t - 1\| = 0. \tag{3.1.1}$$

Clearly, every uniformly continuous semigroup is a C_0 semigroup.

We begin by presenting an example of a uniformly continuous (semi)group. (As we shall see in Theorem 3.1.3, every uniformly continuous (semi)group is of this type.)

Example 3.1.2 *Let $A \in \mathcal{B}(X)$, and define*

$$e^{At} := \sum_{n=0}^{\infty} \frac{(At)^n}{n!}, \qquad t \in \mathbb{R}. \tag{3.1.2}$$

Then e^{At} is a uniformly continuous group on X, and its generator is A. This group satisfies $\|e^{At}\| \le e^{\|A\||t|}$ for all $t \in \mathbb{R}$. In particular, the growth bounds of the semigroups $t \mapsto e^{At}$ and $t \mapsto e^{-At}$ (where $t \ge 0$) are bounded by $\|A\|$ (cf. Definition 2.5.6).

Proof The series in (3.1.2) converges absolutely since

$$\sum_{n=0}^{\infty} \left\| \frac{(At)^n}{n!} \right\| \le \sum_{n=0}^{\infty} \frac{(\|A\||t|)^n}{n!} = e^{\|A\||t|}.$$

This proves that $\|e^{At}\|$ satisfies the given bounds. Clearly $e^{0A} = 1$. Being a power series, the function $t \mapsto e^{At}$ is analytic, hence uniformly continuous for all t. By differentiating the power series (this is permitted since e^{At} is analytic) we find that the generator of e^{At} is A (and the limit in Definition 3.1.1 is uniform). Thus, it only remains to verify the group property $e^{A(s+t)} = e^{As}e^{At}$, which is done as follows:

$$\begin{aligned} e^{A(s+t)} &= \sum_{n=0}^{\infty} \frac{A^n (s+t)^n}{n!} = \sum_{n=0}^{\infty} \frac{A^n}{n!} \sum_{k=0}^{n} \binom{n}{k} s^k t^{n-k} \\ &= \sum_{n=0}^{\infty} \sum_{k=0}^{n} \frac{A^k s^k}{k!} \frac{A^{n-k} t^{n-k}}{(n-k)!} = \sum_{k=0}^{\infty} \frac{A^k s^k}{k!} \sum_{n=k}^{\infty} \frac{A^{n-k} t^{n-k}}{(n-k)!} \\ &= e^{As} e^{At}. \end{aligned}$$

□

Theorem 3.1.3 *Let $\mathfrak{A}$ be a uniformly continuous semigroup. Then the following claims are true:*

(i) *$\mathfrak{A}$ has a bounded generator A and $\mathfrak{A}^t = e^{At}$ for all $t \ge 0$;*
(ii) *$t \mapsto \mathfrak{A}^t$ is analytic and $\frac{d}{dt}\mathfrak{A}^t = A\mathfrak{A}^t = \mathfrak{A}^t A$ for all $t \ge 0$;*
(iii) *$\mathfrak{A}$ can be extended to an analytic group on $\mathbb{R}$ satisfying $\frac{d}{dt}\mathfrak{A}^t = A\mathfrak{A}^t = \mathfrak{A}^t A$ for all $t \in \mathbb{R}$.*

Remark 3.1.4 Actually a slightly stronger result is true: every C_0 semigroup $\mathfrak{A}$ satisfying

$$\limsup_{t\downarrow 0}\|\mathfrak{A}^t - 1\| < 1 \tag{3.1.3}$$

has a bounded generator A and $\mathfrak{A}^t = e^{At}$. Alternatively, every C_0 semigroup $\mathfrak{A}$ for which the operator $\int_0^h \mathfrak{A}^s\,ds$ is invertible for some $h > 0$ has a bounded generator A and $\mathfrak{A}^t = e^{At}$. The proof is essentially the same as the one given below (it uses strong integrals instead of uniform integrals).

Proof of Theorem 3.1.3 (i) For sufficiently small positive h, $\big\|1 - (1/h)\int_0^h \mathfrak{A}^s\,ds\big\| < 1$, hence $(1/h)\int_0^h \mathfrak{A}^s\,ds$ is invertible, and so is $\int_0^h \mathfrak{A}^s\,ds$. By the semigroup property, for $0 < t < h$,

$$\begin{aligned}\frac{1}{t}(\mathfrak{A}^t - 1)\int_0^h \mathfrak{A}^s\,ds &= \frac{1}{t}\Big(\int_0^h \mathfrak{A}^{s+t}\,ds - \int_0^h \mathfrak{A}^s\,ds\Big)\\ &= \frac{1}{t}\Big(\int_t^{t+h} \mathfrak{A}^{s+t}\,ds - \int_0^h \mathfrak{A}^s\,ds\Big)\\ &= \frac{1}{t}\Big(\int_h^{t+h} \mathfrak{A}^{s+t}\,ds - \int_0^t \mathfrak{A}^s\,ds\Big).\end{aligned}$$

Multiply by $\big(\int_0^h \mathfrak{A}^s\,ds\big)^{-1}$ to the right and let $t \downarrow 0$ to get

$$\lim_{t\downarrow 0}\frac{1}{t}(\mathfrak{A}^t - 1) = (\mathfrak{A}^h - 1)\Big(\int_0^h \mathfrak{A}^s\,ds\Big)^{-1}$$

in the uniform operator norm. This shows that $\mathfrak{A}$ has the bounded generator $A = (\mathfrak{A}^h - 1)\big(\int_0^h \mathfrak{A}^s\,ds\big)^{-1}$. By Example 3.1.2, the group e^{At} has the same generator A as $\mathfrak{A}$. By Theorem 3.2.1(vii) below, $\mathfrak{A}^t = e^{At}$ for $t \ge 0$.

(ii)–(iii) See Example 3.1.2 and its proof. □

3.2 The generator of a C_0 semigroup

We now return to the more general class of C_0 semigroups. We already introduced the notion of the *generator* of a C_0 semigroup in Definition 3.1.1. Some basic properties of this generator are listed in the following theorem.

Theorem 3.2.1 *Let $\mathfrak{A}^t$ be a C_0 semigroup on a Banach space X with generator A.*

(i) *For all $x \in X$,*

$$\lim_{h\downarrow 0}\frac{1}{h}\int_t^{t+h} \mathfrak{A}^s x\,ds = \mathfrak{A}^t x.$$

(ii) *For all* $x \in X$ *and* $0 \le s < t < \infty$, $\int_s^t \mathfrak{A}^v x\,dv \in \mathcal{D}(A)$ *and*

$$\mathfrak{A}^t x - \mathfrak{A}^s x = A \int_s^t \mathfrak{A}^v x\,dv.$$

(iii) *For all* $x \in \mathcal{D}(A)$ *and* $t \ge 0$, $\mathfrak{A}^t x \in \mathcal{D}(A)$, $t \mapsto \mathfrak{A}^t x$ *is continuously differentiable in* X, *and*

$$\frac{\mathrm{d}}{\mathrm{d}t}\mathfrak{A}^t x = A\mathfrak{A}^t x = \mathfrak{A}^t A x, \qquad t \ge 0.$$

(iv) *For all* $x \in \mathcal{D}(A)$ *and* $0 \le s \le t < \infty$,

$$\mathfrak{A}^t x - \mathfrak{A}^s x = A \int_s^t \mathfrak{A}^v x\,dv = \int_s^t \mathfrak{A}^v A x\,dv.$$

(v) *For all* $n = 1, 2, 3, \ldots$, *if* $x \in \mathcal{D}(A^n)$, *then* $\mathfrak{A}^t x \in \mathcal{D}(A^n)$ *for all* $t \ge 0$, *the function* $t \mapsto \mathfrak{A}^t x$ *is* n *times continuously differentiable in* X, *and for all* $k = 0, 1, 2, \ldots, n$,

$$\left(\frac{\mathrm{d}}{\mathrm{d}t}\right)^n \mathfrak{A}^t x = A^k \mathfrak{A}^t A^{n-k} x, \qquad t \ge 0.$$

(vi) A *is a closed linear operator and* $\bigcap_{n=1}^{\infty} \mathcal{D}(A^n)$ *is dense in* X. *For each* $x \in \bigcap_{n=1}^{\infty} \mathcal{D}(A^n)$ *the function* $t \mapsto \mathfrak{A}^t x$ *belongs to* $C^\infty(\overline{\mathbb{R}}^+; U)$.

(vii) $\mathfrak{A}$ *is uniquely determined by its generator* A.

Proof (i) This follows from the continuity of $s \mapsto \mathfrak{A}^s x$ (see Lemma 2.2.13(ii)).

(ii) Let $x \in X$ and $h > 0$. Then

$$\begin{aligned}\frac{1}{h}(\mathfrak{A}^h - 1)\int_s^t \mathfrak{A}^v x\,dv &= \frac{1}{h}\int_s^t (\mathfrak{A}^{v+h}x - \mathfrak{A}^v x)\,dv \\ &= \frac{1}{h}\int_t^{t+h} \mathfrak{A}^v x\,dv - \int_s^{s+h} \mathfrak{A}^v x\,dv.\end{aligned}$$

As $h \downarrow 0$ this tends to $\mathfrak{A}^t x - \mathfrak{A}^s x$.

(iii) Let $x \in \mathcal{D}(A)$ and $h > 0$. Then

$$\frac{1}{h}(\mathfrak{A}^h - 1)\mathfrak{A}^t x = \mathfrak{A}^t \frac{1}{h}(\mathfrak{A}^h - 1)x \to \mathfrak{A}^t A x \text{ as } h \downarrow 0.$$

Thus, $\mathfrak{A}^t x \in \mathcal{D}(A)$, and $A\mathfrak{A}^t x = \mathfrak{A}^t A x$ is equal to the right-derivative of $\mathfrak{A}^t x$ at t. To see that it is also a left-derivative we compute

$$\frac{1}{h}(\mathfrak{A}^t x - \mathfrak{A}^{t-h}x) - \mathfrak{A}^t A x = \mathfrak{A}^{t-h}\Big(\frac{1}{h}(\mathfrak{A}^h x - x) - Ax\Big) + (\mathfrak{A}^{t-h} - \mathfrak{A}^t)Ax.$$

This tends to zero because of the uniform boundedness of $\mathfrak{A}^{t-h}$ and the strong continuity of $\mathfrak{A}^t$ (see Lemma 2.2.13).

(iv) We get (iv) by integrating (iii).

(v) This follows from (iii) by induction.

(vi) The linearity of A is trivial. To prove that A is closed we let $x_n \in \mathcal{D}(A)$, $x_n \to x$, and $Ax_n \to y$ in X, and claim that $Ax = y$. By part (iv) with $s = 0$,

$$\mathfrak{A}^t x_n - x_n = \int_0^t \mathfrak{A}^s A x_n \, ds.$$

Both sides converge as $n \to \infty$ (the integrand converges uniformly on $[0, t]$), hence

$$\mathfrak{A}^t x - x = \int_0^t \mathfrak{A}^s y \, ds.$$

Divide by t, let $t \downarrow 0$, and use part (i) to get $Ax = y$.

We still need to show that $\bigcap_{n=1}^{\infty} \mathcal{D}(A^n)$ is dense in X. Pick some real-valued C^∞ function η with compact support in $(0, 1)$ and $\int_0^\infty \eta(s)\, ds = 1$. For each $x \in X$ and $k = 1, 2, 3, \ldots$, we define

$$x_k = k \int_0^1 \eta(ks)\mathfrak{A}^s x \, ds.$$

Then, for each $h > 0$,

$$\begin{aligned}
\frac{1}{h}(\mathfrak{A}^h - 1)x_k &= \frac{1}{h}\int_0^1 \eta(ks)[\mathfrak{A}^{s+h}x - \mathfrak{A}^s x]\, ds \\
&= k \int_0^{1+h} \frac{1}{h}[\eta(k(s-h)) - \eta(ks)]\mathfrak{A}^s x \, ds \\
&\to -k^2 \int_0^1 \dot{\eta}(ks)\mathfrak{A}^s x \, ds \text{ as } h \downarrow 0.
\end{aligned}$$

Thus, $x_k \in \mathcal{D}(A)$ and $Ax_k = -k^2 \int_0^1 \dot{\eta}(ks)\mathfrak{A}^s x$. We can repeat the same argument with η replaced by $\dot{\eta}$, etc., to get $x_k \in \mathcal{D}(A^n)$ for every $n = 1, 2, 3 \ldots$ This means that $x_k \in \bigcap_{n=1}^{\infty} \mathcal{D}(A^n)$.

We claim that $x_k \to x$ as $k \to \infty$, proving the density of $\bigcap_{n=1}^{\infty} \mathcal{D}(A^n)$ in X. To see this we make a change of integration variable to get

$$x_k = \int_0^1 \eta(s)\mathfrak{A}^{s/k} x \, ds.$$

The function $\mathfrak{A}^{s/k}x$ tends uniformly to x on $[0, 1]$, hence the integral tends to $\int_0^\infty \eta(s)x \, ds = x$ as $k \to \infty$.

That $\mathfrak{A}x \in C^\infty(\overline{\mathbb{R}}^+; U)$ whenever $x \in \bigcap_{n=1}^{\infty} \mathcal{D}(A^n)$ follows from (iv).

(vii) Suppose that there is another C_0 semigroup $\mathfrak{A}_1$ with the same generator A. Take $x \in \mathcal{D}(A)$, $t > 0$, and consider the function $s \mapsto \mathfrak{A}^{t-s}\mathfrak{A}_1^s x$, $s \in [0, t]$.

We can use part (iii) and the chain rule to compute its derivative in the form

$$\frac{\mathrm{d}}{\mathrm{d}s}\mathfrak{A}^s\mathfrak{A}_1^{t-s}x = A\mathfrak{A}^s\mathfrak{A}_1^{t-s}x - \mathfrak{A}^s A\mathfrak{A}_1^{t-s}x = \mathfrak{A}^s\mathfrak{A}_1^{t-s}Ax - \mathfrak{A}^s\mathfrak{A}_1^{t-s}Ax = 0.$$

Thus, this function is a constant. Taking $s = 0$ and $s = t$ we get $\mathfrak{A}^t x = \mathfrak{A}_1^t x$ for all $x \in \mathcal{D}(A)$. By the density of $\mathcal{D}(A)$ in X, the same must be true for all $x \in X$. □

To illustrate Definition 3.1.1, let us determine the generators of the shift (semi)groups τ^t, τ_+^t, τ_-^t, $\tau_{[0,T)}^t$, and $\tau_{\mathbb{T}_T}^t$ in Examples 2.3.2 and 2.5.3. The domains of these generators are spaces of the following type:

Definition 3.2.2 Let J be a subinterval of $\mathbb{R}$, $\omega \in \mathbb{R}$, and let U be a Banach space.

(i) A function u belongs to $W_{\mathrm{loc}}^{n,p}(J;U)$ if it is an nth order integral of a function $u^{(n)} \in L_{\mathrm{loc}}^p(J;U)$ (i.e., $u^{(n-1)}(t_2) - u^{(n-1)}(t_1) = \int_{t_1}^{t_2} u^{(n)}(s)\,ds$, etc.).[1] It belongs to $W_\omega^{n,p}(J;U)$ if, in addition, $u^{(k)} \in L_\omega^p(J;U)$ for all $k = 0, 1, 2, \ldots, n$.

(ii) The space $W_{c,\mathrm{loc}}^{n,p}(\mathbb{R};U)$ consists of the functions in $W_{\mathrm{loc}}^{n,p}(\mathbb{R};U)$ whose support is bounded to the left, and the space $W_{\omega,\mathrm{loc}}^{n,p}(\mathbb{R};U)$ consists of the functions u in $W_{\mathrm{loc}}^{n,p}(\mathbb{R};U)$ which satisfy $\pi_- u \in W_\omega^{n,p}(\mathbb{R}^-;U)$.

(iii) The spaces $W_{0,\omega}^{n,p}(J;U)$, $W_{0,\omega,\mathrm{loc}}^{n,p}(\mathbb{R};U)$, $BC_\omega^n(J;U)$, $BC_{\omega,\mathrm{loc}}^n(\mathbb{R};U)$, $BC_{0,\omega}^n(J;U)$, $BC_{0,\omega,\mathrm{loc}}^n(\mathbb{R};U)$, $BUC_\omega^n(J;U)$, $BUC_{\omega,\mathrm{loc}}^n(\mathbb{R};U)$, $Reg_\omega^n(J;U)$, $Reg_{\omega,\mathrm{loc}}^n(\mathbb{R};U)$, $Reg_{0,\omega}^n(J;U)$, and $Reg_{0,\omega,\mathrm{loc}}^n(\mathbb{R};U)$ are defined in an analogous way, with L^p replaced by BC, BC_0, BUC, Reg, or Reg_0.

Example 3.2.3 *The generators of the (semi)groups τ^t, τ_+^t, τ_-^t, $\tau_{[0,T)}^t$, and $\tau_{\mathbb{T}_T}^t$ in Examples 2.3.2 and 2.5.3 are the following:*

(i) *The generator of the bilateral left shift group τ^t on $L_\omega^p(\mathbb{R};U)$ is the differentiation operator $\frac{\mathrm{d}}{\mathrm{d}s}$ with domain $W_\omega^{1,p}(\mathbb{R};U)$, and the generator of the left shift group τ^t on $BUC_\omega(\mathbb{R};U)$ is the differentiation operator $\frac{\mathrm{d}}{\mathrm{d}s}$ with domain $BUC_\omega^1(\mathbb{R};U)$. We denote these generators simply by $\frac{\mathrm{d}}{\mathrm{d}s}$.*

(ii) *The generator of the incoming left shift semigroup τ_+^t on $L_\omega^p(\mathbb{R}^+;U)$ is the differentiation operator $\frac{\mathrm{d}}{\mathrm{d}s}$ with domain $W_\omega^{1,p}(\overline{\mathbb{R}}^+;U)$, and the generator of the left shift semigroup τ_+^t on $BUC_\omega(\overline{\mathbb{R}}^+;U)$ is the differentiation operator $\frac{\mathrm{d}}{\mathrm{d}s}$ with domain $BUC_\omega^1(\overline{\mathbb{R}}^+;U)$. We denote these generators by $\frac{\mathrm{d}}{\mathrm{d}s}_+$.*

[1] Our definition of $W_{\mathrm{loc}}^{n,p}$ implies that the functions in this space are locally absolutely continuous together with their derivatives up to order $n-1$. This is true independently of whether U has the Radon–Nikodym property or not.

(iii) *The generator of the outgoing left shift semigroup τ_-^t on $L^p_\omega(\mathbb{R}^-;U)$ is the differentiation operator $\frac{\mathrm{d}}{\mathrm{d}s}$ with domain $\{u \in W^{1,p}_\omega(\overline{\mathbb{R}}^-;U) \mid u(0)=0\}$, and the generator of the left shift semigroup τ_-^t on $\{u \in BUC_\omega(\overline{\mathbb{R}}^-;U) \mid u(0)=0\}$ is the differentiation operator $\frac{\mathrm{d}}{\mathrm{d}s}$ with domain $\{u \in BUC^1_\omega(\overline{\mathbb{R}}^-;U) \mid u(0)=\dot{u}(0)=0\}$. We denote these generators by $\frac{\mathrm{d}}{\mathrm{d}s}_-$.*

(iv) *The generator of the finite left shift semigroup $\tau^t_{[0,T)}$ on $L^p([0,T);U)$ is the differentiation operator $\frac{\mathrm{d}}{\mathrm{d}s}$ with domain $\{u \in W^{1,p}([0,T];U) \mid u(T)=0\}$, and the generator of the left shift semigroup $\tau^t_{[0,T)}$ on $\{u \in C([0,T];U) \mid u(T)=0\}$ is the differentiation operator $\frac{\mathrm{d}}{\mathrm{d}s}$ with domain $\{u \in C^1([0,T];U) \mid u(T)=\dot{u}(T)=0\}$. We denote these generators by $\frac{\mathrm{d}}{\mathrm{d}s}_{[0,T)}$.*

(v) *The generator of the circular left shift group $\tau^t_{\mathbb{T}_T}$ on $L^p(\mathbb{T}_T;U)$ is the differentiation operator $\frac{\mathrm{d}}{\mathrm{d}s}$ with domain $W^{1,p}(\mathbb{T}_T;U)$ (which can be identified with $\{u \in W^{1,p}([0,T];U) \mid u(T)=u(0)\}$), and the generator of the circular left shift group $\tau^t_{\mathbb{T}_T}$ on $C(\mathbb{T}_T;U)$ is the differentiation operator $\frac{\mathrm{d}}{\mathrm{d}s}$ with domain $C^1(\mathbb{T}_T;U)$ (which can be identified with the set $\{u \in C^1([0,T];U) \mid u(T)=u(0)$ and $\dot{u}(T)=\dot{u}(0)\}$). We denote these generators by $\frac{\mathrm{d}}{\mathrm{d}s}_{\mathbb{T}_T}$.*

Proof The proofs are very similar to each other, so let us only prove, for example, (iii). Since the proof for the L^p-case works in the BUC-case, too, we restrict the discussion to the L^p-case. For simplicity we take $\omega = 0$, but the same argument applies when ω is nonzero.

Suppose that $u \in L^p(\mathbb{R}^-;U)$, and that $\frac{1}{h}(\tau_+^h u - u) \to g$ in $L^p(\mathbb{R}^-;U)$ as $h \downarrow 0$. If we extend u and g to $L^p(\mathbb{R};U)$ by defining them to be zero on $\overline{\mathbb{R}}^+$, then this can be written as $\frac{1}{h}(\tau^h u - u) \to g$ in $L^p(\mathbb{R};U)$ as $h \downarrow 0$.

Fix some $a \in \mathbb{R}$, and for each $t \in \mathbb{R}$, define

$$f(t) = \int_t^{t+a} u(s)\,ds = \int_0^a u(s+t)\,ds.$$

Then

$$\begin{aligned}\frac{1}{h}(\tau^h f - f) &= \int_0^a \frac{1}{h}(\tau^h u(s+t) - u(s+t))\,ds \\ &= \int_t^{t+a} \frac{1}{h}(\tau^h u(s) - u(s))\,ds \\ &\to \int_t^{t+a} g(s)\,ds \text{ as } h \downarrow 0.\end{aligned}$$

On the other hand

$$\begin{aligned}\frac{1}{h}(\tau^h f - f) &= \frac{1}{h}\int_{t+h}^{t+a+h} u(s)\,ds - \frac{1}{h}\int_t^{t+a} u(s)\,ds \\ &= \frac{1}{h}\int_{t+a}^{t+a+h} u(s)\,ds - \frac{1}{h}\int_t^{t+h} u(s)\,ds,\end{aligned}$$

and as $h \downarrow 0$ this tends to $u(t+a) - u(t)$ whenever both t and $t+a$ are Lebesgue points of u. We conclude that for almost all a and t,

$$u(t+a) = u(t) + \int_t^{t+a} g(s)\,ds.$$

By definition, this means that $u \in W^{1,p}(\mathbb{R}; U)$ and that $\dot{u} = g$. Since we extended u to all of $\mathbb{R}$ by defining u to be zero on $\overline{\mathbb{R}}^+$, we have, in addition $u(0) = 0$ (if we redefine u on a set of measure zero to make it continuous everywhere).

To prove the converse claim it suffices to observe that, if $u \in W^{1,p}(\overline{\mathbb{R}}^-; U)$ and $u(0) = 0$, then we can extend u to a function in $W^{1,p}(\mathbb{R}; U)$ by defining u to be zero on $\mathbb{R}^+$, and that

$$\frac{1}{h}(\tau^h u - u)(t) = \frac{1}{h}(u(t+h) - u(t)) = \frac{1}{h}\int_t^{t+h} \dot{u}(s)\,ds,$$

which tends to $\dot{u}$ in $L^p(\mathbb{R}; U)$ as $h \downarrow 0$ (see, e.g., Gripenberg et al. (1990, Lemma 7.4, p. 67)). □

Let us record the following fact for later use:

Lemma 3.2.4 *For $1 \le p < \infty$, $W^{1,p}_\omega(\mathbb{R}; U) \subset BC_{0,\omega}(\mathbb{R}; U)$, i.e., every $u \in W^{1,p}_\omega(\mathbb{R}; U)$ is continuous and $\mathrm{e}^{-\omega t}u(t) \to 0$ as $t \to \pm\infty$.*

Proof The continuity is obvious. The function $u_{-\omega}(t) = \mathrm{e}^{-\omega t}u(t)$ belongs to L^p, and so does its derivative $-\omega u_{-\omega} + \mathrm{e}_{-\omega}\dot{u}$. This implies that $u_{-\omega}(t) \to 0$ as $t \to \infty$. □

By combining Theorem 3.2.1(vi) with Example 3.2.3 we get the major part of the following lemma:

Lemma 3.2.5 *Let $1 \le p < \infty$, $\omega \in \mathbb{R}$, and $n = 0, 1, 2, \ldots$ Then $C_c^\infty(\mathbb{R}; U)$ is dense in $L^p_\omega(\mathbb{R}; U)$, $L^p_{\mathrm{loc}}(\mathbb{R}; U)$, $W^{n,p}(\mathbb{R}; U)$, $W^{n,p}_{\mathrm{loc}}(\mathbb{R}; U)$, $BC_0^n(\mathbb{R}; U)$, and $C^n(\mathbb{R}; U)$.*

Proof It follows from Theorem 3.2.1(vi) and Example 3.2.3 that $\bigcap_{k=1}^\infty W^{k,p}(\mathbb{R}; U)$ is dense in $L^p(\mathbb{R}; U)$ and in $W^{n,p}(\mathbb{R}; U)$. Let u belong to this space. Then $u \in C^\infty$. Choose any $\eta \in C_c^\infty(\mathbb{R}; \mathbb{R})$ satisfying $\eta(t) = 1$ for $|t| \le 1$, and define $u_m(t) = \eta(t/m)u(t)$. Then $u_m \in C_c^\infty(\mathbb{R}; U)$, and $u_m \to u$ in

$L^p(\mathbb{R};U)$ and in $W^{n,p}(\mathbb{R};U)$, proving the density of C_c^∞ in L^p and in $W^{n,p}$. The other claims are proved in a similar manner (see also Lemma 2.3.3). $\square$

Example 3.2.6 *Let $\mathfrak{A}^t$ be a C_0 semigroup on a Banach space X with generator A.*

(i) *For each $\alpha \in \mathbb{C}$, the generator of the exponentially shifted semigroup $e^{\alpha t}\mathfrak{A}^t$, $t \geq 0$ (see Example 2.3.5) is $A + \alpha$.*
(ii) *For each $\lambda > 0$, the generator of the time compressed semigroup $\mathfrak{A}^{\lambda t}$, $t \geq 0$ (see Example 2.3.6) is λA.*
(iii) *For each (boundedly) invertible $E \in \mathcal{B}(X_1; X)$, the generator A_E of the similarity transformed semigroup $\mathfrak{A}_E^t = E^{-1}\mathfrak{A}^t E$, $t \geq 0$ (see Example 2.3.7) is $A_E = E^{-1}AE$, with domain $\mathcal{D}(A_E) = E^{-1}\mathcal{D}(A)$.*

We leave the easy proof to the reader.

Theorem 3.2.1 does not say anything about the spectrum and resolvent set of the generator A. These notions and some related ones are defined as follows:

Definition 3.2.7 Let $A \colon X \supset \mathcal{D}(A) \to X$ be closed, and let $\alpha \in \mathbb{C}$.

(i) α belongs to the *resolvent set* $\rho(A)$ of A if $\alpha - A$ is injective, onto, and has an inverse $(\alpha - A)^{-1} \in \mathcal{B}(X)$. Otherwise α belongs to the *spectrum* $\sigma(A)$ of A.
(ii) α belongs to the *point spectrum* $\sigma_p(A)$, or equivalently, α is an *eigenvalue* of A, if $(\alpha - A)$ is not injective. A vector $x \in X$ satisfying $(\alpha - A)x = 0$ is called an *eigenvector* corresponding to the eigenvalue α.
(iii) α belongs to the *residual spectrum* $\sigma_r(A)$ if $(\alpha - A)$ is injective but its range is not dense in X.
(iv) α belongs to the *continuous spectrum* $\sigma_c(A)$ if $(\alpha - A)$ is injective and has dense range, but the range is not closed.
(v) The *resolvent* of A is the operator-valued function $\alpha \mapsto (\alpha - A)^{-1}$, defined on $\rho(A)$.

By the closed graph theorem, $\sigma(A)$ is the disjoint union of $\sigma_p(A)$, $\sigma_r(A)$, and $\sigma_c(A)$. The different parts of the spectrum need not be closed (see Examples 3.3.1 and 3.3.5), but, as the following lemma shows, the resolvent set is always open, hence the whole spectrum is always closed.

Lemma 3.2.8 *Let A be a (closed) operator $X \supset \mathcal{D}(A) \to X$, with a nonempty resolvent set.*

(i) *For each α and β in the resolvent set of A,*

$$(\alpha - A)^{-1} - (\beta - A)^{-1} = (\beta - \alpha)(\alpha - A)^{-1}(\beta - A)^{-1}. \qquad (3.2.1)$$

In particular, $(\alpha - A)^{-1}(\beta - A)^{-1} = (\beta - A)^{-1}(\alpha - A)^{-1}$.

(ii) *Let* $\alpha \in \rho(A)$ *and denote* $\|(\alpha - A)^{-1}\|$ *by* κ. *Then every* β *in the circle* $|\beta - \alpha| < 1/\kappa$ *belongs to the resolvent set of* A, *and*

$$\|(\beta - A)^{-1}\| \leq \frac{\kappa}{1 - \kappa|\beta - \alpha|}. \tag{3.2.2}$$

(iii) *Let* $\alpha \in \rho(A)$. *Then* $\delta\|(\alpha - A)^{-1}\| \geq 1$, *where* δ *is the distance from* α *to* $\sigma(A)$.

The identity (3.2.1) in (i) is usually called *the resolvent identity*. Note that the closedness of A is a consequence of the fact that A has a nonempty resolvent set.

Proof of Lemma 3.2.8. (i) Multiply the left hand side by $(\alpha - A)$ to the left and by $(\beta - A)$ to the right to get

$$(\alpha - A)\big[(\alpha - A)^{-1} - (\beta - A)^{-1}\big](\beta - A) = \beta - \alpha.$$

(ii) By part (i), for all $\beta \in \mathbb{C}$,

$$(\beta - A) = \big(1 + (\beta - \alpha)(\alpha - A)^{-1}\big)(\alpha - A). \tag{3.2.3}$$

It follows from the contraction mapping principle that if we take $|\beta - \alpha| < 1/\kappa$, then $\big(1 + (\beta - \alpha)(\alpha - A)^{-1}\big)$ is invertible and

$$\Big\|\big(1 + (\beta - \alpha)(\alpha - A)^{-1}\big)^{-1}\Big\| \leq \frac{1}{1 - \kappa|\beta - \alpha|}.$$

This combined with (3.2.3) implies that $\beta \in \rho(A)$ and that (3.2.2) holds.

(iii) This follows from (ii). □

Our next theorem lists some properties of the resolvent $(\lambda - A)^{-1}$ of the generator of a C_0 semigroup. Among others, it shows that the resolvent set of the generator of a semigroup contains a right half-plane.

Theorem 3.2.9 *Let* $\mathfrak{A}^t$ *be a* C_0 *semigroup on a Banach space* X *with generator* A *and growth bound* $\omega_{\mathfrak{A}}$ *(see Definition 2.5.6).*

(i) *Every* $\lambda \in \mathbb{C}^+_{\omega_{\mathfrak{A}}}$ *belongs to the resolvent set of* A, *and*

$$(\lambda - A)^{-(n+1)}x = \frac{1}{n!}\int_0^\infty s^n e^{-\lambda s}\mathfrak{A}^s x\, ds$$

for all $x \in X$, $\lambda \in \mathbb{C}^+_{\omega_{\mathfrak{A}}}$, *and* $n = 0, 1, 2, \ldots$ *In particular,*

$$(\lambda - A)^{-1}x = \int_0^\infty e^{-\lambda s}\mathfrak{A}^s x\, ds$$

for all $x \in X$ *and* $\lambda \in \mathbb{C}^+_{\omega_{\mathfrak{A}}}$.

(ii) *For each* $\omega > \omega_{\mathfrak{A}}$ *there is a finite constant* M *such that*

$$\|(\lambda - A)^{-n}\| \leq M(\Re\lambda - \omega)^{-n}$$

for all $n = 1, 2, 3, \ldots$ *and* $\lambda \in \mathbb{C}^+_\omega$. *In particular,*

$$\|(\lambda - A)^{-1}\| \leq M(\Re\lambda - \omega)^{-1}$$

for all $\lambda \in \mathbb{C}^+_\omega$.

(iii) *For all* $x \in X$, *the following limits exist in the norm of* X:

$$\lim_{\lambda\to+\infty} \lambda(\lambda - A)^{-1}x = x \text{ and } \lim_{\lambda\to+\infty} A(\lambda - A)^{-1}x = 0.$$

(iv) *For all* $t \geq 0$ *and all* $\lambda \in \rho(A)$,

$$(\lambda - A)^{-1}\mathfrak{A}^t = \mathfrak{A}^t(\lambda - A)^{-1}.$$

Proof (i) Define $\mathfrak{A}^t_\lambda = \mathrm{e}^{-\lambda t}\mathfrak{A}^t$ and $A_\lambda = A - \lambda$. Then by Example 3.2.6(i), A_λ is the generator of $\mathfrak{A}_\lambda$. We observe that $\mathfrak{A}_\lambda$ has negative growth bound, i.e., for all $x \in X$, $\mathfrak{A}^t_\lambda x$ tends exponentially to zero as $t \to \infty$. More precisely, for each $\omega_{\mathfrak{A}} < \omega < \Re\lambda$ there is a constant M such that for all $s \geq 0$ (cf. Example 2.3.5),

$$\|\mathrm{e}^{\lambda s}\mathfrak{A}^s\| \leq M\mathrm{e}^{-(\Re\lambda-\omega)s}. \tag{3.2.4}$$

Apply Theorem 3.2.1(ii) with $s = 0$ and $\mathfrak{A}$ and A replaced by $\mathfrak{A}_\lambda$ and A_λ to get

$$\mathfrak{A}^t_\lambda x - x = A_\lambda \int_0^t \mathfrak{A}^s_\lambda x\, ds.$$

Since A_λ is closed, we can let $t \to \infty$ to get

$$x = -A_\lambda \int_0^\infty \mathfrak{A}^s_\lambda x\, ds.$$

On the other hand, if $x \in \mathcal{D}(A)$, then we can do the same thing starting from the identity in Theorem 3.2.1(iv) to get

$$x = -\int_0^\infty \mathfrak{A}^s_\lambda A_\lambda x\, ds.$$

This proves that λ belongs to the resolvent set of A and that

$$(\lambda - A)^{-1}x = \int_0^\infty \mathrm{e}^{-\lambda s}\mathfrak{A}^s x\, ds, \quad x \in X. \tag{3.2.5}$$

To get a similar formula for iterates of $(\lambda - A)^{-1}$ we differentiate this formula with respect to λ. By the resolvent identity in Lemma 3.2.8(i) with $h = \beta - \lambda$,

$$\lim_{h\to 0} \frac{1}{h}\big[(\lambda + h - A)^{-1}x - (\lambda - A)^{-1}x\big] = -(\lambda - A)^{-2}x.$$

The corresponding limit of the right hand side of (3.2.5) is

$$\lim_{h\to 0}\int_0^\infty \frac{1}{h}\big(e^{(\lambda+h)s} - e^{\lambda s}\big)\mathfrak{A}^s x\,ds = \lim_{h\to 0}\int_0^\infty \frac{1}{h}\big(e^{hs} - 1\big)e^{\lambda s}\mathfrak{A}^s x\,ds.$$

As $h \to 0$, $\frac{1}{h}\big(e^{hs} - 1\big) \to s$ uniformly on compact subsets of $\overline{\mathbb{R}}^+$, and

$$\big|\tfrac{1}{h}(e^{hs} - 1)\big| = s\frac{1}{|hs|}\left|\int_0^{hs} e^y\,dy\right| \le s\frac{1}{|hs|}\int_0^{|hs|} e^{|y|}\,dy \le s e^{|hs|}.$$

This combined with (3.2.4) shows that we can use the Lebesgue dominated convergence theorem to move the limit inside the integral to get

$$(\lambda - A)^{-2}x = \int_0^\infty s e^{-\lambda s}\mathfrak{A}^s x\,ds, \qquad x \in X.$$

The same argument can be repeated. Every time we differentiate the right hand side of (3.2.5) the integrand is multiplied by a factor $-s$ (but we can still use the Lebesgue dominated convergence theorem). Thus, to finish the proof of (i) we need to show that

$$\frac{d^n}{d\lambda^n}(\lambda - A)^{-1}x = (-1)^n n!(\lambda - A)^{-(n+1)}x. \tag{3.2.6}$$

To do this we use induction over n, the chain rule, and the fact that the formula is true for $n = 1$, as we have just seen. We leave this computation to the reader.

(ii) Use part (i), (3.2.4), and the fact that (cf. Lemma 4.2.10)

$$\frac{1}{n!}\int_0^\infty s^n e^{-(\Re\lambda-\omega)s}\,ds = (\Re\lambda - \omega)^{-(n+1)}, \qquad \Re\lambda > \omega, \qquad n = 0, 1, 2, \ldots$$

(iii) We observe that the two claims are equivalent to each other since $(\lambda - A)(\lambda - A)^{-1}x = x$. If $x \in \mathcal{D}(A)$, then we can use part (ii) to get

$$\begin{aligned}|\lambda(\lambda - A)^{-1}x - x| &= |A(\lambda - A)^{-1}x|\\ &= |(\lambda - A)^{-1}Ax| \to 0 \text{ as } \lambda \to \infty.\end{aligned}$$

As $\mathcal{D}(A)$ is dense in X and $\limsup_{\lambda\to+\infty}\|\lambda(\lambda - A)^{-1}\| < \infty$ (this, too, follows from part (ii)), it must then be true that $\lambda(\lambda - A)^{-1}x \to x$ for all $x \in X$.

(iv) By Theorem 3.2.1(iii), for all $y \in \mathcal{D}(A)$,

$$\mathfrak{A}^t(\lambda - A)y = (\lambda - A)\mathfrak{A}^t y.$$

Substituting $y = (\lambda - A)^{-1}x$ and applying $(\lambda - A)^{-1}$ to both sides of this identity we find that

$$(\lambda - A)^{-1}\mathfrak{A}^t x = \mathfrak{A}^t(\lambda - A)^{-1}x$$

for all $x \in X$. □

In this proof we used an estimate on $\left|\frac{1}{h}(e^{hs}-1)\right|$ that will be useful later, too, so let us separate this part of the proof into the following slightly more general lemma:

Lemma 3.2.10 *Let* $1 \le p \le \infty$, $\omega \in \mathbb{R}$, *and* $n = 0, 1, 2, \dots$

(i) *For all* $\alpha, \beta \in \mathbb{C}$,

$$|e^\alpha - e^\beta| \le |\alpha - \beta| \max\{e^{\Re\alpha}, e^{\Re\beta}\},$$

$$|e^\alpha - e^\beta - (\alpha-\beta)e^\beta| \le \frac{1}{2}|\alpha-\beta|^2 \max\{e^{\Re\alpha}, e^{\Re\beta}\}.$$

(ii) *The function* $\alpha \mapsto \big(t \mapsto e^{\alpha t},\ t \in \mathbb{R}^-\big)$ *is analytic on the half-plane* $\mathbb{C}^+_\omega$ *in the spaces* $L^p_\omega(\mathbb{R}^-;\mathbb{C})$, $W^{n,p}_\omega(\mathbb{R}^-;\mathbb{C})$, *and* $BC^n_{0,\omega}(\mathbb{R}^-;\mathbb{C})$ *(i.e., it has a complex derivative with respect to* α *in these spaces when* $\Re\alpha > \omega$*). Its derivative is the function* $t \mapsto te^{\alpha t}$, $t \in \mathbb{R}^-$.

(iii) *The function* $\alpha \mapsto \big(t \mapsto e^{\alpha t},\ t > 0\big)$ *is analytic on the half-plane* $\mathbb{C}^-_\omega$ *in the spaces* $L^p_\omega(\mathbb{R}^+;\mathbb{C})$, $W^{n,p}_\omega(\overline{\mathbb{R}}^+;\mathbb{C})$, *and* $BC^n_{0,\omega}(\overline{\mathbb{R}}^+;\mathbb{C})$. *Its derivative is the function* $t \mapsto te^{\alpha t}$, $t > 0$.

Proof (i) Define $f(t) = e^{(\alpha-\beta)t}e^\beta$. Then $\dot f(t) = (\alpha-\beta)e^{(\alpha-\beta)t}e^\beta$ and

$$\begin{aligned}
|e^\alpha - e^\beta| &= |f(1) - f(0)| = \left|\int_0^1 \dot f(s)\,ds\right| \\
&\le |\alpha-\beta| e^{\Re\beta} \int_0^1 \left|e^{(\alpha-\beta)s}\right| ds \\
&\le |\alpha-\beta| e^{\Re\beta} \sup_{0\le s\le 1} e^{\Re(\alpha-\beta)s} \\
&= |\alpha-\beta| e^{\Re\beta} \max\{e^{\Re(\alpha-\beta)}, 1\} \\
&= |\alpha-\beta| \max\{e^{\Re\alpha}, e^{\Re\beta)}\}.
\end{aligned}$$

The similar proof of the second inequality is left to the reader. It can be based on the fact that $\ddot f(t) = (\alpha-\beta)^2 e^{(\alpha-\beta)t}e^\beta$, and that

$$e^\alpha - e^\beta - (\alpha-\beta)e^\beta = f(1) - f(0) - \dot f(0) = \int_0^1 \int_0^s \ddot f(v)\,dv\,ds.$$

(ii) It follows from (i) with α replaced by $(\alpha+h)t$ and β replaced by αt that the function $t \mapsto \frac{1}{h}\big(e^{(\alpha+h)t} - e^{\alpha t}\big) - te^{\alpha t}$ tends to zero as $t \to 0$ (as a complex limit) uniformly for t in each bounded interval. Moreover, combining the growth estimate that (i) gives for this function with the Lebesgue dominated convergence theorem we find that it tends to zero in $L^p_\omega(\mathbb{R}^-;\mathbb{C})$. A similar argument shows that all t-derivatives of this function also tend to zero in $L^p_\omega(\mathbb{R}^-;\mathbb{C})$, i.e., the function itself tends to zero in $W^{n,p}_\omega(\mathbb{R}^-;\mathbb{C})$. The proof of the analyticity in $BC^n_0(\mathbb{R}^-;\mathbb{C})$ is similar (but slightly simpler).

(iii) This proof is completely analogous to the proof of (ii). □

3.3 The spectra of some generators

To get an example of what the spectrum of a generator can look like, let us determine the spectra of the generators of the shift (semi)groups in Examples 2.3.2 and 3.2.3:

Example 3.3.1 *The generators of the (semi)groups τ^t, τ^t_+, τ^t_-, $\tau^t_{[0,T)}$ and $\tau^t_{\mathbb{T}_T}$ in Examples 2.3.2 and 2.5.3 (see Example 3.2.3) have the following spectra:*

(i) *The spectrum of the generator $\frac{d}{ds}$ of the left shift bilateral group τ^t on $L^p_\omega(\mathbb{R};U)$ with $1 \le p < \infty$ or on $BUC_\omega(\mathbb{R};U)$ is equal to the vertical line $\{\Re\lambda = \omega\}$. The whole spectrum is a residual spectrum in the L^1-case, a continuous spectrum in the L^p-case with $1 < p < \infty$, and a point spectrum in the BUC-case.*

(ii) *The spectrum of the generator $\frac{d}{ds}_+$ of the incoming left shift semigroup τ^t_+ on $L^p_\omega(\mathbb{R}^+;U)$ with $1 \le p < \infty$ or on $BUC_\omega(\overline{\mathbb{R}}^+;U)$ is equal to the closed half-plane $\overline{\mathbb{C}}^-_\omega$. The open left half-plane $\mathbb{C}^-_\omega$ belongs to the point spectrum, and the boundary $\{\Re\lambda = \omega\}$ belongs to the continuous spectrum in the L^p-case with $1 \le p < \infty$ and to the point spectrum in the BUC-case.*

(iii) *The spectrum of the generator $\frac{d}{ds}_-$ of the outgoing left shift semigroup τ^t_- on $L^p_\omega(\mathbb{R}^-;U)$ with $1 \le p < \infty$ or on $\{u \in BUC_\omega(\overline{\mathbb{R}}^-;U) \mid u(0) = 0\}$ is equal to the closed half-plane $\overline{\mathbb{C}}^-_\omega$. The open half-plane $\mathbb{C}^-_\omega$ belongs to the residual spectrum, and the boundary $\{\Re\lambda = \omega\}$ belongs to the residual spectrum in the L^1-case and to the continuous spectrum in the other cases.*

(iv) *The spectrum of the generator $\frac{d}{ds}_{[0,T)}$ of the finite left shift semigroup $\tau^t_{[0,T)}$ on $L^p([0,T);U)$ with $1 \le p < \infty$ or on $\{u \in C([0,T];U) \mid u(0) = 0\}$ is empty.*

(v) *The spectrum of the generator $\frac{d}{ds}_{\mathbb{T}_T}$ of the circular left shift group $\tau^t_{\mathbb{T}_T}$ on $L^p(\mathbb{T}_T;U)$ with $1 \le p < \infty$ or on $C(\mathbb{T}_T;U)$ is a pure point spectrum located at $\{2\pi jm/T \mid m = 0, \pm 1, \pm 2, \ldots\}$.*

Proof For simplicity we take $\omega = 0$. The general case can either be reduced to the case $\omega = 0$ with the help of Lemma 2.5.2(ii), or it can be proved directly by a slight modification of the argument below.

(i) As τ^t is a group, both τ^t and τ^{-t} are semigroups, and $\|\tau^t\| = 1$ for all $t \in \mathbb{R}$. It follows from Theorem 3.2.9(i) that every $\lambda \notin j\mathbb{R}$ belongs to the resolvent set of $\frac{d}{ds}$. It remains to show that $j\mathbb{R}$ belongs to the residual spectrum in the L^1-case, to the continuous spectrum in the L^p-case with $1 < p < \infty$, and to the point spectrum in the BUC-case.

Set $\lambda = j\beta$ where $\beta \in \mathbb{R}$, and let $u \in W^{1,p}(\mathbb{R}; U) = \mathcal{D}\left(\frac{d}{ds}\right)$. If $j\beta u - \dot{u} = f$ for some $u \in W^{1,p}(\mathbb{R}; U)$ and $f \in L^p(\mathbb{R}; U)$ then, by the variation of constants formula, for all $T \in \mathbb{R}$,

$$u(t) = e^{j\beta(t-T)}u(T) - \int_T^t e^{j\beta(t-s)} f(s)\, ds.$$

By letting $T \to -\infty$ we get (see Lemma 3.2.4)

$$u(t) = -\lim_{T \to -\infty} \int_T^t e^{j\beta(t-s)} f(s)\, ds.$$

In particular, if $f = 0$ then $u = 0$, i.e., $j\beta - \frac{d}{ds}$ is injective. By letting $t \to +\infty$ we find that

$$\lim_{t \to \infty} \lim_{T \to -\infty} \int_T^t e^{-j\beta s} f(s)\, ds = 0.$$

If $p = 1$, then this implies that the range of $j\beta - \frac{d}{ds}$ is not dense, hence $j\beta \in \sigma_r(\frac{d}{ds})$. If $p > 1$ then it is not true for every $f \in L^p(\mathbb{R}; U)$ that the limits above exist, so the range of $j\beta - \frac{d}{ds}$ is not equal to $L^p(\mathbb{R}; U)$, i.e., $j\omega \in \sigma(\frac{d}{ds})$. On the other hand, if $f \in C_c^\infty(\mathbb{R}; U)$ with $\int_{-\infty}^{\infty} e^{-j\beta s} f(s)\, ds = 0$, and if we define u to be the integral above, then $u \in C_c^\infty(\mathbb{R}; U) \subset W^{1,p}(\mathbb{R}; U)$ and $j\beta y - \dot{u} = f$. The set of functions f of this type is dense in $L^p(\mathbb{R}; U)$ when $1 < p < \infty$. Thus $j\beta - \frac{d}{ds}$ has dense range if $p > 1$, and in this case $j\beta \in \sigma_c(\frac{d}{ds})$.

In the BUC-case the function $e_{j\beta}(t) = e^{j\beta t}$ is an eigenfunction, i.e., $\left(j\beta - \frac{d}{ds}\right)e_{j\beta} = 0$; hence $j\beta \in \sigma_p\left(\frac{d}{ds}\right)$.[2]

(ii) That $\mathbb{C}^+ \subset \rho\left(\frac{d}{ds}_+\right)$ follows from Theorem 3.2.9(i). If $\Re\lambda < 0$ then $\lambda \in \sigma_p\left(\frac{d}{ds}_+\right)$, because then the function $u = e^{\lambda t}$ belongs to $W^{1,p}(\overline{\mathbb{R}}^+; U)$ and $\lambda u - \dot{u} = 0$. The proof that the imaginary axis belongs either to the singular spectrum in the L^p-case or to the point spectrum in the BUC-case is quite similar to the one above, and it is left to the reader (in the L^p-case, let $T \to +\infty$ to get $u(t) = \int_t^\infty e^{j\beta(t-s)} f(s)\, ds$, and see also the footnote about the case $p = 1$).

[2] It is easy to show that the range of $j\beta - \frac{d}{ds}$ is not closed in the L^1-case and BUC-case either. For example, in the L^1-case the range is dense in $\{f \in L^1(\mathbb{R}; U) \mid \int_{\mathbb{R}} f(s)\, ds = 0\}$, but it is not true for every $f \in L^1(\mathbb{R}; U)$ with $\int_{\mathbb{R}} f(s)\, ds = 0$ that the function $u(t) = -\int_{-\infty}^t e^{j\beta(t-s)} f(s)\, ds$ belongs to $L^1(\mathbb{R}; U)$.

(iii) That $\mathbb{C}^+ \subset \lambda \in \rho\left(\frac{\mathrm{d}}{\mathrm{d}s}_-\right)$ follows from Theorem 3.2.9(i). If $\Re\lambda < 0$ or if $\Re\lambda \le 0$ and $p = 1$, then every $f \in \mathcal{R}\left(\lambda - \frac{\mathrm{d}}{\mathrm{d}s}_-\right)$ satisfies

$$\int_{-\infty}^{0} \mathrm{e}^{-\lambda s} f(s)\,ds = 0,$$

hence the range is not dense in this case. We leave the proof of the claim that $\sigma_c = \{\lambda \in \mathbb{C} \mid \Re\lambda = 0\}$ in the other cases to the reader (see the proof of (i)).

(iv) This follows from Theorem 3.2.9(i), since the growth bound of $\tau_{[0,T)}$ is $-\infty$.

(v) For each $m \in \mathbb{Z}$, the derivative of the T-periodic function $\mathrm{e}^{2\pi jmt/T}$ with respect to t is $(2\pi jm/T)\mathrm{e}^{2\pi jmt/T}$, hence $2\pi jm/T$ is an eigenvalue of $\frac{\mathrm{d}}{\mathrm{d}s}_{\mathbb{T}_T}$ with eigenfunction $\mathrm{e}^{2\pi jmt/T}$.

To complete the proof of (v) we have to show that the remaining points λ in the complex plane belong to the resolvent set of $\frac{\mathrm{d}}{\mathrm{d}s}_{\mathbb{T}_T}$. To do this we have to solve the equation $\lambda u - \dot{u} = f$, where, for example, $f \in L^p(\mathbb{T}_T; U)$. By the variation of constants formula, a solution of this equation must satisfy

$$u(s) = \mathrm{e}^{\lambda(s-t)}u(t) - \int_t^s \mathrm{e}^{\lambda(s-v)} f(v)\,dv, \qquad s,\ t \in \mathbb{R}.$$

Taking $s = t + T$, and requiring that $u(t+T) = u(t)$ (in order to ensure T-periodicity of u) we get

$$(1 - \mathrm{e}^{\lambda T})u(t) = -\int_t^{t+T} \mathrm{e}^{\lambda(t+T-v)} f(v)\,dv.$$

The factor on the left hand side is invertible iff λ does not coincide with any of the points $2\pi jm/T$, in which case we get the following formula for the unique T-periodic solution u of $\lambda u - \dot{u} = f$:

$$\begin{aligned} u(t) &= (1 - \mathrm{e}^{-\lambda T})^{-1} \int_t^{t+T} \mathrm{e}^{\lambda(t-v)} f(v)\,dv \\ &= (1 - \mathrm{e}^{-\lambda T})^{-1} \int_0^{T} \mathrm{e}^{-\lambda s} f(t+s)\,ds. \end{aligned}$$

The right-hand side of this formula maps $L^p(\mathbb{T}_T; U)$ into $W^{1,p}(\mathbb{T}_T; U)$ and $C(\mathbb{T}_T; U)$ into $C^1(\mathbb{T}_T; U)$, and by differentiating this formula we find that, indeed, $\lambda u - \dot{u} = f$. □

Example 3.3.2 *The resolvents of the generators $\frac{\mathrm{d}}{\mathrm{d}s}$, $\frac{\mathrm{d}}{\mathrm{d}s}_+$, $\frac{\mathrm{d}}{\mathrm{d}s}_-$, $\frac{\mathrm{d}}{\mathrm{d}s}_{[0,T)}$, and $\frac{\mathrm{d}}{\mathrm{d}s}_{\mathbb{T}_T}$ in Example 3.2.3 can be described as follows:*

(i) *The resolvent $\left(\lambda - \frac{\mathrm{d}}{\mathrm{d}s}\right)^{-1}$ of the generator of the bilateral left shift group τ^t on $L^p_\omega(\mathbb{R}; U)$ and on $BUC_\omega(\mathbb{R}; U)$ maps f into $t \mapsto \int_t^\infty \mathrm{e}^{\lambda(t-s)} f(s)\,ds$,*

$t \in \mathbb{R}$, if $\Re\lambda > \omega$, and it maps f into $t \mapsto -\int_{-\infty}^{t} \mathrm{e}^{\lambda(t-s)} f(s)\,ds$, $t \in \mathbb{R}$, if $\Re\lambda < \omega$.

(ii) *For each $\lambda \in \mathbb{C}^+_\omega$ the resolvent $\left(\lambda - \frac{\mathrm{d}}{\mathrm{d}s}_+\right)^{-1}$ of the generator of the incoming left shift semigroup τ^t_+ on $L^p_\omega(\mathbb{R}^+; U)$ and on $BUC_\omega(\overline{\mathbb{R}}^+; U)$ maps f into $t \mapsto \int_t^\infty \mathrm{e}^{\lambda(t-s)} f(s)\,ds$, $t \ge 0$.*

(iii) *For each $\lambda \in \mathbb{C}^+_\omega$ the resolvent $\left(\lambda - \frac{\mathrm{d}}{\mathrm{d}s}_-\right)^{-1}$ of the generator of the outgoing left shift semigroup τ^t_- on $L^p_\omega(\mathbb{R}^-; U)$ and on $\{u \in BUC_\omega(\overline{\mathbb{R}}^-; U) \mid u(0) = 0\}$ maps f into $t \mapsto \int_t^0 \mathrm{e}^{\lambda(t-s)} f(s)\,ds$, $t \in \overline{\mathbb{R}}^-$.*

(iv) *For each $\lambda \in \mathbb{C}$ the resolvent $\left(\lambda - \frac{\mathrm{d}}{\mathrm{d}s}_{[0,T)}\right)^{-1}$ of the generator of the finite left shift semigroup $\tau^t_{[0,T)}$ on $L^p([0,T); U)$ and on $\{u \in C([0,T]; U) \mid u(T) = 0\}$ maps f into $t \mapsto \int_t^T \mathrm{e}^{\lambda(t-s)} f(s)\,ds$, $t \in [0,T)$.*

(v) *For each $\lambda \in \mathbb{C}$ which is not one of the points $\{2\pi jm/T \mid m = 0, \pm1, \pm2, \ldots\}$ the resolvent $\left(\lambda - \frac{\mathrm{d}}{\mathrm{d}s}_{\mathbb{T}_T}\right)^{-1}$ of the generator of the circular left shift group $\tau^t_{\mathbb{T}_T}$ on $L^p(\mathbb{T}_T; U)$ and on $C(\mathbb{T}_T; U)$ maps f into $t \mapsto (1 - \mathrm{e}^{-\lambda T})^{-1} \int_t^{t+T} \mathrm{e}^{\lambda(t-s)} f(s)\,ds$.*

The proof of this is essentially contained in the proof of Example 3.3.1.

The shift (semi)group examples that we have seen so far have rather exceptional spectra. They play an important role in our theory, but in typical applications one more frequently encounters semigroups of the following type:

Example 3.3.3 *Let $\{\phi_n\}_{n=1}^\infty$ be an orthonormal basis in a separable Hilbert space X, and let $\{\lambda_n\}_{n=1}^\infty$ be a sequence of complex numbers. Then the sum*

$$\mathfrak{A}^t x = \sum_{n=1}^\infty \mathrm{e}^{\lambda_n t} \langle x, \phi_n \rangle \phi_n, \quad x \in X, \quad t \ge 0,$$

converges for each $x \in X$ and $t \ge 0$ and defines a C_0 semigroup if and only if

$$\omega_{\mathfrak{A}} = \sup_{n \ge 0} \Re\lambda_n < \infty.$$

The growth bound of this semigroup is $\omega_{\mathfrak{A}}$, and

$$\|\mathfrak{A}^t\| = \mathrm{e}^{\omega_{\mathfrak{A}} t}, \quad t \ge 0.$$

It is a group if and only if

$$\alpha_{\mathfrak{A}} = \inf_{n \ge 0} \Re\lambda_n > -\infty.$$

in which case

$$\|\mathfrak{A}^t\| = \mathrm{e}^{\alpha_{\mathfrak{A}} t}, \quad t \le 0.$$

In particular, if $\Re\lambda_n = \omega$ *for all n, then* $\mathfrak{A}^t$ *is a group, and*

$$\|\mathfrak{A}^t\| = e^{\omega t}, \quad t \in \mathbb{R}.$$

Proof Clearly, the sum converges always if we choose $x = \phi_n$, in which case $\mathfrak{A}^t \phi_n = e^{\lambda_n t}\phi_n$, and $|\mathfrak{A}^t \phi_n| = e^{\Re\lambda_n t}$. If $\mathfrak{A}^t$ is to be a semigroup, then $\|\mathfrak{A}^t\| \geq e^{\Re\lambda_n t}$ for all n, and by Theorem 2.5.4(i), the number $\omega_{\mathfrak{A}}$ defined above must be finite and less than or equal to the growth bound of $\mathfrak{A}$. If $\mathfrak{A}^t$ is to be a group, then $\mathfrak{A}^{-t}$ is also a semigroup, and the same argument with t replaced by $-t$ shows that necessarily $\alpha_{\mathfrak{A}} > -\infty$ in this case.

Let us suppose that $\omega_{\mathfrak{A}} < \infty$. For each $N = 1, 2, 3, \ldots,$ define

$$\mathfrak{A}^t_N x = \sum_{n=1}^{N} e^{\lambda_n t}\langle x, \phi_n\rangle \phi_n.$$

Then it is easy to show that each $\mathfrak{A}_N$ is a C_0 group (since $\phi_n \perp \phi_k$ when $n \neq k$). For each $t > 0$, the sum converges as $N \to \infty$ because the norm of the tail of the series tends to zero (the sequence ϕ_n is orthonormal):

$$\begin{aligned}
\left|\sum_{n=N+1}^{\infty} e^{\lambda_n t}\langle x, \phi_n\rangle \phi_n\right|^2 &= \sum_{n=N+1}^{\infty} |e^{\lambda_n t}\langle x, \phi_n\rangle \phi_n|^2 \\
&= \sum_{n=N+1}^{\infty} e^{2\Re\lambda_n t}|\langle x, \phi_n\rangle|^2 \\
&\leq e^{2\omega_{\mathfrak{A}} t} \sum_{n=N+1}^{\infty} |\langle x, \phi_n\rangle|^2 \\
&\to 0 \text{ as } N \to \infty.
\end{aligned}$$

Thus $\mathfrak{A}^t \in \mathcal{B}(X)$ (as a strong limit of operators in $\mathcal{B}(X)$). The norm estimate $\|\mathfrak{A}^t\| \leq e^{\omega_{\mathfrak{A}} t}$ follows from the fact that (see the computation above)

$$|\mathfrak{A}^t x|^2 = \left|\sum_{n=1}^{\infty} e^{\lambda_n t}\langle x, \phi_n\rangle \phi_n\right|^2 \leq e^{2\omega_{\mathfrak{A}} t}|x|^2.$$

Moreover, for each x the convergence is uniform in t over bounded intervals since

$$|\mathfrak{A}^t x - \mathfrak{A}^t_N x|^2 \leq e^{2\omega_{\mathfrak{A}} t} \sum_{n=N+1}^{\infty} |\langle x, \phi_n\rangle|^2.$$

This implies that $t \mapsto \mathfrak{A}^t x$ is continuous on $\overline{\mathbb{R}}^+$ for each $x \in X$. Since each $\mathfrak{A}_N$ satisfies $\mathfrak{A}^0_N = 1$ and $\mathfrak{A}^{s+t}_N = \mathfrak{A}^s_N \mathfrak{A}^t_N$, $s, t \geq 0$, the same identities carry over to the limit. We conclude that $\mathfrak{A}$ is a C_0 semigroup.

If $\alpha_{\mathfrak{A}} > -\infty$, then we can repeat the same argument to get convergence also for $t < 0$. □

Definition 3.3.4 A (semi)group of the type described in Example 3.3.3 is called *diagonal*, with *eigenvectors* $\{\phi_n\}_{n=1}^{\infty}$ and *eigenvalues* $\{\lambda_n\}_{n=1}^{\infty}$.

The reason for this terminology is the following:

Example 3.3.5 *The generator A of the (semi)group in Example 3.3.3 is the operator*

$$Ax = \sum_{n=1}^{\infty} \lambda_n \langle x, \phi_n \rangle \phi_n,$$

with domain

$$\mathcal{D}(A) = \left\{ x \in X \;\middle|\; \sum_{n=1}^{\infty} (1 + |\lambda_n|^2) |\langle x, \phi_n \rangle|^2 < \infty \right\}.$$

The spectrum of A is the closure of the set $\{\lambda_n \mid n = 1, 2, 3, \ldots\}$*: every* λ_n *belongs to the point spectrum and cluster points different from all the* λ_n *belong to the continuous spectrum. The resolvent operator is given by*

$$(\alpha - A)^{-1} x = \sum_{n=1}^{\infty} (\alpha - \lambda_n)^{-1} \langle x, \phi_n \rangle \phi_n.$$

Proof Suppose that $\lim_{h \downarrow 0} \frac{1}{h}(\mathfrak{A}^h x - x)$ exists. Taking the inner product with ϕ_n, $n = 1, 2, 3, \ldots$, we get

$$\begin{aligned} \lim_{h \downarrow 0} \frac{1}{h} \langle (\mathfrak{A}^h x - x), \phi_n \rangle &= \lim_{h \downarrow 0} \frac{1}{h} \sum_{k=1}^{\infty} (\mathrm{e}^{\lambda_k h} - 1) \langle x, \phi_k \rangle \langle \phi_k, \phi_n \rangle \\ &= \lim_{h \downarrow 0} \frac{1}{h} (\mathrm{e}^{\lambda_n h} - 1) \langle x, \phi_n \rangle \\ &= \lambda_n \langle x, \phi_n \rangle. \end{aligned}$$

Thus, for all $x \in \mathcal{D}(A)$, we have $Ax = \sum_{n=1}^{\infty} \lambda_n \langle x, \phi_n \rangle \phi_n$. The norm of this vector is finite as is the norm of $x = \sum_{n=1}^{\infty} \langle x, \phi_n \rangle \phi_n$, so we conclude that

$$\mathcal{D}(A) \subset \left\{ x \in X \;\middle|\; \sum_{n=1}^{\infty} (1 + |\lambda_n|^2) |\langle x, \phi_n \rangle|^2 < \infty \right\}.$$

To prove the opposite inclusion, let us suppose that

$$\sum_{n=1}^{\infty} (1 + |\lambda_n|^2) |\langle x, \phi_n \rangle|^2 < \infty.$$

Then the sum $y = \sum_{n=1}^{\infty} \lambda_n \langle x, \phi_n \rangle \phi_n$ converges in X, and for each $h > 0$ we have

$$\left| \frac{1}{h}(\mathfrak{A}^h x - x) - y \right|^2 = \left| \sum_{n=1}^{\infty} \Big(\frac{1}{h}(\mathrm{e}^{\lambda_n h} - 1) - \lambda_n \Big) \langle x, \phi_n \rangle \phi_n \right|^2$$
$$= \sum_{n=1}^{\infty} \left| \frac{1}{h}(\mathrm{e}^{\lambda_n h} - 1) - \lambda_n \right|^2 |\langle x, \phi_n \rangle|^2.$$

Take $h \le 1$. Then, by Lemma 3.2.10, $\frac{1}{h}|\mathrm{e}^{\lambda_n h} - 1| \le |\lambda_n| M_n(h)$, where

$$M_n(h) = \max\{1, \mathrm{e}^{\Re \lambda_n h}\} \le M = \max\{1, \mathrm{e}^{\omega_{\mathfrak{A}}}\},$$

hence

$$\left| \frac{1}{h}(\mathrm{e}^{\lambda_n h} - 1) - \lambda_n \right|^2 |\langle x, \phi_n \rangle|^2 \le (1 + M)^2 |\lambda_n|^2 |\langle x, \phi_n \rangle|^2.$$

Moreover, for each n, $\big(\frac{1}{h}(\mathrm{e}^{\lambda_n h} - 1) - \lambda_n\big) \to 0$ as $h \downarrow 0$. This means that we can use the discrete Lebesgue dominated convergence theorem to conclude $\frac{1}{h}(\mathfrak{A}^h x - x) - y \to 0$ as $h \downarrow 0$, i.e., $x \in \mathcal{D}(A)$.

Obviously every λ_n is an eigenvalue since $\phi_n \in \mathcal{D}(A)$ and $A\phi_n = \lambda_n \phi_n$. The spectrum of A contains therefore at least the closure of $\{\lambda_n \mid n = 1, 2, 3, \dots\}$ (the spectrum is always closed).

If $\inf_{n \ge 0} |\alpha - \lambda_n| > 0$, then there exist two positive constants a and b such that $a(1 + |\lambda_n|) \le |\alpha - \lambda_n| \le b(1 + |\lambda_n|)$, and the sum

$$Bx = \sum_{n=1}^{\infty} (\alpha - \lambda_n)^{-1} \langle x, \phi_n \rangle \phi_n$$

converges for every $x \in X$ and defines an operator $B \in \mathcal{B}(X)$ which maps X onto $\mathcal{D}(A)$. It is easy to show that B is the inverse to $(\alpha - A)$, hence $\alpha \in \rho(A)$. Conversely, suppose that $(\alpha - A)$ has an inverse $(\alpha - A)^{-1}$. Then

$$1 = |\phi_n| = |(\alpha - A)^{-1}(\alpha - A)\phi_n|$$
$$= |(\alpha - A)^{-1}(\alpha - \lambda_n)\phi_n| \le \|(\alpha - A)^{-1}\| |\alpha - \lambda_n|.$$

This shows that every $\alpha \in \rho(A)$ satisfies $\inf_{n \ge 0} |\alpha - \lambda_n| > 0$.

If $\inf_{n \ge 0} |\alpha - \lambda_n| = 0$ but $\alpha \ne \lambda_n$ for all n, then α is not an eigenvalue because $\alpha x - Ax = \sum_{n=1}^{\infty} (\alpha - \lambda_n) \langle x, \phi_n \rangle \phi_n = 0$ only when $\langle x, \phi_n \rangle = 0$ for all n, i.e., $x = 0$. On the other hand, the range of $\alpha - A$ is dense because it contains all finite linear combinations of the base vectors ϕ_n. Thus $\alpha \in \sigma_c(A)$ in this case. □

Example 3.3.3 can be generalized to the case where A is an arbitrary *normal* operator on a Hilbert space X, whose spectrum is contained in some left half-plane. The proofs remain essentially the same, except for the fact that the sums have to be replaced by integrals over a spectral resolution. We refer the reader to Rudin (1973, pp. 301–303) for a precise description of the spectral resolution used in the following theorem (dual operators and semigroups are discussed in Section 3.5).

Example 3.3.6 *Let A be a closed and densely defined normal operator on a Hilbert space X (i.e., $A^*A = AA^*$), and let E be the corresponding spectral resolution of A, so that*

$$\langle Ax, y\rangle_X = \int_{\sigma(A)} \lambda\langle E(d\lambda)x, y\rangle, \quad x \in \mathcal{D}(A), \quad y \in X.$$

Then the following claims are valid.

(i) *For each $n = 1, 2, 3, \ldots$ the domain of A^n is given by*

$$\mathcal{D}\left(A^n\right) = \left\{x \in X \,\middle|\, \int_{\sigma(A)} (1 + |\lambda|^2)^n \langle E(d\lambda)x, x\rangle\right\} < \infty,$$

and

$$\|A^n x\|_X^2 = \left\{x \in X \,\middle|\, \int_{\sigma(A)} |\lambda|^{2n} \langle E(d\lambda)x, x\rangle\right\}.$$

(ii) *For each $\alpha \in \rho(A)$, $0 \le k \le n \in \{1, 2, 3, \ldots\}$, and x, $y \in X$,*

$$\langle A^k(\alpha - A)^{-n}x, y\rangle_X = \int_{\sigma(A)} \lambda^k(\alpha - \lambda)^{-n}\langle E(d\lambda)x, y\rangle.$$

(iii) *A generates a C_0 semigroup $\mathfrak{A}$ on X if and only if the spectrum of A is contained in some left half-plane, i.e.,*

$$\omega_{\mathfrak{A}} = \sup_{\lambda\in\sigma(A)} \Re\lambda < \infty.$$

In this case,

$$\|\mathfrak{A}^t\| = \mathrm{e}^{\omega_{\mathfrak{A}} t}, \quad t \ge 0,$$

and

$$\langle \mathfrak{A}^t x, y\rangle = \int_{\sigma(A)} \mathrm{e}^{\lambda t}\langle E(d\lambda)x, y\rangle, \quad t \ge 0, \quad x \in X, \quad y \in X. \tag{3.3.1}$$

(iv) *A generates a C_0 group $\mathfrak{A}$ on X if and only if $\sigma(A)$ is contained in some vertical strip $\alpha \le \Re\lambda \le \omega$. In this case, if we define*

$$\alpha_{\mathfrak{A}} = \inf\{\Re\lambda \mid \lambda \in \sigma(A)\},$$

then

$$\|\mathfrak{A}^t\| = \mathrm{e}^{\alpha_{\mathfrak{A}} t}, \quad t \le 0,$$

and (3.3.1) holds for all $t \in \mathbb{R}$.

(v) *A C_0 semigroup $\mathfrak{A}$ on X is normal (i.e., $\mathfrak{A}^{*t} = \mathfrak{A}^t$ for all $t \geq 0$) if and only if its generator is normal.*

Proof (i)–(ii) See Rudin (1973, Theorems 12.21, 13.24 and 13.33).

(iii) See Rudin (1973, Theorem 13.37) and the remark following that theorem.

(iv) The proof of this is analogous to the proof of (iii).

(v) See Rudin (1973, Theorem 13.37).

□

Most of the examples of semigroups that we will encounter in this book are either of the type described in Example 2.3.2, 3.3.3, or 3.3.6, or a transformation of these examples of the types listed in Examples 2.3.10–2.3.13.

3.4 Which operators are generators?

There is a celebrated converse to Theorem 3.2.9(i) that gives a complete characterization of the class of operators A that generate C_0 semigroups:

Theorem 3.4.1 (Hille–Yosida) *A linear operator A is the generator of a C_0 semigroup $\mathfrak{A}$ satisfying $\|\mathfrak{A}^t\| \leq M e^{\omega t}$ if and only if the following conditions hold:*

(i) *$\mathcal{D}(A)$ is dense in X;*

(ii) *every real $\lambda > \omega$ belongs to the resolvent set of A, and*

$$\left\|(\lambda - A)^{-n}\right\| \leq \frac{M}{(\lambda - \omega)^n} \text{ for } \lambda > \omega \text{ and } n = 1, 2, 3, \ldots$$

Alternatively, condition (ii) can be replaced by

(ii′) *every real $\lambda > \omega$ belongs to the resolvent set of A, and*

$$\left\|\frac{\partial^n}{\partial \lambda^n}(\lambda - A)^{-1}\right\| \leq \frac{M n!}{(\lambda - \omega)^{n+1}} \text{ for } \lambda > \omega \text{ and } n = 0, 1, 2, \ldots$$

Note that the assumption implies that A must be closed, since it has a nonempty resolvent set.

Proof The necessity of (i) and (ii) follows from Theorems 3.2.1(vi) and 3.2.9 (i)–(ii) (the exact estimate in Theorem 3.2.9(ii) was derived from (3.2.4), which is equivalent to $\|\mathfrak{A}^t\| \leq M e^{\omega t}$). The equivalence of (ii) and (ii′) is a consequence of (3.2.6).

Let us start the proof of the converse claim by observing that the conclusion of Theorem 3.2.9(iii) remains valid, since the proof used only (ii) with $n = 1$

and the density of $\mathcal{D}(A)$ in X. This means that if we define

$$A_\alpha = \alpha A(\alpha - A)^{-1} = \alpha^2(\alpha - A)^{-1} - \alpha,$$

then each $A_\alpha \in \mathcal{B}(X)$, and, for each $x \in \mathcal{D}(A)$, $A_\alpha x \to Ax$ in X as $\alpha \to \infty$. Since A_α is bounded, we can define $\mathfrak{A}^t_\alpha = e^{A_\alpha t}$ as in Example 3.1.2. We claim that for each $x \in X$, the limit $\mathfrak{A}^t x = \lim_{\alpha\to\infty} \mathfrak{A}^t_\alpha x$ exists, uniformly in t on any bounded interval, and that $\mathfrak{A}^t$ is a semigroup with generator A.

Define

$$B_\alpha = A_\alpha + \alpha = \alpha^2(\alpha - A)^{-1}.$$

Then, by (ii), for all $n = 1, 2, 3, \ldots,$

$$\|B^n_\alpha\| \le \frac{M\alpha^{2n}}{(\alpha - \omega)^n}, \tag{3.4.1}$$

and by Theorem 3.2.9(iii) and Example 3.2.6(i),

$$\mathfrak{A}^t_\alpha = e^{-\alpha t} e^{B_\alpha t} = e^{-\alpha t} \sum_{n=0}^{\infty} \frac{B^n_\alpha t^n}{n!}. \tag{3.4.2}$$

Therefore

$$\begin{aligned} \|\mathfrak{A}^t_\alpha\| &\le e^{-\alpha t} \sum_{n=0}^{\infty} \frac{t^n}{n!} \frac{M\alpha^{2n}}{(\alpha - \omega)^n} \\ &= Me^{-\alpha t} e^{(\alpha^2 t)/(\alpha-\omega)} = Me^{(\alpha\omega t)/(\alpha-\omega)}, \qquad t \ge 0. \end{aligned} \tag{3.4.3}$$

This tends to $Me^{\omega t}$ as $\alpha \to \infty$, and the convergence is uniform in t on any bounded interval. Since $(\alpha - A)^{-1}$ and $(\beta - A)^{-1}$ commute (see Lemma 3.2.8(i)), also A_α and A_β commute, i.e., $A_\alpha A_\beta = A_\beta A_\alpha$, and this implies that $\mathfrak{A}^t_\alpha A_\beta = A_\beta \mathfrak{A}^t_\alpha$ for all $\alpha, \beta > \omega$ and $t \in \mathbb{R}$. Thus, for all $x \in X$ and $t \in \mathbb{R}$,

$$\begin{aligned} \mathfrak{A}^t_\alpha x - \mathfrak{A}^t_\beta x &= \int_0^t \frac{d}{ds}\left[\mathfrak{A}^s_\alpha \mathfrak{A}^{t-s}_\beta\right] ds x \\ &= \int_0^t \mathfrak{A}^s_\alpha (A_\alpha - A_\beta)\mathfrak{A}^{t-s}_\beta x\, ds = \int_0^t \mathfrak{A}^s_\alpha \mathfrak{A}^{t-s}_\beta (A_\alpha - A_\beta)x\, ds, \end{aligned}$$

and

$$\begin{aligned} &|\mathfrak{A}^t_\alpha x - \mathfrak{A}^t_\beta x| \\ &\quad\le M^2 \int_0^t e^{(\alpha\omega s)/(\alpha-\omega)} e^{(\beta\omega(t-s))/(\beta-\omega)} |A_\alpha x - A_\beta x|\, ds. \end{aligned} \tag{3.4.4}$$

Let $\alpha, \beta \to \infty$. Then the products of the exponentials tend to $e^{\omega s}e^{\omega(t-s)} = e^{\omega t}$, uniformly in s and t on any bounded interval, and if $x \in \mathcal{D}(A)$, then

$|A_\alpha x - A_\beta x| \to 0$ since both $A_\alpha x \to x$ and $A_\beta x \to x$. Therefore,

$$\lim_{\alpha,\ \beta\to\infty} |\mathfrak{A}^t_\alpha x - \mathfrak{A}^t_\beta x| = 0, \qquad x \in \mathcal{D}(A),$$

uniformly in t on any bounded interval. In other words, $\alpha \mapsto \mathfrak{A}^t_\alpha x$ is a Cauchy family in $C(\overline{\mathbb{R}}^+; X)$, and it has a limit in $C(\overline{\mathbb{R}}^+; X)$. Since we have a uniform bound on the norm of $\mathfrak{A}^t_\alpha x$ for t in each bounded interval (see (3.4.3)) and $\mathcal{D}(A)$ is dense in X, the limit $\lim_{\alpha\to\infty} \mathfrak{A}^t_\alpha x$ must exist in $C(\overline{\mathbb{R}}^+; X)$ for all $x \in X$, uniformly in t on any bounded interval. Let us denote the limit by $\mathfrak{A}^t x$. For each $t \geq 0$ we have $\mathfrak{A}^t \in \mathcal{B}(X)$ (the strong limit of a family of operators in $\mathcal{B}(X)$ belongs to $\mathcal{B}(X)$). By construction $t \mapsto \mathfrak{A}^t x$ is continuous, i.e., $t \mapsto \mathfrak{A}^t$ is strongly continuous. Moreover, $\mathfrak{A}^t$ inherits the semigroup properties $\mathfrak{A}^0 = 1$ and $\mathfrak{A}^{s+t} = \mathfrak{A}^s \mathfrak{A}^t$ from $\mathfrak{A}^t_\alpha$, and it also inherits the bound $\|\mathfrak{A}^t\| \leq M e^{\omega t}$. We conclude that $\mathfrak{A}^t$ is a C_0 semigroup.

The only thing left to be shown is that the generator of $\mathfrak{A}$ is A. Let $x \in \mathcal{D}(A)$. Then by Theorem 3.2.1(iv)

$$\mathfrak{A}^t x - x = \lim_{\alpha\to\infty} (\mathfrak{A}^t_\alpha x - x) = \lim_{\alpha\to\infty} \int_0^t \mathfrak{A}^s_\alpha A_\alpha x\, ds = \int_0^t \mathfrak{A}^s A x\, ds$$

(the integrand converges uniformly on $[0, t]$ to $\mathfrak{A}^s A x$). Divide this by t and let $t \downarrow 0$. This shows that, if we (temporarily) denote the generator of $\mathfrak{A}$ by B, then $\mathcal{D}(A) \subset \mathcal{D}(B)$, and $Bx = Ax$ for all $x \in \mathcal{D}(A)$. In other words, B is an extension of A. But this extension cannot be nontrivial, because if we take some common point α in the resolvent sets of A and B (any $\alpha > \omega$ will do), then

$$X = (\alpha - A)\mathcal{D}(A) = (\alpha - B)\mathcal{D}(A)$$

which implies that

$$\mathcal{D}(B) = (\alpha - B)^{-1} X = (\alpha - A)^{-1} X = \mathcal{D}(A).$$

□

Corollary 3.4.2 *A linear operator A is the generator of a C_0 semigroup $\mathfrak{A}$ satisfying $\|\mathfrak{A}^t\| \leq e^{\omega t}$ if and only if the following conditions hold:*

(i) *$\mathcal{D}(A)$ is dense in X;*

(ii) *every real $\lambda > \omega$ belongs to the resolvent set of A, and*

$$\left\|(\lambda - A)^{-1}\right\| \leq \frac{1}{(\lambda - \omega)} \text{ for } \lambda > \omega.$$

Proof This follows from Theorem 3.4.1 since $\|(\alpha - A)^{-n}\| \leq \|(\alpha - A)^{-1}\|^n$.

□

The case $\omega = 0$ is of special interest:

Definition 3.4.3 By a *bounded semigroup or group* we mean a semigroup or group $\mathfrak{A}$ satisfying $\sup_{t\geq 0}\|\mathfrak{A}^t\| < \infty$ or $\sup_{t\in\mathbb{R}}\|\mathfrak{A}^t\| < \infty$, respectively. By a *contraction semigroup or group* we mean a semigroup or group $\mathfrak{A}$ satisfying $\|\mathfrak{A}^t\| \leq 1$ for all $t \geq 0$ or $t \in \mathbb{R}$, respectively.

Corollary 3.4.4 *Let A be a linear operator $X \supset \mathcal{D}(A) \to X$ with dense domain and let $M < \infty$. Then the following conditions are equivalent:*

(i) *A is the generator of a (bounded) C_0 semigroup $\mathfrak{A}$ satisfying $\|\mathfrak{A}^t\| \leq M$ for all $t \geq 0$;*
(ii) *every positive real λ belongs to the resolvent set of A and*

$$\left\|(\lambda - A)^{-n}\right\| \leq M\lambda^{-n} \text{ for } \lambda > 0 \text{ and } n = 1, 2, 3, \ldots;$$

(iii) *the right half-plane $\mathbb{C}^+$ belongs to the resolvent set of A and*

$$\left\|(\lambda - A)^{-n}\right\| \leq (\Re\lambda)^{-n} \text{ for } \Re\lambda > 0 \text{ and } n = 1, 2, 3, \ldots$$

Proof By Theorem 3.4.1, (i) ⇔ (ii). Obviously (iii) ⇒ (ii). To show that (i) ⇒ (iii) we split $\lambda \in \mathbb{C}$ into $\lambda = \alpha + j\beta$ and apply Theorem 3.4.1 with λ replaced by α, $\mathfrak{A}^t$ replaced by $e^{-j\beta t}\mathfrak{A}^t$ and A replaced by $A - j\beta$. □

Corollary 3.4.5 *Let A be a linear operator $X \supset \mathcal{D}(A) \to X$ with dense domain. Then the following conditions are equivalent*

(i) *A is the generator of a C_0 contraction semigroup;*
(ii) *every positive real λ belongs to the resolvent set of A and*

$$\left\|(\lambda - A)^{-1}\right\| \leq \lambda^{-1} \text{ for } \lambda > 0;$$

(iii) *the right-half plane $\mathbb{C}^+$ belongs to the resolvent set of A, and*

$$\left\|(\lambda - A)^{-1}\right\| \leq (\Re\lambda)^{-1} \text{ for } \Re\lambda > 0.$$

Proof This proof is similar to the proof of Corollary 3.4.4, but we replace Theorem 3.4.1 by Corollary 3.4.2. □

There is also another characterization of the generators of contraction semigroups which is based on *dissipativity*.

Definition 3.4.6 A linear operator $A\colon X \supset \mathcal{D}(A) \to X$ is *dissipative* if for every $x \in \mathcal{D}(A)$ there is a vector $x^* \in X^*$ with $|x^*|^2 = |x|^2 = \langle x^*, x\rangle$ such that $\Re\langle x^*, Ax\rangle \leq 0$ (if X is a Hilbert space, then we take $x^* = x$).[3]

Lemma 3.4.7 *Let $A\colon X \supset \mathcal{D}(A) \to X$ be a linear operator. Then the following conditions are equivalent:*

[3] The dual space X^* is discussed at the beginning of Section 3.5.

(i) *A is dissipative;*
(ii) *$A - j\beta I$ is dissipative for all $\beta \in \mathbb{R}$;*
(iii) *$|(\lambda - A)x| \geq \lambda|x|$ for all $x \in \mathcal{D}(A)$ and all $\lambda > 0$;*
(iv) *$|(\lambda - A)x| \geq \Re\lambda|x|$ for all $x \in \mathcal{D}(A)$ and all $\lambda \in \mathbb{C}^+$.*

Proof (i) ⇒ (ii): This follows from Definition 3.4.6 since, with the notation of that definition, $\Re\langle x^*, j\beta x\rangle = \Re(-j\beta\langle x^*, x\rangle) = \Re(-j\beta|x|^2) = 0$.

(ii) ⇒ (iv): Suppose that (ii) holds. Let $x \in \mathcal{D}(A)$ and $\lambda = \alpha + j\beta$ with $\alpha > 0$ and $\beta \in \mathbb{R}$. Choose some $x^* \in X^*$ with $|x^*|^2 = |x|^2 = \langle x^*, x\rangle$ such that $\Re\langle x^*, Ax\rangle \leq 0$ (by the Hahn–Banach theorem, this is possible). Then

$$\begin{aligned}|\lambda x - Ax||x| &\geq |\langle x^*, \lambda x - Ax\rangle| \geq \Re\langle x^*, \lambda x - Ax\rangle \\ &= \Re\langle x^*, \alpha x\rangle - \Re\langle x^*, (A - j\beta)x\rangle \geq \alpha|x|^2,\end{aligned}$$

and (iv) follows.

(iv) ⇒ (iii): This is obvious.

(iii) ⇒ (i): Let $x \in \mathcal{D}(A)$, and suppose that $\lambda|x| \leq |(\lambda - A)x|$ for all $\lambda > 0$. Choose some $z^*_\lambda \in X^*$ with $|z^*_\lambda| = 1$ such that $\langle z^*_\lambda, (\lambda - A)x\rangle = |(\lambda - A)x|$. Then, for all $\lambda > 0$,

$$\begin{aligned}\lambda|x| &\leq |\lambda x - Ax| = \langle z^*_\lambda, \lambda x - Ax\rangle \\ &= \lambda\Re\langle z^*_\lambda, x\rangle - \Re\langle z^*_\lambda, Ax\rangle \leq \lambda|x| - \Re\langle z^*_\lambda, Ax\rangle.\end{aligned}$$

This implies that $\Re\langle z^*_\lambda, Ax\rangle \leq 0$ and that

$$\lambda\Re\langle z^*_\lambda, x\rangle \geq |\lambda x - Ax| \geq \lambda|x| - |Ax|.$$

For all $\lambda > 0$, let Z^*_λ be the weak* closure of the set $\{z^*_\alpha \mid \alpha \geq \lambda\}$. Then each Z^*_λ is a weak* compact subset of the unit ball in X^*, and for all $z^* \in Z^*_\lambda$ we have

$$\Re\langle z^*, Ax\rangle \leq 0, \quad \Re\langle z^*, x\rangle \geq |x| - \lambda^{-1}|Ax|, \quad |z^*| \leq 1$$

(the functionals $z^* \mapsto \Re\langle z^*, x\rangle$ and $z^* \mapsto \Re\langle z^*, Ax\rangle$ are continuous in the weak* topology). The sets Z^*_λ obviously have the finite intersection property and they are weak* compact, so their intersection $\bigcap_{\lambda>0} Z^*_\lambda$ is nonempty (see, e.g., Rudin (1987, Theorem 2.6, p. 37)). Choose any z^* in this intersection. Then

$$\Re\langle z^*, Ax\rangle \leq 0, \quad \Re\langle z^*, x\rangle \geq |x|, \quad |z^*| \leq 1.$$

The last two inequalities imply that $|z^*| = \langle z^*, x\rangle = |x|$. By taking $x^* = |x|z^*$ in Definition 3.4.6 we find that A is dissipative. □

By using the notion of dissipativity we can add one more condition to the list of equivalent conditions in Corollary 3.4.5.

Theorem 3.4.8 (Lumer–Phillips) *Let A be a linear operator $X \supset \mathcal{D}(A) \to X$ with dense domain. Then the following conditions are equivalent (and they are equivalent to the conditions (ii) and (iii) in Corollary 3.4.5):*

(i) *A is the generator of a C_0 contraction semigroup;*
(iv) *A is dissipative and $\rho(A) \cap \mathbb{C}^+ \neq \emptyset$.*

These conditions are, in particular, true if

(v) *A is closed and densely defined, and both A and A^* are dissipative.*

If X is reflexive, then (v) is equivalent to the other conditions.

Proof (i) $\Rightarrow$ (iv): This follows from Corollary 3.4.5 and Lemma 3.4.7.

(iv) $\Rightarrow$ (i): Suppose that (iv) holds. Then A is closed (since its resolvent set is nonempty). Take some $\lambda = \alpha + j\beta \in \rho(A)$ with $\alpha > 0$ and $\beta \in \mathbb{R}$. If A is dissipative, then we get from Lemma 3.4.7(iv) for all $x \in \mathcal{D}(A)$, $|(\lambda - A)x| \geq \alpha|x|$. This implies that $\|(\lambda - A)^{-1}\| \leq 1/\alpha$. By Lemma 3.2.8, the resolvent set of A contains an open circle with center λ and radius $\alpha = \Re\lambda$. We can repeat this argument with α replaced by first $(3/2)\alpha$, then $(3/2)^2\alpha$, then $(3/2)^3\alpha$, etc., to show that the whole right-half plane belongs to the resolvent set, and that $\|(\lambda - A)^{-1}\| \leq (\Re\lambda)^{-1}$ for all $\lambda \in \mathbb{C}^+$. By Corollary 3.4.5, A is therefore the generator of a C_0 contraction semigroup.

(v) $\Rightarrow$ (iv): By Lemma 3.4.7, $|(1 - A)x| \geq |x|$ for all $x \in \mathcal{D}(A)$. This implies that $1 - A$ is injective and has closed range (see Lemma 9.10.2(iii)). If $\mathcal{R}(1 - A) \neq X$ then, by the Hahn–Banach theorem, there is some nonzero $x^* \in X^*$ such that $\langle x^*, x - Ax = 0\rangle$, or equivalently, $\langle x^*, Ax\rangle = \langle x^*, x\rangle$ for all $x \in \mathcal{D}(A)$. This implies that $x^* \in \mathcal{D}(A^*)$ and that $A^*x^* = x^*$, i.e., $(1 - A^*)x^* = 0$. By Lemma 3.4.7 and the dissipativity of A^*, $|x^*| \leq |(1 - A^*)x^*| = 0$, contradicting our original choice of x^*. Thus $\mathcal{R}(1 - A) = X$. By the closed graph theorem, $(1 - A)^{-1} \in \mathcal{B}(X)$, so $1 \in \rho(A)$, and we have proved that (iv) holds.

If X is reflexive, then A is a generator of a C_0 contraction semigroup if and only if A^* is the generator of a C_0 contraction semigroup (see Theorem 3.5.6(v)), so (v) follows from (iv) in this case. □

In the Hilbert space case there is still another way of characterizing a generator of a contraction semigroup.

Theorem 3.4.9 *Let X be a Hilbert space, and let A be a linear operator $X \supset \mathcal{D}(A) \to X$ with dense domain. Then the following conditions are equivalent:*

(i) *A is the generator of a C_0 contraction semigroup,*
(vi) *there is some $\lambda \in \mathbb{C}^+ \cap \rho(A)$ for which the operator $\mathbf{A}_\lambda = (\overline{\lambda} + A)(\lambda - A)^{-1}$ is a contraction,*
(vii) *all $\lambda \in \mathbb{C}^+$ belong to $\rho(A)$, and $\mathbf{A}_\lambda = (\overline{\lambda} + A)(\lambda - A)^{-1}$ is a contraction,*

If $\mathbf{A}_\lambda$ is defined as in (i) and (ii), then -1 is not an eigenvalue of $\mathbf{A}_\lambda$, and $\mathcal{R}(1+\mathbf{A}_\lambda) = \mathcal{D}(A)$. Conversely, if X is a Hilbert space and if $\mathbf{A}$ is a contraction on X such that -1 is not an eigenvalue of $\mathbf{A}$, then $\mathcal{R}(1+\mathbf{A})$ is dense in X, and, for all $\lambda \in \mathbb{C}^+$, the operator A_λ with $\mathcal{D}(A_\lambda) = \mathcal{R}(1+\mathbf{A})$ defined by

$$A_\lambda x = \lambda x - 2\Re\lambda\,(1+\mathbf{A})^{-1}x, \qquad x \in \mathcal{R}(1+\mathbf{A}),$$

is the generator of a C_0 contraction semigroup on X (and the operator $\mathbf{A}_\lambda$ in (vi) and (vii) corresponding to A_λ is $\mathbf{A}$).

Proof (i) $\Rightarrow$ (vii) $\Rightarrow$ (vi) $\Rightarrow$ (i): Let us denote $\lambda = \alpha + j\beta$ where $\alpha > 0$ and $\beta \in \mathbb{R}$. For all $x \in \mathcal{D}(A)$, if we denote $B = A - j\beta$, then

$$\begin{aligned} |(\lambda - A)x|^2 &= |(\alpha - B)x|^2 = \alpha^2|x|^2 - 2\alpha\Re\langle x, Bx\rangle + |Bx|^2 \\ |(\overline{\lambda} + A)x|^2 &= |(\alpha + B)x|^2 = \alpha^2|x|^2 + 2\alpha\Re\langle x, Bx\rangle + |Bx|^2. \end{aligned}$$

If (i) holds, then by Lemma 3.4.7 and Theorem 3.4.8, $B = A - j\beta$ is dissipative, and we get $|(\overline{\lambda} + A)x| \le |(\lambda - A)x|$ for all $\lambda \in \mathbb{C}^+$ and all $x \in \mathcal{D}(A)$. By Corollary 3.4.5, $\lambda \in \rho(A)$, and by replacing x by $(\lambda - A)^{-1}x$ we find that $|\mathbf{A}_\lambda x| \le |x|$ for all $x \in X$, i.e., $\mathbf{A}_\lambda$ is a contraction. This proves that (i) $\Rightarrow$ (vii). Obviously (vii) $\Rightarrow$ (vi). If (vi) holds, then for that particular value of λ, we have $|\mathbf{A}_\lambda x| \le |x|$ for all $x \in X$, or equivalently, $|(\overline{\lambda} + A)x| \le |(\lambda - A)x|$ for all $x \in \mathcal{D}(A)$. The preceding argument then shows that $B = A - j\beta$ is dissipative, hence so is A, and (i) follows from Theorem 3.4.8. This proves that (i), (vi), and (vii) are equivalent.

Let us next show that -1 cannot be an eigenvalue of $\mathbf{A}_\lambda$ (although $-1 \in \sigma(\mathbf{A}_\lambda)$ whenever A is unbounded) and that $\mathcal{R}(1+\mathbf{A}_\lambda) = \mathcal{D}(A)$. This follows from the (easily verified) identity that

$$1 + \mathbf{A}_\lambda = 2\Re\lambda\,(\lambda - A)^{-1}.$$

Here the right-hand side is injective, hence so is the left-hand side, and the range of the right-hand side is $\mathcal{D}(A)$, hence so is the range of the left-hand side.

It remains to prove the converse part. Let $\mathbf{A}$ be a contraction on X such that -1 is not an eigenvalue of $\mathbf{A}$. Then the operator A_λ is well-defined on $\mathcal{D}(A_\lambda) = \mathcal{R}(1+\mathbf{A})$, and

$$(\lambda 1 - A_\lambda)x = 2\Re\lambda\,(1+\mathbf{A})^{-1}x, \qquad x \in \mathcal{R}(1+\mathbf{A}).$$

This implies that $\lambda - A_\lambda$ is injective, $\mathcal{R}(\lambda - A_\lambda) = X$, and $(\lambda - A_\lambda)^{-1} = 2\Re\lambda\,(1+\mathbf{A})^{-1}$. In particular, $\lambda \in \rho(A_\lambda)$. Arguing as in the proof of the implication (vi) $\Rightarrow$ (i) we find that A_λ is dissipative since $\mathbf{A}$ is a contraction (note that we have the same relationship between A_λ and $\mathbf{A}$ as we had between A and $\mathbf{A}_\lambda$, namely $\mathbf{A} = (\overline{\lambda} + A_\lambda)(\lambda - A_\lambda)^{-1}$). If we knew that $\mathcal{D}(A)$ is dense in X, then we could conclude from Theorem 3.4.8 that A_λ is the generator of a C_0

contraction semigroup. Thus, to complete the proof, the only thing remaining to be verified is that $\mathcal{R}(1 + \mathbf{A})$ is dense in X. This is true if and only if -1 is not an eigenvalue of $\mathbf{A}^*$, so let us prove this statement instead. If $\mathbf{A}^* x = x$ for some $x \in X$, then

$$\langle \mathbf{A}^* x, x \rangle = \langle x, \mathbf{A} x \rangle = \langle x, x \rangle = |x|^2,$$

hence

$$|x - \mathbf{A} x|^2 = |x|^2 - 2\Re \langle x, \mathbf{A} x \rangle + |\mathbf{A} x|^2 = |\mathbf{A} x|^2 - |x|^2 \le 0,$$

and we see that $\mathbf{A} x = x$. This implies that $x = 0$, because -1 was supposed not to be an eigenvalue of $\mathbf{A}$. □

The operator $\mathbf{A}_\lambda$ in Theorem 3.4.9 is called the *Cayley transform* of A with parameter $\alpha \in \mathbb{C}^+$. We shall say much more about this transform in Chapter 11.

3.5 The dual semigroup

Many results in quadratic optimal control rely on the possibility of passing from a system to its dual system. In this section we shall look at the dual of the semigroup $\mathfrak{A}$. The dual of the full system will be discussed in Section 6.2.

In most applications of the duality theory the state space X is a Hilbert space. In this case it is natural to identify the dual X with X itself. This has the effect that the mapping from an operator A on X to its dual A^* becomes conjugate-linear instead of linear, as is the case in the standard Banach space theory. To simplify the passage from the Banach space dual of an operator to the Hilbert space dual we shall throughout use the conjugate-linear dual instead of the ordinary dual of a Banach space.

As usual, we define the dual X^* of the Banach space X to be the space of all bounded linear functionals $x^* \colon X \to \mathbb{C}$. We denote the value of the functional $x^* \in X^*$ acting on the vector $x \in X$ alternatively by

$$x^* x = \langle x, x^* \rangle = \langle x, x^* \rangle_{(X, X^*)}.$$

The norm in X^* is the usual supremum-norm

$$|x^*|_{X^*} := \sup_{|x|_X = 1} |\langle x, x^* \rangle|, \tag{3.5.1}$$

and by the Hahn–Banach theorem, the symmetric relation

$$|x|_X = \sup_{|x^*|_{X^*} = 1} |\langle x, x^* \rangle| \tag{3.5.2}$$

also holds. On this space we use a nonstandard linear structure, defining the sum of two elements x^* and y^* in X^* and the product of a scalar $\lambda \in \mathbb{C}$ and a vector $x^* \in X^*$ by

$$\begin{aligned}\langle x, x^* + y^* \rangle &:= \langle x, x^* \rangle + \langle x, y^* \rangle, && x \in X,\\ \langle x, \lambda x^* \rangle &:= \overline{\lambda} \langle x, x^* \rangle, && x \in X, \quad \lambda \in \mathbb{C}.\end{aligned} \tag{3.5.3}$$

In other words, the mapping $(x, x^*) \mapsto \langle x, x^* \rangle$ is anti-linear (linear in x and conjugate-linear in x^*). All the standard results on the dual of a Banach space and the dual operator remain valid in this conjugate-linear setting, except for the fact that the mapping from an operator A to its dual operator A^* becomes conjugate-linear instead of linear, like in the standard Hilbert space case.

Let A be a closed (unbounded) operator $X \supset \mathcal{D}(A) \to Y$ with dense domain. The domain of the dual A^* of A consists of those $y^* \in Y^*$ for which the linear functional

$$x \mapsto \langle Ax, y^* \rangle_{(Y,Y^*)}, \quad x \in \mathcal{D}(A),$$

can be extended to a bounded linear functional on X. This extension is unique since $\mathcal{D}(A)$ is dense, and it can be written in the form

$$x \mapsto \langle Ax, y^* \rangle_{(Y,Y^*)} = \langle x, x^* \rangle_{(X,X^*)}, \quad x \in \mathcal{D}(A),$$

for some $x^* \in X$. For $y^* \in \mathcal{D}(A^*)$ we define A^* by $A^* y^* = x^*$, where $x^* \in X^*$ is the element above. Thus,

$$\langle Ax, y^* \rangle_{(Y,Y^*)} = \langle x, A^* y^* \rangle_{(X,X^*)}, \quad x \in \mathcal{D}(A), \quad y^* \in \mathcal{D}\big(A^*\big), \tag{3.5.4}$$

and this relationship serves as a definition of A^*.

Lemma 3.5.1 *Let $A\colon X \supset \mathcal{D}(A) \to Y$ be a closed linear operator with dense domain. Then*

(i) *$A^*\colon Y^* \supset \mathcal{D}(A^*) \to U^*$ is a closed linear operator,*
(ii) *if $A \in \mathcal{B}(X; Y)$, then $A^* \in \mathcal{B}(Y^*; X^*)$, and $\|A\| = \|A^*\|$,*
(iii) *$\mathcal{D}(A^*)$ weak*-dense in Y^*,*
(iv) *if Y is reflexive, then $\mathcal{D}(A^*)$ is dense in Y^*.*

Proof (i) It is a routine calculation to show that A^* is linear. Let us show that it is closed. Take some sequence $y_n^* \in \mathcal{D}(A^*)$ such that $y_n^* \to y^* \in Y^*$ and $A^* y_n^* \to x^*$ in X^* as $n \to \infty$. Then, for each $x \in \mathcal{D}(A)$,

$$\langle x, x^* \rangle = \lim_{n\to\infty} \langle x, Ay_n^* \rangle = \lim_{n\to\infty} \langle Ax, y_n^* \rangle = \langle Ax, y^* \rangle.$$

This means that the functional $\langle Ax, y^* \rangle$ can be extended to a bounded linear functional on X, hence $y^* \in \mathcal{D}(A^*)$ and $x^* = A^* y^*$. Thus, A^* is closed.

(ii) If $A \in \mathcal{B}(X;Y)$, then it is clear that $\mathcal{D}(A^*) = Y^*$. Moreover,

$$\begin{aligned}\|A^*\|_{\mathcal{B}(Y^*;X^*)} &= \sup_{|y^*|=1} |A^*y^*|_{X^*} = \sup_{\substack{|x|=1\\|y^*|=1}} |\langle x, A^*y^*\rangle_{(X,X^*)}| \\ &= \sup_{\substack{|x|=1\\|y^*|=1}} |\langle Ax, y^*\rangle_{(X,X^*)}| = \sup_{|x|=1} |Ax|_Y \\ &= \|A\|_{\mathcal{B}(X;Y)}.\end{aligned}$$

(iii) Let $y \in Y$, $y \neq 0$. As A is closed, the set $\{\left[\begin{smallmatrix} Ax \\ x \end{smallmatrix}\right] \mid x \in \mathcal{D}(A)\}$ is a closed subspace of $\left[\begin{smallmatrix} Y \\ X \end{smallmatrix}\right]$, and $\left[\begin{smallmatrix} y \\ 0 \end{smallmatrix}\right]$ certainly does not belong to this subspace. By the Hahn–Banach theorem in $\left[\begin{smallmatrix} Y \\ X \end{smallmatrix}\right]^* = \left[\begin{smallmatrix} Y^* \\ X^* \end{smallmatrix}\right]$, there is some $x_1^* \in X^*$ and $y_1^* \in Y^*$ such that $\langle x, x_1^*\rangle + \langle Ax, y_1^*\rangle = 0$ for all $x \in \mathcal{D}(A)$, but $\langle 0, x_1\rangle - \langle y, y_1^*\rangle \neq 0$. The first equation says that $y_1^* \in \mathcal{D}(A^*)$ (and that $A^*y_1^* = -x_1^*$). Thus, for each nonzero $y \in Y$, it is possible to find some $y^* \in \mathcal{D}(A^*)$ such that $\langle y, y^*\rangle \neq 0$, or equivalently, $Y \ni y = 0$ iff $\langle y, y^*\rangle = 0$ for all $y^* \in \mathcal{D}(A^*)$. This shows that $\mathcal{D}(A^*)$ is weak*-dense in Y^* (apply the Hahn–Banach theorem (Rudin, 1973, Theorem 3.5, p. 59) to the weak*-topology).

(iv) If Y is reflexive, then (iii) implies that $\mathcal{D}(A^*)$ is weakly dense in Y^*, hence dense in Y^* (Rudin 1973, Corollary 3.12(b), p. 65).

□

Lemma 3.5.2 *Let $A: X \supset \mathcal{D}(A) \to Y$ be closed, densely defined, and injective, and suppose that $\mathcal{R}(A) = Y$. Then $A^{-1} \in \mathcal{B}(Y;X)$, and $(A^{-1})^* = (A^*)^{-1}$. We denote this operator by A^{-*}.*

Proof The operator A^{-1} is closed since A is closed, and by the closed graph theorem, it is bounded, i.e., $A^{-1} \in \mathcal{B}(Y;X)$. By Lemma 3.5.1(ii), $(A^{-1})^* \in \mathcal{B}(X^*;Y^*)$. It remains to show that $(A^{-1})^* = (A^*)^{-1}$.

Take some arbitrary $x \in \mathcal{D}(A)$ and $x^* \in X^*$. Then

$$\langle x, x^*\rangle = \langle A^{-1}Ax, x^*\rangle = \langle Ax, (A^{-1})^*x^*\rangle.$$

This implies that $(A^{-1})^*x^* \in \mathcal{D}(A^*)$ and that $A^*(A^{-1})^*x^* = x^*$. Thus, $(A^{-1})^*$ is a left inverse of A^*. If we instead take some arbitrary $x \in X$ and $x^* \in \mathcal{D}(A^*)$, then

$$\langle x, x^*\rangle = \langle AA^{-1}x, x^*\rangle = \langle A^{-1}x, A^*x^*\rangle = \langle x, (A^{-1})^*A^*x^*\rangle.$$

Thus, $(A^{-1})^*$ is also a right inverse of A^*. This means that A^* is invertible, with $(A^*)^{-1} = (A^{-1})^*$. □

Lemma 3.5.3 *Let $A: X \supset \mathcal{D}(A) \to X$ be densely defined, and let $\alpha \in \rho(A)$ (in particular, this means that A is closed). Then $\overline{\alpha} \in \rho(A^*)$, and $((\alpha - A)^*)^{-1} = ((\alpha - A)^{-1})^* = (\alpha - A)^{-*}$.*

Proof By the definition of the dual operator, $(\alpha - A)^* = \overline{\alpha} - A^*$. Therefore Lemma 3.5.3 follows from Lemma 3.5.2, applied to the operator $\alpha - A$. □

Lemma 3.5.4 *Let $A\colon X \supset \mathcal{D}(A) \to Y$ be densely defined, and let $B \in \mathcal{B}(Y;Z)$. Then $(BA)^* = A^*B^*$ (with $\mathcal{D}((BA)^*) = \mathcal{D}(A^*B^*) = \{z^* \in Z^* \mid B^*z \in \mathcal{D}(A^*)\}$).*

Proof Let $x \in \mathcal{D}(A) = \mathcal{D}(BA)$ and $z^* \in \mathcal{D}(A^*B^*) = \{z^* \in Z^* \mid B^*z \in \mathcal{D}(A^*)\}$. Then

$$\langle BAx, z^*\rangle_{(Z,Z^*)} = \langle Ax, B^*z^*\rangle_{(Y,Y^*)} = \langle x, A^*B^*z^*\rangle_{(X,X^*)}.$$

This implies that $z^* \in \mathcal{D}((BA)^*)$, and that $(BA)^*z^* = A^*B^*z^*$. To complete the proof it therefore suffices to show that $\mathcal{D}((BA)^*) \subset \mathcal{D}(A^*B^*)$. Let $z^* \in \mathcal{D}((BA)^*)$. Then, for every $x \in \mathcal{D}(A) = \mathcal{D}(BA)$,

$$\langle Ax, B^*z^*\rangle_{(Y,Y^*)} = \langle BAx, z^*\rangle_{(Z,Z^*)} = \langle x, (BA)^*z^*\rangle_{(X,X^*)}.$$

This implies that $B^*z^* \in \mathcal{D}(A^*)$, and hence $z^* \in \mathcal{D}(A^*B^*)$. □

Lemma 3.5.5 *Let $B \in \mathcal{B}(X;Y)$ be invertible (with an inverse in $\mathcal{B}(Y;X)$), and let $A\colon Y \supset \mathcal{D}(A) \to Z$ be densely defined. Then AB is densely defined (with $\mathcal{D}(AB) = \{x \in X \mid Bx \in \mathcal{D}(A)\}$), and $(AB)^* = B^*A^*$ (with $\mathcal{D}(B^*A^*) = \mathcal{D}(A^*)$).*

Proof The domain of AB is the image under B^{-1} of $\mathcal{D}(A)$ which is dense in Y, and therefore $\mathcal{D}(AB)$ is dense in X (if $x \in X$, and if $y_n \in \mathcal{D}(A)$ and $y_n \to y := Bx$ in y, then $x_n := B^{-1}y_n$ in $\mathcal{D}(AB)$, and $x_n \to B^{-1}y = x$ in X). Thus AB has an adjoint $(AB)^*$.

Let $x \in \mathcal{D}(AB) = \{x \in X \mid Bx \in \mathcal{D}(A)\}$ and $z^* \in \mathcal{D}(A^*)$. Then

$$\langle ABx, z^*\rangle_{(Z,Z^*)} = \langle Bx, A^*z^*\rangle_{(Y,Y^*)} = \langle x, B^*A^*z^*\rangle_{(X,X^*)}.$$

This implies that $z^* \in \mathcal{D}((AB)^*)$, and that $(AB)^*z^* = B^*A^*z^*$. To complete the proof it therefore suffices to show that $\mathcal{D}((AB)^*) \subset \mathcal{D}(A^*)$. Let $z^* \in \mathcal{D}((AB)^*)$. Then, for every $y \in \mathcal{D}(A)$, we have $B^{-1}y \in \mathcal{D}(AB)$, and

$$\begin{aligned}\langle Ay, z^*\rangle_{(Z,Z^*)} &= \langle ABB^{-1}y, z^*\rangle_{(Z,Z^*)} = \langle B^{-1}y, (AB)^*z^*\rangle_{(X,X^*)}\\ &= \langle y, B^{-*}(AB)^*z^*\rangle_{(Y,Y^*)}.\end{aligned}$$

This implies that $z^* \in \mathcal{D}(A^*)$. □

Theorem 3.5.6 *Let $\mathfrak{A}$ be a C_0 semigroup on a Banach space X with generator A.*

(i) *$\mathfrak{A}^{*t} = (\mathfrak{A}^t)^*$, $t \geq 0$, is a locally bounded semigroup on X^* (but it need not be strongly continuous). This semigroup has the same growth bound as $\mathfrak{A}$.*

(ii) *Let* $X^{\odot} = \{x^* \in X^* \mid \lim_{t\downarrow 0} \mathfrak{A}^{*t} x^* = x^*\}$. *Then* $X^{\odot}$ *is a closed subspace of* X^* *which is invariant under* $\mathfrak{A}^*$, *and the restriction* $\mathfrak{A}^{\odot}$ *of* $\mathfrak{A}^*$ *to* $X^{\odot}$ *is a* C_0 *semigroup on* $X^{\odot}$.

(iii) *The generator* $A^{\odot}$ *of the semigroup* $\mathfrak{A}^{\odot}$ *in (ii) is the restriction of* A^* *to* $\mathcal{D}(A^{\odot}) = \{x^* \in \mathcal{D}(A^*) \mid A^* x^* \in X^{\odot}\}$.

(iv) $X^{\odot}$ *is the closure of* $\mathcal{D}(A^*)$ *in* X^*. *Thus,* $\mathcal{D}(A^*) \subset X^{\odot}$ *and* $\mathcal{D}(A^*)$ *is dense in* $X^{\odot}$.

(v) *If* X *is reflexive, then* $X^{\odot} = X^*$, $A^{\odot} = A^*$, *and* $\mathfrak{A}^*$ *is a* C_0 *semigroup on* X^* *with generator* A^*.

(vi) *If* $A \in \mathcal{B}(X)$, *then* $X^{\odot} = X^*$, $A^{\odot} = A^*$, *and* $\mathfrak{A}^*$ *is a* C_0 *semigroup on* X^* *with generator* A^*.

For an example where $X^{\odot} \neq X^*$, see Example 3.5.11 with $p = 1$.

Proof of Theorem 3.5.6 (i) This follows from Lemmas 3.5.1(ii) and 3.5.4.

(ii) The proof of the claim that $\mathfrak{A}^{*t}$ maps $X^{\odot}$ into $X^{\odot}$ is the same as the proof of Lemma 2.2.13(ii).

To show that $X^{\odot}$ is closed we let $x_n^* \in X^{\odot}$, $x_n^* \to x^* \in X^*$. Write

$$\|\mathfrak{A}^{*s} x^* - x^*\| \le \|\mathfrak{A}^{*s} x^* - \mathfrak{A}^{*s} x_n^*\| + \|\mathfrak{A}^{*s} x_n^* - x_n^*\| + \|x_n^* - x^*\|.$$

Given $\epsilon > 0$, we can make $\|\mathfrak{A}^{*s} x^* - \mathfrak{A}^{*s} x_n^*\| + \|x_n^* - x^*\| < \epsilon/2$ for all $0 \le s \le 1$ by choosing n large enough (since $\|\mathfrak{A}^{*s}\| \le M e^{\omega s}$ for some $M > 0$ and $\omega \in \mathbb{R}$). Next we choose $t \le 1$ so small that $\|\mathfrak{A}^{*s} x_n^* - x_n^*\| \le \epsilon/2$ for all $0 \le s \le t$. Then $\|\mathfrak{A}^{*s} x^* - x^*\| \le \epsilon$ for $0 \le s \le t$. This proves that $\lim_{t\downarrow 0} \mathfrak{A}^{*t} x^* = x^*$, hence $x^* \in X^{\odot}$. Thus $X^{\odot}$ is closed in X^*.

Since $X^{\odot}$ is closed in X^*, it is a Banach space with the same norm, and by definition, $\mathfrak{A}^{\odot}$ is a C_0 semigroup on $X^{\odot}$.

(iii) Let $A^{\odot}$ be the generator of $\mathfrak{A}^{\odot}$. Choose some $x \in \mathcal{D}(A)$ and $x^* \in \mathcal{D}(A^{\odot}) \subset X^{\odot} \subset X^*$. Then

$$\begin{aligned}
\langle Ax, x^* \rangle_{(X,X^*)} &= \lim_{t\downarrow 0} \left\langle \frac{1}{t}(\mathfrak{A}^t - 1)x, x^* \right\rangle_{(X,X^*)} \\
&= \lim_{t\downarrow 0} \left\langle x, \frac{1}{t}(\mathfrak{A}^{*t} - 1)x^* \right\rangle_{(X,X^*)} \\
&= \langle x, A^{\odot} x^* \rangle_{(X,X^*)}.
\end{aligned}$$

This implies that $x^* \in \mathcal{D}(A^*)$ and $A^{\odot} x^* = A^* x^*$. In other words, if we let B be the restriction of A^* to $\mathcal{D}(B) = \{x^* \in \mathcal{D}(A^*) \mid A^* x^* \in X^{\odot}\}$, then $A^{\odot} \subset B$, i.e., $\mathcal{D}(A^{\odot}) \subset \mathcal{D}(B)$ and $A^{\odot} X^* = B x^*$ for all $x^* \in \mathcal{D}(A^{\odot})$.

It remains to show that $\mathcal{D}(B) = \mathcal{D}(A^{\odot})$. Choose some $\alpha \in \rho(A^{\odot}) \cap \rho(A^*)$ (by Theorem 3.2.9(i) and Lemma 3.5.3, any α with $\Re\alpha$ large enough will do). Then $\alpha - A^{\odot}$ maps $\mathcal{D}(A^{\odot})$ one-to-one onto $X^{\odot}$, hence $\alpha - B$ maps $\mathcal{D}(B)$

onto $X^{\odot}$, i.e.,

$$X^{\odot} = (\alpha - A^{\odot})\mathcal{D}\left(A^{\odot}\right) = (\alpha - B)\mathcal{D}(B).$$

But $\alpha - B$ is a restriction of $\alpha - A^*$ which is one-to-one on X^*; hence $\alpha - B$ is injective on $\mathcal{D}(B)$, and

$$\mathcal{D}(B) = (\alpha - B)^{-1}X^{\odot} = (\alpha - A^{\odot})^{-1}X^{\odot} = \mathcal{D}(\Lambda).$$

(iv) Let $x^* \in \mathcal{D}(A^*)$. Choose α and M such that $\|\mathfrak{A}^s\| \le M\mathrm{e}^{\alpha s}$ for all $s \ge 0$. Then, for all $x \in X$, all $t \ge 0$, and all real $\alpha > \omega_{\mathfrak{A}}$, by Theorem 3.2.1(ii) and Example 3.2.6(i),

$$\begin{aligned}
\left|\left\langle x, (\mathrm{e}^{-\alpha t}\mathfrak{A}^{*t} - 1)x^*\right\rangle\right| &= \left|\left\langle(\mathrm{e}^{-\alpha t}\mathfrak{A}^t - 1)x, x^*\right\rangle\right| \\
&= \left|\left\langle(\alpha - A)(\alpha - A)^{-1}(\mathrm{e}^{-\alpha t}\mathfrak{A}^t - 1)x, x^*\right\rangle\right| \\
&= \left|\left\langle(\alpha - A)^{-1}(\mathrm{e}^{-\alpha t}\mathfrak{A}^t - 1)x, (\alpha - A)x^*\right\rangle\right| \\
&= \left|\left\langle\int_0^t \mathrm{e}^{-\alpha s}\mathfrak{A}^s x\, ds, (\alpha - A)x^*\right\rangle\right| \\
&\le Mt\|x\|\|(\alpha - A)x^*\|.
\end{aligned}$$

Taking the supremum over all $x \in X$ with $\|x\| = 1$ and using (3.5.1) we get

$$\|(\mathrm{e}^{-\alpha t}\mathfrak{A}^{*t} - 1)x^*\| \le Mt\|(\alpha - A)x^*\| \to 0 \text{ as } t \downarrow 0,$$

which implies that $\lim_{t\downarrow 0}\mathfrak{A}^{*t}x^* = x^*$. This shows that $\mathcal{D}(A^*) \subset X^{\odot}$. That $\mathcal{D}(A^*)$ is dense in $X^{\odot}$ follows from the fact that $\mathcal{D}\left(A^{\odot}\right) \subset \mathcal{D}(A^*)$ and $\mathcal{D}\left(A^{\odot}\right)$ is dense in $X^{\odot}$.

(v)–(vi) These follow from (iv) and Lemma 3.5.1(ii)–(iv). □

Definition 3.5.7 The C_0 semigroup $\mathfrak{A}^{\odot}$ in Theorem 3.5.6 is the dual of the C_0 semigroup $\mathfrak{A}$, $X^{\odot}$ is the $\odot$-dual of X (with respect to $\mathfrak{A}$), and $A^{\odot}$ is the $\odot$-dual of A.

Example 3.5.8 *The dual $\mathfrak{A}^*$ of the diagonal (semi)group $\mathfrak{A}$ in Example 3.3.3 is another diagonal (semi)group where the eigenvectors $\{\phi_n\}_{n=1}^{\infty}$ stay the same but the sequence of eigenvalues $\{\lambda_n\}_{n=1}^{\infty}$ has been replaced by its complex conjugate $\{\overline{\lambda}_n\}_{n=1}^{\infty}$. Thus*

$$\mathfrak{A}^{*t}x = \sum_{n=1}^{\infty} \mathrm{e}^{\overline{\lambda}_n t}\langle x, \phi_n\rangle\phi_n, \quad x \in H, \quad t \ge 0.$$

The dual generator A^* has the same domain as A, and it is given by

$$Ax = \sum_{n=1}^{\infty} \overline{\lambda}_n\langle x, \phi_n\rangle\phi_n, \qquad x \in \mathcal{D}(A).$$

In particular, $\mathfrak{A}^t = \mathfrak{A}^{*t}$ *for all* $t \geq 0$ *and* $A = A^*$ *if and only if all the eigenvalues are real.*

We leave the proof to the reader as an exercise.

Let us next look at the duals of the shift (semi)groups in Examples 2.3.2 and 2.5.3. To do this we need to determine the dual of an L^p-space.

Lemma 3.5.9 *Let* U *be a reflexive Banach space*[4], *let* $1 \leq p < \infty$, $1/p + 1/q = 1$ *(with* $1/\infty = 0$*),* $\omega \in \mathbb{R}$, *and* $J \subset \mathbb{R}$ *(with positive measure).*

(i) *The dual of* $L^p_\omega(J; U)$ *can be identified with* $L^q_{-\omega}(J; U^*)$ *in the sense that every bounded linear functional* f *on* $L^p_\omega(J; U)$ *is of the form*

$$\langle u, f\rangle = \int_J \langle u(t), u^*(t)\rangle_{(U,U^*)}\, dt, \qquad u \in L^p_\omega(J; U),$$

for some $u^* \in L^q_{-\omega}(J; U^*)$. *The norm of the functional* f *is equal to the* $L^q_{-\omega}(J)$*-norm on* u^*.

(ii) $L^p_\omega(J; U)$ *is reflexive iff* $1 < p < \infty$.

Proof For $\omega = 0$ this lemma is contained in Diestel and Uhl (1977, Theorem 1, p. 98 and Corollary 2, p. 100). If f is a bounded linear functional on $L^p_\omega(J; U)$ for some $\omega \neq 0$, then $f_\omega \colon v \mapsto \langle v, f_\omega\rangle = \langle \mathrm{e}_\omega v, f\rangle$ (where $\mathrm{e}_\omega(t) = \mathrm{e}^{\omega t}$) is a bounded linear functional on $L^p(J; U)$, hence this functional has a representation of the form

$$\langle v, f_\omega\rangle = \int_J \langle v(t), u^*_\omega(t)\rangle\, dt$$

for some $u^*_\omega \in L^q(J; U^*)$. Replacing $v \in L^p(J; U)$ by $u = \mathrm{e}_\omega v \in L^q_\omega(J; U)$ and u^*_ω by $u^* = \mathrm{e}_{-\omega} u^*_\omega \in L^q_{-\omega}(J; U^*)$ we get the desired representation

$$\begin{aligned}\langle u, f\rangle = \langle \mathrm{e}_{-\omega} u, f_\omega\rangle &= \int_J \langle \mathrm{e}^{-\omega t} u(t), u^*_\omega(t)\rangle\, dt \\ &= \int_J \langle u(t), \mathrm{e}^{-\omega t} u^*_\omega(t)\rangle\, dt = \int_J \langle u(t), u^*(t)\rangle\, dt.\end{aligned}$$

□

The representation in Lemma 3.5.9 is canonical in the sense that it 'independent of p and ω' in the following sense:

Lemma 3.5.10 *Let* U *be a reflexive Banach space. If* f *is a bounded linear functional on* $L^{p_1}_{\omega_1}(J; U) \cap L^{p_2}_{\omega_2}(J; U)$, *where* $1 \leq p_1 < \infty$, $1 \leq p_2 < \infty$, $\omega_1 \in \mathbb{R}$, *and* $\omega_2 \in \mathbb{R}$, *then we get the same representing function* u^* *for* f *if we use*

[4] In part (i) the reflexivity assumption on U can be weakened to the assumption that U has the Radon–Nikodym property. See Diestel and Uhl (1977, Theorem 1, p. 98).

any combination of p_i and ω_j, $i, j = 1, 2$, in Lemma 3.5.9. In particular, $u^* \in L^{q_1}_{-\omega_1}(J; U^*) \cap L^{q_2}_{-\omega_2}(J; U^*)$, where $1/p_1 + 1/q_1 = 1$ and $1/p_2 + 1/q_2 = 1$.

Proof This follows from the fact that the integral $\int_J \langle u(t), u^*(t)\rangle\, dt$ does not depend on either p or ω (as long as it converges absolutely). □

Example 3.5.11 *Let U be a reflexive Banach space, let $1 \le p < \infty$, $1/p + 1/q = 1$ (with $1/\infty = 0$), and $\omega \in \mathbb{R}$.*

(i) *The dual of the bilateral left shift group τ^t, $t \in \mathbb{R}$, on $L^p_\omega(\mathbb{R}; U)$ is the right shift group τ^{-t}, $t \in \mathbb{R}$, which acts on $L^q_{-\omega}(\mathbb{R}; U^*)$ if $1 < p < \infty$ and on $BUC_{-\omega}(\mathbb{R}; U^*)$ if $p = 1$.*

(ii) *The dual of the incoming left shift semigroup τ^t_+, $t \ge 0$, on $L^p_\omega(\mathbb{R}^+; U)$ is the right shift semigroup*

$$(\tau_+^{-t} u)(t) = (\tau^{-t}\pi_+ u)(s) = \begin{cases} u(s-t), & s > t, \\ 0, & \text{otherwise}, \end{cases}$$

which acts on $L^q_{-\omega}(\mathbb{R}^+; U^*)$ if $1 < p < \infty$ and on $\{u^* \in BUC_{-\omega}(\overline{\mathbb{R}}^+; U^*) \mid u^*(0) = 0\}$ if $p = 1$.

(iii) *The dual of the outgoing left shift semigroup τ^t_-, $t \ge 0$, on $L^p_\omega(\mathbb{R}^-; U)$ is the right shift semigroup*

$$(\tau_-^{-t} u)(s) = (\pi_-\tau^{-t} u)(s) = \begin{cases} u(s-t), & s \le 0, \\ 0, & \text{otherwise}, \end{cases}$$

which acts on $L^q_{-\omega}(\mathbb{R}^-; U^*)$ if $1 < p < \infty$ and on $BUC_{-\omega}(\overline{\mathbb{R}}^-; U^*)$ if $p = 1$.

(iv) *The dual of the finite left shift semigroup $\tau^t_{[0,T)}$, $t \ge 0$, on $L^p_\omega((0,T); U)$ is the right shift semigroup*

$$(\tau_{[0,T)}^{-t} u)(s) = (\pi_{[0,T)}\tau^{-t}\pi_{[0,T)} u)(s) = \begin{cases} u(s-t), & t \le s < T, \\ 0, & \text{otherwise}, \end{cases}$$

which acts on $L^q((0,T); U^*)$ if $1 < p < \infty$ and on $\{u^* \in C([0,T]; U^*) \mid u^*(0) = 0\}$ if $p = 1$.

(v) *The dual of the circular left shift group $\tau^t_{\mathbb{T}_T}$, $t \ge 0$, on $L^p_\omega(\mathbb{T}_T; U)$ is the circular right shift group*

$$(\tau_{\mathbb{T}_T}^{-t} u)(s) = (\tau^{-t} u)(s) = u(s-t),$$

which acts on $L^q(\mathbb{T}_T; U^*)$ if $1 < p < \infty$ and on $C(\mathbb{T}_T; U^*)$ if $p = 1$.

Proof (i) By Lemma 3.5.9, the dual of $L^p_\omega(\mathbb{R};U)$ is $L^q_{-\omega}(\mathbb{R};U^*)$. Let $u \in L^p_\omega(\mathbb{R};U)$ and $u^* \in L^q_{-\omega}(\mathbb{R};U^*)$ and $t \in \mathbb{R}$. Then

$$\begin{aligned}\langle \tau^t u, u^*\rangle &= \int_{-\infty}^{\infty} \langle \tau^t u(s), u^*(s)\rangle\, ds = \int_{-\infty}^{\infty} \langle u(s+t), u^*(s)\rangle\, ds \\ &= \int_{-\infty}^{\infty} \langle u(s), u^*(s-t)\rangle\, ds = \int_{-\infty}^{\infty} \langle u(s), \tau^{-t} u^*(s)\rangle\, ds \\ &= \langle u, \tau^{-t} u^*\rangle.\end{aligned}$$

This shows that $\tau^{*t} = \tau^{-t}$. The rest of (i) follows from Theorem 3.5.6, Definition 3.5.7, and Examples 2.3.2 and 2.5.3.

(ii)–(iv) These follow from (i) and Examples 2.3.2 and 2.5.3. □

The new right shift semigroups that we obtained in Example 3.5.11 are similar to the left shift semigroups that we have encountered earlier. The similarity transform is the *reflection operator* **Я** (in one case combined with a shift), which we define as follows.

Definition 3.5.12 Let $1 \le p \le \infty$, and let U be a Banach space.

(i) For each function $u \in L^p_{\mathrm{loc}}(\mathbb{R};U)$ we define the *reflection* $\mathbf{Я}u$ of u by

$$(\mathbf{Я}u)(s) = u(-s), \qquad s \in \mathbb{R}. \tag{3.5.5}$$

(ii) For each function $u \in Reg_{\mathrm{loc}}(\mathbb{R};U)$ we define the *reflection* $\mathbf{Я}u$ of u by

$$(\mathbf{Я}u)(s) = \lim_{t\downarrow -s} u(t), \qquad s \in \mathbb{R}. \tag{3.5.6}$$

Observe that these two cases are consistent in the sense that in part (ii) we have $(\mathbf{Я}u)(s) = u(-s)$ for all but countably many s.

Lemma 3.5.13 *Let $J \subset \mathbb{R}$, $t \in \mathbb{R}$, $\omega \in \mathbb{R}$, and $1 \le p \le \infty$.*

(i) **Я** *maps $L^p|Reg_\omega(\mathbb{R};U)$ onto $L^p|Reg_{-\omega}(\mathbb{R};U)$, and*
  (a) $\mathbf{Я}^{-1} = \mathbf{Я}$,
  (b) $\mathbf{Я}\tau^t = \tau^{-t}\mathbf{Я}$,
  (c) $\mathbf{Я}\pi_J = \pi_{\mathbf{Я}J}\mathbf{Я}$,[5] *and*
  (d) $\mathbf{Я}^* = \mathbf{Я}$ *(in $L^p_\omega(J;U)$ with reflexive U and $1 \le p < \infty$).*

(ii) $\pi_J^* = \pi_J$ *(in $L^p_\omega(J;U)$ with reflexive U and $1 \le p < \infty$).*

(iii) *The dual of the time compression operator γ_λ (see Example 2.3.6) is the inverse time compression operator $\gamma_{1/\lambda}$ (in $L^p_\omega(J;U)$ with reflexive U and $1 \le p < \infty$).*

[5] In the *Reg*-well-posed case we require χ_J to be right-continuous and define $\mathbf{Я}J$ to be the set whose characteristic function is $\chi_{\mathbf{Я}J}$

(iv) *The right shift (semi)groups in Example 3.5.11 are similar to the corresponding left shift (semi)groups in Examples 2.3.2 and 2.5.3 as follows:*
   (a) $\tau^{\odot} = \mathbf{Я}\tau\mathbf{Я}$;
   (b) $\tau_+^{\odot} = \mathbf{Я}\tau_-\mathbf{Я}$;
   (c) $\tau_-^{\odot} = \mathbf{Я}\tau_+\mathbf{Я}$;
   (d) $\tau_{[0,T)}^{\odot} = \tau^{-T}\mathbf{Я}\tau_{[0,T)}\mathbf{Я}\tau^{T}$;
   (e) $\tau_{\mathbb{T}_T}^{\odot} = \mathbf{Я}\tau_{\mathbb{T}_T}\mathbf{Я}$.

We leave the easy proof to the reader.

3.6 The rigged spaces induced by the generator

In our subsequent theory of $L^p|Reg$-well-posed linear systems we shall need a scale of spaces X_n, $n = 0, \pm 1, \pm 2, \ldots$, which are constructed from X by means of the semigroup generator A. In particular, the spaces X_1 and X_{-1} will be of fundamental importance. To construct these spaces we need not even assume that A generates a C_0 semigroup on X; it is enough if A has a nonempty resolvent set and dense domain.

We begin with the case $n \geq 0$, and define

$$X_0 = X, \qquad X_n = \mathcal{D}\left(A^n\right) \text{ for } n = 1, 2, 3, \ldots$$

Choose an arbitrary number α from the resolvent set of A. Then $(\alpha - A)^{-n}$ maps X one-to-one onto $\mathcal{D}(A^n)$ (this can be proved by induction over n), and we can define a norm in X_n by

$$|x|_n = |x|_{X_n} = \left|(\alpha - A)^n x\right|_X.$$

With this norm each X_n becomes a Banach space, $X_{n+1} \subset X_n$ with a dense injection, and $(A - \alpha)^n$ is an isometric (i.e., norm-preserving) isometry (i.e., bounded linear operator with a bounded inverse) from X_n onto X. If X is a Hilbert space, then so are all the spaces X_n.

If we replace α by some other $\beta \in \rho(A)$, then $(\beta - A)^{-n}$ has the same range as $(\alpha - A)^{-n}$, so if we use β instead of α in the definition of X_n then we still get the same space, but with a different norm. However, the two norms are equivalent since $(\alpha - A)^n(\beta - A)^{-n}$ is an isomorphism (not isometric) on X: for $n = 1$ this follows from the resolvent formula in Lemma 3.2.8(i) which gives

$$(\alpha - A)(\beta - A)^{-1} = 1 + (\alpha - \beta)(\beta - A)^{-1},$$

and by iterating this formula we get the general case. Most of the time the value of $\alpha \in \rho(A)$ which determines the exact norm in X_n is not important.

If A generates a C_0 semigroup $\mathfrak{A}$, then the restriction $\mathfrak{A}_{|X_n}$ of $\mathfrak{A}$ to X_n is a C_0 semigroup on X_n. It follows from Theorem 3.2.1(iii) and Example 3.2.6(i) that $\mathfrak{A}_{|X_n} = (\alpha - A)^{-n}\mathfrak{A}(\alpha - A)^n$, i.e., $\mathfrak{A}_{|X_n}$ and $\mathfrak{A}$ are (isometrically) isometric. Thus, all the important properties of these semigroups are identical. In particular, they all have the same growth bound $\omega_{\mathfrak{A}}$, and the generator of $\mathfrak{A}_n$ is the restriction $A_{|X_{n+1}}$ of A to X_{n+1}. In the sequel we occasionally write (for simplicity) $\mathfrak{A}$ instead of $\mathfrak{A}_{|X_n}$ and A instead of $A_{|X_{n+1}}$ (but we still use the more complicated notions in those cases where the distinction is important).

It is also possible to go in the opposite direction to get spaces X_n with negative index n. This time we first define a sequence of weaker norms in X, namely

$$|x|_{-n} = \left|(\alpha - A)^{-n}x\right|_X \text{ for } n = 1, 2, 3, \ldots,$$

and let X_{-n} be the completion of X with respect to the norm $|\cdot|_{-n}$. Then $(\alpha - A)^n$ has a unique extension to an isometric operator which maps X onto X_{-n}. We denote this operator by $(\alpha - A)^n_{|X}$ and its inverse by $(\alpha - A)^{-n}_{|X_{-n}}$, or sometimes simply by $(\alpha - A)^n$, respectively $(\alpha - A)^{-n}$, if no confusion is likely to arise. In the case $n = 1$ we often write $(\alpha - A_{|X})^{-1}$ instead of $(\alpha - A)^{-1}_{|X_{-1}}$. Thus, for all $n, l = 0, \pm 1, \pm 2, \ldots,$

$$(\alpha - A)^l_{|X_{n+l}} \text{ is an isometry of } X_{n+l} \text{ onto } X_n.$$

If A generates a C_0 semigroup $\mathfrak{A}$ on X, then we can use the formula

$$\mathfrak{A}_{|X_{-n}} = (\alpha - A)^n_{|X}\mathfrak{A}(\alpha - A)^{-n}_{|X_{-n}}$$

to extend (rather than restrict) $\mathfrak{A}$ to a semigroup on each of the spaces X_{-n}. In this way we get a full scale of spaces $X_{n+1} \subset X_n$ for $n = 0, \pm 1, \pm 2, \ldots,$ and a corresponding scale of isometric semigroups $\mathfrak{A}_{|X_{-n}}$. In places where no confusion is likely to arise we abbreviate $\mathfrak{A}_{|X_{-n}}$ to $\mathfrak{A}$. The generator of $\mathfrak{A}_{|X_{-n}}$ is $A_{|X_{-n+1}}$. As in the case of the semigroup itself we sometimes abbreviate $A_{|X_{-n+1}}$ to A.

Above we have defined the norm in X_1 by using the fact that $(\alpha - A)^{-1}$ maps X one-to-one onto X_1 whenever $\alpha \in \rho(A)$. Another commonly used norm in X_1 is the graph norm

$$\|x\|_{X_1} = \left(|x|_X^2 + |Ax|_X^2\right)^{1/2}. \tag{3.6.1}$$

This is the restriction of the norm $\left\|\left[\begin{smallmatrix} x \\ y \end{smallmatrix}\right]\right\| = \left(|x|_X^2 + |y|_X^2\right)^{1/2}$ in $\left[\begin{smallmatrix} X \\ X \end{smallmatrix}\right]$ to the graph $\mathcal{G}(A) = \left\{\left[\begin{smallmatrix} Ax \\ x \end{smallmatrix}\right] \;\middle|\; x \in X\right\}$. This graph is closed since A is closed, so it is a Banach space in itself (or a Hilbert space if X is a Hilbert space). The map which takes $\left[\begin{smallmatrix} Ax \\ x \end{smallmatrix}\right] \in \mathcal{G}(A)$ into x is injective, so we may let $x \in \mathcal{D}(A)$ inherit the norm of $\left[\begin{smallmatrix} Ax \\ x \end{smallmatrix}\right] \in \mathcal{G}(A)$, and this is the norm $\|\cdot\|_{X_1}$ in (3.6.1). This norm is majorized by the earlier introduced norm $|\cdot|_{X_1}$ since

$$|Ax|_X = |(A - \alpha + \alpha)x|_X \leq \|x\|_{X_1} + |\alpha||x|_X,$$

and

$$|x|_X = |(\alpha - A)^{-1}(\alpha - A)x|_X \leq \|(\alpha - A)^{-1}\||x|_{X_1},$$

so by the open mapping theorem, the two norms $|\cdot|_{X_1}$ and $\|\cdot\|_{X_1}$ are equivalent.

A similar norm can be used in $X_n = \mathcal{D}(A^n)$ for $n = 2, 3, \ldots$, namely

$$\|x\|_{X_n} = \left(|x|_X^2 + |A^n x|_X^2\right)^{1/2}. \tag{3.6.2}$$

To prove that this is a norm in X_n we can argue as above: the operator A^n is closed since it is the restriction of $(A^n)_{|X} := A_{|X_{-n+1}} A_{|X_{-n+2}} \cdots A_{|X} \in \mathcal{B}(X; X_{-n})$ to its natural domain $\mathcal{D}(A^n) = \{x \in X \mid (A^n)_{|X} x \in X\}$, and the above norm is the graph norm of A^n on $\mathcal{D}(A^n)$. To show that it is equivalent to the norm $|\cdot|_{X_n}$ we may argue as follows. Take some $\alpha \in \rho(A)$. Then, for each $x \in \mathcal{D}(A^n)$ we have (from the binomial formula)

$$(\alpha - A)^n x = \sum_{k=0}^{n} \binom{n}{k} \alpha^k A^{n-k} x,$$

or equivalently,

$$\begin{aligned} A^n x &= (\alpha - A)^n x - \sum_{k=1}^{n} \binom{n}{k} \alpha^k A^{n-k} x \\ &= \left(1 - \sum_{k=1}^{n} \binom{n}{k} \alpha^k A^{n-k} (\alpha - A)^{-n}\right)(\alpha - A)^n x. \end{aligned}$$

Thus, $|A^n x|_X \leq M|x|_{X_n}$, where M is the norm of the operator $1 - \sum_{k=1}^{n} \binom{n}{k} \alpha^k A^{n-k} (\alpha - A)^{-n} \in \mathcal{B}(X)$, and, of course,

$$|x|_X = |(\alpha - A)^{-n}(\alpha - A)^n x|_X \leq \|(\alpha - A)^{-n}\||x|_{X_n}.$$

This shows that the norm $\|\cdot\|_{X_n}$ is majorized by the norm $|\cdot|_{X_n}$, so by the open mapping theorem, the two norms $|\cdot|_{X_n}$ and $\|\cdot\|_{X_n}$ are equivalent.

Let us illustrate these constructions by looking at Example 3.3.5. In this example we have

$$|x|_{X_n}^2 = \sum_{k=1}^{\infty} |\alpha - \lambda_k|^{2n} |\langle x, \phi_k \rangle|_H^2,$$

where $\alpha \in \rho(A)$, and each X_n is a Hilbert space with the orthogonal basis $\{\phi_n\}_{n=1}^{\infty}$ (it becomes orthonormal if we divide ϕ_k by $|\alpha - \lambda_k|$). For $n \geq 1$ we can alternatively use the equivalent norm

$$|x|_{X_n}^2 = \sum_{k=1}^{\infty}(1 + |\lambda_k|)^{2n}|\langle x, \phi_k\rangle|_H^2;$$

cf. Example 3.3.5.

Remark 3.6.1 This remark explains how the spaces X_n interact with duality. Since $X_{n+1} \subset X_n$ for all $n = 0, \pm 1, \pm 2, \ldots$, with dense embeddings, the duals of these embedding maps are injective (see Lemma 9.10.2(ii)), so they define embeddings $(X_n)^* \subset (X_{n+1})^*$ (which need not be dense). Since $(\alpha - A)^n$ is an isometry of X_{n+l} onto X_l, it follows that $(\alpha - A^*)^n$ is an isometry of $(X_l)^*$ onto $(X_{n+l})^*$ for all $n, l = 0, \pm 1, \pm 2, \ldots$. If X is reflexive, then the embeddings $(X_n)^* \subset (X_{n+1})^*$ are dense, and these spaces are the same as we would get by repeating the argument leading to the definition of the spaces X_n, with X replaced by X^*, A replaced by A^*, and using a different subindex (i.e., $-n$ instead of n). When we discuss the causal and anti-causal dual systems Σ^d and $\Sigma^\dagger$ it is convenient to denote the domain of A^* by X_1^*, and accordingly, in the sequel we use the notation

$$X_{-n}^* := (X^*)_{-n} := (X_n)^*, \qquad n = 0, \pm 1, \pm 2, \ldots$$

In particular,

$$\langle A^n x, x^*\rangle_{(X_l, X_{-l}^*)} = \langle x, A^{*n}x^*\rangle_{(X_{n+l}, X_{-(n+l)}^*)}, \qquad x \in X_{n+l}, \qquad x^* \in X_{-l}^*,$$

where by A^{*n} we mean $A^{*n} := (A^*)^n = (A^n)^*$.

In the Hilbert space case one often uses a slightly different construction, which resembles the one described in Remark 3.6.1. Assume that $W \subset X$ are two Hilbert spaces, with a continuous and dense embedding. Then $(x, y) \mapsto \langle x, y\rangle_X$ is a bounded sesquilinear form on W, and therefore (see, e.g., Kato 1980, pp. 256–257) there is a unique operator $E \in \mathcal{B}(W)$ which is positive and self-adjoint (with respect to the inner product of W) such that

$$\langle x, y\rangle_X = \langle Ex, y\rangle_W = \langle x, Ey\rangle_W = \langle\sqrt{E}x, \sqrt{E}y\rangle_W, \qquad x, y \in W,$$

where $\sqrt{E}$ is the positive self-adjoint square root of E (cf. Lemma A.2.2). For all $x \in W$,

$$|Ex|_X^2 = \langle E\sqrt{E}x, E\sqrt{E}x\rangle_W \le \|E\|_{\mathcal{B}(W)}^2|\sqrt{E}x|_W^2 = \|E\|_{\mathcal{B}(W)}^2|x|_X^2,$$

and this implies that E can be extended to a unique operator in $\mathcal{B}(X)$, which we still denote by the same symbol E. This operator is still self-adjoint in X since $\langle x, Ey\rangle_X = \langle Ex, Ey\rangle_W = \langle Ex, y\rangle_X$ for all $x, y \in W$, and W is dense in X. The space X may be regarded as the completion of W with respect to the norm $|x|_X = |\sqrt{E}x|_W$, and this means that the extended version of $\sqrt{E}$ is an isometric isomorphism of W onto X.

Let V be the completion of X with respect to the norm $|x|_V = |\sqrt{E}x|_X$. By repeating the same argument that we gave above with W replaced by X and X replaced by V we find that E can be extended to a self-adjoint operator in V (which we still denote by the same letter), that $\sqrt{E}$ is an isometric isomorphism of V onto X, and that E is an isometric isomorphism of V onto W. Moreover,

$$\langle x, y\rangle_V = \langle Ex, y\rangle_X = \langle x, Ey\rangle_X = \langle\sqrt{E}x, \sqrt{E}y\rangle_X, \qquad x, y \in X,$$
$$\langle x, y\rangle_V = \langle Ex, y\rangle_X = \langle Ex, Ey\rangle_W, \qquad x, y \in W.$$

The space V can be interpreted as the dual of W with X as pivot space as follows. Every $x \in V$ induces a bounded linear functional on W through the formula

$$\langle x, y\rangle_{(V,W)} = \langle Ex, y\rangle_W,$$

and every bounded linear functional on W is of this type since E maps W one-to-one onto V. This is a norm-preserving mapping of the dual of W onto V, since the norm of the above functional is $|Ex|_W = |x|_V$. That X is a pivot space means that for all $x \in X$ and $y \in W$,

$$\langle x, y\rangle_{(V,W)} = \langle x, y\rangle_X,$$

which is true since both sides are equal to $\langle Ex, y\rangle_W$.

If we apply this procedure (in the Hilbert space case) to the space $X_1 \subset X$ described at the beginning of this section, then we get $V = X^*_{-1}$ and the extended version of E is given by $E = (\alpha - A)^{-1}(\overline{\alpha} - A^*_{|X})^{-1}$ if we use the norm $|x|_1 = |(\alpha - A)x|_X$ in X_1. If we instead use the graph norm $|x|_1^2 = |x|_X^2 + |Ax|_X^2$ in X_1, then the extended version of E is given by $E = (1 + A^*_{|X}A)^{-1}$.

In this book we shall usually identify X with its dual, and identify the dual of W with V as described above. However, occasionally it is important to compute the dual of an operator with respect to the inner product in W or in V instead of computing it with respect to the inner produce in X. Here the following result is helpful.

Proposition 3.6.2 *Let U, Y, and $W \subset X \subset V$ be Hilbert spaces, where the embeddings are continuous and dense, let $E \in \mathcal{B}(V)$ be injective, selfadjoint (with respect to the inner product in V), and suppose that $\sqrt{E}$ maps V isometrically onto X and that $\sqrt{E}_{|X}$ maps X isometrically onto W (the operator E and the space V can be constructed starting from W and X as explained above). We identify U and Y with their duals.*

(i) *Let $B \in \mathcal{B}(U; W)$, let $B' \in \mathcal{B}(W; U)$ be the adjoint of B with respect to the inner product in W, and let $B^* \in \mathcal{B}(X; U)$ be the adjoint of B with* *respect to the inner product in* X *(note that* $B \in \mathcal{B}(U;X)$*). Then* $B^* = B'E_{|X}$*. In particular, this formula can be used to extend* B^* *to* $B'E \in \mathcal{B}(V;U)$*, which is the adjoint of* B *when we identify the dual of* W *with* V*.*

(ii) *Let* $B \in \mathcal{B}(U;X)$*, let* $B^* \in \mathcal{B}(X;U)$ *be the adjoint of* B *with respect to the inner product in* X*, and let* $B'' \in \mathcal{B}(V;U)$ *be the adjoint of* B *with respect to the inner product in* V *(note that* $B \in \mathcal{B}(U;V)$*). Then* $B'' = B^*E$*.*

(iii) *Let* $B \in \mathcal{B}(U;V)$*, let* $B^* \in \mathcal{B}(W;U)$ *be the adjoint of* B *when we identify the dual of* V *with* W *(with* X *as pivot space), and let* $B'' \in \mathcal{B}(V;U)$ *be the adjoint of* B *with respect to the inner product in* V*. Then* $B'' = B^*E$*.*

(iv) *Let* $C \in \mathcal{B}(V;Y)$*, let* $C'' \in \mathcal{B}(Y;V)$ *be the adjoint of* C *with respect to the inner product in* V*, and let* $C^* \in \mathcal{B}(Y;X)$ *be the adjoint of* C *with respect to the inner product in* X *(note that* $C \in \mathcal{B}(X;Y)$*). Then* $C^* = EC''$*. In particular,* $C^* \in \mathcal{B}(Y;W)$*.*

(v) *Let* $C \in \mathcal{B}(X;Y)$*, let* $C^* \in \mathcal{B}(Y;X)$ *be the adjoint of* C *with respect to the inner product in* X*, and let* $C' \in \mathcal{B}(Y;W)$ *be the adjoint of* C *with respect to the inner product in* W *(note that* $C \in \mathcal{B}(W;Y)$*). Then* $C' = EC^*$*.*

(vi) *Let* $C \in \mathcal{B}(W;Y)$*, let* $C' \in \mathcal{B}(Y;W)$ *be the adjoint of* C *with respect to the inner product in* W*, and let* $C^* \in \mathcal{B}(Y;V)$ *be the adjoint of* C *when we identify the dual of* W *with* V *(with* X *as pivot space). Then* $C' = EC^*$*.*

(vii) *Let* $A \in \mathcal{B}(V)$*, and suppose that* X *is invariant under* A*. Let* $A'' \in \mathcal{B}(W)$ *be the adjoint of* A *with respect to the inner product of* V*, and let* $A^*_{|X}$ *be the adjoint of* $A_{|X}$ *with respect to the inner product of* X*. Then* $EA'' = A^*_{|X}E$*. In particular,* W *is invariant under* $A^*_{|X}$*.*

(viii) *Let* $A \in \mathcal{B}(X)$*, and suppose that* W *is invariant under* A*. Let* A^* *be the adjoint of* A *with respect to the inner product of* X*, and let* $A'_{|W}$ *be the adjoint of* $A_{|W}$ *with respect to the inner product of* W*. Then* $A'_{|W}E_{|X} = EA^*$*. In particular,* EX *is invariant under* $A'_{|W}$*.*

(ix) *Let* $A \in \mathcal{B}(W)$*, let* $A' \in \mathcal{B}(W)$ *be the adjoint of* A *with respect to the inner product in* W*, and let* $A^* \in \mathcal{B}(V)$ *be the adjoint of* A *when we identify the dual of* W *with* V*. Then* $A'E = EA^*$*.*

Proof (i) For all $x \in X$ and $u \in U$,

$$\langle u, B^*x\rangle_U = \langle Bu, x\rangle_X = \langle Bu, Ex\rangle_W = \langle u, B'Ex\rangle_U.$$

Thus, $B^* = B'E_{|X}$. If we instead let B^* stand for the adjoint of B when we identify the dual of W by V, then for all $u \in U$ and $x \in V$,

$$\langle B^*x, u\rangle_U = \langle x, Bu\rangle_{(V,W)} = \langle Ex, Bu\rangle_W = \langle B'Ex, u\rangle_U.$$

Thus, $B^* = B'E$.

(ii) Apply (i) with W replaced by X and X replaced by V.

(iii) For all $x \in V$ and $u \in U$,

$$\langle u, B''x\rangle_U = \langle Bu, x\rangle_V = \langle EBu, Ex\rangle_W = \langle Bu, Ex\rangle_{(V,W)} = \langle u, B^*Ex\rangle_U.$$

Thus $B'' = B^*E$.

(iv) For all $x \in X$ and $y \in Y$,

$$\langle x, C^*y\rangle_X = \langle Cx, y\rangle_Y = \langle x, C''y\rangle_V = \langle x, EC''y\rangle_X.$$

Thus $C^* = EC''$.

(v) Apply (iv) with V replaced by X and X replaced by W.

(vi) For all $x \in W$ and $y \in Y$,

$$\langle C'y, x\rangle_W = \langle y, Cx\rangle_Y = \langle C^*y, x\rangle_{(V,W)} = \langle EC^*y, x\rangle_W.$$

Thus, $C' = EC^*$.

(vii) For all $x \in X$ and $y \in V$,

$$\begin{aligned}\langle x, EA''y\rangle_V &= \langle AEx, y\rangle_V = \langle A_{|X}Ex, Ey\rangle_X = \langle x, EA^*_{|X}Ey\rangle_X \\ &= \langle x, A^*_{|X}Ey\rangle_V.\end{aligned}$$

Thus, $EA'' = A^*_{|X}E$ on V. This implies that W is invariant under $A^*_{|X}$.

(viii) Apply (vii) with V replaced by X and X replaced by W.

(ix) For all $x \in V$ and $y \in W$,

$$\begin{aligned}\langle A'Ex, y\rangle_W &= \langle Ex, Ay\rangle_W = \langle x, Ay\rangle_{(V,W)} = \langle A^*x, y\rangle_{(V,W)} \\ &= \langle EA^*x, y\rangle_W.\end{aligned}$$

Thus $A'E = EA^*$. □

3.7 Approximations of the semigroup

The approximation A_α to A that we used in the proof of the Hille–Yosida Theorem 3.4.1 will be quite useful in the sequel, too. For later use, let us record some of the properties of this and some related approximations:

Theorem 3.7.1 *Let A be the generator of a C_0 semigroup on X. Define the space $X_1 = \mathcal{D}(A)$ as in Section 3.6. For all $\alpha \in \rho(A)$ (in particular, for all $\alpha \in \mathbb{C}^+_{\omega_{\mathfrak{A}}}$), define*

$$J_\alpha = \alpha(\alpha - A)^{-1}, \qquad A_\alpha = \alpha A(\alpha - A)^{-1},$$

and for all $h > 0$ and $x \in X$, define

$$J^h x = \frac{1}{h}\int_0^t \mathfrak{A}^s x\,ds, \qquad A^h = \frac{1}{h}(\mathfrak{A}^h - 1)x.$$

Then the following claims are true:

(i) *For all $\alpha \in \rho(A)$ and $h > 0$, $J_\alpha \in \mathcal{B}(X; X_1)$, $J^h \in \mathcal{B}(X; X_1)$, $A_\alpha \in \mathcal{B}(X)$, $A^h \in \mathcal{B}(X)$, and*

$$\begin{aligned} J_\alpha &= \alpha(\alpha - A)^{-1} = 1 + A(\alpha - A)^{-1}, \\ A_\alpha &= AJ_\alpha = \alpha(J_\alpha - 1) = \alpha^2(\alpha - A)^{-1} - \alpha, \\ A^h &= AJ^h = \frac{1}{h}(\mathfrak{A}^h - 1)x = \frac{1}{h}A\int_0^t \mathfrak{A}^s x\,ds, \end{aligned}$$

Moreover, for $\alpha \in \mathbb{C}^+_{\omega_{\mathfrak{A}}}$,

$$J_\alpha x = \alpha\int_0^\infty e^{-\alpha s}\mathfrak{A}^s x\,ds, \qquad x \in X.$$

(ii) *For all $\alpha, \beta \in \rho(A)$ and $h, k, t > 0$, the operators J_α, J_β, J^h, J^k, A_α, A_β, A^h, A^k, and $\mathfrak{A}^t$ commute with each other.*

(iii) *J_α and J^h approximate the identity and A_α and A^h approximate A in the sense that the following limits exist:*

$$\begin{aligned} \lim_{\alpha\to+\infty} J_\alpha x &= \lim_{h\downarrow 0} J^h x = x && \text{in } X \text{ for all } x \in X, \\ \lim_{\alpha\to+\infty} A_\alpha x &= \lim_{h\downarrow 0} A^h x = Ax && \text{in } X \text{ for all } x \in X_1, \\ \lim_{\alpha\to+\infty} \alpha^{-1} J_\alpha x &= \lim_{\alpha\to+\infty} (\alpha - A)^{-1} x = 0 && \text{in } X_1 \text{ for all } x \in X, \\ \lim_{h\downarrow 0} hJ^h x &= \lim_{h\downarrow 0}\int_0^t \mathfrak{A}^s x\,dx = 0 && \text{in } X_1 \text{ for all } x \in X, \\ \lim_{\alpha\to+\infty} \alpha^{-1} A_\alpha x &= \lim_{h\downarrow 0} hA^h x = 0 && \text{in } X \text{ for all } x \in X. \end{aligned}$$

(iv) *$\mathfrak{A}$ is uniformly continuous (hence analytic) iff J_α has a bounded inverse for some $\alpha \in \rho(A)$, or equivalently, iff J^h has a bounded inverse for some $h > 0$.*

Proof (i) Obviously, $J_\alpha \in \mathcal{B}(X; X_1)$, $A_\alpha \in \mathcal{B}(X)$, and $A^h \in \mathcal{B}(X)$. By Theorem 3.2.1(ii), $J^h \in \mathcal{B}(X; X_1)$. The algebraic properties in (i) are easy to verify (see also Theorem 3.2.1(ii)). The integral formula for J_α is found in Theorem 3.2.9(i).

(ii) This is true since $\mathfrak{A}^s$ commutes with $\mathfrak{A}^t$ and with $(\alpha - A)^{-1}$. See also Theorems 3.2.1 and 3.2.9.

(iii) That $\lim_{\alpha\to+\infty} J_\alpha x = \lim_{h\downarrow 0} J^h x = x$ in X for all $x \in X$ follows from Theorems 3.2.1(i) and 3.2.9(iii) and that $\lim_{\alpha\to+\infty} A_\alpha x = \lim_{h\downarrow 0} A^h x = Ax$ in X for all $x \in X_1$ follows from (i), Definition 3.1.1, and Theorem 3.2.9(iii). By Theorem 3.2.9(iii), for all $\beta \in \rho(A)$, $\lim_{\alpha\to+\infty}(\beta - A)(\alpha - A)^{-1}x = 0$ in X for all $x \in X$, and this implies that $\lim_{\alpha\to+\infty} \alpha^{-1} J_\alpha x = 0$ in X_1 for all $x \in X$. To prove that $\lim_{h\downarrow 0} h J^h x = 0$ in X_1 for all $x \in X$ it suffices to observe that, for all $\beta \in \rho(A)$,

$$(\beta - A) h J^h x = (\beta - A)\int_0^h \mathfrak{A}^s x\, ds = \beta \int_0^h \mathfrak{A}^s x\, ds + x - \mathfrak{A}^h x,$$

and here the right-hand side tends to zero in X for every $x \in X$. That $\lim_{\alpha\to+\infty} \alpha^{-1} A_\alpha x = \lim_{h\downarrow 0} h A^h x = 0$ in X for all $x \in X$ follows from (i) and the fact that $\lim_{\alpha\to+\infty} J_\alpha x = \lim_{h\downarrow 0} J^h x = x$ in X for all $x \in X$.

(iv) Obviously $A \in \mathcal{B}(X)$ iff J_α has a bounded inverse. That J^h has a bounded inverse for some $h > 0$ iff $A \in \mathcal{B}(X)$ follows from Example 3.1.2 and Remark 3.1.4. By Theorem 3.1.3, the boundedness of A is equivalent to the uniform continuity of $\mathfrak{A}$. □

Definition 3.7.2 The operators J_α and A_α in Theorem 3.7.1 are called the *Yosida* (or *Abel*) *approximations* of the identity 1 and of A, respectively (with parameter α). The operators J^h and A^h in Theorem 3.7.1 are called the *Cesàro approximations* (of order one) of the identity 1 and of A, respectively (with parameter h).

Theorem 3.7.3 *Let A be the generator of a C_0 semigroup $\mathfrak{A}$ on X, and let $A_\alpha = \alpha A(\alpha - A)^{-1}$ be the Yosida approximation of A. Then for each $x \in X$ and $t \geq 0$, $\lim_{\alpha\to+\infty} e^{A_\alpha t} x = \mathfrak{A}^t x$, and the convergence is uniform in t on any bounded interval.*

The proof of this theorem is contained in the proof of Theorem 3.4.1.

The same result is true if we replace the Yosida approximation by the Cesàro approximation:

Theorem 3.7.4 *Let A be the generator of a C_0 semigroup $\mathfrak{A}$ on X, and let $A_h = \frac{1}{h}(\mathfrak{A}^h - 1)$ be the Cesàro approximation of A. Then for each $x \in X$ and $t \geq 0$, $\lim_{h\downarrow 0} e^{A^h t} x = \mathfrak{A}^t x$, and the convergence is uniform in t on any bounded interval.*

Proof The proof follows the same lines as the proof of Theorem 3.4.1 with A_α replaced by A^h, $\mathfrak{A}^t_\alpha$ replaced by $\mathfrak{A}^t_h = e^{A^h t}$, and B_α replaced by $B^h = A^h + \frac{1}{h} = \frac{1}{h}\mathfrak{A}^h$. We can choose M and ω so that $\|\mathfrak{A}^t\| \leq M e^{\omega t}$ for all $t \geq 0$. Then (3.4.1) is replaced by

$$\|(B^h)^n\| = \left\| \left(\frac{1}{h}\mathfrak{A}^h\right)^n \right\| \le \frac{M e^{\omega h n}}{h^n},$$

and (3.4.3) is replaced by

$$\begin{aligned}\|\mathfrak{A}_h^t\| &\le e^{-t/h} \sum_{n=0}^{\infty} \frac{t^n}{n!} \frac{M e^{\omega h n}}{h^n} \\ &= M e^{-t/h} e^{(t/h) e^{\omega h}} = M e^{t/h(e^{\omega h}-1)}, \qquad t \ge 0.\end{aligned}$$

This tends to $M e^{\omega t}$ as $h \downarrow 0$, uniformly in t on any bounded interval. The new version of estimate (3.4.4) is (for all $h, k > 0$)

$$|\mathfrak{A}_h^t x - \mathfrak{A}_k^t x| \le M^2 \int_0^t e^{s/h(e^{\omega h}-1)} e^{(t-s)/k(e^{\omega k}-1)} |A^h x - A^k x|\, ds,$$

and the remainder of the proof of Theorem 3.4.1 stays the same. □

Theorem 3.7.5 *Let A be the generator of a C_0 semigroup $\mathfrak{A}$ on X. Then, for all $t \ge 0$,*

$$\mathfrak{A}^t x = \lim_{n\to\infty} \left(1 - \frac{t}{n} A\right)^{-n} x, \qquad x \in X,$$

and the convergence is uniform in t on each bounded interval.

Proof By Theorem 3.2.9(i), for all $x \in X$ and $(n-1)/t > \omega_{\mathfrak{A}}$,

$$\begin{aligned}\left(1 - \frac{t}{n} A\right)^{-(n+1)} x &= \left(\frac{n}{t}\right)^{n+1} \left(\frac{n}{t} - A\right)^{-(n+1)} x \\ &= \left(\frac{n}{t}\right)^{n+1} \frac{1}{n!} \int_0^\infty s^n e^{-ns/t} \mathfrak{A}^s x\, ds \\ &= \frac{n^{n+1}}{n!} \int_0^\infty (v e^{-v})^n \mathfrak{A}^{tv} x\, dv.\end{aligned}$$

As $\frac{n^{n+1}}{n!} \int_0^\infty (v e^{-v})^n\, dv = 1$, this implies that

$$\begin{aligned}&\left|\left(1 - \frac{t}{n} A\right)^{-(n+1)} x - \mathfrak{A}^t x\right| \\ &\quad = \left|\frac{n^{n+1}}{n!} \int_0^\infty (v e^{-v})^n (\mathfrak{A}^{tv} x - \mathfrak{A}^t x)\, dv\right| \\ &\quad \le \frac{n^{n+1}}{n!} \int_0^\infty v^n e^{-nv} \left|\mathfrak{A}^{tv} x - \mathfrak{A}^t x\right| dv.\end{aligned}$$

For each $T > 0$, the function $v \mapsto \mathfrak{A}^v$ is uniformly continuous on $[0, T]$. Thus, for every $\epsilon > 0$ it is possible to find a $\delta > 0$ such that $|\mathfrak{A}^{tv} x - \mathfrak{A}^t x| \le \epsilon$ for all $t \in [0, T]$ and $1 - \delta \le v \le 1 + \delta$. We split the integral above into three parts I_1, I_2, and I_3, over the intervals $[0, 1-\delta)$, $[1-\delta, 1+\delta)$, and $[1+\delta, \infty)$, respectively. Then

$$\left|\left(1 - \frac{t}{n}A\right)^{-(n+1)} x - \mathfrak{A}^t x\right| = I_1 + I_2 + I_3.$$

The function $v \mapsto v\mathrm{e}^{-v}$ is increasing on $[0, 1]$, so we can estimate for all $t \in [0, T]$ (choose $M > 0$ and $\omega > 0$ so that $\|\mathfrak{A}^t\| \le M\mathrm{e}^{\omega t} \le M\mathrm{e}^{\omega T}$)

$$\begin{aligned}
I_1 &\le \frac{n^{n+1}((1-\delta)\mathrm{e}^{-(1-\delta)})^n}{n!} \int_0^{1-\delta} \left|\mathfrak{A}^{tv}x - \mathfrak{A}^t x\right| dv \\
&\le 2M\mathrm{e}^{\omega T} \frac{n^{n+1}((1-\delta)\mathrm{e}^{-(1-\delta)})^n}{n!}, \\
I_2 &\le \epsilon \frac{n^{n+1}}{n!} \int_{1-\delta}^{1+\delta} \left(v\mathrm{e}^{-v}\right)^n dv < \epsilon, \\
I_3 &= \frac{n^{n+1}}{n!} \int_{1+\delta}^{\infty} \left(v\mathrm{e}^{-v}\right)^n \left|\mathfrak{A}^{tv}x - \mathfrak{A}^t x\right| dv \\
&\le 2M \frac{n^{n+1}}{n!} \int_{1+\delta}^{\infty} \left(v\mathrm{e}^{-v}\right)^n \mathrm{e}^{\omega T v}\, dv \\
&= 2M \frac{n^{n+1}}{n!} \int_{1+\delta}^{\infty} \left(v\mathrm{e}^{-(1-(1+\omega T)/n)v}\right)^n \mathrm{e}^{-v}\, dv.
\end{aligned}$$

We recall Stirling's formula

$$\lim_{n\to\infty} \frac{n^{n+\frac{1}{2}}}{n!\,\mathrm{e}^n} = \sqrt{2\pi}, \tag{3.7.1}$$

which together with the fact that $(1-\delta)\mathrm{e}^{1-\delta} < 1/\mathrm{e}$ implies that $I_1 \to 0$ as $n \to \infty$. The function $v \mapsto v\mathrm{e}^{-(1-(1+\omega T)/n)v}$ is decreasing for $v \ge (1-(1+\omega T)/n)^{-1}$, so for n large enough, we can estimate I_3 by

$$\begin{aligned}
I_3 &\le 2M \frac{n^{n+1}}{n!} \left((1+\delta)\mathrm{e}^{-(1-(1+\omega T)/n)(1+\delta)}\right)^n \int_{1+\delta}^{\infty} \mathrm{e}^{-v}\, dv \\
&\le 2M \frac{n^{n+1}}{n!} \left((1+\delta)\mathrm{e}^{-(1-(1+\omega T)/n)(1+\delta)}\right)^n.
\end{aligned}$$

Since

$$\lim_{n\to\infty} (1+\delta)\mathrm{e}^{-(1-(1+\omega T)/n)(1+\delta)} = (1+\delta)\mathrm{e}^{-(1+\delta)} < 1/\mathrm{e},$$

we can use Stirling's formula (3.7.1) once more to conclude that I_3 tends to zero as $n \to \infty$. Thus,

$$\lim_{n\to\infty} \left(1 - \frac{t}{n}A\right)^{-(n+1)} x = \mathfrak{A}^t x,$$

uniformly in $t \in [0, T]$. As furthermore

$$\lim_{n\to\infty}\left(1 - \frac{t}{n}A\right)^{-1} x = x$$

uniformly in $t \in [0, T]$ (see Theorem 3.7.1(iii)), this implies that

$$\lim_{n\to\infty}\left(1 - \frac{t}{n}A\right)^{-n} x = \mathfrak{A}^t x,$$

uniformly in t on any bounded interval. □

3.8 The nonhomogeneous Cauchy problem

It is time to study the relationship between the differential equation

$$\begin{aligned} \dot{x}(t) &= Ax(t) + f(t), \qquad t \geq s, \\ x(s) &= x_s, \end{aligned} \tag{3.8.1}$$

and the *variation of constants formula*

$$x(t) = \mathfrak{A}^{t-s}x_s + \int_s^t \mathfrak{A}^{t-v} f(v)\,dv. \tag{3.8.2}$$

It is possible to do this in several different settings, but we choose a setting that is relevant for the full system $\left[\begin{smallmatrix} \mathfrak{A} & \mathfrak{B} \\ \mathfrak{C} & \mathfrak{D} \end{smallmatrix}\right]$. Here the spaces X_n and the extended semigroups $\mathfrak{A}_{|X_n}$ and generators $A_{|X_{n+1}}$ (with $n \leq 0$) introduced in Section 3.6 become important.

Definition 3.8.1 Let $s \in \mathbb{R}$, $x_s \in X$, $n = 0, \pm 1, \pm 2, \ldots$, and $f \in L^1_{\text{loc}}([s, \infty); X_{n-1})$. A function x is a *strong solution of (3.8.1) in* X_n (on the interval $[s, \infty)$) if $x \in C([s, \infty); X_n) \cap W^{1,1}_{\text{loc}}([s, \infty); X_{n-1})$, $x(s) = x_s$, and $\dot{x}(t) = A_{|X_n}x(t) + f(t)$ in X_{n-1} for almost all $t \geq s$. By a *strong solution of (3.8.1)* (without any reference to a space X_n) we mean a strong solution of (3.8.1) in X $(= X_0)$.

Below we shall primarily look for sufficient conditions which imply that we have a strong solution (in X). This means that we must take $x_s \in X$ and $f \in L^1_{\text{loc}}([s, \infty); X_{-1})$, and that (3.8.1) should be interpreted as an equation in X_{-1} (valid for almost all $t \geq s$). Thus, it should really be written in the form (recall that $A_{|X}$ maps $X = X_0$ into X_{-1})

$$\begin{aligned} \dot{x}(t) &= A_{|X}x(t) + f(t), \qquad t \geq s, \\ x(s) &= x_s. \end{aligned} \tag{3.8.3}$$

The integration in (3.8.2) should be carried out in X_{-1}, so that this identity should really be written in the form

$$x(t) = \mathfrak{A}^{t-s}x_s + \int_s^t \mathfrak{A}^{t-v}_{|X_{-1}} f(v)\,dv. \tag{3.8.4}$$

In order for (3.8.1) and (3.8.2) (or more precisely, (3.8.3) and (3.8.4)) to be equivalent we need some sort of smoothness assumptions on f: it should be either smooth in time or smooth in the state space (see parts (iv) and (v) below).

Theorem 3.8.2 *Let $s \in \mathbb{R}$, $x_s \in X$, and $f \in L^1_{\text{loc}}([s,\infty); X_{-1})$.*

(i) *The function x given by* (3.8.4) *is a strong solution of* (3.8.1) *in X_{-1} (hence in X_n for every $n \le -1$).*
(ii) *Equation* (3.8.1) *has at most one strong solution x in X, namely the function x given by* (3.8.4).
(iii) *The function x given by* (3.8.4) *is a strong solution of* (3.8.1) *in X_n for some $n \ge 0$ if and only if $x \in C([s,\infty); X_n)$ and $f \in L^1_{\text{loc}}([s,\infty); X_{n-1})$. (In particular, this implies that $x_s \in X_n$.)*
(iv) *If $f \in L^1_{\text{loc}}([s,\infty); X)$ then the function x given by* (3.8.2) *is a strong solution of* (3.8.1) *in X.*
(v) *If $f \in W^{1,1}_{\text{loc}}([s,\infty); X_{-1})$ then the function x given by* (3.8.4) *is a strong solution of* (3.8.1) *in X, $x \in C^1([s,\infty); X_{-1})$, and $z = \dot{x}$ is a strong solution of the equation*

$$\begin{aligned} \dot{z}(t) &= Az(t) + \dot{f}(t), \qquad t \ge s, \\ z(s) &= Ax_s + f(s) \end{aligned} \tag{3.8.5}$$

in X_{-1}. In particular, $\dot{x}(t) = A_{|X}x(t) + f(t)$ in X_{-1} for all $t \ge s$ (and not just almost all $t \ge s$).
(vi) *If $f = \pi_{[\alpha,\beta)} f_1$, where $s \le \alpha < \beta \le \infty$ and $f_1 \in W^{1,1}_{\text{loc}}([s,\infty); X_{-1})$ then the function x given by* (3.8.4) *is a strong solution of* (3.8.1) *in X.*
(vii) *If f is any finite linear combination of functions of the type presented in (iv)–(vi), then the function x given by* (3.8.4) *is a strong solution of* (3.8.1) *in X.*

Proof (i) Define x by (3.8.4). The term $t \mapsto \mathfrak{A}^{t-s}x_s$ belongs to $C([s,\infty); X) \cap C^1([s,\infty); X_{-1}) \cap C^2([s,\infty); X_{-2})$ and it is a strong solution of (3.8.1) with $f = 0$ in X. Subtracting this term from x we reduce the problem to the case where $x_s = 0$. (The same reduction is valid in the proofs of (ii)–(vii), too.)

That $x \in C([s,\infty); X)$ follows from Proposition 2.3.1 with X replaced by X_{-1}, $\mathfrak{C} = 0$, and $\mathfrak{D} = 0$.

Suppose for the moment that $f \in C([s,\infty); X_{-1})$. Since $A_{|X_{-1}} \in \mathcal{B}(X_{-1}; X_{-2})$, we can then easily justify the following computation for $t \ge 0$

(the double integrals are computed in X_{-1}, and the other integrals in X_{-2} or X_{-1}; see Theorem 3.2.1(ii) for the last step):

$$\begin{aligned}
\int_s^t A_{|X_{-1}} x(v)\,dv &= A_{|X_{-1}} \int_s^t \int_s^v \mathfrak{A}^{v-w}_{|X_{-1}} f(w)\,dw\,dv \\
&= A_{|X_{-1}} \int_s^t \int_w^t \mathfrak{A}^{v-w}_{|X_{-1}} f(w)\,dv\,dw \\
&= A_{|X_{-1}} \int_s^t \int_0^{t-w} \mathfrak{A}^{v}_{|X_{-1}} f(w)\,dv\,dw \\
&= \int_s^t A_{|X_{-1}} \int_0^{t-w} \mathfrak{A}^{v}_{|X_{-1}} f(w)\,dv\,dw \\
&= \int_s^t (\mathfrak{A}^{t-w}_{|X_{-1}} - 1) f(w)\,dw.
\end{aligned}$$

As the set of continuous functions is dense in L^1, the same identity must then be true for all $f \in L^1_{\mathrm{loc}}([s,\infty); X_{-1})$. Rewriting this in terms of the function x in (3.8.4) (with $x_s = 0$) we get

$$x(t) = \int_s^t (A_{|X_{-1}} x(v) + f(v))\,dv.$$

Thus, $x \in W^{1,1}_{\mathrm{loc}}([s,\infty); X_{-2})$ and $\dot{x}(t) = A_{|X_{-1}} x(t) + f(t)$ in X_{-2} for almost all $t \ge s$. Clearly $x(s) = 0$. This implies that x is a strong solution of (3.8.1) in X_{-1} with $x_s = 0$.

(ii) If z is an arbitrary function in $C^1([s,\infty); X)$, then it is easy to show (using Theorem 3.2.1(ii)) that, for each $t > s$, the function $v \mapsto \mathfrak{A}^{t-v} z(v)$ is continuously differentiable in X_{-1}, with derivative $\mathfrak{A}^{t-v}(\dot{z}(v) - A_{|X} z(v))$. Integrating this identity (in X_{n-1}) we get

$$z(t) = \mathfrak{A}^{t-s} z(s) + \int_s^t \mathfrak{A}^{t-v}(\dot{z}(v) - A_{|X} z(v))\,dv.$$

Since $C^1([s,\infty); X)$ is dense in $W^{1,1}_{\mathrm{loc}}([s,\infty); X_{-1}) \cap C([s,\infty); X)$, and since both sides of the above identity depend continuously in X_{-1} on z in the norm of $W^{1,1}_{\mathrm{loc}}([s,\infty); X_{-1}) \cap C([s,\infty); X)$, the same identity must hold for every $z \in W^{1,1}_{\mathrm{loc}}([s,\infty); X_{-1}) \cap C([s,\infty); X)$. In particular, it is true whenever z is a strong solution of (3.8.1) in X_n, in which case we furthermore have $\dot{z}(v) - A_{|X} z(v) = f(v)$ for almost all $v \ge s$. This means that z coincides with the function x given by (3.8.4).

(iii) The necessity of the condition $x \in C([s,\infty); X_n)$ is part of the definition of a strong solution in X_n. The necessity of the condition $f \in L^1_{\mathrm{loc}}([s,\infty); X_{n-1})$ follows from the fact that $f = \dot{x} - A_{|X_n} x$, where $\dot{x} \in L^1_{\mathrm{loc}}([s,\infty); X_{n-1})$ and $A_{|X_n} x \in C([s,\infty); X_{n-1})$.

Conversely, suppose that $x \in C([s,\infty); X_n)$ and that $f \in L^1_{\mathrm{loc}}([s,\infty); X_{n-1})$. By (i), we have $\dot{x} = A_{|X} x + f = A_{|X_n} x + f$ in X_{-2}; in particular, the derivative $\dot{x}$ is computed in X_{-2}. However, the right-hand side of this identity belongs to $L^1_{\mathrm{loc}}([s,\infty);X_{n-1})$, so its integral (which is x) belongs to $W^{1,1}_{\mathrm{loc}}([s,\infty);X_{n-1})$, and the same identity is true a.e. in X_{n-1}. Thus, x is a strong solution in X_n.

(iv)–(vii) In the remainder of the proof we take $x_s = 0$, without loss of generality (see the proof of (i)).

(iv) The proof of (iv) is identical to the proof of (i), with X_{-1} replaced by X.

(v) Since $\mathfrak{A}_{|X_{-1}} \in C^1(\overline{\mathbb{R}}^+;\mathcal{B}(X_{-1};X_{-2}))$ and $f \in C([s,\infty);X_{-1})$, we can differentiate under the integral sign to get (as an identity in X_{-2})

$$\dot{x}(t) = f(t) + A_{|X_{-1}} \int_s^t \mathfrak{A}^{t-v}_{|X_{-1}} f(v)\,dv, \qquad t \geq s.$$

Integrate by parts (or alternatively, write $f(v) = f(s) + \int_s^v \dot{f}(w)\,dw$ and use Fubini's theorem) to show that we can write this (still as an identity in X_{-2}) as

$$\dot{x}(t) = \mathfrak{A}^{t-s}_{|X_{-1}} f(s) + \int_s^t \mathfrak{A}^{t-v}_{|X_{-1}} \dot{f}(v)\,dv, \qquad t \geq s.$$

By (i), the right-hand side of this expression is the strong solution of (3.8.5) in X_{-1}, so from the definition of a strong solution we conclude that $\dot{x} \in C([s,\infty);X_{-1}) \cap W^{1,1}_{\mathrm{loc}}([s,\infty);X_{-2})$. The continuity of $\dot{x}$ in X_{-1} implies that, although we originally computed the derivative $\dot{x}$ of x as a limit in the norm of X_{-2}, this limit actually exists in the norm of X_{-1} (i.e., x is differentiable in the stronger norm of X_{-1}), and that $x \in C^1([s,\infty);X_{-1}) \cap W^{2,1}_{\mathrm{loc}}([s,\infty);X_{-2})$.

We proceed to show that $x \in C([s,\infty);X)$ and that $\dot{x}(t) = A_{|X}x(t) + f(t)$ in X_{-1} for all $t \geq s$. We know from (i) that $\dot{x}(t) = A_{|X_{-1}}x(t) + f(t)$ in X_{-2} for almost all $t \geq s$, and, since both sides are continuous in X_{-2}, we must actually have equality for *all* $t \geq s$. Choose some α in the resolvent set of $A_{|X_{-1}}$ (or equivalently, from the resolvent set of A) and subtract $\alpha x(t)$ from both sides of this identity to get (as an identity in X_{-2})

$$\alpha x(t) - \dot{x}(t) = (\alpha - A_{|X_{-1}})x(t) - f(t),$$

that is

$$x(t) = (\alpha - A_{|X_{-1}})^{-1}\bigl(f(t) + \alpha x(t) - \dot{x}(t)\bigr).$$

As $(\alpha - A_{|X_{-1}})^{-1} \in \mathcal{B}(X_{-1};X)$, and x, $\dot{x}$ and f belong to $C([s,\infty);X_{-1})$, the latter equation shows that $x \in C([s,\infty);X)$, and that $\dot{x}(t) = A_{|X}x(t) + f(t)$ in X_{-1} for all $t \geq s$. Thus, x is a strong solution of (3.8.1) in X.

(vi) Since $\pi_{[\alpha,\beta)} f_1 = \pi_{[\alpha,\infty)} f_1 - \pi_{[\beta,\infty)} f_1$, we can without loss of generality suppose that $\beta = \infty$ (cf. (vii)).

Clearly, the restriction of x to $[s, \alpha)$ is the zero function, so in order to prove the theorem it suffices to show that $\pi_{[\alpha,\infty)}x \in C([\alpha, \infty); X) \cap W^{1,1}_{\mathrm{loc}}([\alpha; \infty); X_{-1})$ and that $x(\alpha) = 0$, because this implies that $x \in C([\alpha, \infty); X) \cap W^{1,1}_{\mathrm{loc}}([\alpha; \infty); X_{-1})$. But this follows from (v) with s replaced by α (and $x_\alpha = 0$).[6]

(vii) This follows from the linearity of (3.8.1) and (3.8.4). □

Sometimes we need more smoothness of a solution than we get from Theorem 3.8.2.

Theorem 3.8.3 *Let $s \in \mathbb{R}$, $x_s \in X$, $f \in W^{2,1}_{\mathrm{loc}}([s, \infty); X_{-1})$, and $A_{|X}x_s + f(s) \in X$. Then the strong solution x of* (3.8.1) *satisfies $x \in C^2([s, \infty); X_{-1}) \cap C^1([s, \infty); X)$, $\dot{x} = z$ is the strong solution of* (3.8.5) *in X, and $\ddot{x} = y$ is the strong solution of*

$$\begin{aligned} \dot{y}(t) &= Ay(t) + \ddot{f}(t), \qquad t \geq s, \\ y(s) &= A\dot{x}(s) + \dot{f}(s) \end{aligned} \tag{3.8.6}$$

in X_{-1}. In particular, $\dot{x} = A_{|X}x + f \in C^1([s, \infty); X_{-1}) \cap C([s, \infty); X)$ and the identities $\dot{x}(t) = A_{|X_{-1}}x(t) + f(t)$ and $\ddot{x}(t) = A_{|X_{-1}}\dot{x}(t) + \dot{f}(t)$ hold X_{-1} for all $t \geq s$.

Proof By Theorem 3.8.2(v), $x \in C([s, \infty); X) \cap C^1([s, \infty); X_{-1})$, and, of course,

$$\dot{x}(t) = A_{|X}x(t) + f(t), \qquad t \geq s.$$

Arguing as in the proof of Theorem 3.8.2(v) (using the density of C^2 in $W^{2,1}$) we can use the extra differentiability assumption on u to show that $x \in C^2([s, \infty); X_{-2})$, and that

$$\ddot{x}(t) = A_{|X_{-1}}\dot{x}(t) + \dot{f}(t), \qquad t \geq s.$$

Let $z = \dot{x}$. Then $z(s) = A_{|X}x_s + f(s) \in X$, and z is the strong solution of the equation (3.8.5) in X_{-1}. However, by Theorem 3.8.2(v), this solution is actually a strong solution in X, i.e., $z \in C([s, \infty); X)$, and it has some additional smoothness, namely $z \in C^1([s, \infty); X_{-1})$. Since $z = \dot{x}$, this means that $x \in C^2([s, \infty); X_{-1}) \cap C^1([s, \infty); X)$, as claimed. □

Above we have only looked at the *local smoothness* of a strong solution of (3.8.1) (or more generally, of the function x defined by the variation of constants formula (3.8.2)). There are also some corresponding *global growth bounds* on the solution and its derivatives.

[6] Although x is continuous, there will be a jump discontinuity in $\dot{x}$ at the cutoff point. Thus, we will not in general have $x \in C^1([s, \infty); X_{-1})$ in this case, but we will still have $x \in Reg^1_{\mathrm{loc}}([s, \infty); X_{-1})$ and $\dot{x} - f \in C([s, \infty); X_{-1})$.

Theorem 3.8.4 *Let A be the generator of a C_0 semigroup $\mathfrak{A}$ with growth bound $\omega_{\mathfrak{A}}$. Let $\omega > \omega_{\mathfrak{A}}$, and let $1 \le p < \infty$. Under the following additional assumptions on the function f in Theorems 3.8.2 and 3.8.3 we get the following additional conclusions about the strong solution x of* (3.8.1) *(and all the listed derivatives exist in the given sense):*

(i) *If $f \in L^p_\omega([s,\infty);X)$, then*

$$\begin{aligned} x &\in BC_{0,\omega}([s,\infty);X) \cap L^p_\omega([s,\infty);X), \\ \dot{x} &\in L^p_\omega([s,\infty);X_{-1}). \end{aligned}$$

(ii) *If $f \in W^{1,p}([s,\infty);X_{-1})$, then*

$$\begin{aligned} x &\in BC_{0,\omega}([s,\infty);X) \cap L^p_\omega([s,\infty);X), \\ \dot{x} &\in BC_{0,\omega}([s,\infty);X_{-1}) \cap L^p_\omega([s,\infty);X_{-1}), \\ \ddot{x} &\in L^p_\omega([s,\infty);X_{-2}). \end{aligned}$$

(iii) *If $f \in W^{2,p}([s,\infty);X_{-1})$, then*

$$\begin{aligned} x &\in BC_{0,\omega}([s,\infty);X) \cap L^p_\omega([s,\infty);X), \\ \dot{x} &\in BC_{0,\omega}([s,\infty);X) \cap L^p_\omega([s,\infty);X), \\ \ddot{x} &\in BC_{0,\omega}([s,\infty);X_{-1}) \cap L^p_\omega([s,\infty);X_{-1}), \\ \dddot{x} &\in L^p_\omega([s,\infty);X_{-2}). \end{aligned}$$

Proof (i) Let Σ be the L^p-well-posed linear system on (X, X, X) described in Proposition 2.3.1 with $B = 1$, $C = 1$, and $D = 0$. Then, according to Theorem 3.8.2(iv), the strong solution x of (3.8.1) can be interpreted as the state trajectory of this system, and furthermore, its output y satisfies $y(t) = x(t)$ for all $t \ge s$. By Theorem 2.5.4, $x \in BC_{0,\omega}([s,\infty);X)$ and $x = y \in L^p_\omega([s,\infty);X)$. This implies that $\dot{x} = A_{|X}x + f \in L^p_\omega([s,\infty);X_{-1})$.

(ii) We again consider the same system as above, but this time on (X_{-1}, X_{-1}, X_{-1}). As above we first conclude that $x \in BC_{0,\omega}([s,\infty);X_{-1}) \cap L^p_\omega([s,\infty);X_{-1})$. We can also apply the same argument with x replaced by $\dot{x}$ (recall that, by Theorem 3.8.2(v), $\dot{x}$ is the strong solution of (3.8.5) in X_{-1}) to get $\dot{x} \in BC_{0,\omega}([s,\infty);X_{-1}) \cap L^p_\omega([s,\infty);X_{-1})$ and $\ddot{x} = A_{|X_{-1}}\dot{x} + \dot{f} \in L^p_\omega([s,\infty);X_{-2})$. Finally, we choose some $\alpha \in \rho(A) = \rho(A_{|X})$ and write the equation $\dot{x} = A_{|X}x + f$ in the form $(\alpha - A_{|X})^{-1}(\alpha x - \dot{x} + f)$ to conclude that $\dot{x} \in BC_{0,\omega}([s,\infty);X) \cap L^p_\omega([s,\infty);X)$.

(iii) Apply (ii) both to the function x itself and to the function $\dot{x}$. □

Another instance where we need a global growth bound on the solution, this time on $\mathbb{R}^-$, is when we want to study the existence and uniqueness of strong solutions of the equation $\dot{x}(t) = Ax(t) + f(t)$ on all of $\mathbb{R}$.

Definition 3.8.5 Let $n = 0, \pm 1, \pm 2, \dots$, and $f \in L^1_{\text{loc}}(\mathbb{R}; X_{n-1})$. A function x is a *strong solution of the equation*

$$\dot{x}(t) = Ax(t) + f(t), \qquad t \in \mathbb{R}, \tag{3.8.7}$$

in X_n (on all of $\mathbb{R}$) if $x \in C(\mathbb{R}; X_n) \cap W^{1,1}_{\text{loc}}(\mathbb{R}; X_{n-1})$, and $\dot{x}(t) = A_{|X_n} x(t) + f(t)$ in X_{n-1} for almost all $t \in \mathbb{R}$. By a *strong solution of (3.8.7)* (without any reference to a space X_n) we mean a strong solution of (3.8.7) in X ($= X_0$).

Without any further conditions we cannot expect a strong solution of (3.8.7) to be unique. For example, if A generates a C_0 group on X, then for every $x_0 \in X$, the function $x(t) = \mathfrak{A}^t x_0$, $t \in \mathbb{R}$, is a strong solution of (3.8.7). We can rule out this case by, e.g., imposing a growth restriction on x at $-\infty$.

Lemma 3.8.6 *Let $\omega \in \mathbb{R}$, and suppose that the semigroup $\mathfrak{A}$ generated by A is ω-bounded (see Definition 2.5.6). Then, for each $f \in L^1_{\text{loc}}(\mathbb{R}; X_{-1})$, the equation* (3.8.7) *can have at most one strong solution x satisfying* $\lim_{t\to-\infty} e^{-\omega t} x(t) = 0$.

If such a solution exists, then we refer to it as the *strong solution of* (3.8.7) *which vanishes at* $-\infty$.

Proof The difference of two strong solutions of (3.8.7) is a strong solution of the equation $\dot{x}(t) = Ax(t)$ on $\mathbb{R}$, so it suffices to show that the only strong solution of (3.8.7) which satisfies $\lim_{t\to-\infty} e^{-\omega t} x(t) = 0$ is the zero solution. Since it is a strong solution on $\mathbb{R}$, it is also a strong solution on $[s, \infty)$ with initial state $x(s)$ for every $s \in \mathbb{R}$, hence by Theorem 3.8.2(iv), $x(t) = \mathfrak{A}^{t-s} x(s)$ for every $t \ge s$. By the ω-boundedness of $\mathfrak{A}$, there is a constant M such that $|x(t)| \le M e^{\omega(t-s)} |x(s)|$, or equivalently, $e^{-\omega t}|x(t)| \le M e^{-\omega s)} |x(s)|$. Let $s \to -\infty$ to conclude that $x(t) = 0$ for all $t \in \mathbb{R}$. □

Theorem 3.8.7 *Let A be the generator of a C_0 semigroup $\mathfrak{A}$ with growth bound $\omega_{\mathfrak{A}}$. Let $\omega > \omega_{\mathfrak{A}}$, and let $1 \le p < \infty$. In all the cases (i)–(iii) listed below the equation* (3.8.7) *has a unique strong solution x satisfying* $\lim_{t\to-\infty} e^{-\omega t} x(t) = 0$, *namely the function*

$$x(t) = \int_{-\infty}^{t} \mathfrak{A}^{t-v} f(v)\, dv, \tag{3.8.8}$$

and this solution has the additional properties listed below.

(i) $f \in L^p_{\omega,\text{loc}}(\mathbb{R}; X)$. *In this case*

$$x \in BC_{0,\omega,\text{loc}}(\mathbb{R}; X) \cap L^p_{\omega,\text{loc}}(\mathbb{R}; X),$$
$$\dot{x} \in L^p_{\omega,\text{loc}}(\mathbb{R}; X_{-1}).$$

(ii) $f \in W^{1,p}_{\omega,\mathrm{loc}}(\mathbb{R}; X_{-1})$. *In this case*

$$x \in BC_{0,\omega,\mathrm{loc}}(\mathbb{R}; X) \cap L^p_{\omega,\mathrm{loc}}(\mathbb{R}; X),$$
$$\dot{x} \in BC_{0,\omega,\mathrm{loc}}(\mathbb{R}; X_{-1}) \cap L^p_{\omega,\mathrm{loc}}(\mathbb{R}; X_{-1}),$$
$$\ddot{x} \in L^p_{\omega,\mathrm{loc}}(\mathbb{R}; X_{-2}).$$

(iii) $f \in W^{2,p}_{\omega,\mathrm{loc}}(\mathbb{R}; X_{-1})$. *In this case*

$$x \in BC_{0,\omega,\mathrm{loc}}(\mathbb{R}; X) \cap L^p_{\omega,\mathrm{loc}}(\mathbb{R}; X),$$
$$\dot{x} \in BC_{0,\omega,\mathrm{loc}}(\mathbb{R}; X) \cap L^p_{\omega,\mathrm{loc}}(\mathbb{R}; X),$$
$$\ddot{x} \in BC_{0,\omega,\mathrm{loc}}(\mathbb{R}; X_{-1}) \cap L^p_{\omega,\mathrm{loc}}(\mathbb{R}; X_{-1}),$$
$$\dddot{x} \in L^p_{\omega,\mathrm{loc}}(\mathbb{R}; X_{-2}).$$

Proof (i) (This proof is very similar to the proof of Theorem 3.8.4.) Let Σ be the L^p-well-posed linear system on (X, X, X) described in Proposition 2.3.1 with $B = 1$, $C = 1$, and $D = 0$. It follows from Theorems 2.5.7 and 3.8.2(iv) and Example 2.5.10 that the function x defined by (3.8.8) is a strong solution of (3.8.7) satisfying $\lim_{t\to-\infty} \mathrm{e}^{-\omega t} x(t) = 0$, hence *the* strong solution satisfying this growth bound. Moreover, by Theorem 2.5.7 and Example 2.5.10, $x \in BC_{0,\omega,\mathrm{loc}}(\mathbb{R}; X)$ and $x = y \in L^p_{\omega,\mathrm{loc}}(\mathbb{R}; X)$. Since $\dot{x} = A_{|X}x + f$, this implies that $\dot{x} \in L^p_{\omega,\mathrm{loc}}(\mathbb{R}; X_{-1})$.

(ii)–(iii) The proofs of (ii)–(iii) are analogous to the proofs of parts (ii)–(iii) of Theorem 3.8.4, and we leave them to the reader. □

Remark 3.8.8 Theorem 3.8.4 remains valid if we replace L^p_ω by $L^\infty_{0,\omega}$ or $Reg_{0,\omega}$ throughout. Theorem 3.8.7 remains valid if we delete the subindex 'loc', or if we replace $L^p_{\omega,\mathrm{loc}}$ by $L^\infty_{0,\omega,\mathrm{loc}}$ or $Reg_{0,\omega,\mathrm{loc}}$ throughout, or if we do both of these operations at the same time. The proofs remain the same.

3.9 Symbolic calculus and fractional powers

In this section we shall develop a basic symbolic calculus for the generators of C_0 semigroups.[7] We shall here consider only two classes of mappings of generators. The first class is the one where the generator A is mapped conformally into $f(A)$ where f is a complex-valued function which is analytic at the spectrum of A (including the point at infinity if A is unbounded). The other class of mapping is the one which gives us the fractional powers of $\gamma - A$ where $\gamma \in \mathbb{C}^+_{\omega_{\mathfrak{A}}}$. In

[7] With some trivial modifications this functional calculus can be applied to any closed operator with a nonempty resolvent set.

Section 3.10 we shall use a similar calculus to construct the semigroup generated by A in a special (analytic) case.

Let us begin with the simplest case where A is bounded. Let Γ be a piecewise continuously differentiable Jordan curve which encircles $\sigma(A)$ counterclockwise, i.e., the index of $\sigma(A)$ with respect to Γ is one. If f is analytic on Γ and inside Γ, then we define $f(A)$ by

$$f(A) = \frac{1}{2\pi j} \oint_\Gamma (\lambda - A)^{-1} f(\lambda)\, d\lambda. \tag{3.9.1}$$

This integral converges in the operator norm topology, e.g., as a Riemann integral (but it can, of course, also be interpreted in the strong sense, where we apply each side to a vector $x \in X$). The definition of $f(A)$ given is standard, and it is found in most books on functional analysis (see, e.g., Rudin 1973, p. 243).

Let us check that the definition (3.9.1) of $f(A)$ coincides with the standard definition in the case where $f(z) = \sum_{k=0}^n a_k z^k$ is a polynomial. In this case we expect to have $f(A) = \sum_{k=0}^n a_k A^k$. By the linearity of the integral in (3.9.1), to prove this it suffices to verify the special case where $f(z) = z^n$ for some $n = 0, 1, 2, \ldots$. In this case we get

$$\begin{aligned}
\frac{1}{2\pi j} \oint_\Gamma \lambda^n (\lambda - A)^{-1}\, d\lambda &= \frac{1}{2\pi j} \oint_\Gamma (\lambda - A + A)^n (\lambda - A)^{-1}\, d\lambda \\
&= \sum_{k=0}^n \binom{n}{k} A^k \frac{1}{2\pi j} \oint_\Gamma (\lambda - A)^{n-k-1}\, d\lambda \\
&= A^n,
\end{aligned}$$

where the last step uses Lemma 3.9.2 below. Thus (3.9.1) is consistent with the standard definition of $f(A)$ in terms of powers of A when f is a polynomial.

If A is unbounded, then (3.9.1) must be slightly modified. In the following discussion, we denote the compactified complex plane $\mathbb{C} \cup \{\infty\}$ by $\overline{\mathbb{C}}$, and we let $\overline{\sigma}(A)$ be the (extended) spectrum of A in $\overline{\mathbb{C}}$, i.e., $\overline{\sigma}(A) = \sigma(A)$ if A is bounded, and $\overline{\sigma}(A) = \sigma(A) \cup \{\infty\}$ if A is unbounded.

Let A be the generator of a C_0 semigroup $\mathfrak{A}$ with growth bound $\omega_{\mathfrak{A}}$. Then we know from Theorem 3.2.9(ii) that $\overline{\sigma}(A) \subset \overline{\mathbb{C}}^-_{\omega_{\mathfrak{A}}} \cup \{\infty\}$ (where we can remove the point at infinity if A is bounded). Let f be a complex-valued function which is analytic on $\overline{\mathbb{C}}^-_{\omega_{\mathfrak{A}}} \cup \{\infty\}$ (f need not be analytic at infinity if A is bounded). We denote the set of points $\lambda \in \overline{\mathbb{C}}$ in which f is not analytic by $\overline{\sigma}(f)$ (this includes the point at infinity if f is not analytic there).

If A and f satisfy the conditions listed in the preceding paragraph, then it is possible to choose a piecewise continuously differentiable Jordan curve Γ in the complex plane which separates $\overline{\sigma}(A)$ from $\overline{\sigma}(f)$, with $\overline{\sigma}(A)$ 'to the left' of Γ and $\overline{\sigma}(f)$ 'to the right' of Γ. If A is bounded, then we can choose

Γ to be a curve encircling $\sigma(A)$ counter-clockwise with $\overline{\sigma}(f)$ on the outside, and if f is analytic at infinity, then we can choose Γ to be a curve encircling $\sigma(f)$ clockwise with $\overline{\sigma}(A)$ on the outside. If both of these conditions hold, then both choices are possible. Unfortunately, they do not produce exactly the same result, so before we try this approach we have to modify (3.9.1) slightly.

Before proceeding further, let us recall two different versions of the *Cauchy formula* for the derivatives of a function.

Lemma 3.9.1 *Let U be a Banach space, and let Γ be a positively oriented piecewise continuously differentiable Jordan curve in $\mathbb{C}$ (i.e., the index of the inside is one).*

(i) *If f is a U-valued function which is analytic on Γ and inside Γ, then, for every λ_0 inside Γ,*

$$\frac{1}{2\pi j}\oint_\Gamma \frac{f(\lambda)}{(\lambda-\lambda_0)^{n+1}}\,d\lambda = \begin{cases} 0, & n<0, \\ 1/(n!)f^{(n)}(\lambda_0), & n\ge 0. \end{cases}$$

(ii) *If instead f is analytic on Γ and outside Γ (including the point at infinity), then for every λ_0 inside Γ,*

$$\frac{1}{2\pi j}\oint_\Gamma (\lambda-\lambda_0)^{n-1} f(\lambda)\,d\lambda = \begin{cases} 0, & n<0, \\ 1/(n!)\,\frac{\mathrm{d}^n}{\mathrm{d}z^n}\, f(\lambda_0+1/z)_{|z=0}, & n\ge 0. \end{cases}$$

Proof (i) In the scalar case this is the standard Cauchy formula for the derivative found in all textbooks (if $n\in\mathbb{Z}_-$ then the integrand is analytic inside Γ, so the result is zero). The operator-valued case can be reduced to the scalar-valued case: if f is $\mathcal{B}(X;Y)$-valued, then we choose arbitrary $x\in X$ and $y^*\in Y^*$ and apply the scalar case to y^*fx.

(ii) We make a change of integration variable from λ to $z=1/(\lambda-\lambda_0)$, $(\lambda-\lambda_0)^{-1}\,d\lambda=-z^{-1}\,dz$. If Γ' is the image of Γ under the mapping $\lambda\mapsto 1/(\lambda-\lambda_0)$, then Γ' is negatively oriented (the outside of Γ is mapped onto the inside of Γ'), and it encircles the origin. Part (i) gives (if we take the negative orientation of Γ' into account)

$$\begin{aligned}\frac{1}{2\pi j}\oint_\Gamma \frac{f(\lambda)}{(\lambda-\lambda_0)^{-n+1}}\,d\lambda &= -\frac{1}{2\pi j}\oint_{\Gamma'}\frac{f(\lambda_0+1/z)}{z^{n+1}}\,dz \\ &= \begin{cases} 0, & n<0, \\ 1/(n!)\,\frac{\mathrm{d}^n}{\mathrm{d}z^n}\, f(\lambda_0+1/z)_{|z=0}, & n\ge 0. \end{cases}\end{aligned}$$

□

Lemma 3.9.2 *Let $A \in \mathcal{B}(X)$, and let Γ be a positively oriented piecewise continuously differentiable Jordan curve which encircles $\rho(A)$. Then,*

$$\frac{1}{2\pi j}\oint_\Gamma (\lambda - A)^{-k}\,d\lambda = \begin{cases} 1, & k = 1, \\ 0, & k \in \mathbb{Z},\ k \neq 1. \end{cases}$$

Proof If $k \in \mathbb{Z}_-$, then the integrand is analytic inside Γ, and the result is zero. If $k \in \mathbb{Z}_+$, then the integrand is analytic outside Γ, including the point at infinity, and the result follows from Lemma 3.9.1(ii) with $n = 1$ and $f(\lambda) = (\lambda - A)^{-k}$. □

By Lemma 3.9.2, if A is bounded and if f is analytic at infinity, then (3.9.1) is equivalent to

$$f(A) = f(\infty) + \frac{1}{2\pi j}\oint_\Gamma (\lambda - A)^{-1}(f(\lambda) - f(\infty))\,d\lambda. \tag{3.9.2}$$

The function inside the integral has a second order zero at infinity, so if we replace Γ by a curve encircling both $\sigma(A)$ and $\sigma(f)$, then it follows from Lemma 3.9.1(ii) (with $n = 1$ and $f(\lambda)$ replaced by $(\lambda - A)^{-1}(f(\lambda) - f(\infty))$ that the resulting integral is zero. Thus, in (2.9.2) we may replace the positively oriented curve Γ which encircles $\sigma(A)$ with $\sigma(f)$ on the outside by a negatively oriented curve which encircles $\sigma(f)$ with $\sigma(A)$ on the outside. If we do so, then $\oint_\Gamma (\lambda - A)^{-1}\,d\lambda = 0$, and (3.9.2) can alternatively be written in the form

$$f(A) = f(\infty) + \frac{1}{2\pi j}\oint_\Gamma (\lambda - A)^{-1} f(\lambda)\,d\lambda. \tag{3.9.3}$$

Here it does not matter if A is bounded or unbounded, as long as Γ and the inside of Γ belong to $\rho(A)$, and f is analytic on Γ and the outside of Γ, including the point at infinity.

From (3.9.3) we immediately conclude the following:

Lemma 3.9.3 *Let A be the generator of a C_0 semigroup $\mathfrak{A}$ with growth rate $\omega_{\mathfrak{A}}$, let f be analytic on $\overline{\mathbb{C}^-_{\omega_{\mathfrak{A}}}} \cup \{\infty\}$, and define $f(A)$ as explained above. Then $f(A) - f(\infty) \in \mathcal{B}(X; X_1)$.*

Proof This follows from (3.9.3): for an arbitrary $\alpha \in \rho(A)$ we have

$$\begin{aligned} f(A) - f(\infty) &= (\alpha - A)^{-1}\frac{1}{2\pi j}\oint_\Gamma (\alpha - A)(\lambda - A)^{-1} f(\lambda)\,d\lambda \\ &= (\alpha - A)^{-1}\frac{1}{2\pi j}\oint_\Gamma \big[(\alpha - \lambda)(\lambda - A)^{-1} - 1\big] f(\lambda)\,d\lambda, \end{aligned}$$

where the integral defines an operator in $\mathcal{B}(X)$. □

As we already mentioned above, the definition of $f(A)$ given in (3.9.1) in the case where A is bounded is standard, but the definition of $f(A)$ in (3.9.3) with unbounded A is less common. However, (3.9.3) can be reduced to (3.9.1) by, e.g., a linear fractional transformation. For example, we can take some $\alpha \in \mathbb{C}^+_{\alpha_\mathfrak{A}} \cap \mathbb{C}^+$, and define $\varphi(\lambda) = 1/(\alpha - \lambda)$. The inverse transformation is $z \mapsto \varphi^{-1}(z) = \alpha - 1/z$. Note that $\alpha \in \rho(A)$, that α is mapped into ∞, and that ∞ is mapped into zero. Let Γ' be the image of Γ under this mapping. If Γ is negatively oriented, then the orientation of Γ' is positive and it encircles the origin (assuming that α lies inside Γ). By changing the integration variable in (3.9.3) we get (note that $d\lambda = z^{-2}\,dz$ and that $1/(2\pi j)\oint_{\Gamma'} z^{-1} f(\alpha - 1/z)\,dz = f(\infty)$)

$$\begin{aligned} f(A) &= f(\infty) + \frac{1}{2\pi j}\oint_{\Gamma'} (\alpha - 1/z - A)^{-1} z^{-2} f(\alpha - 1/z)\,dz \\ &= \frac{1}{2\pi j}\oint_{\Gamma'} \left[1 + (\alpha z - 1 - zA)^{-1}\right] z^{-1} f(\alpha - 1/z)\,dz \\ &= \frac{1}{2\pi j}\oint_{\Gamma'} (\alpha - A)\big(\alpha z - 1 - zA\big)^{-1} f(\alpha - 1/z)\,dz. \end{aligned}$$

Let $B_\alpha = (\alpha - A)^{-1}$ (thus, formally $B_\alpha = \varphi(A)$). Then $B_\alpha \in \mathcal{B}(X)$, and a short algebraic computation shows that

$$(z - B_\alpha)^{-1} = (\alpha - A)\big(\alpha z - 1 - zA\big)^{-1}.$$

Substituting this into the expression for $f(A)$ given above we get

$$f(A) = \frac{1}{2\pi j}\oint_{\Gamma'} (z - B_\alpha)^{-1} f(\alpha - 1/z)\,dz, \qquad B_\alpha = (\alpha - A)^{-1}. \tag{3.9.4}$$

Here Γ' is a positively oriented piecewise continuously differentiable Jordan curve which encircles $\sigma(B_\alpha)$, and the function $z \mapsto f(\alpha - 1/z)$ is analytic on Γ' and inside Γ'. Since we have obtained this from (3.9.3) (which does not depend on α) through a change of integration variable, the right-hand side of (3.9.4) does not depend on α, and it can be used as an alternative definition of $f(A)$.

If f is a rational function whose poles are located in $\mathbb{C}^+_{\omega_\mathfrak{A}}$ and which is analytic at infinity, then there is still another way of defining $f(A)$. Each such function can be written as a constant plus a linear combination of terms of the type $(\alpha_i - \lambda)^{-k_i}$, where each $\alpha_i \in \mathbb{C}^+_{\omega_\mathfrak{A}}$ and $k_i > 0$. It is then natural to define $f(A)$ to be the corresponding linear combination of $(\alpha_i - A)^{-k_i}$. Let us check that this definition is consistent with the one given earlier. To do this it suffices to show that, for all $\alpha \in \mathbb{C}^+_{\omega_\mathfrak{A}}$ and all $k = 1, 2, 3 \ldots,$

$$(\alpha - A)^{-k} = \frac{1}{2\pi j}\oint_{\Gamma} (\lambda - A)^{-1}(\alpha - \lambda)^{-k}\,d\lambda, \tag{3.9.5}$$

where Γ is a negatively oriented piecewise continuously differentiable Jordan curve which encircles α with $\overline{\sigma}(A)$ on the outside. We begin with the case $k = 1$.

Then Lemmas 3.2.8 and 3.9.1 give

$$\begin{aligned}
&\frac{1}{2\pi j}\oint_\Gamma (\lambda - A)^{-1}(\alpha - \lambda)^{-1}\,d\lambda - (\alpha - A)^{-1} \\
&\quad = \frac{1}{2\pi j}\oint_\Gamma [(\lambda - A)^{-1} - (\alpha - A)^{-1}](\alpha - \lambda)^{-1}\,d\lambda \\
&\quad = \frac{1}{2\pi j}\oint_\Gamma (\lambda - A)^{-1}(\alpha - A)^{-1}\,d\lambda \\
&\quad = (\alpha - A)^{-1}\frac{1}{2\pi j}\oint_\Gamma (\lambda - A)^{-1}\,d\lambda = 0.
\end{aligned}$$

The case $k \geq 2$ follows from the case $k = 1$ if we differentiate the special case $k = 1$ of (3.9.5) $k - 1$ times with respect to α.

We shall next look at the related problem of how to define *fractional powers* of $(\gamma - A)$, where A is the generator of a C_0 semigroup $\mathfrak{A}$ and $\gamma > \omega_{\mathfrak{A}}$. This can be done in several different ways, see Pazy (1983). Usually one starts with the *negative* fractional powers of $(\gamma - A)$, and then inverts these to get the positive fractional powers. One method, explained, e.g. in Pazy (1983), is to imitate (3.9.1) with $f(\lambda) = (\gamma - \lambda)^{-\alpha}$, and to let Γ be a path from $\infty e^{-j\epsilon}$ to $\infty e^{j\epsilon}$, where $0 < \epsilon < \pi/2$, passing between $\sigma(A)$ and the interval $[\gamma, \infty)$.[8] Here we shall use a different approach and instead extend the formula for $(\gamma - A)^{-n}$ given in Theorem 3.2.9(i) to fractional values of n.

Definition 3.9.4 Let A be the generator of a C_0 semigroup $\mathfrak{A}$ with growth bound $\omega_{\mathfrak{A}}$. For each $\gamma \in \mathbb{C}^+_{\omega_{\mathfrak{A}}}$ and $\alpha \geq 0$ we define $(\gamma - A)^{-\alpha}$ by

$$\begin{aligned}
(\gamma - A)^0 &= 1, \\
(\gamma - A)^{-\alpha}x &= \frac{1}{\Gamma(\alpha)}\int_0^\infty t^{\alpha-1}e^{-\gamma t}\mathfrak{A}^t x\,dt, \qquad \alpha > 0, \qquad x \in X.
\end{aligned}$$

Lemma 3.9.5 *The operators $(\gamma - A)^{-\alpha}$ introduced in Definition 3.9.4 are bounded linear operators on X, and $\alpha \mapsto (\gamma - A)^{-\alpha}$ is a semigroup, i.e.,*

$$(\gamma - A)^{-(\alpha+\beta)} = (\gamma - A)^{-\alpha}(\gamma - A)^{-\beta}$$

for all $\alpha, \beta > 0$. Moreover, $(\gamma - A)^{-\alpha}$ is injective for all $\alpha \geq 0$.

Proof By assumption, the growth bound of $\mathfrak{A}$ is less than γ, hence the integral used in the definition of $(\gamma - A)^{-\alpha}$ converges absolutely, and it defines an operator in $\mathcal{B}(X)$.

To simplify the notation in our verification of the semigroup property we take $\gamma = 0$ (i.e., we denote $(A - \gamma)$ by A and $e^{-\gamma t}\mathfrak{A}^t$ by $\mathfrak{A}^t$). We take $x \in X$

[8] This method is quite general, and it can be used even in some cases where A is not a generator of a C_0 semigroup.

and make two changes of integration variable to get:

$$
\begin{aligned}
(-A)^{-\alpha}(-A)^{-\beta}x &= \frac{1}{\Gamma(\alpha)\Gamma(\beta)}\int_0^\infty\int_0^\infty s^{\alpha-1}t^{\beta-1}\mathfrak{A}^{s+t}x\,ds\,dt\\
&= \frac{1}{\Gamma(\alpha)\Gamma(\beta)}\int_0^\infty\int_0^\infty s^{\alpha-1}(v-s)^{\beta-1}\mathfrak{A}^{v}x\,dv\,ds\\
&= \frac{1}{\Gamma(\alpha)\Gamma(\beta)}\int_0^\infty\Bigl(\int_0^v s^{\alpha-1}(v-s)^{\beta-1}\,ds\Bigr)\mathfrak{A}^{v}x\,dv\\
&= \frac{1}{\Gamma(\alpha)\Gamma(\beta)}\int_0^1 s^{\alpha-1}(1-s)^{\beta-1}\,ds\int_0^\infty v^{\alpha+\beta-1}\mathfrak{A}^{v}x\,dv\\
&= (-A)^{-(\alpha+\beta)}x;
\end{aligned}
$$

here the last equality follows from Definition 3.9.4 and the fact that the Beta function satisfies (for all $\alpha, \beta > 0$)

$$
B(\alpha,\beta) = \int_0^1 s^{\alpha-1}(1-s)^{\beta-1}\,ds = \frac{\Gamma(\alpha+\beta)}{\Gamma(\alpha)\Gamma(\beta)}.
$$

To show that $(\gamma - A)^{-\alpha}$ is injective we can use the semigroup property in the following way. We choose β so that $\alpha + \beta = n$ is an integer. The operator $(\gamma - A)^{-n} = (\gamma - A)^{-\beta}(\gamma - A)^{-\alpha}$ in injective since $\gamma \in \rho(A)$ (recall that we take $\gamma > \gamma_{\mathfrak{A}}$), hence $(\gamma - A)^{-\alpha}$ in injective. □

The semigroup $\alpha \mapsto (\gamma - A)^{-\alpha}$ is actually a C_0 semigroup (i.e., it is strongly continuous). See Pazy (1983, Corollary 6.5, p. 72).

Since $(\gamma - A)^{-\alpha}$ is injective, it has an inverse defined on its range:

Definition 3.9.6 Let A be the generator of a C_0 semigroup $\mathfrak{A}$ with growth bound $\omega_{\mathfrak{A}}$. For each $\gamma \in \mathbb{C}^+_{\omega_{\mathfrak{A}}}$ and $\alpha \geq 0$ we define $(\gamma - A)^{\alpha}$ to be the inverse of the operator $(\gamma - A)^{-\alpha}$ defined in Definition 3.9.4, with domain $\mathcal{D}((\gamma - A)^{\alpha}) = \mathcal{R}\bigl((\gamma - A)^{-\alpha}\bigr)$.

Lemma 3.9.7 *With the notation of Definitions 3.9.4 and 3.9.6, let $\gamma \in \mathbb{C}^+_{\omega_{\mathfrak{A}}}$ and $\delta \in \mathbb{C}^+_{\omega_{\mathfrak{A}}}$. Then the fractional powers of $(\gamma - A)$ and $(\delta - A)$ have the following properties:*

(i) *$(\gamma - A)^{\alpha} \in \mathcal{B}(X)$ if $\alpha \leq 0$, and $(\gamma - A)^{\alpha}$ is closed if $\alpha > 0$;*
(ii) *$(\delta - A)^{\alpha}(\gamma - A)^{\beta} = (\gamma - A)^{\beta}(\delta - A)^{\alpha}$ if $\alpha \leq 0$ and $\beta \leq 0$;*
(iii) *$\mathcal{D}((\gamma - A)^{\alpha}) \subset \mathcal{D}\bigl((\gamma - A)^{\beta}\bigr)$ if $\alpha \geq \beta$;*
(iv) *$\mathcal{D}((\gamma - A)^{\alpha})$ is dense in X for all $\alpha > 0$ (and equal to X for all $\alpha \leq 0$);*
(v) *$\mathcal{D}((\gamma - A)^{\alpha}) = \mathcal{D}((\delta - A)^{\alpha})$ and $(\delta - A)^{\alpha}(\gamma - A)^{-\alpha} \in \mathcal{B}(X)$ if $\alpha \geq 0$.*

Proof (i) The case $\alpha \leq 0$ is contained in Lemma 3.9.5, and the inverse of a bounded (hence closed) operator is closed.

(ii) Use Fubini's theorem in Definition 3.9.4.

(iii) This is trivial if $\beta \le 0$ or $\alpha = \beta$. Otherwise, by Lemma 3.9.5,

$$(\gamma - A)^{-\alpha} = (\gamma - A)^{-\beta}(\gamma - A)^{-(\alpha-\beta)},$$

hence $\mathcal{R}\big((\gamma - A)^{-\alpha}\big) \subset \mathcal{R}\big((\gamma - A)^{-\beta}\big)$, or equivalently, $\mathcal{D}((\gamma - A)^{\alpha}) \subset \mathcal{D}\big((\gamma - A)^{\beta}\big)$.

(iv) This follows from (iii), since $\mathcal{D}((\gamma - A)^{\alpha})$ contains $\mathcal{D}((\gamma - A)^{n})$ for some positive integer n, and by Theorem 3.2.1(vi), $\mathcal{D}((\gamma - A)^{n}) = \mathcal{D}(A^{n})$ is dense in X.

(v) The boundedness of the operator $(\delta - A)^{\alpha}(\gamma - A)^{-\alpha}$ follows from the closed graph theorem as soon as we have shown that $\mathcal{D}((\gamma - A)^{\alpha}) = \mathcal{D}((\delta - A)^{\alpha})$, or equivalently, that $\mathcal{R}\big((\gamma - A)^{-\alpha}\big) = \mathcal{R}\big((\delta - A)^{-\alpha}\big)$. This is true for integer values of α, so it suffices to consider the case where $0 < \alpha < 1$. Moreover, by (iii), it suffices to show that

$$\mathcal{R}\big((\gamma - A)^{-\alpha} - (\delta - A)^{-\alpha}\big) \subset X_1.$$

By Definition 3.9.4, for all $x \in X$,

$$(\gamma - A)^{-\alpha}x - (\delta - A)^{-\alpha}x = \frac{1}{\Gamma(\alpha)}\int_0^{\infty} t^{\alpha-1}[1 - \mathrm{e}^{-(\delta-\gamma)t}]\mathrm{e}^{-\gamma t}\mathfrak{A}^t x\,dt.$$

Therefore, for $x \in X_1$ we have $[(\gamma - A)^{-\alpha} - (\delta - A)^{-\alpha}]x \in X_1$ and (integrate by parts)

$$[(\gamma - A)^{-\alpha} - (\delta - A)^{-\alpha}]x = \frac{1}{\Gamma(\alpha)}(\gamma - A)^{-1}\int_0^{\infty} \dot{h}(t)\mathrm{e}^{-\gamma t}\mathfrak{A}^t x\,dt,$$

where $h(t) = -t^{\alpha-1}[1 - \mathrm{e}^{-(\delta-\gamma)t}]$. Without loss of generality, suppose that $\delta > \gamma$. Then $\dot{h} \in L^1([0,1]) \cap L^{\infty}([1,\infty)$ and $t \mapsto \mathrm{e}^{-\gamma t}\|\mathfrak{A}^t\| \in L^{\infty}([0,1]) \cap L^1([1,\infty)$, so the integral converges absolutely for all $x \in X$. This implies that $\mathcal{R}\big((\gamma - A)^{-\alpha} - (\delta - A)^{-\alpha}\big) \subset X_1$, as claimed. □

With the fractional powers of $(\gamma - A)$ at our disposal, we can construct a continuous scale of Banach spaces X_α, $\alpha \in \mathbb{R}$, in the same way as we constructed the spaces X_n with integral indices n in Section 3.6. For $\alpha > 0$ we let X_α be the range of $(\gamma - A)^{-\alpha}$ (i.e., the image of X under $(\gamma - A)^{-\alpha}$), with norm

$$|x|_\alpha = |x|_{X_\alpha} = \big|(\gamma - A)^{\alpha}x\big|_X.$$

For $\alpha < 0$ we let X_α be the completion of X with the weaker norm

$$|x|_{-\alpha} = \big|(\gamma - A)^{-\alpha}x\big|_X, \qquad \alpha > 0.$$

All the earlier conclusions listed in Section 3.6 remain valid. In particular, for all $\gamma \in \mathbb{C}^+$, all $\alpha, \beta, \delta \in \mathbb{R}$, and all $t \geq 0$,

$$
\begin{aligned}
&(\gamma - A)^{\beta}_{|X_{\alpha+\beta}} \text{ is an isometry of } X_{\alpha+\beta} \text{ onto } X_\alpha, \\
&(\gamma - A)^{\alpha+\beta}_{|X_\delta} = (\gamma - A)^{\alpha}_{|X_{\delta-\beta}} (\gamma - A)^{\beta}_{|X_\delta}, \\
&(\gamma - A)^{\alpha}_{|X_\delta} \mathfrak{A}^t_{|X_\delta} = \mathfrak{A}^t_{|X_{\delta-\alpha}} (\gamma - A)^{\alpha}_{|X_\delta}.
\end{aligned}
$$

Different choices of γ give identical spaces with equivalent norms, and $(\gamma - A)^\alpha$ commutes with $(\delta - A)^\beta$ for all $\alpha, \beta \in \mathbb{R}$, and all $\gamma, \delta \in \mathbb{C}^+_{\omega_{\mathfrak{A}}}$.

The spaces X_α can be interpreted as *interpolation spaces* between the spaces X_n with integral indices; see Lunardi (1995, Chapters 1,2). The following lemma is related to this fact:

Lemma 3.9.8 *Define the fractional space X_α as above. Then, there is a constant $C > 0$ such that for all $0 < \alpha < 1$, all $x \in X_1 = \mathcal{D}(A)$, and all $\rho > 0$,*

$$
\begin{aligned}
|x|_{X_\alpha} &\leq C\big(\rho^\alpha |x|_X + \rho^{\alpha-1}|x|_{X_1}\big), \\
|x|_{X_\alpha} &\leq 2C|x|_X^{1-\alpha}|x|_{X_1}^{\alpha}.
\end{aligned}
\tag{3.9.6}
$$

The proof of this lemma is based on the following representation of $(\gamma - A)^{-\alpha}$, valid for $0 < \alpha < 1$:

Lemma 3.9.9 *For $0 < \alpha < 1$ the operator $(\gamma - A)^{-\alpha}$ defined in Definition 3.9.4 has the representation*

$$
(\gamma - A)^{-\alpha} = \frac{\sin \pi\alpha}{\pi} \int_0^\infty s^{-\alpha}\big(s + \gamma - A\big)^{-1}\, ds,
$$

where the integral converges absolutely in operator norm.

Proof The absolute convergence in operator norm follows from the Hille–Yosida Theorem 3.4.1 and the assumption that $\gamma \in \mathbb{C}^+_{\omega_{\mathfrak{A}}}$. By using Theorem 3.2.9(i), Fubini's theorem, a change of integration variable $s = v/t$, and the fact that the Gamma-function

$$
\Gamma(\alpha) = \int_0^\infty t^{\alpha-1}\mathrm{e}^{-t}\, dt \tag{3.9.7}
$$

satisfies (for $0 < \alpha < 1$) the reflection formula $\Gamma(\alpha)\Gamma(1-\alpha) = \frac{\sin \pi\alpha}{\pi}$, we get for all $x \in X$,

$$
\begin{aligned}
&\frac{\sin \pi\alpha}{\pi} \int_0^\infty s^{-\alpha}\bigl(s+\gamma-A\bigr)^{-1}x\,ds \\
&= \frac{1}{\Gamma(\alpha)\Gamma(1-\alpha)} \int_0^\infty s^{-\alpha} \int_0^\infty \mathrm{e}^{-(s+\gamma)t}\mathfrak{A}^t x\,dt\,ds \\
&= \frac{1}{\Gamma(\alpha)\Gamma(1-\alpha)} \int_0^\infty \Bigl(\int_0^\infty s^{-\alpha}\mathrm{e}^{-st}\,ds\Bigr)\mathrm{e}^{-\gamma t}\mathfrak{A}^t x\,dt \\
&= \frac{1}{\Gamma(\alpha)\Gamma(1-\alpha)} \Bigl(\int_0^\infty v^{-\alpha}\mathrm{e}^{v}\,dv\Bigr) \int_0^\infty t^{\alpha-1}\mathrm{e}^{-\gamma t}\mathfrak{A}^t x\,dt \\
&= \frac{1}{\Gamma(\alpha)} \int_0^\infty t^{\alpha-1}\mathrm{e}^{-\gamma t}\mathfrak{A}^t x\,dt \\
&= (\gamma - A)^{-\alpha}x.
\end{aligned}
$$

□

Proof of Lemma 3.9.8. Let $\alpha \in (0, 1)$, $\rho > 0$, $x \in X_1$, and recall that $|x|_{X_\alpha} = |(\gamma - A)^\alpha x|_X$ and that $|x|_{X_1}$ is (equivalent to) $|(\gamma - A)x|_X$. Since $0 < \alpha < 1$, we have $0 < 1 - \alpha < 1$, hence by Lemma 3.9.9, Theorem 3.4.1, and the assumption that $\gamma \in \mathbb{C}^+_{\omega_\mathfrak{A}}$ (observe that $\sin \pi(1-\alpha) = \sin \pi\alpha$ and $\sin \pi\alpha \le \pi\alpha$)

$$
\begin{aligned}
|x|_{X_\alpha} &= |(\gamma - A)^{-(1-\alpha)}(\gamma - A)x|_X \\
&\le \frac{\sin \pi(1-\alpha)}{\pi} \int_0^\infty s^{\alpha-1}\bigl|\bigl((s+\gamma)-A\bigr)^{-1}(\gamma - A)x\bigr|\,ds \\
&\le \frac{\sin \pi\alpha}{\pi} \int_0^\rho s^{\alpha-1}\bigl\|\bigl(s+\gamma-A\bigr)^{-1}(\gamma-A)\bigr\|\,|x|_X\,ds \\
&\quad + \frac{\sin \pi(1-\alpha)}{\pi} \int_\rho^\infty s^{\alpha-1}\bigl\|\bigl(s+\gamma-A\bigr)^{-1}\bigr\|\,|x|_{X_1}\,ds \\
&\le |x|_X\alpha \int_0^\rho s^{\alpha-1}\bigl\|1 - s\bigl(s+\gamma-A\bigr)^{-1}\bigr\|\,ds \\
&\quad + |x|_{X_1}(1-\alpha) \int_\rho^\infty s^{\alpha-1}\bigl\|\bigl(s+\gamma-A\bigr)^{-1}\bigr\|\,ds \\
&\le (1+C)|x|_X\alpha \int_0^\rho s^{\alpha-1}\,ds + C|x|_{X_1}(1-\alpha)\int_\rho^\infty s^{\alpha-2}\,ds \\
&= (1+C)|x|_X\rho^\alpha + C|x|_{X_1}\rho^{\alpha-1}.
\end{aligned}
$$

This proves the first inequality (with $C + 1$ instead of C) in (3.9.6). To get the second inequality in (3.9.6) for nonzero x we simply take $\rho = |x|_{X_1}/|x|_X$. □

3.10 Analytic semigroups and sectorial operators

Semigroups obtained from the Cauchy problem for partial differential equations of parabolic type have some extra smoothness properties. They are of the following type.

Definition 3.10.1 Let X be a Banach space, let $0 < \delta \le \pi/2$, and let Δ_δ be the open sector $\Delta_\delta = \{t \in \mathbb{C} \mid t \neq 0,\ |\arg t| < \delta\}$ (see Figure 3.1). The family of operators $\mathfrak{A}^t \in \mathcal{B}(X), t \in \Delta_\delta$, is an *analytic semigroup (with uniformly bounded growth bound ω) in Δ* if the following conditions hold:

(i) $t \mapsto \mathfrak{A}^t$ is analytic in Δ_δ;
(ii) $\mathfrak{A}^0 = 1$ and $\mathfrak{A}^s\mathfrak{A}^t = \mathfrak{A}^{s+t}$ for all $s, t \in \Delta_\delta$;
(iii) there exist constants $M \ge 1$ and $\omega \in \mathbb{R}$ such that

$$\|\mathfrak{A}^t\| \le M e^{\omega t}, \qquad t \in \Delta_\delta;$$

(iv) for all $x \in X$, $\lim_{\substack{t \to 0 \\ t \in \Delta_\delta}} \mathfrak{A}^t x = x$.

A semigroup $\mathfrak{A}$ is *analytic* if it is analytic in some sector Δ of the type described above.

We warn the reader that the sector Δ_δ in Definition 3.10.1 need not be maximal: if $\mathfrak{A}$ is analytic on some sector Δ_δ, then it can often be extended to an analytic semigroup on a larger sector $\Delta_{\delta'}$ with $\delta' > \delta$. If we take the union of all the sectors Δ_δ where $\mathfrak{A}^t$ is analytic (with a uniformly bounded growth bound), then the constants M and ω in (iii) typically deteriorate as we approach the sector boundary.

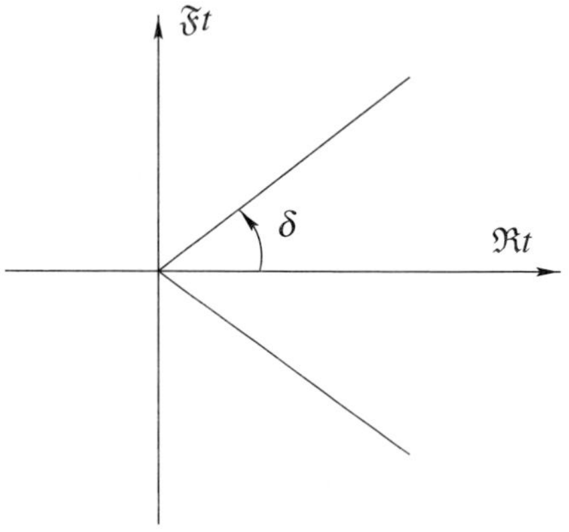

Figure 3.1 The sector Δ_δ

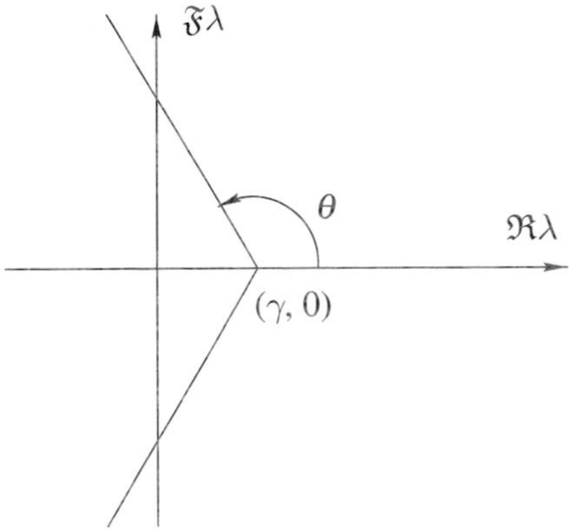

Figure 3.2 The sector $\Sigma_{\theta,\gamma}$

As we shall see in a moment, the generators of the analytic semigroups in Definition 3.10.1 are *sectorial operators*, which are defined as follows.

Definition 3.10.2 For each $\gamma \in \mathbb{R}$ and $\pi/2 < \theta < \pi$, let $\Sigma_{\theta,\gamma}$ be the open sector (see Figure 3.2)

$$\Sigma_{\theta,\gamma} = \{\lambda \in \mathbb{C} \mid \lambda \neq \gamma,\ |\arg(\lambda - \gamma)| < \theta\}.$$

A (closed) densely defined linear operator $X \supset \mathcal{D}(A) \to X$ is *sectorial on* $\Sigma_{\theta,\gamma}$ (with a uniform bound) if the resolvent set of A contains $\Sigma_{\theta,\gamma}$, and if

$$\|(\lambda - A)^{-1}\| \leq \frac{C}{|\lambda - \gamma|}, \qquad \lambda \in \Sigma_{\theta,\gamma}, \tag{3.10.1}$$

for some $C \geq 1$. The operator A is *sectorial* if it is sectorial on some sector $\Sigma_{\theta,\gamma}$ (with $\pi/2 < \theta < \pi$).

Again we warn the reader that the constant θ in Definition 3.10.2 is not maximal: it can always be replaced by a larger constant θ':

Lemma 3.10.3 *If A is sectorial on some sector $\Sigma_{\theta,\gamma}$ with $\pi/2 < \theta < \pi$ then it is also sectorial on some bigger sector $\Sigma_{\theta',\gamma}$ with $\theta' > \theta$. More precisely, this is true for every θ' satisfying $\theta < \theta' < \pi$ and* $\sin(\theta' - \theta) < 1/C$, *where C is the constant in* (3.10.1).

Proof Without loss of generality we can take $\gamma = 0$ (i.e., we replace $A - \gamma I$ by A).

By Lemma 3.2.8(ii), the rays $\{\lambda \neq 0 \mid \arg \lambda = \pm\theta\}$ bounding the sector $\Sigma_{\theta,0}$ belong to the resolvent set of A, and (by continuity) (3.10.1) holds for these λ, too. Let $0 < k < 1$. By (3.10.1) and Lemma 3.2.8(ii), if we take λ close enough to $\alpha \in \Sigma_{\theta,0}$ so that $|\lambda - \alpha| \le \frac{k|\alpha|}{C}$, then $\lambda \in \rho(A)$ and

$$\left\|(\lambda - A)^{-1}\right\| \le \frac{C}{(1-k)|\alpha|}.$$

Choose θ' to satisfy $\theta < \theta' < \pi$ and $\sin(\theta' - \theta) = k/C$. By letting α vary over the rays $\{\lambda \ne 0 \mid \arg\lambda = \pm\theta\}$ and taking $\lambda - \alpha$ orthogonal to λ (and $\Re\lambda < \Re\alpha$) we reach all λ in $\Sigma_{\theta',0} \setminus \Sigma_{\theta,0}$. Moreover, with this choice of λ and α, we have $|\lambda| < |\alpha|$, hence

$$\left\|(\lambda - A)^{-1}\right\| \le \frac{C}{(1-k)\lambda}.$$

Thus, A is sectorial on $\Sigma_{\theta',0}$ with C in (3.10.1) replaced by $C/(1-k)$. □

Lemma 3.10.3 (and its proof) implies that the constant C in (3.10.1) must satisfy $C \ge 1$, because if $C < 1$, then the proof of Lemma 3.10.3 shows that $(\lambda - A)^{-1}$ is a bounded entire function vanishing at infinity, hence identically zero. The optimal constant for $A = \gamma$ is $C = 1$. A similar argument shows that (3.10.1) holds with $\theta = \pi$ if and only if $A = \gamma$.

It is sometimes possible to increase the sector $\Sigma_{\theta,\gamma}$ in which A is sectorial in a different way: we keep θ fixed, but replace γ by $\gamma' < \gamma$. This is possible under the following assumptions:

Lemma 3.10.4 *Let A be a closed linear operator on X, and let $\gamma \in \mathbb{R}$ and $\pi/2 < \theta < \pi$. Then the following conditions are equivalent:*

(i) *A is sectorial on $\Sigma_{\theta,\gamma}$ and $\gamma \in \rho(A)$;*
(ii) *A is sectorial on some sector $\Sigma_{\theta,\gamma'}$ with $\gamma' < \gamma$.*

Proof (i) $\Rightarrow$ (ii): It follows from Lemma 3.10.3 and the assumption $\gamma \in \rho(A)$ that the distance from $\sigma(A)$ to the boundary of the sector Σ_θ, γ is strictly positive, hence $\rho(A) \supset \Sigma_{\theta,\gamma'}$ for some $\gamma' < \gamma$. The norm of the resolvent $(\lambda - A)^{-1}$ is uniformly bounded for all $\lambda \in \Sigma_{\theta,\gamma'}$ satisfying $|\lambda - \gamma| \le |\gamma - \gamma'|$, and for $|\lambda - \gamma| > |\gamma - \gamma'|$ we can estimate

$$\|(\lambda - A)^{-1}\| \le \frac{C}{|\lambda - \gamma|} = \frac{C}{|\lambda - \gamma'|}\frac{|\lambda - \gamma'|}{|\lambda - \gamma|} < \frac{2C}{|\lambda - \gamma'|}.$$

(ii) $\Rightarrow$ (i): This proof is similar to the one above (but slightly simpler). □

Our next result is a preliminary version of Theorem 3.10.6, and we refer the reader to that theorem for a more powerful result.

Theorem 3.10.5 *Let $A\colon X \to X$ be a (densely defined) sectorial operator on the sector $\Sigma_{\theta,\gamma}$. Then A is the generator of a C_0 semigroup $\mathfrak{A}$ satisfying $\|\mathfrak{A}^t\| \le$*

$M e^{\gamma t}$ for some $M \geq 1$ which is continuous in the uniform operator norm on $(0, \infty)$. This semigroup has the representation

$$\mathfrak{A}^t = \frac{1}{2\pi j} \int_\Gamma e^{\lambda t} (\lambda - A)^{-1} \, d\lambda, \tag{3.10.2}$$

where Γ is a smooth curve in $\Sigma_{\theta,\gamma}$ running from $\infty e^{-j\vartheta}$ to $\infty e^{j\vartheta}$, where $\pi/2 < \vartheta < \theta$. For each $t > 0$ the integral converges in the uniform operator topology.

Proof Throughout this proof we take, without loss of generality, $\gamma = 0$. (If $\gamma \neq 0$, then we replace $A - \gamma$ by A and $e^{-\gamma t}\mathfrak{A}^t$ by $\mathfrak{A}^t$; see Examples 2.3.5 and 3.2.6.)

The absolute convergence in operator norm of the integral in (3.10.2) is a consequence of (3.10.1) and the fact that $|e^{e^{j\theta} r}| = e^{\Re e^{j\theta} r} = e^{r \cos\theta}$ for all $r \in \mathbb{R}$ (observe that $\cos\theta < 0$ for $\pi/2 < |\theta| < \pi$). The continuity in the uniform operator topology on $(0, \infty)$ follows from a straightforward estimate (and the Lebesgue dominated convergence theorem). Since the integrand is analytic, we can deform the path of integration without changing the value of the integral to $\Gamma_t = \Gamma_1 \cup \Gamma_2 \cup \Gamma_3$ (see Figure 3.3), where

$$\begin{aligned}
\Gamma_1 &= \{re^{-j\vartheta} \mid \infty > r \geq 1/t\}, \\
\Gamma_2 &= \{t^{-1}e^{j\theta} \mid -\vartheta \leq \theta \leq \vartheta\}, \\
\Gamma_3 &= \{re^{j\vartheta} \mid 1/t \leq r < \infty\}.
\end{aligned}$$

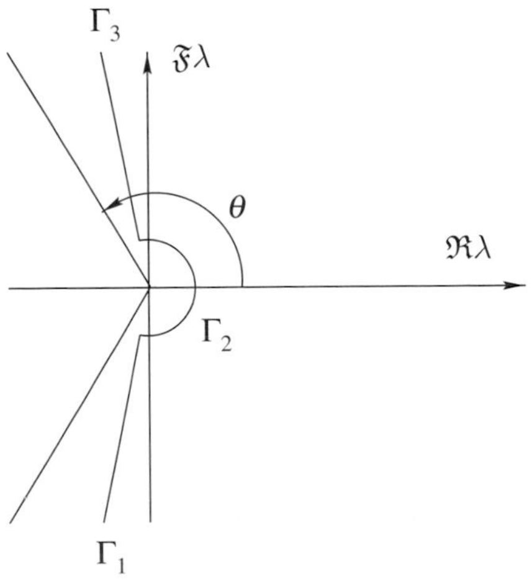

Figure 3.3 The path in the proof of Theorem 3.10.5

Then, by (3.10.1),

$$\begin{aligned}\Big\| \frac{1}{2\pi j} \int_{\Gamma_1} e^{\lambda t} (\lambda - A)^{-1}\, d\lambda \Big\| &\le \frac{1}{2\pi} \int_{\Gamma_1} \| e^{\lambda t} (\lambda - A)^{-1} \|\, d\lambda \\ &\le \frac{C}{2\pi} \int_{1/t}^{\infty} e^{rt\cos\vartheta} r^{-1}\, dr \\ &= \frac{C}{2\pi} \int_{1}^{\infty} e^{s\cos\vartheta} s^{-1}\, ds \\ &= M_1 .\end{aligned}$$

The same estimate is valid for the integral over Γ_3. On Γ_2 we estimate

$$\begin{aligned}\Big\| \frac{1}{2\pi j} \int_{\Gamma_2} e^{\lambda t} (\lambda - A)^{-1}\, d\lambda \Big\| &\le \frac{1}{2\pi} \int_{\Gamma_2} \| e^{\lambda t} (\lambda - A)^{-1} \|\, d\lambda \\ &\le \frac{C}{2\pi} \int_{-\vartheta}^{\vartheta} e^{\cos\theta}\, d\theta \\ &= M_2 .\end{aligned}$$

Together these estimates prove that $\|\mathfrak{A}^t\| \le M$ with $M = 2M_1 + M_2$.

Next we claim that for all $\alpha > 0$,

$$(\alpha - A)^{-1} = \int_0^{\infty} e^{-\alpha t} \mathfrak{A}^t\, dt. \tag{3.10.3}$$

To prove this we choose the curve Γ as we did above, but replace Γ_2 by $\Gamma_2 = \{\epsilon e^{-j\theta} \mid \vartheta \le \theta \le \vartheta\}$, where $0 < \epsilon < \alpha$ is fixed, and adjust Γ_1 and Γ_3 accordingly. We repeat the calculations presented above for this choice of Γ, and find that

$$\frac{1}{2\pi} \int_{\Gamma} \| e^{\lambda t} (\lambda - A)^{-1} \|\, d\lambda \le M(|\log t| + e^{\epsilon t})$$

for some finite M (the logarithm comes from Γ_1 and Γ_3, and the exponent from Γ_2). Thus, with this choice of Γ, if we multiply (3.10.2) by $e^{-\alpha t}$ and integrate over $(0, \infty)$, then the resulting double integral also converges absolutely. By Fubini's theorem, we may change the order of integration to get

$$\int_0^{\infty} e^{-\alpha t} \mathfrak{A}^t\, dt = \frac{1}{2\pi j} \int_{\Gamma} (\alpha - \lambda)^{-1} (\lambda - A)^{-1}\, d\lambda.$$

By (3.10.1), this integral is the limit as $n \to \infty$ of the integral over $\widetilde{\Gamma}_n$, where $\widetilde{\Gamma}_n$ is the closed curve that we get by restricting the variable r in the definition of Γ_1 and Γ_3 to the interval $\epsilon \le r \le n$, and connecting the points $ne^{-i\varphi}$ and $ne^{i\varphi}$ with the arc $\Gamma_4 = \{ne^{-j\theta} \mid \vartheta \ge \theta \ge -\vartheta\}$. However, by the residue theorem, this integral is equal to $(\alpha - A)^{-1}$. This proves (3.10.3).

Equation (3.10.3) together with the bound $\|\mathfrak{A}^t\| \le M$ enables us to repeat the argument that we used in the proof of Theorem 3.2.9(i)–(ii) (starting with formula (3.2.5)) to show that

$$\|(\alpha - A)^{-n}\| \le M\alpha^{-n}$$

for all $n = 1, 2, 3, \ldots$ and $\alpha > 0$. According to Theorem 3.4.1, this implies that A generates a C_0 semigroup $\mathfrak{A}_1$. Thus, to complete the proof it suffices to show that $\mathfrak{A}_1^t = \mathfrak{A}^t$ for all $t > 0$. However, by (3.10.3) and Theorem 3.2.9(i), for every $x^* \in X^*$, $x \in X$, and $\alpha > 0$,

$$\int_0^\infty e^{-\alpha t} x^*(\mathfrak{A}^t x - \mathfrak{A}_1^t x)\, dt = 0.$$

Thus, since the Laplace transform of a (scalar continuous) function determines the function uniquely (see Section 3.12), we must have $x^*\mathfrak{A}^t x = \mathfrak{A}_1^t x$ for all $x^* \in X^*$, $x \in X$, and $t \ge 0$. Thus, $\mathfrak{A}_1^t = \mathfrak{A}^t$ for all $t > 0$. In particular, $\mathfrak{A}$ is strongly continuous. □

As the following theorem shows, the class of semigroups generated by sectorial operators can be characterized in several different ways.

Theorem 3.10.6 *Let A be a closed operator on the Banach space X, and let $\gamma \in \mathbb{R}$. Then the following conditions are equivalent:*

(i) *A is the generator of an analytic semigroup $\mathfrak{A}^t$ with uniformly bounded growth bound γ on a sector $\Delta_\delta = \{t \in \mathbb{C} \mid |\arg t| < \delta\}$ (with $\delta > 0$; see Definition 3.10.1);*

(ii) *every $\lambda \in \mathbb{C}^+_\gamma$ belongs to the resolvent set of A, and there is a constant C such that*

$$\|(\lambda - A)^{-1}\| \le \frac{C}{|\lambda - \gamma|}, \qquad \Re\lambda > \gamma;$$

(iii) *A is sectorial on some sector $\Sigma_{\theta,\gamma}$ (with $\pi/2 < \theta < \pi$; see Definition 3.10.2);*

(iv) *A is the generator of a semigroup $\mathfrak{A}$ which is differentiable on $(0, \infty)$, and there exist finite constants M_0 and M_1 such that*

$$\|\mathfrak{A}^t\| \le M_0 e^{\gamma t}, \qquad \|(A - \gamma)\mathfrak{A}^t\| \le \frac{M_1 e^{\gamma t}}{t}, \qquad t > 0.$$

Proof Throughout this proof we take, without loss of generality, $\gamma = 0$. (If $\gamma \ne 0$, then we replace $A - \gamma$ by A and $e^{-\gamma t}\mathfrak{A}^t$ by $\mathfrak{A}^t$; see Examples 2.3.5 and 3.2.6.)

(i) $\Rightarrow$ (ii): Let $0 < \varphi < \delta$. Let A_φ be the generator of the semigroup $t \mapsto \mathfrak{A}^{e^{j\varphi}t}$. Then, by Theorem 3.2.9(i), every (real) $\alpha > 0$ belongs to the resolvent set of A_φ, and

$$(\alpha - A_\varphi)^{-1}x = \int_0^\infty e^{-\alpha s}\mathfrak{A}^{e^{j\varphi}s}x\, ds.$$

By the estimate that we have on $\mathfrak{A}$ in (i), we can make a change of integration variable from s to $t = e^{j\varphi}s$, $s = e^{-j\varphi}t$ to get

$$\begin{aligned}(\alpha - A_\varphi)^{-1}x &= e^{-j\varphi}\int_0^\infty e^{-\alpha e^{-j\varphi}t}\mathfrak{A}^t x\,ds \\ &= e^{-j\varphi}(\alpha e^{-j\varphi} - A)^{-1} \\ &= (\alpha - e^{j\varphi}A)^{-1},\end{aligned}$$

where the second equality follows from Theorem 3.2.9(i). Thus, $A_\varphi = e^{j\varphi}A$. The estimate in (ii) then follows from (i) and Theorem 3.2.9(ii).

(ii) $\Rightarrow$ (iii): See Lemma 3.10.3.

(iii) $\Rightarrow$ (iv): We know from Theorem 3.10.5 that A generates a C_0 semigroup $\mathfrak{A}$, which has the integral representation (3.10.2). Thus, for all $h > 0$,

$$\frac{1}{h}(\mathfrak{A}^t - 1) = \frac{1}{2\pi j}\int_\Gamma h^{-1}(e^{\lambda h} - 1)e^{\lambda t}(\lambda - A)^{-1}\,d\lambda.$$

The same type of estimates that we developed in the proof of Theorem 3.10.5 (where we split Γ into $\Gamma_1 \cup \Gamma_2 \cup \Gamma_3$) together with Lemma 3.2.10 and the Lebesgue dominated convergence theorem show that this has a limit as $h \downarrow 0$. Thus, $\mathfrak{A}^t$ is differentiable (in operator norm), and

$$A\mathfrak{A}^t = \frac{d}{dt}(\mathfrak{A}^t) = \frac{1}{2\pi j}\int_\Gamma \lambda e^{\lambda t}(\lambda - A)^{-1}\,d\lambda. \tag{3.10.4}$$

As in the proof of Theorem 3.10.5 we again split Γ into $\Gamma_1 \cup \Gamma_2 \cup \Gamma_3$ and estimate. We leave it to the reader to check that the final estimate that we get in this way is of the type

$$\|A\mathfrak{A}^t\| \le M/t,$$

with the same constant $M = 2M_1 + M_2$ as we get in the proof of Theorem 3.10.5.

(iv) $\Rightarrow$ (i): We first claim that, for each $t > 0$, $\mathfrak{A}^t$ maps X into $\bigcap_{n=1}^\infty \mathcal{D}(A^n)$, and prove this as follows. By the differentiability assumption on $\mathfrak{A}^t$, for each $x \in X$, the function $x(t) = \mathfrak{A}^t x$ is differentiable for $t > 0$. By the semigroup property,

$$\dot{x}(t) = \lim_{h\downarrow 0}\frac{1}{h}(\mathfrak{A}^{t+h} - \mathfrak{A}^t)x, = \lim_{h\downarrow 0}\frac{1}{h}(\mathfrak{A}^h - 1)\mathfrak{A}^t x,$$

hence $x(t) = \mathfrak{A}^t x \in \mathcal{D}(A)$, and $\frac{d}{dt}(\mathfrak{A}^t) = A\mathfrak{A}^t$. The same argument shows that

$\mathfrak{A}^t$ maps $\mathcal{D}(A^k)$ into $\mathcal{D}(A^{k+1})$ for all $n = 0, 1, 2, \ldots$ (because $A^k\mathfrak{A}^t x = \mathfrak{A}^t A^k x$ for all $x \in \mathcal{D}(A^k)$; see Theorem 3.2.1(iii)). By the semigroup property, $\mathfrak{A}^t = (\mathfrak{A}^{t/n})^n$, and this together with our earlier observation that $\mathfrak{A}^{t/n}$ maps $\mathcal{D}(A^k)$ into $\mathcal{D}(A^{k+1})$ implies that $\mathfrak{A}^t$ maps X into $\mathcal{D}(A^n)$ for each $t > 0$ and $n = 1, 2, 3, \ldots$ Thus, $\mathfrak{A}^t$ maps X into $\bigcap_{n=1}^{\infty} \mathcal{D}(A^n)$. This together with Theorem 3.2.1(vi) implies that $x(t) = \mathfrak{A}^t x$ is infinitely many times differentiable on $(0, \infty)$, and that

$$x^{(n)}(t) = (\mathfrak{A}^t)^{(n)} = A^n\mathfrak{A}^t x, \qquad t > 0.$$

Since A commutes with $\mathfrak{A}^t$ on $\bigcap_{n=1}^{\infty} \mathcal{D}(A^n)$, and since $\mathfrak{A}^t = (\mathfrak{A}^{t/n})^n$, this implies that

$$(\mathfrak{A}^t)^{(n)} = A^n\mathfrak{A}^t = (A\mathfrak{A}^{t/n})^n, \qquad t > 0. \tag{3.10.5}$$

Observe that the operators $A^n\mathfrak{A}^t$ are closed with $\mathcal{D}(A^n\mathfrak{A}^t) = X$, hence, by the closed graph theorem, they are bounded.

So far we have only used the differentiability of $\mathfrak{A}^t$ and not the norm estimate in (iv). This norm estimate, combined with (3.10.5) and the fact that $\mathrm{e}^n > n^n/n!$ (this is one of the terms in the power series expansion of e^n) gives

$$\left\| \frac{(\mathfrak{A}^t)^{(n)}}{n!} \right\| = \left\| \frac{A^n\mathfrak{A}^t}{n!} \right\| \le \left(\frac{M_1\mathrm{e}}{t} \right)^n, \qquad t > 0, \quad n = 1, 2, 3, \ldots \tag{3.10.6}$$

For each $t > 0$ we can define a formal power series

$$\widetilde{\mathfrak{A}}^z = \sum_{n=0}^{\infty} \frac{(\mathfrak{A}^t)^{(n)}}{n!}(z - t)^n. \tag{3.10.7}$$

The estimate (3.10.6) shows that this power series converges in operator norm for $|z - t| < t/(M_1\mathrm{e})$. Being the limit of a power series, the function $\widetilde{\mathfrak{A}}^z$ is analytic in the circle $|z - t| < t/(M_1\mathrm{e})$.

We claim $\widetilde{\mathfrak{A}}^z = \mathfrak{A}^z$ for real z satisfying $-t/(1 + M_1\mathrm{e}) < z - t < t/(M_1\mathrm{e})$. This is obvious for $z = t$. To see that it is true for other values of z we fix $x \in X$ and use a Taylor expansion of order $N - 1$ with Lagrangian remainder term for $\mathfrak{A}^z$, i.e.,

$$\mathfrak{A}^z x = \sum_{n=0}^{N-1} \frac{(\mathfrak{A}^t)^{(n)}x}{n!}(z - t)^n + \frac{(\mathfrak{A}^\xi)^{(N)}x}{N!}(z - t)^N,$$

where $z < \xi < t$ or $t < \xi < z$, depending on whether $z < t$ or $z > t$. By (3.10.6), the remainder term can be estimated by $(M_1\mathrm{e})^N|z - t|^N(\min\{t, z\})^{-N}$, which tends to zero in the given interval. Thus, $\widetilde{\mathfrak{A}}^z = \mathfrak{A}^z$ for real z satisfying $-t/(1 + M_1\mathrm{e}) < z - t < t/(M_1\mathrm{e})$. Letting $t > 0$ vary we conclude that $\widetilde{\mathfrak{A}}^t$ is an analytic extension of $\mathfrak{A}^t$ to the sector Δ_δ, where $0 < \delta \le \pi/2$ satisfies $\sin\delta = 1/(M_1\mathrm{e})$ (we choose z perpendicular to $z - t$, and assume, without loss of generality, that $M_1 e \geq 1$). The analyticity of $\widetilde{\mathfrak{A}}$ implies that $\widetilde{\mathfrak{A}}$ inherits the semigroup property $\widetilde{\mathfrak{A}}^{s+t} = \widetilde{\mathfrak{A}}^s \widetilde{\mathfrak{A}}^t$ from $\mathfrak{A}$ (two analytic functions which coincide on the real axis coincide everywhere). In each smaller subsector Δ_φ with $\varphi < \delta$ we can choose z and t (with z still perpendicular to $z - t$) so that $(|z-t|M_1 e)/t \leq k < 1$ (uniformly in z and t), and this implies that

$$\|\widetilde{\mathfrak{A}}^z\| \leq \|\mathfrak{A}^t\| + \sum_{n=1}^{\infty} \frac{\|(\mathfrak{A}^t)^{(n)}\|}{n!} |z-t|^n \leq M_0 + k/(1-k).$$

Thus, the extended semigroup is uniformly bounded in each proper subsector Σ_φ of Σ_δ. If we rename φ to δ, then we have shown that $\widetilde{\mathfrak{A}}$ satisfies conditions (i)–(iii) in Definition 3.10.1.

The only thing that remains to be proved is the strong continuity, i.e., condition (iv) in Definition 3.10.1. Because of the density of $\mathcal{D}(A)$ in X and the uniform bound that we have on $\|\widetilde{\mathfrak{A}}^z\|$ for all $z \in \Delta_\delta$, it suffices to show that, for all $x \in \mathcal{D}(A)$,

$$\lim_{\substack{t \to 0 \\ t \in \Delta_\delta}} \mathfrak{A}^t x = x. \tag{3.10.8}$$

By using (3.10.6) and (3.10.7) we can estimate, for all $x \in \mathcal{D}(A)$,

$$\begin{aligned}
|\widetilde{\mathfrak{A}}^z x - x| &\leq |\mathfrak{A}^t x - x| + |\widetilde{\mathfrak{A}}^z x - \mathfrak{A}^t x| \\
&\leq |\mathfrak{A}^t x - x| + \sum_{n=1}^{\infty} \frac{|A^n \mathfrak{A}^t x|}{n!} |z-t|^n \\
&\leq |\mathfrak{A}^t x - x| + |z-t| \sum_{n=1}^{\infty} \frac{\|A^{n-1}\mathfrak{A}^t\| |Ax|}{n(n-1)!} |z-t|^{n-1} \\
&\leq |\mathfrak{A}^t x - x| + |z-t||Ax| \sum_{n=1}^{\infty} \Big(\frac{|z-t|M_1 e}{t}\Big)^{n-1} \\
&\leq |\mathfrak{A}^t x - x| + \frac{|z-t||Ax|}{1-k},
\end{aligned}$$

where the last inequality is true provided we choose z and t to satisfy $(|z - t|M_1 e)/t \leq k < 1$. As before, we take z perpendicular to $z - t$. Let $\varphi = \arg z$. Then $|\varphi| < \delta$, $|z-t|/t = |\sin\varphi| < \sin\delta \leq 1$, and

$$t \cos\delta < t\cos\varphi = |z| \leq t.$$

Thus, with this choice of z and t, the condition $z \to 0$, $z \in \Delta_\delta$, is equivalent to $t \downarrow 0$, and

$$|\widetilde{\mathfrak{A}}^z x - x| \leq |\mathfrak{A}^t x - x| + \frac{t|Ax|}{1-k}.$$

Because of the strong continuity of $\mathfrak{A}^t$, this tends to zero as $t \downarrow 0$. Thus, (3.10.8) holds, and the proof is complete. □

Let us record the following facts, which were established as a part of the proof of Theorem 3.10.6:

Corollary 3.10.7 *If the equivalent conditions listed in Theorem 3.10.6 hold, then, for each $\varphi \in (0, \delta)$, the generator of the semigroup $\mathfrak{A}^{e^{j\varphi}t}$, $t \geq 0$, is $e^{j\varphi}A$, where A is the generator of $\mathfrak{A}$.*

This was established as a part of the proof that (i) ⇒ (ii).

Corollary 3.10.8 *Condition (iv) in Theorem 3.10.6 implies that*

$$\|(A-\gamma)^n \mathfrak{A}^t\| \leq \frac{(nM_1)^n e^{\gamma t}}{t^n}, \qquad t > 0, \quad n = 1, 2, 3, \ldots$$

This follows from (3.10.5) (with A replaced by $A - \gamma$ and $\mathfrak{A}^t$ replaced by $e^{-\gamma t}\mathfrak{A}^t$).

In our applications to well-posed linear systems we are especially interested in the following extension of the estimates in Definition 3.10.2 and Corollary 3.10.8 to fractional powers of $(\gamma - A)$:

Lemma 3.10.9 *Let A be sectorial on some sector $\Sigma_{\theta,\gamma'}$, let $\gamma > \gamma'$, and let $\mathfrak{A}$ be the analytic semigroup generated by A. Then there exist constants M and C such that, for all $0 \leq \alpha \leq 1$,*

$$\begin{aligned} \|(\gamma - A)^\alpha \mathfrak{A}^t\| &\leq M(1 + t^{-\alpha})e^{\gamma' t}, \quad t > 0, \\ \|(\gamma - A)^\alpha (\lambda - A)^{-1}\| &\leq C|\lambda - \gamma'|^{-1}\big(1 + |\lambda - \gamma'|^\alpha\big), \quad \lambda \in \Sigma_{\theta,\gamma'}. \end{aligned}$$

Proof For $\alpha = 0$ and $\alpha = 1$ these estimates follow from Definition 3.10.2 and Theorem 3.10.6(iv). For intermediate values of α we interpolate between these two extreme values by using Lemma 3.9.8 as follows. By that lemma and Theorem 3.10.6(iv), we have for all $x \in X$ and $t > 0$ (recall that $\mathfrak{A}^t x \in \mathcal{D}(A)$ for $t > 0$)

$$\begin{aligned} \|(\gamma - A)^\alpha \mathfrak{A}^t x\| &\leq 2C|\mathfrak{A}^t x|_X^{1-\alpha} |(\gamma' - A)\mathfrak{A}^t x + (\gamma - \gamma')\mathfrak{A}^t x|_X^\alpha \\ &\leq 2CM_0^{1-\alpha}(M_1/t + |\gamma - \gamma'|M_0)^\alpha e^{\gamma' t}|x|_X \\ &\leq C_1(1 + t^{-\alpha})e^{\gamma' t}|x|_X. \end{aligned}$$

To get the second inequality we argue essentially in the same way but replace Theorem 3.10.6(iv) by Definition 3.10.2:

$$\begin{aligned}
&\|(\gamma - A)^{\alpha}(\lambda - A)^{-1}x\| \\
&\quad \le 2C|(\lambda - A)^{-1}x|_X^{1-\alpha}|(\gamma - A)(\lambda - A)^{-1}x|_X^{\alpha} \\
&\quad \le C_1|\lambda - \gamma'|^{\alpha-1}\bigl(1 + |\lambda - \gamma|/|\lambda - \gamma'|\bigr)^{\alpha}|x|_X \\
&\quad \le C_2|\lambda - \gamma'|^{-1}\bigl(1 + |\lambda - \gamma'|^{\alpha}\bigr)|x|_X.
\end{aligned}$$

□

Lemma 3.10.9 enables us to add the following conclusion to Theorem 3.8.2 in the case of an analytic semigroup:

Theorem 3.10.10 *Let $\mathfrak{A}$ be an analytic semigroup on X, and define the spaces X_α, $\alpha \in \mathbb{R}$ as above. Let $s \in \mathbb{R}$, $x_s \in X$, $1 < p \le \infty$, and $f \in L^p_{\text{loc}}([s,\infty); X_{-\alpha})$, with*

$$\alpha < 1 - 1/p.$$

Then the function x given by (3.8.2) *is a strong solution of* (3.8.1) *in X.*

Proof This follows from (3.8.2), Lemma 3.10.9, and Hölder's inequality. □

Let us end this section with a perturbation result. The feedback transform studied in Chapter 7 leads to a perturbation of the original semigroup of the system, so that the generator A of this semigroup is replaced by $A + T$ for some operator T. In the analytic case we are able to allow a fairly large class of perturbations T without destroying the analyticity of the perturbed semigroup.

Theorem 3.10.11 *Let A be the generator of an analytic semigroup on the Banach space X, and define the fractional spaces X_α, $\alpha \in \mathbb{R}$, as in Section 3.9. If $T \in \mathcal{B}(X_\alpha; X_\beta)$ for some $\alpha, \beta \in \mathbb{R}$ with $\alpha - \beta < 1$, then the operator $(A + T)_{|X_\alpha}$ generates an analytic semigroup $\mathfrak{A}^T_{|X_{\alpha-1}}$ on $X_{\alpha-1}$. For $\gamma \in [\alpha - 1, \beta + 1]$, the spaces X_γ are invariant under $\mathfrak{A}^T_{|X_{\alpha-1}}$, and the restriction $\mathfrak{A}^T_{|X_\gamma}$ of $\mathfrak{A}^T_{|X_{\alpha-1}}$ to X_γ is an analytic semigroup on X_γ. The generator of $\mathfrak{A}^T_{|X_\gamma}$ is $(A + T)_{|X_{\gamma+1}}$ if $\gamma \in [\alpha - 1, \beta]$, and it is the part of $A + T$ in X_γ if $\gamma \in (\beta, \beta + 1]$.*[9] *Moreover, if $0 \in [\alpha - 1, \beta + 1]$ (so that $\mathfrak{A}^T_{|X}$ is an analytic semigroup on X) and if we let X^T_α, $\alpha \in \mathbb{R}$, be the analogues of the spaces X_α with A replaced by $A + T$, then $X^T_\gamma = X_\gamma$ for all $\gamma \in [\alpha - 1, \beta + 1]$.*

Proof We begin by studying the special case where $\alpha = 1$ and $0 < \beta \le 1$. By Theorem 3.10.6, every λ in some half-plane $\mathbb{C}^+_\mu$ belongs to the resolvent set of

[9] See Definition 3.14.12 and Theorem 3.14.14.

A, and for all $\lambda \in \mathbb{C}^+_\mu$,

$$\|(\lambda - A)^{-1}\| \le \frac{C}{|\lambda - \mu|}. \tag{3.10.9}$$

We claim that $A + T$ has the same property on some other half-plane $\mathbb{C}^+_\nu$. For all $\lambda \in \mathbb{C}^+_\mu$ we may write $\lambda - A - T$ in the form

$$\lambda - A - T = (\lambda - A)(1 - (\lambda - A)^{-1}T).$$

Here $\lambda - A$ maps X_1 one-to-one onto X, so to show that $\lambda - A - T$ is invertible it suffices to show that $1 - (\lambda - A)^{-1}T$ is invertible in $\mathcal{B}(X_1)$. Fix some $\delta \in \rho(A)$. Then $(\delta - A)^{-\beta}T \in \mathcal{B}(X_1)$, and

$$\begin{aligned}\|(\lambda - A)^{-1}T\|_{\mathcal{B}(X_1)} &= \|(\delta - A)^{\beta}(\lambda - A)^{-1}(\delta - A)^{-\beta}T\|_{\mathcal{B}(X_1)}\\ &\le \|(\delta - A)^{\beta}(\lambda - A)^{-1}\|_{\mathcal{B}(X_1)}\|(\delta - A)^{-\beta}T\|_{\mathcal{B}(X_1)}.\end{aligned}$$

By Lemma 3.10.9 (with X replaced by X_1), we can make the right-hand side less than $\frac{1}{2}$ by choosing $|\lambda|$ large enough, and then it follows from the contraction mapping principle that $1 - (\lambda - A)^{-1}T$ is invertible in $\mathcal{B}(X_1)$ and $\|(1 - (\lambda - A)^{-1}T)^{-1}\|_{\mathcal{B}(X_1)} \le 2$. This proves that, for some sufficiently large $\mu_1 \ge \mu$, every $\mathbb{C}^+_{\mu_1} \subset \rho(A + T)$, and that for all $\lambda \in \mathbb{C}^+_{\mu_1}$, $\lambda - A - T$ maps X_1 one-to-one onto X and

$$\|(\lambda - A - T)^{-1}\|_{\mathcal{B}(X;X_1)} \le 2\|(\lambda - A)^{-1}\|_{\mathcal{B}(X;X_1)}.$$

Fix some $\nu \in \mathbb{C}^+_{\mu_1}$. Then both $\nu - A$ and $\nu - A - T$ are boundedly invertible maps of X_1 onto X, so the above inequality implies the existence of some $C > 0$ such that, for all $\lambda \in \mathbb{C}^+_{\mu_1}$,

$$\|(\nu - A - T)(\lambda - A - T)^{-1}\|_{\mathcal{B}(X)} \le C\|(\nu - A)(\lambda - A)^{-1}\|_{\mathcal{B}(X)}.$$

Equivalently,

$$\|(\nu - \lambda)(\lambda - A - T)^{-1} + 1\|_{\mathcal{B}(X)} \le C\|(\nu - \lambda)(\lambda - A)^{-1} + 1\|_{\mathcal{B}(X)},$$

hence, by the triangle inequality,

$$\|(\lambda - A - T)^{-1}\|_{\mathcal{B}(X)} \le C\|(\lambda - A)^{-1}\|_{\mathcal{B}(X)} + \frac{C+1}{|\lambda - \nu|}. \tag{3.10.10}$$

This, together with (3.10.9) and Theorem 3.10.6, implies that $A + T$ generates an analytic semigroup.

Still assuming that $\alpha = 1$ and $0 < \beta \le 1$, let us show that $X^T_\gamma = X_\gamma$ for all $\gamma \in (0, 1)$ (we already know that this is true for $\gamma = 0$ and $\gamma = 1$). This is equivalent to the claim that $(\nu' - A - T)^{-\gamma}$ has the same range as $(\nu' - A)^{-\gamma}$, where, for example, $\nu' > \nu$, and ν is the same constant as in the proof above.

A short algebraic computation shows that, for $s \ge 0$,

$$(s + \nu' - A)^{-1}(1 + T(s + \nu' - A - T)^{-1}) = (s - A)^{-1},$$

and therefore, by Lemma 3.9.9,

$$\begin{aligned}(\nu' - A - T)^{-\gamma} &- (\nu' - A)^{-\gamma} \\ &= c \int_0^\infty s^{-\gamma}\big(s + \nu' - A\big)^{-1} T\big(s + \nu' - A - T\big)^{-1}\,ds,\end{aligned} \tag{3.10.11}$$

where $c = \frac{\sin \pi\alpha}{\pi}$. Since $T \in \mathcal{B}(X_1; X_\beta)$, we get by a computation similar to the one leading to (3.10.10) that there exist constants $C_1, C_2 > 0$ such that for all $s \ge 0$ (see also (3.10.9)),

$$\big\|T\big(s + \nu' - A - T\big)^{-1}\big\|_{\mathcal{B}(X;X_\beta)} \le C_1 \big\|s(s + \nu' - A)^{-1} + 1\big\|_{\mathcal{B}(X)} \le C_2.$$

If $\gamma \le \beta$, then we can estimate

$$\big\|\big(s + \nu' - A\big)^{-1}\big\|_{\mathcal{B}(X_\beta;X_\gamma)} \le C_3 \big\|\big(s + \nu' - A\big)^{-1}\big\|_{\mathcal{B}(X_\beta)} \le C_4(s + \nu' - \nu)^{-1},$$

and if $\gamma > \beta$ then we get from Lemma 3.10.9 (with $\alpha = \gamma - \beta$)

$$\big\|\big(s + \nu - A\big)^{-1}\big\|_{\mathcal{B}(X_\beta;X_\gamma)} \le C((s + \nu' - \nu)^{-1} + (s + \nu' - \nu)^{\gamma-\beta-1}).$$

In either case we find that the operator norm in $\mathcal{B}(X; X_\gamma)$ of the integrand in (3.10.11) is bounded by a constant times $s^{-\gamma}\big((s + \nu' - \nu)^{-1} + (s + \nu' - \nu)^{\gamma-\beta-1}\big)$, and this implies that the integral converges in $\mathcal{B}(X; X_\gamma)$. Since also $(\nu' - A)^{-\gamma} \in \mathcal{B}(X; X_\gamma)$, this implies that $\mathcal{R}\big((\nu' - A - T)^{-\gamma}\big) \subset X_\gamma$, or in other words, $X_\gamma^T \subset X_\gamma$. To prove the opposite inclusion it suffices to apply the same argument with A replaced by $A + T$ and T replaced by $-T$ (i.e., we interchange the roles of A and $A + T$.

Let us recall what we have proved so far. If $\alpha = 1$ and $0 < \beta \le 1$, then $A + T$ generates an analytic semigroup on X, and $X_\gamma^T = X_\gamma$ for all $\gamma \in [0, 1]$. The restriction $\beta \le 1$ is irrelevant, because if $\beta \ge 1$, then $T \in \mathcal{B}(X_1)$, and we can repeat the same argument with β replaced by one. Thus, the conclusion that we have established so far is valid for all $\beta > 0$ when $\alpha = 1$.

Next we look at another special case, namely the one where $\beta = 0$ and $0 \le \alpha < 1$. The proof is very similar to the one above, so let us only indicate the differences, and leave the details to the reader. The estimate (3.10.9) is still valid. This time we write $\lambda - A - T$ in the form

$$\lambda - A - T = (1 - T(\lambda - A)^{-1})(\lambda - A),$$

where $\lambda - A$ still maps X_1 one-to-one onto X. To prove that $\lambda - A - T$ maps X_1 one-to-one onto X it suffices to show that $1 - T(\lambda - A)^{-1}$ is invertible in $\mathcal{B}(X)$ for sufficiently large $|\lambda|$, and this is done by using the same type of estimates as we saw above. In particular, we find that $\|T(\lambda - A)^{-1}\|_{\mathcal{B}(X)} \le \frac{1}{2}$

and that (3.10.10) holds for $|\lambda|$ large enough. Thus, $A + T$ generates an analytic semigroup on X.

If $\alpha < 0$ and $\beta = 0$, then $T \in \mathcal{B}(X)$, and we may simply replace α by zero in the preceding argument. Thus, $A + T$ generates an analytic semigroup on X whenever $\alpha < 1$ and $\beta = 0$.

Let us finally return to the general case where $\alpha, \beta \in \mathbb{R}$ with $\alpha - \beta < 1$. To deal with this case we replace X by the new state space $X' = X_{\beta-\epsilon}$, where ϵ is an arbitrary number satisfying $0 < \epsilon \le \beta - \alpha + 1$. With respect to this new state space the index α is replaced by $\alpha' = \alpha - \beta + \epsilon \le 1$, β is replaced by $\beta' = \epsilon > 0$, and $T \in \mathcal{B}(X'_{\alpha'}; X'_{\beta'})$. This implies that $T \in \mathcal{B}(X'_1; X'_{\beta'})$ (since $\alpha' \le 1$), so we can apply the first special case where we had $\alpha = 1$ and $\beta > 0$ (replacing X by X'). We conclude that $(A + T)_{|X'_1}$ generates an analytic semigroup $\mathfrak{A}^T_{|X'}$ on $X' = X_{\beta-\epsilon}$ and that $X'^{\,T}_{\gamma'} = X'_{\gamma'}$ for all $\gamma' \in [0, 1]$, where $X'^{\,T}_{\gamma'}$ are the fractional spaces constructed by means of the semigroup $\mathfrak{A}^T_{|X'}$ as described in Section 3.9. Moreover, by restricting $\mathfrak{A}^T_{|X'}$ to $X'^{\,T}_{\gamma'} = X'_{\gamma'}$ we get an analytic semigroup $\mathfrak{A}^T_{|X'_{\gamma'}}$ on $X'_{\gamma'}$ for all $\gamma' \in [0, 1]$. The above ϵ could be any number in $(0, \beta - \alpha + 1]$. In particular, taking $\epsilon = \beta - \alpha + 1$ we get a semigroup $\mathfrak{A}^T_{|X_{\alpha-1}}$ generated by $(A + T)_{|X_\alpha}$ on $X_{\alpha-1}$.

One drawback with the above construction is that the fractional spaces $X'^{\,T}_{\gamma'}$ that we get depend on X', hence on the choice of the parameter ϵ. However, as we saw in Section 3.9, we get the same spaces if we replace X' by any of the spaces $X'^{\,T}_{\gamma'}$, and adjust the index accordingly. Since $X'^{\,T}_{\gamma'} = X'_{\gamma'}$ for all $\gamma' \in [0, 1]$, this means that we may fix one particular value of the parameter ϵ (for example, $\epsilon = \beta - \alpha + 1$ which gives $X' = X_{\alpha-1}$), and base the definition of the fractional spaces induced by $A + T$ on this fixed value of ϵ. After that, by letting ϵ vary in $(0, \beta - \alpha + 1]$ we find that for all $\gamma \in [\alpha - 1, \beta)$, the operator $(A + T)_{|X_{\gamma+1}}$ generates the analytic semigroup $\mathfrak{A}^T_{|X_\gamma}$ on X_γ, where $\mathfrak{A}^T_{|X_\gamma}$ is the restriction of $\mathfrak{A}^T_{|X_{\alpha-1}}$ to X_γ. Moreover, for $\gamma \in [\beta, \beta + 1)$, the restriction $\mathfrak{A}^T_{|X_\gamma}$ of $\mathfrak{A}^T_{|X_{\alpha-1}}$ to X_γ is an analytic semigroup on X_γ. By Theorem 3.14.14, the generator of this semigroup is the part of $A + T$ in X_γ. If $0 \in [\alpha - 1, \beta + 1)$, then we may base our definition of the fractional spaces induced by $A + T$ on the original state space X.

We have now proved most of Theorem 3.10.11. The only open question is whether we can also take $\gamma = \beta$ (instead of $\gamma < \beta$). To deal with this case we replace the state space X by $X' = X_\beta$. Then α is replaced by $\alpha' = \alpha - \beta < 1$ and β is replaced by $\beta' = 0$. By the second of the two special cases that we studied above, $(A + T)_{|X'_1}$ generates an analytic semigroup $\mathfrak{A}^T_{|X_\beta}$ on $X' = X_\beta$ and $X'^{\,T}_1 = X'_1 = X_{\beta+1}$. Thus, $(A + T)_{|X_{\beta+1}}$ generates an analytic semigroup on X_β and $X^T_{\beta+1} = X_{\beta+1}$. Furthermore, the restriction of $\mathfrak{A}^T_{|X_\beta}$ to $X_{\beta+1}$ is an analytic semigroup on this space. □

3.11 Spectrum determined growth

One of the most important properties of a semigroup $\mathfrak{A}$ is its growth bound $\omega_{\mathfrak{A}}$. We defined $\omega_{\mathfrak{A}}$ in Definition 2.5.6 in terms of the behavior of $\|\mathfrak{A}^t\|$ as $t \to \infty$. In practice it is often more convenient to use characterizations of $\omega_{\mathfrak{A}}$ which refer to the generator A of $\mathfrak{A}$ and not directly to $\mathfrak{A}$.

There is one obvious condition that the generator A of $\mathfrak{A}$ must satisfy: by Theorem 3.4.1 the half-plane $\mathbb{C}^+_{\omega_{\mathfrak{A}}}$ belongs to the resolvent set of A. Thus, it is always true that

$$\omega_{\mathfrak{A}} \geq \sup\{\Re\lambda \mid \lambda \in \sigma(A)\}. \tag{3.11.1}$$

If the converse inequality also holds, then we say that $\mathfrak{A}$ has the *spectrum determined growth property*:

Definition 3.11.1 The C_0 semigroup $\mathfrak{A}$ has the *spectrum determined growth property* if

$$\omega_{\mathfrak{A}} = \sup\{\Re\lambda \mid \lambda \in \sigma(A)\}. \tag{3.11.2}$$

Example 3.11.2 *Suppose that the semigroup $\mathfrak{A}$ on X has the spectrum determined growth property. Then so do the following semigroups derived from $\mathfrak{A}$:*

(i) *the exponentially shifted semigroup $t \mapsto e^{\alpha t}\mathfrak{A}^t$ for every $\alpha \in \mathbb{C}$ (see Example 2.3.5);*
(ii) *the time compressed semigroup $t \mapsto \mathfrak{A}^{\lambda t}$ for every $\lambda > 0$ (see Example 2.3.6;*
(iii) *the similarity transformed semigroup $t \mapsto E^{-1}\mathfrak{A}^t E$ for every invertible $E \in \mathcal{B}(X_1; X)$ (see Example 2.3.7).*

The easy proof is left to the reader.

Some other classes of semigroups which have the spectrum determined growth property are the following:

Example 3.11.3 *The following semigroups have the spectrum determined growth property:*

(i) *the left shift semigroups τ, τ_+, τ_-, $\tau_{[0,T)}$, and $\tau_{\mathbb{T}_T}$ in Examples 2.3.2 and 2.5.3;*
(ii) *diagonal semigroups (see Example 3.3.3 and Definition 3.3.4);*
(iii) *analytic semigroups (see Definition 3.10.1).*

Proof (i) See Examples 2.5.3 and 3.3.1.

(ii) See Examples 3.3.3 and 3.3.5.

(iii) We know that (3.11.1) holds for all semigroups, so it suffices to prove the opposite inequality. Choose $\omega \in \mathbb{R}$ so that $\Re\lambda < \omega - \epsilon$ for some $\epsilon > 0$ and all $\omega \in \sigma(A)$. The fact that A is sectorial on some sector $\Sigma_{\theta,\omega'}$ for some $\pi/2 < \theta < \pi$ and some $\omega' \in \mathbb{R}$ then implies that condition (ii) in Theorem 3.10.6 holds. By the same theorem, the growth bound of $\mathfrak{A}$ is at most ω. Since $\epsilon > 0$ is arbitrary we get $\omega_{\mathfrak{A}} \le \sup\{\Re\lambda \mid \lambda \in \sigma(A)\}$. □

Example 3.11.3(iii) is a special case of the following theorem:

Theorem 3.11.4 *A semigroup $\mathfrak{A}^t$ on a Banach space which is continuous in the operator norm on an interval $[t_0, \infty)$, where $t_0 \ge 0$, has the spectrum determined growth property.*

Proof Let $\omega_A = \sup\{\Re\lambda \mid \lambda \in \sigma(A)\}$. By Theorem 2.5.4(i), it suffices to show that, for some $t > 0$, the spectral radius of $\mathfrak{A}^t$ is equal to $e^{\omega_A t}$. However, this follows from the operator norm continuity of $\mathfrak{A}^t$ which implies that $e^{t\sigma(A)} \subset \sigma(\mathfrak{A}^t) \subset \{0\} \cup e^{t\sigma(A)}$; see Davies (1980, Theorems 2.16 and 2.19). □

Corollary 3.11.5 *The semigroup $\mathfrak{A}^t$ on the Banach space X has the spectrum determined growth property (at least) in the following cases:*

(i) *$t \mapsto \mathfrak{A}^t x$ is differentiable on $[t_0, \infty)$ for some $t_0 \ge 0$ and all $x \in X$, or equivalently, $\mathcal{R}(\mathfrak{A}^{t_0}) \subset \mathcal{D}(A)$ for some $t_0 \ge 0$.*
(ii) *$\mathfrak{A}^t$ is compact for all $t \ge t_0$, where $t_0 \ge 0$, or equivalently, $\mathfrak{A}^{t_0}$ is compact for some $t_0 \ge 0$.*

Proof This follows from Theorem 3.11.4 and the fact that in both cases, $t \mapsto \mathfrak{A}^t$ is continuous in the operator norm on $[t_0, \infty)$. See, for example, Pazy (1983, Theorem 3.2 and Lemma 4.2). □

One way of looking at the spectrum determined growth property is to interpret it as a condition that the generator A must not have a spectrum 'at infinity' in the right half-plane $\mathbb{C}^+_{\omega_{\mathfrak{A}}}$. This idea can be made more precise. Theorem 3.4.1 not only implies that the half-plane $\mathbb{C}^+_{\omega_{\mathfrak{A}}}$ belongs to the resolvent set of A, but in fact,

$$\sup_{\Re\lambda \ge \omega_{\mathfrak{A}} + \epsilon} \|(\lambda - A)^{-1}\| < \infty, \qquad \epsilon > 0.$$

As the following theorem shows, this condition can be used to determine $\omega_{\mathfrak{A}}$ whenever the state space is a Hilbert space.

Theorem 3.11.6 *Let A be the generator of a C_0 semigroup $\mathfrak{A}$ on a Hilbert space X. Then*

$$\omega_{\mathfrak{A}} = \inf\Big\{\omega \in \mathbb{R} \;\Big|\; \sup_{\Re\lambda \ge \omega} \|(\lambda - A)^{-1}\| < \infty\Big\}.$$

The proof of this theorem will be given in Section 10.3 (there this theorem is reformulated as Theorem 10.3.7).

This theorem has the following interesting consequence:

Lemma 3.11.7 *Let A be the generator of a C_0 semigroup $\mathfrak{A}$ on a Hilbert space X, and suppose that, for some $\omega \in \mathbb{R}$, some $M > 0$, and some $0 < \epsilon \le 1$, the half-plane $\mathbb{C}^+_\omega$ belongs to the resolvent set of A, and*

$$\|(\lambda - A)^{-1}\| \le M(\Re\lambda - \omega)^{\epsilon-1}, \qquad \Re\lambda > \omega. \tag{3.11.3}$$

Then the growth bound $\omega_{\mathfrak{A}}$ of $\mathfrak{A}$ satisfies

$$\omega_{\mathfrak{A}} \le \omega - \delta,$$

where $\delta > 0$ is given by

$$\delta = \begin{cases} 1/M, & \epsilon = 1, \\ \epsilon(1-\epsilon)^{1/\epsilon - 1} M^{-1/\epsilon}, & 0 < \epsilon < 1. \end{cases}$$

In particular, we conclude that (3.11.3) cannot possibly hold with $\omega = \omega_{\mathfrak{A}}$ (because that would imply $\omega_{\mathfrak{A}} < \omega_{\mathfrak{A}}$). (It cannot hold for $\epsilon > 1$, either.)

Proof The proof is based on Lemma 3.2.8(ii) and Theorem 3.11.6. The case $\epsilon = 1$ follows directly from these two results (the line $\Re\lambda = \omega$ must belong to the resolvent set in this case, and, by continuity, (3.11.3) holds for $\Re\lambda = \omega$, too).

If $0 < \epsilon < 1$, then we take α in Lemma 3.2.8(ii) to lie on the line

$$\Re\alpha = \omega + (1-\epsilon)^{1/\epsilon} M^{-1/\epsilon}$$

(this is the choice that will maximize the constant δ), and take λ to have $\Re\lambda \le \Re\alpha$ and $\Im\lambda = \Im\alpha$. Then, by Lemma 3.2.8(ii), λ will belong to the resolvent set of A if

$$\Re(\alpha - \lambda) = \alpha - \lambda < 1/\|(\alpha - A)^{-1}\|,$$

which by (3.11.3) is true whenever $\Re(\alpha - \lambda) < 1/\kappa$, where

$$\kappa = M^{-1}(\Re\alpha - \omega)^{\epsilon - 1} = (1-\epsilon)^{1/\epsilon - 1} M^{-1/\epsilon}.$$

This can equivalently be rewritten as

$$\Re\alpha - 1/\kappa = \omega - \delta < \Re\lambda \le \omega + (1-\epsilon)^{1/\epsilon} M^{-1/\epsilon}.$$

Moreover, by the same lemma, $\|(\lambda - A)^{-1}\|$ is bounded by

$$\begin{aligned} \|(\lambda - A)^{-1}\| &\le \frac{\|(\lambda - A)^{-1}\|}{1 - \|(\lambda - A)^{-1}\|(\alpha - \lambda)} \\ &\le \frac{\kappa}{1 - \kappa(\alpha - \lambda)} = \frac{1}{\Re\lambda - (\omega - \delta)}. \end{aligned}$$

The conclusion now follows from Theorem 3.11.6. □

There is a related result which connects the growth bound of a semigroup to some L^p-estimates in the time domain:

Theorem 3.11.8 *Let $\mathfrak{A}$ be a C_0 semigroup on the Banach space X and let $\omega \in \mathbb{R}$. Then the following conditions are equivalent:*

(i) $\omega_{\mathfrak{A}} < \omega$;

(ii) $e^{-\omega t}\|\mathfrak{A}^t\| \to 0$ *as* $t \to \infty$;

(iii) $\|\mathfrak{A}^t\| < e^{\omega t}$ *for some* $t > 0$;

(iv) *for all* $x_0 \in X$, *the function* $x(t) = \mathfrak{A}^t x_0$, $t \ge 0$, *belongs to* $L^p_\omega(\mathbb{R}^+; X)$ *for all* $p \in [1, \infty]$;

(v) *for some* $p \in [1, \infty)$ *and all* $x_0 \in X$, *the function* $x(t) = \mathfrak{A}^t x_0$, $t \ge 0$, *belongs to* $L^p_\omega(\mathbb{R}^+; X)$;

(vi) *for all* $q \in [1, \infty]$, *there is a finite constant* M_q *such that, for all* $u \in C_c(\mathbb{R}^-; X)$,

$$\left| \int_{-\infty}^{0} \mathfrak{A}^{-s} u(s)\, ds \right|_X \le M_q \|u\|_{L^q_\omega(\mathbb{R}^-;X)};$$

(vii) *for some* $q \in (1, \infty]$, *some finite constant* M_q, *and all* $u \in C^1_c(\mathbb{R}^-; X_1)$,

$$\left| \int_{-\infty}^{0} \mathfrak{A}^{-s} u(s)\, ds \right|_X \le M_q \|u\|_{L^q_\omega(\mathbb{R}^-;X)};$$

(viii) *for all* $p \in [1, \infty]$, *there is a finite constant* M_p *such that, for all* $u \in C_c(\mathbb{R}; X)$,

$$\left\| t \mapsto \int_{-\infty}^{t} \mathfrak{A}^{t-s} u(s)\, ds \right\|_{L^p_\omega(\mathbb{R};X)} \le M_p \|u\|_{L^p_\omega(\mathbb{R};X)};$$

(ix) *for some* $p \in [1, \infty]$, *some finite constant* M_p, *and all* $u \in C^1_c(\mathbb{R}^+; X)$,

$$\left\| t \mapsto \int_{0}^{t} \mathfrak{A}^{t-s} u(s)\, ds \right\|_{L^p_\omega(\mathbb{R}^+;X)} \le M_p \|u\|_{L^p_\omega(\mathbb{R}^+;X)}.$$

In particular, if $\omega = 0$, *then all the preceding conditions are equivalent to the exponential stability of* $\mathfrak{A}$.

Proof Without loss of generality, we take $\omega = 0$ (see Examples 2.3.5 and 3.2.6).

(i) ⇒ (ii) and (i) ⇒ (iv): These follow from Theorem 2.5.4(i).

(i) ⇒ (vi): Use Theorem 2.5.4(i) and Hölder's inequality.

(i) ⇒ (viii): Apply Theorem 2.5.4(ii) to the system in Proposition 2.3.1 with $B = C = 1$ and $D = 0$.

(ii) ⇒ (iii), (iv) ⇒ (v), (vi) ⇒ (vii), and (viii) ⇒ (ix): These implications are obvious.

(iii) $\Rightarrow$ (i): Clearly, the spectral radius of $\mathfrak{A}^t$ is less than one, and Theorem 2.5.4(i) implies that $\omega_{\mathfrak{A}} < 0$.

(v) $\Rightarrow$ (ii): The operator $x_0 \mapsto x$ is continuous $X \to C(\overline{\mathbb{R}}^+; X)$, hence it is closed $X \to L^p(\mathbb{R}^+; X)$. By the closed graph theorem, there is a constant $M_p > 0$ such that

$$\left(\int_0^\infty |\mathfrak{A}^s x_0|_X^p \, ds\right)^{1/p} \leq M_p |x_0|_X, \qquad x_0 \in X. \tag{3.11.4}$$

By Theorem 2.5.4(i), there are constants $\alpha > 0$ and $M > 0$ such that $\|\mathfrak{A}^t\| \leq M e^{\alpha t}$. Therefore, for all $x_0 \in X$ and $t > 0$,

$$\begin{aligned}\frac{1 - e^{-p\alpha t}}{p\alpha} |\mathfrak{A}^t x_0|_X^p &= \int_0^t e^{-p\alpha s} \, ds |\mathfrak{A}^t x_0|_X^p = \int_0^t e^{-p\alpha s} |\mathfrak{A}^s \mathfrak{A}^{t-s} x_0|_X^p \, ds \\ &\leq \int_0^t \|e^{-\alpha s} \mathfrak{A}^s\|^p |\mathfrak{A}^{t-s} x_0|_X^p \, ds \leq M^p \int_0^t |\mathfrak{A}^{t-s} x_0|_X^p \, ds \\ &\leq M^p M_p^p |x_0|_x^p.\end{aligned}$$

This implies that there is a finite constant M_∞ such that $\|\mathfrak{A}^t\| \leq M_\infty$ for $t \geq 0$. We can now repeat the same computation with $\alpha = 0$ to get

$$\begin{aligned}t \|\mathfrak{A}^t x_0\|_X^p &= \int_0^t |\mathfrak{A}^t x_0|_X^p \, ds \leq \int_0^t \|\mathfrak{A}^s\|^p |\mathfrak{A}^{t-s} x_0|_X^p \, ds \\ &\leq M_\infty^p \int_0^t |\mathfrak{A}^{t-s} x_0|_X^p \, ds \leq M_\infty^p M_p^p |x_0|_x^p,\end{aligned}$$

which implies that

$$\|\mathfrak{A}^t\| \leq (M_p M_\infty) t^{-1/p}.$$

(vii) $\Rightarrow$ (ii): Let $t > 0$. If $q < \infty$ then we can use the density of $C^1([-t, 0); X_1)$ in $L^q([-t, 0); X)$, and if $q = \infty$ then we can use a simple approximation argument and the Lebesgue dominated convergence theorem to show that, for all $t > 0$ and all $u \in C([0, t); X)$,

$$\left| \int_0^t \mathfrak{A}^s u(s) \, ds \right|_X = \left| \int_{-t}^0 \mathfrak{A}^{-s} u(-s) \, ds \right|_X \leq M_q \|u\|_{L^q([0,t);X)}.$$

In particular, taking $u(s) = e^{\alpha s} \mathfrak{A}^{t-s} x_0$, where $\alpha > 0$ and $x_0 \in X$ we get

$$\begin{aligned}\frac{e^{\alpha t} - 1}{\alpha} |\mathfrak{A}^t x_0|_X &= \left| \int_0^t e^{\alpha s} \mathfrak{A}^t x_0 \, ds \right|_X = \left| \int_0^t \mathfrak{A}^s e^{\alpha s} \mathfrak{A}^{t-s} x_0 \, ds \right|_X \\ &\leq e^{\alpha t} M_q K_q |x_0|_X,\end{aligned}$$

where K_q is the (finite) L^q-norm over $\mathbb{R}$ of the function $s \mapsto e^{-\alpha s} \|\mathfrak{A}^s\|$. This implies that $\|\mathfrak{A}^t\| \leq M_1$ for some finite M_1 and all $t \geq 0$. We then repeat the same computation with $\alpha = 0$ to get

$$t|\mathfrak{A}^t x_0|_X = \left|\int_0^t \mathfrak{A}^t x_0\,ds\right|_X = \left|\int_0^t \mathfrak{A}^s\mathfrak{A}^{t-s} x_0\,ds\right|_X$$
$$\le M_q \left\|(s \mapsto \|\mathfrak{A}^s\|)\right\|_{L^q([0,t))}|x_0|_X \le M_q M_1 t^{1/q}|x_0|_X,$$

i.e., $\|\mathfrak{A}^t\| \le M_q M_1 t^{1/q-1}$.

(ix) $\Rightarrow$ (i): First we consider the case $p = \infty$. In (ix) we can replace $C_c^1(\mathbb{R}^+; X)$ by $C_c^1(\mathbb{R}; X)$ if we at the same time replace $L^\infty(\mathbb{R}^+; X)$ by $L^\infty(\mathbb{R}; X)$ and $\int_0^t \mathfrak{A}^{t-s}u(s)\,ds$ by $\int_{-\infty}^t \mathfrak{A}^{t-s}u(s)\,ds$ since the convolution operator $t \mapsto \int_{-\infty}^t \mathfrak{A}^{t-s}u(s)\,ds$ is time-invariant (shift $u \in C_c^1(\mathbb{R}; X)$ to the right until it is supported on $\mathbb{R}^+$, apply the convolution operator, and then shift the result back). The integral $\int_{-\infty}^t \mathfrak{A}^{t-s}u(s)\,ds$ is continuous in t, and by evaluating this integral at zero we conclude that (vii) holds with $q = \infty$. As we saw above, this implies (i).

In the case $p < \infty$ we argue as follows. As $C_c^1(\mathbb{R}^+; X)$ is dense in $L^p(\mathbb{R}^+; X)$, we can weaken the condition $u \in C_c^1(\mathbb{R}^+; X)$ to $u \in L^p(\mathbb{R}^+; X)$, and the same estimate still holds. Let $x_0 \in X$, and define

$$u(t) = \begin{cases} \mathfrak{A}^t x_0, & 0 \le t < 1, \\ 0, & t \ge 1, \end{cases}$$

and $y(t) = \int_0^t \mathfrak{A}^{t-s}u(s)\,ds$. Then

$$y(t) = \begin{cases} t\mathfrak{A}^t x_0, & 0 \le t < 1, \\ \mathfrak{A}^t x_0, & t \ge 1, \end{cases}$$

and by (ix), $y \in L^p(\mathbb{R}^+; X)$. In particular, $t \mapsto \mathfrak{A}^t x_0 \in L^p(\mathbb{R}^+; X)$ for every $x_0 \in X$. Thus (v) is satisfied, hence so is (i). □

3.12 The Laplace transform and the frequency domain

Some of the shift semigroups in Examples 2.3.2 and 2.5.3 and their generators and resolvents described in Examples 3.3.1 and 3.3.2 have simple frequency domain descriptions, which will be presented next. When we say 'frequency domain description' we mean a description given in terms of Laplace transforms of the original functions.

Definition 3.12.1 The *(right-sided) Laplace transform* of a function $u \in L^1_{\mathrm{loc}}(\mathbb{R}^+; U)$ is given by

$$\hat{u}(z) = \int_0^\infty \mathrm{e}^{-zt}u(t)\,dt$$

for all $z \in \mathbb{C}$ for which the integral converges (absolutely). The *left-sided Laplace transform* of a function $u \in L^1_{\mathrm{loc}}(\mathbb{R}^-;U)$ is given by

$$\hat{u}(z) = \int_{-\infty}^{0} \mathrm{e}^{-zt} u(t)\, dt$$

for all $z \in \mathbb{C}$ for which the integral converges (absolutely). The *bilateral Laplace transform* of a function $u \in L^1_{\mathrm{loc}}(\mathbb{R};U)$ is given by

$$\hat{u}(z) = \int_{-\infty}^{\infty} \mathrm{e}^{-zt} u(t)\, dt$$

for all $z \in \mathbb{C}$ for which the integral converges (absolutely). The *finite Laplace transform over the interval* $[0, T]$ of a function $u \in L^1([0, T];U)$ is given by

$$\hat{u}(z) = \int_{0}^{T} \mathrm{e}^{-zt} u(t)\, dt$$

for all $z \in \mathbb{C}$.

The finite Laplace transform is always an entire function. It is easy to see that the domains of definition of the other Laplace transforms, if nonempty, are vertical strips $\{z \in \mathbb{C} \mid \Re z \in J\}$, where J is an interval in $\mathbb{R}$ (open, closed, or semi-closed, bounded or unbounded). The one-sided, left-sided, and finite Laplace transforms can be interpreted as special cases of the bilateral Laplace transform, namely, the case where u vanishes on the complements of $\mathbb{R}^+$, $\mathbb{R}^-$, or $[0, T]$. For the one-sided Laplace transform, either J is empty or the right end-point of J is $+\infty$, and for the left-sided Laplace transform, either J is empty or the left end-point of J is $-\infty$. In the case of the bilateral Laplace transform J may be bounded or unbounded, and it may consist of one point only. In the interiors of their domains all the different Laplace transforms are analytic. For example, if $u \in L^1_c(\mathbb{R};U)$, then the bilateral Laplace transform of u is entire (as in the case of the finite Laplace transform).

All the Laplace transforms listed above determine the original function u uniquely (on its given domain), whenever the domain of the Laplace transform is nonempty. To prove this it suffices to consider the case of the bilateral transform (since the others can be reduced to this one). If $\hat{u}(z)$ is defined for some $z \in \mathbb{C}$, then $u \in L^1_\alpha(\mathbb{C};U)$ where $\alpha = \Re z$. Under some further conditions we can get the following explicit formula for u in terms of $\hat{u}$.

Proposition 3.12.2 *Let $u \in L^1_\alpha(\mathbb{R};U)$, where U is a Banach space, and suppose that the bilateral Laplace transform $\hat{u}$ of u satisfies $\hat{u} \in L^1(\alpha + j\mathbb{R};U)$. Define*

$$v(t) = \frac{1}{2\pi} \int_{-\infty}^{\infty} \mathrm{e}^{(\alpha + j\omega)t} \hat{u}(\alpha + j\omega)\, d\omega, \qquad t \in \mathbb{R}.$$

Then $u = v$ almost everywhere, $v \in BC_{\alpha,0}(\mathbb{R};U)$, and $\hat{u} \in BC_0(\alpha + j\mathbb{R};U)$.

Proof In the scalar case with $\alpha = 0$ this is shown in Rudin (1987, Theorem 9.12, p. 185). We can introduce a general $\alpha \in \mathbb{R}$ by applying the same result with $u(t)$ replaced by $\mathrm{e}^{-\alpha t}u(t)$. The proof of the vector-valued case is identical to the proof of the scalar case given in Rudin (1987). □

It follows from Proposition 3.12.2 that the Laplace transform is injective: if two functions u and v have the same Laplace transform, defined on the same vertical line $\Re z = \alpha$, then they must be equal almost everywhere (apply Proposition 3.12.2 to the difference $u - v$).

In the case of the left-sided or right-sided Laplace transform, the most restrictive condition in Proposition 3.12.2 is the requirement that $\hat{u} \in L^1(\alpha + j\mathbb{R}; U)$. If u is continuous with $u(0) \neq 0$, then Proposition 3.12.2 cannot be applied to the one-sided transforms, due to the fact that $\pi_+ u$ and/or $\pi_- u$ will have a jump discontinuity at zero. The following proposition has been designed to take care of this problem.

Proposition 3.12.3 *Let U be a Banach space.*

(i) *Suppose that $u \in L^1_\alpha(\mathbb{R}^+; U)$ and that there exists some $u_0 \in \mathbb{C}$ such that the function $\omega \mapsto \hat{u}(\alpha + j\omega) - (1 + j\omega)^{-1}u_0$ belongs to $L^1(\mathbb{R}; U)$, where $\hat{u}$ is the (right-sided) Laplace transform of u. Let $\beta \in \mathbb{C}^-_\alpha$, and define for all $t \in \mathbb{R}^+$,*

$$v(t) = \mathrm{e}^{\beta t}u_0 + \frac{1}{2\pi}\int_{-\infty}^{\infty} \mathrm{e}^{(\alpha + j\omega)t}\big[\hat{u}(\alpha + j\omega) - (\alpha + j\omega - \beta)^{-1}u_0\big]\,d\omega.$$

Then v is independent of β, $v \in BC_{\alpha,0}(\mathbb{R}^+; U)$, $v(0+) = u_0$, and $u = v$ almost everywhere on $\mathbb{R}^+$.

(ii) *Suppose that $u \in L^1_\alpha(\mathbb{R}^-; U)$ and that there exists some $u_0 \in \mathbb{C}$ such that the function $\omega \mapsto \hat{u}(\alpha + j\omega) - (1 + j\omega)^{-1}u_0$ belongs to $L^1(\mathbb{R}; U)$, where $\hat{u}$ is the left-sided Laplace transform of u. Let $\beta \in \mathbb{C}^+_\alpha$, and define for all $t \in \mathbb{R}^-$,*

$$v(t) = -\mathrm{e}^{\beta t}u_0 + \frac{1}{2\pi}\int_{-\infty}^{\infty} \mathrm{e}^{(\alpha + j\omega)t}\big[\hat{u}(\alpha + j\omega) - (\alpha + j\omega - \beta)^{-1}u_0\big]\,d\omega.$$

Then v is independent of β, $v \in BC_{\alpha,0}(\mathbb{R}^-; U)$, $v(0-) = -u_0$, and $u = v$ almost everywhere on $\mathbb{R}^+$.

Proof The proof is essentially the same in both cases. First we observe that the function $\omega \mapsto \hat{u}(\alpha + j\omega) + (\alpha + j\omega - \beta)^{-1}u_0$ belongs to $L^1_\alpha(\mathbb{R}; U)$ since the difference

$$\frac{1}{\alpha + j\omega - \beta} - \frac{1}{1 + j\omega} = \frac{1 - \alpha + \beta}{(\alpha + j\omega - \beta)(1 + j\omega)}$$

belongs to $L^1(\mathbb{R})$. We get the conclusion of (i) by applying Proposition 3.12.2 to the function $t \mapsto u(t) - e^{\beta t} u_0$ (whose right-sided Laplace transform is $\hat{u}(\lambda) - (\lambda - \beta)^{-1} u_0$ for $\Re\lambda \geq \alpha$), and we get the conclusion of (ii) by applying Proposition 3.12.2 to the function $t \mapsto u(t) + e^{\beta t} u_0$ (whose left-sided Laplace transform is $\hat{u}(\lambda) - (\lambda - \beta)^{-1} u_0$ for $\Re\lambda \leq \alpha$). □

In the proof of Proposition 3.12.3 given above we observe that, although the Laplace transform is injective in the sense that we explained earlier, it is possible to have two functions, one supported on $\mathbb{R}^+$ and another supported on $\mathbb{R}^-$, which have 'the same' one-sided Laplace transforms in the sense that the two Laplace transforms are analytic continuations of each other. As our following theorem shows, this situation is not at all unusual.

Theorem 3.12.4 *Let U be a Banach space, and let f be a U-valued function which is analytic at infinity with $f(\infty) = 0$. Let Γ be a positively oriented piecewise continuously differentiable Jordan curve such that f is analytic on Γ and outside of Γ, and define*

$$u(t) = \frac{1}{2\pi j} \oint_\Gamma e^{\lambda t} f(\lambda)\, d\lambda.$$

Then u is entire, $u(0) = \lim_{\lambda \to \infty} \lambda f(\lambda)$, $u(t) = O(e^{\beta t})$ as $t \to \infty$ and $u(t) = O(e^{-\alpha t})$ as $t \to -\infty$, where $\beta = \sup\{\Re\lambda \mid \lambda \in \Gamma\}$ and $\alpha = \inf\{\Re\lambda \mid \lambda \in \Gamma\}$. Moreover, the restriction of f to $\mathbb{C}^+_\omega$ is the (right-sided) Laplace transform of $\pi_+ u$, and the restriction of f to $\mathbb{C}^-_\alpha$ is the (left-sided) Laplace transform of $-\pi_- u$.

Proof That u is entire and satisfies the two growth bounds follows immediately from the integral representation. The expression for $u(0)$ follows from Lemma 3.9.1(ii). Thus, it only remains to prove that f, suitably restricted, is the right-sided Laplace transform of $\pi_+ u$ and the left-sided Laplace transform of $-\pi_- u$.

We begin with the case where $f(\lambda) = (\lambda - \gamma)^{-1} u_0$, where γ is encircled by Γ (in particular, $\alpha < \Re\gamma < \beta$). Then a direct inspection shows that the theorem is true with $u(t) = e^{\gamma t} u_0$. Since f is analytic at infinity with $f(\infty) = 0$, it has an expansion of the type $f(\lambda) = u_0/\lambda + O(\lambda^{-2})$ as $\lambda \to \infty$. After subtracting $(\lambda - \gamma)^{-1} u_0$ from f, the new function f satisfies $f(\lambda) = O(\lambda^{-2})$ as $\lambda \to \infty$. Thus, we may, without loss of generality, assume that $f(\lambda) = O(\lambda^{-2})$ as $\lambda \to \infty$. Then $u(0) = 0$. (By subtracting off further terms it is even possible to assume that $f(\lambda) = O(\lambda^{-k})$ as $\lambda \to \infty$ for any finite k.)

Choose some $\beta' > \beta$ and $\alpha' < \alpha$, and define, for all $t \in \mathbb{R}$,

$$v(t) = \frac{1}{2\pi} \int_{-\infty}^{\infty} e^{(\beta'+j\omega)t} f(\beta' + j\omega)\, d\omega,$$

$$w(t) = \frac{1}{2\pi} \int_{-\infty}^{\infty} e^{(\alpha'+j\omega)t} f(\alpha' + j\omega)\, d\omega.$$

We claim that $v(t) = 0$ for $t \leq 0$, that $w(t) = 0$ for $t \geq 0$, and that $v(t) - w(t) = u(t)$ for all $t \in \mathbb{R}$. Suppose that this is true. Then $v = \pi_+ u$, $w = -\pi_- u$, v is the inverse right-sided Laplace transform of the restriction of f to the line $\Re\lambda = \beta'$, and w is the inverse left-sided Laplace transform of the restriction of f to the line $\Re\lambda = \alpha'$. Since the inverse Laplace transform is injective (the proof of this is identical to the proof of the fact that the bilateral Laplace transform is injective), the right-sided Laplace transform of $\pi_+ u$ must coincide with u on $\mathbb{C}^+_{\beta'}$ and the left-sided Laplace transform of $-\pi_- u$ must coincide with u on $\mathbb{C}^-_{\alpha'}$. Thus, it only remains to verify the claim that $v(t) = 0$ for $t \leq 0$, that $w(t) = 0$ for $t \geq 0$, and that $v(t) - w(t) = u(t)$ for all $t \in \mathbb{R}$.

We begin with the claim that $v(t) = 0$ for all $t \leq 0$. For each $R > 0$, by Cauchy's theorem, we have

$$\frac{1}{2\pi j} \oint_{\Gamma_R} e^{\lambda t} f(\lambda)\, d\lambda = 0,$$

where Γ_R is the closed path running from $\beta' - jR$ to $\beta' + jR$ along the line $\Re\lambda = \beta'$, and then back to $\beta' - jR$ along a semi-circle in $\overline{\mathbb{C}}^+_{\beta'}$ centered at β'. If $t \leq 0$, then $e^{\lambda t}$ is bounded on $\mathbb{C}^+_{\beta}$, and we can use the fact that $f(\lambda) = O(\lambda^{-2})$ as $\lambda \to \infty$ to let $R \to \infty$ (the length of the semi-circle is πR, hence this part of the integral tends to zero), and get

$$v(t) = \frac{1}{2\pi} \int_{-\infty}^{\infty} e^{(\beta'+j\omega)t} f(\beta' + j\omega)\, d\omega = \lim_{R\to\infty} \frac{1}{2\pi j} \oint_{\Gamma_R} e^{\lambda t} f(\lambda)\, d\lambda = 0.$$

This proves that $v(t) = 0$ for $t \leq 0$. An analogous proof shows that $w(t) = 0$ for $t \geq 0$.

Finally, let us show that $v(t) - w(t) = u(t)$ for all $t \in \mathbb{R}$. By analyticity, for sufficiently large values of R, we can deform the original path Γ given in the theorem to a rectangular path Γ_R, running from $\beta' - jR$ to $\beta' + jR$ to $\alpha' + jR$ to $\alpha' - jR$, and back to $\beta' - jR$, and get for all $t \in \mathbb{R}$,

$$u(t) = \frac{1}{2\pi j} \oint_{\Gamma_R} e^{\lambda t} f(\lambda)\, d\lambda.$$

Letting again $R \to \infty$ we get

$$\begin{aligned} u(t) &= \lim_{R\to\infty} \frac{1}{2\pi j} \oint_{\Gamma_R} e^{\lambda t} f(\lambda)\, d\lambda \\ &= \frac{1}{2\pi} \int_{-\infty}^{\infty} e^{(\beta'+j\omega)t} f(\beta' + j\omega)\, d\omega \\ &\quad - \frac{1}{2\pi} \int_{-\infty}^{\infty} e^{(\alpha'+j\omega)t} f(\alpha' + j\omega)\, d\omega \\ &= v(t) - w(t). \end{aligned}$$

□

Corollary 3.12.5 *Let U be a Banach space.*

(i) *Suppose that $u \in L^1_\alpha(\mathbb{R}^+; U)$ and that $\hat{u}$ is analytic at infinity. For all $t \in \mathbb{R}^+$, define*

$$v(t) = \frac{1}{2\pi j} \oint_\Gamma e^{\lambda t} \hat{u}(\lambda)\, d\lambda,$$

where Γ is a positively oriented piecewise continuously differentiable Jordan curve which encircles $\sigma(\hat{u})$. Then v is the restriction to $\mathbb{R}^+$ of an entire function, and $u = v$ almost everywhere on $\mathbb{R}^+$.

(ii) *Suppose that $u \in L^1_\alpha(\mathbb{R}^-; U)$ and that $\hat{u}$ is analytic at infinity. For all $t \in \mathbb{R}^-$, define*

$$v(t) = \frac{1}{2\pi j} \oint_\Gamma e^{\lambda t} \hat{u}(\lambda)\, d\lambda,$$

where Γ is a negatively oriented piecewise continuously differentiable Jordan curve which encircles $\sigma(\hat{u})$. Then v is the restriction to $\mathbb{R}^-$ of an entire function, and $u = v$ almost everywhere on $\mathbb{R}^-$.

Proof This follows from Theorem 3.12.4. □

By combining the symbolic calculus developed in Section 3.9 with Corollary 3.12.5 we get the following result.

Theorem 3.12.6 *Let A be the generator of a C_0 semigroup $\mathfrak{A}$ on X with growth bound $\omega_{\mathfrak{A}}$. Let $u \in L^1_\omega(\mathbb{R}^-; \mathbb{C})$ for some $\omega > \omega_{\mathfrak{A}}$, and suppose that the left-sided Laplace transform $\hat{u}$ of u is analytic at infinity. Then $\hat{u}(A) \in \mathcal{B}(X; X_1)$ and*

$$\hat{u}(A)x = \int_{-\infty}^{0} u(t)\mathfrak{A}^{-t} x\, dt, \qquad x \in X.$$

Proof The assumptions imply that $\hat{u}$ is analytic on $\mathbb{C}^-_\omega \cup \infty$ with $\hat{u}(\infty) = 0$. In particular, by Lemma 3.9.3, $\hat{u}(A) \in \mathcal{B}(X; X_1)$.

Let Γ be a negatively oriented piecewise continuously differentiable Jordan curve which encircles $\sigma(\hat{u})$ with $\overline{\sigma}(A)$ on the outside. Then Theorem 3.2.9, Corollary 3.12.5, Fubini's theorem, and (3.9.3) give, for all $x \in X$,

$$\begin{aligned}\int_{-\infty}^{0} u(t)\mathfrak{A}^{-t}x\,dt &= \frac{1}{2\pi j}\int_{-\infty}^{0}\oint_{\Gamma} \mathrm{e}^{\lambda t}\hat{u}(\lambda)\,d\lambda\,\mathfrak{A}^{-t}x\,dt \\ &= \frac{1}{2\pi j}\oint_{\Gamma}\hat{u}(\lambda)\int_{-\infty}^{0}\mathrm{e}^{\lambda t}\mathfrak{A}^{-t}x\,dt\,d\lambda \\ &= \frac{1}{2\pi j}\oint_{\Gamma}\hat{u}(\lambda)\int_{0}^{\infty}\mathrm{e}^{-\lambda t}\mathfrak{A}^{t}x\,dt\,d\lambda \\ &= \frac{1}{2\pi j}\oint_{\Gamma}\hat{u}(\lambda)(\lambda - A)^{-1}x\,d\lambda \\ &= \hat{u}(A)x.\end{aligned}$$

□

Corollary 3.12.7 *Let A be the generator of a C_0 semigroup $\mathfrak{A}$ on X with growth bound $\omega_{\mathfrak{A}}$. Let $u \in L^1_{-\omega}(\mathbb{R}^+;\mathbb{C})$ for some $\omega > \omega_{\mathfrak{A}}$, and suppose that the right-sided Laplace transform $\hat{u}$ of u is analytic at infinity. Then $\hat{u}(-A) \in \mathcal{B}(X;X_1)$ and*

$$\hat{u}(-A)x = \int_0^\infty u(t)\mathfrak{A}^t x\,dt, \qquad x \in X. \tag{3.12.1}$$

Proof This follows from Theorem 3.12.6 through a change of integration variables (i.e., we apply Theorem 3.12.6 to the function $t \mapsto u(-t)$). □

It is possible to use (3.12.1) as a *definition* of $\hat{u}(-A)$, and in this way it is possible to extend the functional calculus presented in the first half of Section 3.9 to a larger class of functions of A, namely those that correspond to functions $u \in L^1_{\mathrm{loc}}(\mathbb{R}^+)$ satisfying

$$\int_0^\infty |u(t)|\|\mathfrak{A}^t\|\,dt < \infty \tag{3.12.2}$$

(thus, u should be integrable with respect to the weight function $\|\mathfrak{A}\|$). One such example was the definition of the fractional powers of $\gamma - A$ given in the second half of Section 3.9.

Theorem 3.12.8 *Let A be the generator of a C_0 semigroup $\mathfrak{A}$ on X with growth bound $\omega_{\mathfrak{A}}$. Let $u, v \in L^1_{\omega}(\mathbb{R}^-;\mathbb{C})$ for some $\omega > \omega_{\mathfrak{A}}$, and suppose that the left-sided Laplace transforms of u and v are analytic at infinity. Define $w(t) = \int_t^0 u(t-s)v(s)\,ds$ for all those $t \in \mathbb{R}^-$ for which the integral converges. Then $w \in L^1_{\omega}(\mathbb{R}^-;\mathbb{C})$ (in particular, w is defined for almost all t), $\widehat{w}(\lambda) = \hat{u}(\lambda)\hat{v}(\lambda)$*

for $\lambda \in \mathbb{C}^-_\omega$, *and*

$$\widehat{w}(A) = \hat{u}(A)\hat{v}(A).$$

Proof The function $(s,t) \mapsto u(t-s)v(s)$ is measurable (see, e.g., Hewitt and Stromberg (1965, pp. 396–397)). By Fubini's theorem and a change of integration variable,

$$\begin{aligned}\int_{-\infty}^0 e^{\omega t}|w(t)|\,dt &\le \int_{-\infty}^0 \int_t^0 |e^{\omega(t-s)}u(t-s)||e^{\omega s}v(s)|\,ds\,dt \\ &= \int_{-\infty}^0 \int_{-\infty}^s |e^{\omega(t-s)}u(t-s)|\,dt\,|e^{\omega s}v(s)|\;ds \\ &= \int_{-\infty}^0 |e^{\omega t}u(t)|\,dt \int_{-\infty}^0 |e^{\omega s}v(s)|\,ds < \infty.\end{aligned}$$

This proves that $w \in L^1_\omega(\mathbb{R}^-;\mathbb{C})$. Using Fubini's theorem once more we get for all $\lambda \in \mathbb{C}^-_\omega$,

$$\begin{aligned}\widehat{w}(\lambda) &= \int_{-\infty}^0 e^{-\lambda t}w(t)\,dt \\ &= \int_{-\infty}^0 \int_t^0 e^{-\lambda(t-s)}u(t-s)e^{-\lambda s}v(s)\,ds\,dt \\ &= \int_{-\infty}^0 \int_{-\infty}^s e^{-\lambda(t-s)}u(t-s)\,dt\;e^{-\lambda s}v(s)\;ds \\ &= \int_{-\infty}^0 e^{-\lambda t}u(t)\,dt \int_{-\infty}^0 e^{-\lambda s}v(s)\,ds \\ &= \hat{u}(\lambda)\hat{v}(\lambda).\end{aligned}$$

In particular, $\widehat{w}$ is analytic at infinity. To show that $\widehat{u}(A) = \hat{u}(A)\hat{v}(A)$ it suffices to repeat the computation above with $e^{-\lambda t}$ replaced by $\mathfrak{A}^{-t}$ (and use Theorem 3.12.6). □

Theorem 3.12.9 *Let A be the generator of a C_0 semigroup $\mathfrak{A}$ on X with growth bound $\omega_{\mathfrak{A}}$. Let $u \in L^1_\omega(\mathbb{R}^-;\mathbb{C})$ and $v \in L^1_\omega(\mathbb{R}^-;X)$ for some $\omega > \omega_{\mathfrak{A}}$, and suppose that the left-sided Laplace transforms of u and v are analytic at infinity. Define $w(t) = \int_t^0 u(t-s)v(s)\,ds$ for all those $t \in \mathbb{R}^-$ for which the integral converges. Then $w \in L^1_\omega(\mathbb{R}^-;X)$ (in particular, w is defined for almost all t), $\widehat{w}(\lambda) = \hat{u}(\lambda)\hat{v}(\lambda)$ for $\lambda \in \mathbb{C}^-_\omega$, and*

$$\int_{-\infty}^0 \mathfrak{A}^{-t}w(t)\,dt = \hat{u}(A)\int_{-\infty}^0 \mathfrak{A}^{-t}v(t)\,dt.$$

Proof The proof of this theorem is the same as the proof of Theorem 3.12.8. □

Lemma 3.12.10 *Let $\beta \in \mathbb{R}$ and $\alpha > 0$, and define*

$$f(t) = \frac{1}{\Gamma(\alpha)} e^{\beta t} t^{\alpha - 1}, \qquad t \in \mathbb{R}^+.$$

Then the (right-sided) Laplace transform of f is given by

$$\hat{f}(\lambda) = (\lambda - \beta)^{-\alpha}, \qquad \lambda \in \mathbb{C}^+_\beta.$$

We leave the proof to the reader (one possibility is to make a (complex) change of variable from t to $u = t/(\lambda - \beta)$ in (3.9.7)).

If U is a Hilbert space and $p = 2$, then it is possible to say much more about the behavior of the various Laplace transforms. Some results of this type are described in Section 10.3.

3.13 Shift semigroups in the frequency domain

The following proposition describes how the shift semigroups introduced in Examples 2.3.2 and 2.5.3 behave in the terms of Laplace transforms.

Proposition 3.13.1 *Let U be a Banach space, let $1 \le p \le \infty$, let $\omega \in \mathbb{R}$, and let $T > 0$.*

(i) *Let τ^t be the bilateral shift $(\tau^t u)(s) = u(s + t)$ for all $s, t \in \mathbb{R}$ and all $u \in L^1_{\mathrm{loc}}(\mathbb{R}; U)$. Then, for all $t \in \mathbb{R}$ and all $u \in L^1_{\mathrm{loc}}(\mathbb{R}; U)$, the bilateral Laplace transforms $\widehat{\tau^t u}$ and $\hat{u}$ of $\tau^t u$, respectively u, have the same domain (possibly empty), and for all z in this common domain,*

$$\widehat{\tau^t u}(z) = e^{zt} \hat{u}(z).$$

(ii) *Let $\tau^t_+ = \pi_+ \tau^t$ for all $t \ge 0$, where τ^t is the bilateral shift defined in (i). Then, for all $t \ge 0$ and all $u \in L^1_{\mathrm{loc}}(\mathbb{R}^+; U)$, the (one-sided) Laplace transforms $\widehat{\tau^t_+ u}$ and $\hat{u}$ of $\tau^t_+ u$, respectively u, have the same domain (possibly empty), and for all z in this common domain,*

$$\widehat{\tau^t_+ u}(z) = e^{zt} \int_t^\infty e^{-zs} u(s)\, ds.$$

In particular, if $u \in L^p_\omega(\mathbb{R}^+; U)$, then $\widehat{\tau^t u}$ and $\hat{u}$ are defined at least for all $z \in \mathbb{C}^+_\omega$ and the above formula holds.

(iii) *Let $\tau^t_- = \tau^t \pi_-$ for all $t \ge 0$, where τ^t is the bilateral shift defined in (i). Then, for all $t \ge 0$ and all $u \in L^1_{\mathrm{loc}}(\mathbb{R}^-; U)$, the left-sided Laplace transforms $\widehat{\tau^t_- u}$ and $\hat{u}$ of $\tau^t_- u$, respectively u, have the same domain (possibly empty), and for all z in this common domain,*

$$\widehat{\tau^t_- u}(z) = e^{zt} \hat{u}(z).$$

In particular, if $u \in L^p_\omega(\mathbb{R}^-;U)$*, then* $\widehat{\tau^t u}$ *and* $\hat{u}$ *are defined at least for all* $z \in \mathbb{C}^-_\omega$ *and the above formula holds.*

(iv) *Let* $\tau^t_{[0,T)}$*,* $t \geq 0$*, be the finite left shift introduced in Example 2.3.2(iv). Then, for all* $u \in L^1([0,T];U)$*, the finite Laplace transform of* $\tau^t_{[0,T)}u$ *over* $[0,T]$ *is given by (for all* $z \in \mathbb{C}$*)*

$$\widehat{\tau^t_{[0,T)}u}(z) = \begin{cases} e^{zt}\int_t^T e^{-zs}u(s)\,ds, & 0 \leq t \leq T, \\ 0, & \textit{otherwise}. \end{cases}$$

(v) *Let* $\tau^t_{\mathbb{T}_T}$*,* $t \in \mathbb{R}$*, be the circular left shift introduced in Example 2.3.2(v). Then, for all* $u \in L^1(\mathbb{T}_T;U)$*, the finite Laplace transform of* $\tau^t_{\mathbb{T}_T}u$ *over* $[0,T]$ *is given by (for all* $z \in \mathbb{C}$*)*

$$\widehat{\tau^t_{\mathbb{T}_T}u}(z) = e^{zt}\int_t^{t+T} e^{-zs}u(s)\,ds.$$

In particular, this Laplace transform is a T*-periodic function of* t*.*

Proof The proof is the same in all cases: it suffices to make a change of variable $v = s + t$ in the integral defining the Laplace transform of the shifted u (and adjust the bounds of integration appropriately):

$$\int e^{-zs}u(s+t)\,ds = e^{zt}\int e^{-zv}u(v)\,dv.$$

□

From Proposition 3.13.1 we observe that especially the bilateral shift τ^t and the outgoing shift τ^t_- have simple Laplace transform descriptions, whereas the descriptions of the other shifts are less transparent. Fortunately, all the generators and the resolvents of the various shifts have simple descriptions.

Proposition 3.13.2 *The generators of the (semi)groups* τ^t*,* τ^t_+*,* τ^t_-*,* $\tau^t_{[0,T)}$*, and* $\tau^t_{\mathbb{T}_T}$ *in Examples 2.3.2 and 2.5.3 (see Example 3.2.3), have the following descriptions in terms of Laplace transforms:*

(i) *For all* u *in the domain of the generator* $\frac{d}{ds}$ *of the bilateral left shift group* τ^t *on* $L^1_\omega(\mathbb{R};U)$*, the bilateral Laplace transforms of* u *and* $\dot{u} = \frac{d}{ds}u$ *are defined (at least) on the vertical line* $\Re z = \omega$*, and*

$$\widehat{\dot{u}}(z) = z\hat{u}(z), \qquad \Re z = \omega.$$

(ii) *For all* u *in the domain of the generator* $\frac{d}{ds}_+$ *of the incoming left shift semigroup* τ^t_+ *on* $L^p_\omega(\mathbb{R}^+;U)$ *or on* $BUC_\omega(\mathbb{R}^+;U)$*, the Laplace transforms of* u *and* $\dot{u} = \frac{d}{ds}_+u$ *are defined (at least) on the half-plane* $\mathbb{C}^+_\omega$*, and*

$$\widehat{\dot{u}}(z) = z\hat{u}(z) - u(0), \qquad \Re z > \omega.$$

(iii) *For all u in the domain of the generator $\frac{d}{ds}_-$ of the outgoing left shift semigroup τ_-^t on $L^p_\omega(\mathbb{R}^-;U)$ or on $\{u \in BUC_\omega(\overline{\mathbb{R}}^-;U) \mid u(0)=0\}$, the left-sided Laplace transforms of u and $\dot{u} = \frac{d}{ds}_+ u$ are defined (at least) on the half-plane $\mathbb{C}^-_\omega$, and*

$$\widehat{\dot{u}}(z) = z\hat{u}(z), \qquad \Re z < \omega.$$

(iv) *For all u in the domain of the generator $\frac{d}{ds}_{[0,T)}$ of the finite left shift semigroup $\tau^t_{[0,T)}$ on $L^p([0,T);U)$ or on $\{u \in C([0,T];U) \mid u(T)=0\}$, the finite Laplace transforms over $[0,T]$ of u and $\dot{u} = \frac{d}{ds}_{[0,T)} u$ are related as follows:*

$$\widehat{\dot{u}}(z) = z\hat{u}(z) - u(0), \qquad z \in \mathbb{C}.$$

(v) *For all u in the domain of the generator $\frac{d}{ds}_{\mathbb{T}_T}$ of the circular left shift group $\tau^t_{\mathbb{T}_T}$ on $L^p(\mathbb{T}_T;U)$ or on $C(\mathbb{T}_T;U)$, the finite Laplace transforms over $[0,T]$ of u and $\dot{u} = \frac{d}{ds}_{\mathbb{T}_T} u$ are related as follows:*

$$\widehat{\dot{u}}(z) = z\hat{u}(z) + (e^{-zT} - 1)u(0), \qquad z \in \mathbb{C}.$$

Proof All the proofs are identical: it suffices to integrate by parts in the integral defining the particular Laplace transform, and to take into account possible nonzero boundary terms (i.e., the terms $-u(0)$ and $(e^{-zT} - 1)u(0)$ in (ii), (iv) and (v)). □

Proposition 3.13.3 *The resolvents of the generators $\frac{d}{ds}$, $\frac{d}{ds}_+$, $\frac{d}{ds}_-$, $\frac{d}{ds}_{[0,T)}$, and $\frac{d}{ds}_{\mathbb{T}_T}$ in Examples 3.2.3 and 2.5.3 (see Example 3.3.2), have the following descriptions in terms of Laplace transforms:*

(i) *For all $f \in L^1_\omega(\mathbb{R};U)$ and all λ with $\Re\lambda \neq \omega$, the bilateral Laplace transforms of f and $u = \left(\lambda - \frac{d}{ds}\right)^{-1} f$ are defined (at least) on the vertical line $\Re z = \omega$, and*

$$\hat{u}(z) = (\lambda - z)^{-1}\hat{f}(z), \qquad \Re z = \omega.$$

(ii) *For all $f \in L^p_\omega(\mathbb{R}^+;U)$ or $f \in BUC_\omega(\mathbb{R}^+;U)$ and all $\lambda \in \mathbb{C}^+_\omega$, the Laplace transforms of f and $u = \left(\lambda - \frac{d}{ds}_+\right)^{-1} f$ are defined (at least) on the half-plane $\mathbb{C}^+_\omega$, and*

$$\hat{u}(z) = (\lambda - z)^{-1}(\hat{f}(z) - \hat{f}(\lambda)), \qquad \Re z > \omega.$$

(iii) *For all $f \in L^p_\omega(\mathbb{R}^-;U)$ or $f \in BUC_\omega(\mathbb{R}^-;U)$ with $f(0)=0$ and all λ with $\Re\lambda > \omega$, the left-sided Laplace transforms of f and $u = \left(\lambda - \frac{d}{ds}_-\right)^{-1} f$ are defined (at least) on the half-plane $\mathbb{C}^-_\omega$, and*

$$\hat{u}(z) = (\lambda - z)^{-1}\hat{f}(z), \qquad \Re z < \omega.$$

(iv) *For all* $f \in L^p_\omega([0,T];U)$ *or* $f \in C_\omega([0,T];U)$ *with* $f(T) = 0$ *and all* $\lambda \in \mathbb{C}$, *the finite Laplace transforms over* $[0,T]$ *of* f *and* $u = \big(\lambda - \frac{\mathrm{d}}{\mathrm{d}s}{}_{[0,T)}\big)^{-1} f$ *are related as follows:*

$$\hat{u}(z) = (\lambda - z)^{-1}(\hat{f}(z) - \hat{f}(\lambda)), \qquad z \neq \lambda.$$

(v) *For all* $f \in L^p(\mathbb{T}_T;U)$ *or* $f \in C(\mathbb{T}_T;U)$ *and all* $\lambda \notin \{2\pi jm/T \mid m = 0, \pm 1, \pm 2, \ldots\}$, *the finite Laplace transforms over* $[0,T]$ *of* f *and* $u = \big(\lambda - \frac{\mathrm{d}}{\mathrm{d}s}{}_{\mathbb{T}_T}\big)^{-1} f$ *are related as follows:*

$$\hat{u}(z) = (\lambda - z)^{-1}(\hat{f}(z) - (1 - \mathrm{e}^{-zT})(1 - \mathrm{e}^{-\lambda T})^{-1}\hat{f}(\lambda)), \qquad z \neq \lambda.$$

Proof Throughout this proof the convergence of the indicated Laplace transforms is obvious, so let us only concentrate on the formulas relating $\hat{u}(z)$ to $\hat{f}(z)$.

(i) This follows from Proposition 3.13.2(i), which given $\hat{f}(z) = \lambda\hat{u}(z) - \widehat{\dot{u}}(z) = (\lambda - z)\hat{u}(z)$.

(ii) By Proposition 3.13.2(ii), $\hat{f}(z) = \lambda\hat{u}(z) - \widehat{\dot{u}}(z) = (\lambda - z)\hat{u}(z) + u(0)$. Taking $z = \lambda$ we get $u(0) = \hat{f}(\lambda)$, and so

$$(\lambda - z)\hat{u}(z) = \hat{f}(z) - \hat{f}(\lambda).$$

(iii)–(iv) These proofs are identical to the proofs of (i) and (ii), respectively.

(v) By Proposition 3.13.2(v),

$$\hat{f}(z) = \lambda\hat{u}(z) - \widehat{\dot{u}}(z) = (\lambda - z)\hat{u}(z) + (1 - \mathrm{e}^{-zT})u(0).$$

Taking $z = \lambda$ we get $u(0) = (1 - \mathrm{e}^{-\lambda T})^{-1}\hat{f}(\lambda)$, and so

$$(\lambda - z)\hat{u}(z) = \hat{f}(z) - (1 - \mathrm{e}^{-zT})(1 - \mathrm{e}^{-\lambda T})^{-1}\hat{f}(\lambda).$$

□

3.14 Invariant subspaces and spectral projections

We begin by introducing some terminology.

Definition 3.14.1 We say that the Banach space X is the *direct sum* of Y and Z and write either $X = Y \dotplus Z$ or $X = \left[\begin{smallmatrix} Y \\ Z \end{smallmatrix}\right]$ if Y and Z are closed subspaces of X and every $x \in X$ has a unique representation of the form $x = y + z$ where $y \in Y$ and $z \in Z$. A subspace Y of X is *complemented* if X is the direct sum of Y and some other subspace Z.

Instead of writing $x = y + z$ (corresponding to the notation $X = Y \dotplus Z$) as we did above we shall also use the alternative notation $x = \left[\begin{smallmatrix} y \\ z \end{smallmatrix}\right]$ (corresponding to the notation $X = \left[\begin{smallmatrix} Y \\ Z \end{smallmatrix}\right]$), and we identify $y \in Y$ with $\left[\begin{smallmatrix} y \\ 0 \end{smallmatrix}\right] \in X$. Note that every closed subspace Y of a Hilbert space X is complemented: we may take the complement of $Y \subset X$ to be $Z = Y^{\perp}$.

Lemma 3.14.2 *Let X be a Banach space.*

(i) *If π is an arbitrary projection on X (i.e., $\pi \in \mathcal{B}(X)$ and $\pi = \pi^2$), then $X = \mathcal{N}(\pi) \dotplus \mathcal{R}(\pi)$. (In particular, the range of a projection operator is closed.)*

(ii) *Conversely, to each splitting of X into a direct sum $X = Y \dotplus Z$ there is a unique projection π such that $Y = \mathcal{N}(\pi)$ and $Z = \mathcal{R}(\pi)$. We call π the* projection of X onto Z along Y.

Proof (i) Clearly, every $x \in X$ can be split into $x = \pi x + (1 - \pi)x$ where $\pi x \in \mathcal{R}(\pi)$ and $(1 - \pi)x \in \mathcal{N}(\pi)$ (since $\pi(1 - \pi)x = \pi x - \pi^2 x = 0$). Obviously, if $x \in \mathcal{N}(1 - \pi)$, then $x = \pi x$, so $x \in \mathcal{R}(\pi)$, and conversely, if $x \in \mathcal{R}(\pi)$, then $x = \pi y$ for some $y \in X$, so $\pi x = \pi^2 y = \pi y = x$, i.e., $x \in \mathcal{N}(1 - \pi)$. This means that $\mathcal{R}(\pi) = \mathcal{N}(1 - \pi)$, hence $\mathcal{R}(\pi)$ is closed.

(ii) For each $x \in X$, split x (uniquely) into $x = y + z$ where $y \in Y$ and $z \in Z$. Define $\pi x = y$. Then $\pi x = x$ iff $x \in Y$ and $\pi x = 0$ iff $x \in Z$. It is easy to see that the operator π defined in this way is linear and closed, and that it satisfies $\pi = \pi^2$. By the closed graph theorem, $\pi \in \mathcal{B}(X)$. □

Definition 3.14.3 Let X be a Banach space, and let $A \colon X \supset \mathcal{D}(A) \to X$ be a linear operator.

(i) A subspace Y of X is an *invariant* subspace of A if $Ax \in Y$ for every $x \in \mathcal{D}(A) \cap Y$.

(ii) A pair of subspaces Y and Z of X are *reducing subspaces* of A if $X = Y \dotplus Z$, every $x \in \mathcal{D}(A)$ is of the form $x = y + z$ where $y \in \mathcal{D}(A) \cap Y$ and $z \in \mathcal{D}(A) \cap Z$, and both Y and Z are invariant subspaces of A.

(iii) By an invariant subspace of a C_0 semigroup $\mathfrak{A}$ on X we mean a subspace Y which is an invariant subspace of $\mathfrak{A}^t$ for every $t \geq 0$.

(iv) By a pair of reducing subspaces of a C_0 semigroup $\mathfrak{A}$ on X we mean a pair of subspaces which are reducing subspaces of $\mathfrak{A}^t$ for every $t \geq 0$.

Theorem 3.14.4 *Let $\mathfrak{A}$ be a C_0 semigroup on a Banach space X with generator A and growth bound $\omega_{\mathfrak{A}}$, and let Y be a closed subspace of X. Denote the component of $\rho(A)$ which contains an interval $[\omega, \infty)$ by $\rho_\infty(A)$ (by Theorem 3.2.9(i) such an interval always exists). Then the following conditions are equivalent.*

(i) *Y is an invariant subspace of* $\mathfrak{A}$.
(ii) *Y is an invariant subspace of* $(\lambda - A)^{-1}$ *for some* $\lambda \in \rho_\infty(A)$.
(iii) *Y is an invariant subspace of* $(\lambda - A)^{-1}$ *for all* $\lambda \in \rho_\infty(A)$.
(iv) *Y is an invariant subspace of A and* $\rho(A_{|\mathcal{D}(A)\cap Y}) \cap \rho_\infty(A) \neq \emptyset$.

If these equivalent conditions hold, then it is also true that

(v) $\mathfrak{A}_{|Y}$ *is a C_0 semigroup on Y whose generator is* $A_{|\mathcal{D}(A)\cap Y}$.

Here it is important that Y is *closed*. For example, $\mathcal{D}(A)$ is an invariant subspace of $\mathfrak{A}$ but not of A.

Proof (i) $\Rightarrow$ (ii): If (i) holds, then by Theorem 3.2.9(i), for all $\lambda \in \omega_{\mathfrak{A}}^+$ and all $x \in Y$,

$$(\lambda - A)^{-1}x = \int_0^\infty e^{-\lambda s}\mathfrak{A}^s x \in Y.$$

(ii) $\Rightarrow$ (iii): Take an arbitrary $x^* \in Y^\perp$, where

$$Y^\perp = \{x^* \in X^* \mid \langle x, x^*\rangle_{(X,X^*)} = 0 \text{ for all } x \in Y\}.$$

Fix $x \in Y$, and take some $\lambda_0 \in \rho_\infty(A)$ such that Y is invariant under $(\lambda_0 - A)^{-1}$. Then $(\lambda_0 - A)^{-n}x \in Y$ for all $n = 1, 2, 3, \ldots$, hence $\langle(\lambda_0 - A)^{-n}x, x^*\rangle_{(X,X^*)} = 0$ for all $n = 1, 2, 3, \ldots$ Define $f(\lambda) = \langle(\lambda - A)^{-n}x, x^*\rangle_{(X,X^*)}$. Then f is analytic in $\rho_\infty(A)$, and all its derivatives vanish at the point λ_0 (see (3.2.6)). Therefore $f(\lambda) = \langle(\lambda - A)^{-n}x, x^*\rangle_{(X,X^*)} = 0$ for all $\lambda \in \rho_\infty(A)$. Taking the intersection over all $x^* \in Y^\perp$ we find that $(\lambda - A)^{-n}x \in Y$.

(iii) $\Rightarrow$ (i): If (iii) holds, then by Theorem 3.7.5, for all $t \ge 0$ and all $x \in Y$,

$$\mathfrak{A}^t x = \lim_{n\to\infty}\left(1 - \frac{t}{n}A\right)^{-n} x \in Y.$$

(i) $\Rightarrow$ (iv): For all $x \in \mathcal{D}(A) \cap Y$,

$$Ax = \lim_{h\downarrow 0}\frac{1}{h}(\mathfrak{A}^h - 1)x \in Y.$$

That $\rho(A_{|\mathcal{D}(A)\cap Y}) \cap \rho_\infty(A) \neq \emptyset$ follows from (v) and Theorem 3.2.9(i).

(i) $\Rightarrow$ (v): Trivially, $\mathfrak{A}_{|Y}$ is a C_0 semigroup on Y (since Y is invariant, and we use the same norm in Y as in X). Let us denote its generator by $\widetilde{A}$. We have just seen that $x \in \mathcal{D}(\widetilde{A})$ whenever $x \in \mathcal{D}(A) \cap Y$, and that $\widetilde{A}x = Ax$ for all $x \in \mathcal{D}(A) \cap Y$. Conversely, if $x \in \mathcal{D}(\widetilde{A})$, then $x \in Y$ and $\frac{1}{h}(\mathfrak{A}^h - 1)x$ has a limit in Y as $h \downarrow 0$, so $x \in \mathcal{D}(A) \cap Y$.

(iv) $\Rightarrow$ (ii): Assume (iv). Let $\lambda \in \rho(A_{|\mathcal{D}(A)\cap Y}) \cap \rho_\infty(A)$. Then $\lambda - A$ maps $\mathcal{D}(A) \cap Y$ one-to-one onto Y (since $\lambda \in \rho(A_{|\mathcal{D}(A)\cap Y})$ and $(\lambda - A_{|\mathcal{D}(A)\cap Y})y = (\lambda - A)y$ for every $y \in \mathcal{D}(A) \cap Y$). This implies that $(\lambda - A)^{-1}$ maps Y one-to-one onto $\mathcal{D}(A) \cap Y$. In particular, Y is invariant under $(\lambda - A)^{-1}$. □

Invariance is preserved under the symbolic calculus described at the beginning of Section 3.9.

Lemma 3.14.5 *Let $A \in \mathcal{B}(X)$, let Γ be a piecewise continuously differentiable Jordan curve which encircles $\sigma(A)$ counter-clockwise, and let f be analytic on Γ and inside Γ. Define $f(A)$ by* (3.9.1)*. Then the following claims are true.*

(i) *If Y is a closed invariant subspace of A, then Y is also invariant under $f(A)$.*
(ii) *If Y and Z are a pair of reducing subspaces of A, then they are also reducing for $f(A)$.*

Proof This follows from (3.9.1), the fact that Y and Z are closed, and Theorem 3.14.4. □

In the decomposition of systems into smaller parts we shall encounter closed invariant subspaces contained in the domain of the generator.

Theorem 3.14.6 *Let $A\colon X \supset \mathcal{D}(A) \to X$ be a closed linear operator, and let Y be an invariant subspace of A which is contained in $\mathcal{D}(A)$ and closed in X. Then $Y \subset \bigcap_{n=1}^{\infty} \mathcal{D}(A^n)$, $A_{|Y} \in \mathcal{B}(Y)$, and $(A^n)_{|Y} = (A_{|Y})^n$ for all $n = 1, 2, 3, \ldots$. If A is the generator of a C_0 semigroup $\mathfrak{A}$ on X, then Y is invariant under $\mathfrak{A}$ and $\mathfrak{A}_{|Y}$ is a uniformly continuous semigroup on Y whose generator is $A_{|Y}$.*

Proof The operator $A_{|Y}$ is closed since A is closed. Its domain is all of Y, and therefore, by the closed graph theorem, $A_{|Y} \in \mathcal{B}(Y)$.

Let $x \in Y$. Then $x \in \mathcal{D}(A)$ (since $Y \subset \mathcal{D}(A)$), and $Ax \in Y$ (since Y is invariant). Repeating the same argument with x replaced by Ax we find that $Ax \in \mathcal{D}(A)$, i.e, $x \in \mathcal{D}\left(A^2\right)$, and that $(A^2)_{|Y} = (A_{|Y})^2$. Continuing in the same way we find that $Y \subset \bigcap_{n=1}^{\infty} \mathcal{D}(A^n)$ and that $(A^n)_{|Y} = (A_{|Y})^n$ for all $n = 1, 2, 3, \ldots$

That Y is invariant under $\mathfrak{A}$ follows from Theorem 3.14.4 (every λ with $|\lambda| > \|A_{|Y}\|$ belongs to the resolvent set of $A_{|Y}$). That $\mathfrak{A}_{|Y}$ is uniformly continuous follows from the fact that its generator $A_{|Y}$ is bounded. □

We now turn our attention to subspaces which are *reducing* and not just invariant.

Lemma 3.14.7 *Let $X = Y \dotplus Z$, let π be the projection of X onto Y along Z, and let $A\colon X \supset \mathcal{D}(A) \to X$ be a linear operator. Then Y and Z is a pair of reducing subspaces of A if and only if π maps $\mathcal{D}(A)$ into $\mathcal{D}(A)$ and $\pi Ax = A\pi x$ for all $x \in \mathcal{D}(A)$.*

Proof Assume that π maps $\mathcal{D}(A)$ into itself and that $\pi Ax = A\pi x$ for all $x \in \mathcal{D}(A)$. Then every $x \in \mathcal{D}(A)$ can be split into $x = \pi x + (1-\pi)x$ where $\pi x \in \mathcal{D}(A) \cap Y$ and $(1-\pi)x \in \mathcal{D}(A) \cap Z$. Moreover, for all $y \in \mathcal{D}(A) \cap Y$, $\pi Ay = A\pi y = Ay$, hence $Ay \in Y$, and for all $z \in \mathcal{D}(A) \cap Z$, $\pi Az = A\pi z = 0$, hence $Az \in Z$. Thus both Y and Z are invariant subspaces of A, and Y and Z is a pair of reducing subspaces of A.

Conversely, suppose that Y and Z is a pair of reducing subspaces of A. Then π maps $\mathcal{D}(A)$ into $\mathcal{D}(A)$ (since every $x \in X$ has a unique representation $x = y + z$ with $y \in \mathcal{D}(A) \cap Y$ and $z \in \mathcal{D}(A) \cap Z$, and $\pi x = y$). Moreover, for all $x \in \mathcal{D}(A)$, $\pi A(1-\pi)x = 0$ and $(1-\pi)A\pi = 0$ (since $1-\pi$ is the complementary projection of X onto Z along Y), and

$$\pi A = \pi A\pi = A\pi.$$

□

Theorem 3.14.8 *Let $X = Y \dotplus Z$, let π be the projection of X onto Y along Z, and let $A\colon X \supset \mathcal{D}(A) \to X$ be a linear operator with a nonempty resolvent set. Then the following conditions are equivalent:*

(i) *Y and Z are reducing subspaces of A,*
(ii) *Y and Z are reducing subspaces of $(\lambda - A)^{-1}$ for some $\lambda \in \rho(A)$,*
(iii) *Y and Z are reducing subspaces of $(\lambda - A)^{-1}$ for all $\lambda \in \rho(A)$.*

If, in addition, A is the generator of a C_0 semigroup $\mathfrak{A}$, then (i)–(iii) are equivalent to

(iv) *Y and Z are reducing subspaces of $\mathfrak{A}$.*

Proof Let π denote the projection of X onto Y along Z.

(i) ⇒ (iii): Assume (i). By Lemma 3.14.7, $(\lambda - A)\pi x = \pi(\lambda - A)x$ for all $\lambda \in \mathbb{C}$ and all $x \in \mathcal{D}(A)$. In particular, if we take $\lambda \in \rho(A)$, then we can apply $(\lambda - \widetilde{A})^{-1}$ to both sides of this identity and replace x by $(\lambda - A)^{-1}x$ to get for all $x \in X$,

$$\pi(\lambda - A)^{-1}x = (\lambda - A)^{-1}\pi x.$$

By Lemma 3.14.7, this implies (iii).

(iii) ⇒ (ii): This is trivial.

(ii) $\Rightarrow$ (i): Assume (ii). Then, for all $x \in \mathcal{D}(A)$,

$$\pi x = (\lambda - A)^{-1}\pi(\lambda - A)x.$$

This implies that π maps $\mathcal{D}(A)$ into itself. Applying $(\lambda - A)$ to both sides of this identity we get $(\lambda - A)\pi x = \pi(\lambda - A)x$, or equivalently, $\pi Ax = A\pi x$. By Lemma 3.14.7, this implies (i).

(iv) $\Rightarrow$ (ii): This follows from the fact that (i) implies (ii) in Theorem 3.14.4.

(iii) $\Rightarrow$ (iv): If (iii) holds, then by Theorem 3.7.5, for all $t \geq 0$ and all $x \in X$,

$$\mathfrak{A}^t \pi x = \lim_{n\to\infty}\Big(1 - \frac{t}{n}A\Big)^{-n}\pi x = \pi \lim_{n\to\infty}\Big(1 - \frac{t}{n}A\Big)^{-n} x = \pi\mathfrak{A}^t x.$$

By Lemma 3.14.7, this implies (iv). □

Corollary 3.14.9 *If the equivalent conditions (i)–(iii) in Theorem 3.14.8 hold, then $\rho(A) = \rho(A_{|Y}) \cap \rho(A_{|Z})$, and for all $\lambda \in \rho(A)$,*

$$\begin{aligned}(\lambda - A_{|Y})^{-1} &= (\lambda - A)^{-1}_{|Y}, \quad (\lambda - A_{|Z})^{-1} = (\lambda - A)^{-1}_{|Z},\\ (\lambda - A)^{-1} &= (\lambda - A_{|Y})^{-1}\pi + (\lambda - A_{|Z})^{-1}(1 - \pi),\end{aligned} \tag{3.14.1}$$

where π is the projection of X onto Y along Z.

Proof Let $\lambda \in \mathbb{C}$. Then $\lambda \in \rho(A)$ if and only if the equation $(\lambda - A)x = w$ has a unique solution x for every $w \in X$, and x depends continuously on w. By projecting this equation onto Y and Z (i.e., we multiply the equation by π and $1 - \pi$, where π is the projection onto Y along Z, and recall that π commutes with A) we get the two independent equations

$$(\lambda - A_{|Y})y = \pi w, \qquad (\lambda - A_{|Y})z = (1 - \pi)w,$$

where $y = \pi x \in Y$ and $z = x - y \in Z$. The original equation is solvable if and only if both of these equations are solvable, and this implies that $\rho(A) = \rho(A_+) \cap \rho(A_-)$. Furthermore,

$$\begin{aligned}x &= (\lambda - A)^{-1}w = y + z\\ &= (\lambda - A_{|Y})^{-1}\pi w + (\lambda - A_{|Y})^{-1}(1 - \pi)w,\end{aligned}$$

which gives us (3.14.1). □

One common way to construct invariant subspaces of a semigroup is to use a *spectral projection*. This is possible whenever the spectrum of the generator is not connected.

Theorem 3.14.10 *Let $A \colon X \supset \mathcal{D}(A) \to X$ be a densely defined linear operator with a nonempty resolvent set $\rho(A)$. Let Γ be a positively oriented piecewise continuously differentiable Jordan curve contained in $\rho(A)$ which separates*

$\sigma(A)$ into two nontrivial parts $\sigma(A) = \sigma_+(A) \cup \sigma_-(A)$, where $\sigma_+(A)$ lies inside Γ and $\sigma_-(A)$ lies outside Γ (in particular, $\infty \notin \sigma_+(A)$). Then the operator $\pi \in \mathcal{B}(X)$ defined by

$$\pi = \frac{1}{2\pi j} \oint_\Gamma (\lambda - A)^{-1}\, d\lambda. \tag{3.14.2}$$

is a projection which maps X into $\mathcal{D}(A)$. Denote $X_+ = \mathcal{R}(\pi)$ and $X_- = \mathcal{N}(\pi)$, $A_+ = A_{|X_+}$ and $A_- = A_{|X_-}$. Then the following claims are true.

(i) *$X = X_+ \dot{+} X_-$, and X_+ and X_- are reducing subspaces of A and of $(\lambda - A)^{-1}$ for all $\lambda \in \rho(A)$.*
(ii) *$X_+ \subset \bigcap_{n=1}^\infty \mathcal{D}(A^n)$, $A_+ \in \mathcal{B}(X_+)$, and for all $n = 1, 2, 3, \ldots$, $(A_+)^n = A^n_{|X_+}$ and $(A_-)^n = A^n_{|\mathcal{D}(A^n) \cap X_-}$.*
(iii) *$\sigma(A_+) = \sigma_+(A)$ and $\sigma(A_-) = \sigma_-(A)$.*
(iv) *If A is the generator of a C_0 semigroup $\mathfrak{A}$, then A_+ is the generator of the norm-continuous semigroup $\mathfrak{A}_+ = \mathfrak{A}_{|X_+}$ and A_- is the generator of the C_0 semigroup $\mathfrak{A}_- = \mathfrak{A}_{|X_-}$.*

The projection π constructed above is often referred to as the *Riesz projection* corresponding to the part $\sigma_+(A)$ of $\sigma(A)$.

Proof As in Section 3.6, let us denote $\mathcal{D}(A)$ by X_1. Then the function $\lambda \mapsto (\lambda - A)^{-1}$ is bounded and (uniformly) continuous on Γ with values in $\mathcal{B}(X; X_1)$, so $\pi \in \mathcal{B}(X; X_1)$.

Let $\mu \in \rho(A)$. Then $(\mu - A)^{-1}$ is analytic on Γ, so

$$\begin{aligned}(\mu - A)^{-1}\pi &= (\mu - A)^{-1} \frac{1}{2\pi j} \oint_\Gamma (\lambda - A)^{-1}\, d\lambda \\ &= \frac{1}{2\pi j} \oint_\Gamma (\mu - A)^{-1}(\lambda - A)^{-1}\, d\lambda = \pi(\mu - A)^{-1},\end{aligned}$$

since $(\mu - A)^{-1}$ and $(\lambda - A)^{-1}$ commute. Thus, π commutes with $(\mu - A)^{-1}$. Furthermore, by the resolvent identity (3.2.1),

$$\begin{aligned}(\mu - A)^{-1}\pi &= \frac{1}{2\pi j} \oint_\Gamma \frac{(\mu - A)^{-1} - (\lambda - A)^{-1}}{\lambda - \mu}\, d\lambda \\ &= \frac{(\mu - A)^{-1}}{2\pi j} \oint_\Gamma (\lambda - \mu)^{-1}\, d\lambda - \frac{1}{2\pi j} \oint_\Gamma \frac{(\lambda - A)^{-1}}{\lambda - \mu}\, d\lambda \\ &= \begin{cases} (\mu - A)^{-1} - \frac{1}{2\pi j} \oint_\Gamma \frac{(\lambda - A)^{-1}}{\lambda - \mu}\, d\lambda, & \text{if } \mu \text{ lies inside } \Gamma, \\ -\frac{1}{2\pi j} \oint_\Gamma \frac{(\lambda - A)^{-1}}{\lambda - \mu}\, d\lambda, & \text{if } \mu \text{ lies outside } \Gamma. \end{cases}\end{aligned} \tag{3.14.3}$$

We next show that π is a projection. If we perturb the path Γ in such a way that the new path Γ_1 lies outside Γ, and so that both Γ_1 and Γ together with the area in between lie in $\rho(A)$, then, by the analyticity of the resolvent $(\lambda - A)^{-1}$ in this area,

$$\pi = \frac{1}{2\pi j} \oint_{\Gamma_1} (\mu - A)^{-1}\, d\mu.$$

Therefore, by (3.14.2), (3.14.3) (note that the index of λ with respect to Γ_1 is one)

$$\begin{aligned}
\pi^2 &= \frac{1}{2\pi j} \oint_{\Gamma_1} (\mu - A)^{-1} \pi\, d\mu \\
&= -\frac{1}{2\pi j} \oint_{\Gamma_1} \frac{1}{2\pi j} \oint_{\Gamma} \frac{(\lambda - A)^{-1}}{\lambda - \mu}\, d\lambda\, d\mu \\
&= -\frac{1}{2\pi j} \oint_{\Gamma} (\lambda - A)^{-1} \frac{1}{2\pi j} \oint_{\Gamma_1} \frac{d\mu}{\lambda - \mu}\, d\lambda \\
&= \frac{1}{2\pi j} \oint_{\Gamma} (\lambda - A)^{-1}\, d\lambda = \pi.
\end{aligned}$$

Thus $\pi^2 = \pi$, and we have proved that π is a projection. By Lemma 3.14.2, $X = X_+ \dotplus X_-$. (In particular, X_+ and X_- are closed.)

We now proceed to verify properties (i)–(iv).

(i) We know from the argument above that π commutes with $(\lambda - A)^{-1}$ for all $\lambda \in \rho(A)$, and we get (i) from Lemma 3.14.7 (applied to $(\lambda - A)^{-1}$) and Theorem 3.14.8.

(ii) This follows from Theorem 3.14.6.

(iii) By Corollary 3.14.9, $\rho(A) = \rho(A_+) \cap \rho(A_-)$. Thus, to prove (iii) it suffices to show that A_+ has no spectrum outside Γ and that A_- has no spectrum inside Γ.

Let $\mu \in \rho(A)$ lie outside Γ. Then $\mu \in \rho(A_+)$, and by (3.14.3) and Corollary 3.14.9, for all $x \in X_+$,

$$(\mu - A_+)^{-1} x = -\frac{1}{2\pi j} \oint_{\Gamma} \frac{(\lambda - A)^{-1}}{\lambda - \mu}\, x\, d\lambda.$$

Thus, in particular,

$$\left\|(\mu - A_+)^{-1}\right\| \le \frac{1}{2\pi} \oint_{\Gamma} |\lambda - \mu|^{-1} \left\|(\lambda - A)^{-1}\right\| |d\lambda|.$$

We know that neither A nor A_+ has any spectrum in a neighborhood of Γ (since $\Gamma \in \rho(A)$), and away from Γ the right-hand side of the above inequality is bounded, uniformly in μ. This implies that A_+ cannot have any spectrum outside Γ, because by Lemma 3.2.8(iii), $\left\|(\mu - A_+)^{-1}\right\| \to \infty$ as μ approaches a point in $\sigma(A_+)$. In the same way it can be shown that A_- cannot have any spectrum inside Γ: note that by (3.14.3),

$$(\mu - A_-)^{-1}x = \frac{1}{2\pi j}\oint_\Gamma \frac{(\lambda - A)^{-1}}{\lambda - \mu} x\,d\lambda$$

for all $x \in X_-$ and all $\mu \in \rho(A)$ which lie inside Γ.

(iv) See Theorems 3.14.6 and 3.14.8. □

In the case of a *normal* semigroup it is possible to use a different type of spectral projection, which does not require the spectrum of the generator to be disconnected.

Theorem 3.14.11 *Let A be a closed and densely defined normal operator on a Hilbert space X (i.e., $A^*A = AA^*$), and let E be the corresponding spectral resolution of A, so that*

$$\langle Ax, y\rangle_X = \int_{\sigma(A)} \lambda \langle E(d\lambda)x, y\rangle, \quad x \in \mathcal{D}(A), \quad y \in X.$$

Let F be a bounded Borel set in $\mathbb{C}$, and let $\pi = E(F)$, i.e.,

$$\langle \pi x, y\rangle_X = \int_{\sigma(A)\cap F} \langle E(d\lambda)x, y\rangle, \quad x \in X, \quad y \in X.$$

Then π is an orthogonal projection which maps X into $\mathcal{D}(A)$. Denote $X_+ = \mathcal{R}(\pi)$ and $X_- = \mathcal{N}(\pi)$, $A_+ = A_{|X_+}$ and $A_- = A_{|X_-}$. Then the claims (i), (ii), and (iv) in Theorem 3.14.10 hold, and the claim (iii) is replaced by

(iii′) $\sigma(A_+) = \overline{\sigma(A)\cap F}$, *and* $\sigma(A_-) = \overline{\sigma(A)\setminus F}$.

We leave the proof of this theorem to the reader. (That π is a self-adjoint projection follows from the definition of a spectral resolution, and the rest follows either directly from the properties of a spectral resolution or from an argument similar to the one used in the proof of Theorem 3.14.10. Note, in particular, that $\mathcal{R}(\pi) \subset \mathcal{D}(A)$ since F is bounded.)

So far we have primarily looked at *closed* invariant subspaces. There is also another class of subspaces that play an important role in the theory, namely invariant Banach spaces which are *continuously embedded* in the state space. An example of this is the spaces X_α with $\alpha > 0$ introduced in Sections 3.6 and 3.9. These spaces are typically invariant under the semigroup $\mathfrak{A}$, but not under its generator A.

Definition 3.14.12 Let $A\colon X \supset \mathcal{D}(A) \to A$ be a linear operator, and let Y be a subspace of X (not necessarily dense). By the *part of A in Y* we mean the operator $\widetilde{A}$ which is the restriction of A to

$$\mathcal{D}(\widetilde{A}) = \{x \in \mathcal{D}(A) \cap Y \mid Ax \in Y\}.$$

Definition 3.14.13 Let $\mathfrak{A}$ be a C_0 semigroup on the Banach space X with generator A, and let Y be another Banach space embedded in X (not necessarily densely). We call Y *A-admissible* if Y is invariant under $\mathfrak{A}$ and the restriction of $\mathfrak{A}$ to Y is strongly continuous in the norm of Y (i.e, $\mathfrak{A}_{|Y}$ is a C_0 semigroup on Y).

Theorem 3.14.14 *Let $\mathfrak{A}$ be a C_0 semigroup on the Banach space X with generator A and growth bound $\omega_{\mathfrak{A}}$, and let Y be another Banach space embedded in X (not necessarily densely). Then Y is A-admissible if and only if the following two conditions hold:*

(i) *Y is an invariant subspace of $(\lambda - A)^{-1}$ for all real $\lambda > \omega_{\mathfrak{A}}$,*
(ii) *the part of A in Y is the generator of a C_0 semigroup on Y.*

When these conditions hold, then the generator of $\mathfrak{A}_{|Y}$ is the part of A in Y.

Proof Assume that Y is A-admissible. Then Theorem 3.2.9(i) (applied to $\mathfrak{A}_{|Y}$) implies (i) (and it is even true that Y is an invariant subspace of $(\lambda - A)^{-1}$ for all real $\lambda \in \mathbb{C}^+_{\omega_{\mathfrak{A}}}$). Denote the generator of $\mathfrak{A}_{|Y}$ by A_1, and denote the part of A in Y by $\widetilde{A}$. Since the norm in Y is stronger than the norm in X, it follows easily that $\mathcal{D}(A_1) \subset \mathcal{D}(A) \cap Y$, and that for $x \in \mathcal{D}(A_1)$, $Ax = A_1x \in Y$. Thus, $\widetilde{A}$ is an extension of A_1. On the other hand, if $x \in \mathcal{D}(\widetilde{A})$, then $Ax \in Y$, and each term in the identity

$$\mathfrak{A}^t x - x = \int_0^t \mathfrak{A}^s Ax\, ds, \qquad t \geq 0,$$

belongs to Y. Dividing by t and letting $t \downarrow 0$ we find that $x \in \mathcal{D}(A_1)$. Thus, $A_1 = \widetilde{A}$, and $\widetilde{A}$ is the generator of the C_0 semigroup $\mathfrak{A}_{|Y}$ on Y.

Conversely, suppose that (i) and (ii) hold. Denote the C_0 semigroup generated by $\widetilde{A}$ by $\widetilde{\mathfrak{A}}$. For all $x \in \mathcal{D}(\widetilde{A})$ and all $\lambda > \omega_{\mathfrak{A}}$ we have (since $\lambda \in \rho(A)$)

$$(\lambda - A)^{-1}(\lambda - \widetilde{A})x = (\lambda - A)^{-1}(\lambda - A)x = x,$$

and for all $y \in Y$ we have (because of (ii))

$$(\lambda - \widetilde{A})(\lambda - A)^{-1}x = (\lambda - A)(\lambda - A)^{-1}x = x.$$

Thus, $(\lambda - \widetilde{A})$ maps $\mathcal{D}(\widetilde{A})$ one-to-one onto Y, and $(\lambda - \widetilde{A})^{-1}$ is the restriction of $(\lambda - A)^{-1}$ to Y. Fix $t > 0$, and choose n so large that $n/t > \omega_{\mathfrak{A}}$. Then, for all $\lambda > \omega_{\mathfrak{A}}$ and all $y \in Y$,

$$\left(1 - \frac{t}{n}A\right)^{-n} y = \left(1 - \frac{t}{n}\widetilde{A}\right)^{-n} y.$$

Let $n \to \infty$. By Theorem 3.7.5, the left-hand side tends to $\mathfrak{A}^t y$ in X and the right-hand side tends to $\widetilde{\mathfrak{A}}^t y$ in Y, hence in X. Thus, $\widetilde{\mathfrak{A}} = \mathfrak{A}_{|Y}$. This implies both that Y is invariant under $\mathfrak{A}$, and that $\mathfrak{A}_{|Y}$ is strongly continuous. □

Let us end this section by proving the following extension of Theorem 3.14.8 (to get that theorem we take $\widetilde{X} = X$, $E = \pi$, $\widetilde{A} = A$, and $\widetilde{\mathfrak{A}} = \mathfrak{A}$).

Theorem 3.14.15 *Let A be the generator of a C_0 semigroup $\mathfrak{A}$ on X, let $\widetilde{A}$ be the generator of a C_0 semigroup $\widetilde{\mathfrak{A}}$ on $\widetilde{X}$, and let $E \in \mathcal{B}(X; \widetilde{X})$. Then the following conditions are equivalent.*

(i) $E\mathfrak{A}^t = \widetilde{\mathfrak{A}}^t E$ *for all* $t \geq 0$.
(ii) $E(\lambda - A)^{-1} = (\lambda - \widetilde{A})^{-1}E$ *for some* $\lambda \in \rho(A) \cap \rho(\widetilde{A})$.
(iii) $E(\lambda - A)^{-1} = (\lambda - \widetilde{A})^{-1}E$ *for all* $\lambda \in \rho(A) \cap \rho(\widetilde{A})$.
(iv) E *maps* $\mathcal{D}(A)$ *into* $\mathcal{D}(\widetilde{A})$, *and* $EAx = \widetilde{A}Ex$ *for all* $x \in \mathcal{D}(A)$.

If, in addition, $\mathfrak{A}$ *and* $\widetilde{\mathfrak{A}}^t$ *are groups, then these conditions are further equivalent to*

(v) $E\mathfrak{A}^t = \widetilde{\mathfrak{A}}^t E$ *for all* $t \in R$.

Proof (i) $\Rightarrow$ (ii): If (i) holds, then by Theorem 3.2.9(i), for all $\lambda > \max\{\omega_{\mathfrak{A}}, \omega_{\widetilde{\mathfrak{A}}}\}$ and all $x \in X$,

$$E(\lambda - A)^{-1}x = \int_0^\infty e^{-\lambda s} E\mathfrak{A}^s x = \int_0^\infty e^{-\lambda s}\widetilde{\mathfrak{A}}^s Ex = (\lambda - \widetilde{A})^{-1}Ex.$$

(ii) $\Rightarrow$ (iv): By (ii), for all $x \in \mathcal{D}(A)$,

$$Ex = (\lambda - \widetilde{A})^{-1}E(\lambda - A)x.$$

This implies that E maps $\mathcal{D}(A)$ into $\mathcal{D}(\widetilde{A})$. Applying $(\lambda - \widetilde{A})$ to both sides of this identity we get $(\lambda - \widetilde{A})Ex = E(\lambda - A)x$, or equivalently, $EAx = \widetilde{A}Ex$.

(iv) $\Rightarrow$ (iii): If (iv) holds, then $(\lambda - \widetilde{A})Ex = E(\lambda - A)x$ for all $\lambda \in \mathbb{C}$ and all $x \in \mathcal{D}(A)$. In particular, if we take $\lambda \in \rho(A) \cap \rho(\widetilde{A})$, then we can apply $(\lambda - \widetilde{A})^{-1}$ to both sides of this identity and replace x by $(\lambda - A)^{-1}x$ to get for all $x \in X$,

$$E(\lambda - A)^{-1}x = (\lambda - \widetilde{A})^{-1}Ex.$$

(iii) $\Rightarrow$ (i): If (iii) holds, then by Theorem 3.7.5, for all $t \geq 0$ and all $x \in X$,

$$\mathfrak{A}^t Ex = \lim_{n\to\infty}\Big(1 - \frac{t}{n}A\Big)^{-n} Ex = E \lim_{n\to\infty}\Big(1 - \frac{t}{n}\widetilde{A}\Big)^{-n} x = E\widetilde{\mathfrak{A}}^t x.$$

(v) We get $E\mathfrak{A}^{-t} = \widetilde{\mathfrak{A}}^{-t}E$ for all $t \geq 0$ by multiplying the identity in (i) by $\widetilde{\mathfrak{A}}^{-t}$ to the left and by $\mathfrak{A}^{-t}$ to the right.

□

3.15 Comments

By now, most of the results in this chapter are classic. We refer the reader to Davies (1980), Dunford and Schwartz (1958, 1963, 1971), Goldstein (1985), Lunardi (1995), Nagel (1986), Hille and Phillips (1957), Pazy (1983), and Yosida (1974) for the history and theory of C_0 semigroups beyond what we have presented here.

Sections 3.2–3.3 The generators of the shift semigroups and their resolvents have been studied in, e.g., Hille and Phillips (1957, Sections 19.2–19.4). Diagonal semigroups appear frequently in the theory of *parabolic* and *hyperbolic* partial differential equations. Sometimes the basis of eigenvectors of the generator is not orthonormal, but instead a *Riesz basis* in the sense of Curtain and Zwart (1995, Definition 2.3.1). However, a Riesz basis can be transformed into an orthonormal basis by means of a similarity transformation of the type described in Example 2.3.7. This makes it easy to extend the theory for diagonal semigroups presented in Examples 3.3.3 and 3.3.5 to semigroups whose generator has a set of eigenvectors which are a Riesz basis for the state space. These types of semigroups are studied in some detail by Curtain and Zwart (1995).

Sections 3.4, 3.7, and 3.8 See, for example, Pazy (1983), for the history of the Hille–Yosida and Lumer–Phillips theorems, the different approximation theorems, and the Cauchy problem.

Section 3.5 Our presentation of the dual semigroup follows roughly Hille and Phillips (1957, Chapter 14), except for the fact that we use the conjugate-linear dual instead of the linear dual.

Section 3.6 Rigged spaces of the type discussed in Section 3.6 are part of the traditional semigroup formulation of partial differential equations of *parabolic* and *hyperbolic* type. See, for example, Lions (1971), Lunardi (1995), and Lasiecka and Triggiani (2000a, b). In the reflexive case the space X_{-1} is usually defined to be the dual of the domain of the adjoint operator (see Remark 3.6.1). The spaces X_1 and X_{-1} have been an important part of the theory of well-posed linear systems since Helton (1976). Spaces X_α of fractional order are introduced in Section 3.9.

Section 3.9 This section is based in part on Dunford and Schwartz (1958, Section VII.9), Rudin (1973, Chapter 10), and Pazy (1983, Section 2.6).

Section 3.10 The results of this section are fairly standard. Analytic semigroups most frequently arise from the solution of *parabolic* or heavily damped *hyperbolic* partial differential equations. Theorem 3.10.11 has been modeled after Lunardi (1995, Proposition 2.4.1) and Mikkola (2002, Lemma 9.4.3). For further results about analytic semigroups we refer the reader to standard text books, such as Goldstein (1985, Sections 1.5, 2.4 and 2.5), Hille and Phillips (1957, Chapter 17), Lunardi (1995), Lasiecka and Triggiani (2000a, b), and Pazy (1983, Sections 2.5, 7.2 and 7.3).

Section 3.11 The proof of Theorem 3.11.4 is based on Davies (1980, Theorem 2.19), which goes back to Hille and Phillips (1957, Theorem 16.4.1, p. 460). This theorem is also found in Nagel (1986, p. 87). The reduction of the spectrum determined growth property to the corresponding spectral inclusion that we use in the proof of Theorem 3.11.4 is classic; see, e.g., Slemrod (1976, pp. 783–784), Triggiani (1975, p. 387), Zabczyk (1975), or Nagel (1986, p. 83). Corollary 3.11.5(i) is due to Triggiani (1975, pp. 387–388) and Corollary 3.11.5(ii) is due to Zabczyk (1975). Theorem 3.11.6 was proved independently by Herbst (1983), Huang (1985), and Prüss (1984). Lemma 3.11.7 has been modeled after G. Weiss (1988b, Theorem 3.4). The implication (v) $\Rightarrow$ (i) in Theorem 3.11.8 is often called *Datko's theorem* after Datko (1970, p. 615) who proves this implication in the Hilbert space case with $p = 2$. The general version of the same implication was proved by Pazy (1972) (see Pazy (1983, Theorem 4.1 and p. 259)). Our proof follows the one given by Pritchard and Zabczyk (1981, Theorem 3.4). In the reflexive case the implication (vii) $\Rightarrow$ (i) can be reduced to the implication (v) $\Rightarrow$ (i) through duality, but we prefer to give a direct proof, which is valid even in the non reflexive case (the non reflexive part of the implication (vii) $\Rightarrow$ (i) may be new). Our proof of the implication (ix) $\Rightarrow$ (i) in Theorem 3.11.8 has been modeled after G. Weiss (1988b, Theorem 4.2) (we have not been able to find this explicit implication in the existing literature). Examples of semigroups which do *not have the spectrum determined growth property* are given in Curtain and Zwart (1995, Example 5.1.4 and Exercise 5.6), Greiner *et al.* (1981, Example 4.2), Davies (1980, Theorem 2.17), Hille and Phillips (1957, p. 665), and Zabczyk (1975). For additional results about the spectrum determined growth property, and more generally, the spectral mapping property, we refer the reader to van Neerven (1996), which is devoted to this very question.

Section 3.12 The frequency domain plays a very important role in many texts, including this one. Most mathematical texts use the Fourier transform instead of the Laplace transform, and they replace the right half-plane by the upper half-plane. However, we prefer to stick to the engineering tradition at this point. It is possible to develop a symbolic calculus for generators of semigroups based on the representation of $\hat{u}(A)$ in Theorem 3.12.6 (this is done in Dunford and Schwartz (1958, Section VIII.2) in the case where A generates a group rather than a semigroup), and our definition of $(\gamma - A)^{-\alpha}$ given in Section 3.9 is based on a special case of that calculus.

Section 3.13 Most books on operator theory *define* the shift semigroups using the characterization given in Propositions 3.13.1 and 3.13.2 (and they completely ignore the time-domain versions of these semigroups). This is especially true in the Hilbert space L^2-well-posed case. In our context the time-domain versions are more natural to work with.

Section 3.14 The results in this section are classic (but not always that easy to find in the literature). Our presentation follows loosely Curtain and Zwart (1995) and Pazy (1983).

4

The generators of a well-posed linear system

In Chapter 2 we introduced the notion of a well-posed linear system by describing the algebraic and continuity properties of the mappings from the initial state and the input function to the final state and the output function. Here we develop an alternative approach which replaces the algebraic 'integral level' description of a system by a 'differential level' description involving one differential equation and one algebraic equation. We also introduce the transfer function of a well-posed linear system. Furthermore, by using the 'differential level' we are able to introduce an even more general class of systems which need not be well-posed.

4.1 Introduction

We started out in Section 1.1 with a system of the type

$$\begin{aligned} \dot{x}(t) &= Ax(t) + Bu(t), \\ y(t) &= Cx(t) + Du(t), \qquad t \geq s, \\ x(s) &= x_s. \end{aligned} \tag{4.1.1}$$

where A is the generator of a semigroup $\mathfrak{A}$ on X, $B \in \mathcal{B}(U;X)$, $C \in \mathcal{B}(X;Y)$, and $D \in \mathcal{B}(U;Y)$. Then we constructed the operators $\mathfrak{B}$, $\mathfrak{C}$, and $\mathfrak{D}$, that together determine both a *Reg*-well-posed and an L^p-well-posed linear system for all p, $1 \leq p \leq \infty$ (see Proposition 2.3.1). Here we investigate the converse question: Given a $L^p|Reg$-well-posed linear system $\Sigma = \left[\begin{smallmatrix} \mathfrak{A} & \mathfrak{B} \\ \mathfrak{C} & \mathfrak{D} \end{smallmatrix}\right]$, is it possible to find operators B, C and D (possibly unbounded) in such a way that the operators $\mathfrak{B}$, $\mathfrak{C}$, and $\mathfrak{D}$ can be reconstructed from $\mathfrak{A}$, B, C and D by the use of the formulas

$$
\begin{aligned}
\mathfrak{B}u &= \int_{-\infty}^{0} \mathfrak{A}^{-v} Bu(v)\, dv, \\
\mathfrak{C}x &= \left(t \mapsto C\mathfrak{A}^t x, \qquad t \geq 0\right), \\
\mathfrak{D}u &= \left(t \mapsto \int_{-\infty}^{t} C\mathfrak{A}^{t-v} Bu(v)\, dv + Du(t), \qquad t \in \mathbb{R}\right),
\end{aligned}
\tag{4.1.2}
$$

and such that the *state trajectory* x and the *output function* y in some sense are given by the variation of constants formula

$$
\begin{aligned}
x(t) &= \mathfrak{A}^{t-s} x_s + \int_s^t \mathfrak{A}^{t-v} Bu(v)\, dv, & t \geq s, \\
y(t) &= C\mathfrak{A}^{t-s} x_s + C \int_s^t \mathfrak{A}^{t-v} Bu(v)\, dv + Du(t), & t \geq s?
\end{aligned}
\tag{4.1.3}
$$

The major question presented above about the existence of generating operators B, C, and D can be split into several minor ones.

We have already discussed the generator A of the semigroup $\mathfrak{A}$ in Chapter 3. For this we did not need any of the other operators $\mathfrak{B}$, $\mathfrak{C}$, and $\mathfrak{D}$ of the system. The same separation of a system into independent components can be continued. For example, the relationship between B and $\mathfrak{B}$ does not involve the operators $\mathfrak{C}$ and $\mathfrak{D}$, so it should be possible to ignore $\mathfrak{C}$ and $\mathfrak{D}$ in the study of 'how $\mathfrak{B}$ is generated by B'. Similarly, $\mathfrak{B}$ and $\mathfrak{D}$ are not needed in the study of the relationship between $\mathfrak{C}$ and C. For this reason we introduce the following more detailed version of Definition 2.2.1:

Definition 4.1.1 Let U, X, and Y be Banach spaces, let $\mathfrak{A}$ be a C_0 semigroup on X, and let $1 \leq p \leq \infty$.

(i) A continuous operator $\mathfrak{B}\colon L^p_c(\mathbb{R}^-; U) \to X$ is a *L^p-well-posed input map* for $\mathfrak{A}$ with input space U if $\mathfrak{A}^t\mathfrak{B}u = \mathfrak{B}\tau^t_- u$ for all $u \in L^p_c(\mathbb{R}^-; U)$ and $t \geq 0$.

(ii) A continuous operator $\mathfrak{B}\colon Reg_c(\mathbb{R}^-; U) \to X$ is a *Reg-well-posed input map* for $\mathfrak{A}$ with input space U if $\mathfrak{A}^t\mathfrak{B}u = \mathfrak{B}\tau^t_- u$ for all $u \in Reg_c(\mathbb{R}^-; U)$ and $t \geq 0$.

(iii) A continuous operator $\mathfrak{C}\colon X \to L^p_{\text{loc}}(\mathbb{R}^+; Y)$ is a *L^p-well-posed output map* for $\mathfrak{A}$ with output space Y if $\mathfrak{C}\mathfrak{A}^t x = \tau^t_+ \mathfrak{C}x$ for all $x \in X$ and $t \geq 0$.

(iv) A continuous operator $\mathfrak{C}\colon X \to Reg_{\text{loc}}(\overline{\mathbb{R}}^+; Y)$ is a *Reg-well-posed output map* for $\mathfrak{A}$ with output space Y if $\mathfrak{C}\mathfrak{A}^t x = \tau^t_+ \mathfrak{C}x$ for all $x \in X$ and $t \geq 0$.

(v) By a $L^p|Reg$-well-posed input or output map of $\mathfrak{A}$ we mean an input or output map that is either *Reg*-well-posed or L^p-well-posed for some p, $1 \leq p \leq \infty$. (Thus, the case $p = \infty$ is included, except when explicitly excluded.)

(vi) By a well-posed input or output map of $\mathfrak{A}$ we mean an input or output map that is either *Reg*-well-posed or L^p-well-posed for some p, $1 \le p < \infty$. (Thus, the case $p = \infty$ is excluded.)

Note that the relevant parts of Definition 2.2.6, Definition 2.2.7, Lemma 2.2.8, Theorem 2.2.11, Theorem 2.2.12, Theorem 2.4.1, Lemma 2.4.3, Theorem 2.4.4, Theorem 2.5.4, and Theorem 2.5.7 still apply. In the L^p-case, for every $\omega > \omega_{\mathfrak{A}}$, $\mathfrak{C}$ maps X continuously into $L^p_\omega(\mathbb{R}^+; Y)$, and $\mathfrak{B}$ can be extended to a continuous operator $L^p_\omega(\mathbb{R}^-; U) \to X$. The same statement is true with L^p replaced by Reg_0, too.

4.2 The control operator

Let us now return to the question of how to find generators B and C for $\mathfrak{B}$ and $\mathfrak{C}$. We begin with $\mathfrak{B}$. Here we notice for the first time a significant difference between the cases of L^∞-well-posedness and *Reg*-well-posedness (see Section 3.6 for the notations $\mathfrak{A}_{|X_{-1}}$ and $A_{|X}$):

Theorem 4.2.1 *Let U and X be Banach spaces, let $\mathfrak{A}$ be a C_0 semigroup on X with growth bound $\omega_{\mathfrak{A}}$, let $\mathfrak{B}$ be a $L^p|Reg$-well-posed input map for $\mathfrak{A}$ with input space U, and let $\omega > \omega_{\mathfrak{A}}$. Then the following claims are true:*

(i) *There is a unique operator $B \in \mathcal{B}(U; X_{-1})$ such that*

$$\mathfrak{B}u = \int_{-\infty}^{0} \mathfrak{A}^{-s}_{|X_{-1}} Bu(s)\, ds; \tag{4.2.1}$$

this equation is valid as an equation in X_{-1} for all $u \in L^p_\omega(\mathbb{R}^-; U)$ if $\mathfrak{B}$ is L^p-well-posed for some $p < \infty$, and for all $u \in Reg_{0,\omega}(\mathbb{R}^-; U)$ if $\mathfrak{B}$ is L^∞-well-posed or Reg-well-posed.

(ii) *If $\mathfrak{B}$ is L^p-well-posed for some $p < \infty$, then $\mathfrak{B}$ is uniquely determined by B and $\mathfrak{A}$, and if $\mathfrak{B}$ is L^∞-well-posed or Reg-well-posed then the restriction of $\mathfrak{B}$ to $Reg_{0,\omega}(\mathbb{R}^-; U)$ is uniquely determined by B and $\mathfrak{A}$.*

(iii) *Define $e_{\lambda,n}(s) = (1/n!)(-s)^n e^{\lambda s}$. Then, for all $u \in U$, $\lambda \in \mathbb{C}_{\omega_{\mathfrak{A}}}$, and $n = 0, 1, 2, \ldots,$*

$$(\lambda - A)^{-(n+1)}_{|X_{-1}} Bu = (1/n!) \int_0^\infty s^n e^{-\lambda s} \mathfrak{A}^s_{|X_{-1}} Bu\, ds = \mathfrak{B}(e_{\lambda,n} u).$$

In particular,

$$(\lambda - A_{|X})^{-1} Bu = \int_0^\infty e^{-\lambda s} \mathfrak{A}^s_{|X_{-1}} Bu\, ds = \mathfrak{B}(e_{\lambda,0} u).$$

Proof (i) Fix some $\alpha \in \mathbb{C}^+_{\omega_{\mathfrak{A}}}$, and let $e_\alpha(s) = e^{\alpha s}$, $s \in \mathbb{R}$. Then $\tau^t e_\alpha = e^{\alpha t} e_\alpha$ for all $t \in \mathbb{R}$ and $e_\alpha \in L^p_{\omega,\mathrm{loc}}(\mathbb{R}; \mathbb{R}) \cap BC_{0,\omega,\mathrm{loc}}(\mathbb{R}; \mathbb{R})$ for all $\omega_{\mathfrak{A}} < \omega < \Re\alpha$ and

$n = 1, 2, 3, \ldots$ We define

$$Bu = (\alpha - A_{|X})\mathfrak{B}(e_\alpha u), \qquad u \in U,$$

where $A_{|X}$ is the generator of $\mathfrak{A}_{|X_{-1}}$ (that the above integral is well-defined follows from Theorem 2.5.4). The operator $u \mapsto \mathfrak{B}(e_\alpha u)$ is continuous $U \to X$, hence $B \in \mathcal{B}(U; X_{-1})$.

We claim that (4.2.1) holds for this operator B. We begin the proof of (4.2.1) with the case where u is replaced by $\pi_{(-\infty,t)}(e_\alpha u)$ for some fixed $u \in U$ and $t \le 0$. Then (4.2.1) becomes

$$\mathfrak{B}\pi_{(-\infty,t)}(e_\alpha u) = \int_{-\infty}^{0} \mathfrak{A}^{-s}_{|X_{-1}} B\pi_{(-\infty,t)}(e_\alpha(s)u)\,ds.$$

To see that this identity is true for all $t \le 0$ we can use Definition 4.1.1(i), Theorem 3.2.9(i), and Lemma 2.2.9 to get

$$\begin{aligned}
\int_{-\infty}^{t} \mathfrak{A}^{-s}_{|X_{-1}} B(e_\alpha(s)u)\,ds &= \int_{0}^{\infty} e^{\alpha(t-v)}\mathfrak{A}^{v-t}_{|X_{-1}} Bu\,dv \\
&= e^{\alpha t}\mathfrak{A}^{-t}_{|X_{-1}} \int_{0}^{\infty} e^{-\alpha v}\mathfrak{A}^{v}_{|X_{-1}} Bu\,dv \\
&= e^{\alpha t}\mathfrak{A}^{-t}_{|X_{-1}}(\alpha - A_{|X})^{-1} Bu \\
&= e^{\alpha t}\mathfrak{A}^{-t}\mathfrak{B}(e_\alpha u) \\
&= e^{\alpha t}\mathfrak{B}\tau^{-t}\pi_{-}(e_\alpha u) \\
&= e^{\alpha t}\mathfrak{B}\pi_{(-\infty,t)}\tau^{-t}(e_\alpha u) \\
&= \mathfrak{B}\pi_{(-\infty,t)}(e_\alpha u).
\end{aligned}$$

Thus, (4.2.1) holds when u is replaced by $\pi_{(-\infty,t)}(e_\alpha u)$ for some fixed $u \in U$ and $t \le 0$.

By linearity, (4.2.1) must then also hold for all functions of the type $\pi_{[t_1,t_2)}(e_\alpha u) = \pi_{(-\infty,t_2)}(e_\alpha u) - \pi_{(-\infty,t_2)}(e_\alpha u)$, for all $u \in U$ and $-\infty < t_1 < t_2 \le 0$, hence for all linear combinations of such functions. This class of functions is dense in $L^p_\omega(\mathbb{R}^-; U)$ if $p < \infty$ and in $Reg_{0,\omega}(\mathbb{R}^-; U)$ if $p = \infty$, so by the continuity of both sides of (4.2.1) with respect to convergence in $L^p_\omega(\mathbb{R}^-; U)$, this formula must be true even for this larger class of functions.

The uniqueness of B follows from part (iii).

(ii) This follows from (i).

(iii) Use (4.2.1) and Theorem 3.2.9(i). □

Definition 4.2.2 The operator B in Theorem 4.2.1 is called *the control operator* (induced by $\mathfrak{A}$ and $\mathfrak{B}$).

Corollary 4.2.3 *In the setting of Theorem 4.2.1, the input operators $\mathfrak{B}^t$ and $\mathfrak{B}^t_s$, $-\infty < s < t < \infty$, introduced in Definition 2.2.6 have the representations*

$$\mathfrak{B}^t u = \int_{-\infty}^{t} \mathfrak{A}^{t-s}_{|X_{-1}} Bu(s)\,ds, \tag{4.2.2}$$

$$\mathfrak{B}^t_s u = \int_{s}^{t} \mathfrak{A}^{t-v}_{|X_{-1}} Bu(v)\,dv. \tag{4.2.3}$$

The representation (4.2.2) *is valid for all $u \in L^p_\omega((-\infty, t); U)$ if $\mathfrak{B}$ is L^p-well-posed for some $p < \infty$, and for all $u \in Reg_{0,\omega}((-\infty, t); U)$ if $\mathfrak{B}$ is L^∞-well-posed or Reg-well-posed; here $\omega > \omega_{\mathfrak{A}}$. The representation* (4.2.2) *is valid for all $u \in L^p([s, t); U)$ if $\mathfrak{B}$ is L^p-well-posed for some $p < \infty$, and for all $u \in Reg([s, t); U)$ if $\mathfrak{B}$ is L^∞-well-posed or Reg-well-posed. (In both cases the integrals are computed in X_{-1}, but the final results belongs to X.)*

Proof This follows immediately from Definition 2.2.6 and Theorem 4.2.1(i). □

There is a simple converse to Theorem 4.2.1(i):

Theorem 4.2.4 *Let $\mathfrak{A}$ be a C_0 semigroup on X, and let $1 \le p < \infty$. Then B is the control operator of a $L^p|Reg$-well-posed input map for $\mathfrak{A}$ with input space U if and only if $B \in \mathcal{B}(U; X^{-1})$ and the map $\mathfrak{B}\colon L^1_c(\mathbb{R}^-; U) \to X_{-1}$ defined by*

$$\mathfrak{B}u = \int_{-\infty}^{0} \mathfrak{A}^{-s}_{|X_{-1}} Bu(s)\,ds$$

maps $L^p|Reg_c(\mathbb{R}^-; U)$ into X.

Proof The necessity follows from Theorem 4.2.1(i). For the converse claim we can use the closed graph theorem to get the continuity of $\mathfrak{B}$ as an operator with values in X. The intertwining property in Definition 4.1.1(i)–(ii) is proved as follows:

$$\begin{aligned}\mathfrak{B}\tau^t_- u &= \int_{-\infty}^{-t} \mathfrak{A}^{-s}_{|X_{-1}} Bu(s+t)\,ds = \int_{-\infty}^{0} \mathfrak{A}^{-s+t}_{|X_{-1}} Bu(s)\,ds \\ &= \mathfrak{A}^t \int_{-\infty}^{0} \mathfrak{A}^{-s}_{|X_{-1}} Bu(s)\,ds = \mathfrak{A}^t \mathfrak{B}u.\end{aligned}$$

□

Example 4.2.5 *Let A be the semigroup generator and B the control operator of the delay line in Example 2.3.4. Then, for each $\lambda \in \mathbb{C}$ and $u \in U$, the function $(\lambda - A_{|X})^{-1}Bu \in W^{1,p}([0, T])$ is given by*

$$\big((\lambda - A_{|X})^{-1}Bu\big)(t) = e^{\lambda(t-T)}u, \qquad 0 \le t \le T.$$

In particular, B can be (formally) interpreted as the distribution-valued operator $Bu = u\delta_T$, where δ_T represents the Dirac delta (a unit point mass) at T.

We prefer to use only a *formal* interpretation of B as a delta-function, and avoid the introduction of the notion of a U-valued distribution or measure.

Proof By Theorem 4.2.1(iii) with $n = 1$,

$$(\lambda - A_{|X})^{-1}Bu = \mathfrak{B}(\mathrm{e}_\lambda u),$$

where $\mathrm{e}_\lambda(t) = \mathrm{e}^{\lambda t}$ for all $t \le 0$. By the definition of $\mathfrak{B}$ in Example 2.3.4, the right hand side is the restriction of the function $t \mapsto \mathrm{e}^{\lambda(t-T)}u$ to the interval $[0, T]$. By Example 3.3.2(iv), for each $f \in L^p|Reg([0, T); U)$, $(\lambda - A)^{-1}f$ is the restriction of the function $t \mapsto \int_t^T \mathrm{e}^{\lambda(t-s)} f(s)\,ds$ to the interval $[0, T]$. We get formally the same result by taking $f = u\delta_T$. □

The same argument can be used to find the control operators of the realizations in Example 2.6.5:

Example 4.2.6 *In parts (i)–(iv) below we let A be the semigroup generator and B the control operator of the realizations listed in Example 2.6.5(i)–(iv), respectively.*

(i) *In the exactly controllable shift realization in Example 2.6.5(i), for each $\lambda \in \mathbb{C}^+_\omega$ and $u \in U$, the function $(\lambda - A_{|X})^{-1}Bu \in L^p_\omega(\mathbb{R}^-; U)$ is given by*

$$(\lambda - A_{|X})^{-1}Bu = \mathrm{e}^{\lambda t}u, \qquad t \le 0.$$

Thus, the control operator B for this system can be (formally) interpreted as the distribution-valued operator $Bu = u\delta_0$, where δ_0 represents the Dirac delta at zero.

(ii) *In the exactly observable shift realization in Example 2.6.5(ii), for each $\lambda \in \mathbb{C}^+_\omega$ and $u \in U$, the function $(\lambda - A_{|X})^{-1}Bu \in L^p_\omega(\mathbb{R}^+; Y)$ is given by*

$$(\lambda - A_{|X})^{-1}Bu = \pi_+\mathfrak{D}\pi_-\mathrm{e}_\lambda u,$$

where $\mathrm{e}_\lambda(t) = \mathrm{e}^{\lambda t}$ for $t \in \mathbb{R}$.

(iii) *In the bilateral input shift realization in Example 2.6.5(iii), for each $\lambda \in \mathbb{C}^+_\omega$ and $u \in U$, the function $(\lambda - A_{|X})^{-1}Bu \in L^p_\omega(\mathbb{R}; U)$ is given by*

$$(\lambda - A_{|X})^{-1}Bu = \begin{cases} 0, & t \ge 0, \\ \mathrm{e}^{\lambda t}u, & t < 0. \end{cases}$$

Thus, the control operator B for this system can be (formally) interpreted as the distribution-valued operator $Bu = u\delta_0$, where δ_0 represents the Dirac delta at zero.

(iv) *In the bilateral output shift realization in Example 2.6.5(iv), for each* $\lambda \in \mathbb{C}^+_\omega$ *and* $u \in U$*, the function* $(\lambda - A_{|X})^{-1}Bu \in L^p_\omega(\mathbb{R}; Y)$ *is given by*

$$\big((\lambda - A_{|X})^{-1}Bu\big)(t) = \begin{cases} \big(\pi_+\mathfrak{D}\pi_-\mathrm{e}_\lambda u\big)(t), & t \geq 0, \\ \widehat{\mathfrak{D}}(\lambda)\mathrm{e}^{\lambda t}u, & t < 0, \end{cases}$$

where $\mathrm{e}_\lambda(t) = \mathrm{e}^{\lambda t}$ *for* $t \in \mathbb{R}$.

The proof is the same as the proof of Example 4.2.5.

The case $p = 1$ is special in the following sense:

Theorem 4.2.7 *If* $p = 1$ *and* X *is reflexive, then the operator* B *in Theorem 4.2.1 maps* U *continuously into* X*, i.e.,* $B \in \mathcal{B}(U; X)$.

Proof For each $u \in U$, define

$$u_n(t) = \begin{cases} nu, & -1/n \leq t < 0, \\ 0, & \text{otherwise.} \end{cases}$$

Then $u_n \in L^1(\mathbb{R}^-, U)$, and $\|u_n\|_{L^1} = |u|_U$. In particular, this norm is bounded uniformly in n. Therefore, by the continuity of $\mathfrak{B}$, the norm of $\mathfrak{B}u_n$ is bounded uniformly in n. If X is reflexive, then the unit ball of X is weakly sequentially compact, hence there is some subsequence $\mathfrak{B}u_{n_k}$ which converges to a limit in X. However, by using the representation formula for $\mathfrak{B}$ given in Theorem 4.2.1(i) we find that the whole sequence converges to Bu in X_{-1}. Thus, B maps U into X. The continuity of $B\colon U \to X$ follows from the closed graph theorem (B is closed from U to X since it is continuous from U to X_{-1}). □

We remark that the assumption that X is reflexive cannot be removed from Theorem 4.2.7. For example, the control operator of the shift line in Example 2.3.4 has a distribution interpretation as the operator which maps $u \in U$ into $u\delta_T$, where δ_T represents a unit point mass at the point T (see Example 4.2.5). This is a distribution which belongs to X_{-1} in all L^p-cases with $1 \leq p < \infty$, but it does not belong to X even when $p = 1$. See also Example 4.5.13.

Corollary 4.2.8 *Let* $\mathfrak{A}$ *be a* C_0 *semigroup on a reflexive state space* X*. Then* B *is the control operator of an* L^1*-well-posed input map for* $\mathfrak{A}$ *with input space* U *if and only if* $B \in \mathcal{B}(U; X)$.

Proof The necessity follows from Theorem 4.2.7, and the proof of the sufficiency is trivial (cf. Example 2.5.10). □

From Theorem 4.2.1(iii) it is possible to derive the following growth estimates involving B and the resolvent of A.

Proposition 4.2.9 *In the setting of Theorem 4.2.1, for each $\omega > \omega_{\mathfrak{A}}$ there is a finite constant M such that*

$$\left\|(\lambda - A_{|X})^{-n} B\right\| \leq \frac{M}{n^{(1-1/p)/2}(\Re\lambda - \omega)^{n-1+1/p}}$$

for all $n = 1, 2, 3, \ldots$ and all $\lambda \in \mathbb{C}^+_\omega$ in the case where $\mathfrak{B}$ is L^p-well-posed with $1 \leq p < \infty$, and such that

$$\left\|(\lambda - A_{|X})^{-n} B\right\| \leq \frac{M}{\sqrt{n}\,(\Re\lambda - \omega)^{n-1}}$$

for all $n = 1, 2, 3, \ldots$ and all $\lambda \in \mathbb{C}^+_\omega$ in the case where $\mathfrak{B}$ is L^∞-well-posed or Reg-well-posed. In particular, taking $n = 1$ we get

$$\left\|(\lambda - A_{|X})^{-1} B\right\| \leq \frac{M}{(\Re\lambda - \omega)^{1/p}}$$

in the case where $\mathfrak{B}$ is L^p-well-posed with $1 \leq p < \infty$, and

$$\left\|(\lambda - A_{|X})^{-1} B\right\| \leq M$$

in the case where $\mathfrak{B}$ is L^∞-well-posed or Reg-well-posed.

Proof By Theorems 4.2.1(iii) and 2.5.4(ii), there is a finite constant M such that

$$\left|(\lambda - A_{|X})^{-(n+1)} Bu\right|_X = |\mathfrak{B}(\mathrm{e}_{\lambda,n}u)|_X \leq M \|e_{\lambda,n}\|_{L^p_\omega(\mathbb{R}^-)} |u|_U$$

for all $u \in U$ and all $\lambda \in \mathbb{C}^+_\omega$. The desired estimate then follows from the estimate on $\|e_{\lambda,n}\|_{L^p_\omega(\mathbb{R}^-)}$ that we get from Lemma 4.2.10 below. □

Lemma 4.2.10 *Let $\alpha > 0$, $n \geq 0$. Then there is a constant K such that*

$$\begin{aligned} \frac{1}{\Gamma(n+1)}\left(\int_0^\infty \left(s^n \mathrm{e}^{-\alpha s}\right)^p ds\right)^{1/p} &= \frac{\Gamma(pn+1)^{1/p}}{\Gamma(n+1)(\alpha p)^{(n+1/p)}} \\ &\leq \frac{K}{n^{(1-1/p)/2}\alpha^{(n+1/p)}} \end{aligned}$$

for $1 \leq p < \infty$, and

$$\sup_{s\geq 0} \frac{s^n \mathrm{e}^{-\alpha s}}{\Gamma(n+1)} = \frac{n^n \mathrm{e}^{-n}}{\Gamma(n+1)\alpha^n} \leq \frac{K}{\sqrt{n}\,\alpha^n}.$$

We leave the proof of this lemma to the reader (the integral can be reduced to the standard integral defining the Γ-function by a change of integration variable, and the inequality follows from Stirling's formula (3.7.1)).

The following theorem presents two different extensions of Theorem 4.2.1(iii).

Theorem 4.2.11 *We use the same setting as in Theorem 4.2.1 (in particular, we take $\omega > \omega_{\mathfrak{A}}$).*

(i) *Let $v \in L^p_\omega(\mathbb{R}^-;\mathbb{C})$ if $\mathfrak{B}$ is L^p-well-posed for some $p < \infty$ and let $v \in Reg_{0,\omega}(\mathbb{R}^-;\mathbb{C})$ if $\mathfrak{B}$ is L^∞-well-posed or Reg-well-posed. In addition, suppose that the left-sided Laplace transform $\hat{v}$ of v is analytic at infinity. Then, for all $u_0 \in U$,*

$$\mathfrak{B}(vu_0) = \hat{v}(A_{|X})Bu_0.$$

(ii) *Let $v \in L^1_\omega(\mathbb{R}^-;\mathbb{C})$, and let $u \in L^p_\omega(\mathbb{R}^-;U)$ if $\mathfrak{B}$ is L^p-well-posed for some $p < \infty$ and $u \in Reg_{0,\omega}(\mathbb{R}^-;U)$ if $\mathfrak{B}$ is L^∞-well-posed or Reg-well-posed. In addition, suppose that the left-sided Laplace transforms of v and u are analytic at infinity. Define $w(t) = \int_t^0 v(t-s)u(s)\,ds$ for all those $t \in \mathbb{R}^-$ for which the integral converges. Then $w \in L^p|Reg_\omega(\mathbb{R}^-;U)$, and*

$$\mathfrak{B}w = \hat{v}(A)\mathfrak{B}u.$$

Proof Note that, in both parts of the theorem, $v \in L^1_{\omega'}(\mathbb{R}^-;\mathbb{C})$ and $u \in L^1_{\omega'}(\mathbb{R}^-;U)$ for every $\omega' \in (\omega_{\mathfrak{A}}, \omega)$. Part (i) then follows from Theorems 3.12.6 and 4.2.1(i), and part (ii) from Theorems 3.12.9, 4.2.1(i), and A.3.4. □

4.3 Differential representations of the state

As the following theorem shows, the state trajectory $x(t) = \mathfrak{A}^{t-s}x_s + \mathfrak{B}^t_s u$ is the strong solution of the appropriate differential equation:

Theorem 4.3.1 *Let U and X be Banach spaces, let $\mathfrak{A}$ be a C_0 semigroup on X, and let $\mathfrak{B}$ be a $L^p|Reg$-well-posed input map for $\mathfrak{A}$ with input space U, and let B be the corresponding control operator (see Definition 4.2.2). Then the following claims are true:*

(i) *If $\mathfrak{B}$ is L^p-well-posed for some $p < \infty$, then for each $s \in \mathbb{R}$, $x_s \in X$, and $u \in L^p_{\mathrm{loc}}([s,\infty);U)$, the function*

$$x(t) = \mathfrak{A}^{t-s}x_s + \mathfrak{B}^t_s u = \mathfrak{A}^{t-s}x_s + \int_s^t \mathfrak{A}^{t-v}_{|X_{-1}} Bu(v)\,dv \tag{4.3.1}$$

is the unique strong solution of the equation

$$\begin{aligned} \dot{x}(t) &= Ax(t) + Bu(t), \qquad t \ge s, \\ x(s) &= x_s. \end{aligned} \tag{4.3.2}$$

Moreover, if $\omega > \omega_{\mathfrak{A}}$, *and if* $u \in L^p_\omega([s,\infty);U)$, *then* $x \in BC_{0,\omega}([s,\infty);X)$.

(ii) *If* $\mathfrak{B}$ *is* L^∞*-well-posed or Reg-well-posed, then for each* $s \in \mathbb{R}$, $x_s \in X$, *and* $u \in Reg_{\text{loc}}([s,\infty);U)$, *the function x defined in* (4.3.1) *is the unique strong solution of the equation* (4.3.2) *(i.e., x(t) is a continuous function of t in X). Moreover, if* $\omega > \omega_{\mathfrak{A}}$, *and if* $u \in Reg_{0,\omega}([s,\infty);U)$, *then* $x \in BC_{0,\omega}([s,\infty);X)$.

Proof (i) This follows from Corollary 4.2.3 and Theorems 2.2.12, 2.5.4(iii), and 3.8.2(ii),(iii).

(ii) As in the proof of (i), most of this follows from Theorems 4.2.1(i), 2.5.4(iv), and 3.8.2(ii),(iii), but this time we need a separate argument showing that the solution is continuous in t since we cannot appeal to Theorem 2.2.12. This argument goes as follows.

The state x consists of two parts,

$$x(t) = \mathfrak{A}^t x_s + \mathfrak{B}^t_s u = \mathfrak{A}^t x_s + \mathfrak{B}\tau^t \pi_{[s,\infty)} u.$$

The first term is continuous and has the right growth bound at infinity, so it can be ignored. We can also ignore the projection $\pi_{[s,\infty)}$ by simply replacing $\pi_{[s,\infty)}u$ by u. Thus, it suffices to show that $\mathfrak{B}^t u = \mathfrak{B}\tau^t u$ is a continuous function of t for every $u \in Reg_{c,\text{loc}}(\mathbb{R};U)$. To do this we use a boot-strap argument, starting from the case where u is continuous.

If u is continuous, then $\mathfrak{B}^t u = \mathfrak{B}\tau^t u$ is continuous, since $\tau^t u$ depends continuously on t in $C_c(\mathbb{R};U)$.

Suppose next that u is continuous on $(-\infty, t_1)$ and vanishes on $[t_1,\infty)$. Then $u = \pi_{(-\infty,t_1)}u$. Clearly $\mathfrak{B}^t u = \mathfrak{B}\tau^t u$ is continuous on $(-\infty,t_1)$ and has the left-hand limit $\mathfrak{B}\tau^{t_1}u$ at t_1 (because $\mathfrak{B}\tau^t u = \mathfrak{B}\pi_-\tau^t u$ and the function $\pi_-\tau^t u$ depends continuously on t in $C_c(\mathbb{R}^-;U)$ for $t \le t_1$). For $t \ge t_1$ we have

$$\mathfrak{B}\pi_-\tau^t u = \mathfrak{B}\pi_-\tau^{t-t_1}\pi_-\tau^{t_1}u = \mathfrak{A}^{t-t_1}\mathfrak{B}\tau^{t_1}u.$$

This term is continuous on $[t_1,\infty)$ and has the value $\mathfrak{B}\tau^{t_1}u$ at t_1. Thus, also for this class of functions $\mathfrak{B}^t u$ is continuous in t.

If u is continuous on $(-\infty,t_1)$ and $[t_1,\infty)$, then we can write u as the sum of two functions, one of which is continuous on $\mathbb{R}$, and the other continuous on $(-\infty,t_1)$ and zero on $[t_1,\infty)$. Thus, by linearity, $\mathfrak{B}^t u$ is a continuous function of t.

If u has a finite number of discontinuities, then u can be written as a finite sum of functions with only one discontinuity, and, by linearity, the function $\mathfrak{B}^t u$ is continuous in X.

If u is an arbitrary function in $Reg_{c,\text{loc}}(\mathbb{R};U)$, then it can be written as a locally uniformly convergent limit of functions u_n with a finite number of discontinuities. But then $\mathfrak{B}^t u_n$ converges locally uniformly to $\mathfrak{B}^t u$, hence the limit function must be continuous.

In particular, we observe that

$$\lim_{t\downarrow s} \mathfrak{B}_s^t u = \lim_{t\uparrow s} \mathfrak{B}\tau^t \pi_{[s,\infty)} u = \lim_{t\uparrow s} \mathfrak{B}\pi_- \tau^t \pi_{[s,\infty)} u = 0,$$

hence the state $x(t)$ of Σ satisfies $x(s) = x_s$. □

It is also possible to prove a similar result for the state trajectory of Σ corresponding to the initial time $-\infty$ (this type of state trajectory was introduced in Definition 2.5.8).

Theorem 4.3.2 *Let U and X be Banach spaces, let $\mathfrak{A}$ be a C_0 semigroup on X, let $\mathfrak{B}$ be a $L^p|Reg$-well-posed input map for $\mathfrak{A}$ with input space U, and let $\omega > \omega_{\mathfrak{A}}$. Let B be the corresponding control operator. Then the following claims are true:*

(i) *Let $\mathfrak{B}$ be L^p-well-posed for some $p < \infty$ and let $u \in L^p_{\omega,\mathrm{loc}}(\mathbb{R}; U)$. Then the function*

$$x(t) = \mathfrak{B}^t u = \int_{-\infty}^t \mathfrak{A}^{t-s}_{|X_{-1}} Bu(s)\,ds, \qquad t \in \mathbb{R},$$

is the unique strong solution of the equation

$$\dot{x}(t) = Ax(t) + Bu(t), \qquad t \in \mathbb{R}$$

satisfying $BC_{0,\omega,\mathrm{loc}}(\mathbb{R}; X)$. If $u \in L^p_\omega(\mathbb{R}; U)$, then $x \in BC_{0,\omega}(\mathbb{R}; X)$.

(ii) *If instead $\mathfrak{B}$ is L^∞-well-posed or Reg-well-posed, then (i) remains true if we replace the conditions $u \in L^p_{\omega,\mathrm{loc}}(\mathbb{R}; U)$ and $u \in L^p_\omega(\mathbb{R}; U)$ by $u \in Reg_{0,\omega,\mathrm{loc}}(\mathbb{R}; U)$, respectively $u \in Reg_{0,\omega}(\mathbb{R}; U)$.*

Proof The proof is the same in both cases, so let us only prove, e.g., (ii). That x is a strong solution of the equation $\dot{x}(t) = Ax(t) + u(t)$ follows from Theorems 2.5.7 and 4.3.1. If the support of u is bounded to the left, then Theorem 4.3.1(ii) gives $x \in BC_{c,\mathrm{loc}}(\mathbb{R}; X) \subset BC_{0,\omega,\mathrm{loc}}(\mathbb{R}; X)$. The continuous dependence of $x \in BC_{0,\omega,\mathrm{loc}}(\mathbb{R}; X)$ on $u \in Reg_{0,\omega,\mathrm{loc}}(\mathbb{R}; U)$ (with support bounded to the left) follows from Example 2.5.3 and Theorem 2.5.4(ii). This set of functions u is dense in $Reg_{0,\omega,\mathrm{loc}}(\mathbb{R}; U)$, hence $x \in BC_{0,\omega,\mathrm{loc}}(\mathbb{R}; X)$ whenever $u \in Reg_{0,\omega,\mathrm{loc}}(\mathbb{R}; U)$. The uniqueness of x follows from Lemma 3.8.6. □

By analogy to Definition 2.5.8 we introduce the following terminology:

Definition 4.3.3 We call the function $t \mapsto x(t) = \mathfrak{B}^t u$ in Theorem 4.3.2 the *state trajectory* of the pair $\left[\mathfrak{A} \,\middle|\, \mathfrak{B}\right]$ with initial time $-\infty$ and input u.

With the help of Theorem 4.3.1 we can prove that if we have a semigroup $\mathfrak{A}$ and an $L^p|Reg$-well-posed input map $\mathfrak{B}$ for $\mathfrak{A}$, then it is always possible to embed these operators in a $L^p|Reg$-well-posed system:

Theorem 4.3.4 *Let $\mathfrak{A}$ be a C_0 semigroup on X, and let $\mathfrak{B}$ be a $L^p|Reg$-well-posed input map for $\mathfrak{A}$ with input space U. Let $C \in \mathcal{B}(X;Y)$ and $D \in \mathcal{B}(U;Y)$. For each $x \in X$ and $u \in L^p|Reg_{c,\mathrm{loc}}(\mathbb{R};U)$, define*

$$(\mathfrak{C}x)(t) = \begin{cases} C\mathfrak{A}^t x, & t \geq 0, \\ 0, & t < 0, \end{cases} \qquad (\mathfrak{D}u)(t) = C\mathfrak{B}^t u + Du(t).$$

Then $\left[\begin{smallmatrix} \mathfrak{A} & \mathfrak{B} \\ \mathfrak{C} & \mathfrak{D} \end{smallmatrix}\right]$ is a Reg-well-posed linear system if $\mathfrak{B}$ is L^∞-well-posed or Reg-well-posed, and it is an L^q-well-posed linear system for all q, $p \leq q < \infty$ if $\mathfrak{B}$ is L^p-well-posed for some $p < \infty$. (The observation operator of this system is C and the feedthrough operator is D; see Definitions 4.4.3 and 4.5.11, respectively.)

Proof Let us begin by inspecting the algebraic properties required by Definition 2.2.1. Obviously $\tau_+^t \mathfrak{C}x = \pi_+(s \mapsto C\mathfrak{A}^{s+t}x) = \pi_+(s \mapsto C\mathfrak{A}^s\mathfrak{A}^t x) = \mathfrak{C}\mathfrak{A}^t x$. The time-invariance and causality of $\mathfrak{D}$ are also obvious. Thus, it only remains to compute the Hankel operator of $\mathfrak{D}$. For all $t > 0$ and all $u \in L^p|Reg_{c,\mathrm{loc}}(\overline{\mathbb{R}}^+;U)$, we have

$$(\pi_+\mathfrak{D}\pi_- u)(t) = \pi_+ C\mathfrak{B}^t \pi_- u = \pi_+ C\mathfrak{B}\tau_-^t u = \pi_+ C\mathfrak{A}^t\mathfrak{B}u.$$

Thus, $\pi_+\mathfrak{D}\pi_- = \mathfrak{C}\mathfrak{B}$ as required.

Obviously, $\mathfrak{C}$ is continuous $X \to Reg_{\mathrm{loc}}(\overline{\mathbb{R}}^+;Y)$ and $X \to L^p_{\mathrm{loc}}(\mathbb{R}^+;Y)$. That $\mathfrak{D}$ is continuous $L^p_{c,\mathrm{loc}}(\mathbb{R};U) \to L^p_{c,\mathrm{loc}}(\mathbb{R};Y)$ (or $Reg_{c,\mathrm{loc}}(\mathbb{R};U) \to Reg_{c,\mathrm{loc}}(\mathbb{R};Y)$ in the L^∞-well-posed and Reg-well-posed case) follows from Theorem 4.3.1. The additional claim about the L^q-well-posedness follows from Theorem 2.4.4. □

As the following lemma shows, the input map converts smoothness in time to smoothness in the state space.

Lemma 4.3.5 *Let U and X be Banach spaces, let $\mathfrak{A}$ be a C_0 semigroup on X, let $\mathfrak{B}$ be a $L^p|Reg$-well-posed input map for $\mathfrak{A}$ with input space U, let $\omega > \omega_{\mathfrak{A}}$, and let $n = 1, 2, 3, \ldots$ Then the following claims are true:*

(i) *If $\mathfrak{B}$ is L^p-well-posed for some $p < \infty$, then $\mathfrak{B}$ maps the set $\{u \in W^{n,p}_\omega(\overline{\mathbb{R}}^-;U) \mid u(0) = \dot u(0) = \cdots = u^{(n-1)}(0) = 0\}$ continuously into $X_n = \mathcal{D}(A^n)$, and $\mathfrak{B}\dot u = A\mathfrak{B}u$ for all $u \in W^{1,p}_\omega(\overline{\mathbb{R}}^-;U)$ with $u(0) = 0$.*

(ii) *If $\mathfrak{B}$ is L^∞-well-posed or Reg-well-posed, then $\mathfrak{B}$ maps the set $\{u \in Reg^n_{0,\omega}(\overline{\mathbb{R}}^-;U) \mid u(0) = \dot u(0) = \cdots = u^{(n-1)}(0) = 0\}$ continuously* into $X_n = \mathcal{D}(A^n)$, and $\mathfrak{B}\dot{u} = A\mathfrak{B}u$ for all $u \in Reg^1_{0,\omega}(\overline{\mathbb{R}}^-;U)$ with $u(0) = 0$.

(iii) *If* $\mathfrak{B}$ *is* L^p*-well-posed for some* $p < \infty$ *and* $u \in W^{1,p}_\omega(\overline{\mathbb{R}}^-;U)$*, then* $A_{|X}\mathfrak{B}u + Bu(0) \in X$ *and* $\mathfrak{B}\dot{u} = A_{|X}\mathfrak{B}u + Bu(0)$.

(iv) *If* $\mathfrak{B}$ *is* L^∞*-well-posed or Reg-well-posed and* $u \in Reg^1_{0,\omega}(\overline{\mathbb{R}}^-;U)$*, then* $A_{|X}\mathfrak{B}u + Bu(0) \in X$ *and* $\mathfrak{B}\dot{u} = A_{|X}\mathfrak{B}u + Bu(0)$.

Proof In the proofs of both (i) and (iii) it suffices to prove the case $n = 1$, which can then be iterated to give the general case.

(i) If $u \in W^{1,p}_\omega(\overline{\mathbb{R}}^-;U)$ with $u(0) = 0$, then, by Example 3.2.3(iii), the function $t \mapsto \mathfrak{B}^t_0 u = \mathfrak{B}\tau^t_- u$ has a right derivative at zero. We differentiate this function, and use the identity $\mathfrak{B}\tau^t_- u = \mathfrak{A}^t\mathfrak{B}u$ from Definition 2.2.1 to conclude that $\mathfrak{B}u \in X_1 = \mathcal{D}(A)$, and that $A\mathfrak{B}u = \mathfrak{B}\dot{u}$.

(ii) The proof is otherwise the same as the proof of (i), but we have to work harder to show that $t \mapsto \mathfrak{B}^t_0 u = \mathfrak{B}\tau^t_- u$ has a right derivative at zero when $u \in Reg^1_{0,\omega}(\overline{\mathbb{R}}^-;U)$ and $u(0) = 0$. To do this we use the representation formula in Theorem 4.2.1(i), which gives for $h > 0$ (define $u(t) = 0$ for $t > 0$)

$$\begin{aligned}\frac{1}{h}\mathfrak{B}(\tau^h u - u) &= \frac{1}{h}\int_{-\infty}^0 \mathfrak{A}^{-s}B(u(s+h) - u(s))\,ds\\ &= \frac{1}{h}\int_{-\infty}^0\int_0^h \mathfrak{A}^{-s}B\dot{u}(s+v)\,ds\,dv\\ &= \frac{1}{h}\int_0^h\int_{-\infty}^0 \mathfrak{A}^{-s}B\dot{u}(s+v)\,ds\,dv\\ &= \frac{1}{h}\int_0^h \mathfrak{B}\tau^v\dot{u}\,dv.\end{aligned}$$

By Theorem 4.3.2, $\mathfrak{B}\tau^v\dot{u}$ is continuous in v, so the integral tends to $(\mathfrak{B}\dot{u})(0)$ as $h \downarrow 0$.

(iii) Let $\alpha > \omega$, and define $\mathrm{e}_\alpha(t) = \mathrm{e}^{\alpha t}$, $t \in \mathbb{R}$. Define $v = u - \mathrm{e}_\alpha u(0)$. Then (i) with $n = 1$ applies to v, and we get $\mathfrak{B}(u - \mathrm{e}_\alpha u(0)) \in X_1$ and

$$\mathfrak{B}(\dot{u} - \alpha\mathrm{e}_\alpha u(0)) = A\mathfrak{B}(u - \mathrm{e}_\alpha u(0)).$$

This, together with the fact that $\mathfrak{B}\mathrm{e}_\alpha u(0) = (\alpha - A_{|X})^{-1}Bu(0)$ (see Theorem 4.2.1(iii)) gives

$$\begin{aligned}\mathfrak{B}\dot{u} &= A_{|X}\mathfrak{B}u - A_{|X}(\alpha - A_{|X})^{-1}Bu(0) + \alpha(\alpha - A_{|X})^{-1}Bu(0)\\ &= A_{|X}\mathfrak{B}u + Bu(0).\end{aligned}$$

(iv) The proof of (iv) is identical to the proof of (iii). □

By using Lemma 4.3.5 we can strengthen the conclusion of Theorem 4.3.2 in the case where u has additional smoothness.

Theorem 4.3.6 *Let U and X be Banach spaces, let $\mathfrak{A}$ be a C_0 semigroup on X, let $\mathfrak{B}$ be a $L^p|Reg$-well-posed input map for $\mathfrak{A}$ with input space U, and let $\omega > \omega_{\mathfrak{A}}$. Then the following claims are true:*

(i) *Let $\mathfrak{B}$ be L^p-well-posed for some $p < \infty$, let $u \in W^{1,p}_{\omega,\mathrm{loc}}(\mathbb{R}; U)$, and let x be the unique strong solution of*

$$\dot{x}(t) = Ax(t) + Bu(t), \qquad t \in \mathbb{R},$$

in $BC_{0,\omega,\mathrm{loc}}(\mathbb{R}; X)$ (cf. Theorem 4.3.2). Then $x \in BC^1_{0,\omega,\mathrm{loc}}(\mathbb{R}; X)$ and $x(t) = \mathfrak{B}^t u$ and $\dot{x}(t) = \mathfrak{B}^t \dot{u}$, $t \in \mathbb{R}$. In particular, $\dot{x} = A_{|X} x + Bu \in BC_{0,\omega,\mathrm{loc}}(\mathbb{R}; X)$, and $\dot{x}$ is the unique strong solution of

$$\ddot{x}(t) = A\dot{x}(t) + B\dot{u}(t), \qquad t \in \mathbb{R},$$

in $BC_{0,\omega,\mathrm{loc}}(\mathbb{R}; X)$. If $u \in W^{1,p}_{\omega}(\mathbb{R}; U)$, then $x \in BC^1_{0,\omega}(\mathbb{R}; X)$, and $A_{|X} x + Bu \in BC_{0,\omega}(\mathbb{R}; X)$.

(ii) *If instead $\mathfrak{B}$ is L^∞-well-posed or Reg-well-posed, then the conclusion of (i) remains valid if we replace the conditions $u \in W^{1,p}_{\omega,\mathrm{loc}}(\mathbb{R}; U)$ and $u \in W^{1,p}_{\omega}(\mathbb{R}; U)$ by $u \in Reg^1_{0,\omega,\mathrm{loc}}(\mathbb{R}; U)$ respectively $u \in Reg^1_{0,\omega}(\mathbb{R}; U)$.*

Proof Most of this follows from Theorem 4.3.2. The only significant new claims are that $\dot{x} \in BC_{0,\omega,\mathrm{loc}}(\mathbb{R}; X)$ and that $\dot{x}(t) = \mathfrak{B}^t \dot{u}$ for all $t \in \mathbb{R}$. If this is true, then $\dot{x} = A_{|X} x + Bu \in BC_{0,\omega,\mathrm{loc}}(\mathbb{R}; X)$. Thus, it remains to show that $\dot{x} \in BC_{0,\omega,\mathrm{loc}}(\mathbb{R}; X)$ and that $\dot{x}(t) = \mathfrak{B}^t \dot{u}$, $t \in \mathbb{R}$.

(i) The differentiability of x follows from the fact that $\frac{1}{h}(\tau^{t+h} - \tau^t)u \to \dot{u}$ in $L^p_{\omega,\mathrm{loc}}(\mathbb{R}; U)$ as $h \to 0$ (see Example 3.2.3), hence x is continuously differentiable and $\dot{x}(t) = \mathfrak{B}\tau^t \dot{u} = \mathfrak{B}^t \dot{u}$ for all $t \geq 0$. This together with Theorem 4.3.2 implies that $\dot{x} \in BC_{0,\omega,\mathrm{loc}}(\mathbb{R}; X)$.

(ii) Apply the argument given in the proof of Lemma 4.3.5 with u replaced by $\tau^t u$ to show that $\dot{x}(t) = \mathfrak{B}\tau^t \dot{u}$, and proceed as above. □

By using Lemma 4.3.5 and Theorem 4.3.6 we can likewise strengthen the conclusion of Theorem 4.3.1 in the case where u has additional smoothness.

Theorem 4.3.7 *Let U and X be Banach spaces, let $\mathfrak{A}$ be a C_0 semigroup on X, let $\mathfrak{B}$ be a $L^p|Reg$-well-posed input map for $\mathfrak{A}$ with input space U, and let $\omega > \omega_{\mathfrak{A}}$.*

(i) *If $\mathfrak{B}$ is L^p-well-posed for some $p < \infty$, then for each $s \in \mathbb{R}$, $x_s \in X$, and $u \in W^{1,p}_{\mathrm{loc}}([s, \infty); U)$ satisfying $A_{|X} x_s + Bu(s) \in X$, the function*

$$x(t) = \mathfrak{A}^{t-s} x_s + \mathfrak{B}^t_s u = \mathfrak{A}^{t-s} x_s + \int_s^t \mathfrak{A}^{t-v}_{|X_{-1}} Bu(v)\, dv \qquad (4.3.1)$$

is continuously differentiable in X with respect to t on $[s,\infty)$, and

$$\begin{aligned}\dot{x}(t) &= A_{|X}x(t) + Bu(t), \qquad t \ge s,\\ x(s) &= x_s.\end{aligned} \tag{4.3.2}$$

Moreover,

$$\begin{aligned}\dot{x}(t) &= \mathfrak{A}^{t-s}[A_{|X}x_s + Bu(s)] + \mathfrak{B}_s^t\dot{u}\\ &= \mathfrak{A}^{t-s}[A_{|X}x_s + Bu(s)] + \int_s^t \mathfrak{A}_{|X_{-1}}^{t-v}B\dot{u}(v)\,dv, \qquad t \ge s,\end{aligned} \tag{4.3.3}$$

$\dot{x} = A_{|X}x + Bu \in C([s,\infty);X)$, and $\dot{x}$ is the unique strong solution of

$$\begin{aligned}\ddot{x}(t) &= A\dot{x}(t) + B\dot{u}(t), \qquad t \ge s,\\ \dot{x}(s) &= A_{|X}x_s + Bu(s).\end{aligned} \tag{4.3.4}$$

If $u \in W^{1,p}_\omega([s,\infty);U)$, then $x \in BC^1_{0,\omega}([s,\infty);X)$, and $A_{|X}x + Bu \in BC_{0,\omega}([s,\infty);X)$.

(ii) *If instead $\mathfrak{B}$ is L^∞-well-posed or Reg-well-posed, then the conclusion of (i) remains valid if we replace the conditions $u \in W^{1,p}_{\omega,\mathrm{loc}}([s,\infty);U)$ and $u \in W^{1,p}_\omega([s,\infty);U)$ by $u \in Reg^1_{0,\omega,\mathrm{loc}}([s,\infty);U)$, respectively $u \in Reg^1_{0,\omega}([s,\infty);U)$.*

Proof Most of this follows from Theorem 4.3.1, and arguing as in the proof of Theorem 4.3.6 we find that the only new thing which we have to prove is that x is continuously differentiable in X and that $\dot{x}$ is the strong solution of (4.3.4). Without loss of generality, let us take $s = 0$, because otherwise we may replace s by 0, u by $\tau^{-s}u$, y by $\tau^{-s}y$, and $x(t)$ by $x(t-s)$.

Take some $\alpha \in \mathbb{C}^+_{\omega_{\mathfrak{A}}}$ (where $\omega_{\mathfrak{A}}$ is the growth bound of $\mathfrak{A}$), and define $\mathrm{e}_\alpha(s) = \mathrm{e}^{\alpha s}$, $s \in \mathbb{R}$, and

$$v = \pi_+u + \pi_-\mathrm{e}_\alpha u(0).$$

Then $v \in W^{1,p}_{\omega,\mathrm{loc}}(\mathbb{R};U)$ in case (i), $v \in Reg^1_{0,\omega,\mathrm{loc}}(\mathbb{R};U)$ in case (ii), and

$$\begin{aligned}x(t) &= \mathfrak{A}^t x_0 + \mathfrak{B}_0^t u\\ &= \mathfrak{A}^t x_0 - \mathfrak{B}\tau_-^t \mathrm{e}_\alpha u(0) + \mathfrak{B}^t v\\ &= \mathfrak{A}^t(x_0 - \mathfrak{B}\mathrm{e}_\alpha u(0)) + \mathfrak{B}^t v\\ &= x_1(t) + x_2(t).\end{aligned}$$

Recall that, by Theorem 4.2.1(iii),

$$\mathfrak{B}\mathrm{e}_\alpha u_0 = (\alpha - A_{|X})^{-1}Bu_0$$

for all $u_0 \in U$. The assumptions $x_0 \in X$ and $A_{|X}x_0 + Bu(0) \in X$ imply $(\alpha - A_{|X})x_0 - Bu(0) \in X$, hence $x_0 - (\alpha - A_{|X})^{-1}Bu(0) = x_0 - \mathfrak{B}\mathrm{e}_\alpha u(0) \in X_1$.

Thus, $x_1(t) = \mathfrak{A}^t(x_0 - \mathfrak{B}e_\alpha u(0))$ is continuously differentiable in X with $\dot{x}_1 = Ax_1$. By Theorem 4.3.6, also x_2 is continuously differentiable in X with $\dot{x}_2 = A_{|X}x_2 + Bu$. Thus x is continuously differentiable in X and $\dot{x} = A_{|X}x + Bu$ (in X). □

Corollary 4.3.8 *Let U and X be Banach spaces, let $\mathfrak{A}$ be a C_0 semigroup on X, and let $\mathfrak{B}$ be a $L^p|Reg$-well-posed input map for $\mathfrak{A}$ with input space U. Let $s \in \mathbb{R}$, $x_s \in X$, and $u_s \in U$.*

(i) *If $\mathfrak{B}$ is L^p-well-posed for some $p < \infty$, then the following conditions are equivalent:*
  (a) $A_{|X}x_s + Bu_s \in X$;
  (b) *for each $u \in W^{1,p}_{\mathrm{loc}}([s,\infty);U)$ satisfying $u(s) = u_s$ the strong solution x of (4.3.2) is continuously differentiable in X;*
  (c) *for some $u \in W^{1,p}_{\mathrm{loc}}([s,\infty);U)$ satisfying $u(s) = u_s$ the strong solution x of (4.3.2) is continuously differentiable in X.*

(ii) *If $\mathfrak{B}$ is L^∞-well-posed or Reg-well-posed, then the following conditions are equivalent:*
  (a) $A_{|X}x_s + Bu_s \in X$;
  (b) *for each $u \in Reg^1_{\mathrm{loc}}([s,\infty);U)$ satisfying $u(s) = u_s$ the strong solution x of (4.3.2) is continuously differentiable in X;*
  (c) *for some $u \in Reg^1_{\mathrm{loc}}([s,\infty);U)$ satisfying $u(s) = u_s$ the strong solution x of (4.3.2) is continuously differentiable in X.*

Proof The proof is the same in both cases. By Theorem 4.3.7, (a) $\Rightarrow$ (b), and obviously (b) $\Rightarrow$ (c). If the strong solution x of (4.3.2) is continuously differentiable in X, then $\dot{x}(s) = A_{|X}x(s) + Bu(s) = A_{|X}x_s + Bu_s \in X$, so (c) $\Rightarrow$ (a). □

Occasionally we shall need a version of Corollary 4.3.8 where we do not know that $\mathfrak{B}$ is $L^p|Reg$-well-posed, but only that the control operator B satisfies $B \in \mathcal{B}(U;X_{-1})$. As the following theorem shows, the conclusion of Corollary 4.3.8 remains valid in this case, too, provided we increase the regularity of the input function slightly.

Theorem 4.3.9 *Let U and X be Banach spaces, let $\mathfrak{A}$ be a C_0 semigroup on X, and let $B \in \mathcal{B}(U;X_{-1})$. Let $s \in \mathbb{R}$, $x_s \in X$, and $u_s \in U$. Then the following conditions are equivalent:*

(i) $A_{|X}x_s + Bu_s \in X$;
(ii) *for each $u \in W^{2,1}_{\mathrm{loc}}([s,\infty);U)$ satisfying $u(s) = u_s$ the strong solution x of (4.3.2) is continuously differentiable in X;*
(iii) *for some $u \in W^{2,1}_{\mathrm{loc}}([s,\infty);U)$ satisfying $u(s) = u_s$ the strong solution x of (4.3.2) is continuously differentiable in X.*

Proof That (i) $\Rightarrow$ (ii) follows from Theorem 3.8.3. Obviously, (ii) $\Rightarrow$ (iii). If (iii) holds, then $\dot{x}(s) = A_{|X}x(s) + Bu(s) = A_{|X}x_s + Bu_s \in X$; hence (iii) $\Rightarrow$ (i). $\square$

It is possible to reformulate parts of the conclusions of Theorems 4.3.6, 4.3.7 and 4.3.9 by introducing some additional spaces, namely

$$\mathcal{D}(A\&B) = \left\{\left[\begin{smallmatrix} w \\ u \end{smallmatrix}\right] \in \left[\begin{smallmatrix} X \\ U \end{smallmatrix}\right] \;\middle|\; A_{|X}w + Bu \in X\right\}, \tag{4.3.5}$$

$$(X + BU)_1 = \{w \in X \mid A_{|X}w + Bu \in X \text{ for some } u \in U\}. \tag{4.3.6}$$

The former of the above spaces becomes important in Section 4.6 in connection with the definition of a system node. The latter is important in the theory of compatible systems and boundary control systems presented in Chapter 5.

Lemma 4.3.10 *Let A be a densely defined operator on X with a nonempty resolvent set, and let $B \in \mathcal{B}(U; X_{-1})$. Define $\mathcal{D}(A\&B)$ by (4.3.5). Then $\mathcal{D}(A\&B)$ is dense in $\left[\begin{smallmatrix} X \\ U \end{smallmatrix}\right]$, and the restriction $A\&B$ of $\left[A_{|X} \;\; B\right] : \left[\begin{smallmatrix} X \\ U \end{smallmatrix}\right] \to X_{-1}$ to $\mathcal{D}(A\&B)$ is closed (as an unbounded operator from $\left[\begin{smallmatrix} X \\ U \end{smallmatrix}\right]$ to X). The domain $\mathcal{D}(A\&B)$ of $A\&B$ is a Banach space with the graph norm*

$$\left|\left[\begin{smallmatrix} w \\ u \end{smallmatrix}\right]\right|_{\mathcal{D}(A\&B)} = \left(|A_{|X}w + Bu|_X^2 + |w|_X^2 + |u|_U^2\right)^{1/2}, \tag{4.3.7}$$

and it is a Hilbert space if both X and U are Hilbert spaces.

Proof We begin by showing that $\mathcal{D}(A\&B)$ is dense in $\left[\begin{smallmatrix} X \\ U \end{smallmatrix}\right]$. Take some arbitrary $\left[\begin{smallmatrix} x \\ u \end{smallmatrix}\right] \in \left[\begin{smallmatrix} X \\ U \end{smallmatrix}\right]$, and fix some $\alpha \in \rho(A_{|X}) = \rho(A)$. Then $x - (\alpha - A_{|X})^{-1}Bu \in X$, so we can find a sequence $w_n \in X_1$ which converges to $x - (\alpha - A_{|X})^{-1}Bu$ in X. Define $x_n = w_n + (\alpha - A_{|X})^{-1}Bu$. Then $x_n \to x \in X$ and

$$\begin{aligned} A_{|X}x_n + Bu &= A_{|X}w_n + A_{|X}(\alpha - A_{|X})^{-1}Bu + Bu \\ &= A_{|X}w_n + \alpha(\alpha - A_{|X})^{-1}Bu \in X, \end{aligned}$$

so $\left[\begin{smallmatrix} x_n \\ u \end{smallmatrix}\right] \in \mathcal{D}(A\&B)$. This proves the density of $\mathcal{D}(A\&B)$ in $\left[\begin{smallmatrix} X \\ U \end{smallmatrix}\right]$.

The closedness of $A\&B$ follows from the continuity of $\left[A_{|X} \;\; B\right] : \left[\begin{smallmatrix} X \\ U \end{smallmatrix}\right] \to X_{-1}$: if $\left[\begin{smallmatrix} x_n \\ u_n \end{smallmatrix}\right] \to \left[\begin{smallmatrix} x \\ u \end{smallmatrix}\right]$ in $\left[\begin{smallmatrix} X \\ U \end{smallmatrix}\right]$ and $A_{|X}x_n + Bu_n \to z$ in X, then we have both $A_{|X}x_n + Bu_n \to z$ and $A_{|X}x_n + Bu_n \to A_{|X}x + Bu$ in X_{-1}, so $z = A_{|X}x + Bu$. $\square$

Theorem 4.3.11 (i) *The pair of functions $\left[\begin{smallmatrix} x \\ u \end{smallmatrix}\right]$ in Theorem 4.3.6(i)–(ii) belongs to $BC_{0,\omega,\mathrm{loc}}(\mathbb{R}; \mathcal{D}(A\&B))$. If $u \in W_\omega^{1,p}(\mathbb{R}; U)$, respectively $u \in Reg_{0,\omega}^1(\mathbb{R}; U)$, then $\left[\begin{smallmatrix} x \\ u \end{smallmatrix}\right] \in BC_{0,\omega}(\mathbb{R}; \mathcal{D}(A\&B))$.*

(ii) *Also the pair of functions $\left[\begin{smallmatrix} x \\ u \end{smallmatrix}\right]$ in Theorem 4.3.7(i)–(ii) belongs to $C([s,\infty); \mathcal{D}(A\&B))$. If $u \in W_\omega^{1,p}([s,\infty); U)$, respectively $u \in Reg_{0,\omega}^1([s,\infty);)$, then $\left[\begin{smallmatrix} x \\ u \end{smallmatrix}\right] \in BC_{0,\omega}([s,\infty); \mathcal{D}(A\&B))$.*

(iii) *The pair of functions* $\left[\begin{smallmatrix} x \\ u \end{smallmatrix}\right]$ *in Theorem 4.3.9 belongs to* $C([s,\infty);\mathcal{D}(A\&B))$ *as well. If* $u \in W^{2,1}_{\omega}([s,\infty);U)$*, then* $\left[\begin{smallmatrix} x \\ u \end{smallmatrix}\right] \in BC_{0,\omega}([s,\infty);\mathcal{D}(A\&B))$.

Proof This follows from Lemma 4.3.10 and the listed theorems. (A pair of functions $\left[\begin{smallmatrix} x \\ u \end{smallmatrix}\right]$ is continuous in $\mathcal{D}(A\&B)$ if and only if x and $A_{|X}x + Bu$ are continuous in X and u is continuous in U. Similar statements are valid for the needed estimates of the size of the norm of $\left[\begin{smallmatrix} x \\ u \end{smallmatrix}\right]$ in $\mathcal{D}(A\&B)$.) □

Lemma 4.3.12 *Let A be a densely defined operator on X with a nonempty resolvent set, let $\alpha \in \rho(A)$, and let $B \in \mathcal{B}(U;X_{-1})$.*

(i) *Define*

$$X + BU = \{z \in X_{-1} \mid z = x + Bu \text{ for some } x \in X \text{ and } u \in U\}.$$

Then $X + BU$ is a Banach space with the norm

$$|z|_{X+BU} = \inf_{x+Bu=z} \left(|x|_X^2 + |u|_U^2\right)^{1/2},$$

satisfying $X \subset X + BU \subset X_{-1}$, and it is a Hilbert space if both X and U are Hilbert spaces. The operator $\begin{bmatrix} 1 & B \end{bmatrix}$ maps $\left[\begin{smallmatrix} X \\ U \end{smallmatrix}\right]$ continuously onto $X + BU$.

(ii) *Define*

$$(X + BU)_1 = (\alpha - A_{|X})^{-1}(X + BU).$$

Then $(X + BU)_1$ is a Banach space with the norm

$$\begin{aligned} |w|_{(X+BU)_1} &= |(\alpha - A_{|X})w|_{X+BU} \\ &= \inf_{(\alpha - A_{|X})^{-1}(x+Bu)=w} \left(|x|_X^2 + |u|_U^2\right)^{1/2}, \end{aligned}$$

satisfying $X_1 \subset (X + BU)_1 \subset X$, and it is a Hilbert space if both X and U are Hilbert spaces. The operator $(\alpha - A_{|X})^{-1}\begin{bmatrix} 1 & B \end{bmatrix}$ maps $\left[\begin{smallmatrix} X \\ U \end{smallmatrix}\right]$ continuously into $(X + BU)_1$.

(iii) *The space $(X + BU)_1$ in (ii) can alternatively be defined as*

$$(X + BU)_1 = \{w \in X \mid A_{|X}w + Bu \in X \text{ for some } u \in U\}.$$

An equivalent norm in $(X + BU)_1$ is given by

$$\|w\|_{(X+BU)_1} = \inf_{A_{|X}w+Bu\in X} \left(|w|_X^2 + |A_{|X}w + Bu|_X^2 + |u|_U^2\right)^{1/2}.$$

See Sections 5.1 and 5.2 for additional results on the spaces $X + BU$ and $(X + BU)_1$. In particular, although the embeddings $X + BU \subset X_{-1}$ and $(X + BU)_1 \subset X$ are dense, the embeddings $X \subset X + BU$ and $X_1 \subset (X + BU)_1$ need not be dense.

The proof of Lemma 4.3.12 is based on the following construction:

Lemma 4.3.13 *Let $E \in \mathcal{B}(U;Y)$, where U and Y are Banach spaces. For each $y \in \mathcal{R}(E)$, let $|y|_Z = \inf\{|u|_U \mid y = Eu\}$. This defines a norm on $\mathcal{R}(E)$ that makes $\mathcal{R}(E)$ a Banach space Z, and this norm is (not necessarily strictly) stronger than the norm of Y. The operator E is a contraction from U into Z, and it induces an isometric isomorphism of the quotient $U/\mathcal{N}(E)$ onto Z. If U is a Hilbert space, then so is Z, and then E is an isometric isomorphism of $\mathcal{N}(E)^\perp$ onto Z.*

See Section 9.1 for a further discussion of the quotient space $U/\mathcal{N}(E)$ and the factorization of E through $U/\mathcal{N}(E)$.

Proof of Lemma 4.3.13 Fix some $y \in \mathcal{R}(E)$. Then the set of all $u \in U$ satisfying $Eu = y$ is an equivalence class in the quotient space $W = U/\mathcal{N}(E)$ (the difference of any two elements in this set belong to $\mathcal{N}(E)$). The norm in W of this equivalence class is $\inf_{v \in U; Ev=y} |v|_U$, which is the same number which we denoted by $|y|_Z$ above. The space W is a Banach space, and it is a Hilbert space if U is a Hilbert space (and in the latter case we can identify W with $\mathcal{N}(E)^\perp$). We factor E into $E = \left[\begin{smallmatrix} F \\ \pi_W \end{smallmatrix}\right]$, where π_W is the (continuous and open) quotient map of U onto W (see Definition 9.1.5). Then $F \in \mathcal{B}(W;Y)$, F is injective, and $\mathcal{R}(F) = \mathcal{R}(E)$. We can therefore let a point in $\mathcal{R}(E) = \mathcal{R}(F)$ inherit the norm of its preimage in W, i.e., if $y = Fw$, then we define (as we did above) $|y|_Z = |w|_W$. This is a norm in $\mathcal{R}(E)$, and we denote $\mathcal{R}(E)$ equipped with this norm by Z. Then Z is a Banach space which is an isometrically isomorphic image of $W = U/\mathcal{N}(E)$ under F, and Z is a Hilbert space whenever U is a Hilbert space. The continuity of F implies that the norm of Z is stronger than the norm of Y. □

Proof of Lemma 4.3.12 (i) We get (i) from Lemma 4.3.13 by replacing U by $\left[\begin{smallmatrix} X \\ U \end{smallmatrix}\right]$, Y by X_{-1}, and E by $\begin{bmatrix} 1 & B \end{bmatrix} \in \mathcal{B}\left(\left[\begin{smallmatrix} X \\ U \end{smallmatrix}\right]; X_{-1}\right)$.

(ii) We get (ii) from Lemma 4.3.13 by replacing U by $\left[\begin{smallmatrix} X \\ U \end{smallmatrix}\right]$, Y by X, and E by $(\alpha - A_{|X})^{-1} \begin{bmatrix} 1 & B \end{bmatrix} \in \mathcal{B}\left(\left[\begin{smallmatrix} X \\ U \end{smallmatrix}\right]; X\right)$. Alternatively, we may apply the same construction that we used in the definition of the spaces X_n in Section 3.6.

(iii) Suppose that $w \in X$, $u \in U$, and $A_{|X}w + Bu \in X$. Then $x := \alpha w - (A_{|X}w + Bu) \in X$ and $(\alpha - A_{|X})w = x + Bu \in X + BU$, so $w = (\alpha - A_{|X})^{-1}(x + Bu) \in (X + BU)_1$. Conversely, if $w \in (X + BU)_1$, then there exist $x \in X$ and $u \in U$ such that $w = (\alpha - A_{|X})^{-1}(x + Bu)$, or equivalently, $(\alpha - A_{|X})w = x + Bu$. This implies that $A_{|X}w + Bu = \alpha w - x \in X$. This establishes the alternative description of $(X + BU)_1$ given in (iii).

To prove that $\|\cdot\|_{(X+BU)_1}$ is a norm we use Lemma 4.3.13 with U replaced by $\mathcal{D}(A\&B)$, Y replaced by X, and E replaced by the bounded linear operator which maps $\left[\begin{smallmatrix} w \\ u \end{smallmatrix}\right] \in \mathcal{D}(A\&B)$ into $w \in X$.

Let us finally prove that the two norms on $(X+BU)_1$ are equivalent. In the definition of both norms we minimize over the same set of vectors $u \in U$: the vector x in the definition of $|w|_{(X+BU)_1}$ is uniquely determined by $w \in (X+BU)_1$ and $u \in U$ since $x = (\alpha - A_{|X})w - Bu$, and for all $w \in (X+BU)_1$ and $u \in U$, with x defined in this way, $A_{|X}w + Bu \in X$ iff $x \in X$. Thus, it suffices to show that there is a finite constant K such that, for all $u \in U$ with $(\alpha - A_{|X})^{-1}(x+Bu) = w$, we have estimates of the type

$$\begin{aligned} |x|_X &\le K(|w|_X + |A_{|X}w + Bu|_x), \\ |w|_X &\le K(|x|_X + |u|_U), \\ |A_{|X}w + Bu|_X &\le K(|x|_X + |u|_U). \end{aligned}$$

But this follows from the assumed continuity properties of $(\alpha - A_{|X})^{-1}$ and B and the facts that

$$\begin{aligned} x &= \alpha w - (A_{|X}w + Bu), \\ w &= (\alpha - A_{|X})^{-1}(x + Bu), \\ Aw + Bu &= -x + \alpha(\alpha - A_{|X})^{-1}(x + Bu). \end{aligned}$$

□

Theorem 4.3.14 (i) *The pair of functions $\left[\begin{smallmatrix} x \\ u \end{smallmatrix}\right]$ in Theorem 4.3.6(i)–(ii) belongs to $BC_{0,\omega,\mathrm{loc}}(\mathbb{R}; (X+BU)_1)$. If $u \in W^{1,p}_\omega(\mathbb{R}; U)$, respectively $u \in Reg^1_{0,\omega}(\mathbb{R}; U)$, then $\left[\begin{smallmatrix} x \\ u \end{smallmatrix}\right] \in BC_{0,\omega}(\mathbb{R}; (X+BU)_1)$.*

(ii) *Also the pair of functions $\left[\begin{smallmatrix} x \\ u \end{smallmatrix}\right]$ in Theorem 4.3.7(i)–(ii) belongs to $C([s,\infty); (X+BU)_1)$. If $u \in W^{1,p}_\omega([s,\infty); U)$, respectively $u \in Reg^1_{0,\omega}([s,\infty);)$, then $\left[\begin{smallmatrix} x \\ u \end{smallmatrix}\right] \in BC_{0,\omega}([s,\infty); (X+BU)_1)$.*

(iii) *The pair of functions $\left[\begin{smallmatrix} x \\ u \end{smallmatrix}\right]$ in Theorem 4.3.9 belongs to $C([s,\infty); (X+BU)_1)$ as well. If $u \in W^{2,1}_\omega([s,\infty); U)$, then $\left[\begin{smallmatrix} x \\ u \end{smallmatrix}\right] \in BC_{0,\omega}([s,\infty); (X+BU)_1)$.*

Proof This follows from Theorem 4.3.7 and part (iii) of Lemma 4.3.12. □

4.4 The observation operator

We proceed with a study of the output map $\mathfrak{C}$, and start with a preliminary result which resembles Lemma 4.3.5. It says that the output map converts smoothness in the state space into smoothness in time.

Lemma 4.4.1 *Let X and Y be Banach spaces, let $\mathfrak{A}$ be a C_0 semigroup on X with growth bound $\omega_{\mathfrak{A}}$, let $\mathfrak{C}$ be a $L^p|Reg$-well-posed output map for $\mathfrak{A}$ with* *output space Y, let $\omega > \omega_{\mathfrak{A}}$, and let $n = 1, 2, 3, \ldots$. Then the following claims are true:*

(i) *If $\mathfrak{C}$ is L^p-well-posed for some $p < \infty$, then $\mathfrak{C}$ maps $X_n = \mathcal{D}(A^n)$ continuously into $W^{n,p}_\omega(\overline{\mathbb{R}}^+; Y) \subset BC^{(n-1)}_{0,\omega}(\overline{\mathbb{R}}^+; Y)$, and $(\mathfrak{C}x)' = \mathfrak{C}Ax$ for $x \in X_1$.*

(ii) *If $\mathfrak{C}$ is L^∞-well-posed or Reg-well-posed, then $\mathfrak{C}$ maps X continuously into $BC_{0,\omega}(\overline{\mathbb{R}}^+; Y)$ and $X_n = \mathcal{D}(A^n)$ continuously into $BC^n_{0,\omega}(\overline{\mathbb{R}}^+; Y)$, and $(\mathfrak{C}x)' = \mathfrak{C}Ax$ for $x \in X_1$.*

Proof (i) Let $x \in X_1$. Then, by Theorem 3.2.1(iii), $\mathfrak{A}^t x$ is continuously differentiable in X. Since $\tau^t_+ \mathfrak{C}x = \mathfrak{C}\mathfrak{A}^t x$, this implies that the function $t \mapsto \tau^t_+ \mathfrak{C}x$ is continuously differentiable in $L^p|Reg_\omega(\overline{\mathbb{R}}^+; Y)$. In the L^p-case with $p < \infty$ we find that $\mathfrak{C}x \in W^{1,p}_\omega(\overline{\mathbb{R}}^+; Y)$ (see Example 3.2.3(ii)); in particular, $\mathfrak{C}x \in BC_{0,\omega}(\overline{\mathbb{R}}^+; Y)$ (see Lemma 3.2.4). Moreover, by differentiating both sides of the identity $\tau^t_+ \mathfrak{C}x = \mathfrak{C}\mathfrak{A}^t x$ with respect to t and then setting $t = 0$ we get $(\mathfrak{C}x)' = \mathfrak{C}Ax$ for $x \in X_1$. By repeating the same argument with x replaced by Ax, etc. we find that $\mathfrak{C}x \in W^{n,p}_\omega(\overline{\mathbb{R}}^+; Y)$ whenever $x \in X_n$.

(ii) If $\mathfrak{C}$ is L^∞-well-posed or *Reg*-well-posed, then the identity $\tau^t_+ \mathfrak{C}x = \mathfrak{C}\mathfrak{A}^t x$ combined with the strong continuity of $\mathfrak{A}$ and Examples 2.3.2(ii) and 2.5.3 shows that $\mathfrak{C}$ maps X continuously into $BC_\omega(\overline{\mathbb{R}}^+; Y)$. As this is true also when we replace ω by ω' with $\omega_{\mathfrak{A}} < \omega' < \omega$, we must in fact have $\mathfrak{C}x \in BC_{0,\omega}(\overline{\mathbb{R}}^+; Y)$. But then the same argument that we used to prove (i) shows that $\mathfrak{C}$ also maps X_n continuously into $\mathfrak{C}x \in BC^n_{0,\omega}(\overline{\mathbb{R}}^+; Y)$ and that $(\mathfrak{C}x)' = \mathfrak{C}Ax$. □

With the help of Lemma 4.4.1 it is easy to prove the following basic result on the existence of an observation operator.

Theorem 4.4.2 *Let X and Y be Banach spaces, let $\mathfrak{A}$ be a C_0 semigroup on X with growth bound $\omega_{\mathfrak{A}}$, let $\mathfrak{C}$ be a $L^p|Reg$-well-posed output map for $\mathfrak{A}$ with output space Y, and let $\omega > \omega_{\mathfrak{A}}$. Then the following claims are true:*

(i) *There exists a unique operator $C \in \mathcal{B}(X_1; Y)$ such that*

$$(\mathfrak{C}x)(t) = C\mathfrak{A}^t x \tag{4.4.1}$$

for all $x \in X_1$ and all $t \geq 0$.

(ii) *If $\mathfrak{C}$ is L^∞-well-posed or Reg-well-posed, then the operator C in (i) can be extended to a unique operator $C_{|X}$ in $\mathcal{B}(X; Y)$, and* (4.4.1) *holds for all $x \in X$ and all $t \geq 0$ with C replaced by $C_{|X}$.*

(iii) *The operator $\mathfrak{C}$ is uniquely determined by $\mathfrak{A}$ and C.*

(iv) *For all* $x \in X$, $\lambda \in \mathbb{C}^+_{\omega_{\mathfrak{A}}}$, *and* $n = 0, 1, 2, \ldots$,

$$C(\lambda - A)^{-(n+1)}x = \frac{1}{n!}\int_0^\infty s^n e^{-\lambda s}(\mathfrak{C}x)(s)\,ds.$$

In particular,

$$C(\lambda - A)^{-1}x = \int_0^\infty e^{-\lambda s}(\mathfrak{C}x)(s)\,ds.$$

Thus, in terms of the Laplace transform introduced in Definition 3.12.1, Theorem 4.4.2(iii) implies that $\widehat{(\mathfrak{C}x_0)}(\lambda) = C(\lambda - A)^{-1}x_0$ for all $\lambda \in \mathbb{C}^+_{\omega_{\mathfrak{A}}}$ and all $x_0 \in X$.

Proof (i)–(ii) For each $x \in X_1$ (or $x \in X$ if $\mathfrak{A}$ is L^∞-well-posed or *Reg*-well-posed) we define $Cx = (\mathfrak{C}x)(0)$. Then, by Lemma 4.4.1, $C \in \mathcal{B}(X_1; Y)$ (or $C \in \mathcal{B}(X; Y)$), and, for all $t \geq 0$,

$$(\mathfrak{C}x)(t) = (\tau_+^t \mathfrak{C}x)(0) = (\mathfrak{C}\mathfrak{A}^t x)(0) = C\mathfrak{A}^t x.$$

The uniqueness of C follows from the fact that, necessarily, $Cx = (\mathfrak{C}x)(0)$ for $x \in X_1$.

(iii) This follows from (i) and the fact that X_1 is dense in X.

(iv) By part (i) and Theorem 3.2.9(i), the identity in (iv) is true if $x \in X_1$. As X_1 is dense in X and both sides of this identity depend continuously on $x \in X$, the same identities must then be true for all $x \in X$. □

Definition 4.4.3

(i) The operator C in Theorem 4.4.2 is called *the observation operator* (induced by $\mathfrak{A}$ and $\mathfrak{C}$).
(ii) The observation operator C is *bounded* if it can be extended to an operator $C_{|X}$ in $\mathcal{B}(X; Y)$.

The output map $\mathfrak{C}$ has the following representation in terms of the semigroup $\mathfrak{A}$ and the observation operator C:

Theorem 4.4.4 *Let* $\mathfrak{A}$ *be a* C_0 *semigroup on the Banach space* X, *and let* $\mathfrak{C}$ *be a* $L^p|Reg$*-well-posed output map for* $\mathfrak{A}$ *with output space* Y. *Then, for each* $x \in X$, *the function* $t \mapsto \int_0^t \mathfrak{A}^s x\,ds$, $t \geq 0$, *is continuous with values in* X_1, *the function* $t \mapsto C\int_0^t \mathfrak{A}^s x\,ds$, $t \geq 0$, *belongs to* $W^{1,p}_{\mathrm{loc}}(\overline{\mathbb{R}}^+; Y)$ *if* $\mathfrak{C}$ *is* L^p*-well-posed for some* $p < \infty$ *and to* $C^1(\overline{\mathbb{R}}^+; Y)$ *if* $\mathfrak{C}$ *is* L^∞*-well-posed or Reg-well-posed, and*

$$(\mathfrak{C}x)(t) = \frac{d}{dt} C\int_0^t \mathfrak{A}^s x\,ds \quad \textit{for almost all } t \geq 0. \tag{4.4.2}$$

Proof By Theorem 3.2.1(ii), $\int_0^t \mathfrak{A}^s x\,ds \in X_1$ for all $t \ge 0$, and, for all $\alpha \in \rho(A)$ and $t \ge 0$,

$$\begin{aligned}\int_0^t \mathfrak{A}^s x\,ds &= (\alpha - A)^{-1}(\alpha - A)\int_0^t \mathfrak{A}^s x\,ds \\ &= (\alpha - A)^{-1}\left(\alpha\int_0^t \mathfrak{A}^s x\,ds - \mathfrak{A}^t x + x\right).\end{aligned} \tag{4.4.3}$$

This implies that x_1 is continuous with values in X_1, as claimed.

In the L^∞-well-posed case and the *Reg*-well-posed case the remainder of the conclusion of Theorem 4.4.4 can be proved as follows. We first use Theorem 4.4.2(ii) and integrate (4.4.1) with C replaced by $C_{|X}$ to get $\int_0^t (\mathfrak{C}x)(s)\,ds = \int_0^t C_{|X}\mathfrak{A}^s x\,ds = C_{|X}\int_0^t \mathfrak{A}^s x\,ds$ for $t \ge 0$. The restriction of $C_{|X}$ to X_1 is C, so this implies that the function $t \mapsto C\int_0^t \mathfrak{A}^s x\,ds$ is continuously differentiable, and that (4.4.2) holds.

In the L^p-well-posed case with $p < \infty$ the function $\mathfrak{C}x$ belongs to $L^p_{\mathrm{loc}}(\overline{\mathbb{R}}^+; Y)$, so the function $t \mapsto \int_0^t (\mathfrak{C}x)(s)\,ds$ belongs to $W^{1,p}_{\mathrm{loc}}(\overline{\mathbb{R}}^+; Y)$. Thus, to complete the proof it suffices to show that $\int_0^t (\mathfrak{C}x)(s)\,ds = C\int_0^t \mathfrak{A}^s x\,ds$ for $t \ge 0$. Indeed, it follows from Theorem 4.4.2(i) that this identity is true whenever $x \in X_1$, because then $(\mathfrak{C}x)(s) = C\mathfrak{A}^s x$, and hence $\int_0^t (\mathfrak{C}x)(s)\,ds = \int_0^t C\mathfrak{A}^s x\,ds = C\int_0^t \mathfrak{A}^s x\,ds$. If $x \notin X_1$, then we approximate x by a sequence $x_n \to x$ in X, with each $x_n \in X_1$. By the continuity of $\mathfrak{C}$ from X to $L^p_{\mathrm{loc}}(\mathbb{R}^+; U)$, the restriction of $\mathfrak{C}x_n$ to $[0, t)$ converges to $\mathfrak{C}x$ in $L^p([0, t); Y)$, so $\int_0^t (\mathfrak{C}x_n)(s)\,ds \to \int_0^t (\mathfrak{C}x)(s)\,ds$ as $n \to \infty$. It follows from (4.4.3) that $\int_0^t \mathfrak{A}^s x_n\,ds \to \int_0^t \mathfrak{A}^s x\,ds$ in X_1, hence $C\int_0^t \mathfrak{A}^s x_n\,ds \to C\int_0^t \mathfrak{A}^s x\,ds$ in Y as $n \to \infty$. Thus, also in this case $\int_0^t (\mathfrak{C}x)(s)\,ds = C\int_0^t \mathfrak{A}^s x\,ds$ for all $t \ge 0$. □

Example 4.4.5 *The observation operator of the delay line in Example 2.3.4 is the point evaluation operator* $Cx = x(0)$.

Example 4.4.6 *The observation operators of the exactly controllable shift realization in Example 2.6.5(i) and the bilateral input shift realization in Example 2.6.5(iii) are the operators* $Cu = (\mathfrak{D}u)(0)$. *The observation operators of the exactly observable shift realization in Example 2.6.5(ii) and the bilateral output shift realization in Example 2.6.5(iv) are the point evaluation operators* $Cy = y(0)$.

This follows from Theorem 4.4.2(i) and the definition of $\mathfrak{A}$ and $\mathfrak{C}$ in Example 2.6.5.

There is also an easy converse to Theorem 4.4.2:

Theorem 4.4.7 *Let $\mathfrak{A}$ be a C_0 semigroup on X, and let $1 \le p < \infty$. Then*

(i) *C is the observation operator of an L^p-well-posed output map for $\mathfrak{A}$ with output space Y if and only if $C \in \mathcal{B}(X_1; Y)$ and the map $\mathfrak{C}\colon X_1 \to C(\overline{\mathbb{R}}^+; Y)$ defined by*

$$(\mathfrak{C}x)(t) = C\mathfrak{A}^t x, \qquad t \ge 0,$$

can be extended to a continuous map $X \to L^p_{\mathrm{loc}}(\mathbb{R}^+; Y)$.

(ii) *C is the observation operator of an L^∞-well-posed or Reg-well-posed output map for $\mathfrak{A}$ with output space Y if and only if $C \in \mathcal{B}(X; Y)$.*

We leave the easy proof to the reader (the necessity part is contained in Theorem 4.4.2(i)–(ii)).

The following theorem is the analogue of Theorem 4.3.4 for output maps.

Theorem 4.4.8 *Let $\mathfrak{A}$ be a C_0 semigroup on X, and let $\mathfrak{C}$ be a $L^p|$Reg-well-posed output map for $\mathfrak{A}$ with input space Y. Let $B \in \mathcal{B}(U; X)$ and $D \in \mathcal{B}(U; Y)$. For each $u \in L^1_c(\mathbb{R}^-; U)$, define*

$$\mathfrak{B}u = \int_{-\infty}^{0} \mathfrak{A}^{-s} Bu(s)\, ds,$$

and let $\mathfrak{D}$ be the unique continuous time-invariant causal operator which maps $L^1_{c,\mathrm{loc}}(\mathbb{R}; U)$ continuously into $L^p_{c,\mathrm{loc}}(\mathbb{R}; Y)$ and satisfies

$$(\mathfrak{D}(uv))(t) = \int_{-\infty}^{t} (\mathfrak{C}Bu)(t-s)v(s)\, ds + Duv(t) \tag{4.4.4}$$

for all $u \in U$ and $v \in L^1_{c,\mathrm{loc}}(\mathbb{R}; \mathbb{C})$ and almost all $t \in \mathbb{R}$ (see Theorem A.3.8). Then $\left[\begin{smallmatrix} \mathfrak{A} & \mathfrak{B} \\ \mathfrak{C} & \mathfrak{D} \end{smallmatrix}\right]$ is a well-posed linear system on (Y, X, U) for all q, $1 \le q \le p$, if the original system is L^p-well-posed, and for all q, $1 \le q \le \infty$, if the original system is Reg-well-posed. It is, in addition, Reg-well-posed if $\mathfrak{C}$ is L^∞-well-posed or Reg-well-posed. (The control operator of this system is B and the feedthrough operator is D; see Definitions 4.2.2 and 4.5.11, respectively.)

Proof The proof of the claim that $\mathfrak{B}$ is an L^1-well-posed input map for $\mathfrak{A}$ is the same as in Proposition 2.3.1, and we leave this proof to the reader. For each simple function u with compact support (simple means that u takes only finitely many values) it follows from (4.4.4) and the linearity of $\mathfrak{D}$ that $\tau^t\mathfrak{D}u = \mathfrak{D}\tau^t u$ for all $t \in \mathbb{R}$, and that $\pi_-\mathfrak{D}\pi_+ u = 0$. The set of simple functions is dense in $L^1_{c,\mathrm{loc}}(\mathbb{R}; U)$, so the same identities must be true for all $u \in L^1_{c,\mathrm{loc}}(\mathbb{R}; U)$. By the same argument, to show that the Hankel operator of $\mathfrak{D}$ is $\mathfrak{C}\mathfrak{B}$ it suffices to show that for all $u \in U$, $v \in L^1_{c,\mathrm{loc}}(\mathbb{R}; \mathbb{C})$, and almost all $t \ge 0$, it is true that

$$(\mathfrak{D}(u\pi_- v))(t) = (\mathfrak{C}\mathfrak{B}(u\pi_- v))(t).$$

By linearity and density, it suffices to show that this is true when we take v to be of the form $v = \pi_{(-\infty,s)}e_\alpha$ where $\alpha > \omega_{\mathfrak{A}}$ and $s \le 0$. First, take $v = \pi_- e_\alpha$. Then, for almost all $t \ge 0$, By Definition 4.1.1(iii)–(iv) and Theorem 3.2.9(i),

$$\begin{aligned}(\mathfrak{D}(u\pi_- e_\alpha))(t) &= \int_{-\infty}^{0} (\mathfrak{C}Bu)(t-s)e^{\alpha s}\,ds \\ &= \int_0^\infty (\mathfrak{C}Bu)(t+s)e^{-\alpha s}\,ds \\ &= \int_0^\infty \tau_+^t(\mathfrak{C}Bu)(s)e^{-\alpha s}\,ds \\ &= \int_0^\infty (\mathfrak{C}\mathfrak{A}^t Bu)(s)e^{-\alpha s}\,ds \\ &= C(\alpha - A)^{-1}\mathfrak{A}^t Bu.\end{aligned}$$

By the time-invariance of $\mathfrak{D}$, this means that for all $s \le 0$ and almost all $t \ge 0$ (see also Lemmas 2.2.9 and 2.5.2(ii) and Theorem 3.2.9(i))

$$\begin{aligned}(\mathfrak{D}(u\pi_{(-\infty,s)}e_\alpha))(t) &= e^{\alpha s}(\mathfrak{D}(u\tau^{-s}\pi_- e_\alpha))(t) \\ &= e^{\alpha s}\tau^{-s}(\mathfrak{D}(u\pi_- e_\alpha))(t) \\ &= e^{\alpha s}C(\alpha - A)^{-1}\mathfrak{A}^{t-s}Bu \\ &= e^{\alpha s}C\mathfrak{A}^t\mathfrak{A}^{-s}(\alpha - A)^{-1}Bu \\ &= e^{\alpha s}C\mathfrak{A}^t\mathfrak{A}^{-s}\mathfrak{B}(ue_\alpha) \\ &= e^{\alpha s}(\mathfrak{C}(\mathfrak{B}(u\pi_-\tau^{-s}e_\alpha)))(t) \\ &= (\mathfrak{C}(\mathfrak{B}(u\pi_{(-\infty,s)}e_\alpha)))(t).\end{aligned}$$

Thus, the claim is true for this class of functions. As we already observed above, by linearity and density, $\mathfrak{D} = \mathfrak{C}\mathfrak{B}$. □

From Theorem 4.4.2(iv) it is possible to derive the following growth estimates involving C and the resolvent of A.

Proposition 4.4.9 *In the setting of Theorem 4.4.2, for each $\omega > \omega_{\mathfrak{A}}$ there is a finite constant M such that*

$$\left\|C(\lambda - A)^{-n}\right\| \le \frac{M}{n^{1/2p}(\Re\lambda - \omega)^{n-1/p}}$$

for all $n = 1, 2, 3, \ldots$ and all $\lambda \in \mathbb{C}_\omega^+$ in the case where $\mathfrak{C}$ is L^p-well-posed with $1 \le p < \infty$, and such that

$$\left\|C(\lambda - A)^{-n}\right\| \le \frac{M}{(\Re\lambda - \omega)^n}$$

for all $n = 1, 2, 3, \dots$ *and all* $\lambda \in \mathbb{C}^+_\omega$ *in the case where* $\mathfrak{C}$ *is* L^∞*-well-posed or Reg-well-posed. In particular, taking* $n = 1$ *we get*

$$\left\|C(\lambda - A)^{-1}\right\| \le \frac{M}{(\Re\lambda - \omega)^{1-1/p}}$$

in the case where $\mathfrak{B}$ *is* L^p*-well-posed with* $1 \le p < \infty$*, and*

$$\left\|C(\lambda - A)^{-1}\right\| \le \frac{M}{\Re\lambda - \omega}$$

in the case where $\mathfrak{B}$ *is* L^∞*-well-posed or Reg-well-posed.*

Proof In the L^∞-case and *Reg*-case this follows from Theorems 3.2.9(ii) and 4.4.2(ii). By Theorem 2.5.4(ii), there is a constant $M < \infty$ such that

$$\left\|e_{-\omega}\mathfrak{C}x\right\|_{L^p(\mathbb{R}^+)} \le M|x|,$$

where $e_{-\omega}(s) = e^{-\omega s}$, $s > 0$. By Theorem 4.4.2(ii), for all $n = 0, 1, 2, \dots,$

$$|C(\lambda - A)^{-(n+1)}x| \le \frac{1}{n!}\int_0^\infty s^n e^{-(\Re\lambda-\omega)s}\left|e^{-\omega s}(\mathfrak{C}x)(s)\right|.$$

The given estimate then follows from Hölder's inequality and Lemma 4.2.10. □

The following theorem is an extension of Theorem 4.4.2(iv).

Theorem 4.4.10 *We use the same setting as in Theorem 4.4.2 (in particular, we take* $\omega > \omega_{\mathfrak{A}}$*). Define* q *by* $1/q + 1/p = 1$ *if* $\mathfrak{C}$ *is* L^p*-well-posed with* $p < \infty$*, and let* $q = 1$ *if* $\mathfrak{C}$ *is* L^∞*-well-posed or Reg-well-posed. Let* $v \in L^q_{-\omega}(\mathbb{R}^+; \mathbb{C})$*, and suppose that the (right-sided) Laplace transform* $\hat{v}$ *of* v *is analytic at infinity. Then, for all* $x_0 \in X$*,*

$$\int_0^\infty v(s)(\mathfrak{C}x_0)(s)\,ds = C\hat{v}(-A)x_0. \qquad (4.4.5)$$

Proof It suffices to prove this in the case where $x_0 \in X_1$, because X_1 is dense in X and both sides of (4.4.5) depend continuously on $x_0 \in X$ (cf. Lemma 3.5.9, and observe that $\hat{v}(-A) \in \mathcal{B}(X; X_1)$ since $\hat{v}(\infty) = 0$). For $x_0 \in X_1$ formula (4.4.5) follows from Corollary 3.12.7 and Theorem 4.4.2(i) (note that $v \in L^1_{-\omega'}(\mathbb{R}^+; \mathbb{C})$ for all $\omega' \in (\omega_{\mathfrak{A}}, \omega)$). □

4.5 The feedthrough operator

We have now developed representations for $\mathfrak{A}$ (the semigroup generated by A), $\mathfrak{B}$ (the input map generated by A and B), and $\mathfrak{C}$ (the output map generated by A and C). We are still facing the most difficult part of the major question that we posed at the beginning of this chapter, namely the development of a representation for the input/output map $\mathfrak{D}$. There are some special cases where such a representation is easy to obtain, so let us begin with these cases and return to the more general case later. However, before that, let us prove the following smoothness result for $\mathfrak{D}$, which is analogous to those given in Lemmas 4.3.5 and 4.4.1 for $\mathfrak{B}$ and $\mathfrak{C}$.

Lemma 4.5.1 *Let $\Sigma = \left[\begin{smallmatrix}\mathfrak{A} & \mathfrak{B}\\ \mathfrak{C} & \mathfrak{D}\end{smallmatrix}\right]$ be a $L^p|Reg$-well-posed linear system on (Y, X, U) with growth bound $\omega_{\mathfrak{A}}$, let $\omega > \omega_{\mathfrak{A}}$, and let $n = 0, 1, 2, \ldots$*

(i) *If Σ is L^p-well-posed for some $p < \infty$, then $\mathfrak{D}$ maps $W^{n,p}_{\omega}(\mathbb{R}; U)$ and $W^{n,p}_{\omega,\mathrm{loc}}(\mathbb{R}; U)$ continuously into $W^{n,p}_{\omega}(\mathbb{R}; Y)$, respectively $W^{n,p}_{\omega,\mathrm{loc}}(\mathbb{R}; Y)$.*
(ii) *If Σ is L^∞-well-posed or Reg-well-posed, then $\mathfrak{D}$ maps $BUC^n_{0,\omega}(\mathbb{R}; U)$ and $BUC^n_{0,\omega,\mathrm{loc}}(\mathbb{R}; U)$ continuously into $BUC^n_{0,\omega}(\mathbb{R}; Y)$, respectively $BUC^n_{0,\omega,\mathrm{loc}}(\mathbb{R}; Y)$.*

In both cases, $(\mathfrak{D}u)^{(k)} = \mathfrak{D}u^{(k)}$ for $0 \le k \le n$.

Proof As long as we are only interested in the values of $\mathfrak{D}$ on some interval $(-\infty, M]$ we may without loss of generality assume that $u \in W^{n,p}_{\omega}(\mathbb{R}; U)$ or $u \in BUC^n_{0,\omega}(\mathbb{R}; U)$ (because of the causality of $\mathfrak{D}$). The claims (i) and (ii) then the follow from Examples 2.5.3 and 3.2.3 since

$$\frac{1}{h}(\tau^h \mathfrak{D}u - u) = \mathfrak{D}\frac{1}{h}(\tau^h u - u),$$

and $\frac{1}{h}(\tau^h u - u) \to \dot{u}$ in $L^p_{\omega}(\mathbb{R}; U)$ or $BUC_{\omega}(\mathbb{R}; U)$ as $h \to 0$ whenever $u \in W^{1,p}_{\omega}(\mathbb{R}; U)$ or $u \in BUC^1_{\omega}(\mathbb{R}; U)$. □

Our first 'easy' representation theorem for the input/output operator $\mathfrak{D}$ deals with the case where the observation operator is bounded:

Theorem 4.5.2 *Let $\Sigma = \left[\begin{smallmatrix}\mathfrak{A} & \mathfrak{B}\\ \mathfrak{C} & \mathfrak{D}\end{smallmatrix}\right]$ be a $L^p|Reg$-well-posed linear system on (Y, X, U) with growth bound $\omega_{\mathfrak{A}}$, let $\omega > \omega_{\mathfrak{A}}$, and suppose that the observation operator C is bounded, i.e., it can be extended to an operator $C_{|X}$ in $\mathcal{B}(X; Y)$ (this is true, for example, if Σ is L^∞-well-posed or Reg-well-posed). Then there is a unique*[1] *operator $D \in \mathcal{B}(U; Y)$ such that*

$$(\mathfrak{D}u)(t) = C_{|X}\mathfrak{B}^t u + Du(t) = C_{|X}\int_{-\infty}^{t} \mathfrak{A}^{t-s}_{|X_{-1}} Bu(s)\,ds + Du(t), \qquad t \in \mathbb{R}; \tag{4.5.1}$$

[1] For the uniqueness of D it is important that we have extended C to the whole space X, and not just to some other intermediate space W lying between X_1 and X, e.g., to the space $(X + BU)_1$ described in Lemma 4.3.12. In the latter situation D may depend on the space W and the particular extension of C to W, which may not be unique. This question is discussed at length in Chapter 5.

this equation is valid a.e. for all $u \in L^p_{\omega,\mathrm{loc}}(\mathbb{R};U)$ if Σ is L^p-well-posed for some $p < \infty$, and for all $u \in Reg_{0,\omega,\mathrm{loc}}(\mathbb{R};U)$ if Σ is L^∞-well-posed or Reg-well-posed.

Compare this result to Theorem 4.3.4.

In the proof of this theorem we use the following lemma:

Lemma 4.5.3 *Let $\Sigma = \left[\begin{smallmatrix} \mathfrak{A} & \mathfrak{B} \\ \mathfrak{C} & \mathfrak{D} \end{smallmatrix}\right]$ be a $L^p|Reg$-well-posed linear system on (Y, X, U) with growth bound $\omega_{\mathfrak{A}}$. For each $\lambda \in C^+_{\omega_{\mathfrak{A}}}$ we define $\mathrm{e}_\lambda(t) = \mathrm{e}^{\lambda t}$ for $t \in \mathbb{R}$. Then, for each $u_0 \in U$, there is a $y_0 \in Y$ such that the function $\mathfrak{D}(\mathrm{e}_\lambda u)$ is (a.e.) equal to $\mathrm{e}_\lambda y_0$. Thus, we may write $y_0 = (\mathfrak{D}(\mathrm{e}_\lambda u))(0)$. The mapping $u_0 \mapsto y_0 = (\mathfrak{D}(\mathrm{e}_\lambda u))(0)$ is a bounded linear operator $U \to Y$.*

Proof of Lemma 4.5.3 By Lemma 4.5.1, $\mathfrak{D}(\mathrm{e}_\lambda u)$ is continuous. As $\mathfrak{D}$ commutes with translations, we have

$$\mathfrak{D}(\mathrm{e}_\lambda u)(t) = (\tau^t \mathfrak{D}(\mathrm{e}_\lambda u))(0) = (\mathfrak{D}\tau^t(\mathrm{e}_\lambda u))(0) = \mathrm{e}^{\lambda t}(\mathfrak{D}(\mathrm{e}_\lambda u))(0).$$

That the mapping $u \mapsto (\mathfrak{D}(\mathrm{e}_\lambda u))(0)$ is linear and continuous follows from Theorem 2.5.4 (since the norm in $L^p_\omega(\mathbb{R}^-;Y)$ of $t \mapsto \mathrm{e}^{\lambda t}(\mathfrak{D}(\mathrm{e}_\lambda u))(0)$ is a constant times $|(\mathfrak{D}(\mathrm{e}_\lambda u))(0)|_Y$). □

Proof of Theorem 4.5.2 We begin by observing that C is bounded if Σ is L^∞-well-posed or *Reg*-well-posed; see Theorem 4.4.7(ii). Moreover, the two different formulas for $\mathfrak{D}$ in (4.5.1) are equivalent because of Theorem 4.2.1 (and a change of integration variable).

Next we prove uniqueness of D. Fix $\alpha > \omega$, and apply (4.5.1) to the function $\mathrm{e}_\alpha u$, where $u \in U$ and $\mathrm{e}_\alpha(t) = \mathrm{e}^{\alpha t}$ for $t \in \mathbb{R}$. Then, by Theorem 4.2.1(iii) and Lemmas 4.5.1 and 4.5.3, for all $t \in \mathbb{R}$ and $u \in U$,

$$\mathrm{e}^{\alpha t}(\mathfrak{D}(\mathrm{e}_\alpha u))(0) = \mathrm{e}^{\alpha t} C_{|X}(\alpha - A_{|X})^{-1} B u + \mathrm{e}^{\alpha t} D u.$$

Taking $t = 0$ and solving this equation for D we get

$$Du = (\mathfrak{D}(\mathrm{e}_\alpha u))(0) - C_{|X}(\alpha - A_{|X})^{-1} B u, \qquad u \in U.$$

Thus, D is unique. It is also clear that if we use the formula above to define D, then $D \in \mathcal{B}(U;Y)$. We claim that (4.5.1) holds for this operator D.

We first want to check that (4.5.1) holds when u is replaced by $\pi_{(-\infty,s)}\mathrm{e}_\alpha u$ for some fixed $u \in U$ and $s \in \mathbb{R}$. For $t < s$ the left hand side of (4.5.1) becomes in this case

$$\begin{aligned}(\mathfrak{D}(\pi_{(-\infty,s)}\mathrm{e}_\alpha u))(t) &= (\mathfrak{D}(\mathrm{e}_\alpha u))(t) = \mathrm{e}^{\alpha t}(\mathfrak{D}(\mathrm{e}_\alpha u))(0) \\ &= \mathrm{e}^{\alpha t}(C_{|X}(\alpha - A_{|X})^{-1} B u + D u),\end{aligned}$$

whereas the right hand side is

$$\begin{aligned} C_{|X}\mathfrak{B}\tau^t(\pi_{(-\infty,s)}\mathrm{e}_\alpha u) + D(\pi_{(-\infty,s)}\mathrm{e}_\alpha u)(t) &= \mathrm{e}^{\alpha t} C_{|X}\mathfrak{B}(\mathrm{e}_\alpha u) + D(\mathrm{e}_\alpha u)(t) \\ &= \mathrm{e}^{\alpha t}(C_{|X}(\alpha - A_{|X})^{-1}Bu + Du). \end{aligned}$$

Thus, the two functions agree for $t < s$. For $t \geq s$ we make a similar computation: the left hand side is then given by

$$\begin{aligned} (\pi_{[s,\infty)}\mathfrak{D}(\pi_{(-\infty,s)}\mathrm{e}_\alpha u))(t) &= (\tau^{-s}\pi_+\mathfrak{D}\pi_-\tau^s(\mathrm{e}_\alpha u))(t) = (\tau^{-s}\mathfrak{C}\mathfrak{B}\tau^s(\mathrm{e}_\alpha u))(t) \\ &= \mathrm{e}^{\alpha s}(\tau^{-s}\mathfrak{C}\mathfrak{B}(\mathrm{e}_\alpha u))(t) = \mathrm{e}^{\alpha s}(\mathfrak{C}\mathfrak{B}(\mathrm{e}_\alpha u))(t-s) \\ &= \mathrm{e}^{\alpha s} C_{|X}\mathfrak{A}^{t-s}\mathfrak{B}(\mathrm{e}_\alpha u), \end{aligned}$$

whereas the right hand side is

$$\begin{aligned} C_{|X}\mathfrak{B}\tau^t(\pi_{(-\infty,s)}\mathrm{e}_\alpha u) + D(\pi_{(-\infty,s)}\mathrm{e}_\alpha u)(t) &= C_{|X}\mathfrak{B}\tau^{t-s}\pi_-(\tau^s\mathrm{e}_\alpha u) \\ &= \mathrm{e}^{\alpha s} C_{|X}\mathfrak{B}\tau^{t-s}\pi_-(\mathrm{e}_\alpha u) = \mathrm{e}^{\alpha s} C_{|X}\mathfrak{A}^{t-s}\mathfrak{B}(\mathrm{e}_\alpha u). \end{aligned}$$

Thus, the two functions agree on all of $\mathbb{R}$.

The proof is completed in the same way as the proof of Theorem 4.2.1(i) (use linearity, density, and continuity). □

Theorem 4.5.2 has several important consequence, some of which will be presented next.

Theorem 4.5.4 *If* $\left[\begin{smallmatrix}\mathfrak{A} & \mathfrak{B} \\ \mathfrak{C} & \mathfrak{D}\end{smallmatrix}\right]$ *is an* L^∞*-well-posed linear system on* (Y, X, U)*, and if we restrict the domains of* $\mathfrak{B}$ *and* $\mathfrak{D}$ *to* $Reg_c(\mathbb{R}^-; U)$ *and* $Reg_{c,\mathrm{loc}}(\mathbb{R}; U)$*, respectively, then the resulting system is Reg-well-posed.*

Proof By Theorems 4.4.2 and 4.5.2, $\mathfrak{C}$ maps X continuously into $BC_\omega(\overline{\mathbb{R}}^+; U)$, $C_{|X} \in \mathcal{B}(X; Y)$, and the representation formula (4.5.1) is valid. It follows from this formula that if u is piecewise constant and right-continuous, then $\mathfrak{D}u$ is right-continuous and has a left hand limit at each point. If $u \in Reg_{c,\mathrm{loc}}(\mathbb{R}; \mathrm{loc})$ is arbitrary, then we can choose a sequence u_n of functions of this type that converge locally uniformly to u. But then $\mathfrak{D}u_n$ converges locally uniformly to $\mathfrak{D}u$, so $\mathfrak{D}u$ must be right-continuous and have a left hand limit at each point. □

Corollary 4.5.5 *Let* $\Sigma = \left[\begin{smallmatrix}\mathfrak{A} & \mathfrak{B} \\ \mathfrak{C} & \mathfrak{D}\end{smallmatrix}\right]$ *be a* $L^p|Reg$*-well-posed linear system on* (Y, X, U) *with growth bound* $\omega_\mathfrak{A}$*, let* $\omega > \omega_\mathfrak{A}$*, and let* $s \in \mathbb{R}$ *and* $x_s \in X$*.*

(i) *If* Σ *is* L^p*-well-posed for some* $p < \infty$*, and if the observation operator* C *is bounded, then for each* $u \in L^p_{\mathrm{loc}}([s, \infty); U)$*, the output* y *of* Σ *with initial time* $s \in \mathbb{R}$*, initial value* $x_s \in X$ *and input* u *is given by*

$$y(t) = C_{|X}x(t) + Du(t), \qquad t \geq s, \tag{4.5.2}$$

where x is the strong solution of the equation

$$\begin{aligned} \dot{x}(t) &= Ax(t) + Bu(t), \qquad t \ge s, \\ x(s) &= x_s. \end{aligned} \tag{4.5.3}$$

Moreover, if $u \in L^p_\omega([s,\infty);U)$, *then* $x \in BC_{0,\omega}([s,\infty);X)$ *and* $y - Du \in BC_{0,\omega}([s,\infty);Y)$. *In particular,* Σ *can be extended to an* L^∞*-well-posed linear system on* (Y,X,U).

(ii) *If* Σ *is* L^∞*-well-posed or Reg-well-posed, then for each* $u \in Reg_{\mathrm{loc}}([s,\infty);U)$, *the output* y *of* Σ *with initial time* $s \in \mathbb{R}$, *initial value* $x_s \in X$, *and input* u *is given by* (4.5.2) *where* x *is the strong solution of* (4.5.3). *Moreover, if* $u \in Reg_{0,\omega}([s,\infty);U)$, *then* $x \in BC_{0,\omega}([s,\infty);X)$ *and* $y - Du \in BC_{0,\omega}([s,\infty);Y)$.

Proof Combine Theorems 4.2.1, 4.3.1, and 4.5.2. □

Corollary 4.5.6 *Let* $\Sigma = \left[\begin{smallmatrix} \mathfrak{A} & \mathfrak{B} \\ \mathfrak{C} & \mathfrak{D} \end{smallmatrix}\right]$ *be a* $L^p|Reg$*-well-posed linear system on* (Y,X,U) *with growth bound* $\omega_{\mathfrak{A}}$, *and let* $\omega > \omega_{\mathfrak{A}}$.

(i) *If* Σ *is* L^p*-well-posed for some* $p < \infty$, *and if the observation operator* C *is bounded, then the operator* $u \mapsto \mathfrak{D}u - Du$ *maps* $L^p_\omega(\mathbb{R};U)$ *continuously into* $BC_{0,\omega}(\mathbb{R};Y)$.

(ii) *If* Σ *is* L^∞*-well-posed or Reg-well-posed, then the operator* $u \mapsto \mathfrak{D}u - Du$ *maps* $Reg_{0,\omega}(\mathbb{R};U)$ *continuously into* $BC_{0,\omega}(\mathbb{R};Y)$.

Proof Combine Theorem 4.3.2 with Theorem 4.5.2. □

Example 4.5.7 *The input/output map* $u \mapsto \tau^{-T}u$ *of the delay line in Example 2.3.4 is not the input/output map of a Reg-well-posed or* L^∞*-well-posed linear system.*

This follows from Corollary 4.5.6, because there is no operator $D \in \mathcal{B}(U;Y)$ that would make $\tau^{-T}u - Du$ continuous for all $u \in Reg_{0,\omega}(\mathbb{R};U)$.

Theorem 4.5.8 *Every static operator* $\mathfrak{D} \in TIC_{\mathrm{loc}}(U;Y)$ *(cf. Definitions 2.6.1 and 2.6.2) has a representation of the type*

$$(\mathfrak{D}u)(t) = Du(t), \qquad t \in \mathbb{R},$$

for some $D \in \mathcal{B}(U;Y)$. *This representation is valid for all* $u \in L^p_{c,\mathrm{loc}}(\mathbb{R};U)$ *and almost all* $t \in \mathbb{R}$ *if* $\mathfrak{D} \in TIC^p_{\mathrm{loc}}(U;Y)$ *with* $p < \infty$, *and for all* $u \in Reg_{c,\mathrm{loc}}(\mathbb{R};U)$ *and all* $t \in \mathbb{R}$ *if* $\mathfrak{D} \in TIC^\infty_{\mathrm{loc}}(U;Y)$ *or* $\mathfrak{D} \in TIC^{Reg}_{\mathrm{loc}}(U;Y)$. *(Thus,* $\mathfrak{D}$ *can be extended to a bounded operator* $L^p_{\mathrm{loc}}(\mathbb{R};U) \to L^p_{\mathrm{loc}}(\mathbb{R};Y)$ *with the same representation.)*

Proof Trivially, $\mathfrak{D}$ can be realized as the input/output map of a well-posed linear system of the form $\left[\begin{smallmatrix} \mathfrak{A} & 0 \\ 0 & \mathfrak{D} \end{smallmatrix}\right]$, where $\mathfrak{A}$ is an arbitrary C_0 semigroup on an arbitrary state space. The result then follows from Theorem 4.5.2. □

By using Corollary 4.5.6 we can extend Lemma 4.5.1 in the following way.

Proposition 4.5.9 *Let* $\Sigma = \left[\begin{smallmatrix} \mathfrak{A} & \mathfrak{B} \\ \mathfrak{C} & \mathfrak{D} \end{smallmatrix}\right]$ *be an* L^∞*-well-posed or Reg-well-posed linear system on* (Y, X, U) *with growth bound* $\omega_{\mathfrak{A}}$*, let* $\omega > \omega_{\mathfrak{A}}$*, let* D *be the feedthrough operator of* Σ *(see Theorem 4.5.2), and let* $n = 0, 1, 2, \ldots$ *Then*

(i) $\mathfrak{D}$ *maps* $Reg^n_{0,\omega}(\mathbb{R}; U)$ *and* $Reg^n_{0,\omega,\mathrm{loc}}(\mathbb{R}; U)$ *continuously into* $Reg^n_{0,\omega}(\mathbb{R}; Y)$*, respectively* $Reg^n_{0,\omega,\mathrm{loc}}(\mathbb{R}; Y)$*, and*
(ii) $u \mapsto \mathfrak{D}u - Du$ *maps* $Reg^n_{0,\omega}(\mathbb{R}; U)$ *and* $Reg^n_{0,\omega,\mathrm{loc}}(\mathbb{R}; U)$ *continuously into* $BC^n_{0,\omega}(\mathbb{R}; Y)$*, respectively* $BC^n_{0,\omega,\mathrm{loc}}(\mathbb{R}; Y)$*.*

Moreover, $(\mathfrak{D}u)^{(k)} = \mathfrak{D}u^{(k)}$ *for* $0 \le k \le n$*.*

Proof Clearly, (i) follows from (ii). Case (ii) with $n = 0$ is contained in Corollary 4.5.6. If $n > 0$ then we use the representation

$$(\mathfrak{D}u)(t) - Du(t) = C_{|X} \int_{-\infty}^{t} \mathfrak{A}^{t-s} Bu(s)\,ds$$

found in Theorem 4.5.2. By the boundedness of $C_{|X}$, it suffices to show that the operator

$$(\mathfrak{F}u)(t) = \int_{-\infty}^{t} \mathfrak{A}^{t-s} Bu(s)\,ds$$

maps $Reg^n_{0,\omega}(\mathbb{R}; U)$ continuously into $BC^n_{0,\omega}(\mathbb{R}; X)$. This operator is the input/output map of an L^1-well-posed system on (X_{-1}, X_{-1}, U) with semigroup $\mathfrak{A}$, bounded control operator B, bounded observation operator 1, and zero feedthrough; cf. Proposition 2.3.1. By Theorem 4.3.4, the same system is *Reg*-well-posed on (X, X, U). If $u \in Reg^n_{0,\omega}(\mathbb{R}; U)$, then $u \in W^{1,n}_{\beta,\mathrm{loc}}(\mathbb{R}; U)$ for all $\beta \in (\omega_{\mathfrak{A}}, \omega)$, and we conclude from part (i) that $\mathfrak{F}u \in W^{1,n}_{\beta,\mathrm{loc}}(\mathbb{R}; X_{-1})$ and that $(\mathfrak{F}u)^{(k)} = \mathfrak{F}u^{(k)}$ for all $0 \le k \le n$. But $u^{(k)} \in Reg_{0,\omega}(\mathbb{R}; U)$, and by the case $n = 0$ proved above, we have $(\mathfrak{F}u)^{(k)} = \mathfrak{F}u^{(k)} \in BC_{0,\omega}(\mathbb{R}; X)$. This shows that $\mathfrak{F}$ maps $Reg^n_{0,\omega}(\mathbb{R}; U)$ continuously into $BC^n_{0,\omega}(\mathbb{R}; X)$, hence $C\mathfrak{F}$ maps $Reg^n_{0,\omega}(\mathbb{R}; U)$ continuously into $BC^n_{0,\omega}(\mathbb{R}; Y)$. □

The other relatively easy case where we can construct a feedthrough operator D is the one where the control operator B is bounded.

Theorem 4.5.10 *Let* $\Sigma = \left[\begin{smallmatrix} \mathfrak{A} & \mathfrak{B} \\ \mathfrak{C} & \mathfrak{D} \end{smallmatrix}\right]$ *be a* $L^p|Reg$*-well-posed linear system on* (Y, X, U) *with growth bound* $\omega_{\mathfrak{A}}$*, let* $\omega > \omega_{\mathfrak{A}}$*, and suppose that the control operator is bounded, i.e.,* $B \in \mathcal{B}(U; X)$ *(this is, in particular, true if* X *is reflexive and* Σ *is* L^1*-well-posed). Then there is a unique operator* $D \in \mathcal{B}(U; Y)$ *such that*

$$(\mathfrak{D}u)(t) = C\mathfrak{B}^t u + Du(t) = C \int_{-\infty}^{t} \mathfrak{A}^{t-s} Bu(s)\,ds + Du(t), \qquad t \in \mathbb{R}; \tag{4.5.4}$$

this equation is valid a.e. for all u that belong piecewise to $W^{1,1}_{\omega,\text{loc}}(\mathbb{R};U)$ (i.e., there should exist $-\infty < t_0 < t_1 < t_2 < \dots$, with $t_n \to \infty$ as $n \to \infty$, such that $\pi_{(-\infty,t_0)}u \in W^{1,1}_{\omega}((-\infty,t_0);U)$ and $\pi_{[t_{n-1},t_n)}u \in W^{1,1}([t_{n-1},t_n);U)$ for $n \ge 1$). Thus, for such u, the output y of Σ with initial time $s \in \mathbb{R}$, initial value $x_s \in X$, and input u is given by

$$y(t) = Cx(t) + Du(t), \qquad t \ge s, \tag{4.5.5}$$

where x is the strong solution of (4.5.3).

Proof The major part of the proof of this theorem is identical to the proof of Theorem 4.5.2 with one small correction: we replace X_{-1} by X and X by X_1 throughout. Then, with the exception of the last two lines, all the statements in the proof of Theorem 4.5.2 remain valid. Observe, in particular, that $\mathfrak{B}(\mathrm{e}_\alpha u) = (\alpha - A)^{-1}Bu \in X_1$ in this case.

As in the proof of Theorem 4.5.2, we conclude that (4.5.4) is valid for all finite linear combination of functions of the type $\mathrm{e}_\alpha u$ and $\pi_{[s,\infty)}(\mathrm{e}_\alpha u)$. By subtracting such a linear combination from u we can assume that $u \in W^{1,1}_{\omega,\text{loc}}(\mathbb{R};U)$ (i.e, we can remove the possible discontinuities of u).

We observe that both sides of (4.5.4) with u replaced by $\pi_{[s,\infty)}(\mathrm{e}_\alpha u)$ are infinitely many times differentiable with respect to α, as long as $\alpha < \omega$ (see Lemma 3.2.10). Differentiation with respect to α replaces $\mathrm{e}^{\alpha t}$ by $t\mathrm{e}^{\alpha t}$, and we conclude that (4.5.4) is valid for functions u of the type $\pi_{[s,\infty)}(\mathrm{e}_\alpha P)$ where $s \in \mathbb{R}$ is arbitrary and P is an arbitrary polynomial with vector coefficients in U. This class of functions is dense in $W^{1,1}_{\omega,\text{loc}}(\mathbb{R};U)$, so (4.5.4) must hold for all functions u that belong piecewise to $W^{1,1}_{\omega,\text{loc}}(\mathbb{R};U)$. □

Definition 4.5.11 The operator D in Theorems 4.5.2 and 4.5.10 is called the *feedthrough operator* of Σ.

The input/output map has the following interesting representation in the case where the control operator is bounded (compare this theorem to Theorem 4.4.8):

Theorem 4.5.12 *Let $\Sigma = \left[\begin{smallmatrix}\mathfrak{A} & \mathfrak{B} \\ \mathfrak{C} & \mathfrak{D}\end{smallmatrix}\right]$ be a $L^p|Reg$-well-posed linear system on (Y,X,U) with growth bound $\omega_{\mathfrak{A}}$, let $\omega > \omega_{\mathfrak{A}}$, and suppose that the control operator is bounded, i.e., $B \in \mathcal{B}(U;X)$ (this is, in particular, true if X is reflexive and Σ is L^1-well-posed). Then*

$$(\mathfrak{D}(uv))(t) = \int_{-\infty}^{t} (\mathfrak{C}Bu)(t-s)v(s)\,ds + Duv(t) \tag{4.4.4}$$

for all $u \in U$ and $v \in L^p|Reg_{\omega,\text{loc}}(\mathbb{R})$; here D is the operator in Theorem 4.5.2. Moreover, $\mathfrak{B}$ can be extended to a continuous mapping $L^1_\omega(\mathbb{R}^-;U) \to X$ and $\mathfrak{D}$ can be extended to a continuous mapping $L^1_{\omega,\text{loc}}(\mathbb{R};U) \to L^p_{\omega,\text{loc}}(\mathbb{R};Y)$. Thus, Σ can be extended to an L^1-well-posed linear system on (Y,X,U).

Proof Obviously, if B is bounded, then we can use the representation of $\mathfrak{B}$ given in Theorem 4.2.1(i) to extend $\mathfrak{B}$ to an operator in $\mathcal{B}(L^1_\omega(\mathbb{R}^-;U);X)$. Thus, it suffices to prove (4.4.4), because the rest of the claims follow from Theorem 4.4.8. Moreover, by continuity, it suffices to prove (4.4.4) for $v \in W^{1,p}_{c,\mathrm{loc}}(\mathbb{R})$.

Let $u \in U$ and $v \in W^{1,p}_{c,\mathrm{loc}}(\mathbb{R};\mathbb{R})$. Then, by Theorem 3.8.2(v) with X replaced by X_1, $\int_{-\infty}^t \mathfrak{A}^{t-s}Buv(s)\,ds \in X_1$. By Theorem 3.7.1(iii) with X replaced by X_1, if we let $J_\alpha = \alpha(\alpha - A)^{-1}$ be the Yosida approximation of the identity, then $J_\alpha \in \mathcal{B}(X;X_1)$, J_α commutes with $\mathfrak{A}^s$ for all $s \geq 0$, and

$$J_\alpha \int_{-\infty}^t \mathfrak{A}^{t-s}Buv(s)\,ds \to \int_{-\infty}^t \mathfrak{A}^{t-s}Buv(s)\,ds$$

in X_1 as $\alpha \to \infty$. Thus, by Theorem 4.5.10,

$$\begin{aligned}
(\mathfrak{D}uv)(t) &= C\int_{-\infty}^t \mathfrak{A}^{t-s}Buv(s)\,ds + Duv(t)\\
&= \lim_{\alpha\to\infty} CJ_\alpha \int_{-\infty}^t \mathfrak{A}^{t-s}Buv(s)\,ds + Duv(t)\\
&= \lim_{\alpha\to\infty} \int_{-\infty}^t CJ_\alpha\mathfrak{A}^{t-s}Buv(s)\,ds + Duv(t)\\
&= \lim_{\alpha\to\infty} \int_{-\infty}^t C\mathfrak{A}^{t-s}J_\alpha Buv(s)\,ds + Duv(t)\\
&= \lim_{\alpha\to\infty} \int_{-\infty}^t (\mathfrak{C}J_\alpha Bu)(t-s)v(s)\,ds + Duv(t)\\
&= \int_{-\infty}^t (\mathfrak{C}Bu)(t-s)v(s)\,ds + Duv(t),
\end{aligned}$$

where in the last line we have used the Lebesgue dominated convergence theorem and the fact that $J_\alpha Bu \to Bu$ in X as $\alpha \to \infty$ (see Theorem 3.7.1(iii)). □

Example 4.5.13 *The input/output map $u \mapsto \tau^{-T}u$ of the delay line in Example 2.3.4 is not the input/output map of an L^1-well-posed linear system with reflexive state space.*

This follows from Theorems 4.2.7 and 4.5.12, since this input/output map does not have a representation of the type (4.4.4).

4.6 The transfer function and the system node

Let us begin by giving a name to the operator $u \mapsto (\mathfrak{D}(\mathrm{e}_z u))(0)$ that we encountered in Lemma 4.5.3.

Definition 4.6.1 The operator-valued function $z \mapsto \big(u \mapsto (\mathfrak{D}(e_z u))(0)\big)$ in Lemma 4.5.3, defined on $\mathbb{C}^+_{\omega_{\mathfrak{A}}}$, is called the *transfer function* of Σ, and it is denoted by $\widehat{\mathfrak{D}}(z)$.[2]

Lemma 4.6.2 *The transfer function $\widehat{\mathfrak{D}}$ is an analytic $\mathcal{B}(U;Y)$-valued function on $\mathbb{C}^+_{\omega_{\mathfrak{A}}}$, and, for each $\omega > \omega_{\mathfrak{A}}$, it is uniformly bounded in the closed half-plane $\overline{\mathbb{C}}^+_{\omega}$.*

Proof Choose some $\omega_{\mathfrak{A}} < \alpha < \omega$. Then the family $e_z u$ is bounded in $W^{1,p}_{\alpha}((-\infty, 1]; U)$ for every p, $1 \le p \le \infty$, and in $BC_{0,\omega}(\mathbb{R};\mathbb{C})$, uniformly over all $z \in \overline{\mathbb{C}}^+_{\omega}$ and all $u \in U$ with $|u| \le 1$. Therefore $\widehat{\mathfrak{D}}(z)$ is uniformly bounded on the half-plane $\overline{\mathbb{C}}^+_{\omega}$ as claimed. The analyticity of $\widehat{\mathfrak{D}}u(z)$ for each $u \in U$ follows from the fact that $e_z u$ has a complex derivative with respect to z (see Lemma 3.2.10). Thus $\widehat{\mathfrak{D}}$ is analytic in the strong topology, hence in the uniform operator topology (see Hille and Phillips 1957, Theorem 3.10.1). □

In the situation discussed in Theorems 4.5.2 and 4.5.10 the transfer function has the following representation in terms of the the control, observation, and feedthrough operators:

Theorem 4.6.3 *Let $\Sigma = \left[\begin{smallmatrix} \mathfrak{A} & \mathfrak{B} \\ \mathfrak{C} & \mathfrak{D} \end{smallmatrix}\right]$ be a $L^p|Reg$-well-posed linear system.*

(i) *If the observation operator C of Σ is bounded, then the transfer function $\widehat{\mathfrak{D}}$ is given by*

$$\widehat{\mathfrak{D}}(z) = C_{|X}(z - A_{|X})^{-1}B + D, \qquad \Re z > \omega_{\mathfrak{A}}. \tag{4.6.1}$$

In particular, this is true if Σ is L^∞-well-posed or Reg-well-posed.

(ii) *If the control operator B is bounded, then the transfer function $\widehat{\mathfrak{D}}$ is given by*

$$\widehat{\mathfrak{D}}(z) = C(z - A)^{-1}B + D, \qquad \Re z > \omega_{\mathfrak{A}}. \tag{4.6.2}$$

In particular, this is true if Σ is L^1-well-posed and X is reflexive.

This follows from Theorems 4.2.1(iii), 4.5.2, and 4.5.10.

[2] In Definition 4.7.4 we extend the domain of $\widehat{\mathfrak{D}}$ to the whole resolvent set of A. See also Lemma 4.7.5.

If neither B nor C is bounded, then the representation formula for $\widehat{\mathfrak{D}}$ becomes more complicated. One way to prove this more general formula is to first prove a representation formula for the output y, where neither B nor C is assumed to be bounded. This representation formula contains a new operator, the *combined observation/feedthrough operator* $C\&D$, which is defined a follows.

Definition 4.6.4 Let $\Sigma = \left[\begin{smallmatrix} \mathfrak{A} & \mathfrak{B} \\ \mathfrak{C} & \mathfrak{D} \end{smallmatrix}\right]$ be a $L^p|Reg$-well-posed linear system on (Y, X, U) with semigroup generator A, control operator B, observation operator C, and growth bound $\omega_{\mathfrak{A}}$. The *combined observation/feedthrough operator* $C\&D\colon \left[\begin{smallmatrix} X \\ U \end{smallmatrix}\right] \supset \mathcal{D}(C\&D) \to Y$ and the *system node* $S\colon \left[\begin{smallmatrix} X \\ U \end{smallmatrix}\right] \supset \mathcal{D}(S) \to \left[\begin{smallmatrix} X \\ Y \end{smallmatrix}\right]$ of Σ are defined by

$$C\&D\begin{bmatrix} x \\ u \end{bmatrix} = C\big[x - (\alpha - A_{|X})^{-1}Bu\big] + \widehat{\mathfrak{D}}(\alpha)u, \tag{4.6.3}$$

$$S = \begin{bmatrix} A\&B \\ C\&D \end{bmatrix}, \tag{4.6.4}$$

where α is an arbitrary point in $\mathbb{C}^+_{\omega_{\mathfrak{A}}}$,[3] and $A\&B$ is the restriction of $\big[A_{|X}\ B\big] : \left[\begin{smallmatrix} X \\ U \end{smallmatrix}\right] \to X_{-1}$ to $\mathcal{D}(A\&B)$. The three operators $A\&B$, $C\&D$ and S have the same domain (described in Lemma 4.3.10), namely

$$\begin{aligned} \mathcal{D}(S) &= \mathcal{D}(A\&B) = \mathcal{D}(C\&D) \\ &= \big\{\left[\begin{smallmatrix} x \\ u \end{smallmatrix}\right] \in \left[\begin{smallmatrix} X \\ U \end{smallmatrix}\right] \bigm| A_{|X}x + Bu \in X\big\}. \end{aligned} \tag{4.6.5}$$

The motivation for the notation $C\&D$ that we have introduced above is that in the case where the observation operator C is bounded (i.e., it has a bounded extension to X), the operator $C\&D$ is the restriction to $\mathcal{D}(C\&D)$ of the operator $\big[C\quad D\big] : \left[\begin{smallmatrix} X \\ U \end{smallmatrix}\right] \to Y$. This follows from Theorems 4.5.2 and 4.6.3. (In particular, at least in this case $C\&D$ does not depend on α; see Theorem 4.6.7 for the general statement.)

Theorem 4.6.5 *Let $\Sigma = \left[\begin{smallmatrix} \mathfrak{A} & \mathfrak{B} \\ \mathfrak{C} & \mathfrak{D} \end{smallmatrix}\right]$ be an $L^p|Reg$-well-posed linear system with semigroup generator A, control operator B, observation operator C, transfer function $\widehat{\mathfrak{D}}$, and growth bound $\omega_{\mathfrak{A}}$. Let $\omega > \omega_{\mathfrak{A}}$ and $\Re\alpha > \omega_{\mathfrak{A}}$. Define $C\&D$ and S as in Definition 4.6.4 (using the particular value of α specified above).*

(i) *Let Σ be L^p-well-posed for some p, $1 \le p < \infty$, let $u \in W^{1,p}_{\omega,\mathrm{loc}}(\mathbb{R}; U)$, and let $x(t) = \mathfrak{B}^t u$, $t \in \mathbb{R}$, and $y = \mathfrak{D}u$ be the state trajectory and output function of Σ with initial time $-\infty$ and input function u (cf. Definition 2.5.8). Then $x \in BC^1_{0,\omega,\mathrm{loc}}(\mathbb{R}; X)$, $\left[\begin{smallmatrix} x \\ u \end{smallmatrix}\right] \in BC_{0,\omega,\mathrm{loc}}(\mathbb{R}; \mathcal{D}(S))$,*

[3] We shall see in Theorem 4.6.7 that $C\&D$ does not depend on α.

$y \in W^{1,p}_{\omega,\mathrm{loc}}(\mathbb{R}; Y)$ *(in particular, y is continuous), and*

$$\begin{bmatrix} \dot{x}(t) \\ y(t) \end{bmatrix} = S \begin{bmatrix} x(t) \\ u(t) \end{bmatrix}, \qquad t \in \mathbb{R}. \tag{4.6.6}$$

If $u \in W^{1,p}_{\omega}(\mathbb{R}; U)$*, then* $x \in BC^1_{0,\omega}(\mathbb{R}; X)$, $\left[\begin{smallmatrix} x \\ u \end{smallmatrix}\right] \in BC_{0,\omega}(\mathbb{R}; \mathcal{D}(S))$*, and* $y \in W^{1,p}_{\omega}(\mathbb{R}; Y)$.

(ii) *Let* Σ *be* L^∞*-well-posed or Reg-well-posed, and denote the feedthrough operator of* Σ *by* D *(cf. Theorem 4.5.2). Let* $u \in Reg^1_{0,\omega,\mathrm{loc}}(\mathbb{R}; U)$*, and define x and y as in (i). Then* $x \in BC^1_{0,\omega,\mathrm{loc}}(\mathbb{R}; X)$, $\left[\begin{smallmatrix} x \\ u \end{smallmatrix}\right] \in BC_{0,\omega,\mathrm{loc}}(\mathbb{R}; \mathcal{D}(S))$, $y - Du \in BC^1_{0,\omega,\mathrm{loc}}(\mathbb{R}; Y)$*, and* (4.6.6) *holds. In this case we can alternatively write* (4.6.6) *in the form*

$$\begin{bmatrix} \dot{x}(t) \\ y(t) \end{bmatrix} = \begin{bmatrix} A_{|X} & B \\ C_{|X} & D \end{bmatrix} \begin{bmatrix} x(t) \\ u(t) \end{bmatrix}, \qquad t \in \mathbb{R}, \tag{4.6.7}$$

where $C_{|X}$ *stands for the unique extension of* C *to* $\mathcal{B}(X; Y)$*. If* $u \in Reg^1_{0,\omega}(\mathbb{R}; U)$*, then* $x \in BC^1_{0,\omega}(\mathbb{R}; X)$, $\left[\begin{smallmatrix} x \\ u \end{smallmatrix}\right] \in BC_{0,\omega}(\mathbb{R}; \mathcal{D}(S))$*, and* $y - Du \in BC^1_{0,\omega}(\mathbb{R}; Y)$.

Proof The claims concerning the differentiability and growth rate of the function x are taken from Theorem 4.3.6, and also the claim that $\dot{x}(t) = A\&B \left[\begin{smallmatrix} x(t) \\ u(t) \end{smallmatrix}\right] = A_{|X}x(t) + Bu(t)$ for all $t \in \mathbb{R}$ is found in the same place. The claims about the continuity and growth rate of $\left[\begin{smallmatrix} x \\ u \end{smallmatrix}\right]$ are found in Theorem 4.3.11. By Lemma 4.5.1, $\mathfrak{D}u \in W^{1,p}_{\omega,\mathrm{loc}}(\mathbb{R}; Y)$ or $\mathfrak{D}u \in W^{1,p}_{\omega}(\mathbb{R}; Y)$ (depending on the assumption on u) in case (i), and by Proposition 4.5.9(ii), $y - Du \in BC^1_{0,\omega,\mathrm{loc}}(\mathbb{R}; Y)$ or $y - Du \in BC^1_{0,\omega}(\mathbb{R}; Y)$ in case (ii) (again depending on the assumption on u). The bottom half of (4.6.7) is a rewritten version of (4.5.2). Thus, it only remains to prove that $y(t) = C\&D \left[\begin{smallmatrix} x(t) \\ u(t) \end{smallmatrix}\right]$ for all $t \in \mathbb{R}$ in case (i). In this proof we may, without loss of generality, take $t = 0$: to get a nonzero t we simply replace u by $\tau^t u$ and $x(0)$ by $x(t)$ in the argument below.

Define $e_\alpha(s) = e^{\alpha s}$ for $s \in \mathbb{R}$. The function $\pi_-(u - e_\alpha u(0))$ belongs to $W^{1,p}_{\omega,\mathrm{loc}}(\mathbb{R}; U)$, hence by Lemma 4.5.1, $\mathfrak{D}\pi_-(u - e_\alpha u(0)) \in W^{1,p}_{\omega,\mathrm{loc}}(\mathbb{R}; Y)$. By causality, this function coincides with $\mathfrak{D}(u - e_\alpha u(0))$ on $\mathbb{R}^-$, so necessarily

$$\big(\mathfrak{D}\pi_-(u - e_\alpha u(0))\big)(0) = \big(\mathfrak{D}(u - e_\alpha u(0))\big)(0) = y(0) - \widehat{\mathfrak{D}}(\alpha)u(0).$$

On the other hand, by Definition 2.2.1,

$$\pi_+\mathfrak{D}\pi_-(u - e_\alpha u(0)) = \mathfrak{C}\mathfrak{B}(u - e_\alpha u(0)),$$

where, by Lemma 4.3.5, $\mathfrak{B}(u - e_\alpha u(0)) \in X_1$. By Theorems 4.4.2(i) and 4.2.1(i),(iii), for all $s \geq 0$,

$$\begin{aligned}\big(\mathfrak{D}\pi_-(u - e_\alpha u)\big)(s) &= \big(\mathfrak{C}\mathfrak{B}(u - e_\alpha u(0))\big)(s)\\ &= C\mathfrak{A}^s\mathfrak{B}(u - e_\alpha u(0))\\ &= C\mathfrak{A}^s\big(x(0) - (\alpha - A)^{-1}Bu(0)\big).\end{aligned}$$

Taking $s = 0$ we get

$$y(0) = C\big(x(0) - (\alpha - A)^{-1}Bu(0)\big) + \widehat{\mathfrak{D}}(\alpha)u(0) = C\&D\begin{bmatrix} x(0)\\ u(0)\end{bmatrix}.$$

□

Corollary 4.6.6 *A well-posed linear system is determined uniquely by its semigroup generator A, its control operator B, its observation operator C, and its transfer function $\widehat{\mathfrak{D}}$, evaluated at one point $\alpha \in \mathbb{C}^+_{\omega_\mathfrak{A}}$ (where $\omega_\mathfrak{A}$ is the growth bound of the semigroup).*

Proof See Definition 2.2.4 and Theorems 3.2.1(vii), 4.2.1(ii), 4.4.2(iii), 4.5.2, and 4.6.5 (the space $W^{1,p}_{\omega,\mathrm{loc}}(\mathbb{R}; U)$ is dense in $L^p_{\omega,\mathrm{loc}}(\mathbb{R}; U)$ for $p < \infty$). □

Another consequence of Theorem 4.6.5 is the following representation formula for the transfer function of a system.

Theorem 4.6.7 *Let $\Sigma = \left[\begin{smallmatrix}\mathfrak{A} & \mathfrak{B}\\ \mathfrak{C} & \mathfrak{D}\end{smallmatrix}\right]$ be an $L^p|Reg$-well-posed linear system with semigroup generator A, control operator B, observation operator C, transfer function $\widehat{\mathfrak{D}}$, and growth bound $\omega_\mathfrak{A}$. Let $\omega > \omega_\mathfrak{A}$ and $\Re\alpha > \omega_\mathfrak{A}$. Define $C\&D$ and S as in Definition 4.6.4 (using the particular value of α specified above). Then,*

$$\widehat{\mathfrak{D}}(z) = C\&D\begin{bmatrix}(z - A_{|X})^{-1}B\\ 1\end{bmatrix}, \qquad \Re z > \omega_\mathfrak{A}. \tag{4.6.8}$$

Moreover, for all $\alpha \in \mathbb{C}^+_{\omega_\mathfrak{A}}$ and $z \in \mathbb{C}^+_{\omega_\mathfrak{A}}$,

$$\begin{aligned}\widehat{\mathfrak{D}}(z) - \widehat{\mathfrak{D}}(\alpha) &= C[(z - A_{|X})^{-1} - (\alpha - A_{|X})^{-1}]B\\ &= (\alpha - z)C(z - A)^{-1}(\alpha - A_{|X})^{-1}B,\end{aligned} \tag{4.6.9}$$

and $C\&D$ does not depend on the particular value of α which is used in its definition.

Proof We apply Theorem 4.6.5 to the function $u = e_z v$, where $v \in U$, $e_z(t) = e^{zt}$ for $t \in \mathbb{R}$ and $\Re z > \omega_\mathfrak{A}$, and evaluate y at zero. By Definition 4.6.1, $y(0) = \widehat{\mathfrak{D}}(z)v$. On the other hand, by Theorem 4.6.5, $y(0) = C\&D\left[\begin{smallmatrix}x(0)\\ u(0)\end{smallmatrix}\right]$, where (by Theorem 4.2.1(iii)) $x(0) = (z - A_{|X})^{-1}Bv$ and $u(0) = v$. This proves (4.6.8).

Formula (4.6.9) follows from (4.6.3), (4.6.8) and the resolvent identity (3.2.1). That $C\&D$ does not depend on the value used in its definition follows from (4.6.9), which shows that for all $\left[\begin{smallmatrix} x \\ u \end{smallmatrix}\right] \in \mathcal{D}(S)$ we have

$$C[x - (\alpha - A_{|X})^{-1}Bu] + \widehat{\mathfrak{D}}(\alpha) = C[x - (z - A_{|X})^{-1}Bu] + \widehat{\mathfrak{D}}(z).$$

□

Corollary 4.6.8 *Let $\Sigma = \left[\begin{smallmatrix} \mathfrak{A} & \mathfrak{B} \\ \mathfrak{C} & \mathfrak{D} \end{smallmatrix}\right]$ be a $L^p|Reg$-well-posed linear system on (Y, X, U) with semigroup generator A, control operator B, and observation operator C. Then the operator-valued function*

$$(\lambda - \alpha)C(\lambda - A)^{-1}(\alpha - A_{|X})^{-1}B$$

is analytic both in λ for $\lambda \in \mathbb{C}^+_{\omega_\mathfrak{A}}$ and in α for $\alpha \in \mathbb{C}^+_{\omega_\mathfrak{A}}$, and for each $\omega > \omega_\mathfrak{A}$ it is uniformly bounded in $\Re\lambda \geq \omega$ and $\Re\alpha \geq \omega$.

This follows from Lemma 4.6.2 and Theorem 4.6.7.

At this point we ask the reader to recall Definition 3.12.1 of the Laplace transform of a function.

Theorem 4.6.9 *Let $\Sigma = \left[\begin{smallmatrix} \mathfrak{A} & \mathfrak{B} \\ \mathfrak{C} & \mathfrak{D} \end{smallmatrix}\right]$ be a $L^p|Reg$-well-posed linear system on (Y, X, U) with growth bound $\omega_\mathfrak{A}$, let $\omega > \omega_\mathfrak{A}$, and let $u \in L^p_\omega(\mathbb{R}^+; U)$ if Σ is L^p-well-posed for some $1 \leq p < \infty$, and $u \in Reg_{0,\omega}(\overline{\mathbb{R}}^+; U)$ if Σ is L^∞-well-posed or Reg-well-posed. Let x be the state trajectory and let y be the output of Σ with initial time zero, initial state x_0, and input function u. Then $x \in BC_{0,\omega}(\mathbb{R}^+; X)$, $y \in L^p|Reg_{0,\omega}(\mathbb{R}^+; Y)$, and the Laplace transforms of x and y satisfy*

$$\begin{aligned} \hat{x}(z) &= (z - A)^{-1}x_0 + (z - A_{|X})^{-1}B\hat{u}(z), \\ \hat{y}(z) &= C(z - A)^{-1}x_0 + \widehat{\mathfrak{D}}(z)\hat{u}(z), \qquad z \in \mathbb{C}_\omega. \end{aligned} \tag{4.6.10}$$

Proof Both the solution x and the output y can be split into two parts, $x = x_1 + x_2$, and $y = y_1 + y_2$, where

$$\begin{aligned} x_1(t) &= \mathfrak{A}^t x_0, \qquad & x_2(t) &= \mathfrak{B}^t_0 u, \qquad t \geq 0, \\ y_1 &= \mathfrak{C}x_0, & y_2 &= \mathfrak{D}\pi_+ u. \end{aligned}$$

It follows from Theorem 2.5.4 that both x_1 and x_2 belong to $BC_{0,\omega}(\mathbb{R}^+; X)$, and that both y_1 and y_1 belong to $L^p|Reg_{0,\omega}(\mathbb{R}^+; Y)$. In particular, the Laplace transforms of all these functions converge absolutely in the half-plane $\mathbb{C}^+_\omega$. Let $z \in \mathbb{C}^+_\omega$. By Theorem 3.2.9(i), $\hat{x}_1(z) = (z - A)^{-1}x_0$, and by Theorem 4.4.2(iv),

$\hat{y}_1(z) = C(z-A)^{-1}x_0$. Using the representation (4.2.3) for $\mathfrak{B}_0^t$ and Theorem 3.2.9(i) (with X replaced by X_{-1}) we get (after two changes of the order of integration and a change of integration variable)

$$\begin{aligned}
\hat{x}_2(z) &= \int_0^\infty e^{-zt} \int_0^t \mathfrak{A}_{|X_{-1}}^{t-v} Bu(v)\,dv\,dt \\
&= \int_0^\infty \int_v^\infty e^{-zt}\,\mathfrak{A}_{|X_{-1}}^{t-v} Bu(v)\,dt\,dv \\
&= \int_0^\infty \int_0^\infty e^{-z(s+v)}\,\mathfrak{A}_{|X_{-1}}^{s} Bu(v)\,ds\,dv \\
&= \int_0^\infty e^{-zs}\,\mathfrak{A}_{|X_{-1}}^{s} B \int_0^\infty e^{-zv}u(v)\,dv\,ds \\
&= (z-A_{|X})^{-1}B\hat{u}(z).
\end{aligned}$$

Thus, it only remains to prove that

$$\widehat{\mathfrak{D}\pi_+ u}(z) = \widehat{\mathfrak{D}}(z)\hat{u}(z), \qquad z \in \mathbb{C}_\omega^+. \tag{4.6.11}$$

We first investigate the validity of (4.6.11) for functions u of the type $u = \pi_{[0,s)}e_\alpha v$, where $0 < s \le \infty$, $\omega_{\mathfrak{A}} < \alpha < \omega$, $v \in U$ are fixed, and $e_\alpha(t) = e^{\alpha t}$, $t \in \mathbb{R}$. Then, for $\Re z > \omega > \alpha$,

$$\hat{u}(z) = \int_0^s e^{(\alpha-z)t}\,dt\,v = (z-\alpha)^{-1}(1-e^{(\alpha-z)s})v.$$

To compute $\mathfrak{D}u$ we write u in the form

$$u = \pi_{(-\infty,s)}e_\alpha v - \pi_{(-\infty,0)}e_\alpha v = u_1 - u_2.$$

Let (cf. Theorem 4.2.1(iii))

$$x = \mathfrak{B}(e_\alpha v) = (\alpha - A_{|X})^{-1}Bv.$$

Then,

$$\begin{aligned}
\pi_{[0,s)}\mathfrak{D}u_1 &= \pi_{[0,s)}e_\alpha\widehat{\mathfrak{D}}(\alpha)v, \\
\pi_{[s,\infty)}\mathfrak{D}u_1 &= \tau^{-s}\pi_+\mathfrak{D}\pi_-\tau^s u_1 = \tau^{-s}\mathfrak{C}\mathfrak{B}(\tau^s e_\alpha v) \\
&= e^{\alpha s}\tau^{-s}\mathfrak{C}\mathfrak{B}(e_\alpha v) = e^{\alpha s}\tau^{-s}\mathfrak{C}x, \\
\pi_{[0,\infty)}\mathfrak{D}u_2 &= \mathfrak{C}\mathfrak{B}u_2 = \mathfrak{C}x.
\end{aligned}$$

We multiply this by the function e_{-z}, integrate over $\mathbb{R}^+$, and use Theorems

4.4.2(iv) and 4.6.7 to get

$$
\begin{aligned}
(\widehat{\mathfrak{D}\pi_+ u})(z) &= \int_0^s e^{(\alpha-z)t}\widehat{\mathfrak{D}}(\alpha)v\,dt \\
&\quad + \int_s^\infty e^{\alpha s}e^{-zt}(\mathfrak{C}x)(t-s)\,dt - \int_0^\infty e^{-zt}(\mathfrak{C}x)(t) \\
&= (z-\alpha)^{-1}(1-e^{(\alpha-z)s})\widehat{\mathfrak{D}}(\alpha)v \\
&\quad - (1-e^{(\alpha-z)s})\int_0^\infty e^{-zt}(\mathfrak{C}x)(t)\,dt \\
&= (z-\alpha)^{-1}(1-e^{(\alpha-z)s})\widehat{\mathfrak{D}}(\alpha)v \\
&\quad - (1-e^{(\alpha-z)s})C(z-A)^{-1}(\alpha-A_{|X})^{-1}Bv \\
&= (1-e^{(\alpha-z)s})(z-\alpha)^{-1}\widehat{\mathfrak{D}}(z)v \\
&= \widehat{\mathfrak{D}}(z)\hat{u}(z).
\end{aligned}
$$

Thus, (4.6.11) holds for this class of functions u.

The proof is completed using the standard density argument (cf. the proof of Theorem 4.2.1(i)). □

Corollary 4.6.10 *A causal time-invariant operator which either maps $L^p_\omega(\mathbb{R};U)$ continuously into $L^p_\omega(\mathbb{R};Y)$ for some $\omega \in \mathbb{R}$ and p, $1 \le p < \infty$, or satisfies the assumptions of Theorem 2.6.7, has a transfer function $\widehat{\mathfrak{D}}$ with the following properties:*

(i) *$\widehat{\mathfrak{D}}(z)u = (\mathfrak{D}(e_z u))(0)$ for all $u \in U$ and $z \in \mathbb{C}^+_\omega$. Here $e_z(t) = e^{zt}$ for all $t \in \mathbb{R}$.*

(ii) *$\widehat{\mathfrak{D}}$ is an analytic $\mathcal{B}(U;Y)$-valued function on $\mathbb{C}^+_\omega$, and, for each $\alpha > \omega$ it is bounded in the closed half-plane $\overline{\mathbb{C}}^+_\alpha$.*

(iii) *If $u \in L^p|Reg_\omega(\overline{\mathbb{R}}^+;U)$, then the Laplace transform of $\mathfrak{D}\pi_+ u$ is given by*

$$
(\widehat{\mathfrak{D}\pi_+ u})(z) = \widehat{\mathfrak{D}}(z)\hat{u}(z), \qquad z \in \mathbb{C}^+_\omega. \tag{4.6.12}
$$

(iv) *$\mathfrak{D}$ is determined uniquely by $\widehat{\mathfrak{D}}$.*

Proof (i)–(ii) This follows from Definition 4.6.1, Lemma 4.6.2 and Theorems 2.6.6, 2.6.7, and 4.6.9.

(iii) Since the Laplace transform is injective, (ii) implies that u and $\widehat{\mathfrak{D}}$ determine $\mathfrak{D}u$ uniquely whenever u vanishes on $\mathbb{R}^-$. By the time-invariance of $\mathfrak{D}$, this implies that u and $\widehat{\mathfrak{D}}$ determine $\mathfrak{D}u$ uniquely whenever the support of u is bounded from the left. This set of functions is dense in the domain of $\mathfrak{D}$, so $\widehat{\mathfrak{D}}$ determines $\mathfrak{D}$ uniquely. □

Theorem 4.6.5 can also be used to develop the following formula for the output function of a $L^p|Reg$-well-posed linear system for a restricted class of input functions and initial states.

Theorem 4.6.11 *Let $\Sigma = \left[\begin{smallmatrix} \mathfrak{A} & \mathfrak{B} \\ \mathfrak{C} & \mathfrak{D} \end{smallmatrix}\right]$ be a $L^p|Reg$-well-posed linear system with system node S and growth bound $\omega_{\mathfrak{A}}$.*

(i) *Let Σ be L^p-well-posed for some $p < \infty$, $s \in \mathbb{R}$, $x_s \in X$, $u \in W^{1,p}_{\mathrm{loc}}([s,\infty);U)$, and let $\left[\begin{smallmatrix} x_s \\ u(s) \end{smallmatrix}\right] \in \mathcal{D}(S)$ (i.e., $A_{|X}x_s + Bu(s) \in X$). Let x and y be the state trajectory and output function of Σ with initial time s, initial state x_s, and input function u. Then $x \in C^1([s,\infty);X)$, $\left[\begin{smallmatrix} x \\ u \end{smallmatrix}\right] \in C([s,\infty);\mathcal{D}(S))$, $y \in W^{1,p}_{\mathrm{loc}}([s,\infty);Y)$ (in particular, y is continuous), and*

$$\begin{bmatrix} \dot{x}(t) \\ y(t) \end{bmatrix} = S \begin{bmatrix} x(t) \\ u(t) \end{bmatrix}, \qquad t \ge s. \tag{4.6.13}$$

If $u \in W^{1,p}_{\omega}([s,\infty);U)$ for some $\omega > \omega_{\mathfrak{A}}$, then $x \in BC^1_{0,\omega}([s,\infty);X)$, $\left[\begin{smallmatrix} x \\ u \end{smallmatrix}\right] \in BC_{0,\omega}([s,\infty);\mathcal{D}(S))$, and $y \in W^{1,p}_{\omega}([s,\infty);Y)$.

(ii) *If Σ is L^∞-well-posed or Reg-well-posed, then all the claims in (i) are valid if we replace the condition $u \in W^{1,p}_{\mathrm{loc}}([s,\infty);U)$ by $u \in Reg^1_{\mathrm{loc}}([s,\infty);U)$, the condition $u \in W^{1,p}_{\omega}([s,\infty);U)$ by $u \in Reg^1_{0,\omega}([s,\infty);U)$, the condition $y \in W^{1,p}_{\mathrm{loc}}([s,\infty);Y)$ by $y - Du \in C^1([s,\infty);Y)$, and the condition $y \in W^{1,p}_{\omega}([s,\infty);Y)$ by $y - Du \in BC^1_{0,\omega}([s,\infty);Y)$. In addition, in this case 4.6.13 can be written in the form*

$$\begin{bmatrix} \dot{x}(t) \\ y(t) \end{bmatrix} = \begin{bmatrix} A_{|X} & B \\ C_{|X} & D \end{bmatrix} \begin{bmatrix} x(t) \\ u(t) \end{bmatrix}, \qquad t \ge s, \tag{4.6.14}$$

where $C_{|X}$ stands for the unique extension of C to $\mathcal{B}(X;Y)$.

Proof All the claims concerning the state x are taken from Theorem 4.3.7, which also contains the top half of (4.6.13) (i.e., the equation $\dot{x}(t) = A_{|X}x(t) + Bu(t)$ for $t \ge s$). The representation (4.6.14) is taken from Corollary 4.5.5. Together with Theorem 4.6.3 this implies (4.6.13) in case (ii). The claim that $y - Du \in C^1([s,\infty);Y)$ in case (ii) follows from the boundedness of C and the fact that $y - Du = Cx$ where $x \in C^1([s,\infty);X)$. Thus, it remains to show that $y \in W^{1,p}_{\mathrm{loc}}([s,\infty);Y)$ and that $y(t) = C\&D \left[\begin{smallmatrix} x(t) \\ u(t) \end{smallmatrix}\right]$ for $t \ge s$ in case (i).

For simplicity we take $s = 0$ (i.e., we replace s by 0, u by $\tau^{-s}u$, y by $\tau^{-s}y$, and $x(t)$ by $x(t-s)$).

According to Definition 2.2.7 with $s = 0$,

$$x(t) = \mathfrak{A}^t x_0 + \mathfrak{B}_{t0} u, \qquad t \geq 0,$$
$$y = \mathfrak{C} x_0 + \mathfrak{D}_0 u.$$

Fix some α with $\Re\alpha > \omega > \omega_{\mathfrak{A}}$, let $e_\alpha(t) = e^{\alpha t}$, $t \in \mathbb{R}$, and define

$$v = \pi_+ u + \pi_- e_\alpha u(0).$$

Then $v \in W^{1,p}_{\omega,\mathrm{loc}}(\mathbb{R}, U)$. As in the proof of Theorem 4.3.7 we write x in the form $x(t) = x_1(t) + x_2(t)$ where $x_1(t) = \mathfrak{A}^t(x_0 - \mathfrak{B} e_\alpha u(0)) \in X_1$ for all $t \geq 0$ and $x_2(t) = \mathfrak{B}^t v$. We split y accordingly into

$$\begin{aligned} y &= \mathfrak{C} x_0 + \mathfrak{D}\pi_+ u \\ &= \mathfrak{C} x_0 + \mathfrak{D} v - \pi_+ \mathfrak{D} \pi_- e_\alpha u(0) \\ &= \mathfrak{C}(x_0 - \mathfrak{B} e_\alpha u(0)) + \mathfrak{D} v \\ &= y_1 + y_2. \end{aligned}$$

By Lemma 4.4.1 and Theorem 4.4.2(i), $y_1 \in W^{1,p}_{\mathrm{loc}}(\overline{\mathbb{R}}^+; Y)$ and $y_1(t) = C x_1(t)$ for all $t \geq 0$. By Theorem 4.6.5, also $y_2 \in W^{1,p}_{\mathrm{loc}}(\overline{\mathbb{R}}^+; Y)$, and the representation for y_2 given there (with u replaced by v) is valid. This, together with Theorem 4.2.1(iii), gives for all $t \geq 0$,

$$\begin{aligned} y(t) &= C x_1(t) + C\mathfrak{B}\big(\tau^t v - e_\alpha v(t)\big) + \widehat{\mathfrak{D}}(\alpha) v(t) \\ &= C x_1(t) + C\big(x_2(t) - \mathfrak{B} e_\alpha u(t)\big) + \widehat{\mathfrak{D}}(\alpha) u(t) \\ &= C\big(x(t) - \mathfrak{B} e_\alpha u(t)\big) + \widehat{\mathfrak{D}}(\alpha) u(t) \\ &= C\big[x(t) - (\alpha - A_{|X})^{-1} B u(t)\big] + \widehat{\mathfrak{D}}(\alpha) u(t) \\ &= C\&D \begin{bmatrix} x(t) \\ u(t) \end{bmatrix}. \end{aligned}$$

□

It is also possible to develop a representation for the output of a well-posed linear system in the general case, which is based on the representation described in Theorem 4.6.11.

Theorem 4.6.12 *Let $\Sigma = \left[\begin{smallmatrix} \mathfrak{A} & \mathfrak{B} \\ \mathfrak{C} & \mathfrak{D} \end{smallmatrix}\right]$ be a $L^p|Reg$-well-posed linear system with system node $S = \left[\begin{smallmatrix} A\&B \\ C\&D \end{smallmatrix}\right]$ and control operator B. Let x and y be the state trajectory and output function of Σ with initial time $s \in \mathbb{R}$, initial state $x_s \in X$, and input function u, where $u \in L^p_{\mathrm{loc}}([s, \infty); U)$ if Σ is L^p-well-posed for some $p < \infty$ and $u \in Reg_{\mathrm{loc}}([s, \infty); U)$ if Σ is L^∞-well-posed or Reg-well-posed. Let*

$$\begin{bmatrix} x_1(t) \\ u_1(t) \end{bmatrix} = \int_s^t \begin{bmatrix} x(v) \\ u(v) \end{bmatrix} dv, \qquad t \geq s,$$

be the integral of $\left[\begin{smallmatrix} x \\ u \end{smallmatrix}\right]$. *Then*

$$x(t) = \mathfrak{A}^{t-s} x_s + \int_s^t \mathfrak{A}^{t-v}_{|X_{-1}} Bu(v)\,dv, \qquad t \ge s, \tag{4.6.15}$$

$\left[\begin{smallmatrix} x_1 \\ u_1 \end{smallmatrix}\right] \in W^{1,p}_{\mathrm{loc}}([s,\infty); \mathcal{D}(S))$ *if* Σ *is* L^p*-well-posed for some* $p < \infty$ *and* $\left[\begin{smallmatrix} x_1 \\ u_1 \end{smallmatrix}\right] \in Reg^1_{\mathrm{loc}}([s,\infty); \mathcal{D}(S))$ *if* Σ *is* L^∞*-well-posed or Reg-well-posed, and*

$$y(t) = \frac{\mathrm{d}}{\mathrm{d}t} C\&D \begin{bmatrix} x_1(t) \\ u_1(t) \end{bmatrix} \text{ for almost all } t \ge s. \tag{4.6.16}$$

See Theorem 4.7.11 for a related result.

Proof That x is given by (4.6.15) follows from (4.3.1).

By the linearity of the problem, it suffices to prove the theorem in the two cases where either $x_s = 0$ or $u = 0$. If $u = 0$, then the conclusion follows immediately from Definition 2.2.7 and Theorem 4.4.4, since $\left[\begin{smallmatrix} X_1 \\ 0 \end{smallmatrix}\right] \subset \mathcal{D}(S)$ and $C\&D\left[\begin{smallmatrix} x \\ 0 \end{smallmatrix}\right] = Cx$ for every $x \in X_1$. Thus, in the sequel we may assume that $x_s = 0$.

If $x_s = 0$, then (according to Definition 2.2.7) $y = \mathfrak{D}\pi_{[s,\infty)}u$. Define $y_1(t) = \int_s^t y(v)\,dv = \int_s^t (\mathfrak{D}\pi_{[s,\infty)}u)(v)\,dv$ for $t \ge s$. Then $y_1 = \mathfrak{D}\pi_{[s,\infty)}u_1$ since these functions coincide at the point s, and, according to Proposition 4.5.9, they have the same derivative. Thus, y_1 is the output function of Σ with initial time s, initial state zero and input function u_1. The corresponding state trajectory is x_1 (this follows from the variation of constants formula (4.6.15)). By Theorem 4.6.11, $\left[\begin{smallmatrix} x_1 \\ u_1 \end{smallmatrix}\right] \in W^{1,p}_{\mathrm{loc}}([s,\infty); \mathcal{D}(S))$ or $\left[\begin{smallmatrix} x_1 \\ u_1 \end{smallmatrix}\right] \in Reg^1_{\mathrm{loc}}([s,\infty); \mathcal{D}(S))$ (depending on which case of well-posedness we consider), and $y_1(t) = C\&D\left[\begin{smallmatrix} x_1(t) \\ u_1(t) \end{smallmatrix}\right]$ for all $t \ge s$. Differentiating this equation we get (4.6.16). □

It is possible to reformulate Theorems 4.6.5 and 4.6.11 by making a change of variable, and replacing the state x by $z = x - (\alpha - A_{|X})^{-1}Bu$. (This particular combination of x and u appears in the definition of $C\&D$, too.)

Corollary 4.6.13 *Let* $\Sigma = \left[\begin{smallmatrix} \mathfrak{A} & \mathfrak{B} \\ \mathfrak{C} & \mathfrak{D} \end{smallmatrix}\right]$ *be an* $L^p|Reg$*-well-posed linear system with semigroup generator* A*, control operator* B*, observation operator* C*, transfer function* $\widehat{\mathfrak{D}}$*, and growth bound* $\omega_{\mathfrak{A}}$*. Let* $\omega > \omega_{\mathfrak{A}}$ *and* $\alpha \in \mathbb{C}^+_{\omega_{\mathfrak{A}}}$.

(i) *Let* Σ *be* L^p*-well-posed for some* p, $1 \le p < \infty$*. Then, for each* $u \in W^{1,p}_{\omega,\mathrm{loc}}(\mathbb{R}; U)$*, the function* $\mathfrak{D}u$ *belongs to* $W^{1,p}_{\omega,\mathrm{loc}}(\mathbb{R}; Y)$ *(in particular, it is continuous), and it is given by*

$$y(t) = Cz(t) + \widehat{\mathfrak{D}}(\alpha)u(t), \qquad t \in \mathbb{R}, \tag{4.6.17}$$

where z is the unique strong solution (with initial time $-\infty$) in $BC_{0,\omega,\mathrm{loc}}(\mathbb{R};X_1)$ of the equation

$$\dot{z}(t) = Az(t) + (\alpha - A_{|X})^{-1}B(\alpha u - \dot{u}), \qquad t \in \mathbb{R}. \tag{4.6.18}$$

If $u \in W^{1,p}_{\omega}(\mathbb{R};U)$, then $z \in BC_{0,\omega}(\mathbb{R};X_1)$.

(ii) *If Σ is L^p-well-posed for some $p < \infty$, then for each $s \in \mathbb{R}$, $x_s \in X$, and $u \in W^{1,p}_{\mathrm{loc}}([s,\infty);U)$ satisfying $A_{|X}x_s + Bu(s) \in X$, the output y of Σ with initial time s, initial value x_s and input u satisfies $y \in W^{1,p}_{\mathrm{loc}}([s,\infty);Y)$, and it is given by*

$$y(t) = Cz(t) + \widehat{\mathfrak{D}}(\alpha)u(t), \qquad t \geq s, \tag{4.6.19}$$

where z is the strong solution in X_1 of

$$\begin{aligned} \dot{z}(t) &= Az(t) + (\alpha - A_{|X})^{-1}B(\alpha u - \dot{u}), \qquad t \geq s, \\ z(s) &= x_s - (\alpha - A_{|X})^{-1}Bu(s) \\ &= (\alpha - A_{|X})^{-1}\big(\alpha x_s - (A_{|X}x_s + Bu(s))\big). \end{aligned} \tag{4.6.20}$$

If $u \in W^{1,p}_{\omega}([s,\infty);U)$, then $z \in BC_{0,\omega}([s,\infty);X_1)$.

(iii) *If instead Σ is L^∞-well-posed or Reg-well-posed, then the conclusions of (i) and (ii) remain valid if we replace $W^{1,p}_{\omega,\mathrm{loc}}$ by $Reg^1_{0,\omega,\mathrm{loc}}$, $W^{1,p}_{\omega}$ by $Reg^1_{0,\omega}$, and $W^{1,p}_{\mathrm{loc}}$ by Reg^1_{loc} throughout.*

Proof (i) By Theorem 4.6.5, if we define $z(t) := \mathfrak{B}\tau^t u - (\alpha - A_{|X})^{-1}Bu(t)$ for $t \in \mathbb{R}$, then $z \in BC^1_{0,\omega,\mathrm{loc}}(\mathbb{R};X)$ and (4.6.17) holds. The control operator $(\alpha - A_{|X})^{-1}B$ in (4.6.18) is bounded, so by Proposition 2.3.1 and Theorem 4.3.2(i), equation (4.6.18) has a unique strong solution in $BC_{0,\omega,\mathrm{loc}}(\mathbb{R};X)$. Thus, (i) follows as soon as we show that z is a strong solution in X of (4.6.18). This is easy: recall that (by Theorem 4.3.2) $x := \mathfrak{B}\tau u$ is a strong solution in X of $\dot{x} = Ax + Bu$ and that $z = x - (\alpha - A_{|X})^{-1}Bu$, hence z is a strong solution of the equation

$$\begin{aligned} \dot{z} &= \dot{x} - (\alpha - A_{|X})^{-1}B\dot{u} \\ &= A_{|X}\big(z + (\alpha - A_{|X})^{-1}Bu\big) + Bu - (\alpha - A_{|X})^{-1}B\dot{u} \\ &= A_{|X}z + (\alpha - A_{|X})^{-1}(\alpha u - \dot{u}). \end{aligned}$$

(ii)–(iii) The proofs of (ii)–(iii) are essentially the same as the proof of (i) given above, with Theorem 4.3.2(i) replaced by Theorem 4.3.1 and Theorem 4.6.5 replaced by Theorem 4.6.11. □

There is another, equivalent version of Corollary 4.6.13 which is based on a slightly different change of state variable:

Corollary 4.6.14 *Let* $\Sigma = \left[\begin{smallmatrix} \mathfrak{A} & \mathfrak{B} \\ \mathfrak{C} & \mathfrak{D} \end{smallmatrix}\right]$ *be an* $L^p|Reg$*-well-posed linear system with semigroup generator* A*, control operator* B*, observation operator* C*, transfer function* $\widehat{\mathfrak{D}}$*, and growth bound* $\omega_{\mathfrak{A}}$*. Let* $\omega > \omega_{\mathfrak{A}}$ *and* $\alpha \in \mathbb{C}^+_{\omega_{\mathfrak{A}}}$*.*

(i) *Let* Σ *be* L^p*-well-posed for some* p*,* $1 \le p < \infty$*. Then, for each* $u \in W^{1,p}_{\omega,\mathrm{loc}}(\mathbb{R}; U)$*, the function* $\mathfrak{D}u$ *belongs to* $W^{1,p}_{\omega,\mathrm{loc}}(\mathbb{R}; Y)$ *(in particular, it is continuous), and it is given by*

$$y(t) = C(\alpha - A)^{-1} w(t) + \widehat{\mathfrak{D}}(\alpha) u(t), \quad t \in \mathbb{R}, \tag{4.6.21}$$

where w *is the unique strong solution (with initial time* $-\infty$*) in* $BC_{0,\omega,\mathrm{loc}}(\mathbb{R}; X)$ *of the equation*

$$\dot{w}(t) = A w(t) + B(\alpha u - \dot{u}), \quad t \in \mathbb{R}. \tag{4.6.22}$$

If $u \in W^{1,p}_{\omega}(\mathbb{R}; U)$*, then* $w \in BC_{0,\omega}(\mathbb{R}; X)$*.*

(ii) *If* Σ *is* L^p*-well-posed for some* $p < \infty$*, then for each* $s \in \mathbb{R}$*,* $x_s \in X$*, and* $u \in W^{1,p}_{\mathrm{loc}}([s,\infty); U)$ *satisfying* $A_{|X} x_s + B u(s) \in X$*, the output* y *of* Σ *with initial time* s*, initial value* x_s *and input* u *satisfies* $y \in W^{1,p}_{\mathrm{loc}}([s,\infty); Y)$*, and it is given by*

$$y(t) = C(\alpha - A)^{-1} w(t) + \widehat{\mathfrak{D}}(\alpha) u(t), \quad t \ge s, \tag{4.6.23}$$

where w *is the strong solution in* X *of*

$$\begin{aligned} \dot{w}(t) &= A w(t) + B(\alpha u - \dot{u}), \quad t \ge s, \\ w(s) &= \alpha x_s - A_{|X} x_s - B u(s). \end{aligned} \tag{4.6.24}$$

If $u \in W^{1,p}_{\omega}([s,\infty); U)$*, then* $w \in BC_{0,\omega}([s,\infty); X)$*.*

(iii) *If instead* Σ *is* L^∞*-well-posed or Reg-well-posed, then the conclusions of (i) and (ii) remain valid if we replace* $W^{1,p}_{\omega,\mathrm{loc}}$ *by* $Reg^1_{0,\omega,\mathrm{loc}}$*,* $W^{1,p}_{\omega}$ *by* $Reg^1_{0,\omega}$*, and* $W^{1,p}_{\mathrm{loc}}$ *by* Reg^1_{loc} *throughout.*

Proof This follows from Corollary 4.6.13 since $w = (\alpha - A)z = \alpha x - A_{|X} x - Bu$ is a strong solution of (4.6.22) or (4.6.24) in X if and only if z is a strong solution of (4.6.18) or (4.6.20) in X_1. □

4.7 Operator nodes

In the preceding section we arrived at the notion of system node induced by a $L^p|Reg$-well-posed linear system. This node is an unbounded operator S of the type $S = \left[\begin{smallmatrix} A\&B \\ C\&D \end{smallmatrix}\right] : \left[\begin{smallmatrix} X \\ U \end{smallmatrix}\right] \supset \mathcal{D}(S) \to \left[\begin{smallmatrix} X \\ Y \end{smallmatrix}\right]$, where $A\&B$ is the restriction of $\left[\begin{smallmatrix} A_{|X} & B \end{smallmatrix}\right] : \left[\begin{smallmatrix} X \\ U \end{smallmatrix}\right] \to X_{-1}$ to $\mathcal{D}(S) = \left\{\left[\begin{smallmatrix} x \\ u \end{smallmatrix}\right] \in \left[\begin{smallmatrix} X \\ U \end{smallmatrix}\right] \mid A_{|X} x + Bu \in X\right\}$, and $C\&D$ is the combined observation/feedthrough operator. In this section we give some necessary conditions and some sufficient conditions for an unbounded operator $S\colon \begin{bmatrix} X \\ U \end{bmatrix} \supset \mathcal{D}(S) \to \begin{bmatrix} X \\ Y \end{bmatrix}$ to be the system node of a $L^p|Reg$-well-posed linear system. We begin by listing some necessary conditions.

Proposition 4.7.1 *Let $S\colon \begin{bmatrix} X \\ U \end{bmatrix} \supset \mathcal{D}(S) \to \begin{bmatrix} X \\ Y \end{bmatrix}$ be the system node of an $L^p|Reg$-well-posed linear system on (Y, X, U). Denote the two complementary projections of $\begin{bmatrix} X \\ Y \end{bmatrix}$ onto X respectively Y by π_X and π_Y, and write S in the form $S = \begin{bmatrix} A\&B \\ C\&D \end{bmatrix}$, where $A\&B = \pi_X S$ and $C\&D = \pi_Y S$ (with $\mathcal{D}(A\&B) = \mathcal{D}(C\&D) = \mathcal{D}(S)$). Then S has the following properties.*

(i) *S is closed (as an operator $\begin{bmatrix} X \\ U \end{bmatrix} \to \begin{bmatrix} X \\ Y \end{bmatrix}$).*
(ii) *The operator $A\colon X \supset \mathcal{D}(A) \to X$ defined by $Ax = A\&B \begin{bmatrix} x \\ 0 \end{bmatrix}$ on $\mathcal{D}(A) = \left\{x \in X \mid \begin{bmatrix} x \\ 0 \end{bmatrix} \in \mathcal{D}(S)\right\}$ is the generator of a C_0 semigroup on X.*
(iii) *The operator $A\&B$ (with $\mathcal{D}(A\&B) = \mathcal{D}(S)$) can be extended to an operator $\begin{bmatrix} A_X & B \end{bmatrix} \in \mathcal{B}(\begin{bmatrix} X \\ U \end{bmatrix}; X_{-1})$, where X_{-1} is defined as in Section 3.6.*
(iv) *$\mathcal{D}(S) = \left\{\begin{bmatrix} x \\ u \end{bmatrix} \in \begin{bmatrix} X \\ U \end{bmatrix} \mid A_{|X}x + Bu \in X\right\}$.*

Proof The decomposition $S = \begin{bmatrix} A\&B \\ C\&D \end{bmatrix}$ is part of the definition of S given in Definition 4.6.4 (and the same decomposition can be carried out for all possible linear operators $S\colon \begin{bmatrix} X \\ U \end{bmatrix} \supset \mathcal{D}(S) \to \begin{bmatrix} X \\ Y \end{bmatrix}$). The claims (ii)–(iv) are part of Definition 4.6.4.

It remains to show that S is closed. Recall from Lemma 4.3.10 that the operator $A\&B$ is closed. Let $\begin{bmatrix} w_n \\ u_n \end{bmatrix} \to \begin{bmatrix} w \\ u \end{bmatrix}$ in $\begin{bmatrix} X \\ U \end{bmatrix}$, and suppose that $\begin{bmatrix} x_n \\ y_n \end{bmatrix} = S\begin{bmatrix} w_n \\ u_n \end{bmatrix} \to \begin{bmatrix} x \\ y \end{bmatrix}$ in $\begin{bmatrix} X \\ Y \end{bmatrix}$. Equivalently, $x_n = A\&B \begin{bmatrix} w_n \\ u_n \end{bmatrix} = A_{|X}x_n + Bu_n \to x$ in X and $y_n = C\&D \begin{bmatrix} w_n \\ u_n \end{bmatrix} \to y$ in Y. Since $A\&B$ is closed, this implies that $A\&B \begin{bmatrix} w \\ u \end{bmatrix} = x$, or equivalently, that $A_{|X}w + Bu = x$. Fix some $\alpha \in \mathbb{C}^+_{\omega\mathfrak{A}}$ (= the growth bound of the system). Then $w_n - (\alpha - A_{|X})^{-1}Bu_n = (\alpha - A_{|X})^{-1}[\alpha w_n - (A_{|X}w_n + Bu_n)] \to (\alpha - A_{|X})^{-1}[\alpha w - A_{|X}w + Bu] = w - (\alpha - A_{|X})^{-1}Bu$ in X_1, so $y_n = C\&D \begin{bmatrix} w_n \\ u_n \end{bmatrix} = C[w_n - (\alpha - A_{|X})^{-1}Bu_n] + \widehat{\mathfrak{D}}(\alpha)u_n \to C[w - (\alpha - A_{|X})^{-1}Bu] + \widehat{\mathfrak{D}}(\alpha)u = C\&D \begin{bmatrix} w \\ u \end{bmatrix}$ in Y. Thus, $y = C\&D \begin{bmatrix} w \\ u \end{bmatrix}$, and this proves that S is closed. □

The conditions (i)–(iv) in Proposition 4.7.1 contain no direct reference to the $L^p|Reg$-well-posedness of the underlying system, apart from the requirement that A generates a C_0 semigroup. This makes it possible to introduce the following class of operators, which we call *operator nodes*.

Definition 4.7.2 By an *operator node* on a triple of Banach spaces (Y, X, U) we mean an operator $S\colon \begin{bmatrix} X \\ U \end{bmatrix} \supset \mathcal{D}(S) \to \begin{bmatrix} X \\ Y \end{bmatrix}$ which satisfies the following conditions. Denote the two complementary projections of $\begin{bmatrix} X \\ Y \end{bmatrix}$ onto X, respectively Y, by π_X and π_Y, and write S in the form $S = \begin{bmatrix} A\&B \\ C\&D \end{bmatrix}$, where $A\&B = \pi_X S$ and $C\&D = \pi_Y S$ (with $\mathcal{D}(A\&B) = \mathcal{D}(C\&D) = \mathcal{D}(S)$). Then we require that

(i) S is closed (as an operator $\begin{bmatrix} X \\ U \end{bmatrix} \to \begin{bmatrix} X \\ Y \end{bmatrix}$).

(ii) The operator $A\colon X \supset \mathcal{D}(A) \to X$ defined by $Ax = A\&B \left[\begin{smallmatrix} x \\ 0 \end{smallmatrix}\right]$ on $\mathcal{D}(A) = \left\{x \in X \mid \left[\begin{smallmatrix} x \\ 0 \end{smallmatrix}\right] \in \mathcal{D}(S)\right\}$ has a nonempty resolvent set, and $\mathcal{D}(A)$ is dense in X.

(iii) The operator $A\&B$ (with $\mathcal{D}(A\&B) = \mathcal{D}(S)$) can be extended to an operator $\left[A_X \;\; B\right] \in \mathcal{B}(\left[\begin{smallmatrix} X \\ U \end{smallmatrix}\right]; X_{-1})$, where X_{-1} is defined as in Section 3.6.

(iv) $\mathcal{D}(S) = \left\{\left[\begin{smallmatrix} x \\ u \end{smallmatrix}\right] \in \left[\begin{smallmatrix} X \\ U \end{smallmatrix}\right] \mid A_{|X}x + Bu \in X\right\}$.

If, in addition, A is the generator of a C_0 semigroup, then we call S a *system node*. Finally, we call a system node $L^p|Reg$-well-posed if it is the system node of a $L^p|Reg$-well-posed linear system (as in Definition 4.6.4).

Indeed, by Proposition 4.7.1, the system node of a $L^p|Reg$-well-posed linear system is a system node in the sense of Definition 4.7.2.

Lemma 4.7.3 *An operator node S on (Y, X, U) has the following additional properties (in addition to those listed in Definition 4.7.2):*

(v) *The operator $A\&B$ is closed (as an operator $\left[\begin{smallmatrix} X \\ U \end{smallmatrix}\right] \to X$).*

(vi) *For every $u \in U$, the set $\left\{x \in X \mid \left[\begin{smallmatrix} x \\ u \end{smallmatrix}\right] \in \mathcal{D}(S)\right\}$ is dense in X. (Thus, in particular, $\mathcal{D}(S)$ is dense in $\left[\begin{smallmatrix} X \\ U \end{smallmatrix}\right]$.)*

(vii) *The operator $C\&D$ is continuous from $\mathcal{D}(S)$ with the graph norm to Y, and the operator C defined by $Cx = C\&D \left[\begin{smallmatrix} x \\ 0 \end{smallmatrix}\right]$ for all $x \in X_1 = \mathcal{D}(A)$ is continuous from X_1 to Y.*

(viii) *For every $\alpha \in \rho(A) = \rho(A_{|X})$, the operator $\left[\begin{smallmatrix} 1 & -(\alpha - A_{|X})^{-1}B \\ 0 & 1 \end{smallmatrix}\right]$ maps $\mathcal{D}(S)$ one-to-one onto $\left[\begin{smallmatrix} X_1 \\ U \end{smallmatrix}\right]$ and it is bounded and invertible on $\left[\begin{smallmatrix} X \\ U \end{smallmatrix}\right]$. The inverse of this operator (which maps $\left[\begin{smallmatrix} X_1 \\ U \end{smallmatrix}\right]$ one-to-one onto $\mathcal{D}(S)$ and $\left[\begin{smallmatrix} X \\ U \end{smallmatrix}\right]$ one-to-one onto itself) is $\left[\begin{smallmatrix} 1 & (\alpha - A_{|X})^{-1}B \\ 0 & 1 \end{smallmatrix}\right]$.*

(ix) *The graph norm in $\mathcal{D}(S)$ is equivalent to the graph norm in $\mathcal{D}(A\&B)$ described in Lemma 4.3.10, and it is also equivalent to the norm*

$$\left\|\left[\begin{smallmatrix} x \\ u \end{smallmatrix}\right]\right\| := \left(|x - (\alpha - A_{|X})^{-1}Bu|_{X_1}^2 + |u|_U^2\right)^{1/2},$$

for each $\alpha \in \rho(A)$.

Proof (v)–(vi) For the proofs of (v) and (vi) we refer the reader to the proof of Lemma 4.3.10.

(vii) Every closed operator is continuous from its domain with the graph norm into its range space, so S is continuous from $\mathcal{D}(S)$ to $\left[\begin{smallmatrix} X \\ Y \end{smallmatrix}\right]$. Therefore also $C\&D$ is continuous from $\mathcal{D}(S)$ to Y. The space $\left[\begin{smallmatrix} X_1 \\ 0 \end{smallmatrix}\right]$ is a closed subspace of $\mathcal{D}(S)$, and C is the restriction of $C\&D$ to this subspace, so C is continuous from X_1 to Y.

(viii) As $(\alpha - A_{|X})^{-1}B \in \mathcal{B}(U; X)$, the operator $E_\alpha := \left[\begin{smallmatrix} 1 & -(\alpha - A_{|X})^{-1}B \\ 0 & 1 \end{smallmatrix}\right]$ maps $\left[\begin{smallmatrix} X \\ U \end{smallmatrix}\right]$ one-to-one onto itself, with inverse $E_\alpha^{-1} = \left[\begin{smallmatrix} 1 & (\alpha - A_{|X})^{-1}B \\ 0 & 1 \end{smallmatrix}\right]$ (and both

E_α and E_α^{-1} are bounded). If $\left[\begin{smallmatrix} x \\ u \end{smallmatrix}\right] \in \mathcal{D}(S)$, then $A_{|X}x + Bu \in X$ and

$$x - (\alpha - A_{|X})^{-1}Bu = (\alpha - A_{|X})^{-1}[\alpha x - (A_{|X}x + Bu)] \in X_1,$$

so E_α maps $\mathcal{D}(S)$ into $\left[\begin{smallmatrix} X_1 \\ U \end{smallmatrix}\right]$. Conversely, if $\left[\begin{smallmatrix} x \\ u \end{smallmatrix}\right] \in \left[\begin{smallmatrix} X_1 \\ U \end{smallmatrix}\right]$, then

$$A_{|X}[x + (\alpha - A_{|X})^{-1}Bu] + Bu = Ax + \alpha(\alpha - A_{|X})^{-1}Bu \in X,$$

so E_α^{-1} maps $\left[\begin{smallmatrix} X_1 \\ U \end{smallmatrix}\right]$ into $\mathcal{D}(S)$.

(ix) The graph norm in $\mathcal{D}(S)$ is given by

$$\left|\left[\begin{smallmatrix} x \\ u \end{smallmatrix}\right]\right| := \left(\left|A\&B\left[\begin{smallmatrix} x \\ u \end{smallmatrix}\right]\right|_X^2 + \left|C\&D\left[\begin{smallmatrix} x \\ u \end{smallmatrix}\right]\right|_Y^2 + |x|_X^2 + |u|_U^2\right)^{1/2},$$

so it appears to be stronger than the graph norm in $\mathcal{D}(A\&B)$ (which is otherwise the same, but the term involving $C\&D$ is missing). However, since $\mathcal{D}(S) = \mathcal{D}(A\&B)$, it follows from the open mapping theorem that these two norms are equivalent. The third equivalent norm given in (ix) is the norm of the image of $\left[\begin{smallmatrix} x \\ u \end{smallmatrix}\right]$ under E_α. The norm of this image is equivalent to the norm of $\left[\begin{smallmatrix} x \\ u \end{smallmatrix}\right]$ in $\mathcal{D}(S)$ since E_α maps $\mathcal{D}(S)$ one-to-one onto $\left[\begin{smallmatrix} X_1 \\ U \end{smallmatrix}\right]$ (and E_α is closed as an operator from $\mathcal{D}(S)$ to $\left[\begin{smallmatrix} X_1 \\ U \end{smallmatrix}\right]$ since it is bounded as an operator from $\left[\begin{smallmatrix} X \\ U \end{smallmatrix}\right]$ to itself). □

Each operator node has a main operator, a control operator, an observation operator, and a transfer function:

Definition 4.7.4 Let $S = \left[\begin{smallmatrix} A\&B \\ C\&D \end{smallmatrix}\right]$ be an operator node.

(i) The operator A in Definition 4.6.4 is called the *main operator* of S. If S is a system node, then we shall also refer to A as the *semigroup generator* of S.
(ii) The operator B in Definition 4.6.4 is called the *control operator* of S.
(iii) The operator $C\colon X_1$ defined by $Cx = C\&D\left[\begin{smallmatrix} x \\ 0 \end{smallmatrix}\right]$ is called the *observation operator* of S.
(iv) The *transfer function* of S is the operator-valued function

$$\widehat{\mathfrak{D}}(z) = C\&D\begin{bmatrix}(z - A_{|X})^{-1}B \\ 1\end{bmatrix}, \qquad z \in \rho(A). \tag{4.7.1}$$

Lemma 4.7.5 *Let $S = \left[\begin{smallmatrix} A\&B \\ C\&D \end{smallmatrix}\right]$ be an operator node with main operator A.*

(i) *The transfer function of S is analytic on $\rho(A)$, and for all $\alpha \in \rho(A)$ and $z \in \rho(A)$,*

$$\begin{aligned}\widehat{\mathfrak{D}}(z) - \widehat{\mathfrak{D}}(\alpha) &= C[(z - A_{|X})^{-1} - (\alpha - A_{|X})^{-1}]B \\ &= (\alpha - z)C(z - A)^{-1}(\alpha - A_{|X})^{-1}B.\end{aligned} \tag{4.7.2}$$

(ii) *For all* $\alpha \in \rho(A)$ *and all* $\left[\begin{smallmatrix} x \\ u \end{smallmatrix}\right] \in \mathcal{D}(S)$,

$$C\&D \begin{bmatrix} x \\ u \end{bmatrix} = C\big[x - (\alpha - A_{|X})^{-1} B u\big] + \widehat{\mathfrak{D}}(\alpha) u. \tag{4.7.3}$$

(iii) *Suppose that* S *is a* $L^p|Reg$*-well-posed system node, and denote the corresponding* $L^p|Reg$*-well-posed linear system by* Σ*. Then the semigroup generator* A*, the control operator* B*, and the observation operator* C *coincide with the corresponding operators of* Σ*. The transfer function* $\widehat{\mathfrak{D}}$ *of* S *is an extension to* $\rho(A)$ *of the transfer function of* Σ *(which is defined in Definition 4.6.1 only on the half-plane* $\mathbb{C}_{\omega_{\mathfrak{A}}}$*, where* $\omega_{\mathfrak{A}}$ *is the growth bound of the semigroup generated by* A*).*

In the sequel we shall throughout use this extended version of $\widehat{\mathfrak{D}}$ also in the non-well-posed case, and *regard* $\widehat{\mathfrak{D}}$ *as a function defined on all of* $\rho(A)$ *by* (4.7.1).

Proof of Lemma 4.7.5 (i) By Lemma 4.7.3(viii), for each $z \in \rho(A_{|X}) = \rho(A)$, the operator $\left[\begin{smallmatrix} (z - A_{|X})^{-1} B \\ 1 \end{smallmatrix}\right]$ maps U into $\mathcal{D}(S)$. Furthermore, it is an analytic function of z. This implies that $\widehat{\mathfrak{D}}$ is well-defined and analytic on $\rho(A)$. Formula (4.7.2) follows from (4.7.1) and the resolvent identity (3.2.1).

(ii) By (4.7.1), for all $\alpha \in \rho(A)$ and all $\left[\begin{smallmatrix} x \\ u \end{smallmatrix}\right] \in \mathcal{D}(S)$,

$$\begin{aligned} C\&D \begin{bmatrix} x \\ u \end{bmatrix} - \widehat{\mathfrak{D}}(\alpha) u &= C\&D \begin{bmatrix} x - (\alpha - A_{|X})^{-1} B u \\ 0 \end{bmatrix} \\ &= C\big[x - (\alpha - A_{|X})^{-1} B u\big]. \end{aligned}$$

(iii) That the semigroup generator A, the control operator B, and the observation operator C coincide with the corresponding operators of Σ follows from Definitions 4.6.4, 4.7.2, and 4.7.4. That $\widehat{\mathfrak{D}}$ is an extension in the $L^p|Reg$-well-posed case of the transfer function defined in Definition 4.6.1 follows from Theorem 4.6.7. □

Above we started with an operator S which satisfies the conditions (i)–(iv) listed in Definition 4.6.4, and from this operator we extracted the main operator A, the control operator B, the observation operator C, and the transfer function $\widehat{\mathfrak{D}}$. Conversely, it is possible to start with A, B, C, and the transfer function $\widehat{\mathfrak{D}}$ evaluated at one point in $\rho(A)$, and from this set of data construct a system node as follows.

Lemma 4.7.6 *Let* A *be a densely defined operator on a Banach space* X *with a nonempty resolvent set, let* $\alpha \in \rho(A)$*, let* X_{-1} *be the completion of* X *under the norm* $|x|_{X_{-1}} := |(\alpha - A)^{-1} x|_X$*, and let* $X_1 = \mathcal{D}(A)$*, with norm* $|x|_{X_1} := |(\alpha - A) x|_X$*. Denote the extension of* A *to an operator in* $\mathcal{B}(X; X_{-1})$ *by* $A_{|X}$*.*

Let $B \in \mathcal{B}(U; X_{-1})$, let $C \in \mathcal{B}(X_1; Y)$, and let $D \in \mathcal{B}(U; Y)$, where U and Y are two more Banach spaces. Define $S\colon \left[\begin{smallmatrix} X \\ U \end{smallmatrix}\right] \supset \mathcal{D}(S) \to \left[\begin{smallmatrix} X \\ Y \end{smallmatrix}\right]$ by $S := \left[\begin{smallmatrix} A\&B \\ C\&D \end{smallmatrix}\right]$, where

$$\begin{aligned} \mathcal{D}(S) : &= \mathcal{D}(A\&B) = \mathcal{D}(C\&D) \\ &= \left\{ \begin{bmatrix} x \\ u \end{bmatrix} \in \begin{bmatrix} X \\ U \end{bmatrix} \;\middle|\; A_{|X}x + Bu \in X \right\}, \\ A\&B : &= \begin{bmatrix} A_{|X} & B \end{bmatrix}_{|\mathcal{D}(S)}, \\ C\&D \begin{bmatrix} x \\ u \end{bmatrix} : &= C(x - (\alpha - A_{|X})^{-1}Bu) + Du, \qquad \begin{bmatrix} x \\ u \end{bmatrix} \in \mathcal{D}(S). \end{aligned}$$

Then S is an operator node on (Y, X, U). The main operator of S is A, the observation operator of S is B, the control operator of S is C, and the transfer function of S satisfies $\widehat{\mathfrak{D}}(\alpha) = D$.

Proof Most of this is obvious. The only thing which needs to be checked is that S is closed, but this proof is essentially the same as the proof of Proposition 4.7.1(i) with $\widehat{\mathfrak{D}}(\alpha)$ replaced by D (the only properties of C and $\widehat{\mathfrak{D}}(\alpha)$ that were used in that proof were the facts that $C \in \mathcal{B}(X_1; Y)$ and $\widehat{\mathfrak{D}}(\alpha) \in \mathcal{B}(U; Y)$). □

We have now two different characterizations of operator nodes, the original one in Definition 4.7.2, and the alternative one in Lemma 4.7.6. There is still a third way of characterizing an operator node which has the advantage that it contains no explicit reference to the rigged spaces X_1 and X_{-1}, to the control operator, to the observation operator, and to the transfer function.

Lemma 4.7.7 *Let U, X and Y be Banach spaces and let $S\colon \mathcal{D}(S) \to \left[\begin{smallmatrix} X \\ Y \end{smallmatrix}\right]$ be a linear operator with $\mathcal{D}(S) \subset \left[\begin{smallmatrix} X \\ U \end{smallmatrix}\right]$. We decompose S into $S = \left[\begin{smallmatrix} A\&B \\ C\&D \end{smallmatrix}\right]$, where $A\&B\colon \mathcal{D}(S) \to X$ and $C\&D\colon \mathcal{D}(S) \to Y$. We denote $\mathcal{D}(A) = \left\{x \in X \mid \left[\begin{smallmatrix} x \\ 0 \end{smallmatrix}\right] \in \mathcal{D}(S)\right\}$, and we define $A\colon \mathcal{D}(A) \to X$ by $Ax = A\&B \left[\begin{smallmatrix} x \\ 0 \end{smallmatrix}\right]$. Then S is an operator node on (Y, X, U) if and only if the following conditions hold:*

(i) *S is closed as an operator $\left[\begin{smallmatrix} X \\ U \end{smallmatrix}\right]$ to $\left[\begin{smallmatrix} X \\ Y \end{smallmatrix}\right]$ (with domain $\mathcal{D}(S)$).*
(ii) *$A\&B$ is closed as an operator $\left[\begin{smallmatrix} X \\ U \end{smallmatrix}\right]$ to X (with domain $\mathcal{D}(S)$).*
(iii) *A has a nonempty resolvent set, and $\mathcal{D}(A)$ is dense in X.*
(iv) *For every $u \in U$ there exists an $x \in X$ with $\left[\begin{smallmatrix} x \\ u \end{smallmatrix}\right] \in \mathcal{D}(S)$.*

Proof The necessity of (i)–(iv) is more or less obvious: (i) and (iii) are part of Definition 4.7.2, (ii) is part (v) of Lemma 4.7.3, and (iv) follows from part (vi) of the same lemma.

Next we turn to the sufficiency part. For this we have to construct the operator B in Definition 4.6.4. Take $u \in U$ and let $x \in X$ be as in (ii). We define

$Bu = A\&B\left[\begin{smallmatrix} x \\ u \end{smallmatrix}\right] - A_{|X}x$. It is easy to see that Bu does not depend on the choice of x and that B is linear.

To prove the boundedness of B (from U to X_{-1}), we show that it is closed. Take a sequence u_n in U such that $u_n \to u$ and $Bu_n \to z$ in X_{-1}. We have to show that $Bu = z$. Take $\alpha \in \rho(A)$ and define $x_n = (\alpha - A_{|X})^{-1}Bu_n$. Then $A_{|X}x_n + Bu_n = \alpha(\alpha - A_{|X})^{-1}Bu_n \in X$, so $\left[\begin{smallmatrix} x_n \\ u_n \end{smallmatrix}\right] \in \mathcal{D}(S)$. Moreover, $\left[\begin{smallmatrix} x_n \\ u_n \end{smallmatrix}\right] \to \left[\begin{smallmatrix} (\alpha - A_{|X})^{-1}z \\ u \end{smallmatrix}\right]$ in $\left[\begin{smallmatrix} X \\ U \end{smallmatrix}\right]$ and $A\&B\left[\begin{smallmatrix} x_n \\ u_n \end{smallmatrix}\right] = A_{|X}x_n + Bu_n \to \alpha(\alpha - A_{|X})^{-1}z$ in X. By the closedness of $A\&B$,

$$A\&B\begin{bmatrix} (\alpha - A_{|X})^{-1}z \\ u \end{bmatrix} = \alpha(\alpha - A_{|X})^{-1}z.$$

This implies that $Bu = A\&B\left[\begin{smallmatrix} (\alpha - A_{|X})^{-1}z \\ u \end{smallmatrix}\right] - A_{|X}(\alpha - A_{|X})^{-1}z = z$. By the closed graph theorem, B is bounded.

We still have to show that $\mathcal{D}(S) = V$, where $V := \left\{\left[\begin{smallmatrix} x \\ u \end{smallmatrix}\right] \in \left[\begin{smallmatrix} X \\ U \end{smallmatrix}\right] \mid A_{|X}x + Bu \in X\right\}$. By construction, $A_{|X}x + Bu = A\&B\left[\begin{smallmatrix} x \\ u \end{smallmatrix}\right] \in X$ for all $\left[\begin{smallmatrix} x \\ u \end{smallmatrix}\right] \in \mathcal{D}(S)$. Thus $\mathcal{D}(S) \subset V$. Conversely, suppose that $\left[\begin{smallmatrix} x \\ u \end{smallmatrix}\right] \in V$. Then, by condition (iv), there exists $x_0 \in X$ such that $\left[\begin{smallmatrix} x_0 \\ u \end{smallmatrix}\right] \in \mathcal{D}(S)$, hence $A_{|X}x_0 + Bu \in X$. This implies that $A_{|X}(x - x_0) \in X$, hence $\left[\begin{smallmatrix} x - x_0 \\ 0 \end{smallmatrix}\right] \in \mathcal{D}(S)$. This gives $\left[\begin{smallmatrix} x \\ u \end{smallmatrix}\right] \in \mathcal{D}(S)$. Thus $\mathcal{D}(S) = V$. □

Everything which we have said so far in this section applies to general operator nodes. In the case of a system node (where the main operator generates a C_0 semigroup) we can say something more: it induces a 'dynamical system' of a certain type:

Lemma 4.7.8 *Let S be a system node on (Y, X, U). Then, for each $s \in \mathbb{R}$, $x_s \in X$ and $u \in W^{2,1}_{\mathrm{loc}}([s, \infty); U)$ with $\left[\begin{smallmatrix} x_s \\ u(s) \end{smallmatrix}\right] \in \mathcal{D}(S)$, the equation*

$$\begin{bmatrix} \dot{x}(t) \\ y(t) \end{bmatrix} = S\begin{bmatrix} x(t) \\ u(t) \end{bmatrix}, \quad t \geq s, \quad x(s) = x_s, \tag{4.7.4}$$

has a unique solution $\left[\begin{smallmatrix} x \\ y \end{smallmatrix}\right]$ satisfying $x \in C^1([s, \infty); X)$, $\left[\begin{smallmatrix} x \\ u \end{smallmatrix}\right] \in C([s, \infty); \mathcal{D}(S))$, and $y \in C([s, \infty); Y)$.

Proof By Theorem 4.3.9, the equation $\dot{x}(t) = Ax(t) + Bu(t)$ has a unique strong solution on $[s, \infty)$ with $x(s) = x_s$, this solution is continuously differentiable in X, and $\dot{x}(t) = A_{|X}x(t) + Bu(t)$ for all $t \geq s$. In particular, x, u, and $A_{|X}x + Bu$ are continuous in X, U, and X, respectively, and this implies that $\left[\begin{smallmatrix} x \\ u \end{smallmatrix}\right]$ is continuous with values in $\mathcal{D}(S) = \mathcal{D}(A\&B)$. Since $C\&D \in \mathcal{B}(\mathcal{D}(S); Y)$, also y is continuous with values in Y. □

Occasionally we shall also need to discuss solutions of (4.7.4) with a less smooth input function $u \in L^1_{\mathrm{loc}}([s, \infty); U)$ and an arbitrary initial state

$x_s \in X$. The definition of the state trajectory x is straightforward: by Theorem 3.8.2(i), for every $u \in L^1_{\mathrm{loc}}([s,\infty);U)$ and $x_s \in X$, the equation $\dot{x}(t) = A_{|X_{-1}}x(t) + Bu(t)$ has a unique strong solution in X_{-1} on $[s,\infty)$ with initial state $x(s) = x_s$, which is given by the variation of constants formula

$$x(t) = \mathfrak{A}^{t-s}x_s + \int_s^t \mathfrak{A}^{t-v}_{|X_{-1}} Bu(v)\,dv, \qquad t \ge s. \tag{4.7.5}$$

However, the definition of the output y is less straightforward. One possibility is to interpret the output as a distribution. By integrating the equation $y(t) = C\&D\left[\begin{smallmatrix} x(t) \\ u(t) \end{smallmatrix}\right]$ (which is part of (4.7.4)) two times and then differentiating the result two times we get

$$y(t) = \frac{\mathrm{d}^2}{\mathrm{d}t^2}\, C\&D \int_s^t (t-v)\begin{bmatrix} x(v) \\ u(v) \end{bmatrix} dv, \qquad t \ge s. \tag{4.7.6}$$

As the following lemma shows, the function to which we apply $C\&D$ in the formula above does belong to $\mathcal{D}(C\&D) = \mathcal{D}(S)$, so if we interpret the second order derivative in the distribution sense, then this formula defines a Y-valued distribution (of order at most two).

Lemma 4.7.9 *Let S be a system node on (Y, X, U). Let $s \in \mathbb{R}$, $x_s \in X$, and $u \in L^1_{\mathrm{loc}}([s,\infty);U)$, define x by* (4.7.5)*, and let*

$$\begin{bmatrix} x_2(t) \\ u_2(t) \end{bmatrix} = \int_s^t (t-v)\begin{bmatrix} x(v) \\ u(v) \end{bmatrix} dv, \qquad t \ge s,$$

be the second order integral of $\left[\begin{smallmatrix} x \\ u \end{smallmatrix}\right]$. Then $x_2 \in C^1([s,\infty);X)$ and $\left[\begin{smallmatrix} x_2 \\ u_2 \end{smallmatrix}\right] \in C([s,\infty);\mathcal{D}(S))$.

Proof Because of the linearity of the system, it suffices to prove the lemma separately in the two cases where either $x_s = 0$ or $u = 0$. We begin with the latter. If $u = 0$, then already the first order integral $x_1(t) = \int_s^t x(v)\,dv$ is continuous with values in X_1 (as shown at the very beginning of the proof of Theorem 4.4.4), so we even have $x_2 \in C^1([s,\infty);X_1)$ and $\left[\begin{smallmatrix} x_2 \\ 0 \end{smallmatrix}\right] \in C^1\big([s,\infty);\left[\begin{smallmatrix} X_1 \\ 0 \end{smallmatrix}\right]\big) \subset C^1([s,\infty);\mathcal{D}(S))$ in this case.

It remains to study the case where $x_s = 0$ but $u \ne 0$. By integrating (4.7.5) (with $x_s = 0$) two times (and simplifying the resulting double integral) we find that

$$x_2(t) = \int_s^t \mathfrak{A}^{t-v}_{|X_{-1}} Bu_2(s)\,ds.$$

Thus, x_2 is the strong solution of the equation $\dot{x}_2(t) = A_{|X}x_2(t) + u_2(t)$, $t \ge s$, with initial state $x_2(s) = 0$. By Lemma 4.7.8, $x_2 \in C^1([s,\infty);X)$ and $\left[\begin{smallmatrix} x_2 \\ u_2 \end{smallmatrix}\right] \in C([s,\infty);\mathcal{D}(S))$. Thus, the conclusion of Lemma 4.7.9 holds in this case, too. □

Definition 4.7.10 Let S be a system node on (Y, X, U), let $s \in \mathbb{R}$, $x_s \in X$, and let $u \in L^1_{\mathrm{loc}}([s,\infty);U)$. Then the *state trajectory* x of S with initial time s, initial state x_s, and input function u is the function x given by (4.7.5), and the *output* y is the distribution given by (4.7.6), where the second order derivative on the right-hand side is interpreted in the distribution sense (by Lemma 4.7.9, the function to which this derivative is applied is a continuous Y-valued function).

Note that, by Theorem 4.6.12, in the well-posed case this state trajectory and output function coincide with the state trajectory and output function of the corresponding well-posed linear system.

It is of interest to know some reasonable conditions under which the output is a function rather than a distribution (even in the non-well-posed case).

Theorem 4.7.11 *Let* $S = \left[\begin{smallmatrix} A\&B \\ C\&D \end{smallmatrix}\right]$ *be a system node on* (Y, X, U)*. Let* $s \in \mathbb{R}$*,* $x_s \in X$*, and* $u \in L^1_{\mathrm{loc}}([s,\infty);U)$*, and let* x *and* y *be the state trajectory and output of* S *with initial time* s*, initial state* x_s*, and input function* u*. If* $x \in W^{1,1}_{\mathrm{loc}}([s,\infty);X)$*, then* $\left[\begin{smallmatrix} x \\ u \end{smallmatrix}\right] \in L^1_{\mathrm{loc}}([s,\infty);\mathcal{D}(S))$*,* $y \in L^1_{\mathrm{loc}}([s,\infty);Y)$*, and* $\left[\begin{smallmatrix} x \\ y \end{smallmatrix}\right]$ *is the unique solution with the above properties of the equation*

$$\begin{bmatrix} \dot{x}(t) \\ y(t) \end{bmatrix} = S \begin{bmatrix} x(t) \\ u(t) \end{bmatrix} \text{ for almost all } t \geq s, \quad x(s) = x_s. \tag{4.7.7}$$

If $u \in C([s,\infty);U)$ *and* $x \in C^1([s,\infty);X)$*, then* $\left[\begin{smallmatrix} x \\ u \end{smallmatrix}\right] \in C([s,\infty);\mathcal{D}(S))$*,* $y \in C([s,\infty);Y)$*, and the equation* (4.7.7) *holds for all* $t \geq s$*.*

Proof Throughout this proof we assume that $u \in L^1_{\mathrm{loc}}([s,\infty);U)$ and $x \in W^{1,1}_{\mathrm{loc}}([s,\infty);X)$.

Let us begin with the uniqueness claim. If $\left[\begin{smallmatrix} x \\ u \end{smallmatrix}\right] \in L^1_{\mathrm{loc}}([s,\infty);\mathcal{D}(S))$ and (4.7.7) holds, then x is a strong solution (in X) of the equation $\dot{x}(t) = A_{|X}x(t) + Bu(t)$ on $[s,\infty)$ with initial state x_s $\left(\text{recall that } A\&B = \left[A_{|X} \;\; B\right]_{\mathcal{D}(S)}\right)$. This fact determines x uniquely in terms of x_s and u; see Theorem 3.8.2(ii). Once we know that x is unique, formula (4.7.7) defines y uniquely.

The state trajectory x is, as always, the strong solution in X_{-1} of the equation $\dot{x}(t) = Ax(t) + Bu(t)$ on $[s,\infty)$ with initial state x_s. Since we now assume that $x \in W^{1,1}_{\mathrm{loc}}([s,\infty);X)$, this equation implies that $\dot{x} = A_{|X}x + Bu \in L^1_{\mathrm{loc}}([s,\infty);X)$, and hence $\left[\begin{smallmatrix} x \\ u \end{smallmatrix}\right] \in L^1_{\mathrm{loc}}([s,\infty);\mathcal{D}(S))$ and $\dot{x}(t) = A\&B \left[\begin{smallmatrix} x(t) \\ u(t) \end{smallmatrix}\right]$ for almost all $t \geq s$. We can now interpret the distribution derivative on the right-hand side of (4.7.6) as an ordinary derivative, and carrying out the differentiation we find that $y \in L^1_{\mathrm{loc}}([s,\infty);Y)$, and that $y(t) = C\&D \left[\begin{smallmatrix} x(t) \\ u(t) \end{smallmatrix}\right]$ for almost all $t \geq s$. The proof of the last claim (where u is continuous and x continuously differentiable) is similar. □

The transfer function of a general system node has more or less the same interpretation as the transfer function of an $L^p|Reg$-well-posed linear system.

Lemma 4.7.12 *Let $S = \left[\begin{smallmatrix} A\&B \\ C\&D \end{smallmatrix}\right]$ be a system node on (Y, X, U), with semigroup generator A, the control operator B, the observation operator C, and the transfer function $\widehat{\mathfrak{D}}$. Denote the growth bound of the semigroup $\mathfrak{A}$ generated by A by $\omega_{\mathfrak{A}}$. Let $\omega > \omega_{\mathfrak{A}}$, and suppose that the input function u in Lemma 4.7.8 satisfies $W^{2,1}_{\omega}(\overline{\mathbb{R}}^+; U)$. Then the state trajectory x and the output function y in Lemma 4.7.8 satisfy $x \in BC^1_{0,\omega}(\overline{\mathbb{R}}^+; X) \cap W^{1,1}_{\omega}(\overline{\mathbb{R}}^+; X)$, $\left[\begin{smallmatrix} x \\ u \end{smallmatrix}\right] \in BC_{0,\omega}(\overline{\mathbb{R}}^+; \mathcal{D}(S)) \cap L^1_{\omega}(\mathbb{R}^+; \mathcal{D}(S))$, and $y \in BC_{0,\omega}(\overline{\mathbb{R}}^+; Y) \cap L^1_{\omega}(\mathbb{R}^+; Y)$. In this case the Laplace transforms $\hat{x}$ of x and $\hat{y}$ of y are given (in terms of the Laplace transform $\hat{u}$ of u) by* (4.6.10).

Proof That $x \in BC^1_{0,\omega}(\overline{\mathbb{R}}^+; X) \cap W^{1,1}_{\omega}(\overline{\mathbb{R}}^+; X)$ follows from Theorem 3.8.4. This implies (as in Theorem 4.3.11) that $\left[\begin{smallmatrix} x \\ u \end{smallmatrix}\right] \in BC_{0,\omega}(\overline{\mathbb{R}}^+; \mathcal{D}(S)) \cap L^1_{\omega}(\mathbb{R}^+; \mathcal{D}(S))$ (where $\mathcal{D}(S) = \mathcal{D}(A\&B)$). Since $y = C\&D\left[\begin{smallmatrix} x \\ u \end{smallmatrix}\right]$, this furthermore implies that $y \in BC_{0,\omega}(\overline{\mathbb{R}}^+; Y) \cap L^1_{\omega}(\mathbb{R}^+; Y)$. We can therefore take Laplace transforms in (4.7.4) to get for all $z \in \overline{\mathbb{C}}^+_{\omega}$,

$$\begin{bmatrix} z\hat{x}(z) - x(0) \\ \hat{y}(z) \end{bmatrix} = \begin{bmatrix} A\&B \\ C\&D \end{bmatrix} \begin{bmatrix} \hat{x}(z) \\ \hat{u}(z) \end{bmatrix}.$$

Since $A\&B = \begin{bmatrix} A_{|X} & B \end{bmatrix}$, we can write the first of the above equations in the form

$$z\hat{x}(z) - x(0) = A_{|X}\hat{x}(z) + B\hat{u}(z),$$

which we can solve for $\hat{x}(z)$ to get the first equation in (4.6.10) (note that $(z - A_{|X})^{-1}_{|X} = (z - A)^{-1}$). Substituting this into the second of the two equation above we get

$$\begin{aligned} \hat{y}(z) &= C\&D \begin{bmatrix} (z - A)^{-1}x(0) + (z - A_{|X})^{-1}B\hat{u}(z) \\ \hat{u}(z) \end{bmatrix} \\ &= C(z - A)^{-1}x(0) + C\&D \begin{bmatrix} (z - A_{|X})^{-1}B \\ 1 \end{bmatrix} \hat{u}(z) \\ &= C(z - A)^{-1}x(0) + \widehat{\mathfrak{D}}(z)\hat{u}(z). \end{aligned}$$

□

Among all system nodes, the class of L^p-well-posed system nodes with $1 \le p < \infty$ can be characterized in the following way:

Theorem 4.7.13 *Let S be a system node on three Banach spaces (Y, X, U), and $1 \le p < \infty$. Then S is L^p-well-posed if and only if, for some (hence all)*

$T > 0$, there is a constant M such that all solutions $\left[\begin{smallmatrix} x \\ y \end{smallmatrix}\right]$ of the type described in Lemma 4.7.8 with $s = 0$ satisfy

$$|x(T)|_X + \|y\|_{L^p([0,T);Y)} \le M\big(|x(0)|_X + \|u\|_{L^p([0,T);U)}\big). \tag{4.7.8}$$

Proof The necessity of (4.7.8) follows from Definitions 2.2.1, 2.2.6, 2.2.7, and Theorem 4.6.11(i). To prove the converse we must construct the input map $\mathfrak{B}$, the output map $\mathfrak{C}$, and the input/output map $\mathfrak{D}$ of the corresponding L^p-well-posed linear system (the existence of the semigroup $\mathfrak{A}$ is part of the assumption of a system node). In this construction we use the fact that if (4.7.8) holds, then (by the time-invariance of the equation), it is also true that, for all initial times $s \in \mathbb{R}$, the solutions $\left[\begin{smallmatrix} x \\ y \end{smallmatrix}\right]$ of the type described in Lemma 4.7.8 satisfy

$$|x(s+T)|_X + \|y\|_{L^p([s,s+T);Y)} \le M\big(|x(s)|_X + \|u\|_{L^p([s,s+T);U)}\big). \tag{4.7.9}$$

We begin with the input map $\mathfrak{B}$. If we replace the state space X by X_{-1}, then B is a bounded control operator, and it follows from Proposition 2.3.1 (with arbitrary choices of C and D) that

$$\mathfrak{B}u := \int_{\infty}^{0} \mathfrak{A}^{-v}_{|X_{-1}} Bu(v)\, dv$$

is an L^1-well-posed input map for $\mathfrak{A}$. In particular, $\mathfrak{B}$ satisfies the intertwining condition $\mathfrak{A}^t_{|X_{-1}}\mathfrak{B}u = \mathfrak{B}\tau^t_- u$ (with values in X_{-1}) for all $u \in L^1_c(\mathbb{R}^-;U)$ and $t \ge 0$. If $u \in W^{2,1}_{\mathrm{loc}}([-T,\infty);U)$ and $u(-T) = 0$, then the solution x mentioned in Lemma 4.7.8 with $s = -T$ and $x(-T) = 0$ is given by

$$x(t) = \int_{-T}^{t} \mathfrak{A}^{t-v}_{|X_{-1}} Bu(v)\, dv = \mathfrak{B}^t_{-T}u, \qquad t \ge -T,$$

and by (4.7.9), for this class of functions, $|\mathfrak{B}^0_{-T}u|_X = |x(0)|_X \le M\|u\|_{L^p([-T,0);U)}$. The set of functions u in $W^{2,1}([-T,0);U)$ with $u(-T) = 0$ is dense in $L^p([-T,0);U)$, and this implies that $\mathfrak{B}^0_{-T} \in \mathcal{B}(L^p([-T,0);U);X)$, or equivalently, $\mathfrak{B}\pi_{[-T,0)} \in \mathcal{B}(L^p([-T,0);U);X)$. By Lemma 2.4.3, this implies that $\mathfrak{B} \in \mathcal{B}(L^p_c(\mathbb{R}^-;U);X)$, and thus $\mathfrak{B}$ is an L^p-well-posed input map for $\mathfrak{A}$.

Next we define $\mathfrak{C}x$ for all $x \in X_1$ by

$$(\mathfrak{C}x)(t) = C\mathfrak{A}^t x, \qquad t \ge 0, \qquad x \in X_1.$$

Then $\mathfrak{C}$ maps X_1 into $C(\overline{\mathbb{R}}^+;Y)$ since $C \in \mathcal{B}(X_1;Y)$ and $t \mapsto \mathfrak{A}^t x$ is continuous in X_1 whenever $x \in X_1$, and it is easy to show that $\mathfrak{C}$ satisfies the intertwining condition $\mathfrak{C}\mathfrak{A}^t x = \tau^t_+\mathfrak{C}x$ for all $x \in X_1$ and all $t \ge 0$. If we take $u = 0$ and $x_0 \in X_1$ in Lemma 4.7.8, then $y = \mathfrak{C}x_0$, and so, by (4.7.8), $\|\mathfrak{C}x_0\|_{L^p([0,T);Y)} \le M|x(0)|_X$. This, together with Lemma 2.4.3, implies that $\mathfrak{C}$ can be extended to a continuous map from X to $L^p_{\mathrm{loc}}(\mathbb{R}^+;Y)$. By continuity (and since X_1 is dense in X), also the extended map satisfies the intertwining condition $\mathfrak{C}\mathfrak{A}^t = \tau_+^t\mathfrak{C}$ for all $t \geq 0$, so $\mathfrak{C}$ is an L^p-well-posed output map for $\mathfrak{A}$.

Finally, we construct the input/output map $\mathfrak{D}u$. We first consider the class of input functions $u \in W^{2,1}_{c,\text{loc}}(\mathbb{R}; U)$ which vanish on $(-\infty, s)$ for some $s \in \mathbb{R}$. The corresponding state trajectory x and output function y in Lemma 4.7.8 with $x(s) = 0$ are given by

$$x(t) = \mathfrak{B}^t u, \qquad y(t) = C\&D \begin{bmatrix} x(t) \\ u(t) \end{bmatrix}, \qquad t \geq 0,$$

so for this class of inputs it is natural to define $\mathfrak{D}u$ by

$$(\mathfrak{D}u)(t) = C\&D \begin{bmatrix} \mathfrak{B}^t u \\ u(t) \end{bmatrix}, \qquad t \in \mathbb{R}$$

(note, in particular, that $\left[\begin{smallmatrix} x \\ u \end{smallmatrix}\right] \in C(\mathbb{R}; \mathcal{D}(S))$ if we define $x(t) = 0$ for $t < s$). On its present domain, $\mathfrak{D}$ is clearly time-invariant. Let us look at the special case where $u(0) = \dot{u}(0) = 0$. Since $\left[\begin{smallmatrix} x(0) \\ u(0) \end{smallmatrix}\right] \in \mathcal{D}(S)$ and $u(0) = 0$, this implies that $x(0) = \mathfrak{B}u \in X_1$. Moreover, $\pi_- u \in W^{2,1}_c(\mathbb{R}; U)$, so we may apply the same argument with u replaced by $\pi_- u$. This gives for all $t \geq 0$,

$$\begin{aligned} (\mathfrak{D}\pi_- u)(t) &= C\&D \begin{bmatrix} \mathfrak{B}\tau^t \pi_- u \\ 0 \end{bmatrix} = C \int_{-\infty}^0 \mathfrak{A}^{t-v}_{|X_{-1}} Bu(v)\, dv \\ &= C\mathfrak{A}^t_{|X_{-1}} \int_{-\infty}^0 \mathfrak{A}^{-v}_{|X_{-1}} Bu(v)\, dv \\ &= C\mathfrak{A}^t \mathfrak{B}u = (\mathfrak{C}(\mathfrak{B}u))(t). \end{aligned}$$

Thus, $\pi_+\mathfrak{D}\pi_- u = \mathfrak{C}\mathfrak{B}u$ whenever $u \in W^{2,1}_{c,\text{loc}}(\mathbb{R}; U)$ and $u(0) = \dot{u}(0) = 0$. A similar (but simpler) computation shows that $\pi_-\mathfrak{D}\pi_+ u = 0$ for the same set of functions u.

So far we have not yet used (4.7.8) in our construction of $\mathfrak{D}$ (so the same construction is valid without this extra condition). Suppose now that $u \in W^{2,1}_{c,\text{loc}}(\mathbb{R}; U)$ vanishes on $\mathbb{R}^-$. Then, by (4.7.8),

$$\|\mathfrak{D}u\|_{L^p([0,T);Y)} \leq \|u\|_{L^p([0,T);U)}.$$

The set of functions $u \in W^{2,1}([0, T); U)$ which satisfy $u(0) = \dot{u}(0) = 0$ is dense in $L^p([0, T); U)$, so the preceding inequality implies that $\pi_{[0,T)}\mathfrak{D}\pi_{[0,T)}$ can be extended to an operator in $\mathcal{B}(L^p([0, T); U); L^p([0, T); Y))$. It follows from Lemma 2.4.3 that $\mathfrak{D}$ can be extended to a continuous operator from $L^p_{c,\text{loc}}([0, T); U)$ to $L^p_{c,\text{loc}}([0, T); Y)$, and (by continuity and density) the extended operator is still time-invariant and causal, and its Hankel operator is $\mathfrak{C}\mathfrak{B}$.

We have now constructed an L^p-well-posed linear system $\Sigma = \left[\begin{smallmatrix} \mathfrak{A} & \mathfrak{B} \\ \mathfrak{C} & \mathfrak{D} \end{smallmatrix}\right]$. The semigroup generator A, the control operator B, and the observation operator C of Σ are the same as the corresponding operators of the given system node S. It is also true that S and Σ have the same transfer function, hence they have the same observation/feedthrough operator $C\&D$, so S is the system node of Σ, and hence L^p-well-posed. □

The preceding theorem can also be stated in a slightly different form:

Theorem 4.7.14 *Let S be a system node on three Banach spaces (Y, X, U) with semigroup $\mathfrak{A}$, control operator B, and observation operator C, and let $1 \le p < \infty$. Then S is L^p-well-posed if and only if the following conditions hold:*

(i) *For all $u \in L^p_c(\mathbb{R}^-; U)$,*

$$\mathfrak{B}u = \int_{-\infty}^{0} \mathfrak{A}^{-s} Bu(s)\, ds \in X$$

(although the integral is computed in X_{-1}).

(ii) *The operator $\mathfrak{C}\colon X_1 \to C(\overline{\mathbb{R}}^+; Y)$ defined by*

$$(\mathfrak{C}x)(t) = C\mathfrak{A}^t x, \qquad t \ge 0,$$

can be extended to a continuous operator $X \to L^p_{\mathrm{loc}}(\mathbb{R}^+; Y)$.

(iii) *The operator $\mathfrak{D}\colon W^{2,1}_{c,\mathrm{loc}}(\mathbb{R}; U) \to C_c(\mathbb{R}; Y)$ defined by*

$$(\mathfrak{D}u)(t) = C\&D \begin{bmatrix} \mathfrak{B}^t u \\ u(t) \end{bmatrix}, \qquad t \in \mathbb{R},$$

can be extended to a continuous operator $L^p_{c,\mathrm{loc}}(\mathbb{R}; U) \to L^p_{c,\mathrm{loc}}(\mathbb{R}; Y)$.

Proof The necessity of the given conditions follows from Theorems 4.2.4 and 4.4.7, and the fact that $(\mathfrak{D}u)(t) = C\&D \left[\begin{smallmatrix} \mathfrak{B}^t u \\ u(t) \end{smallmatrix}\right]$, $t \in \mathbb{R}$, whenever $u \in W^{2,1}_{c,\mathrm{loc}}(\mathbb{R}; U)$ (as noticed in the proof of Theorem 4.7.13). The proof of the sufficiency is a simplified version of the proof of Theorem 4.7.13 (use Theorems 4.2.4 and 4.4.7 to conclude that $\mathfrak{B}$ and $\mathfrak{C}$ are L^p-well-posed input and output maps). □

Sometimes it is more convenient to use this theorem in a form which has no explicit reference to the system $\left[\begin{smallmatrix} \mathfrak{A} & \mathfrak{B} \\ \mathfrak{C} & \mathfrak{D} \end{smallmatrix}\right]$ induced by the node $S = \left[\begin{smallmatrix} A\&B \\ C\&D \end{smallmatrix}\right]$.

Theorem 4.7.15 *Let S be an operator node (see Definition 4.7.2) on three Banach spaces (Y, X, U) with main operator A, control operator B, and observation operator C, and let $1 \le p < \infty$. Then S is an L^p-well-posed system node if and only if the following conditions hold:*

(i) *For all* $x_0 \in X$ *and all* $u \in W^{2,1}_{\mathrm{loc}}(\mathbb{R}^+; U)$, *the equation*

$$\begin{aligned} \dot{x}(t) &= A_{|X} x(t) + Bu(t), \qquad t \geq 0, \\ x(0) &= x_0, \end{aligned} \tag{4.7.10}$$

has a unique strong solution x *in the sense of Definition 3.8.1 (i.e.,* $x \in C(\mathbb{R}^+; X) \cap W^{1,1}_{\mathrm{loc}}(\mathbb{R}^+; X_{-1})$, $x(0) = x_0$, *and* $\dot{x}(t) = A_{|X} x(t) + Bu(t)$ *for almost all* $t \geq 0$*).*

(ii) *For each* $T > 0$ *there is a constant* $K_T < \infty$ *so that, for all* $x_0 \in X$ *and* $u \in W^{2,1}_{\mathrm{loc}}(\mathbb{R}^+; U)$ *with* $\left[\begin{smallmatrix} x_0 \\ u(0) \end{smallmatrix}\right] \in \mathcal{D}(S)$, *if we let* x *be as in (i) and define* $y(t) = C\&D \left[\begin{smallmatrix} x(t) \\ u(t) \end{smallmatrix}\right]$, $t \geq 0$, *then*

$$|x(T)| + \|y\|_{L^p([0,T);Y)} \leq K_T \big(\|x_0\|_X + \|u\|_{L^p([0,T);U)}\big).$$

Proof The necessity of the given conditions follows from Lemma 4.7.8 and Theorem 4.7.13, and so does the sufficiency if we can show that (i) and (ii) imply that A is the generator of a C_0 semigroup.

Assume (i) and (ii). If for all $t \geq 0$ and $x_0 \in X$ we define $\mathfrak{A}^t x_0 = x(t)$, where x is the strong solution of (4.7.10) with $u = 0$, then $\mathfrak{A}^t$ is a strongly continuous family of operators in $\mathcal{B}(X)$ satisfying $\mathfrak{A}^0 = 1$. It is also easy to show that $\mathfrak{A}$ has the semigroup property $\mathfrak{A}^{s+t} = \mathfrak{A}^s \mathfrak{A}^t$ for all $s, t \geq 0$ (by the uniqueness of the solution of (4.7.10), the solution at time $t > 0$ corresponding to the initial value $x(s)$ is equal to $x(s+t)$). Thus, $\mathfrak{A}$ is a C_0 semigroup.

We denote the generator of $\mathfrak{A}$ by A_1, and claim that $A_1 = A$. Suppose first that $x_0 \in \mathcal{D}(A_1)$. Then we know (see Theorem 3.2.1(iii)) that $x(t) = \mathfrak{A}^t x_0$ is continuously differentiable in X. In particular, $A_1 x_0 = \dot{x}(0) = Ax_0 \in X$. Choose some $\alpha \in \rho(A)$. Then $(\alpha - A)x_0 \in X$, and since $(\alpha - A)^{-1}$ maps X onto $\mathcal{D}(A)$, we get $x_0 \in \mathcal{D}(A)$. Thus, $\mathcal{D}(A_1) \subset \mathcal{D}(A)$, and $A_1 x_0 = Ax_0$ for all $x_0 \in \mathcal{D}(A_1)$. Conversely, suppose that $x_0 \in \mathcal{D}(A)$. Define $z_0 = (\alpha - A)x_0$ where $\alpha \in \rho(A)$, and let $z(t) = \mathfrak{A}^t z_0$. Then, by (i), $z \in C(\mathbb{R}^+; X) \cap W^{1,1}_{\mathrm{loc}}(\mathbb{R}^+; X_{-1})$, $z(0) = z_0$, and $\dot{z}(t) = Az(t)$ for almost all $t \geq 0$. But $Az \in C(\mathbb{R}^+; X_{-1})$, so we have, in fact, $z \in C(\mathbb{R}^+; X) \cap C^1(\mathbb{R}^+; X_{-1})$, and $\dot{z}(t) = Az(t)$ (in X_{-1}) for all $t \geq 0$. Define $x(t) = (\alpha - A)^{-1} z(t)$ for $t \geq 0$. Then $x \in C(\mathbb{R}^+; X_1) \cap C^1(\mathbb{R}^+; X)$, $x(0) = x_0$, and

$$\begin{aligned} \dot{x}(t) &= (\alpha - A)^{-1} \dot{z}(t) = (\alpha - A)^{-1} Az(t) \\ &= A(\alpha - A)^{-1} z(t) = Ax(t) \end{aligned}$$

for all $t \geq 0$. Thus, by (i), $x(t) = \mathfrak{A}^t x_0$. Since x is continuously differentiable, we have (by Definition 3.1.1), $x_0 \in \mathcal{D}(A_1)$. This means that $A = A_1$, i.e, A is the generator of $\mathfrak{A}$. □

Above we have exclusively looked at L^p-well-posedness with $p < \infty$. Similar results are true in the *Reg*-well-posed case. For simplicity, we only give the analogue of Theorem 4.7.13.

Theorem 4.7.16 *Let $S = \left[\begin{smallmatrix} A\&B \\ C\&D \end{smallmatrix}\right]$ be a system node on three Banach spaces (Y, X, U). Then S is Reg-well-posed if and only if, for some (hence all) $T > 0$, there is a constant M such that all solutions $\left[\begin{smallmatrix} x \\ y \end{smallmatrix}\right]$ of the type described in Lemma 4.7.8 with $s = 0$ satisfy*

$$|x(T)|_X + \|y\|_{Reg([0,T];Y)} \le M\big(|x(0)|_X + \|u\|_{Reg([0,T];U)}\big). \tag{4.7.11}$$

Proof The necessity of (4.7.11) is easily proved (as in the L^p-case with $p < \infty$), so let us concentrate on the converse part. We begin by showing that the observation operator C of S can be extended to an operator $C_{|X} \in \mathcal{B}(X;Y)$. To see this, we take $s = 0$, $x_0 \in X_1$ and $u = 0$ in Lemma 4.7.8 to get the continuous output $y(t) = C\mathfrak{A}^t x_0$. By (4.7.11), $\sup_{0\le t\le T} |y(t)|_Y \le M|x(0)|_X$, so, in particular (take $t = 0$), $|Cx_0|_Y \le M|x(0)|_X$. This, together with the density of X_1 in X, implies that C has a (unique) extension to an operator $C_{|X} \in \mathcal{B}(X;Y)$. From this and (4.7.3) we further conclude that $C\&D$ has a (unique) extension to an operator $\left[C_{|X}\ \ D\right] \in \mathcal{B}(\left[\begin{smallmatrix} X \\ U \end{smallmatrix}\right]; Y)$. This means that, as far as the output is concerned, we are in the situation described in Theorem 4.3.4: if we can show that the operator $\mathfrak{B}u = \int_{-\infty}^0 \mathfrak{A}_{|X_{-1}}^{-s} Bu(s)\,ds$ is a *Reg*-well-posed input map for $\mathfrak{A}$, then by applying that theorem we get a *Reg*-well-posed system whose system node is the given node S (here $\mathfrak{A}$ is the semigroup generated by the main operator A of the node, and $B \in \mathcal{B}(U; X_{-1})$ is the control operator).

Arguing in the same way as we did at the beginning of the proof of Theorem 4.7.13, we find that (4.7.9) holds with L^p replaced by *Reg*, and that it suffices to show that the operator $\mathfrak{B}^0_{-T} : u \mapsto \int_{-T}^t \mathfrak{A}_{|X_{-1}}^{t-v} Bu(v)\,dv$ is a bounded linear operator from $Reg([-T,0];U)$ to X. We know that this operator is bounded from $L^1([-T,0];U)$ to X_{-1}, and by (4.7.9) with L^p replaced by *Reg*, $s = -T$, and $x(s) = 0$,

$$|\mathfrak{B}^0_{-T}u|_X \le M\|u\|_{Reg([-T,0];U)}, \tag{4.7.12}$$

whenever $u \in W^{2,1}([-T,0];U)$ with $u(-T) = 0$. This is the right inequality, but we need it for all $u \in Reg([-T,0];U)$, and not just the subset of u described above.

By the density of $W^{2,1}([-T,0];U)$ in $C([-T,0];U)$, if (4.7.12) holds for all $u \in W^{2,1}([-T,0];U)$ with $u(-T) = 0$, then it must also hold for all $u \in C([-T,0];U)$ with $u(-T) = 0$. To show that it also holds for discontinuous functions in $Reg([-T,0];U)$ (even without the limitation $u(-T) = 0$) we first study how $\mathfrak{B}^0_{-T}$ behaves when it is applied to a specific class of discontinuous 'triangular' functions. For each $n = 1, 2, 3, \ldots$, we define

$$\eta_n(s) = \begin{cases} 1 - ns, & 0 \le s \le 1/n, \\ 0, & \text{otherwise.} \end{cases}$$

We fix some $r \in [-T, 0)$ and $u \in U$, and study how $\mathfrak{B}^0_{-T}(\tau^r \eta_n u)$ behaves as $n \to \infty$. Clearly, $\tau^r \eta_n u \to 0$ in $L^1([-T, 0); U)$ as $n \to \infty$, so we know that $\mathfrak{B}^0_{-T}(\tau^r \eta_n u) \to 0$ in X_{-1} as $n \to \infty$. We claim that we have in fact convergence in X, and not just in X_{-1}. To prove this, we may, without loss of generality, take $n > -1/r$ (so that $(\tau^r \eta_n)(s) = 0$ for $s \ge 0$). Then, after a change of integration variable we get

$$\mathfrak{B}^0_{-T}(\tau^r \eta_n u) = \mathfrak{A}^{-(r+1/n)}_{|X_{-1}} \int_0^{1/n} ns\, \mathfrak{A}^s_{|X_{-1}} Bu\, ds.$$

To show that this tends to zero in X it suffices to show that the integral (belongs to X and) tends to zero in X. Let us denote this integral by x_n. Then $x_n \to 0$ in X_{-1} (since $\int_0^{1/n} ns\, ds \to 0$ as $n \to \infty$). Fix some $\alpha \in \rho(A)$, multiply x_n by $(\alpha - A_{|X_{-1}})$, and integrate by parts to show that

$$(\alpha - A_{|X_{-1}})x_n = \alpha x_n + n \int_0^{1/n} \mathfrak{A}^s_{|X_{-1}} Bu\, ds - \mathfrak{A}^{1/n}_{|X_{-1}} Bu.$$

Here the right hand side (belongs to X_{-1} and) tends to zero in X_{-1} as $n \to \infty$, so by multiplying both sides by $(\alpha - A_{|X})^{-1}$ we find that x_n (belongs to X and) tends to zero in X as $n \to \infty$. This proves our earlier claim that $\mathfrak{B}^0_{-T}(\tau^r \eta_n u)$ (belongs to X and) tends to zero in X as $n \to \infty$.

Now let us consider the case where u is a step function, vanishing on $(-\infty, -T)$, with a finite number of jump discontinuities at the points $-T \le t_1 < t_2 < \cdots < t_k < t_{k+1} = 0$ (we include the point $-T$ in this set if $u(-T) \ne 0$, and we always include the point zero). Define

$$u_n = u - \sum_{j=0}^{k} \tau^{t_k} \eta_n [u(t_k) - u(t_k-)].$$

Then each u_n is continuous on $[-T, 0]$ with $u_n(-T) = 0$, so we may apply (4.7.12) to u_n. Thus

$$|\mathfrak{B}^0_{-T} u_n|_X \le M \|u_n\|_{Reg([-T,0];U)}.$$

For $n > \max_{1 \le j \le k} 1/(t_{j+1} - t_j)$, u_n is a (continuous) piecewise linear function where at most one of the correction terms $\eta_n[u(t_k) - u(t_k-)]$ is nonzero at each point, hence its values 'interpolates' the values of the original function u, and therefore $\|u_n\|_{Reg([-T,0];U)} = \|u\|_{Reg([-T,0];U)}$. Thus, we can continue the above inequality to get

$$|\mathfrak{B}^0_{-T} u_n|_X \le M \|u\|_{Reg([-T,0];U)}.$$

By our earlier argument (applied to each of the correction terms separately), $\mathfrak{B}^0_{-T}(u - u_n)$ (belongs to X and) tends to zero in X as $n \to \infty$. We conclude that (4.7.12) holds whenever u is a step function with at most a finite number of discontinuities in $[-T, 0]$. This set of functions is dense in $Reg([-T, 0]; U)$, and therefore (4.7.12) must hold for all $u \in Reg([-T, 0]; U)$. □

As a part of the proof of Theorem 4.7.16 we essentially proved the following result.

Lemma 4.7.17 *Let* $\Sigma = \left[\begin{smallmatrix} \mathfrak{A} & \mathfrak{B} \\ \mathfrak{C} & \mathfrak{D} \end{smallmatrix}\right]$ *be a Reg-well-posed linear system on* (Y, X, U). *Then, for each* $u \in Reg_c(\mathbb{R}^-; U)$ *and* $\epsilon > 0$ *there is a function* $v \in BC_c(\overline{\mathbb{R}}^-; U)$ *such that* $\|\mathfrak{B}(u - v)\|_X \le \epsilon$.

In other words, the space $BC_c(\overline{\mathbb{R}}^-; U)$ is dense in $Reg_c(\mathbb{R}^-; U)$ in the semi-norm induced by $\mathfrak{B}$.

Proof Since *Reg* is the closure of the space of step functions (see Section A.1) and $\mathfrak{B}$ is continuous, we can find a compactly supported step function w such that $\|\mathfrak{B}(u - w)\|_X \le \epsilon/2$. By the argument used in the preceding proof, we can find a compactly supported continuous function v such that $\|\mathfrak{B}(w - v)\|_X \le \epsilon/2$. Then $\|\mathfrak{B}(w - v)\|_X \le \epsilon$. □

There are certain algebraic formulas involving the transfer function of an operator node that we shall need from time to time.

Lemma 4.7.18 *Let* $S = \left[\begin{smallmatrix} A\&B \\ C\&D \end{smallmatrix}\right]$ *be an operator node, with main operator* A, *control operator* B, *observation operator* C, *and transfer function* $\widehat{\mathfrak{D}}$, *and let* $\alpha \in \rho(A) = \rho(A_{|X})$. *Then the following claims are true.*

(i) *The operator* $\left[\begin{smallmatrix} \alpha & 0 \\ 0 & 1 \end{smallmatrix}\right] - \left[\begin{smallmatrix} A\&B \\ 0 & 0 \end{smallmatrix}\right]$ *maps* $\mathcal{D}(S)$ *one-to-one onto* $\left[\begin{smallmatrix} X \\ U \end{smallmatrix}\right]$, *its inverse is* $\left[\begin{smallmatrix} (\alpha - A)^{-1} & (\alpha - A_{|X})^{-1}B \\ 0 & 1 \end{smallmatrix}\right]$, *and the following operator identities hold in* $\mathcal{B}(\left[\begin{smallmatrix} X \\ U \end{smallmatrix}\right]; \left[\begin{smallmatrix} X \\ Y \end{smallmatrix}\right])$:

$$\begin{bmatrix} (\alpha - A)^{-1} & (\alpha - A_{|X})^{-1}B \\ C(\alpha - A)^{-1} & \widehat{\mathfrak{D}}(\alpha) \end{bmatrix} \tag{4.7.13}$$

$$= \begin{bmatrix} 1 & 0 \\ C\&D \end{bmatrix} \left(\begin{bmatrix} \alpha & 0 \\ 0 & 1 \end{bmatrix} - \begin{bmatrix} A\&B \\ 0 \quad 0 \end{bmatrix} \right)^{-1} \tag{4.7.14}$$

$$= \begin{bmatrix} 1 & 0 \\ C\&D \end{bmatrix} \begin{bmatrix} (\alpha - A)^{-1} & (\alpha - A_{|X})^{-1}B \\ 0 & 1 \end{bmatrix}. \tag{4.7.15}$$

(ii) *The operator* $\left[\begin{smallmatrix} 1 & -(\alpha - A_{|X})^{-1}B \\ 0 & 1 \end{smallmatrix}\right]$ *maps* $\mathcal{D}(S)$ *one-to-one onto* $\left[\begin{smallmatrix} X_1 \\ U \end{smallmatrix}\right]$, *its inverse is* $\left[\begin{smallmatrix} 1 & (\alpha - A_{|X})^{-1}B \\ 0 & 1 \end{smallmatrix}\right]$, *and the following operator identities hold in* $\mathcal{B}(\left[\begin{smallmatrix} X_1 \\ U \end{smallmatrix}\right]; \left[\begin{smallmatrix} X \\ Y \end{smallmatrix}\right])$:

$$\begin{bmatrix} A\&B \\ C\&D \end{bmatrix} \begin{bmatrix} 1 & (\alpha - A_{|X})^{-1}B \\ 0 & 1 \end{bmatrix} = \begin{bmatrix} A & \alpha(\alpha - A_{|X})^{-1}B \\ C & \widehat{\mathfrak{D}}(\alpha) \end{bmatrix}, \tag{4.7.16}$$

$$\begin{bmatrix} 1 & 0 \\ C\&D \end{bmatrix} \begin{bmatrix} 1 & (\alpha - A_{|X})^{-1}B \\ 0 & 1 \end{bmatrix} = \begin{bmatrix} 1 & (\alpha - A_{|X})^{-1}B \\ C & \widehat{\mathfrak{D}}(\alpha) \end{bmatrix}. \tag{4.7.17}$$

(iii) *The following operator identities hold in* $\mathcal{B}(\mathcal{D}(S); \left[\begin{smallmatrix} X \\ Y \end{smallmatrix}\right])$:

$$\begin{aligned}
\left(\begin{bmatrix} \alpha & 0 \\ 0 & 0 \end{bmatrix} - \begin{bmatrix} A\&B \\ C\&D \end{bmatrix}\right) &= \begin{bmatrix} \alpha - A & 0 \\ -C & -\widehat{\mathfrak{D}}(\alpha) \end{bmatrix} \begin{bmatrix} 1 & -(\alpha - A_{|X})^{-1}B \\ 0 & 1 \end{bmatrix} \\
&= \begin{bmatrix} 1 & 0 \\ -C(\alpha - A)^{-1} & 1 \end{bmatrix} \begin{bmatrix} \alpha - A & 0 \\ 0 & -\widehat{\mathfrak{D}}(\alpha) \end{bmatrix} \begin{bmatrix} 1 & -(\alpha - A_{|X})^{-1}B \\ 0 & 1 \end{bmatrix} \\
&= \begin{bmatrix} 1 & 0 \\ -C(\alpha - A)^{-1} & 1 \end{bmatrix} \left(\begin{bmatrix} \alpha & 0 \\ 0 & -\widehat{\mathfrak{D}}(\alpha) \end{bmatrix} - \begin{bmatrix} A\&B \\ 0 \quad 0 \end{bmatrix}\right).
\end{aligned}$$

(iv) *The operator* $\left[\begin{smallmatrix} \alpha & 0 \\ 0 & 0 \end{smallmatrix}\right] - S$ *is invertible if and only if* $\widehat{\mathfrak{D}}(\alpha)$ *is invertible, or equivalently, if and only if* $\left[\begin{smallmatrix} \alpha - A & 0 \\ -C & -\widehat{\mathfrak{D}}(\alpha) \end{smallmatrix}\right]$ *is invertible, or equivalently, if and only if* $\left[\begin{smallmatrix} \alpha & 0 \\ 0 & -\widehat{\mathfrak{D}}(\alpha) \end{smallmatrix}\right] - \left[\begin{smallmatrix} A\&B \\ 0 \quad 0 \end{smallmatrix}\right]$ *is invertible, in which case*

$$\begin{aligned}
\left(\begin{bmatrix} \alpha & 0 \\ 0 & 0 \end{bmatrix} - \begin{bmatrix} A\&B \\ C\&D \end{bmatrix}\right)^{-1} &= \begin{bmatrix} 1 & (\alpha - A_{|X})^{-1}B \\ 0 & 1 \end{bmatrix} \begin{bmatrix} \alpha - A & 0 \\ -C & -\widehat{\mathfrak{D}}(\alpha) \end{bmatrix}^{-1} \\
&= \begin{bmatrix} 1 & (\alpha - A_{|X})^{-1}B \\ 0 & 1 \end{bmatrix} \begin{bmatrix} \alpha - A & 0 \\ 0 & -\widehat{\mathfrak{D}}(\alpha) \end{bmatrix}^{-1} \begin{bmatrix} 1 & 0 \\ C(\alpha - A)^{-1} & 1 \end{bmatrix} \\
&= \left(\begin{bmatrix} \alpha & 0 \\ 0 & -\widehat{\mathfrak{D}}(\alpha) \end{bmatrix} - \begin{bmatrix} A\&B \\ 0 \quad 0 \end{bmatrix}\right)^{-1} \begin{bmatrix} 1 & 0 \\ C(\alpha - A)^{-1} & 1 \end{bmatrix} \\
&= \begin{bmatrix} (\alpha - A)^{-1} & 0 \\ 0 & 0 \end{bmatrix} - \begin{bmatrix} (\alpha - A_{|X})^{-1}B \\ 1 \end{bmatrix} [\widehat{\mathfrak{D}}(\alpha)]^{-1} \begin{bmatrix} C(\alpha - A)^{-1} & 1 \end{bmatrix}.
\end{aligned}$$

Proof The proofs of these identities are easy algebraic computations which use Lemma 4.7.3, Lemma 4.7.5, and Definition 4.7.4. □

4.8 Examples of generators

Here we compute the system nodes and the transfer functions of most of the different systems that we have encountered so far.

Example 4.8.1 *The system node S of the delay line in Example 2.3.4 (with $1 \le p < \infty$) is given by*

$$S\begin{bmatrix} x \\ u \end{bmatrix} = \begin{bmatrix} \dot{x} \\ x(0) \end{bmatrix}$$

for all $\left[\begin{smallmatrix} x \\ u \end{smallmatrix}\right] \in \mathcal{D}(S) = \left\{ \left[\begin{smallmatrix} x \\ u \end{smallmatrix}\right] \in \left[\begin{smallmatrix} W^{1,p}([0,T];U) \\ U \end{smallmatrix}\right] \mid x(T) = u \right\}$. *In particular, the observation operator is $Cx = x(0)$ for all $x \in \mathcal{D}(\tau_T) = \{u \in W^{1,p}([0,T];U) \mid u(T) = 0\}$. The transfer function is $\widehat{\mathfrak{D}}(z) = \mathrm{e}^{-Tz}$, $z \in \mathbb{C}$.*

Proof We begin with the transfer function $\widehat{\mathfrak{D}}$. The input/output map $\mathfrak{D}$ is the delay operator $\mathfrak{D}u = \tau^{-T}u$; hence by Definition 4.6.1, $\widehat{\mathfrak{D}}(z) = \mathrm{e}^{-Tz}$.

Next we try to identify $\mathcal{D}(S)$. We have $\left[\begin{smallmatrix} x \\ u \end{smallmatrix}\right] \in \mathcal{D}(S)$ iff $A_{|X}x + Bu \in X$, or equivalently, iff $(\alpha - A_{|X})^{-1}(A_{|X}x + Bu) \in \mathcal{D}(A)$ for $\alpha \in \rho(A)$. Define $z = (\alpha - A_{|X})^{-1}(A_{|X}x + Bu)$. Then

$$z = -x + \alpha(\alpha - A)^{-1}x + (\alpha - A_{|X})^{-1}Bu,$$

which by Examples 3.3.2(iv) and 4.2.5 can be written in the form (for almost all $t \in [0, T)$)

$$z(t) = -x(t) + \alpha \int_t^T \mathrm{e}^{\alpha(t-s)}x(s)\,ds + \mathrm{e}^{\alpha(t-T)}u.$$

The condition $z \in \mathcal{D}(A)$ means that $z \in W^{1,p}([0,T];U)$ and $z(T) = 0$, and by the above identity, this is equivalent to $x \in W^{1,p}([0,T];U)$ and $x(T) = u$.

To compute $S\left[\begin{smallmatrix} x \\ u \end{smallmatrix}\right]$ for $\left[\begin{smallmatrix} x \\ u \end{smallmatrix}\right] \in \mathcal{D}(S)$ we argue as follows. For all $\left[\begin{smallmatrix} x \\ u \end{smallmatrix}\right] \in \mathcal{D}(S)$ and all $\alpha \in \rho(A) = \rho(A_{|X})$, we have by Lemmas 4.7.3(viii) and 4.7.18(ii),

$$\begin{aligned} S\begin{bmatrix} x \\ u \end{bmatrix} &= S\begin{bmatrix} 1 & (\alpha - A_{|X})^{-1}B \\ 0 & 1 \end{bmatrix}\begin{bmatrix} 1 & -(\alpha - A_{|X})^{-1}B \\ 0 & 1 \end{bmatrix}\begin{bmatrix} x \\ u \end{bmatrix} \\ &= \begin{bmatrix} A & \alpha(\alpha - A_{|X})^{-1}B \\ C & \widehat{\mathfrak{D}}(\alpha) \end{bmatrix}\begin{bmatrix} x - (\alpha - A_{|X})^{-1}Bu \\ u \end{bmatrix}. \end{aligned}$$

Here $x - (\alpha - A_{|X})^{-1}Bu \in X_1$, and by Example 4.2.5, it is equal to the function $t \mapsto x(t) + \mathrm{e}^{\alpha(t-T)}u$. By Examples 3.2.3(iv) and 4.4.5,

$$\begin{bmatrix} A \\ C \end{bmatrix}\left(x - (\alpha - A_{|X})^{-1}Bu\right) = \begin{bmatrix} t \mapsto \dot{x}(t) - \alpha\mathrm{e}^{\alpha(t-T)}u \\ x(0) - \mathrm{e}^{-\alpha T}u \end{bmatrix},$$

and by Example 4.2.5,

$$\begin{bmatrix} \alpha(\alpha - A_{|X})^{-1}B \\ \widehat{\mathfrak{D}}(\alpha) \end{bmatrix} u = \begin{bmatrix} t \mapsto \alpha e^{\alpha(t-T)}u \\ e^{-\alpha T}u \end{bmatrix}.$$

Adding these two terms we get $S\left[\begin{smallmatrix} x \\ u \end{smallmatrix}\right] = \left[\begin{smallmatrix} \dot{x} \\ x(0) \end{smallmatrix}\right]$. □

Example 4.8.2 *Let $\Sigma = \left[\begin{smallmatrix} \mathfrak{A} & \mathfrak{B} \\ \mathfrak{C} & \mathfrak{D} \end{smallmatrix}\right]$ be an $L^p|Reg$-well-posed linear system on (Y, X, U) with system node $S = \left[\begin{smallmatrix} A\&B \\ C\&D \end{smallmatrix}\right]$.*

(i) *For each $\alpha \in \mathbb{C}$, the system node S_α and the transfer function $\widehat{\mathfrak{D}}_\alpha$ of the exponentially shifted system Σ_α in Example 2.3.5 are*

$$S_\alpha = \begin{bmatrix} A\&B \\ C\&D \end{bmatrix} + \begin{bmatrix} \alpha & 0 \\ 0 & 0 \end{bmatrix}, \qquad \widehat{\mathfrak{D}}_\alpha(z) = \widehat{\mathfrak{D}}(z - \alpha).$$

(ii) *For each $\lambda > 0$, the system node S_λ and the transfer function $\widehat{\mathfrak{D}}_\lambda$ of the time compressed system Σ_λ in Example 2.3.6 are*

$$S_\lambda = \begin{bmatrix} \lambda[A\&B] \\ C\&D \end{bmatrix}, \qquad \widehat{\mathfrak{D}}_\lambda(z) = \widehat{\mathfrak{D}}(z/\lambda).$$

(iii) *For each (boundedly) invertible $E \in \mathcal{B}(X_1; X)$, the system node S_E and the transfer function $\widehat{\mathfrak{D}}_E$ of the similarity transformed system Σ_E in Example 2.3.7 are*

$$S_E = \begin{bmatrix} E^{-1} & 0 \\ 0 & 1 \end{bmatrix} \begin{bmatrix} A\&B \\ C\&D \end{bmatrix} \begin{bmatrix} E & 0 \\ 0 & 1 \end{bmatrix}, \qquad \widehat{\mathfrak{D}}_E(z) = \widehat{\mathfrak{D}}(z).$$

Once more we leave the proof to the reader. We also leave the (easy) formulations and the proofs of the corresponding results for Examples 2.3.10–2.3.13 to the reader (see Example 5.1.17).

The generator of the Lax–Phillips model can be characterized as follows (for simplicity we restrict our attention to the case $1 \le p < \infty$):

Theorem 4.8.3 *Let $1 \le p < \infty$ and $\omega \in \mathbb{R}$, let $\Sigma = \left[\begin{smallmatrix} \mathfrak{A} & \mathfrak{B} \\ \mathfrak{C} & \mathfrak{D} \end{smallmatrix}\right]$ be an L^p-well-posed linear system on (Y, X, U) with system node $S = \left[\begin{smallmatrix} A\&B \\ C\&D \end{smallmatrix}\right]$, and let T be the generator of the corresponding Lax–Phillips model $\mathfrak{T}$ of type L^p_ω defined in Definition 2.7.2.*

(i) *The domain of T consists of all the vectors $\left[\begin{smallmatrix} y_0 \\ x_0 \\ u_0 \end{smallmatrix}\right] \in \left[\begin{smallmatrix} W^{1,p}_\omega(\overline{\mathbb{R}}^-;Y) \\ X \\ W^{1,p}_\omega(\overline{\mathbb{R}}^+;U) \end{smallmatrix}\right]$ which satisfy $\left[\begin{smallmatrix} x_0 \\ u_0(0) \end{smallmatrix}\right] \in \mathcal{D}(S)$ and $y_0(0) = C\&D\left[\begin{smallmatrix} x_0 \\ u_0(0) \end{smallmatrix}\right]$, and on its domain T is given by*

$$T\begin{bmatrix} y_0 \\ x_0 \\ u_0 \end{bmatrix} = \begin{bmatrix} \dot{y}_0 \\ A\&B\left[\begin{smallmatrix} x_0 \\ u_0(0) \end{smallmatrix}\right] \\ \dot{u}_0 \end{bmatrix}.$$

Thus, the following three conditions are equivalent (see Definition 3.12.1 for the definitions of the Laplace transforms $\hat{u}$ and $\hat{u}_0$ of u and u_0, and of the left-sided Laplace transforms $\hat{y}$ and $\hat{y}_0$ of y and y_0):

(a) $\begin{bmatrix} y_0 \\ x_0 \\ u_0 \end{bmatrix} \in \mathcal{D}(T)$ *and* $\begin{bmatrix} y \\ x \\ u \end{bmatrix} = T \begin{bmatrix} y_0 \\ x_0 \\ u_0 \end{bmatrix}$;

(b) $y_0 \in W^{1,p}_\omega(\overline{\mathbb{R}}^-;Y)$, $x_0 \in X$, $u_0 \in W^{1,p}_\omega(\overline{\mathbb{R}}^+;U)$, $\begin{bmatrix} x_0 \\ u_0(0) \end{bmatrix} \in \mathcal{D}(S)$ *and*

$$\begin{bmatrix} x \\ y_0(0) \end{bmatrix} = S \begin{bmatrix} x_0 \\ u_0(0) \end{bmatrix}, \qquad \begin{bmatrix} y \\ u \end{bmatrix} = \begin{bmatrix} \dot{y}_0 \\ \dot{u}_0 \end{bmatrix};$$

(c) $y_0 \in W^{1,p}_\omega(\overline{\mathbb{R}}^-;Y)$, $x_0 \in X$, $u_0 \in W^{1,p}_\omega(\overline{\mathbb{R}}^+;U)$, $\begin{bmatrix} x_0 \\ u_0(0) \end{bmatrix} \in \mathcal{D}(S)$ *and*

$$\begin{aligned} &\begin{bmatrix} x \\ y_0(0) \end{bmatrix} = S \begin{bmatrix} x_0 \\ u_0(0) \end{bmatrix}, \\ &\hat{y}(z) = z\hat{y}_0(z) - y_0(0), \qquad \Re z < \omega, \\ &\hat{u}(z) = z\hat{u}_0(z) + u_0(0), \qquad \Re z > \omega. \end{aligned}$$

(ii) *The spectrum of T contains the vertical line $\{\Re\alpha = \omega\}$. A point $\alpha \in \mathbb{C}^+_\omega$ belongs to the spectrum of T iff it belongs to the spectrum of A, and a point $\alpha \in \mathbb{C}^-_\omega$ belongs to the spectrum of T iff $\begin{bmatrix} \alpha & 0 \\ 0 & 0 \end{bmatrix} - S$ is not invertible.*

(iii) *Let $\alpha \in \rho(T) \cap \mathbb{C}^+_\omega$ and let $\begin{bmatrix} y \\ x \\ u \end{bmatrix} \in \begin{bmatrix} L^2_\omega(\mathbb{R}^-;Y) \\ X \\ L^2_\omega(\mathbb{R}^+;U) \end{bmatrix}$. Then the following three conditions are equivalent:*

(a) $\begin{bmatrix} y_0 \\ x_0 \\ u_0 \end{bmatrix} = (\alpha - T)^{-1} \begin{bmatrix} y \\ x \\ u \end{bmatrix}$;

(b)
$$\begin{cases} \begin{bmatrix} x_0 \\ y_0(0) \end{bmatrix} = \begin{bmatrix} (\alpha - A)^{-1} & (\alpha - A_{|X})^{-1}B \\ C(\alpha - A)^{-1} & \widehat{\mathfrak{D}}(\alpha) \end{bmatrix} \begin{bmatrix} x \\ \hat{u}(\alpha) \end{bmatrix}, \\ y_0(t) = e^{\alpha t} y_0(0) + \int_t^0 e^{\alpha(t-s)} y(s)\,ds, \qquad t \le 0, \\ u_0(t) = \int_t^\infty e^{\alpha(t-s)} u(s)\,ds, \qquad t \ge 0; \end{cases}$$

(c)
$$\begin{cases} \begin{bmatrix} x_0 \\ y_0(0) \end{bmatrix} = \begin{bmatrix} (\alpha - A)^{-1} & (\alpha - A_{|X})^{-1}B \\ C(\alpha - A)^{-1} & \widehat{\mathfrak{D}}(\alpha) \end{bmatrix} \begin{bmatrix} x \\ \hat{u}(\alpha) \end{bmatrix}, \\ \hat{y}_0(z) = \dfrac{\hat{y}(z) + y_0(0)}{\alpha - z}, \qquad \Re z < \omega, \\ \hat{u}_0(z) = \dfrac{\hat{u}(z) - \hat{u}(\alpha)}{\alpha - z}, \qquad \Re z > \omega. \end{cases}$$

(iv) *Let $\alpha \in \rho(T) \cap \mathbb{C}^-_\omega$ and let $\begin{bmatrix} y \\ x \\ u \end{bmatrix} \in \begin{bmatrix} L^2_\omega(\mathbb{R}^-;Y) \\ X \\ L^2_\omega(\mathbb{R}^+;U) \end{bmatrix}$. Then the following three* *conditions are equivalent:*

(a) $\begin{bmatrix} y_0 \\ x_0 \\ u_0 \end{bmatrix} = (\alpha - T)^{-1} \begin{bmatrix} y \\ x \\ u \end{bmatrix}$;

(b)
$$\begin{cases} \begin{bmatrix} x_0 \\ u_0(0) \end{bmatrix} = \left(\begin{bmatrix} \alpha & 0 \\ 0 & 0 \end{bmatrix} - S \right)^{-1} \begin{bmatrix} x \\ \hat{y}(\alpha) \end{bmatrix}, \\ y_0(t) = -\int_{-\infty}^t e^{\alpha(t-s)} y(s)\,ds, \qquad t \le 0 \\ u_0(t) = e^{\alpha t} u_0(0) - \int_0^t e^{\alpha(t-s)} u(s)\,ds, \qquad t \ge 0; \end{cases}$$

(c)
$$\begin{cases} \begin{bmatrix} x_0 \\ u_0(0) \end{bmatrix} = \left(\begin{bmatrix} \alpha & 0 \\ 0 & 0 \end{bmatrix} - S \right)^{-1} \begin{bmatrix} x \\ \hat{y}(\alpha) \end{bmatrix}, \\ \hat{y}_0(z) = \dfrac{\hat{y}(z) - \hat{y}(\alpha)}{\alpha - z}, \qquad \Re z < \omega, \\ \hat{u}_0(z) = \dfrac{\hat{u}(z) + u_0(0)}{\alpha - z}, \qquad \Re z > \omega. \end{cases}$$

Proof (i) Let $\begin{bmatrix} y_0 \\ x_0 \\ u_0 \end{bmatrix} \in \begin{bmatrix} Y \\ X \\ U \end{bmatrix}$, and let $\begin{bmatrix} y_t \\ x_t \\ u_t \end{bmatrix} = \mathfrak{T}^t \begin{bmatrix} y_0 \\ x_0 \\ u_0 \end{bmatrix}$. By the definition of the Lax–Phillips model given in Definition 2.7.2 and by Example 3.2.3(ii), the limit $\lim_{t\downarrow 0} \frac{1}{t}(u_t - u_0)$ exists in $L^p(\mathbb{R}^+; U)$ if and only if $u_0 \in W^{1,p}(\overline{\mathbb{R}}^+; U)$, in which case this limit is equal to $\dot{u}$. Suppose that this is true. Then, by Corollary 4.3.8, x_t is continuously differentiable in X if and only if $A_{|X} x_0 + B u_0(0) \in X$. Suppose that this is the case. Then, according to Theorem 4.6.11, the output z of Σ belongs to $W^{1,p}_{\mathrm{loc}}(\overline{\mathbb{R}}^+; Y)$ and satisfies $z(t) = C\&D \begin{bmatrix} x_t \\ u_0(t) \end{bmatrix} = C\&D \begin{bmatrix} x_t \\ u_t(0) \end{bmatrix}$ for $t \ge 0$. The y-component of the Lax–Phillips semigroup consists of two parts:

$$y_t(s) = \begin{cases} y_0(t+s), & s < -t, \\ z(s+t), & -t \le s < 0. \end{cases}$$

This implies (cf. Example 3.2.3 and its proof) that the limit $\lim_{t\downarrow 0} \frac{1}{t}(y_t - y_0)$ exists in $L^p_\omega(\mathbb{R}^-; Y)$ if and only if $y_0 \in W^{1,p}_\omega(\overline{\mathbb{R}}^-; Y)$ and $y_0(0) = z(0) = C\&D \begin{bmatrix} x_0 \\ u_0(0) \end{bmatrix}$, and that this limit is equal to $\dot{y}$ in this case. To get the formulas involving the Laplace transforms of u and u_0 and the left-sided Laplace transforms of y and y_0 it suffices to integrate by parts in the definitions of the Laplace transforms.

(ii)–(iv) Let $\alpha \in \mathbb{C}$, $\begin{bmatrix} y_0 \\ x_0 \\ u_0 \end{bmatrix} \in \mathcal{D}(T)$, and $(\alpha - T)\begin{bmatrix} y_0 \\ x_0 \\ u_0 \end{bmatrix} = \begin{bmatrix} y \\ x \\ u \end{bmatrix}$. By (i), this is equivalent to the following conditions: $y_0 \in W^{1,p}_\omega(\overline{\mathbb{R}}^-; Y)$, $u_0 \in W^{1,p}_\omega(\overline{\mathbb{R}}^+; U)$,

$\left[\begin{smallmatrix} x_0 \\ u_0(0) \end{smallmatrix}\right] \in \mathcal{D}(S)$, and

$$\begin{aligned} \alpha y_0 - \dot{y}_0 &= y, \\ \left(\begin{bmatrix} \alpha & 0 \\ C\&D \end{bmatrix} - \begin{bmatrix} A\&B \\ 0 \quad 0 \end{bmatrix}\right) \begin{bmatrix} x_0 \\ u_0(0) \end{bmatrix} &= \begin{bmatrix} x \\ y_0(0) \end{bmatrix}, \\ \alpha u_0 - \dot{u}_0 &= u. \end{aligned} \tag{4.8.1}$$

By Examples 3.2.3(ii) and 3.3.1(ii), if $\Re\alpha = \omega$, then the range of the operator $u_0 \mapsto \alpha u_0 - \dot{u}_0$ in the last equation in (4.8.1) is not all of $L^p_\omega(\mathbb{R}^+; U)$. Thus, the line $\{\Re\alpha = \omega\}$ belongs to the spectrum of T (and we have proved the first statement in (ii)).

(ii)&(iii) If $\Re\alpha > \omega$, then the the last equation in (4.8.1) has the unique solution u_0 given in the last equation in (iii)(b). Observe that $u_0(0) = \hat{u}(\alpha)$. We substitute this value in the equation $\alpha x_0 - A\&B \left[\begin{smallmatrix} x_0 \\ u_0(0) \end{smallmatrix}\right] = x$ (which is part of the middle equation in (4.8.1)), and observe that the resulting equation has a unique solution x_0 for all $x \in X$ if and only if $\alpha \in \rho(A) = \rho(A_{|X})$. When this is the case, we can solve $\left[\begin{smallmatrix} x_0 \\ y_0(0) \end{smallmatrix}\right]$ in terms of $\left[\begin{smallmatrix} x \\ u_0(0) \end{smallmatrix}\right]$ from the middle equation in (4.8.1), and the result is the one given in formula (iii)(b) (see also Lemma 4.7.18(i)). Once $y_0(0)$ is known we can use the variation of constants formula in the first equation in (4.8.1) to get the second equation in (iii)(b). Thus, we conclude that, for $\Re\alpha > \omega$, $\alpha \in \rho(T)$ if and only if $\alpha \in \rho(A)$, and that (iii) holds. (We leave the proof of the fact that (iii)(b) $\Leftrightarrow$ (iii)(c) to the reader.)

(ii)&(iv) If $\Re\alpha < \omega$, then the first equation in (4.8.1) has the unique solution y_0 given in the second equation in (iv)(b). Observe that $y_0(0) = -\hat{y}(\alpha)$. The middle equation in (4.8.1) has a unique solution $\left[\begin{smallmatrix} x_0 \\ u_0(0) \end{smallmatrix}\right]$ in terms of $\left[\begin{smallmatrix} x \\ \hat{y}(\alpha) \end{smallmatrix}\right]$ if and only if $\left[\begin{smallmatrix} \alpha & 0 \\ 0 & 0 \end{smallmatrix}\right] - S$ is invertible, and the solution is, in that case, given by the first equation in (iv)(b). Once $u_0(0)$ is known we can use the variation of constants formula in the last equation in (4.8.1) to get the last equation in (iii)(b). Thus, we conclude that, for $\Re\alpha < \omega$, $\alpha \in \rho(T)$ if and only if $\left[\begin{smallmatrix} \alpha & 0 \\ 0 & 0 \end{smallmatrix}\right] - S$ is invertible, and that (iv) holds. (We again leave the proof of the fact that (iv)(b) $\Leftrightarrow$ (iv)(c) to the reader.) □

4.9 Diagonal and normal systems

By a *diagonal system* we understand an $L^p|Reg$-well-posed linear system on (Y, X, U) with a diagonal or normal semigroup on the Hilbert space X (which is separable in the diagonal case). We discussed this type of semigroup in Section 3.3 (see Definition 3.3.4 and Example 3.3.6), but we have not yet looked at full systems of this type.

We begin with the notationally simpler diagonal case. Our first task is to develop representation formulas in terms of the spectral resolution of the semigroup generator for the input map, output map, and input/output map.

We introduce the same notation as in Example 3.3.3. We let X be a separable Hilbert space spanned by the orthonormal basis $\{\phi_n\}_{n=1}^{\infty}$. To simplify some of the formulas below we introduce the notation

$$y^*x = \langle x, y\rangle_X$$

(thus, y^* can be interpreted as the bounded linear operator in X^* which maps x in to $\langle x, y\rangle$). Then every vector $x \in X$ can be written in the form

$$x = \sum_{n=1}^{\infty} \phi_n\phi_n^*x = \sum_{n=1}^{\infty} x_n\phi_n, \qquad x_n = \phi_n^*x = \langle x, \phi_n\rangle_X,$$

where $\phi_n\phi_n^*$ is the orthogonal projection onto the one-dimensional subspace spanned by ϕ_n. We call $\{x_n\}_{n=1}^{\infty}$ the *coordinates* of x in the basis $\{\phi_n\}_{n=1}^{\infty}$. The inner product of two arbitrary vectors in X can be written in the form

$$\langle x, y\rangle_X = y^*x = \sum_{n=1}^{\infty} x_n\overline{y}_n, \qquad x_n = \phi_n^*x, \qquad y_n = \phi_n^*y,$$

We let

$$Ax = \sum_{n=1}^{\infty} \lambda_n\phi_n\phi_n^*x$$

be the generator of a diagonal semigroup $\mathfrak{A}$ with eigenvalues λ_n and eigenvectors ϕ_n (see Examples 3.3.3 and 3.3.5), and let $X_1 = \mathcal{D}(A)$. By Example 3.3.5, the vectors $x \in X_1$ are characterized by the fact that their coordinates x_n in the basis ϕ_n satisfy

$$\|x\|_{X_1}^2 = \sum_{n=1}^{\infty}(1 + |\lambda_n|^2)|x_n|^2 < \infty, \qquad x_n = \phi_n^*x, \qquad x \in X_1.$$

Let us define the operator $\Lambda \in \mathcal{B}(X_1; X)$ by

$$\Lambda x = \sum_{n=1}^{\infty}(1 + |\lambda_n|^2)^{1/2}\phi_n\phi_n^*x = \sum_{n=1}^{\infty}(1 + |\lambda_n|^2)^{1/2}x_n\phi_n. \tag{4.9.1}$$

Then Λ is closed and self-adjoint as an operator in X, and it has a bounded inverse

$$\Lambda^{-1}x = \sum_{n=1}^{\infty}(1 + |\lambda_n|^2)^{-1/2}\phi_n\phi_n^*x = \sum_{n=1}^{\infty}(1 + |\lambda_n|^2)^{-1/2}x_n\phi_n.$$

Obviously, Λ is an isometric isomorphism of X_1 onto X, i.e., it maps X_1 onto X, and

$$\|\Lambda x\|_X = \|x\|_{X_1}, \quad x \in X_1.$$

Thus X_1 is a Hilbert space with inner product

$$\langle x, y\rangle_{X_1} = \langle \Lambda x, \Lambda y\rangle_X = \sum_{n=1}^{\infty}(1 + |\lambda_n|^2)x_n\overline{y}_n, \quad x_n = \phi_n^* x, \quad y_n = \phi_n^* y.$$

In particular,

$$\|\phi_n\|_{X_1} = (1 + |\lambda_n|^2)^{1/2},$$

and $\{\phi_n\}_{n=1}^{\infty}$ is an orthogonal (but not orthonormal) basis in X_1. The standard formula for the expansion of a vector in terms of this basis in X_1 is

$$\begin{aligned} x &= \sum_{n=1}^{\infty} \frac{\langle x, \phi_n\rangle_{X_1}}{\|\phi_n\|_{X_1}^2}\phi_n = \sum_{n=1}^{\infty} \frac{(1 + |\lambda_n|^2)\langle x, \phi_n\rangle_X}{(1 + |\lambda_n|^2)}\phi_n \\ &= \sum_{n=1}^{\infty}\langle x, \phi_n\rangle_X \phi_n = \sum_{n=1}^{\infty} \phi_n\phi_n^* x. \end{aligned}$$

Thus, we get the same result if we expand $x \in X_1$ with respect to the orthonormal basis $\{\phi_n\}_{n=1}^{\infty}$ in X, or with respect to the same orthogonal (but not orthonormal) basis in X_1.

We proceed to construct the space X_{-1} in the same way as we did in Section 3.6, but with $\alpha - A$ replaced by Λ, i.e., we let X_{-1} be the completion of X with the weaker norm

$$\|x\|_{X_{-1}}^2 = \|\Lambda^{-1}x\|_X^2 = \sum_{n=1}^{\infty}(1 + |\lambda_n|^2)^{-1}|x_n|^2, \quad x_n = \phi_n^* x.$$

This norm is induced by the inner product

$$\langle x, y\rangle_{X_{-1}} = \langle \Lambda^{-1}x, \Lambda^{-1}y\rangle_X = \sum_{n=1}^{\infty}(1 + \lambda_n^2)^{-1}x_n\overline{y}_n.$$

In particular, X_{-1} is a Hilbert space, and Λ can be extended to an isometric isomorphism of X onto X_{-1}. The sequence $\{\phi_n\}_{n=1}^{\infty}$ is still an orthogonal basis in X_1, but it is not orthonormal since

$$\|\phi_n\|_{X_{-1}} = (1 + |\lambda_n|^2)^{-1/2}.$$

Thus, vectors $x \in X_{-1}$ are characterized by the fact that they can be written in the form $x = \sum_{n=1}^{\infty} x_n\phi_n$, where

$$\|x\|_{X_{-1}}^2 = \sum_{n=1}^{\infty}(1 + |\lambda_n|^2)^{-1}|x_n|^2 < \infty, \quad x \in X_{-1}.$$

Here the coefficients x_n are given by

$$x_n = \frac{\langle x, \phi_n \rangle_{X_{-1}}}{\|\phi_n\|^2_{X_{-1}}} = (1 + |\lambda_n|^2)\langle x, \phi_n \rangle_{X_{-1}},$$

and they are equal to $\phi_n^* x$ whenever $x \in X$.

As X is a Hilbert space, we can identify X^* with X (every $y^* \in X^*$ is of the form $x \mapsto y^* x = \langle x, y \rangle_X = \sum_{n=1}^{\infty} x_n \overline{y}_n$ for some $y \in X$). It is also possible to identify the dual of X_{-1} with X_{-1} and the dual of X_1 with X_1, but it is more convenient to identify the dual of X_{-1} with X_1, arguing as follows. Every $y^* \in (X_{-1})^*$ is of the form $x \mapsto \langle x, y_1 \rangle_{X_{-1}}$ for some $y_1 \in X_{-1}$. For $x \in X$ we can write this as (since Λ^{-1} is self-adjoint in X)

$$\langle x, y \rangle_{X_{-1}} = \langle \Lambda^{-1} x, \Lambda^{-1} y_1 \rangle_X = \langle x, \Lambda^{-2} y_1 \rangle_X = \langle x, y \rangle_X,$$

where $y = \Lambda^{-2} y_1 \in X_1$. Thus, to every $y^* \in (X_{-1})^*$ there corresponds a unique $y \in X_1$ such that

$$y^* x = \langle x, y \rangle_X, \quad x \in X,$$

and this induces a duality pairing

$$y^* x = \langle x, y \rangle_{(X_{-1}, X_1)} = \sum_{n=1}^{\infty} x_n \overline{y}_n, \quad x \in X_{-1}, \quad y \in X_1$$

between X_{-1} and X_1. The use of the same expression $y^* x$ to represent both $\langle x, y \rangle_X$ and $\langle x, y \rangle_{(X_{-1}, X_1)}$, depending on the context, is possible since the actual formula for its computation, $y^* x = \sum_{n=1}^{\infty} x_n \overline{y}_n$, is the same in both cases. (The construction presented above is comparable to the one in Remark 3.6.1 if we identify X^* with X and use the fact that Λ is self-adjoint, hence $\mathcal{D}(\Lambda) = \mathcal{D}(\Lambda)^*$.)

If the diagonal semigroup $\mathfrak{A}$ is part of an $L^p|Reg$-well-posed linear system $\left[\begin{smallmatrix} \mathfrak{A} & \mathfrak{B} \\ \mathfrak{C} & \mathfrak{D} \end{smallmatrix}\right]$ on (Y, X, U), then we can develop representations based on the eigenvectors and eigenvalues of the generator A of $\mathfrak{A}$ for all the different operators appearing in the theory. In our representation for the input operator $\mathfrak{B}$ we need the notion of a *left-sided Laplace transform*, introduced in Definition 3.12.1.

Theorem 4.9.1 *Let $\Sigma = \left[\begin{smallmatrix} \mathfrak{A} & \mathfrak{B} \\ \mathfrak{C} & \mathfrak{D} \end{smallmatrix}\right]$ be an $L^p|Reg$-well-posed linear system on (Y, X, U), where $\mathfrak{A}$ is a diagonal semigroup on the separable Hilbert space X generated by the operator A with eigenvalues λ_n and eigenvectors ϕ_n, $n = 1, 2, 3, \ldots$ Let B be the control operator, C the observation operator, $C\&D$ the combined observation/feedthrough operator, and $\widehat{\mathfrak{D}}$ the transfer function of Σ, and define $C_n = C\phi_n$ and $B_n = \phi_n^* B$, $n = 1, 2, 3, \ldots$ Denote the growth bound of $\mathfrak{A}$ by $\omega_{\mathfrak{A}}$, and let $\omega > \omega_{\mathfrak{A}}$ and $\alpha \in \rho(A)$. Then the following representation formulas are valid (all the sums converge in the strong topology in the given spaces):*

(i) $Ax = \sum_{n=1}^{\infty} \lambda_n \phi_n x_n$, where $x_n = \phi_n^* x$ and the sum converges in X for every $x \in X_1$;

(ii) $(\alpha - A)^{-1} x = \sum_{n=1}^{\infty} (\alpha - \lambda_n)^{-1} \phi_n x_n$, where $x_n = \phi_n^* x$ and the sum converges in X_1 for every $x \in X$;

(iii) $Bu = \sum_{n=1}^{\infty} \phi_n B_n u$, where the sum converges in X_{-1} for every $u \in U$;

(iv) $Cx = \sum_{n=1}^{\infty} C_n x_n$, where the sum converges in Y for every $x \in X_1$;

(v) $C\&D \left[\begin{smallmatrix} x \\ u \end{smallmatrix}\right] = \sum_{n=1}^{\infty} C_n [x_n - (\alpha - \lambda_n)^{-1} B_n u] + \widehat{\mathfrak{D}}(\alpha) u$, where $x_n = \phi_n^* x$ and the sum converges in Y for every $\left[\begin{smallmatrix} x \\ u \end{smallmatrix}\right] \in \mathcal{D}(C\&D)$ and $\alpha \in \mathbb{C}^+_{\omega_{\mathfrak{A}}}$;

(vi) $\widehat{\mathfrak{D}}(z) u = \sum_{n=1}^{\infty} C_n \big((z - \lambda_n)^{-1} - (\alpha - \lambda_n)^{-1} \big) B_n u + \widehat{\mathfrak{D}}(\alpha) u = \sum_{n=1}^{\infty} C_n (\alpha - \lambda_n)^{-1} (\alpha - z)(z - \lambda_n)^{-1} B_n u + \widehat{\mathfrak{D}}(\alpha) u$, where the sums converge in Y for every $u \in U$ and $z \in \rho(A)$;

(vii) $\mathfrak{A}^t x = \sum_{n=1}^{\infty} \mathrm{e}^{\lambda_n t} \phi_n x_n$ where $x_n = \phi_n^* x$ and the sum converges in X for every $x \in X$, uniformly in t on any bounded interval;

(viii) $\mathfrak{B} u = \sum_{n=1}^{\infty} \phi_n B_n \hat{u}(\lambda_n)$, where $\hat{u}$ is the left-sided Laplace transform of u, and the sum converges in X for every $u \in L^p|Reg_\omega(\mathbb{R}^-; U)$ with $\omega > \omega_{\mathfrak{A}}$;

(ix) $(\mathfrak{C} x)(t) = \sum_{n=1}^{\infty} \mathrm{e}^{\lambda_n t} C_n x_n$, where $x_n = \phi_n^* x$, and the sum converges in $L^p|Reg_\omega(\overline{\mathbb{R}}^+; Y)$ for every $x \in X$ and in $BC_{0,\omega}(\overline{\mathbb{R}}^+; Y)$ for every $x \in X_1$;

(x) $(\mathfrak{D} u)(t) = \sum_{n=1}^{\infty} C_n B_n \Big(\int_{-\infty}^{t} \mathrm{e}^{(\lambda_n (t-s)} u(s)\, ds - (\alpha - \lambda_n)^{-1} u(t) \Big) + \widehat{\mathfrak{D}}(\alpha) u(t) = \sum_{n=1}^{\infty} C_n B_n \int_{-\infty}^{0} \mathrm{e}^{-\lambda_n s} \big(u(s + t) - \mathrm{e}^{\alpha s} u(t) \big)\, ds + \widehat{\mathfrak{D}}(\alpha) u(t)$, where the sums converge in $BC_{\omega,\mathrm{loc}}(\mathbb{R}; Y)$ for all $u \in W^{1,p}_{\omega,\mathrm{loc}}(\mathbb{R}; U)$ if Σ is L^p-well-posed with $p < \infty$, and for all $u \in BC_{\omega,\mathrm{loc}}(\mathbb{R}; U)$ if Σ is Reg-well-posed.

Further size estimates on the sequences C_n and B_n (which together with the eigenvectors ϕ_n determine B and C) will be given in Section 10.6.

Proof (i)–(ii) See Examples 3.3.3 and 3.3.5.

(iii) For each $x \in X_{-1}$, the sum $\sum_{n=1}^{\infty} \phi_n \phi_n^* x$ converges to x in X_{-1}. We apply this to $x = Bu$ to get, for all $u \in U$,

$$Bu = \sum_{n=1}^{\infty} \phi_n \phi_n^* Bu = \sum_{n=1}^{\infty} \phi_n B_n u,$$

where $B_n = \phi_n^* B \in U^*$.

(iv) For each $x \in X_1$, the sum $\sum_{n=1}^{\infty} \phi_n \phi_n^* x$ converges to x in X_1. This, together with the continuity of $C \in \mathcal{B}(X_1; Y)$, implies that for each $x \in X_1$,

$$Cx = C \sum_{n=1}^{\infty} \phi_n \phi_n^* x = \sum_{n=1}^{\infty} C \phi_n \phi_n^* x = \sum_{n=1}^{\infty} C_n x_n,$$

where $C_n = C\phi_n \in Y$, $x_n = \phi_n^* x$, and the sum converges in Y.

(v) Apply (ii)–(iv) to (4.7.3).

(vi) This follows from (ii)–(iv) and Theorem 4.6.9.

(vii) See Example 3.3.3.

(viii) For each $x \in X$, the sum $\sum_{n=1}^{\infty} \phi_n \phi_n^* x$ converges to x in X. By Theorem 4.2.1(i), if $u \in L^p|Reg_{\omega}(\mathbb{R}^-; U)$, then $x = \mathfrak{B}u = \int_{-\infty}^{0} \mathfrak{A}^{-s} Bu(s)\, ds \in X$, so the preceding expansion is valid in this case. This combined with (iii) and (vii) gives

$$\begin{aligned}
\mathfrak{B}u &= \sum_{n=1}^{\infty} \phi_n \phi_n^* \int_{-\infty}^{0} \mathfrak{A}^{-s} Bu(s)\, ds = \sum_{n=1}^{\infty} \phi_n \int_{-\infty}^{0} \phi_n^* \mathfrak{A}^{-s} Bu(s)\, ds \\
&= \sum_{n=1}^{\infty} \phi_n \int_{-\infty}^{0} \mathrm{e}^{-\lambda_n s} \phi_n^* Bu(s)\, ds = \sum_{n=1}^{\infty} \phi_n B_n \int_{-\infty}^{0} \mathrm{e}^{-\lambda_n s} u(s)\, ds \\
&= \sum_{n=1}^{\infty} \phi_n B_n \hat{u}(\lambda_n),
\end{aligned}$$

where $B_n = \phi_n^* B$, $\hat{u}$ is the left-sided Laplace transform of u, and the sum still converges in X (the operator $\phi_n^* \in X^*$ is bounded and can be moved inside the integral, and the operator $B_n = \phi_n^* B \in U^*$ is bounded and can be moved out of the integral).

(ix) This follows from (iv), Theorems 2.5.4(ii),(iv) and 4.4.2, and the fact that the sum $x = \sum_{n=1}^{\infty} \phi_n \phi_n^* x$ converges in X whenever $x \in X$ and in X_1 whenever $x \in X_1$.

(x) Use (ii)–(iv) and Theorems 4.5.2 and 4.6.5.

□

Let us now turn to the more general case of a *normal* semigroup. All the results of Theorem 4.9.1 remain valid, modulo the fact that the sums must be replaced by integrals over the spectrum of A. In particular, the identity $x = \sum_{n=1}^{\infty} \phi_n \phi_n^* x$ is replaced by $x = \int_{\sigma(A)} E(d\lambda)x$, and the inner products in X_1, X, and X_{-1} are given by

$$\begin{aligned}
\langle x, y\rangle_{X_1} &= \int_{\sigma(A)} (1 + |\lambda|)^2 \langle E(d\lambda)x, y\rangle, \\
\langle x, y\rangle_{X} &= \int_{\sigma(A)} \langle E(d\lambda)x, y\rangle, \\
\langle x, y\rangle_{X_{-1}} &= \int_{\sigma(A)} (1 + |\lambda|)^{-2} \langle E(d\lambda)x, y\rangle.
\end{aligned}$$

Theorem 4.9.2 *Let* $\Sigma = \left[\begin{smallmatrix} \mathfrak{A} & \mathfrak{B} \\ \mathfrak{C} & \mathfrak{D} \end{smallmatrix}\right]$ *be an* $L^p|Reg$*-well-posed linear system on* (Y, X, U)*, where* $\mathfrak{A}$ *is a normal semigroup on the Hilbert space* X *generated by the operator* A *with spectral resolution* E *(cf. Example 3.3.6). Let* B *be the control operator,* C *the observation operator,* $C\&D$ *the combined observation/*

feedthrough operator, and $\widehat{\mathfrak{D}}$ *the transfer function of* Σ. *Denote the growth bound of* $\mathfrak{A}$ *by* $\omega_{\mathfrak{A}}$, *and let* $\omega > \omega_{\mathfrak{A}}$ *and* $\alpha \in \rho(A)$. *Then the following representation formulas are valid (all the integrals converge in the strong topology in the given spaces as limits* $\lim_{r\to\infty}\int_{\sigma(A)\cap D(r)}$, *where* $D(r) = \{\lambda \in \mathbb{C} \mid |\lambda| \le r\}$*):*

(i) $Ax = \int_{\sigma(A)} \lambda E(d\lambda)x$, *where the integral converges in* X *for every* $x \in X_1$;

(ii) $(\alpha - A)^{-1}x = \int_{\sigma(A)}(\alpha - \lambda)^{-1}E(d\lambda)x$, *where the integral converges in* X_1 *for every* $x \in X$;

(iii) $Bu = \int_{\sigma(A)} E(d\lambda)Bu$, *where the integral converges in* X_{-1} *for every* $u \in U$;

(iv) $Cx = \int_{\sigma(A)} CE(d\lambda)x$, *where the integral converges in* Y *for every* $x \in X_1$;

(v) $C\&D\left[\begin{smallmatrix} x \\ u \end{smallmatrix}\right] = \int_{\sigma(A)} CE(d\lambda)(x - (\alpha - \lambda)^{-1}Bu) + \widehat{\mathfrak{D}}(\alpha)u$, *where the integral converges in* Y *for every* $\left[\begin{smallmatrix} x \\ u \end{smallmatrix}\right] \in \mathcal{D}(C\&D)$;

(vi) $\widehat{\mathfrak{D}}(z)u = \int_{\sigma(A)} C\big((z-\lambda)^{-1} - (\alpha - \lambda)^{-1}\big)E(d\lambda)Bu + \widehat{\mathfrak{D}}(\alpha)u = \int_{\sigma(A)} C(\alpha - \lambda)^{-1}(\alpha - z)(z - \lambda)^{-1}E(d\lambda)Bu + \widehat{\mathfrak{D}}(\alpha)u$, *where the integrals converge in* Y *for every* $u \in U$ *and every* $z \in \rho(A)$;

(vii) $\mathfrak{A}^t x = \int_{\sigma(A)} e^{\lambda t} E(d\lambda)x$ *where the integral converges in* X *for every* $x \in X$, *uniformly in* t *on any bounded interval;*

(viii) $\mathfrak{B}u = \int_{\sigma(A)} E(d\lambda)B\hat{u}(\lambda)$, *where* $\hat{u}$ *is the left-sided Laplace transform of* u, *and the integral converges in* X *for every* $u \in L^p|Reg_\omega(\mathbb{R}^-;U)$;

(xi) $(\mathfrak{C}x)(t) = \int_{\sigma(A)} e^{\lambda t} CE(d\lambda)x$, *where the integral converges in* $L^p|Reg_\omega(\overline{\mathbb{R}}^+;Y)$ *for every* $x \in X$ *and in* $BC_{0,\omega}(\overline{\mathbb{R}}^+;Y)$ *for every* $x \in X_1$;

(x) $(\mathfrak{D}u)(t) = \int_{\sigma(A)} CE(d\lambda)B\Big(\int_{-\infty}^t e^{(\lambda(t-s)}u(s)\,ds - (\alpha - \lambda)^{-1}u(t)\Big) + \widehat{\mathfrak{D}}(\alpha)u(t) = \int_{\sigma(A)} CE(d\lambda)B\int_{-\infty}^0 e^{-\lambda s}\big(u(s+t) - e^{\alpha s}u(t)\big)\,ds + \widehat{\mathfrak{D}}(\alpha)u(t)$, *where the integrals converge in* $BC_{\omega,\text{loc}}(\mathbb{R};Y)$ *for all* $u \in W^{1,p}_{\omega,\text{loc}}(\mathbb{R};U)$ *if* Σ *is* L^p*-well-posed with* $p < \infty$, *and for all* $u \in BC_{\omega,\text{loc}}(\mathbb{R};U)$ *if* Σ *is Reg-well-posed.*

As we mentioned above, modulo the replacement of all sums by integrals, the proof of this theorem is identical to the proof of Theorem 4.9.1. We therefore leave this proof to the reader. (The needed integration theory is found in, e.g., Rudin (1973, Chapters 12–13).)

4.10 Decompositions of systems

In many cases an $L^p|Reg$-well-posed linear system $\Sigma = \left[\begin{smallmatrix} \mathfrak{A} & \mathfrak{B} \\ \mathfrak{C} & \mathfrak{D} \end{smallmatrix}\right]$ or a system node $S = \left[\begin{smallmatrix} A\&B \\ C\&D \end{smallmatrix}\right]$ on (Y, X, U) has a structure which makes it possible to split into smaller parts, so that the whole system can be regarded as one of the connections of its separate parts described in Examples 2.3.11–2.3.13. One such case appears when the control operator B is not injective, and $\mathcal{N}(B)$ is complemented in the input space U (which is always true if U is a Hilbert space).

Proposition 4.10.1 *Let $S := \left[\begin{smallmatrix} A\&B \\ C\&D \end{smallmatrix}\right]$ be an operator node on (Y, X, U) with control operator B, and suppose that $\mathcal{N}(B)$ is complemented in U. Decompose U into $U = \left[\begin{smallmatrix} U_1 \\ U_0 \end{smallmatrix}\right]$, where $U_0 = \mathcal{N}(B)$ and U_1 is a complementing subspace (if U is a Hilbert space then we can take $U_1 = U_0^\perp$). This decomposition induces a (unique) decomposition of S into*

$$S = \begin{bmatrix} [A\&B]_1 & 0 \\ [C\&D]_1 & D_0 \end{bmatrix} : \begin{bmatrix} \mathcal{D}(S_1) \\ U_0 \end{bmatrix} \to \begin{bmatrix} X \\ Y \end{bmatrix}, \tag{4.10.1}$$

where $S_1 := \left[\begin{smallmatrix} [A\&B]_1 \\ [C\&D]_1 \end{smallmatrix}\right]$ is an operator node on (Y, X, U_1) and $D_0 \in \mathcal{B}(U_0; Y)$. The node S_1 has the same main operator A and the same observation operator C as S. Its control operator B_1 is injective, and it is the restriction of B to U_1, so that $B = \begin{bmatrix} B_1 & 0 \end{bmatrix}$. The transfer function $\widehat{\mathfrak{D}}$ of S has the decomposition

$$\widehat{\mathfrak{D}}(z) = \begin{bmatrix} \widehat{\mathfrak{D}}_1(z) & D_0 \end{bmatrix}, \qquad z \in \rho(A),$$

where $\widehat{\mathfrak{D}}_1$ is the transfer function of S_1. We have $\left[\begin{smallmatrix} x \\ u_1 \\ u_0 \end{smallmatrix}\right] \in \mathcal{D}(S)$ if and only if $\left[\begin{smallmatrix} x \\ u_1 \end{smallmatrix}\right] \in \mathcal{D}(S_1)$. If S is the system node of an $L^p|Reg$-well-posed linear system $\Sigma = \left[\begin{smallmatrix} \mathfrak{A} & \mathfrak{B} \\ \mathfrak{C} & \mathfrak{D} \end{smallmatrix}\right]$, then this system is decomposed into

$$\Sigma = \left[\begin{array}{c|cc} \mathfrak{A} & \mathfrak{B}_1 & 0 \\ \hline \mathfrak{C} & \mathfrak{D}_1 & D_0 \end{array}\right], \tag{4.10.2}$$

where $\Sigma_1 = \left[\begin{smallmatrix} \mathfrak{A} & \mathfrak{B}_1 \\ \mathfrak{C} & \mathfrak{D}_1 \end{smallmatrix}\right]$ is the $L^p|Reg$-well-posed linear system on (Y, X, U_1) induced by the system node S_1.

Thus, S may be interpreted as the sum junction (cf. Example 2.3.11) of S_1 and the static system D_0 (whose state space has dimension zero).

Proof Both $\left[\begin{smallmatrix} X \\ U_1 \end{smallmatrix}\right]$ and $\left[\begin{smallmatrix} 0 \\ U_0 \end{smallmatrix}\right]$ are closed subspaces of $\left[\begin{smallmatrix} X \\ U \end{smallmatrix}\right]$, and this implies that the restrictions of S to each of these subspaces is closed. Let us denote these restrictions by S_1, respectively S_0, with domains $\mathcal{D}(S_1) = \mathcal{D}(S) \cap \left[\begin{smallmatrix} X \\ U_1 \end{smallmatrix}\right]$ and $\mathcal{D}(S_0) = \mathcal{D}(S) \cap \left[\begin{smallmatrix} 0 \\ U_0 \end{smallmatrix}\right]$. Since $\mathcal{D}(S) = \left\{\left[\begin{smallmatrix} x \\ u \end{smallmatrix}\right] \in \left[\begin{smallmatrix} X \\ U \end{smallmatrix}\right] \;\middle|\; A_{|X}x + Bu \in X\right\}$ and since $Bu = 0$ for all $u \in U_0$, it is clear that $\left[\begin{smallmatrix} x \\ u_1 \\ u_0 \end{smallmatrix}\right] \in \mathcal{D}(S)$ if and only if $\left[\begin{smallmatrix} x \\ u_1 \end{smallmatrix}\right] \in \mathcal{D}(S_1)$. Thus, $\mathcal{D}(S_0) = \left[\begin{smallmatrix} 0 \\ U_0 \end{smallmatrix}\right]$, and by the closed graph theorem, $S_0 \left[\begin{smallmatrix} 0 \\ u_0 \end{smallmatrix}\right] = \left[\begin{smallmatrix} 0 \\ D_0 \end{smallmatrix}\right] u_0$ for some $D_0 \in \mathcal{B}(U_0; Y)$ and all $u \in U_0$. It is also clear from Definition 4.6.4 that S_1 is an operator node on (Y, X, U_1). The proofs of the remaining claims are easy, and they are left to the reader. □

Another case where it is possible to reduce the system is when the observation operator does not have dense range but the closure of the range is complemented in the output space Y.

Proposition 4.10.2 *Let $S := \left[\begin{smallmatrix} A\&B \\ C\&D \end{smallmatrix}\right]$ be an operator node on (Y, X, U) with observation operator C, and suppose that $\overline{\mathcal{R}(C)}$ is complemented in Y. Decompose Y into $Y = \left[\begin{smallmatrix} Y_1 \\ Y_0 \end{smallmatrix}\right]$, where $Y_1 = \overline{\mathcal{R}(C)}$ and Y_0 is a complementing subspace (if Y is a Hilbert space then we can take $Y_0 = Y_1^\perp$). This decomposition induces a (unique) decomposition of S into*

$$S = \begin{bmatrix} A\&B \\ [C\&D]_1 \\ 0 \;\; D_0 \end{bmatrix} : \begin{bmatrix} \mathcal{D}(S) \\ U \end{bmatrix} \to \begin{bmatrix} X \\ Y_1 \\ Y_0 \end{bmatrix}, \tag{4.10.3}$$

where $S_1 := \left[\begin{smallmatrix} A\&B \\ [C\&D]_1 \end{smallmatrix}\right]$ is an operator node on (Y_1, X, U) and $D_0 \in \mathcal{B}(U; Y_0)$. The node S_1 has the same domain as S, and its observation operator C_1 is C, with range space Y_1, so that C_1 has dense range. The transfer function $\widehat{\mathfrak{D}}$ of S has the decomposition

$$\widehat{\mathfrak{D}}(z) = \begin{bmatrix} \widehat{\mathfrak{D}}_1(z) \\ D_0 \end{bmatrix}, \qquad z \in \rho(A),$$

where $\widehat{\mathfrak{D}}_1$ is the transfer function of S_1. If S is the system node of an $L^p|Reg$-well-posed linear system $\Sigma = \left[\begin{smallmatrix} \mathfrak{A} & \mathfrak{B} \\ \mathfrak{C} & \mathfrak{D} \end{smallmatrix}\right]$, then this system is decomposed into

$$\Sigma = \left[\begin{array}{c|c} \mathfrak{A} & \mathfrak{B} \\ \hline \mathfrak{C} & \mathfrak{D}_1 \\ 0 & D_0 \end{array}\right], \tag{4.10.4}$$

where $\Sigma_1 = \left[\begin{smallmatrix} \mathfrak{A} & \mathfrak{B}_1 \\ \mathfrak{C} & \mathfrak{D} \end{smallmatrix}\right]$ is the $L^p|Reg$-well-posed linear system on (Y_1, X, U) induced by the system node S_1.

Thus, S may be interpreted as the T-junction (cf. Example 2.3.12) of S_1 and the static system D_0 (whose state space has dimension zero).

Proof We split S into $\left[\begin{smallmatrix} A\&B \\ [C\&D]_1 \\ [C\&D]_0 \end{smallmatrix}\right]$ in accordance with the splitting of the range space $\left[\begin{smallmatrix} X \\ Y_1 \\ Y_0 \end{smallmatrix}\right]$. Then $[C\&D]_1 \in \mathcal{B}(\mathcal{D}(S); Y_1)$, $[C\&D]_0 \in \mathcal{B}(\mathcal{D}(S); Y_0)$, and both $\left[\begin{smallmatrix} A\&B \\ [C\&D]_1 \end{smallmatrix}\right]$ and $\left[\begin{smallmatrix} A\&B \\ [C\&D]_0 \end{smallmatrix}\right]$ are operator nodes, with the same domain as S and output space Y_1, respectively Y_0. We get the observation operators of these operator nodes by splitting the original observation operator C into $\left[\begin{smallmatrix} C_1 \\ C_0 \end{smallmatrix}\right]$. Since $Y_1 = \overline{\mathcal{R}(C)}$ and Y_0 is a complementing subspace, we have $\overline{\mathcal{R}(C_1)} = Y_1$ and $C_0 = 0$. The latter fact implies that $[C\&D]_0$ has a (unique) extension to an operator $\left[0 \;\; D_0\right] \in \mathcal{B}\left(\left[\begin{smallmatrix} X \\ U_1 \end{smallmatrix}\right]; Y_0\right)$. The (easy) proofs of the remaining claims are left to the reader. $\square$

We can also combine the two preceding propositions into one.

Proposition 4.10.3 *Let $S := \left[\begin{smallmatrix} A\&B \\ C\&D \end{smallmatrix}\right]$ be an operator node on (Y, X, U) with control operator B and observation operator C, and suppose that $\mathcal{N}(B)$ is complemented in U and that $\overline{\mathcal{R}(C)}$ is complemented in Y. Decompose U and Y into $U = \left[\begin{smallmatrix} U_1 \\ U_0 \end{smallmatrix}\right]$ and $Y = \left[\begin{smallmatrix} Y_1 \\ Y_0 \end{smallmatrix}\right]$, where $U_0 = \mathcal{N}(B)$, $Y_1 = \overline{\mathcal{R}(C)}$, and U_1 and Y_0 are complementing subspaces. These decompositions induce a (unique) decomposition of S into*

$$S = \begin{bmatrix} [A\&B]_1 & 0 \\ [C\&D]_1 & D_{10} \\ 0 \; D_{01} & D_{00} \end{bmatrix} : \begin{bmatrix} \mathcal{D}(S_1) \\ U_0 \end{bmatrix} \to \begin{bmatrix} X \\ Y_1 \\ Y_0 \end{bmatrix}, \tag{4.10.5}$$

where $S_1 := \left[\begin{smallmatrix} [A\&B]_1 \\ [C\&D]_1 \end{smallmatrix}\right]$ is an operator node on (Y_1, X, U_1), $D_{01} \in \mathcal{B}(U_1; Y_0)$, $D_{10} \in \mathcal{B}(U_0; Y_1)$, and $D_{00} \in \mathcal{B}(U_0; Y_0)$. The node S_1 has the same main operator A as S. Its control operator B_1 is injective, and it is the restriction of B to U_1, so that $B = \begin{bmatrix} B_1 & 0 \end{bmatrix}$. Its observation operator C_1 is C with range space Y_1. The transfer function $\widehat{\mathfrak{D}}$ of S has the decomposition

$$\widehat{\mathfrak{D}}(z) = \begin{bmatrix} \widehat{\mathfrak{D}}_1(z) & D_{10} \\ D_{01} & D_{00} \end{bmatrix}, \qquad z \in \rho(A),$$

where $\widehat{\mathfrak{D}}_1$ is the transfer function of S_1. We have $\left[\begin{smallmatrix} x \\ u_1 \\ u_0 \end{smallmatrix}\right] \in \mathcal{D}(S)$ if and only if $\left[\begin{smallmatrix} x \\ u_1 \end{smallmatrix}\right] \in \mathcal{D}(S_1)$.

Proof Combine Propositions 4.10.1 and 4.10.2. □

So far we have only studied decompositions of systems which are based on the decomposition of the input and output spaces. A more intriguing question is to what extent it is possible to split the state space into two parts, leaving the input and output spaces intact, and to write the whole system as a parallel connection of two subsystems (cf. Example 2.3.13). For this to be possible we need to split the state space X into two complementary reducing subspaces $X = \left[\begin{smallmatrix} X_+ \\ X_- \end{smallmatrix}\right]$. One way is to use the spectral projections described in Theorems 3.14.10 and 3.14.11. More generally, let us assume that we have an arbitrary decomposition of X into two reducing subspaces X_+ and X_- of the semigroup generator A, and that $X_+ \subset \mathcal{D}(A)$.

Theorem 4.10.4 *Let $\Sigma = \left[\begin{smallmatrix} \mathfrak{A} & \mathfrak{B} \\ \mathfrak{C} & \mathfrak{D} \end{smallmatrix}\right]$ be an $L^p|Reg$-well-posed linear system on (Y, X, U) with system node $S = \left[\begin{smallmatrix} A\&B \\ C\&D \end{smallmatrix}\right]$, semigroup generator A, control operator B, observation operator C, and transfer function $\widehat{\mathfrak{D}}$. Let X_+ and X_- be a pair of reducing subspaces of A with $X_+ \subset \mathcal{D}(A)$. Then Σ is the parallel* *connection (see Example 2.3.13)*

$$\Sigma = \left[\begin{array}{cc|c} \mathfrak{A}_+ & 0 & \mathfrak{B}_+ \\ 0 & \mathfrak{A}_- & \mathfrak{B}_- \\ \hline \mathfrak{C}_+ & \mathfrak{C}_- & \mathfrak{D}_+ + \mathfrak{D}_- \end{array}\right]$$

of the two systems $\Sigma_+ = \left[\begin{array}{c|c} \mathfrak{A}_+ & \mathfrak{B}_+ \\ \hline \mathfrak{C}_+ & \mathfrak{D}_+ \end{array}\right]$ *and* $\Sigma_- = \left[\begin{array}{c|c} \mathfrak{A}_- & \mathfrak{B}_- \\ \hline \mathfrak{C}_- & \mathfrak{D}_- \end{array}\right]$ *on* (Y, X_+, U)*, respectively* (Y, X_-, U)*, which are defined as follows. Let* π *be the projection of* X *onto* X_+ *along* X_-*. Then* π *can be extended to a projection operator* $\pi_{|X_{-1}}$ *in* X_{-1} *with* $\mathcal{R}\left(\pi_{|X_{-1}}\right) = \mathcal{R}(\pi) = X_+$*. The system* Σ_+ *is generated by the system node*

$$S_+ = \begin{bmatrix} A_+ & B_+ \\ C_+ & 0 \end{bmatrix} = \begin{bmatrix} A_{|X_+} & \pi_{|X_{-1}} B \\ C_{|X_+} & 0 \end{bmatrix} \tag{4.10.6}$$

where $A_+ \in \mathcal{B}(X_+)$*,* $B_+ \in \mathcal{B}(U; X_+)$*,* $C_+ \in \mathcal{B}(X_+; Y)$*, and*

$$\widehat{\mathfrak{D}}_+(z) = C_+(z - A_+)^{-1} B_+, \qquad z \in \rho(A_+) \supset \rho(A). \tag{4.10.7}$$

In particular, Σ_+ *is Reg-well-posed and* L^p*-well-posed for all* $p \in [1, \infty]$*, and* $\mathfrak{A}_+ = \mathfrak{A}_{|X_+}$*,* $\mathfrak{B}_+ = \pi \mathfrak{B}$*, and* $\mathfrak{C}_+ = \mathfrak{C}_{|X_+}$*. The system* Σ_- *is* $L^p|$*Reg-well-posed, and it is given by*

$$\left[\begin{array}{c|c} \mathfrak{A}_- & \mathfrak{B}_- \\ \hline \mathfrak{C}_- & \mathfrak{D}_- \end{array}\right] = \left[\begin{array}{c|c} \mathfrak{A}_{|X_-} & (1 - \pi)\mathfrak{B}_- \\ \hline \mathfrak{C}_{|X_-} & \mathfrak{D} - \mathfrak{D}_+ \end{array}\right]. \tag{4.10.8}$$

Its system node is

$$S_- = \begin{bmatrix} (1 - \pi) & 0 \\ 0 & 1 \end{bmatrix} \begin{bmatrix} A\&B \\ C\&D \end{bmatrix}_{|\mathcal{D}(S_-)}. \tag{4.10.9}$$

with domain

$$\mathcal{D}(S_-) = \mathcal{D}(S) \cap \begin{bmatrix} X_- \\ U \end{bmatrix}.$$

The semigroup generator of S_- *is* $A_- = A_{|X_-}$*, the control operator is* $B_- = (1 - \pi_{|X_{-1}})B$*, the observation operator is* $C_- = C_{|X_-}$*, and the transfer function* $\widehat{\mathfrak{D}}_-$ *satisfies*

$$\widehat{\mathfrak{D}}_-(z) = \widehat{\mathfrak{D}}(z) - \widehat{\mathfrak{D}}_+(z), \qquad z \in \rho(A). \tag{4.10.10}$$

Thus, we may write Σ *and* S *in the block matrix forms*

$$\Sigma = \left[\begin{array}{cc|c} \mathfrak{A}_{|X_+} & 0 & \pi\mathfrak{B} \\ 0 & \mathfrak{A}_{|X_-} & (1-\pi)\mathfrak{B} \\ \hline \mathfrak{C}_{|X_+} & \mathfrak{C}_{|X_-} & \mathfrak{D} \end{array}\right], \tag{4.10.11}$$

$$S = \begin{bmatrix} A_{|X_+} & 0 \; B_+ \\ 0 & [A\&B]_- \\ C_{|X_+} & [C\&D]_- \end{bmatrix}, \tag{4.10.12}$$

where $B_+ = \pi_{|X_{-1}} B$ *and* $\begin{bmatrix} [A\&B]_- \\ [C\&D]_- \end{bmatrix} = S_-$.

Proof Since X_+ and X_- are reducing subspaces of A, they are also reducing subspaces of $\mathfrak{A}$, and we immediately get the decomposition of Σ given in (4.10.11). Let X_{-1-} be the analogue of the space X_{-1} defined in Section 3.6 with X replaced by X_- and A replaced by A_-. Then X_+ and X_{-1-} are reducing subspaces of $A_{|X_{-1}}$. Let $\pi_{|X_{-1}}$ be the projection onto X_+ along X_{-1-}. Then π is the restriction of $\pi_{|X_{-1}}$ to X. The decomposition of X_{-1} into $\begin{bmatrix} X_+ \\ X_{-1-} \end{bmatrix}$ induces the corresponding decomposition of S given in (4.10.12).

Define S_+ by (4.10.6). By Theorem 3.14.6, $A_+ \in \mathcal{B}(X_+)$. The operator C_+ is a closed operator defined on all of X_+, so by the closed graph theorem, $C_+ \in \mathcal{B}(X_+; Y)$. Since $\pi_{|X_{-1}} \in \mathcal{B}(X_{-1}; X_+)$, we also have $B_+ \in \mathcal{B}(U; X_+)$. Thus, S_+ generates a system $\Sigma_+ = \left[\begin{array}{c|c} \mathfrak{A}_+ & \mathfrak{B}_+ \\ \hline \mathfrak{C}_+ & \mathfrak{D}_+ \end{array}\right]$ whose semigroup $\mathfrak{A}_+$ is uniformly continuous, and which is both *Reg*-well-posed and L^p-well-posed for all $p \in [1, \infty]$ (see Proposition 2.3.1). It is easy to see that $\mathfrak{A}_+ = \mathfrak{A}_{|X_+}$, $\mathfrak{B}_+ = \pi\mathfrak{B}$, and $\mathfrak{C}_+ = \mathfrak{C}_{|X_+}$, and of course, the transfer function is given by (4.10.7).

Define Σ_- by (4.10.8). We claim that Σ_- is an $L^p|Reg$-well-posed linear system. This is not difficult to prove. Indeed, the four operators in Σ_- have the right continuity properties (since both Σ and Σ_+ are $L^p|Reg$-well-posed). Most of the algebraic conditions in Definitions 2.2.1 and 2.2.3 are obviously satisfied: $\mathfrak{A}_-$ is a C_0 semigroup on X_-, and the two intertwining conditions $\mathfrak{A}_-^t\mathfrak{B}_- = \mathfrak{B}\tau_-^t$ and $\mathfrak{C}_-\mathfrak{A}_-^t = \tau_+^t\mathfrak{C}$ hold for all $t \geq 0$. That the Hankel operator of $\mathfrak{D}_-$ is $\mathfrak{C}_-\mathfrak{B}_-$ is also easy to show:

$$\pi_+\mathfrak{D}_-\pi_- = \pi_+\mathfrak{D}\pi_- - \pi_+\mathfrak{D}_+\pi_- = \mathfrak{C}\mathfrak{B} - \mathfrak{C}_+\mathfrak{B}_+ = \mathfrak{C}\mathfrak{B} - \mathfrak{C}\pi\mathfrak{B} = \mathfrak{C}_-\mathfrak{B}_-.$$

We conclude that Σ_- is an $L^p|Reg$-well-posed linear system. The remaining claims about S_-, A_-, B_-, C_-, and $\widehat{\mathfrak{D}}_-$ are also easy to verify. □

If the space X_+ is obtained by a Riesz projection of the type described in Theorem 3.14.10, then we can say something more about the transfer function $\widehat{\mathfrak{D}}_+$ in Theorem 4.10.4.

Theorem 4.10.5 *Let $\Sigma = \left[\begin{smallmatrix} \mathfrak{A} & \mathfrak{B} \\ \mathfrak{C} & \mathfrak{D} \end{smallmatrix}\right]$ be an $L^p|Reg$-well-posed linear system on (Y, X, U) with semigroup generator A, control operator B, observation operator C, and transfer function $\widehat{\mathfrak{D}}$. Let Γ be a positively oriented piecewise continuously differentiable Jordan curve contained in $\rho(A)$ which separates $\sigma(A)$ into two nontrivial parts $\sigma(A) = \sigma_+(A) \cup \sigma_-(A)$, where $\sigma_+(A)$ lies inside Γ and $\sigma_-(A)$ lies outside Γ, and define the Riesz projection π as in Theorem 3.14.10 with $X_+ = \mathcal{R}(\pi)$ and $X_- = \mathcal{N}(\pi)$. Decompose Σ into Σ_+ and Σ_- as in Theorem 4.10.4, and let λ_0 be a point inside Γ. Then $\widehat{\mathfrak{D}}_+$ is analytic at infinity, and it has a expansion*

$$\widehat{\mathfrak{D}}_+(\lambda) = \sum_{n=1}^{\infty} D_{-n}(\lambda - \lambda_0)^{-n}, \tag{4.10.13}$$

valid in a neighborhood of infinity, where $D_{-n} \in \mathcal{B}(U; Y)$ is given by

$$D_{-n} = \frac{1}{2\pi j} \oint_{\Gamma} (\lambda - \lambda_0)^{n-1} \widehat{\mathfrak{D}}(\lambda)\, d\lambda = C_+(A_+ - \lambda_0)^{n-1} B_+. \tag{4.10.14}$$

In particular, $\widehat{\mathfrak{D}}_+$ depends only on Γ and on the transfer function $\widehat{\mathfrak{D}}$, and not on the particular realization. If $\sigma_+(A)$ consists of the single point λ_0, then the above expansion is valid in all of $\mathbb{C} \setminus \{\lambda_0\}$, and the coefficients D_{-n} coincide with the coefficients with negative index in the Laurent series of $\widehat{\mathfrak{D}}$ at λ_0.

Proof The function $\widehat{\mathfrak{D}}_+(\lambda_0 + \lambda) = C_+(\lambda_0 + \lambda - A)^{-1} B_+$ is analytic at infinity and vanishes at infinity since $A_+ \in \mathcal{B}(X_+)$. It therefore has a Taylor expansion

$$\widehat{\mathfrak{D}}(\lambda_0 + 1/z) = \sum_{n=1}^{\infty} D_{-n} z^n,$$

valid in a neighborhood of zero. After a change of variable (replace z by $1/(\lambda - \lambda_0)$) this becomes (4.10.13). The Taylor coefficients D_{-n} can be computed by means of a Cauchy integral, and after the same change of variable that we made above (cf. Lemma 3.9.1(ii) and its proof) we get

$$\begin{aligned} D_{-n} &= \frac{1}{2\pi j} \oint_{\Gamma_1} (\lambda - \lambda_0)^{n-1} \widehat{\mathfrak{D}}_+(\lambda)\, d\lambda \\ &= \frac{1}{2\pi j} \oint_{\Gamma_1} (\lambda - \lambda_0)^{n-1} C_+(\lambda - A_+)^{-1} B_+\, d\lambda \\ &= C_+(A_+ - \lambda_0)^{n-1} B_+, \end{aligned}$$

where the last line follows from the preceding line and Lemma 4.10.6. Since both $(\lambda - \lambda_0)^{n-1}$ and $\widehat{\mathfrak{D}}_-$ are analytic inside Γ, we have

$$\frac{1}{2\pi j}\oint_{\Gamma_1} \lambda^{n-1}\widehat{\mathfrak{D}}_-(\lambda)\, d\lambda = 0,$$

and therefore also the first identity in (4.10.14) holds (recall that $\widehat{\mathfrak{D}}(z) = \widehat{\mathfrak{D}}_+(z) + \widehat{\mathfrak{D}}_-(z)$ for all $z \in \rho(A)$).

If $\sigma_+(A)$ consists of the single point λ_0, then $\widehat{\mathfrak{D}}_+$ is analytic on $\mathbb{C} \setminus \{\lambda_0\}$, and therefore the expansion (4.10.13) holds in $\mathbb{C} \setminus \{\lambda_0\}$, and it coincides with the Laurent expansion of $\widehat{\mathfrak{D}}_+$ at λ_0, and also with the negative part of the Laurent expansion of $\widehat{\mathfrak{D}}$ at λ_0. □

In the above proof we used the following lemma.

Lemma 4.10.6 *Let $A \in \mathcal{B}(X)$ and let $\lambda_0 \in \mathbb{C}$. Then $(\lambda - A)^{-1}$ is analytic at infinity, and it has an expansion*

$$(\lambda - A)^{-1} = \sum_{n=1}^{\infty} A_{-n}(\lambda - \lambda_0)^{-n}, \tag{4.10.15}$$

valid in a neighborhood of infinity, where $A_{-n} \in \mathcal{B}(U; Y)$ is given by

$$A_{-n} = \frac{1}{2\pi j}\oint_{\Gamma} (\lambda - \lambda_0)^{n-1}(\lambda - A)^{-1}\, d\lambda = (A - \lambda_0)^{n-1}; \tag{4.10.16}$$

here Γ is a positively oriented piecewise continuously differentiable Jordan curve which encircles both λ_0 and $\sigma(A)$. In particular, $A_{|X} = 1$. If $\sigma(A)$ consists of the single point λ_0, then the above expansion is valid in all of $\mathbb{C} \setminus \{\lambda_0\}$, and the coefficients A_{-n} coincide with the coefficients with negative index in the Laurent expansion of $(\lambda - A)^{-1}$ at λ_0.

Proof That $(\lambda - A)^{-1}$ has an expansion of the type (4.10.15) with the coefficients A_{-n} given by the first of the two expression in (4.10.16) is proved in the same way as in the proof of Theorem 4.10.5 (take $\sigma_-(A) = \emptyset$, $B = 1$, and $C = 1$). The alternative formula for the coefficients follows from (3.9.1) (see also the discussion following (3.9.1)). □

4.11 Comments

Most of the results in this chapter are (independently) due to Salamon (1987, 1989), Smuljan (1986) (these two authors deal with the Hilbert space case with $p = 2$) and G.Weiss (1989a, b, 1991a, 1994a). Some of the results for *Reg*-well-posed linear systems appear to be new. Our proofs follow mainly those given by Salamon (1989).

Section 4.2 The main part of Theorem 4.2.1 is found in Salamon (1989, Theorem 3.1), Smuljan (1986, Theorem 2.1), and G. Weiss (1989a, Theorem 3.9). The representation of a *Reg*-well-posed input map appears to be new. An example is given in G. Weiss (1991a, Section 3) which shows that the representation in Theorem 4.2.1 need not be valid for inputs in L^∞_{loc}. Theorem 4.2.7 is due to G. Weiss (1989a, Theorem 4.8). Our proof of this theorem is simpler than the one given by G. Weiss (1989a). The estimates in Proposition 4.2.9 with $n > 1$ may be new (the case $n = 1$ is well-known). Also Theorem 4.2.11 appears to be new.

Section 4.3 Part (i) of Theorem 4.3.1 (in the Hilbert space case with $p = 2$) is due to Salamon (1989, Lemma 2.3) (and it can also be found in the PDE literature), but part (ii) appears to be new. The space $(X + BU)_1 = (\alpha - A)^{-1}(X + BU)$ appears in Salamon (1987, Section 2.2) in the case where B is injective, and in the form presented here in Staffans (1997). It can also be found in the PDE literature.

Section 4.4 The main part of Theorem 4.4.2 is due to Salamon (1987, Theorem 3.1), Smuljan (1986, Theorem 3.2), and G. Weiss (1989b, Theorem 3.3). Our proof follows the one given by Salamon (1987). Theorems 4.4.2(ii) and 4.4.7(ii) are due to G. Weiss (1989b, Proposition 6.5). Theorem 4.4.8 may be new, and also the size estimates in Proposition 4.4.9 with $n > 1$ may be new. Theorem 4.4.10 is apparently new.

Section 4.5 Theorem 4.5.2 is a simple corollary of the more general results in G. Weiss (1994a) (which will be discussed in Chapter 5). Lemma 4.5.3 is found in Lax and Phillips (1973, pp. 187–188) and G. Weiss (1991a, Theorem 2.3). Theorem 4.5.4 and Corollaries 4.5.5 and 4.5.6 appear to be new. Theorem 4.5.8 is, of course, well-known, but its proof is new.

Section 4.6 Our definition of a transfer function is the same as the one used by Lax and Phillips (1973, pp. 187–188). Theorem 4.6.3 can be deduced from, e.g., Helton (1976) or G. Weiss (1989c, Proposition 4.7). The combined observation/feedthrough operator appears explicitly in the work by Arov and Nudelman (1996) and Smuljan (1986), and less explicitly in the works by Curtain and Weiss (1989) and Salamon (1987). The representation formula in Theorem 4.6.11 is found in Helton (1976, p. 148) and Salamon (1987, formula (2.1;2)) in a rather implicit way, and more explicitly in Arov and Nudelman (1996, formula (2.1)), Curtain and Weiss (1989, formula (4.1)), Salamon (1989, formula (2.1)), and Šmuljan (1986, formulas (4.6)). Different versions of Theorem 4.6.9 are found in Arov and Nudelman (1996, formula (2.9)), Curtain and Weiss (1989, formula (4.3)), Helton (1976, formula (2.4)), and Salamon (1987, formula (2.3)), and Smuljan (1986, formula (5.6)). The Hilbert space case with $p = 2$ of the representation in Theorem 4.6.9 is well-known. The L^p-case with $1 \le p < \infty$

was proved by G. Weiss (1991a, Theorem 2.3), but the *Reg*-well-posed case may be new.

Section 4.7 Operator and system nodes of the type that we consider here seem to have appeared for the first time in Smuljan (1986), and in a less explicit form in Salamon (1987). Our notation $C\&D\left[\begin{smallmatrix} x \\ u \end{smallmatrix}\right]$ corresponds to Smuljan's notation $N\langle x, u\rangle$ and Salamon's notation $\left(x - (\alpha - A)^{-1}Bu\right) + \widehat{\mathfrak{D}}(\alpha)u$. Compare this to formula (4.6.3). System nodes have since then been used in Arov and Nudelman (1996), Malinen *et al.* (2003), (Staffans, 2001a, 2002a, b, c), Staffans and Weiss (2002, 2004). Many of the results presented in this section are found in Malinen *et al.* (2003).

Section 4.8 Theorem 4.8.3 (in a Hilbert space setting with $p = 2$) is found in Staffans and Weiss (2002, Theorem 6.3 and Proposition 6.4). Related results are found in Lax and Phillips (1973, Theorems 6.1 and 6.2), Grabowski and Callier (1996, Theorem 2.2), Engel (1998, Theorem 2), and Smuljan (1986, Section 4).

Section 4.9 Diagonal systems are often used as examples illustrating more general results, and they also arise naturally from partial differential equations on a bounded domain. A system whose generator has a Riesz basis of eigenvectors can be transformed into a diagonal system through a similarity transform (see Example 2.3.7 and our earlier comments about Section 3.3). Most of Theorem 4.9.2 is contained (more or less explicitly) in Rudin (1973, Chapter 13) (see, in particular, Rudin 1973, Theorems 13.33 and 13.37).

Section 4.10 Proposition 4.10.3 is a minor extension of Malinen *et al.* (2003, Proposition 2.5). Theorems 4.10.4 and 4.10.5 may be new (at least formally), but similar decompositions have, of course, been used before, e.g., in the finite-dimensional case.

5

Compatible and regular systems

Compatibility of a system has to do with the question of to what extent the system node $S = \left[\begin{smallmatrix} A\&B \\ C\&D \end{smallmatrix}\right]$ in the differential/algebraic equation $\left[\begin{smallmatrix} \dot{x} \\ y \end{smallmatrix}\right] = S \left[\begin{smallmatrix} x \\ u \end{smallmatrix}\right]$ can be extended to an operator of the type $\left[\begin{smallmatrix} A_{|W} & B \\ C_{|W} & D \end{smallmatrix}\right]$, defined on $\left[\begin{smallmatrix} W \\ U \end{smallmatrix}\right]$, where $\mathcal{D}(S) \subset \left[\begin{smallmatrix} W \\ U \end{smallmatrix}\right] \subset \left[\begin{smallmatrix} X \\ U \end{smallmatrix}\right]$. Typical boundary control systems are compatible. Regularity is related to the behavior of the transfer function at $+\infty$.

5.1 Compatible systems

The system nodes and operator nodes treated in Sections 4.6 and 4.7 have a natural decomposition $S = \left[\begin{smallmatrix} A\&B \\ C\&D \end{smallmatrix}\right]$, where $A\&B$ maps $\mathcal{D}(S)$ into the state space X and $C\&D$ maps $\mathcal{D}(S)$ into the output space Y. If the control operator B is bounded, i.e., if $B \in \mathcal{B}(U; X)$, then $\mathcal{D}(S) = \left[\begin{smallmatrix} X_1 \\ U \end{smallmatrix}\right]$, and S further decomposes into $S = \left[\begin{smallmatrix} A & B \\ C & D \end{smallmatrix}\right]$, where $A \in \mathcal{B}(X_1; X)$, $B \in \mathcal{B}(U; X)$, $C \in \mathcal{B}(X_1; Y)$, and $D \in \mathcal{B}(U; Y)$. In this case many of the formulas in Sections 4.6 and 4.7 simplify, and they start to look exactly like they would look in the classical finite-dimensional case.

The same simplification of S into $S = \left[\begin{smallmatrix} A & B \\ C & D \end{smallmatrix}\right]$ would be possible whenever $\mathcal{D}(S)$ is of the form $\mathcal{D}(S) = \left[\begin{smallmatrix} W \\ U \end{smallmatrix}\right]$ for some arbitrary subspace W of X. Unfortunately, such a decomposition is never valid if B is unbounded, i.e., $B \notin \mathcal{B}(U; X)$; this follows from (4.6.5). Therefore, we have to settle for a weaker condition in this case: Is it maybe possible to extend the operator node S in such a way that the extended operator is defined on $\left[\begin{smallmatrix} W \\ U \end{smallmatrix}\right]$, where $\mathcal{D}(S) \subset \left[\begin{smallmatrix} W \\ U \end{smallmatrix}\right] \subset \left[\begin{smallmatrix} X \\ U \end{smallmatrix}\right]$?

We have already seen a partial answer to the question above: the definition of an operator node contains the requirement that $A\&B$ always can be extended to an operator $\left[A_{|X} \;\; B\right]$ on $\left[\begin{smallmatrix} X \\ U \end{smallmatrix}\right]$, at the expense of also extending the range space of this operator so that it becomes X_{-1} instead of the original X. Unfortunately, a similar extension of the range space of the operator $C\&D$ is in most cases not allowed, since this space is the (often finite-dimensional) output space of the whole system.[1] Still, it may be possible to extend $C\&D$ to a bounded linear operator on some larger space $\left[\begin{smallmatrix} W \\ U \end{smallmatrix}\right]$ without changing the output space Y, where W is a Banach (or Hilbert) space. We must have $X_1 \subset W$ since $\left[\begin{smallmatrix} X_1 \\ 0 \end{smallmatrix}\right] \subset \mathcal{D}(S)$, and it is natural to require that $W \subset X$. The extension of $C\&D$ can then be written in the form $\begin{bmatrix} C_{|W} & D \end{bmatrix}$, and we get $S = \left[\begin{smallmatrix} A_{|X} & B \\ C_{|W} & D \end{smallmatrix}\right]_{|\mathcal{D}(S)}$.

Before looking further into the question of the existence of this extension of S, let us study some of its consequences. We start with a definition.

Definition 5.1.1 Let $S = \left[\begin{smallmatrix} A\&B \\ C\&D \end{smallmatrix}\right]$ be an operator node on (Y, X, U) with main operator A, control operator B, and observation operator C. Then S is *compatible* if there exist a Banach space W, with $X_1 \subset W \subset X$ (with continuous embeddings), and an operator $C_{|W} \in \mathcal{B}(W; Y)$ such that

(i) $C_{|W}x = Cx$ for all $x \in X_1$, and
(ii) $\mathcal{R}\left((\alpha - A_{|X})^{-1}B\right) \subset W$ for some $\alpha \in \rho(A)$.

We call W a *compatible extension* of X_1 and $C_{|W}$ a *compatible extension* of the observation operator C, or alternatively, an *extended observation operator*. By a compatible $L^p|Reg$-well-posed linear system Σ we mean a system whose system node is compatible.

Note that the space W in (i) is always dense in X since X_1 is dense in X, but that X_1 need not be dense in W. See Remarks 5.1.13 and 5.1.14. Therefore, *there may exist more than one extension $C_{|W}$ of C to W.*

Lemma 5.1.2 *The following operator nodes (and systems) are compatible:*[2]

(i) *Operator nodes with a bounded observation operator; in particular, nodes that are L^∞-well-posed or Reg-well-posed.*
(ii) *Operator nodes with a bounded control operator; in particular, L^1-well-posed system nodes with a reflexive state space.*
(iii) *System nodes whose control operator B and observation operator C satisfy $B \in \mathcal{B}(U; X_{\alpha-1})$ and $C \in \mathcal{B}(X_\alpha; Y)$, where $\alpha \in [0, 1]$ and X_α is defined as in Section 3.9 (U is the input space and Y the output space).*
(iv) *The delay line in Example 2.3.4.*

Proof Take $W = X$ in (i) and $W = X_1$ in (ii), and see Theorems 4.2.7 and 4.4.2(ii). Note that both of these are special cases of (iii). That (iii) holds is obvious from the standard properties of the spaces X_α (see Section 3.9). In (iv) we can take, for example, $W = W^{1,p}([0, T]; U)$ (see Example 4.8.1). □

[1] From an input/output point of view, the spaces U and Y are determined by the original problem, whereas the state space may vary from one realization to another.

[2] These systems are even *regular* whenever they are $L^p|Reg$-well-posed: see Lemma 5.7.1.

For further examples of compatible systems, see Theorems 5.1.12 and 5.2.3.

Lemma 5.1.3 *Let A be a densely defined operator on X with a nonempty resolvent set, and let W be a Banach space satisfying $X_1 \subset W \subset X$, and let $\alpha \in \rho(A)$. For each $k = 0, \pm 1, \pm 2, \dots$, define*

$$W_k = (\alpha - A)_{|W}^{-k} W = \{(\alpha - A)_{|W}^{-k} x \mid x \in W\}.$$

Then $X_{k+1} \subset W_k \subset X_k$, W_k is a Banach space with the norm

$$|x|_{W_k} = |(\alpha - A)_{|W_k}^{k} x|_W,$$

W_k is a Hilbert spaces if W is a Hilbert space, W_k is dense in X_k, and X_{k+1} is dense in W_k if and only if X_1 is dense in W. Moreover, for each $\alpha \in \rho(A)$ and k, $l = 0, \pm 1, \pm 2, \dots$, $(\alpha - A)_{|W_k}^{l} \in \mathcal{B}(W_k; W_{k-l})$. Moreover, all possible choices of $\alpha \in \rho(A)$ give the same spaces W_k.

Proof The proof of this lemma is virtually identical to the argument that we gave in Section 3.6 in our construction of the spaces X_k. □

The most important of these spaces (apart from W itself) will be $W_{-1} = (\alpha - A)_{|W} W$, $X \subset W_{-1} \subset X_{-1}$.

At the beginning of this section we were looking for an extension of the observation/feedthrough operator $C\&D$ to a larger domain of the type $\left[\begin{smallmatrix} W \\ U \end{smallmatrix}\right]$, for some Banach space W, $X_1 \subset W \subset X$, but then we continued to instead discuss extensions of the observation operator C. Obviously, if we first extend $C\&D$ to an operator on $\left[\begin{smallmatrix} W \\ U \end{smallmatrix}\right]$ and then restrict this extended operator to $\left[\begin{smallmatrix} W \\ 0 \end{smallmatrix}\right]$, then we get an extension of C. The converse statement is also true in the following sense.

Lemma 5.1.4 *Let $S = \left[\begin{smallmatrix} A\&B \\ C\&D \end{smallmatrix}\right]$ be a compatible operator node on (Y, X, U), with main operator A, control operator B, observation operator C, and extended observation operator $C_{|W}$ defined on W, $X_1 \subset W \subset X$. Let $W_{-1} = (\alpha - A)_{|W} W$ where $\alpha \in \rho(A)$ (cf. Lemma 5.1.3). Then the following claims hold.*

(i) $\mathcal{D}(S) \subset \left[\begin{smallmatrix} W \\ U \end{smallmatrix}\right] \subset \left[\begin{smallmatrix} X \\ U \end{smallmatrix}\right]$, *and*

$$S = \begin{bmatrix} A_{|W} & B \\ C_{|W} & D \end{bmatrix}_{|\mathcal{D}(S)} \tag{5.1.1}$$

where $\left[\begin{smallmatrix} A_{|W} & B \\ C_{|W} & D \end{smallmatrix}\right] \in \mathcal{B}(\left[\begin{smallmatrix} W \\ U \end{smallmatrix}\right]; \left[\begin{smallmatrix} W_{-1} \\ Y \end{smallmatrix}\right])$, *and*

$$D = \widehat{\mathfrak{D}}(\alpha) - C_{|W}(\alpha - A_{|X})^{-1} B, \quad \alpha \in \rho(A), \tag{5.1.2}$$

belongs to $\mathcal{B}(U; Y)$ and is independent of $\alpha \in \rho(A)$. In particular, $(\alpha - A_{|X})^{-1} B \in \mathcal{B}(U; W)$ for all $\alpha \in \rho(A)$, and D is determined uniquely by S and $C_{|W}$.

(ii) *The transfer function of S can be written in the form*

$$\widehat{\mathfrak{D}}(z) = C_{|W}(z - A_{|X})^{-1}B + D, \qquad z \in \rho(A). \tag{5.1.3}$$

(iii) *S is determined uniquely by A, B, $C_{|W}$, and D. If X_1 is dense in W then $C_{|W}$ and D are determined uniquely by S.*

Proof (i) That A maps W into W_{-1} and that B maps U into W_{-1} follow from Definition 5.1.1 and Lemma 5.1.3. The boundedness of these operators follows from the closed graph theorem (they are continuous with values in X_{-1}, hence closed as operators with values in W_{-1}). Thus in particular, $(\alpha - A_{|X})^{-1}B \in \mathcal{B}(U; W)$. Fix some $\alpha \in \rho(A)$, and define D by (5.1.2). Then $\begin{bmatrix} C_{|W} & D \end{bmatrix} \in \mathcal{B}(\left[\begin{smallmatrix} W \\ U \end{smallmatrix}\right]; Y)$, and by (4.7.3), $C\&D = \begin{bmatrix} C_{|W} & D \end{bmatrix}_{|\mathcal{D}(S)}$. That D is independent of α follows from (4.7.2).

(ii) This is a rewritten version of (5.1.2).

(iii) That S is determined uniquely by A, B, $C_{|W}$, and D follows from (5.1.1). Clearly, if X_1 is dense in W then $C_{|W}$ is determined uniquely by C (which is determined uniquely by S, and D is then determined uniquely by (5.1.2). □

Definition 5.1.5 We call the operator $\left[\begin{smallmatrix} A_{|W} & B \\ C_{|W} & D \end{smallmatrix}\right] \in \mathcal{B}(\left[\begin{smallmatrix} W \\ U \end{smallmatrix}\right]; \left[\begin{smallmatrix} W_{-1} \\ Y \end{smallmatrix}\right])$ in Lemma 5.1.4(i) a *compatible extension* of the operator node S, or alternatively, an *extended operator node*. We call D the *feedthrough operator induced by S and $C_{|W}$*.

Remark 5.1.6 Clearly, an extended operator node $\left[\begin{smallmatrix} A_{|W} & B \\ C_{|W} & D \end{smallmatrix}\right]$ determines the original operator node S uniquely through (5.1.1), but the converse is not true unless X_1 is dense in W. For example, in the delay line in Example 5.1.2(iv), if we take $W = W^{1,p}([0, T]; U)$, then we can take D to be an arbitrary operator in $\mathcal{B}(U)$ and let $C_{|W}\left[\begin{smallmatrix} x \\ u \end{smallmatrix}\right] = x(0) - D_W x(T)$. See also Lemma 5.1.10 and Remark 5.2.9.

For compatible $L^p|Reg$-well-posed linear systems we can in practice replace the system node S by its extension $\left[\begin{smallmatrix} A_{|W} & B \\ C_{|W} & D \end{smallmatrix}\right]$.

Corollary 5.1.7 *Let $S = \left[\begin{smallmatrix} A\&B \\ C\&D \end{smallmatrix}\right]$ be a compatible system node on (Y, X, U), and let $\left[\begin{smallmatrix} A_{|W} & B \\ C_{|W} & D \end{smallmatrix}\right] \in \mathcal{B}(\left[\begin{smallmatrix} W \\ U \end{smallmatrix}\right]; \left[\begin{smallmatrix} W_{-1} \\ Y \end{smallmatrix}\right])$ be a compatible extension of S. Then the following claims are true.*

(i) *We can replace the equation* (4.7.4) *in Lemma 4.7.8 by*

$$\begin{bmatrix} \dot{x}(t) \\ y(t) \end{bmatrix} = \begin{bmatrix} A_{|W} & B \\ C_{|W} & D \end{bmatrix} \begin{bmatrix} x(t) \\ u(t) \end{bmatrix}, \qquad t \geq s, \quad x(s) = x_s. \tag{5.1.4}$$

(ii) *If S is $L^p|Reg$-well-posed, then we can replace equations* (4.6.6) *and* (4.6.7) *in Theorem 4.6.5 by*

$$\begin{bmatrix} \dot{x}(t) \\ y(t) \end{bmatrix} = \begin{bmatrix} A_{|W} & B \\ C_{|W} & D \end{bmatrix} \begin{bmatrix} x(t) \\ u(t) \end{bmatrix}, \qquad t \in \mathbb{R}, \tag{5.1.5}$$

and we can replace formulas (4.6.13) *and* (4.6.14) *in Theorem 4.6.11 by* (5.1.4).

Proof This is obvious. □

For each compatible operator node, there is a canonical *minimal* space W, namely the space $(X + BU)_1$ introduced in Lemma 4.3.12(ii)–(iii):

Theorem 5.1.8 *Let $S = \left[\begin{smallmatrix} A\&B \\ C\&D \end{smallmatrix}\right]$ be a compatible operator node on (Y, X, U) with a compatible extension $\left[\begin{smallmatrix} A_{|W} & B \\ C_{|W} & D \end{smallmatrix}\right] \in \mathcal{B}(\left[\begin{smallmatrix} W \\ U \end{smallmatrix}\right]; \left[\begin{smallmatrix} W_{-1} \\ Y \end{smallmatrix}\right])$. Let $\alpha \in \rho(A)$, and define (cf. Lemma 4.3.12)*

$$(X + BU)_1 = (\alpha - A_{|X})^{-1}(X + BU).$$

Then $X_1 \subset (X + BU)_1 \subset W$, and if we let $C_{|(X+BU)_1}$ be the restriction of $C_{|W}$ to $(X + BU)_1$, then $C_{|(X+BU)_1}$ is a compatible extension of C.

Proof By Definition 5.1.1, W contains both $X_1 = (\alpha - A)^{-1}X$ and $(\alpha - A_{|X})^{-1}BU$, hence $(X + BU)_1 = (\alpha - A_{|X})^{-1}(X + BU) \subset W$ (and the embedding is, of course, continuous). That $X_1 \subset (X + BU)_1$ follows from the inclusion $X \subset X + BU$. Thus, the conditions of Definition 5.1.1 are satisfied if we replace W by $(X + BU)_1$. □

Corollary 5.1.9 *A operator node $S = \left[\begin{smallmatrix} A\&B \\ C\&D \end{smallmatrix}\right]$ with main operator A, control operator B, and observation operator C is compatible if and only if C can be extended to a bounded linear operator $C_{|(X+BU)_1} \in \mathcal{B}((X + BU)_1; Y)$.*

This follows from Theorem 5.1.8.

We saw in Lemma 5.1.4(i) that the operator D in that lemma is determined uniquely by A, B, $C_{|W}$, and $\widehat{\mathfrak{D}}$. In the case where $W = (X + BU)_1$ it is also true that $C_{|W}$ is determined uniquely by A, B, C, D, and $\widehat{\mathfrak{D}}$:

Lemma 5.1.10 *In the case where $W = (X + BU)_1$ we can add the following conclusion to Lemma 5.1.4(i):*

(iv) *The operator $C_{|(X+BU)_1}$ is determined uniquely by S and D.*

Proof Every $w \in (X + BU)_1$ can be written in the form $w = x_1 + (\alpha - A_{|X})^{-1}Bu$, where $x_1 \in X_1$, $u \in U$, and $\alpha \in \rho(A)$. Thus, $C_{|(X+BU)_1}w = Cx_1 + C_{|(X+BU)_1}(\alpha - A_{|X})^{-1}Bu$, and to show that $C_{|(X+BU)_1}$ is unique it suffices to show that $C_{|(X+BU)_1}$ is unique on $\mathcal{R}\big((\alpha - A_{|X})^{-1}B\big)$. However, this follows from the fact that, according to (5.1.2), $C_{|(X+BU)_1}(\alpha - A_{|X})^{-1}B = \widehat{\mathfrak{D}}(\alpha) - D$. □

Let us now return to the question of exactly what type of operator nodes are known to be compatible. We begin with a crucial lemma.

Lemma 5.1.11 *Let Σ be a $L^p|Reg$-well-posed linear system Σ on (Y, X, U) with control operator B and observation operator C. Then C is continuous with respect to the (weaker) norm in X_1 induced by $(X + BU)_1$, i.e., there is a constant M such that*

$$|Cx|_Y \le M|x|_{(X+BU)_1}, \quad x \in X_1. \tag{5.1.6}$$

Thus, C has a (unique) extension to a bounded linear operator from the closure of X_1 in $(X + BU)_1$ to Y.

Proof Let $x \in X_1$, and let $\alpha \in \rho(A)$. By Lemma 4.3.12, there exist $z \in X_1$ and $u \in U$ with

$$|z|_{X_1} + |u|_U \le 2|x|_{(X+BU)_1}$$

such that $x = z + (\alpha - A_{|X})^{-1}Bu$. As both $x \in X_1$ and $z \in X_1$, this implies that $Bu \in X$. By Theorem 3.7.1(iii),

$$Bu = \lim_{\lambda\to+\infty} \lambda(\lambda - A)^{-1}Bu$$

(in X). This, combined with (4.7.2), gives

$$\begin{aligned} C(\alpha - A)^{-1}Bu &= \lim_{\lambda\to+\infty} C(\alpha - A)^{-1}(\lambda - \alpha)(\lambda - A)^{-1}Bu \\ &= \widehat{\mathfrak{D}}(\alpha)u - \lim_{\lambda\to+\infty} \widehat{\mathfrak{D}}(\lambda)u. \end{aligned}$$

If we denote $M = \sup_{\Re\lambda\ge\Re\alpha}\|\widehat{\mathfrak{D}}(\lambda)\|$, then $M < \infty$ (see Corollary 4.6.8), and $|C(\alpha - A)^{-1}Bu|_Y \le 2M|u|_U$. Thus

$$\begin{aligned} |Cx|_Y &\le |Cz|_Y + |C(\alpha - A)^{-1}Bu|_Y \\ &\le \|C\|_{\mathcal{B}(X_1;Y)}|z|_{X_1} + 2M|u|_U \\ &\le 2(\|C\|_{\mathcal{B}(X_1;Y)} + 2M)|x|_{(X+BU)_1}. \end{aligned}$$

□

With the aid of Lemma 5.1.11 we are now ready to prove the following result, which shows that most $L^p|Reg$-well-posed linear systems are compatible.

Theorem 5.1.12 *Let Σ be a $L^p|Reg$-well-posed linear system on (Y, X, U) with control operator B. Then Σ is compatible in (at least) the following cases:*

(i) *the closure of* X *in* $X + BU$ *is complemented in* $X + BU$*;*
(ii) X *and* U *are Hilbert spaces;*
(iii) *at least one of the spaces* X, U, *or* Y *is finite-dimensional.*

Proof (i) Write $X + BU$ in the form $X + BU = \overline{X} \dotplus Z$, where $\overline{X}$ is the closure of X in $X + BU$ and Z is a complementing subspace to $\overline{X}$. Then $(X + BU)_1 = (\alpha - A_{|X})^{-1}(\overline{X} + Z) = W_1 \dotplus W_2$, with $W_1 = (\alpha - A_{|X})^{-1}\overline{X}$ and $W_2 = (\alpha - A_{|X})^{-1}Z$. By Lemma 5.1.11, C can be extended to an operator $C_{W_1} \in \mathcal{B}(W_1; Y)$. For each $w = w_1 + w_2 \in W_1 \dotplus W_2$ we can define, e.g., $C_{|(X+BU)_1} w = C_{W_1} w_1$. Then $C_{|(X+BU)_1} \in \mathcal{B}((X + BU)_1; Y)$ is an extension of $C \in \mathcal{B}(X_1; Y)$, and by Corollary 5.1.9, Σ is compatible.

(ii) If both X and U are Hilbert spaces, then $X + BU$ is a Hilbert space, and every closed subspace of a Hilbert space is complemented.

(iii) If X is finite-dimensional, then $X_{-1} = X$, hence $X = X + BU$ and (i) holds. If U is finite-dimensional, then the co-dimension of X in $X + BU$ is finite and X is complemented. If Y is finite-dimensional, then we can use the Hahn–Banach theorem to extend C from the closure of X_1 in $(X + BU)_1$ to all of $(X + BU)_1$. □

Remark 5.1.13 It is clear that the extension of C to $(X + BU)_1$ is unique if and only if X_1 is dense in $(X + BU)_1$, or equivalently, if and only if X is dense in $X + BU$. It is easy to give examples where these inclusions are not dense. For example, in the delay line described in Lemma 5.1.2(iv) we have

$$X_1 = \{x \in W^{1,p}([0, T]; U) \mid x(T) = 0\}$$

and

$$(X + BU)_1 = W^{1,p}([0, T]; U).$$

Observe that X is closed in $X + BU$ and X_1 is closed in $(X + BU)_1$ in this case, and that this is a Hilbert space example if $p = 2$ and U is a Hilbert space.

Remark 5.1.14 As a complement to Remark 5.1.13, let us give another example where X is dense in $X + BU$ and X_1 is dense in the space $(X + BU)_1$. We let A generate an exponentially stable analytic semigroup on the Banach space X, let $U = X$, $B = A_{|X}^{1/4}$, and $\mathfrak{B}u = \int_{-\infty}^{0} \mathfrak{A}_{|X_{-1}}^{-s} Bu(s)\, ds$. The exponential stability of $\mathfrak{A}$ implies that A is invertible, that $A_{|X}^{\alpha}$ maps X onto $X_{-\alpha}$ for all $\alpha \in \mathbb{R}$, and that we can choose the norm in $X_{-\alpha}$ so that $\| A_{|X}^{\alpha} x\|_{X_{-\alpha}} = \|x\|_X$. By Theorem 5.7.3, $\mathfrak{B}$ is an L^2-admissible input map for A with control operator B, and $X + BU = \{z = x + A_{|X}^{1/4} u \mid x,\ u \in X\}$. Clearly $X + BU = X_{-1/4}$ (since $A_{|X}^{1/4}$ maps X onto $X_{-1/4}$). The norm in $X + BU$ is given by

$$|z|_{X+BU} = \inf_{x + A_{|X}^{1/4} u = z} \left(|x|_X^2 + |u|_X^2\right)^{1/2}.$$

Choosing $x = 0$ and $z = A_{|X}^{1/4}u$ we observe that $\|z\|_{X+BU} \le \|A_{|X}^{-1/4}z\|_X = \|z\|_{X_{-1/4}}$. This and the closed graph theorem imply that the norms in $X + BU$ and $X_{-1/4}$ are equivalent. As X is dense in $X_{-1/4}$, this means that X is also dense in $X + BU$.

Remark 5.1.15 If $C\colon X \to Y$ is closable, i.e., if the closure of the graph of C in $\left[\begin{smallmatrix} Y \\ X \end{smallmatrix}\right]$ is the graph of an operator $X \to Y$ (this is equivalent to the condition that Cx_n cannot tend to a nonzero limit in Y if $x_n \in X_1$ and $x_n \to 0$ in X), then this closure of C (defined on its domain equipped with the graph norm) is the *maximally defined* extension on C that can be used in Definition 5.1.1, i.e., the domain of the closure of C contains every possible space W. However, C will not, in general, be closable, and even if it is, then $C_{|W}$ need not be the restriction of W of the maximal extension of C. In particular, the observation operator for the delay line is not closable.

We end this section with some comments on how compatibility is preserved and how the generators change under some standard transformations. We leave the easy proofs to the reader.

Example 5.1.16 *Let* $\Sigma = \left[\begin{smallmatrix} \mathfrak{A} & \mathfrak{B} \\ \mathfrak{C} & \mathfrak{D} \end{smallmatrix}\right]$ *be a compatible* $L^p|Reg$*-well-posed linear system on* (Y, X, U) *with extended system node* $\left[\begin{smallmatrix} A_{|W} & B \\ C_{|W} & D \end{smallmatrix}\right] \in \mathcal{B}(\left[\begin{smallmatrix} W \\ U \end{smallmatrix}\right]; \left[\begin{smallmatrix} W_{-1} \\ Y \end{smallmatrix}\right])$.

(i) *For each* $\alpha \in \mathbb{C}$*, the exponentially shifted system* Σ_α *in Example 2.3.5 is compatible, with extended system node* $\left[\begin{smallmatrix} A_{|W}+\alpha & B \\ C_{|W} & D \end{smallmatrix}\right]$.

(ii) *For each* $\lambda > 0$*, the time compressed system* Σ_λ *in Example 2.3.6 is compatible, with extended system node* $\left[\begin{smallmatrix} \lambda A_{|W} & \lambda B \\ C_{|W} & D \end{smallmatrix}\right]$.

(iii) *For each (boundedly) invertible* $E \in \mathcal{B}(X_1; X)$*, the similarity transformed system* Σ_E *in Example 2.3.7 is compatible, with extended system node* $\left[\begin{smallmatrix} E^{-1}AE & E^{-1}B \\ C_{|W}E & D \end{smallmatrix}\right]$.

Example 5.1.17 *Let* Σ_1 *and* Σ_2 *be two compatible systems (with matching input and output spaces) which are* $L^p|Reg$*-well-posed in the same sense, with extended system nodes* $\left[\begin{smallmatrix} A_{1|W_1} & B_1 \\ C_{|W_1} & D_1 \end{smallmatrix}\right]$*, respectively* $\left[\begin{smallmatrix} A_{2|W_2} & B_2 \\ C_{|W_2} & D_2 \end{smallmatrix}\right]$*. Then the following systems derived from these two systems are also compatible, and they have the following extended system nodes:*

(i) *The cross-product of* Σ_1 *and* Σ_2 *in Example 2.3.10 with extended system node*

$$\begin{bmatrix} A_{|W} & B \\ C_{|W} & D \end{bmatrix} = \left[\begin{array}{cc|cc} A_{1|W_1} & 0 & B_1 & 0 \\ 0 & A_{2|W_2} & 0 & B_2, \\ \hline C_{|W_1} & 0 & D_1 & 0 \\ 0 & C_{|W_2} & 0 & D_2 \end{array}\right].$$

(ii) *The sum junction of* Σ_1 *and* Σ_2 *in Example 2.3.11 with extended system node*

$$\begin{bmatrix} A_{|W} & B \\ C_{|W} & D \end{bmatrix} = \left[\begin{array}{cc|cc} A_{1|W_1} & 0 & B_1 & 0 \\ 0 & A_{2|W_2} & 0 & B_2, \\ \hline C_{|W_1} & C_{|W_2} & D_1 & D_1 \end{array}\right].$$

(iii) *The T-junction of* Σ_1 *and* Σ_2 *in Example 2.3.12 with extended system node*

$$\begin{bmatrix} A_{|W} & B \\ C_{|W} & D \end{bmatrix} = \left[\begin{array}{cc|c} A_{1|W_1} & 0 & B_1 \\ 0 & A_{2|W_2} & B_2, \\ \hline C_{|W_1} & 0 & D_1 \\ 0 & C_{|W_2} & D_2 \end{array}\right].$$

(iv) *The parallel connection of* Σ_1 *and* Σ_2 *in Example 2.3.13 with extended system node*

$$\begin{bmatrix} A_{|W} & B \\ C_{|W} & D \end{bmatrix} = \left[\begin{array}{cc|c} A_{1|W_1} & 0 & B_1 \\ 0 & A_{2|W_2} & B_2, \\ \hline C_{|W_1} & C_{|W_2} & D_1 + D_2 \end{array}\right].$$

5.2 Boundary control systems

In this section we take a closer look at so-called *boundary control systems.*

Definition 5.2.1 Let $S = \left[\begin{smallmatrix} A\&B \\ C\&D \end{smallmatrix}\right]$ be an operator node on (Y, X, U) with control operator $B \in \mathcal{B}(U; X_{-1})$. We call S a *boundary control node* if B is injective and $\mathcal{R}(B) \cap X = 0$. If, in addition, S is an $L^p|Reg$-well-posed system node, then we call the corresponding system a *boundary control system.*

Here the requirement that B should be injective is not important, but it simplifies some of the formulas below. It can be removed completely if U is a Hilbert space, and it can be replaced by the assumption that the null space of B is complemented. See Proposition 4.10.1 and Theorem 5.2.16.

The following lemma gives a number of equivalent conditions for an operator node to be a boundary control node:

Lemma 5.2.2 *Let* $S = \left[\begin{smallmatrix} A\&B \\ C\&D \end{smallmatrix}\right]$ *be an operator node on* (Y, X, U) *with main operator* A *and control operator* B*. Suppose that* B *is injective, and let* $\alpha \in \rho(A)$*. Then the following conditions are equivalent:*

(i) S *is a boundary control node;*
(ii) $\mathcal{R}(B) \cap X = 0$*;*
(iii) $X + BU$ *is the direct sum* $X \dotplus BU$ *of* X *and* BU *(i.e., both* X *and* BU *are closed in* $X + BU$ *and* $X \cap BU = 0$*);*

(iv) $\mathcal{R}\left((\alpha - A_{|X})^{-1}B\right) \cap X_1 = 0$;
(v) $(X + BU)_1$ *is the direct sum of* X_1 *and* $(\alpha - A_{|X})^{-1}BU$.

When these conditions hold, there exists a finite positive constant and M *such that, for all* $x \in X$, $x_1 \in X_1$, *and* $u \in U$

$$M^{-1}\big(|x|_X^2 + |u|_U^2\big) \le |x + Bu|_{X+BU}^2 \le |x|_X^2 + |u|_U^2,$$
$$M^{-1}\big(|x_1|_{X_1}^2 + |u|_U^2\big) \le |x_1 + (\alpha - A_{|X})^{-1}Bu|_{(X+BU)_1}^2 \le M\big(|x_1|_{X_1}^2 + |u|_U^2\big).$$

For the major part of the conclusion of this lemma it is not really important that B is injective (we may replace U by $U/\mathcal{N}(B)$).

Proof (i) $\Leftrightarrow$ (ii). See Definition 5.2.1.

(ii) $\Leftrightarrow$ (iv). This is true since $(\alpha - A)^{-1}$ maps X one-to-one onto X_1.

(iii) $\Leftrightarrow$ (v). This is true since $(\alpha - A)^{-1}$ is an isomorphism of X onto X_1 and $(\alpha - A)_{|X+BU}^{-1}$ is an isomorphism of $X + BU$ onto W.

(ii) $\Rightarrow$ (iii). Since we assume that $\mathcal{R}(B) \cap X = 0$, every $z \in X + BU$ has a unique representation $z = x + Bu$, where $x \in X$ and $u \in U$. This implies that (see Lemma 4.3.12), for all $x \in X$, $|x|_{X+BU} = |x|_X$, and for all $z \in BU$, $|z|_{X+BU} = |u|_U$ where $z = Bu$. This means that both X and BU are closed in $X + BU$. As moreover $X \cap BU = 0$, we find that $X + BU = X \dotplus BU$.

(iii) $\Rightarrow$ (ii). This is trivial.

We have now proved the equivalence of (i)–(v). It still remains to prove the additional claims. The inequality $|x + Bu|_{X+BU}^2 \le |x|_X^2 + |u|_U^2$ is built into the definition of the norm in $X + BU$, and the converse inequality $\big(|x|_X^2 + |u|_U^2\big) \le M|x + Bu|_{X+BU}^2$ follows from a general result on complemented subspaces (see, e.g., Rudin (1973, Theorem 5.16) and recall that $|Bu|_{X+BU} = |u|_U$). The corresponding norm inequality in $(X + BU)_1$ follows from this and the fact that $(\alpha - A)^{-1}$ is an isomorphism of X onto X_1 and that $(\alpha - A_{|X})^{-1}$ is an isomorphism of $X + BU$ onto $(X + BU)_1$. □

Theorem 5.2.3 *Every boundary control node is* compatible.

Proof See Theorem 5.1.12(i) and Lemma 5.2.2(iii). □

Example 5.2.4 *The exactly controllable shift realization in Example 2.6.5(i) and the bilateral input shift realization in Example 2.6.5(iii) are boundary control systems. In particular, they are* compatible.

Proof Combine Theorem 5.2.3 with Example 4.2.6 and Lemma 5.2.2. □

Corollary 5.2.5 *Every* $\mathfrak{D} \in TIC_\omega^p(U;Y)$, *where* $1 \le p < \infty$, $\omega \in R$, *and* U *and* Y *are Banach spaces, has a compatible realization.*

Proof See Examples 2.6.5 and 5.2.4. □

Boundary control nodes are not only *compatible*: it is furthermore possible to replace the extended operator node $\left[\begin{smallmatrix} A_{|W} & B \\ C_{|W} & D \end{smallmatrix}\right]$ by another set of 'generating operators' $(\Delta, \Gamma, C_{|(X+BU)_1}, D)$. Here Δ and Γ are unique, D is a free parameter in $\mathcal{B}(U;Y)$ which we interpret as a feedthrough operator, and $C_{|(X+BU)_1}$ is the corresponding (unique) compatible extension of C to the space $(X+BU)_1$. In particular, if we require $D=0$, then all of Δ, Γ, and $C_{|(X+BU)_1}$ are unique.

Theorem 5.2.6 *Let $S = \left[\begin{smallmatrix} A\&B \\ C\&D \end{smallmatrix}\right]$ be a boundary control node on (Y, X, U) with main operator A, control operator B, observation operator C, and transfer function $\widehat{\mathfrak{D}}$. Let $D \in \mathcal{B}(U;Y)$, $\alpha \in \rho(A)$, and let $(X+BU)_1 = (\alpha - A_{|X})^{-1}(X+BU)$.*

(i) *There exist unique operators $\Delta \in \mathcal{B}((X+BU)_1;X)$, $\Gamma \in \mathcal{B}((X+BU)_1;U)$, and $C_{|(X+BU)_1} \in \mathcal{B}((X+BU)_1;Y)$ satisfying*

$$\Delta = A_{|X} + B\Gamma, \qquad C_{|(X+BU)_1} = C\&D\begin{bmatrix} 1 \\ \Gamma \end{bmatrix} - D\Gamma \tag{5.2.1}$$

(in particular, the operator $\left[\begin{smallmatrix} 1 \\ \Gamma \end{smallmatrix}\right]$ maps $(X+BU)_1$ into $\mathcal{D}(S)$). The operator Δ is an extension of the semigroup generator $A \in \mathcal{B}(X_1;X)$, and $\mathcal{N}(\Gamma) = X_1$. Moreover, the following two conditions are equivalent:

(a) *$x \in X$, $w \in X$, $u \in U$, and $x = A_{|X}w + Bu$;*

(b) *$w \in (X+BU)_1$, $x = \Delta w$, and $u = \Gamma w$.*

(ii) *When the operators Δ, Γ, and $C_{|(X+BU)_1}$ are applied to $(\alpha - A_{|X})^{-1}B$ we get the following results:*

$$\begin{aligned} \Delta(\alpha - A_{|X})^{-1}B &= \alpha(\alpha - A_{|X})^{-1}B, \\ \Gamma(\alpha - A_{|X})^{-1}B &= 1, \\ C_{|(X+BU)_1}(\alpha - A_{|X})^{-1}B &= \widehat{\mathfrak{D}}(\alpha) - D. \end{aligned} \tag{5.2.2}$$

Thus, if we define $Q_\alpha := (\alpha - A_{|X})^{-1}B\Gamma$, then Q_α is the projection of $(X+BU)_1$ onto $(\alpha - A_{|X})^{-1}BU$ with null space X_1, and

$$\begin{aligned} \Delta Q_\alpha &= \alpha Q_\alpha, \\ \Gamma Q_\alpha &= \Gamma, \\ C_{|(X+BU)_1} Q_\alpha &= \big(\widehat{\mathfrak{D}}(\alpha) - D\big)\Gamma. \end{aligned} \tag{5.2.3}$$

(iii) *The operator $C_{|(X+BU)_1}$ can alternatively be written in the form*

$$C_{|(X+BU)_1} = C(\alpha - A)^{-1}(\alpha - \Delta) + (\widehat{\mathfrak{D}}(\alpha) - D)\Gamma$$

(where the right hand side is independent of $\alpha \in \rho(A)$), and it is a

compatible extension of the observation operator $C \in \mathcal{B}(X_1; Y)$ with corresponding feedthrough operator D, i.e.,

$$C\&D = \begin{bmatrix} C_{|(X+BU)_1} & D \end{bmatrix}_{|\mathcal{D}(S)}.$$

Proof (i) For each $w \in (X + BU)_1 \subset X$, $(\alpha - A_{|X})w \in X + BU$, hence $A_{|X}w \in X + BU$. By Lemma 5.2.2 (with u replaced by $-u$), $A_{|X}w$ has a unique representation $A_{|X}w = x - Bu$ where $x \in X$ and $u \in U$. In other words, for each $w \in (X + BU)_1$, the requirements

$$x = A_{|X}w + Bu, \qquad x \in X, \qquad u \in U,$$

determine x and u uniquely. Moreover, since $A_{|X}$ maps $(X + BU)_1$ continuously into $X + BU$, and since the pair $\left[\begin{smallmatrix} x \\ u \end{smallmatrix}\right] \in \left[\begin{smallmatrix} X \\ U \end{smallmatrix}\right]$ depends continuously on $A_{|X}w \in X + BU$ (see Lemma 5.2.2), we find that $\left[\begin{smallmatrix} x \\ u \end{smallmatrix}\right] \in \left[\begin{smallmatrix} X \\ U \end{smallmatrix}\right]$ depends continuously on $w \in (X + BU)_1$. Thus, if we define

$$\Delta w = x, \qquad \Gamma w = u,$$

then $\Delta \in \mathcal{B}((X + BU)_1; X)$, $\Gamma \in \mathcal{B}((X + BU)_1, U)$, and $\Delta = A + B\Gamma$. The uniqueness of Δ and Γ follows from the fact that the decomposition of Aw into $x - Bu$ is unique. To see that $\left[\begin{smallmatrix} 1 \\ \Gamma \end{smallmatrix}\right]$ maps $(X + BU)_1$ into $\mathcal{D}(S)$ it suffices to observe that $\begin{bmatrix} A & B \end{bmatrix} \left[\begin{smallmatrix} 1 \\ \Gamma \end{smallmatrix}\right] = A + B\Gamma = \Delta$ maps $(x + BU)_1$ into X ($\mathcal{D}(S)$ is defined in Definition 4.6.4). Clearly the second equation in (5.2.1) then defines a unique $C_{|(X+BU)_1} \in \mathcal{B}((X + BU)_1; Y)$.

If $w \in X_1$ then $x = Aw$ and $u = 0$, hence Δ is an extension of $A \in \mathcal{B}(X_1; X)$ and $X_1 \subset \mathcal{N}(\Gamma)$. Conversely, if $u = \Gamma w = 0$, then $A_{|X}w = x \in X$, hence $w \in X_1$. This shows that $\mathcal{N}(\Gamma) = X_1$.

That (a) $\Rightarrow$ (b) was built into the definition of Δ and Γ that we gave above (recall that, by Lemma 4.3.12, if $w \in X$, $u \in U$, and $A_{|X}w + Bu \in X$, then $w \in (X - BU)_1$). The converse implication follows from (5.2.1).

(ii) Continuing with the same notation as in the proof of (i), if $u \in U$ and we let $w := (\alpha - A_{|X})^{-1}Bu$, then $x = \alpha w$, and we have $x = \Delta w = \alpha w = \alpha(\alpha - A_{|X})^{-1}Bu$ and $u = \Gamma w = \Gamma(\alpha - A_{|X})^{-1}Bu$. This gives us the first two identities in (5.2.2). This together with (4.7.1) and (5.2.1) gives

$$\begin{aligned}
\widehat{\mathfrak{D}}(\alpha) &= C\&D \begin{bmatrix} (\alpha - A_{|X})^{-1}B \\ 1 \end{bmatrix} \\
&= C\&D \begin{bmatrix} (\alpha - A_{|X})^{-1}B \\ \Gamma(\alpha - A_{|X})^{-1}B \end{bmatrix} \\
&= C\&D \begin{bmatrix} 1 \\ \Gamma \end{bmatrix} (\alpha - A_{|X})^{-1}B \\
&= \big(C_{|(X+BU)_1} + D\Gamma\big)(\alpha - A_{|X})^{-1}B \\
&= C_{|(X+BU)_1}(\alpha - A_{|X})^{-1}B + D.
\end{aligned}$$

Clearly (5.2.3) follows from (5.2.2).

(iii) The alternative formula for $C_{|(X+BU)_1}$ follows from (4.7.3). For each $w \in X_1$, $\Gamma w = 0$ and

$$C_{|(X+BU)_1} w = C\&D \begin{bmatrix} w \\ \Gamma w \end{bmatrix} = C\&D \begin{bmatrix} w \\ 0 \end{bmatrix} = Cw,$$

so $C_{|(X+BU)_1}$ is a (compatible) extension of $C \in \mathcal{B}(X_1; Y)$. By (5.1.2) and (5.2.2), the corresponding feedthrough operator is

$$\widehat{\mathfrak{D}}(\alpha) - C_{|(X+BU)_1}(\alpha - A_{|X})^{-1} B = D.$$

□

Corollary 5.2.7 *Let $S = \left[\begin{smallmatrix} A\&B \\ C\&D \end{smallmatrix}\right]$ be a boundary control node on (Y, X, U), with main operator A, control operator B, observation operator C, and transfer function $\widehat{\mathfrak{D}}$. Let $D \in \mathcal{B}(U; Y)$, $\alpha \in \rho(A)$, and define $C_{|(X+BU)_1}$ as in Theorem 5.2.6. The the following claims are true.*

(i) *The transfer function $\widehat{\mathfrak{D}}$ is given by*

$$\widehat{\mathfrak{D}}(z) = C_{|(X+BU)_1}(z - A_{|X})^{-1} B + D, \qquad z \in \rho(A).$$

(ii) *If S is a system node, then we can replace the equation* (4.7.4) *in Lemma 4.7.8 by*

$$\begin{bmatrix} \dot{x}(t) \\ y(t) \end{bmatrix} = \begin{bmatrix} A_{|(X+BU)_1} & B \\ C_{|(X+BU)_1} & D \end{bmatrix} \begin{bmatrix} x(t) \\ u(t) \end{bmatrix}, \quad t \ge s, \quad x(s) = x_s. \tag{5.2.4}$$

(iii) *If S is an $L^p|Reg$-well-posed system node then we can replace equations* (4.6.6) *and* (4.6.7) *in Theorem 4.6.5 by*

$$\begin{bmatrix} \dot{x}(t) \\ y(t) \end{bmatrix} = \begin{bmatrix} A_{|(X+BU)_1} & B \\ C_{|(X+BU)_1} & D \end{bmatrix} \begin{bmatrix} x(t) \\ u(t) \end{bmatrix}, \qquad t \in \mathbb{R}, \tag{5.2.5}$$

and we can replace formulas (4.6.13) *and* (4.6.14) *in Theorem 4.6.11 by* (5.2.4).

Proof This follows from Lemma 5.1.4, Corollary 5.1.7, and Theorem 5.2.6.

□

In the above corollary we only used a small part of Theorem 5.2.6, namely the fact that $\begin{bmatrix} C_{|(X+BU)_1} & D \end{bmatrix}$ is a compatible extension of $C\&D$. We can also use the operators Δ and Γ to get a different representation formula for the state trajectory x in Lemma 4.7.8 and Theorems 4.6.5 and 4.6.11.

Corollary 5.2.8 *Let $S = \left[\begin{smallmatrix} A\&B \\ C\&D \end{smallmatrix}\right]$ be a system node on (Y, X, U) which is also a boundary control node, with semigroup generator A, control operator B, observation operator C, and transfer function $\widehat{\mathfrak{D}}$. Let $D \in \mathcal{B}(U; Y)$, $\alpha \in \rho(A)$, and define Δ, Γ, and $C_{|(X+BU)_1}$ as in Theorem 5.2.6. Then the following claims are true.*

(i) *The state trajectory x and output function y in Lemma 4.7.8 satisfy*

$$\begin{aligned} \dot{x}(t) &= \Delta x(t), \qquad \Gamma x(t) = u(t), && t \ge s, \\ y(t) &= C_{|(X+BU)_1} x(t) + Du(t), && t \ge s, \\ x(s) &= x_s. \end{aligned} \tag{5.2.6}$$

(ii) *If S is $L^p|Reg$-well-posed then the state trajectory x and output function y in Theorem 4.6.5 satisfy*

$$\begin{aligned} \dot{x}(t) &= \Delta x(t), \qquad \Gamma x(t) = u(t), && t \in \mathbb{R}, \\ y(t) &= C_{|(X+BU)_1} x(t) + Du(t), && t \in \mathbb{R}, \end{aligned} \tag{5.2.7}$$

and the state trajectory x and output function y in Theorem 4.6.11 satisfy (5.2.6).

Proof This follows from Theorem 5.2.6 and Corollary 5.2.7. □

Remark 5.2.9 Especially in the L^∞-well-posed and *Reg*-well-posed cases it may be confusing that we have several different extensions of C, namely the unique extension to an operator $C_{|X} \in \mathcal{B}(X;Y)$ given by Theorem 4.4.2(ii), and the extensions given in Theorem 5.2.6 to a family of operators in $\mathcal{B}((X + BU)_1;Y)$ which are parametrized by the operator D. It is not true, in general, that $C_{|(X+BU)_1} \in \mathcal{B}((X + BU)_1;Y)$ in Theorem 5.2.6 is a restriction of $C_{|X}$ to $(X + BU)_1$: this is true if and only if we choose D in Theorem 5.2.6 to be the feedthrough operator D defined in Theorem 4.5.2. The same comment applies to the case where the system is regular in the sense of Definition 5.6.3.

Remark 5.2.10 Continuing the preceding remark, although the operator $\Delta \in \mathcal{B}((X + BU)_1;X)$ is an extension of the generator $A \in \mathcal{B}(X_1;X)$, it is *not* the restriction to $(X + BU)_1$ of the extension $A_{|X}$ of A to $\mathcal{B}(X;X_{-1})$ that we constructed in Section 3.6. This follows from the fact that $(\alpha - A)$ maps X_1 *onto* X for $\alpha \in \rho(A)$, whereas $(\alpha - \Delta)$ maps all of $(X + BU)_1$ into X. In particular, $(\alpha - A)$ is injective but $(\alpha - \Delta)$ is not.

Remark 5.2.11 The operators Δ and Γ in Theorem 5.2.6 can be interpreted in the following way. The operator Δ is an 'abstract partial differential operator' with an insufficient set of boundary conditions, so that Δ does not itself generate a semigroup, i.e., the solution x of (5.2.6) is not unique if we remove the 'abstract boundary condition' $\Gamma x(t) = u(t)$. Often Γ is referred to as a 'trace operator'. It follows from (5.2.2) that $x := (\alpha - A_{|X})^{-1}Bu$ is the unique (static) solution in $(X + BU)_1$ of the 'abstract elliptic problem'

$$\Delta x = \alpha x, \qquad \Gamma x = u. \tag{5.2.8}$$

Boundary control systems arise in at least two different ways: in realization theory (see Example 5.2.4), and in the theory of partial differential equations. In the latter case one often starts by studying the abstract elliptic problem (5.2.8). In this setting the given data are the spaces U, Y, and $W \subset X$ (where $W := (X + BU)_1$) and the operators $\Delta \in \mathcal{B}(W; X)$, $\Gamma \in \mathcal{B}(W; U)$, $C_{|W} \in \mathcal{B}(W; Y)$, and $D \in \mathcal{B}(U; Y)$ (where the last operator is usually taken to be zero). These operators typically have the following properties:

Corollary 5.2.12 *The Banach spaces U, X, Y, and $W := (X + BU)_1$ and the operators $\Delta \in \mathcal{B}(W; X)$, $\Gamma \in \mathcal{B}(W; U)$, and $C_{|W} \in \mathcal{B}(W; Y)$ constructed in Theorem 5.2.6 have the following properties:*

(i) *W is densely and continuously embedded in X;*
(ii) *$\mathcal{N}(\Gamma)$ is dense in X;*
(iii) *for some $\alpha \in \mathbb{C}$, $\alpha - \Delta$ maps $\mathcal{N}(\Gamma)$ one-to-one onto X;*
(iv) *Γ is right-invertible (i.e., $\Gamma\Gamma^{-1}_{\mathrm{right}} = 1$ for some $\Gamma^{-1}_{\mathrm{right}} \in \mathcal{B}(U; W)$).*

Proof Claim (i) is contained in Lemma 4.3.12(i), and claim (ii) is part of Theorem 5.2.6(i). Clearly $\mathcal{N}(\Gamma)$ is dense in X since $\mathcal{N}(\Gamma) = X_1$, and that Γ is right-invertible follows from (5.2.2). The restriction of Δ to $\mathcal{N}(\Gamma) = X_1$ is equal to the main operator A of the node, and (iii) holds for all $\alpha \in \rho(A)$. □

It turns out that the necessary conditions on W, Δ, Γ, and $C_{|W}$ listed above are also *sufficient* for these operators to determine a boundary control node.

Theorem 5.2.13 *Let U, X, Y, and W be Banach spaces, and suppose that $\Delta \in \mathcal{B}(W; X)$, $\Gamma \in \mathcal{B}(W; U)$, $C_{|W} \in \mathcal{B}(W; Y)$, and $D \in \mathcal{B}(U; Y)$ satisfy conditions (i)–(iv) in Corollary 5.2.12. Then there is a unique boundary control node $S = \left[\begin{smallmatrix} A\&B \\ C\&D \end{smallmatrix}\right]$ such that the operators Δ, Γ, and $C_{|W}$ constructed in Theorem 5.2.6 coincide with the given ones, and $W = (X + BU)_1$ (possibly with a different but equivalent norm). This operator node can be constructed in the following way.*

(i) *The main operator A of S is given by $A := \Delta_{|\mathcal{N}(\Gamma)}$. It is closed in X, and the constant α in condition (iii) in Corollary 5.2.12 belongs to its resolvent set.*
(ii) *The spaces $X_1 \subset X \subset X_{-1}$ are constructed as in Section 3.6, with α given by condition (iii) in Corollary 5.2.12, and A is extended to an operator $A_{|X} \in \mathcal{B}(X; X_{-1})$. In particular, $X_1 = \mathcal{N}(\Gamma)$, and the norm in X_1 is equivalent to the norm inherited from W.*
(iii) *$B = (\Delta - A_{|X})\Gamma^{-1}_{\mathrm{right}} \in \mathcal{B}(U; X_{-1})$, where $\Gamma^{-1}_{\mathrm{right}} \in \mathcal{B}(U; W)$ is an arbitrary right-inverse of Γ (the result is independent of the choice of $\Gamma^{-1}_{\mathrm{right}}$).*

(iv) $\mathcal{D}(S) := \left\{ \left[\begin{smallmatrix} w \\ u \end{smallmatrix} \right] \in \left[\begin{smallmatrix} W \\ U \end{smallmatrix} \right] \mid u = \Gamma w \right\}$.

(v) $S := \left[\begin{smallmatrix} A_{|X} & B \\ C_{|W} & D \end{smallmatrix} \right]_{|\mathcal{D}(S)}$.

Proof Let us begin by establishing the uniqueness of S. In Theorem 5.2.6 we had $X_1 = \mathcal{N}(\Gamma)$, so A must be given by (i), and the constant α in condition (iii) in Corollary 5.2.12 must belong to its resolvent set. The formula given in (iii) for B is obtained from (5.2.1). By part (iv) of Definition 4.7.2 and by the equivalence of (a) and (b) in part (i) of Theorem 5.2.6, $\mathcal{D}(S)$ must be given by (iv). Finally (v) holds for all compatible operator nodes. Thus, the node S is unique (if it exists). In particular, the operator B defined in (iii) does not depend on the particular choice of $\Gamma_{\text{right}}^{-1}$.

We continue with the existence part of the proof. To show that A is closed in X we let $X_1' := \mathcal{N}(\Gamma)$ with the norm inherited from W. By condition (iii) in Corollary 5.2.12, $\alpha - A$ maps X_1' one-to-one continuously onto X. By the closed graph theorem, the inverse is bounded from X to X_1', and hence (by the continuity of the inclusion $W \subset X$), from X to itself. Therefore $\alpha - A$ is closed, hence so is A, and $\alpha \in \rho(A)$. That the norm in $\mathcal{N}(\Gamma)$ inherited from W is equivalent to the norm defined in Section 3.6 (i.e., the norm induced by $\alpha - A$ from X) follows from the fact that $\alpha - A$ is a bounded operator from X_1' to X with a bounded inverse.

We are now in a position where we can define $A_{|X}$, B, and S as described in (ii)–(v). It remains to show that S is a boundary control node, that $W = (X + BU)_1$, and that the operators Δ, Γ, and $C_{|W}$ constructed in Theorem 5.2.6 coincide with the given ones.

We begin with the claim that $W = (X + BU)_1$. Trivially, $X_1 = \mathcal{N}(\Gamma)$ is a closed subspace of W, so to show that $(X + BU)_1 = X_1 + (\alpha - A_{|X})^{-1} BU \subset W$ with a continuous inclusion it suffices to show that $(\alpha - A_{|X})^{-1} B$ maps U continuously into W. But this follows from the fact that

$$
\begin{aligned}
(\alpha - A_{|X})^{-1} B &= (\alpha - A_{|X})^{-1} (\Delta - A_{|X}) \Gamma_{\text{right}}^{-1} \\
&= \Gamma_{\text{right}}^{-1} + (\alpha - A_{|X})^{-1} (\Delta + \alpha) \Gamma_{\text{right}}^{-1},
\end{aligned}
\tag{5.2.9}
$$

where $\Gamma_{\text{right}}^{-1} \in \mathcal{B}(U; W)$, $\Delta + \alpha \in \mathcal{B}(W; X)$, and $(\alpha - A_{|X})_{|X}^{-1} = (\alpha - A)^{-1} \in \mathcal{B}(X; X_1)$. In particular, since the last term belongs to $X_1 = \mathcal{N}(\Gamma)$, we have, in addition,

$$
\Gamma (\alpha - A_{|X})^{-1} B = 1. \tag{5.2.10}
$$

To prove the converse inclusion $W \subset (X + BU)_1$ we take an arbitrary $w \in W$, and define $u = \Gamma w$ and $x = w - (\alpha - A_{|X})^{-1} Bu$. Recall that $\Gamma \in \mathcal{B}(W; U)$ and that $(\alpha - A_{|X})^{-1} Bu \in \mathcal{B}(U; W)$, hence the mapping $w \mapsto \left[\begin{smallmatrix} x \\ u \end{smallmatrix} \right]$ is continuous from W to $\left[\begin{smallmatrix} W \\ U \end{smallmatrix} \right]$. Actually, the range of the mapping $w \mapsto x$ belongs to

$X_1 = \mathcal{N}(\Gamma)$, since

$$\Gamma x = \Gamma w - \Gamma(\alpha - A_{|X})^{-1} B u = u - u = 0.$$

For each $w \in W$ we can find a pair $\left[\begin{smallmatrix} x \\ u \end{smallmatrix}\right] \in \left[\begin{smallmatrix} X_1 \\ U \end{smallmatrix}\right]$ which depends continuously on w such that $w = x + (\alpha - A_{|X})^{-1} B u$. This implies that $W = (X + BU)_1$, with equivalent norms.

We next show that $\mathcal{D}(S)$ defined in (iv) satisfies

$$\mathcal{D}(S) = \left\{ \left[\begin{smallmatrix} w \\ u \end{smallmatrix}\right] \in \left[\begin{smallmatrix} X \\ U \end{smallmatrix}\right] \;\middle|\; A_{|X} w + B u \in X \right\}. \tag{5.2.11}$$

Let $\alpha \in \rho(A)$. Each of the lines in the following list is equivalent to both the preceding and the following line (recall that $(\alpha - A_{|X})^{-1}$ maps X onto-to-one onto X_1, that $(\alpha - A_{|X})^{-1} B$ maps U into W, that $X_1 = \mathcal{N}(\Gamma)$, and that $\Gamma(\alpha - A_{|X})^{-1} B = 1$):

$$\begin{aligned}
&\left[\begin{smallmatrix} w \\ u \end{smallmatrix}\right] \in \left[\begin{smallmatrix} X \\ U \end{smallmatrix}\right] \text{ and } A_{|X} w + B u \in X, \\
&\left[\begin{smallmatrix} w \\ u \end{smallmatrix}\right] \in \left[\begin{smallmatrix} X \\ U \end{smallmatrix}\right] \text{ and } (\alpha - A_{|X})^{-1}(A_{|X} w + B u) \in X_1, \\
&\left[\begin{smallmatrix} w \\ u \end{smallmatrix}\right] \in \left[\begin{smallmatrix} X \\ U \end{smallmatrix}\right] \text{ and } -w + (\alpha - A_{|X})^{-1} w + (\alpha - A_{|X})^{-1} B u \in X_1, \\
&\left[\begin{smallmatrix} w \\ u \end{smallmatrix}\right] \in \left[\begin{smallmatrix} W \\ U \end{smallmatrix}\right] \text{ and } -w + (\alpha - A_{|X})^{-1} B u \in X_1, \\
&\left[\begin{smallmatrix} w \\ u \end{smallmatrix}\right] \in \left[\begin{smallmatrix} W \\ U \end{smallmatrix}\right] \text{ and } -\Gamma w + \Gamma(\alpha - A_{|X})^{-1} B u = 0, \\
&\left[\begin{smallmatrix} w \\ u \end{smallmatrix}\right] \in \left[\begin{smallmatrix} W \\ U \end{smallmatrix}\right] \text{ and } u = \Gamma w.
\end{aligned}$$

This proves (5.2.11).

By (5.2.11), the operator S defined in (v) is the restriction of the operator $\left[\begin{smallmatrix} A_{|X} & B \\ C_{|W} & D \end{smallmatrix}\right] \in \mathcal{B}(\left[\begin{smallmatrix} W \\ U \end{smallmatrix}\right]; \left[\begin{smallmatrix} X_{-1} \\ Y \end{smallmatrix}\right])$ to the domain $\left\{ \left[\begin{smallmatrix} x \\ u \end{smallmatrix}\right] \in \left[\begin{smallmatrix} W \\ U \end{smallmatrix}\right] \mid S \left[\begin{smallmatrix} x \\ u \end{smallmatrix}\right] \in \left[\begin{smallmatrix} X \\ U \end{smallmatrix}\right] \right\}$. As can easily be seen, this implies that S is closed. According to Definition 4.7.2, S is an operator node. By (5.2.10), B is injective. If for some $u \in U$ we have $Bu \in X$, then $(\alpha - A_{|X})^{-1} B u \in X_1 = \mathcal{N}(\Gamma)$, and hence $u = \Gamma(\alpha - A_{|X})^{-1} B u = 0$. Thus, S satisfies all the requirements of a boundary control node listed in Definition 5.2.1.

It remains to show that the operators Δ, Γ, and $C_{|W}$ constructed from the node S in Theorem 5.2.6 coincide with the given ones. By (5.2.11), $w \in W$ and $u = \Gamma u$ if and only if $A_{|X} w + B u \in X$. This was the property that we used to define Γ in Theorem 5.2.6, so the given operator Γ coincides with the operator Γ constructed in Theorem 5.2.6. That the operator Δ also coincides with the one constructed in Theorem 5.2.6 follows from the fact that they are both the restriction of $A_{|X} + B\Gamma$ to $(X + BU)_1$ (see (5.2.1) and the definition of B in (iii)). Finally, in both cases $C_{|W}$ is the compatible extension of C to W corresponding to the given feedthrough operator D, and according to Lemma 5.1.10, this determines $C_{|W}$ uniquely. □

A boundary control node can be regarded as a special case of the following more general class of operator nodes:

Definition 5.2.14 Let $S = \left[\begin{smallmatrix} A\&B \\ C\&D \end{smallmatrix}\right]$ be an operator node on (Y, X, U) with control operator $B \in \mathcal{B}(U; X_{-1})$. We call S a *mixed boundary/distributed control node* if the inverse image under B of $\mathcal{R}(B) \cap X$ is complemented in U. (If U is a Hilbert space, then this is true if and only if the inverse image under B of $\mathcal{R}(B) \cap X$ is closed in U.) If, in addition, S is an $L^p|Reg$-well-posed system node, then we call the corresponding system a *mixed boundary/distributed control system.*

Clearly, every operator node with a finite-dimensional input space is a mixed boundary/distributed control node. An example of a system which is *not* a mixed boundary/distributed control system is given in Remark 5.1.14.

It follows from Definition 5.2.14 that S is a mixed boundary/distributed control node on (Y, X, U) if and only if we can split U into $U = \left[\begin{smallmatrix} U_1 \\ U_2 \end{smallmatrix}\right]$ in such a way that, if we split B accordingly into $B = \begin{bmatrix} B_1 & B_2 \end{bmatrix}$ (so that $B \left[\begin{smallmatrix} u_1 \\ u_2 \end{smallmatrix}\right] = \begin{bmatrix} B_1 & B_2 \end{bmatrix} \left[\begin{smallmatrix} u_1 \\ u_2 \end{smallmatrix}\right] = B_1 u_1 + B_2 u_2$), then $\mathcal{R}(B_1) \cap X = 0$ and $B_2 u_2 \in X$ for all $u_2 \in U$ (we define U_2 to be the inverse image under B of $\mathcal{R}(B) \cap X$, and U_1 to be a complementing subspace). Both $\left[\begin{smallmatrix} X \\ U_1 \\ 0 \end{smallmatrix}\right]$ and $\left[\begin{smallmatrix} 0 \\ 0 \\ U_0 \end{smallmatrix}\right]$ are closed subspaces of $\left[\begin{smallmatrix} X \\ U_1 \\ U_0 \end{smallmatrix}\right]$, and this implies that the restriction of S to each of these subspaces is closed. Let us denote these restrictions by S_1 respectively S_2. As $B \left[\begin{smallmatrix} u_1 \\ u_2 \end{smallmatrix}\right] = \begin{bmatrix} B_1 & B_2 \end{bmatrix} \left[\begin{smallmatrix} u_1 \\ u_2 \end{smallmatrix}\right] = B_1 u_1 + B_2 u_2$ where $B_2 u_2 \in X$, it is clear that $\left[\begin{smallmatrix} x \\ u_1 \\ u_2 \end{smallmatrix}\right] \in \mathcal{D}(S)$ if and only if $A_{|X_{-1}} x + B_1 u_1 \in X$. This implies that $\mathcal{D}(S_1) = \left\{ \left[\begin{smallmatrix} x \\ u_1 \end{smallmatrix}\right] \in \left[\begin{smallmatrix} X \\ U_1 \end{smallmatrix}\right] \mid A_{|X_{-1}} x + B_1 u_1 \in X \right\}$ and $\mathcal{D}(S_2) = U_2$. In particular, it follows from Definition 4.7.2 that S_1 is an operator node on (Y, X, U_1), with main operator A, control operator B_1, and observation operator C. Furthermore, by the closed graph theorem, S_2 is of the form $S_2 = \left[\begin{smallmatrix} B_2 \\ D_2 \end{smallmatrix}\right]$ with $B_2 \in \mathcal{B}(U_2 : X)$ and $D_2 \in \mathcal{B}(U_2; Y)$. Splitting S_1 vertically into $S_1 = \left[\begin{smallmatrix} [A\&B]_1 \\ [C\&D]_1 \end{smallmatrix}\right]$, we conclude that S has the representation $S = \left[\begin{smallmatrix} [A\&B]_1 & B_2 \\ [C\&D]_1 & D_2 \end{smallmatrix}\right]$. By construction, S_1 is a boundary control node on (Y, X, U_1). If S is a system node then so is S_1, and if S is $L^p|Reg$-well-posed, then so is S_1.

The following result is a generalization of Theorem 5.2.3:

Theorem 5.2.15 *Every mixed boundary/distributed control system is compatible.*

Proof With the notation introduced above, by Theorem 5.2.3, S_1 is compatible, and since $B_2 \in \mathcal{B}(U_2; X)$, this implies that also S is compatible. □

The transfer function, the state trajectory, and the output function of a mixed boundary/distributed control systems can be represented in the following way:

Theorem 5.2.16 *Let* $S = \left[\begin{smallmatrix} A\&B \\ C\&D \end{smallmatrix}\right] = \left[\begin{smallmatrix} [A\&B]_1 & B_2 \\ [C\&D]_1 & D_2 \end{smallmatrix}\right]$ *be a mixed boundary/distributed control node on* $(Y, X, U) = \left(Y, X, \left[\begin{smallmatrix} U_1 \\ U_2 \end{smallmatrix}\right]\right)$ *(where we use the notations introduced after Definition 5.2.14). Let* $D_1 \in \mathcal{B}(U_1; Y)$*, and define* Δ*,* Γ*, and* $C_{|(X+BU)_1}$ *as in Theorem 5.2.6 with* U *replaced by* U_1*,* S *replaced by* $S_1 = \left[\begin{smallmatrix} [A\&B]_1 \\ [C\&D]_1 \end{smallmatrix}\right]$*,* B *replaced by the control operator* B_1 *of* Σ_1*, and* D *replaced by* D_1*. Then the following claims are true.*

(i) *The transfer function* $\widehat{\mathfrak{D}}$ *of* S *is given by*

$$\begin{aligned}\widehat{\mathfrak{D}}(z)\left[\begin{smallmatrix} u_1 \\ u_2 \end{smallmatrix}\right] &= C_{|(X+BU)_1}(z - A_{|X})^{-1}(B_1 u_1 + B_2 u_2) \\ &\quad + D_1 u_1 + D_2 u_2, \qquad z \in \rho(A).\end{aligned}$$

(ii) *If* S *is a system node, then the state trajectory* x *and output function* y *in Lemma 4.7.8 satisfy*

$$\begin{aligned} \dot{x}(t) &= \Delta x(t) + B_2 u_2(t), && t \geq s, \\ \Gamma x(t) &= u_1(t), && t \geq s, \\ y(t) &= C_{|(X+BU)_1} x(t) + D_1 u_1(t) + D_2 u_2(t), && t \geq s, \\ x(s) &= x_s. \end{aligned} \tag{5.2.12}$$

(iii) *If* S *is a* $L^p|Reg$*-well-posed system node, then the state trajectory* x *and output function* y *in Theorem 4.6.5 satisfy*

$$\begin{aligned} \dot{x}(t) &= \Delta x(t) + B_2 u_2(t), && t \in \mathbb{R}, \\ \Gamma x(t) &= u_1(t), && t \in \mathbb{R}, \\ y(t) &= C_{|(X+BU)_1} x(t) + D_1 u_1(t) + D_2 u_2(t), && t \in \mathbb{R}, \end{aligned} \tag{5.2.13}$$

and the state trajectory x *and output function* y *in Theorem 4.6.11 satisfy* (5.2.12).

Proof (i) This is the formula for the transfer function of a compatible node given in (5.1.3), with the $X = (X + BU)_1$, substitutions $B = \left[B_1 \ B_2\right]$ and $D = \left[D_1 \ D_2\right]$.

(ii) Let x_1 and y_1 be the state trajectory and output function that we get by applying Lemma 4.7.8 with S replaced by S_1, u replaced by u_1, and $x_1(s) = x_s$, and let x_2 and y_2 be the state trajectory and output function that we get by applying Lemma 4.7.8 with S replaced by $S_2 = \left[\begin{smallmatrix} A & B_2 \\ C & D_2 \end{smallmatrix}\right]$, u replaced by u_2, and $x_2(s) = 0$. Then $x = x_1 + x_2$, $y = y_1 + y_2$, and we get (ii) by applying Corollary 5.2.8(i) to x_1 and y_1 and Corollary 5.1.7(i) to x_2 and y_2 (recall that $\Delta_{|X_1} = A$, and that the restriction of $C_{|(X+BU)_1}$ to X_1 is C).

(iii) This proof is similar to the one above. □

5.3 Approximations of the identity in the state space

In Section 5.1 we introduced the notion of a *compatible system*, which is based on an extension of the observation operator C. The purpose of this section is to define some specific extensions by means of summability methods. These extensions have the advantage that they are determined *uniquely* by the system, even if X_1 is not dense in $(X + BU)_1$, and that (in some of the cases) X_1 is dense in their domains. They have the drawback that their domains are not Hilbert spaces, even when U, X, and Y are Hilbert spaces. We shall also extend the combined observation/feedthrough operator $C\&D$ by the same method.

The extensions of C that we present below are all based on the same idea: we first approximate C by CJ^h or CJ_α where J^h and J_α are approximations of the identity operator on X, and then let $h \downarrow 0$ or $\alpha \to \infty$. The approximations of the identity that we use are various Cesàro and Yosida approximations of the semigroup $\mathfrak{A}$ at zero. We already defined the Cesàro approximation J^h of order one and the Yosida approximation J_α in Definition 3.7.2, but in addition to these approximations we shall need Cesàro approximations of order different from one, defined as follows. (For completeness we repeat the definition of the Yosida approximation from Definition 3.7.2.)

Definition 5.3.1 The *Yosida* (or *Abel*) *approximation* J_α with parameter $\alpha \in \rho(A)$ of the identity on X is given by

$$J_\alpha x = \alpha(\alpha - A)^{-1}x, \qquad x \in X.$$

The *Cesàro approximation* $J^{\beta,h}$ *of order* $\beta > 0$ with parameter $h > 0$ of the identity on X is given by

$$J^{\beta,h}x = \beta h^{-\beta} \int_0^h (h - s)^{\beta-1}\mathfrak{A}^s x\, ds, \qquad x \in X.$$

The *Cesàro approximation* $J^{0,h}$ *of order zero* with parameter $h > 0$ of the identity on X is given by

$$J^{0,h}x = \mathfrak{A}^h x, \qquad x \in X.$$

By the *Cesàro approximation* without any reference to its order we mean the Cesàro approximation of order one.

In particular, $\mathfrak{A}$ is strongly continuous if and only if the zero order Cesàro approximation of the identity tends strongly to 1 as $h \downarrow 0$ (this is the reason for the abbreviation C_0 semigroup for a strongly continuous semigroup), and $J^{1,h}$ coincides with the operator J^h defined in Definition 3.7.2.

The Yosida and Cesàro approximations of the identity defined above have the following properties (for completeness we repeat part of Theorem 3.7.1):

Theorem 5.3.2 *Let $\alpha \in \rho(A)$, γ, $\beta \geq 0$, h, k, and $t > 0$.*

(i) $J_\alpha \in \mathcal{B}(X; X_1)$, $J^{\beta,h} \in \mathcal{B}(X)$, $J^{\beta,h} \in \mathcal{B}(X; X_1)$ *for* $\beta \geq 1$, *and*

$$\begin{aligned} AJ_\alpha &= \alpha(J_\alpha - 1), \\ AJ^{\beta,h} &= \frac{\beta}{h}\big(J^{\beta-1,h} - 1\big), && \beta \geq 1, \\ \frac{\partial}{\partial h} J^{\beta,h} &= AJ^{\beta,h} + \frac{\beta}{h}\big(1 - J^{\beta,h}\big), && \beta \geq 1, \\ \frac{\partial}{\partial h} J^{\beta,h} &= \frac{\beta}{h}\big(J^{\beta-1,h} - J^{\beta,h}\big), && \beta \geq 1. \end{aligned}$$

Moreover, for $\alpha \in \mathbb{C}^+_{\omega_\mathfrak{A}}$,

$$J_\alpha x = \alpha \int_0^\infty e^{-\alpha s} \mathfrak{A}^s x \, ds, \qquad x \in X.$$

(ii) *The operators* J_α, $J^{\beta,h}$, $J^{\gamma,k}$, *and* $\mathfrak{A}^t$ *commute with each other and with* A *(regarded as an operator in* $\mathcal{B}(X_1; X)$*).*

(iii) $J^{\beta,h}$ *and* J_α *approximate the identity on* X *in the sense that*

$$\lim_{\alpha \to +\infty} J_\alpha x = \lim_{h \downarrow 0} J^{\beta,h} x = x$$

in X *(for all* $x \in X$*). (Here* $\alpha \to +\infty$ *along the positive real axis.)*

(iv) *For all* $x \in X$,

$$J_\alpha x = \frac{\alpha^{\beta+1}}{\Gamma(\beta+1)} \int_0^\infty e^{-\alpha v} v^\beta J^{\beta,v} x \, dv,$$

where

$$\Gamma(\beta) = \int_0^\infty s^{\beta-1} e^{-s} \, ds$$

and

$$\frac{\alpha^{\beta+1}}{\Gamma(\beta+1)} \int_0^\infty e^{-\alpha v} v^\beta \, dv = 1.$$

(v) *For all* $\gamma > 0$ *and all* $x \in X$,

$$J^{\beta+\gamma,h} x = \frac{\gamma h^{-(\beta+\gamma)}}{B(\beta+1, \gamma+1)} \int_0^h (h - v)^{\gamma-1} v^\beta J^{\beta,v} x \, dv,$$

where

$$B(\beta, \gamma) = \int_0^1 (1-s)^{\beta-1} s^{\gamma-1}\, ds = \frac{\Gamma(\beta)\Gamma(\gamma)}{\Gamma(\beta+\gamma)} \tag{5.3.1}$$

is the Beta-function, and

$$\frac{\gamma h^{-(\beta+\gamma)}}{B(\beta+1, \gamma+1)} \int_0^h (h-v)^{\gamma-1} v^{\beta}\, dv = 1. \tag{5.3.2}$$

(vi) *For all $\beta \in (0,1)$ and all $x \in X$,*

$$\mathfrak{A}^t J^{\beta,h} x = \left(\frac{t+h}{h}\right)^{\beta} J^{\beta,t+h} x - h^{-\beta} \int_0^t a^{h,\beta}(t-s) s^{\beta} J^{\beta,s} x\, ds,$$

where

$$a^{h,\beta}(s) = \frac{\beta}{\Gamma(\beta)\Gamma(1-\beta)} \int_0^h (s+h-w)^{-\beta-1} w^{\beta-1}\, dw,$$

is nonnegative on $\overline{\mathbb{R}}^+$, $\int_0^\infty a^{h,\beta}(s)\, ds = 1$, and

$$\int_0^t a^{\beta,h}(t-s) s^{\beta}\, ds = (t+h)^{\beta} - h^{\beta}, \qquad t > 0, \tag{5.3.3}$$

$$\int_0^t a^{\beta,h}(t-s) s^{\beta-1}\, ds = (t+h)^{\beta-1}, \qquad t > 0. \tag{5.3.4}$$

(vii) *For all $x \in X$,*

$$\mathfrak{A}^t J^{1,h} x = \frac{t+h}{h} J^{1,t+h} x - \frac{t}{h} J^{1,t} x.$$

(viii) $J^{\beta,h} J^{\gamma,k} \in \mathcal{B}(X; X_1)$ *if $\beta + \gamma \geq 1$.*

Proof (i) The claims about J_α are contained in Theorem 3.7.1(i), and so are the claims about $J^{\beta,h}$ for $\beta = 1$. Obviously $J^{\beta,h} \in \mathcal{B}(X; X)$. To get the two formulas of $\frac{\partial}{\partial h} J^{\beta,h}$ we differentiate the definition of $J^{\beta,h}$, first in its original form and then in the equivalent form

$$J^{\beta,h} = \beta h^{-\beta} \int_0^h s^{\beta-1} \mathfrak{A}^{h-s} x\, ds.$$

By combining these two formulas we get the formula for $A J^{\beta,h}$, which in turn implies that $J^{\beta,h} \in \mathcal{B}(X; X_1)$ for $\beta \geq 1$.

(ii) This is true since $\mathfrak{A}^s$ commutes with $\mathfrak{A}^t$ and with $(\alpha - A)^{-1}$.

(iii) See Theorem 3.2.9(iii) (or Theorem 3.7.1(iii)) for the Yosida approximation. Trivially, $J^{0,h}x = \mathfrak{A}^h x \to x$ as $h \downarrow 0$ (by the strong continuity of $\mathfrak{A}$). For $\beta > 0$ we have $\beta h^{-\beta} \int_0^h (h-s)^{\beta-1}\, ds = 1$, hence

$$\begin{aligned}
|J^{\beta,h}x - x|_X &= \left|\beta h^{-\beta} \int_0^h (h-s)^{\beta-1}(\mathfrak{A}^s x - x)\, ds\right|_X \\
&\le \beta h^{-\beta} \int_0^h (h-s)^{\beta-1}|\mathfrak{A}^s x - x|_X\, ds \\
&\le \sup_{0<s<h} |\mathfrak{A}^s x - x|_X \to 0 \text{ as } h \downarrow 0.
\end{aligned}$$

(iv) This is trivial if $\beta = 0$. Otherwise we write out $J^{\beta,v}x$ as an integral, change the order of integration, make the substitution $v = s + t/\alpha$, and use the definition of the Gamma function

$$\Gamma(\beta) = \int_0^\infty \mathrm{e}^{-t} t^{\beta-1}\, dt.$$

(v) This is trivial if $\beta = 0$. Otherwise we write out $J^{\beta,v}x$ as an integral, change the order of integration, make the substitution $v = s + (h-s)t$ in the inner integral, and use the fact that

$$\Gamma(\beta+1) = \beta\Gamma(\beta).$$

(vi) Clearly, by a change of integration variable

$$\begin{aligned}
\mathfrak{A}^t J^{\beta,h}x &= J^{\beta,h}\mathfrak{A}^t x \\
&= \beta h^{-\beta} \int_0^h (h-s)^{\beta-1}\mathfrak{A}^{s+t}x\, ds \\
&= \beta h^{-\beta} \int_t^{t+h} (h+t-v)^{\beta-1}\mathfrak{A}^v x\, dv \\
&\quad - \beta h^{-\beta} \int_0^t (t+h-v)^{\beta-1}\mathfrak{A}^v x\, dv \\
&= \left(\frac{t+h}{h}\right)^\beta J^{\beta,t+h}x - \beta h^{-\beta} \int_0^t (t+h-v)^{\beta-1}\mathfrak{A}^v x\, dv.
\end{aligned}$$

Thus, the claim in (vi) is equivalent to the claim that (for simplicity we have denoted $a^{h,\beta}$ by a)

$$\begin{aligned}
\beta \int_0^t &(t+h-v)^{\beta-1}\mathfrak{A}^v x\, dv \\
&= \int_0^t a(t-s) s^\beta J^{\beta,s}\, ds \\
&= \beta \int_0^t a(t-s) \int_0^s (s-v)^{\beta-1}\mathfrak{A}^v x\, dv\, ds \\
&= \beta \int_0^t \Big(\int_v^t a(t-s)(s-v)^{\beta-1}\, ds \Big) \mathfrak{A}^v x\, dv \\
&= \beta \int_0^t \Big(\int_0^{t-v} a(t-v-s) s^{\beta-1}\, ds \Big) \mathfrak{A}^v x\, dv,
\end{aligned}$$

which is implied by (5.3.4). The identity (5.3.4) can be derived from (5.3.3): we rewrite (5.3.4) in the form

$$\int_0^t a^{\beta,h}(s)(t-s)^\beta\, ds = (t+h)^\beta - h^\beta, \qquad t > 0,$$

and differentiate with respect to t. To prove (5.3.3) we can, e.g., first integrate by parts, use the fact that

$$\int_0^t (t-s)^{\beta-1} s^{-\beta}\, ds = B(\beta, 1-\beta) = \Gamma(\beta)\Gamma(1-\beta), \qquad t > 0,$$

observe that

$$\int_0^s a^{h,\beta}(v)\, dv = \frac{1}{\Gamma(\beta)\Gamma(1-\beta)} \int_0^s (s-w)^{-\beta}(w+h)^{\beta-1}\, dw,$$

and change the order of integration. To prove that $\int_0^\infty a^{h,\beta}(s)\, ds = 1$ it suffices to change the order of the two integrals.

(vii) The proof of this is a (much) simplified version of the proof of (vi).

(viii) This follows from (i) if $\beta \ge 1$ or $\gamma \ge 1$. If both $\beta < 1$ and $\gamma < 1$ then we can apply the operator A to the identity

$$\begin{aligned}
J^{\beta,h} J^{\gamma,k} x = B(\beta+1, \gamma+1) \\
\times \Big[&(h+k)^{\beta+\gamma} J^{\beta+\gamma,h+k} x \\
&- \int_0^k a^{h,\beta}(k-v) v^{\beta+\gamma} J^{\beta+\gamma,v} x\, dv \\
&- \int_0^h a^{k,\gamma}(h-v) v^{\beta+\gamma} J^{\beta+\gamma,v} x\, dv \Big],
\end{aligned} \tag{5.3.5}$$

where the notation is the same as in (vi), and use (i). We leave the verification of this identity (based on (vi)) to the reader. □

By Theorems 3.7.1(i) and 5.3.2(vii), we can approximate C by $C^{\beta,h} = CJ^{\beta,h}$ or $C_\alpha = CJ_\alpha$ for $\beta \geq 1$, $h > 0$, and $\alpha \in \rho(A)$ (compare this to the approximations of A defined in Definition 3.7.2). All these approximations belong to $\mathcal{B}(X_1; Y)$, and even to $\mathcal{B}(X; Y)$ in the case of C_α and $C^{\beta,h}$ with $h \geq 1$. Actually, it is possible to define $C^{\beta,h}$ in a different way which produces an operator in $\mathcal{B}(X; Y)$ for a larger range of β. This method is based on the following fact.

Lemma 5.3.3 *Let $1 \leq p \leq \infty$, and let $\mathfrak{C}\colon X \to L^p_{\mathrm{loc}}(\mathbb{R}^+; Y)$ be an L^p-well-posed output map for the C_0 semigroup $\mathfrak{A}$, and let C be the corresponding control operator. Then, for all $\beta > 0$ and $h > 0$,*

$$\begin{aligned} CJ^{\beta,h}x &= \beta h^{-\beta} C \int_0^h (h-s)^{\beta-1}\mathfrak{A}^s x\, ds \\ &= \beta h^{-\beta} \int_0^h (h-s)^{\beta-1}(\mathfrak{C}x)(s)\, ds, \qquad x \in X_1. \end{aligned}$$

Proof We get this formula by moving C inside the integral (this is permitted since $C \in \mathcal{B}(X_1; Y)$) and using the fact that $(\mathfrak{C}x)(s) = C\mathfrak{A}^s x$ for all $x \in X_1$ and (almost) all $s \geq 0$. □

By Hölder's inequality, the second integral characterization of $CJ^{\beta,h}$ given in Lemma 5.3.3 defines an operator in $\mathcal{B}(X; Y)$ for all $\beta > 1/p$ if $p > 1$, and for all $\beta \geq 1$ if $p = 1$. Therefore, we prefer to use this as the definition of $C^{\beta,h}$.

Definition 5.3.4 Let X and Y be Banach spaces, let $\mathfrak{A}$ be a C_0 semigroup on X with growth bound $\omega_{\mathfrak{A}}$, and let $\mathfrak{C}$ be an L^p-well-posed output map for $\mathfrak{A}$ with output space Y, with $1 \leq p \leq \infty$.

(i) For each $\alpha \in \rho(A)$, the *Yosida* (or *Abel*) *approximation* $C_\alpha \in \mathcal{B}(X; Y)$ of C with parameter α is given by

$$C_\alpha x = CJ_\alpha X = \alpha C(\alpha - A)^{-1}x, \qquad x \in X.$$

(ii) For each $h > 0$ and $\beta > 1/p$ if $p > 1$, $\beta \geq 1$ if $p = 1$, the *Cesàro approximation* $C^{\beta,h} \in \mathcal{B}(X; Y)$ of C of order β of C with parameter h is given by

$$C^{\beta,h}x = \beta h^{-\beta} \int_0^h (h-s)^{\beta-1}(\mathfrak{C}x)(s)\, ds, \qquad x \in X.$$

These approximations of C inherit many properties from the corresponding approximations of the identity.

Lemma 5.3.5 *Let X and Y be Banach spaces, let $\mathfrak{A}$ be a C_0 semigroup on X with growth bound $\omega_{\mathfrak{A}}$, and let $\mathfrak{C}$ be an L^p-well-posed output map for $\mathfrak{A}$*

with output space Y, with $1 \le p \le \infty$. Let $\alpha \in \rho(A)$, $h > 0$, and let $\beta > 1/p$ if $p > 1$, $\beta \ge 1$ if $p = 1$.

(i) *For all $\alpha \in \mathbb{C}^+_{\omega_{\mathfrak{A}}}$, C_α is given by*

$$C_\alpha x = \alpha \int_0^\infty e^{-\alpha s} (\mathfrak{C}x)(s)\, ds, \qquad x \in X.$$

(ii) *There exists a constant M (which depends on $\mathfrak{C}$, β, and ω but not on h and α) such that*

$$\begin{aligned} \|C_\alpha\| &\le (\alpha - \omega)^{1/p} M, && \alpha \ge \omega + 1 > \omega_{\mathfrak{A}} + 1, \\ \|C^{\beta,h}\| &\le h^{-1/p} M, && 0 < h \le 1. \end{aligned}$$

(iii) *The following identities are valid for all $x \in X$ and almost all $t > 0$:*

$$\begin{aligned} C_\alpha J^{\beta,h} x &= C^{\beta,h} J_\alpha x, \\ (\mathfrak{C}(J_\alpha x))(t) &= C_\alpha \mathfrak{A}^t x, \\ (\mathfrak{C}(J^{\beta,h} x))(t) &= C^{\beta,h} \mathfrak{A}^t x. \end{aligned}$$

(iv) *For all $x \in X$,*

$$C_\alpha x = \frac{\alpha^{\beta+1}}{\Gamma(\beta+1)} \int_0^\infty e^{-\alpha v} v^\beta C^{\beta,v} x\, dv,$$

where

$$\frac{\alpha^{\beta+1}}{\Gamma(\beta+1)} \int_0^\infty e^{-\alpha v} v^\beta\, dv = 1.$$

(v) *For all $\gamma > 0$ and all $x \in X$,*

$$C^{\beta+\gamma,h} x = \frac{\gamma h^{-(\beta+\gamma)}}{B(\beta+1, \gamma+1)} \int_0^h (h - v)^{\gamma-1} v^\beta C^{\beta,v} x\, dv,$$

where

$$\frac{\gamma h^{-(\beta+\gamma)}}{B(\beta+1, \gamma+1)} \int_0^h (h - v)^{\gamma-1} v^\beta\, dv = 1. \tag{5.3.2}$$

(vi) *For all $\beta \in (0, 1)$, all $x \in X$, and almost all $t > 0$,*

$$\mathfrak{C}(J^{\beta,h} x)(t) = \left(\frac{t+h}{h}\right)^\beta C^{\beta,t+h} x - h^{-\beta} \int_0^t a^{h,\beta}(t - s) s^\beta C^{\beta,s} x\, ds,$$

where

$$a^{h,\beta}(s) = \frac{\beta}{\Gamma(\beta)\Gamma(1-\beta)} \int_0^h (s + h - w)^{-\beta-1} w^{\beta-1}\, dw,$$

is nonnegative on $\overline{\mathbb{R}}^+$, $\int_0^\infty a^{h,\beta}(s)\,ds = 1$, *and*

$$\int_0^t a^{\beta,h}(t-s)s^\beta\,ds = (t+h)^\beta - h^\beta, \qquad t>0, \tag{5.3.3}$$

$$\int_0^t a^{\beta,h}(t-s)s^{\beta-1}\,ds = (t+h)^{\beta-1}, \qquad t>0. \tag{5.3.4}$$

(vii) *For all* $x \in X$ *and almost all* $t > 0$,

$$(\mathfrak{C}J^{1,h}x)(t) = \frac{t+h}{h}C^{1,t+h}x - \frac{t}{h}C^{1,t}x.$$

Proof The integral formula for C_α in (i) follows from the corresponding integral formula for J_α (see, e.g., Theorem 5.3.2(i)). To get the inequalities in (ii) we use Hölder's inequality and some simple estimates (the estimate for C_α can also be reduced to the one in Proposition 4.4.9). To get the remaining claims we multiply the formulas in Theorem 5.3.2 by $C \in \mathcal{B}(X_1;Y)$, and use the density of X_1 in X and the continuity of all the operators to get the result for all $x \in X$ (instead of $x \in X_1$). □

5.4 Extended observation operators

After the preliminary considerations in Section 5.3, we are now ready to study how the Yosida and Cesàro approximations C_α and $C^{\beta,h}$ behave as $\alpha \to +\infty$ (along the real axis) or $h \downarrow 0$.

Definition 5.4.1 Let X and Y be Banach spaces, let $\mathfrak{A}$ be a C_0 semigroup on X and let $\mathfrak{C}$ be an L^p-well-posed output map for $\mathfrak{A}$ with output space Y, with $1 \le p \le \infty$. Let $\beta > 1/p$ if $p > 1$, $\beta \ge 1$ if $p = 1$.

(i) The weak *Yosida* (or *Abel*) *extension* $\widetilde{C}_w$ of C is the operator

$$\widetilde{C}_w x = \lim_{\alpha\to+\infty} C_\alpha x = \lim_{\alpha\to+\infty} \alpha C(\alpha - A)^{-1}x,$$

defined for those $x \in X$ for which this limit exists in the weak sense. (Here $\alpha \to +\infty$ along the positive real axis.)

(ii) The *weak Cesàro extension* $\overline{C}^\beta_w$ of C of order β is the operator

$$\overline{C}^\beta_w x = \lim_{h\downarrow 0} C^{\beta,h}x = \lim_{h\downarrow 0} \beta h^{-\beta}\int_0^h (h-s)^{\beta-1}(\mathfrak{C}x)(s)\,ds,$$

defined for those $x \in X$ for which this limit exists in the weak sense.

(iii) The *strong Yosida* and *Cesàro extensions* $\widetilde{C}_s$ and $\overline{C}^\beta_s$ of C are defined in the same way as the weak extensions $\widetilde{C}_w$ and $\overline{C}^\beta_w$, but with weak limits replaced by strong limits.

(iv) By $\widetilde{C}$ we mean either $\widetilde{C}_w$ or $\widetilde{C}_s$, and by $\overline{C}^\beta$ we mean either $\overline{C}^\beta_w$ or $\overline{C}^\beta_s$, depending on the context.

These extensions of C will be used in two different ways. In Theorem 5.4.8 we extend the formula

$$(\mathfrak{C}x)(t) = C\mathfrak{A}^t x,$$

valid for all $x \in X_1$ and (almost) all $t > 0$, to all $x \in X$ by replacing C by either $\overline{C}^\beta$ or $\widetilde{C}$. In Section 5.6 we use these extensions as *compatible extensions* of C to which we apply the theory developed in Section 5.1 (and at the same time extend this theory slightly).

The domains of $\overline{C}^\beta$ and $\widetilde{C}$ can be made into Banach spaces as follows:

Definition 5.4.2 Make the same assumptions as in Definition 5.4.1, and denote the growth bound of $\mathfrak{A}$ by $\omega_{\mathfrak{A}}$.

(i) We define a norm $|\cdot|_{\widetilde{C}}$ on $\mathcal{D}(\widetilde{C})$ as follows: If $p = \infty$, then $|x|_{\widetilde{C}} = |x|_X$, and if $p < \infty$ then

$$\begin{aligned} |x|_{\widetilde{C}} &= |x|_X + \sup_{\alpha > 1+\omega_{\mathfrak{A}}} |C_\alpha x|_Y \\ &= |x|_X + \sup_{\alpha > 1+\omega_{\mathfrak{A}}} \left| \alpha \int_0^\infty \mathrm{e}^{-\alpha s} (\mathfrak{C}x)(s)\, ds \right|_Y \\ &= |x|_X + \sup_{\alpha > 1+\omega_{\mathfrak{A}}} \left| \alpha C(\alpha - A)^{-1} x \right|, \\ & x \in \mathcal{D}(\widetilde{C}_w) \text{ or } x \in \mathcal{D}(\widetilde{C}_s). \end{aligned}$$

(ii) We define a norm $|\cdot|_{\overline{C}^\beta}$ on $\mathcal{D}\left(\overline{C}^\beta\right)$ as follows. If $p = \infty$, then $|x|_{\overline{C}^\beta} = |x|_X$, and if $p < \infty$ then

$$\begin{aligned} |x|_{\overline{C}^\beta} &= |x|_X + \sup_{0<h<1} |C^{\beta,h} x|_Y \\ &= |x|_X + \sup_{0<h<1} \left| \beta h^{-\beta} \int_0^h (h-s)^{\beta-1} (\mathfrak{C}x)(s)\, ds \right|_Y, \\ & x \in \mathcal{D}\left(\overline{C}^\beta_w\right) \text{ or } x \in \mathcal{D}\left(\overline{C}^\beta_s\right). \end{aligned}$$

The main properties of the extensions $\widetilde{C}$ and $\overline{C}^\beta$ are listed in the following theorem.

Theorem 5.4.3 *Let X and Y be Banach spaces, let $\mathfrak{A}$ be a C_0 semigroup on X, and let $\mathfrak{C}$ be an L^p-well-posed output map for $\mathfrak{A}$ with output space Y and observation operator C. Let $\beta \geq 1$ if $p = 1$ or $\beta > 1/p$ if $p > 1$, and let $\gamma \geq \beta$. Then, both the weak and the strong versions of the following claims are true:*

(i) *If C is bounded then $\mathcal{D}(\widetilde{C}) = \mathcal{D}\left(\overline{C}^\beta\right) = X$ and $\widetilde{C}x = \overline{C}^\beta x = Cx$ for all $x \in X$. This is, in particular, true when $\mathfrak{C}$ is L^∞-well-posed.*

(ii) *The norms defined in Definition 5.4.2 make the domains of the operators $\overline{C}^\beta$ and $\widetilde{C}$ (both the weak and the strong versions) into Banach spaces, and these operators are continuous from their domains equipped with these norms into Y. Moreover,*

$$C \subset \overline{C}^\beta \subset \overline{C}^\gamma \subset \widetilde{C}$$

(i.e., $\widetilde{C}$ is an extension of $\overline{C}^\gamma$, which is an extension of $\overline{C}^\beta$, which is an extension of $\overline{C}^\gamma$, which is an extension of C), and

$$X_1 = \mathcal{D}(C) \subset \mathcal{D}\left(\overline{C}^\beta\right) \subset \mathcal{D}(\overline{C}^\gamma) \subset \mathcal{D}(\widetilde{C}) \subset X$$

with continuous embeddings. The embeddings $\mathcal{D}\left(\overline{C}^\beta\right) \subset X$ and $\mathcal{D}(\widetilde{C}) \subset X$ are always dense, and the embedding $X_1 \subset \mathcal{D}\left(\overline{C}_s^\beta\right)$ (i.e., the strong version of $\overline{C}^\beta$) is dense if $\beta \le 1$.[3] *The domains of the strong versions of these extensions are closed subspaces of the domains of the corresponding weak version. (Thus, the embeddings of X_1 in $\mathcal{D}\left(\overline{C}_w^\beta\right)$ and $\mathcal{D}(\widetilde{C}_w)$ cannot be dense unless $\overline{C}_w^\beta = \overline{C}_s^\beta$ and $\widetilde{C}_w = \widetilde{C}_s$.)*

(iii) *If $p < \infty$ and*

$$\lim_{h\downarrow 0} \frac{1}{h} \int_0^h |y - (\mathfrak{C}x)(s)|^p \, ds = 0,$$

then $x \in \mathcal{D}\left(\overline{C}^\beta\right) \subset \mathcal{D}(\widetilde{C})$ and

$$y = \overline{C}^\beta x = \widetilde{C}x.$$

(iv) *If $p < \infty$, $\beta > 1/p$, and*

$$\limsup_{h\downarrow 0} \frac{1}{h} \int_0^h |(\mathfrak{C}x)(s)|^p \, ds < \infty,$$

then $x \in \mathcal{D}(\widetilde{C})$ if and only if $x \in \mathcal{D}\left(\overline{C}^\beta\right)$.

The proof of this theorem is rather long. It is based on a number of auxiliary lemmas.

Lemma 5.4.4 *Let $1 \le p < \infty$, $u \in L^p_{\mathrm{loc}}(\mathbb{R}^+; U)$, $v \in U$, and $h > 0$. Then*

$$\begin{aligned}\left(\frac{1}{h}\int_0^h |u(s)|^p \, ds\right)^{1/p} - |v| &\le \left(\frac{1}{h}\int_0^h |u(s) - v|^p \, ds\right)^{1/p} \\ &\le \left(\frac{1}{h}\int_0^h |u(s)|^p \, ds\right)^{1/p} + |v|.\end{aligned}$$

[3] We do not know if the embeddings $X_1 \subset \mathcal{D}\left(\overline{C}_s^\beta\right)$ with $\beta > 1$ and $X_1 \subset \mathcal{D}(\widetilde{C}_s)$ are always dense. See also Remark 5.4.7.

Thus,

$$\limsup_{h\downarrow 0}\frac{1}{h}\int_0^h |u(s)|^p\,ds < \infty$$

if and only if

$$\limsup_{h\downarrow 0}\frac{1}{h}\int_0^h |u(s)-v|^p\,ds < \infty.$$

Proof This follows from the triangle inequality and the fact that $1/h\int_0^h |v|^p\,ds = |v|^p$. □

Lemma 5.4.5 *Let* $1 \le p < \infty$, $u \in L^p_{\mathrm{loc}}(\mathbb{R}^+;U)$, $v \in U$, $h > 0$, *and* $1/p + 1/q = 1$.

(i) *Let* $\beta \ge 1$ *if* $p = 1$ *and* $\beta > 1/p$ *if* $p > 1$. *Then, for all* $\gamma \ge \beta$,

$$\left|\gamma h^{-\gamma}\int_0^h (h-s)^{\gamma-1}u(s)\,ds - v\right| \le \sup_{0<t<h}\left|\beta t^{-\beta}\int_0^t (t-s)^{\beta-1}u(s)\,ds - v\right|.$$

(ii) *Let* $\beta \ge 1$ *if* $p = 1$ *and* $\beta > 1/p$ *if* $p > 1$. *Then*

$$\left|\beta h^{-\beta}\int_0^h (h-s)^{\beta-1}u(s)\,ds - v\right| \le K_\beta\left(\frac{1}{h}\int_0^h |u(s)-v|^p\,ds\right)^{1/p},$$

where $K_\beta = \begin{cases}\beta, & p = 1,\ \beta \ge 1,\\ \beta/(1+\beta q - q)^{1/q}, & p > 1,\ \beta > 1/p.\end{cases}$

(iii) *For all* $\beta > 0$,

$$\left|\beta h^{-\beta}\int_0^h (h-s)^{\beta-1}u(s)\,ds - v\right| \le \operatorname*{ess\,sup}_{0<s<h}|u(s)-v|.$$

(iv) *If* $u \in L^p_\omega(\mathbb{R}^+;U)$ *for some* $\omega \ge 0$, *then*

$$\left|\alpha\int_0^\infty \mathrm{e}^{-\alpha s}u(s)\,ds - v\right| \le \frac{\alpha^{\beta+1}}{\Gamma(\beta+1)(\alpha-\omega)^{\beta+1}}\int_0^\infty s^\beta \mathrm{e}^{-s}|K(s/(\alpha-\omega))|\,ds$$

for all $\alpha > \omega$, *and* $\beta \ge 1$ *if* $p = 1$, $\beta > 1/p$ *if* $p > 1$; *here*

$$K(t) = \mathrm{e}^{-\omega t}\beta t^{-\beta}\int_0^t (t-s)^{\beta-1}(u(s)-v)\,ds.$$

(v) *If $u \in L^p_\omega(\mathbb{R}^+; U)$ for some $\omega \ge 0$, $\beta \ge 1$ if $p = 1$, $\beta > 1/p$ if $p > 1$, and $\gamma \ge \beta$, then*

$$\begin{aligned}
&\lim_{h\downarrow 0} \operatorname*{ess\,sup}_{0<s<h} |u(s) - v| = 0 \\
&\Rightarrow \lim_{h\downarrow 0} \frac{1}{h} \int_0^h |u(s) - v|^p \, ds = 0 \\
&\Rightarrow v = \lim_{h\downarrow 0} \beta t^{-\beta} \int_0^t (t-s)^{\beta-1} u(s)\, ds \\
&\Rightarrow v = \lim_{h\downarrow 0} \gamma t^{-\gamma} \int_0^t (t-s)^{\gamma-1} u(s)\, ds \\
&\Rightarrow v = \lim_{\alpha\to+\infty} \alpha \int_0^\infty \mathrm{e}^{-\alpha s} u(s)\, ds.
\end{aligned}$$

(vi) *If $u \in L^p_\omega(\mathbb{R}^+; U)$ for some $\omega \ge 0$, $\beta > 1/p$, and*

$$\limsup_{h\downarrow 0} \frac{1}{h} \int_0^h |u(s)|^p \, ds < \infty,$$

then

$$v = \lim_{\alpha\to+\infty} \alpha \int_0^\infty \mathrm{e}^{-\alpha s} u(s)\, ds,$$

if and only if

$$v = \lim_{h\downarrow 0} \beta t^{-\beta} \int_0^t (t-s)^{\beta-1} u(s)\, ds.$$

Proof We may, without loss of generality, throughout the proof take $v = 0$, i.e., we move v inside the integrals and replace $u - v$ by u; note that

$$\beta h^{-\beta} \int_0^h (h-s)^{\beta-1}\, ds = \alpha \int_0^\infty \mathrm{e}^{-\alpha s}\, ds = 1.$$

(i) We change the order of integration in the right hand side of the formula below to get (this argument is identical to the proof of Theorem 5.3.2(v); note that the integrals are well-defined because of the restriction on β)

$$\begin{aligned}
\gamma h^{-\gamma} &\int_0^h (h-s)^{\gamma-1} u(s)\, ds \\
&= \frac{\Gamma(\gamma+1) h^{-\gamma}}{\Gamma(\gamma-\beta)\Gamma(\beta+1)} \\
&\quad \times \int_0^h v^\beta (h-v)^{\gamma-\beta-1} \left(\beta v^{-\beta} \int_0^v (v-s)^{\beta-1} u(s)\, ds \right) dv.
\end{aligned}$$

We take the norm of both sides, move the norm inside the integral on the right hand side, estimate the term inside the parentheses by its supremum over $(0, h)$, and (5.3.2) to get (i).

(ii) If $p = 1$ then we simply move the norm inside the integral, and estimate $|h - s|^{\beta-1}$ by $h^{\beta-1}$. For $p > 1$ we use Hölder's inequality as follows:

$$\begin{aligned}
&\left|\beta h^{-\beta} \int_0^h (h-s)^{\beta-1} u(s)\, ds\right| \\
&\quad \le \beta h^{-\beta} \int_0^h (h-s)^{\beta-1} |u(s)|\, ds \\
&\quad \le \beta h^{-\beta} \left(\int_0^h (h-s)^{q\beta-q} |u(s)|\, ds\right)^{1/q} \left(\int_0^h |u(s)|^p\, ds\right)^{1/p} \\
&\quad = \beta h^{-\beta} \frac{h^{\beta-1+1/q}}{(1+\beta q - q)^{1/q}} \left(\int_0^h |u(s)|^p\, ds\right)^{1/p} \\
&\quad = \frac{\beta}{(1+\beta q - q)^{1/q}} \left(\frac{1}{h}\int_0^h |u(s)|^p\, ds\right)^{1/p}.
\end{aligned}$$

(iii) This follows from Hölder's inequality.

(iv) We first observe that the following identities hold (to get the first identity we change the order of integration; to get the last we make a change of integration variable)

$$\begin{aligned}
&\alpha \int_0^\infty \mathrm{e}^{-\alpha s} u(s)\, ds \\
&\quad = \frac{\alpha^{\beta+1}}{\Gamma(\beta+1)} \int_0^\infty t^\beta \mathrm{e}^{-\alpha t} \beta t^{-\beta} \int_0^t (t-s)^{\beta-1} u(s)\, ds\, dt \\
&\quad = \frac{\alpha^{\beta+1}}{\Gamma(\beta+1)} \int_0^\infty t^\beta \mathrm{e}^{-(\alpha-\omega)t} K(t)\, dt \\
&\quad = \frac{\alpha^{\beta+1}}{\Gamma(\beta+1)(\alpha-\omega)^{\beta+1}} \int_0^\infty s^\beta \mathrm{e}^{-s} K(s/(\alpha-\omega))\, ds
\end{aligned}$$

By taking the norm of the first and last terms and moving the norm inside the integral in the last term we get (iv).

(v) The first implication is obvious, the second follows from (ii), and the third from (i). To get the last implication we observe that the function K in (iv) is bounded and $\lim_{s\downarrow 0} K(s) = 0$, hence the result follows from (iv) and the Lebesgue dominated convergence theorem.

(vi) One half of (vi) follows from (v). To prove the opposite direction we may assume that $u \in L^\infty(\mathbb{R}^+; U)$ if $p = \infty$ and that

$$\sup_{0<h<\infty} \frac{1}{h} \int_0^h |u(s)|^p\, ds < \infty \text{ if } p < \infty,$$

because otherwise we replace $u(s)$ by $\mathrm{e}^{-(\omega+1)s}u(s)$; this does not change the two limits since

$$\begin{aligned}\lim_{\alpha\to+\infty}\alpha\int_0^\infty \mathrm{e}^{-\alpha s}u(s)\,ds &= \lim_{\alpha\to+\infty}(\alpha-\omega-1)\int_0^\infty \mathrm{e}^{-\alpha s}u(s)\,ds\\ &= \lim_{\alpha\to+\infty}\alpha\int_0^\infty \mathrm{e}^{-\alpha s}\mathrm{e}^{-(\omega+1)s}u(s)\,ds\end{aligned}$$

and, since by (ii) and (iii),

$$\lim_{h\downarrow 0}\beta t^{-\beta}\int_0^t (t-s)^{\beta-1}\big(1-\mathrm{e}^{-(\omega+1)s}\big)u(s)\,ds = 0.$$

The result then follows from (ii), (iii), and Pitt's tauberian theorem (Hille and Phillips, 1957, Theorem 18.3.3, see also the remark in Hille and Phillips 1957, p. 508). □

Proof of Theorem 5.4.3 In this proof, whenever we apply Lemma 5.4.5 we replace u by $\mathfrak{C}x$ or $\tau^t\mathfrak{C}x$, and let v be an appropriate mean at zero of $\mathfrak{C}x$ or $\tau^t\mathfrak{C}x$.

(i) This follows from Theorem 4.4.2(ii), Lemma 5.4.5(v), and the fact that strong convergence implies weak convergence.

(ii) The inclusions of the operators and their domains follow from Theorem 4.4.2(i), Lemma 5.4.5(v), and the fact that strong convergence implies weak convergence.

To prove that the spaces are Banach spaces we have to prove completeness. The proofs are very similar in all the cases, so let us just treat, for example, $\mathcal{D}\left(\overline{C}^\beta_w\right)$.

Let x_n be a Cauchy sequence in the norm $|\cdot|_{\overline{C}^\beta}$. Then x_n has a limit x in X, hence $\mathfrak{C}x_n\to\mathfrak{C}x$ in $L^p_\omega(\mathbb{R}^+;Y)$ for all $\omega>\omega_{\mathfrak{A}}$. In particular, for all $h\in(0,1)$,

$$\int_0^h (h-s)^{\beta-1}(\mathfrak{C}x_n)(s)\,ds \to \int_0^h (h-s)^{\beta-1}(\mathfrak{C}x)(s)\,ds.$$

Moreover, since $h\mapsto \beta h^{-\beta}\int_0^h(h-s)^{\beta-1}(\mathfrak{C}x_n)(s)\,ds$ is a Cauchy sequence in $L^\infty((0,1);Y)$, the limit must satisfy

$$\sup_{0<h<1}\left|\beta h^{-\beta}\int_0^h (h-s)^{\beta-1}(\mathfrak{C}x)(s)\,ds\right|.$$

For each $y^*\in Y^*$, the sequence $h\mapsto y^*\beta h^{-\beta}\int_0^h(h-s)^{\beta-1}(\mathfrak{C}x_n)(s)\,ds$ is a Cauchy sequence in $C([0,1];\mathbb{C})$, and therefore the limit $h\mapsto y^*\beta h^{-\beta}\int_0^h(h-s)^{\beta-1}(\mathfrak{C}x)(s)\,ds$ belongs to $C([0,1];\mathbb{C})$. This proves that the weak limit

$$\lim_{h\downarrow 0}\beta h^{-\beta}\int_0^h (h-s)^{\beta-1}(\mathfrak{C}x)(s)\,ds$$

exists, hence $x \in \mathcal{D}\left(\overline{C}^{\beta}_{w}\right)$. Moreover, because of the uniform weak convergence,

$$\begin{aligned}\overline{C}^{\beta}_{w}x &= \lim_{h\downarrow 0}\lim_{n\to\infty} \beta h^{-\beta}\int_0^h (h-s)^{\beta-1}(\mathfrak{C}x_n)(s)\,ds \\ &= \lim_{n\to\infty}\lim_{h\downarrow 0} \beta h^{-\beta}\int_0^h (h-s)^{\beta-1}(\mathfrak{C}x_n)(s)\,ds = \lim_{n\to\infty}\overline{C}^{\beta}_{w}x_n.\end{aligned}$$

Thus, $\mathcal{D}\left(\overline{C}^{\beta}_{w}\right)$ is complete, and $\overline{C}^{\beta}_{w}$ is continuous $\mathcal{D}\left(\overline{C}^{\beta}_{w}\right) \to Y$.

It is obvious that the domains of the strong versions of these extensions are closed subspaces of the domains of the corresponding weak version, since the norms are the same.

The continuity of the given embeddings are either obvious or follow immediately from Lemma 5.3.5 and Lemma 5.4.5, with one minor exception, namely the embedding $\mathcal{D}(\overline{C}^{\gamma}) \subset \mathcal{D}(\widetilde{C})$. The continuity of this embedding follows from the closed graph theorem since the embedding $\mathcal{D}(\overline{C}^{\gamma}) \subset X$ is continuous, hence the embedding $\mathcal{D}(\overline{C}^{\gamma}) \subset \mathcal{D}(\widetilde{C})$ is closed and thus continuous.

The density of the embeddings $\mathcal{D}\left(\overline{C}^{\beta}\right) \subset X$ and $\mathcal{D}(\widetilde{C}) \subset X$ follow from the density of the embedding $X_1 \subset X$.

To show that X_1 is dense in $\mathcal{D}\left(\overline{C}^{\beta}\right)$ when $\beta \le 1$ it suffices to show that, for example, $J_\alpha x \to x$ in $\mathcal{D}\left(\overline{C}^{\beta}\right)$ as $\alpha \to +\infty$. We know that $J_\alpha x \to x$ in X (see Theorem 5.3.2(iii)), so to show convergence in $\mathcal{D}\left(\overline{C}^{\beta}\right)$ it is (necessary and) sufficient to show that $C^{\beta,h}J_\alpha x \to C^{\beta,h}x$ as $\alpha \to +\infty$, uniformly in $h \in [0,1]$. Let $x \in \mathcal{D}\left(\overline{C}^{\beta}\right)$ and $\epsilon > 0$. By Lemma 5.4.6 below, there is a $\delta > 0$ and $a > 0$ such that

$$|C^{\beta,h}J_\alpha x - \overline{C}^{\beta}x|_Y \le \epsilon/2, \qquad 0 < h < \delta, \qquad \alpha > a,$$

hence

$$|C^{\beta,h}J_\alpha x - C^{\beta,h}x|_Y \le \epsilon, \qquad 0 < h < \delta, \qquad \alpha > a.$$

For $h \ge \delta$ we can estimate (cf. Lemma 5.3.5)

$$|C^{\beta,h}J_\alpha x - C^{\beta,h}x|_Y \le \|C^{\beta,h}\||J_\alpha x - x|_X \le h^{-1/p}M|J_\alpha x - x|_X,$$

which tends to zero as $\alpha \to +\infty$, uniformly in $h \in [\delta, 1]$. Thus, $C^{\beta,h}J_\alpha x \to C^{\beta,h}x$ as $\alpha \to +\infty$, uniformly in $h \in [0,1]$, and we have shown that $J_\alpha x \to x$ in $\mathcal{D}\left(\overline{C}^{\beta}\right)$.

(iii)–(iv) These follow from Lemma 5.4.5(v)–(vi). □

Lemma 5.4.6 *Let X and Y be Banach spaces, let $\mathfrak{A}$ be a C_0 semigroup on X, and let $\mathfrak{C}$ be an L^p-well-posed output map for $\mathfrak{A}$ with output space Y and observation operator C. Let $\beta = 1$ if $p = 1$ or $1/p < \beta \le 1$ if $p > 1$, and let*

$\gamma \ge 0$. *Then, for each* $x \in \mathcal{D}\left(\overline{C}_s^\beta\right)$,

$$\overline{C}^\beta x = \lim_{h\downarrow 0, k\downarrow 0} C^{\beta,h} J^{\beta+\gamma,k} x = \lim_{h\downarrow 0, \alpha\to+\infty} C^{\beta,h} J_\alpha x, \tag{5.4.1}$$

where the limits are strong limits in Y. *If, instead,* $x \in \mathcal{D}\left(\overline{C}_w^\beta\right)$, *then the same statement is true if we replace the strong limits by weak limits.*

Proof We claim that, in order to prove (5.4.1), it suffices to prove a special case of (5.4.1), namely

$$\overline{C}^\beta x = \lim_{h\downarrow 0, k\downarrow 0} C^{\beta,h} J^{\beta,k} x. \tag{5.4.2}$$

Suppose that (5.4.2) holds. Then $|C^{\beta,h} J^{\beta,k} x - \overline{C}^\beta x|_Y \to 0$ as $h \downarrow 0$ and $k \downarrow 0$, and by Theorem 5.3.2(v),

$$\begin{aligned}
&|C^{\beta,h} J^{\beta+\gamma,k} x - \overline{C}^\beta x|_Y \\
&\quad\le \frac{\gamma k^{-(\beta+\gamma)}}{B(\beta+1,\gamma+1)} \int_0^k (k-v)^{\gamma-1} v^\beta |C^{\beta,h} J^{\beta,v} x - \overline{C}^\beta x|_Y \, dv \\
&\quad\le \sup_{0<v<k} |C^{\beta,h} J^{\beta,k} x - \overline{C}^\beta x|_Y .
\end{aligned}$$

A similar (slightly longer) computation (cf. Theorem 5.3.2(iv) and Lemma 5.4.5(iv)) shows that (5.4.2) implies that $\overline{C}^\beta x = \lim_{h\downarrow 0,\alpha\to+\infty} C^{\beta,h} J_\alpha x$. Thus, it suffices to prove (5.4.2).

The proof of (5.4.2) is slightly different depending on whether $\beta < 1$ or $\beta = 1$. Below we treat only the case $\beta < 1$, and leave the easier case $\beta = 1$ to the reader (replace Lemma 5.3.5(vi) by Lemma 5.3.5(vii)).

By Lemma 5.3.5(vi),

$$\begin{aligned}
&C^{\beta,h} J^{\beta,k} x - \overline{C}^\beta x \\
&\quad = \beta k^{-\beta} \int_0^k (k-s)^{\beta-1} (1+s/h)^\beta \left(C^{\beta,s+h} x - \overline{C}^\beta x\right) ds \\
&\quad\quad - \beta k^{-\beta} \int_0^k (k-s)^{\beta-1} h^{-\beta} \int_0^s a^{h,\beta}(s-v) v^\beta \left(C^{\beta,v} x - \overline{C}^\beta x\right) dv\, ds.
\end{aligned}$$

Let $\delta > 0$. We split the region $h > 0$, $k > 0$, $h + k < \delta$ in two parts, depending on whether $k \le h$ or $k > h$, and, by symmetry (recall that $C^{\beta,h} J^{\beta,k} = C^{\beta,k} J^{\beta,h}$), we may assume that $k \le h$. Then

$$|C^{\beta,h} J^{\beta,k} x - \overline{C}^\beta x|_Y \le I_1 + I_2$$

where

$$I_1 \le \beta k^{-\beta} \int_0^k (k-s)^{\beta-1}\bigl(1+s/h\bigr)^{\beta} |C^{\beta,s+h}x - \overline{C}^{\beta}x|_Y \, ds$$
$$\le \sup_{0<s<k} \bigl(1+s/h\bigr)^{\beta} |C^{\beta,s+h}x - \overline{C}^{\beta}x|_Y$$
$$\le 2^{\beta} \sup_{0<s<\delta} |C^{\beta,v}x - \overline{C}^{\beta}x|_Y \to 0 \text{ as } \delta \downarrow 0$$

and

$$I_2 \le \beta k^{-\beta} \int_0^k (k-s)^{\beta-1} h^{-\beta} \int_0^s a^{h,\beta}(s-v) v^{\beta} \bigl|C^{\beta,v}x - \overline{C}^{\beta}x\bigr|_Y \, dv \, ds$$
$$\le \sup_{0<s<k} h^{-\beta} \int_0^s a^{h,\beta}(s-v) v^{\beta} \bigl|C^{\beta,v}x - \overline{C}^{\beta}x\bigr|_Y \, dv$$
$$\le \sup_{0<v<k} h^{-\beta} v^{\beta} \bigl|C^{\beta,v}x - \overline{C}^{\beta}x\bigr|_Y$$
$$\le \sup_{0<v<\delta} \bigl|C^{\beta,v}x - \overline{C}^{\beta}x\bigr|_Y \to 0 \text{ as } \delta \downarrow 0.$$

This proves (5.4.2).

The proof for the weak case is identical to the one above, except for the fact that we replace $\mathfrak{C}$ by $y^*\mathfrak{C}$ and C by y^*C where $y^* \in Y^*$ is arbitrary. □

Remark 5.4.7 If $\beta \le 1$, then X_1 is dense in $\mathcal{D}\left(\overline{C}^{\beta}_w\right)$ in the topology induced by the seminorms

$$\|x\|_X + \sup_{0<h<1} |y^* C^{\beta,h} x|_Y, \qquad y^* \in Y^*.$$

The proof is the same as in the strong case.

The following theorem extends the formula

$$(\mathfrak{C}x)(t) = C\mathfrak{A}^t x,$$

valid for all $x \in X_1$ and (almost) all $t > 0$, to all $x \in X$ by replacing C by either $\overline{C}^{\beta}$ or $\widetilde{C}$.

Theorem 5.4.8 *Let X and Y be Banach spaces, let $\mathfrak{A}$ be a C_0 semigroup on X, and let $\mathfrak{C}$ be an L^p-well-posed output map for $\mathfrak{A}$ with output space Y and observation operator C. Let $\beta \ge 1$ if $p = 1$ or $\beta > 1/p$ if $p > 1$, and let $\gamma > \beta$. Then, both the weak and the strong versions of the following claims are true:*

(i) *If C is bounded then $(\mathfrak{C}x)(t) = C\mathfrak{A}^t x$ for all $x \in X$ and (almost) all $t \ge 0$. This is, in particular, true when $\mathfrak{C}$ is L^∞-well-posed.*

(ii) *If $p < \infty$, then $\mathfrak{A}^t x \in \mathcal{D}\left(\overline{C}^\beta\right)$ if and only if the limit $\lim_{h\downarrow 0} \beta h^{-\beta} \int_0^h (h-s)^{\beta-1}(\mathfrak{C}x)(t+s)\,ds$ exists, and in this case*

$$\overline{C}^\beta \mathfrak{A}^t x = \lim_{h\downarrow 0} \beta h^{-\beta} \int_0^h (h-s)^{\beta-1}(\mathfrak{C}x)(t+s)\,ds.$$

(iii) *If $p < \infty$, then $\mathfrak{A}^t x \in \mathcal{D}(\widetilde{C})$ if and only if the limit $\lim_{\alpha\to+\infty} \alpha \int_0^\infty \mathrm{e}^{-\alpha s}(\mathfrak{C}x)(t+s)\,ds$ exists, and in this case*

$$\widetilde{C}\mathfrak{A}^t x = \lim_{\alpha\to+\infty} \alpha \int_0^\infty \mathrm{e}^{-\alpha s}(\mathfrak{C}x)(t+s)\,ds.$$

(iv) *If $p < \infty$, $\beta > 1/p$, and*

$$\limsup_{h\downarrow 0} \frac{1}{h} \int_0^h |(\mathfrak{C}x)(t+s)|^p\,ds < \infty,$$

then $\mathfrak{A}^t x \in \mathcal{D}(\widetilde{C})$ if and only if $\mathfrak{A}^t x \in \mathcal{D}\left(\overline{C}^\beta\right)$ (and $\widetilde{C}\mathfrak{A}^t x = \overline{C}^\beta \mathfrak{A}^t x$).

(v) *If t is a right L^p Lebesgue point of $\mathfrak{C}x$ (see Definition 5.4.9), then $(\mathfrak{A}^t x)(t) \in \mathcal{D}\left(\overline{C}_s^\beta\right)$ and*

$$(\mathfrak{C}x)(t) = \overline{C}_s^\beta \mathfrak{A}^t x = \widetilde{C}_s \mathfrak{A}^t x. \tag{5.4.3}$$

In particular, $(\mathfrak{A}^t x)(t) \in \mathcal{D}\left(\overline{C}_s^\beta\right)$ and (5.4.3) holds for almost all $t \geq 0$.

Proof (i) See Theorem 4.4.2(ii).

(ii)–(iii) Combine the definitions of $\overline{C}^\beta$ and $\widetilde{C}$ with the fact that $\pi_+\tau^t\mathfrak{C}x = \mathfrak{C}\mathfrak{A}^t x$ for all $x \in X$ and $t \geq 0$.

(iv) Combine (ii) and (iii) with Theorem 5.4.3(iv).

(v) Combine (ii) and (iii) with Definition 5.4.9 and Theorem 5.4.3(iii). □

The preceding theorem used the following definition and the following auxiliary lemma.

Definition 5.4.9 The point $t \in \mathbb{R}$ is called a *right L^p Lebesgue point*, $1 \leq p < \infty$, of $u \in L^p_{\mathrm{loc}}(\mathbb{R}; U)$ if

$$\lim_{h\downarrow 0} \frac{1}{h} \int_0^h |u(t+s) - u(t)|_U^p\,ds = 0.$$

It is an L^∞ *Lebesgue point* of $u \in L^\infty_{\mathrm{loc}}(\mathbb{R}; U)$ if

$$\lim_{h\downarrow 0} \operatorname*{ess\,sup}_{0<s<h} |u(t+h) - u(t)|_U = 0.$$

Left L^p Lebesgue points are defined in the same way, with the interval $(0, h)$ replaced by the interval $(-h, 0)$. A point t is a (*double sided*) *L^p Lebesgue point* if it is both a right and a left Lebesgue point.

Note that, by Hölder's inequality, if t is an L^p Lebesgue point of u for some $p > 1$, then it is an L^r Lebesgue point for all $r \in [1, p]$.

Lemma 5.4.10 *Let $u \in L^p_{\mathrm{loc}}(\mathbb{R}; U)$ for some p, $1 \le p < \infty$. Then almost all $t \in \mathbb{R}$ are Lebesgue points of u.*

Proof After redefining u on a set of measure zero we may assume that the range of u is separable-valued, i.e., there is a countable set $\{v_n \mid n \in \mathbb{Z}^+\}$ which is dense in $\{u(t) \mid t \in \mathbb{R}\}$. For any $n \in \mathbb{Z}^+$, the function $\phi_n(t) = |u(t) - v_n|^p$ belongs to L^1_{loc}, so

$$\phi_n(t) = \lim_{h \to 0} \frac{1}{2h} \int_{-h}^{h} \phi_n(t+s)\, ds \tag{5.4.4}$$

for almost all $t \in \mathbb{R}$. Thus, there is a set E of measure zero such that (5.4.4) holds for all $n \in \mathbb{Z}^+$ if $t \notin E$. We claim that all $t \notin E$ are L^p Lebesgue points of u. Fix $t \notin E$. By the triangle inequality,

$$\begin{aligned}
&\left| \frac{1}{2h} \int_{-h}^{h} |u(t+s) - u(t)|^p\, ds \right|^{1/p} \\
&\quad \le \left| \frac{1}{2h} \int_{-h}^{h} |u(t+s) - v_n|^p\, ds \right|^{1/p} + |v_n - u(t)| \\
&\quad = \left[\left| \frac{1}{2h} \int_{-h}^{h} |u(t+s) - v_n|^p\, ds \right|^{1/p} - (\phi_n(t))^{1/p} \right] + 2|v_n - u(t)|.
\end{aligned}$$

For each $\epsilon > 0$ we can first choose n so large that $2|v_n - u(t)| < \epsilon/2$, and then choose h so small that the term inside the square bracket is less than $\epsilon/2$. Thus, the left hand side is less than ϵ. This shows that t is an L^p Lebesgue point of u. □

5.5 Extended observation/feedthrough operators

By using Theorems 5.4.3 and 5.4.8 we are able to extend the output formula

$$y(t) = (C\&D) \begin{bmatrix} x(t) \\ u(t) \end{bmatrix}, \qquad t \ge s,$$

in Theorem 4.6.11(i) for an L^p-well-posed linear system to arbitrary initial states $x_s \in X$ and inputs $u \in L^p_{\mathrm{loc}}([s, \infty); U)$. (The corresponding result for the L^∞-well-posed and *Reg*-well-posed cases is found in Corollary 4.5.5.) This requires an extension of the combined observation/feedthrough operator $C\&D$.

Definition 5.5.1 Let $\Sigma = \left[\begin{array}{c|c} \mathfrak{A} & \mathfrak{B} \\ \hline \mathfrak{C} & \mathfrak{D} \end{array}\right]$ be a $L^p|Reg$-well-posed linear system on (Y, X, U) with system node $S = \left[\begin{smallmatrix} A\&B \\ C\&D \end{smallmatrix}\right]$, semigroup generator A, control operator B, and observation operator C. Let $\alpha \in \rho(A)$, and let $\beta \ge 1$ if $p = 1$, $\beta > 1/p$ if $p > 1$.

(i) The *weak Yosida* (or *Abel*) *extension* $\widetilde{C\&D}_w$ of $C\&D$ is the operator

$$\widetilde{C\&D}_w \begin{bmatrix} x \\ u \end{bmatrix} = \widetilde{C}_w\big[x - (\alpha - A_{|X})^{-1}Bu\big] + \widehat{\mathfrak{D}}(\alpha)u.$$

with domain

$$\mathcal{D}(\widetilde{C\&D}_w) = \Big\{\begin{bmatrix} x \\ u \end{bmatrix} \in \begin{bmatrix} X \\ U \end{bmatrix} \;\Big|\; x - (\alpha - A_{|X})^{-1}Bu \in \mathcal{D}(\widetilde{C}_w)\Big\}.$$

(ii) The extensions $\overline{C\&D}^\beta_w$, $\widetilde{C\&D}_s$, and $\overline{C\&D}^\beta_s$ are defined in the same way, with $\widetilde{C}_w$ replaced by $\overline{C}^\beta_w$, $\widetilde{C}_s$, or $\overline{C}^\beta_s$, respectively.

(iii) By $\widetilde{C\&D}$ we mean either $\widetilde{C\&D}_w$ or $\widetilde{C\&D}_s$, and by $\overline{C\&D}^\beta$ we mean either $\overline{C\&D}^\beta_w$ or $\overline{C\&D}^\beta_s$, depending on the context.

Lemma 5.5.2 *The operators $\widetilde{C\&D}_w$, $\overline{C\&D}^\beta_w$, $\widetilde{C\&D}_s$, and $\overline{C\&D}^\beta_s$ in Definition 5.5.1 have the following properties.*

(i) *$\mathcal{D}(\widetilde{C\&D}_w)$ is a Banach space with norm*

$$\Big|\begin{bmatrix} x \\ u \end{bmatrix}\Big|_{\mathcal{D}(\widetilde{C\&D}_w)} = \Big\{|x|^2_X + |u|^2_U + |x - (\alpha - A_{|X})^{-1}Bu|^2_{\mathcal{D}(\widetilde{C}_w)}\Big\}^{1/2}.$$

(ii) *The operator $\widetilde{C\&D}_w$ belongs to $\mathcal{B}(\mathcal{D}(\widetilde{C\&D}_w); Y)$, and it does not depend on $\alpha \in \rho(A)$ (although the norm in $\mathcal{D}(\widetilde{C\&D}_w)$ does depend on α).*

(iii) *The same claims remain true for the operators $\overline{C\&D}^\beta_s$, $\widetilde{C\&D}_w$, and $\widetilde{C\&D}_s$ if we replace $\widetilde{C}_w$ by $\overline{C}^\beta_w$, $\widetilde{C}_s$, or $\overline{C}^\beta_s$, respectively.*

Proof (i) The operator $\big[1 \;\; (\alpha - A_{|X})^{-1}B\big]$ is a bounded operator $\begin{bmatrix} X \\ U \end{bmatrix}$ to X, hence a closed operator from $\mathcal{D}(\widetilde{C\&D}_w)$ to $\mathcal{D}(\widetilde{C}_w)$. We have defined the norm in $\mathcal{D}(\widetilde{C\&D}_w)$ to be the graph norm of this operator, and therefore it is a Banach space.

(ii) Clearly $\widetilde{C\&D}_w \in \mathcal{B}(\widetilde{C\&D}_w; Y)$. That this operator does not depend on the value of α follows from (4.7.2) and the fact that the restriction of $\widetilde{C\&D}_w$ to X_1 is C.

(iii) The proof stays the same if we replace $\widetilde{C}_w$ by $\overline{C}^\beta_w$, $\widetilde{C}_s$, or $\overline{C}^\beta_s$, respectively. □

We are now ready to present the following representation of the input/output map of an L^p-well-posed linear system.

Theorem 5.5.3 *Let $\Sigma = \left[\begin{smallmatrix} \mathfrak{A} & \mathfrak{B} \\ \mathfrak{C} & \mathfrak{D} \end{smallmatrix}\right]$ be an L^p-well-posed, $1 \le p < \infty$, linear system with growth bound $\omega_{\mathfrak{A}}$, and let $\omega > \omega_{\mathfrak{A}}$. Let $u \in L^p_{\omega,\mathrm{loc}}(\mathbb{R}; U)$, and let $x(t) = \mathfrak{B}^t u$, $t \in \mathbb{R}$, and $y = \mathfrak{D}u$ be the state trajectory and output function of*

Σ with initial time $-\infty$ and input function u (cf. Definition 2.5.8). Let $\beta \geq 1$ if $p = 1$, and $\beta > 1/p$ if $p > 1$.

(i) *If t is a right L^p Lebesgue point of u, then $\left[\begin{smallmatrix} x(t) \\ u(t) \end{smallmatrix}\right] \in \mathcal{D}\left(\overline{C\&D^\beta}\right)$ (the strong or the weak version; cf. Definition 5.4.1) if and only if the limit*

$$\overline{y}^\beta(t) = \lim_{h\downarrow 0} \beta/h^\beta \int_0^h (h-s)^{\beta-1} y(t+s)\,ds, \tag{5.5.1}$$

exists (in the strong or weak sense), in which case this limit is equal to

$$\overline{y}^\beta(t) = \overline{C\&D}^\beta \begin{bmatrix} x(t) \\ u(t) \end{bmatrix}. \tag{5.5.2}$$

In particular,

$$y(t) = \overline{C\&D}^\beta \begin{bmatrix} x(t) \\ u(t) \end{bmatrix}, \tag{5.5.3}$$

whenever t is a right L^p Lebesgue point of both u and y, and hence, for almost all $t \in \mathbb{R}$.

(ii) *If $u \in L^p_\omega(\mathbb{R}; U)$, then the conclusion of part (i) remains true if we replace $\overline{C\&D}^\beta$ by $\widetilde{C\&D}$ and the limit in (5.5.1) by*

$$\widetilde{y}(t) = \lim_{\alpha\to+\infty} \alpha \int_0^\infty e^{-\alpha s} y(t+s)\,ds. \tag{5.5.4}$$

The proof of this theorem uses the following lemma.

Lemma 5.5.4 *Let $1 \leq p < \infty$, and let $\mathfrak{D} \in TIC^p_\omega(U;Y)$ (see Definition 2.6.2). Let $0 < h < 1$, $\beta \geq 1$ if $p = 1$, and $\beta > 1/p$ if $p > 1$. Then there exist constants K_1 and K_2 such that*

$$\begin{aligned} &\left| \beta/h^\beta \int_0^h (h-s)^{\beta-1} (\mathfrak{D}\pi_+ u)(s)\,ds \right| \\ &\qquad \leq K_1 \left(\frac{1}{h} \int_0^h |(\mathfrak{D}\pi_+ u)(s)|^p\,ds \right)^{1/p} \\ &\qquad \leq K_2 \left(\frac{1}{h} \int_0^h |u(s)|^p\,ds \right)^{1/p} \end{aligned}$$

for all $u \in L^p_{\mathrm{loc}}(\mathbb{R}^+; U)$. In particular, if the last term in this chain of inequalities tends to zero as $h \downarrow 0$, then so do the other two.

Proof Define $\mathbf{D} = \pi_{[0,1)} \mathfrak{D} \pi_{[0,1)}$. Then

$$\left(\int_0^h |(\mathfrak{D}\pi_+ u)(s)|^p\,ds \right)^{1/p} \leq \|\mathbf{D}\| \left(\int_0^h |u(s)|^p\,ds \right)^{1/p},$$

and the result follows from Lemma 5.4.5(ii) with $v = 0$ and u replaced by $\mathfrak{D}\pi_+ u$. □

Proof of Theorem 5.5.3 (i) Let us first assume that $t = 0$ and $u(0) = 0$, and that zero is a right L^p Lebesgue point of u. We write $\pi_+ y = \pi_+ \mathfrak{D} u$ in the form $\pi_+ y = \pi_+ \mathfrak{D}(\pi_- u + \pi_+ u) = \mathfrak{C}\mathfrak{B} u + \mathfrak{D}\pi_+ u$, and integrate this identity to get

$$\begin{aligned} &\beta/h^\beta \int_0^h (h-s)^{\beta-1} y(s)\,ds \\ &\quad = \beta/h^\beta \int_0^h (h-s)^{\beta-1}\big((\mathfrak{C}\mathfrak{B}u)(s) + (\mathfrak{D}\pi_+ u)(s)\big)(s)\,ds. \end{aligned}$$

Let $h \downarrow 0$ and use Definitions 5.4.9 and 5.4.1 and Lemma 5.5.4 to get

$$\begin{aligned} \overline{y}^\beta(0) &= \lim_{h\downarrow 0} \beta/h^\beta \int_0^h (h-s)^{\beta-1} y(s)\,ds \\ &= \lim_{h\downarrow 0} \beta/h^\beta \int_0^h (h-s)^{\beta-1} (\mathfrak{C}\mathfrak{B}u)(s)\,ds \\ &= \overline{C}^\beta \mathfrak{B} u = \overline{C}^\beta x(0). \end{aligned}$$

In the case where $t = 0$ and $u(t) = 0$ this proves that the limit in (5.5.1) exists if and only if $x(0) \in \mathcal{D}\left(\overline{C}^\beta\right)$, and in this case $y(0) = \overline{C}^\beta x(0)$. Equivalently, since $u(0) = 0$, the limit in (5.5.1) exists if and only if $\left[\begin{smallmatrix} x(0) \\ u(0) \end{smallmatrix}\right] = \left[\begin{smallmatrix} x(0) \\ 0 \end{smallmatrix}\right] \in \mathcal{D}\left(\overline{C\&D}^\beta\right)$, and in this case (5.5.2) holds.

If $t \neq 0$ or $u(t) \neq 0$ then we apply the same argument to the function $v = \tau^t u - \mathrm{e}_\alpha u(t)$, where $\alpha \in \mathbb{C}^+_{\omega_{\mathfrak{A}}}$ and $\mathrm{e}_\alpha(s) = \mathrm{e}^{\alpha s}$ for $s \in \mathbb{R}$. Then $\mathfrak{B}u$ is replaced by $\mathfrak{B}^t u - (\alpha - A_{|X})^{-1} B u(t) = x(t) - (\alpha - A_{|X})^{-1} B u(t)$ (see Definition 2.2.6 and Theorem 4.2.1(i),(iii)) and y is replaced by $\tau^t y - \widehat{\mathfrak{D}}(\alpha) u(t)$ (see Definition 4.6.1). Thus, by the preceding argument, if t is a right L^p Lebesgue point of u, then the limit in (5.5.1) exists if and only if $x(t) - (\alpha - A_{|X})^{-1} B u(t) \in \mathcal{D}\left(\overline{C}^\beta\right)$ and

$$\begin{aligned} \overline{y}^\beta(t) &= [\overline{C}^\beta x(t) - (\alpha - A_{|X})^{-1} B u(t)] + \widehat{\mathfrak{D}}(\alpha) u(t) \\ &= \overline{C\&D}^\beta \begin{bmatrix} x(t) \\ u(t) \end{bmatrix}. \end{aligned}$$

If, furthermore, t is a right L^p Lebesgue point of y, then from the second implication in Lemma 5.4.5(v) with u replaced by y, we get $\overline{y}^\beta(t) = y(t)$, hence (5.5.3) holds. By Lemma 5.4.10, the set of points which are Lebesgue points of both u and y is dense in $\mathbb{R}$.

(ii) The proof stays essentially the same (only the exact references change slightly) if we replace $\overline{C\&D}^\beta$ by $\widetilde{C\&D}$, and we leave this part of the proof to the reader. (The stronger assumption that $u \in L^p_\omega(\mathbb{R}; U)$ is needed in Lemma

5.4.5(v), and it guarantees that the integral in (5.5.4) converges for α large enough.) □

A similar representation is valid for the output of an L^p-well-posed linear system with a finite initial time.

Theorem 5.5.5 *Let* $\Sigma = \left[\begin{smallmatrix} \mathfrak{A} & \mathfrak{B} \\ \mathfrak{C} & \mathfrak{D} \end{smallmatrix}\right]$ *be an* L^p*-well-posed,* $1 \le p < \infty$*, linear system with system node* $S = \left[\begin{smallmatrix} A\&B \\ C\&D \end{smallmatrix}\right]$*. Let* $s \in \mathbb{R}$*,* $x_s \in X$*,* $u \in L^p_{\mathrm{loc}}([s,\infty);U)$*, and let* x *and* y *be the state trajectory and output function of* Σ *with initial time* s*, initial state* x_s*, and input function* u *(cf. Definition 2.2.6). Let* $\beta \ge 1$ *if* $p = 1$*, and* $\beta > 1/p$ *if* $p > 1$*.*

(i) *For almost all* $t \in [s,\infty)$*,*

$$y(t) = \overline{C\&D}^{\beta}_{s} \begin{bmatrix} x(t) \\ u(t) \end{bmatrix}. \tag{5.5.5}$$

More precisely, (5.5.5) *holds at every point* t *which is a right* L^p *Lebesgue point of both* u *and* y*.*

(ii) *The statement (i) is also true if we replace* $\overline{C\&D}^{\beta}_{s}$ *by* $\overline{C\&D}^{\beta}_{w}$*,* $\widetilde{C\&D}_{s}$*, or* $\widetilde{C\&D}_{w}$*.*

Proof This follows from Definition 2.2.7 and Theorems 5.4.8(v), and 5.5.3. (Recall that $\overline{C}^{\beta}_{w}$, $\widetilde{C}_{s}$, or $\widetilde{C}_{w}$ are extensions of $\overline{C}^{\beta}_{s}$ and that $\overline{C\&D}^{\beta}_{w}$, $\widetilde{C\&D}_{s}$, or $\widetilde{C\&D}_{w}$ are extensions of $\overline{C\&D}^{\beta}_{s}$.) □

5.6 Regular systems

The notion of a *regular* system is an important special case of the notion of a compatible $L^p|Reg$-well-posed system. To define this notion we first define what we mean by the regularity of an input/output map:

Definition 5.6.1 Let $\omega \in \mathbb{R}$, let $\mathfrak{D} \in TIC_{\omega}(U;Y)$ satisfy the assumptions of Corollary 4.6.10 (i.e., $\mathfrak{D}$ has a well-posed realization), and denote the transfer function of $\mathfrak{D}$ by $\widehat{\mathfrak{D}}$.

(i) $\mathfrak{D}$ is *weakly regular* if the (weak) limit $\lim_{\alpha\to+\infty} y^*\widehat{\mathfrak{D}}(\alpha)u$ exists for all $u \in U$ and $y^* \in Y^*$.
(ii) $\mathfrak{D}$ is *strongly regular* if the (strong) limit $\lim_{\alpha\to+\infty} \widehat{\mathfrak{D}}(\alpha)u$ exists in Y for all $u \in U$.
(iii) $\mathfrak{D}$ is *uniformly regular* if the (uniform) $\lim_{\alpha\to+\infty} \widehat{\mathfrak{D}}(\alpha)$ exists in $\mathcal{B}(U;Y)$.

In each case the operator D which maps $u \in U$ into

$$Du = \lim_{\alpha \to +\infty} \widehat{\mathfrak{D}}(\alpha) u$$

is called the *feedthrough operator* of $\mathfrak{D}$.

Lemma 5.6.2 *The feedthrough operator D is bounded, i.e., $D \in \mathcal{B}(U; Y)$.*

Proof This follows from the uniform boundedness principle since the operators $\widehat{\mathfrak{D}}(\alpha)$ are bounded. □

Definition 5.6.3 By a *weakly*, *strongly*, or *uniformly regular* linear system $\Sigma = \left[\begin{smallmatrix} \mathfrak{A} & \mathfrak{B} \\ \mathfrak{C} & \mathfrak{D} \end{smallmatrix}\right]$ on (Y, X, U) we mean a $L^p|Reg$-well-posed system whose input/output map $\mathfrak{D}$ is weakly, strongly, or uniformly regular, and by the *feedthrough operator* of Σ we mean the feedthrough operator of $\mathfrak{D}$.

Several examples of weakly or strongly regular systems are given in Section 5.7. (Most of the systems appearing in practice seem to be regular.)

Theorem 5.6.4 *Every weakly regular system is* compatible. *In particular, for weakly regular systems the conclusion of Corollary 5.1.7 is valid with $W = \mathcal{D}(\widetilde{C}_w)$ and $C_{|W} = \widetilde{C}_w$, with D given by the feedthrough operator of $\widehat{\mathfrak{D}}$ (defined in Definition 5.6.1).*

Proof In the L^∞-case and *Reg*-case this follows from Lemma 5.1.2(i)–(ii), and in the L^p-case with $1 \le p < \infty$ from the equivalence of (i) and (iv) in Theorem 5.6.5 stated below. □

The following theorem lists a number of equivalent characterizations of the regularity of a system.

Theorem 5.6.5 *Let $\Sigma = \left[\begin{smallmatrix} \mathfrak{A} & \mathfrak{B} \\ \mathfrak{C} & \mathfrak{D} \end{smallmatrix}\right]$ be an L^p-well-posed, $1 \le p < \infty$, linear system with growth bound $\omega_{\mathfrak{A}}$, semigroup generator A, control operator B, observation operator C, and transfer function $\widehat{\mathfrak{D}}$. Let $\beta > 1/p$, let $\omega > \omega_{\mathfrak{A}}$, and define the different extensions of C as in Definition 5.4.1. Then both the strong and the weak versions of the following conditions are equivalent, and the strong versions of these conditions imply the weak versions. (In these conditions the text within square brackets gives different ways of computing the feedthrough operator D, and the conditions are still equivalent if this text is omitted.)*

(i) *Σ is regular [with feedthrough operator D].*
(ii) *The limit*

$$[Du =] \lim_{\alpha \to +\infty} \widehat{\mathfrak{D}}(\alpha) u$$

exists for all $u \in U$.

(iii) *The limit*

$$[Du =] \lim_{h\downarrow 0} \beta h^{-\beta} \int_0^h (h-s)^{\beta-1}(\mathfrak{D}1_+u)(s)\,ds$$

exists for each $u \in U$. *Here* $(1_+u)(t) = u$ *for* $t \geq 0$ *and* $(1_+u)(t) = 0$ *for* $t < 0$.

(iii′) *The limit*

$$[Du =] \lim_{\alpha\to+\infty} \alpha \int_0^\infty \mathrm{e}^{-\alpha t}(\mathfrak{D}1_+u)(s)\,ds$$

exists for each $u \in U$.

(iv) *For some* $\alpha \in \rho(A)$, $\mathcal{R}\left((\alpha - A_{|X})^{-1}B\right) \subset \mathcal{D}\left(\overline{C}^\beta\right)$ *[and* $D = \widehat{\mathfrak{D}}(\alpha) - \overline{C}^\beta(\alpha - A_{|X})^{-1}B$*]*.

(iv′) *For some* $\alpha \in \rho(A)$, $\mathcal{R}\left((\alpha - A_{|X})^{-1}B\right) \subset \mathcal{D}(\widetilde{C})$ *[and* $D = \widehat{\mathfrak{D}}(\alpha) - \widetilde{C}(\alpha - A_{|X})^{-1}B$*]*.

(v) *For all* $\alpha \in \rho(A)$, $\mathcal{R}\left((\alpha - A_{|X})^{-1}B\right) \subset \mathcal{D}\left(\overline{C}^\beta\right)$ *[and* $D = \widehat{\mathfrak{D}}(\alpha) - \overline{C}^\beta(\alpha - A_{|X})^{-1}B$*]*.

(v′) *For all* $\alpha \in \rho(A)$, $\mathcal{R}\left((\alpha - A_{|X})^{-1}B\right) \subset \mathcal{D}(\widetilde{C})$ *[and* $D = \widehat{\mathfrak{D}}(\alpha) - \widetilde{C}(\alpha - A_{|X})^{-1}B$*]*.

(vi) *For all* $u \in L^p_{\omega,\mathrm{loc}}(\mathbb{R}; U)$, *it is true for almost all* $t \in \mathbb{R}$ *that the state trajectory* x *of* Σ *with initial time* $-\infty$ *and input function* u *satisfies* $x(t) \in \mathcal{D}\left(\overline{C}^\beta\right)$ *[and the corresponding output* $y(t)$ *is given by* $y(t) = \overline{C}^\beta x(t) + Du(t)$*]*.

(vi′) *For all* $u \in L^p_{\omega,\mathrm{loc}}(\mathbb{R}; U)$, *it is true for almost all* $t \in \mathbb{R}$ *that the state trajectory* x *of* Σ *with initial time* $-\infty$ *and input function* u *satisfies* $x(t) \in \mathcal{D}(\widetilde{C})$ *[and the corresponding output* $y(t)$ *is given by* $y(t) = \widetilde{C}x(t) + Du(t)$*]*.

(vii) *For all* $s \in \mathbb{R}$, $x_s \in X$ *and all* $u \in L^p_{\mathrm{loc}}(\mathbb{R}^+; U)$, *it is true for almost all* $t \in [s, \infty)$ *that the state trajectory* x *of* Σ *with initial state* x_s, *input function* u, *and initial time* s *satisfies* $x(t) \in \mathcal{D}\left(\overline{C}^\beta\right)$ *[and the corresponding output* $y(t)$ *is given by* $y(t) = \overline{C}^\beta x(t) + Du(t)$*]*.

(vii′) *For all* $s \in \mathbb{R}$, $x_s \in X$ *and all* $u \in L^p_{\mathrm{loc}}(\mathbb{R}^+; U)$, *it is true for almost all* $t \in [s, \infty)$ *that the state trajectory* x *of* Σ *with initial state* x_s, *input function* u, *and initial time* s *satisfies* $x(t) \in \mathcal{D}(\widetilde{C})$ *[and the corresponding output* $y(t)$ *is given by* $y(t) = \widetilde{C}x(t) + Du(t)$*]*.

Proof By Definition 5.6.3, (i) $\Leftrightarrow$ (ii). The plan is to show that (ii) $\Leftrightarrow$ (iii′), that (iii) $\Leftrightarrow$ (iii′), that (iii) $\Leftrightarrow$ (iv) $\Leftrightarrow$ (v) and (iii′) $\Leftrightarrow$ (iv′) $\Leftrightarrow$ (v′), that (iv) $\Leftrightarrow$ (vi) and (iv′) $\Leftrightarrow$ (vi′), and that (vi) $\Leftrightarrow$ (vii) and (vi′) $\Leftrightarrow$ (vii′). This implies that all the listed conditions are equivalent.

(ii) ⇔ (iii′): This follows from Theorem 4.6.9 (the Laplace transform of 1_+u is u/α).

(iii) ⇒ (iii′): This follows from the last implication in Lemma 5.4.5(v) with u replaced by $\mathfrak{D}1_+u$.

(iii′) ⇒ (iii): This follows from the second inequality in Lemma 5.5.4 and Lemma 5.4.5(vi) (with u replaced by $\mathfrak{D}1_+u$).

(iii) ⇔ (iv) ⇔ (v): Define $x(t) = \mathfrak{B}^t 1_+u$ and $y = \mathfrak{D}1_+u$. Then zero is a right Lebesgue point of 1_+u and $x(0) = 0$, so by Theorem 5.5.3 with $t = 0$, $\left[\begin{smallmatrix} 0 \\ u \end{smallmatrix}\right] \in \mathcal{D}\left(\overline{C\&D}^\beta\right)$ if and only if the limit in (iii) exists. By Definition 5.5.1 and Lemma 5.5.2, $\left[\begin{smallmatrix} 0 \\ u \end{smallmatrix}\right] \in \mathcal{D}\left(\overline{C\&D}^\beta\right)$ if and only if $(\alpha - A_{|X})^{-1}Bu \in \mathcal{D}\left(\overline{C}^\beta\right)$ for some $\alpha \in \rho(A)$, or equivalently, for all $\alpha \in \rho(A)$. Moreover, if we denote the limit in (iii) by Du, then $Du = \overline{C\&D}^\beta \left[\begin{smallmatrix} 0 \\ u \end{smallmatrix}\right] = -\overline{C}^\beta(\alpha - A_{|X})^{-1}Bu + \widehat{\mathfrak{D}}(\alpha)u$ in this case.

(iii′) ⇔ (iv′) ⇔ (v′): This proof is the same as the one above, except that we replace the Cesàro means and extensions by the Yosida (or Abel) means and extensions.

(iv) ⇒ (vi) and (iv′) ⇒ (vi′): This follows from Theorem 5.5.3, since $\overline{C\&D}^\beta \left[\begin{smallmatrix} x \\ u \end{smallmatrix}\right] = \overline{C}^\beta\left[x - (\alpha - A_{|X})^{-1}Bu\right] + \widehat{\mathfrak{D}}(\alpha)u$ and $(\alpha - A_{|X})^{-1}Bu \in \mathcal{D}\left(\overline{C}^\beta\right)$ for all $u \in U$ (and the corresponding result is also true with $\overline{C\&D}^\beta$ and $\overline{C}^\beta$ replaced by $\widetilde{C\&D}$, respectively $\widetilde{C}$).

(vi) ⇒ (iv) and (vi′) ⇒ (iv′): Let $\alpha \in \mathbb{C}^+_{\omega_\mathfrak{A}}$, and let $u(t) = e^{\alpha t}u_0$, $t \in \mathbb{R}$, where $u_0 \in U$. Then, by Theorem 4.2.1(iii), the corresponding state of Σ with initial time $-\infty$ is given by $x(t) = e^{\alpha t}(\alpha - A_{|X})^{-1}Bu_0$, $t \in \mathbb{R}$. Clearly, this implies that if $x(t) \in \mathcal{D}\left(\overline{C}^\beta\right)$ for some $t \in \mathbb{R}$, then $(\alpha - A_{|X})^{-1}Bu_0 \in \mathcal{D}\left(\overline{C}^\beta\right)$. In particular, this is true if (vi) holds. The corresponding output is given by (see Lemma 4.5.3, and Definition 4.6.1) $y(t) = e^{\alpha t}\widehat{\mathfrak{D}}(\alpha)u_0$ for almost all $t \in \mathbb{R}$, which should be compared to the formula that we get from (vi), namely

$$y(t) = \overline{C}^\beta x(t) + Du(t) = e^{\alpha t}\left(\overline{C}^\beta(\alpha - A_{|X})^{-1}B + D\right)u_0.$$

Thus, if we let D stand for the operator in (vi), then $D = \widehat{\mathfrak{D}}(\alpha) - \overline{C}^\beta(\alpha - A_{|X})^{-1}B$, as (iv) says. The proof that (vi′) ⇒ (iv′) is analogous.

(vi) ⇒ (vii) and (vi′) ⇒ (vii′): This follows from Theorems 5.4.3(ii), 5.4.8(v) and 5.5.3 (in the last theorem we replace u by $\pi_{[s,\infty)}u$).

(vii) ⇒ (vi) and (vii′) ⇒ (vi′): Let $u \in L^p_{\omega,\mathrm{loc}}(\mathbb{R}; U)$, and define $x(t) = \mathfrak{B}^t u$ and $y = \mathfrak{D}u$. Fix an arbitrary $s \in \mathbb{R}$. Then, by Theorem 2.5.7, $x(t)$ for $t \geq s$ and $\pi_{[s,\infty)}y$ can be interpreted as the state and output of Σ with initial time s, initial value $x(s)$, and input u. Thus, by (vii), $x(t) \in \mathcal{D}\left(\overline{C}^\beta\right)$ and the output $y(t)$ is given by $y(t) = \overline{C}^\beta x(t) + Du(t)$ for almost all $t \geq s$. Since s is arbitrary, this proves (vi). (The proof that (vii′) ⇒ (vi′) is analogous.) □

In the case $p = 1$ we can add the following conclusion:

Theorem 5.6.6 *Every L^1-well-posed linear system is weakly regular, and hence, the weak versions of the equivalent conditions listed in Theorem 5.6.5 hold. In addition, it is possible to take $\beta = 1$ in this case.*

Proof To prove this theorem it suffices to show that the weak version of condition (iii) in Theorem 5.6.5 always holds with $\beta = 1$ whenever $p = 1$. (The assumption that $\beta > 1/p$ with $p = 1$ was needed only in the proof of the implication (iii′) ⇒ (iii), which was based on Lemma 5.4.5(vi). See also Lemma 5.4.5(v).)

Without loss of generality, suppose that the growth bound of Σ is negative (see Example 3.2.6(i)). Fix $y^* \in Y^*$ and $u \in U$. Then $v \mapsto y^*\mathfrak{D}(uv)$ is a causal time-invariant continuous mapping $L^1(\mathbb{R};\mathbb{C}) \to L^1(\mathbb{R};C)$. Let us denote this mapping by a. We claim that this operator is a convolution operator with a kernel which is a measure of finite total variation on $\overline{\mathbb{R}}^+$. This can be seen as follows. The adjoint operator a^* is an anti-causal time-invariant continuous mapping $L^\infty(\mathbb{R};\mathbb{C}) \to L^\infty(\mathbb{R};C)$. In particular, since $\tau^h u \to u$ in $L^\infty(\mathbb{R};C)$ whenever $u \in BC_0(\mathbb{R};\mathbb{C})$ (see Example 2.3.2(i)), this means that a^* maps $BC_0(\mathbb{R};\mathbb{C})$ into $C(\mathbb{R};\mathbb{C})$. Thus we can evaluate the resulting function at zero, and conclude that the mapping $u \mapsto (a^*u)(0)$ belongs to the dual of $BC_0(\mathbb{R};\mathbb{C})$. By the Riesz representation theorem, there is a measure μ on $\overline{\mathbb{R}}^+$ with finite total variation such that

$$(a^*u)(0) = \int_{\overline{\mathbb{R}}^+} u(s)\mu(ds)$$

(μ must be supported on $\overline{\mathbb{R}}^+$ since a^* is anti-causal). This together with the time-invariance of a^* implies that

$$(a^*u)(t) = \int_{\overline{\mathbb{R}}^+} u(t+s)\mu(ds),$$

which in turn implies that a^* maps $BC_0(\mathbb{R};\mathbb{C})$ into itself (see, e.g., Gripenberg *et al.* (1990, Theorem 6.1(iii), p. 97)). The adjoint a^{**} of a^*, which maps the dual of $BC_0(\mathbb{R};\mathbb{C})$ into itself, i.e., the space of measures with finite total variation on $\mathbb{R}$ into itself, is then convolution with the complex conjugate of μ. By restricting a^{**} to $L^1(\mathbb{R};\mathbb{C})$ we get the original operator a. This argument proves our claim that a is a convolution operator of the form

$$(au)(t) = (\overline{\mu} * u)(t) = \int_{\overline{\mathbb{R}}^+} u(t-s)\overline{\mu}(ds).$$

If we apply this to the Heaviside function 1_+, then we simply get

$$(a1_+)(t) = \overline{\mu}([0,t]), \qquad t \geq 0,$$

which tends to $\overline{\mu}(\{0\})$ as $t \downarrow 0$. Thus, Theorem 5.6.5(iii) holds in the weak sense with $\beta = 1$, and Σ is weakly regular. □

It is sometimes possible to prove regularity of an input/output map $\mathfrak{D}$ (or of a system whose input/output map is $\mathfrak{D}$) by appealing directly to time-domain properties of $\mathfrak{D}$ which imply part (iii) of Theorem 5.6.5. The following theorem gives some examples of this type. In addition, it shows that we can sometimes say more about the behavior of $\widehat{\mathfrak{D}}$ at infinity.

Theorem 5.6.7 *Let $\omega \in \mathbb{R}$, and let $\mathfrak{D} \in TIC_\omega(U;Y)$ have a well-posed realization. Then, in all the following cases $\mathfrak{D}$ is strongly regular with a zero feedthrough operator:*

(i) *$\mathfrak{D}$ is an operator of the type constructed in Theorem A.3.5.*
(ii) *$\|\mathfrak{D}\pi_+ u\|_{L^p|Reg([0,t);Y)} \le k(t)\|u\|_{L^p|Reg([0,t);U)}$ for all $u \in L^p|Reg_{\mathrm{loc}}(\overline{\mathbb{R}}^+;U)$, where $k(t) \to 0$ as $t \downarrow 0$. In this case $\mathfrak{D}$ is even uniformly regular.*
(iii) *$\mathfrak{D}$ is a convolution operator of the type given in Theorem A.3.4. In this case $\mathfrak{D}$ is even uniformly regular.*

Moreover, we have the following additional conclusions about $\widehat{\mathfrak{D}}$:

(i′) *In case (i), let $\omega' = \omega$ if $p = 1$ and $\omega' > \omega$ if $p > 1$. Then, for each $u \in U$, $\widehat{\mathfrak{D}}(\alpha)u \to 0$ as $|\alpha|$ tends to infinity in the half-plane $\Re\alpha \ge \omega'$.*
(ii′) *In case (ii), $\|\widehat{\mathfrak{D}}(\alpha)\|_{\mathcal{B}(U;Y)} \to 0$ as $\Re\alpha \to +\infty$, uniformly in $\Im\alpha$.*
(iii′) *In case (iii) the conclusions of (i′) and (ii′) hold (with the same choice of ω'). If, in addition, the function A in Theorem A.3.4 is measurable in the norm of $\mathcal{B}(X;Y)$, then $\|\widehat{\mathfrak{D}}(\alpha)\|_{\mathcal{B}(U;Y)} \to 0$ as $|\alpha|$ tends to infinity in the half-plane $\Re\alpha \ge \omega'$.*

Proof (i) In this case the function $\mathfrak{D}1_+u$ in Theorem 5.6.5(iii) is given by

$$(\mathfrak{D}1_+u)(t) = \int_0^t (\mathfrak{C}u)(s)\,ds.$$

This function is continuous with a limit zero at zero, so the condition in Theorem 5.6.5(iii) holds with $D = 0$ (cf. Lemma 5.4.5).

(ii) Again with reference to Theorem 5.6.5(iii), since $\|1_+u\|_{L^p([0,t);U)} = t^{1/p}\|u\|_U$ (where we take $p = \infty$ in the *Reg*-well-posed case), we get

$$t^{-1/p}\|\mathfrak{D}1_+u\|_{L^p|Reg([0,t);Y)} \le k(t), \qquad t > 0.$$

This together with Lemma 5.4.5(ii) and Theorem 5.6.5(iii) implies regularity with $D = 0$. The estimate above is uniform in u with $\|u\|_U \le 1$, and by checking the constants in our proof of the implication (iii) ⇒ (ii) in Theorem 5.6.5 we find that the limit in (ii) is uniform in u (see, in particular, the proof of Lemma

5.4.5(v) with u replaced by $\mathfrak{D}1_+u$ and $v = 0$, and move the norm inside the integral in the definition of $K(t)$ in Lemma 5.4.5(iv)). Thus $\mathfrak{D}$ is uniformly regular.

(iii) Both (i) and (ii) can be applied in this case.

(i′) This follows from Lemma 5.6.10(i) below, applied to the function $s \mapsto e^{-\omega' s}(\mathfrak{C}u)(s)$, which belongs to $L^1(\mathbb{R}^+; Y)$.

(ii′) Use the same argument as in the proof of (ii), but replace the function $\mathfrak{D}1_+u$ (where $u \in U$) by $\mathfrak{D}\pi_+v$, where $v(t) = e^{j\beta}(t)u$ and $\beta \in \mathbb{R}$. By the same argument that we gave above, the limit in Theorem 5.6.5(iii) is uniform both in $u \in U$ with $\|u\|_U \leq 1$ and $\beta \in \mathbb{R}$.

(iii′) We can apply both (i′) and (ii′) in this case. To prove the last claim we apply (i′) with X replaced by $\mathbb{C}$ and Y replaced by $\mathcal{B}(X; Y)$; this is possible since $A \in L^1(\mathbb{R}^+; \mathcal{B}(X; Y))$ in this case. □

By Theorem 5.6.4, every weakly regular system is compatible. A partial converse to this theorem is also true:

Theorem 5.6.8 *Let Σ be a compatible $L^p|Reg$-well-posed linear system on (Y, X, U), with extended observation operator $C_{|W} \in \mathcal{B}(W; Y)$, where W is a compatible extension of X_1.*

(i) *Σ is weakly regular if either of the following conditions hold:*
  (a) *The (weak) limit* $\lim_{\alpha\to+\infty} w^*\alpha(\alpha - A)^{-1}w = w^*Ew$ *exists for all $w^* \in W^*$ and $w \in W$ (and some $E \in \mathcal{B}(W)$). If $E = 1$ then X_1 is weakly dense in W, $W \subset \mathcal{D}(\widetilde{C}_w)$, and $C_{|W}$ is the restriction of $\widetilde{C}_w$ to W.*
  (b) *X_1 is weakly dense in W, and* $\limsup_{\alpha\to+\infty} \|\alpha(\alpha - A)^{-1}\|_{\mathcal{B}(W)} < \infty$. *In this case $W \subset \mathcal{D}(\widetilde{C}_w)$ and $C_{|W}$ is the restriction of $\widetilde{C}_w$ to W.*

(ii) *Σ is strongly regular if any one of the following conditions hold:*
  (a) *The (strong) limit* $\lim_{\alpha\to+\infty} \alpha(\alpha - A)^{-1}w = Ew$ *exists in W for all $w \in W$ (and some $E \in \mathcal{B}(W)$). If $E = 1$, then X_1 is dense in W, $W \subset \mathcal{D}(\widetilde{C}_s)$, and $C_{|W}$ is the restriction of $\widetilde{C}_s$ to W.*
  (b) *X_1 is dense in W, and* $\limsup_{\alpha\to+\infty} \|\alpha(\alpha - A)^{-1}\|_{\mathcal{B}(W)} < \infty$. *In this case X_1 is dense in W, $W \subset \mathcal{D}(\widetilde{C}_s)$, and $C_{|W}$ is the restriction of $\widetilde{C}_s$ to W.*
  (c) *W is invariant under $\mathfrak{A}$ and the restriction of $\mathfrak{A}$ to W is a C_0 semigroup on W. In this case X_1 is dense in W, $W \subset \mathcal{D}(\widetilde{C}_s)$, and $C_{|W}$ is the restriction of $\widetilde{C}_s$ to W.*

(iii) *Σ is uniformly regular if the (uniform) limit* $\lim_{\alpha\to+\infty} \alpha(\alpha - A)^{-1}$ *exists in $\mathcal{B}(W)$.*

Proof The proofs of cases (i)–(iii) are very similar to each other, so let us prove only one of them, namely (ii).

(ii)(a) For each $\beta \in \mathbb{C}$,

$$(\alpha - A)^{-1}(\beta - A) = (\alpha - A)^{-1}\big[\beta - \alpha + (\alpha - A)\big] = 1 + (\beta - \alpha)(\alpha - A)^{-1},$$

which tends strongly to $1 - E$ as $\alpha \to +\infty$. Thus, for each $\beta \in \rho(A)$ and $x \in W_{-1}$,

$$(\alpha - A)^{-1}x = (\alpha - A)^{-1}(\beta - A)(\beta - A)^{-1}x \to (1 - E)(\beta - A)^{-1}x$$

in W as $\alpha \to +\infty$. This implies that, for each $u \in U$,

$$\widehat{\mathfrak{D}}(\alpha)u = C_{|W}(\alpha - A)^{-1}Bu + Du \to C_{|W}(1 - E)(\beta - A)^{-1}Bu + Du$$

in Y as $\alpha \to +\infty$. This proves the strong regularity of Σ (and, at the same time we find that the feedthrough operator of Σ is $C_{|W}(1 - E)(\beta - A)^{-1}Bu + D)$. If $E = 1$, then every $w \in W$ is the limit in W of $\alpha(\alpha - A)^{-1}w$ as $\alpha \to +\infty$, hence X_1 is dense in W (recall that $\alpha(\alpha - A)^{-1}w \in W_1 \subset X_1$). This further implies that, for each $w \in W$, $\alpha C(\alpha - A)^{-1}w = \alpha C_w(\alpha - A)^{-1}w \to C_{|W}w$ in Y as $\alpha \to +\infty$, hence $w \in \mathcal{D}(\widetilde{C}_s)$ and $\widetilde{C}_s w = C_{|W}w$. (The feedthrough operator of Σ is D in this case.)

(ii)(b) By Theorem 5.3.2(iii), for each $x \in X_1$, $\alpha(\alpha - A)^{-1}x \to x$ in X_1 as $\alpha \to +\infty$, hence $\alpha(\alpha - A)^{-1}x \to x$ in W as $\alpha \to +\infty$. This together with the condition $\limsup_{\alpha\to+\infty}\|\alpha(\alpha - A)^{-1}\|_{\mathcal{B}(W)} < \infty$ implies that, for every $w \in W$, $\alpha(\alpha - A)^{-1}w \to w$ in W as $\alpha \to +\infty$. Thus (b) follows from (a) (with $E = 1$).

(ii)(c) This follows from (b) and Theorems 3.2.1(vi) and 3.2.9(ii) with X replaced by W. □

Corollary 5.6.9 *Let Σ be a compatible $L^p|Reg$-well-posed linear system on (Y, X, U), with extended observation operator $C_{|W} \in \mathcal{B}(W; Y)$, where $W = X_\alpha$ for some $\alpha \in [0, 1]$ is a compatible extension of X_1 (the spaces X_α are defined as in Section 3.9). Then Σ is regular, $W \subset \mathcal{D}(\widetilde{C}_s)$, and $C_{|W}$ is the restriction of $\widetilde{C}_s$ to W.*

Proof This follows from Theorem 5.6.8(ii)(c) (see also the discussion in Section 3.9). □

In proof of Theorem 5.6.7(i′) we used the following lemma.

Lemma 5.6.10 (Riemann–Lebesgue Lemma)

(i) *If $u \in L^1(\mathbb{R}^+; U)$, then $\hat{u}(z) \to 0$ as $|z| \to \infty$ in the closed right half-plane $\overline{\mathbb{C}}^+$.*

(ii) *If $u \in L^1(\mathbb{R}; U)$, then $\hat{u}(z) \to 0$ as $|z| \to \infty$ along the imaginary axis $j\mathbb{R}$.*

Proof (i) Let us first suppose that $u \in W_0^{1,1}(\overline{\mathbb{R}}^+; U)$. Then, by Proposition 3.13.2(i),(ii), the Laplace transform of u satisfies $z\hat{u}(z) = \widehat{\dot{u}}(z)$ in $\overline{\mathbb{C}}^+$, where $|\widehat{\dot{u}}(z)| \le \|\dot{u}\|_{L^1(\mathbb{R}^+;U)}$ for all z in $\overline{\mathbb{C}}^+$. In particular, $\hat{u}(z) = O(|z|^{-1})$ as $|z| \to \infty$ in $\overline{\mathbb{C}}^+$. Thus, the conclusion of part (i) of Lemma 5.6.10 holds whenever $u \in W_0^{1,1}(\overline{\mathbb{R}}^+; U)$.

Given an arbitrary $u \in L^1(\mathbb{R}^+; U)$ and $\epsilon > 0$ we can approximate u by a function $v \in W_0^{1,1}(\overline{\mathbb{R}}^+; U)$ so that $\|u - v\|_{L^1(\mathbb{R}^+;U)} \le \epsilon$ (note that $W_0^{1,1}(\overline{\mathbb{R}}^+; U)$ is dense in $L^1(\mathbb{R}^+; U)$ since it is the domain of the generator of the outgoing shift τ_+^* on $L^1(\mathbb{R}^+; U)$). We then have $|\hat{u}(z) - \hat{v}(z)| \le \epsilon$ for all $z \in \overline{\mathbb{C}}^+$, and hence $\limsup_{|z|\to\infty, z\in\overline{\mathbb{C}}^+} |\hat{u}(z)| \le \epsilon$. Here ϵ can be made arbitrarily small. Thus, $\lim_{|z|\to\infty, z\in\overline{\mathbb{C}}^+} \hat{u}(z) = 0$.

(ii) The proof of part (ii) is very similar to the one above (use the bilateral shift τ on $L^1(\mathbb{R}; U)$ instead of τ_+^*), and it is left to the reader. □

5.7 Examples of regular systems

In this section we present some examples of regular systems.

Lemma 5.7.1 *The following $L^p|Reg$-well-posed linear systems are strongly regular:*

- (i) *Systems with a bounded observation operator; in particular, systems that are L^∞-well-posed or Reg-well-posed.*
- (ii) *Systems with a bounded control operator; in particular, L^1-well-posed systems with a reflexive state space.*
- (iii) *Systems whose control operator B and observation operator C satisfy $B \in \mathcal{B}(U; X_{\alpha-1})$ and $C \in \mathcal{B}(X_\alpha; Y)$, where $\alpha \in [0, 1]$ and X_α is defined as in Section 3.9 (U is the input space and Y the output space).*
- (iv) *The delay line in Example 2.3.4.*

Proof Both (i) and (ii) are special cases of (iii), and (iii) is a reformulation of Corollary 5.6.9. That the delay line is regular follows from Example 4.8.1. □

Note that, by Theorem 5.6.6, L^1-well-posed linear systems are always weakly regular, even when the state space is not reflexive.

Example 5.7.2 *If the system Σ in Example 5.1.16 is weakly, strongly, or uniformly regular, then so are the systems in Example 5.1.16(i)–(iii) derived from Σ. If the systems Σ_1, and Σ_2 in Example 5.1.17 are weakly, strongly, or uniformly* *regular, then so are the systems in Example 5.1.17(i)–(iv) derived from Σ_1 and Σ_2.*

This is obvious.

Theorem 5.7.3 *Let $\mathfrak{A}$ be an analytic semigroup on X with generator A (cf. Definition 3.10.1), let $B \in \mathcal{B}(U; X_\beta)$, $C \in \mathcal{B}(X_\alpha; Y)$, and $D \in \mathcal{B}(U; Y)$, where $\beta, \alpha \in \mathbb{R}$ satisfy $\alpha - \beta < 1$, and the spaces X_β and X_α are defined as in Section 3.10. Define $\mathfrak{B}$, $\mathfrak{C}$, and $\mathfrak{D}$ by*

$$\begin{aligned}\mathfrak{B}u &= \int_{-\infty}^{0} \mathfrak{A}^{-s}_{|X_{\alpha-1}} Bu(s)\,ds,\\ \mathfrak{C}x &= \bigl(t \mapsto C\mathfrak{A}^t x, \qquad t \ge 0\bigr),\\ \mathfrak{D}u &= \left(t \mapsto \int_{-\infty}^{t} C\mathfrak{A}^{t-s}_{|X_{\alpha-1}} Bu(s)\,ds + Du(t), \qquad t \in \mathbb{R}\right),\end{aligned}$$

and let $\Sigma = \left[\begin{smallmatrix}\mathfrak{A} & \mathfrak{B}\\ \mathfrak{C} & \mathfrak{D}\end{smallmatrix}\right]$. Then

(i) *Σ is an L^1-well-posed linear system on (Y, X_γ, U) for all $\gamma \in (\alpha - 1, \beta]$.*
(ii) *Σ is an L^p-well-posed linear system, $1 < p < \infty$, on (Y, X_γ, U) for all $\gamma \in (\alpha - 1/p, \beta + 1 - 1/p)$.*
(iii) *Σ is an L^∞-well-posed and Reg-well-posed linear system on (Y, X_γ, U) for all $\gamma \in [\alpha, \beta + 1)$.*

In all cases the systems described above are uniformly regular, with an extended system node $\left[\begin{smallmatrix}A_{|X_\alpha} & B\\ C & D\end{smallmatrix}\right] \in \mathcal{B}(\left[\begin{smallmatrix}X_\alpha\\ U\end{smallmatrix}\right]; \left[\begin{smallmatrix}X_{\alpha-1}\\ Y\end{smallmatrix}\right])$. The function $t \mapsto C\mathfrak{A}^t_{|X_{\alpha-1}} B$ appearing in the definition of $\mathfrak{D}$ belongs to $L^1_{\text{loc}}\bigl(\overline{\mathbb{R}}^+, \mathcal{B}(U;Y)\bigr) \cap C\bigl(\mathbb{R}^+, \mathcal{B}(U;Y)\bigr)$; in particular, it is measurable in the operator-norm. The transfer function $\widehat{\mathfrak{D}}(\lambda) = C(\lambda - A_{|X_\alpha})^{-1}B + D$ satisfies

$$\begin{aligned}&\limsup_{\substack{|\lambda|\to\infty\\ |\arg\lambda|\le\theta}} \bigl(|\lambda|^\epsilon \|\widehat{\mathfrak{D}}(\lambda) - D\|\bigr)\\ &\quad= \limsup_{\substack{|\lambda|\to\infty\\ |\arg\lambda|\le\theta}} \bigl(|\lambda|^\epsilon \|C(\lambda - A_{|X_\alpha})^{-1}B\|\bigr) < \infty\end{aligned} \tag{5.7.1}$$

for some $\theta > \pi/2$; here $\epsilon = \min\{1, \beta - \alpha + 1\} > 0$. In particular

$$\lim_{\substack{|\lambda|\to\infty\\ \Re\lambda\ge 0}} \|\widehat{\mathfrak{D}}(\lambda) - D\| = 0.$$

Proof We claim that it suffices to prove the theorem under the extra assumption that $\beta \le \alpha$. If $\beta > \alpha$, then the assumption of the theorem holds with both β and α replaced by γ for every $\gamma \in [\alpha, \beta]$. We can then apply this theorem with

β and α replaced by γ, and by letting γ vary in the interval $[\alpha, \beta]$ we get the desired conclusion.

Let $1/p + 1/q = 1$ (with $1/\infty = 0$), let $\beta \le \alpha < \beta + 1$, and let γ satisfy the inequalities in (i)–(iii). If $\gamma < \beta$, then it follows from Proposition 2.3.1 that $\mathfrak{B}$ is an L^1-well-posed input map for $\mathfrak{A}$; hence it is an L^p-well-posed input map for every $p \ge 1$. If $\gamma > \alpha$, then, by Proposition 2.3.1, $\mathfrak{C}$ is an L^∞-well-posed output map for $\mathfrak{A}$; hence it is an L^p-well-posed output map for every $p \ge 1$. The assumptions $B \in \mathcal{B}(U; X_\beta)$ and $C \in \mathcal{B}(X_\alpha; Y)$ combined with Lemma 3.10.9 implies that there are constants $M \ge 1$ and $\omega \in \mathbb{R}$ such that (in the two first inequalities we assume that $\beta \le \gamma \le \alpha$)

$$\|\mathfrak{A}^t_{|X_{\alpha-1}} B\|_{\mathcal{B}(U;X_\gamma)} \le M t^{-(\gamma-\beta)} e^{\omega t}, \qquad t > 0,$$

$$\|C\mathfrak{A}^t_{|X_\gamma}\|_{\mathcal{B}(X_\gamma;Y)} \le M t^{-(\alpha-\gamma)} e^{\omega t}, \qquad t > 0,$$

$$\|C\mathfrak{A}^t_{|X_{\alpha-1}} B\|_{\mathcal{B}(U;Y)} \le M t^{-(\alpha-\beta)} e^{\omega t}, \qquad t > 0.$$

The first inequality combined with Hölder's inequality (and the listed restrictions on γ) implies that $\mathfrak{B}$ maps $L^p|Reg_c(\mathbb{R}^-; U)$ continuously into X_γ, hence $\mathfrak{B}$ is an $L^p|Reg$-well-posed input map for $\mathfrak{A}$. The second inequality shows that $\mathfrak{C}$ is an $L^p|Reg$-well-posed output map for $\mathfrak{A}$. The third inequality combined with Theorem A.3.7 implies that $\mathfrak{D}$ maps $L^p|Reg_{c,\mathrm{loc}}(U; Y)$ continuously into itself. To show that $\pi_+\mathfrak{D}\pi_- = \mathfrak{C}\mathfrak{B}$ we move $C\mathfrak{A}^t_{|X_\gamma}$ out of the integral (this is permitted since $C\mathfrak{A}^t_{|X_\gamma} \in \mathcal{B}(X_\gamma; Y)$ and the integral converges absolutely for almost all t) to get, for (almost) all $t > 0$,

$$(\mathfrak{D}u)(t) = C\mathfrak{A}^t_{|X_\gamma} \int_{-\infty}^0 \mathfrak{A}^{-s}_{|X_{\alpha-1}} Bu(s)\, ds = (\mathfrak{C}\mathfrak{B}u)(t).$$

Thus, $\Sigma = \left[\begin{smallmatrix} \mathfrak{A} & \mathfrak{B} \\ \mathfrak{C} & \mathfrak{D} \end{smallmatrix}\right]$ is an $L^p|Reg$-well-posed linear system on $(Y, X_\gamma; U)$.

To prove condition (5.7.1) it suffices to pick any sector $\Sigma_{\theta',\omega}$ on which A is sectorial, let $\pi/2 < \theta < \theta'$, and use Lemma 3.10.9. □

Example 5.7.4 *Let $Y = l^2(\mathbb{Z}^+; \mathbb{C})$. We let $\mathbb{C}^+ = \{z \in \mathbb{C} \mid \Re z > 0\}$, $a > 1$, and define $f : \mathbb{C}^+ \to \mathbb{C}$ and $g : \mathbb{C}^+ \to Y$ by*

$$f(s) = \frac{s}{(1+s)^2}, \qquad s \in \mathbb{C}^+,$$

$$g(s) = \left[f(s)\ f(s/a)\ f(s/a^2)\ f(s/a^3) \cdots \right], \qquad s \in \mathbb{C}^+.$$

Then g is the transfer function of an L^2-well-posed linear system which is weakly regular but not strongly regular. The dual of this system is strongly regular but *not uniformly regular. Moreover, if we choose a to be large enough, then the limit* $\lim_{\lambda\to+\infty}\|g(\lambda)\|_Y$ *does not exist.*

Proof Clearly

$$|f(s)| \le \min\{|s|^{-1}, |s|\}, \qquad s \in \mathbb{C}^+, \tag{5.7.2}$$

since $|(1+s)| \ge \max\{|s|, 1\}$ for all $s \in \mathbb{C}^+$. Using (5.7.2) we can show that g is bounded on C_0. Indeed, for all $s \in \mathbb{C}^+$,

$$\begin{aligned}
\|g(s)\|^2 &= \sum_{k=0}^{\infty} \left|f(a^{-k}s)\right|^2 \le \sum_{k=0}^{\infty} \min\{a^{2k}|s|^{-2}, a^{-2k}|s|^2\} \\
&\le \sum_{k=-\infty}^{\infty} \min\{a^{2k}|s|^{-2}, a^{-2k}|s|^2\} \\
&\le \sum_{k=-\infty}^{\infty} \min\{e^{2(k\log a-\log|s|)}, e^{-2(k\log a-\log|s|)}\} \\
&= \sum_{k=-\infty}^{\infty} e^{-2\left|\log|s|-k\log a\right|}.
\end{aligned}$$

The function $h(\gamma) = \sum_{k=-\infty}^{\infty} e^{-2|\gamma-k\log a|}$ is periodic in γ with period $\log a$. Elementary computations show that its maximum value is $h_{\max} = (1 + a^{-2})/(1 - a^{-2})$, at $\gamma = k\log a$, and its minimum value is $h_{\min} = 2/(a - a^{-1})$, at $\gamma = (k + 1/2)\log a$, where $k \in \mathbb{Z}$. In particular, this shows that g is bounded on $\mathbb{C}^+$.

Let us show that g is analytic. For any $y = \{y_k\}_{k=0}^{\infty} \in Y$ we have

$$\langle g(s), y\rangle = \sum_{k=1}^{\infty} f(a^{-k}s)\bar{y}_k.$$

Using the estimate $|f(s)| \le |s|$ (see (5.7.2)), we see that the partial sums of the above series converge uniformly on compact subsets of $\mathbb{C}^+$. These partial sums are analytic, so $\langle g, y\rangle$ is analytic for any $y \in Y$. It follows that g is weakly analytic, hence analytic.

Let us next show that $\lim_{\lambda\to+\infty}\langle g(\lambda), y\rangle = 0$ for all $y \in Y$, i.e., g is weakly regular. Let $y = \{y_k\}_{k=0}^{\infty}$ and $\epsilon > 0$. We can find $n \in \mathbb{Z}^+$ such that

$$\sum_{k=n}^{\infty} |y_k|^2 \le \frac{\epsilon^2}{h_{max}}.$$

Then, for $\lambda \in (0, \infty)$, by (5.7.2), the Schwarz inequality and the estimate $g(\lambda) \leq h_{max}$ obtained earlier,

$$
\begin{aligned}
|\langle g(\lambda), y\rangle| &\leq \sum_{k=0}^{\infty} |f(a^{-k}\lambda) y_k| \\
&\leq \sum_{k=0}^{n-1} |f(a^{-k}\lambda) y_k| + \Big(\sum_{k=n}^{\infty} |f(a^{-k}\lambda)|^2\Big)^{1/2} \Big(\sum_{k=n}^{\infty} |y_k|^2\Big)^{1/2} \\
&\leq \sum_{k=0}^{n-1} a^k \lambda^{-2} |y_k| + (h_{max})^{\frac{1}{2}} \Big(\sum_{k=n}^{\infty} |y_k|^2\Big)^{1/2} \\
&\leq \sum_{k=0}^{n-1} a^k \lambda^{-2} |y_k| + \epsilon.
\end{aligned}
$$

For sufficiently large λ the sum is less than 2ϵ. Thus, $\langle g(\lambda), y\rangle \to 0$ as $\lambda \to \infty$. This proves that g is weakly regular.

Let us show that g is not strongly regular. Strong regularity of g would mean that $\lim_{\lambda\to+\infty} g(\lambda) = 0$ in Y, since the feedthrough operator of g is zero. But this is not true, since for all positive integers k, $\|g(a^k)\| > |f(1)| = 1/4$.

By Theorem 10.3.5, g is the transfer function of some operator $\mathfrak{D} \in TIC^2(\mathbb{C}; Y)$. Let $\Sigma = \left[\begin{smallmatrix} \mathfrak{A} & \mathfrak{B} \\ \mathfrak{C} & \mathfrak{D} \end{smallmatrix}\right]$ be an arbitrary L^2-well-posed realization of this transfer function. Then Σ is a weakly regular system with feedthrough operator $D = 0$, but Σ is not strongly regular. By Theorem 6.2.15(ii)–(iii), the causal dual system is weakly regular but not uniformly regular. As the output space of the dual system is $\mathbb{C}$, weak regularity is equivalent to strong regularity in this case.

We can also show that for certain a, $\lim_{\lambda\to+\infty} \|g(\lambda)\|_Y$ does not exist. (In fact, this is true for any $a > 1$, but it is easiest to prove it for large a.) On one hand, we observed above that $\|g(a^k)\| > |f(1)| = 1/4$ for all positive integers k. On the other hand, we have $\|g(a^{k+1/2})\| \leq h_{min}^{1/2} = \big(2/(a - a^{-1})\big)^{1/2}$ for all positive integers k. For a large enough, the latter value is smaller than the former, so $\lim_{\lambda\to+\infty} \|g(\lambda)\|_Y$ does not exist (at least) for these values of a. □

5.8 Comments

Section 5.1 The notion of a *compatible system* was introduced by Helton (1976, p. 148) and adapted by Fuhrmann (1981), but it did not attract a great deal of attention at that time. (Helton's definition differs slightly from ours in the sense that he does not require $C_{|W}$ to be continuous with respect to the norm in W.) The version of compatibility that we use was independently rediscovered by

Mikkola while he was working on his thesis (Mikkola, 2002 which also contains a number of additional 'regularity' notions). Helton also introduces a stronger version of a uniformly regular system in Helton (1976, p. 155) (he requires $\widehat{\mathfrak{D}}(z)$ to tend to D in operator norm as $\Re z \to \infty$, uniformly in $\Im z$.) Lemma 5.1.11 was proved only recently in Staffans and Weiss (2002) (in the Hilbert space case with $p = 2$). The example in Remark 5.1.14 is due to George Weiss (private communication).

Section 5.2 Boundary control problems for partial differential equations are the most important single source of interesting well-posed linear systems, and the literature about these problems is huge. The formalism where the basic building blocks in the construction of the solution are Δ and Γ was introduced by Fattorini (1968), who studies a mixed boundary/interior control problem of the type presented in Theorem 5.2.16 with the output equal to the state, and gives several examples of such boundary control systems. Since then work on the boundary control problem for partial differential equations has been carried on by, for example, Avalos and Lasiecka (1996), Avalos *et al.* (1999), Barbu (1998), Bensoussan *et al.* (1992), Bradley and Horn (1995), Chang and Lasiecka (1986), Curtain and Ichikawa (1996), Curtain and Salamon (1986), Da Prato and Ichikawa (1985), Da Prato *et al.* (1986), Delfour *et al.* (1986), Desch *et al.* (1985), Flandoli (1984, 1986, 1987, 1993), Hansen and Zhang (1997), Hendrickson and Lasiecka (1993, 1995), Ho and Russell (1983), Horn (1992a, b, 1994, 1996, 1998a, b), Horn and Lasiecka (1994), Komornik (1997), Lagnese (1977, 1978, 1980, 1983a, b, 1989, 1995), Lagnese and Lions (1988), Lasiecka (1980, 1992), Lasiecka *et al.* (1995), Lasiecka *et al.* (1997), Lasiecka and Triggiani (1981, 1983a, b, c, 1986, 1987a, b, c, 1988, 1989a, b, 1990a, b, 1991b, c, d, 1992a, b, c, 1993a, b, 2000a, b), Lions (1971), Lions and Magenes (1972), McMillan and Triggiani (1994a, b, c), Morgül (1990, 1994), Pandolfi (1989, 1998), Russell (1971, 1973a, b, 1978), and Triggiani (1979, 1980a, b, 1988, 1989, 1991, 1992, 1993a, b, 1994a, b, 1997). Boundary control systems also arise naturally in the theory of control of delay equation; see, e.g., Bensoussan *et al.* (1992), Delfour and Karrakchou (1987), Pandolfi (1990, 1995), and Salamon (1984, 1987). Our presentation in Section 5.2 has been modeled after Salamon (1987, Section 2.2) (some of the details were worked out by Jarmo Malinen).

Section 5.3 The summability results presented in this section are classic. This section can be regarded as an expanded version of Hille and Phillips (1957, Sections 10.6 and 18.2).

Section 5.4 The strong Yosida and Cesàro extensions (of order one) of C were introduced in G. Weiss (1989b, Section 4) and G. Weiss (1994a, Definition 5.6).

He calls the former the 'Λ-extension' and the latter the 'Lebesgue' extension.[4] The Cesàro extensions of order different from one are new. Theorems 5.4.3 and 5.4.8 and Lemmas 5.4.5, 5.4.6, and 5.4.10 (with $\beta = \gamma = 1$) is due to G. Weiss (1989b, Proposition 4.3), G. Weiss (1994a, Remark 5.7 and Lemma 6.1), and G. Weiss (1994b, Section 5). G. Weiss (1994a, Theorem 5.2) proves Lemma 5.4.5(iv) directly (for $p > 1$ and $\beta = 1$) without using Pitt's tauberian theorem as we do.

Section 5.5 Theorem 5.5.3 is formally new, but the main part of this theorem is mentioned in G. Weiss (1989c) without proof. Theorem 5.5.5 is an expanded version of Staffans and Weiss (2002, Theorem 3.2).

Section 5.6 Uniform regularity was introduced by Ober and Montgomery-Smith (1990) and Ober and Wu (1996), strong regularity by G. Weiss (1989c), and weak regularity by M. Weiss and G. Weiss (1997) and Staffans and G. Weiss (2002). The strong version of Theorem 5.6.5 with $p > 1$ is due to G. Weiss (1994a, Theorem 5.8), and the weak version is due to Staffans and Weiss (2002). The case $p = 1$ appears to be new. For the proof of this theorem it is important that we define regularity in the case $p = 1$ in a slightly different way from that which G. Weiss (1994a, Remark 5.9) does: we use the Yosida extension of C whereas G. Weiss uses the Cesàro extension of order one (these produce the same class of regular systems for $p > 1$, but we do not know if this is true when $p = 1$). A strong version of Theorem 5.6.6 was announced in G. Weiss (1994a) without proof. All the different type of regularity (weak, strong, or uniform) are *realization independent*, i.e., they depend only on the input/output map of the system and its transfer function. On the contrary, the compatibility property appears to be realization dependent.

Section 5.7 Theorem 5.7.3 (which is part of the folklore in the literature for parabolic partial differential equations) is found in, e.g., Staffans (1998e) and Staffans and Weiss (1998). Example 5.7.4 is taken from Staffans and Weiss (2004).

[4] In his early work G. Weiss uses a nonstandard definition of 'Lebesgue point.' In that terminology, a function u has a Lebesgue point at t if the Cesàro mean of order one of u exists at t and is equal to $u(t)$. Most authors define a Lebesgue point in the same way as we do in Definition 5.4.9 with $p = 1$.

6

Anti-causal, dual, and inverted systems

In this chapter we introduce various transformations on a system. An anti-causal system evolves in the backward time direction. To get the flow-inverted system we interchange the roles of the input and the output. Time-inversion means that we reverse the direction of time. To get a time-flow-inverted system we perform both of these transformations at the same time. Both well-posed and non-well-posed versions of these transformations are given.

6.1 Anti-causal systems

Up to now we have only considered *causal* systems which are well-posed in the *forward time direction*, i.e., we have always chosen the initial time to be smaller than the final time. It is possible to develop a completely analogous theory for *anti-causal* systems which are well-posed in the *backward time direction*. These systems appear naturally, e.g., when we want to pass from a system to its dual.

To get an anti-causal system it suffices to take a causal system and reverse the time direction as follows.

Definition 6.1.1 Let U, X, and Y be Banach spaces, and let $1 \le p \le \infty$. An *anti-causal $L^p|Reg$-well-posed linear system* Σ on (Y, X, U) consists of a quadruple $\Sigma = \left[\begin{smallmatrix} \mathfrak{A} & \mathfrak{B} \\ \mathfrak{C} & \mathfrak{D} \end{smallmatrix}\right]$ satisfying the following conditions:

(i) the operator family $\widetilde{\mathfrak{A}}^t = \mathfrak{A}_{-t}$, $t \ge 0$, is a C_0 semigroup on X;
(ii) $\mathfrak{B}\colon L^p|Reg_c(\overline{\mathbb{R}}^+; U) \to X$ satisfies $\mathfrak{A}_s \mathfrak{B} u = \mathfrak{B} \tau^s \pi_+ u$ for all $u \in L^p|Reg_c(\overline{\mathbb{R}}^+; U)$ and all $s \le 0$;
(iii) $\mathfrak{C}\colon X \to L^p|Reg_{\mathrm{loc}}(\mathbb{R}^-; Y)$ satisfies $\mathfrak{C}\mathfrak{A}_s x = \pi_- \tau^s \mathfrak{C} x$ for all $x \in X$ and all $s \le 0$;

(iv) $\mathfrak{D}\colon L^p|Reg_{\mathrm{loc},c}(\mathbb{R};U) \to L^p|Reg_{\mathrm{loc},c}(\mathbb{R};Y)$ satisfies $\tau^s\mathfrak{D}u = \mathfrak{D}\tau^s u$, $\pi_+\mathfrak{D}\pi_- u = 0$, and $\pi_-\mathfrak{D}\pi_+ u = \mathfrak{C}\mathfrak{B}u$ for all $u \in L^p|Reg_{\mathrm{loc},c}(\mathbb{R};U)$ and all $s \in \mathbb{R}$.

The different components of Σ are given the same names as in the causal case, with the prefix 'backward'. Thus, $\mathfrak{A}$ is the *backward semigroup*, $\mathfrak{B}$ is the *backward input map*, $\mathfrak{C}$ is the *backward output map*, and $\mathfrak{D}$ is the *backward input/output map*.

Here $L^p|Reg_{\mathrm{loc},c}$ and $Reg_{\mathrm{loc},c}$ stand for the subspace of functions in $L^p|Reg_{\mathrm{loc}}$ whose support is bounded to the right.

By comparing Definition 6.1.1 with Definitions 2.2.1 and 2.2.3 we find the following obvious connection (see also Definition 3.5.12):

Lemma 6.1.2 *The quadruple $\left[\begin{smallmatrix}\mathfrak{A} & \mathfrak{B}\\ \mathfrak{C} & \mathfrak{D}\end{smallmatrix}\right]$ is an anti-causal $L^p|Reg$-well-posed linear system on (Y, X, U) if and only if the reflected quadruple*

$$\left[\begin{array}{c|c}\widetilde{\mathfrak{A}}^t & \widetilde{\mathfrak{B}}\\ \hline \widetilde{\mathfrak{C}} & \widetilde{\mathfrak{D}}\end{array}\right] = \left[\begin{array}{c|c}\mathfrak{A}_{-t} & \mathfrak{C}\mathbf{Я}\\ \hline \mathbf{Я}\mathfrak{B} & \mathbf{Я}\mathfrak{D}\mathbf{Я}\end{array}\right]$$

is an $L^p|Reg$-well-posed linear system on $\mathbb{R}$.

Definition 6.1.3 Let U, X, and Y be Banach spaces, and let $\Sigma = \left[\begin{smallmatrix}\mathfrak{A} & \mathfrak{B}\\ \mathfrak{C} & \mathfrak{D}\end{smallmatrix}\right]$ be an anti-causal $L^p|Reg$-well-posed linear system on (Y, X, U). For each $t \in \mathbb{R}$, $x^t \in X$, $s \le t$, and $u \in L^p|Reg_{\mathrm{loc}}((-\infty, t); U)$ we define *the backward state trajectory* x and *backward output function* y of Σ with initial time t, initial state x^t, and input function u by

$$\begin{aligned} x(s) &= \mathfrak{A}_{s-t}x^t + \mathfrak{B}\tau^s\pi_{[s,t)}u, \qquad s \le t,\\ y &= \tau^{-t}\mathfrak{C}x^t + \mathfrak{D}\pi_{(-\infty,t)}u. \end{aligned} \tag{6.1.1}$$

In particular, if the initial time t is zero and the initial state x_0, then

$$\begin{aligned} x(s) &= \mathfrak{A}_s x_0 + \mathfrak{B}\tau^s\pi_{[s,0)}u, \qquad s \le 0,\\ y &= \mathfrak{C}x_0 + \mathfrak{D}\pi_{(-\infty,0]}u. \end{aligned} \tag{6.1.2}$$

The definition of the backward state trajectory and backward output function in Definition 6.1.3 is motivated in part by the following result.

Lemma 6.1.4 *Let x and y be the backward state trajectory and backward output function of the anti-causal system Σ with initial state x_0, initial time t, and input function u (cf. Definition 6.1.3). Let $\widetilde{\Sigma}$ be the corresponding causal system described in Lemma 6.1.2, and let $s \in \mathbb{R}$. Define $\tilde{x}$ and $\tilde{y}$ by $\tilde{x}(v) = x(s + t - v)$, $v \ge s$, and $\tilde{y} = \mathbf{Я}_{(s+t)/2}y$. Then x and y are the state trajectory and output function of $\widetilde{\Sigma}$ with initial state x_0, initial time s, and input function $\tilde{u} = \mathbf{Я}_{(s+t)/2}u$. Conversely, if $\tilde{x}$ and $\tilde{y}$ are the state trajectory and output function of the causal* *system $\widetilde{\Sigma}$ with initial state x_0, initial time s, and input function $\tilde{u}$, then the functions x and y defined by $x(v) = \tilde{x}(s+t-v)$, $v \le t$, and $y = \mathbf{Я}_{(s+t)/2}\tilde{y}$ are the backward state trajectory and backward output function of Σ with initial state x_0, initial time t, and input function $u = \mathbf{Я}_{(s+t)/2}\tilde{u}$.*

We leave the easy (algebraic) proof to the reader.

We define the system node of an anti-causal system in the following way.

Definition 6.1.5 Let $\Sigma = \left[\begin{smallmatrix}\mathfrak{A} & \mathfrak{B}\\ \mathfrak{C} & \mathfrak{D}\end{smallmatrix}\right]$ be an anti-causal $L^p|Reg$-well-posed linear system on (Y, X, U). The *system node* S of Σ is the operator

$$S = \begin{bmatrix} A\&B \\ C\&D \end{bmatrix} = \begin{bmatrix} -\widetilde{A\&B} \\ \widetilde{C\&D} \end{bmatrix} = \begin{bmatrix} -1 & 0 \\ 0 & 1 \end{bmatrix} \widetilde{S},$$

where $\widetilde{S} = \begin{bmatrix} \widetilde{A\&B} \\ \widetilde{C\&D} \end{bmatrix}$ is the system node of the corresponding causal system described in Lemma 6.1.2.

Observe that the domain of S in this definition is the same as the domain of $\widetilde{S}$.

The reason for this definition is the following result.

Theorem 6.1.6 *Let $\Sigma = \left[\begin{smallmatrix}\mathfrak{A} & \mathfrak{B}\\ \mathfrak{C} & \mathfrak{D}\end{smallmatrix}\right]$ be an anti-causal $L^p|Reg$-well-posed linear system on (Y, X, U) with system node S.*

(i) *If Σ is L^p-well-posed for some $p < \infty$ then, for each $t \in \mathbb{R}$, $u \in W^{1,p}_{\mathrm{loc}}((-\infty, t]; U)$, and $x^t \in X$ satisfying $\left[\begin{smallmatrix} x^t \\ u(t) \end{smallmatrix}\right] \in \mathcal{D}(S)$, the backward state trajectory x and the backward output function y of Σ with initial time t, initial value x^t and input function u satisfy $x \in C^1((-\infty, t]$, $\left[\begin{smallmatrix} x \\ u \end{smallmatrix}\right] \in C((-\infty, t]; \mathcal{D}(S))$, $y \in W^{1,p}_{\mathrm{loc}}((-\infty, t]; Y)$, and*

$$\begin{bmatrix} \dot{x}(s) \\ y(s) \end{bmatrix} = S \begin{bmatrix} x(s) \\ u(s) \end{bmatrix}, \qquad s \le t. \tag{6.1.3}$$

(ii) *The same claims are true if Σ is L^∞-well-posed or Reg-well-posed and $u \in Reg^1_{\mathrm{loc}}((-\infty, t]; U)$, provided we replace the claim $y \in W^{1,p}_{\mathrm{loc}}((-\infty, t]; Y)$ by $y - Du \in C^1((-\infty, t]; Y)$, where D is the feedthrough operator of Σ. In this case $y(s) = C_{|X}x(s) + Du(s)$ for all $s \le t$, where $C_{|X}$ stands for the unique extension of C to $\mathcal{B}(X; Y)$.*

Proof This follows from Lemma 6.1.4 and Theorem 4.6.11. □

Clearly, we could add a number of conclusions to this theorem by transforming the rest of Theorem 4.6.11, too, into an anti-causal setting.

For later use, let us develop the theory of anti-causal systems a little further, along the same lines as we developed the theory for causal systems in Section 2.2. We begin with the following analogue of Definition 2.2.6.

Definition 6.1.7 Let $\Sigma = \left[\begin{smallmatrix} \mathfrak{A} & \mathfrak{B} \\ \mathfrak{C} & \mathfrak{D} \end{smallmatrix}\right]$ be an anti-causal $L^p|Reg$-well-posed linear system on (Y, X, U).

(i) We interpret $\mathfrak{B}$ as an operator $L^p|Reg_{\mathrm{loc},c}(\mathbb{R}; U) \to X$ by defining $\mathfrak{B}u = \mathfrak{B}\pi_+ u$ for all $u \in L^p|Reg_{\mathrm{loc},c}(\mathbb{R}; U)$.

(ii) We interpret $\mathfrak{C}$ as an operator $X \to L^p|Reg_{\mathrm{loc},c}(\mathbb{R}; Y)$ by defining $\mathfrak{C}x = \pi_-\mathfrak{C}x$ for all $x \in X$.

(iii) The *backward state transition map* $\mathfrak{A}_s^t$, the *backward input map* $\mathfrak{B}_s^t$, the *backward output map* $\mathfrak{C}_s^t$, and the *backward input/output map* $\mathfrak{D}_s^t$ with initial time $t \in \mathbb{R}$ and final time $s \le t$ are defined by

$$\begin{bmatrix} \mathfrak{A}_s^t & \mathfrak{B}_s^t \\ \mathfrak{C}_s^t & \mathfrak{D}_s^t \end{bmatrix} := \begin{bmatrix} \mathfrak{A}_{s-t} & \mathfrak{B}\tau^s\pi_{[s,t)} \\ \pi_{[s,t)}\tau^{-t}\mathfrak{C} & \pi_{[s,t)}\mathfrak{D}\pi_{[s,t)} \end{bmatrix}, \qquad s \le t.$$

(iv) The *backward input map* $\mathfrak{B}_s$ and the *backward input/output map* $\mathfrak{D}_s$ with final time $s \in \mathbb{R}$ (and initial time $+\infty$) are defined by

$$\mathfrak{B}_s = \mathfrak{B}_s^\infty := \mathfrak{B}\tau^s, \qquad \mathfrak{D}_s = \mathfrak{D}_s^\infty := \pi_{[s,\infty)}\mathfrak{D}, \qquad t \in \mathbb{R}.$$

(v) The *backward output map* $\mathfrak{C}^t$ and the *backward input/output map* $\mathfrak{D}^t$ with initial time $t \in \mathbb{R}$ (and final time $-\infty$) are defined by

$$\mathfrak{C}^t = \mathfrak{C}_{-\infty}^t := \tau^{-t}\mathfrak{C}, \qquad \mathfrak{D}^t = \mathfrak{D}_{-\infty}^t := \mathfrak{D}\pi_{(-\infty,t)}, \qquad t \in \mathbb{R}.$$

We can reinterpret the algebraic conditions listed in Definition 6.1.1 in terms of the operators introduced in Definition 6.1.7 as follows.

Lemma 6.1.8 *Let $\Sigma = \left[\begin{smallmatrix} \mathfrak{A} & \mathfrak{B} \\ \mathfrak{C} & \mathfrak{D} \end{smallmatrix}\right]$ be a anti-causal $L^p|Reg$-well-posed linear system on (Y, X, U), and let $\widetilde{\Sigma} = \left[\begin{smallmatrix} \widetilde{\mathfrak{A}} & \widetilde{\mathfrak{B}} \\ \widetilde{\mathfrak{C}} & \widetilde{\mathfrak{D}} \end{smallmatrix}\right]$ be the corresponding causal L^p-well-posed linear system (cf. Lemma 6.1.2). For each $h \in \mathbb{R}$, define (see also Definition 3.5.12)*

$$\boldsymbol{\mathcal{R}}_h = \tau^{-h}\boldsymbol{\mathcal{R}}\tau^h (= \boldsymbol{\mathcal{R}}\tau^{2h} = \tau^{-2h}\boldsymbol{\mathcal{R}}). \tag{6.1.4}$$

Then the (backward) maps introduced in Definition 6.1.7 (applied to the anti-causal system Σ) and the (forward) maps introduced in Definition 2.2.6 (applied to the causal system $\widetilde{\Sigma}$) are related to each other in the following way for all $s \le t$

$$\begin{aligned} \begin{bmatrix} \mathfrak{A}_s^t & \mathfrak{B}_s^t \\ \mathfrak{C}_s^t & \mathfrak{D}_s^t \end{bmatrix} &= \begin{bmatrix} \widetilde{\mathfrak{A}}_s^t & \widetilde{\mathfrak{B}}_s^t \boldsymbol{\mathcal{R}}_{(s+t)/2} \\ \boldsymbol{\mathcal{R}}_{(s+t)/2}\widetilde{\mathfrak{C}}_s^t & \boldsymbol{\mathcal{R}}_{(s+t)/2}\widetilde{\mathfrak{D}}_s^t \boldsymbol{\mathcal{R}}_{(s+t)/2} \end{bmatrix}, \\ \begin{bmatrix} \mathfrak{B}_s \\ \mathfrak{D}_s \end{bmatrix} &= \begin{bmatrix} \widetilde{\mathfrak{B}}^s \boldsymbol{\mathcal{R}}_s \\ \boldsymbol{\mathcal{R}}_s \widetilde{\mathfrak{D}}^s \boldsymbol{\mathcal{R}}_s \end{bmatrix}, \\ \begin{bmatrix} \mathfrak{C}^t & \mathfrak{D}^t \end{bmatrix} &= \begin{bmatrix} \boldsymbol{\mathcal{R}}_t \widetilde{\mathfrak{C}}_t & \boldsymbol{\mathcal{R}}_t \widetilde{\mathfrak{D}}_t \boldsymbol{\mathcal{R}}_t \end{bmatrix}. \end{aligned} \tag{6.1.5}$$

We leave the straightforward algebraic proof to the reader. Observe that $\mathbf{Я}_h$ is a reflection of the real axis around the point h which in the *Reg*-case has been modified so that it maps Reg_{loc} onto itself. In particular,

$$\begin{aligned}\mathbf{Я}_{(s+t)/2}\chi_{(-\infty,t)} &= \chi_{[s,\infty)},\\ \mathbf{Я}_{(s+t)/2}\chi_{[s,t)} &= \chi_{[s,t))},\\ \mathbf{Я}_{(s+t)/2}\chi_{[s,\infty)} &= \chi_{(-\infty,t))}.\end{aligned}$$

Lemma 6.1.9 *Let* $\Sigma = \left[\begin{smallmatrix}\mathfrak{A} & \mathfrak{B}\\ \mathfrak{C} & \mathfrak{D}\end{smallmatrix}\right]$ *be an* $L^p|Reg$*-well-posed linear system on* (Y, X, U)*. Then the operators* $\mathfrak{A}^t_s$, $\mathfrak{B}^t_s$, $\mathfrak{C}^t_s$, *and* $\mathfrak{D}^t_s$ *introduced in Definition 6.1.7 have the following algebraic properties:*

(i) *The initial condition* (2.2.3) *holds for all* $t \in \mathbb{R}$.
(ii) *The causality condition* (2.2.4) *holds for all* $s \le t$.
(iii) *The time-invariance condition* (2.2.5) *holds for all* $s \le t$ *and* $h \in \mathbb{R}$,.
(iv) *For all* $s \le r \le t$*, the following composition condition holds:*

$$\begin{aligned}\left[\begin{array}{c|c}\mathfrak{A}^t_s & \mathfrak{B}^t_s\\ \hline \mathfrak{C}^t_s & \mathfrak{D}^t_s\end{array}\right] &= \left[\begin{array}{c|cc}\mathfrak{A}^r_s & \mathfrak{B}^r_s & 0\\ \hline \mathfrak{C}^r_s & \mathfrak{D}^r_s & 1\end{array}\right]\left[\begin{array}{c|c}\mathfrak{A}^t_r & \mathfrak{B}^t_r\\ \hline 0 & 1\\ \mathfrak{C}^t_r & \mathfrak{D}^t_r\end{array}\right]\\ &= \left[\begin{array}{c|c}\mathfrak{A}^r_s\mathfrak{A}^t_r & \mathfrak{B}^r_s + \mathfrak{A}^r_s\mathfrak{B}^t_r\\ \hline \mathfrak{C}^r_s\mathfrak{A}^t_r + \mathfrak{C}^t_r & \mathfrak{D}^r_s + \mathfrak{C}^r_s\mathfrak{B}^t_r + \mathfrak{D}^t_r\end{array}\right].\end{aligned} \tag{6.1.6}$$

The proof of this lemma is analogous to the proof of Lemma 2.2.8, and we leave it to the reader.

Our following theorem is an anti-causal version of Theorem 2.2.14.

Theorem 6.1.10 *Let* U, X, *and* Y *be Banach spaces, and let* $\mathfrak{A}^t_s\colon X \to X$, $\mathfrak{B}^t_s\colon L^p|Reg_{\mathrm{loc},c}(\mathbb{R}; U) \to X$, $\mathfrak{C}^t_s\colon X \to L^p|Reg_{\mathrm{loc},c}(\mathbb{R}; Y)$*, and* $\mathfrak{D}^t_s\colon L^p|Reg_{\mathrm{loc},c}(\mathbb{R}; U) \to L^p|Reg_{\mathrm{loc},c}(\mathbb{R}; Y)$ *be four families of bounded linear operators indexed by* $-\infty < s \le t < \infty$*. Suppose that* $\mathfrak{A}^0_0 x = \lim_{s\uparrow 0}\mathfrak{A}^0_s x = x$ *for all* $x \in X$*, that for all* $s \le t$ *and* $h \in \mathbb{R}$ *time-invariance condition* (2.2.5) *holds, and that for all* $s \le r \le t$,

$$\begin{aligned}\left[\begin{array}{c|c}\mathfrak{A}^t_s & \mathfrak{B}^t_s\\ \hline \mathfrak{C}^t_s & \mathfrak{D}^t_s\end{array}\right] &= \begin{bmatrix}1 & 0 & 0\\ 0 & \pi_{[s,r)} & \pi_{[r,t)}\end{bmatrix}\left[\begin{array}{c|cc}\mathfrak{A}^r_s & \mathfrak{B}^r_s & 0\\ \hline \mathfrak{C}^r_s & \mathfrak{D}^r_s & 0\\ 0 & 0 & 1\end{array}\right]\\ &\quad\times\left[\begin{array}{c|cc}\mathfrak{A}^t_r & 0 & \mathfrak{B}^t_r\\ \hline 0 & 1 & 0\\ \mathfrak{C}^t_r & 0 & \mathfrak{D}^t_r\end{array}\right]\begin{bmatrix}1 & 0\\ 0 & \pi_{[s,r)}\\ 0 & \pi_{[r,t)}\end{bmatrix}.\end{aligned} \tag{6.1.7}$$

Then we get an anti-causal $L^p|Reg$-well-posed linear system $\Sigma = \left[\begin{smallmatrix}\mathfrak{A} & \mathfrak{B}\\ \mathfrak{C} & \mathfrak{D}\end{smallmatrix}\right]$ by defining $\mathfrak{A}_s = \mathfrak{A}_s^0$ for $s \le 0$ and, for all $x \in X$ and $u \in L^p|Reg_{\mathrm{loc},c}(\mathbb{R}; U)$,

$$\mathfrak{B}u = \lim_{t\to\infty} \mathfrak{B}_0^t u, \qquad \mathfrak{C}x = \lim_{s\to-\infty} \mathfrak{C}_s^0 x, \qquad \mathfrak{D}u = \lim_{\substack{s\to-\infty\\ t\to\infty}} \mathfrak{D}_s^t u \quad (6.1.8)$$

(in particular, these limits exist in X, $L^p|Reg_{\mathrm{loc}}(\mathbb{R}^+; Y)$, respectively $L^p|Reg_{\mathrm{loc},c}(\mathbb{R}; Y)$). Moreover, the given operator families $\mathfrak{A}_s^t$, $\mathfrak{B}_s^t$, $\mathfrak{C}_s^t$, and $\mathfrak{D}_s^t$ are identical to those derived from Σ as described in Definition 6.1.7.

Observe that the only significant difference in the assumptions of Theorems 2.2.14 and 6.1.10 is that two of the factors in (6.1.7) have changed places compared to (2.2.10). We leave this proof, too, to the reader (it is the same as the proof of Theorem 2.2.14, with the positive and negative time directions interchanged).

6.2 The dual system

The theory developed in Sections 3.5 and 6.1 makes it possible to introduce the dual of an L^p-well-posed linear system.

Theorem 6.2.1 *Let $1 \le p < \infty$, and let $\Sigma = \left[\begin{smallmatrix}\mathfrak{A} & \mathfrak{B}\\ \mathfrak{C} & \mathfrak{D}\end{smallmatrix}\right]$ be an L^p-well-posed linear system on the reflexive Banach spaces (Y, X, U). Define $\Sigma^\dagger = \left[\begin{smallmatrix}\mathfrak{A}^\dagger & \mathfrak{B}^\dagger\\ \mathfrak{C}^\dagger & \mathfrak{D}^\dagger\end{smallmatrix}\right]$, where, for all $t \le 0$,*

$$\begin{aligned} (\mathfrak{A}^\dagger)_t &= (\mathfrak{A}^*)^{-t}, & \mathfrak{B}^\dagger &= \mathfrak{C}^* \\ \mathfrak{C}^\dagger &= \mathfrak{B}^* & \mathfrak{D}^\dagger &= \mathfrak{D}^*. \end{aligned} \qquad (6.2.1)$$

Then $\Sigma^\dagger$ is an anti-causal L^q-well-posed linear system on (U^*, X^*, Y^*), where $1/p + 1/q = 1$ (and $1/\infty = 0$).

Here the formula (6.2.1) requires some explanation. We fix some $\omega > \omega_{\mathfrak{A}}$, where $\omega_{\mathfrak{A}}$ is the growth bound of $\mathfrak{A}$. Then, by Theorem 2.5.4,

$$\begin{aligned} &\mathfrak{B} \in \mathcal{B}(L^p_\omega(\mathbb{R}^-; U); X), \\ &\mathfrak{C} \in \mathcal{B}(X; L^p_\omega(\mathbb{R}^+; Y)), \\ &\mathfrak{D} \in \mathcal{B}(L^p_\omega(\mathbb{R}; U); L^p_\omega(\mathbb{R}; Y)). \end{aligned}$$

We can therefore compute the duals of these operators as in Lemma 3.5.9, and get

$$\begin{aligned} \mathfrak{B}^\dagger &= \mathfrak{C}^* \in \mathcal{B}(L^q_{-\omega}(\mathbb{R}^+; Y^*); X^*), \\ \mathfrak{C}^\dagger &= \mathfrak{B}^* \in \mathcal{B}(X^*; L^q_{-\omega}(\mathbb{R}^-; U^*)), \\ \mathfrak{D}^\dagger &= \mathfrak{D}^* \in \mathcal{B}(L^q_{-\omega}(\mathbb{R}; Y^*); L^q_{-\omega}(\mathbb{R}; U^*)). \end{aligned} \qquad (6.2.2)$$

By restricting the domain of $\mathfrak{C}^*$ we can interpret $\mathfrak{C}^*$ as an operator $L^q_c(\mathbb{R}^+; Y^*) \to X^*$, and the operator $\mathfrak{B}^*$ can be interpreted as an operator $X^* \to L^q_{\text{loc}}(\mathbb{R}^-; U^*)$. As part of the proof of Theorem 6.3.1 we show that $\mathfrak{D}^*$ is anti-causal and time-invariant, and this fact can be used to restrict and extend this operator to an operator mapping $L^q_{\text{loc},c}(\mathbb{R}; Y^*) \to L^q_{\text{loc},c}(\mathbb{R}; U^*)$, as required by Definition 6.1.1. The resulting operators do not depend on the value of $\omega \geq \omega_{\mathfrak{A}}$ (we prove this as part of the proof of Theorem 6.2.1).

Proof of Theorem 6.2.1 By Theorem 3.5.6, $\mathfrak{A}^*$ is a C_0 semigroup on X^*. As indicated in the discussion above, the operators $\mathfrak{C}^*$, $\mathfrak{B}^*$, and $\mathfrak{D}^*$ have the right continuity properties required from a backward input map, output map, and input/output map of an anti-causal L^q-well-posed linear system. The algebraic conditions in Definition 6.1.1 with $\left[\begin{smallmatrix} \mathfrak{A} & \mathfrak{B} \\ \mathfrak{C} & \mathfrak{D} \end{smallmatrix}\right]$ replaced by $\left[\begin{smallmatrix} \mathfrak{A}^\dagger & \mathfrak{B}^\dagger \\ \mathfrak{C}^\dagger & \mathfrak{D}^\dagger \end{smallmatrix}\right]$ follow directly from the corresponding algebraic conditions in Definition 2.2.1, since $(\tau^t)^* = \tau^{-t}$, $\pi_-^* = \pi_-$, and $\pi_+^* = \pi_+$. Thus, the only thing which needs to be checked is that the resulting system is independent of the constant ω in (6.2.2).

To show that $\mathfrak{B}^\dagger$ does not depend on ω it suffices to show that $\mathfrak{B}^\dagger$ applied to y^* does not depend on ω in the case where $y^* \in L^q_c(\mathbb{R}^+; Y^*)$. A sufficient condition for this to be true is that, for all $x \in X$ and $y^* \in L^q_c(\mathbb{R}^+; Y^*)$, the value of $\langle x, \mathfrak{B}^\dagger y^* \rangle_{(X,X^*)}$ does not depend on ω. However, this is true since

$$\begin{aligned}\langle x, \mathfrak{B}^\dagger y^* \rangle_{(X,X^*)} &= \langle x, \mathfrak{C}^* y^* \rangle_{(X,X^*)} = \langle \mathfrak{C}x, y^* \rangle_{(L^p_\omega(\mathbb{R}^+;Y), L^q_{-\omega}(\mathbb{R}^+;Y^*))} \\ &= \int_0^\infty \langle (\mathfrak{C}x)(s), y^*(s) \rangle_{(Y,Y^*)},\end{aligned}$$

and the integral does not depend on ω. To show that $\mathfrak{C}^\dagger$ does depend on ω it suffices to show that, for all $x^* \in X^*$ and all $u \in L^p_c(\mathbb{R}^-; U)$, the value of $\int_{-\infty}^0 \langle u(s), (\mathfrak{C}^\dagger x^*)(s) \rangle\, ds$ does not depend on ω, and this is true since

$$\begin{aligned}\int_{-\infty}^0 \langle u(s), (\mathfrak{C}^\dagger x^*)(s) \rangle\, ds &= \int_{-\infty}^0 \langle u(s), (\mathfrak{B}^* x^*)(s) \rangle\, ds \\ &= \langle u, \mathfrak{B}^* x^* \rangle_{(L^p_\omega(\mathbb{R}^-;U), L^q_{-\omega}(\mathbb{R}^-;U^*))} \\ &= \langle \mathfrak{B}u, x^* \rangle_{(X,X^*)}.\end{aligned}$$

By the anti-causality and time-invariance of $\mathfrak{D}^\dagger$, to show that this operator does not depend on ω it suffices to know, for all compactly supported $u \in L^p_c(\mathbb{R}; U)$ and $y^* \in L^q_c(\mathbb{R}; Y^*)$ the value of $\int_{-\infty}^\infty \langle u(s), (\mathfrak{D}^\dagger y^*)(s) \rangle\, ds$ does not depend on

ω, and this is true since

$$\begin{aligned}\int_{-\infty}^{\infty} \langle u(s), (\mathfrak{D}^{\dagger} y^*)(s)\rangle\, ds &= \int_{-\infty}^{\infty} \langle u(s), (\mathfrak{D}^* y^*)(s)\rangle\, ds \\ &= \langle u, \mathfrak{D}^* y^*\rangle_{(L^p_\omega(\mathbb{R};U), L^q_{-\omega}(\mathbb{R},U^*))} \\ &= \langle \mathfrak{D} u, y^*\rangle_{(L^p_\omega(\mathbb{R};Y), L^q_{-\omega}(\mathbb{R},Y^*))} \\ &= \int_{-\infty}^{\infty} \langle (\mathfrak{D} u)(s), y^*(s)\rangle\, ds.\end{aligned}$$

□

Definition 6.2.2 The system $\Sigma^{\dagger}$ in Theorem 6.2.1 is called the *anti-causal dual* of Σ.

By reversing the direction of time in Theorem 6.2.1 we get the following causal system.

Theorem 6.2.3 *Let $1 \le p < \infty$, and let $\Sigma = \left[\begin{smallmatrix} \mathfrak{A} & \mathfrak{B} \\ \mathfrak{C} & \mathfrak{D}\end{smallmatrix}\right]$ be an L^p-well-posed linear system on the reflexive Banach spaces (Y, X, U). Define*

$$\Sigma^d = \left[\begin{array}{c|c} \mathfrak{A}^d & \mathfrak{B}^d \\ \hline \mathfrak{C}^d & \mathfrak{D}^d \end{array}\right] = \left[\begin{array}{c|c} \mathfrak{A}^* & \mathfrak{C}^* \mathbf{Я} \\ \hline \mathbf{Я}\mathfrak{B}^* & \mathbf{Я}\mathfrak{D}^*\mathbf{Я} \end{array}\right] \tag{6.2.3}$$

where $\mathfrak{A}^{*t} = (\mathfrak{A}^t)^*$ for all $t \ge 0$. Then Σ^d is an L^q-well-posed linear system on (U^*, X^*, Y^*), where $1/p + 1/q = 1$ (and $1/\infty = 0$).

Proof This follows from Lemma 6.1.2 and Theorem 6.2.1. □

Definition 6.2.4 The system Σ^d in Theorem 6.2.3 is called the *causal dual* of Σ.

Example 6.2.5 *Let $\Sigma = \left[\begin{smallmatrix} \mathfrak{A} & \mathfrak{B} \\ \mathfrak{C} & \mathfrak{D}\end{smallmatrix}\right]$ be an L^p-well-posed linear system on the reflexive Banach spaces (Y, X, U), with $1 \le p < \infty$.*

(i) *For each $\alpha \in \mathbb{C}$, the dual $\mathfrak{A}^d_\alpha$ of the exponentially shifted system Σ_α in Example 2.3.5 is*

$$\Sigma^d_\alpha = \left[\begin{array}{c|c} \mathfrak{A}^d_\alpha & \mathfrak{B}^d_\alpha \\ \hline \mathfrak{C}^d_\alpha & \mathfrak{D}^d_\alpha \end{array}\right] = \left[\begin{array}{c|c} e_{\overline{\alpha}}\mathfrak{A} & \mathfrak{C}^*\mathbf{Я}e_{-\overline{\alpha}} \\ \hline e_{\overline{\alpha}}\mathbf{Я}\mathfrak{B}^* & e_{\overline{\alpha}}\mathbf{Я}\mathfrak{D}^*\mathbf{Я}e_{-\overline{\alpha}} \end{array}\right].$$

(ii) *For each $\lambda > 0$, the dual $\mathfrak{A}^d_\lambda$ of the time compressed system Σ_λ in Example 2.3.6 is*

$$\Sigma^d_\lambda = \left[\begin{array}{c|c} \mathfrak{A}^d_\lambda & \mathfrak{B}^d_\lambda \\ \hline \mathfrak{C}^d_\lambda & \mathfrak{D}^d_\lambda \end{array}\right] = \left[\begin{array}{c|c} \mathfrak{A}^*_\lambda & \mathfrak{C}^*\mathbf{Я}\gamma_{1/\lambda} \\ \hline \gamma_\lambda\mathbf{Я}\mathfrak{B}^* & \gamma_\lambda\mathbf{Я}\mathfrak{D}^*\mathbf{Я}\gamma_{1/\lambda} \end{array}\right].$$

(iii) *For each (boundedly) invertible* $E \in \mathcal{B}(X_1; X)$, *the dual* $\mathfrak{A}_E^d$ *of the similarity transformed system* Σ_E *in Example 2.3.7 is*

$$\Sigma_E^d = \left[\begin{array}{c|c} \mathfrak{A}_E^d & \mathfrak{B}_E^d \\ \hline \mathfrak{C}_E^d & \mathfrak{D}_E^d \end{array}\right] = \left[\begin{array}{c|c} E^*\mathfrak{A}^*E^{-*} & E^*\mathfrak{C}^*\mathbf{Я} \\ \hline \mathbf{Я}\mathfrak{B}^*E^{-*} & \mathbf{Я}\mathfrak{D}^*\mathbf{Я} \end{array}\right].$$

This follows from Examples 2.3.5–2.3.7 and Theorem 6.2.3.

Example 6.2.6 *We consider the systems in Examples 2.3.10–2.3.13 in the* L^p*-setting with* $1 \le p < \infty$ *and with reflexive input spaces, state spaces and output spaces.*

(i) *The dual* $\mathfrak{A}^d$ *of the cross-product of the systems* Σ_1 *and* Σ_2 *in Example 2.3.10 is the cross-product of the duals* Σ_1^d *and* Σ_2^d.
(ii) *The dual* $\mathfrak{A}^d$ *of the sum junction of the systems* Σ_1 *and* Σ_2 *in Example 2.3.11 is the T-junction of the duals* Σ_1^d *and* Σ_2^d.
(iii) *The dual* $\mathfrak{A}^d$ *of the T-junction of the systems* Σ_1 *and* Σ_2 *in Example 2.3.12 is the sum junction of the duals* Σ_1^d *and* Σ_2^d.
(iv) *The dual* $\mathfrak{A}^d$ *of the parallel connection of the systems* Σ_1 *and* Σ_2 *in Example 2.3.13 is the parallel connection of the duals* Σ_1^d *and* Σ_2^d.

This follows from Examples 2.3.10–2.3.13 and Theorem 6.2.3.

Example 6.2.7 *The dual of the delay line example 2.3.4 in* L^p *with* $1 < p < \infty$ *and reflexive* U *is a similarity transformed version of the same delay line in* L^q, $1/p + 1/q = 1$, *with the similarity transformation given in Lemma 3.5.13(iv)(d). (The signal enters at the left end of the line, and leaves at the right end.)*

This follows from Example 2.3.4, Theorem 6.2.3, and Lemma 3.5.13.

Example 6.2.8 *Let* $\mathfrak{D} \in TIC_\omega^p(U; Y)$ *where* $1 < p < \infty$, $\omega \in \mathbb{R}$ *and* U *and* Y *are reflexive Banach spaces. Then the exactly controllable shift realization of* $\mathfrak{D}$ *in Example 2.6.5(i) and the exactly observable shift realization of* $\mathfrak{D}^d$ *in Example 2.6.5(ii) are duals of each other. Also the bilateral input shift realization of* $\mathfrak{D}$ *in Example 2.6.5(iii) and the bilateral output shift realization of* $\mathfrak{D}^d$ *in Example 2.6.5(iv) are duals of each other.*

This is obvious (see Lemma 3.5.13).

The formal relationships between the finite time input/state/output maps $\left[\begin{array}{c|c} \mathfrak{A}_s^t & \mathfrak{B}_s^t \\ \hline \mathfrak{C}_s^t & \mathfrak{D}_s^t \end{array}\right]$ and the corresponding maps for the anti-causal dual system are simpler than the relationships with the causal dual system, due to the absence of reflection operators. Specifically, we have the following result.

Lemma 6.2.9 *Let $1 \le p < \infty$, and let $\Sigma = \left[\begin{smallmatrix} \mathfrak{A} & \mathfrak{B} \\ \mathfrak{C} & \mathfrak{D} \end{smallmatrix}\right]$ be an L^p-well-posed linear system on the reflexive Banach spaces (Y, X, U), and let $\Sigma^\dagger$ be the corresponding anti-causal dual system. Then the (forward) maps introduced in Definition 2.2.6 (applied to the original system Σ) and the (backward) maps introduced in Definition 6.1.7 (applied to $\Sigma^\dagger$) are related in the following way for all $-\infty < s < t < \infty$,*

$$\left[\begin{array}{c|c} (\mathfrak{A}^\dagger)_s^t & (\mathfrak{B}^\dagger)_s^t \\ \hline (\mathfrak{C}^\dagger)_s^t & (\mathfrak{D}^\dagger)_s^t \end{array}\right] = \left[\begin{array}{c|c} (\mathfrak{A}_s^t)^* & (\mathfrak{C}_s^t)^* \\ \hline (\mathfrak{B}_s^t)^* & (\mathfrak{D}_s^t)^* \end{array}\right],$$

$$\left[\begin{array}{c} (\mathfrak{B}^\dagger)_s \\ \hline (\mathfrak{D}^\dagger)_s \end{array}\right] = \left[\begin{array}{c} (\mathfrak{C}^s)^* \\ \hline (\mathfrak{D}^s)^* \end{array}\right], \tag{6.2.4}$$

$$\left[\, (\mathfrak{C}^\dagger)^t \,\middle|\, (\mathfrak{D}^\dagger)^t \,\right] = \left[\, (\mathfrak{B}_t)^* \,\middle|\, (\mathfrak{D}_t)^* \,\right].$$

Proof This follows immediately from Definitions 2.2.6, 6.1.7, and 6.2.2. □

By combining this result with Lemma 6.1.8 we get the corresponding relationships between the original system Σ and the causal dual system Σ^d.

Theorem 6.2.10 *Let $1 \le p < \infty$, $1/p + 1/q = 1$, $s < t$, $x_s \in X$, $x_t^* \in X^*$, $u \in L^p((s,t);U)$, and $y^* \in L^q((s,t);Y^*)$. Let x and y be the state trajectory and output function (restricted to (s,t)) of $\Sigma = \left[\begin{smallmatrix} \mathfrak{A} & \mathfrak{B} \\ \mathfrak{C} & \mathfrak{D} \end{smallmatrix}\right]$ with initial time s, initial state x_s, and input function u, and let x^* and u^* be the backward state trajectory and output function (restricted to (s,t)) of the anti-causal dual system $\Sigma^\dagger$ with initial time t, initial state x_t^*, and input function y^*. Then*

$$\begin{aligned} &\langle x(t), x_t^* \rangle_{(X,X^*)} + \int_s^t \langle y(r), y^*(r) \rangle_{(Y,Y^*)}\, dr \\ &\quad = \langle x_s, x^*(s) \rangle_{(X,X^*)} + \int_s^t \langle u(r), u^*(r) \rangle_{(U,U^*)}\, dr. \end{aligned}$$

Of course, it is also possible to write this identity by using the causal dual of Σ instead, at the expense of having to add a number of reflection operators $\mathbf{Я}$, leading to a more complicated formula.

Proof We observe that $u = \pi_{[s,t)} u$ and $y^* = \pi_{[s,t)} y^*$, and use Definitions 2.2.7 and 6.1.3 and Lemma 6.2.9 to get

$$\begin{aligned} \langle x(t), x_t^* \rangle + \langle y, y^* \rangle &= \langle \mathfrak{A}_s^t x_s + \mathfrak{B}_s^t u, x_t^* \rangle + \langle \mathfrak{C}_s^t x_s + \mathfrak{D}_s^t u, y^* \rangle \\ &= \langle \mathfrak{A}_s^t x_s, x_t^* \rangle + \langle \mathfrak{C}_s^t x_s, y^* \rangle + \langle \mathfrak{B}_s^t u, x_t^* \rangle + \langle \mathfrak{D}_s^t u, y^* \rangle \\ &= \langle x_s, (\mathfrak{A}_s^t)^* x_t^* + (\mathfrak{C}_s^t)^* y^* \rangle + \langle u, (\mathfrak{B}_s^t)^* x_t^* + (\mathfrak{D}_s^t)^* y^* \rangle \\ &= \langle x_s, x^*(s) \rangle + \langle u, u^* \rangle. \end{aligned}$$

□

It is possible to formulate versions of Theorem 6.2.10 where either $s = -\infty$ or $t = +\infty$. For simplicity we only give the former result, and leave the formulation of the latter to the reader.

Corollary 6.2.11 *Let $\Sigma = \left[\begin{smallmatrix} \mathfrak{A} & \mathfrak{B} \\ \mathfrak{C} & \mathfrak{D} \end{smallmatrix}\right]$ be an L^p-well-posed linear system on (Y, X, U) with $1 \le p < \infty$, and let $\omega > \omega_{\mathfrak{A}}$. Let $1/p + 1/q = 1$, $x_t^* \in X^*$, $u \in L^p_\omega((-\infty, t); U)$, and $y^* \in L^q_{-\omega}((-\infty, t); Y^*)$. Let x and y be the state trajectory and output function of Σ with initial time $-\infty$, initial state zero, and input function u, and let x^* and u^* be the backward state trajectory and output function of the anti-causal dual system $\Sigma^\dagger$ with initial time t, initial state x_t^*, and input function y^*. Then*

$$\begin{aligned} \langle x(t), x_t^* \rangle_{(X,X^*)} &+ \int_{-\infty}^t \langle y(s), y^*(s) \rangle_{(Y,Y^*)}\, ds \\ &= \int_{-\infty}^t \langle u(s), u^*(s) \rangle_{(U,U^*)}\, ds. \end{aligned}$$

To prove this it suffices to let $s \to -\infty$ in Theorem 6.2.10.

As our following theorem shows, there is a simple connection between the causal dual of Σ and the adjoint of the Lax–Phillips model induced by Σ.

Theorem 6.2.12 *Let $1 \le p < \infty$, $1/p + 1/q = 1$, let $\Sigma = \left[\begin{smallmatrix} \mathfrak{A} & \mathfrak{B} \\ \mathfrak{C} & \mathfrak{D} \end{smallmatrix}\right]$ be an L^p-well-posed linear system on the reflexive Banach spaces (Y, X, U), and let Σ^d be the causal dual of Σ. Let $\omega \in \mathbb{R}$, and let $\mathfrak{T}$ be the Lax–Phillips model of type L^p_ω induced by Σ. Then the Lax–Phillips model $\mathfrak{T}^d$ of type L^q_ω induced by Σ^d is given by*

$$\mathfrak{T}^d = \begin{bmatrix} 0 & 0 & \mathbf{Я} \\ 0 & 1 & 0 \\ \mathbf{Я} & 0 & 0 \end{bmatrix} \mathfrak{T}^* \begin{bmatrix} 0 & 0 & \mathbf{Я} \\ 0 & 1 & 0 \\ \mathbf{Я} & 0 & 0 \end{bmatrix}. \tag{6.2.5}$$

Proof The state space of $\mathfrak{T}^d$, which is $\left[\begin{smallmatrix} L^q_\omega(\mathbb{R}^-; U^*) \\ X \end{smallmatrix}\right] L^q_\omega(\mathbb{R}^+; Y^*)$, coincides with the state space of the semigroup of the right hand side of (6.2.5) since we identify the duals of $L^p_\omega(\mathbb{R}^-; Y)$ and $L^p_\omega(\mathbb{R}^+; U)$ with $L^q_{-\omega}(\mathbb{R}^-; Y^*)$, respectively $L^q_{-\omega}(\mathbb{R}^+; U^*)$, and since $\mathbf{Я}$ maps $L^q_{-\omega}(\mathbb{R}^-; Y^*)$ and $L^q_{-\omega}(\mathbb{R}^+; U^*)$ onto $L^q_\omega(\mathbb{R}^+; Y^*)$, respectively $L^q_\omega(\mathbb{R}^-; U^*)$. The rest of the proof is a simple algebraic computation based on Definitions 2.2.6 and 2.7.4, Example 3.5.11, and

Lemma 3.5.13:

$$\mathfrak{T}^d = \begin{bmatrix} \tau^t\pi_- & \pi_-\tau^t\boldsymbol{Я}\mathfrak{B}^* & \pi_-\tau^t\boldsymbol{Я}\mathfrak{D}^*\boldsymbol{Я}\pi_+ \\ 0 & \mathfrak{A}^{*t} & \mathfrak{C}^*\boldsymbol{Я}\tau^t\pi_+ \\ 0 & 0 & \pi_+\tau^t \end{bmatrix}$$

$$= \begin{bmatrix} \boldsymbol{Я}\tau^{-t}\pi_+\boldsymbol{Я} & \boldsymbol{Я}\pi_+\tau^{-t}\mathfrak{B}^* & \boldsymbol{Я}\pi_+\mathfrak{D}^*\tau^{-t}\pi_-\boldsymbol{Я} \\ 0 & \mathfrak{A}^{*t} & \mathfrak{C}^*\tau^{-t}\pi_-\boldsymbol{Я} \\ 0 & 0 & \boldsymbol{Я}\pi_-\tau^{-t}\boldsymbol{Я} \end{bmatrix}$$

$$= \begin{bmatrix} 0 & 0 & \boldsymbol{Я} \\ 0 & 1 & 0 \\ \boldsymbol{Я} & 0 & 0 \end{bmatrix} \begin{bmatrix} \pi_-\tau^{-t} & 0 & 0 \\ \mathfrak{C}^*\tau^{-t}\pi_- & \mathfrak{A}^{*t} & 0 \\ \pi_+\mathfrak{D}^*\tau^{-t}\pi_- & \pi_+\tau^{-t}\mathfrak{B}^* & \tau^{-t}\pi_+ \end{bmatrix} \begin{bmatrix} 0 & 0 & \boldsymbol{Я} \\ 0 & 1 & 0 \\ \boldsymbol{Я} & 0 & 0 \end{bmatrix}$$

$$= \begin{bmatrix} 0 & 0 & \boldsymbol{Я} \\ 0 & 1 & 0 \\ \boldsymbol{Я} & 0 & 0 \end{bmatrix} \begin{bmatrix} \tau^t\pi_- & \pi_-\tau^t\mathfrak{C} & \pi_-\tau^t\mathfrak{D}\pi_+ \\ 0 & \mathfrak{A}^t & \mathfrak{B}\tau^t\pi_+ \\ 0 & 0 & \pi_+\tau^t \end{bmatrix}^* \begin{bmatrix} 0 & 0 & \boldsymbol{Я} \\ 0 & 1 & 0 \\ \boldsymbol{Я} & 0 & 0 \end{bmatrix}$$

$$= \begin{bmatrix} 0 & 0 & \boldsymbol{Я} \\ 0 & 1 & 0 \\ \boldsymbol{Я} & 0 & 0 \end{bmatrix} \mathfrak{T}^{*t} \begin{bmatrix} 0 & 0 & \boldsymbol{Я} \\ 0 & 1 & 0 \\ \boldsymbol{Я} & 0 & 0 \end{bmatrix}.$$

□

Theorem 6.2.13 *Let $1 \le p < \infty$, and let $\Sigma = \left[\begin{smallmatrix} \mathfrak{A} & \mathfrak{B} \\ \mathfrak{C} & \mathfrak{D} \end{smallmatrix}\right]$ be an L^p-well-posed linear system on the reflexive Banach spaces (Y, X, U), with system node $S = \left[\begin{smallmatrix} A\&B \\ C\&D \end{smallmatrix}\right]$, semigroup generator A, control operator B, observation operator C, transfer function $\widehat{\mathfrak{D}}$, and growth bound $\omega_{\mathfrak{A}}$. We denote the corresponding operators for the causal dual $\Sigma^d = \left[\begin{smallmatrix} \mathfrak{A}^d & \mathfrak{B}^d \\ \mathfrak{C}^d & \mathfrak{D}^d \end{smallmatrix}\right]$ by the same letters and the superscript d. Then $\omega_{\mathfrak{A}^d} = \omega_{\mathfrak{A}}$, and*

$$S^d = \begin{bmatrix} [A\&B]^d \\ [C\&D]^d \end{bmatrix} = \begin{bmatrix} A\&B \\ C\&D \end{bmatrix}^* = S^*.$$

Moreover, $A^d = A^*$, $B^d = C^*$ and $C^d = B^*$, and the transfer functions of Σ and Σ^d are related by

$$\widehat{\mathfrak{D}}^d(z) = \widehat{\mathfrak{D}}(\overline{z})^*, \qquad \overline{z} \in \sigma(A). \tag{6.2.6}$$

The proof of Theorem 6.2.13 is partially based on the following lemma, which is also of independent interest.

Lemma 6.2.14 *Let S be an operator node on (Y, X, U), where X is reflexive. We denote the main operator of S by A, the control operator by B, the observation operator by C, and the transfer function $\widehat{\mathfrak{D}}$. Let X^*_{-1} be the analogue of X_{-1} constructed in Section 3.6 with X replaced by X^* and A replaced by A^**

(cf. Remark 3.6.1). Then the (unbounded) adjoint of S is given by

$$S^* = \begin{bmatrix} [A\&B]^d \\ [C\&D]^d \end{bmatrix} : \begin{bmatrix} X^* \\ Y^* \end{bmatrix} \supset \mathcal{D}(S^*) \to \begin{bmatrix} X^* \\ U^* \end{bmatrix},$$

where $[A\&B]^d = \begin{bmatrix} A^*_{|X^*} & C^* \end{bmatrix}_{|\mathcal{D}(S^*)}$ *and*

$$\begin{aligned} \mathcal{D}(S^*) &:= \left\{ \begin{bmatrix} x^* \\ y^* \end{bmatrix} \in \begin{bmatrix} X^* \\ Y^* \end{bmatrix} \;\middle|\; A^*_{|X^*}x^* + C^*y^* \in X^* \right\}, \\ [C\&D]^d \begin{bmatrix} x^* \\ y^* \end{bmatrix} &:= B^*\left(x^* - (\overline{\alpha} - A^*_{|X^*})^{-1}C^*y^*\right) + \widehat{\mathfrak{D}}(\alpha)^*y^*, \end{aligned} \tag{6.2.7}$$

where α is an arbitrary number in $\rho(A) = \rho(A_{|X_{-1}})$ (the resulting operator $[C\&D]^d$ is independent of α). In particular, S^* is an operator node (in the sense of Definition 4.7.2), and if S is a system node, then so is S^*. The main operator of S^* is A^*, the control operator is C^*, the observation operator is B^*, and the transfer function is $z \mapsto (\widehat{\mathfrak{D}}(\overline{z}))^*$, $z \in \rho(A^*)$.

Proof Let $\alpha \in \rho(A)$, and define

$$E_\alpha := \begin{bmatrix} 1 & (\alpha - A_{|X})^{-1}B \\ 0 & 1 \end{bmatrix}.$$

This operator is bounded and invertible on $\left[\begin{smallmatrix} X \\ U \end{smallmatrix}\right]$, and it maps $\left[\begin{smallmatrix} X_1 \\ U \end{smallmatrix}\right]$ one-to-one onto $\mathcal{D}(S)$ (see Lemma 4.7.3(viii)). Therefore, by Lemma 3.5.5,

$$S^* = \left(SE_\alpha E_\alpha^{-1}\right)^* = E_\alpha^{-*}(SE_\alpha)^*.$$

In particular, $\mathcal{D}(S^*) = \mathcal{D}((SE_\alpha)^*)$. The first factor above is the adjoint of $E_\alpha^{-1} = \left[\begin{smallmatrix} 1 & -(\alpha - A_{|X})^{-1}B \\ 0 & 1 \end{smallmatrix}\right]$, and it is given by $E_\alpha^{-*} = \left[\begin{smallmatrix} 1 & 0 \\ -B^*(\overline{\alpha} - A^*)^{-1} & 1 \end{smallmatrix}\right]$. By (4.7.16),

$$SE_\alpha = \begin{bmatrix} A & \alpha(\alpha - A_{|X})^{-1}B \\ C & \widehat{\mathfrak{D}}(\alpha) \end{bmatrix}.$$

This operator can be interpreted in two ways: on one hand, it is an unbounded operator from $\left[\begin{smallmatrix} X \\ U \end{smallmatrix}\right]$ to $\left[\begin{smallmatrix} X \\ Y \end{smallmatrix}\right]$ with domain $\left[\begin{smallmatrix} X_1 \\ U \end{smallmatrix}\right]$. On the other hand, it can also be interpreted as a bounded operator from $\left[\begin{smallmatrix} X_1 \\ U \end{smallmatrix}\right]$ to $\left[\begin{smallmatrix} X \\ Y \end{smallmatrix}\right]$. The adjoint of the latter operator is a bounded operator from $\left[\begin{smallmatrix} X^* \\ Y^* \end{smallmatrix}\right]$ to $\left[\begin{smallmatrix} X^*_{-1} \\ U^* \end{smallmatrix}\right]$ (recall that we identify the dual of X_1 with X^*_{-1}; see Remark 3.6.1), and it is given by $\left[\begin{smallmatrix} A^*_{|X^*} & C^* \\ \overline{\alpha}B^*(\overline{\alpha} - A^*)^{-1} & (\widehat{\mathfrak{D}}(\alpha))^* \end{smallmatrix}\right]$.

This implies, in particular, that for all $\begin{bmatrix} x \\ u \end{bmatrix} \in \begin{bmatrix} X_1 \\ U \end{bmatrix}$ and all $\begin{bmatrix} x^* \\ y^* \end{bmatrix} \in \begin{bmatrix} X^* \\ Y^* \end{bmatrix}$,

$$\begin{aligned}
&\left\langle \begin{bmatrix} x^* \\ y^* \end{bmatrix}, SE_\alpha \begin{bmatrix} x \\ u \end{bmatrix} \right\rangle_{\left(\left[\begin{smallmatrix} X^* \\ Y^* \end{smallmatrix}\right], \left[\begin{smallmatrix} X \\ Y \end{smallmatrix}\right]\right)} \\
&= \left\langle \begin{bmatrix} A^*_{|X^*} & C^* \\ \overline{\alpha} B^*(\overline{\alpha} - A^*)^{-1} & (\widehat{\mathfrak{D}}(\alpha))^* \end{bmatrix} \begin{bmatrix} x^* \\ y^* \end{bmatrix}, \begin{bmatrix} x \\ u \end{bmatrix} \right\rangle_{\left(\left[\begin{smallmatrix} X^*_{-1} \\ Y^* \end{smallmatrix}\right], \left[\begin{smallmatrix} X_1 \\ Y \end{smallmatrix}\right]\right)} \\
&= \langle A^*_{|X^*} x^* + C^* y^*, x \rangle_{(X^*_{-1}; X_1)} \\
&\quad + \langle \overline{\alpha} B^*(\overline{\alpha} - A^*)^{-1} x^* + (\widehat{\mathfrak{D}}(\alpha))^* y^*, y \rangle_{(Y^*;Y)}.
\end{aligned}$$

By the definition of the domain of the adjoint of an unbounded operator, if we interpret $SE_\alpha \colon \begin{bmatrix} X \\ U \end{bmatrix} \supset \mathcal{D}(SE_\alpha) \to \begin{bmatrix} X \\ U \end{bmatrix}$ as an unbounded operator, then $\begin{bmatrix} x^* \\ y^* \end{bmatrix} \in \mathcal{D}((SE_\alpha)^*)$ if and only if the above expression, regarded as a function of $\begin{bmatrix} x \\ u \end{bmatrix} \in \begin{bmatrix} X_1 \\ U \end{bmatrix}$, can be extended to a bounded linear functional on $\begin{bmatrix} X \\ Y \end{bmatrix}$. Clearly, this is true if and only if $A^*_{|X^*} x^* + C^* y^* \in X^*$. Thus, $\mathcal{D}((SE_\alpha)^*) = \mathcal{D}(S^*) = \left\{ \begin{bmatrix} x^* \\ y^* \end{bmatrix} \in \begin{bmatrix} X^* \\ Y^* \end{bmatrix} \;\middle|\; A^*_{|X^*} x^* + C^* y^* \in X^* \right\}$. Furthermore, the same computation shows that

$$(SE_\alpha)^* = \begin{bmatrix} A^*_{|X^*} & C^* \\ \overline{\alpha} B^*(\overline{\alpha} - A^*)^{-1} & (\widehat{\mathfrak{D}}(\alpha))^* \end{bmatrix}_{|\mathcal{D}(S^*)}.$$

Multiplying this identity by $E_\alpha^{-*} = \begin{bmatrix} 1 & 0 \\ -B^*(\overline{\alpha} - A^*)^{-1} & 1 \end{bmatrix}$ to the left we get for all $\begin{bmatrix} x^* \\ y^* \end{bmatrix} \in \mathcal{D}(S^*)$,

$$\begin{aligned}
S^* \begin{bmatrix} x^* \\ y^* \end{bmatrix} &= E_\alpha^{-*}(SE_\alpha)^* \begin{bmatrix} x^* \\ y^* \end{bmatrix} \\
&= \begin{bmatrix} 1 & 0 \\ -B^*(\overline{\alpha} - A^*)^{-1} & 1 \end{bmatrix} \begin{bmatrix} A^*_{|X^*} & C^* \\ B^*\overline{\alpha}(\overline{\alpha} - A^*)^{-1} & \widehat{\mathfrak{D}}(\alpha)^* \end{bmatrix} \begin{bmatrix} x^* \\ y^* \end{bmatrix} \\
&= \begin{bmatrix} 1 & 0 \\ -B^*(\overline{\alpha} - A^*)^{-1} & 1 \end{bmatrix} \begin{bmatrix} A^*_{|X^*} x^* + C^* y^* \\ B^*\overline{\alpha}(\overline{\alpha} - A^*)^{-1} x^* + \widehat{\mathfrak{D}}(\alpha)^* y^* \end{bmatrix} \\
&= \begin{bmatrix} A^*_{|X^*} x^* + C^* y^* \\ B^*(\overline{\alpha} - A^*)^{-1}[(\overline{\alpha} - A^*_{|X^*}) x^* + C^* y^*] + \widehat{\mathfrak{D}}(\alpha)^* y^* \end{bmatrix} \\
&= \begin{bmatrix} A^*_{|X^*} x^* + C^* y^* \\ B^*(x - (\overline{\alpha} - A^*_{|X^*})^{-1} C^* y^*) + \widehat{\mathfrak{D}}(\alpha)^* y^* \end{bmatrix} = \begin{bmatrix} [A\&B]^d \\ [C\&D]^d \end{bmatrix} \begin{bmatrix} x^* \\ y^* \end{bmatrix},
\end{aligned}$$

which gives equation (6.2.7).

Checking Definition 4.7.2 we find that S^* is an operator node on (U^*, X^*, Y^*) (that $\mathcal{D}(A^*)$ is dense in X^* follows from Lemma 3.5.1(iv)). If A is the generator of a C_0 semigroup, then so is A^*, so S^* is a system node whenever S is so. The claims concerning the main operator, the control operator, and the observation operator have already been established. If we denote the transfer function of S^* by $\widehat{\mathfrak{D}}^d$, then, by taking $x^* = (\overline{\alpha} - A^*_{|X^*})^{-1} C^* y^*$ in (6.2.7) we get for all

$\overline{\alpha} \in \rho(A^*)$ and $y^* \in Y^*$,

$$\widehat{\mathfrak{D}}^d(\overline{\alpha})y^* = [C\&D]^d \begin{bmatrix} (\overline{\alpha} - A^*_{|X^*})^{-1}C^*y^* \\ y^* \end{bmatrix} = \widehat{\mathfrak{D}}(\alpha)^* y^*.$$

□

Proof of Theorem 6.2.13 Fix some $h > 0$, and let $\left[\begin{smallmatrix} x_0 \\ u_0 \end{smallmatrix}\right] \in \mathcal{D}(S)$, $\left[\begin{smallmatrix} x_0^d \\ y_0^d \end{smallmatrix}\right] \in \mathcal{D}\left(S^d\right)$, and $u \in W^{1,p}([0,h];U)$ with $u(0) = u_0$. If $p > 1$, then we take $y^d \in W^{1,q}([0,h];Y)$, and if $p = \infty$, then we take $y \in Reg^1([0,h];Y)$. In both cases we require, in addition, that $y^d(0) = y_0^d$. Let x be the state trajectory and let y be the output function (restricted to $[0,h]$) of Σ with initial state x_0, initial time 0, and input function u, and let x^d be the state trajectory and let u^d be the output function (restricted to $[0,h]$) of Σ^d with initial state x_0^d, initial time 0, and input function y^d. Then, by Theorem 4.6.11, for all $t \in [0,h]$, $\left[\begin{smallmatrix} x(t) \\ u(t) \end{smallmatrix}\right] \in \mathcal{D}(S)$, $\left[\begin{smallmatrix} x^d(t) \\ u^d(t) \end{smallmatrix}\right] \in \mathcal{D}\left(S^d\right)$, and

$$\begin{bmatrix} \dot{x}(t) \\ y(t) \end{bmatrix} = S \begin{bmatrix} x(t) \\ u(t) \end{bmatrix}, \qquad \begin{bmatrix} (x^d)'(t) \\ u^d(t) \end{bmatrix} = S^d \begin{bmatrix} x^d(t) \\ y^d(t) \end{bmatrix}, \qquad t \in [0,h].$$

This combined with Theorem 6.2.10 and Lemma 6.1.4 (with $s = 0$ and $t = h$) implies that

$$\begin{aligned} &\langle x_0, x^d(h) - x_0^d \rangle + \int_0^h \left\langle u(t), [C\&D]^d \begin{bmatrix} x^d(h-t) \\ y^d(h-t) \end{bmatrix} \right\rangle dt \\ &= \langle x(h) - x_0, x_0^d \rangle + \int_0^h \left\langle C\&D \begin{bmatrix} x(t) \\ u(t) \end{bmatrix}, y^d(h-t) \right\rangle d\sigma. \end{aligned}$$

Divide by h and let $h \downarrow 0$ to get

$$\left\langle S \begin{bmatrix} x_0 \\ u_0 \end{bmatrix}, \begin{bmatrix} x_0^d \\ y_0^d \end{bmatrix} \right\rangle = \left\langle \begin{bmatrix} x_0 \\ u_0 \end{bmatrix}, S^d \begin{bmatrix} x_0^d \\ y_0^d \end{bmatrix} \right\rangle. \tag{6.2.8}$$

This being true for all $\left[\begin{smallmatrix} x_0 \\ u_0 \end{smallmatrix}\right] \in \mathcal{D}(S)$ and $\left[\begin{smallmatrix} x_0^d \\ y_0^d \end{smallmatrix}\right] \in \mathcal{D}\left(S^d\right)$, we conclude that $\mathcal{D}\left(S^d\right) \subset \mathcal{D}(S^*)$, and that S^d is the restriction of S^* to $\mathcal{D}\left(S^d\right)$.

In Theorem 6.2.13 we introduced S^d as the system node of the dual system Σ^d, and by Lemma 6.2.14, also S^* is a system node on the same triple of spaces (U^*, X^*, Y^*). These two nodes have the same main operator A^*; see Theorem 3.5.6(v) and Lemma 6.2.14. They must also have the same control operator C^*, since S^d is a restriction of S^*, and since the extended operator $\left[A^*_{|X^*} \;\; C^*\right]$ is determined uniquely by its restriction to $\mathcal{D}\left(S^d\right)$ (which is dense in $\left[\begin{smallmatrix} X^* \\ Y^* \end{smallmatrix}\right]$).

However, this implies that

$$\mathcal{D}\left(S^d\right) = \left\{\left[\begin{smallmatrix} x^* \\ y^* \end{smallmatrix}\right] \in \left[\begin{smallmatrix} X^* \\ Y^* \end{smallmatrix}\right] \mid A^*_{|X^*}x^* + C^*y^* \in X^*\right\} = \mathcal{D}\left(S^*\right).$$

The remaining claims follow from Lemma 6.2.14. □

Theorem 6.2.15 *Let $1 \le p < \infty$, and let Σ be a* compatible *L^p-well-posed linear system on the reflexive Banach spaces (Y, X, U), with semigroup generator A, control operator B, extended observation operator $C_{|W}$ defined on $W \subset X$, and corresponding feedthrough operator D. We denote the corresponding operators for the causal dual system Σ^d by the same letters with a superscript d.*

(i) *If X_1 is dense in W, then Σ^d is compatible with extended observation operator $C^d_{|V^*} = B^*_{|V^*}: V^* \to U^*$, where V^* is the dual of the space $V = W_{-1}$, and $B^*_{|V^*}$ is the adjoint of the operator $B: U \to V$. The corresponding feedthrough operator D^d is given by $D^d = D^*$.*
(ii) *If $1 < p < \infty$, then Σ is weakly regular if and only if Σ^d is weakly regular, and the feedthrough operators D and D^d satisfy $D^d = D^*$.*
(iii) *Σ is uniformly regular if and only if Σ^d is uniformly regular, and the feedthrough operators D and D^d satisfy $D^d = D^*$.*

Proof (i) We begin with some comments on the assumption that X_1 is dense in W. If this is true, then X_{k+1} is dense in W_k for all $k = 0, \pm 1, \pm 2, \ldots$ (see Lemma 5.1.3). By Lemma 9.10.2(ii), the adjoints of the embeddings $X_{k+1} \subset W_k \subset X_k$ are then injective, and they define embeddings $X^*_{-k} \subset (W_k)^* \subset X^*_{-(k+1)}$ (where $X^*_{-(k+1)}$ is the dual of X_{k+1}, cf. Remark 3.6.1). In particular $X^*_1 \subset V^* \subset X^*$.

By Definition 5.1.1, the compatibility of Σ (together with the closed graph theorem) implies that $B \in \mathcal{B}(U; V)$ and $C_{|W} \in \mathcal{B}(W; Y)$. Therefore $C^d_{|V^*} = B^*_{|V^*} \in \mathcal{B}(V^*; U^*)$ and $B^d = C^* \in \mathcal{B}(Y^*; W^*) = \mathcal{B}(Y^*; V^*_{-1})$, and hence Σ^d is compatible.

To prove that $D^d = D^*$ we fix some $\alpha \in \rho(A^d)$ and make the following computation (cf. (5.1.2) and (6.2.6)):

$$\begin{aligned} D^d &= \widehat{\mathfrak{D}}^d(\alpha) - C^d_{|V^*}(\alpha - A^d_{|X})^{-1}B^d \\ &= \widehat{\mathfrak{D}}^*(\overline{\alpha}) - B^*_{|V^*}(\alpha - A^*_{|V})^{-1}C^* \\ &= \left(\widehat{\mathfrak{D}}(\overline{\alpha}) - C_{|W}(\overline{\alpha} - A_{|X})^{-1}B\right)^* = D^*. \end{aligned}$$

(ii) Weak regularity of Σ means that for all $u \in U$ and all $y^* \in Y^*$,

$$\lim_{\alpha \to +\infty} \langle \widehat{\mathfrak{D}}(\alpha)u, y^* \rangle_{(Y,Y^*)} = \langle Du, y^* \rangle_{(Y,Y^*)}$$

for $D \in \mathcal{B}(U; Y)$, whereas weak regularity of Σ^d means that for all $u \in U^{**} = U$,

$$\lim_{\alpha\to+\infty} \langle u, \widehat{\mathfrak{D}}^d(\alpha)y^*\rangle_{(U,U^*)} = \langle u, D^d y^*\rangle_{(U,U^*)}$$

for some $D^d \in \mathcal{B}(Y^*; U^*)$. But

$$\langle \widehat{\mathfrak{D}}(\alpha)u, y^*\rangle_{(Y,Y^*)} = \langle u, \widehat{\mathfrak{D}}^*(\alpha)y^*\rangle_{(U,U^*)} = \langle u, \widehat{\mathfrak{D}}^d(\alpha)y^*\rangle_{(U,U^*)},$$

so if one of these limits exists, then so does the other, and the two limits coincide. Furthermore, also in this case $D^d = D^*$ because

$$\langle Du, y^*\rangle_{(Y,Y^*)} = \langle u, D^* y^*\rangle_{(U,U^*)} = \langle u, D^d y^*\rangle_{(U,U^*)}.$$

(iii) This follows from the fact that $\widehat{\mathfrak{D}}(\alpha) \to D$ in $\mathcal{B}(U; Y)$ iff $\widehat{\mathfrak{D}}^d(\alpha) = \widehat{\mathfrak{D}}^*(\alpha)$ to D^* in $\mathcal{B}(Y^*; U^*)$ since

$$\|\widehat{\mathfrak{D}}(\alpha) - D\| = \|\widehat{\mathfrak{D}}^*(\alpha) - D^*\|;$$

cf. Lemma 3.5.1. □

Let us end this section by showing that the more general dual system nodes discussed in Lemma 6.2.14 satisfy an appropriate version of Theorem 6.2.10.

Lemma 6.2.16 *Let S be a system node on (Y, X, U), where X is reflexive, let $x_0 \in X$ and $u \in W^{2,1}_{\mathrm{loc}}(\overline{\mathbb{R}}^+; U)$ with $\left[\begin{smallmatrix} x_0 \\ u(0)\end{smallmatrix}\right] \in \mathcal{D}(S)$, and let x and y be the corresponding state trajectory and output function of S given by Lemma 4.7.8. Let S^* be the dual system node on (U^*, X^*, Y^*) (see Lemma 6.2.14), let $x_0^d \in X^*$, and $y^d \in W^{2,1}_{\mathrm{loc}}(\overline{\mathbb{R}}^+; Y^*)$ with $\left[\begin{smallmatrix} x_0^d \\ y^d(0)\end{smallmatrix}\right] \in \mathcal{D}(S^*)$, and let x^d and u^d be the corresponding state trajectory and output function of S^* given by Lemma 4.7.8. Then, for all $t > 0$,*

$$\begin{aligned}\langle x(t), x_0^d\rangle_{(X,X^*)} &+ \int_0^t \langle y(s), y^*(t-s)\rangle_{(Y,Y^*)}\,ds \\ &= \langle x_0, x^d(t)\rangle_{(X,X^*)} + \int_0^t \langle u(s), u^*(t-s)\rangle_{(U,U^*)}\,ds.\end{aligned} \tag{6.2.9}$$

Proof By Lemma 4.7.8, for all $s \in [0, t]$,

$$\begin{aligned}\left\langle \begin{bmatrix} \dot{x}(s) \\ y(s)\end{bmatrix}, \begin{bmatrix} x^d(t-s) \\ y^d(t-s)\end{bmatrix}\right\rangle &= \left\langle S\begin{bmatrix} x(s) \\ u(s)\end{bmatrix}, \begin{bmatrix} x^d(t-s) \\ y^d(t-s)\end{bmatrix}\right\rangle \\ &= \left\langle \begin{bmatrix} x(s) \\ u(s)\end{bmatrix}, S^*\begin{bmatrix} x^d(t-s) \\ y^d(t-s)\end{bmatrix}\right\rangle \\ &= \left\langle \begin{bmatrix} x(s) \\ u(s)\end{bmatrix}, \begin{bmatrix} (x^d)'(t-s) \\ u^d(t-s)\end{bmatrix}\right\rangle,\end{aligned}$$

or equivalently,

$$\frac{\mathrm{d}}{\mathrm{d}s}\langle x(s), x^d(t-s)\rangle_{(X,X^*)} + \langle y(s), y^d(t-s)\rangle_{(Y,Y^*)}$$
$$= \langle u(s), u^d(t-s)\rangle_{(U,U^*)}.$$

Integrating this identity over $[0, t]$ we get (6.2.9). □

6.3 Flow-inversion

The idea behind flow-inversion is the following.[1] We start with a well-posed linear system $\Sigma = \left[\begin{smallmatrix}\mathfrak{A} & \mathfrak{B}\\ \mathfrak{C} & \mathfrak{D}\end{smallmatrix}\right]$ on (Y, X, U). Let x be the state trajectory and let y be the output function of this system with initial time zero, initial state x_0, and input function $u \in L^p|Reg_{\mathrm{loc}}(\overline{\mathbb{R}}^+; U)$, i.e.,

$$\begin{aligned} x(t) &= \mathfrak{A}_0^t x_0 + \mathfrak{B}_0^t u = \mathfrak{A}^t x_0 + \mathfrak{B}\tau^t \pi_+ u, \qquad t \ge 0,\\ y &= \mathfrak{C}_0 x_0 + \mathfrak{D}_0 u = \mathfrak{C} x_0 + \mathfrak{D}\pi_+ u. \end{aligned} \tag{6.3.1}$$

If $\mathfrak{D}$ has an inverse in $TIC_{\mathrm{loc}}(Y; U)$, then we can solve both $x(t)$ and u in terms of x_0 and y to get

$$\begin{aligned} \begin{bmatrix} x(t)\\ \pi_+ u\end{bmatrix} &= \begin{bmatrix} \mathfrak{A}^t & \mathfrak{B}\tau^t\\ 0 & 1\end{bmatrix}\begin{bmatrix} x_0\\ \pi_+ u\end{bmatrix}\\ &= \begin{bmatrix} \mathfrak{A}^t & \mathfrak{B}\tau^t\\ 0 & 1\end{bmatrix}\begin{bmatrix} 1 & 0\\ \mathfrak{C} & \mathfrak{D}\end{bmatrix}^{-1}\begin{bmatrix} x_0\\ \pi_+ y\end{bmatrix}\\ &= \begin{bmatrix} \mathfrak{A}^t - \mathfrak{B}\tau^t\mathfrak{D}^{-1}\mathfrak{C} & \mathfrak{B}\mathfrak{D}^{-1}\tau^t\\ -\mathfrak{D}^{-1}\mathfrak{C} & \mathfrak{D}^{-1}\end{bmatrix}\begin{bmatrix} x_0\\ \pi_+ y\end{bmatrix}, \qquad t \ge 0. \end{aligned} \tag{6.3.2}$$

This set of equations is of the same nature as (6.3.1), and it suggests that it may be possible to interpret x and u as the state trajectory and output function of another well-posed linear system with initial time zero, initial state x_0, and input function y. Indeed, this is true if we exclude the L^∞-well-posed case, and restrict our attention to *well-posed linear systems*, i.e, systems that are either L^p-well-posed for some $p < \infty$ or *Reg*-well-posed (cf. Definition 2.2.4).[2]

[1] The results presented in this section are equivalent to the output feedback results presented in Chapter 7 in the sense that the flow-inverted system can be interpreted as an output feedback connection (see Remark 7.1.10), and an output feedback connection can be interpreted as a flow-inverse (see Remark 7.2.3). Therefore, to avoid undue repetition we leave some of the proofs to the reader.

[2] It is possible to include the case $p = \infty$, too, by adding the requirement that the input/output map $\mathfrak{D}$ is invertible both in $TIC^\infty_{\mathrm{loc}}$ and in TIC^{Reg}_{loc}; this is needed for the strong continuity of the

Theorem 6.3.1 *Let $\Sigma = \left[\begin{smallmatrix}\mathfrak{A} & \mathfrak{B}\\ \mathfrak{C} & \mathfrak{D}\end{smallmatrix}\right]$ be a well-posed linear system on (Y, X, U), and suppose that $\mathfrak{D}$ has an inverse in $TIC_{\mathrm{loc}}(Y; U)$. Then the system*

$$\begin{aligned}
\Sigma_\times = \begin{bmatrix} \mathfrak{A}_\times & \mathfrak{B}_\times\tau \\ \mathfrak{C}_\times & \mathfrak{D}_\times \end{bmatrix} &= \begin{bmatrix} \mathfrak{A} & \mathfrak{B}\tau \\ 0 & 1 \end{bmatrix} \begin{bmatrix} 1 & 0 \\ \mathfrak{C} & \mathfrak{D} \end{bmatrix}^{-1} \\
&= \begin{bmatrix} 1 & -\mathfrak{B}\tau \\ 0 & \mathfrak{D} \end{bmatrix}^{-1} \begin{bmatrix} \mathfrak{A} & 0 \\ -\mathfrak{C} & 1 \end{bmatrix} \\
&= \begin{bmatrix} \mathfrak{A} - \mathfrak{B}\tau\mathfrak{D}^{-1}\mathfrak{C} & \mathfrak{B}\mathfrak{D}^{-1}\tau \\ -\mathfrak{D}^{-1}\mathfrak{C} & \mathfrak{D}^{-1} \end{bmatrix} \\
&= \begin{bmatrix} \mathfrak{A} & 0 \\ 0 & 0 \end{bmatrix} + \begin{bmatrix} \mathfrak{B}\tau \\ 1 \end{bmatrix} \mathfrak{D}^{-1} \begin{bmatrix} -\mathfrak{C} & 1 \end{bmatrix}
\end{aligned} \tag{6.3.3}$$

is a linear system on (U, X, Y) which is well-posed in the same sense as Σ. If x and y are the state trajectory and output function of Σ with initial time $s \in \mathbb{R}$, initial state $x_s \in X$, and input function $u \in L^p|Reg_{\mathrm{loc}}([s,\infty); U)$, then x and u are the state trajectory and output function of $\Sigma_\times$ with the same initial time s, the same initial state x_s, and input function y. In particular, for $s = 0$ we have

$$\begin{bmatrix} x(t) \\ \pi_+ u \end{bmatrix} = \begin{bmatrix} \mathfrak{A}_\times^t & \mathfrak{B}_\times\tau^t \\ \mathfrak{C}_\times & \mathfrak{D}_\times \end{bmatrix} \begin{bmatrix} x_0 \\ y \end{bmatrix}, \qquad t \ge 0. \tag{6.3.4}$$

The proof of Theorem 6.3.1 is based on the following formula for the Hankel operator of the causal inverse of an operator in $TIC_\alpha(U)$:

Lemma 6.3.2 *Let $\mathfrak{D} \in TIC_{\mathrm{loc}}(U; Y)$ have an inverse in $TIC_{\mathrm{loc}}(Y; U)$. Then*

$$\pi_+\mathfrak{D}^{-1}\pi_- = -\mathfrak{D}^{-1}\pi_+\mathfrak{D}\pi_-\mathfrak{D}^{-1}.$$

Proof This follows from the causality of $\mathfrak{D}$ and $\mathfrak{D}^{-1}$ as follows:

$$\begin{aligned}
0 = \mathfrak{D}^{-1}\pi_+\pi_- &= \mathfrak{D}^{-1}\pi_+\mathfrak{D}(\pi_+ + \pi_-)\mathfrak{D}^{-1}\pi_- \\
&= \mathfrak{D}^{-1}\mathfrak{D}\pi_+\mathfrak{D}^{-1}\pi_- + \mathfrak{D}^{-1}\pi_+\mathfrak{D}\pi_-\mathfrak{D}^{-1} \\
&= \pi_+\mathfrak{D}^{-1}\pi_- + \mathfrak{D}^{-1}\pi_+\mathfrak{D}\pi_-\mathfrak{D}^{-1}.
\end{aligned}$$

□

Proof of Theorem 6.3.1 We begin by observing that all the different formulas for components of $\Sigma_\times$ are equivalent; this follows from Lemma A.4.2 and some easy algebraic manipulations. Moreover, if $\Sigma_\times$ is a well-posed linear system,

closed-loop semigroup. It is not clear if the latter condition is implied by the former or not. If it is true, then the L^∞ case could be treated in the same way as the other cases.

then these formulas show that the state trajectory x and the output function u of $\Sigma_\times$ with initial time zero, initial value x_0, and input function y, are given by (6.3.4). This together with a time shift implies that x and u are the state trajectory and output function of $\Sigma_\times$ with initial time s, initial state x_s, and input function y if and only if x and y are the state trajectory and output function of Σ with initial time $s \in \mathbb{R}$, initial state x, and input function u.

The continuity of the operators in $\Sigma_\times$ is obvious. The strong continuity of $\mathfrak{A}_\times$ follows from the strong continuity of τ in $L^p_{c,\text{loc}}(\mathbb{R};U)$ (see Example 2.5.3) for $p < \infty$ or from the strong continuity of $\mathfrak{B}\tau$ from $Reg_{c,\text{loc}}(\mathbb{R};U)$ to X (see Theorem 4.3.1(ii)).[3] Thus, to complete the proof it suffices to check that the algebraic properties in Definition 2.2.1 hold.

The key ingredient in the proof of the algebraic properties is the formula

$$\pi_+\mathfrak{D}^{-1}\pi_- = -\mathfrak{D}^{-1}\mathfrak{C}\mathfrak{B}\mathfrak{D}^{-1}, \tag{6.3.5}$$

which follows from Lemma 6.3.2 and the fact that $\pi_+\mathfrak{D}\pi_- = \mathfrak{C}\mathfrak{B}$.

(i) Clearly $\mathfrak{A}^0_\times = \mathfrak{A}^0 = 1$. To show that $\mathfrak{A}_\times$ is a semigroup we use Definition 2.2.1(i)–(iv), Lemma A.4.1, the causality and time-invariance of $\mathfrak{D}^{-1}$, and (7.1.4) to compute

$$\begin{aligned}
\mathfrak{A}^s_\times\mathfrak{A}^t_\times &= \big(\mathfrak{A}^s - \mathfrak{B}\tau^s\mathfrak{D}^{-1}\mathfrak{C}\big)\big(\mathfrak{A}^t - \mathfrak{B}\mathfrak{D}^{-1}\tau^t\mathfrak{C}\big)\\
&= \mathfrak{A}^s\mathfrak{A}^t - \mathfrak{A}^s\mathfrak{B}\mathfrak{D}^{-1}\tau^t\mathfrak{C} - \mathfrak{B}\tau^s\mathfrak{D}^{-1}\mathfrak{C}\mathfrak{A}^t\\
&\quad + \mathfrak{B}\tau^s\mathfrak{D}^{-1}\mathfrak{C}\mathfrak{B}\mathfrak{D}^{-1}\tau^t\mathfrak{C}\\
&= \mathfrak{A}^{s+t} - \mathfrak{B}\tau^s\pi_-\mathfrak{D}^{-1}\pi_-\tau^t\mathfrak{C}\\
&\quad - \mathfrak{B}\tau^s\pi_+\mathfrak{D}^{-1}\pi_+\tau^t\mathfrak{C}\\
&\quad - \mathfrak{B}\tau^s\pi_+\mathfrak{D}^{-1}\pi_-\tau^t\mathfrak{C}\\
&= \mathfrak{A}^{s+t} - \mathfrak{B}\tau^s\tau^t\mathfrak{D}^{-1}\mathfrak{C}\\
&= \mathfrak{A}^{s+t}_\times.
\end{aligned}$$

(ii) The proof of the identity $\mathfrak{A}^t_\times\mathfrak{B}_\times = \mathfrak{B}_\times\tau^t\pi_-$ in Definition 2.2.1(ii) uses the same ingredients:

$$\begin{aligned}
\mathfrak{A}^t_\times\mathfrak{B}_\times &= \big(\mathfrak{A}^t - \mathfrak{B}\tau^t\mathfrak{D}^{-1}\mathfrak{C}\big)\mathfrak{B}\mathfrak{D}^{-1}\\
&= \mathfrak{A}^t\mathfrak{B}\mathfrak{D}^{-1} - \mathfrak{B}\tau^t\mathfrak{D}^{-1}\mathfrak{C}\mathfrak{B}\mathfrak{D}^{-1}\\
&= \mathfrak{B}\tau^t\pi_-\mathfrak{D}^{-1}\pi_- + \mathfrak{B}\tau^t\pi_+\mathfrak{D}^{-1}\pi_-\\
&= \mathfrak{B}\mathfrak{D}^{-1}\tau^t\pi_- = \mathfrak{B}_\times\tau^t\pi_-.
\end{aligned}$$

[3] If it is true that $\mathfrak{D}^{-1}$ maps $Reg_{\text{loc}}(\overline{\mathbb{R}}^+;Y)$ into $Reg_{\text{loc}}(\overline{\mathbb{R}}^+;U)$ also in the L^∞-case (which seems plausible), then the closed-loop semigroup is strongly continuous also in the L^∞-case, and we can remove the restriction $p < \infty$ imposed on the system at the beginning of this chapter.

(iii) Also the proof of the identity $\mathfrak{C}_\times \mathfrak{A}^t_\times = \pi_+ \tau^t \mathfrak{C}_\times$ in Definition 2.2.1(iii) is similar:

$$\begin{aligned}\mathfrak{C}_\times \mathfrak{A}^t_\times &= \mathfrak{D}^{-1}\mathfrak{C}(\mathfrak{A}^t + \mathfrak{B}\mathfrak{D}^{-1}\tau^t\mathfrak{C})\\ &= \mathfrak{D}^{-1}\mathfrak{C}\mathfrak{A}^t + \mathfrak{D}^{-1}\mathfrak{C}\mathfrak{B}\mathfrak{D}^{-1}\tau^t\mathfrak{C}\\ &= \pi_+\mathfrak{D}^{-1}\pi_+\tau^t\mathfrak{C} + \pi_+\mathfrak{D}^{-1}\pi_-\tau^t\mathfrak{C}\\ &= \pi_+\tau^t\mathfrak{D}^{-1}\mathfrak{C} = \pi_+\tau^t\mathfrak{C}_\times.\end{aligned}$$

(iv) The time-invariance and causality of $\mathfrak{D}_\times = \mathfrak{D}^{-1}$ are part of the assumption of the theorem, and by (6.3.5), the Hankel operator of $\mathfrak{D}_\times$ is

$$\pi_+\mathfrak{D}_\times\pi_- = -\mathfrak{D}^{-1}\mathfrak{C}\mathfrak{B}\mathfrak{D}^{-1} = \mathfrak{C}_\times\mathfrak{B}_\times.$$

□

Definition 6.3.3 We say that Σ is *flow-invertible* if its input/output map is invertible in TIC_{loc}, and we call the system $\Sigma_\times$ in Theorem 6.3.1 the *flow-inverse* of Σ.

Remark 6.3.4 In the classical system

$$\begin{aligned}\dot{x}(t) &= Ax(t) + Bu(t),\\ y(t) &= Cx(t) + Du(t), \qquad t \geq 0,\\ x(0) &= x_0,\end{aligned} \tag{6.3.6}$$

it is possible to interpret y as the input function and u as the output function if and only if D is invertible. In this case the flow-inverted system $\Sigma_\times$ is again a classical system with generators

$$\begin{aligned}\begin{bmatrix} A_\times & B_\times \\ C_\times & D_\times \end{bmatrix} &= \begin{bmatrix} A & B \\ 0 & 1 \end{bmatrix}\begin{bmatrix} 1 & 0 \\ C & D \end{bmatrix}^{-1}\\ &= \begin{bmatrix} 1 & -B \\ 0 & D \end{bmatrix}^{-1}\begin{bmatrix} A & 0 \\ -C & 1 \end{bmatrix}\\ &= \begin{bmatrix} A - BD^{-1}C & BD^{-1} \\ -D^{-1}C & D^{-1} \end{bmatrix}\\ &= \begin{bmatrix} A & 0 \\ 0 & 0 \end{bmatrix} + \begin{bmatrix} B \\ 1 \end{bmatrix} D^{-1} \begin{bmatrix} -C & 1 \end{bmatrix}.\end{aligned} \tag{6.3.7}$$

Observe the striking similarity between this formula and the one given in Theorem 6.3.1. Operator node versions of this result are given in Theorems 6.3.6 and 6.3.16.

The crucial assumption in Theorem 6.3.1 is, of course, that $\mathfrak{D}$ has an inverse in $TIC_{\mathrm{loc}}(Y; Y)$. This assumption can be characterized in many different ways:

Theorem 6.3.5 *Let* $\Sigma = \left[\begin{smallmatrix}\mathfrak{A} & \mathfrak{B}\\ \mathfrak{C} & \mathfrak{D}\end{smallmatrix}\right]$ *be a well-posed linear system on* (Y, X, U) *with growth bound* $\omega_{\mathfrak{A}}$. *Then the following conditions are equivalent:*

(i) Σ *is flow-invertible.*
(ii) *For some* $T > 0$, *the operator* $\pi_{[0,T)}\mathfrak{D}\pi_{[0,T)}$ *has a bounded inverse defined on* $L^p|Reg([0, T); Y)$.
(iii) *For all* s *and* t, $-\infty < s < t < \infty$, *the operator* $\pi_{[s,t)}\mathfrak{D}\pi_{[s,t)}$ *has a bounded inverse defined on* $L^p|Reg([s, t); Y)$.
(iv) *The operator* $\mathfrak{D}$ *has an inverse in* $TIC_{\mathrm{loc}}(Y; U)$.
(v) *The operator* $\mathfrak{D}$ *has an inverse in* $TIC_\alpha(Y; U)$ *for some* $\alpha > \omega_{\mathfrak{A}}$.
(vi) *In the set of equations (6.3.1) we can interpret* $x(t)$ *and* π_+u *as the state trajectory and output function of a well-posed linear system with initial state* $x_0 \in X$ *and input function* $y \in L^p|Reg_{\mathrm{loc}}(\overline{\mathbb{R}}^+; Y)$.

Proof (i) $\Leftrightarrow$ (iv): See Definition 6.3.3.

(i) $\Rightarrow$ (vi): This follows from Theorem 6.3.1.

(vi) $\Rightarrow$ (v): By taking $x_0 = 0$ in (6.3.1) we get

$$y = \mathfrak{D}\pi_+u.$$

This equation determines u uniquely and continuously in $L^p|Reg_\alpha(\overline{\mathbb{R}}^+; U)$ for some $\alpha \in \mathbb{R}$ as a function of $v \in L^p|Reg_\alpha(\overline{\mathbb{R}}^+; U)$ (if and) only if $\mathfrak{D}$ has an inverse in $TIC_\alpha(Y; U)$.

(v) $\Rightarrow$ (ii): If (v) holds, then $\pi_{[0,T)}\mathfrak{D}^{-1}\pi_{[0,T)}$ is the inverse to $\pi_{[0,T)}\mathfrak{D}\pi_{[0,T)}$ in $\mathcal{B}(L^p|Reg([0, T); U))$ (for all $T > 0$).

(ii) $\Rightarrow$ (iii): Let (ii) hold. With the notation introduced in Definition 2.2.6, this means that $\mathfrak{D}_0^T$ has an inverse in $\mathcal{B}(L^p|Reg([0, T); U))$. This and the time-invariance of $\mathfrak{D}$ imply that $\pi_{[s,s+T)}\mathfrak{D}\pi_{[s,s+T)} = \mathfrak{D}_s^{s+T}$ has an inverse in $\mathcal{B}(L^p|Reg([s, s+T); U))$ for all $s \in \mathbb{R}$ (cf. Lemma 2.2.8(iii)).

The last identity in Lemma 2.2.8(iv) can be written in block matrix form as (for all $s \le r \le t$)

$$\mathfrak{D}_s^t = \begin{bmatrix}\pi_{[s,r)} & \pi_{[r,t)}\end{bmatrix}\begin{bmatrix}\mathfrak{D}_s^r & 0\\ \mathfrak{C}_r^t\mathfrak{B}_s^r & \mathfrak{D}_r^t\end{bmatrix}\begin{bmatrix}\pi_{[s,r)}\\ \pi_{[r,t)}\end{bmatrix},$$

hence,

$$\pi_{[s,t)}\mathfrak{D}_s^t = \begin{bmatrix}\pi_{[s,r)} & \pi_{[r,t)}\end{bmatrix}\begin{bmatrix}\mathfrak{D}_s^r & 0\\ -\mathfrak{C}_r^t\mathfrak{B}_s^r & \mathfrak{D}_r^t\end{bmatrix}\begin{bmatrix}\pi_{[s,r)}\\ \pi_{[r,t)}\end{bmatrix}. \tag{6.3.8}$$

In particular, $\pi_{[s,r)}\mathfrak{D}_s^t = \mathfrak{D}_{[s,r)}$ and $\mathfrak{D}_s^t\pi_{[r,t)} = \mathfrak{D}_r^t$. If we here replace t by $s + T$ and r by t, with $s < t < s + T$, then the invertibility of $\mathfrak{D}_s^{s+T}$ implies that $\mathfrak{D}_s^t$ maps $L^p|Reg([s, t); U)$ onto itself, and that $\mathfrak{D}_t^{s+T}$ is injective on $L^p|Reg([t, s + T); U)$. The latter condition implies that $\mathfrak{D}_s^t$ is injective on $L^p|Reg([s, t); U)$ (use the time-invariance and replace $2s + T - t$ by t; note that $s < t < s + T$

iff $s < 2s + T - t < s + T$). Thus, for all t satisfying $0 < t - s \le T$, $\mathfrak{D}_s^t$ maps $L^p|Reg([s,t);U)$ one-to-one onto itself. By the closed graph theorem, $\mathfrak{D}_s^t$ has a continuous inverse.

We still need to remove the restriction $t - s < T$. Clearly, to do this it is enough to show that if $\mathfrak{D}_0^T$ is invertible, then so is $\mathfrak{D}_0^{2T}$ (because then we can first iterate this argument to show that $\mathfrak{D}_0^{2^n T}$ is invertible for all positive integers n, and then replace T by $2^n T$ in the argument given above to show that $\mathfrak{D}_0^t$ is invertible whenever $0 < t - s < 2^n T$). However, the invertibility of $\mathfrak{D}_0^{2T}$ follows from (6.3.8) with $s = 0$, $r = T$, and $t = 2T$, since all the block matrices on the right hand side are invertible (cf. Lemma A.4.2(i)).

(iii) $\Rightarrow$ (iv): Trivially, (iii) implies that $\mathfrak{D}$ maps $L^p_{c,\mathrm{loc}}(\mathbb{R};U)$ or $Reg_{c,\mathrm{loc}}(\mathbb{R};U)$ one-to-one onto itself, and that the inverse is continuous. The time-invariance of the inverse is also trivial. □

Let us next compute the system node of the flow-inverted system.

Theorem 6.3.6 *Let $\Sigma = \left[\begin{smallmatrix} \mathfrak{A} & \mathfrak{B} \\ \mathfrak{C} & \mathfrak{D} \end{smallmatrix}\right]$ be a well-posed flow-invertible linear system on (Y, X, U) with system node $S = \left[\begin{smallmatrix} A\&B \\ C\&D \end{smallmatrix}\right]$. Denote the system node of the flow-inverted system by $S_\times = \left[\begin{smallmatrix} [A\&B]_\times \\ [C\&D]_\times \end{smallmatrix}\right]$. Then the operator $\left[\begin{smallmatrix} 1 & 0 \\ C\&D \end{smallmatrix}\right]$ maps $\mathcal{D}(S)$ continuously onto $\mathcal{D}(S_\times)$, its inverse is $\left[\begin{smallmatrix} 1 & 0 \\ [C\&D]_\times \end{smallmatrix}\right]$, and*

$$\begin{bmatrix} A\&B \\ C\&D \end{bmatrix} = \begin{bmatrix} [A\&B]_\times \\ 0 \quad 1 \end{bmatrix} \begin{bmatrix} 1 \quad 0 \\ [C\&D]_\times \end{bmatrix}^{-1} \quad \text{(on } \mathcal{D}(S)\text{)}, \tag{6.3.9}$$

$$\begin{bmatrix} [A\&B]_\times \\ [C\&D]_\times \end{bmatrix} = \begin{bmatrix} A\&B \\ 0 \quad 1 \end{bmatrix} \begin{bmatrix} 1 \quad 0 \\ C\&D \end{bmatrix}^{-1} \quad \text{(on } \mathcal{D}(S_\times)\text{)}. \tag{6.3.10}$$

Proof Let $\left[\begin{smallmatrix} x_0 \\ u_0 \end{smallmatrix}\right] \in \mathcal{D}(S)$, i.e., $x_0 \in X$, $u_0 \in U$, and $A_{|X_{-1}} x_0 + B u_0 \in X$. Define

$$y_0 = C\&D \begin{bmatrix} x_0 \\ u_0 \end{bmatrix}.$$

Choose an arbitrary $u \in C^1(\overline{\mathbb{R}}^+;U)$ with $u(0) = u_0$. Let x and y be the state trajectory and output function of Σ with initial time zero, initial state x_0, and input function u. Then, by Theorem 4.6.11, x is continuously differentiable in X, $y \in W^{1,p}_{\mathrm{loc}}(\overline{\mathbb{R}}^+;Y)$ (or $y \in C^1(\overline{\mathbb{R}}^+;Y)$ in the *Reg*-well-posed case), and for all $t \ge 0$,

$$\begin{bmatrix} \dot{x}(t) \\ y(t) \end{bmatrix} = S \begin{bmatrix} x(t) \\ u(t) \end{bmatrix}, \qquad \begin{bmatrix} x(t) \\ y(t) \end{bmatrix} = \begin{bmatrix} 1 \quad 0 \\ C\&D \end{bmatrix} \begin{bmatrix} x(t) \\ u(t) \end{bmatrix}. \tag{6.3.11}$$

In particular, $y(0) = y_0$. On the other hand, we can also consider the system $\Sigma_\times$ with initial time zero, initial state x_0, and input function y. By Theorem 6.3.1, the state trajectory and output function of this system are x and u, where

x and u are the same functions as above. By Theorem 4.3.7, Corollary 4.3.8, and Theorem 4.6.11, $\left[\begin{smallmatrix} x_0 \\ y(0) \end{smallmatrix}\right] = \left[\begin{smallmatrix} x_0 \\ y_0 \end{smallmatrix}\right] \in \mathcal{D}(S_\times)$, and, for all $t \geq 0$,

$$\begin{bmatrix} \dot{x}(t) \\ u(t) \end{bmatrix} = S_\times \begin{bmatrix} x(t) \\ y(t) \end{bmatrix}, \qquad \begin{bmatrix} x(t) \\ u(t) \end{bmatrix} = \begin{bmatrix} 1 & 0 \\ [C\&D]_\times \end{bmatrix} \begin{bmatrix} x(t) \\ y(t) \end{bmatrix}. \tag{6.3.12}$$

In particular, taking $t = 0$ in (6.3.11) and (6.3.12) we find that $\left[\begin{smallmatrix} 1 & 0 \\ C\&D \end{smallmatrix}\right]$ maps $\mathcal{D}(S)$ into $\mathcal{D}(S_\times)$, that it has a left inverse $\left[\begin{smallmatrix} 1 & 0 \\ [C\&D]_\times \end{smallmatrix}\right]$, and that

$$S \begin{bmatrix} x_0 \\ u_0 \end{bmatrix} = \begin{bmatrix} \dot{x}(0) \\ y_0 \end{bmatrix} = \begin{bmatrix} [A\&B]_\times \\ 0 \quad 1 \end{bmatrix} \begin{bmatrix} x_0 \\ y_0 \end{bmatrix} = \begin{bmatrix} [A\&B]_\times \\ 0 \quad 1 \end{bmatrix} \begin{bmatrix} 1 & 0 \\ C\&D \end{bmatrix} \begin{bmatrix} x_0 \\ y_0 \end{bmatrix}.$$

Thus, $S = \left[\begin{smallmatrix} [A\&B]_\times \\ 0 \quad 1 \end{smallmatrix}\right] \left[\begin{smallmatrix} 1 & 0 \\ C\&D \end{smallmatrix}\right]$. By interchanging the roles of Σ and $\Sigma_\times$ we find that $\left[\begin{smallmatrix} 1 & 0 \\ [C\&D]_\times \end{smallmatrix}\right]$ is also a right inverse of $\left[\begin{smallmatrix} 1 & 0 \\ C\&D \end{smallmatrix}\right]$, and that $S_\times = \left[\begin{smallmatrix} A\&B \\ 0 \quad 1 \end{smallmatrix}\right] \left[\begin{smallmatrix} 1 & 0 \\ [C\&D]_\times \end{smallmatrix}\right]$. □

Motivated by Theorem 6.3.6 we extend Definition 6.3.3 to arbitrary operator nodes as follows.

Definition 6.3.7 Let $S = \left[\begin{smallmatrix} A\&B \\ C\&D \end{smallmatrix}\right]$ be an operator node on (Y, X, U). We call this operator node *flow-invertible* if there exists an operator node $S_\times = \left[\begin{smallmatrix} [A\&B]_\times \\ [C\&D]_\times \end{smallmatrix}\right]$ on (U, X, Y) which together with S satisfies the following conditions: the operator $\left[\begin{smallmatrix} 1 & 0 \\ C\&D \end{smallmatrix}\right]$ maps $\mathcal{D}(S)$ continuously onto $\mathcal{D}(S_\times)$, its inverse is $\left[\begin{smallmatrix} 1 & 0 \\ [C\&D]_\times \end{smallmatrix}\right]$, and (6.3.9)–(6.3.10) hold. In this case we call S and $S_\times$ *flow-inverses* of each other.

Obviously, the flow-inverse of a node S is unique (when it exists). Furthermore, by Theorem 6.3.6, if a well-posed linear system is flow-invertible, then its system node is flow-invertible, and the flow-inverted operator node is well-posed. As we shall see in Corollary 6.3.15, the converse statement is also true, at least in the L^p-well-posed case with $p < \infty$. However, before proving this converse statement, let us first look at some easy consequences of Definition 6.3.7.

Lemma 6.3.8 *Let $S = \left[\begin{smallmatrix} A\&B \\ C\&D \end{smallmatrix}\right]$ be a flow-invertible operator node on (Y, X, U), with main operator A, control operator B, observation operator C, and transfer function $\widehat{\mathfrak{D}}$, and let $S_\times = \left[\begin{smallmatrix} [A\&B]_\times \\ [C\&D]_\times \end{smallmatrix}\right]$ be its flow-inverse, with main operator $A_\times$, control operator $B_\times$, observation operator $C_\times$, and transfer function $\widehat{\mathfrak{D}}_\times$. Then the following claims are true.*

(i) *$\alpha \in \rho(A_\times)$ if and only if $\left[\begin{smallmatrix} \alpha & 0 \\ 0 & 0 \end{smallmatrix}\right] - S$ is invertible, in which case*

$$\left(\begin{bmatrix} \alpha & 0 \\ 0 & 0 \end{bmatrix} - S\right)^{-1} = \begin{bmatrix} (\alpha - A_\times)^{-1} & -(\alpha - A_{\times|X})^{-1} B_\times \\ C_\times(\alpha - A_\times)^{-1} & -\widehat{\mathfrak{D}}_\times(\alpha) \end{bmatrix}. \tag{6.3.13}$$

(ii) *$\alpha \in \rho(A)$ if and only if $\left[\begin{smallmatrix} \alpha & 0 \\ 0 & j0 \end{smallmatrix}\right] - S_\times$ is invertible, in which case*

$$\left(\begin{bmatrix} \alpha & 0 \\ 0 & 0 \end{bmatrix} - S_\times\right)^{-1} = \begin{bmatrix} (\alpha - A)^{-1} & -(\alpha - A_{|X})^{-1}B \\ C(\alpha - A)^{-1} & -\widehat{\mathfrak{D}}(\alpha) \end{bmatrix}. \qquad (6.3.14)$$

(iii) *If $\alpha \in \rho(A)$, then $\alpha \in \rho(A_\times)$ if and only if $\widehat{\mathfrak{D}}(\alpha)$ is invertible. In this case*

$$\begin{aligned} &\begin{bmatrix} (\alpha - A_\times)^{-1} & -(\alpha - A_{\times|X})^{-1}B_\times \\ C_\times(\alpha - A_\times)^{-1} & -\widehat{\mathfrak{D}}_\times(\alpha) \end{bmatrix} = \begin{bmatrix} (\alpha - A)^{-1} & 0 \\ 0 & 0 \end{bmatrix} \\ &\qquad - \begin{bmatrix} (\alpha - A_{|X})^{-1}B \\ 1 \end{bmatrix} [\widehat{\mathfrak{D}}(\alpha)]^{-1} \begin{bmatrix} C(\alpha - A)^{-1} & 1 \end{bmatrix}. \end{aligned} \qquad (6.3.15)$$

In particular, in this case $\widehat{\mathfrak{D}}_\times(\alpha) = \widehat{\mathfrak{D}}^{-1}(\alpha)$, and U is isomorphic to Y (hence they have the same dimension).

(iv) *If $\alpha \in \rho(A_\times)$, then $\alpha \in \rho(A)$ if and only if $\widehat{\mathfrak{D}}_\times(\alpha)$ is invertible. In this case*

$$\begin{aligned} &\begin{bmatrix} (\alpha - A)^{-1} & -(\alpha - A_{|X})^{-1}B \\ C(\alpha - A)^{-1} & -\widehat{\mathfrak{D}}(\alpha) \end{bmatrix} = \begin{bmatrix} (\alpha - A_\times)^{-1} & 0 \\ 0 & 0 \end{bmatrix} \\ &\qquad - \begin{bmatrix} (\alpha - A_{\times|X})^{-1}B_\times \\ 1 \end{bmatrix} [\widehat{\mathfrak{D}}_\times(\alpha)]^{-1} \begin{bmatrix} C_\times(\alpha - A_\times)^{-1} & 1 \end{bmatrix}. \end{aligned} \qquad (6.3.16)$$

The right-hand side of formulas (6.3.15) and (6.3.16) can be written in several equivalent forms. See Lemma 4.7.18.

Proof (i) By (6.3.9) (note that $\begin{bmatrix} \alpha & 0 \end{bmatrix} \left[\begin{smallmatrix} 1 & 0 \\ [C\&D]_\times \end{smallmatrix}\right] = \begin{bmatrix} \alpha & 0 \end{bmatrix}$),

$$\begin{bmatrix} \alpha & 0 \\ 0 & 0 \end{bmatrix} - S = \left(\begin{bmatrix} \alpha & 0 \\ 0 & -1 \end{bmatrix} - \begin{bmatrix} [A\&B]_\times \\ 0 \end{bmatrix}\right) \begin{bmatrix} 1 & 0 \\ [C\&D]_\times \end{bmatrix}^{-1}. \qquad (6.3.17)$$

The second factor on the right-hand side is a bounded bijection between $\mathcal{D}(S)$ and $\mathcal{D}(S_\times)$, and the first factor on the right-hand side has a bounded inverse defined on $\left[\begin{smallmatrix} X \\ U \end{smallmatrix}\right]$ if and only if $\alpha \in \rho(A_\times)$ (see Lemma 4.7.18(i)). Thus, $\left[\begin{smallmatrix} \alpha & 0 \\ 0 & 0 \end{smallmatrix}\right] - S$ is invertible if and only if $\alpha \in \rho(A_\times)$. Formula (6.3.13) follows from Lemma 4.7.18(i) with S replaced by $S_\times$.

(ii) We get (ii) from (i) by interchanging S and $S_\times$.

(iii) This follows from (i) and Lemma 4.7.18(iv).

(iv) We get (iv) from (iii) by interchanging S and $S_\times$. □

Our following theorem lists a number of equivalent conditions for two operator nodes to be flow-inverses of each other.

Theorem 6.3.9 *Let $S = \left[\begin{smallmatrix} A\&B \\ C\&D \end{smallmatrix}\right]$ be an operator node on (Y, X, U), with main operator A, control operator B, observation operator C, and transfer function $\widehat{\mathfrak{D}}$, and let $S_\times = \left[\begin{smallmatrix} [A\&B]_\times \\ [C\&D]_\times \end{smallmatrix}\right]$ be an operator node on (U, X, Y), with main* *operator* $A_\times$, *control operator* $B_\times$, *observation operator* $C_\times$, *and transfer function* $\widehat{\mathfrak{D}}_\times$. *Then the following conditions are equivalent:*

(i) *S and $S_\times$ are flow-inverses of each other.*
(ii) *The operator* $\left[\begin{smallmatrix}1 & 0\\ [C\&D]_\times\end{smallmatrix}\right]$ *maps* $\mathcal{D}(S_\times)$ *one-to-one onto* $\mathcal{D}(S)$, *and* (6.3.9) *holds.*
(iii) *For all* $\alpha \in \rho(A_\times)$, *the operator* $\left[\begin{smallmatrix}\alpha & 0\\ 0 & 0\end{smallmatrix}\right] - S$ *maps* $\mathcal{D}(S)$ *one-to-one onto* $\left[\begin{smallmatrix}X\\ Y\end{smallmatrix}\right]$, *and its (bounded) inverse is given by* (6.3.13).
(iv) *For some* $\alpha \in \rho(A_\times)$, *the operator* $\left[\begin{smallmatrix}\alpha & 0\\ 0 & 0\end{smallmatrix}\right] - S$ *maps* $\mathcal{D}(S)$ *one-to-one onto* $\left[\begin{smallmatrix}X\\ Y\end{smallmatrix}\right]$ *and* (6.3.13) *holds.*
(v) *The operator* $\left[\begin{smallmatrix}1 & 0\\ C\&D\end{smallmatrix}\right]$ *maps* $\mathcal{D}(S)$ *one-to-one onto* $\mathcal{D}(S_\times)$, *and* (6.3.10) *holds.*
(vi) *For all* $\alpha \in \rho(A)$, *the operator* $\left[\begin{smallmatrix}\alpha & 0\\ 0 & 0\end{smallmatrix}\right] - S_\times$ *maps* $\mathcal{D}(S_\times)$ *one-to-one onto* $\left[\begin{smallmatrix}X\\ Y\end{smallmatrix}\right]$, *and its (bounded) inverse is given by* (6.3.14).
(vii) *For some* $\alpha \in \rho(A)$, *the operator* $\left[\begin{smallmatrix}\alpha & 0\\ 0 & 0\end{smallmatrix}\right] - S_\times$ *maps* $\mathcal{D}(S_\times)$ *one-to-one onto* $\left[\begin{smallmatrix}X\\ Y\end{smallmatrix}\right]$ *and* (6.3.14) *holds.*

When these equivalent conditions hold, then $\left[\begin{smallmatrix}1\\ C\end{smallmatrix}\right]$ *maps* $\mathcal{D}(A)$ *into* $\mathcal{D}(S_\times)$, $\left[\begin{smallmatrix}1\\ C_\times\end{smallmatrix}\right]$ *maps* $\mathcal{D}(A_\times)$ *into* $\mathcal{D}(S)$, *and*

$$\begin{aligned} A &= A_{\times|\mathcal{D}(A)} + B_\times C, & A_\times &= A_{|\mathcal{D}(A_\times)} + BC_\times, \\ 0 &= [C\&D]_\times \begin{bmatrix}1\\ C\end{bmatrix}, & 0 &= C\&D \begin{bmatrix}1\\ C_\times\end{bmatrix}. \end{aligned} \tag{6.3.18}$$

Proof It suffices to prove that (i)–(iv) are equivalent, because the remaining equivalences follow from the facts that (i) is symmetric with respect to S and $S_\times$, and that we get (v)–(vii) by interchanging S and $S_\times$ in (ii)–(iv). Also observe that (6.3.18), which is equivalent to

$$\begin{bmatrix}[A\&B]_\times\\ [C\&D]_\times\end{bmatrix}\begin{bmatrix}1\\ C\end{bmatrix} = \begin{bmatrix}A\\ 0\end{bmatrix}, \quad \begin{bmatrix}A\&B\\ C\&D\end{bmatrix}\begin{bmatrix}1\\ C_\times\end{bmatrix} = \begin{bmatrix}A_\times\\ 0\end{bmatrix}, \tag{6.3.19}$$

follows from (i) and (6.3.9)–(6.3.10) since $\left[\begin{smallmatrix}\mathcal{D}(A)\\ 0\end{smallmatrix}\right] \in \mathcal{D}(S)$ and $\left[\begin{smallmatrix}\mathcal{D}(A_\times)\\ 0\end{smallmatrix}\right] \in \mathcal{D}(S_\times)$.

(i) $\Rightarrow$ (ii): This is obvious (see Definition 6.3.7).

(ii) $\Rightarrow$ (i): Suppose that (ii) holds. Then $\left[\begin{smallmatrix}1 & 0\\ C\&D\end{smallmatrix}\right]\left[\begin{smallmatrix}1 & 0\\ [C\&D]_\times\end{smallmatrix}\right] = \left[\begin{smallmatrix}1 & 0\\ 0 & 1\end{smallmatrix}\right]$ on $\mathcal{D}(S_\times)$ (since, by assumption, $C\&D\left[\begin{smallmatrix}1 & 0\\ [C\&D]_\times\end{smallmatrix}\right] = \left[\begin{smallmatrix}0 & 1\end{smallmatrix}\right]$, and we always have $\left[\begin{smallmatrix}1 & 0\end{smallmatrix}\right]\left[\begin{smallmatrix}1 & 0\\ [C\&D]_\times\end{smallmatrix}\right] = \left[\begin{smallmatrix}1 & 0\end{smallmatrix}\right]$). Thus, $\left[\begin{smallmatrix}1 & 0\\ C\&D\end{smallmatrix}\right]$ is a left-inverse of $\left[\begin{smallmatrix}1 & 0\\ [C\&D]_\times\end{smallmatrix}\right]$. However, as (by assumption) $\left[\begin{smallmatrix}1 & 0\\ [C\&D]_\times\end{smallmatrix}\right]$ is both injective and onto, it is invertible, so the left inverse is also a right inverse, i.e., the inverse of $\left[\begin{smallmatrix}1 & 0\\ [C\&D]_\times\end{smallmatrix}\right]$ is $\left[\begin{smallmatrix}1 & 0\\ C\&D\end{smallmatrix}\right]$. The identity (6.3.10) can equivalently be written in the form $\left[\begin{smallmatrix}[A\&B]_\times\\ [C\&D]_\times\end{smallmatrix}\right] = \left[\begin{smallmatrix}A\&B\\ 0 & 1\end{smallmatrix}\right]\left[\begin{smallmatrix}1 & 0\\ [C\&D]_\times\end{smallmatrix}\right]$. The top part $[A\&B]_\times = A\&B\left[\begin{smallmatrix}1 & 0\\ [C\&D]_\times\end{smallmatrix}\right]$ of this identity is contained in (6.3.9), and the bottom part $[C\&D]_\times = \begin{bmatrix} 0 & 1 \end{bmatrix} \left[\begin{smallmatrix} 1 & 0 \\ [C\&D]_\times \end{smallmatrix}\right]$ is always valid. We conclude that (ii) $\Rightarrow$ (i).

(i) $\Rightarrow$ (iii): See Lemma 6.3.8(i).

(iii) $\Rightarrow$ (iv): This is obvious.

(iv) $\Rightarrow$ (ii): By Lemma 4.7.18(i) with S replaced by $S_\times$, (6.3.13) implies

$$\left(\begin{bmatrix} \alpha & 0 \\ 0 & 0 \end{bmatrix} - S\right)^{-1} \begin{bmatrix} 1 & 0 \\ 0 & -1 \end{bmatrix} = \begin{bmatrix} 1 \quad 0 \\ [C\&D]_\times \end{bmatrix} \begin{bmatrix} \alpha - A_{|X_\times} & -B_\times \\ 0 & -1 \end{bmatrix}^{-1}.$$

The second factor on the right-hand side maps $\left[\begin{smallmatrix} X \\ U \end{smallmatrix}\right]$ one-to-one onto $\mathcal{D}(S_\times)$, and the left-hand side maps $\left[\begin{smallmatrix} X \\ U \end{smallmatrix}\right]$ one-to-one onto $\mathcal{D}(S)$. Thus, $\left[\begin{smallmatrix} 1 & 0 \\ [C\&D]_\times \end{smallmatrix}\right]$ maps $\mathcal{D}(S_\times)$ one-to-one onto $\mathcal{D}(S)$. Inverting this equation we get (6.3.17). As we noticed in the proof of Lemma 6.3.8(i), formula (6.3.17) is equivalent to (6.3.9). □

With the help of Theorem 6.3.9 we are now able to give a necessary and sufficient condition for the flow-invertibility of an operator node.

Theorem 6.3.10 *An operator node $S = \left[\begin{smallmatrix} A\&B \\ C\&D \end{smallmatrix}\right]$ on (Y, X, U) is flow-invertible if and only if the following condition holds. For some $\alpha \in \mathbb{C}$, the operator $\left[\begin{smallmatrix} \alpha & 0 \\ 0 & 0 \end{smallmatrix}\right] - S$ maps $\mathcal{D}(S)$ one-to-one onto $\left[\begin{smallmatrix} X \\ Y \end{smallmatrix}\right]$, and if we denote its inverse by*

$$\begin{bmatrix} M_{11}(\alpha) & M_{12}(\alpha) \\ M_{21}(\alpha) & M_{22}(\alpha) \end{bmatrix} := \left(\begin{bmatrix} \alpha & 0 \\ 0 & 0 \end{bmatrix} - S\right)^{-1}, \tag{6.3.20}$$

then $M_{11}(\alpha)$ is injective and has dense range. In this case the main operator, the control operator, the observation operator, and the transfer function (evaluated at α) of $S_\times$ are given by

$$\begin{aligned} A_\times &= \alpha - M_{11}^{-1}(\alpha), & B_\times &= -(\alpha - A_{\times|X})M_{12}(\alpha), \\ C_\times &= M_{21}(\alpha)(\alpha - A_\times), & \widehat{\mathfrak{D}}(\alpha) &= -M_{22}(\alpha). \end{aligned} \tag{6.3.21}$$

In particular, $\alpha \in \rho(A_\times)$.

Proof The necessity of the given conditions follows from Lemma 6.3.8(i) (since (6.3.20)–(6.3.21) are equivalent to (6.3.13)) . Conversely, suppose that the operator $\left[\begin{smallmatrix} \alpha & 0 \\ 0 & 0 \end{smallmatrix}\right] - S$ maps $\mathcal{D}(S)$ one-to-one onto $\left[\begin{smallmatrix} X \\ Y \end{smallmatrix}\right]$, and that $M_{11}(\alpha)$ is injective and has dense range. Then $A_\times$ defined in (6.3.21) is densely defined, and $\alpha \in \rho(A_\times)$. By Lemma 4.7.6, the equations listed in (6.3.21) define a unique operator node $S_\times$. By Theorem 6.3.9(iv), S and $S_\times$ are flow-inverses of each other. □

In most cases we are able to add four more conditions which are equivalent to those listed in Theorem 6.3.9:

Theorem 6.3.11 *Make the same assumptions and introduce the same notation as in Theorem 6.3.9. In addition, suppose that $\rho(A) \cap \rho(A_\times) \neq \emptyset$ (this is, in* *particular, true if both S and $S_\times$ are system nodes). Then the conditions (i)–(vii) listed in Theorem 6.3.9 are equivalent to each one of the following conditions:*

(viii) *For all $\alpha \in \rho(A) \cap \rho(A_\times)$, $\widehat{\mathfrak{D}}(\alpha)$ is invertible and* (6.3.15) *holds.*
(ix) *For some $\alpha \in \rho(A) \cap \rho(A_\times)$, $\widehat{\mathfrak{D}}(\alpha)$ is invertible and* (6.3.15) *holds.*
(x) *For all $\alpha \in \rho(A) \cap \rho(A_\times)$, $\widehat{\mathfrak{D}}_\times(\alpha)$ is invertible and* (6.3.16) *holds.*
(xi) *For some $\alpha \in \rho(A) \cap \rho(A_\times)$, $\widehat{\mathfrak{D}}_\times(\alpha)$ is invertible and* (6.3.16) *holds.*

Proof As in the proof of Theorem 6.3.9 it suffices to show that (viii) and (ix) are equivalent to the conditions listed in Theorem 6.3.9, because we get (x) and (xi) by interchanging S and $S_\times$ in (viii) and (ix). Below, when we refer to (i)–(vii) we mean the conditions listed in Theorem 6.3.9.

(iii) $\Rightarrow$ (viii): This follows from Lemma 4.7.18(iv) (applied to all $\alpha \in \rho(A) \cap \rho(A_\times)$).

(viii) $\Rightarrow$ (ix): This is obvious.

(ix) $\Rightarrow$ (iv): This, too, follows from Lemma 4.7.18(iv) (applied to a particular $\alpha \in \rho(A) \cap \rho(A_\times)$). □

The right-hand side of formulas (6.3.15) and (6.3.16) can be written in several equivalent forms. See Lemma 4.7.18.

Corollary 6.3.12 *Under the assumption of Theorem 6.3.11, for all $\alpha \in \rho(A) \cap \rho(A_\times)$, we have $\widehat{\mathfrak{D}}_\times(\alpha) = [\widehat{\mathfrak{D}}(\alpha)]^{-1}$. In particular, U and Y are isomorphic (hence they have the same dimension). Moreover, the operator $(\alpha - A_\times)^{-1}(\alpha - A)$ which maps $\mathcal{D}(A)$ onto $\mathcal{D}(A_\times)$ is given by*

$$\begin{aligned}(\alpha - A_\times)^{-1}(\alpha - A) &= 1 - (\alpha - A_{|X})^{-1} B \widehat{\mathfrak{D}}_\times(\alpha) C \\ &= 1 - (\alpha - A_{\times|X})^{-1} B_\times C\end{aligned}$$

and its inverse $(\alpha - A)^{-1}(\alpha - A_\times)$ is given by

$$\begin{aligned}(\alpha - A)^{-1}(\alpha - A_\times) &= 1 + (\alpha - A_{\times|X})^{-1} B_\times \widehat{\mathfrak{D}}(\alpha) C_\times \\ &= 1 + (\alpha - A_{|X})^{-1} B C_\times.\end{aligned}$$

Proof The above formulas are part of the conclusion of Theorem 6.3.11. That U and Y must be isomorphic follows from the fact that $\widehat{\mathfrak{D}}(\alpha)$ is a continuously invertible bijection of U onto Y. □

By using Theorem 6.3.11 we can make the following addition to Theorem 6.3.10.

Theorem 6.3.13 *The operator node $S = \left[\begin{smallmatrix} A\&B \\ C\&D \end{smallmatrix}\right]$ on (Y, X, U), with main operator A, control operator B, observation operator C, and transfer function $\widehat{\mathfrak{D}}$ is flow-invertible if the following condition holds. For some $\alpha \in \mathbb{C}$, $\widehat{\mathfrak{D}}(\alpha)$ is* *invertible, and the operator*

$$1 - (\alpha - A_{|X})^{-1} B[\widehat{\mathfrak{D}}(\alpha)]^{-1} C$$

maps $\mathcal{D}(A)$ *one-to-one onto a dense subset of* X*. When these conditions hold, then the flow-inverted operator node* $S_\times$ *is determined by* (6.3.15) *and Lemma 4.7.6.*

The proof is analogous to the proof of Theorem 6.3.10, and we leave it to the reader (replace (6.3.13) by (6.3.15) and Theorem 6.3.9 by Theorem 6.3.11). Note that the condition given in this theorem is sufficient *but not necessary* for flow-invertibility (see Example 6.5.10 which is flow-invertible, but has $\rho(A) \cap \rho(A_\times) = \emptyset$). It is also necessary if we require, in addition, that both S and $S_\times$ are system nodes, because then $\rho(A) \cap \rho(A_\times) \neq \emptyset$ (see Lemma 6.3.8(iii)).

The original idea behind the flow-inversion of a well-posed linear system was to interchange the roles of the input and output (see Theorem 6.3.1). A similar interpretation is valid for the flow-inversion of system nodes, too.

Theorem 6.3.14 *Let* $S = \left[\begin{smallmatrix} A\&B \\ C\&D \end{smallmatrix}\right]$ *be a flow-invertible system node on* (Y, X, U)*, whose flow-inverse* $S_\times$ *is also a system node (on* (U, X, Y)*). Let* x *and* y *be the state trajectory and output function of* S *with initial time* $s \in \mathbb{R}$*, initial state* $x_s \in X$*, and input function* $u \in L^1_{\mathrm{loc}}([s, \infty); U)$*, and suppose that* $x \in W^{1,1}_{\mathrm{loc}}([s, \infty); X)$*. Then* $y \in L^1_{\mathrm{loc}}([s, \infty); Y)$*, and* x *and* u *are the state trajectory and output function of* $S_\times$ *with initial time* s*, initial state* x_s *and input function* y*.*

Proof By Theorem 4.7.11, $\left[\begin{smallmatrix} x \\ u \end{smallmatrix}\right] \in L^1_{\mathrm{loc}}([s, \infty); \mathcal{D}(S))$, $y \in L^1_{\mathrm{loc}}([s, \infty); Y)$, and $\left[\begin{smallmatrix} x \\ y \end{smallmatrix}\right]$ is the unique solution with the above properties of the equation

$$\begin{bmatrix} \dot{x}(t) \\ y(t) \end{bmatrix} = S \begin{bmatrix} x(t) \\ u(t) \end{bmatrix} \text{ for almost all } t \geq s, \quad x(s) = x_s.$$

Since $\left[\begin{smallmatrix} 1 & 0 \\ \multicolumn{2}{c}{C\&D} \end{smallmatrix}\right]$ maps $\mathcal{D}(S)$ continuously onto $\mathcal{D}(S_\times)$, this implies that $\left[\begin{smallmatrix} x \\ y \end{smallmatrix}\right] = \left[\begin{smallmatrix} 1 & 0 \\ \multicolumn{2}{c}{C\&D} \end{smallmatrix}\right] \left[\begin{smallmatrix} x \\ u \end{smallmatrix}\right] \in L^1_{\mathrm{loc}}([s, \infty); \mathcal{D}(S_\times))$. Moreover, since $\left[\begin{smallmatrix} 1 & 0 \\ \multicolumn{2}{c}{C\&D} \end{smallmatrix}\right]^{-1} = \left[\begin{smallmatrix} 1 & 0 \\ \multicolumn{2}{c}{[C\&D]_\times} \end{smallmatrix}\right]$, we have for almost all $t \geq s$,

$$\begin{aligned} \begin{bmatrix} \dot{x}(t) \\ u(t) \end{bmatrix} &= \begin{bmatrix} A\&B \\ 0 \quad 1 \end{bmatrix} \begin{bmatrix} x(t) \\ u(t) \end{bmatrix} = \begin{bmatrix} A\&B \\ 0 \quad 1 \end{bmatrix} \begin{bmatrix} 1 \quad 0 \\ [C\&D]_\times \end{bmatrix} \begin{bmatrix} 1 \quad 0 \\ C\&D \end{bmatrix} \begin{bmatrix} x(t) \\ u(t) \end{bmatrix} \\ &= \begin{bmatrix} [A\&B]_\times \\ [C\&D]_\times \end{bmatrix} \begin{bmatrix} x(t) \\ y(t) \end{bmatrix}. \end{aligned}$$

By Theorem 4.7.11, this implies that x and u are the state and output function of $S_\times$ with initial time s, initial state x_s, and input function y. □

With the help of Theorem 6.3.14 we are now ready to prove the following result, that we announced after Definition 6.3.7.

Corollary 6.3.15 *Let* $\Sigma = \left[\begin{smallmatrix} \mathfrak{A} & \mathfrak{B} \\ \mathfrak{C} & \mathfrak{D} \end{smallmatrix}\right]$ *be an* L^p*-well-posed linear system on* (Y, X, U) *with* $1 \le p < \infty$*. Then* Σ *is flow-invertible (in the sense of Definition 6.3.3) if and only if its system node* S *is flow-invertible (in the sense of Definition 6.3.7) and the flow-inverted operator node* $S_\times$ *is an* L^p*-well-posed system node.*

Proof As we already observed after Definition 6.3.7, the necessity of the condition above on S for Σ to be flow-invertible is clear from Theorem 6.3.6. Conversely, suppose that S is flow-invertible, and that $S_\times$ is a well-posed system node. Denote the system induced by $S_\times$ by $\Sigma_\times$. Then it follows from Lemma 4.7.8 and Theorem 6.3.14 that for all $x_0 \in X$ and $u \in W^{2,1}_{\mathrm{loc}}(\mathbb{R}^+; U)$ with $\left[\begin{smallmatrix} x_0 \\ u(0) \end{smallmatrix}\right] \in \mathcal{D}(S)$, and for all $t \ge 0$,

$$\begin{aligned}\begin{bmatrix} \mathfrak{A}^t & \mathfrak{B}\tau^t \\ 0 & 1 \end{bmatrix}\begin{bmatrix} x_0 \\ \pi_+ u \end{bmatrix} &= \begin{bmatrix} x(t) \\ \pi_+ u \end{bmatrix} = \begin{bmatrix} \mathfrak{A}^t_\times & \mathfrak{B}_\times\tau^t \\ \mathfrak{C}_\times & \mathfrak{D}_\times \end{bmatrix}\begin{bmatrix} x_0 \\ \pi_+ y \end{bmatrix} \\ &= \begin{bmatrix} \mathfrak{A}^t_\times & \mathfrak{B}_\times\tau^t \\ \mathfrak{C}_\times & \mathfrak{D}_\times \end{bmatrix}\begin{bmatrix} 1 & 0 \\ \mathfrak{C} & \mathfrak{D} \end{bmatrix}\begin{bmatrix} x_0 \\ \pi_+ u \end{bmatrix}.\end{aligned}$$

This set of data is dense in $\left[\begin{smallmatrix} X \\ L^p(\mathbb{R}^+;U) \end{smallmatrix}\right]$, so the same identity must be true for all $x_0 \in X$ and $u \in L^p(\mathbb{R}^+; U)$. In particular, this implies that $\mathfrak{D}_\times\pi_+$ is a left-inverse of $\mathfrak{D}\pi_+$. A similar argument with Σ interchanged with $\Sigma_\times$ shows that $\mathfrak{D}_\times\pi_+$ is also a right inverse of $\mathfrak{D}\pi_+$. By Theorem 6.3.5, Σ is flow-invertible. □

Our next theorem shows that compatibility is preserved under flow-inversion in most cases.

Theorem 6.3.16 *Let* $S = \left[\begin{smallmatrix} A\&B \\ C\&D \end{smallmatrix}\right]$ *be a compatible operator node on* (Y, X, U)*, and let* $\left[\begin{smallmatrix} A_{|W} & B \\ C_{|W} & D \end{smallmatrix}\right] \in \mathcal{B}(\left[\begin{smallmatrix} W \\ U \end{smallmatrix}\right]; \left[\begin{smallmatrix} W_{-1} \\ Y \end{smallmatrix}\right])$ *be a compatible extension of* S *(here* $X_1 \subset W \subset X$ *and* W_{-1} *is defined as in Lemma 5.1.3). Suppose that* S *is flow-invertible. Denote the flow-inverted operator node by* $S_\times = \left[\begin{smallmatrix} [A\&B]_\times \\ [C\&D]_\times \end{smallmatrix}\right]$*, let* $X_{\times 1}$ *and* $X_{\times -1}$ *be the analogues of* X_1 *and* X_{-1} *for* $S_\times$*, and let* $W_{\times -1}$ *be the analogue of* W_{-1} *for* $S_\times$ *(i.e.,* $W_{\times -1} = (\alpha - A_\times)_{|W} W$ *for some* $\alpha \in \rho(A_\times)$*).*

(i) *If* D *has a left inverse* $D^{-1}_{\mathrm{left}} \in \mathcal{B}(Y; U)$*, then* $X_{\times 1} \subset W$ *and* $S_\times$ *is compatible with extended observation operator* $C_{\times|W}\colon W \to U$ *and corresponding feedthrough operator* $D_\times$ *given by*

$$\begin{aligned} C_{\times|W} &= -D^{-1}_{\mathrm{left}} C_{|W}, \\ D_\times &= D^{-1}_{\mathrm{left}}, \end{aligned} \tag{6.3.22}$$

and the main operator $A_\times$ *of* $S_\times$ *is given by*

$$A_\times = \left(A_{|X} - B D^{-1}_{\mathrm{left}} C_{|W}\right)_{|X_{\times 1}}.$$

In this case the space W_{-1} can be identified with a closed subspace of $W_{\times-1}$, so that $X \subset W_{-1} \subset X_{-1} \cap X_{\times-1}$. With this identification,

$$A_{|W} = A_{\times|W} + B_\times C_{|W}, \qquad B = B_\times D$$

(where by $A_{|W}$ and $A_{\times|W}$ we mean the restrictions of $A_{|X}$ and $A_{\times|X}$ to W).

(ii) *If D is invertible, then $W_{-1} = W_{\times-1}$, $A_\times W \subset W_{-1}$, $B_\times U \subset W_{-1}$, and the operator $\begin{bmatrix} A_{\times|W} & B_\times \\ C_{\times|W} & D_\times \end{bmatrix} \in \mathcal{B}(\begin{bmatrix} W \\ U \end{bmatrix}; \begin{bmatrix} W_{-1} \\ Y \end{bmatrix})$ defined by*

$$\begin{aligned}\begin{bmatrix} A_{\times|W} & B_\times \\ C_{\times|W} & D_\times \end{bmatrix} &= \begin{bmatrix} A_{|W} - BD^{-1}C_{|W} & BD^{-1} \\ -D^{-1}C_{|W} & D^{-1} \end{bmatrix} \\ &= \begin{bmatrix} A_{|W} & 0 \\ 0 & 0 \end{bmatrix} + \begin{bmatrix} B \\ 1 \end{bmatrix} D^{-1} \begin{bmatrix} -C_{|W} & 1 \end{bmatrix} \\ &= \begin{bmatrix} A_{|W} & 0 \\ 0 & 0 \end{bmatrix} + \begin{bmatrix} B \\ 1 \end{bmatrix} \begin{bmatrix} C_{\times|W} & 1 \end{bmatrix} \\ &= \begin{bmatrix} A_{|W} & 0 \\ 0 & 0 \end{bmatrix} + \begin{bmatrix} B_\times \\ 1 \end{bmatrix} \begin{bmatrix} -C_{|W} & 1 \end{bmatrix}\end{aligned}$$

is a compatible extension of $S_\times$.

Proof (i) Take $\begin{bmatrix} x \\ y \end{bmatrix} \in \mathcal{D}(S_\times)$, and define $u = [C\&D]_\times \begin{bmatrix} x \\ y \end{bmatrix}$. Then $\begin{bmatrix} x \\ u \end{bmatrix} \in \mathcal{D}(S)$ and $y = C\&D \begin{bmatrix} x \\ u \end{bmatrix} = C_{|W}x + Du$. Multiplying the above identity by D_{left}^{-1} to the left we get for all $\begin{bmatrix} x \\ y \end{bmatrix} \in \mathcal{D}(S_\times)$,

$$u = [C\&D]_\times \begin{bmatrix} x \\ y \end{bmatrix} = -D_{\mathrm{left}}^{-1} C_{|W}x + D_{\mathrm{left}}^{-1} y.$$

The right-hand side is defined (and continuous) on all of $W \times Y$. By (6.3.15), for all $y \in Y$ and all $\alpha \in \rho(A) \cap \rho(A_\times)$,

$$(\alpha - A_\times)^{-1} B_\times y = (\alpha - A)^{-1} B \widehat{\mathfrak{D}}_\times(\alpha) y \in W,$$

so $\mathcal{R}(B_\times) \in W_{\times-1}$. This implies that $\begin{bmatrix} A_{\times|W} & B_\times \\ C_{\times|W} & D_\times \end{bmatrix}$ is a compatible extension of $S_\times$, with $C_{\times|W} = -D_{\mathrm{left}}^{-1} C_{|W}$ and $D_\times = D_{\mathrm{left}}^{-1}$. By (6.3.18), for all $x \in X_{\times 1}$, we have $A_\times x = (A_{|X} + BC_\times)x = (A_{|X} - BD_{\mathrm{left}}^{-1} C_{|W})x$, as claimed.

Next we construct an embedding operator $J\colon W_{-1} \to W_{\times-1}$. This operator is required to be injective, and its restriction to X should be the identity operator. We define

$$\begin{aligned} J &= (\alpha - A_{\times|W} - B_\times C_{|W})(\alpha - A_{|W})^{-1}, \\ J_\times &= (\alpha - A_{|W} - BC_{\times|W})(\alpha - A_{\times|W})^{-1}. \end{aligned} \tag{6.3.23}$$

The compatibility of S and $S_\times$ implies that $J \in \mathcal{B}(W_{-1}; W_{\times-1})$ and $J_\times \in \mathcal{B}(W_{\times-1}; W_{-1})$ and by (6.3.18), both J and $J_\times$ reduce to the identity operator on X.

We claim that $J_\times \in \mathcal{B}(W_{\times-1}; W_{-1})$ is a left inverse of $J \in \mathcal{B}(W_{-1}; W_{\times-1})$, or equivalently, that $(\alpha - A_{|W})^{-1} J_\times J(\alpha - A)_{|W}$ is the identity on W. To see that this is the case we use (6.3.23), (6.3.22), (6.3.15), and (5.1.3) (in this order) to compute

$$\begin{aligned}
&(\alpha - A_{|W})^{-1} J_\times J(\alpha - A)_{|W} \\
&\quad = (\alpha - A_{|W})^{-1}(\alpha - A_{|W} - BC_{\times|W}) \\
&\qquad \times (\alpha - A_{\times|W})^{-1}(\alpha - A_{\times|W} - B_\times C_{|W}) \\
&\quad = (1 - (\alpha - A_{|W})^{-1}BC_{\times|W})(1 - (\alpha - A_{\times|W})^{-1}B_\times C_{|W}) \\
&\quad = (1 + (\alpha - A_{|W})^{-1}BD_{\text{left}}^{-1}C_{|W})(1 - (\alpha - A_{|W})^{-1}B\widehat{\mathfrak{D}}^{-1}(\alpha)C_{|W}) \\
&\quad = 1 + (\alpha - A_{|W})^{-1}B\big[D_{\text{left}}^{-1} - \widehat{\mathfrak{D}}^{-1}(\alpha) - D_{\text{left}}^{-1}C_{|W}(\alpha - A_{|W})^{-1}B\widehat{\mathfrak{D}}^{-1}(\alpha)\big]C_{|W} \\
&\quad = 1 + (\alpha - A_{|W})^{-1}BD_{\text{left}}^{-1}\big[\widehat{\mathfrak{D}}(\alpha) - D - C_{|W}(\alpha - A_{|W})^{-1}B\big]\widehat{\mathfrak{D}}^{-1}(\alpha)C_{|W} \\
&\quad = 1.
\end{aligned}$$

This implies that the operator J is injective; hence it defines a (not necessarily dense) embedding of W_{-1} into $W_{\times-1}$. In the sequel we shall identify W_{-1} with the range of J. That W_{-1} is closed in $W_{\times-1}$ follows from the fact that J has a bounded left inverse.

The identification of W_{-1} with a subspace of $W_{\times-1}$ means that the embedding operator $J = (\alpha - A_{\times|W} - B_\times C_{|W})(\alpha - A_{|W})^{-1}$ becomes the identity on W_{-1}, and hence, with this identification, $(\alpha - A)_{|W} = (\alpha - A_{\times|W} - B_\times C_{|W})$, or equivalently,

$$A_{|W} = A_{\times|W} + B_\times C_{|W}.$$

The remaining identity $B = B_\times D$ can be verified as follows. By (6.3.15) and the fact that $A_{\times|W} = A_{|W} - B_\times C_{|W}$,

$$\begin{aligned}
B_\times \widehat{\mathfrak{D}}(\alpha) &= (\alpha - A_{\times|W})(\alpha - A_{|W})^{-1}B \\
&= (\alpha - A_{|W} + B_\times C_{|W})(\alpha - A_{|W})^{-1}B \\
&= B + B_\times C_{|W}(\alpha - A_{|W})^{-1}B \\
&= B + B_\times(\widehat{\mathfrak{D}}(\alpha) - D) \\
&= B_\times \widehat{\mathfrak{D}}(\alpha) + B - B_\times D.
\end{aligned}$$

Thus $B = B_\times D$.

(ii) Part (ii) follows from part (i) if we interchange S and $S_\times$. (This will also interchange W_{-1} with $W_{\times-1}$ and J with $J_\times$.) □

Let us finally look at the flow-inversion of a regular system.

Theorem 6.3.17 *Let* $\Sigma = \left[\begin{smallmatrix} \mathfrak{A} & \mathfrak{B} \\ \mathfrak{C} & \mathfrak{D} \end{smallmatrix}\right]$ *be a weakly regular ($L^p|Reg$-well-posed) flow-invertible linear system on* (Y, X, U) *with system node* $S = \left[\begin{smallmatrix} A\&B \\ C\&D \end{smallmatrix}\right]$, *semigroup generator A, control operator B, and observation operator C. We define the extensions $\widetilde{C}_w$ and $\widetilde{C}_s$ of the observation operator C as in Definition 5.4.1, and let D be the corresponding feedthrough operator (see Theorem 5.6.5).*

(i) *The system node S is compatible with extended system node $\left[\begin{smallmatrix} A_{|W} & B \\ C_{|W} & D \end{smallmatrix}\right]$ where $W = \mathcal{D}(\widetilde{C}_w)$ and $C_{|W} = \widetilde{C}_w$, hence parts (i) and (ii) of Theorem 6.3.16 apply whenever D is left invertible or invertible, respectively. In particular, if D is left-invertible, then the flow-inverted system $\Sigma_\times$ is compatible.*

(ii) *If S is strongly regular, then the operator D is coercive (see Definition 9.10.1). If, in addition, the closure of the range of D is complemented in Y (this is, in particular, true if Y is a Hilbert space), then D is left invertible, $\mathcal{D}\left(\widetilde{C}_{\times s}\right)$ is a closed subspace of $\mathcal{D}(\widetilde{C}_s)$, and Theorem 6.3.16(i) applies with $W = \mathcal{D}(\widetilde{C}_s)$ and $C_{|W} = \widetilde{C}_{\times s}$ (in particular, the flow-inverted system $\Sigma_\times$ is* compatible*).*

(iii) *Suppose that Σ is strongly regular. Then the flow-inverted system is strongly regular iff D is invertible. In this case $\mathcal{D}\left(\widetilde{C}_{\times s}\right) = \mathcal{D}(\widetilde{C}_s)$, and Theorem 6.3.16(ii) applies with $W = \mathcal{D}(\widetilde{C}_s)$ and $C_{|W} = \widetilde{C}_s$.*

(iv) *In the Reg-well-posed case both the original and the flow-inverted system are strongly regular, D is invertible, and Theorem 6.3.16(ii) applies with $W = \mathcal{D}(\widetilde{C}_s)$ and $C_{|W} = \widetilde{C}_s$.*

(v) *In the L^1-well-posed case with a reflexive state space X both the original and the flow-inverted system are strongly regular, D is invertible, and Theorem 6.3.16(ii) applies with $W = \mathcal{D}(\widetilde{C}_s)$ and $C_{|W} = \widetilde{C}_s$.*

(vi) *If Σ is uniformly regular then the flow-inverted system $\Sigma_\times$ is uniformly regular, D is invertible, and Theorem 6.3.16(ii) applies with $W = \mathcal{D}(\widetilde{C}_s)$ and $C_{|W} = \widetilde{C}_s$.*

Proof (i) See Theorem 5.6.4.

(ii) Let $u \in U$. Then, by the strong regularity of Σ,

$$\lim_{\alpha\to+\infty}\left[\widehat{\mathfrak{D}}(\alpha)u - Du\right] = 0.$$

Multiplying the function inside the limit by the bounded function $\widehat{\mathfrak{D}}^{-1}(\alpha)$ (see Lemma 4.6.2 and Corollary 6.3.12) to the left we get

$$u = \lim_{\alpha\to+\infty} \widehat{\mathfrak{D}}^{-1}(\alpha)Du.$$

This implies

$$|u|_U \le \limsup_{\alpha\to+\infty}\|\widehat{\mathfrak{D}}^{-1}(\alpha)\|\,|Du|_U,$$

or equivalently,

$$|Du|_U \geq L|u|_U,$$

where $L = \left(\limsup_{\alpha\to+\infty}\|\widehat{\mathfrak{D}}^{-1}(\alpha)\|\right)^{-1} > 0$. Thus, D is coercive (see Definition 9.10.1), and it has a bounded left inverse defined on $\mathcal{R}(D)$. If $\overline{\mathcal{R}(D)}$ is complemented then this left inverse can be extended to all of Y (we extend it to $\overline{\mathcal{R}(D)}$ by continuity and define it to be zero on the complement). By Theorem 5.6.4, Σ is compatible with $W = \mathcal{D}(\widetilde{C}_s)$ and $C_{|W} = \widetilde{C}_s$, and Theorem 7.5.1(i) applies.

Next we intend to show that $\mathcal{D}\left(\widetilde{C}_{\times s}\right) \subset \mathcal{D}(\widetilde{C}_s)$, and that, on its domain, $\widetilde{C}_{\times s} = D_{\text{left}}^{-1}\widetilde{C}_s$. Let $x_0 \in \mathcal{D}\left(\widetilde{C}_{\times s}\right)$, i.e., suppose that the strong limit

$$\begin{aligned}\widetilde{C}_{\times s}x_0 &= \lim_{\alpha\to+\infty} \alpha C_\times(\alpha - A_\times)^{-1}x_0\\ &= \lim_{\alpha\to+\infty} \alpha\widehat{\mathfrak{D}}^{-1}(\alpha)C(\alpha - A)^{-1}x_0\end{aligned}$$

exists; to get the second equality we have used Theorem 6.3.11. Subtract $\widetilde{C}_{\times s}x_0$ from both sides of this equation and multiply by the bounded function $\widehat{\mathfrak{D}}(\alpha)$ to get

$$\begin{aligned}0 &= \lim_{\alpha\to+\infty}\left[\alpha C(\alpha - A)^{-1}x_0 - \widehat{\mathfrak{D}}(\alpha)\widetilde{C}_{\times s}x_0\right]\\ &= \lim_{\alpha\to+\infty} \alpha C(\alpha - A)^{-1}x_0 - D\widetilde{C}_{\times s}x_0.\end{aligned}$$

This shows that $\mathcal{D}\left(\widetilde{C}_{\times s}\right) \subset \mathcal{D}(\widetilde{C}_s)$, and that, on its domain, $\widetilde{C}_{\times s} = D_{\text{left}}^{-1}\widetilde{C}_s$.

To show that $\mathcal{D}\left(\widetilde{C}_{\times s}\right)$ is a closed subspace of $\mathcal{D}(\widetilde{C}_s)$ we must show that the norms on these two spaces are equivalent. These norms are given by

$$\begin{aligned}|x|_{\widetilde{C}_s} &= |x|_X + \sup_{\alpha>1+\omega_{\mathfrak{A}}} \left|\alpha C(\alpha - A)^{-1}x\right|,\\ |x|_{\widetilde{C}_{\times s}} &= |x|_X + \sup_{\alpha>1+\omega_{\mathfrak{A}_\times}} \left|\alpha C_\times(\alpha - A_\times)^{-1}x\right|\\ &= |x|_X + \sup_{\alpha>1+\omega_{\mathfrak{A}_\times}} \left|\widehat{\mathfrak{D}}^{-1}(\alpha)\alpha C(\alpha - A)^{-1}x\right|.\end{aligned}$$

Indeed, these norms are equivalent since we can always dominate the suprema over $(1 + \min\{\omega_{\mathfrak{A}}, 1 + \omega_{\mathfrak{A}_\times}\}, \max\{\omega_{\mathfrak{A}}, \omega_{\mathfrak{A}_\times}\})$ by a constant times $|x|_X$, and both $\widehat{\mathfrak{D}}(\alpha)$ and $\widehat{\mathfrak{D}}^{-1}(\alpha)$ are uniformly bounded on $(1 + \max\{\omega_{\mathfrak{A}}, 1 + \omega_{\mathfrak{A}_\times}\}, \infty)$.

(iii) If D is invertible then, for each $u \in U$,

$$\lim_{\alpha\to+\infty}\left[u - \widehat{\mathfrak{D}}(\alpha)D^{-1}u\right] = 0.$$

Multiplying the function inside the limit by the bounded function $\widehat{\mathfrak{D}}_\times(\alpha) = \widehat{\mathfrak{D}}^{-1}(\alpha)$ to the left we get

$$0 = \lim_{\alpha\to+\infty}\left[\widehat{\mathfrak{D}}_\times(\alpha)u - D^{-1}u\right] = \lim_{\alpha\to+\infty}\widehat{\mathfrak{D}}_\times(\alpha)u - D^{-1}u.$$

Thus, the closed-loop system is strongly regular (and its feedthrough operator is the expected D^{-1}).

Conversely, suppose that the closed-loop system is strongly regular with feedthrough operator $D_\times$. We know from (ii) that D is left invertible and that $D_\times D = 1$. If we interchange Σ and $\Sigma_\times$, then the same argument shows that $D_\times$ is left invertible and that $DD_\times = 1$. Thus D and $D_\times$ are invertible and $D_\times = D^{-1}$.

We know from (ii) that $\mathcal{D}\left(\widetilde{C}_{\times s}\right) \subset \mathcal{D}(\widetilde{C}_s)$. To prove the converse inclusion it suffices to interchange Σ and $\Sigma_\times$ and apply (ii).

(iv)–(v) See (iii) and Lemma 5.7.1.

(vi) We know from (ii) that D is coercive, and by applying (ii) to the dual system Σ^d (see Theorem 6.2.15(iii)) we find that also D^* is coercive. By Lemma 9.10.2(iii)–(iv), D is invertible. Repeating the computation at the beginning of the proof of (iii) with strong limits replaced by limits in the operator norm (and u replaced by the identity operator) we find that $\mathfrak{D}_\times(\alpha)$ tends to D^{-1} in operator norm as $\alpha \to +\infty$. □

In the case of a system with an analytic semigroup of the type studied in Theorem 5.7.3 the question of the flow-invertibility of the system is particularly easy:

Example 6.3.18 *Let $\Sigma = \left[\begin{smallmatrix}\mathfrak{A} & \mathfrak{B}\\ \mathfrak{C} & \mathfrak{D}\end{smallmatrix}\right]$ be one of the systems considered in Theorem 5.7.3. In all the different cases considered there, Σ is flow-invertible if and only if D is invertible, in which case the closed-loop system is of the same type as Σ. More precisely, the closed-loop semigroup $\mathfrak{A}_\times$ can be extended to an analytic semigroup on $\mathfrak{A}_{\times|X_{\alpha-1}}$ on $X_{\alpha-1}$. For all $\gamma \in [\alpha-1, \beta+1]$, the spaces X_γ are invariant under $\mathfrak{A}_{\times|X_{\alpha-1}}$, and the restriction $\mathfrak{A}_{\times|X_\gamma}$ of $\mathfrak{A}_{\times|X_{\alpha-1}}$ to X_γ is an analytic semigroup on X_γ. The generator of $\mathfrak{A}_{\times|X_\gamma}$ is $(A - BD^{-1}C)_{|X_{\gamma+1}}$ if $\gamma \in [\alpha-1, \beta]$, and it is the part of $A - BD^{-1}C$ in X_γ if $\gamma \in (\beta, \beta+1]$. If we define $X_{\times\alpha-1} = X_{\alpha-1}$, and let $X_{\times\gamma}$ be the fractional order space with index $\gamma-\alpha-1$ constructed by means of the semigroup $\mathfrak{A}_{\times|X_{\alpha-1}}$ on $X_{\alpha-1}$, then $X_{\times\gamma} = X_\gamma$ for all $\gamma \in [\alpha-1, \beta+1]$. The closed-loop observation operator is $C_\times = (1-DK)^{-1}C$, the closed-loop control operator is $B_\times = B(1-KD)^{-1}$, and the closed-loop feedthrough operator is $D_\times = D(1-KD)^{-1}$.*

Proof Let us begin by showing that Σ is flow-invertible if and only if D is invertible. Suppose that Σ is flow-invertible. Then, by Theorem 6.3.17(vi), D

is invertible (and the flow-inverted system is uniformly regular). Conversely, if D is invertible, then for all $T > 0$ and almost all $t \in [0, T]$,

$$D^{-1}\pi_{[0,t)}(\mathfrak{D}\pi_{[0,t)}u)(t) = u(t) + \int_0^t D^{-1}C\mathfrak{A}^{t-s}_{|X_{\alpha-1}} Bu(s)\,ds.$$

The operator-norm of the integral term is dominated by the L^1-norm of $t \mapsto D^{-1}C\mathfrak{A}^t_{|X_{\alpha-1}} B$ (see, e.g., Theorems A.3.4(i) and A.3.7(i)), and by choosing T small enough we can make this norm less than $\frac{1}{2}$. By the contraction mapping principle, the right-hand side is invertible in $\mathcal{B}(L^p|Reg([0, T); \mathcal{B}(U)))$, hence $\pi_{[0,t)}\mathfrak{D}\pi_{[0,t)}$ is invertible in $\mathcal{B}(L^p|Reg([0, T); \mathcal{B}(U; Y)))$. It then follows from Theorem 6.3.5 that Σ is flow-invertible.

The formulas for the generators of the flow-inverted system are found in Theorems 6.3.17(vi) and 6.3.16(ii). All the extra claims about the closed-loop semigroup can be derived from Theorem 3.10.11. □

As the following theorem says, flow-inversion almost commutes with duality:

Theorem 6.3.19 *Let (Y, X, U) be reflexive Banach spaces, let $1 < p < \infty$, let Σ be an L^p-well-posed linear system on (Y, X, U), and let S be an operator node on (Y, X, U) (not necessarily the one induced by Σ).*

(i) *The system Σ is flow-invertible if and only if the causal dual system Σ^d is flow-invertible, in which case the flow-inverted systems satisfy $(\Sigma^d)_\times = \left[\begin{smallmatrix} 1 & 0 \\ 0 & -1 \end{smallmatrix}\right](\Sigma_\times)^d\left[\begin{smallmatrix} 1 & 0 \\ 0 & -1 \end{smallmatrix}\right]$.*
(ii) *The operator node S is flow-invertible if and only if S^* is flow-invertible, in which case the flow-inverted operator nodes satisfy $(S^*)_\times = \left[\begin{smallmatrix} 1 & 0 \\ 0 & -1 \end{smallmatrix}\right](S_\times)^*\left[\begin{smallmatrix} 1 & 0 \\ 0 & -1 \end{smallmatrix}\right]$.*
(iii) *If S is flow-invertible and both S and S^* are strongly regular with feedthrough operators D, respectively D^*, then D is invertible. In this case both $S_\times$ and $(S^*)_\times$ are strongly regular.*

Proof (i) It follows from Definition 6.3.3, Lemmas 3.5.2 and 3.5.13, and Theorems 6.2.3 and 6.3.5 that Σ is flow-invertible if and only if Σ^d is flow-invertible. That $(\Sigma^d)_\times = \left[\begin{smallmatrix} 1 & 0 \\ 0 & -1 \end{smallmatrix}\right](\Sigma_\times)^d\left[\begin{smallmatrix} 1 & 0 \\ 0 & -1 \end{smallmatrix}\right]$ follows from (6.2.3) and (6.3.3) (recall that, by Lemma 3.5.13 and Example 3.5.11, $\boldsymbol{Я}^* = \boldsymbol{Я}$, $\tau^t\boldsymbol{Я} = \boldsymbol{Я}\tau^{-t}$, and $(\tau^t)^* = \tau^{-t}$).

(ii) By Theorem 6.3.9, the operator node S is flow-invertible with flow-inverse $S_\times$ if and only if (6.3.13) holds (for some $\alpha \in \rho(A_\times)$). This identity holds if and only if the dual identity holds which we get by taking the adjoint of each side. The adjoint of the left-hand side of (6.3.13) is $\left(\left[\begin{smallmatrix} \overline{\alpha} & 0 \\ 0 & 0 \end{smallmatrix}\right] - S^*\right)^{-1}$, i.e., it is equal to the left-hand side of (6.3.13) with α replaced by $\overline{\alpha}$ and S replaced by S^*. The adjoint of the right-hand side of (6.3.13) is

$$\begin{bmatrix} (\overline{\alpha} - (A_\times)^*)^{-1} & (\overline{\alpha} - A^*_{\times|X^*})^{-1}(C_\times)^* \\ -(B_\times)^*(\overline{\alpha} - (A_\times)^*)^{-1} & -(\widehat{\mathfrak{D}}_\times(\alpha))^* \end{bmatrix},$$

and by Lemma 6.2.14, this is equal to the right-hand side of (6.3.13) with α replaced by $\overline{\alpha}$ and $S_\times$ replaced by $\left[\begin{smallmatrix} 1 & 0 \\ 0 & -1 \end{smallmatrix}\right](S_\times)^*\left[\begin{smallmatrix} 1 & 0 \\ 0 & -1 \end{smallmatrix}\right]$. Thus, S is flow-invertible with flow-inverse $S_\times$ if and only if S^* is flow-invertible with flow-inverse $\left[\begin{smallmatrix} 1 & 0 \\ 0 & -1 \end{smallmatrix}\right](S_\times)^*\left[\begin{smallmatrix} 1 & 0 \\ 0 & -1 \end{smallmatrix}\right]$.

(iii) The regularity of Σ and Σ^d combined with (ii) and Theorem 6.3.17(ii) implies that both D and D^* are coercive. By Lemma 9.10.2(iii)–(iv), D is invertible. The strong regularity of $S_\times$ and $(S^*)_\times$ follow from Theorem 6.3.17(ii). □

6.4 Time-inversion

The systems that we have studied so far have been either causal and propagate information in the forward time direction, or anti-causal and propagate information in the backward time direction. In the finite-dimensional case this distinction is not necessary: a linear system with finite-dimensional state space can be solved both forward in time and backward in time. Some infinite-dimensional systems have the same property. We shall refer to this property as *time-invertibility*. Whereas flow-invertibility is a property of the input/output map of the system, time-invertibility is a property of the semigroup.

Let x and y be the state trajectory and output function of an $L^p|Reg$-well-posed linear system $\Sigma = \left[\begin{smallmatrix} \mathfrak{A} & \mathfrak{B} \\ \mathfrak{C} & \mathfrak{D} \end{smallmatrix}\right]$ on (Y, X, U) with initial time $s \in \mathbb{R}$, initial state $x(s) \in X$, and input function $u \in L^p|Reg_{\mathrm{loc}}([s, \infty, U)$. Then, for all $t \geq s$, by Definitions 2.2.6 and 2.2.7,

$$\begin{aligned} x(t) &= \mathfrak{A}^{t-s}x(s) + \mathfrak{B}^t_s u, \\ \pi_{[s,t)}y &= \mathfrak{C}^t_s x(s) + \mathfrak{D}^t_s u, \qquad t \geq s. \end{aligned} \tag{6.4.1}$$

We fix $t > s$ and try to interpret t as the initial time and s as the final time, still regarding u to be the input function and y the output function. Since $x(t)$ and $\pi_{[s,t)}y$ depend only on $x(s)$ and the restriction of u to $[s, t)$, we can, without loss of generality, assume that u vanishes outside of $[s, t)$, i.e., $u = \pi_{[s,t)}u$. Clearly, we can solve $x(s)$ and $\pi_{[s,t)}y$ from (6.4.1) for all possible choices of $x(t) \in X$ and $u \in L^p|Reg([s, t); U)$ if and only if $\mathfrak{A}^{t-s}$ is invertible. When this is the case, we denote $(\mathfrak{A}^{t-s})^{-1}$ by $\mathfrak{A}^{s-t}$ and get from (6.4.1),

$$\begin{aligned}\begin{bmatrix} x(s) \\ \pi_{[s,t)}y \end{bmatrix} &= \begin{bmatrix} 1 & 0 \\ \mathfrak{C}_s^t & \mathfrak{D}_s^t \end{bmatrix}\begin{bmatrix} x(s) \\ \pi_{[s,t)}u \end{bmatrix} \\ &= \begin{bmatrix} 1 & 0 \\ \mathfrak{C}_s^t & \mathfrak{D}_s^t \end{bmatrix}\begin{bmatrix} \mathfrak{A}^{t-s} & \mathfrak{B}_s^t \\ 0 & 1 \end{bmatrix}^{-1}\begin{bmatrix} x(t) \\ \pi_{[s,t)}u \end{bmatrix} \\ &= \begin{bmatrix} \mathfrak{A}^{s-t} & -\mathfrak{A}^{s-t}\mathfrak{B}_s^t \\ \mathfrak{C}_s^t\mathfrak{A}^{s-t} & \mathfrak{D}_s^t - \mathfrak{C}_s^t\mathfrak{A}^{s-t}\mathfrak{B}_s^t \end{bmatrix}\begin{bmatrix} x(t) \\ \pi_{[s,t)}u \end{bmatrix}, \qquad t \ge s. \end{aligned} \tag{6.4.2}$$

This equation resembles (6.4.1), but it cannot correspond to a causal system since the initial time t is bigger than the final time s. However, as the following theorem says, it can be interpreted as the input/state/output relationship of an anti-causal system.

Theorem 6.4.1 *Let $\Sigma = \left[\begin{smallmatrix} \mathfrak{A} & \mathfrak{B} \\ \mathfrak{C} & \mathfrak{D} \end{smallmatrix}\right]$ be an $L^p|Reg$-well-posed linear system on (Y, X, U), and suppose that $\mathfrak{A}^t$ is invertible for some $t > 0$. Then $\mathfrak{A}^t$ is invertible for all $t \ge 0$ and $\mathfrak{A}$ becomes a strongly continuous group if we define $\mathfrak{A}^t = (\mathfrak{A}^{-t})^{-1}$ for $t < 0$. For all $-\infty < s \le t < \infty$ we define*

$$\begin{bmatrix} (\mathfrak{A}^a)_s^t & (\mathfrak{B}^a)_s^t \\ (\mathfrak{C}^a)_s^t & (\mathfrak{D}^a)_s^t \end{bmatrix} = \begin{bmatrix} 1 & 0 \\ \mathfrak{C}_s^t & \mathfrak{D}_s^t \end{bmatrix}\begin{bmatrix} \mathfrak{A}^{t-s} & \mathfrak{B}_s^t \\ 0 & 1 \end{bmatrix}^{-1}. \tag{6.4.3}$$

Then $\Sigma^a = \left[\begin{smallmatrix} \mathfrak{A}^a & \mathfrak{B}^a \\ \mathfrak{C}^a & \mathfrak{D}^a \end{smallmatrix}\right]$ is an anti-causal $L^p|Reg$-well-posed linear system, where $\mathfrak{A}_s^a = (\mathfrak{A}_s^a)^0 = \mathfrak{A}^s$ for $s \le 0$ and

$$\mathfrak{B}^a = \lim_{t\to\infty}(\mathfrak{B}^a)_0^t, \qquad \mathfrak{C}^a = \lim_{s\to-\infty}(\mathfrak{C}^a)_s^0, \qquad \mathfrak{D}^a = \lim_{\substack{s\to-\infty \\ t\to\infty}}(\mathfrak{D}^a)_s^t. \tag{6.4.4}$$

Moreover, if x and y are the state trajectory and output function of Σ with initial time s, initial state $x(s)$, and input function $u \in L^p|Reg_{\mathrm{loc}}([s,\infty); U)$, then, for all $t \ge s$,

$$\begin{bmatrix} x(s) \\ \pi_{[s,t)}y \end{bmatrix} = \begin{bmatrix} (\mathfrak{A}^a)_s^t & (\mathfrak{B}^a)_s^t \\ (\mathfrak{C}^a)_s^t & (\mathfrak{D}^a)_s^t \end{bmatrix}\begin{bmatrix} x(t) \\ \pi_{[s,t)}u \end{bmatrix}. \tag{6.4.5}$$

The intuitive interpretation of this theorem is that Σ^a is the 'same' system as Σ, but it propagates information in the backward time direction instead of the forward time direction. Observe, in particular, that $\mathfrak{A}_t^a = \mathfrak{A}^t$ for all $t \le 0$.

Proof of Theorem 6.4.1 We begin with the proof of the fact that $\mathfrak{A}^t$ is invertible for all $t > 0$ if it is invertible for one $t > 0$. Assume that $\mathfrak{A}^{t_0}$ is invertible. Then for all $0 \le t \le t_0$, $\mathfrak{A}^{t_0} = \mathfrak{A}^{t_0-t}\mathfrak{A}^t = \mathfrak{A}^t\mathfrak{A}^{t_0-t}$. This implies that $\mathfrak{A}^t$ is both injective and onto, hence invertible. By the closed graph theorem, the inverse is continuous. To remove the condition $t \le t_0$, it suffices to observe that $\mathfrak{A}^{2t_0} = \mathfrak{A}^{t_0}\mathfrak{A}^{t_0}$ is invertible, hence $\mathfrak{A}^{4t_0} = \mathfrak{A}^{2t_0}\mathfrak{A}^{2t_0}$ is invertible, etc.

Since $\mathfrak{A}^t$ is invertible for all $t \ge 0$, we can define $\mathfrak{A}^t$ for $t < 0$ by $\mathfrak{A}^t = (\mathfrak{A}^{-t})^{-1}$, and the semigroup property $\mathfrak{A}^{s+t} = \mathfrak{A}^s\mathfrak{A}^t$ is then valid for all $s, t \in \mathbb{R}$, as is easily shown. The strong left-continuity at zero of this group follows from the strong left-continuity of $\mathfrak{A}^t$ at $t = 1$ (see Lemma 2.2.13) and the fact that $\mathfrak{A}^t = \mathfrak{A}^{t+1}\mathfrak{A}^{-1}$.

In particular, we observe that $t \mapsto \mathfrak{A}^a_{-t} = \mathfrak{A}^{-t}$ is a C_0 semigroup, as required by Definition 6.1.1.

The remainder of the proof of the fact that Σ^a is an anti-causal $L^p|Reg$-well-posed linear system appeals to Theorem 6.1.10. To be able to apply this theorem, we must first show that the time-invariance condition (2.2.5) and the composition condition (6.1.7) hold when we replace $\mathfrak{A}^t_s$, $\mathfrak{B}^t_s$, $\mathfrak{C}^t_s$, and $\mathfrak{D}^t_s$ by $(\mathfrak{A}^a)^t_s$, $(\mathfrak{B}^a)^t_s$, $(\mathfrak{C}^a)^t_s$, and $(\mathfrak{D}^a)^t_s$. The time invariance condition for $(\mathfrak{A}^a)^t_s$, $(\mathfrak{B}^a)^t_s$, $(\mathfrak{C}^a)^t_s$, and $(\mathfrak{D}^a)^t_s$ follows from (2.2.5) and (6.4.3), so this only leaves the composition property.

Let x and y be the state trajectory and output function of Σ with initial time s, initial state $x(s)$, and input function $u \in L^p|Reg_{\mathrm{loc}}([s,\infty);U)$. Then, by (6.4.2) and (6.4.3), $x(t)$ and $\pi_{[s,t)}y$ are given by (6.4.5) for all $t \ge s$. In particular, for all $t \ge r \ge s$,

$$\begin{aligned}
\begin{bmatrix} x(s) \\ \pi_{[s,r)}y \\ \pi_{[r,t)}y \end{bmatrix} &= \begin{bmatrix} 1 & 0 & 0 \\ \mathfrak{C}^r_s & \mathfrak{D}^r_s & 0 \\ 0 & 0 & 1 \end{bmatrix} \begin{bmatrix} x(s) \\ \pi_{[s,r)}u \\ \pi_{[r,t)}y \end{bmatrix}, \\
\begin{bmatrix} x(r) \\ \pi_{[s,r)}u \\ \pi_{[r,t)}y \end{bmatrix} &= \begin{bmatrix} \mathfrak{A}^{r-s} & \mathfrak{B}^r_s & 0 \\ 0 & 1 & 0 \\ 0 & 0 & 1 \end{bmatrix} \begin{bmatrix} x(s) \\ \pi_{[s,r)}u \\ \pi_{[r,t)}y \end{bmatrix}, \\
\begin{bmatrix} x(r) \\ \pi_{[s,r)}u \\ \pi_{[r,t)}y \end{bmatrix} &= \begin{bmatrix} 1 & 0 & 0 \\ 0 & 1 & 0 \\ \mathfrak{C}^t_r & 0 & \mathfrak{D}^t_r \end{bmatrix} \begin{bmatrix} x(r) \\ \pi_{[s,r)}u \\ \pi_{[r,t)}u \end{bmatrix}, \\
\begin{bmatrix} x(t) \\ \pi_{[s,r)}u \\ \pi_{[r,t)}u \end{bmatrix} &= \begin{bmatrix} \mathfrak{A}^{t-r} & 0 & \mathfrak{B}^t_r \\ 0 & 1 & 0 \\ 0 & 0 & 1 \end{bmatrix} \begin{bmatrix} x(r) \\ \pi_{[s,r)}u \\ \pi_{[r,t)}u \end{bmatrix}.
\end{aligned} \tag{6.4.6}$$

We now interpret $\left[\begin{smallmatrix} x(t) \\ \pi_{[s,t)}u \end{smallmatrix}\right]$ as the initial data, and solve successively (working backwards)

$$\begin{bmatrix} x(r) \\ \pi_{[s,r)}u \\ \pi_{[r,t)}u \end{bmatrix}, \quad \begin{bmatrix} x(r) \\ \pi_{[s,r)}u \\ \pi_{[r,t)}y \end{bmatrix}, \quad \begin{bmatrix} x(s) \\ \pi_{[s,r)}u \\ \pi_{[r,t)}y \end{bmatrix}, \quad \begin{bmatrix} x(s) \\ \pi_{[s,r)}y \\ \pi_{[r,t)}y \end{bmatrix}, \quad \begin{bmatrix} x(s) \\ \pi_{[s,t)}y \end{bmatrix}$$

to get (the inverses below exist since $\mathfrak{A}^{r-s}$ and $\mathfrak{A}^{t-r}$ are invertible)

$$
\begin{aligned}
\left[\begin{array}{c|c} (\mathfrak{A}^a)_s^t & (\mathfrak{B}^a)_s^t \\ \hline (\mathfrak{C}^a)_s^t & (\mathfrak{D}^a)_s^t \end{array}\right]
&= \begin{bmatrix} 1 & 0 & 0 \\ 0 & \pi_{[s,r)} & \pi_{[r,t)} \end{bmatrix}
\begin{bmatrix} 1 & 0 & 0 \\ \mathfrak{C}_s^r & \mathfrak{D}_s^r & 0 \\ 0 & 0 & 1 \end{bmatrix}
\begin{bmatrix} \mathfrak{A}^{r-s} & \mathfrak{B}_s^r & 0 \\ 0 & 1 & 0 \\ 0 & 0 & 1 \end{bmatrix}^{-1} \\
&\quad \times \begin{bmatrix} 1 & 0 & 0 \\ 0 & 1 & 0 \\ \mathfrak{C}_r^t & 0 & \mathfrak{D}_r^t \end{bmatrix}
\begin{bmatrix} \mathfrak{A}^{t-r} & 0 & \mathfrak{B}_r^t \\ 0 & 1 & 0 \\ 0 & 0 & 1 \end{bmatrix}^{-1}
\begin{bmatrix} 1 & 0 \\ 0 & \pi_{[s,r)} \\ 0 & \pi_{[r,t)} \end{bmatrix} \\
&= \begin{bmatrix} 1 & 0 & 0 \\ 0 & \pi_{[s,r)} & \pi_{[r,t)} \end{bmatrix}
\left[\begin{array}{c|cc} (\mathfrak{A}^a)_s^r & (\mathfrak{B}^a)_s^r & 0 \\ \hline (\mathfrak{C}^a)_s^r & (\mathfrak{D}^a)_s^r & 0 \\ 0 & 0 & 1 \end{array}\right] \\
&\quad \times \left[\begin{array}{c|cc} (\mathfrak{A}^a)_r^t & 0 & (\mathfrak{B}^a)_r^t \\ \hline 0 & 1 & 0 \\ (\mathfrak{C}^a)_r^t & 0 & (\mathfrak{D}^a)_r^t \end{array}\right]
\begin{bmatrix} 1 & 0 \\ 0 & \pi_{[s,r)} \\ 0 & \pi_{[r,t)} \end{bmatrix},
\end{aligned}
$$

which is exactly (6.1.7) with $\mathfrak{A}_s^t$, $\mathfrak{B}_s^t$, $\mathfrak{C}_s^t$, and $\mathfrak{D}_s^t$ replaced by $(\mathfrak{A}^a)_s^t$, $(\mathfrak{B}^a)_s^t$, $(\mathfrak{C}^a)_s^t$, and $(\mathfrak{D}^a)_s^t$.

By Theorem 6.1.10, the system Σ^a is an anti-causal $L^p|Reg$-well-posed linear system. □

By reversing the direction of time in Theorem 6.4.1 we get a new causal system:

Definition 6.4.2 Let $\Sigma = \left[\begin{smallmatrix} \mathfrak{A} & \mathfrak{B} \\ \mathfrak{C} & \mathfrak{D} \end{smallmatrix}\right]$ be an $L^p|Reg$-well-posed linear system on (Y, X, U). We call Σ *time-invertible* if $\mathfrak{A}^t$ is invertible for some $t > 0$. When this is the case, then we call the system Σ^a in Theorem 6.4.1 the *backward system* and the system $\Sigma^{\text{Я}}$ that we get from Σ^a by reflecting the time axis, i.e.,

$$
\Sigma^{\text{Я}} = \left[\begin{array}{c|c} (\mathfrak{A}^{\text{Я}})^t & \mathfrak{B}^{\text{Я}} \\ \hline \mathfrak{C}^{\text{Я}} & \mathfrak{D}^{\text{Я}} \end{array}\right] = \left[\begin{array}{c|c} (\mathfrak{A}^a)_{-t} & \mathfrak{B}^a \mathbf{Я} \\ \hline \mathbf{Я}\mathfrak{C}^a & \mathbf{Я}\mathfrak{D}^a\mathbf{Я} \end{array}\right],
$$

the *time-inverted system.*

Theorem 6.4.3 *Let $\Sigma = \left[\begin{smallmatrix} \mathfrak{A} & \mathfrak{B} \\ \mathfrak{C} & \mathfrak{D} \end{smallmatrix}\right]$ be a time-invertible well-posed linear system on (Y, X, U) with system node $S = \left[\begin{smallmatrix} A\&B \\ C\&D \end{smallmatrix}\right]$. Then the backward system Σ^a in Theorem 6.4.1 has the same system node S as Σ, and the system node $S^{\text{Я}}$ of the time-inverted system $\Sigma^{\text{Я}}$ is given by $\left[\begin{smallmatrix} [A\&B]^{\text{Я}} \\ [C\&D]^{\text{Я}} \end{smallmatrix}\right] = \left[\begin{smallmatrix} -A\&B \\ C\&D \end{smallmatrix}\right]$. In particular, $\mathcal{D}(S^{\text{Я}}) = \mathcal{D}(S)$ and $\Sigma^{\text{Я}}$ is compatible if and only if Σ is compatible.*

Proof Let $\left[\begin{smallmatrix} x_0 \\ u_0 \end{smallmatrix}\right] \in \mathcal{D}(S)$, i.e., $x_0 \in X$, $u_0 \in U$, and $A_{|X}x_0 + Bu_0 \in X$. Choose an arbitrary $u \in C^1(\mathbb{R}^+; U)$ satisfying $u(0) = u_0$, and let x and y be the state trajectory and output function of Σ with initial time zero, initial state x_0, and input function u. By Theorem 4.6.11, $x \in C^1(\overline{\mathbb{R}}^+; X)$, $y \in W^{1,p}_{\mathrm{loc}}(\overline{\mathbb{R}}^+; Y)$ (or $y \in C^1(\overline{\mathbb{R}}^+; Y)$ in the *Reg*-well-posed case), $\left[\begin{smallmatrix} x(t) \\ u(t) \end{smallmatrix}\right] \in \mathcal{D}(S)$ for all $t \geq 0$, and

$$\begin{bmatrix} \dot{x}(t) \\ y(t) \end{bmatrix} = S \begin{bmatrix} x(t) \\ u(t) \end{bmatrix}, \qquad t \geq 0.$$

Define $\tilde{x}(t) = x(-t)$ for $t \leq 0$, $\tilde{u} = \text{Я}u$, and $\tilde{y} = \text{Я}y$. Then obviously $x \in C^1((\overline{\mathbb{R}}^-; X)$, $u \in W^{1,p}(\overline{\mathbb{R}}^-; Y)$, $y \in W^{1,p}(\overline{\mathbb{R}}^-; Y)$, $\left[\begin{smallmatrix} \tilde{x}(t) \\ \tilde{u}(t) \end{smallmatrix}\right] \in \mathcal{D}(S)$ for all $t \leq 0$, and

$$\begin{bmatrix} -\dot{\tilde{x}}(t) \\ \tilde{y}(t) \end{bmatrix} = S \begin{bmatrix} x(t) \\ u(t) \end{bmatrix}, \qquad t \leq 0. \tag{6.4.7}$$

On the other hand, by Theorem 6.4.1 and Definition 6.4.2, the restriction of $\tilde{x}$ and $\tilde{y}$ to $[-1, 0]$ are the state trajectory and output function of $\Sigma^{\text{Я}}$ with initial time -1, initial state $\tilde{x}(-1)$, and input function $\pi_{[-1,0]}\tilde{u}$. Therefore, by Theorem 4.6.11,

$$\begin{bmatrix} \dot{\tilde{x}}(t) \\ \tilde{y}(t) \end{bmatrix} = S^{\text{Я}} \begin{bmatrix} x(t) \\ u(t) \end{bmatrix}, \qquad -1 \leq t \leq 0.$$

Taking $t = 0$ we find that, for all $\left[\begin{smallmatrix} x_0 \\ u_0 \end{smallmatrix}\right] \in \mathcal{D}(S)$,

$$\begin{bmatrix} [A\&B]^{\text{Я}} \\ [C\&D]^{\text{Я}} \end{bmatrix} \begin{bmatrix} x_0 \\ u_0 \end{bmatrix} = \begin{bmatrix} -A\&B \\ C\&D \end{bmatrix}.$$

This shows that $\mathcal{D}(S) \subset \mathcal{D}(S^{\text{Я}})$, and that $\left[\begin{smallmatrix} [A\&B]^{\text{Я}} \\ [C\&D]^{\text{Я}} \end{smallmatrix}\right] = \left[\begin{smallmatrix} -A\&B \\ C\&D \end{smallmatrix}\right]$ on $\mathcal{D}(S)$. To get the opposite inclusion $\mathcal{D}(S^{\text{Я}}) \subset \mathcal{D}(S)$ we interchange Σ and $\Sigma^{\text{Я}}$ with each other and repeat the same argument. That Σ^a has the same system node as Σ follows from Definition 6.1.5.

Since $\Sigma^{\text{Я}}$ has the same observation/feedthrough node as Σ, and since $B^{\text{Я}} = -B$, it is obvious that $\Sigma^{\text{Я}}$ is compatible if and only if Σ is compatible. □

The notion of time-inversion can easily be extended to an arbitrary operator node.

Definition 6.4.4 Let $S = \left[\begin{smallmatrix} A\&B \\ C\&D \end{smallmatrix}\right]$ be an operator node on (Y, X, U). Then we call the operator node $S^{\text{Я}} = \left[\begin{smallmatrix} -A\&B \\ C\&D \end{smallmatrix}\right]$ the time-inverse of S.

Indeed, by Definition 4.7.2, $\left[\begin{smallmatrix} -A\&B \\ C\&D \end{smallmatrix}\right]$ is an operator node on (Y, X, U) whenever S is so.

There is a very simple relationship between the transfer function of a given operator node and the transfer function of its time-inverse.

Lemma 6.4.5 *Let $S = \left[\begin{smallmatrix} A\&B \\ C\&D \end{smallmatrix}\right]$ be an operator node on (Y, X, U) with main operator A and transfer function $\widehat{\mathfrak{D}}$. Then the main operator $A^{\text{Я}}$, the control* *operator* $B^{\text{Я}}$*, and the transfer function* $\widehat{\mathfrak{D}}^{\text{Я}}$ *of the time-inverted operator node* $S^{\text{Я}} = \left[\begin{smallmatrix} -A\&B \\ C\&D \end{smallmatrix}\right]$ *are given by*

$$A^{\text{Я}} = -A, \quad B^{\text{Я}} = -B, \quad \widehat{\mathfrak{D}}^{\text{Я}}(z) = \widehat{\mathfrak{D}}(-z), \quad -z \in \rho(A).$$

In particular, $\rho(A^{\text{Я}}) = -\rho(A)$.

Proof Obviously $A^{\text{Я}} = -A$ and $B^{\text{Я}} = -B$. This together with (4.7.1) implies that $\rho(A^{\text{Я}}) = -\rho(A)$ and that $\widehat{\mathfrak{D}}^{\text{Я}}(z) = \widehat{\mathfrak{D}}(-z)$ for $z \in \rho(A^{\text{Я}})$. □

As the following lemma shows, time-inversion almost commutes with duality (in the same sense as flow-inversion almost commutes with duality):

Theorem 6.4.6 *Let* (Y, X, U) *be reflexive Banach spaces, let* $1 < p < \infty$*, let* Σ *be an* L^p*-well-posed linear system on* (Y, X, U)*, and let* S *be an operator node on* (Y, X, U) *(not necessarily the one induced by* Σ*).*

(i) *The system* Σ *is time-invertible if and only if the causal dual system* Σ^d *is time-invertible, in which case the time-inverted systems satisfy* $(\Sigma^d)^{\text{Я}} = \left[\begin{smallmatrix} -1 & 0 \\ 0 & 1 \end{smallmatrix}\right] (\Sigma^{\text{Я}})^d \left[\begin{smallmatrix} -1 & 0 \\ 0 & 1 \end{smallmatrix}\right]$.
(ii) *The time-inverse* $S^{\text{Я}}$ *of* S *and the time-inverse* $(S^*)^{\text{Я}}$ *of the dual operator node* S^* *are related by* $(S^*)^{\text{Я}} = \left[\begin{smallmatrix} -1 & 0 \\ 0 & 1 \end{smallmatrix}\right] (S^{\text{Я}})^* \left[\begin{smallmatrix} -1 & 0 \\ 0 & 1 \end{smallmatrix}\right]$.

We leave the easy proof to the reader.

We end this section with a number of examples which illustrate various things that can happen when a well-posed linear system is time-inverted. (Some of these examples also refer to the notion of time-flow-inversion, which will be discussed in the next section.)

Example 6.4.7 *The circular shift* Σ *in Example 7.7.1 is time-invertible. Both this system and the corresponding time-inverted system are regular. The space* $W = (X + BU)_1$ *introduced in Lemma 4.3.12 is given by* $W = W^{1,p}([0, T]; U)$*. On this space, the 'forward extension' of* C *(i.e., the extension of* C *with respect to the original system) is given by* $\overline{C}x = x(0)$ *and the corresponding 'forward feedthrough operator' is* $\overline{D} = 0$*, whereas the 'backward extension' of* C *(i.e., the extension of* C *with respect to the time-inverted system) is given by* $\widetilde{C}x = x(0) - x(T)$ *and the corresponding 'backward feedthrough operator' is* $\widetilde{D}u = u$*. In particular, the forward and backward extensions are different.*

Proof In this proof we use the same notation as in Example 7.7.1.

Clearly, by the construction in Lemma 4.3.12 and by the description of $\mathcal{D}(S)$ given in Example 7.7.1, $W = W^{1,p}([0, T]; U)$. By the same example and

Definition 5.6.1, the forward and backward feedthrough operators are given by

$$\overline{D} = \lim_{\lambda\to+\infty} \widehat{\mathfrak{D}}(\lambda) = 0, \qquad \widetilde{D} = \lim_{\lambda\to-\infty} \widehat{\mathfrak{D}}(\lambda) = 1.$$

For all $(x, u) \in \mathcal{D}(S)$ we have $C\&D\left[\begin{smallmatrix} x \\ u \end{smallmatrix}\right] = x(0)$ (see Example 7.7.1), and by Theorem 5.6.4,

$$\overline{C}x = \overline{C}x + \overline{D}u = C\&D\begin{bmatrix} x \\ u \end{bmatrix} = x(0) = \widetilde{C}x + \widetilde{D}u = \widetilde{C}x + u.$$

For each $x \in W$ we can choose a unique $u \in U$ so that $\left[\begin{smallmatrix} x \\ u \end{smallmatrix}\right] \in \mathcal{D}(S)$, namely $u = x(T)$. Therefore we must have $\overline{C}x = x(0)$ and $\widetilde{C}x = x(0) - x(T)$. □

Example 6.4.8 *There exists a regular time-invertible L^2-well-posed linear system on $(\mathbb{C}, X, \mathbb{C})$ (where X is a Hilbert space) whose input/output map is zero, such that the corresponding time-inverted system is not regular (and its input/output map is nonzero). In particular, weak, strong, and uniform regularity are not preserved under time-inversion in general.*

Proof Define $\widehat{\mathfrak{D}}(z) = g^*(\overline{z})g(z)$, where g is the function defined in Example 5.7.4. Then $\widehat{\mathfrak{D}}$ is a bounded analytic function in the half-plane $\Re z \geq 0$, and $\lim_{\lambda\to+\infty} \widehat{\mathfrak{D}}(\lambda)$ does not exist, if we choose a to be large enough. By Theorem 10.3.5, $\widehat{\mathfrak{D}}$ is the transfer function of some $\mathfrak{D} \in TIC(\mathbb{C})$. Let

$$\Sigma = \left[\begin{array}{c|c} \mathfrak{A} & \mathfrak{B} \\ \hline \mathfrak{C} & \mathfrak{D} \end{array}\right] = \left[\begin{array}{c|c} \tau & \pi_- \\ \hline \pi_+\mathfrak{D} & \mathfrak{D} \end{array}\right]$$

be the bilateral input shift realization of $\mathfrak{D}$ on $(\mathbb{C}, L^2(\mathbb{C}); \mathbb{C})$ introduced in Example 2.6.5(iii). This system is not regular (in this case weak, strong, and uniform regularity are equivalent since both the input and the output space are one-dimensional). It is time-invertible since τ is a group. Let us compute the anti-causal time-inverted input/output map $(\mathfrak{D}^a)_s^t$ in Theorem 6.4.1. By Definition 2.2.6(iii), (6.4.2), and (6.4.3),

$$\begin{aligned}
(\mathfrak{D}^a)_s^t &= \mathfrak{D}_s^t - \pi_{[s,t)}\tau^{-s}\pi_+\mathfrak{D}\tau^{s-t}\pi_-\tau^t\pi_{[s,t)} \\
&= \mathfrak{D}_s^t - \pi_{[s,t)}\tau^{-s}\mathfrak{D}\tau^{s-t}\tau^t\pi_{[s,t)} \\
&= \mathfrak{D}_s^t - \pi_{[s,t)}\mathfrak{D}\pi_{[s,t)} = 0.
\end{aligned}$$

Thus, the input/output map of the time-inverted system is zero, and the time-inverted system is regular. We get the system described in Example 6.4.8 by interchanging Σ and $\Sigma^{\text{Я}}$. □

Note that in this example the spectrum of the generator (which is the imaginary axis) separates the open right half-plane $\mathbb{C}^+$ from the open left half-plane

$\mathbb{C}^-$, and the restrictions of $\widehat{\mathfrak{D}}$ to $\mathbb{C}^+$, respectively $\mathbb{C}^-$, *are not analytic continuations of each other*. The restriction of $\widehat{\mathfrak{D}}$ to $\mathbb{C}^-$ is zero, and so is the restriction of $\widehat{\mathfrak{D}}^{\mathfrak{R}}$ to $\mathbb{C}^+$, whereas the restriction of $\widehat{\mathfrak{D}}$ to $\mathbb{C}^+$ and the restriction of $\widehat{\mathfrak{D}}^{\mathfrak{R}}$ to $\mathbb{C}^-$ are nonzero.

Our following example is of a similar nature, but it is even more striking:

Example 6.4.9 *Let U and Y be Hilbert spaces, let $\widehat{\mathfrak{D}}_+$ be an arbitrary bounded analytic $\mathcal{B}(U;Y)$-valued function on the open right half-plane $\mathbb{C}^+$, and let $\widehat{\mathfrak{D}}_-$ be an arbitrary bounded analytic $\mathcal{B}(U;Y)$-valued function on the open left half-plane $\mathbb{C}^-$. Then there is a stable time-invertible L^2-well-posed linear system Σ on $(Y, L^2(\mathbb{R};U), U)$ whose transfer function $\widehat{\mathfrak{D}}$ is defined on $\mathbb{C} \setminus j\mathbb{R}$, and satisfies*

$$\widehat{\mathfrak{D}}(z) = \begin{cases} \widehat{\mathfrak{D}}_+(z), & \Re z > 0, \\ \widehat{\mathfrak{D}}_-(z), & \Re z < 0. \end{cases}$$

The semigroup of this system is the left-shift group τ on $L^2(\mathbb{R};U)$, and its spectrum separates $\mathbb{C}^+$ from $\mathbb{C}^-$.

Here 'stable' means that Σ is ω-bounded with $\omega = 0$. See Definition 8.1.1.

Proof Our construction of the realization of the given transfer function is a slight modification of the bilateral input shift realization described in Example 2.6.5(iii); another possibility would have been to modify the bilateral output shift realization described in Example 2.6.5(iv).

We take the state space of Σ to be $X = L^2(\mathbb{R};U)$ and we take the semigroup $\mathfrak{A}$ of Σ to be the bilateral left shift group τ on X described in Example 2.3.2(i). The generator $\frac{d}{ds}$ of this group is described in Example 3.2.3. The space $X_1 = \mathcal{D}\left(\frac{d}{ds}\right)$ is given by $X_1 = W^{1,2}_\omega(\mathbb{R};U)$, and the resolvent set of $\frac{d}{ds}$ is $\mathbb{C} - j\mathbb{R}$.

We let $\mathfrak{D}_+ \in TIC^2(U;Y)$ be the causal time-invariant operator from $L^2(\mathbb{R};U)$ to $L^2(\mathbb{R};Y)$ whose transfer function, restricted to $\mathbb{C}^+$, is $\widehat{\mathfrak{D}}_+$ (see Theorem 10.3.5), and we let $\mathfrak{D}_-$ be the causal time-invariant operator from $L^2(\mathbb{R};U)$ to $L^2(\mathbb{R};Y)$ whose transfer function, restricted to $\mathbb{C}^-$, is $\widehat{\mathfrak{D}}_-$ in the following sense: we require the transfer function of the causal time-invariant operator $\boldsymbol{\mathfrak{R}}\mathfrak{D}_-\boldsymbol{\mathfrak{R}}$ to be the function $z \mapsto \widehat{\mathfrak{D}}_-(-z)$ for $z \in \mathbb{C}^+$. Both of these operators map $X_1 = W^{1,2}_\omega(\mathbb{R};U)$ into $W^{1,2}_\omega(\mathbb{R};Y)$ since they intertwine the left-shifts on $L^2(\mathbb{R};U)$ and on $L^2(\mathbb{R};Y)$ (see Theorem 3.14.15).

We define the system Σ by

$$\Sigma = \left[\begin{array}{c|c} \mathfrak{A} & \mathfrak{B} \\ \hline \mathfrak{C} & \mathfrak{D} \end{array}\right] = \left[\begin{array}{c|c} \tau & \pi_- \\ \hline \pi_+(\mathfrak{D}_+ - \mathfrak{D}_-) & \mathfrak{D}_+ \end{array}\right].$$

Indeed, it is easy to see that this is an L^2-well-posed linear system. If $\mathfrak{D}_- = 0$, then this system is identical to the bilateral input realization described in

Example 2.6.5(iii). In particular, it has the same control operator B as that realization (see Example 4.2.6). Its observation operator is different, namely

$$Cu = (\mathfrak{D}_+ u)(0) - (\mathfrak{D}_- u)(0), \qquad u \in W^{1,2}_\omega(\mathbb{R}; U).$$

The system Σ is time-invertible, since $\mathfrak{A}$ is a group. Let us first compute the anti-causal time-inverted system Σ^a described in Theorem 6.4.1. Using (6.4.3), we get (as for any time-inverted system)

$$\left[\begin{array}{c|c} (\mathfrak{A}^a)^t_s & (\mathfrak{B}^a)^t_s \\ \hline (\mathfrak{C}^a)^t_s & (\mathfrak{D}^a)^t_s \end{array}\right] = \left[\begin{array}{c|c} \mathfrak{A}^{s-t} & -\mathfrak{A}^{s-t}\mathfrak{B}^t_s \\ \hline \mathfrak{C}^t_s\mathfrak{A}^{s-t} & (\mathfrak{D}^t_s - \mathfrak{C}^t_s\mathfrak{A}^{s-t}\mathfrak{B}^t_s) \end{array}\right],$$

which in this particular case simplifies into

$$\left[\begin{array}{c|c} (\mathfrak{A}^a)^t_s & (\mathfrak{B}^a)^t_s \\ \hline (\mathfrak{C}^a)^t_s & (\mathfrak{D}^a)^t_s \end{array}\right] = \left[\begin{array}{c|c} \tau^{s-t} & -\tau^s\pi_{[s,t)} \\ \hline \pi_{[s,t)}(\mathfrak{D}_+ - \mathfrak{D}_-)\tau^{-t} & \pi_{[s,t)}\mathfrak{D}_-\pi_{[s,t)} \end{array}\right].$$

Letting $s \to -\infty$ and $t \to +\infty$ as described in Theorem 6.1.10 we get for all $t \in \mathbb{R}$,

$$\left[\begin{array}{c|c} (\mathfrak{A}^a)_t & \mathfrak{B}^a \\ \hline \mathfrak{C}^a & \mathfrak{D}^a \end{array}\right] = \left[\begin{array}{c|c} \tau^{-t} & -\pi_+ \\ \hline \pi_-(\mathfrak{D}_+ - \mathfrak{D}_-) & \mathfrak{D}_- \end{array}\right].$$

In particular, the input/output of the time-inverted system $\Sigma^{Я}$ is $\mathfrak{D}^{Я} = \mathbf{Я}\mathfrak{D}_-\mathbf{Я}$, and this implies that the restriction of the transfer function $\widehat{\mathfrak{D}}^{Я}$ of $\Sigma^{Я}$ to $\mathbb{C}^+$ is the function $z \mapsto \widehat{\mathfrak{D}}_-(-z)$. Equivalently, the restriction of the transfer function of the original system Σ to $\mathbb{C}^-$ is $\widehat{\mathfrak{D}}_-$. □

Example 6.4.10 *There is a stable L^2-well-posed time-invertible single input single output system Σ such that neither Σ nor the time-inverted system $\Sigma^{Я}$ is regular. The spectrum of the semigroup generator A is contained in the imaginary axis $j\mathbb{R}$ and it does not separate the open right half-plane from the open left half-plane.*

This example is based on Example 6.4.9, with $U = Y = \mathbb{C}$ and with specific choices of $\widehat{\mathfrak{D}}_-$ and $\widehat{\mathfrak{D}}_+$, but using a smaller state space, after factoring out a redundant subspace of $L^2(\mathbb{R})$.

Proof Let $E = (-\infty, -1] \cup [1, \infty)$ and put $\Omega = \mathbb{C} \setminus jE$. Thus, Ω contains the open left and right half-planes and also a connecting bridge between them. For $s \in \Omega$, $s^2 + 1$ is not a real number in $\overline{\mathbb{R}^-}$. Since the logarithm function can be defined to be analytic on $\mathbb{C} \setminus \overline{\mathbb{R}^-}$ and such that $\log z$ is real for $z > 0$, we can define

$$\widehat{\mathfrak{D}}(s) = \cos\log(s^2 + 1), \qquad s \in \Omega. \tag{6.4.8}$$

Then $\widehat{\mathfrak{D}}$ is a bounded analytic function on Ω (its nontangential limits on jE are different, depending if we come from the right or from the left). Moreover, $\widehat{\mathfrak{D}}$ does not have limits as $s \to +\infty$ or $s \to -\infty$ along the real axis.

We now use Example 6.4.9 to construct a time-invertible realization Σ of $\widehat{\mathfrak{D}}$ (which is not yet the final realization). We simply take $\widehat{\mathfrak{D}}_+$ to be the restriction of $\widehat{\mathfrak{D}}$ to $\mathbb{C}^+$, and $\widehat{\mathfrak{D}}_-$ to be the restriction of $\widehat{\mathfrak{D}}$ to $\mathbb{C}^-$. According to Example 6.4.9 we get a realization Σ whose state space is $L^2(\mathbb{R};\mathbb{C})$, whose main group is τ^t, with generator $\frac{\mathrm{d}}{\mathrm{d}s}$ and $\mathcal{D}\left(\frac{\mathrm{d}}{\mathrm{d}s}\right) = W^{1,2}_{\omega}(\mathbb{R};\mathbb{C})$. The input map is π_-, and the output map is $\pi_+(\mathfrak{D}_+ - \mathfrak{D}_-)$.

Since $\widehat{\mathfrak{D}}_+(j\omega) - \widehat{\mathfrak{D}}_-(j\omega) = 0$ if (and only if) $\omega \in [-1, 1]$, the space X_0 of all band-limited functions in $L^2(\mathbb{R};\mathbb{C})$ whose spectrum is confined to $[-1, 1]$ (i.e., their bilateral Laplace transforms vanish on jE) is an unobservable subspace for Σ (this means that X_0 is an invariant subspace for $\mathfrak{A}$ and $X_0 \subset \mathcal{N}(\mathfrak{C})$). Moreover, X_0 is invariant also for the adjoint semigroup $\mathfrak{A}^*$. We factor out X_0, obtaining a reduced system Σ whose state space is the orthogonal complement of X_0 (see Theorem 9.1.9(ii)). Thus, the state space of Σ is

$$X = \{x \in L^2(\mathbb{R}\mathbb{C}) \mid \hat{x} \in L^2(jE;\mathbb{C})\}.$$

Thus, the functions in X contain 'only high frequencies.' The semigroup $\mathfrak{A}$ of the reduced system Σ is the restriction of τ to X. The spectrum of the generator A of this semigroup is jE, and so $\rho(A) = \Omega$ is connected. Denoting the orthogonal projection of $L^2(\mathbb{R};\mathbb{C})$ onto X by π_X, we find that the input map is $\mathfrak{B} = \pi_X\pi_-$. The output map of Σ is $\mathfrak{C} = \pi_+(\mathfrak{D}_+ - \mathfrak{D}_-)_{|X}$. The extended input/output map of the reduced system Σ coincides with that of Σ, and therefore the restriction of the transfer function $\widehat{\mathfrak{D}}$ of Σ to $\mathbb{C}^+$ is $\widehat{\mathfrak{D}}_+$. As Ω is connected, the transfer function of Σ must be equal to the (analytic) function $\widehat{\mathfrak{D}}$ defined in (6.4.8) on all of Ω.

Finally, we make an interesting observation. In this example $\widehat{\mathfrak{D}}_+(s) = \widehat{\mathfrak{D}}_-(-s)$ for all $s \in \mathbb{C}^+$, and Ω is invariant under a 180° rotation of the complex plane. In particular, $\mathfrak{D}_- = \mathbf{\mathfrak{R}}\mathfrak{D}\mathbf{\mathfrak{R}}$. This implies that the realization Σ constructed as in Example 6.4.9 has the property that $\Sigma^{\mathfrak{R}}$ is unitarily similar to Σ, with similarity operator $-\mathbf{\mathfrak{R}}$. Since $\mathbf{\mathfrak{R}}$ commutes with π_X, the reduced system Σ has the same property: $\Sigma^{\mathfrak{R}}$ is unitarily similar to Σ, with similarity operator $-\mathbf{\mathfrak{R}}$. □

Remark 6.4.11 The construction presented in Example 6.4.10 can be extended into a general procedure to construct time-invertible systems with $\rho(A)$ connected, whose transfer functions behave in a specified way at $+\infty$ and $-\infty$. In all the cases we may choose the group $\mathfrak{A}$ to be the same left-shift as in Example 6.4.10, and we only vary the transfer function $\widehat{\mathfrak{D}}$. We start with an arbitrary H^∞ function φ in the unit disk $\mathbb{D}^-$. Then we use a conformal map η to map the unit disk onto the region Ω in Example 6.4.10 in such a way that 1 is mapped onto $+\infty$ and -1 is mapped onto $-\infty$. After that we define $\widehat{\mathfrak{D}}(s) = \varphi(\eta^{-1}(s))$ and realize $\widehat{\mathfrak{D}}$ in the same way as we did in Example 6.4.10. By an appropriate choice of φ we can adjust the behavior of $\widehat{\mathfrak{D}}$ at $\pm\infty$ (and also at any point of jE). For example, if φ is bounded away from zero at ± 1 but does not have limits (taken along the real axis) at these points, then the system that we get is time-invertible, flow-invertible, and time-flow-invertible, but neither the system itself nor any of the inverted systems is regular.

6.5 Time-flow-inversion

Suppose that we have a well-posed time-invertible system Σ for which the time-inverted system is flow-invertible, or alternatively, suppose that Σ is flow-invertible and that the flow-inverted system is time-invertible. In both cases the final result will be the same in the following sense. Let x and y be the state trajectory and output function of $\Sigma = \left[\begin{smallmatrix} \mathfrak{A} & \mathfrak{B} \\ \mathfrak{C} & \mathfrak{D} \end{smallmatrix}\right]$ on (Y, X, U) with initial time $s \in \mathbb{R}$, initial state $x(s) \in X$, and input function $u \in L^p|Reg_{\text{loc}}([s, \infty, U)$. Then, for all $t \geq s$, (6.4.1) holds. Time-inversion of this system means that we interpret $x(t)$ and u as the given data and solve for $x(s)$ and y, and flow-inversion of the resulting system means that we interpret $x(t)$ and $\pi_{[s,t)}y$ as initial data and solve for $x(s)$ and $\pi_{[s,t)}u$. If we do the inversions in the opposite order we still end up with the same interpretation.

Actually, the final result, namely the interpretation of $x(t)$ and $\pi_{[s,t)}y$ as the given data and $x(s)$ and $\pi_{[s,t)}u$ as the derived data is possible under weaker assumptions than those indicated above: the original system Σ need neither be time-invertible nor flow-invertible. The only important property is that the operator $\Sigma_s^t = \left[\begin{smallmatrix} \mathfrak{A}_s^t & \mathfrak{B}_s^t \\ \mathfrak{C}_s^t & \mathfrak{D}_s^t \end{smallmatrix}\right]$ on the right-hand side of (6.4.1) is invertible from $\left[\begin{smallmatrix} X \\ L^p|Reg([s,t);U) \end{smallmatrix}\right]$ to $\left[\begin{smallmatrix} X \\ L^p|Reg([s,t);Y) \end{smallmatrix}\right]$ for all $s < t$. In this case we call the system *time-flow-invertible*. Solving (6.4.1) for $x(s)$ and $\pi_{[s,t)}u$ we get

$$\begin{bmatrix} x(s) \\ \pi_{[s,t)}u \end{bmatrix} = \begin{bmatrix} \mathfrak{A}_s^t & \mathfrak{B}_s^t \\ \mathfrak{C}_s^t & \mathfrak{D}_s^t \end{bmatrix}^{-1} \begin{bmatrix} x(t) \\ \pi_{[s,t)}y \end{bmatrix}, \qquad t \geq s. \tag{6.5.1}$$

By now it should come as no surprise that this can be interpreted as a well-posed linear system, which must be anti-causal since the initial time t is bigger than the final time s.

Theorem 6.5.1 *Let $\Sigma = \left[\begin{smallmatrix} \mathfrak{A} & \mathfrak{B} \\ \mathfrak{C} & \mathfrak{D} \end{smallmatrix}\right]$ be a well-posed linear system on (Y, X, U), and suppose that $\Sigma_s^t = \left[\begin{smallmatrix} \mathfrak{A}_s^t & \mathfrak{B}_s^t \\ \mathfrak{C}_s^t & \mathfrak{D}_s^t \end{smallmatrix}\right]$ is invertible as an operator from $\left[\begin{smallmatrix} X \\ L^p|Reg([s,t);U) \end{smallmatrix}\right]$ to $\left[\begin{smallmatrix} X \\ L^p|Reg([s,t);Y) \end{smallmatrix}\right]$ for some $-\infty < s < t < \infty$. Then Σ_s^t is invertible between these spaces for all $-\infty < s \leq t < \infty$ (note that Σ_t^t is the identity operator on*

$\left[\begin{smallmatrix} X \\ 0 \end{smallmatrix}\right]$). *For all* $-\infty < s \le t < \infty$ *we define*

$$(\Sigma^b)_s^t = \left[\begin{array}{c|c} (\mathfrak{A}^b)_s^t & (\mathfrak{B}^b)_s^t \\ \hline (\mathfrak{C}^b)_s^t & (\mathfrak{D}^b)_s^t \end{array}\right] = \left[\begin{array}{c|c} \mathfrak{A}_s^t & \mathfrak{B}_s^t \\ \hline \mathfrak{C}_s^t & \mathfrak{D}_s^t \end{array}\right]^{-1}, \tag{6.5.2}$$

and extend $(\mathfrak{C}^b)_s^t$ *and* $(\mathfrak{D}^b)_s^t$ *to* $L^p|Reg_{\mathrm{loc}}(\mathbb{R};U)$ *by requiring* $(\mathfrak{C}^b)_s^t = (\mathfrak{C}^b)_s^t \pi_{[s,t)}$ *and* $(\mathfrak{D}^b)_s^t = (\mathfrak{D}^b)_s^t \pi_{[s,t)}$. *Then* $\Sigma^b = \left[\begin{array}{c|c} \mathfrak{A}^b & \mathfrak{B}^b \\ \hline \mathfrak{C}^b & \mathfrak{D}^b \end{array}\right]$ *is an anti-causal* $L^p|Reg$*-well-posed linear system, where* $\mathfrak{A}_s^b = (\mathfrak{A}_s^b)^0 = \mathfrak{A}^s$ *for* $s \le 0$ *and*

$$\mathfrak{B}^b = \lim_{t\to\infty} (\mathfrak{B}^b)_0^t, \qquad \mathfrak{C}^b = \lim_{s\to-\infty} (\mathfrak{C}^b)_s^0, \qquad \mathfrak{D}^b = \lim_{\substack{s\to-\infty \\ t\to\infty}} (\mathfrak{D}^b)_s^t. \tag{6.5.3}$$

Moreover, if x *and* y *are the state trajectory and output function of* Σ *with initial time* s*, initial state* $x(s)$*, and input function* $u \in L^p|Reg_{\mathrm{loc}}([s,\infty);U)$*, then, for all* $t \ge s$*,*

$$\begin{bmatrix} x(s) \\ \pi_{[s,t)}u \end{bmatrix} = \left[\begin{array}{c|c} (\mathfrak{A}^b)_s^t & (\mathfrak{B}^b)_s^t \\ \hline (\mathfrak{C}^b)_s^t & (\mathfrak{D}^b)_s^t \end{array}\right] \begin{bmatrix} x(t) \\ \pi_{[s,t)}y \end{bmatrix}. \tag{6.5.4}$$

Proof of Theorem 6.5.1 We begin with the proof of the fact that Σ_s^t is invertible for all $s < t$ if it is invertible for some $s < t$. Assume that $\Sigma_{s_0}^{t_0}$ is invertible from $\left[\begin{smallmatrix} X \\ L^p|Reg([s_0,t_0);U) \end{smallmatrix}\right]$ to $\left[\begin{smallmatrix} X \\ L^p|Reg([s_0,t_0);Y) \end{smallmatrix}\right]$ where $s_0 < t_0$. For every $r \in [s_0, t_0]$, it follows from (2.2.10) that $\Sigma_{s_0}^r$ is injective and that $\Sigma_r^{t_0}$ is onto. Using the time invariance (2.2.5) we observe that $\Sigma_{s_0}^{s_0+t_0-r}$ is onto, and replacing $s_0 + t_0 - r$ by r we find that $\Sigma_{s_0}^r$ is onto. Being both injective and onto, $\Sigma_{s_0}^r$ is invertible. By the closed graph theorem, the inverse is continuous. We use the time-invariance (2.2.5) once more to conclude that Σ_s^t is invertible whenever $t - s \le T$ where $T = t_0 - s_0$. To remove the condition $t - s \le T$ we observe that, by (2.2.10) with $s = 0$, $r = T$, and $t = 2T$, Σ_0^{2T} is invertible, hence Σ_0^{4T} is invertible, etc.

Thus, Σ_s^t is invertible for all $s < t$, and $(\Sigma^b)_s^t$ is well-defined. It follows from (6.5.1) and (6.5.2) that (6.5.4) holds.

Next we claim that (6.1.7) holds with Σ_s^t replaced by $(\Sigma^b)_s^t$. However, this follows immediately from (6.5.2) and the fact that Σ_s^t satisfies (2.2.10) (see Lemma 2.2.8).

By definition, $(\mathfrak{A}^b)_0^0 x = x$ for all $x \in X$. Let us show that also $\lim_{t\uparrow 0} (\mathfrak{A}^b)_t^0 x = x$ for all $x \in X$. Take an arbitrary $x_0 \in X$, and define

$$\begin{bmatrix} x(-1) \\ \pi_{[-1,0)}u \end{bmatrix} = \begin{bmatrix} (\mathfrak{A}^b)_{-1}^0 \\ (\mathfrak{C}^b)_{-1}^0 \end{bmatrix} x_0,$$

$$x(t) = \mathfrak{A}_{-1}^t x(-1) + \mathfrak{B}_{-1}^t u, \qquad -1 < t \le 0.$$

Then $x(t)$ is the state trajectory of Σ with initial time -1, initial state $x(-1)$, and input function u. By (6.5.2), $x(0) = x_0$ and the corresponding output function vanishes on $[0, 1)$. This combined with (6.5.4) shows that $x(t) = (\mathfrak{A}^b)^0_t x_0$ for all $t \in [-1, 0]$. By Theorems 2.2.12 and 2.3.3, $\lim_{t\uparrow 0} x(t) = \lim_{t\uparrow 0}(\mathfrak{A}^b)^0_t x_0 = x_0$.

By Theorem 6.1.10, Σ^b is an anti-causal $L^p|Reg$-well-posed linear system. □

It is often convenient to replace the anti-causal system Σ^b in Theorem 6.5.1 by the corresponding causal system.

Definition 6.5.2 Let $\Sigma = \left[\begin{array}{c|c}\mathfrak{A} & \mathfrak{B}\\ \hline \mathfrak{C} & \mathfrak{D}\end{array}\right]$ be a well-posed linear system on (Y, X, U). We call Σ *time-flow-invertible* if $\Sigma^t_s = \left[\begin{array}{c|c}\mathfrak{A}^t_s & \mathfrak{B}^t_s\\ \hline \mathfrak{C}^t_s & \mathfrak{D}^t_s\end{array}\right]$ is invertible for some $s < t$. When this is the case, then we call the system $\Sigma^{\mathbf{Я}}_{\times}$ that we get from the system Σ^b in Theorem 6.5.1 by reflecting the time axis, i.e.,

$$\Sigma^{\mathbf{Я}}_{\times} = \left[\begin{array}{c|c}(\mathfrak{A}^{\mathbf{Я}}_{\times})^t & \mathfrak{B}^{\mathbf{Я}}_{\times}\\ \hline \mathfrak{C}^{\mathbf{Я}}_{\times} & \mathfrak{D}^{\mathbf{Я}}_{\times}\end{array}\right] = \left[\begin{array}{c|c}(\mathfrak{A}^b)_{-t} & \mathfrak{B}^b\mathbf{Я}\\ \hline \mathbf{Я}\mathfrak{C}^b & \mathbf{Я}\mathfrak{D}^b\mathbf{Я}\end{array}\right],$$

the *time-flow-inverted system*.

As our following theorem shows, there is a simple connection between the time-flow-inversion of Σ and the inverse of the Lax–Phillips model induced by Σ.

Theorem 6.5.3 *Let $\Sigma = \left[\begin{array}{c|c}\mathfrak{A} & \mathfrak{B}\\ \hline \mathfrak{C} & \mathfrak{D}\end{array}\right]$ be an $L^p|Reg$-well-posed linear system on (Y, X, U), let $\omega \in \mathbb{R}$, and let $\mathfrak{T}$ be the Lax–Phillips model of type $L^p|Reg_\omega$ induced by Σ. Then Σ is time-flow-invertible if and only if $\mathfrak{T}$ is invertible, in which case the Lax–Phillips model $\mathfrak{T}^{\mathbf{Я}}_{\times}$ of type $L^p|Reg_{-\omega}$ induced by $\Sigma^{\mathbf{Я}}_{\times}$ is given by*

$$\mathfrak{T}^{\mathbf{Я}}_{\times} = \begin{bmatrix} 0 & 0 & \mathbf{Я}\\ 0 & 1 & 0\\ \mathbf{Я} & 0 & 0\end{bmatrix} \mathfrak{T}^{-1} \begin{bmatrix} 0 & 0 & \mathbf{Я}\\ 0 & 1 & 0\\ \mathbf{Я} & 0 & 0\end{bmatrix}. \tag{6.5.5}$$

Proof Clearly, as $\mathbf{Я}$ maps $L^p|Reg_\omega(\mathbb{R}^-; Y)$ and $L^p|Reg_\omega(\mathbb{R}^+; U)$ onto $L^p|Reg_{-\omega}(\mathbb{R}^+; Y)$, respectively $L^p|Reg_{-\omega}(\mathbb{R}^-; U)$, the state space of the semigroup on the right hand side of (6.5.5) coincides with the state space of $\mathfrak{T}^{\mathbf{Я}}_{\times}$.

Let $\begin{bmatrix} y_0\\ x_0\\ u_0\end{bmatrix} \in \begin{bmatrix} L^p|Reg_\omega(\mathbb{R}^-;Y)\\ X\\ L^p|Reg_\omega(\mathbb{R}^+;U)\end{bmatrix}$, let $t > 0$, and let $\begin{bmatrix} y_t\\ x_t\\ u_t\end{bmatrix} = \mathfrak{T}^t \begin{bmatrix} y_0\\ x_0\\ u_0\end{bmatrix}$. This means explicitly that

$$\begin{aligned}
\pi_{(-\infty,-t)} y_t &= \tau^t y_0,\\
\begin{bmatrix} x_t\\ \pi_{[0,t)}\tau^{-t} y_t\end{bmatrix} &= \begin{bmatrix} \mathfrak{A}^t & \mathfrak{B}^t_0\\ \mathfrak{C}^t_0 & \mathfrak{D}^t_0\end{bmatrix} \begin{bmatrix} x_0\\ \pi_{[0,t)} u_0\end{bmatrix},\\
u_t &= \tau^t \pi_{[t,\infty)} u_0.
\end{aligned}$$

The first and last equations define invertible mappings of $L^p|Reg(\mathbb{R}^-;Y)$ and $L^p|Reg([t,\infty);U)$ onto $L^p|Reg((-\infty,-t);Y)$, respectively $L^p|Reg(\mathbb{R}^+;U)$, so $\mathfrak{T}$ is invertible if and only if Σ is time-flow-invertible. In this case we get (see also Lemma 6.1.8)

$$y_0 = \tau^{-t}\pi_{(-\infty,-t)}y_t,$$
$$\begin{bmatrix} x_0 \\ \text{Я}_{t/2}\pi_{[0,t)}u_0 \end{bmatrix} = \begin{bmatrix} (\mathfrak{A}^{\text{я}})^t_s & (\mathfrak{B}^{\text{я}})^t_s \\ (\mathfrak{C}^{\text{я}})^t_s & (\mathfrak{D}^{\text{я}})^t_s \end{bmatrix} \begin{bmatrix} x_t \\ \text{Я}_{t/2}\pi_{[0,t)}\tau^{-t}y_t \end{bmatrix},$$
$$\pi_{[t,\infty)}u_0 = \tau^{-t}u_t,$$

or equivalently,

$$\pi_{(-\infty,-t)}\text{Я}u_0 = \tau^t\text{Я}u_t,$$
$$\begin{bmatrix} x_0 \\ \tau^{-t}\pi_{[-t,0)}\text{Я}u_0 \end{bmatrix} = \begin{bmatrix} (\mathfrak{A}^{\text{я}})^t_s & (\mathfrak{B}^{\text{я}})^t_s \\ (\mathfrak{C}^{\text{я}})^t_s & (\mathfrak{D}^{\text{я}})^t_s \end{bmatrix} \begin{bmatrix} x_t \\ \pi_{[0,t)}\text{Я}y_t \end{bmatrix},$$
$$\text{Я}y_0 = \tau^t\pi_{[t,\infty)}\text{Я}y_t.$$

This is just another way of writing the formula which we get by applying both sides of (6.5.5) to $\left[\begin{smallmatrix} \text{я}u_t \\ x_t \\ \text{я}y_t \end{smallmatrix}\right]$. □

Next we investigate the system node of the time-flow-inverted system.

Theorem 6.5.4 *Let* $\Sigma = \left[\begin{smallmatrix} \mathfrak{A} & \mathfrak{B} \\ \mathfrak{C} & \mathfrak{D} \end{smallmatrix}\right]$ *be a time-flow-invertible well-posed linear system on* (Y, X, U) *with system node* $S = \left[\begin{smallmatrix} A\&B \\ C\&D \end{smallmatrix}\right]$. *Then the operator* $\left[\begin{smallmatrix} 1 & 0 \\ C\&D \end{smallmatrix}\right]$ *maps* $\mathcal{D}(S)$ *continuously onto* $\mathcal{D}\left(S^{\text{я}}_{\times}\right)$, *its inverse is* $\left[\begin{smallmatrix} 1 & 0 \\ [C\&D]^{\text{я}}_{\times} \end{smallmatrix}\right]$, *and*

$$\begin{bmatrix} A\&B \\ C\&D \end{bmatrix} = \begin{bmatrix} -[A\&B]^{\text{я}}_{\times} \\ 0 \quad 1 \end{bmatrix} \begin{bmatrix} 1 \quad 0 \\ [C\&D]^{\text{я}}_{\times} \end{bmatrix}^{-1} \quad (on\ \mathcal{D}(S)), \tag{6.5.6}$$

$$\begin{bmatrix} [A\&B]^{\text{я}}_{\times} \\ [C\&D]^{\text{я}}_{\times} \end{bmatrix} = \begin{bmatrix} -A\&B \\ 0 \quad 1 \end{bmatrix} \begin{bmatrix} 1 \quad 0 \\ C\&D \end{bmatrix}^{-1} \quad (on\ \mathcal{D}\left(S^{\text{я}}_{\times}\right)). \tag{6.5.7}$$

As we shall see, the proof of this theorem is a combination of the proofs of Theorems 6.3.6 and 6.4.3.

Proof We begin the proof exactly in the same way as the proof of Theorem 6.4.3, up to formula (6.4.7). In addition, we observe that

$$\begin{bmatrix} x_0 \\ y_0 \end{bmatrix} = \begin{bmatrix} 1 \quad 0 \\ C\&D \end{bmatrix} \begin{bmatrix} x_0 \\ u_0 \end{bmatrix}.$$

By Theorem 6.5.1 and Definition 6.5.2, the restriction of $\tilde{x}$ and $\tilde{u}$ to $[-1, 0]$ are the state trajectory and output function of $\Sigma^{\text{я}}_{\times}$ with initial time -1, initial state $\tilde{x}(-1)$, and input function $\pi_{[-1,0]}\tilde{y}$. Therefore, by Theorem 4.6.11,

$$\begin{bmatrix}\dot{\tilde{x}}(t)\\ \tilde{u}(t)\end{bmatrix} = S^{\text{я}}_{\times}\begin{bmatrix}\tilde{x}(t)\\ \tilde{y}(t)\end{bmatrix}, \quad \begin{bmatrix}\tilde{x}(t)\\ \tilde{u}(t)\end{bmatrix} = \begin{bmatrix}1 & 0\\ [C\&D]^{\text{я}}_{\times}\end{bmatrix}\begin{bmatrix}\tilde{x}(t)\\ \tilde{y}(t)\end{bmatrix}, \quad -1 \le t \le 0.$$

Taking $t = 0$ we find that, for all $\left[\begin{smallmatrix}x_0\\ u_0\end{smallmatrix}\right] \in \mathcal{D}(S)$, we have $\left[\begin{smallmatrix}x_0\\ y_0\end{smallmatrix}\right] = \left[\begin{smallmatrix}1 & 0\\ C\&D\end{smallmatrix}\right]\left[\begin{smallmatrix}x_0\\ u_0\end{smallmatrix}\right] \in \mathcal{D}\left(S^{\text{я}}_{\times}\right)$, and

$$\begin{bmatrix}[A\&B]^{\text{я}}_{\times}\\ [C\&D]^{\text{я}}_{\times}\end{bmatrix}\begin{bmatrix}x_0\\ y_0\end{bmatrix} = \begin{bmatrix}[A\&B]^{\text{я}}_{\times}\\ [C\&D]^{\text{я}}_{\times}\end{bmatrix}\begin{bmatrix}1 & 0\\ C\&D\end{bmatrix}\begin{bmatrix}x_0\\ u_0\end{bmatrix} = \begin{bmatrix}-A\&B\\ C\&D\end{bmatrix}\begin{bmatrix}x_0\\ u_0\end{bmatrix}.$$

By interchanging the roles of Σ and $\Sigma^{\text{я}}_{\times}$ (and, at the same time, interchanging the roles of u and y), we can prove in the same way that, for all $\left[\begin{smallmatrix}x_0\\ y_0\end{smallmatrix}\right] \in \mathcal{D}\left(S^{\text{я}}_{\times}\right)$, we have $\left[\begin{smallmatrix}x_0\\ y_0\end{smallmatrix}\right] = \left[\begin{smallmatrix}1 & 0\\ [C\&D]^{\text{я}}_{\times}\end{smallmatrix}\right]\left[\begin{smallmatrix}x_0\\ y_0\end{smallmatrix}\right] \in \mathcal{D}(S)$, and that

$$\begin{bmatrix}A\&B\\ C\&D\end{bmatrix}\begin{bmatrix}x_0\\ u_0\end{bmatrix} = \begin{bmatrix}A\&B\\ C\&D\end{bmatrix}\begin{bmatrix}1 & 0\\ [C\&D]^{\text{я}}_{\times}\end{bmatrix}\begin{bmatrix}x_0\\ y_0\end{bmatrix} = \begin{bmatrix}-[A\&B]^{\text{я}}_{\times}\\ [C\&D]^{\text{я}}_{\times}\end{bmatrix}\begin{bmatrix}x_0\\ y_0\end{bmatrix}.$$

These two identities show that $\left[\begin{smallmatrix}1 & 0\\ C\&D\end{smallmatrix}\right]$ maps $\mathcal{D}(S)$ onto $\mathcal{D}\left(S^{\text{я}}_{\times}\right)$ with inverse $\left[\begin{smallmatrix}1 & 0\\ [C\&D]^{\text{я}}_{\times}\end{smallmatrix}\right]$, and that all the other identities listed in Theorem 6.5.4 hold. □

Motivated by Theorem 6.5.4 we extend Definition 6.5.2 to arbitrary operator nodes as follows.

Definition 6.5.5 Let $S = \left[\begin{smallmatrix}A\&B\\ C\&D\end{smallmatrix}\right]$ be an operator node on (Y, X, U). We call this operator node *time-flow-invertible* if there exists an operator node $S^{\text{я}}_{\times} = \left[\begin{smallmatrix}[A\&B]^{\text{я}}_{\times}\\ [C\&D]^{\text{я}}_{\times}\end{smallmatrix}\right]$ on (U, X, Y) which together with S satisfies the following conditions: the operator $\left[\begin{smallmatrix}1 & 0\\ C\&D\end{smallmatrix}\right]$ maps $\mathcal{D}(S)$ continuously onto $\mathcal{D}\left(S^{\text{я}}_{\times}\right)$, its inverse is $\left[\begin{smallmatrix}1 & 0\\ [C\&D]^{\text{я}}_{\times}\end{smallmatrix}\right]$, and (6.5.6)–(6.5.7) hold. In this case we call S and $S^{\text{я}}_{\times}$ *time-flow-inverses* of each other.

Formulas involving the operator nodes S and $S^{\text{я}}_{\times}$ can readily be obtained from our earlier formulas for time-inversion and flow-inversion with the help of the following lemma.

Lemma 6.5.6 *Let $S = \left[\begin{smallmatrix}A\&B\\ C\&D\end{smallmatrix}\right]$ be an operator node on (Y, X, U). Then S is time-flow-invertible if and only if S is time-flow-invertible, in which case the time-flow-inverse $S^{\text{я}}_{\times}$ of S is given by $S^{\text{я}}_{\times} = (S_{\times})^{\text{я}} = (S^{\text{я}})_{\times}$ (where $(S_{\times})^{\text{я}}$ is the time-inverse of the flow-inverse of S and $(S^{\text{я}})_{\times}$ is the flow-inverse of the time-inverse of S).*

Proof This is obvious (see Definitions 6.3.7, 6.4.4, and 6.5.5). □

Lemma 6.5.7 *Let $S = \left[\begin{smallmatrix}A\&B\\ C\&D\end{smallmatrix}\right]$ be a time-flow-invertible operator node on (Y, X, U), with main operator A, control operator B, observation operator*

C, and transfer function $\mathfrak{D}$*, and let* $S^{\text{я}}_{\times} = \begin{bmatrix} [A\&B]^{\text{я}}_{\times} \\ [C\&D]^{\text{я}}_{\times} \end{bmatrix}$ *be its time-flow-inverse, with main operator* $A^{\text{я}}_{\times}$*, control operator* $B^{\text{я}}_{\times}$*, observation operator* $C^{\text{я}}_{\times}$*, and transfer function* $\mathfrak{D}^{\text{я}}_{\times}$*. Then the following claims are true.*

(i) $\alpha \in \rho(A^{\text{я}}_{\times})$ *if and only if* $\left[\begin{smallmatrix} \alpha & 0 \\ 0 & 0 \end{smallmatrix}\right] + S$ *is invertible, in which case*

$$\left(\begin{bmatrix} \alpha & 0 \\ 0 & 0 \end{bmatrix} + S\right)^{-1} = \begin{bmatrix} (\alpha - A^{\text{я}}_{\times})^{-1} & (\alpha - A^{\text{я}}_{\times|X})^{-1} B^{\text{я}}_{\times} \\ C^{\text{я}}_{\times}(\alpha - A^{\text{я}}_{\times})^{-1} & \widehat{\mathfrak{D}}^{\text{я}}_{\times}(\alpha) \end{bmatrix}. \tag{6.5.8}$$

(ii) $\alpha \in \rho(A)$ *if and only if* $\left[\begin{smallmatrix} \alpha & 0 \\ 0 & 0 \end{smallmatrix}\right] + S^{\text{я}}_{\times}$ *is invertible, in which case*

$$\left(\begin{bmatrix} \alpha & 0 \\ 0 & 0 \end{bmatrix} + S^{\text{я}}_{\times}\right)^{-1} = \begin{bmatrix} (\alpha - A)^{-1} & (\alpha - A_{|X})^{-1} B \\ C(\alpha - A)^{-1} & \widehat{\mathfrak{D}}(\alpha) \end{bmatrix}. \tag{6.5.9}$$

(iii) *If* $\alpha \in \rho(-A)$*, then* $\alpha \in \rho(A^{\text{я}}_{\times})$ *if and only if* $\widehat{\mathfrak{D}}(-\alpha)$ *is invertible. In this case*

$$\begin{aligned} \begin{bmatrix} (\alpha - A^{\text{я}}_{\times})^{-1} & (\alpha - A^{\text{я}}_{\times|X})^{-1} B^{\text{я}}_{\times} \\ C^{\text{я}}_{\times}(\alpha - A^{\text{я}}_{\times})^{-1} & \widehat{\mathfrak{D}}^{\text{я}}_{\times}(\alpha) \end{bmatrix} &= \begin{bmatrix} (\alpha + A)^{-1} & 0 \\ 0 & 0 \end{bmatrix} \\ &+ \begin{bmatrix} -(\alpha + A_{|X})^{-1} B \\ 1 \end{bmatrix} [\widehat{\mathfrak{D}}(-\alpha)]^{-1} \begin{bmatrix} -C(\alpha + A)^{-1} & 1 \end{bmatrix}. \end{aligned} \tag{6.5.10}$$

In particular, in this case $\widehat{\mathfrak{D}}^{\text{я}}_{\times}(\alpha) = \widehat{\mathfrak{D}}^{-1}(-\alpha)$*, and U is isomorphic to Y (hence they have the same dimension).*

(iv) *If* $\alpha \in \rho(-A^{\text{я}}_{\times})$*, then* $\alpha \in \rho(A)$ *if and only if* $\widehat{\mathfrak{D}}^{\text{я}}_{\times}(-\alpha)$ *is invertible. In this case*

$$\begin{aligned} \begin{bmatrix} (\alpha - A)^{-1} & (\alpha - A_{|X})^{-1} B \\ C(\alpha - A)^{-1} & \widehat{\mathfrak{D}}(\alpha) \end{bmatrix} &= \begin{bmatrix} (\alpha + A^{\text{я}}_{\times})^{-1} & 0 \\ 0 & 0 \end{bmatrix} \\ &+ \begin{bmatrix} -(\alpha + A^{\text{я}}_{\times|X})^{-1} B^{\text{я}}_{\times} \\ 1 \end{bmatrix} [\widehat{\mathfrak{D}}^{\text{я}}_{\times}(-\alpha)]^{-1} \begin{bmatrix} -C^{\text{я}}_{\times}(\alpha + A^{\text{я}}_{\times})^{-1} & 1 \end{bmatrix}. \end{aligned} \tag{6.5.11}$$

Proof See Lemmas 6.3.8, 6.4.5, and 6.5.6. □

Corollary 6.5.8 *Let* $S = \left[\begin{smallmatrix} A\&B \\ C\&D \end{smallmatrix}\right]$ *be an operator node on* (Y, X, U)*, with main operator A, control operator B, observation operator C, and transfer function* $\mathfrak{D}$*, and let* $S^{\text{я}}_{\times} = \begin{bmatrix} [A\&B]^{\text{я}}_{\times} \\ [C\&D]^{\text{я}}_{\times} \end{bmatrix}$ *be an operator node on* (U, X, Y)*, with main operator* $A^{\text{я}}_{\times}$*, control operator* $B^{\text{я}}_{\times}$*, observation operator* $C^{\text{я}}_{\times}$*, and transfer function* $\mathfrak{D}^{\text{я}}_{\times}$*. Then the following conditions are equivalent:*

(i) *S and* $S^{\text{я}}_{\times}$ *are time-flow-inverses of each other.*

(ii) *The operator* $\begin{bmatrix} 1 \quad 0 \\ [C\&D]^{\text{я}}_{\times} \end{bmatrix}$ *maps* $\mathcal{D}\left(S^{\text{я}}_{\times}\right)$ *one-to-one onto* $\mathcal{D}(S)$*, and (6.5.6) holds.*

(iii) *For all* $\alpha \in \rho(A^{\text{я}}_{\times})$, *the operator* $\left[\begin{smallmatrix}\alpha & 0\\ 0 & 0\end{smallmatrix}\right] + S$ *maps* $\mathcal{D}(S)$ *one-to-one onto* $\left[\begin{smallmatrix}X\\ Y\end{smallmatrix}\right]$, *and its (bounded) inverse is given by* (6.5.8).

(iv) *For some* $\alpha \in \rho(A^{\text{я}}_{\times})$, *the operator* $\left[\begin{smallmatrix}\alpha & 0\\ 0 & 0\end{smallmatrix}\right] + S$ *maps* $\mathcal{D}(S)$ *one-to-one onto* $\left[\begin{smallmatrix}X\\ Y\end{smallmatrix}\right]$ *and* (6.5.8) *holds.*

(v) *The operator* $\left[\begin{smallmatrix}1 & 0\\ C\&D\end{smallmatrix}\right]$ *maps* $\mathcal{D}(S)$ *one-to-one onto* $\mathcal{D}\left(S^{\text{я}}_{\times}\right)$, *and* (6.5.7) *holds.*

(vi) *For all* $\alpha \in \rho(A)$, *the operator* $\left[\begin{smallmatrix}\alpha & 0\\ 0 & 0\end{smallmatrix}\right] + S^{\text{я}}_{\times}$ *maps* $\mathcal{D}\left(S^{\text{я}}_{\times}\right)$ *one-to-one onto* $\left[\begin{smallmatrix}X\\ Y\end{smallmatrix}\right]$, *and its (bounded) inverse is given by* (6.5.9).

(vii) *For some* $\alpha \in \rho(A)$, *the operator* $\left[\begin{smallmatrix}\alpha & 0\\ 0 & 0\end{smallmatrix}\right] + S^{\text{я}}_{\times}$ *maps* $\mathcal{D}\left(S^{\text{я}}_{\times}\right)$ *one-to-one onto* $\left[\begin{smallmatrix}X\\ Y\end{smallmatrix}\right]$ *and* (6.5.9) *holds.*

When these equivalent conditions hold, then $\left[\begin{smallmatrix}1\\ C\end{smallmatrix}\right]$ *maps* $\mathcal{D}(A)$ *into* $\mathcal{D}\left(S^{\text{я}}_{\times}\right)$, $\left[\begin{smallmatrix}1\\ C^{\text{я}}_{\times}\end{smallmatrix}\right]$ *maps* $\mathcal{D}\left(A^{\text{я}}_{\times}\right)$ *into* $\mathcal{D}(S)$, *and*

$$\begin{aligned} A &= -A^{\text{я}}_{\times|\mathcal{D}(A)} - B^{\text{я}}_{\times}C, & A^{\text{я}}_{\times} &= -A_{|\mathcal{D}(A^{\text{я}}_{\times})} - BC^{\text{я}}_{\times}, \\ 0 &= [C\&D]^{\text{я}}_{\times}\begin{bmatrix}1\\ C\end{bmatrix}, & 0 &= C\&D\begin{bmatrix}1\\ C^{\text{я}}_{\times}\end{bmatrix}. \end{aligned} \tag{6.5.12}$$

Proof See Theorem 6.3.9 and Lemma 6.5.6. □

Corollary 6.5.9 *Make the same assumptions and introduce the same notation as in Corollary 6.5.8. In addition, suppose that* $\rho(-A) \cap \rho(A^{\text{я}}_{\times}) \neq \emptyset$. *Then the conditions (i)–(vii) listed in Corollary 6.5.8 are equivalent to each one of the following conditions:*

(viii) *For all* $\alpha \in \rho(-A) \cap \rho(A^{\text{я}}_{\times})$, $\widehat{\mathfrak{D}}(-\alpha)$ *is invertible and* (6.5.10) *holds.*

(ix) *For some* $\alpha \in \rho(-A) \cap \rho(A^{\text{я}}_{\times})$, $\widehat{\mathfrak{D}}(\alpha)$ *is invertible and* (6.5.10) *holds.*

(x) *For all* $\alpha \in \rho(A) \cap \rho(-A^{\text{я}}_{\times})$, $\widehat{\mathfrak{D}}^{\text{я}}_{\times}(-\alpha)$ *is invertible and* (6.5.11) *holds.*

(xi) *For some* $\alpha \in \rho(A) \cap \rho(-A^{\text{я}}_{\times})$, $\widehat{\mathfrak{D}}^{\text{я}}_{\times}(-\alpha)$ *is invertible and* (6.5.11) *holds.*

In particular, when these conditions hold, then U and Y are isomorphic (hence they have the same dimension).

Proof This follows from Theorem 6.3.11 and Lemma 6.5.6 (see also Corollary 6.3.12). □

In the finite-dimensional case (i.e., the case where the state space is finite-dimensional) the condition $\rho(-A) \cap \rho(A^{\text{я}}_{\times}) \neq \emptyset$ is redundant, and the system is time-flow-invertible if and only if $\widehat{\mathfrak{D}}(\infty)$ is invertible. In particular, a finite-dimensional system cannot be time-flow-invertible unless the dimension of the input space is the same as the dimension of the output space. However, as the following example shows, this condition is no longer necessary in the infinite-dimensional case. In this example the input and output spaces are of different dimension, and Corollary 6.5.9 does not apply due to the fact that we here have $\rho(-A) \cap \rho(A^{\text{я}}_{\times}) = \emptyset$ (since A and $A^{\text{я}}_{\times}$ have the same spectrum, namely the closed left half-plane $\overline{\mathbb{C}^-}$).

Example 6.5.10 *Let $U = 0$, $Y = \mathbb{C}$, let $\mathfrak{A}^t = \tau^t_+$ be the left-shift on $X = L^2(\mathbb{R}^+;\mathbb{C})$, and let $\mathfrak{C} = 1$. Then $\Sigma = \left[\begin{smallmatrix}\mathfrak{A}\\ \mathfrak{C}\end{smallmatrix}\right]$ is a time-flow-invertible system on $(Y;X;U)$ (with no input) whose time-flow-inverted system $\Sigma^{\text{я}}_{\times}$ coincides with its causal dual system $\Sigma^d = \left[\mathfrak{A}^d \middle| \mathfrak{B}^d\right]$ (which has no output). In particular, $(\mathfrak{A}^{\text{я}}_{\times})^t = (\mathfrak{A}^d)^t = (\tau^t_+)^*$ is the right-shift on $L^2(\mathbb{R}^+;\mathbb{C})$ and $\mathfrak{B}^{\text{я}}_{\times} = 1$.*

Proof This system is conservative, meaning that the operator $\Sigma^t_0 = \left[\begin{smallmatrix}\mathfrak{A}^t\\ \mathfrak{C}^t_0\end{smallmatrix}\right]$ is unitary from X to $\left[\begin{smallmatrix}X\\ L^2(0,t)\end{smallmatrix}\right]$ for all $t \geq 0$, hence invertible. The inverse of a unitary operator coincides with its adjoint; hence the time-flow-inverted system coincides with the causal dual system in this case. □

The system described above is time-flow-invertible but neither time-invertible (since the semigroup is not a group) nor flow-invertible (since the input and output dimensions are not the same). Note that both the system itself and the time-flow-inverted system are regular.

Finally, we observe that time-flow-inversion commutes with duality:

Theorem 6.5.11 *Let (Y, X, U) be reflexive Banach spaces, let $1 < p < \infty$, let Σ be an L^p-well-posed linear system on (Y, X, U), and let S be an operator node on (Y, X, U) (not necessarily the one induced by Σ).*

(i) *The system Σ is time-flow-invertible if and only if the causal dual system Σ^d is time-flow-invertible, in which case the time-flow-inverted systems satisfy $(\Sigma^d)^{\text{я}} = (\Sigma^{\text{я}})^d$.*
(ii) *The operator node S is time-flow-invertible if and only if the dual operator node S^* is time-flow-invertible, in which case the time-flow-inverted operator nodes satisfy $(S^*)^{\text{я}} = (S^{\text{я}})^*$.*

Thus, time-flow-inversion commutes with the duality transformation.

We leave the easy proof to the reader.

Remark 6.5.12 It is easy to find examples of well-posed systems Σ where none, or any one but not the other two, or any two but not the third one, or all three of the 'inverted' systems $\Sigma^{\text{я}}$, $\Sigma_{\times}$, and $\Sigma^{\text{я}}_{\times}$ exist. Indeed, all combinations are possible, as can be seen by comparing the different conditions for the existence of the different inverses. By inspecting (6.3.3), (6.4.3), and (6.5.2) we can draw some additional conclusions. For example, if Σ is both time-invertible and flow-invertible, then both the time-inverted and flow-inverted systems are time-flow-invertible, and they are time-flow-inverses of each other. Similar statements are true when Σ is both time-invertible and time-flow-invertible, or both flow-invertible and time-flow-invertible. Finally, if all three inverses $\Sigma^{\mathfrak{R}}$, $\Sigma_\times$, and $\Sigma_\times^{\mathfrak{R}}$ exist, then they are all time-invertible, flow-invertible, and time-flow-invertible, and a combination of any two of the inversions gives the third.

6.6 Partial flow-inversion

In Section 6.3 we introduced the operation of interchanging the input and the output of a system with each other, and called it flow-inversion. Here we shall look at a more general situation where we split both the input and the output into two parts, and only the second part of the input is interchanged with the second part of the output (the first parts of the input and output are not changed). We call this *partial flow-inversion*. (Another commonly used name is the *chain scattering transformation*.)

Let $\Sigma = \left[\begin{array}{c|cc} \mathfrak{A} & \mathfrak{B}_1 & \mathfrak{B}_2 \\ \hline \mathfrak{C}_1 & \mathfrak{D}_{11} & \mathfrak{D}_{12} \\ \mathfrak{C}_2 & \mathfrak{D}_{21} & \mathfrak{D}_{22} \end{array}\right]$ be a well-posed linear system on $\left(\left[\begin{smallmatrix} Y_1 \\ Y_2 \end{smallmatrix}\right], X, \left[\begin{smallmatrix} U_1 \\ U_2 \end{smallmatrix}\right]\right)$. Let x be the state trajectory and let $\left[\begin{smallmatrix} y_1 \\ y_2 \end{smallmatrix}\right]$ be the output function of this system with initial time zero, initial state x_0, and input function $\left[\begin{smallmatrix} u_1 \\ u_2 \end{smallmatrix}\right] \in L^p|Reg_{\text{loc}}(\overline{\mathbb{R}}^+; \left[\begin{smallmatrix} U_1 \\ U_2 \end{smallmatrix}\right])$, i.e.,

$$
\begin{aligned}
x(t) &= \mathfrak{A}^t x_0 + \mathfrak{B}_1 \tau^t \pi_+ u_1 + \mathfrak{B}_2 \tau^t \pi_+ u_2, \qquad t \geq 0, \\
y_1 &= \mathfrak{C}_1 x_0 + \mathfrak{D}_{11} \pi_+ u_1 + \mathfrak{D}_{12} \pi_+ u_2, \\
y_2 &= \mathfrak{C}_2 x_0 + \mathfrak{D}_{21} \pi_+ u_1 + \mathfrak{D}_{22} \pi_+ u_2,
\end{aligned} \tag{6.6.1}
$$

If $\mathfrak{D}_{22}$ has an inverse in $TIC_{\text{loc}}(Y_2; U_2)$, then we can solve for $x(t)$, y_1 and u_2 in terms of x_0, u_1, and y_2 to get for all $t \geq 0$,

$$
\begin{aligned}
\begin{bmatrix} x(t) \\ \pi_+ y_1 \\ \pi_+ u_2 \end{bmatrix} &= \begin{bmatrix} \mathfrak{A}^t & \mathfrak{B}_1 \tau^t & \mathfrak{B}_2 \tau^t \\ \mathfrak{C}_1 & \mathfrak{D}_{11} & \mathfrak{D}_{12} \\ 0 & 0 & 1 \end{bmatrix} \begin{bmatrix} x_0 \\ \pi_+ u_1 \\ \pi_+ u_2 \end{bmatrix} \\
&= \begin{bmatrix} \mathfrak{A}^t & \mathfrak{B}_1 \tau^t & \mathfrak{B}_2 \tau^t \\ \mathfrak{C}_1 & \mathfrak{D}_{11} & \mathfrak{D}_{12} \\ 0 & 0 & 1 \end{bmatrix} \begin{bmatrix} 1 & 0 & 0 \\ 0 & 1 & 0 \\ \mathfrak{C}_2 & \mathfrak{D}_{21} & \mathfrak{D}_{22} \end{bmatrix}^{-1} \begin{bmatrix} x_0 \\ \pi_+ u_1 \\ \pi_+ y_2 \end{bmatrix}.
\end{aligned} \tag{6.6.2}
$$

It should not come as a great surprise that this formula can be interpreted as a new well-posed linear system.

Theorem 6.6.1 *Let* $\Sigma = \left[\begin{array}{c|cc} \mathfrak{A} & \mathfrak{B}_1 & \mathfrak{B}_2 \\ \hline \mathfrak{C}_1 & \mathfrak{D}_{11} & \mathfrak{D}_{12} \\ \mathfrak{C}_2 & \mathfrak{D}_{21} & \mathfrak{D}_{22} \end{array}\right]$ *be a well-posed linear system on* $\left(\left[\begin{smallmatrix} Y_1 \\ Y_1 \end{smallmatrix}\right], X, \left[\begin{smallmatrix} U_1 \\ U_2 \end{smallmatrix}\right]\right)$, *and suppose that* $\mathfrak{D}_{22}$ *has an inverse in* $TIC_{\text{loc}}(Y_2; U_2)$. *Then* *the system*

$$
\begin{aligned}
\Sigma^{\curvearrowleft} &= \left[\begin{array}{c|cc} \mathfrak{A}^{\curvearrowleft} & \mathfrak{B}_1^{\curvearrowleft}\tau & \mathfrak{B}_2^{\curvearrowleft}\tau \\ \hline \mathfrak{C}_1^{\curvearrowleft} & \mathfrak{D}_{11}^{\curvearrowleft} & \mathfrak{D}_{12}^{\curvearrowleft} \\ \mathfrak{C}_2^{\curvearrowleft} & \mathfrak{D}_{21}^{\curvearrowleft} & \mathfrak{D}_{22}^{\curvearrowleft} \end{array}\right] \\
&= \left[\begin{array}{c|cc} \mathfrak{A} & \mathfrak{B}_1\tau & \mathfrak{B}_2\tau \\ \hline \mathfrak{C}_1 & \mathfrak{D}_{11} & \mathfrak{D}_{12} \\ 0 & 0 & 1 \end{array}\right] \left[\begin{array}{c|cc} 1 & 0 & 0 \\ \hline 0 & 1 & 0 \\ \mathfrak{C}_2 & \mathfrak{D}_{21} & \mathfrak{D}_{22} \end{array}\right]^{-1} \\
&= \left[\begin{array}{c|cc} 1 & 0 & -\mathfrak{B}_2\tau \\ \hline 0 & 1 & -\mathfrak{D}_{12} \\ 0 & 0 & \mathfrak{D}_{22} \end{array}\right]^{-1} \left[\begin{array}{c|cc} \mathfrak{A} & \mathfrak{B}_1\tau & 0 \\ \hline \mathfrak{C}_1 & \mathfrak{D}_{11} & 0 \\ -\mathfrak{C}_2 & -\mathfrak{D}_{21} & 1 \end{array}\right] \\
&= \left[\begin{array}{c|cc} \mathfrak{A}^t & \mathfrak{B}_1\tau^t & 0 \\ \hline \mathfrak{C}_1 & \mathfrak{D}_{11} & 0 \\ 0 & 0 & 0 \end{array}\right] + \left[\begin{array}{c} \mathfrak{B}_2\tau^t \\ \hline \mathfrak{D}_{12} \\ 1 \end{array}\right] \mathfrak{D}_{22}^{-1} \left[\begin{array}{c|cc} -\mathfrak{C}_2 & -\mathfrak{D}_{21} & 1 \end{array}\right]
\end{aligned}
\qquad (6.6.3)
$$

is a linear system on $\left(\left[\begin{smallmatrix} Y_1 \\ U_2 \end{smallmatrix}\right], X, \left[\begin{smallmatrix} U_1 \\ Y_2 \end{smallmatrix}\right]\right)$ *which is well-posed in the same sense as* Σ. *If* x *and* $\left[\begin{smallmatrix} y_1 \\ y_2 \end{smallmatrix}\right]$ *are the state trajectory and output function of* Σ *with initial time* $s \in \mathbb{R}$, *initial state* $x_s \in X$, *and input function* $\left[\begin{smallmatrix} u_1 \\ u_2 \end{smallmatrix}\right] \in L^p|Reg_{\mathrm{loc}}([s,\infty);\left[\begin{smallmatrix} U_1 \\ U_2 \end{smallmatrix}\right])$, *then* x *and* $\left[\begin{smallmatrix} y_1 \\ u_2 \end{smallmatrix}\right]$ *are the state trajectory and output function of* $\Sigma^{\curvearrowleft}$ *with the same initial time* s, *the same initial state* x_s, *and input function* $\left[\begin{smallmatrix} u_1 \\ y_2 \end{smallmatrix}\right]$. *In particular, for* $s = 0$ *we have*

$$
\begin{bmatrix} x(t) \\ \pi_+ y_1 \\ \pi_+ u_2 \end{bmatrix} = \begin{bmatrix} \mathfrak{A}^{\curvearrowleft} & \mathfrak{B}_1^{\curvearrowleft}\tau & \mathfrak{B}_2^{\curvearrowleft}\tau \\ \mathfrak{C}_1^{\curvearrowleft} & \mathfrak{D}_{11}^{\curvearrowleft} & \mathfrak{D}_{12}^{\curvearrowleft} \\ \mathfrak{C}_2^{\curvearrowleft} & \mathfrak{D}_{21}^{\curvearrowleft} & \mathfrak{D}_{22}^{\curvearrowleft} \end{bmatrix} \begin{bmatrix} x(t) \\ \pi_+ u_1 \\ \pi_+ y_2 \end{bmatrix}, \qquad t \ge 0. \qquad (6.6.4)
$$

We leave the proof of this theorem to the reader. The most obvious (and easiest) way is to imitate the proof of Theorem 6.3.1. Another possibility is to absorb the first input u_1 and the first output y_1 into the semigroup, i.e., to write $\left[\begin{smallmatrix} \mathfrak{A} & \mathfrak{B}_1 \\ \mathfrak{C}_1 & \mathfrak{D}_{11} \end{smallmatrix}\right]$ as a Lax–Phillips semigroup, and to then apply Theorem 6.3.1 to the new system whose state is $\left[\begin{smallmatrix} y_1 \\ x \\ u_1 \end{smallmatrix}\right]$, with input function u_2 and output function y_2.

Definition 6.6.2 We say that the $L^p|Reg$-well-posed linear system $\Sigma = \left[\begin{smallmatrix} \mathfrak{A} & \mathfrak{B}_1 & \mathfrak{B}_2 \\ \mathfrak{C}_1 & \mathfrak{D}_{11} & \mathfrak{D}_{12} \\ \mathfrak{C}_2 & \mathfrak{D}_{21} & \mathfrak{D}_{22} \end{smallmatrix}\right]$ is *partially flow-invertible* if $\mathfrak{D}_{22}$ is invertible in TIC_{loc}, and we call the system $\Sigma^{\curvearrowleft}$ in Theorem 6.6.1 the *partial flow-inverse* of Σ.

Corollary 6.6.3 *The $L^p|Reg$-well-posed linear system* $\left[\begin{array}{c|cc}\mathfrak{A} & \mathfrak{B}_1 & \mathfrak{B}_2 \\ \hline \mathfrak{C}_1 & \mathfrak{D}_{11} & \mathfrak{D}_{12} \\ \mathfrak{C}_2 & \mathfrak{D}_{21} & \mathfrak{D}_{22}\end{array}\right]$ *is partially flow-invertible if and only if the system* $\left[\begin{array}{c|c}\mathfrak{A} & \mathfrak{B}_2 \\ \hline \mathfrak{C}_2 & \mathfrak{D}_{22}\end{array}\right]$ *is flow-invertible, and, with the notation of Theorems 6.3.1 and 6.6.1,*

$$\left[\begin{array}{c|c}\mathfrak{A}_\times & \mathfrak{B}_\times \\ \hline \mathfrak{C}_{\times 2} & \mathfrak{D}_{\times 22}\end{array}\right] = \left[\begin{array}{c|c}\mathfrak{A}^{\curvearrowleft} & \mathfrak{B}^{\curvearrowleft} \\ \hline \mathfrak{C}_2^{\curvearrowleft} & \mathfrak{D}_{22}^{\curvearrowleft}\end{array}\right].$$

This follows from Definitions 6.3.3 and 6.6.2.

Remark 6.6.4 In the classical system

$$\begin{aligned}\dot{x}(t) &= Ax(t) + B_1u_1(t) + B_2u_2(t), \\ y_1(t) &= C_1x(t) + D_{11}u_1(t) + D_{12}u_2(t), \\ y_2(t) &= C_2x(t) + D_{21}u_1(t) + D_{22}u_2(t), \qquad t \ge 0, \\ x(0) &= x_0,\end{aligned} \tag{6.6.5}$$

it is possible to interpret $\left[\begin{smallmatrix}u_1 \\ y_2\end{smallmatrix}\right]$ as the input function and $\left[\begin{smallmatrix}y_1 \\ u_2\end{smallmatrix}\right]$ as the output function if and only if D_{22} is invertible. In this case the partially flow-inverted system $\Sigma^{\curvearrowleft}$ is again a classical system with generators

$$\begin{aligned}\begin{bmatrix}A^{\curvearrowleft} & B_1^{\curvearrowleft} & B_2^{\curvearrowleft} \\ C_1^{\curvearrowleft} & D_{11}^{\curvearrowleft} & D_{12}^{\curvearrowleft} \\ C_2^{\curvearrowleft} & D_{21}^{\curvearrowleft} & D_{22}^{\curvearrowleft}\end{bmatrix} &= \begin{bmatrix}A & B_1 & B_2 \\ C_1 & D_{11} & D_{12} \\ 0 & 0 & 1\end{bmatrix}\begin{bmatrix}1 & 0 & 0 \\ 0 & 1 & 0 \\ C_2 & D_{21} & D_{22}\end{bmatrix}^{-1} \\ &= \begin{bmatrix}1 & 0 & -B_2 \\ 0 & 1 & -D_{12} \\ 0 & 0 & D_{22}\end{bmatrix}^{-1}\begin{bmatrix}A & B_1 & 0 \\ C_1 & D_{11} & 0 \\ -C_2 & -D_{21} & 1\end{bmatrix} \\ &= \begin{bmatrix}A & B_1 & 0 \\ C_1 & D_{11} & 0 \\ 0 & 0 & 0\end{bmatrix} + \begin{bmatrix}B_2 \\ D_{12} \\ 1\end{bmatrix} D_{22}^{-1}\begin{bmatrix}-C_2 & -D_{21} & 1\end{bmatrix}.\end{aligned} \tag{6.6.6}$$

Observe the striking similarity between this formula and the one given in Theorem 6.6.1. Operator node versions of this result are given in Theorems 6.6.5 and 6.6.17.

Let us next compute the system node of the partially flow-inverted system.

Theorem 6.6.5 *Let* $\Sigma = \left[\begin{array}{c|cc}\mathfrak{A} & \mathfrak{B}_1 & \mathfrak{B}_2 \\ \hline \mathfrak{C}_1 & \mathfrak{D}_{11} & \mathfrak{D}_{12} \\ \mathfrak{C}_2 & \mathfrak{D}_{21} & \mathfrak{D}_{22}\end{array}\right]$ *be a well-posed partially flow-invertible linear system on* $\left(\left[\begin{smallmatrix}Y_1 \\ Y_2\end{smallmatrix}\right], X, \left[\begin{smallmatrix}U_1 \\ U_2\end{smallmatrix}\right]\right)$ *with system node* $S = \left[\begin{smallmatrix}A\&B \\ [C\&D]_1 \\ [C\&D]_2\end{smallmatrix}\right]$. *Denote the system node of the partially flow-inverted system by* $S^{\curvearrowleft} =$

$\begin{bmatrix} [A\&B]^{\curvearrowleft} \\ [C\&D]_1^{\curvearrowleft} \\ [C\&D]_2^{\curvearrowleft} \end{bmatrix}$. *Then the operator* $\begin{bmatrix} 1 & 0 & 0 \\ 0 & 1 & 0 \\ & [C\&D]_2 & \end{bmatrix}$ *maps* $\mathcal{D}(S)$ *continuously onto* $\mathcal{D}(S^{\curvearrowleft})$, *its inverse is* $\begin{bmatrix} 1 & 0 & 0 \\ 0 & 1 & 0 \\ & [C\&D]_2^{\curvearrowleft} & \end{bmatrix}$, *and*

$$S = \begin{bmatrix} [A\&B]^{\curvearrowleft} \\ [C\&D]_1^{\curvearrowleft} \\ 0\;0\;1 \end{bmatrix} \begin{bmatrix} 1\;0\;0 \\ 0\;1\;0 \\ [C\&D]_2^{\curvearrowleft} \end{bmatrix}^{-1} \quad (\text{on } \mathcal{D}(S)), \tag{6.6.7}$$

$$S^{\curvearrowleft} = \begin{bmatrix} A\&B \\ [C\&D]_1 \\ 0\;0\;1 \end{bmatrix} \begin{bmatrix} 1\;0\;0 \\ 0\;1\;0 \\ [C\&D]_2 \end{bmatrix}^{-1} \quad (\text{on } \mathcal{D}(S^{\curvearrowleft})). \tag{6.6.8}$$

We leave this proof, too, to the reader (it is very similar to the proof of Theorem 6.3.6).

Motivated by Theorem 6.6.5 we extend Definition 6.6.2 to arbitrary operator nodes as follows.

Definition 6.6.6 Let $S = \begin{bmatrix} A\&B \\ [C\&D]_1 \\ [C\&D]_2 \end{bmatrix}$ be an operator node on $\left(\begin{bmatrix} Y_1 \\ Y_2 \end{bmatrix}, X, \begin{bmatrix} U_1 \\ U_2 \end{bmatrix}\right)$. We call this operator node *partially flow-invertible* if there exists an operator node $S^{\curvearrowleft} = \begin{bmatrix} [A\&B]^{\curvearrowleft} \\ [C\&D]_1^{\curvearrowleft} \\ [C\&D]_2^{\curvearrowleft} \end{bmatrix}$ on $\left(\begin{bmatrix} Y_1 \\ U_2 \end{bmatrix}, X, \begin{bmatrix} U_1 \\ Y_2 \end{bmatrix}\right)$ which together with S satisfies the following conditions: the operator $\begin{bmatrix} 1 & 0 & 0 \\ 0 & 1 & 0 \\ & [C\&D]_2 & \end{bmatrix}$ maps $\mathcal{D}(S)$ continuously onto $\mathcal{D}(S^{\curvearrowleft})$, its inverse is $\begin{bmatrix} 1 & 0 & 0 \\ 0 & 1 & 0 \\ & [C\&D]_2^{\curvearrowleft} & \end{bmatrix}$, and (6.6.7)–(6.6.8) hold. In this case we call S and $S^{\curvearrowleft}$ *partial flow-inverses* of each other.

We again first look at some immediate consequences of this lemma.

Lemma 6.6.7 *Let* $S = \begin{bmatrix} A\&B \\ [C\&D]_1 \\ [C\&D]_2 \end{bmatrix}$ *be an operator node on* $\left(\begin{bmatrix} Y_1 \\ Y_2 \end{bmatrix}, X, \begin{bmatrix} U_1 \\ U_2 \end{bmatrix}\right)$, *with main operator* A, *control operator* $B = \begin{bmatrix} B_1 & B_2 \end{bmatrix}$, *observation operator* $C = \begin{bmatrix} C_1 \\ C_2 \end{bmatrix}$, *and transfer function* $\widehat{\mathfrak{D}} = \begin{bmatrix} \widehat{\mathfrak{D}}_1 \\ \widehat{\mathfrak{D}}_2 \end{bmatrix} = \begin{bmatrix} \widehat{\mathfrak{D}}_{11} & \widehat{\mathfrak{D}}_{12} \\ \widehat{\mathfrak{D}}_{21} & \widehat{\mathfrak{D}}_{22} \end{bmatrix}$, *and let* $S^{\curvearrowleft} = \begin{bmatrix} [A\&B]^{\curvearrowleft} \\ [C\&D]_1^{\curvearrowleft} \\ [C\&D]_2^{\curvearrowleft} \end{bmatrix}$ *be its flow-inverse, with main operator* $A^{\curvearrowleft}$, *control operator* $B^{\curvearrowleft} = \begin{bmatrix} B_1^{\curvearrowleft} & B_2^{\curvearrowleft} \end{bmatrix}$, *observation operator* $C^{\curvearrowleft} = \begin{bmatrix} C_1^{\curvearrowleft} \\ C_2^{\curvearrowleft} \end{bmatrix}$, *and transfer function* $\widehat{\mathfrak{D}}^{\curvearrowleft} = \begin{bmatrix} \widehat{\mathfrak{D}}_1^{\curvearrowleft} \\ \widehat{\mathfrak{D}}_2^{\curvearrowleft} \end{bmatrix} = \begin{bmatrix} \widehat{\mathfrak{D}}_{11}^{\curvearrowleft} & \widehat{\mathfrak{D}}_{12}^{\curvearrowleft} \\ \widehat{\mathfrak{D}}_{21}^{\curvearrowleft} & \widehat{\mathfrak{D}}_{22}^{\curvearrowleft} \end{bmatrix}$. *Then the following claims are true.*

(i) $\alpha \in \rho(A^{\circlearrowleft})$ *if and only if* $\left[\begin{smallmatrix} \alpha & 0 & 0 \\ 0 & 1 & 0 \\ 0 & 0 & 0 \end{smallmatrix}\right] - \left[\begin{smallmatrix} A\&B \\ 0 \\ [C\&D]_2 \end{smallmatrix}\right]$ *is invertible, in which case*

$$
\begin{bmatrix} 1\,0\,0 \\ [C\&D]_1 \\ 0\,0\,1 \end{bmatrix} \left(\begin{bmatrix} \alpha & 0 & 0 \\ 0 & 1 & 0 \\ 0 & 0 & 0 \end{bmatrix} - \begin{bmatrix} A\&B \\ 0 \\ [C\&D]_2 \end{bmatrix} \right)^{-1}
= \begin{bmatrix} (\alpha - A^{\circlearrowleft})^{-1} & (\alpha - A^{\circlearrowleft}_{|X})^{-1} B^{\circlearrowleft}_1 & -(\alpha - A^{\circlearrowleft}_{|X})^{-1} B^{\circlearrowleft}_2 \\ C^{\circlearrowleft}_1 (\alpha - A^{\circlearrowleft})^{-1} & \widehat{\mathfrak{D}}^{\circlearrowleft}_{11}(\alpha) & -\widehat{\mathfrak{D}}^{\circlearrowleft}_{12}(\alpha) \\ C^{\circlearrowleft}_2 (\alpha - A^{\circlearrowleft})^{-1} & \widehat{\mathfrak{D}}^{\circlearrowleft}_{21}(\alpha) & -\widehat{\mathfrak{D}}^{\circlearrowleft}_{22}(\alpha) \end{bmatrix}. \tag{6.6.9}
$$

(ii) $\alpha \in \rho(A)$ *if and only if* $\left[\begin{smallmatrix} \alpha & 0 & 0 \\ 0 & 1 & 0 \\ 0 & 0 & 0 \end{smallmatrix}\right] - \left[\begin{smallmatrix} [A\&B]^{\circlearrowleft} \\ 0 \\ [C\&D]^{\circlearrowleft}_2 \end{smallmatrix}\right]$ *is invertible, in which case*

$$
\begin{bmatrix} 1\,0\,0 \\ [C\&D]^{\circlearrowleft}_1 \\ 0\,0\,1 \end{bmatrix} \left(\begin{bmatrix} \alpha & 0 & 0 \\ 0 & 1 & 0 \\ 0 & 0 & 0 \end{bmatrix} - \begin{bmatrix} [A\&B]^{\circlearrowleft} \\ 0 \\ [C\&D]^{\circlearrowleft}_2 \end{bmatrix} \right)^{-1}
= \begin{bmatrix} (\alpha - A)^{-1} & (\alpha - A_{|X})^{-1} B_1 & -(\alpha - A_{|X})^{-1} B_2 \\ C_1 (\alpha - A)^{-1} & \widehat{\mathfrak{D}}_{11}(\alpha) & -\widehat{\mathfrak{D}}_{12}(\alpha) \\ C_2 (\alpha - A)^{-1} & \widehat{\mathfrak{D}}_{21}(\alpha) & -\widehat{\mathfrak{D}}_{22}(\alpha) \end{bmatrix}. \tag{6.6.10}
$$

(iii) *If* $\alpha \in \rho(A)$, *then* $\alpha \in \rho(A^{\circlearrowleft})$ *if and only if* $\widehat{\mathfrak{D}}_{22}(\alpha)$ *is invertible. In this case*

$$
\begin{aligned}
&\begin{bmatrix} (\alpha - A^{\circlearrowleft})^{-1} & (\alpha - A^{\circlearrowleft}_{|X})^{-1} B^{\circlearrowleft}_1 & -(\alpha - A^{\circlearrowleft}_{|X})^{-1} B^{\circlearrowleft}_2 \\ C^{\circlearrowleft}_1 (\alpha - A^{\circlearrowleft})^{-1} & \widehat{\mathfrak{D}}^{\circlearrowleft}_{11}(\alpha) & -\widehat{\mathfrak{D}}^{\circlearrowleft}_{12}(\alpha) \\ C^{\circlearrowleft}_2 (\alpha - A^{\circlearrowleft})^{-1} & \widehat{\mathfrak{D}}^{\circlearrowleft}_{21}(\alpha) & -\widehat{\mathfrak{D}}^{\circlearrowleft}_{22}(\alpha) \end{bmatrix} \\
&= \begin{bmatrix} (\alpha - A)^{-1} & (\alpha - A_{|X})^{-1} B_1 & 0 \\ C_1 (\alpha - A)^{-1} & \widehat{\mathfrak{D}}_{11}(\alpha) & 0 \\ 0 & 0 & 0 \end{bmatrix} \\
&\quad - \begin{bmatrix} (\alpha - A_{|X})^{-1} B_2 \\ \widehat{\mathfrak{D}}_{12}(\alpha) \\ 1 \end{bmatrix} [\widehat{\mathfrak{D}}_{22}(\alpha)]^{-1} \begin{bmatrix} C_2 (\alpha - A)^{-1} & \widehat{\mathfrak{D}}_{21}(\alpha) & 1 \end{bmatrix}.
\end{aligned} \tag{6.6.11}
$$

(iv) *If $\alpha \in \rho(A^{\curvearrowleft})$, then $\alpha \in \rho(A)$ if and only if $\widehat{\mathfrak{D}}^{\curvearrowleft}_{22}(\alpha)$ is invertible. In this case*

$$
\begin{aligned}
&\begin{bmatrix} (\alpha - A)^{-1} & (\alpha - A_{|X})^{-1}B_1 & -(\alpha - A_{|X})^{-1}B_2 \\ C_1(\alpha - A)^{-1} & \widehat{\mathfrak{D}}_{11}(\alpha) & -\widehat{\mathfrak{D}}_{12}(\alpha) \\ C_2(\alpha - A)^{-1} & \widehat{\mathfrak{D}}_{21}(\alpha) & -\widehat{\mathfrak{D}}_{22}(\alpha) \end{bmatrix} \\
&\quad = \begin{bmatrix} (\alpha - A^{\curvearrowleft})^{-1} & (\alpha - A^{\curvearrowleft}_{|X})^{-1}B^{\curvearrowleft}_1 & 0 \\ C^{\curvearrowleft}_1(\alpha - A^{\curvearrowleft})^{-1} & \widehat{\mathfrak{D}}^{\curvearrowleft}_{11}(\alpha) & 0 \\ 0 & 0 & 0 \end{bmatrix} \\
&\quad\quad - \begin{bmatrix} (\alpha - A^{\curvearrowleft}_{|X})^{-1}B^{\curvearrowleft}_2 \\ \widehat{\mathfrak{D}}^{\curvearrowleft}_{12}(\alpha) \\ 1 \end{bmatrix} [\widehat{\mathfrak{D}}^{\curvearrowleft}_{22}(\alpha)]^{-1} \begin{bmatrix} C^{\curvearrowleft}_2(\alpha - A^{\curvearrowleft})^{-1} & \widehat{\mathfrak{D}}^{\curvearrowleft}_{21}(\alpha) & 1 \end{bmatrix}.
\end{aligned}
\tag{6.6.12}
$$

Proof (i) By (6.6.7),

$$
\begin{aligned}
&\begin{bmatrix} \alpha & 0 & 0 \\ 0 & 1 & 0 \\ 0 & 0 & 0 \end{bmatrix} - \begin{bmatrix} A\&B \\ 0 \\ [C\&D]_2 \end{bmatrix} \\
&\quad = \left(\begin{bmatrix} \alpha & 0 & 0 \\ 0 & 1 & 0 \\ 0 & 0 & -1 \end{bmatrix} - \begin{bmatrix} [A\&B]^{\curvearrowleft} \\ 0 \\ 0 \end{bmatrix} \right) \begin{bmatrix} 1\,0\,0 \\ 0\,1\,0 \\ [C\&D]^{\curvearrowleft}_2 \end{bmatrix}^{-1}.
\end{aligned}
\tag{6.6.13}
$$

The second factor on the right-hand side is a bounded bijection between $\mathcal{D}(S)$ and $\mathcal{D}(S^{\curvearrowleft})$, and the first factor on the right-hand side has a bounded inverse defined on $\left[\begin{smallmatrix} X \\ U_1 \\ U_2 \end{smallmatrix}\right]$ if and only if $\alpha \in \rho(A^{\curvearrowleft})$ (see Lemma 4.7.18(i)). Thus, $\left[\begin{smallmatrix} \alpha & 0 & 0 \\ 0 & 1 & 0 \\ 0 & 0 & 0 \end{smallmatrix}\right] - \left[\begin{smallmatrix} A\&B \\ 0 \\ [C\&D]_2 \end{smallmatrix}\right]$ is invertible if and only if $\alpha \in \rho(A^{\curvearrowleft})$. Inverting (6.6.13) we get

$$
\begin{aligned}
&\begin{bmatrix} 1\,0\,0 \\ [C\&D]_1 \\ 0\,0\,1 \end{bmatrix} \left(\begin{bmatrix} \alpha & 0 & 0 \\ 0 & 1 & 0 \\ 0 & 0 & 0 \end{bmatrix} - \begin{bmatrix} A\&B \\ 0 \\ [C\&D]_2 \end{bmatrix} \right)^{-1} \\
&\quad = \begin{bmatrix} 1\,0\,0 \\ [C\&D]_1 \\ 0\,0\,1 \end{bmatrix} \begin{bmatrix} 1\,0\,0 \\ 0\,1\,0 \\ [C\&D]^{\curvearrowleft}_2 \end{bmatrix} \begin{bmatrix} \alpha - A^{\curvearrowleft}_{|X} & -B^{\curvearrowleft}_1 & -B^{\curvearrowleft}_1 \\ 0 & 1 & 0 \\ 0 & 0 & -1 \end{bmatrix}^{-1},
\end{aligned}
$$

where, according to (6.6.8), $\left[\begin{smallmatrix} 1\,0\,0 \\ [C\&D]_1 \\ 0\,0\,1 \end{smallmatrix}\right] \left[\begin{smallmatrix} 1\,0\,0 \\ 0\,1\,0 \\ [C\&D]^{\curvearrowleft}_2 \end{smallmatrix}\right] = \left[\begin{smallmatrix} 1\,0\,0 \\ [C\&D]^{\curvearrowleft}_1 \\ [C\&D]^{\curvearrowleft}_2 \end{smallmatrix}\right]$. We get formula (6.6.9) from Lemma 4.7.18(i) with S replaced by $S^{\curvearrowleft}$.

(ii) We get (ii) from (i) by interchanging S and $S^{\curvearrowleft}$.

(iii) Assume that $\alpha \in \rho(A)$. Then (cf. Lemma 4.7.18(iii))

$$\left(\begin{bmatrix}\alpha & 0 & 0\\ 0 & 1 & 0\\ 0 & 0 & 0\end{bmatrix} - \begin{bmatrix} A\&B \\ 0 \\ [C\&D]_2\end{bmatrix}\right)\left(\begin{bmatrix}\alpha & 0 & 0\\ 0 & 1 & 0\\ 0 & 0 & 1\end{bmatrix} - \begin{bmatrix} A\&B \\ 0 \\ 0\end{bmatrix}\right)^{-1}$$
$$= \begin{bmatrix} 1 & 0 & 0 \\ 0 & 1 & 0 \\ -C_2(\alpha - A)^{-1} & -\widehat{\mathfrak{D}}_{21}(\alpha) & -\widehat{\mathfrak{D}}_{21}(\alpha)\end{bmatrix},$$

where the second factor on the left-hand side maps $\left[\begin{smallmatrix} X \\ U_1 \\ U_2\end{smallmatrix}\right]$ one-to-one onto $\mathcal{D}(S)$. Thus, it follows from (i) that $\alpha \in \rho(A^{\curvearrowleft})$ if and only if $\widehat{\mathfrak{D}}_{22}(\alpha)$ is invertible in this case. We leave the computation which shows that the left-hand side of (6.6.9) is equal to the right-hand side of (6.6.11) to the reader (cf. Lemma 4.7.18).

(iv) We get (iv) from (iii) by interchanging S and $S^{\curvearrowleft}$. □

Our following theorem lists a number of equivalent conditions for two operator nodes to be partial flow-inverses of each other.

Theorem 6.6.8 *Let* $S = \left[\begin{smallmatrix} A\&B \\ [C\&D]_1 \\ [C\&D]_2\end{smallmatrix}\right]$ *be an operator node on* $\left(\left[\begin{smallmatrix} Y_1 \\ Y_2\end{smallmatrix}\right], X, \left[\begin{smallmatrix} U_1 \\ U_2\end{smallmatrix}\right]\right)$, *with main operator* A, *control operator* $B = \begin{bmatrix} B_1 & B_2\end{bmatrix}$, *observation operator* $C = \left[\begin{smallmatrix} C_1 \\ C_2\end{smallmatrix}\right]$, *and transfer function* $\widehat{\mathfrak{D}} = \left[\begin{smallmatrix} \widehat{\mathfrak{D}}_{11} & \widehat{\mathfrak{D}}_{12} \\ \widehat{\mathfrak{D}}_{21} & \widehat{\mathfrak{D}}_{22}\end{smallmatrix}\right]$, *and let* $S^{\curvearrowleft} = \left[\begin{smallmatrix} [A\&B]^{\curvearrowleft} \\ [C\&D]_1^{\curvearrowleft} \\ [C\&D]_2^{\curvearrowleft}\end{smallmatrix}\right]$ *be an operator node on* $\left(\left[\begin{smallmatrix} Y_1 \\ U_2\end{smallmatrix}\right], X, \left[\begin{smallmatrix} U_1 \\ Y_2\end{smallmatrix}\right]\right)$, *with main operator* $A^{\curvearrowleft}$, *control operator* $B^{\curvearrowleft} = \begin{bmatrix} B_1^{\curvearrowleft} & B_2^{\curvearrowleft}\end{bmatrix}$, *observation operator* $C^{\curvearrowleft} = \left[\begin{smallmatrix} C_1^{\curvearrowleft} \\ C_2^{\curvearrowleft}\end{smallmatrix}\right]$, *and transfer function* $\widehat{\mathfrak{D}}^{\curvearrowleft} = \left[\begin{smallmatrix} \widehat{\mathfrak{D}}_{11}^{\curvearrowleft} & \widehat{\mathfrak{D}}_{12}^{\curvearrowleft} \\ \widehat{\mathfrak{D}}_{21}^{\curvearrowleft} & \widehat{\mathfrak{D}}_{22}^{\curvearrowleft}\end{smallmatrix}\right]$. *Then the following conditions are equivalent:*

(i) *S and $S^{\curvearrowleft}$ are partial flow-inverses of each other.*

(ii) *The operator* $\left[\begin{smallmatrix} 1\ 0\ 0 \\ 0\ 1\ 0 \\ [C\&D]_2^{\curvearrowleft}\end{smallmatrix}\right]$ *maps* $\mathcal{D}(S^{\curvearrowleft})$ *one-to-one onto* $\mathcal{D}(S)$ *and (6.6.7) holds.*

(iii) *For all* $\alpha \in \rho(A^{\curvearrowleft})$, *the operator* $\left[\begin{smallmatrix} \alpha & 0 & 0 \\ 0 & 1 & 0 \\ 0 & 0 & 0\end{smallmatrix}\right] - \left[\begin{smallmatrix} A\&B \\ 0 \\ [C\&D]_2\end{smallmatrix}\right]$ *maps* $\mathcal{D}(S)$ *one-to-one onto* $\left[\begin{smallmatrix} X \\ U_1 \\ Y_2\end{smallmatrix}\right]$ *and (6.6.9) holds.*

(iv) *For some* $\alpha \in \rho(A^{\curvearrowleft})$, *the operator* $\left[\begin{smallmatrix} \alpha & 0 & 0 \\ 0 & 1 & 0 \\ 0 & 0 & 0\end{smallmatrix}\right] - \left[\begin{smallmatrix} A\&B \\ 0 \\ [C\&D]_2\end{smallmatrix}\right]$ *maps* $\mathcal{D}(S)$ *one-to-one onto* $\left[\begin{smallmatrix} X \\ U_1 \\ Y_2\end{smallmatrix}\right]$ *and (6.6.9) holds.*

(v) *The operator* $\left[\begin{smallmatrix} 1\ 0\ 0 \\ 0\ 1\ 0 \\ [C\&D]_2\end{smallmatrix}\right]$ *maps* $\mathcal{D}(S)$ *one-to-one onto* $\mathcal{D}(S^{\curvearrowleft})$ *and (6.6.8) holds.*

(vi) *For all* $\alpha \in \rho(A)$, *the operator* $\left[\begin{smallmatrix} \alpha & 0 & 0 \\ 0 & 1 & 0 \\ 0 & 0 & 0\end{smallmatrix}\right] - \left[\begin{smallmatrix} [A\&B]^{\curvearrowleft} \\ 0 \\ [C\&D]_2^{\curvearrowleft}\end{smallmatrix}\right]$ *maps* $\mathcal{D}(S^{\curvearrowleft})$ *one-to-one onto* $\left[\begin{smallmatrix} X \\ U_1 \\ U_2\end{smallmatrix}\right]$ *and (6.6.10) holds.*

(vii) *For some* $\alpha \in \rho(A)$, *the operator* $\begin{bmatrix} \alpha & 0 & 0 \\ 0 & 1 & 0 \\ 0 & 0 & 0 \end{bmatrix} - \begin{bmatrix} [A\&B]^{\curvearrowleft} \\ 0 \\ [C\&D]_2^{\curvearrowleft} \end{bmatrix}$ *maps* $\mathcal{D}(S^{\curvearrowleft})$ *one-to-one onto* $\begin{bmatrix} X \\ U_1 \\ U_2 \end{bmatrix}$ *and* (6.6.10) *holds.*

When these equivalent conditions hold, then $\begin{bmatrix} 1 \\ 0 \\ C_2 \end{bmatrix}$ *maps* $\mathcal{D}(A)$ *into* $\mathcal{D}(S^{\curvearrowleft})$, $\begin{bmatrix} 1 \\ 0 \\ C_2^{\curvearrowleft} \end{bmatrix}$ *maps* $\mathcal{D}(A^{\curvearrowleft})$ *into* $\mathcal{D}(S)$, *and*

$$\begin{aligned} A &= A^{\curvearrowleft}_{|\mathcal{D}(A)} + B_2^{\curvearrowleft} C_2, & A^{\curvearrowleft} &= A_{|\mathcal{D}(A^{\curvearrowleft})} + B_2 C_2^{\curvearrowleft}, \\ \begin{bmatrix} C_1 \\ 0 \end{bmatrix} &= \begin{bmatrix} [C\&D]_1^{\curvearrowleft} \\ [C\&D]_2^{\curvearrowleft} \end{bmatrix} \begin{bmatrix} 1 \\ 0 \\ C_2 \end{bmatrix}, & \begin{bmatrix} C_1^{\curvearrowleft} \\ 0 \end{bmatrix} &= \begin{bmatrix} [C\&D]_1 \\ [C\&D]_2 \end{bmatrix} \begin{bmatrix} 1 \\ 0 \\ C_2^{\curvearrowleft} \end{bmatrix}. \end{aligned} \tag{6.6.14}$$

Proof It suffices to prove that (i)–(iv) are equivalent, because the remaining equivalences follow from the facts that (i) is symmetric with respect to S and $S^{\curvearrowleft}$, and that we get (v)–(vii) by interchanging S and $S^{\curvearrowleft}$ in (ii)–(iv). Also observe that (6.6.14), which is equivalent to

$$\begin{bmatrix} [A\&B]^{\curvearrowleft} \\ [C\&D]_1^{\curvearrowleft} \\ [C\&D]_2^{\curvearrowleft} \end{bmatrix} \begin{bmatrix} 1 \\ 0 \\ C_2 \end{bmatrix} = \begin{bmatrix} A \\ C_1 \\ 0 \end{bmatrix}, \quad \begin{bmatrix} A\&B \\ [C\&D]_1 \\ [C\&D]_2 \end{bmatrix} \begin{bmatrix} 1 \\ 0 \\ C_2^{\curvearrowleft} \end{bmatrix} = \begin{bmatrix} A^{\curvearrowleft} \\ C_1^{\curvearrowleft} \\ 0 \end{bmatrix}, \tag{6.6.15}$$

follows from (i) and (6.6.7)–(6.6.8) since $\begin{bmatrix} \mathcal{D}(A) \\ 0 \\ 0 \end{bmatrix} \in \mathcal{D}(S)$ and $\begin{bmatrix} \mathcal{D}(A^{\curvearrowleft}) \\ 0 \\ 0 \end{bmatrix} \in \mathcal{D}(S^{\curvearrowleft})$.

(i) $\Rightarrow$ (ii): This is obvious (see Definition 6.6.6).

(ii) $\Rightarrow$ (i): This proof is very similar to the proof of the corresponding implication in Theorem 6.3.9.

(i) $\Rightarrow$ (iii): See Lemma 6.6.7(i).

(iii) $\Rightarrow$ (iv): This is obvious.

(iv) $\Rightarrow$ (ii): Clearly (6.6.9) implies that

$$\begin{aligned} &\left(\begin{bmatrix} \alpha & 0 & 0 \\ 0 & 1 & 0 \\ 0 & 0 & 0 \end{bmatrix} - \begin{bmatrix} A\&B \\ 0 \\ [C\&D]_2 \end{bmatrix} \right)^{-1} \begin{bmatrix} 1 & 0 & 0 \\ 0 & 1 & 0 \\ 0 & 0 & -1 \end{bmatrix} \\ &= \begin{bmatrix} (\alpha - A^{\curvearrowleft})^{-1} & (\alpha - A^{\curvearrowleft}_{|X})^{-1} B_1^{\curvearrowleft} & (\alpha - A^{\curvearrowleft}_{|X})^{-1} B_2^{\curvearrowleft} \\ Z_0 & Z_1 & Z_2 \\ C_2^{\curvearrowleft} (\alpha - A^{\curvearrowleft})^{-1} & \widehat{\mathfrak{D}}_{21}^{\curvearrowleft}(\alpha) & \widehat{\mathfrak{D}}_{22}^{\curvearrowleft}(\alpha) \end{bmatrix}, \end{aligned}$$

where $Z_0 \in \mathcal{B}(X)$, $Z_1 \in \mathcal{B}(U_1; X)$, and $Z_2 \in \mathcal{B}(Y_2; X)$. Multiplying this identity by $\begin{bmatrix} 0 & 1 & 0 \end{bmatrix}$ to the left we get $u_1 = Z_0 x + Z_1 u_1 + Z_3 y_2$ for all $\begin{bmatrix} x \\ u_1 \\ y_2 \end{bmatrix} \in \begin{bmatrix} X \\ U_1 \\ Y_2 \end{bmatrix}$.

Thus, $Z_0 = 0$, $Z_1 = 1$, and $Z_2 = 0$, and we find that

$$
\begin{aligned}
&\left(\begin{bmatrix} \alpha & 0 & 0 \\ 0 & 1 & 0 \\ 0 & 0 & 0 \end{bmatrix} - \begin{bmatrix} A\&B \\ 0 \\ [C\&D]_2 \end{bmatrix}\right)^{-1} \begin{bmatrix} 1 & 0 & 0 \\ 0 & 1 & 0 \\ 0 & 0 & -1 \end{bmatrix} \\
&= \begin{bmatrix} (\alpha - A^{\curvearrowleft})^{-1} & (\alpha - A^{\curvearrowleft}_{|X})^{-1} B_1^{\curvearrowleft} & (\alpha - A^{\curvearrowleft}_{|X})^{-1} B_2^{\curvearrowleft} \\ 0 & 1 & 0 \\ C_2^{\curvearrowleft}(\alpha - A^{\curvearrowleft})^{-1} & \widehat{\mathfrak{D}}_{21}^{\curvearrowleft}(\alpha) & \widehat{\mathfrak{D}}_{22}^{\curvearrowleft}(\alpha) \end{bmatrix}.
\end{aligned} \tag{6.6.16}
$$

We can factor the operator on the right-hand side (cf. Lemma 4.7.18) to get

$$
\begin{aligned}
&\left(\begin{bmatrix} \alpha & 0 & 0 \\ 0 & 1 & 0 \\ 0 & 0 & 0 \end{bmatrix} - \begin{bmatrix} A\&B \\ 0 \\ [C\&D]_2 \end{bmatrix}\right)^{-1} \begin{bmatrix} 1 & 0 & 0 \\ 0 & 1 & 0 \\ 0 & 0 & -1 \end{bmatrix} \\
&= \begin{bmatrix} 1\,0\,0 \\ 0\,1\,0 \\ [C\&D]_2^{\curvearrowleft} \end{bmatrix} \left(\begin{bmatrix} \alpha & 0 & 0 \\ 0 & 1 & 0 \\ 0 & 0 & 1 \end{bmatrix} - \begin{bmatrix} [A\&B]^{\curvearrowleft} \\ 0 \\ 0 \end{bmatrix}\right)^{-1}.
\end{aligned} \tag{6.6.17}
$$

Here the second factor on the right-hand side maps $\left[\begin{smallmatrix} X \\ U_1 \\ Y_2 \end{smallmatrix}\right]$ one-to-one onto $\mathcal{D}(S^{\curvearrowleft})$, whereas the left-hand side maps $\left[\begin{smallmatrix} X \\ U_1 \\ Y_2 \end{smallmatrix}\right]$ one-to-one onto $\mathcal{D}(S)$. Thus $\left[\begin{smallmatrix} 1\,0\,0 \\ 0\,1\,0 \\ [C\&D]_2^{\curvearrowleft} \end{smallmatrix}\right]$ maps $\mathcal{D}(S^{\curvearrowleft})$ one-to-one onto $\mathcal{D}(S)$.

It follows from (6.6.9) and (6.6.16) that

$$
\begin{aligned}
&\left(\begin{bmatrix} \alpha & 0 & 0 \\ 0 & 0 & 0 \\ 0 & 0 & 0 \end{bmatrix} - S\right) \\
&\quad\times \begin{bmatrix} (\alpha - A^{\curvearrowleft})^{-1} & (\alpha - A^{\curvearrowleft}_{|X})^{-1} B_1^{\curvearrowleft} & (\alpha - A^{\curvearrowleft}_{|X})^{-1} B_2^{\curvearrowleft} \\ 0 & 1 & 0 \\ C_2^{\curvearrowleft}(\alpha - A^{\curvearrowleft})^{-1} & \widehat{\mathfrak{D}}_{21}^{\curvearrowleft}(\alpha) & \widehat{\mathfrak{D}}_{22}^{\curvearrowleft}(\alpha) \end{bmatrix} \\
&\quad= \begin{bmatrix} 1 & 0 & 0 \\ -C_1^{\curvearrowleft}(\alpha - A^{\curvearrowleft})^{-1} & -\widehat{\mathfrak{D}}_{21}^{\curvearrowleft}(\alpha) & -\widehat{\mathfrak{D}}_{22}^{\curvearrowleft}(\alpha) \\ 0 & 0 & -1 \end{bmatrix}.
\end{aligned}
$$

This equation can be simplified: we multiply it by $\left[\begin{smallmatrix} \alpha & 0 & 0 \\ 0 & 1 & 0 \\ 0 & 0 & 1 \end{smallmatrix}\right] - \left[\begin{smallmatrix} [A\&B]^{\curvearrowleft} \\ 0 \\ 0 \end{smallmatrix}\right]$ from the right to get

$$
\left(\begin{bmatrix} \alpha & 0 & 0 \\ 0 & 0 & 0 \\ 0 & 0 & 0 \end{bmatrix} - S\right) \begin{bmatrix} 1\,0\,0 \\ 0\,1\,0 \\ [C\&D]_2^{\curvearrowleft} \end{bmatrix} = \begin{bmatrix} \alpha & 0 & 0 \\ 0 & 0 & 0 \\ 0 & 0 & 0 \end{bmatrix} - \begin{bmatrix} [A\&B]^{\curvearrowleft} \\ [C\&D]_1^{\curvearrowleft} \\ 0\,0\,1 \end{bmatrix},
$$

which can easily be seen to be equivalent to (6.6.7). □

In most cases we are able to add four more conditions which are equivalent to those listed in Theorem 6.6.8:

Theorem 6.6.9 *Make the same assumptions and introduce the same notation as in Theorem 6.6.8. In addition, suppose that $\rho(A) \cap \rho(A^{\curvearrowleft}) \neq \emptyset$ (this is, in particular, true if both S and $S^{\curvearrowleft}$ are system nodes). Then the conditions (i)–(vii) listed in Theorem 6.6.8 are equivalent to each one of the following conditions:*

(viii) *For all $\alpha \in \rho(A) \cap \rho(A^{\curvearrowleft})$, $\widehat{\mathfrak{D}}_{22}(\alpha)$ is invertible and* (6.6.11) *holds.*
(ix) *For some $\alpha \in \rho(A) \cap \rho(A^{\curvearrowleft})$, $\widehat{\mathfrak{D}}_{22}(\alpha)$ is invertible and* (6.6.11) *holds.*
(x) *For all $\alpha \in \rho(A) \cap \rho(A^{\curvearrowleft})$, $\widehat{\mathfrak{D}}^{\curvearrowleft}_{22}(\alpha)$ is invertible and* (6.6.12) *holds.*
(xi) *For some $\alpha \in \rho(A) \cap \rho(A^{\curvearrowleft})$, $\widehat{\mathfrak{D}}^{\curvearrowleft}(\alpha)$ is invertible and* (6.6.12) *holds.*

We leave the easy proof to the reader (it is similar to the proof of Theorem 6.3.11, and it consists primarily of a number of algebraic manipulations).

We remark that the right-hand side of formulas (6.6.11) and (6.6.12) can be written in several equivalent forms, similar to those appearing in Lemma 4.7.18(iv).

Corollary 6.6.10 *Under the assumption of Theorem 6.6.9, for all $\alpha \in \rho(A) \cap \rho(A^{\curvearrowleft})$, we have $\widehat{\mathfrak{D}}^{\curvearrowleft}_{22}(\alpha) = [\widehat{\mathfrak{D}}_{22}(\alpha)]^{-1}$. In particular, U_2 and Y_2 are isomorphic (hence they have the same dimension). Moreover, the operator $(\alpha - A^{\curvearrowleft})^{-1}(\alpha - A)$ which maps $\mathcal{D}(A)$ onto $\mathcal{D}(A^{\curvearrowleft})$ is given by*

$$\begin{aligned}(\alpha - A^{\curvearrowleft})^{-1}(\alpha - A) &= 1 - (\alpha - A_{|X})^{-1} B_2 \widehat{\mathfrak{D}}^{\curvearrowleft}_{22}(\alpha) C_2 \\ &= 1 - (\alpha - A^{\curvearrowleft}_{|X})^{-1} B^{\curvearrowleft}_2 C_2\end{aligned}$$

and its inverse $(\alpha - A)^{-1}(\alpha - A^{\curvearrowleft})$ is given by

$$\begin{aligned}(\alpha - A)^{-1}(\alpha - A^{\curvearrowleft}) &= 1 + (\alpha - A^{\curvearrowleft}_{|X})^{-1} B^{\curvearrowleft}_2 \widehat{\mathfrak{D}}_{22}(\alpha) C^{\curvearrowleft}_2 \\ &= 1 + (\alpha - A_{|X})^{-1} B_2 C^{\curvearrowleft}_2.\end{aligned}$$

Proof The above formulas are part of the conclusion of Theorem 6.6.9. That U_2 and Y_2 must be isomorphic follows from the fact that $\widehat{\mathfrak{D}}_{22}(\alpha)$ is a continuously invertible bijection of U_2 onto Y_2. □

With the help of Theorem 6.6.8 we are able to give a necessary and sufficient condition for the partial flow-invertibility of an operator node.

Theorem 6.6.11 *An operator node $S = \left[\begin{smallmatrix} A\&B \\ [C\&D]_1 \\ [C\&D]_2 \end{smallmatrix}\right]$ on $\left(\left[\begin{smallmatrix} Y_1 \\ Y_2 \end{smallmatrix}\right], X, \left[\begin{smallmatrix} U_1 \\ U_2 \end{smallmatrix}\right]\right)$ is partially flow-invertible if and only if the following condition holds. For some $\alpha \in \mathbb{C}$, the operator $\left[\begin{smallmatrix} \alpha & 0 & 0 \\ 0 & 1 & 0 \\ 0 & 0 & 0 \end{smallmatrix}\right] - \left[\begin{smallmatrix} A\&B \\ 0 \\ [C\&D]_2 \end{smallmatrix}\right]$ maps $\mathcal{D}(S)$ one-to-one onto $\left[\begin{smallmatrix} X \\ U_1 \\ Y_2 \end{smallmatrix}\right]$, and if we denote its inverse by $\left[\begin{smallmatrix} M_{11}(\alpha) & M_{12}(\alpha) & M_{13}(\alpha) \\ M_{21}(\alpha) & M_{22}(\alpha) & M_{23}(\alpha) \\ M_{31}(\alpha) & M_{32}(\alpha) & M_{33}(\alpha) \end{smallmatrix}\right]$, then $M_{11}(\alpha)$ is injective* *and has dense range. In this case the partially flow-inverted operator node* $S^{\curvearrowleft}$ *can be recovered from* (6.6.9) *and Lemma 4.7.6.*

We leave the straightforward proof to the reader (see the proof of Theorem 6.3.10).

It is possible to sharpen Theorem 6.6.11 by proving an analogue of Corollary 6.6.3 for arbitrary operator nodes. In order to do this we need to study the 'reduced' operator node that we get by 'removing inactive inputs and outputs'. This removal is done in the following way.

Definition 6.6.12 Let $S = \left[\begin{smallmatrix} A\&B \\ [C\&D]_1 \\ [C\&D]_2 \end{smallmatrix}\right]$ be an operator node on $\left(\left[\begin{smallmatrix} Y_1 \\ Y_2 \end{smallmatrix}\right], X, \left[\begin{smallmatrix} U_1 \\ U_2 \end{smallmatrix}\right]\right)$, with main operator A, control operator $B = \begin{bmatrix} B_1 & B_2 \end{bmatrix}$, observation operator $C = \left[\begin{smallmatrix} C_1 \\ C_2 \end{smallmatrix}\right]$, and transfer function $\widehat{\mathfrak{D}} = \left[\begin{smallmatrix} \widehat{\mathfrak{D}}_1 \\ \widehat{\mathfrak{D}}_2 \end{smallmatrix}\right] = \left[\begin{smallmatrix} \widehat{\mathfrak{D}}_{11} & \widehat{\mathfrak{D}}_{12} \\ \widehat{\mathfrak{D}}_{21} & \widehat{\mathfrak{D}}_{22} \end{smallmatrix}\right]$. By the *restriction of S to (Y_2, X, U_2)* we mean the operator node whose whose main operator is A, control operator B_2, observation operator C_2, and transfer function $\widehat{\mathfrak{D}}_{22}$.

It is not difficult to check that this is, indeed, an operator node on (Y_2, X, U_2) (see Lemma 4.7.6). Moreover, if we denote this node by S_0, then

$$\mathcal{D}(S_0) = \left\{ \left[\begin{smallmatrix} x \\ u_2 \end{smallmatrix}\right] \,\Big|\, \left[\begin{smallmatrix} x \\ 0 \\ u_2 \end{smallmatrix}\right] \in \mathcal{D}(S) \right\}. \tag{6.6.18}$$

Lemma 6.6.13 *Let* $S = \left[\begin{smallmatrix} A\&B \\ [C\&D]_1 \\ [C\&D]_2 \end{smallmatrix}\right]$ *be an operator node on* $\left(\left[\begin{smallmatrix} Y_1 \\ Y_2 \end{smallmatrix}\right], X, \left[\begin{smallmatrix} U_1 \\ U_2 \end{smallmatrix}\right]\right)$, *and let* $S_0 = \left[\begin{smallmatrix} [A\&B]_0 \\ [C\&D]_0 \end{smallmatrix}\right]$ *be the restriction of S to (Y_2, X, U_2). Then the following two conditions are equivalent, for all $\alpha \in \mathbb{C}$:*

(i) $\left[\begin{smallmatrix} \alpha & 0 \\ 0 & 0 \end{smallmatrix}\right] - S_0$ *maps $\mathcal{D}(S_0)$ one-to-one onto* $\left[\begin{smallmatrix} X \\ Y_2 \end{smallmatrix}\right]$;

(ii) $\left[\begin{smallmatrix} \alpha & 0 & 0 \\ 0 & 1 & 0 \\ 0 & 0 & 0 \end{smallmatrix}\right] - \left[\begin{smallmatrix} A\&B \\ 0 \\ [C\&D]_2 \end{smallmatrix}\right]$ *maps $\mathcal{D}(S)$ one-to-one onto* $\left[\begin{smallmatrix} X \\ U_1 \\ Y_2 \end{smallmatrix}\right]$.

Proof Let us denote the common main operator of S and S_0 by A. Fix any $\beta \in \rho(A)$. Then $\left(\left[\begin{smallmatrix} \beta & 0 \\ 0 & 1 \end{smallmatrix}\right] - \left[\begin{smallmatrix} [A\&B]_0 \\ 0 \quad 0 \end{smallmatrix}\right]\right)^{-1}$ maps $\left[\begin{smallmatrix} X \\ Y_2 \end{smallmatrix}\right]$ one-to-one onto $\mathcal{D}(S_0)$ and $\left(\left[\begin{smallmatrix} \beta & 0 & 0 \\ 0 & 1 & 0 \\ 0 & 0 & 1 \end{smallmatrix}\right] - \left[\begin{smallmatrix} A\&B \\ 0 \\ 0 \end{smallmatrix}\right]\right)^{-1}$ maps $\left[\begin{smallmatrix} X \\ U_1 \\ Y_2 \end{smallmatrix}\right]$ one-to-one onto $\mathcal{D}(S)$ (see Lemma 4.7.18). Thus, to prove the equivalence of (i) and (ii) it suffices to show that $\left(\left[\begin{smallmatrix} \alpha & 0 \\ 0 & 0 \end{smallmatrix}\right] - S_0\right)\left(\left[\begin{smallmatrix} \beta & 0 \\ 0 & 1 \end{smallmatrix}\right] - \left[\begin{smallmatrix} [A\&B]_0 \\ 0 \quad 0 \end{smallmatrix}\right]\right)^{-1}$ is invertible on $\left[\begin{smallmatrix} X \\ Y_2 \end{smallmatrix}\right]$ if and only if $\left(\left[\begin{smallmatrix} \alpha & 0 & 0 \\ 0 & 1 & 0 \\ 0 & 0 & 0 \end{smallmatrix}\right] - \left[\begin{smallmatrix} A\&B \\ 0 \\ [C\&D]_2 \end{smallmatrix}\right]\right)\left(\left[\begin{smallmatrix} \beta & 0 & 0 \\ 0 & 1 & 0 \\ 0 & 0 & 1 \end{smallmatrix}\right] - \left[\begin{smallmatrix} A\&B \\ 0 \\ 0 \end{smallmatrix}\right]\right)^{-1}$ is invertible on $\left[\begin{smallmatrix} X \\ U_1 \\ Y_2 \end{smallmatrix}\right]$. The first of the two operators above is explicitly given by (in obvious notation)

$$\begin{bmatrix} (\alpha - A)(\beta - A)^{-1} & (\alpha - \beta)(\beta - A_{|X})^{-1}B_2 \\ -C(\beta - A)^{-1} & -\widehat{\mathfrak{D}}_{22}(\beta) \end{bmatrix}, \tag{6.6.19}$$

and the latter is given by

$$\begin{bmatrix} (\alpha - A)(\beta - A)^{-1} & (\alpha - \beta)(\beta - A_{|X})^{-1}B_1 & (\alpha - \beta)(\beta - A_{|X})^{-1}B_2 \\ 0 & 1 & 0 \\ -C(\beta - A)^{-1} & -\widehat{\mathfrak{D}}_{21}(\beta) & -\widehat{\mathfrak{D}}_{22}(\beta) \end{bmatrix}. \tag{6.6.20}$$

By permuting second and third row and column we can make this operator upper block triangular, where the first diagonal block is the operator in (6.6.19) and the second diagonal block is the identity operator. It follows from Lemma A.4.2(i) that the operator in (6.6.20) is invertible if and only if the one in (6.6.19) is invertible. □

Corollary 6.6.14 *Let $S = \left[\begin{smallmatrix} A\&B \\ [C\&D]_1 \\ [C\&D]_2 \end{smallmatrix}\right]$ be an operator node on $\left(\left[\begin{smallmatrix} Y_1 \\ Y_2 \end{smallmatrix}\right], X, \left[\begin{smallmatrix} U_1 \\ U_2 \end{smallmatrix}\right]\right)$, and let $S_0 = \left[\begin{smallmatrix} [A\&B]_0 \\ [C\&D]_0 \end{smallmatrix}\right]$ be the restriction of S to (Y_2, X, U_2). Then S is partially flow-invertible if and only if S_0 is flow-invertible. Moreover, if we denote the partial flow-inverse of S by $S^{\curvearrowleft}$ and the flow-inverse of S by $S_\times$, then (in an obvious notation)*

$$\begin{aligned} A^{\curvearrowleft} &= A_{0\times}, & B_2^{\curvearrowleft} &= B_{0\times}, \\ C_2^{\curvearrowleft} &= C_{0\times}, & \widehat{\mathfrak{D}}_{22}^{\curvearrowleft} &= \widehat{\mathfrak{D}}_{0\times}. \end{aligned}$$

Proof See Theorems 6.3.10 and 6.6.11, Lemma 6.6.13, and the identities (6.3.13) and (6.6.9).

□

Our following theorem is an extension of Theorem 6.3.14.

Theorem 6.6.15 *Let $S = \left[\begin{smallmatrix} A\&B \\ [C\&D]_1 \\ [C\&D]_2 \end{smallmatrix}\right]$ be a flow-invertible system node on $\left(\left[\begin{smallmatrix} Y_1 \\ Y_2 \end{smallmatrix}\right], X, \left[\begin{smallmatrix} U_1 \\ U_2 \end{smallmatrix}\right]\right)$, whose flow-inverse $S^{\curvearrowleft}$ is also a system node (on $\left(\left[\begin{smallmatrix} Y_1 \\ U_2 \end{smallmatrix}\right], X, \left[\begin{smallmatrix} U_1 \\ Y_2 \end{smallmatrix}\right]\right)$). Let x and $y = \left[\begin{smallmatrix} y_1 \\ y_2 \end{smallmatrix}\right]$ be the state trajectory and output function of S with initial time $s \in \mathbb{R}$, initial state $x_s \in X$, and input function $u = \left[\begin{smallmatrix} u_1 \\ u_2 \end{smallmatrix}\right] \in L^1_{\mathrm{loc}}\left([s,\infty); \left[\begin{smallmatrix} U_1 \\ U_2 \end{smallmatrix}\right]\right)$, and suppose that $x \in W^{1,1}_{\mathrm{loc}}([s,\infty); X)$. Then $y \in L^1_{\mathrm{loc}}\left([s,\infty); \left[\begin{smallmatrix} Y_1 \\ Y_2 \end{smallmatrix}\right]\right)$, and x and $\left[\begin{smallmatrix} y_1 \\ u_2 \end{smallmatrix}\right]$ are the state trajectory and output function of $S^{\curvearrowleft}$ with initial time s, initial state x_s and input function $\left[\begin{smallmatrix} u_1 \\ y_2 \end{smallmatrix}\right]$.*

The proof is very similar to the proof of Theorem 6.3.14, and we leave it to the reader.

Corollary 6.6.16 *Let $\Sigma = \left[\begin{smallmatrix} A\&B \\ [C\&D]_1 \\ [C\&D]_2 \end{smallmatrix}\right]$ be an L^p-well-posed linear system on $\left(\left[\begin{smallmatrix} Y_1 \\ Y_2 \end{smallmatrix}\right], X, \left[\begin{smallmatrix} U_1 \\ U_2 \end{smallmatrix}\right]\right)$ with $1 \le p < \infty$. Then Σ is partially flow-invertible (in the sense of Definition 6.6.2) if and only if its system node S is partially flow-invertible (in the sense of Definition 6.6.6) and the partially flow-inverted operator node $S^{\curvearrowleft}$ is an L^p-well-posed system node.*

The proof is analogous to the proof of Corollary 6.3.15 (and it is left to the reader).

Our next theorem is an extension of Theorem 6.3.16.

Theorem 6.6.17 *Let* $S = \begin{bmatrix} A\&B \\ [C\&D]_1 \\ [C\&D]_2 \end{bmatrix}$ *be a compatible operator node on* $\left(\begin{bmatrix} Y_1 \\ Y_2 \end{bmatrix}, X, \begin{bmatrix} U_1 \\ U_2 \end{bmatrix}\right)$, *and let* $\begin{bmatrix} A_{|W} & B_1 & B_2 \\ [C_1]_{|W} & D_{11} & D_{12} \\ [C_2]_{|W} & D_{21} & D_{22} \end{bmatrix} \in \mathcal{B}\left(\begin{bmatrix} W \\ U_1 \\ U_2 \end{bmatrix}; \begin{bmatrix} W_{-1} \\ Y_1 \\ Y_2 \end{bmatrix}\right)$ *be a compatible extension of S (here* $X_1 \subset W \subset X$ *and* W_{-1} *is defined as in Lemma 5.1.3). Suppose that S is partially flow-invertible and that* D_{22} *is invertible. Denote the partially flow-inverted operator node by* $S^{\curvearrowleft} = \begin{bmatrix} [A\&B]^{\curvearrowleft} \\ [C\&D]_1^{\curvearrowleft} \\ [C\&D]_2^{\curvearrowleft} \end{bmatrix}$, *let* $X_1^{\curvearrowleft}$ *and* $X_{-1}^{\curvearrowleft}$ *be the analogues of* X_1 *and* X_{-1} *for* $S^{\curvearrowleft}$, *and let* $W_{-1}^{\curvearrowleft}$ *be the analogue of* W_{-1} *for* $S^{\curvearrowleft}$ *(i.e.,* $W_{-1}^{\curvearrowleft} = (\alpha - A^{\curvearrowleft})_{|W} W$ *for some* $\alpha \in \rho(A^{\curvearrowleft})$*). Then* $W_{-1}^{\curvearrowleft}$ *can be identified with* W_{-1}, *and* $S^{\curvearrowleft}$ *is compatible over W with the compatible extension*

$$\begin{aligned}
\begin{bmatrix} [A^{\curvearrowleft}]_{|W} & B_1^{\curvearrowleft} & B_2^{\curvearrowleft} \\ [C_1^{\curvearrowleft}]_{|W} & D_{11}^{\curvearrowleft} & D_{12}^{\curvearrowleft} \\ [C_2^{\curvearrowleft}]_{|W} & D_{21}^{\curvearrowleft} & D_{22}^{\curvearrowleft} \end{bmatrix} &= \begin{bmatrix} A_{|W} & B_1 & B_2 \\ [C_1]_{|W} & D_{11} & D_{12} \\ 0 & 0 & 1 \end{bmatrix} \begin{bmatrix} 1 & 0 & 0 \\ 0 & 1 & 0 \\ [C_2]_{|W} & D_{21} & D_{22} \end{bmatrix}^{-1} \\
&= \begin{bmatrix} 1 & 0 & -B_2 \\ 0 & 1 & -D_{12} \\ 0 & 0 & D_{22} \end{bmatrix}^{-1} \begin{bmatrix} A_{|W} & B_1 & 0 \\ [C_1]_{|W} & D_{11} & 0 \\ -[C_2]_{|W} & -D_{21} & 1 \end{bmatrix} \\
&= \begin{bmatrix} A & B_1 & 0 \\ [C_1]_{|W} & D_{11} & 0 \\ 0 & 0 & 0 \end{bmatrix} \\
&\quad + \begin{bmatrix} B_2 \\ D_{12} \\ 1 \end{bmatrix} D_{22}^{-1} \begin{bmatrix} -[C_2]_{|W} & -D_{21} & 1 \end{bmatrix}.
\end{aligned} \tag{6.6.21}$$

We leave the formulation and proof of the corresponding result where D_{22} is only *left-invertible* to the reader (see part (i) of Theorem 6.3.16).

Proof By (6.6.8),

$$\begin{bmatrix} [C\&D]_1^{\curvearrowleft} \\ [C\&D]_2^{\curvearrowleft} \end{bmatrix} = \begin{bmatrix} [C\&D]_1 \\ 0\,0\,1 \end{bmatrix} \begin{bmatrix} 1\,0\,0 \\ 0\,1\,0 \\ [C\&D]_2 \end{bmatrix}^{-1}.$$

This operator has an obvious extension to an operator in $\mathcal{B}\left(\begin{bmatrix} W \\ U_1 \\ U_2 \end{bmatrix}; \begin{bmatrix} Y_1 \\ Y_2 \end{bmatrix}\right)$, namely

$$\begin{bmatrix} [C_1]_{|W} & D_{11} & D_{12} \\ 0 & 0 & 1 \end{bmatrix} \begin{bmatrix} 1 & 0 & 0 \\ 0 & 1 & 0 \\ [C_2]_{|W} & D_{21} & D_{22} \end{bmatrix}^{-1}.$$

Thus, $S^{\curvearrowleft}$ is compatible over W. We identify $W_{-1}^{\curvearrowleft}$ with W_{-1} in the same way as in Theorem 6.3.16, applied to the operator node that we get by dropping the first input and the first output of S (the flow-inverse of this operator node is the one that we get from $S^{\curvearrowleft}$ by again dropping the first input and the first output). With this identification, the operator given in (6.6.21) is an obvious extension to an operator in $\mathcal{B}\left(\left[\begin{smallmatrix} W \\ U_1 \\ U_2 \end{smallmatrix}\right]; \left[\begin{smallmatrix} W_{-1} \\ Y_1 \\ Y_2 \end{smallmatrix}\right]\right)$ of $\left[\begin{smallmatrix} [A\&B]^{\curvearrowleft} \\ [C\&D]_1^{\curvearrowleft} \\ [C\&D]_2^{\curvearrowleft} \end{smallmatrix}\right] = \left[\begin{smallmatrix} A\&B \\ [C\&D]_1 \\ 0\,0\,1 \end{smallmatrix}\right]\left[\begin{smallmatrix} 1\,0\,0 \\ 0\,1\,0 \\ [C\&D]_2 \end{smallmatrix}\right]^{-1}$. □

Let us finally look at the partial flow-inversion of a regular system.

Theorem 6.6.18 *Let* $\Sigma = \left[\begin{smallmatrix} \mathfrak{A} & \mathfrak{B}_1 & \mathfrak{B}_2 \\ \mathfrak{C}_1 & \mathfrak{D}_{11} & \mathfrak{D}_{12} \\ \mathfrak{C}_2 & \mathfrak{D}_{21} & \mathfrak{D}_{22} \end{smallmatrix}\right]$ *be a weakly regular ($L^p|Reg$-well-posed) partially flow-invertible linear system on* $\left(\left[\begin{smallmatrix} Y_1 \\ Y_1 \end{smallmatrix}\right], X, \left[\begin{smallmatrix} U_1 \\ U_2 \end{smallmatrix}\right]\right)$ *with system node* $S = \left[\begin{smallmatrix} A\&B \\ [C\&D]_1 \\ [C\&D]_2 \end{smallmatrix}\right]$, *semigroup generator A, control operator* $B = \begin{bmatrix} B_1 & B_2 \end{bmatrix}$, *and observation operator* $C = \left[\begin{smallmatrix} C_1 \\ C_2 \end{smallmatrix}\right]$. *We define the extensions $\widetilde{C}_w$ and $\widetilde{C}_s$ of the observation operator C as in Definition 5.4.1, and let* $D = \left[\begin{smallmatrix} D_{11} & D_{12} \\ D_{21} & D_{22} \end{smallmatrix}\right]$ *be the corresponding feedthrough operator (see Theorem 5.6.5).*

(i) *If D_{22} is invertible, then Theorem 6.6.17 applies with $W = \mathcal{D}(\widetilde{C}_w)$ and $C_{|W} = \widetilde{C}_w$ (in particular, both Σ and the partially flow-inverted system $\Sigma^{\curvearrowleft}$ are compatible).*

(ii) *If Σ is strongly regular, then the partially flow-inverted system is strongly regular iff D_{22} is invertible. In this case $\mathcal{D}(\widetilde{C}_{\underset{\sim}{s}}^{\curvearrowleft}) = \mathcal{D}(\widetilde{C}_s)$, and Theorem 6.6.17 applies with $W = \mathcal{D}(\widetilde{C}_s)$ and $C_{|W} = \widetilde{C}_s$.*

(iii) *In the Reg-well-posed case both the original and the partially flow-inverted system are strongly regular, D_{22} is invertible, and Theorem 6.6.17 applies with $W = \mathcal{D}(\widetilde{C}_s)$ and $C_{|W} = \widetilde{C}_s$.*

(iv) *In the L^1-well-posed case with a reflexive state space X both the original and the flow-inverted system are strongly regular, D_{22} is invertible, and Theorem 6.6.17 applies with $W = \mathcal{D}(\widetilde{C}_s)$ and $C_{|W} = \widetilde{C}_s$.*

(v) *If Σ is uniformly regular then the partially flow-inverted system $\Sigma^{\curvearrowleft}$ is uniformly regular, D_{22} is invertible, and Theorem 6.6.17 applies with $W = \mathcal{D}(\widetilde{C}_s)$ and $C_{|W} = \widetilde{C}_s$.*

Once more we leave the proof to the reader (it resembles the proof of Theorem 6.3.17).

The following analogue of Theorem 6.3.19 is true as well.

Theorem 6.6.19 *Let* $\left(\left[\begin{smallmatrix} Y_1 \\ Y_2 \end{smallmatrix}\right], X, \left[\begin{smallmatrix} U_1 \\ U_2 \end{smallmatrix}\right]\right)$ *be reflexive Banach spaces, let* $1 < p < \infty$, *let Σ be an L^p-well-posed linear system on* $\left(\left[\begin{smallmatrix} Y_1 \\ Y_2 \end{smallmatrix}\right], X, \left[\begin{smallmatrix} U_1 \\ U_2 \end{smallmatrix}\right]\right)$, *and let S be an operator node on* $\left(\left[\begin{smallmatrix} Y_1 \\ Y_2 \end{smallmatrix}\right], X, \left[\begin{smallmatrix} U_1 \\ U_2 \end{smallmatrix}\right]\right)$ *(not necessarily the one induced by Σ).*

(i) *The system* Σ *is partially flow-invertible if and only if the causal dual system* Σ^d *is partially flow-invertible, in which case the partially flow-inverted systems satisfy* $(\Sigma^d)^{\curvearrowleft} = \begin{bmatrix} 1 & 0 & 0 \\ 0 & 1 & 0 \\ 0 & 0 & -1 \end{bmatrix} (\Sigma^{\curvearrowleft})^d \begin{bmatrix} 1 & 0 & 0 \\ 0 & 1 & 0 \\ 0 & 0 & -1 \end{bmatrix}$.

(ii) *The operator node* S *is partially flow-invertible if and only if* S^* *is partially flow-invertible, in which case the partially flow-inverted operator nodes satisfy* $(S^*)^{\curvearrowleft} = \begin{bmatrix} 1 & 0 & 0 \\ 0 & 1 & 0 \\ 0 & 0 & -1 \end{bmatrix} (S^{\curvearrowleft})^* \begin{bmatrix} 1 & 0 & 0 \\ 0 & 1 & 0 \\ 0 & 0 & -1 \end{bmatrix}$.

(iii) *If* S *is partially flow-invertible and both* S *and* S^* *are strongly regular with feedthrough operators* $D = \begin{bmatrix} D_{11} & D_{12} \\ D_{21} & D_{22} \end{bmatrix}$ *respectively* $D^* = \begin{bmatrix} D_{11}^* & D_{21}^* \\ D_{12}^* & D_{22}^* \end{bmatrix}$, *then* D_{22} *is invertible. In this case both* $S^{\curvearrowleft}$ *and* $(S^*)^{\curvearrowleft}$ *are strongly regular.*

Proof (i) The proof of part (i) is similar to the corresponding part of the proof of Theorem 6.3.19, and it is left to the reader.

(ii) The proof of part (ii) is also similar to the corresponding part of the proof of Theorem 6.3.19, but it is significantly longer (too long to be given here in detail). If we base it on (6.6.9) which is the partial flow-inversion analogue of (6.3.13), then the right-hand side does not cause any problems: the adjoint of the right-hand side of (6.6.9) is equal to the right-hand side of (6.6.9) with α replaced by $\overline{\alpha}$ and $S^{\curvearrowleft}$ replaced by $\begin{bmatrix} 1 & 0 & 0 \\ 0 & 1 & 0 \\ 0 & 0 & -1 \end{bmatrix} (S^{\curvearrowleft})^* \begin{bmatrix} 1 & 0 & 0 \\ 0 & 1 & 0 \\ 0 & 0 & -1 \end{bmatrix}$. However, this time we cannot appeal directly to Lemma 6.2.14 to compute the adjoint of the operator on the left-hand side of (6.6.9). This can be done approximately in the same way as in the proof of Lemma 6.2.14, if we start by multiplying both $\begin{bmatrix} 1\,0\,0 \\ [C\&D]_1 \\ 0\,0\,1 \end{bmatrix}$ and $\left(\begin{bmatrix} \alpha & 0 & 0 \\ 0 & 1 & 0 \\ 0 & 0 & 0 \end{bmatrix} - \begin{bmatrix} A\&B \\ 0 \\ [C\&D]_2 \end{bmatrix} \right)$ to the right by $\begin{bmatrix} (\beta-A)^{-1} & (\beta-A_{|X})^{-1}B_1 & (\beta-A_{|X})^{-1}B_2 \\ 0 & 1 & 0 \\ 0 & 0 & 1 \end{bmatrix}$, where $\beta \in \rho(A)$. Unfortunately, this leads to a lengthy computation. Therefore, let us only prove part (ii) in the special case where $\rho(A) \cap \rho(A^{\curvearrowleft}) \neq \emptyset$, so that Theorem 6.6.9 applies. In this case, instead of computing the adjoint of the left-hand side of (6.6.9), we may compute the adjoint of the right-hand side of (6.6.11), which is easy. (We leave the proof of the general case to the reader.)

(iii) The proof of part (iii) is similar to the corresponding part of the proof of Theorem 6.3.19, and it is left to the reader. □

6.7 Comments

The main part of the results presented in this chapter are taken from Staffans and Weiss (2004) (the well-posed case) and from Malinen *et al.* (2003) and Staffans (2001a, 2002a, c) (for general system or operator nodes).

Section 6.1 Anti-causal systems have been used especially in connection with the anti-causal dual; see, e.g., Salamon (1987) and Staffans (1997, 1998a). Our presentation has been modeled after Staffans (1997, 1998a) and Staffans and Weiss (2004).

Section 6.2 Most of the duality theory has been part of the folklore in well-posed linear systems and in partial differential equations for quite some time. The traditional definition of a dual system is based on the generating operators of the system. Our treatment of the dual of an L^p-well-posed system is an extension of the corresponding discussions in Staffans (1997) and Staffans (1998a). Slightly different but equivalent versions (in the Hilbert space case with $p = 2$) are given in Arov and Nudelman (1996) and Staffans and Weiss (2004). Theorem 6.2.10 is found in Salamon (1987, Theorem 3.3) and in Staffans (1998a, Lemma 2.15) (in the Hilbert space case with $p = 2$). Theorem 6.2.13 is proved in Staffans and Weiss (2004) (in the Hilbert space case with $p = 2$). Lemma 6.2.14 is taken from Malinen *et al.* (2003). An early version of this lemma appears in Šmuljan (1986). Theorem 6.2.15(ii) is due to Staffans and Weiss (2004), and Theorem 6.2.15(iii) to Ober and Montgomery-Smith (1990, Section 6).

Section 6.3 Flow-inversion does not change the relationships between the different signals appearing in the system (only the grouping of 'external' signals into inputs and outputs changes), so in Willems' (1991) behavioral sense the system stays the same. The main part of the results on flow-inversion are taken from Staffans and Weiss (2004) (the well-posed case) and from Staffans (2001a, 2002a, c) (for general system or operator nodes). In particular, one direction of Lemma 6.3.8(i),(iii) was proved in Staffans and Weiss (2004, Corollary 5.3) (in the well-posed setting). The other direction was used implicitly in Staffans and Weiss (2004, Theorem 5.4), and it was explicitly formulated and proved (independently) by Kalle Mikkola in the spring of 2003 (in a the feedback setting; cf. Lemma 7.4.4).

As we already mentioned earlier, flow-inversion is very closely connected to output feedback. See Chapter 7 for additional comments on various feedback results.

Section 6.4 Time-inversion has been used in the study of optimal control of hyperbolic partial differential equations for quite some time; see, e.g., Flandoli *et al.* (1988, Section 2.4). This notion was formalized (in the well-posed case) in Staffans and Weiss (2004). The examples in this section are taken from Staffans and Weiss (2004).

Section 6.5 Again, the first formal study of time-flow-inversion (in the well-posed case) seems to be Staffans and Weiss (2004). In the conservative case (which will be discussed in Chapter 11) the time-flow-inversion coincides with the duality transformation, and many of the results of this section were known from before in this form (see Sz.-Nagy and Foiaş (1970) and the comments in Chapter 11).

Section 6.6 Partial flow-inversion plays an important role in the theory of doubly coprime factorizations presented in Section 8.3, and also in the standard H^∞ control theory. There this transformation is often referred to as a *chain scattering transformation*. See, for example, Staffans (1998d).

7
Feedback

Feedback plays a central role in control theory. We describe different types of feedback, such as output feedback, state feedback, and output injection, both in the well-posed case and in the non-well-posed case. Feedback and partial feedback can be seen as special cases of flow-inversion or partial flow-inversion, and the converse is also true.

7.1 Static output feedback

The notions of stabilization and detection deal with the possibility of stabilizing a well-posed linear system by the use of either a state feedback or an output injection. Therefore, before we can study these notions, we must first look at different kinds of feedback connections.

We start by presenting the most basic type of feedback, namely static output feedback. It is possible to regard this type of feedback as a special case of flow-inversion, which was discussed in Section 6.3 (see Remark 7.2.3), and conversely, it is possible to regard flow-inversion as a special case of static output feedback (see Remark 7.1.10). In particular, this means that we for technical reasons have to exclude the L^∞-well-posed case, and restrict the discussion in this chapter to *well-posed linear systems*, i.e, systems that are either L^p-well-posed for some $p < \infty$ or *Reg*-well-posed (cf. Definition 2.2.4).[1]

Figure 7.1 contains a diagram describing static output feedback. In that diagram K is a bounded linear operator from the output space into the input

[1] It is possible to include the case $p = \infty$, too, by adding the requirement that the return difference (the operator $(1 - \mathfrak{D}K)$ in Theorem 7.1.2, etc.) is invertible both in $TIC^\infty_{\mathrm{loc}}$ and in TIC^{Reg}_{loc}; this is needed for the strong continuity of the closed-loop semigroup. It is not clear if the latter condition is implied by the former or not. If it is true, then the L^∞ case could be treated in the same way as the other cases.

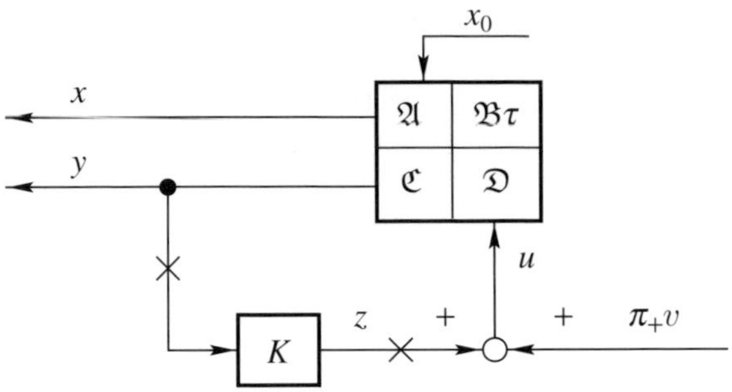

Figure 7.1 Static output feedback connection

space. Thus, if we consider this feedback configuration with initial time zero, initial value x_0, and input v, we find that the state $x(t)$ at time $t \geq 0$, the output y, the effective input u, and the feedback signal z satisfy the equations

$$\begin{aligned} x(t) &= \mathfrak{A}^t x_0 + \mathfrak{B}\tau^t u, \\ y &= \mathfrak{C}x_0 + \mathfrak{D}u, \\ u &= z + \pi_+ v, \\ z &= Ky, \end{aligned} \tag{7.1.1}$$

which can be formally solved as

$$\begin{aligned} x(t) &= \big(\mathfrak{A}^t + \mathfrak{B}\tau^t K (1 - \mathfrak{D}K)^{-1} \mathfrak{C}\big)x_0 + \mathfrak{B}(1 - K\mathfrak{D})^{-1} \tau^t \pi_+ v, \\ y &= (1 - \mathfrak{D}K)^{-1} (\mathfrak{C}x_0 + \mathfrak{D}\pi_+ v), \\ u &= (1 - K\mathfrak{D})^{-1} (K\mathfrak{C}x_0 + \pi_+ v), \\ z &= (1 - K\mathfrak{D})^{-1} K (\mathfrak{C}x_0 + \mathfrak{D}\pi_+ v). \end{aligned} \tag{7.1.2}$$

We say that the feedback operator K is admissible whenever these equations are valid.

For the notions used in the following definition, see Definitions 2.2.4 and 2.6.2. See also Remark 2.2.5.

Definition 7.1.1 Let $\Sigma = \left[\begin{array}{c|c} \mathfrak{A} & \mathfrak{B} \\ \hline \mathfrak{C} & \mathfrak{D} \end{array}\right]$ be a well-posed linear system on (Y, X, U). Then $K \in \mathcal{B}(Y; U)$ is called an *admissible (static) output feedback operator for* Σ if the operator $1 - K\mathfrak{D}$ has an inverse in $TIC_{\text{loc}}(U)$.

It is a fundamental fact that x and y in (7.1.2) can be interpreted as the state and output of another well-posed linear system with initial state x_0 and input u if (and only if; see Theorem 7.1.8) K is an admissible output feedback operator:

Theorem 7.1.2 *Let $\Sigma = \left[\begin{smallmatrix} \mathfrak{A} & \mathfrak{B} \\ \mathfrak{C} & \mathfrak{D} \end{smallmatrix}\right]$ be a well-posed linear system on (Y, X, U), and let $K \in \mathcal{B}(Y; U)$ be an admissible output feedback operator for Σ. Then the system*

$$\Sigma^K = \left[\begin{array}{c|c} \mathfrak{A}^K & \mathfrak{B}^K \tau \\ \hline \mathfrak{C}^K & \mathfrak{D}^K \end{array}\right] = \left[\begin{array}{c|c} \mathfrak{A} + \mathfrak{B}\tau K (1 - \mathfrak{D}K)^{-1} \mathfrak{C} & \mathfrak{B}(1 - K\mathfrak{D})^{-1} \tau \\ \hline (1 - \mathfrak{D}K)^{-1} \mathfrak{C} & \mathfrak{D}(1 - K\mathfrak{D})^{-1} \end{array}\right] \tag{7.1.3}$$

is another well-posed linear system on (Y, X, U). We call this system the closed-loop system *with (output) feedback operator K. The state $x(t)$ at time t and the output y of Σ^K with initial time zero, initial value x_0, and input v, are given by* (7.1.2).

It is possible to prove this theorem by reducing it to a special case of Theorem 6.3.1 (see Remark 7.2.3), but for the convenience of the reader we have chosen to give an independent proof, analogous to that of Theorem 6.3.1.

Proof of Theorem 7.1.2 Clearly, if Σ^K is a well-posed linear system, then the state $x(t)$ at time t and the output y of Σ^K with initial time zero, initial value x_0, and input v, are given by (7.1.2).

The continuity of the operators in Σ^K is obvious. The strong continuity of $\mathfrak{A}^K$ follows from the strong continuity of τ in $L^p_{c,\mathrm{loc}}(\mathbb{R}; U)$ (see Example 2.5.3) for $p < \infty$ or from the strong continuity of $\mathfrak{B}\tau$ from $Reg_{c,\mathrm{loc}}(\mathbb{R}; U)$ to X (see Theorem 4.3.1(ii)).[2] Thus, to complete the proof it suffices to check that the algebraic properties in Definition 2.2.1 hold.

The key ingredients in the proof of the algebraic properties are the formulas

$$\begin{aligned} \pi_+(1 - K\mathfrak{D})^{-1}\pi_- &= K(1 - \mathfrak{D}K)^{-1}\mathfrak{C}\mathfrak{B}(1 - K\mathfrak{D})^{-1}, \\ \pi_+(1 - \mathfrak{D}K)^{-1}\pi_- &= (1 - \mathfrak{D}K)^{-1}\mathfrak{C}\mathfrak{B}(1 - K\mathfrak{D})^{-1}K, \end{aligned} \tag{7.1.4}$$

which follow from Lemmas 6.3.2, A.4.1, and the fact that $\pi_+\mathfrak{D}\pi_- = \mathfrak{C}\mathfrak{B}$.

(i) Clearly $(\mathfrak{A}^K)^0 = \mathfrak{A}^0 = 1$. To show that $\mathfrak{A}^K$ is a semigroup we use Definition 2.2.1(i)–(iv), Lemma A.4.1, the causality and time-invariance of

[2] If it is true that $(1 - \mathfrak{D}K)^{-1}$ maps $Reg_{\mathrm{loc}}(\overline{\mathbb{R}}^+; Y)$ into itself also in the L^∞-case (which seems plausible), then the closed-loop semigroup is strongly continuous also in the L^∞-case, and we can remove the restriction $p < \infty$ imposed on the system at the beginning of this chapter.

$(1-K\mathfrak{D})^{-1}$, and (7.1.4) to compute

$$\begin{aligned}
(\mathfrak{A}^K)^s(\mathfrak{A}^K)^t &= \big(\mathfrak{A}^s+\mathfrak{B}\tau^s K(1-\mathfrak{D}K)^{-1}\mathfrak{C}\big)\big(\mathfrak{A}^t+\mathfrak{B}K(1-\mathfrak{D}K)^{-1}\tau^t\mathfrak{C}\big)\\
&= \mathfrak{A}^s\mathfrak{A}^t+\mathfrak{A}^s\mathfrak{B}K(1-\mathfrak{D}K)^{-1}\tau^t\mathfrak{C}\\
&\quad +\mathfrak{B}\tau^s K(1-\mathfrak{D}K)^{-1}\mathfrak{C}\mathfrak{A}^t\\
&\quad +\mathfrak{B}\tau^s K(1-\mathfrak{D}K)^{-1}\mathfrak{C}\mathfrak{B}(1-K\mathfrak{D})^{-1}K\tau^t\mathfrak{C}\\
&= \mathfrak{A}^{s+t}+\mathfrak{B}\tau^s\pi_-K(1-\mathfrak{D}K)^{-1}\pi_-\tau^t\mathfrak{C}\\
&\quad +\mathfrak{B}\tau^s\pi_+K(1-\mathfrak{D}K)^{-1}\pi_+\tau^t\mathfrak{C}\\
&\quad +\mathfrak{B}\tau^s\pi_+K(1-\mathfrak{D}K)^{-1}\pi_-\tau^t\mathfrak{C}\\
&= \mathfrak{A}^{s+t}+\mathfrak{B}\tau^s\tau^t K(1-\mathfrak{D}K)^{-1}\mathfrak{C}\\
&= (\mathfrak{A}^K)^{s+t}.
\end{aligned}$$

(ii) The proof of the identity $(\mathfrak{A}^K)^t\mathfrak{B}^K=\mathfrak{B}^K\tau^t\pi_-$ in Definition 2.2.1(ii) uses the same ingredients:

$$\begin{aligned}
(\mathfrak{A}^K)^t\mathfrak{B}^K &= \big(\mathfrak{A}^t+\mathfrak{B}\tau^t K(1-\mathfrak{D}K)^{-1}\mathfrak{C}\big)\mathfrak{B}(1-K\mathfrak{D})^{-1}\\
&= \mathfrak{A}^t\mathfrak{B}(1-K\mathfrak{D})^{-1}+\mathfrak{B}\tau^t K(1-\mathfrak{D}K)^{-1}\mathfrak{C}\mathfrak{B}(1-K\mathfrak{D})^{-1}\\
&= \mathfrak{B}\tau^t\pi_-(1-K\mathfrak{D})^{-1}\pi_-+\mathfrak{B}\tau^t\pi_+(1-\mathfrak{D}K)^{-1}\pi_-\\
&= \mathfrak{B}(1-K\mathfrak{D})^{-1}\tau^t\pi_-=\mathfrak{B}^K\tau^t\pi_-
\end{aligned}$$

(iii) Also the proof of the identity $\mathfrak{C}^K(\mathfrak{A}^K)^t=\pi_+\tau^t\mathfrak{C}^K$ in Definition 2.2.1(iii) is similar:

$$\begin{aligned}
\mathfrak{C}^K(\mathfrak{A}^K)^t &= (1-\mathfrak{D}K)^{-1}\mathfrak{C}\big(\mathfrak{A}^t+\mathfrak{B}(1-K\mathfrak{D})^{-1}K\tau^t\mathfrak{C}\big)\\
&= (1-\mathfrak{D}K)^{-1}\mathfrak{C}\mathfrak{A}^t+(1-\mathfrak{D}K)^{-1}\mathfrak{C}\mathfrak{B}(1-K\mathfrak{D})^{-1}K\tau^t\mathfrak{C}\\
&= \pi_+(1-\mathfrak{D}K)^{-1}\pi_+\tau^t\mathfrak{C}+\pi_+(1-\mathfrak{D}K)^{-1}\pi_-\tau^t\mathfrak{C}\\
&= \pi_+\tau^t(1-\mathfrak{D}K)^{-1}\mathfrak{C}=\pi_+\tau^t\mathfrak{C}^K.
\end{aligned}$$

(iv) The time-invariance and causality of $\mathfrak{D}^K=\mathfrak{D}(1-K\mathfrak{D})^{-1}$ follow from the time-invariance and causality of $\mathfrak{D}$ and $(1-K\mathfrak{D})^{-1}$. Thus, to verify part (iv) of Definition 2.2.1 it suffices to compute the Hankel operator of $\mathfrak{D}^K$ as follows:

$$\begin{aligned}
\pi_+\mathfrak{D}^K\pi_- &= \pi_+\mathfrak{D}(\pi_++\pi_-)(1-K\mathfrak{D})^{-1}\pi_-\\
&= \mathfrak{D}\pi_+(1-K\mathfrak{D})^{-1}\pi_-+\pi_+\mathfrak{D}\pi_-(1-K\mathfrak{D})^{-1}\\
&= \mathfrak{D}K(1-\mathfrak{D}K)^{-1}\mathfrak{C}\mathfrak{B}(1-K\mathfrak{D})^{-1}+\mathfrak{C}\mathfrak{B}(1-K\mathfrak{D})^{-1}\\
&= \big(1+(1-1+\mathfrak{D}K)(1-\mathfrak{D}K)^{-1}\big)\mathfrak{C}\mathfrak{B}(1-K\mathfrak{D})^{-1}\\
&= (1-\mathfrak{D}K)^{-1}\mathfrak{C}\mathfrak{B}(1-K\mathfrak{D})^{-1}\\
&= \mathfrak{C}^K\mathfrak{B}^K.
\end{aligned}$$

□

Remark 7.1.3 The system Σ^K defined in (7.1.3) can be written in many different equivalent block matrix forms, namely

$$
\begin{aligned}
\Sigma^K &= \left[\begin{array}{c|c} \mathfrak{A}^K & \mathfrak{B}^K\tau \\ \hline \mathfrak{C}^K & \mathfrak{D}^K \end{array}\right] \\
&= \left[\begin{array}{c|c} \mathfrak{A} + \mathfrak{B}\tau K (1 - \mathfrak{D}K)^{-1}\mathfrak{C} & \mathfrak{B}(1 - K\mathfrak{D})^{-1}\tau \\ \hline (1 - \mathfrak{D}K)^{-1}\mathfrak{C} & \mathfrak{D}(1 - K\mathfrak{D})^{-1} \end{array}\right] \\
&= \left[\begin{array}{c|c} \mathfrak{A} & \mathfrak{B}\tau \\ \hline \mathfrak{C} & \mathfrak{D} \end{array}\right] \left[\begin{array}{c|c} 1 & 0 \\ \hline -K\mathfrak{C} & 1 - K\mathfrak{D} \end{array}\right]^{-1} \\
&= \left[\begin{array}{c|c} 1 & -\mathfrak{B}K\tau \\ \hline 0 & 1 - \mathfrak{D}K \end{array}\right]^{-1} \left[\begin{array}{c|c} \mathfrak{A} & \mathfrak{B}\tau \\ \hline \mathfrak{C} & \mathfrak{D} \end{array}\right] \\
&= \left[\begin{array}{c|c} \mathfrak{A} & \mathfrak{B}\tau \\ \hline \mathfrak{C} & \mathfrak{D} \end{array}\right] + \left[\begin{array}{c} \mathfrak{B}\tau \\ \hline \mathfrak{D} \end{array}\right] K (1 - \mathfrak{D}K)^{-1} \left[\begin{array}{c|c} \mathfrak{C} & \mathfrak{D} \end{array}\right] \\
&= \left[\begin{array}{c|c} \mathfrak{A} & \mathfrak{B}\tau \\ \hline \mathfrak{C} & \mathfrak{D} \end{array}\right] + \left[\begin{array}{c} \mathfrak{B}\tau \\ \hline \mathfrak{D} \end{array}\right] K \left[\begin{array}{c|c} \mathfrak{C}^K & \mathfrak{D}^K \end{array}\right] \\
&= \left[\begin{array}{c|c} \mathfrak{A} & \mathfrak{B}\tau \\ \hline \mathfrak{C} & \mathfrak{D} \end{array}\right] + \left[\begin{array}{c} \mathfrak{B}^K\tau \\ \hline \mathfrak{D}^K \end{array}\right] K \left[\begin{array}{c|c} \mathfrak{C} & \mathfrak{D} \end{array}\right].
\end{aligned}
\tag{7.1.5}
$$

We leave the simple algebraic proofs of these identities to the reader. It is possible to give direct interpretations of these block matrix forms of the closed-loop system in terms of the diagram in Figure 7.1. The factor $\left[\begin{array}{c|c} 1 & 0 \\ \hline -K\mathfrak{C} & 1-K\mathfrak{D} \end{array}\right]^{-1}$ in the first block matrix form is simply the mapping from $\left[\begin{smallmatrix} x_0 \\ v \end{smallmatrix}\right]$ to $\left[\begin{smallmatrix} x_0 \\ u \end{smallmatrix}\right]$. The factor $\left[\begin{array}{c|c} \mathfrak{A} & \mathfrak{B}\tau \\ \hline \mathfrak{C} & \mathfrak{D} \end{array}\right]$ in the second block matrix form is the (open-loop) mapping from $\left[\begin{smallmatrix} x_0 \\ v \end{smallmatrix}\right]$ to $\left[\begin{smallmatrix} x(t) \\ y \end{smallmatrix}\right]$ if we cut the loop at either of the two loop break points (marked with crosses), and the factor $\left[\begin{array}{c|c} 1 & -\mathfrak{B}K\tau \\ \hline 0 & 1-\mathfrak{D}K \end{array}\right]^{-1}$ is the correction term which is introduced when the loop is closed. In the next block matrix form the first term $\left[\begin{array}{c|c} \mathfrak{A} & \mathfrak{B}\tau \\ \hline \mathfrak{C} & \mathfrak{D} \end{array}\right]$ on the right hand side is the system response from $\left[\begin{smallmatrix} x_0 \\ \pi_+ v \end{smallmatrix}\right]$ to $\left[\begin{smallmatrix} x \\ y \end{smallmatrix}\right]$ when the loop is cut at the break point located at the input to K. In the second term, $\left[\begin{array}{c|c} \mathfrak{C} & \mathfrak{D} \end{array}\right]$ is the open-loop mapping from $\left[\begin{smallmatrix} x_0 \\ \pi_+ v \end{smallmatrix}\right]$ to this break point, $(1 - \mathfrak{D}K)^{-1}$ is the closed-loop response to an external signal entering at the break point (the closed-loop system sensitivity to an external signal entering at this point), and $\left[\begin{array}{c} \mathfrak{B}\tau \\ \hline \mathfrak{D} \end{array}\right] K$ is the open-loop mapping from the break point to $\left[\begin{smallmatrix} x \\ y \end{smallmatrix}\right]$. The last two block matrix forms have similar interpretations.

Remark 7.1.4 If in the classical system

$$\begin{aligned}\dot{x}(t) &= Ax(t) + Bu(t),\\ y(t) &= Cx(t) + Du(t), \qquad t \ge 0,\\ x(0) &= x_0,\end{aligned} \tag{7.1.6}$$

we replace u by $u = Ky + v$, then we get a new well-defined system of the same type iff $1 - DK$ is invertible, or equivalently, iff $1 - KD$ is invertible. In the new system the operators $\left[\begin{array}{c|c} A & B \\ \hline C & D \end{array}\right]$ have been replaced by

$$\begin{aligned}
\left[\begin{array}{c|c} A^K & B^K \\ \hline C^K & D^K \end{array}\right] &= \left[\begin{array}{c|c} A + BK(1 - DK)^{-1}C & B(1 - KD)^{-1} \\ \hline (1 - DK)^{-1}C & D(1 - KD)^{-1} \end{array}\right] \\
&= \left[\begin{array}{c|c} A & B \\ \hline C & D \end{array}\right] \left[\begin{array}{c|c} 1 & 0 \\ \hline -KC & 1 - KD \end{array}\right]^{-1} \\
&= \left[\begin{array}{c|c} 1 & -BK \\ \hline 0 & 1 - DK \end{array}\right]^{-1} \left[\begin{array}{c|c} A & B \\ \hline C & D \end{array}\right] \\
&= \left[\begin{array}{c|c} A & B \\ \hline C & D \end{array}\right] + \left[\begin{array}{c} B \\ \hline D \end{array}\right] K(1 - DK)^{-1} \left[\begin{array}{c|c} C & D \end{array}\right] \\
&= \left[\begin{array}{c|c} A & B \\ \hline C & D \end{array}\right] + \left[\begin{array}{c} B \\ \hline D \end{array}\right] K \left[\begin{array}{c|c} C^K & D^K \end{array}\right] \\
&= \left[\begin{array}{c|c} A & B \\ \hline C & D \end{array}\right] + \left[\begin{array}{c} B^K \\ \hline D^K \end{array}\right] K \left[\begin{array}{c|c} C & D \end{array}\right].
\end{aligned} \tag{7.1.7}$$

Observe the striking similarity between this formula and (7.1.5). (Usually the feedthrough operator D is taken to be zero, in which case this formula simplifies significantly and the invertibility condition on $1 - DK$ drops out.) This remark is actually valid in a much more general setting; see Remark 7.7.4.

Definition 7.1.5 With reference to Figure 7.1, we call $K\mathfrak{D}$ the *input loop gain operator*, $\mathfrak{D}K$ the *output loop gain operator*, $1 - K\mathfrak{D}$ the *input return difference operator*, $1 - \mathfrak{D}K$ the *output return difference operator*, $(1 - K\mathfrak{D})^{-1}$ the *input sensitivity operator*, $(1 - \mathfrak{D}K)^{-1}$ the *output sensitivity operator*, $K\mathfrak{D}(1 - K\mathfrak{D})^{-1}$ the *input complementary sensitivity operator*, and $\mathfrak{D}K(1 - \mathfrak{D}K)^{-1}$ the *output complementary sensitivity operator*.

Remark 7.1.6 It is possible, without loss of generality, to assume that $Y = U$ and that $K = 1$ in Theorem 7.1.2 by modifying the system Σ; see Theorem

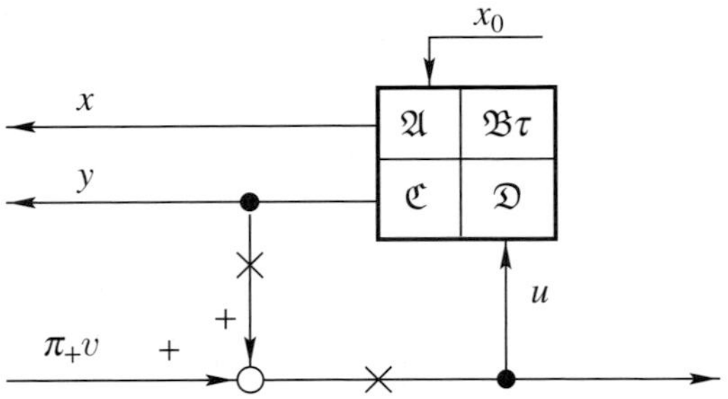

Figure 7.2 Positive identity feedback

7.2.1 and Figure 7.7. We shall refer to this case as (positive) *identity feedback*. In this case the formulas for the closed-loop system drawn in Figure 7.2 with input v and outputs $\left[\begin{smallmatrix} y \\ u \end{smallmatrix}\right]$ become

$$\left[\begin{array}{c|c} \mathfrak{A} + \mathfrak{B}\tau\,(1-\mathfrak{D})^{-1}\,\mathfrak{C} & \mathfrak{B}\,(1-\mathfrak{D})^{-1}\,\tau \\ \hline (1-\mathfrak{D})^{-1}\,\mathfrak{C} & \mathfrak{D}\,(1-\mathfrak{D})^{-1} \\ (1-\mathfrak{D})^{-1}\,\mathfrak{C} & (1-\mathfrak{D})^{-1} \end{array}\right]$$

$$= \left[\begin{array}{c|c} \mathfrak{A} & \mathfrak{B}\tau \\ \hline \mathfrak{C} & \mathfrak{D} \\ 0 & 1 \end{array}\right] \left[\begin{array}{c|c} 1 & 0 \\ \hline -\mathfrak{C} & 1-\mathfrak{D} \end{array}\right]^{-1}$$

$$= \left[\begin{array}{c|cc} 1 & -\mathfrak{B}\tau & 0 \\ \hline 0 & 1-\mathfrak{D} & 0 \\ 0 & -1 & 1 \end{array}\right]^{-1} \left[\begin{array}{c|c} \mathfrak{A} & \mathfrak{B}\tau \\ \hline \mathfrak{C} & \mathfrak{D} \\ 0 & 1 \end{array}\right]$$

$$= \left[\begin{array}{c|c} \mathfrak{A} & \mathfrak{B}\tau \\ \hline \mathfrak{C} & \mathfrak{D} \\ 0 & 1 \end{array}\right] + \left[\begin{array}{c} \mathfrak{B}\tau \\ \hline \mathfrak{D} \\ 1 \end{array}\right] (1-\mathfrak{D})^{-1} \left[\begin{array}{c|c} \mathfrak{C} & \mathfrak{D} \end{array}\right]$$

$$= \left[\begin{array}{c|c} \mathfrak{A} & 0 \\ \hline 0 & -1 \\ 0 & 0 \end{array}\right] + \left[\begin{array}{c} \mathfrak{B}\tau \\ \hline 1 \\ 1 \end{array}\right] (1-\mathfrak{D})^{-1} \left[\begin{array}{c|c} \mathfrak{C} & 1 \end{array}\right].$$

The interpretations of these block matrix forms in terms of Figure 7.2 are similar to those given in Remark 7.1.3. If we instead use *negative identity feedback* as drawn in Figure 7.3, then we get the analogous closed-loop

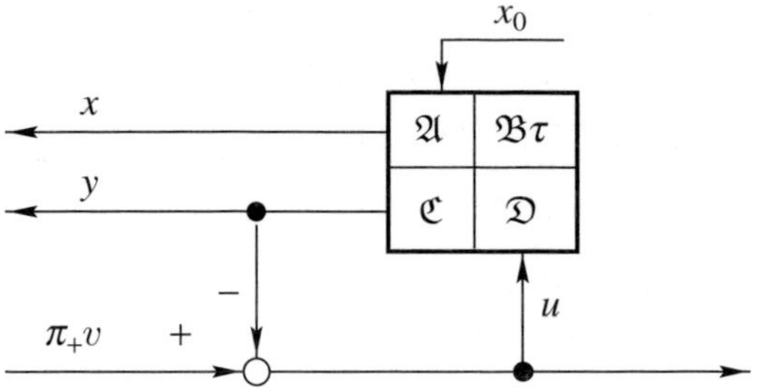

Figure 7.3 Negative identity feedback

system

$$
\begin{aligned}
&\left[\begin{array}{c|c} \mathfrak{A} - \mathfrak{B}\tau\,(1+\mathfrak{D})^{-1}\,\mathfrak{C} & \mathfrak{B}\,(1+\mathfrak{D})^{-1}\,\tau \\ \hline (1+\mathfrak{D})^{-1}\,\mathfrak{C} & \mathfrak{D}\,(1+\mathfrak{D})^{-1} \\ -\,(1+\mathfrak{D})^{-1}\,\mathfrak{C} & (1+\mathfrak{D})^{-1} \end{array}\right] \\
&\quad = \left[\begin{array}{c|c} \mathfrak{A} & \mathfrak{B}\tau \\ \hline \mathfrak{C} & \mathfrak{D} \\ 0 & 1 \end{array}\right] \left[\begin{array}{c|c} 1 & 0 \\ \hline \mathfrak{C} & 1+\mathfrak{D} \end{array}\right]^{-1} \\
&\quad = \left[\begin{array}{c|cc} 1 & \mathfrak{B}\tau & 0 \\ \hline 0 & 1+\mathfrak{D} & 0 \\ 0 & 1 & 1 \end{array}\right]^{-1} \left[\begin{array}{c|c} \mathfrak{A} & \mathfrak{B}\tau \\ \hline \mathfrak{C} & \mathfrak{D} \\ 0 & 1 \end{array}\right] \\
&\quad = \left[\begin{array}{c|c} \mathfrak{A} & \mathfrak{B}\tau \\ \hline \mathfrak{C} & \mathfrak{D} \\ 0 & 1 \end{array}\right] - \left[\begin{array}{c} \mathfrak{B}\tau \\ \hline \mathfrak{D} \\ 1 \end{array}\right] (1+\mathfrak{D})^{-1} \left[\begin{array}{c|c} \mathfrak{C} & \mathfrak{D} \end{array}\right] \\
&\quad = \left[\begin{array}{c|c} \mathfrak{A} & 0 \\ \hline 0 & 1 \\ 0 & 0 \end{array}\right] - \left[\begin{array}{c} \mathfrak{B}\tau \\ \hline -1 \\ 1 \end{array}\right] (1+\mathfrak{D})^{-1} \left[\begin{array}{c|c} \mathfrak{C} & -1 \end{array}\right].
\end{aligned}
$$

Repeated feedbacks behave in the expected way:

Lemma 7.1.7 *Let $K \in \mathcal{B}(Y;U)$ be an admissible output feedback operator for Σ. Then $L \in \mathcal{B}(Y;U)$ is an admissible output feedback operator for the closed-loop system Σ^K iff $L+K$ is an admissible output feedback operator for Σ, and $\Sigma^{L+K} = (\Sigma^K)^L$. In particular, $-K$ is always an admissible feedback operator for Σ^K, and $(\Sigma^K)^{-K} = \Sigma$.*

Proof That $L \in \mathcal{B}(Y;U)$ is an admissible output feedback operator for the closed-loop system Σ^K iff $K+L$ is an admissible output feedback operator for Σ follows from the fact that

$$\begin{aligned}1 - L\mathfrak{D}^K &= 1 - L\mathfrak{D}(1 - K\mathfrak{D})^{-1} = (1 - K\mathfrak{D} - L\mathfrak{D})(1 - K\mathfrak{D})^{-1}\\ &= \big(1 - (L + K)\mathfrak{D}\big)(1 - K\mathfrak{D})^{-1}\end{aligned}$$

is invertible in $TIC_{\mathrm{loc}}(U)$ if and only if $\big(1 - (L + K)\mathfrak{D}\big)$ is invertible in $TIC_{\mathrm{loc}}(U)$. Moreover, this computation and another similar one where $1 - L\mathfrak{D}^K$ is replaced by $1 - \mathfrak{D}^K L$ shows that

$$\begin{aligned}(1 - L\mathfrak{D}^K)^{-1} &= (1 - K\mathfrak{D})\big(1 - (L + K)\mathfrak{D}\big)^{-1},\\ (1 - \mathfrak{D}^K L)^{-1} &= \big(1 - \mathfrak{D}(L + K)\big)^{-1}(1 - \mathfrak{D}K).\end{aligned}$$

Using these facts in (7.1.5) we get $\Sigma^{L+K} = \big(\Sigma^K\big)^L$. □

As the following theorem shows, there are several different ways to characterize the admissibility of a feedback operator:

Theorem 7.1.8 *Let $\Sigma = \left[\begin{smallmatrix}\mathfrak{A} & \mathfrak{B}\\ \mathfrak{C} & \mathfrak{D}\end{smallmatrix}\right]$ be a well-posed linear system on (Y, X, U) with growth bound $\omega_{\mathfrak{A}}$, and let $K \in \mathcal{B}(Y; U)$. Then the following conditions are equivalent:*

(i) *K is an admissible output feedback operator for Σ.*
(ii) *For some $T > 0$, the operator $\pi_{[0,T)}(1 - K\mathfrak{D})\pi_{[0,T)}$ has an inverse in $\mathcal{B}(L^p|Reg([0, T); U))$.*
(ii′) *For some $T > 0$, the operator $\pi_{[0,T)}(1 - \mathfrak{D}K)\pi_{[0,T)}$ has an inverse in $\mathcal{B}(L^p|Reg([0, T); Y))$.*
(iii) *For all s and t, $-\infty < s < t < \infty$, the operator $\pi_{[s,t)}(1 - K\mathfrak{D})\pi_{[s,t)}$ has an inverse in $\mathcal{B}(L^p|Reg([s, t); U))$.*
(iii′) *For all s and t, $-\infty < s < t < \infty$, the operator $\pi_{[s,t)}(1 - \mathfrak{D}K)\pi_{[s,t)}$ has an inverse in $\mathcal{B}(L^p|Reg([s, t); U))$ for all s and t.*
(iv) *The operator $1 - K\mathfrak{D}$ has an inverse in $TIC_{\mathrm{loc}}(U)$.*
(iv′) *The operator $1 - \mathfrak{D}K$ has an inverse in $TIC_{\mathrm{loc}}(Y)$.*
(v) *The operator $1 - K\mathfrak{D}$ has an inverse in $TIC_\alpha(U)$ for some $\alpha \geq \omega_{\mathfrak{A}}$.*
(v′) *The operator $1 - \mathfrak{D}K$ has an inverse in $TIC_\alpha(Y)$ for some $\alpha \geq \omega_{\mathfrak{A}}$.*
(vi) *The functions x and y in the diagram in Figure 7.1 (i.e., the set of equations (7.1.1)) can be interpreted as the state and output of a well-posed linear system with initial state $x_0 \in X$ and input $v \in L^p|Reg_{\mathrm{loc}}(\overline{\mathbb{R}}^+; U)$.*
(vii) *There is some $\alpha \geq \omega_{\mathfrak{A}}$ for which the diagram in Figure 7.1 (i.e., the set of equations (7.1.1)) with $x_0 = 0$ defines a continuous linear mapping from the external input $v \in L^p|Reg_\alpha(\overline{\mathbb{R}}^+; U)$ to the internal input $u \in L^p|Reg_\alpha(\overline{\mathbb{R}}^+; U)$.*

Proof That (ii) ⇔ (ii′), (iii) ⇔ (iii′), (iv) ⇔ (iv′), and (v) ⇔ (v′) follows from Lemma A.4.1. The proof of the equivalence of (i)–(vi) is a copy of the proof of the corresponding equivalence in Theorem 6.3.5, with $\mathfrak{D}$ replaced by $1 - K\mathfrak{D}$. Thus, the only additional thing which we have to prove is that (vii) is equivalent to the other conditions.

(vi) ⇒ (vii): By (7.1.1), $u = Ky + \pi_+ v$. Thus, if (vi) holds, then it follows from Theorem 2.5.4 that the mapping $v \mapsto u$ is continuous from $L^p|Reg_\alpha(\overline{\mathbb{R}}^+; U)$ to itself. Thus (vii) holds.

(vii) ⇒ (v): By taking $x_0 = 0$ and eliminating the variable z from (7.1.1) we get

$$(1 - K\mathfrak{D})u = \pi_+ v.$$

This equation determines u uniquely and continuously in $L^p|Reg_\alpha(\overline{\mathbb{R}}^+; U)$ for some $\alpha \in \mathbb{R}$ as a function of $v \in L^p|Reg_\alpha(\overline{\mathbb{R}}^+; U)$ if and only if $1 - K\mathfrak{D}$ has an inverse in $TIC_\alpha(U)$. □

Corollary 7.1.9 *Let $\Sigma = \left[\begin{smallmatrix} \mathfrak{A} & \mathfrak{B} \\ \mathfrak{C} & \mathfrak{D} \end{smallmatrix}\right]$ be a well-posed linear system on (Y, X, U), and let $K \in \mathcal{B}(Y; U)$. Then K is an admissible output feedback operator for Σ if any one of the following conditions hold:*

(i) $\|\pi_{[0,T)}(K\mathfrak{D})\pi_{[0,T)}\|_{\mathcal{B}(L^p|Reg([0,T);U))} < 1$ *for some* $T > 0$;
(ii) $\|\pi_{[0,T)}(\mathfrak{D}K)\pi_{[0,T)}\|_{\mathcal{B}(L^p|Reg([0,T);Y))} < 1$ *for some* $T > 0$;
(iii) $\lim_{T\downarrow 0}\|\pi_{[0,T)}\mathfrak{D}\pi_{[0,T)}\|_{\mathcal{B}(L^p|Reg([0,T);U);L^p|Reg([0,T);Y))} = 0$;
(iv) *$\mathfrak{D}$ is a convolution operator of the type given in Theorem A.3.7(i);*
(v) *$p = 1$ and $\mathfrak{D}$ is a convolution operator of the type given in Theorem A.3.8.*

Proof (i)–(ii) In these cases condition (ii) or (ii′) in Theorem 7.1.8 is satisfied.
(iii) This is a special case of both (i) and (ii).
(iv)–(v) These two cases are contained in (iii). □

Remark 7.1.10 Corollary 6.3.12 enables us to interpret a flow-inversion as a feedback connection whenever $\rho(A) \cap \rho(A_\times) \neq \emptyset$; see Figure 7.4. In this figure

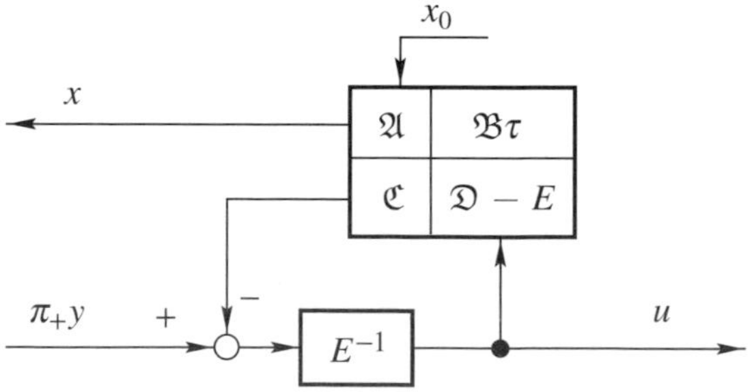

Figure 7.4 Flow-inversion

E is an arbitrary continuous linear operator which maps U onto Y, and the result is independent of the choice of E. For example, we may take $E = \widehat{\mathfrak{D}}(\alpha)$ with $\alpha \in \rho(A) \cap \rho(A_\times)$.

7.2 Additional feedback connections

It is possible to add some inputs and outputs to the basic output feedback connection described in the preceding section. This does not have any significant impact on the theory, as long as the feedback loop stays the same. For example, let us consider the feedback connection drawn in Figure 7.5. The difference compared to Figure 7.1 is that we have added another input labeled w and regarded the internal signal z as an additional output. An almost equivalent connection is drawn in Figure 7.6. Here the inputs are the same as in Figure 7.5, but the outputs are y_1 and u instead of y and z. The appropriate equations describing these systems are

$$\begin{aligned} x(t) &= \mathfrak{A}^t x_0 + \mathfrak{B}\tau^t u, \\ y &= \mathfrak{C}x_0 + \mathfrak{D}u, \\ z &= Ky + K\pi_+ w, \\ y_1 &= y + \pi_+ w, \\ u &= z + \pi_+ v, \end{aligned} \tag{7.2.1}$$

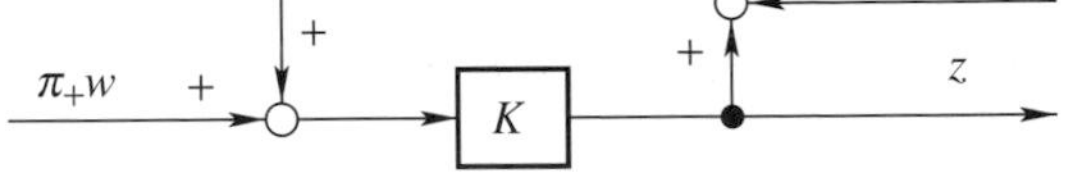

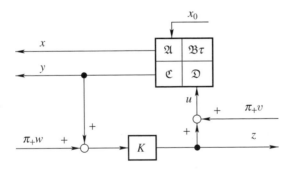

Figure 7.5 Another static output feedback

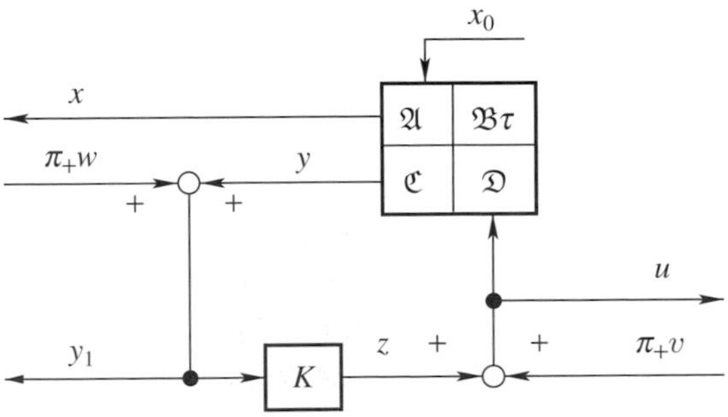

Figure 7.6 A third static output feedback

which can be formally solved as

$$
\begin{aligned}
x(t) &= \big(\mathfrak{A}^t + \mathfrak{B}\tau^t K (1-\mathfrak{D}K)^{-1}\mathfrak{C}\big)x_0 + \mathfrak{B}(1-K\mathfrak{D})^{-1}K\tau^t\pi_+ w,\\
&\quad + \mathfrak{B}(1-K\mathfrak{D})^{-1}\tau^t\pi_+ v,\\
y &= (1-\mathfrak{D}K)^{-1}(\mathfrak{C}x_0 + \mathfrak{D}K\pi_+ w + \mathfrak{D}\pi_+ v),\\
z &= (1-K\mathfrak{D})^{-1}K(\mathfrak{C}x_0 + \pi_+ w + \mathfrak{D}\pi_+ v),\\
y_1 &= (1-\mathfrak{D}K)^{-1}(\mathfrak{C}x_0 + \pi_+ w + \mathfrak{D}\pi_+ v),\\
u &= (1-K\mathfrak{D})^{-1}(K\mathfrak{C}x_0 + K\pi_+ w + \pi_+ v).
\end{aligned}
\tag{7.2.3}
$$

This leads us to the following theorem.

Theorem 7.2.1 *Let* $\Sigma = \left[\begin{smallmatrix}\mathfrak{A} & \mathfrak{B}\\ \mathfrak{C} & \mathfrak{D}\end{smallmatrix}\right]$ *be a well-posed linear system on* (Y, X, U), *and let* $K \in \mathcal{B}(Y; U)$ *be an admissible output feedback operator for* Σ.

(i) *The system*

$$
\left[\begin{array}{c|cc}
\mathfrak{A} + \mathfrak{B}\tau K(1-\mathfrak{D}K)^{-1}\mathfrak{C} & \mathfrak{B}(1-K\mathfrak{D})^{-1}K\tau & \mathfrak{B}(1-K\mathfrak{D})^{-1}\tau\\ \hline
(1-\mathfrak{D}K)^{-1}\mathfrak{C} & \mathfrak{D}(1-K\mathfrak{D})^{-1}K & \mathfrak{D}(1-K\mathfrak{D})^{-1}\\
K(1-\mathfrak{D}K)^{-1}\mathfrak{C} & (1-K\mathfrak{D})^{-1}K & K(1-\mathfrak{D}K)^{-1}\mathfrak{D}
\end{array}\right]
$$

$$
= \left[\begin{array}{c|cc}
\mathfrak{A} & 0 & \mathfrak{B}\tau\\ \hline
\mathfrak{C} & 0 & \mathfrak{D}\\
0 & K & 0
\end{array}\right]
\left[\begin{array}{c|cc}
1 & 0 & 0\\ \hline
-\mathfrak{C} & 1 & -\mathfrak{D}\\
0 & -K & 1
\end{array}\right]^{-1}
$$

$$
= \left[\begin{array}{c|cc}
1 & 0 & -\mathfrak{B}\tau\\ \hline
0 & 1 & -\mathfrak{D}\\
0 & -K & 1
\end{array}\right]^{-1}
\left[\begin{array}{c|cc}
\mathfrak{A} & 0 & \mathfrak{B}\tau\\ \hline
\mathfrak{C} & 0 & \mathfrak{D}\\
0 & K & 0
\end{array}\right]
$$

$$
= \left[\begin{array}{c|cc}
\mathfrak{A} & 0 & \mathfrak{B}\tau\\ \hline
\mathfrak{C} & 0 & \mathfrak{D}\\
0 & K & 0
\end{array}\right]
+ \left[\begin{array}{cc}
0 & \mathfrak{B}\tau\\ \hline
0 & \mathfrak{D}\\
K & 0
\end{array}\right]
\begin{bmatrix} 1 & -\mathfrak{D}\\ -K & 1\end{bmatrix}^{-1}
\left[\begin{array}{c|cc}
\mathfrak{C} & 0 & \mathfrak{D}\\
0 & K & 0
\end{array}\right]
$$

$$
= \left[\begin{array}{c|cc}
\mathfrak{A} & 0 & 0\\ \hline
0 & -1 & 0\\
0 & 0 & -1
\end{array}\right]
+ \left[\begin{array}{cc}
0 & \mathfrak{B}\tau\\ \hline
1 & 0\\
0 & 1
\end{array}\right]
\begin{bmatrix} 1 & -\mathfrak{D}\\ -K & 1\end{bmatrix}^{-1}
\left[\begin{array}{c|cc}
\mathfrak{C} & 1 & 0\\
0 & 0 & 1
\end{array}\right]
$$

where

$$
\begin{bmatrix} 1 & -\mathfrak{D}\\ -K & 1\end{bmatrix}^{-1}
= \begin{bmatrix} (1-\mathfrak{D}K)^{-1} & \mathfrak{D}(1-K\mathfrak{D})^{-1}\\ (1-K\mathfrak{D})^{-1}K & (1-K\mathfrak{D})^{-1}\end{bmatrix}
$$

is another well-posed linear system on (Y, X, U). *The state* $x(t)$ *at time* t *and the outputs* y *and* z *of this system with initial time zero, initial value* x_0, *and inputs* w *and* v *are given by* (7.2.2), *and it is the closed-loop system corresponding to the feedback drawn in Figure 7.5.*

(ii) *The system*

$$
\left[\begin{array}{c|cc}
\mathfrak{A} + \mathfrak{B}\tau K (1-\mathfrak{D}K)^{-1}\mathfrak{C} & \mathfrak{B}(1-K\mathfrak{D})^{-1}K\tau & \mathfrak{B}(1-K\mathfrak{D})^{-1}\tau \\
\hline
(1-\mathfrak{D}K)^{-1}\mathfrak{C} & (1-\mathfrak{D}K)^{-1} & \mathfrak{D}(1-K\mathfrak{D})^{-1} \\
K(1-\mathfrak{D}K)^{-1}\mathfrak{C} & (1-K\mathfrak{D})^{-1}K & (1-K\mathfrak{D})^{-1}
\end{array}\right]
$$

$$
= \left[\begin{array}{c|cc}
\mathfrak{A} & 0 & \mathfrak{B}\tau \\
\hline
0 & 1 & 0 \\
0 & 0 & 1
\end{array}\right]
\left[\begin{array}{c|cc}
1 & 0 & 0 \\
\hline
-\mathfrak{C} & 1 & -\mathfrak{D} \\
0 & -K & 1
\end{array}\right]^{-1}
$$

$$
= \left[\begin{array}{c|cc}
1 & 0 & -\mathfrak{B}\tau \\
\hline
0 & 1 & -\mathfrak{D} \\
0 & -K & 1
\end{array}\right]^{-1}
\left[\begin{array}{c|cc}
\mathfrak{A} & 0 & 0 \\
\hline
\mathfrak{C} & 1 & 0 \\
0 & 0 & 1
\end{array}\right]
$$

$$
= \left[\begin{array}{c|cc}
\mathfrak{A} & 0 & \mathfrak{B}\tau \\
\hline
\mathfrak{C} & 1 & \mathfrak{D} \\
0 & K & 1
\end{array}\right]
+ \left[\begin{array}{cc}
0 & \mathfrak{B}\tau \\
\hline
0 & \mathfrak{D} \\
K & 0
\end{array}\right]
\begin{bmatrix}
1 & -\mathfrak{D} \\
-K & 1
\end{bmatrix}^{-1}
\left[\begin{array}{c|cc}
\mathfrak{C} & 0 & \mathfrak{D} \\
0 & K & 0
\end{array}\right]
$$

$$
= \left[\begin{array}{c|cc}
\mathfrak{A} & 0 & 0 \\
\hline
0 & 0 & 0 \\
0 & 0 & 0
\end{array}\right]
+ \left[\begin{array}{cc}
0 & \mathfrak{B}\tau \\
\hline
1 & 0 \\
0 & 1
\end{array}\right]
\begin{bmatrix}
1 & -\mathfrak{D} \\
-K & 1
\end{bmatrix}^{-1}
\left[\begin{array}{c|cc}
\mathfrak{C} & 1 & 0 \\
0 & 0 & 1
\end{array}\right]
$$

is another well-posed linear system on (Y, X, U). *The state* $x(t)$ *at time* t *and the outputs* y_1 *and* u *of this system with initial time zero, initial value* x_0, *and inputs* w *and* v *are given by* (7.2.3), *and it is the closed-loop system corresponding to the feedback drawn in Figure 7.6.*

Proof The simplest way to prove (i) is to observe that Figure 7.5 is equivalent to (a part of) Figure 7.7, which is an identity output feedback connection for

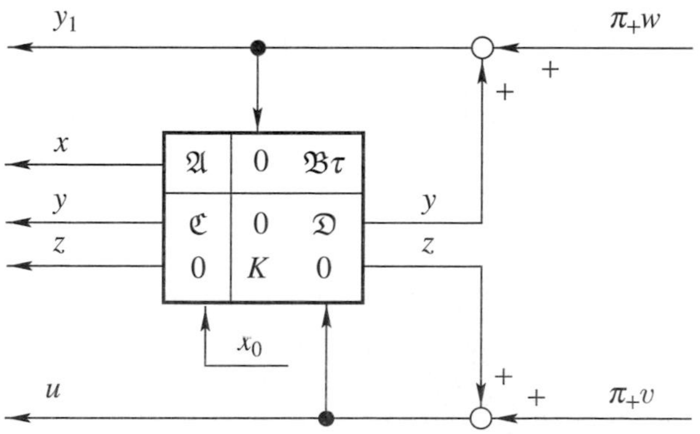

Figure 7.7 Output feedback in block form

the system

$$\begin{bmatrix} \mathfrak{A} & 0 & \mathfrak{B}\tau \\ \hline \mathfrak{C} & 0 & \mathfrak{D} \\ 0 & K & 0 \end{bmatrix},$$

cf. Remark 7.1.6. This identity output feedback operator is admissible iff $\left[\begin{smallmatrix} 1 & -\mathfrak{D} \\ -K & 1 \end{smallmatrix}\right]$ is invertible in $TIC_{\mathrm{loc}}(\left[\begin{smallmatrix} Y \\ U \end{smallmatrix}\right])$, and by Lemma A.4.1, this condition is equivalent to the admissibility of K as an output feedback operator for the original system. To prove (ii) it suffices to observe that the only difference between the system drawn in Figure 7.6 and the one drawn in Figure 7.5 is that there is an extra feedthrough term $\left[\begin{smallmatrix} 1 & 0 \\ 0 & 1 \end{smallmatrix}\right]$ from the inputs to the outputs. □

Observe that Figure 7.7 has been drawn in such a way that it contains all of Figures 7.1, 7.5, and 7.6. To get Figure 7.1 we delete the input w and the outputs y_1, z, and u. To get Figure 7.5 we delete the outputs y_1 and u. To get Figure 7.6 we delete the outputs y and z.

The preceding result enable us to add some additional equivalent conditions to the list in Theorem 7.1.8.

Corollary 7.2.2 *Let $\Sigma = \left[\begin{smallmatrix} \mathfrak{A} & \mathfrak{B} \\ \mathfrak{C} & \mathfrak{D} \end{smallmatrix}\right]$ be a well-posed linear system on (Y, X, U) with growth bound $\omega_{\mathfrak{A}}$, and let $K \in \mathcal{B}(Y; U)$. Then the following conditions are equivalent to those listed in Theorem 7.1.8:*

(ii″) *The operator $\pi_{[0,T)} \left[\begin{smallmatrix} 1 & -\mathfrak{D} \\ -K & 1 \end{smallmatrix}\right] \pi_{[0,T)}$ has an inverse in $\mathcal{B}\big(L^p|Reg([0,T); \left[\begin{smallmatrix} Y \\ U \end{smallmatrix}\right])\big)$ for some $T > 0$.*

(iii″) *The operator $\pi_{[s,t)} \left[\begin{smallmatrix} 1 & -\mathfrak{D} \\ -K & 1 \end{smallmatrix}\right] \pi_{[s,t)}$ has an inverse in $\mathcal{B}\big(L^p|Reg([s,t); \left[\begin{smallmatrix} Y \\ U \end{smallmatrix}\right])\big)$ for all s and t, $-\infty < s < t < \infty$.*

(iv″) *The operator $\left[\begin{smallmatrix} 1 & -\mathfrak{D} \\ -K & 1 \end{smallmatrix}\right]$ has an inverse in $TIC_{\mathrm{loc}}(\left[\begin{smallmatrix} Y \\ U \end{smallmatrix}\right])$.*

(v″) *The operator $\left[\begin{smallmatrix} 1 & -\mathfrak{D} \\ -K & 1 \end{smallmatrix}\right]$ has an inverse in $TIC_\alpha(\left[\begin{smallmatrix} Y \\ U \end{smallmatrix}\right])$ for some $\alpha \ge \omega_{\mathfrak{A}}$.*

(vi″) *The functions x and $\left[\begin{smallmatrix} y \\ z \end{smallmatrix}\right]$ in the diagrams in Figures 7.5 and 7.7 can be interpreted as the state and output of a well-posed linear system with initial state $x_0 \in X$ and input $\left[\begin{smallmatrix} w \\ v \end{smallmatrix}\right] \in L^p|Reg_{\mathrm{loc}}(\overline{\mathbb{R}}^+; \left[\begin{smallmatrix} Y \\ U \end{smallmatrix}\right])$.*

(vi‴) *The functions x and $\left[\begin{smallmatrix} y_1 \\ u \end{smallmatrix}\right]$ in the diagrams in Figures 7.6 and 7.7 can be interpreted as the state and output of a well-posed linear system with initial state $x_0 \in X$ and input $\left[\begin{smallmatrix} w \\ v \end{smallmatrix}\right] \in L^p|Reg_{\mathrm{loc}}(\overline{\mathbb{R}}^+; \left[\begin{smallmatrix} Y \\ U \end{smallmatrix}\right])$.*

This follows from Theorem 7.2.1 and its proof.

Remark 7.2.3 The diagrams in Figure 7.6 and 7.7 have the interesting property that it is possible to recover the extended version of the original system $\left[\begin{smallmatrix} \mathfrak{A} & \mathfrak{B} \\ \mathfrak{C} & \mathfrak{D} \end{smallmatrix}\right]$ drawn in Figure 7.8 by reversing the directions of two of the lines, i.e, by interchanging the inputs $\left[\begin{smallmatrix} w \\ v \end{smallmatrix}\right]$ with the outputs $\left[\begin{smallmatrix} y_1 \\ u \end{smallmatrix}\right]$. This is a typical case of partial flow-inversion. Actually, it suffices to change the direction of one of the

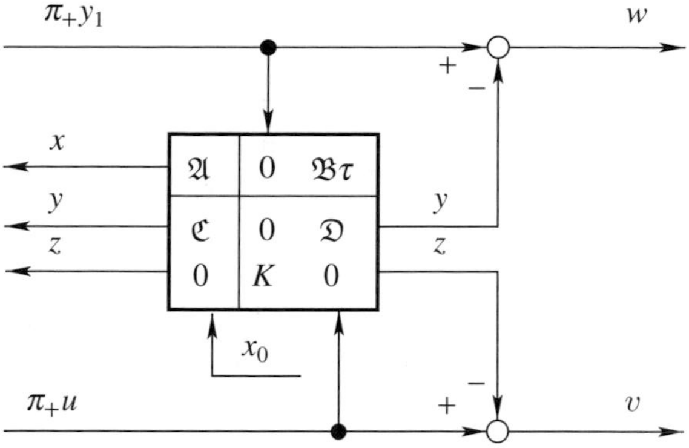

Figure 7.8 Cancellation of static output feedback

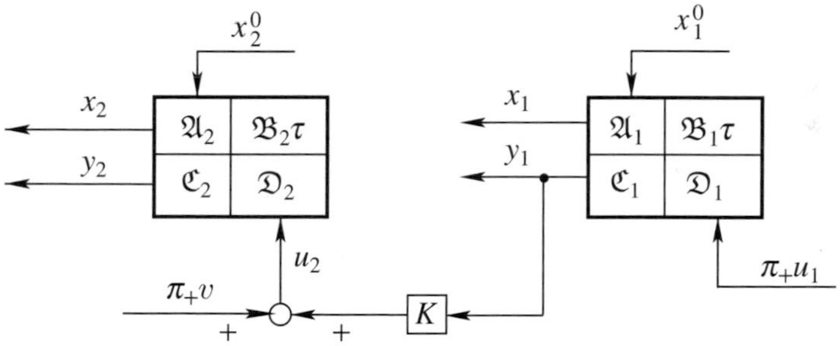

Figure 7.9 Cascade connection through K

two lines to get (slightly different) extended versions of the original system. This is maybe most easily seen from Figure 7.6.

Example 7.2.4 *Let* $\Sigma_1 = \left[\begin{array}{c|c} \mathfrak{A}_1 & \mathfrak{B}_1 \\ \hline \mathfrak{C}_1 & \mathfrak{D}_1 \end{array}\right]$ *be a well-posed linear system on* (Y_1, X_1, U_1)*, let* $\Sigma_2 = \left[\begin{array}{c|c} \mathfrak{A}_2 & \mathfrak{B}_2 \\ \hline \mathfrak{C}_2 & \mathfrak{D}_2 \end{array}\right]$ *be another linear system on* (Y_2, X_2, U_2) *which is well-posed in the same sense, and let* $K \in \mathcal{B}(Y_1; X_2)$*. Define* $U = \left[\begin{smallmatrix} U_1 \\ U_2 \end{smallmatrix}\right]$, $X = \left[\begin{smallmatrix} X_1 \\ X_2 \end{smallmatrix}\right]$, $Y = \left[\begin{smallmatrix} Y_1 \\ Y_2 \end{smallmatrix}\right]$, *and*

$$\Sigma = \left[\begin{array}{c|c} \mathfrak{A} & \mathfrak{B}\tau \\ \hline \mathfrak{C} & \mathfrak{D} \end{array}\right] = \left[\begin{array}{cc|cc} \mathfrak{A}_1 & 0 & \mathfrak{B}_1\tau & 0 \\ \mathfrak{B}_2\tau K\mathfrak{C}_1 & \mathfrak{A}_2 & \mathfrak{B}_2 K\mathfrak{D}_1\tau & \mathfrak{B}_2\tau \\ \hline \mathfrak{C}_1 & 0 & \mathfrak{D}_1 & 0 \\ \mathfrak{D}_2 K\mathfrak{C}_1 & \mathfrak{C}_2 & \mathfrak{D}_2 K\mathfrak{D}_1 & \mathfrak{D}_2 \end{array}\right].$$

Then $\left[\begin{array}{c|c} \mathfrak{A} & \mathfrak{B} \\ \hline \mathfrak{C} & \mathfrak{D} \end{array}\right]$ *is a linear system on* (Y, X, U) *which is well-posed in the same sense as* Σ_1 *and* Σ_2*. See the equivalent Figures 7.9 and 7.10. We call* Σ *the* cascade connection *of* Σ_1 *and* Σ_2 *through* K*.*

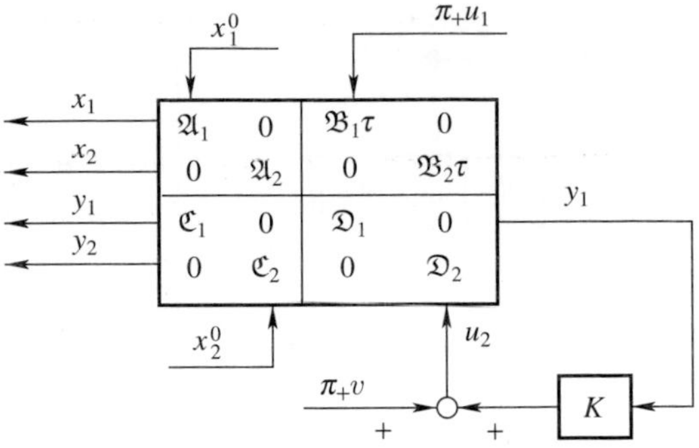

Figure 7.10 Cascade connection in block form

Proof This is an output feedback connection for the cross-product of Σ_1 and Σ_2 described in Example 2.3.10 with feedback operator $\left[\begin{smallmatrix} 0 & 0 \\ K & 0 \end{smallmatrix}\right]$. Since

$$\begin{bmatrix} 1 & 0 \\ 0 & 1 \end{bmatrix} - \begin{bmatrix} 0 & 0 \\ K & 0 \end{bmatrix} \begin{bmatrix} \mathfrak{D}_1 & 0 \\ 0 & \mathfrak{D}_2 \end{bmatrix} = \begin{bmatrix} 1 & 0 \\ -K\mathfrak{D}_1 & 1 \end{bmatrix}$$

is always invertible (the inverse is $\left[\begin{smallmatrix} 1 & 0 \\ K\mathfrak{D}_1 & 1 \end{smallmatrix}\right]$), this feedback connection is always well-posed. The explicit formulas for the closed-loop system follow from Theorem 7.1.2 and (7.1.5). □

Example 7.2.5 *Let* $\Sigma_1 = \left[\begin{array}{c|c} \mathfrak{A}_1 & \mathfrak{B}_1 \\ \hline \mathfrak{C}_1 & \mathfrak{D}_1 \end{array}\right]$ *be a well-posed linear system on* (Y, X_1, U)*, and let* $\Sigma_2 = \left[\begin{array}{c|c} \mathfrak{A}_2 & \mathfrak{B}_2 \\ \hline \mathfrak{C}_2 & \mathfrak{D}_2 \end{array}\right]$ *be another linear system on* (U, X_2, Y) *which is well-posed in the same sense. Suppose that* $1 - \mathfrak{D}_2\mathfrak{D}_1$ *is invertible in* $TIC_{\mathrm{loc}}(U)$*. Define* $X = \left[\begin{smallmatrix} X_1 \\ X_2 \end{smallmatrix}\right]$ *and*

$$\Sigma = \left[\begin{array}{c|c} \mathfrak{A} & \mathfrak{B}\tau \\ \hline \mathfrak{C} & \mathfrak{D} \end{array}\right] = \left[\begin{array}{cc|cc} \mathfrak{A}_1 & 0 & 0 & \mathfrak{B}_1\tau \\ 0 & \mathfrak{A}_2 & \mathfrak{B}_2\tau & 0 \\ \hline \mathfrak{C}_1 & 0 & 0 & \mathfrak{D}_1 \\ 0 & \mathfrak{C}_2 & \mathfrak{D}_2 & 0 \end{array}\right]$$

$$+ \left[\begin{array}{cc} 0 & \mathfrak{B}_1\tau \\ \mathfrak{B}_2\tau & 0 \\ \hline 0 & \mathfrak{D}_1 \\ \mathfrak{D}_2 & 0 \end{array}\right] \begin{bmatrix} 1 & -\mathfrak{D}_1 \\ -\mathfrak{D}_2 & 1 \end{bmatrix}^{-1} \left[\begin{array}{cc|cc} \mathfrak{C}_1 & 0 & 0 & \mathfrak{D}_1 \\ 0 & \mathfrak{C}_2 & \mathfrak{D}_2 & 0 \end{array}\right]$$

$$= \left[\begin{array}{cc|cc} \mathfrak{A}_1 & 0 & 0 & 0 \\ 0 & \mathfrak{A}_2 & 0 & 0 \\ \hline 0 & 0 & -1 & 0 \\ 0 & 0 & 0 & -1 \end{array}\right]$$

$$+ \left[\begin{array}{cc} 0 & \mathfrak{B}_1\tau \\ \mathfrak{B}_2\tau & 0 \\ \hline 1 & 0 \\ 0 & 1 \end{array}\right] \begin{bmatrix} 1 & -\mathfrak{D}_1 \\ -\mathfrak{D}_2 & 1 \end{bmatrix}^{-1} \left[\begin{array}{cc|cc} \mathfrak{C}_1 & 0 & 1 & 0 \\ 0 & \mathfrak{C}_2 & 0 & 1 \end{array}\right],$$

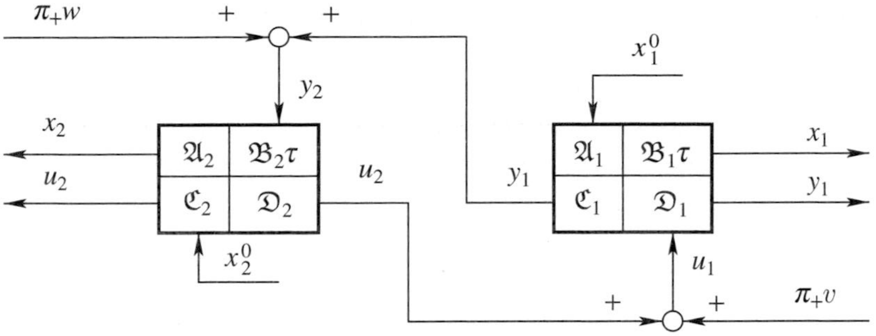

Figure 7.11 Dynamic feedback

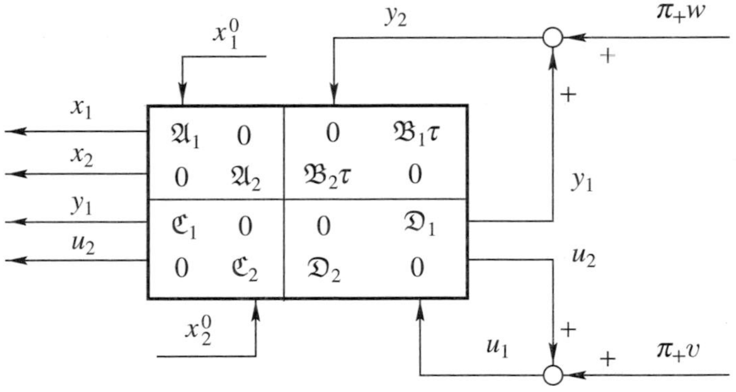

Figure 7.12 Dynamic feedback in block form

where

$$\begin{bmatrix} 1 & -\mathfrak{D}_1 \\ -\mathfrak{D}_2 & 1 \end{bmatrix}^{-1} = \begin{bmatrix} (1-\mathfrak{D}_1\mathfrak{D}_2)^{-1} & \mathfrak{D}_1(1-\mathfrak{D}_2\mathfrak{D}_1)^{-1} \\ (1-\mathfrak{D}_2\mathfrak{D}_1)^{-1}\mathfrak{D}_2 & (1-\mathfrak{D}_2\mathfrak{D}_1)^{-1} \end{bmatrix}.$$

Then $\left[\begin{smallmatrix} \mathfrak{A} & \mathfrak{B} \\ \mathfrak{C} & \mathfrak{D} \end{smallmatrix}\right]$ *is a linear system on* $\left(\left[\begin{smallmatrix} Y \\ U \end{smallmatrix}\right], X, \left[\begin{smallmatrix} Y \\ U \end{smallmatrix}\right]\right)$ *which is well-posed in the same sense as* Σ_1 *and* Σ_2*. See the equivalent Figures 7.11 and 7.12. We call* Σ *the* dynamic feedback connection *of* Σ_1 *and* Σ_2*.*

Proof This can be interpreted as an output feedback connection for the cross-product of Σ_1 and Σ_2 described in Example 2.3.10 with feedback operator $\left[\begin{smallmatrix} 0 & 1 \\ 1 & 0 \end{smallmatrix}\right]$. However, we can simplify the computations slightly if we reorder the two inputs u_1 and y_2 into $\left[\begin{smallmatrix} y_2 \\ u_1 \end{smallmatrix}\right]$ (and keep the order of the outputs as $\left[\begin{smallmatrix} y_1 \\ u_2 \end{smallmatrix}\right]$. Then we get an identity feedback, and the result follows from Remark 7.1.6. Observe that this identity feedback is admissible iff $\left[\begin{smallmatrix} 1 & -\mathfrak{D}_1 \\ -\mathfrak{D}_2 & 1 \end{smallmatrix}\right]$ is invertible in $TIC_{\mathrm{loc}}(\left[\begin{smallmatrix} Y \\ U \end{smallmatrix}\right])$, which by Lemma A.4.1 is true iff $1-\mathfrak{D}_2\mathfrak{D}_1$ is invertible in $TIC_{\mathrm{loc}}(U)$, or equivalently, iff $1-\mathfrak{D}_1\mathfrak{D}_2$ is invertible in $TIC_{\mathrm{loc}}(Y)$. □

In many applications the notion of a *partial feedback* plays a very important role. By this we mean the following. The inputs and outputs of the system are grouped into two parts, an *active* and an *inactive* part, out of which only the active part is allowed to be part of a feedback loop. This is the same situation that we encountered in partial flow-inversion.

Lemma 7.2.6 *Let* $\Sigma = \left[\begin{array}{c|c} \mathfrak{A} & \mathfrak{B} \\ \hline \mathfrak{C} & \mathfrak{D} \end{array}\right] = \left[\begin{array}{c|cc} \mathfrak{A} & \mathfrak{B}_1 & \mathfrak{B}_2 \\ \hline \mathfrak{C}_1 & \mathfrak{D}_{11} & \mathfrak{D}_{12} \\ \mathfrak{C}_2 & \mathfrak{D}_{21} & \mathfrak{D}_{22} \end{array}\right]$ *be a well-posed linear system on* $\left(\left[\begin{smallmatrix} Y_1 \\ Y_2 \end{smallmatrix}\right], X, \left[\begin{smallmatrix} U_1 \\ U_2 \end{smallmatrix}\right]\right)$*, and let* $K \in \mathcal{B}(Y_2; U_2)$*.*

(i) *The operator* $\left[\begin{smallmatrix} 0 & 0 \\ 0 & K \end{smallmatrix}\right]$ *is an admissible feedback operator for* Σ *if and only if* K *is an admissible feedback operator for the (reduced) system* $\left[\begin{array}{c|c} \mathfrak{A} & \mathfrak{B}_2 \\ \hline \mathfrak{C}_2 & \mathfrak{D}_{22} \end{array}\right]$ *(see Figure 7.13). The closed-loop system, which for simplicity we denote by* Σ^K*, is given by*

$$
\begin{aligned}
\Sigma^K &= \left[\begin{array}{c|cc} \mathfrak{A}^K & \mathfrak{B}_1^K\tau & \mathfrak{B}_2^K\tau \\ \hline \mathfrak{C}_1^K & \mathfrak{D}_{11}^K & \mathfrak{D}_{12}^K \\ \mathfrak{C}_2^K & \mathfrak{D}_{21}^K & \mathfrak{D}_{22}^K \end{array}\right] \\
&= \left[\begin{array}{c|cc} \mathfrak{A} & \mathfrak{B}_1\tau & \mathfrak{B}_2\tau \\ \hline \mathfrak{C}_1 & \mathfrak{D}_{11} & \mathfrak{D}_{12} \\ \mathfrak{C}_2 & \mathfrak{D}_{21} & \mathfrak{D}_{22} \end{array}\right] \left[\begin{array}{c|cc} 1 & 0 & 0 \\ \hline 0 & 1 & 0 \\ -K\mathfrak{C}_2 & -K\mathfrak{D}_{21} & 1 - K\mathfrak{D}_{22} \end{array}\right]^{-1} \\
&= \left[\begin{array}{c|cc} 1 & 0 & -\mathfrak{B}_2 K\tau \\ \hline 0 & 1 & -\mathfrak{D}_{12}K \\ 0 & 0 & 1 - \mathfrak{D}_{22}K \end{array}\right]^{-1} \left[\begin{array}{c|cc} \mathfrak{A} & \mathfrak{B}_1\tau & \mathfrak{B}_2\tau \\ \hline \mathfrak{C}_1 & \mathfrak{D}_{11} & \mathfrak{D}_{12} \\ \mathfrak{C}_2 & \mathfrak{D}_{21} & \mathfrak{D}_{22} \end{array}\right] \\
&= \left[\begin{array}{c|cc} \mathfrak{A} & \mathfrak{B}_1\tau & \mathfrak{B}_2\tau \\ \hline \mathfrak{C}_1 & \mathfrak{D}_{11} & \mathfrak{D}_{12} \\ \mathfrak{C}_2 & \mathfrak{D}_{21} & \mathfrak{D}_{22} \end{array}\right] + \left[\begin{array}{c} \mathfrak{B}_2\tau \\ \hline \mathfrak{D}_{12} \\ \mathfrak{D}_{22} \end{array}\right] K (1 - \mathfrak{D}_{22}K)^{-1} \left[\begin{array}{c|cc} \mathfrak{C}_2 & \mathfrak{D}_{21} & \mathfrak{D}_{22} \end{array}\right] \\
&= \left[\begin{array}{c|cc} \mathfrak{A} & \mathfrak{B}_1\tau & \mathfrak{B}_2\tau \\ \hline \mathfrak{C}_1 & \mathfrak{D}_{11} & \mathfrak{D}_{12} \\ \mathfrak{C}_2 & \mathfrak{D}_{21} & \mathfrak{D}_{22} \end{array}\right] + \left[\begin{array}{c} \mathfrak{B}_2\tau \\ \hline \mathfrak{D}_{12} \\ \mathfrak{D}_{22} \end{array}\right] K \left[\begin{array}{c|cc} \mathfrak{C}_2^K & \mathfrak{D}_{21}^K & \mathfrak{D}_{22}^K \end{array}\right] \\
&= \left[\begin{array}{c|cc} \mathfrak{A} & \mathfrak{B}_1\tau & \mathfrak{B}_2\tau \\ \hline \mathfrak{C}_1 & \mathfrak{D}_{11} & \mathfrak{D}_{12} \\ \mathfrak{C}_2 & \mathfrak{D}_{21} & \mathfrak{D}_{22} \end{array}\right] + \left[\begin{array}{c} \mathfrak{B}_2^K\tau \\ \hline \mathfrak{D}_{12}^K \\ \mathfrak{D}_{22}^K \end{array}\right] K \left[\begin{array}{c|cc} \mathfrak{C}_2 & \mathfrak{D}_{21} & \mathfrak{D}_{22} \end{array}\right].
\end{aligned}
$$

We call Σ^K *the* closed-loop system with partial (output) feedback operator K.

(ii) *In the case of* partial positive identity feedback *where* $Y_2 = U_2$ *and* $K = 1$ *it is possible to write the closed-loop system in the alternative*

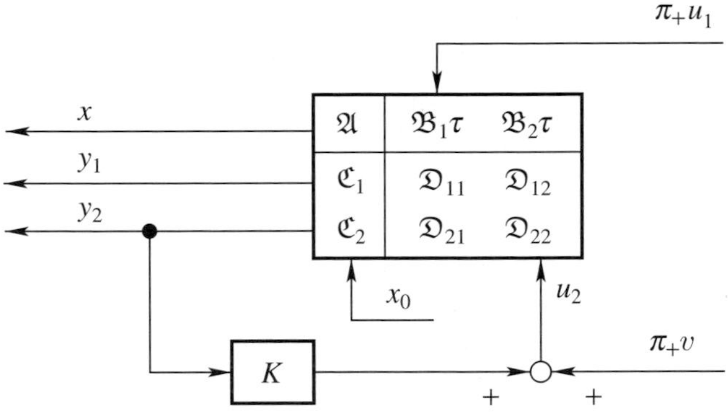

Figure 7.13 Partial feedback

form

$$
\begin{aligned}
\Sigma^{\left[\begin{smallmatrix}0&0\\0&1\end{smallmatrix}\right]} &= \left[\begin{array}{c|cc} \mathfrak{A}^K & \mathfrak{B}_1^K\tau & \mathfrak{B}_2^K\tau \\ \hline \mathfrak{C}_1^K & \mathfrak{D}_{11}^K & \mathfrak{D}_{12}^K \\ \mathfrak{C}_2^K & \mathfrak{D}_{21}^K & \mathfrak{D}_{22}^K \end{array}\right] \\
&= \left[\begin{array}{c|cc} \mathfrak{A} & \mathfrak{B}_1\tau & \mathfrak{B}_2\tau \\ \hline \mathfrak{C}_1 & \mathfrak{D}_{11} & \mathfrak{D}_{12} \\ \mathfrak{C}_2 & \mathfrak{D}_{21} & \mathfrak{D}_{22} \end{array}\right] + \left[\begin{array}{c} \mathfrak{B}_2\tau \\ \hline \mathfrak{D}_{12} \\ \mathfrak{D}_{22} \end{array}\right] (1-\mathfrak{D}_{22})^{-1} \left[\begin{array}{c|cc} \mathfrak{C}_2 & \mathfrak{D}_{21} & \mathfrak{D}_{22} \end{array}\right] \\
&= \left[\begin{array}{c|cc} \mathfrak{A} & \mathfrak{B}_1\tau & 0 \\ \hline \mathfrak{C}_1 & \mathfrak{D}_{11} & 0 \\ 0 & 0 & -1 \end{array}\right] + \left[\begin{array}{c} \mathfrak{B}_2\tau \\ \hline \mathfrak{D}_{12} \\ 1 \end{array}\right] (1-\mathfrak{D}_{22})^{-1} \left[\begin{array}{c|cc} \mathfrak{C}_2 & \mathfrak{D}_{21} & 1 \end{array}\right].
\end{aligned}
$$

(iii) *In the case of* partial negative identity feedback *where* $Y_2 = U_2$ *and* $K = -1$ *it is possible to write the closed-loop system in the alternative form*

$$
\begin{aligned}
\Sigma^{\left[\begin{smallmatrix}0&0\\0&-1\end{smallmatrix}\right]} &= \left[\begin{array}{c|cc} \mathfrak{A}^K & \mathfrak{B}_1^K\tau & \mathfrak{B}_2^K\tau \\ \hline \mathfrak{C}_1^K & \mathfrak{D}_{11}^K & \mathfrak{D}_{12}^K \\ \mathfrak{C}_2^K & \mathfrak{D}_{21}^K & \mathfrak{D}_{22}^K \end{array}\right] \\
&= \left[\begin{array}{c|cc} \mathfrak{A} & \mathfrak{B}_1\tau & \mathfrak{B}_2\tau \\ \hline \mathfrak{C}_1 & \mathfrak{D}_{11} & \mathfrak{D}_{12} \\ \mathfrak{C}_2 & \mathfrak{D}_{21} & \mathfrak{D}_{22} \end{array}\right] - \left[\begin{array}{c} \mathfrak{B}_2\tau \\ \hline \mathfrak{D}_{12} \\ \mathfrak{D}_{22} \end{array}\right] (1+\mathfrak{D}_{22})^{-1} \left[\begin{array}{c|cc} \mathfrak{C}_2 & \mathfrak{D}_{21} & \mathfrak{D}_{22} \end{array}\right] \\
&= \left[\begin{array}{c|cc} \mathfrak{A} & \mathfrak{B}_1\tau & 0 \\ \hline \mathfrak{C}_1 & \mathfrak{D}_{11} & 0 \\ 0 & 0 & 1 \end{array}\right] - \left[\begin{array}{c} \mathfrak{B}_2\tau \\ \hline \mathfrak{D}_{12} \\ -1 \end{array}\right] (1+\mathfrak{D}_{22})^{-1} \left[\begin{array}{c|cc} \mathfrak{C}_2 & \mathfrak{D}_{21} & -1 \end{array}\right].
\end{aligned}
$$

Proof By Lemma A.4.2(i)–(ii), $\left[\begin{smallmatrix}1&0\\0&1\end{smallmatrix}\right] - \left[\begin{smallmatrix}0&0\\0&K\end{smallmatrix}\right]\left[\begin{smallmatrix}\mathfrak{D}_{11}&\mathfrak{D}_{21}\\\mathfrak{D}_{21}&\mathfrak{D}_{22}\end{smallmatrix}\right] = \left[\begin{smallmatrix}1&0\\-K\mathfrak{D}_{21}&1-K\mathfrak{D}_{22}\end{smallmatrix}\right]$ is invertible in $TIC_{\mathrm{loc}}(\left[\begin{smallmatrix}U_1\\U_2\end{smallmatrix}\right])$ if and only if $1 - K\mathfrak{D}_{22}$ is invertible in $TIC_{\mathrm{loc}}(U_2)$. Thus, Lemma 7.2.6 follows from Theorems 7.1.2 and 7.1.8 (and some simple algebra; cf. Theorem 6.6.1). □

The cascade connection in Example 7.2.4 is a simple example of partial positive identity feedback. This connection is always well-posed since the identity operator is a well-posed output feedback operator for the reduced system $\left[\begin{array}{cc|c}\mathfrak{A}_1&0&0\\0&\mathfrak{A}_2&\mathfrak{B}_2\\\hline\mathfrak{C}_1&0&0\end{array}\right]$.

7.3 State feedback and output injection

The notion of a *state feedback* can be regarded as a special case of the partial identity feedback described in Lemma 7.2.6(ii). Intuitively, a state feedback means that an additional output is created, and this output is then fed back into the input, as shown in Figure 7.14. In this figure the original system is represented by $\left[\begin{array}{c|c}\mathfrak{A}&\mathfrak{B}\\\hline\mathfrak{C}&\mathfrak{D}\end{array}\right]$. We find two additional components, namely a new observability map $\mathfrak{K}$ (from the initial state to the new output) and a new input/output map $\mathfrak{F}$ (from the original input to the new output). The pair $\left[\,\mathfrak{K}\,\middle|\,\mathfrak{F}\,\right]$ is *admissible* if the resulting system is well-posed, i.e., if $\left[0\ \ 1\right]$ is an admissible output feedback operator for the extended system:

Definition 7.3.1 Let $\Sigma = \left[\begin{array}{c|c}\mathfrak{A}&\mathfrak{B}\\\hline\mathfrak{C}&\mathfrak{D}\end{array}\right]$ be a well-posed linear system on (Y, X, U). The pair $\left[\,\mathfrak{K}\,\middle|\,\mathfrak{F}\,\right]$ is an *admissible state feedback pair* for Σ if the extended system $\left[\begin{array}{c|c}\mathfrak{A}&\mathfrak{B}\\\hline\mathfrak{C}&\mathfrak{D}\\\mathfrak{K}&\mathfrak{F}\end{array}\right]$ is a well-posed linear system on $\left(\left[\begin{smallmatrix}Y\\U\end{smallmatrix}\right], X, U\right)$ and $\left[0\ \ 1\right]$ is an admissible output feedback operator for this extended system.

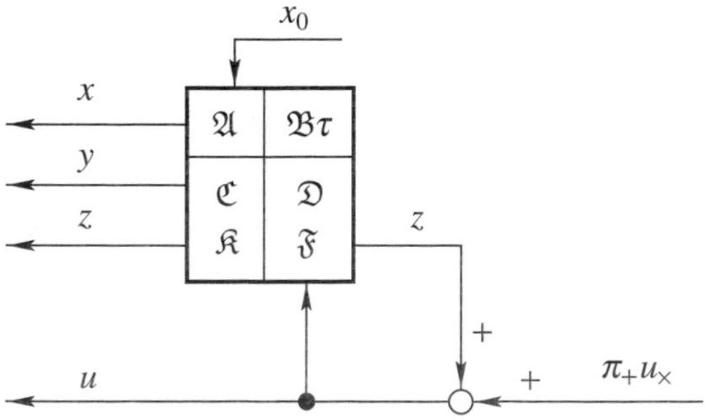

Figure 7.14 State feedback

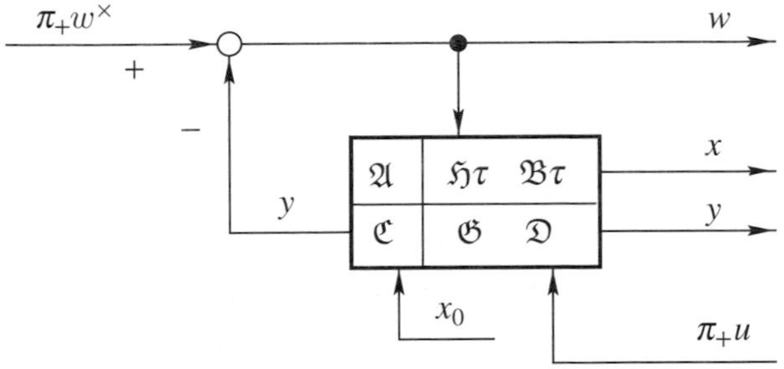

Figure 7.15 Output injection

The notion of an *output injection* is analogous. In this case a new input is created, into which we feed the negative of the original output y plus a new perturbation $w^\times$, as shown in Figure 7.15. The original system is still represented by $\left[\begin{smallmatrix}\mathfrak{A} & \mathfrak{B}\\ \mathfrak{C} & \mathfrak{D}\end{smallmatrix}\right]$. Observe that this time we use *negative* feedback instead of positive feedback.[3] In this figure we find a new controllability map $\mathfrak{H}$ (from the new input to the state) and a new input/output map $\mathfrak{G}$ (from the new input to the original output). The pair $\left[\begin{smallmatrix}\mathfrak{H}\\ \mathfrak{G}\end{smallmatrix}\right]$ is *admissible* if the resulting system is well-posed:

Definition 7.3.2 Let $\Sigma = \left[\begin{smallmatrix}\mathfrak{A} & \mathfrak{B}\\ \mathfrak{C} & \mathfrak{D}\end{smallmatrix}\right]$ be a well-posed linear system on (Y, X, U). The pair $\left[\begin{smallmatrix}\mathfrak{H}\\ \mathfrak{G}\end{smallmatrix}\right]$ is an *admissible output injection pair* for Σ if the extended system $\left[\begin{smallmatrix}\mathfrak{A} & \mathfrak{H} & \mathfrak{B}\\ \mathfrak{C} & \mathfrak{G} & \mathfrak{D}\end{smallmatrix}\right]$ is a well-posed linear system on $\left(Y, X, \left[\begin{smallmatrix}Y\\ U\end{smallmatrix}\right]\right)$, and $\left[\begin{smallmatrix}-1\\ 0\end{smallmatrix}\right]$ is an admissible output feedback operator for this extended system. See Figure 7.15.

Lemma 7.3.3 *Let* $\Sigma = \left[\begin{smallmatrix}\mathfrak{A} & \mathfrak{B}\\ \mathfrak{C} & \mathfrak{D}\end{smallmatrix}\right]$ *be a well-posed linear system on* (Y, X, U).

(i) *The following conditions are equivalent:*
  (a) *the pair* $\left[\mathfrak{K} \,|\, \mathfrak{F}\right]$ *is an admissible state feedback pair for* Σ;
  (b) *the extended system* $\left[\begin{smallmatrix}\mathfrak{A} & \mathfrak{B}\\ \mathfrak{C} & \mathfrak{D}\\ \mathfrak{K} & \mathfrak{F}\end{smallmatrix}\right]$ *is a well-posed linear system on* $\left(\left[\begin{smallmatrix}Y\\ U\end{smallmatrix}\right], X, U\right)$, *and* $1 - \mathfrak{F}$ *has an inverse in* $TIC_{\mathrm{loc}}(U)$.

  When these conditions hold, then the closed-loop state feedback system $\Sigma_\times$ *with input* $u_\times$ *and outputs* y, *and* z *that we get by using the output feedback operator* $\left[0\ 1\right]$ *in the extended system (see Figure 7.14) is*

[3] Our use of positive feedback in the state feedback loop and negative feedback in the output injection loop is motivated by the fact that in this way we get the 'standard' distribution of minus signs in the doubly coprime factorization (8.3.3). Positive or negative feedback in both the state feedback loop and the output injection loop leads to one minus sign in each of the two factors in (8.3.3).

given by

$$\Sigma_\times = \left[\begin{array}{c|c} \mathfrak{A}_\times & \mathfrak{B}_\times\tau \\ \hline \mathfrak{C}_\times & \mathfrak{D}_\times \\ \mathfrak{K}_\times & \mathfrak{F}_\times \end{array}\right]$$
$$= \left[\begin{array}{c|c} \mathfrak{A} + \mathfrak{B}\tau(1-\mathfrak{F})^{-1}\mathfrak{K} & \mathfrak{B}(1-\mathfrak{F})^{-1}\tau \\ \hline \mathfrak{C} + \mathfrak{D}(1-\mathfrak{F})^{-1}\mathfrak{K} & \mathfrak{D}(1-\mathfrak{F})^{-1} \\ (1-\mathfrak{F})^{-1}\mathfrak{K} & \mathfrak{F}(1-\mathfrak{F})^{-1} \end{array}\right]$$
$$= \left[\begin{array}{c|c} \mathfrak{A} & \mathfrak{B}\tau \\ \hline \mathfrak{C} & \mathfrak{D} \\ \mathfrak{K} & \mathfrak{F} \end{array}\right] \left[\begin{array}{c|c} 1 & 0 \\ \hline -\mathfrak{K} & 1-\mathfrak{F} \end{array}\right]^{-1}$$
$$= \left[\begin{array}{c|c} \mathfrak{A} & \mathfrak{B}\tau \\ \hline \mathfrak{C} & \mathfrak{D} \\ \mathfrak{K} & \mathfrak{F} \end{array}\right] + \left[\begin{array}{c} \mathfrak{B}\tau \\ \hline \mathfrak{D} \\ \mathfrak{F} \end{array}\right] (1-\mathfrak{F})^{-1} \left[\begin{array}{c|c} \mathfrak{K} & \mathfrak{F} \end{array}\right]$$
$$= \left[\begin{array}{c|c} \mathfrak{A} & 0 \\ \hline \mathfrak{C} & 0 \\ 0 & -1 \end{array}\right] + \left[\begin{array}{c} \mathfrak{B}\tau \\ \hline \mathfrak{D} \\ 1 \end{array}\right] (1-\mathfrak{F})^{-1} \left[\begin{array}{c|c} \mathfrak{K} & 1 \end{array}\right].$$

(ii) *The following conditions are also equivalent:*
   (a) *the pair* $\left[\begin{smallmatrix} \mathfrak{H} \\ \mathfrak{G} \end{smallmatrix}\right]$ *is an admissible output injection pair for* Σ;
   (b) *the extended system* $\left[\begin{array}{c|cc} \mathfrak{A} & \mathfrak{H} & \mathfrak{B} \\ \hline \mathfrak{C} & \mathfrak{G} & \mathfrak{D} \end{array}\right]$ *is a well-posed linear system on* $(Y, X, \left[\begin{smallmatrix} Y \\ U \end{smallmatrix}\right])$, *and* $1 + \mathfrak{G}$ *has an inverse in* $TIC_{\mathrm{loc}}(Y)$.

When these conditions hold, then the closed-loop output injection system $\Sigma^\times$ *with inputs* $w^\times$ *and* u *and output* y *that we get by using the output feedback operator* $\left[\begin{smallmatrix} 1 \\ 0 \end{smallmatrix}\right]$ *in the extended system (see Figure 7.14) is given by*

$$\Sigma^\times = \left[\begin{array}{c|cc} \mathfrak{A}^\times & \mathfrak{H}^\times\tau & \mathfrak{B}^\times\tau \\ \hline \mathfrak{C}^\times & \mathfrak{G}^\times & \mathfrak{D}^\times \end{array}\right]$$
$$= \left[\begin{array}{c|cc} \mathfrak{A} - \mathfrak{H}\tau(1+\mathfrak{G})^{-1}\mathfrak{C} & \mathfrak{H}(1+\mathfrak{G})^{-1}\tau & \mathfrak{B}\tau - \mathfrak{H}(1+\mathfrak{G})^{-1}\mathfrak{D}\tau \\ \hline (1+\mathfrak{G})^{-1}\mathfrak{C} & (1+\mathfrak{G})^{-1}\mathfrak{G} & (1+\mathfrak{G})^{-1}\mathfrak{D} \end{array}\right]$$
$$= \left[\begin{array}{c|c} 1 & \mathfrak{H}\tau \\ \hline \mathfrak{C} & 1+\mathfrak{G} \end{array}\right]^{-1} \left[\begin{array}{c|cc} \mathfrak{A} & \mathfrak{H}\tau & \mathfrak{B}\tau \\ \hline \mathfrak{C} & \mathfrak{G} & \mathfrak{D} \end{array}\right]$$
$$= \left[\begin{array}{c|cc} \mathfrak{A} & \mathfrak{H}\tau & \mathfrak{B}\tau \\ \hline \mathfrak{C} & \mathfrak{G} & \mathfrak{D} \end{array}\right] - \left[\begin{array}{c} \mathfrak{H}\tau \\ \hline \mathfrak{G} \end{array}\right] (1+\mathfrak{G})^{-1} \left[\begin{array}{c|cc} \mathfrak{C} & \mathfrak{G} & \mathfrak{D} \end{array}\right]$$
$$= \left[\begin{array}{c|cc} \mathfrak{A} & 0 & \mathfrak{B}\tau \\ \hline 0 & 1 & 0 \end{array}\right] - \left[\begin{array}{c} \mathfrak{H}\tau \\ \hline -1 \end{array}\right] (1+\mathfrak{G})^{-1} \left[\begin{array}{c|cc} \mathfrak{C} & -1 & \mathfrak{D} \end{array}\right].$$

Proof See Lemma 7.2.6(ii)–(iii). □

Remark 7.3.4 We shall frequently regard the signal u in Figure 7.14 (i.e., the input to the open-loop system) as an additional output of the closed-loop state feedback system. This output has the same observability map $(1-\mathfrak{F})^{-1}\mathfrak{K}$ as the output z, and its input/output map, given by $(1-\mathfrak{F})^{-1}$, differs from the input/output map from $u_\times$ to z by an identity operator (see Theorem 7.2.1 and Lemma 7.3.3). Similar remarks apply to the signals w in Figure 7.15, $u_\times$ and $w^\times$ in Figure 8.1, u and $w^\times$ in Figure 8.2, $u_\times$ and w in Figure 8.3, etc. See also Remark 7.1.6.

7.4 The closed-loop generators

Let us next look at the closed-loop system node and combined observation/feedthrough operator.

Theorem 7.4.1 *Let* $\Sigma = \left[\begin{smallmatrix} \mathfrak{A} & \mathfrak{B} \\ \mathfrak{C} & \mathfrak{D} \end{smallmatrix}\right]$ *be a well-posed linear system on* (Y, X, U)*, and let* $K \in \mathcal{B}(Y; U)$ *be an admissible output feedback operator for* Σ*. Let* $S = \left[\begin{smallmatrix} A\&B \\ C\&D \end{smallmatrix}\right]$ *and* $S^K = \left[\begin{smallmatrix} [A\&B]^K \\ [C\&D]^K \end{smallmatrix}\right]$ *be the open- and closed-loop system nodes. Then the operator*

$$M = \begin{bmatrix} 1 & 0 \\ 0 & 1 \end{bmatrix} - \begin{bmatrix} 0 \\ K[C\&D] \end{bmatrix} \tag{7.4.1}$$

maps $\mathcal{D}(S)$ *continuously onto* $\mathcal{D}(S^K)$*. Its inverse is given by*

$$M^{-1} = \begin{bmatrix} 1 & 0 \\ 0 & 1 \end{bmatrix} + \begin{bmatrix} 0 \\ K[C\&D]^K \end{bmatrix}, \tag{7.4.2}$$

and

$$S^K = SM^{-1}. \tag{7.4.3}$$

It is possible to derive this theorem from Theorem 6.6.5, but for completeness we give a separate proof.

Proof Let $\left[\begin{smallmatrix} x_0 \\ u_0 \end{smallmatrix}\right] \in \mathcal{D}(S)$, i.e., $x_0 \in X$, $u_0 \in U$, and $A_{|X_{-1}}x_0 + Bu_0 \in X$. Define

$$v_0 = u_0 - K[C\&D]\begin{bmatrix} x_0 \\ u_0 \end{bmatrix}.$$

Choose an arbitrary $u \in C^1(\overline{\mathbb{R}}^+; U)$ with $u(0) = u_0$. Let x, y, and v be the state and the two outputs of the system drawn in Figure 7.16 with initial time zero, initial state x_0, and input u. Then, by Theorem 4.6.11, x is continuously differentiable in X, $y \in W^{1,p}_{\mathrm{loc}}(\overline{\mathbb{R}}^+; Y)$, $v \in W^{1,p}_{\mathrm{loc}}(\overline{\mathbb{R}}^+; U)$ (or $y \in C^1(\overline{\mathbb{R}}^+; Y)$,

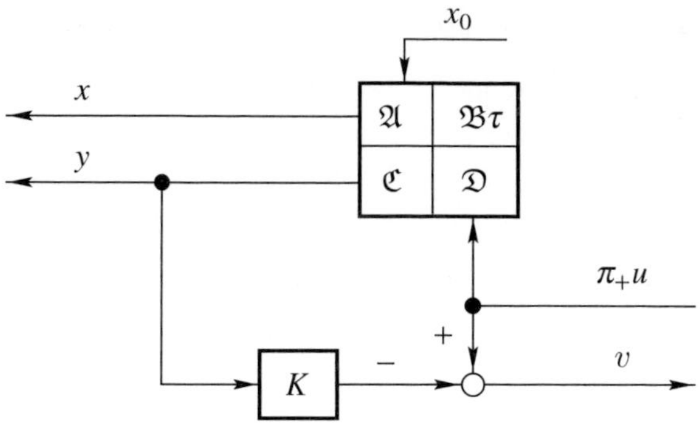

Figure 7.16 Original system with one extra output

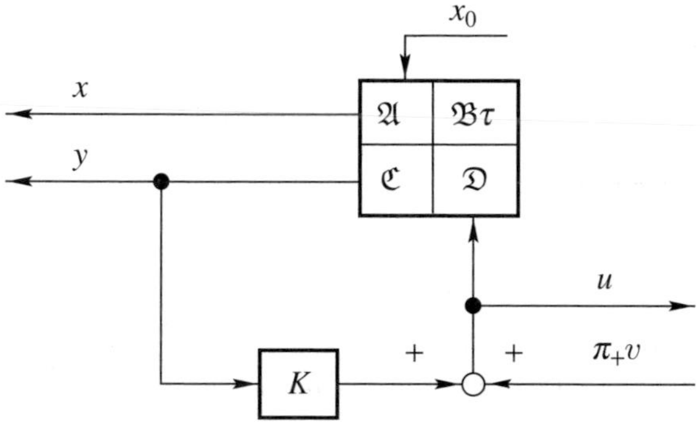

Figure 7.17 Closed-loop system with one extra output

$v \in C^1(\overline{\mathbb{R}}^+; U)$ in the *Reg*-well-posed case), and for all $t \geq 0$,

$$\begin{bmatrix} \dot{x}(t) \\ y(t) \end{bmatrix} = S \begin{bmatrix} x(t) \\ u(t) \end{bmatrix}, \qquad \begin{bmatrix} x(t) \\ v(t) \end{bmatrix} = \begin{bmatrix} x(t) \\ u(t) - Ky(t) \end{bmatrix} = M \begin{bmatrix} x(t) \\ u(t) \end{bmatrix}. \tag{7.4.4}$$

In particular, $\left[\begin{smallmatrix} x_0 \\ v_0 \end{smallmatrix}\right] = M \left[\begin{smallmatrix} x_0 \\ u_0 \end{smallmatrix}\right]$. On the other hand, we can also consider the system drawn in Figure 7.17 with initial time zero, initial state x_0, and input v, where v is the same function as above. This is a special case of the closed-loop system studied in Theorem 7.2.1(ii). Clearly, the relationships between u and v is exactly the same in both diagrams (although u is the input in the first figure, and v is the input in the second). Therefore, the two states labeled x and also the two outputs labeled y in Figure 7.16 and Figure 7.17 will be identical, and the output u in Figure 7.17 will be identical to the input u in Figure 7.16. By Theorem 4.3.7, Corollary 4.3.8, and Theorem 4.6.11, $\left[\begin{smallmatrix} x_0 \\ v(0) \end{smallmatrix}\right] = \left[\begin{smallmatrix} x_0 \\ v_0 \end{smallmatrix}\right] \in \mathcal{D}\left(S^K\right)$, and, for all $t \geq 0$,

$$\begin{aligned} \begin{bmatrix} \dot{x}(t) \\ y(t) \end{bmatrix} &= S^K \begin{bmatrix} x(t) \\ v(t) \end{bmatrix}, \\ \begin{bmatrix} x(t) \\ u(t) \end{bmatrix} &= \begin{bmatrix} x(t) \\ v(t) + Ky(t) \end{bmatrix} = \left(\begin{bmatrix} 1 & 0 \\ 0 & 1 \end{bmatrix} + \begin{bmatrix} 0 \\ K[C\&D]^K \end{bmatrix} \right) \begin{bmatrix} x(t) \\ v(t) \end{bmatrix}. \end{aligned} \tag{7.4.5}$$

In particular, taking $t = 0$ in (7.4.4) and (7.4.5) we get

$$S \begin{bmatrix} x_0 \\ u_0 \end{bmatrix} = S^K \begin{bmatrix} x_0 \\ v_0 \end{bmatrix}, \quad \begin{bmatrix} x_0 \\ u_0 \end{bmatrix} = \left(\begin{bmatrix} 1 & 0 \\ 0 & 1 \end{bmatrix} + \begin{bmatrix} 0 \\ K[C\&D]^K \end{bmatrix} \right) \begin{bmatrix} x_0 \\ v_0 \end{bmatrix}.$$

This shows that the operator defined by the right hand side of (7.4.2) is a left inverse to M, and that (7.4.1)–(7.4.3) are valid if M is invertible. To show that the operator in (7.4.2) is also a right inverse of M we argue as above, but start by first fixing $\left[\begin{smallmatrix} x_0 \\ v_0 \end{smallmatrix} \right] \in \mathcal{D}\left(S^K\right)$, defining u_0 by

$$u_0 = v_0 + K[C\&D]^K \begin{bmatrix} x_0 \\ v_0 \end{bmatrix},$$

and interchanging the role of Figures 7.16 and 7.17. □

Motivated by Theorem 7.4.1 we extend Definition 7.1.1 to arbitrary operator nodes as follows.

Definition 7.4.2 Let $S = \left[\begin{smallmatrix} A\&B \\ C\&D \end{smallmatrix} \right]$ be an operator node on (Y, X, U). Then $K \in \mathcal{B}(Y; U)$ is an *admissible output feedback operator for the operator node* S if there exists another operator node $S^K = \left[\begin{smallmatrix} [A\&B]^K \\ [C\&D]^K \end{smallmatrix} \right]$ on (Y, X, U) which together with S satisfies the following conditions: the operator

$$M = \begin{bmatrix} 1 & 0 \\ 0 & 1 \end{bmatrix} - \begin{bmatrix} 0 \\ K[C\&D] \end{bmatrix} \tag{7.4.1}$$

maps $\mathcal{D}(S)$ continuously onto $\mathcal{D}\left(S^K\right)$, its inverse is given by

$$M^{-1} = \begin{bmatrix} 1 & 0 \\ 0 & 1 \end{bmatrix} + \begin{bmatrix} 0 \\ K[C\&D]^K \end{bmatrix}, \tag{7.4.2}$$

and

$$S^K = SM^{-1}. \tag{7.4.3}$$

In this case we call S^K the *closed-loop node* with output feedback operator K. If both S and S^K are system nodes, then we call K *system node admissible*.

Remark 7.4.3 It is obvious that if K is an admissible output feedback for the operator node S, then $-K$ is an admissible output feedback for the closed-loop operator node S^K. Moreover, the closed-loop operator node that we get from S^K by using $-K$ as the feedback operator is the original node S.

Definition 7.4.2 has some immediate consequences.

Lemma 7.4.4 *Let $S = \left[\begin{smallmatrix} A\&B \\ C\&D \end{smallmatrix}\right]$ be an operator node on (Y, X, U), with main operator A, control operator B, observation operator C, and transfer function $\widehat{\mathfrak{D}}$. Let $K \in \mathcal{B}(Y;U)$ be an admissible output feedback operator for S, and let $S^K = \left[\begin{smallmatrix} [A\&B]^K \\ [C\&D]^K \end{smallmatrix}\right]$ be the closed-loop operator node with main operator A^K, control operator B^K, observation operator C^K, and transfer function $\widehat{\mathfrak{D}}^K$. Then the following claims are true.*

(i) *$\alpha \in \rho(A^K)$ if and only if $\left(\left[\begin{smallmatrix} \alpha & 0 \\ 0 & 1 \end{smallmatrix}\right] - \left[\begin{smallmatrix} A\&B \\ K[C\&D] \end{smallmatrix}\right]\right)$ is invertible, in which case*

$$\begin{bmatrix} 1 & 0 \\ C\&D \end{bmatrix}\left(\begin{bmatrix} \alpha & 0 \\ 0 & 1 \end{bmatrix} - \begin{bmatrix} A\&B \\ K[C\&D] \end{bmatrix}\right)^{-1} = \begin{bmatrix} (\alpha - A^K)^{-1} & (\alpha - A^K_{|X})^{-1}B^K \\ C^K(\alpha - A^K)^{-1} & \widehat{\mathfrak{D}}^K(\alpha) \end{bmatrix}. \tag{7.4.6}$$

(ii) *$\alpha \in \rho(A)$ if and only if $\left(\left[\begin{smallmatrix} \alpha & 0 \\ 0 & 1 \end{smallmatrix}\right] + \left[\begin{smallmatrix} -[A\&B]^K \\ K[C\&D]^K \end{smallmatrix}\right]\right)$ is invertible, in which case*

$$\begin{bmatrix} 1 & 0 \\ [C\&D]^K \end{bmatrix}\left(\begin{bmatrix} \alpha & 0 \\ 0 & 1 \end{bmatrix} + \begin{bmatrix} -[A\&B]^K \\ K[C\&D]^K \end{bmatrix}\right)^{-1} = \begin{bmatrix} (\alpha - A)^{-1} & (\alpha - A_{|X})^{-1}B \\ C(\alpha - A)^{-1} & \widehat{\mathfrak{D}}(\alpha) \end{bmatrix}. \tag{7.4.7}$$

(iii) *If $\alpha \in \rho(A)$, then $\alpha \in \rho(A^K)$ if and only if $1 - K\widehat{\mathfrak{D}}(\alpha)$ is invertible. In this case*

$$\begin{aligned} &\begin{bmatrix} (\alpha - A^K)^{-1} & (\alpha - A^K_{|X})^{-1}B^K \\ C^K(\alpha - A^K)^{-1} & \widehat{\mathfrak{D}}^K(\alpha) \end{bmatrix} \\ &\quad = \begin{bmatrix} (\alpha - A)^{-1} & (\alpha - A_{|X})^{-1}B \\ C(\alpha - A)^{-1} & \widehat{\mathfrak{D}}(\alpha) \end{bmatrix} \\ &\qquad + \begin{bmatrix} (\alpha - A_{|X})^{-1}B \\ \widehat{\mathfrak{D}}(\alpha) \end{bmatrix}(1 - K\widehat{\mathfrak{D}}(\alpha))^{-1}K\begin{bmatrix} C(\alpha - A)^{-1} & \widehat{\mathfrak{D}}(\alpha) \end{bmatrix}. \end{aligned} \tag{7.4.8}$$

(iv) *If* $\alpha \in \rho(A^K)$, *then* $\alpha \in \rho(A)$ *if and only if* $1 + K\widehat{\mathfrak{D}}^K(\alpha)$ *is invertible. In this case*

$$\begin{aligned}
&\begin{bmatrix} (\alpha - A)^{-1} & (\alpha - A_{|X})^{-1}B \\ C(\alpha - A)^{-1} & \widehat{\mathfrak{D}}(\alpha) \end{bmatrix} \\
&\quad = \begin{bmatrix} (\alpha - A^K)^{-1} & (\alpha - A^K_{|X})^{-1}B^K \\ C^K(\alpha - A^K)^{-1} & \widehat{\mathfrak{D}}^K(\alpha) \end{bmatrix} \\
&\quad\quad - \begin{bmatrix} (\alpha - A^K_{|X})^{-1}B^K \\ \widehat{\mathfrak{D}}^K(\alpha) \end{bmatrix} (1 + K\widehat{\mathfrak{D}}^K(\alpha))^{-1}K \\
&\quad\quad \times \begin{bmatrix} C^K(\alpha - A^K)^{-1} & \widehat{\mathfrak{D}}^K(\alpha) \end{bmatrix}.
\end{aligned} \tag{7.4.9}$$

Proof We shall derive this lemma from Lemmas 6.3.8 and 6.6.7, applied to the operator nodes $S_0 = \left[\begin{smallmatrix} A\&B \\ [C\&D]_2 \end{smallmatrix}\right]$, respectively $S_2 = \left[\begin{smallmatrix} A\&B \\ C\&D \\ [C\&D]_2 \end{smallmatrix}\right]$, where $[C\&D]_2 = \begin{bmatrix} 0 & 1 \end{bmatrix} - K[C\&D]$. It is not difficult to check that the assumption that K is an admissible output feedback operator for S implies that S_0 is flow-invertible, with the flow-inverse $S_{0\times} = \left[\begin{smallmatrix} [A\&B]^K \\ [C\&D]^K_2 \end{smallmatrix}\right]$, where $[C\&D]^K_2 = \begin{bmatrix} 0 & 1 \end{bmatrix} + K[C\&D]^K$. The operator node S_2 can be interpreted as a node of the type treated in Lemma 6.6.7 if we take $U_1 = 0$ and $U_2 = U$ (so that the first input u_1 is absent). By Corollary 6.6.14, S_2 is partially flow invertible since S_0 is flow-invertible. Let us denote the partial flow-inverse of S_2 by $S_2^{\curvearrowleft}$. Comparing Definitions 6.6.6 and 7.4.2 to each other we find that, in the same way as we obtain S from S_2 by dropping the second output, we obtain S^K from $S_2^{\curvearrowleft}$ by dropping the second output. In particular, $[A\&B]^K = [A\&B]_2^{\curvearrowleft}$. The conclusion of Lemma 7.4.4 now follows from Lemma 6.6.7. (The operator $\left[\begin{smallmatrix} \alpha & 0 & 0 \\ 0 & 1 & 0 \\ 0 & 0 & 0 \end{smallmatrix}\right] - \left[\begin{smallmatrix} A\&B \\ 0 \\ [C\&D]_2 \end{smallmatrix}\right]$ in Lemma 6.6.7 collapses into $\left(\left[\begin{smallmatrix} \alpha & 0 \\ 0 & 0 \end{smallmatrix}\right] - S_0\right)$ since the first input is absent. The sign changes in some of the formulas are due to the fact that $\left(\left[\begin{smallmatrix} \alpha & 0 \\ 0 & 1 \end{smallmatrix}\right] - \left[\begin{smallmatrix} A\&B \\ K[C\&D] \end{smallmatrix}\right]\right) = \left[\begin{smallmatrix} 1 & 0 \\ 0 & -1 \end{smallmatrix}\right]\left(\left[\begin{smallmatrix} \alpha & 0 \\ 0 & 0 \end{smallmatrix}\right] - S_0\right)$ and that $\left(\left[\begin{smallmatrix} \alpha & 0 \\ 0 & 1 \end{smallmatrix}\right] + \left[\begin{smallmatrix} -[A\&B]^K \\ K[C\&D]^K \end{smallmatrix}\right]\right) = \left[\begin{smallmatrix} 1 & 0 \\ 0 & -1 \end{smallmatrix}\right]\left(\left[\begin{smallmatrix} \alpha & 0 \\ 0 & 0 \end{smallmatrix}\right] - S_{0\times}\right)$.) □

The following theorem is the feedback version of Theorem 6.3.9.

Theorem 7.4.5 *Let* $S = \left[\begin{smallmatrix} A\&B \\ C\&D \end{smallmatrix}\right]$ *be an operator node on* (Y, X, U), *with main operator* A, *control operator* B, *observation operator* C, *and transfer function* $\mathfrak{D}$, *let* $K \in \mathcal{B}(Y; U)$, *and let* $S^K = \left[\begin{smallmatrix} [A\&B]^K \\ [C\&D]^K \end{smallmatrix}\right]$ *be an operator node on* (Y, X, U), *with main operator* A^K, *control operator* B^K, *observation operator* C^K, *and transfer function* $\mathfrak{D}^K$. *Let* M *be the operator defined in* (7.4.1), *and denote the operator on the right hand side of* (7.4.2) *by* M^K. *Then the following conditions are equivalent:*

(i) K *is an admissible output feedback operator for* S, *and the closed-loop system node is* S^K.
(ii) *The operator* M^K *maps* $\mathcal{D}(S^K)$ *one-to-one onto* $\mathcal{D}(S)$, *and* $S^K = SM^K$.

(iii) *For all* $\alpha \in \rho(A^K)$ *the operator* $\left(\left[\begin{smallmatrix}\alpha & 0\\ 0 & 1\end{smallmatrix}\right] - \left[\begin{smallmatrix}A\&B\\ K[C\&D]\end{smallmatrix}\right]\right)$ *maps* $\mathcal{D}(S)$ *one-to-one onto* $\left[\begin{smallmatrix}X\\ U\end{smallmatrix}\right]$ *and (7.4.6) holds.*

(iv) *For some* $\alpha \in \rho(A^K)$ *the operator* $\left(\left[\begin{smallmatrix}\alpha & 0\\ 0 & 1\end{smallmatrix}\right] - \left[\begin{smallmatrix}A\&B\\ K[C\&D]\end{smallmatrix}\right]\right)$ *maps* $\mathcal{D}(S)$ *one-to-one onto* $\left[\begin{smallmatrix}X\\ U\end{smallmatrix}\right]$ *and (7.4.6) holds.*

(v) *The operator* M *maps* $\mathcal{D}(S)$ *one-to-one onto* $\mathcal{D}(S^K)$, *and* $S = S^K M$.

(vi) *For all* $\alpha \in \rho(A)$ *the operator* $\left(\left[\begin{smallmatrix}\alpha & 0\\ 0 & 1\end{smallmatrix}\right] + \left[\begin{smallmatrix}-[A\&B]^K\\ K[C\&D]^K\end{smallmatrix}\right]\right)$ *maps* $\mathcal{D}(S^K)$ *one-to-one onto* $\left[\begin{smallmatrix}X\\ U\end{smallmatrix}\right]$ *and (7.4.7) holds.*

(vii) *For some* $\alpha \in \rho(A)$, *the operator* $\left(\left[\begin{smallmatrix}\alpha & 0\\ 0 & 1\end{smallmatrix}\right] + \left[\begin{smallmatrix}-[A\&B]^K\\ K[C\&D]^K\end{smallmatrix}\right]\right)$ *maps* $\mathcal{D}(S^K)$ *one-to-one onto* $\left[\begin{smallmatrix}X\\ U\end{smallmatrix}\right]$ *and (7.4.7) holds.*

When these equivalent conditions hold, then $\left[\begin{smallmatrix}1\\ -KC\end{smallmatrix}\right]$ *maps* $\mathcal{D}(A)$ *into* $\mathcal{D}(S^K)$, $\left[\begin{smallmatrix}1\\ KC^K\end{smallmatrix}\right]$ *maps* $\mathcal{D}(A^K)$ *into* $\mathcal{D}(S)$, *and*

$$\begin{aligned} A &= A^K_{|\mathcal{D}(A)} - B^K KC, & A^K &= A_{|\mathcal{D}(A^K)} + BKC^K, \\ C &= [C\&D]^K \begin{bmatrix} 1 \\ -KC \end{bmatrix}, & C^K &= C\&D \begin{bmatrix} 1 \\ KC^K \end{bmatrix}. \end{aligned} \tag{7.4.10}$$

Proof As in the proof of Lemma 7.4.4, let us define the two auxiliary operator nodes $S_0 = \left[\begin{smallmatrix}A\&B\\ [C\&D]_2\end{smallmatrix}\right]$, respectively $S_2 = \left[\begin{smallmatrix}A\&B\\ C\&D\\ [C\&D]_2\end{smallmatrix}\right]$, where $[C\&D]_2 = \left[0\ 1\right] - K[C\&D]$. Arguing in the same way as in the proof of that lemma we find that S_0 is flow-invertible whenever K is an admissible feedback operator for S, that S_2 is partially flow-invertible whenever S_0 is flow-invertible, and that K is an admissible feedback operator for S whenever S_2 is partially flow-invertible. Thus, these three conditions are equivalent. This argument reduces Theorem 7.4.5 to a special case of Theorems 6.3.9 and 6.6.8. We leave the details to the reader (e.g., show that each of the conditions (i)–(iv) in these three theorems are equivalent to each other). □

With the help of Theorem 7.4.5 we are able to give a necessary and sufficient condition for the admissibility of a feedback operator.

Theorem 7.4.6 *Let* $S = \left[\begin{smallmatrix}A\&B\\ C\&D\end{smallmatrix}\right]$ *be an operator node on* (Y, X, U), *and let* $K \in \mathcal{B}(Y; U)$. *Then* K *is an admissible output feedback operator for* S *if and only if the following condition holds. For some* $\alpha \in \mathbb{C}$, *the operator* $\left(\left[\begin{smallmatrix}\alpha & 0\\ 0 & 1\end{smallmatrix}\right] - \left[\begin{smallmatrix}A\&B\\ K[C\&D]\end{smallmatrix}\right]\right)$ *maps* $\mathcal{D}(S)$ *one-to-one onto* $\left[\begin{smallmatrix}X\\ U\end{smallmatrix}\right]$, *and if we denote*

$$\begin{bmatrix} M_{11}(\alpha) & M_{12}(\alpha) \\ M_{21}(\alpha) & M_{22}(\alpha) \end{bmatrix} := \begin{bmatrix} 1 & 0 \\ C\&D \end{bmatrix} \left(\begin{bmatrix} \alpha & 0 \\ 0 & 1 \end{bmatrix} - \begin{bmatrix} A\&B \\ K[C\&D] \end{bmatrix} \right)^{-1}, \tag{7.4.11}$$

then $M_{11}(\alpha)$ *is injective and has dense range. In this case the main operator, the control operator, the observation operator, and the transfer function (evaluated* *at α) of the closed-loop operator node S^K are given by*

$$\begin{aligned} A^K &= \alpha - M_{11}^{-1}(\alpha), & B^K &= (\alpha - A^K_{|X})M_{12}(\alpha), \\ C^K &= M_{21}(\alpha)(\alpha - A^K), & \widehat{\mathfrak{D}}(\alpha) &= M_{22}(\alpha). \end{aligned} \tag{7.4.12}$$

In particular, $\alpha \in \rho(A^K)$.

The proof is analogous to the proof of Theorem 6.3.10, and we leave it to the reader.

In most cases we are able to add four more conditions which are equivalent to those listed in Theorem 7.4.5:

Theorem 7.4.7 *Make the same assumptions and introduce the same notation as in Theorem 7.4.5. In addition, suppose that $\rho(A) \cap \rho(A^K) \neq \emptyset$ (this is, in particular, true if both S and S^K are system nodes). Then the conditions (i)–(vii) listed in Theorem 7.4.5 are equivalent to each one of the following conditions:*

(viii) *For all $\alpha \in \rho(A) \cap \rho(A^K)$, $1 - K\widehat{\mathfrak{D}}(\alpha)$ is invertible and* (7.4.8) *holds.*
(ix) *For some $\alpha \in \rho(A) \cap \rho(A^K)$, $1 - K\widehat{\mathfrak{D}}(\alpha)$ is invertible and* (7.4.8) *holds.*
(x) *For all $\alpha \in \rho(A) \cap \rho(A^K)$, $1 + K\widehat{\mathfrak{D}}^K(\alpha)$ is invertible and* (7.4.9) *holds.*
(xi) *For some $\alpha \in \rho(A) \cap \rho(A^K)$, $1 + K\widehat{\mathfrak{D}}^K(\alpha)$ is invertible and* (7.4.9) *holds.*

We leave the easy proof to the reader (see the proof of Theorem 6.3.11).

We remark that the right-hand sides of formulas (7.4.8) and (7.4.9) can be written in several equivalent forms, similar to those appearing in Lemma 4.7.18(iv).

Corollary 7.4.8 *Under the assumption of Theorem 7.4.7, for all $\alpha \in \rho(A) \cap \rho(A^K)$, we have $1 + K\widehat{\mathfrak{D}}^K(\alpha) = (1 - K\widehat{\mathfrak{D}}(\alpha))^{-1}$ and $1 + \widehat{\mathfrak{D}}^K(\alpha)K = (1 - \widehat{\mathfrak{D}}(\alpha)K)^{-1}$. Moreover, the operator $(\alpha - A^K)^{-1}(\alpha - A)$ which maps $\mathcal{D}(A)$ onto $\mathcal{D}\left(A^K\right)$ is given by*

$$\begin{aligned} (\alpha - A^K)^{-1}(\alpha - A) &= 1 + (\alpha - A_{|X})^{-1}B(1 - K\widehat{\mathfrak{D}}(\alpha))^{-1}KC \\ &= 1 + (\alpha - A^K_{|X})^{-1}B^K KC \end{aligned}$$

and its inverse $(\alpha - A)^{-1}(\alpha - A^K)$ is given by

$$\begin{aligned} (\alpha - A)^{-1}(\alpha - A^K) &= 1 - (\alpha - A^K_{|X})^{-1}B^K(1 - K\widehat{\mathfrak{D}}(\alpha))KC^K \\ &= 1 - (\alpha - A_{|X})^{-1}BKC^K. \end{aligned}$$

Proof The above formulas are part of the conclusion of Theorem 7.4.7 (see also Lemma A.4.2). □

By using Theorem 7.4.7 we can make the following addition to Theorem 7.4.6.

Theorem 7.4.9 *Let $S = \left[\begin{smallmatrix} A\&B \\ C\&D \end{smallmatrix}\right]$ be an operator node on (Y, X, U), with main operator A, control operator B, observation operator C, and transfer function $\widehat{\mathfrak{D}}$. An operator $K \in \mathcal{B}(Y; U)$ is an admissible output feedback operator for S if the following condition holds: For some $\alpha \in \mathbb{C}$, $1 - K\widehat{\mathfrak{D}}(\alpha)$ is invertible, and the operator*

$$1 - (\alpha - A_{|X})^{-1} B[1 - K\widehat{\mathfrak{D}}(\alpha)]^{-1} KC$$

maps $\mathcal{D}(A)$ one-to-one onto a dense subset of X. When these conditions hold, then the closed-loop operator node S^K is determined by (7.4.8) *and Lemma 4.7.6.*

The proof is analogous to the proof of Theorem 6.3.10, and we leave it to the reader (replace (6.3.13) by (7.4.6) and Theorem 6.3.9 by Theorem 7.4.7). Note that the condition given in this theorem is sufficient *but not necessary* for the admissibility of K. However, it is necessary if we require, in addition, that both S and S^K are system nodes, because then $\rho(A) \cap \rho(A^K) \neq \emptyset$ (see Lemma 7.4.4(iii)).

The original idea that we used to introduce the notion of (partial) output feedback for a well-posed linear system was to feed a part of the output back into the input. A similar interpretation is valid for the output feedback of system nodes, too.

Theorem 7.4.10 *Let $S = \left[\begin{smallmatrix} A\&B \\ C\&D \end{smallmatrix}\right]$ be a system node on (Y, X, U), let $K \in \mathcal{B}(Y; U)$ be a system node admissible output feedback operator for S, and denote the closed-loop system by $S^K = \left[\begin{smallmatrix} [A\&B]^K \\ [C\&D]^K \end{smallmatrix}\right]$. Let x and y be the state trajectory and output of S with initial time $s \in \mathbb{R}$, initial state $x_s \in X$, and input function $u \in L^1_{\mathrm{loc}}([s, \infty); U)$, and suppose that $x \in W^{1,1}_{\mathrm{loc}}([s, \infty); X)$. Then $y \in L^1_{\mathrm{loc}}([s, \infty); Y)$, and x and y are the state trajectory and output of S^K with initial time s, initial state x_s and input function $v = y - Ku$.*

Proof By Theorem 4.7.11, $\left[\begin{smallmatrix} x \\ u \end{smallmatrix}\right] \in L^1_{\mathrm{loc}}([s, \infty); \mathcal{D}(S))$, $y \in L^1_{\mathrm{loc}}([s, \infty); Y)$, and $\left[\begin{smallmatrix} x \\ y \end{smallmatrix}\right]$ is the unique solution with the above properties of the equation

$$\begin{bmatrix} \dot{x}(t) \\ y(t) \end{bmatrix} = S \begin{bmatrix} x(t) \\ u(t) \end{bmatrix} \quad \text{for almost all } t \geq s, \qquad x(s) = x_s.$$

Let M be the operator defined in (7.4.1), and let $v = u - Ky$. Then M maps $\mathcal{D}(S)$ continuously onto $\mathcal{D}(S^K)$, $SM^{-1} = S^K$, and $M \left[\begin{smallmatrix} x \\ u \end{smallmatrix}\right] = \left[\begin{smallmatrix} x \\ v \end{smallmatrix}\right]$. Therefore

$\left[\begin{smallmatrix} x \\ v \end{smallmatrix}\right] \in L^1_{\mathrm{loc}}\big([s,\infty);\mathcal{D}\left(S^K\right)\big)$, and for almost all $t \geq s$,

$$\begin{aligned}\begin{bmatrix}\dot{x}(t)\\ y(t)\end{bmatrix} &= S\begin{bmatrix}x(t)\\ u(t)\end{bmatrix} = SM^{-1}M\begin{bmatrix}x(t)\\ u(t)\end{bmatrix}\\ &= S^K\begin{bmatrix}x(t)\\ v(t)\end{bmatrix}.\end{aligned}$$

By Theorem 4.7.11, this implies that x and u are the state and output function of S^K with initial time s, initial state x_s, and input function y. □

The following corollary is a state feedback version of Corollary 6.3.15.

Corollary 7.4.11 *Let $\Sigma = \left[\begin{smallmatrix} \mathfrak{A} & \mathfrak{B} \\ \mathfrak{C} & \mathfrak{D} \end{smallmatrix}\right]$ be an L^p-well-posed linear system on (Y, X, U) with $1 \leq p < \infty$, and let $K \in \mathcal{B}(Y;U)$. Then K is an admissible output feedback operator for Σ (in the sense of Definition 7.1.1) if and only if K is a system node admissible output feedback operator for its system node S (in the sense of Definition 7.4.2) and the closed-loop system node S^K is L^p-well-posed.*

Proof By Theorem 7.4.1, the above condition on S for K to be an admissible feedback operator for Σ is necessary. Conversely, suppose that K is an admissible feedback operator for Σ, and that S^K is a well-posed system node. Denote the system induced by S^K by Σ^K. Then it follows from Lemma 4.7.8 and Theorem 7.4.10 that for all $x_0 \in X$ and $u \in W^{2,1}_{\mathrm{loc}}(\mathbb{R}^+;U)$ with $\left[\begin{smallmatrix} x_0 \\ u(0) \end{smallmatrix}\right] \in \mathcal{D}(S)$, and for all $t \geq 0$,

$$\begin{aligned}\begin{bmatrix}\mathfrak{A}^t & \mathfrak{B}\tau^t\\ 0 & 1\end{bmatrix}\begin{bmatrix}x_0\\ \pi_+u\end{bmatrix} &= \begin{bmatrix}x(t)\\ \pi_+u\end{bmatrix} = \begin{bmatrix}(\mathfrak{A}^K)^t & \mathfrak{B}^K\tau^t\\ K\mathfrak{C}^K & 1+K\mathfrak{D}^K\end{bmatrix}\begin{bmatrix}x_0\\ \pi_+v\end{bmatrix}\\ &= \begin{bmatrix}(\mathfrak{A}^K)^t & \mathfrak{B}^K\tau^t\\ K\mathfrak{C}^K & 1+K\mathfrak{D}^K\end{bmatrix}\begin{bmatrix}1 & 0\\ -K\mathfrak{C} & 1-K\mathfrak{D}\end{bmatrix}\begin{bmatrix}x_0\\ \pi_+u\end{bmatrix}.\end{aligned}$$

This set of data is dense in $\left[\begin{smallmatrix} X \\ L^p(\mathbb{R}^+;U) \end{smallmatrix}\right]$, so the same identity must be true for all $x_0 \in X$ and $u \in L^p(\mathbb{R}^+;U)$. In particular, this implies that $1 + K\mathfrak{D}^K\pi_+$ is a left-inverse of $1 - K\mathfrak{D}\pi_+$. A similar argument with Σ interchanged with Σ^K shows that $1 + K\mathfrak{D}^K\pi_+$ is also a right inverse of $1 - K\mathfrak{D}\pi_+$. By Theorem 7.1.8, K is an admissible feedback operator for Σ. □

7.5 Regularity of the closed-loop system

If the original operator node S is compatible or regular, then it is possible to say something more about the closed-loop generators. We begin with the compatible case.

Theorem 7.5.1 *Let $S = \left[\begin{smallmatrix} A\&B \\ C\&D \end{smallmatrix}\right]$ be a compatible operator node on (Y, X, U), and let $\left[\begin{smallmatrix} A_{|W} & B \\ C_{|W} & D \end{smallmatrix}\right] \in \mathcal{B}(\left[\begin{smallmatrix} W \\ U \end{smallmatrix}\right]; \left[\begin{smallmatrix} W_{-1} \\ Y \end{smallmatrix}\right])$ be a compatible extension of S (here $X_1 \subset W \subset X$ and W_{-1} is defined as in Lemma 5.1.3). Let $K \in \mathcal{B}(Y; U)$ be an admissible output feedback operator for S. Denote the closed-loop operator node by $S^K = \left[\begin{smallmatrix} [A\&B]_\times \\ [C\&D]_\times \end{smallmatrix}\right]$, let X_1^K and X_{-1}^K be the analogues of X_1 and X_{-1} for S^K, and let W_{-1}^K be the analogue of W_{-1} for S^K (i.e., $W_{-1}^K = (\alpha - A^K)_{|W} W$ for some $\alpha \in \rho(A^K)$).*

(i) *If $1 - DK$ has a left inverse $(1 - DK)_{\text{left}}^{-1} \in \mathcal{B}(Y)$ (or equivalently, $1 - KD$ has a left inverse $(1 - KD)_{\text{left}}^{-1} \in \mathcal{B}(U)$), then $X_1^K \subset W$ and S^K is compatible with the extended observation operator $C_{|W}^K : W \to U$ and corresponding feedthrough operator D^K given by*

$$\begin{aligned} C_{|W}^K &= (1 - DK)_{\text{left}}^{-1} C_{|W}, \\ D^K &= (1 - DK)_{\text{left}}^{-1}, \end{aligned} \tag{7.5.1}$$

and the the main operator A^K of S^K is given by

$$A^K = \left(A_{|X} + BKC_{|W}^K\right)_{|X_1^K}.$$

In this case the space W_{-1} can be identified with a closed subspace of W_{-1}^K, so that $X \subset W_{-1} \subset X_{-1} \cap X_{-1}^K$. With this identification,

$$A_{|W} = A_{|W}^K - B^K K C_{|W}, \qquad B = B^K(1 + KD)$$

(where by $A_{|W}$ and $A_{|W}^K$ we mean the restrictions of $A_{|X}$ and $A_{|X}^K$ to W).

(ii) *If $1 - DK$ is invertible (or equivalently, $1 - KD$ is invertible), then $W_{-1} = W_{-1}^K$, $A^K W \subset W_{-1}$, $B^K U \subset W_{-1}$, and the operator $\left[\begin{smallmatrix} A_{|W}^K & B^K \\ C_{|W}^K & D^K \end{smallmatrix}\right] \in \mathcal{B}(\left[\begin{smallmatrix} W \\ U \end{smallmatrix}\right]; \left[\begin{smallmatrix} W_{-1} \\ Y \end{smallmatrix}\right])$ defined by*

$$\begin{aligned} \begin{bmatrix} A_{|W}^K & B^K \\ C_{|W}^K & D^K \end{bmatrix} &= \begin{bmatrix} A_{|W} + BK(1 - DK)^{-1} C_{|W} & B(1 - KD)^{-1} \\ (1 - DK)^{-1} C_{|W} & D(1 - KD)^{-1} \end{bmatrix} \\ &= \begin{bmatrix} A_{|W} & B \\ C_{|W} & D \end{bmatrix} + \begin{bmatrix} B \\ D \end{bmatrix} K(1 - DK)^{-1} \begin{bmatrix} C_{|W} & D \end{bmatrix} \\ &= \begin{bmatrix} A_{|W} & B \\ C_{|W} & D \end{bmatrix} + \begin{bmatrix} B \\ D \end{bmatrix} K \begin{bmatrix} C_{|W}^K & D^K \end{bmatrix} \\ &= \begin{bmatrix} A_{|W} & B \\ C_{|W} & D \end{bmatrix} + \begin{bmatrix} B^K \\ D^K \end{bmatrix} K \begin{bmatrix} C_{|W} & D \end{bmatrix} \end{aligned}$$

is a compatible extension of S^K.

Proof This can be regarded as a special case of Theorem 6.6.17 (or Theorem 6.3.16). See the proof of Theorem 7.4.5 and the equivalent Figures 7.16 and 7.17. □

Remark 7.5.2 The (left) invertibility of $1 - DK$ in Theorem 7.5.1 is *not a necessary condition* for the compatibility of the closed-loop system. This has to do with the possible nonuniqueness of D. For example, in the case of a boundary control system the operator D can be chosen in an arbitrary way (by adjusting the extended observation operator C_W; see Theorem 5.2.6). In particular, in a boundary control system we can always choose D in such a way that $1 - DK$ is not invertible, as soon as $K \neq 0$.

Next we look at the different regular cases.

Theorem 7.5.3 *Let $\Sigma = \left[\begin{smallmatrix} \mathfrak{A} & \mathfrak{B} \\ \mathfrak{C} & \mathfrak{D} \end{smallmatrix}\right]$ be a weakly regular (L^p|Reg-well-posed) linear system on (Y, X, U) with system node $S = \left[\begin{smallmatrix} A\&B \\ C\&D \end{smallmatrix}\right]$, semigroup generator A, control operator B, and observation operator C. We define the extensions $\widetilde{C}_w$ and $\widetilde{C}_s$ of the observation operator C as in Definition 5.4.1, and let D be the corresponding feedthrough operator (see Theorem 5.6.5). Let $K \in \mathcal{B}(Y;U)$ be an admissible output feedback operator for Σ.*

(i) *The system node S is compatible with the extended system node $\left[\begin{smallmatrix} A_{|W} & B \\ C_{|W} & D \end{smallmatrix}\right]$ where $W = \mathcal{D}(\widetilde{C}_w)$ and $C_{|W} = \widetilde{C}_w$, hence parts (i) and (ii) of Theorem 7.5.1 apply whenever $1 - DK$ is left invertible or invertible, respectively. In particular, if $1 - DK$ if left-invertible, then the closed-loop system Σ^K is compatible.*

(ii) *If S is strongly regular, then the operators $1 - KD$ and $1 - DK$ are coercive (see Definition 9.10.1). If, in addition, the closure of the range of $1 - KD$ is complemented in U or the closure of the range of $1 - DK$ is complemented in Y (this is, in particular, true if U or Y is a Hilbert space), then $1 - KD$ is left invertible, $\mathcal{D}\left(\widetilde{C}_s^K\right)$ is a closed subspace of $\mathcal{D}(\widetilde{C}_s)$, and Theorem 7.5.1(i) applies with $W = \mathcal{D}(\widetilde{C}_s)$ and $C_{|W} = \widetilde{C}_s^K$ (in particular, the closed-loop system Σ^K* is compatible*).*

(iii) *Suppose that Σ is strongly regular. Then the closed-loop system is strongly regular iff $1 - KD$ is invertible.*[4] *In this case $\mathcal{D}\left(\widetilde{C}_s^K\right) = \mathcal{D}(\widetilde{C}_s)$, and Theorem 7.5.1(ii) applies with $W = \mathcal{D}(\widetilde{C}_s)$ and $C_{|W} = \widetilde{C}_s$.*

(iv) *In the Reg-well-posed case both the original and the closed-loop system are strongly regular, $1 - KD$ is invertible, and Theorem 7.5.1(ii) applies with $W = \mathcal{D}(\widetilde{C}_s)$ and $C_{|W} = \widetilde{C}_s$.*

(v) *In the L^1-well-posed case with a reflexive state space X both the original and the closed-loop system are strongly regular, $1 - KD$ is invertible, and Theorem 7.5.1(ii) applies with $W = \mathcal{D}(\widetilde{C}_s)$ and $C_{|W} = \widetilde{C}_s$.*

(vi) *If Σ is uniformly regular then the closed-loop system Σ^K is uniformly regular, $1 - KD$ is invertible, and Theorem 7.5.1(ii) applies with $W = \mathcal{D}(\widetilde{C}_s)$ and $C_{|W} = \widetilde{C}_s$.*

[4] In the L^p-well-posed strongly regular case it is not known if $(1 - KD)$ is always invertible, i.e., if the closed-loop system is always strongly regular.

Proof This can be regarded as a special case of Theorem 6.6.18 (or Theorem 6.3.17). See the proof of Theorem 7.4.5 and the equivalent Figures 7.16 and 7.17. □

Remark 7.5.4 By using Example 5.7.4 we can show that weak regularity is not preserved in general under feedback. Since a cascade connection can be interpreted as a special case of feedback, it suffices to show that weak regularity is not preserved under cascade connections. Let Σ be the system in Example 5.7.4, and let Σ^d be its (causal) dual. Look at the cascade of Σ and Σ^d. Each of these is weakly regular. However, the transfer function of the cascade is the scalar function $g^*(\overline{s})g(s)$, which for real values of s is equal to $\|g(s)\|^2$. This transfer function is not regular since $\lim_{\alpha\to+\infty}\|g(\alpha)\|^2$ does not exist.

7.6 The dual of the closed-loop system

As the following theorem says, output feedback commutes with duality:

Theorem 7.6.1 *Let (Y, X, U) be reflexive Banach spaces, let $1 < p < \infty$, let Σ be an L^p-well-posed linear system on (Y, X, U), and let S be an operator node on (Y, X, U) (not necessarily the one induced by Σ). Let $K \in \mathcal{B}(Y; U)$.*

(i) *K is an admissible output feedback operator for Σ if and only if K^* is an admissible output feedback operator for the causal dual system Σ^d, in which case the closed-loop systems satisfy $(\Sigma^d)^{K^*} = (\Sigma^K)^d$. In particular, the generators and the transfer functions of these two systems are also identical.*

(ii) *K is an admissible output feedback operator for the operator node S if and only if S^* is an admissible output feedback operator for the dual operator node S^*, in which case the closed-loop operator nodes satisfy $(S^*)^{K^*} = (S^K)^*$.*

(iii) *If K is an admissible output feedback operator for S and both S and S^* are strongly regular with feedthrough operators D, respectively D^*, then $1 - KD$ is invertible. In this case both S^K and $(S^*)^K$ are strongly regular.*

Proof This can be regarded as a special case of Theorem 6.6.19. See the proof of Theorem 7.4.5 and the equivalent Figures 7.16 and 7.17. □

7.7 Examples

In this section we present some feedback examples.

Example 7.7.1 (Circular shift from delay line) *Let Σ be the delay line described in Example 2.3.4 with $1 \le p < \infty$. The system node of this delay line is given by*

$$S\begin{bmatrix} x \\ u \end{bmatrix} = \begin{bmatrix} A\&B \\ C\&D \end{bmatrix}\begin{bmatrix} x \\ u \end{bmatrix} = \begin{bmatrix} \dot{x} \\ x(0) \end{bmatrix} \tag{7.7.1}$$

for all $\left[\begin{smallmatrix} x \\ u \end{smallmatrix}\right] \in \mathcal{D}(S) = \left\{\left[\begin{smallmatrix} x \\ u \end{smallmatrix}\right] \in \left[\begin{smallmatrix} W^{1,p}([0,T]) \\ \mathbb{C} \end{smallmatrix}\right] \mid x(T) + u = 0\right\}$, and the transfer function is $\widehat{\mathfrak{D}}(z) = \mathrm{e}^{-Tz}$. The identity operator on $\mathbb{C}$ is an admissible feedback operator for this system, and the generators of the corresponding closed-loop system Σ^1 are given by

$$\begin{bmatrix} [A\&B]^1 \\ [C\&D]^1 \end{bmatrix}\begin{bmatrix} x \\ u \end{bmatrix} = \begin{bmatrix} \dot{x} \\ x(0) \end{bmatrix} \tag{7.7.2}$$

for all $\left[\begin{smallmatrix} x \\ u \end{smallmatrix}\right] \in \mathcal{D}\left([C\&D]^1\right) = \left\{\left[\begin{smallmatrix} x \\ u \end{smallmatrix}\right] \in \left[\begin{smallmatrix} W^{1,p}([0,T]) \\ \mathbb{C} \end{smallmatrix}\right] \mid x(T) + u = x(0)\right\}$. Thus, the closed-loop semigroup $\mathfrak{A}^1$ is the circular shift $\tau_{\mathbb{T}_T}$, the closed-loop observation operator C^1 is given by $C^1 x = x(0)$ (for $x \in \mathcal{D}\left(A^1\right) = \{x \in W^{1,p}([0,T]) \mid x(T) = x(0)\}$), and the transfer function is

$$\widehat{\mathfrak{D}}^1(z) = \frac{\mathrm{e}^{-Tz}}{1 - \mathrm{e}^{-Tz}}.$$

Proof The description of the generators of the delay line is taken from Example 4.8.1. By, for example, Theorem 7.1.8 with T replaced by $T/2$, the identity operator is an admissible feedback operator (and it is easy to show that $(1 - \mathfrak{D})^{-1} = \sum_{n=0}^{\infty} \tau^{-nT}$). The delay line Σ is regular with feedthrough operator $D = 0$, so by Theorem 7.5.3, the closed-loop system is also regular, and the formulas in Theorems 7.4.5, 7.4.7, 7.5.1, and 7.5.3 apply. These formulas tell us that the generators and the transfer function of the closed-loop system are those given above. Comparing the closed-loop semigroup generator to that of the circular shift in Example 3.2.3(v) we realize that they are the same, hence the closed-loop semigroup is the circular shift. □

Sometimes it is possible to use the specific structure of a feedback system to simplify the study of the compatibility or regularity of the closed-loop system.

Example 7.7.2 *Let $\Sigma = \left[\begin{smallmatrix} \mathfrak{A} & \mathfrak{B} \\ \mathfrak{C} & \mathfrak{D} \end{smallmatrix}\right]$ be the cascade connection of two well-posed linear systems $\Sigma_1 = \left[\begin{smallmatrix} \mathfrak{A}_1 & \mathfrak{B}_1 \\ \mathfrak{C}_1 & \mathfrak{D}_1 \end{smallmatrix}\right]$ and $\Sigma_2 = \left[\begin{smallmatrix} \mathfrak{A}_2 & \mathfrak{B}_2 \\ \mathfrak{C}_2 & \mathfrak{D}_2 \end{smallmatrix}\right]$ through $K \in \mathcal{B}(Y_1; X_2)$ presented in Example 7.2.4. Suppose that both Σ_1 and Σ_2 are compatible, with extended system nodes $\left[\begin{smallmatrix} [A_1]_{|W_1} & B_1 \\ [C_1]_{|W_1} & D_1 \end{smallmatrix}\right]$, respectively $\left[\begin{smallmatrix} [A_2]_{|W_2} & B_2 \\ [C_2]_{|W_2} & D_2 \end{smallmatrix}\right]$, where*

$\mathcal{D}(A_1) \subset W_1 \subset X_1$ and $\mathcal{D}(A_2) \subset W_2 \subset X_2$. Then the following claims are true:

(i) *Σ is compatible, with extended system node*

$$\left[\begin{array}{c|c} A_{|W} & B \\ \hline C_{|W} & D \end{array}\right] = \left[\begin{array}{cc|cc} [A_1]_{|W_1} & 0 & B_1 & 0 \\ B_2K[C_1]_{|W_1} & [A_2]_{|W_2} & B_2KD_1 & B_2 \\ \hline [C_1]_{|W_1} & 0 & D_1 & 0 \\ D_2K[C_1]_{|W_1} & [A_2]_{|W_2} & D_2KD_1 & D_2 \end{array}\right],$$

where $W = \left[\begin{smallmatrix} W_1 \\ W_2 \end{smallmatrix}\right]$. In particular,

$$\mathcal{D}(A) = \left\{ \begin{bmatrix} x_1 \\ x_2 \end{bmatrix} \in \begin{bmatrix} X_1 \\ X_2 \end{bmatrix} \middle| \; x_1 \in \mathcal{D}(A_1) \text{ and } B_2KC_1x_1 + [A_2]_{|W_2}x_2 \in X_2 \right\}.$$

(ii) *If both Σ_1 and Σ_2 are strongly or uniformly regular, then so is Σ.*

By (i), if both Σ_1 and Σ_2 are weakly regular, then Σ is compatible (but it need not be weakly regular).

Proof This follows from Examples 5.7.2 and 7.2.4 and Theorems 7.5.1 and 7.5.3 (the operator $\left[\begin{smallmatrix} 1 & 0 \\ -KD_1 & 1 \end{smallmatrix}\right]$ is always invertible, with inverse $\left[\begin{smallmatrix} 1 & 0 \\ KD_1 & 1 \end{smallmatrix}\right]$). $\square$

Example 7.7.2 can be regarded as a special case of the following more general result:

Lemma 7.7.3 *Consider the partial feedback for the well-posed linear system $\Sigma = \left[\begin{smallmatrix} \mathfrak{A} & \mathfrak{B} \\ \mathfrak{C} & \mathfrak{D} \end{smallmatrix}\right] = \left[\begin{smallmatrix} \mathfrak{A} & \mathfrak{B}_1 & \mathfrak{B}_2 \\ \mathfrak{C}_1 & \mathfrak{D}_{11} & \mathfrak{D}_{12} \\ \mathfrak{C}_2 & \mathfrak{D}_{21} & \mathfrak{D}_{22} \end{smallmatrix}\right]$ on $\left(\left[\begin{smallmatrix} Y_1 \\ Y_2 \end{smallmatrix}\right], X, \left[\begin{smallmatrix} U_1 \\ U_2 \end{smallmatrix}\right]\right)$ with admissible feedback operator $K \in \mathcal{B}(Y_2; X_2)$ described in Lemma 7.2.6. Suppose that Σ is compatible with extended system node $\left[\begin{smallmatrix} A_{|W} & B_1 & B_2 \\ [C_1]_{|W} & D_{11} & D_{12} \\ [C_2]_{|W} & D_{21} & D_{22} \end{smallmatrix}\right]$ (where $\mathcal{D}(A) \subset W \subset X$). Then the following claims are true:*

(i) *If $1 - KD_{22}$ is left-invertible, then the closed-loop system is compatible over the same space W (and appropriate versions of the formulas in Theorem 7.5.1(i) apply).*

(ii) *If* $1 - KD_{22}$ *is invertible, then the closed-loop system has the extended system node*

$$
\left[\begin{array}{c|cc} A^K_{|W} & B^K_1 & B^K_2 \\ \hline [C_1]^K_{|W} & D^K_{11} & D^K_{12} \\ [C_2]^K_{|W} & D^K_{21} & D^K_{22} \end{array}\right]
= \left[\begin{array}{c|cc} A_{|W} & B_1 & B_2 \\ \hline [C_1]_{|W} & D_{11} & D_{12} \\ [C_2]_{|W} & D_{21} & D_{22} \end{array}\right]
+ \left[\begin{array}{c} B_2 \\ \hline D_{12} \\ D_{22} \end{array}\right] K(1 - D_{22}K)^{-1} \left[\begin{array}{c|cc} [C_2]_{|W} & D_{21} & D_{22} \end{array}\right]
$$

$$
= \left[\begin{array}{c|cc} A_{|W} & B_1 & B_2 \\ \hline [C_1]_{|W} & D_{11} & D_{12} \\ [C_2]_{|W} & D_{21} & D_{22} \end{array}\right]
+ \left[\begin{array}{c} B_2 \\ \hline D_{12} \\ D_{22} \end{array}\right] K \left[\begin{array}{c|cc} [C_2]^K_{|W} & D^K_{21} & D^K_{22} \end{array}\right]
$$

$$
= \left[\begin{array}{c|cc} A_{|W} & B_1 & B_2 \\ \hline [C_1]_{|W} & D_{11} & D_{12} \\ [C_2]_{|W} & D_{21} & D_{22} \end{array}\right]
+ \left[\begin{array}{c} B^K_2 \\ \hline D^K_{12} \\ D^K_{22} \end{array}\right] K \left[\begin{array}{c|cc} [C_2]_{|W} & D_{21} & D_{22} \end{array}\right].
$$

Here $A^K_{|W}$ *maps* W *continuously into* $(\alpha - A)^{-1}X$ *for all* $\alpha \in \mathbb{C}$ *with sufficiently large real part.*

(iii) *Suppose that* Σ *is strongly regular. Then the closed-loop system is strongly regular iff* $1 - KD_{22}$ *is invertible. In this case* $\mathcal{D}(\widetilde{C}^K_s) = \mathcal{D}(\widetilde{C}_s)$*, and the formulas in (ii) apply with* $W = \mathcal{D}(\widetilde{C}_s)$*.*

(iv) *In the Reg-well-posed case both the open and closed-loop system are strongly regular,* $1 - KD_{22}$ *is invertible, and the formulas in (ii) apply with* $W = \mathcal{D}(\widetilde{C}_s)$*.*

(v) *In the* L^1*-well-posed case with reflexive state space* X *both the open and closed-loop system are strongly regular,* $1 - KD_{22}$ *is invertible, and the formulas in (ii) apply with* $W = \mathcal{D}(\widetilde{C}_s)$*.*

(vi) *If* Σ *is uniformly regular then the closed-loop system* Σ^K *is uniformly regular,* $1 - KD$ *is invertible, and the formulas in (ii) apply with* $W = \mathcal{D}(\widetilde{C}_s)$*.*

Proof This follows from Example 5.7.2, Lemma 7.2.6, and Theorems 7.5.1 and 7.5.3 (the feedthrough operator of $\left[\begin{smallmatrix} \mathfrak{D}_{11} & \mathfrak{D}_{12} \\ \mathfrak{D}_{21} & \mathfrak{D}_{22} \end{smallmatrix}\right]$ is $\left[\begin{smallmatrix} D_{11} & D_{12} \\ D_{21} & D_{22} \end{smallmatrix}\right]$, and $\left[\begin{smallmatrix} 1 & 0 \\ 0 & 1 \end{smallmatrix}\right] - \left[\begin{smallmatrix} 0 & 0 \\ 0 & K \end{smallmatrix}\right]\left[\begin{smallmatrix} D_{11} & D_{21} \\ D_{21} & D_{22} \end{smallmatrix}\right] = \left[\begin{smallmatrix} 1 & 0 \\ -KD_{21} & 1-KD_{22} \end{smallmatrix}\right]$ is invertible if and only if $1 - KD_{22}$ is invertible). $\square$

Remark 7.7.4 The reader should compare the formulas for the closed-loop systems in Theorem 7.1.2, Example 7.2.4, and Lemma 7.2.6 with the corresponding formulas for the generators of these systems in the compatible case, given in Theorem 7.5.1(ii), Example 7.7.2, and Lemma 7.7.3, respectively. The analogy is striking! We invite the reader to use the same principle to formulate and prove the corresponding formulas for the generators of all the other feedback systems discussed in Sections 7.2 and 7.3. In the literature it is often assumed that all the feedthrough operators that appear inside a feedback loop vanish, and in this case the needed invertibility conditions become vacuous and the formulas for the generators simplify slightly. Thus, in these cases the closed-loop systems are always compatible or strongly or uniformly regular whenever the original systems are compatible or strongly or uniformly regular, provided that the feedback is admissible. In this sense compatibility and strong and uniform regularity are preserved under (static output) feedback.

Our final example concerns feedback in a system with an analytic semigroup of the type studied in Theorem 5.7.3.

Example 7.7.5 *Let* $\Sigma = \left[\begin{smallmatrix} \mathfrak{A} & \mathfrak{B} \\ \mathfrak{C} & \mathfrak{D} \end{smallmatrix}\right]$ *be one of the systems considered in Theorem 5.7.3, and let* $K \in \mathcal{B}(Y;U)$. *In all the different cases considered there,* K *is an admissible output feedback operator for* Σ *if and only if* $1 - DK$ *is invertible, in which case the closed-loop system is of the same type as* Σ. *More precisely, the closed-loop semigroup* $\mathfrak{A}^K$ *can be extended to an analytic semigroup on* $\mathfrak{A}^K_{|X_{\alpha-1}}$ *on* $X_{\alpha-1}$. *For all* $\gamma \in [\alpha-1, \beta+1]$, *the spaces* X_γ *are invariant under* $\mathfrak{A}^K_{|X_{\alpha-1}}$, *and the restriction* $\mathfrak{A}^K_{|X_\gamma}$ *of* $\mathfrak{A}^K_{|X_{\alpha-1}}$ *to* X_γ *is an analytic semigroup on* X_γ. *The generator of* $\mathfrak{A}^K_{|X_\gamma}$ *is* $(A + BK(1-DK)^{-1}C)_{|X_{\gamma+1}}$ *if* $\gamma \in [\alpha-1, \beta]$, *and it is the part of* $A + BK(1-DK)^{-1}C$ *in* X_γ *if* $\gamma \in (\beta, \beta+1]$. *If we define* $X^K_{\alpha-1} = X_{\alpha-1}$, *and let* X^K_γ *be the fractional order space with index* $\gamma - \alpha - 1$ *constructed by means of the semigroup* $\mathfrak{A}^K_{|X_{\alpha-1}}$ *on* $X_{\alpha-1}$, *then* $X^K_\gamma = X_\gamma$ *for all* $\gamma \in [\alpha-1, \beta+1]$. *The closed-loop observation operator is* $C^K = (1 - DK)^{-1}C$, *the closed-loop control operator is* $B^K = B(1-KD)^{-1}$, *and the closed-loop feedthrough operator is* $D^K = D(1-KD)^{-1}$.

The proof of this is similar to the proof of Example 6.3.18, and it is left to the reader.

7.8 Comments

Feedback plays a central role in many different areas in mathematics and control theory. For example, many aspects of the theory of Volterra integral and functional equation (see, e.g., Gripenberg *et al.* (1990)) can be formulated as feedback problems. The PDE literature is full of different feedback notions and results (see, e.g., Lasiecka and Triggiani 2000a,b), many of which fit into the setting of this chapter. Below we comment only on those results in the literature which are most closely connected to the theory presented here.

Section 7.1 Theorem 7.1.2 is due to G. Weiss (1994b, Theorem 6.1) in the Hilbert space case with $p = 2$. The part of this theorem which says that $\mathfrak{A}^K$ is a C_0 semigroup is in Salamon (1987, Theorem 4.2). Our proof has been modeled after Malinen (2000). Lemma 7.1.7 is due to G. Weiss (1994b, Remark 6.4). The equivalence (ii) $\Leftrightarrow$ (iii) in Theorem 7.1.8 was proved by Salamon (1987, Lemma 4.1). That condition (vii) in Theorem 7.1.8 implies that K is an admissible feedback operator for Σ was pointed out by Staffans (1998a, Lemma 3.5(i)). A particular *Reg*-well-posed version of Corollary 7.1.9 is found in Desch *et al.* (1985, Theorem 2.1 and Corollary 2).

Section 7.2 The results in this section are straightforward extensions of the corresponding classical results. The idea of recovering an extended version of the original system from the feedback systems in Figures 7.6 and 7.7 by partial flow inversion (i.e., by reversing the direction of some of the arrows and changing some signs in the summation junctions) was used extensively in Curtain, Weiss, and Weiss (1996) in a slightly different context. The same idea can be applied to the flow-inversion in Figure 7.1. The cascade connection in Example 7.2.4 has been modeled after Weiss and Curtain (1997, Lemma 5.1).

Section 7.3 This sections is based mainly on Staffans (1998a) and to some extent also on Weiss and Curtain (1997) (the first version of Staffans (1998a) was written independently of Curtain, Weiss and Weiss (1996, 2001) and Weiss and Curtain (1997), but there was some later cross-influence; see Staffans (1998a)). State feedback with a bounded feedback operator K (and output injection with a bounded injection operator H) has been studied by Morris (1994).

Section 7.4 This sections is based mainly on Staffans (2002c) and Staffans and Weiss (2004). Theorem 7.4.5 is (formally) new. In the L^2-well-posed Hilbert space case some parts of Theorem 7.4.7 are found in G. Weiss (1994b, Proposition 3.7) and also in G. Weiss (1994b, formulas (6.14), (7.3), and (7.21)). Non-well-posed feedbacks have also been studied in, e.g., Curtain *et al.* (2001), Flandoli *et al.* (1988), Lasiecka and Triggiani (2000a, b), Mikkola (2002), Weiss and Curtain (1997), and Weiss and Rebarber (2001).

Section 7.5 Theorem 7.5.1 is (formally) new. The claims related to A^K and $C^K_{|W}$ in the well-posed version of part (i) were discovered by Mikkola in 1997 while he was working on his thesis (see the comments that Mikkola makes regarding the history of Mikkola (2002, Proposition 6.6.18)), and we were later able to add the claims involving B^K. The main part of Theorem 7.5.3 is due to Weiss

(1994a, Theorems 7.2 and 7.8). Our proof of this theorem is based on Theorem 7.5.1, and it differs from G. Weiss's proof in the respect that we use a different (and more general) method of embedding W_{-1} into W_{-1}^K. In particular, we are able to conclude that W_{-1} is *closed* in W_{-1}^K, and that the closed-loop system is compatible whenever $1 - DK$ is left-invertible. (The latter fact was first proved by Mikkola in an unpublished manuscript from 1997.) Remark 7.5.4 is taken from Staffans and Weiss (2002).

Section 7.6 The main part of Theorem 7.6.1 was announced and used in Staffans (1997) and Weiss and Weiss (1997). A formal proof was finally published in Staffans and Weiss (2004).

Section 7.7 The regular Hilbert space L^2-well-posed version of Example 7.7.2 is due to Weiss and Curtain (1997, Lemma 5.1). Lemma 7.7.3 and Remark 7.7.4 are used (implicitly) in Staffans (1998d).

8
Stabilization and detection

A system is stabilizable if it is possible to make it stable by using state feedback. It is detectable if it is possible to make it stable by using output injection. The stability notion that we use implies well-posedness, but it is weaker than exponential stability. For simplicity we require also the open-loop system to be well-posed.

8.1 Stability

So far we have more or less ignored the question of the *stability* of a well-posed linear system, but this will change from now on. The word 'stable' is usually used as a synonym for 'exponentially stable.' We shall not use this interpretation here, since there are many infinite-dimensional systems that are not exponentially stable, but they are still stable in a weaker sense.[1] Note, however, that stability still implies well-posedness in the setting that we use.

Sometimes it is important to to know that a specific part of a system is stable, although the rest of the system may be unstable, and for this reason we introduce the following terminology:

Definition 8.1.1 Let Σ be a well-posed linear system $\Sigma = \left[\begin{smallmatrix} \mathfrak{A} & \mathfrak{B} \\ \mathfrak{C} & \mathfrak{D} \end{smallmatrix}\right]$ on (Y, X, U).

(i) With reference to Definition 2.5.6, we say that
  (a) Σ is *state/state bounded* if $\mathfrak{A}$ is bounded, i.e., $\mathfrak{A}$ is ω-bounded with $\omega = 0$;
  (b) Σ is *input/state bounded* if $\mathfrak{B}$ is bounded, i.e., $\mathfrak{B}$ is ω-bounded with $\omega = 0$;

[1] For simplicity we treat only the case where the exponential growth bound ω in Definition 2.5.6 is zero, and leave the obvious extension to the case $\omega \neq 0$ to the reader.

  (c) Σ is *state/output bounded* if $\mathfrak{C}$ is bounded, i.e., $\mathfrak{C}$ is ω-bounded with $\omega = 0$;
  (d) Σ is *input/output bounded* or *input/output stable* if $\mathfrak{D}$ is bounded, i.e., $\mathfrak{D}$ is ω-bounded with $\omega = 0$;
  (e) Σ is *input bounded* or *input stable* if conditions (b) and (d) above hold;
  (f) Σ is *output bounded* or *output stable* if conditions (c) and (d) above hold;
  (g) Σ is *bounded* or *stable* if conditions (a)–(d) above hold.

(ii) We say that
  (a) Σ is *strongly state/state stable* if $\mathfrak{A}$ is strongly stable, i.e., for every $x \in X$, $\mathfrak{A}^t x \to 0$ in X as $t \to \infty$;
  (b) Σ is *strongly input/state stable* if $\mathfrak{B}$ is strongly stable, i.e., $\mathfrak{B}$ is ω-bounded with $\omega = 0$ and, for every $u \in L^p|Reg_0(\mathbb{R}; U)$, $\mathfrak{B}\tau^t u \to 0$ in X as $t \to \infty$;
  (c) Σ is *strongly state/output stable* if $\mathfrak{C}$ is strongly stable, i.e., $\mathfrak{C}$ is ω-bounded with $\omega = 0$, and, in the *Reg*-well-posed case we require, in addition, that $\mathfrak{C}$ maps X into $BC_0(\overline{\mathbb{R}}^+; Y)$;
  (d) Σ is *strongly input/output stable* if $\mathfrak{D}$ is strongly stable, i.e., $\mathfrak{D}$ is ω-bounded with $\omega = 0$, and, in the *Reg*-well-posed case we require, in addition, that $\mathfrak{D}$ maps $Reg_0(\mathbb{R}; U)$ into $Reg_0(\mathbb{R}; Y)$;
  (e) Σ is *strongly input stable* if conditions (b) and (d) above hold;
  (f) Σ is *strongly output stable* if conditions (c) and (d) above hold;
  (g) Σ is *strongly stable* if conditions (a)–(d) above hold.

(iii) Σ is *weakly stable* if conditions (a)–(d) in (ii) hold with strong convergence replaced by weak convergence (thus, in (c) and (d) we use the weak versions of $Reg_0(\overline{\mathbb{R}}^+; Y)$ and $Reg_0(\mathbb{R}; Y)$).

(iv) By *exponential stability* of Σ or one of its components we mean that Σ or the corresponding component is ω-bounded for some $\omega < 0$.

The different stability notions defined in Definition 8.1.1 are far from independent of each other. The most basic relationships are given in the following lemma:

Lemma 8.1.2 *Let* $\Sigma = \left[\begin{array}{c|c} \mathfrak{A} & \mathfrak{B} \\ \hline \mathfrak{C} & \mathfrak{D} \end{array}\right]$ *be a well-posed linear system on* (Y, X, U).

(i) *For each component of* Σ, *exponential stability implies strong stability, strong stability implies weak stability, and weak stability implies boundedness.*

(ii) *Exponential stability of* $\mathfrak{A}$ *implies exponential stability of* $\mathfrak{B}$, $\mathfrak{C}$, *and* $\mathfrak{D}$, *and exponential stability of either* $\mathfrak{B}$ *or* $\mathfrak{C}$ *implies exponential stability* *of $\mathfrak{D}$. In particular, Σ is exponentially stable if and only if $\mathfrak{A}$ is exponentially stable.*

(iii) *If $\mathfrak{A}$ and $\mathfrak{B}$ are bounded, then $\mathfrak{B}$ is weakly or strongly stable if and only if the restriction of $\mathfrak{A}$ to the closure of the range of $\mathfrak{B}$ (i.e., the reachable subspace; see Definition 9.1.2) is weakly or strongly stable.*

(iv) *Weak or strong stability of $\mathfrak{A}$ together with boundedness of $\mathfrak{C}$ implies that $\mathfrak{C}$ is weakly or strongly stable.*

(v) *Weak or strong stability of $\mathfrak{B}$ together with boundedness of $\mathfrak{D}$ implies that $\mathfrak{D}$ is weakly or strongly stable.*

(vi) *If $\mathfrak{B}$, $\mathfrak{C}$, and $\mathfrak{D}$ are bounded, then $\mathfrak{D}$ is weakly or strongly stable if and only if the restriction of $\mathfrak{C}$ to the closure of the range of $\mathfrak{B}$ (i.e., the reachable subspace; see Definition 9.1.2) is weakly or strongly stable.*

(vii) *Σ is weakly or strongly stable if and only if it is stable and $\mathfrak{A}$ is weakly or strongly stable.*

(viii) *In the L^1-well-posed case, boundedness of $\mathfrak{A}$ implies boundedness of $\mathfrak{B}$ and boundedness of $\mathfrak{C}$ implies boundedness of $\mathfrak{D}$. In particular, Σ is bounded if and only if both $\mathfrak{A}$ and $\mathfrak{C}$ are bounded, and Σ is weakly or strongly stable if and only if $\mathfrak{A}$ is weakly or strongly stable and $\mathfrak{C}$ is bounded.*

(ix) *In the Reg-well-posed case, boundedness of $\mathfrak{A}$ implies boundedness of $\mathfrak{C}$, and boundedness of $\mathfrak{B}$ implies boundedness of $\mathfrak{D}$. In particular, Σ is bounded if and only if both $\mathfrak{A}$ and $\mathfrak{B}$ are bounded, and Σ is weakly or strongly stable if and only if $\mathfrak{A}$ is weakly or strongly stable and $\mathfrak{B}$ is bounded.*

Observe that, in most of the statements above we assume something about $\mathfrak{A}$ and say something about $\mathfrak{B}$ or $\mathfrak{C}$, or we assume something about $\mathfrak{B}$ or $\mathfrak{C}$ and say something about $\mathfrak{D}$. To go in the opposite direction we need stabilizability or detectability conditions of some kind (see also (iii) and (vi)). These will be discussed in the next section. Condition (ii) provides a partial explanation of why exponential stability is in widespread use.

Proof of Lemma 8.1.2 (i) Most of this is obvious. The only non-obvious parts are the claims that exponential stability of $\mathfrak{B}$ implies strong stability, and that exponential stability of $\mathfrak{D}$ implies strong stability in the *Reg*-well-posed case. The former of these two claims is true because, for every $u \in L^p|Reg_0(\mathbb{R}, U)$ and every $\omega < 0$, $\tau^t\pi_-u \to 0$ in $L^p|Reg_\omega(\mathbb{R}^-; U)$ as $t \to \infty$, and $\mathfrak{B}$ in bounded on $L^p|Reg_\omega(\mathbb{R}^-; U)$ for some $\omega < 0$. To prove the latter claim we first use Example 2.6.5 to get an exponentially stable realization of $\mathfrak{D}$, and then we apply part (v).

(ii) That exponential stability of $\mathfrak{A}$ implies exponential stability of $\mathfrak{B}$, $\mathfrak{C}$, and $\mathfrak{D}$ follows from Theorem 2.5.4(ii). If $\mathfrak{B}$ is exponentially stable, then $\mathfrak{D}$ satisfies condition (iii) in Theorem 2.6.6 for some $\omega < 0$, and the realization of

$\mathfrak{D}$ given in Example 2.6.5(i) is exponentially stable. This means that also $\mathfrak{D}$ is exponentially stable. If $\mathfrak{C}$ is exponentially stable, then $\mathfrak{D}$ satisfies condition (iv) in Theorem 2.6.6 for some $\omega < 0$, and the realization of $\mathfrak{D}$ given in Example 2.6.5(ii) is exponentially stable. Thus, again $\mathfrak{D}$ is exponentially stable.

(iii) Suppose that the restriction of $\mathfrak{A}$ to $\mathcal{R}(\mathfrak{B})$ is weakly or strongly stable, and let $u \in L^p|Reg_0(\mathbb{R}; U)$. For each $t \in \mathbb{R}$ we can split $x(t) = \mathfrak{B}\tau^t u$ into

$$x(t) = \mathfrak{B}\tau^{t-T}(\pi_+ + \pi_-)\tau^T u = \mathfrak{B}\tau^t\pi_{[T,\infty)}u + \mathfrak{A}^{t-T}\mathfrak{B}\tau^T u.$$

Here the first term tends to zero in norm as $T \to \infty$, uniformly in $t \geq T$, and the second term tends weakly or strongly to zero as $t \to \infty$ and T is fixed. Thus, $x(t) \to 0$ weakly or strongly as $t \to \infty$. This shows that $\mathfrak{B}$ is weakly or strongly stable. Conversely, suppose that $\mathfrak{B}$ is strongly stable. Let $x_0 \in \mathcal{R}(\mathfrak{B})$, and choose some $u \in L^p|Reg_0(\mathbb{R}^-; U)$ such that $x_0 = \mathfrak{B}u$. Then $\mathfrak{A}^t x_0 = \mathfrak{A}^t\mathfrak{B}u = \mathfrak{B}\tau^t\pi_- u \to 0$ weakly or strongly as $t \to \infty$. Thus, $\mathfrak{A}^t x_0 \to 0$ weakly or strongly for all $x_0 \in \mathcal{R}(\mathfrak{B})$. By continuity and the boundedness of $\mathfrak{A}$, the same statement is also true for all $x_0 \in \overline{\mathcal{R}(\mathfrak{B})}$.

(iv), (v), and (ix): The claims (iv) and (v) are vacuous in the L^p-well-posed case with $p < \infty$, so it suffices to consider the *Reg*-well-posed case. In this case (iv), (v) and (ix) follow from the boundedness of the observation operator C (see Theorem 4.4.2(i)–(ii)) and the representation formula for $\mathfrak{D}$ given in Theorem 4.5.2.

(vi) This proof is similar to the proof of (iii) (only the *Reg*-well-posed case needs a proof).

(vii) See (iii)–(v).

(viii) This follows from the representation formulas

$$\mathfrak{B} = \sum_{n=0}^{\infty} \mathfrak{A}^{nT}\mathfrak{B}\pi_{[-T,0)}\tau^{-nT},$$

$$\mathfrak{D} = \sum_{n=-\infty}^{\infty} \tau^{-nT}\left(\mathfrak{C}\mathfrak{B}\pi_{[-T,0)} + \pi_{[0,T)}\mathfrak{D}\pi_{[0,T)}\right)\tau^{nT}$$

in Lemma 2.4.3, since $\|\tau^{-nT}\| = \|\tau^{nT}\| = 1$, $\|\mathfrak{B}\pi_{[-T,0)}\| < \infty$, $\|\mathfrak{D}\pi_{[0,T)}\| < \infty$, and $\|u\|_{L_1(\mathbb{R};U)} = \sum_{n=-\infty}^{\infty}\|\pi_{[nT,(n+1)T)}u\|_{L_1(\mathbb{R};U)}$ for every $u \in L^1(\mathbb{R}; U)$. □

As our next lemma shows, the stability of a system is reflected in its frequency domain behavior.

Lemma 8.1.3 *Let* $\Sigma = \left[\begin{smallmatrix}\mathfrak{A} & \mathfrak{B}\\ \mathfrak{C} & \mathfrak{D}\end{smallmatrix}\right]$ *be a well-posed linear system with semigroup generator* A*, control operator* B*, observation operator* C*, and transfer function* $\widehat{\mathfrak{D}}$.

(i) *Σ is state/state bounded if and only if the right half-plane $\Re\lambda > 0$ belongs to the resolvent set of A and there is a constant $M < \infty$ such that*

$$\|(\lambda - A)^{-n}\| \le M(\Re\lambda)^{-n}$$

for $\Re\lambda > 0$ and $n = 1, 2, 3, \ldots$

(ii) *If Σ is state/output bounded, then, for each $n = 1, 2, 3, \ldots$, $C(\lambda - A)^{-n}$ has an analytic extension to the right half-plane $\Re\lambda > 0$ and there is a constant $M < \infty$ such that the extended functions satisfy*

$$\|C(\lambda - A)^{-n}\| \le Mn^{-1/2p}(\Re\lambda)^{1/p-n}$$

for $\Re\lambda > 0$ and $n = 1, 2, 3, \ldots$ if Σ is L^p-well-posed and

$$\|C(\lambda - A)^{-n}\| \le M(\Re\lambda)^{-n}$$

for $\Re\lambda > 0$ and $n = 1, 2, 3, \ldots$ if Σ is Reg-well-posed.

(iii) *If Σ is input/state bounded, then, for each $n = 1, 2, 3, \ldots$, $(\lambda - A)^{-1}B$ has an analytic extension to the right half-plane $\Re\lambda > 0$, and there is a constant $M < \infty$ such that the extended functions satisfy*

$$\|(\lambda - A)^{-n}B\| \le Mn^{(1/p-1)/2}(\Re\lambda)^{1-1/p-n}$$

for $\Re\lambda > 0$ and $n = 1, 2, 3, \ldots$ if Σ is L^p-well-posed and

$$\|(\lambda - A)^{-n}B\| \le Mn^{-1/2}(\Re\lambda)^{1-n}$$

for $\Re\lambda > 0$ and $n = 1, 2, 3, \ldots$ if Σ is Reg-well-posed.

(iv) *If Σ is input/output bounded, then the transfer function $\widehat{\mathfrak{D}}$ has a bounded analytic extension to the right half-plane $\Re\lambda > 0$.*

The converse of (iv) is true in the L^2-well-posed Hilbert space case (see Theorem 10.3.5). Some partial converses to (ii) and (iii) are also true in the same setting (see Theorems 10.6.1 and 10.6.2).

Proof (i) This is a restatement of Corollary 3.4.4.

(ii) Let $x_0 \in X$. As $\mathfrak{C}$ is ω-bounded with $\omega = 0$, the Laplace transform of $\mathfrak{C}x_0$ is an analytic function in the right half-plane $\Re\lambda > 0$. By (3.2.6) and Theorem 4.4.2(iv), for $\Re\lambda > \omega_{\mathfrak{A}}$ and $n = 0, 1, 2, \ldots$,

$$C(\lambda - A)^{-(n+1)}x_0 = \frac{(-1)^n}{n!}\frac{\partial^n}{\partial\lambda^n}(\widehat{\mathfrak{C}x_0})(\lambda).$$

Thus, the functions on the left-hand side can be analytically extended to the right half-plane $\Re\lambda > 0$. To derive the given norm estimates we apply Hölder's inequality and Lemma 4.2.10 in the same way as we did in the proof of Proposition 4.4.9.

(iii) The proof of (iii) is similar to the proof of (ii), and we leave it to the reader (replace Theorem 4.4.2 by Theorem 4.2.1).

(iv) Without loss of generality, we may assume that $\omega_{\mathfrak{A}} \le 0$ (choose a different realization of $\mathfrak{D}$, if necessary). By Lemma 4.5.3 and Definition 4.6.1, for all $u \in U$ and all λ in the right half-plane $\Re\lambda > 0$,

$$\mathfrak{D}(e_\lambda u) = e_\lambda \widehat{\mathfrak{D}}(\lambda) u$$

where $e_\lambda(t) = e^{\lambda t}$. In particular, by first taking the $L^p|Reg$-norm over $(-\infty, 0]$ and then dividing both sides by $\|e_\lambda\|_{L^p|Reg((-\infty;0])}$ we get

$$\|\widehat{\mathfrak{D}}(\lambda) u\|_Y \le \|\mathfrak{D}\|_{TIC(U;Y)} \|u\|_U, \qquad \Re\lambda > 0.$$

□

There is also a version of Lemma 8.1.3 which deals with *exponential stability* instead of boundedness:

Corollary 8.1.4 *Let* $\Sigma = \left[\begin{smallmatrix} \mathfrak{A} & \mathfrak{B} \\ \mathfrak{C} & \mathfrak{D} \end{smallmatrix}\right]$ *be a well-posed linear system with semigroup generator A, control operator B, observation operator C, and transfer function* $\widehat{\mathfrak{D}}$.

(i) Σ *is exponentially state/state stable if and only if some right half-plane* $\Re\lambda > -\epsilon$ *with* $\epsilon > 0$ *belongs to the resolvent set of A and there is a constant* $M < \infty$ *such that*

$$\left\|(\lambda - A)^{-n}\right\| \le M(\Re\lambda + \epsilon)^{-n}$$

for $\Re\lambda > -\epsilon$ *and* $n = 1, 2, 3, \dots$

(ii) *If* Σ *is exponentially state/output stable, then, for each* $n = 1, 2, 3, \dots,$ $C(\lambda - A)^{-n}$ *has an analytic extension to some right half-plane* $\Re\lambda > -\epsilon$ *with* $\epsilon > 0$*, and there is a constant* $M < \infty$ *such that the extended functions satisfy*

$$\|C(\lambda - A)^{-n}\| \le M n^{-1/2p} (\Re\lambda + \epsilon)^{1/p-n}$$

for $\Re\lambda > -\epsilon$ *and* $n = 1, 2, 3, \dots$ *if* Σ *is* L^p*-well-posed and*

$$\|C(\lambda - A)^{-n}\| \le M(\Re\lambda + \epsilon)^{-n}$$

for $\Re\lambda > -\epsilon$ *and* $n = 1, 2, 3, \dots$ *if* Σ *is Reg-well-posed.*

(iii) *If* Σ *is exponentially input/state stable, then, for each* $n = 1, 2, 3, \dots,$ $(\lambda - A)^{-1} B$ *has an analytic extension to some right half-plane* $\Re\lambda > -\epsilon$ *with* $\epsilon > 0$*, and there is a constant* $M < \infty$ *such that the extended functions satisfy*

$$\|(\lambda - A)^{-n} B\| \le M n^{(1/p-1)/2} (\Re\lambda + \epsilon)^{1-1/p-n}$$

for $\Re\lambda > -\epsilon$ *and* $n = 1, 2, 3, \dots$ *if* Σ *is* L^p*-well-posed and*

$$\|(\lambda - A)^{-n} B\| \leq M n^{-1/2} (\Re\lambda + \epsilon)^{1-n}$$

for $\Re\lambda > -\epsilon$ *and* $n = 1, 2, 3, \dots$ *if* Σ *is Reg-well-posed.*

(iv) *If* Σ *is exponentially input/output stable, then the transfer function* $\widehat{\mathfrak{D}}$ *has a bounded analytic extension to some right half-plane* $\Re\lambda > -\epsilon$ *with* $\epsilon > 0$.

This follows from Lemma 8.1.3 by an exponential shift; see Example 4.8.2(i).

Next we observe that in some cases exponential stability is equivalent to input/state or state/output boundedness:

Theorem 8.1.5 *Let* Σ *be an* $L^p|Reg$*-well-posed linear system on* (Y, X, U) *with control operator* B *and observation operator* C.

(i) *If* Σ *is* L^p*-well-posed for some* $p < \infty$, *and if* $\|Cx\|_Y \geq \epsilon \|x\|_X$ *for some* $\epsilon > 0$ *and all* $x \in X_1$, *then* Σ *is exponentially stable if and only if it is state/output bounded.*

(ii) *If* Σ *is Reg-well-posed or* L^p*-well-posed for some* $p > 1$, *and if* $\mathcal{N}(B)$ *is complemented in* U *and* $\mathcal{R}(B) \supset X$, *then* Σ *is exponentially stable if and only if it is input/state bounded.*

Proof By Lemma 8.1.2(i), exponential stability always implies both input/state and state/output boundedness, so it suffices to prove the converse claims.

(i) Suppose that Σ is state/output bounded. Then there is a constant K such that $\|\mathfrak{C}x\|_{L^p(\mathbb{R}^+;Y)} \leq K\|x\|_X$ for all $x \in X$. Under the additional coercivity condition on C, this together with Theorem 4.4.2(i) implies that, for all $x \in X_1$, $\|t \mapsto \mathfrak{A}^t x\|_{L^p(\mathbb{R}^+;Y)} \leq K/\epsilon \|x_x\|$ for some $\epsilon > 0$. As X_1 is dense in X, the same inequality must hold for all $x \in X$, too. By Theorem 3.11.8(v), this implies the exponential stability of $\mathfrak{A}$.

(ii) Without loss of generality we can assume that B is injective (see Proposition 4.10.1). By the closed graph theorem and the assumption that $\mathcal{R}(B) \supset X$, B^{-1} maps X continuously into U. This implies that the system that we get by replacing U by X and B by the identity operator is also input/state bounded. Thus, condition (vii) in Theorem 3.11.8 is satisfied with $\omega = 0$, and $\mathfrak{A}$ is exponentially stable. □

Our next result concerns the asymptotic behavior of the state and output of a system which is strongly stable in the appropriate sense.

Theorem 8.1.6 *Let* $\Sigma = \left[\begin{smallmatrix} \mathfrak{A} & \mathfrak{B} \\ \mathfrak{C} & \mathfrak{D} \end{smallmatrix}\right]$ *be a well-posed linear system on* (Y, X, U).

(i) *If* $\mathfrak{A}$ *is strongly stable and* $\mathfrak{B}$ *is bounded, then, for all* $s \in \mathbb{R}$, *all* $x_s \in X$, *and all* $u \in L^p|Reg_0([s, \infty); U)$, *the state* x *of* Σ *with initial time* s, *initial* *state x_s and input u (see Definition 2.2.7) satisfies $x \in BC_0([s,\infty);X)$. If, in addition, $\mathfrak{C}$ and $\mathfrak{D}$ are bounded, then the corresponding output y satisfies $y \in L^p|Reg_0([s,\infty);Y)$, and, in the Reg-well-posed case, $y - Du \in BC_0([s,\infty);Y)$ (where D is the feedthrough operator of $\mathfrak{D}$).*

(ii) *If $\mathfrak{B}$ is strongly stable, then, for each $u \in L^p|Reg_0(\mathbb{R};U)$, the equation $\dot{x}(t) = Ax(t) + Bu(t)$, $t \in \mathbb{R}$, has a solution in $BC_0(\mathbb{R};X)$, namely $x = \mathfrak{B}\tau u$. This solution is unique in $BC_0(\mathbb{R};X)$ if $\mathfrak{A}$ is bounded. If $\mathfrak{B}$ is strongly stable and $\mathfrak{D}$ is bounded, then the corresponding output $y = \mathfrak{D}u$ satisfies $y \in L^p|Reg_0(\mathbb{R};U)$, and, in the Reg-well-posed case, $y - Du \in BC_0(\mathbb{R};Y)$ (where D is the feedthrough operator of $\mathfrak{D}$).*

Proof (i) The state x is given by $x(t) = \mathfrak{A}^{t-s}x_s + \mathfrak{B}\tau^t\pi_{[s,\infty)}u$ for $t \ge s$. This tends to zero as $t \to \infty$ because of the strong stability of $\mathfrak{A}$ and $\mathfrak{B}$ (see Definition 8.1.1 and Lemma 8.1.2(iii)). The claim that $y \in L^p|Reg_0([s,\infty);Y)$ can be deduced from Lemma 8.1.2(vii) and the representation $y = \tau^{-s}\mathfrak{C}X_S + \mathfrak{D}\pi_{[s,\infty)}u$. That $y - Du$ is continuous in the regular case follows from the fact that $y(t) = Cx(t) + Du(t)$ for all $t \ge s$ where $C \in \mathcal{B}(X;Y)$; see Theorem 4.4.2(ii) and Corollary 4.5.5.

(ii) Most of the proof of (ii) is identical to the proof of Theorem 4.3.2, i.e., by repeating that argument we find that $x = \mathfrak{B}\tau u$ is a strong solution of the equation $\dot{x}(t) = Ax(t) + Bu(t)$, that $x(t) \to 0$ as $t \to -\infty$, and that this equation has a unique strong solution in $BC_0(\mathbb{R};X)$ if $\mathfrak{A}$ is bounded. By Lemma 8.1.2(iii), $x(t) \to 0$ as $t \to \infty$. Trivially, $y = \mathfrak{D}u \in L^p(\mathbb{R};Y)$ in the L^p-case when $\mathfrak{D}$ is bounded, and by Lemma 8.1.2(iii),(v), $y \in Reg_0(\mathbb{R};Y)$ in the *Reg*-well-posed case. In the regular case, if u has compact support, then, by applying (i), we find that $y - Du$ is continuous. By the boundedness of $\mathfrak{D}$, the same must then be true for arbitrary $u \in Reg_0(\mathbb{R};U)$ since $Reg_c(\mathbb{R};U)$ is dense in $Reg_0(\mathbb{R};U)$ and $BC_0(\mathbb{R};Y)$ is closed in $Reg_0(\mathbb{R};Y)$. □

Finally, let us investigate what type of stability is preserved under output feedback:

Theorem 8.1.7 *Let $\Sigma = \left[\begin{smallmatrix} \mathfrak{A} & \mathfrak{B} \\ \mathfrak{C} & \mathfrak{D} \end{smallmatrix}\right]$ be a well-posed linear system on (Y, X, U), let $K \in \mathcal{B}(Y;U)$ be an admissible output feedback operator for Σ, and denote the closed-loop system by $\Sigma^K = \left[\begin{smallmatrix} \mathfrak{A}^K & \mathfrak{B}^K \\ \mathfrak{C}^K & \mathfrak{D}^K \end{smallmatrix}\right]$.*

(i) *If $\mathfrak{B}$ and $K\mathfrak{C}^K$ are bounded, then the following claims are true:*
  (a) *$\mathfrak{A}^K$ is bounded if and only if $\mathfrak{A}$ is bounded.*
  (b) *If both $\mathfrak{A}$ and $K\mathfrak{C}^K$ are strongly stable (the latter condition is redundant in the L^p-case with $p < \infty$) then $\mathfrak{A}^K$ is strongly stable.*
  (c) *In the L^p-well-posed case with $p < \infty$, if $\mathfrak{A}$ is exponentially stable, then so is $\mathfrak{A}^K$.*

(ii) *If* $\mathfrak{C}$ *and* $\mathfrak{B}^K K$ *are bounded, then the following claims are true:*
   (a) $\mathfrak{A}^K$ *is bounded if and only if* $\mathfrak{A}$ *is bounded.*
   (b) *If both* $\mathfrak{A}$ *and* $\mathfrak{B}^K K$ *are strongly stable then* $\mathfrak{A}^K$ *is strongly stable.*
   (c) *In the Reg-well-posed case and the* L^p*-well-posed case with* $p > 1$*, if* $\mathfrak{A}$ *is exponentially stable, then so is* $\mathfrak{A}^K$*.*

(iii) *If both* $\mathfrak{B}$ *and* $K\mathfrak{D}^K$ *are bounded, or strongly stable, or exponentially stable, then so is* $\mathfrak{B}^K$*.*

(iv) *If both* $\mathfrak{C}$ *and* $\mathfrak{D}^K K$ *are bounded, or strongly stable, or exponentially stable, then so is* $\mathfrak{C}^K$*.*

(v) *If* Σ *is bounded, then the the following conditions are equivalent:*
   (a) Σ^K *is bounded;*
   (b) Σ^K *is input/output bounded;*
   (c) *the operator* $1 - K\mathfrak{D}$ *has in inverse in* $TIC(U)$ *(where TIC stands for* TIC_ω *with* $\omega = 0$*);*
   (d) *the operator* $1 - \mathfrak{D}K$ *has an inverse in* $TIC(Y)$*;*
   (e) *the operator* $\left[\begin{smallmatrix} 1 & -\mathfrak{D} \\ -K & 1 \end{smallmatrix}\right]$ *has an inverse in* $TIC(\left[\begin{smallmatrix} Y \\ U \end{smallmatrix}\right])$*;*
   (f) *the diagram in Figure 7.1 (i.e., the set of equations (7.1.1)) with* $x_0 = 0$ *defines a continuous linear mapping from the external input* $v \in L^p|Reg(\overline{\mathbb{R}}^+; U)$ *to the internal input* $u \in L^p|Reg(\overline{\mathbb{R}}^+; U)$*.*

   When these conditions hold, then
   (g) *in the* L^p*-well-posed case with* $p < \infty$*,* Σ^K *is strongly stable if and only if* Σ *is strongly stable,*
   (h) *in the Reg-well-posed case,* Σ^K *is strongly stable if* Σ *is strongly stable and at least one of the components of* Σ^K *is strongly stable, and*
   (j) Σ^K *is exponentially stable if and only if* Σ *is exponentially stable.*

Proof (i) By (7.1.5), $\mathfrak{A}^K = \mathfrak{A} + \mathfrak{B}\tau K\mathfrak{C}^K$, hence (a) holds. To prove (b) it suffices to show that, for every $x_0 \in X$, $\mathfrak{B}\tau^t K\mathfrak{C}^K x_0 \to 0$ as $t \to \infty$. But this follows from Theorem 8.1.6(ii) with u replaced by $K\mathfrak{C}^K x_0 \in L^p|Reg_0(\overline{\mathbb{R}}^+, U)$. By Theorem 3.11.8(v), to prove (c) it suffices to show that $\mathfrak{B}\tau K\mathfrak{C}^K x_0 \in L^p(\mathbb{R}^+; X)$ for all $x_0 \in X$. By the boundedness of $K\mathfrak{C}^K$, $K\mathfrak{C}^K x_0 \in L^p(\mathbb{R}^+; U)$. We can interpret $\mathfrak{B}\tau$ as the input/output map of the system with generating operators $\left[\begin{smallmatrix} A & B \\ 1 & 0 \end{smallmatrix}\right]$ (see Theorem 4.3.4). This system is exponentially stable, and by Theorem 2.5.4, $\mathfrak{B}\tau$ maps $L^p(\mathbb{R}^+; U)$ into $L^p(\mathbb{R}^+; X)$.

(ii) The proof of (ii) is very similar to the proof of (i), but this time we use the formula $\mathfrak{A}^K = \mathfrak{A} + \mathfrak{B}^K K\tau\mathfrak{C}$ in (7.1.5). The claim (a) follows immediately. In case (b), by Lemma 8.1.2(iv), for each $x_0 \in S$, $\mathfrak{C}x_0 \in L^p|Reg_0(\overline{\mathbb{R}}^+; Y)$, and by Theorem 8.1.6(ii), $\mathfrak{B}^K K\tau^t\mathfrak{C}x_0 \to 0$ as $t \to \infty$. To prove (c) we shall appeal to Theorem 3.11.8(vii). Let $u \in C_c^1(\mathbb{R}; X_1)$, and consider the system with generating operators $\left[\begin{smallmatrix} A & 1 \\ C & 0 \end{smallmatrix}\right]$ (see Theorem 4.3.4). The exponential stability of

$\mathfrak{A}$ implies that this system is exponentially stable, hence its input/output map is bounded. If we apply this input/output map to u, then, by Theorem 4.5.10, we get the function

$$y(t) = C \int_{-\infty}^{t} \mathfrak{A}^{t-s} u(s)\, ds$$

and there is a constant $M_1 < \infty$ such that $\|y\|_{L^p|Reg_0(\mathbb{R};Y)} \le M_1 \|u\|_{L^p|Reg_0(\mathbb{R};X)}$. To this function we apply $\mathfrak{B}^K K$ and get

$$x = \int_{-\infty}^{0} (\mathfrak{A}^K)^{-t} B^K K C \int_{-\infty}^{t} \mathfrak{A}^{t-s} u(s)\, ds\, dt.$$

The boundedness of $\mathfrak{B}^K K$ implies that there is a constant M_2 such that $\|x\|_X \le M_2 \|y\|_{L^p|Reg_0(\mathbb{R};Y)} \le M_2 M_1 \|u\|_{L^p|Reg_0(\mathbb{R};X)}$. On the other hand, we can use Fubini's theorem to write x in the form (note that $B^K K C \in \mathcal{B}(X_1; X^K_{-1})$)

$$\begin{aligned} x &= \int_{-\infty}^{0} \int_{-s}^{0} (\mathfrak{A}^K)^{-t} B^K K C \mathfrak{A}^{t-s} u(s)\, dt\, ds \\ &= \int_{-\infty}^{0} \int_{-s}^{0} (\mathfrak{A}^K)^{-t} B^K K (\tau^{-s} \mathfrak{C} u(s))(t)\, dt\, ds \\ &= \int_{-\infty}^{0} \mathfrak{B}^K K \tau^{-s} \mathfrak{C} u(s)\, ds \\ &= \int_{-\infty}^{0} ((\mathfrak{A}^K)^{-s} - \mathfrak{A}^{-s}) u(s)\, ds. \end{aligned}$$

By Theorem 3.11.8, $\mathfrak{A}^K$ is exponentially stable.

(iii)–(iv) This is true because $\mathfrak{B}^K = \mathfrak{B} + \mathfrak{B} K \mathfrak{D}^K$ and $\mathfrak{C}^K = \mathfrak{C} + \mathfrak{D}^K K \mathfrak{C}$.

(v) Trivially, (a) $\Rightarrow$ (b), and we get the converse implication from, for example, (i), (iii), and (iv). To prove that (b) $\Leftrightarrow$ (c) it suffices to observe that $1 + K\mathfrak{D}^K = (1 - K\mathfrak{D})^{-1}$. The equivalence of (c), (d), and (e) is proved in Lemma A.4.1. Finally, (f) $\Leftrightarrow$ (c) since the map from $\pi_+ v$ to u in Figure 7.1 is $(1 - K\mathfrak{D})^{-1}$. The claim (g) follows from (i)(b) and Lemma 8.1.2(vii), the claim (h) from Lemma 8.1.2(vii), (i)(b), (ii)(b), and (iii), and the claim (j) from (i)(c) and (ii)(c). □

Corollary 8.1.8 *Suppose that $\mathfrak{D} \in TIC(U)$ has an inverse in $TIC(U)$ (and in the Reg-well-posed case suppose, in addition, that $\mathfrak{D}$ has a well-posed realization; cf. Theorem 2.6.7). Then $\mathfrak{D}$ is exponentially stable if and only if $\mathfrak{D}^{-1}$ is exponentially stable.*

Proof We let $\mathfrak{D}$ be exponentially stable and choose an exponentially stable realization $\left[\begin{smallmatrix} \mathfrak{A} & \mathfrak{B} \\ \mathfrak{C} & \mathfrak{D} \end{smallmatrix}\right]$ of $\mathfrak{D} - 1$ (cf. Theorems 2.6.6 and 2.6.7). To this system we can apply negative identity feedback (cf. Remark 7.1.10) to get a realization of $\mathfrak{D}^{-1}$. By Theorem 8.1.7(v)(j), this realization is exponentially stable. In particular, $\mathfrak{D}^{-1}$ is exponentially stable. The converse implication is proved in the same way. □

8.2 Stabilizability and detectability

The notion of (closed-loop) stabilization is based on different types of feedback, i.e., (static or dynamic) output feedback, state feedback, and output injection. In addition we shall also study a case where we at the same time want to add both a state feedback pair $\left[\mathfrak{K} \,\middle|\, \mathfrak{F}\right]$ and an output injection pair $\left[\frac{\mathfrak{H}}{\mathfrak{G}}\right]$ to a given system $\left[\begin{smallmatrix}\mathfrak{A} & \mathfrak{B} \\ \mathfrak{C} & \mathfrak{D}\end{smallmatrix}\right]$. If we try to write a figure similar to Figures 7.14 and 7.15, we immediately observe that we need one more input/output map $\mathfrak{E}$ (from the output injection input to the state feedback output); see Figure 8.1. This operator need not always exist (as a bounded operator), and this forces us to introduce still another definition:

Definition 8.2.1 Let $\Sigma = \left[\begin{smallmatrix}\mathfrak{A} & \mathfrak{B} \\ \mathfrak{C} & \mathfrak{D}\end{smallmatrix}\right]$ be a well-posed linear system on (Y, X, U). The pairs $\left[\mathfrak{K} \,\middle|\, \mathfrak{F}\right]$ and $\left[\frac{\mathfrak{H}}{\mathfrak{G}}\right]$ are called *jointly admissible state feedback and output injection pairs* for Σ if $\left[\mathfrak{K} \,\middle|\, \mathfrak{F}\right]$ is an admissible state feedback pair for Σ, $\left[\frac{\mathfrak{H}}{\mathfrak{G}}\right]$ is an admissible output injection pair for Σ, and in addition, there exists a operator $\mathfrak{E}$, called the *interaction operator*, such that the combined extended system $\Sigma_{\text{ext}} = \left[\begin{smallmatrix}\mathfrak{A} & \mathfrak{H} & \mathfrak{B} \\ \mathfrak{C} & \mathfrak{G} & \mathfrak{D} \\ \mathfrak{K} & \mathfrak{E} & \mathfrak{F}\end{smallmatrix}\right]$ is a well-posed linear system on $\left(\left[\begin{smallmatrix}Y \\ X\end{smallmatrix}\right], X, \left[\begin{smallmatrix}Y \\ U\end{smallmatrix}\right]\right)$.

Lemma 8.2.2 *Let $\Sigma = \left[\begin{smallmatrix}\mathfrak{A} & \mathfrak{B} \\ \mathfrak{C} & \mathfrak{D}\end{smallmatrix}\right]$ be a well-posed linear system on (Y, X, U). Then the following conditions are equivalent:*

(i) *The pairs $\left[\mathfrak{K} \,\middle|\, \mathfrak{F}\right]$ and $\left[\frac{\mathfrak{H}}{\mathfrak{G}}\right]$ are jointly admissible state feedback and output injection pairs with interaction operator $\mathfrak{E}$.*

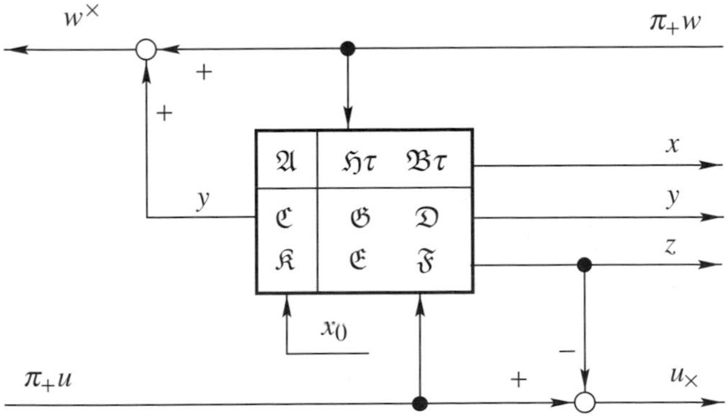

Figure 8.1 The extended system

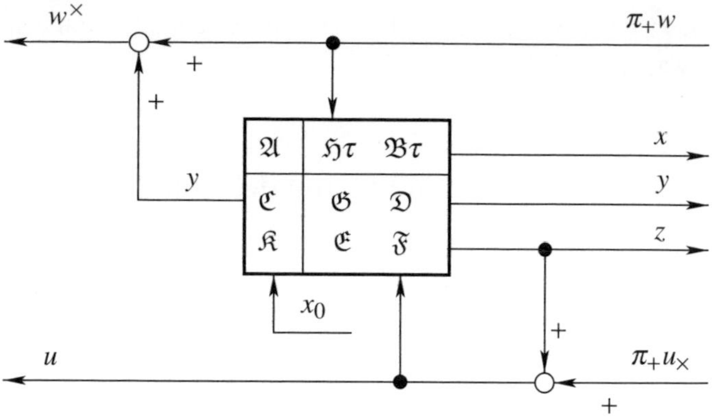

Figure 8.2 Right coprime factor

(ii) *The system* Σ_{ext} *in Definition 8.2.1 is a well-posed linear system on* $\left(\left[\begin{smallmatrix}Y\\U\end{smallmatrix}\right], X, \left[\begin{smallmatrix}Y\\U\end{smallmatrix}\right]\right)$*, and both* $\left[\begin{smallmatrix}0&0\\0&1\end{smallmatrix}\right]$ *and* $\left[\begin{smallmatrix}-1&0\\0&0\end{smallmatrix}\right]$ *are admissible output feedback operators for* Σ_{ext}.

(iii) *The system* Σ_{ext} *in Definition 8.2.1 is a well-posed linear system on* $\left(\left[\begin{smallmatrix}Y\\U\end{smallmatrix}\right], X, \left[\begin{smallmatrix}Y\\U\end{smallmatrix}\right]\right)$*, and* $1-\mathfrak{F}$ *and* $1+\mathfrak{G}$ *have inverses in* $TIC_{\text{loc}}(U)$*, respectively* $TIC_{\text{loc}}(Y)$.

Suppose that these conditions hold. Then the closed-loop system $\Sigma_\times$ *that we get by using* $\left[\begin{smallmatrix}0&0\\0&1\end{smallmatrix}\right]$ *as an output feedback operator for* Σ_{ext} *with inputs* $u_\times$ *and* w *and outputs* y *and* z *(see Figure 8.2[2]) is given by*

$$\Sigma_\times = \left[\begin{array}{c|cc} \mathfrak{A}_\times & \mathfrak{H}_\times\tau & \mathfrak{B}_\times\tau \\ \hline \mathfrak{C}_\times & \mathfrak{G}_\times & \mathfrak{D}_\times \\ \mathfrak{K}_\times & \mathfrak{E}_\times & \mathfrak{F}_\times \end{array}\right]$$

$$= \left[\begin{array}{c|cc} \mathfrak{A}+\mathfrak{B}\tau(1-\mathfrak{F})^{-1}\mathfrak{K} & \mathfrak{H}\tau+\mathfrak{B}(1-\mathfrak{F})^{-1}\mathfrak{E}\tau & \mathfrak{B}(1-\mathfrak{F})^{-1}\tau \\ \hline \mathfrak{C}+\mathfrak{D}(1-\mathfrak{F})^{-1}\mathfrak{K} & \mathfrak{G}+\mathfrak{D}(1-\mathfrak{F})^{-1}\mathfrak{E} & \mathfrak{D}(1-\mathfrak{F})^{-1} \\ (1-\mathfrak{F})^{-1}\mathfrak{K} & (1-\mathfrak{F})^{-1}\mathfrak{E} & (1-\mathfrak{F})^{-1}\mathfrak{F} \end{array}\right]$$

$$= \left[\begin{array}{c|cc} \mathfrak{A} & \mathfrak{H}\tau & \mathfrak{B}\tau \\ \hline \mathfrak{C} & \mathfrak{G} & \mathfrak{D} \\ \mathfrak{K} & \mathfrak{E} & \mathfrak{F} \end{array}\right] + \left[\begin{array}{c} \mathfrak{B}\tau \\ \hline \mathfrak{D} \\ \mathfrak{F} \end{array}\right](1-\mathfrak{F})^{-1}\left[\begin{array}{c|cc} \mathfrak{K} & \mathfrak{E} & \mathfrak{F} \end{array}\right]$$

$$= \left[\begin{array}{c|cc} \mathfrak{A} & \mathfrak{H}\tau & 0 \\ \hline \mathfrak{C} & \mathfrak{G} & 0 \\ 0 & 0 & -1 \end{array}\right] + \left[\begin{array}{c} \mathfrak{B}\tau \\ \hline \mathfrak{D} \\ 1 \end{array}\right](1-\mathfrak{F})^{-1}\left[\begin{array}{c|cc} \mathfrak{K} & \mathfrak{E} & 1 \end{array}\right],$$

[2] See the proof of Theorem 8.4.1 for an explanation of the captions of Figures 8.2 and 8.3.

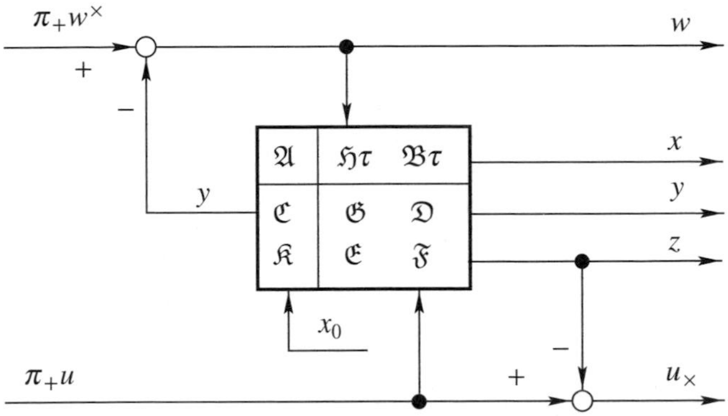

Figure 8.3 Left coprime factor

and the closed-loop system $\Sigma^\times$ that we get by using $\left[\begin{smallmatrix}1 & 0\\ 0 & 0\end{smallmatrix}\right]$ as an output feedback operator for Σ_{ext} with inputs u and $w^\times$ and outputs y and z (see Figure 8.3) is given by

$$
\begin{aligned}
\Sigma^\times &= \left[\begin{array}{c|cc} \mathfrak{A}^\times & \mathfrak{H}^\times\tau & \mathfrak{B}^\times\tau \\ \hline \mathfrak{C}^\times & \mathfrak{G}^\times & \mathfrak{D}^\times \\ \mathfrak{K}^\times & \mathfrak{E}^\times & \mathfrak{F}^\times \end{array}\right] \\
&= \left[\begin{array}{c|cc} \mathfrak{A} - \mathfrak{H}\tau\,(1+\mathfrak{G})^{-1}\mathfrak{C} & \mathfrak{H}\,(1+\mathfrak{G})^{-1}\tau & \mathfrak{B}\tau - \mathfrak{H}\,(1+\mathfrak{G})^{-1}\mathfrak{D}\tau \\ \hline (1+\mathfrak{G})^{-1}\mathfrak{C} & \mathfrak{G}\,(1+\mathfrak{G})^{-1} & (1+\mathfrak{G})^{-1}\mathfrak{D} \\ \mathfrak{K} - \mathfrak{E}\,(1+\mathfrak{G})^{-1}\mathfrak{C} & \mathfrak{E}\,(1+\mathfrak{G})^{-1} & \mathfrak{F} - \mathfrak{E}\,(1+\mathfrak{G})^{-1}\mathfrak{D} \end{array}\right] \\
&= \left[\begin{array}{c|cc} \mathfrak{A} & \mathfrak{H}\tau & \mathfrak{B}\tau \\ \hline \mathfrak{C} & \mathfrak{G} & \mathfrak{D} \\ \mathfrak{K} & \mathfrak{E} & \mathfrak{F} \end{array}\right] - \left[\begin{array}{c} \mathfrak{H}\tau \\ \hline \mathfrak{G} \\ \mathfrak{E} \end{array}\right] (1+\mathfrak{G})^{-1} \left[\begin{array}{c|cc} \mathfrak{C} & \mathfrak{G} & \mathfrak{D} \end{array}\right] \\
&= \left[\begin{array}{c|cc} \mathfrak{A} & 0 & \mathfrak{B}\tau \\ \hline 0 & 1 & 0 \\ \mathfrak{K} & 0 & \mathfrak{F} \end{array}\right] - \left[\begin{array}{c} \mathfrak{H}\tau \\ \hline -1 \\ \mathfrak{E} \end{array}\right] (1+\mathfrak{G})^{-1} \left[\begin{array}{c|cc} \mathfrak{C} & -1 & \mathfrak{D} \end{array}\right].
\end{aligned}
$$

Proof See Lemma 7.2.6(ii)–(iii). □

Remark 8.2.3 According to Lemma 7.1.7, it is possible to recover the extended system Σ_{ext} from either of the systems $\Sigma_\times$ or $\Sigma^\times$ by using negative feedback. For example,the feedback connection drawn in Figure 8.4 is equivalent to Σ_{ext}.

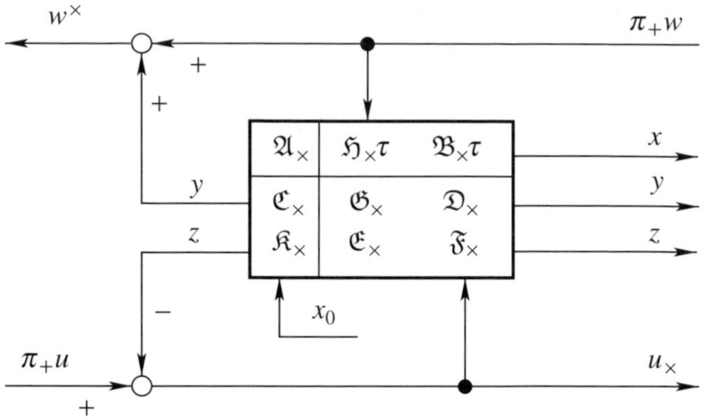

Figure 8.4 Cancellation of state feedback

We are now ready to define the most basic types of stabilizability and detectability (coprime stabilization and detection and dynamic stabilization will be introduced later in Definitions 8.4.5 and 8.5.1, respectively).[3]

Definition 8.2.4 Let Σ be a well-posed linear system on (Y, X, U).

(i) An operator $K \in \mathcal{B}(Y; U)$ is called a *stabilizing output feedback operator* for Σ if K is an admissible output feedback operator for Σ and the resulting closed-loop system Σ^K in Theorem 7.1.2 is bounded (see Definition 8.1.1). The system Σ is (closed-loop) *stabilizable by static output feedback* if there exists a stabilizing output feedback operator $K \in \mathcal{B}(Y; U)$ for Σ.

(ii) A pair $[\mathfrak{K} | \mathfrak{F}]$ is called a *stabilizing state feedback pair* for Σ if $[\mathfrak{K} | \mathfrak{F}]$ is an admissible state feedback pair for Σ and the resulting closed-loop system $\Sigma_\times$ in Lemma 7.3.3 is bounded. The system Σ is (closed-loop) *stabilizable* if there exists a stabilizing state feedback pair for Σ.

(iii) A pair $\left[\frac{\mathfrak{H}}{\mathfrak{G}}\right]$ is called a *detecting output injection pair* for Σ if $\left[\frac{\mathfrak{H}}{\mathfrak{G}}\right]$ is an admissible output injection pair for Σ and the resulting closed-loop system $\Sigma^\times$ in Lemma 7.3.3 is bounded. The system Σ is (closed-loop) *detectable* if there exists a detecting output injection pair for Σ.

(iv) The pairs $[\mathfrak{K} | \mathfrak{F}]$ and $\left[\frac{\mathfrak{H}}{\mathfrak{G}}\right]$ are called *jointly stabilizing and detecting state feedback and output injection pairs* for Σ if they are jointly admissible state feedback and output injection pairs with some interaction operator $\mathfrak{E}$, and both the closed-loop systems $\Sigma_\times$ and $\Sigma^\times$ in Lemma 8.2.2 are bounded. The system Σ is (closed-loop) *jointly*

[3] For simplicity we shall throughout assume that *the original system is well-posed*. It is also possible to develop a more general theory based on the notion of feedback for arbitrary operator nodes.

stabilizable and detectable if there exist some jointly stabilizing and detecting state feedback and output injection pairs.

(v) To these definitions we add one of the words '*weakly*', '*strongly*', or '*exponentially*' whenever the closed-loop system is stable in the corresponding sense (see Definition 8.1.1).

(vi) If to these definitions we add one or several of the qualifiers '*state/state*', '*input/state*', '*state/output*', '*input/output*', '*input*', or '*output*', then we mean that only the corresponding part of the closed-loop system has to be bounded or stable in the appropriate sense (recall that '*input*' means both 'input/state' and 'input/output' and that '*output*' means both 'state/output' and 'input/output').

Of course, we could also introduce ω-stabilizing versions of these notions by requiring that the closed-loop systems (or the appropriate parts of the closed-loop systems) are ω-bounded. See Definition 2.5.6.

Remark 8.2.5 Definition 8.2.4 deserves a comment. In the case of static output feedback we do not add any new inputs or outputs to the system, and part (i) of Definition 8.2.4 is straightforward. In the other parts of Definition 8.2.4 we do not just require the original inputs and outputs of the system to be stabilized, but also the added inputs and outputs. For example, in the case of state feedback, it is not enough if $\mathfrak{A}_\times$, $\mathfrak{B}_\times$, $\mathfrak{C}_\times$, and $\mathfrak{D}_\times$ are bounded or strongly stable, but also $\mathfrak{K}_\times$ and $\mathfrak{F}_\times$ should be bounded or strongly stable (see Figure 7.14). In the case of *exponential* stabilization this is not a problem, since exponential stability of the closed-loop system is equivalent to exponential stability of its semigroup; thus any added inputs or outputs are automatically stabilized.

We observe the following basic relationships between the different methods of stabilization:

Lemma 8.2.6

(i) *Exponential stabilizability implies strong stabilizability, strong stabilizability implies weak stabilizability, and weak stabilizability implies stabilizability.*

(ii) *Exponential detectability implies strong detectability, strong detectability implies weak detectability, and weak detectability implies detectability.*

(iii) *Every system which is stabilizable by static output feedback in any of the senses described in Definition 8.2.4(v),(vi) is jointly stabilizable and detectable in the same sense.*

(iv) *Every system which is jointly stabilizable and detectable in any of the senses described in Definition 8.2.4(v),(vi) is both stabilizable and detectable in the same sense.*

The converse to (iii) is not true (even in the finite-dimensional situation). We do not know if the converse to (iv) is true or not.

Proof (i)–(ii) See Lemma 8.1.2(i).

(iii) Take $\left[\mathfrak{K}\middle|\mathfrak{F}\right] = K\left[\mathfrak{C}\middle|\mathfrak{D}\right]$, $\left[\begin{smallmatrix}\mathfrak{H}\\ \mathfrak{G}\end{smallmatrix}\right] = -\left[\begin{smallmatrix}\mathfrak{B}\\ \mathfrak{D}\end{smallmatrix}\right]K$, and $\mathfrak{E} = K - K\mathfrak{D}K$ (or $\mathfrak{E} = -K\mathfrak{D}K$).

(iv) This is obvious. □

The properties of a system of being stabilizable or detectable are preserved under static output feedback:

Lemma 8.2.7 *All the different notions of stabilizability and detectability listed in Definition 8.2.4(i)–(vi) are preserved under (admissible) static output feedback, i.e., the closed-loop system is stabilizable or detectable in exactly the same sense as the original system.*

Proof In the case of output stabilizability this follows from Lemma 7.1.7: if K is an arbitrary admissible output feedback operator for Σ and K_1 is a stabilizing output feedback for Σ, then $K_1 - K$ is a stabilizing output feedback for the system Σ^K in Theorem 7.1.2. The proof of the fact that the different versions of (state feedback) stabilizability and (output injection) detectability are preserved under static output feedback is left to the reader (it is a simplified version of the argument below, where either the output injection column or the state feedback row is ignored).

Let K be an admissible output feedback operator, and let $\left[\mathfrak{K}\middle|\mathfrak{F}\right]$ and $\left[\begin{smallmatrix}\mathfrak{H}\\ \mathfrak{G}\end{smallmatrix}\right]$ be a jointly stabilizing and detecting state feedback and output injection pair for Σ with interaction operator $\mathfrak{E}$. Consider the extended system

$$\Sigma_{\mathrm{ext}} = \left[\begin{array}{c|cc} \mathfrak{A} & \mathfrak{H} & \mathfrak{B} \\ \hline \mathfrak{C} & \mathfrak{G} & \mathfrak{D} \\ \mathfrak{K} & \mathfrak{E} & \mathfrak{F} \end{array}\right].$$

By Lemma 7.2.6, all the operators $\left[\begin{smallmatrix}0 & 0\\ K & 0\end{smallmatrix}\right]$, $\left[\begin{smallmatrix}0 & 0\\ 0 & 1\end{smallmatrix}\right]$, and $\left[\begin{smallmatrix}-1 & 0\\ 0 & 0\end{smallmatrix}\right]$ are admissible output feedback operators for Σ_{ext}, and the latter two are stabilizing. Let us denote the system that we get by using $\left[\begin{smallmatrix}0 & 0\\ K & 0\end{smallmatrix}\right]$ as an output feedback operator by

$$\Sigma^K = \left[\begin{array}{c|cc} \mathfrak{A}^K & \mathfrak{H}^K & \mathfrak{B}^K \\ \hline \mathfrak{C}^K & \mathfrak{G}^K & \mathfrak{D}^K \\ \mathfrak{K}^K & \mathfrak{E}^K & \mathfrak{F}^K \end{array}\right].$$

This is an extended version of the system Σ^K in Theorem 7.1.2. As in Lemma 8.2.2, we denote the systems that we get by using either $\left[\begin{smallmatrix}0 & 0\\ 0 & 1\end{smallmatrix}\right]$ or $\left[\begin{smallmatrix}-1 & 0\\ 0 & 0\end{smallmatrix}\right]$ as output feedback operator by $\Sigma_\times$ and $\Sigma^\times$, respectively. By Lemma 7.1.7, we can recover $\Sigma_\times$ from Σ^K by using the output feedback operator $\left[\begin{smallmatrix}0 & 0\\ -K & 1\end{smallmatrix}\right]$, and we can recover $\Sigma^\times$ from Σ^K by using the output feedback operator $-\left[\begin{smallmatrix} 1 & 0 \\ K & 0 \end{smallmatrix}\right]$. However, to use $\left[\begin{smallmatrix} 0 & 0 \\ -K & 1 \end{smallmatrix}\right]$ as an output feedback operator for Σ^K is almost the same thing as using the output feedback operator $\left[\begin{smallmatrix} 0 & 0 \\ 0 & 1 \end{smallmatrix}\right]$ for the modified system

$$\left[\begin{array}{c|cc} \mathfrak{A}^K & \mathfrak{H}^K & \mathfrak{B}^K \\ \hline \mathfrak{C}^K & \mathfrak{G}^K & \mathfrak{D}^K \\ \mathfrak{K}^K - K\mathfrak{C}^K & \mathfrak{E}^K - K\mathfrak{G}^K & \mathfrak{F}^K - K\mathfrak{D}^K \end{array}\right]; \tag{8.2.1}$$

this leads to the stable closed-loop system

$$\left[\begin{array}{c|cc} \mathfrak{A}_\times & \mathfrak{H}_\times & \mathfrak{B}_\times \\ \hline \mathfrak{C}_\times & \mathfrak{G}_\times & \mathfrak{D}_\times \\ \mathfrak{K}_\times - K\mathfrak{C}_\times & \mathfrak{E}_\times - K\mathfrak{G}_\times & \mathfrak{F}_\times - K\mathfrak{D}_\times \end{array}\right].$$

Let us replace Σ^K by (8.2.1). We can still use $-\left[\begin{smallmatrix} 1 & 0 \\ K & 0 \end{smallmatrix}\right]$ as an output feedback operator for this modified system, and it is almost the same thing as using the output feedback operator $\left[\begin{smallmatrix} -1 & 0 \\ 0 & 0 \end{smallmatrix}\right]$ for the further modified system

$$\left[\begin{array}{c|cc} \mathfrak{A}^K & \mathfrak{H}^K + \mathfrak{B}^K K & \mathfrak{B}^K \\ \hline \mathfrak{C}^K & \mathfrak{G}^K + \mathfrak{D}^K K & \mathfrak{D}^K \\ \mathfrak{K}^K - K\mathfrak{C}^K & \mathfrak{E}^K - K\mathfrak{G}^K + \mathfrak{F}^K K - K\mathfrak{D}^K K & \mathfrak{F}^K - K\mathfrak{D}^K \end{array}\right];$$

the result is the stable closed-loop system

$$\left[\begin{array}{c|cc} \mathfrak{A}^\times & \mathfrak{H}^\times + \mathfrak{B}^\times K & \mathfrak{B}^\times \\ \hline \mathfrak{C}^\times & \mathfrak{G}^\times + \mathfrak{D}^\times K & \mathfrak{D}^\times \\ \mathfrak{K}^\times - K\mathfrak{C}^\times & \mathfrak{E}^\times - K\mathfrak{G}^\times + \mathfrak{F}^\times K - K\mathfrak{D}^\times K & \mathfrak{F}^\times - K\mathfrak{D}^\times \end{array}\right].$$

This modification does not affect the stability of the system that we get by using $\left[\begin{smallmatrix} -1 & 0 \\ 0 & 0 \end{smallmatrix}\right]$ as an output feedback operator (the result is analogous to the one above). Thus, $\left[\mathfrak{K}^K - K\mathfrak{C}^K \middle| \mathfrak{F}^K - K\mathfrak{D}^K\right]$ and $\left[\begin{smallmatrix} \mathfrak{H}^K + \mathfrak{B}^K K \\ \hline \mathfrak{G}^\times + \mathfrak{D}^\times K \end{smallmatrix}\right]$ are jointly stabilizing and detecting state feedback and output injection pairs for the output feedback system $\left[\begin{array}{c|c} \mathfrak{A}^K & \mathfrak{B}^K \\ \hline \mathfrak{C}^K & \mathfrak{D}^K \end{array}\right]$ with interaction operator $\mathfrak{E}^K - K\mathfrak{G}^K + \mathfrak{F}^K K - K\mathfrak{D}^K K$. Explicitly, these operators are given by

$$\begin{aligned} &\begin{bmatrix} \mathfrak{G}^K + \mathfrak{D}^K K & \mathfrak{D}^K \\ \mathfrak{E}^K - K\mathfrak{G}^K + \mathfrak{F}^K K - K\mathfrak{D}^K K & \mathfrak{F}^K - K\mathfrak{D}^K \end{bmatrix} \\ &= \begin{bmatrix} (1-\mathfrak{D}K)^{-1}(\mathfrak{G}+\mathfrak{D}K) & \mathfrak{D}(1-K\mathfrak{D})^{-1} \\ \mathfrak{E}-(1-K\mathfrak{D})^{-1}K\mathfrak{G}+\mathfrak{F}(1-K\mathfrak{D})^{-1}K-K\mathfrak{D}(1-K\mathfrak{D})^{-1}K & (\mathfrak{F}-K\mathfrak{D})(1-K\mathfrak{D})^{-1} \end{bmatrix}. \end{aligned}$$

□

We have the following *necessary* conditions for stabilizability and detectability:

Lemma 8.2.8 *Let* $\Sigma = \left[\begin{smallmatrix} \mathfrak{A} & \mathfrak{B} \\ \mathfrak{C} & \mathfrak{D} \end{smallmatrix}\right]$ *be a well-posed linear system with semigroup generator* A*, control operator* B*, and observation operator* C*.*

(i) *If* Σ *is state/state stabilizable, then*

$$\mathcal{R}\left(\begin{bmatrix}(\lambda - A) & B\end{bmatrix}\right) \supset X, \qquad \Re\lambda > 0. \tag{8.2.2}$$

If Σ *is exponentially stabilizable, then* (8.2.2) *holds for all* λ *in some right half-plane* $\Re\lambda > -\epsilon$ *where* $\epsilon > 0$.

(ii) *If* Σ *is state/state detectable, then*

$$\mathcal{N}\left(\begin{bmatrix}(\lambda - A) \\ C\end{bmatrix}\right) = 0. \qquad \Re\lambda > 0. \tag{8.2.3}$$

If Σ *is exponentially detectable, then* (8.2.3) *holds for all* λ *in some half-plane* $\Re\lambda > -\epsilon$ *where* $\epsilon > 0$.

In the finite-dimensional case conditions (8.2.2) and (8.2.3) are usually referred to as the *Hautus rank conditions*. See also Lemmas 9.6.6 and 9.6.9.

Proof We introduce the same notation as in Lemma 7.3.3. In addition, we denote the state feedback semigroup generator, control operator, and observation operator by $A_\times$, $B_\times$, and $\left[\begin{smallmatrix} C_\times \\ K_\times \end{smallmatrix}\right]$, and we denote the output injection semigroup generator, control operator, and observation operator by $A^\times$ and $\begin{bmatrix}H^\times & B^\times\end{bmatrix}$, and $C^\times$.

(i) Choose some state/state stabilizing state feedback pair $\left[\begin{smallmatrix} \mathfrak{K} & \mathfrak{F} \end{smallmatrix}\right]$. By Lemma 7.3.3 and Corollary 7.4.8,

$$(\lambda - A_\times)^{-1} = (\lambda - A)^{-1} + (\lambda - A)^{-1} B K_\times (\lambda - A_\times)^{-1}$$

for $\Re\lambda > \max\{\omega_{\mathfrak{A}}, \omega_{\mathfrak{A}_\times}\}$ (where $\omega_{\mathfrak{A}_\times} \le 0$ since $\left[\begin{smallmatrix} \mathfrak{K} & \mathfrak{F} \end{smallmatrix}\right]$ is state/state stabilizing). Multiply by $(\lambda - A)$ to the left and reorder the terms to get

$$1 = \begin{bmatrix}(\lambda - A) & B\end{bmatrix} \begin{bmatrix} 1 \\ -K_\times \end{bmatrix} (\lambda - A_\times)^{-1}$$

for $\Re\lambda > \max\{\omega_{\mathfrak{A}}, \omega_{\mathfrak{A}_\times}\}$; this should be interpreted as an operator identity in $\mathcal{B}(X; X_{-1})$. Both sides are analytic functions of λ in $\Re\lambda > \omega_{\mathfrak{A}_\times}$, so the same identity extends to the half-plane $\Re\lambda > \omega_{\mathfrak{A}_\times}$. This implies that $\mathcal{R}\left(\begin{bmatrix}(\lambda - A) & B\end{bmatrix}\right) \supset X$ for these λ.

(ii) This proof is similar to the proof of (i), but this time we use the identity

$$1 = (\lambda - A^\times)^{-1} \begin{bmatrix}1 & H^\times\end{bmatrix} \begin{bmatrix}(\lambda - A) \\ C\end{bmatrix},$$

valid for $\Re\lambda > \max\{\omega_{\mathfrak{A}}, \omega_{\mathfrak{A}^\times}\}$. □

In the case where either the input dimension or the output dimension of the system is finite, then we get some further necessary conditions for stabilizability and detectability:

Lemma 8.2.9 *Let* $\Sigma = \left[\begin{smallmatrix} \mathfrak{A} & \mathfrak{B} \\ \mathfrak{C} & \mathfrak{D} \end{smallmatrix}\right]$ *be a well-posed linear system on* (Y, X, U) *with semigroup generator* A*, control operator* B*, observation operator* C*, and transfer function* $\widehat{\mathfrak{D}}$*.*

(i) *If* U *is finite-dimensional, then stabilizability implies meromorphic extendibility in the following sense:*
   (a) *If* Σ *is state/state stabilizable then the resolvent* $(\lambda - A)^{-1}$ *is meromorphic in the right half-plane* $\Re\lambda > 0$.
   (b) *If* Σ *is input/output stabilizable then* $\widehat{\mathfrak{D}}$ *has a meromorphic extension to the right half-plane* $\Re\lambda > 0$.
   (c) *If* Σ *is input/state and input/output stabilizable then* $(\lambda - A)^{-1}B$ *has a meromorphic extension to the right half-plane* $\Re\lambda > 0$.
   (d) *If* Σ *is state/output and input/output stabilizable then* $C(\lambda - A)^{-1}$ *has a meromorphic extension to the right half-plane* $\Re\lambda > 0$.
   (e) *If we replace 'stabilizable' by 'exponentially stabilizable' in (a)–(d) above, then the conclusions of (a)–(d) are valid in some right half-plane* $\Re\lambda > -\epsilon$ *where* $\epsilon > 0$.

(ii) *If* Y *is finite-dimensional, then claims (a)–(e) above remain true if we replace 'stabilizable' by 'detectable'.*

Observe that, by (i)(a) and (ii)(a), a system where the semigroup is one of the shift semigroups in Example 2.3.2(i)–(iii) can never be exponentially stabilizable or detectable if the input, respectively output, space is finite dimensional, because their resolvents are not meromorphic in any half-plane $\Re\lambda > -\epsilon$ with $\epsilon > 0$ (see Example 3.3.1(i)–(iii)).

Proof We introduce the same notation as in the proof of Lemma 8.2.8. To prove (i) it suffices to observe that, by Lemma 7.3.3 and Theorem 7.4.7, for $\Re\lambda > \max\{\omega_{\mathfrak{A}}, \omega_{\mathfrak{A}_\times}\}$,

$$\begin{aligned}
(\lambda - A)^{-1} &= (\lambda - A_\times)^{-1} \\
&\qquad - (\lambda - A_\times)^{-1} B_\times (1 + \widehat{\mathfrak{F}}_\times(\lambda))^{-1} K_\times (\lambda - A_\times)^{-1}, \\
(\lambda - A)^{-1} B &= (\lambda - A_\times)^{-1} B_\times (1 + \widehat{\mathfrak{F}}_\times(\lambda))^{-1}, \\
C(\lambda - A)^{-1} &= \left(C_\times - \widehat{\mathfrak{D}}_\times(\lambda)(1 + \widehat{\mathfrak{F}}_\times(\lambda))^{-1} K_\times\right)(\lambda - A_\times)^{-1}, \\
\widehat{\mathfrak{D}}(\lambda) &= \widehat{\mathfrak{D}}_\times(\lambda)(1 + \widehat{\mathfrak{F}}_\times(\lambda))^{-1},
\end{aligned}$$

where the right-hand side is meromorphic in $\Re\lambda > \omega_{\mathfrak{A}_\times}$.

(ii) The proof is similar to the one above, but it uses the facts that, for $\Re\lambda > \max\{\omega_{\mathfrak{A}}, \omega_{\mathfrak{A}^\times}\}$,

$$\begin{aligned}
(\lambda - A)^{-1} &= (\lambda - A^\times)^{-1} \\
&\quad + (\lambda - A^\times)^{-1} H^\times (1 + \widehat{\mathfrak{G}}^\times(\lambda))^{-1} C^\times (\lambda - A^\times)^{-1}, \\
(\lambda - A)^{-1} B &= (\lambda - A^\times)^{-1} \left(B^\times + H^\times (1 - \widehat{\mathfrak{G}}^\times(\lambda))^{-1} \widehat{\mathfrak{D}}^\times(\lambda) \right), \\
C(\lambda - A)^{-1} &= (1 - \widehat{\mathfrak{G}}^\times(\lambda))^{-1} C^\times (\lambda - A^\times)^{-1}, \\
\widehat{\mathfrak{D}}(\lambda) &= (1 - \widehat{\mathfrak{G}}^\times(\lambda))^{-1} \widehat{\mathfrak{D}}^\times(\lambda),
\end{aligned}$$

where the right-hand sides are meromorphic in $\Re\lambda > \omega_{\mathfrak{A}^\times}$. □

As the following proposition shows, most of the different transformations on systems that we have studied preserve stabilizability and detectability (with the obvious exception of the exponential shift in Example 2.3.5):

Proposition 8.2.10 *Let Σ, Σ_1, and Σ_2 be well-posed linear systems.*

(i) *The following systems derived from Σ have exactly the same stabilizability and detectability properties (listed in Definition 8.2.4(i)–(vi)) as the original system Σ:*
   (a) *the time compressed system Σ_λ in Example 2.3.6;*
   (b) *the similarity transformed system Σ_E in Example 2.3.7;*
   (c) *the closed-loop system with output feedback operator K in Theorem 7.1.2;*
   (d) *the extended closed-loop system with output feedback operator K in Theorem 7.2.1;*
   (e) *the closed-loop system with partial output feedback operator K in Lemma 7.2.6.*

(ii) *If both Σ_1 and Σ_2 are stabilizable or detectable in one of the senses listed in Definition 8.2.4(i)–(vi), then the following systems derived from these two are stabilizable or detectable in the same sense:*
   (a) *the cross-product of Σ_1 and Σ_2 presented in Example 2.3.10;*
   (b) *the cascade connection of Σ_1 and Σ_2 through K presented in Example 7.2.4;*
   (c) *the dynamic feedback connection of Σ_1 and Σ_2 presented in Example 7.2.5.*

In the finite-dimensional case a converse to (ii) is true: if the cross-product, or the cascade connection, or the dynamic feedback connection of Σ_1 and Σ_2 is stabilizable and detectable, then so are both Σ_1 and Σ_2. It was conjectured by Weiss and Curtain (1997) that the same converse claim is true in the infinite-dimensional situation, too.

Proof of Proposition 8.2.10 (i) We leave the easy proof of (a), (b), and (d) to the reader. The claims (c) and (e) follow from Lemma 8.2.7.

(ii) The straightforward proof of (a) is left to the reader. The claims (b) and (c) follow from (a) and Lemma 8.2.7. □

Note that some of the transformed systems listed in Section 2.3 are not mentioned in Proposition 8.2.10, namely the sum junction (Example 2.3.11), the T-junction (Example 2.3.12), and the parallel connection (Example 2.3.13). It is easy to see that these transformed systems do not necessarily inherit the stabilizability and dectectability properties of the two subsystems. For example, the finite-dimensional system on $(\mathbb{C}, \mathbb{C}^2, \mathbb{C})$ generated by the system node $\left[\begin{array}{cc|c} 1 & 0 & 1 \\ 0 & 1 & 1 \\ \hline 1 & 1 & 0 \end{array}\right]$ is neither stabilizable nor detectable, but it is the parallel connection of two copies of the system generated by $\left[\begin{array}{c|c} 1 & 1 \\ \hline 1 & 0 \end{array}\right]$, which is both stabilizable and detectable.[4]

By Theorem 8.1.7(v)(j), if a system is bounded but not exponentially stable, then it is impossible to make it exponentially stable by use of admissible output feedback. An analogous result is true for strong stability in the L^p-case with $p < \infty$. As the following theorem shows, a very similar result is true for state feedbacks and output injections:

Theorem 8.2.11 *Let Σ be a well-posed linear system.*

(i) *Σ is bounded if and only if Σ is stabilizable, detectable (not necessarily jointly), and input/output bounded.*

(ii) *The following conditions are equivalent:*
   (a) *Σ is strongly stable;*
   (b) *Σ is bounded, strongly input/state stable, and strongly stabilizable.*
   In the L^p-well-posed case with $p < \infty$ these conditions are equivalent to
   (c) *Σ is bounded and strongly detectable,*
   and in the Reg-well-posed case they are equivalent to
   (d) *Σ is bounded, strongly state/output stable, and strongly detectable.*

(iii) *In the Reg-well-posed case and the L^p-well-posed case with $p > 1$ the following conditions are equivalent:*
   (a) *Σ is exponentially stable;*
   (b) *Σ is bounded and exponentially stabilizable.*
   In the L^p-well-posed case with $p < \infty$ condition (a) is equivalent to
   (c) *Σ is bounded and exponentially detectable.*

[4] A related question will be discussed in Section 9.7. Note that in the above example the point spectra of the semigroup generators of the two subsystems have a nonempty intersection.

Proof Since boundedness implies stabilizability and detectability and strong or exponential stability implies strong or exponential stabilizability and detectability, it suffices to prove the converse claims.

(i) We introduce the same notation as in Lemma 7.3.3. The observability map $\mathfrak{C}$ is given by $\mathfrak{C} = \mathfrak{C}_\times - \mathfrak{D}\mathfrak{K}_\times$, hence $\mathfrak{C}$ is bounded whenever $\mathfrak{C}_\times$, $\mathfrak{D}$, and $\mathfrak{K}_\times$ are so. The input map $\mathfrak{B}$ is given by $\mathfrak{B} = \mathfrak{B}^\times - \mathfrak{H}^\times\mathfrak{D}$, hence $\mathfrak{B}$ is bounded whenever $\mathfrak{B}^\times$, $\mathfrak{H}^\times$, and $\mathfrak{D}$ are so. Finally, $\mathfrak{A} = \mathfrak{A}_\times - \mathfrak{B}\tau\mathfrak{K}_\times$, hence $\mathfrak{A}$ is bounded whenever $\mathfrak{A}_\times$, $\mathfrak{B}$, and $\mathfrak{K}_\times$ are so.

(ii) We use Lemma 7.1.7 to interpret Σ as a feedback connection of $\Sigma_\times$ (cf. Figure 8.4), and apply Theorem 8.1.7(ii)(b) with the replacements $\mathfrak{A} \to \mathfrak{A}_\times$, $K = \begin{bmatrix} 0 & 1 \end{bmatrix}$, $\mathfrak{A}^K \to \mathfrak{A}$, $\mathfrak{C} \to \begin{bmatrix} \mathfrak{C}_\times \\ \mathfrak{K}_\times \end{bmatrix}$, and $\mathfrak{B}^K \to \mathfrak{B}$. This gives us the implication (b) $\Rightarrow$ (a). To get the two remaining implications (c) $\Rightarrow$ (a) and (d) $\Rightarrow$ (a) we instead interpret Σ as a feedback connection of $\Sigma^\times$, and apply Theorem 8.1.7(i)(b) with the replacements $\mathfrak{A} \to \mathfrak{A}^\times$, $K = \begin{bmatrix} 1 \\ 0 \end{bmatrix}$, $\mathfrak{A}^K \to \mathfrak{A}$, $\mathfrak{B} \to \begin{bmatrix} \mathfrak{H}^\times & \mathfrak{B}^\times \end{bmatrix}$, and $\mathfrak{C}^K \to \mathfrak{C}$. (Note that, by Lemma 8.1.2(ii),(vii), strong or exponential stability of $\mathfrak{A}$ together with the boundedness of Σ implies strong or exponential stability of Σ.)

(iii) The proof is the same as the proof of (ii), except that this time we use parts (ii)(c) and (i)(c) of Theorem 8.1.7. □

Note how in this proof we used the convention that not only is the original system stabilized under state feedback or output injection, but also the added inputs and outputs should be stabilized (cf. Remark 8.2.5). However, *we did not require any added inputs* $\left[\begin{smallmatrix} \mathfrak{H} \\ \mathfrak{G} \end{smallmatrix}\right]$ *or outputs* $\left[\begin{smallmatrix} \mathfrak{K} & \mathfrak{F} \end{smallmatrix}\right]$ *to be* open-loop *stable*, or strongly stable, or exponentially stable (this is not true, in general, even in the finite-dimensional case).

A number of additional conclusions could be added to Theorem 8.2.11, such as the following: an input/output bounded system Σ is output stable if and only if it is right output stabilizable, it is input stable if and only if it is input stabilizable, etc.

Let us end this section by observing that, in the reflexive case, stabilizability and detectability are dual to each other in the following sense:

Lemma 8.2.12 *Let Σ be an L^p-well-posed with $1 < p < \infty$ on the reflexive Banach spaces (Y, X, U), and let Σ^d represent the causal dual system on (U^*, X^*, Y^*) (see Theorem 6.2.3).*

(i) *The following relationships hold between the output feedback stabilizability of Σ and Σ^d:*
   (a) *Σ is state/state stabilizable, or weakly stabilizable, or exponentially stabilizable by output feedback if and only if Σ^d has the same property;*

(b) Σ *is input/state stabilizable or exponentially stabilizable by output feedback if and only if* Σ^d *is state/output stabilizable in the same sense;*

(c) Σ *is input/output stabilizable or exponentially stabilizable by output feedback if and only if* Σ^d *has the same property;*

(d) Σ *is stabilizable, or weakly stabilizable, or exponentially stabilizable by output feedback if and only if* Σ^d *has the same property.*

(ii) *The following relationships hold between the (state feedback) stabilizability of* Σ *and the (output injection) detectability of* Σ^d*:*

(a) Σ *is state/state stabilizable, or weakly stabilizable, or exponentially stabilizable if and only if* Σ^d *is detectable in the same sense;*

(b) Σ *is input/state stabilizable or exponentially stabilizable if and only if* Σ^d *is state/output detectable in the same sense;*

(c) Σ *is state/output stabilizable or exponentially stabilizable if and only if* Σ^d *is input/state detectable in the same sense;*

(d) Σ *is input/output stabilizable or exponentially stabilizable if and only if* Σ^d *detectable in the same sense;*

(e) Σ *is stabilizable, or weakly stabilizable, or exponentially stabilizable by output feedback if and only if* Σ^d *detectable in the same sense.*

(iii) *All the claims in (i) remain true if we replace 'stabilizable by output feedback' by 'jointly stabilizable and detectable' throughout.*

Proof All of this follows directly from Theorem 6.2.3, Theorem 7.6.1, and Definition 8.2.4. □

Observe that strong stabilization and detection are missing in the list above, due to the fact that strong stability is not preserved under duality. For example, if $\mathfrak{D}$ is bounded, then the exactly observable shift realization of $\mathfrak{D}$ in Example 2.6.5 is strongly stable, but its adjoint, the exactly controllable shift realization of $\mathfrak{D}^d$ (see Example 6.2.8), is not.

As the reader can easily check, much of Theorem 8.1.7 remains valid in the case $p = 1$.

8.3 Coprime fractions and factorizations

The notions of stabilizability and detectability introduced in Definition 8.2.4 are closely related to the notions of *coprime fractions*. To introduce this notion we must first define what we mean by 'coprime', which we interpret in the Bezout sense.

Definition 8.3.1

(i) The operators $\mathfrak{N} \in TIC(U;Y)$ and $\mathfrak{M} \in TIC(U;Z)$ are *right coprime* if there exist operators $\widetilde{\mathfrak{Y}} \in TIC(Y;U)$ and $\widetilde{\mathfrak{X}} \in TIC(Z;U)$ that together with $\mathfrak{N}$ and $\mathfrak{M}$ satisfy the *right Bezout identity*

$$\widetilde{\mathfrak{X}}\mathfrak{M} - \widetilde{\mathfrak{Y}}\mathfrak{N} = 1 \tag{8.3.1}$$

in $TIC(U)$. If the same conditions are true with TIC replaced by TIC_α for some $\alpha < 0$, then we call $\mathfrak{N}$ and $\mathfrak{M}$ *exponentially right coprime*.[5]

(ii) The operators $\widetilde{\mathfrak{N}} \in TIC(U;Y)$ and $\widetilde{\mathfrak{M}} \in TIC(Z;Y)$ are *left coprime* if there exist operators $\mathfrak{Y} \in TIC(Y;U)$ and $\mathfrak{X} \in TIC(Y;Z)$ that together with $\widetilde{\mathfrak{N}}$ and $\widetilde{\mathfrak{M}}$ satisfy the *left Bezout identity*

$$\widetilde{\mathfrak{M}}\mathfrak{X} - \widetilde{\mathfrak{N}}\mathfrak{Y} = 1 \tag{8.3.2}$$

in $TIC(Y)$. If the same conditions are true with TIC replaced by TIC_α for some $\alpha < 0$, then we call $\mathfrak{N}$ and $\mathfrak{M}$ *exponentially left coprime*.[6]

Thus, $\mathfrak{N}$ and $\mathfrak{M}$ are right coprime iff $\left[\begin{smallmatrix}\mathfrak{N}\\ \mathfrak{M}\end{smallmatrix}\right]$ has a left inverse in $TIC\left(\left[\begin{smallmatrix}Y\\ Z\end{smallmatrix}\right];U\right)$, and $\widetilde{\mathfrak{N}}$ and $\widetilde{\mathfrak{M}}$ are left coprime iff $\left[\widetilde{\mathfrak{N}}\ \widetilde{\mathfrak{M}}\right]$ has a right inverse in $TIC\left(Y;\left[\begin{smallmatrix}U\\ Z\end{smallmatrix}\right]\right)$.

Definition 8.3.2 Let $\mathfrak{D} \in TIC_\omega(U;Y)$ and $\mathfrak{Q} \in TIC_\omega(Y;U)$ for some $\omega \geq 0$.

(i) We call $\mathfrak{N}\mathfrak{M}^{-1}$ a *right coprime fraction* of $\mathfrak{D}$ with *numerator* $\mathfrak{N}$ and *denominator* $\mathfrak{M}$ if $\mathfrak{N} \in TIC(U;Y)$ and $\mathfrak{M} \in TIC(U)$ are right coprime, $\mathfrak{M}$ has an inverse in $TIC_\beta(U)$ for some $\beta \geq 0$, and $\mathfrak{D} = \mathfrak{N}\mathfrak{M}^{-1}$ in $TIC_\gamma(U;Y)$ where $\gamma = \max\{\omega,\beta\}$.

(ii) We call $\widetilde{\mathfrak{M}}^{-1}\widetilde{\mathfrak{N}}$ a *left coprime fraction* of $\mathfrak{D}$ with *numerator* $\widetilde{\mathfrak{N}}$ and *denominator* $\widetilde{\mathfrak{M}}$ if $\widetilde{\mathfrak{M}} \in TIC(Y)$ and $\widetilde{\mathfrak{N}} \in TIC(U;Y)$ are left coprime, $\widetilde{\mathfrak{M}}$ has an inverse in $TIC_\beta(Y)$ for some $\beta \geq 0$, and $\mathfrak{D} = \widetilde{\mathfrak{M}}^{-1}\widetilde{\mathfrak{N}}$ in $TIC_\gamma(U;Y)$ where $\gamma = \max\{\omega,\beta\}$.

(iii) A *doubly coprime factorization* of $\mathfrak{D}$ consists of eight operators in TIC (of the appropriate dimensions) satisfying

$$\begin{bmatrix}\widetilde{\mathfrak{M}} & -\widetilde{\mathfrak{N}}\\ -\widetilde{\mathfrak{Y}} & \widetilde{\mathfrak{X}}\end{bmatrix}\begin{bmatrix}\mathfrak{X} & \mathfrak{N}\\ \mathfrak{Y} & \mathfrak{M}\end{bmatrix} = \begin{bmatrix}\mathfrak{X} & \mathfrak{N}\\ \mathfrak{Y} & \mathfrak{M}\end{bmatrix}\begin{bmatrix}\widetilde{\mathfrak{M}} & -\widetilde{\mathfrak{N}}\\ -\widetilde{\mathfrak{Y}} & \widetilde{\mathfrak{X}}\end{bmatrix} = 1 \tag{8.3.3}$$

in $TIC\left(\left[\begin{smallmatrix}Y\\ U\end{smallmatrix}\right]\right)$, and, in addition, we require that $\mathfrak{N}\mathfrak{M}^{-1}$ is a right and $\widetilde{\mathfrak{M}}^{-1}\widetilde{\mathfrak{N}}$ a left coprime fraction of $\mathfrak{D}$.

[5] In the *Reg*-well-posed case we shall throughout require all the operators appearing in this definition to have *Reg*-well-posed realizations. See Theorem 2.6.7.

[6] We could, of course, have written the two Bezout identities in the form $\widetilde{\mathfrak{X}}\mathfrak{M} + \widetilde{\mathfrak{Y}}\mathfrak{N} = 1$ and $\widetilde{\mathfrak{X}}\mathfrak{M} + \widetilde{\mathfrak{Y}}\mathfrak{N} = 1$. The reason for the extra minus sign is that we want these equations to be part of the doubly coprime factorization (8.3.3).

(iv) A *joint doubly coprime factorization* of $\mathfrak{D}$ and $\mathfrak{Q}$ is a doubly coprime factorization (8.3.3) of $\mathfrak{D}$ where, in addition, $\mathfrak{Y}\mathfrak{X}^{-1}$ is a right and $\widetilde{\mathfrak{X}}^{-1}\widetilde{\mathfrak{Y}}$ a left coprime fraction of $\mathfrak{Q}$.

(v) If all the involved operators (except $\mathfrak{D}$, $\mathfrak{Q}$, $\mathfrak{M}^{-1}$, $\widetilde{\mathfrak{M}}^{-1}$, $\mathfrak{X}^{-1}$, and $\widetilde{\mathfrak{X}}^{-1}$) belong to TIC_α for some $\alpha < 0$, then we call these fractions or factorizations *exponentially right or left or doubly coprime*.

In (i) and (ii), it follows from Lemma 2.6.8 that $TIC \subset TIC_\omega \subset TIC_\gamma$ and that $TIC \subset TIC_\alpha \subset TIC_\gamma$. However, as the following lemma shows, we can always, without loss of generality, take $\omega = \beta = \gamma$. This lemma also clarifies some of the other basic relationships between the different parts of Definition 8.3.2.

Lemma 8.3.3 *We make the same assumptions and introduce the same notation as in Definition 8.3.2.*

(i) *If $\mathfrak{D}$ has a right coprime fraction $\mathfrak{D} = \mathfrak{N}\mathfrak{M}^{-1}$, then, for each $\gamma \geq 0$, $\mathfrak{D} \in TIC_\gamma(U;Y)$ if and only if $\mathfrak{M}^{-1} \in TIC_\gamma(U)$. Thus, we may, without loss of generality, take $\omega = \beta = \gamma$ in Definition 8.3.2(i).*

(ii) *If $\mathfrak{D}$ has a left coprime fraction $\mathfrak{D} = \widetilde{\mathfrak{M}}^{-1}\widetilde{\mathfrak{N}}$, then, for each $\gamma \geq 0$, $\mathfrak{D} \in TIC_\gamma(U;Y)$ if and only if $\widetilde{\mathfrak{M}}^{-1} \in TIC_\gamma(Y)$. Thus, we may, without loss of generality, take $\omega = \beta = \gamma$ in Definition 8.3.2(ii).*

(iii) *If $\mathfrak{D}$ has both a right coprime fraction $\mathfrak{N}\mathfrak{M}^{-1}$ and a left coprime fraction $\widetilde{\mathfrak{N}}^{-1}\widetilde{\mathfrak{M}}$, then, for each $\gamma \geq 0$, $\mathfrak{M}^{-1} \in TIC_\gamma(U)$ if and only if $\widetilde{\mathfrak{M}}^{-1} \in TIC_\gamma(Y)$, and this is true if and only if $\mathfrak{D} \in TIC_\gamma(U;Y)$.*

(iv) *In parts (iii) and (iv) of Definition 8.3.2, it suffices to require that $\mathfrak{N}\mathfrak{M}^{-1}$ is a right coprime fraction of $\mathfrak{D}$ or $\widetilde{\mathfrak{M}}^{-1}\mathfrak{N}$ is a right coprime fraction of $\mathfrak{D}$ because these two conditions are equivalent to each other whenever* (8.3.3) *holds.*

(v) *In part (iii) of Definition 8.3.2, for each $\gamma \geq 0$, the operator $\mathfrak{X}$ is invertible in $TIC_\gamma(Y)$ if and only if $\widetilde{\mathfrak{X}}$ is invertible in $TIC_\gamma(U)$.*

(vi) *In part (iv) of Definition 8.3.2, it suffices to require that $\mathfrak{Y}\mathfrak{X}^{-1}$ is a right coprime fraction of $\mathfrak{Q}$ or $\widetilde{\mathfrak{X}}^{-1}\mathfrak{Y}$ is a right coprime fraction of $\mathfrak{Q}$ because these two conditions are equivalent to each other whenever* (8.3.3) *holds.*

(vii) *Suppose that $\mathfrak{N}\mathfrak{M}^{-1}$ is a right coprime fraction of $\mathfrak{D} \in TIC_\omega(U;Y)$, that $\widetilde{\mathfrak{X}}^{-1}\widetilde{\mathfrak{Y}}$ is a left coprime fraction of $\mathfrak{Q} \in TIC_\omega(Y;U)$, and that the numerators and denominators satisfy the right Bezout identity $\widetilde{\mathfrak{X}}\mathfrak{M} - \widetilde{\mathfrak{Y}}\mathfrak{N} = 1$. Then $\mathfrak{M} - \mathfrak{Q}\mathfrak{N}$ and $\widetilde{\mathfrak{X}} - \widetilde{\mathfrak{Y}}\mathfrak{D}$ are invertible in $TIC_\omega(U)$ and*

$$\widetilde{\mathfrak{X}} = (\mathfrak{M} - \mathfrak{Q}\mathfrak{N})^{-1}, \qquad \widetilde{\mathfrak{Y}} = (\mathfrak{M} - \mathfrak{Q}\mathfrak{N})^{-1}\mathfrak{Q},$$
$$\mathfrak{M} = (\widetilde{\mathfrak{X}} - \widetilde{\mathfrak{Y}}\mathfrak{D})^{-1}, \qquad \mathfrak{N} = \mathfrak{D}(\widetilde{\mathfrak{X}} - \widetilde{\mathfrak{Y}}\mathfrak{D})^{-1}.$$

In particular, $\widetilde{\mathfrak{X}}$ and $\widetilde{\mathfrak{Y}}$ are determined uniquely by $\mathfrak{Q}$, $\mathfrak{M}$ and $\mathfrak{N}$, and $\mathfrak{M}$ and $\mathfrak{N}$ are determined uniquely by $\mathfrak{D}$, $\widetilde{\mathfrak{X}}$ and $\widetilde{\mathfrak{Y}}$.

Proof (i)–(ii) Let ω and β be the constants in parts (i) and (ii) of Definition 8.3.2. Clearly $\mathfrak{D} \in TIC_\beta$ both in (i) and (ii) since $\mathfrak{D} = \mathfrak{N}\mathfrak{M}^{-1} \in TIC_\beta$ or $\mathfrak{D} = \widetilde{\mathfrak{M}}^{-1}\widetilde{\mathfrak{N}} \in TIC_\beta$. Conversely, multiplying the Bezout identity $\widetilde{\mathfrak{X}}\mathfrak{M} - \widetilde{\mathfrak{Y}}\mathfrak{N} = 1$ by $\mathfrak{M}$ to the right we get

$$\mathfrak{M}^{-1} = \widetilde{\mathfrak{X}} - \widetilde{\mathfrak{Y}}\mathfrak{D}$$

in TIC_β. Here the right-hand side belongs to TIC_ω, hence $\mathfrak{M}^{-1} \in TIC_\omega$ (cf. Lemmas 2.6.4 and 2.6.8). The claim $\widetilde{\mathfrak{M}}^{-1} \in TIC_\omega$ in (ii) is proved analogously.

(iii) This follows from (i)–(ii).

(iv) By Lemma A.4.2(iv), $\mathfrak{M}$ is invertible in $TIC_\omega(U)$ if and only if $\widetilde{\mathfrak{M}}$ is invertible in $TIC_\omega(Y)$. Equation (8.3.3) implies that $\widetilde{\mathfrak{M}}\mathfrak{N} = \widetilde{\mathfrak{N}}\mathfrak{M}$, hence $\mathfrak{N}\mathfrak{M}^{-1} = \widetilde{\mathfrak{M}}^{-1}\widetilde{\mathfrak{N}}$.

(v)–(vi) See the proof of (iv).

(vii) Substitute $\widetilde{\mathfrak{Y}} = \widetilde{\mathfrak{X}}\mathfrak{Q}$ in the right Bezout identity to get $\widetilde{\mathfrak{X}}(\mathfrak{M} - \mathfrak{Q}\mathfrak{N}) = 1$, and then multiply by $\widetilde{\mathfrak{X}}^{-1}$ to the left to get $\widetilde{\mathfrak{X}}^{-1} = (\mathfrak{M} - \mathfrak{Q}\mathfrak{N})$. This combined with (i) implies that $\mathfrak{M} - \mathfrak{Q}\mathfrak{N}$ is invertible in $TIC_\omega(U)$ and that $\widetilde{\mathfrak{X}} = (\mathfrak{M} - \mathfrak{Q}\mathfrak{N})^{-1}$ and $\widetilde{\mathfrak{Y}} = \widetilde{\mathfrak{X}}\mathfrak{Q} = (\mathfrak{M} - \mathfrak{Q}\mathfrak{N})^{-1}\mathfrak{Q}$. The other claims are proved in an analogous way. □

Above we have defined coprime fractions and factorizations in the operator algebra TIC. These notions have natural extensions to the frequency domain.

Definition 8.3.4 Let U and Y be Banach spaces and let $\omega \in \mathbb{R}$. The space $H^\infty_\omega(U;Y)$ consists of all bounded analytic $\mathcal{B}(U;Y)$-valued functions φ on $\Re z > \omega$ with

$$\|\varphi\|_{H^\infty_\omega(U;Y)} = \sup_{\Re z > \omega} \|\varphi(z)\|_{\mathcal{B}(U;Y)}.$$

In the case $\omega = 0$ we abbreviate $H^\infty_\omega(U;Y)$ to $H^\infty(U;Y)$.

We recall the following result which connects $TIC(U;Y)$ to $H^\infty(U;Y)$:

Lemma 8.3.5 *Let $\mathfrak{D} \in TIC_\omega(U;Y)$ and $\mathfrak{E} \in TIC_\omega(Y;Z)$ for some $\omega \in \mathbb{R}$, and denote the transfer functions of these operators by $\widehat{\mathfrak{D}}$ and $\widehat{\mathfrak{E}}$. Then $\widehat{\mathfrak{D}} \in H^\infty_\omega(U;Y)$, $\widehat{\mathfrak{E}} \in H^\infty_\omega(Y;Z)$, $\|\widehat{\mathfrak{D}}\|_{H^\infty_\omega(U;Y)} \le \|\mathfrak{D}\|_{TIC_\omega(U;Y)}$, $\|\widehat{\mathfrak{E}}\|_{H^\infty_\omega(Y;Z)} \le \|\mathfrak{E}\|_{TIC_\omega(Y;Z)}$, and the transfer function of $\mathfrak{E}\mathfrak{D}$ is $\widehat{\mathfrak{E}}\widehat{\mathfrak{D}}$. In particular, if $\mathfrak{D} \in TIC(U;Y)$ then $\widehat{\mathfrak{D}} \in H^\infty(U;Y)$ and $\|\widehat{\mathfrak{D}}\|_{H^\infty}(U;Y) \le \|\mathfrak{D}\|_{TIC(U;Y)}$.*

Proof See Corollary 4.6.10 and the proof of Lemma 8.1.3(iv). □

A converse to this lemma is true in the L^2-well-posed Hilbert space case (and the norm inequalities become equalities); see Theorem 10.3.5.

Definition 8.3.6 We define the notions '*right coprime*', '*left coprime*', '*coprime fraction*' and '*doubly coprime factorization*' in H^∞ in the same way as in Definitions 8.3.1 and 8.3.2 with *TIC* replaced by H^∞.

Remark 8.3.7 In the sequel we shall formulate our results almost exclusively in the time domain, i.e., we work with the algebra *TIC* of input/output maps. However, all the results carry over to the frequency domain; we simply replace the input/output maps by their transfer functions, replace *TIC* by H^∞, and use Lemma 8.3.5 and Definition 8.3.4. In particular, as soon as we have a coprime fraction or a doubly coprime factorization in the time domain, then we get the corresponding coprime fraction or a doubly coprime factorization in the frequency domain. Also Lemma 8.3.3 remains true in this setting (since the analytic continuation of an analytic function is unique). We shall occasionally apply a result originally formulated in the time domain setting to the frequency domain case without specific warning. However, it may not always be possible to go in the opposite direction (except in the L^2-well-posed Hilbert space case) since a coprime condition in H^∞ is weaker than the corresponding coprime condition in *TIC* (the H^∞-factors in (8.3.1), (8.3.2) and (8.3.3) need not be the transfer functions of input/output maps in *TIC*).

As a consequence of Remark 8.3.7 we get the following *necessary* conditions for two operators in *TIC* to be right or left coprime:

Lemma 8.3.8 *Let* $\mathfrak{N} \in TIC(U;Y)$, $\mathfrak{M} \in TIC(U)$, $\widetilde{\mathfrak{N}} \in TIC(U;Y)$, $\widetilde{\mathfrak{M}} \in TIC(Y)$, $\mathfrak{D} \in TIC_\omega(U;Y)$ *(for some* $\omega \geq 0$*), and denote the corresponding transfer functions by* $\widehat{\mathfrak{N}}$, $\widehat{\mathfrak{M}}$, $\widehat{\widetilde{\mathfrak{N}}}$, $\widehat{\widetilde{\mathfrak{M}}}$, *and* $\widehat{\mathfrak{D}}$.

(i) *If* $\mathfrak{N}$ *and* $\mathfrak{M}$ *are right coprime (in TIC) then* $\begin{bmatrix} \widehat{\mathfrak{N}} \\ \widehat{\mathfrak{M}} \end{bmatrix}$ *has a left inverse in* $H^\infty(\left[\begin{smallmatrix} Y \\ U \end{smallmatrix}\right]; U)$.
(ii) *If* $\widetilde{\mathfrak{N}}$ *and* $\widetilde{\mathfrak{M}}$ *are left coprime (in TIC) then* $\begin{bmatrix} \widehat{\widetilde{\mathfrak{N}}} & \widehat{\widetilde{\mathfrak{M}}} \end{bmatrix}$ *has a right inverse in* $H^\infty(Y; \left[\begin{smallmatrix} U \\ Y \end{smallmatrix}\right])$.
(iii) *If* $\mathfrak{N}\mathfrak{M}^{-1}$ *is a right coprime fraction of* $\mathfrak{D}$ *(in TIC), then* $\widehat{\mathfrak{N}}\widehat{\mathfrak{M}}^{-1}$ *is a right coprime fraction of* $\widehat{\mathfrak{D}}$ *(in* H^∞*).*
(iv) *If* $\widetilde{\mathfrak{M}}^{-1}\widetilde{\mathfrak{N}}$ *is a left coprime fraction of* $\mathfrak{D}$ *(in TIC), then* $\widehat{\widetilde{\mathfrak{M}}}^{-1}\widehat{\widetilde{\mathfrak{N}}}$ *is a left coprime fraction of* $\widehat{\mathfrak{D}}$ *(in* H^∞*).*
(v) *The transfer functions of a doubly coprime factorization of* $\mathfrak{D}$ *(in TIC) define a doubly coprime factorization of* $\widehat{\mathfrak{D}}$ *(in* H^∞*).*

This follows from Remark 8.3.7.

A coprime fraction is unique, up to a unit:

Lemma 8.3.9 *Let $\mathfrak{D} \in TIC_\omega(U;Y)$ for some $\omega \geq 0$.*

(i) *Let $\mathfrak{N}\mathfrak{M}^{-1}$ be a right coprime fraction of $\mathfrak{D}$.*
  (a) *If $\mathfrak{D} = \mathfrak{N}_1\mathfrak{M}_1^{-1}$ where $\mathfrak{N}_1 \in TIC(U;Y)$, $\mathfrak{M}_1 \in TIC(U)$, and $\mathfrak{M}_1$ has an inverse in $TIC_\omega(U)$ (but $\mathfrak{N}_1$ and $\mathfrak{M}_1$ are not necessarily right coprime), then $\mathfrak{N}_1 = \mathfrak{N}\mathfrak{U}$ and $\mathfrak{M}_1 = \mathfrak{M}\mathfrak{U}$, where $\mathfrak{U} \in TIC(U)$ and $\mathfrak{U} = \mathfrak{M}^{-1}\mathfrak{M}_1$ in $TIC_\omega(U)$.*
  (b) *The set of all right coprime fractions of $\mathfrak{D}$ can be parametrized in the form $(\mathfrak{N}\mathfrak{U})(\mathfrak{M}\mathfrak{U})^{-1}$, where $\mathfrak{U}$ is an invertible operator in $TIC(U)$.*

(ii) *Let $\widetilde{\mathfrak{M}}^{-1}\widetilde{\mathfrak{N}}$ be a left coprime fraction of $\mathfrak{D}$.*
  (a) *If $\mathfrak{D} = \widetilde{\mathfrak{M}}_1^{-1}\widetilde{\mathfrak{N}}_1$ where $\widetilde{\mathfrak{N}}_1 \in TIC(U;Y)$, $\widetilde{\mathfrak{M}}_1 \in TIC(Y)$, and $\widetilde{\mathfrak{M}}_1$ has an inverse in $TIC_\omega(Y)$ (but $\widetilde{\mathfrak{N}}_1$ and $\widetilde{\mathfrak{M}}_1$ are not necessarily right coprime), then $\widetilde{\mathfrak{N}}_1 = \mathfrak{U}\widetilde{\mathfrak{N}}$ and $\widetilde{\mathfrak{M}}_1 = \mathfrak{U}\widetilde{\mathfrak{M}}$, where $\mathfrak{U} \in TIC(Y)$ and $\mathfrak{U} = \widetilde{\mathfrak{M}}_1\widetilde{\mathfrak{M}}^{-1}$ in $TIC_\omega(Y)$.*
  (b) *Then the set of all left coprime fractions of $\mathfrak{D}$ can be parametrized in the form $(\widetilde{\mathfrak{U}}\widetilde{\mathfrak{M}})^{-1}(\widetilde{\mathfrak{U}}\widetilde{\mathfrak{N}})$, where $\widetilde{\mathfrak{U}}$ is an invertible operator in $TIC(Y)$.*

Proof (i)(a) Choose some operators $\widetilde{\mathfrak{X}} \in TIC(U)$ and $\widetilde{\mathfrak{Y}} \in TIC(Y;U)$ which together with $\mathfrak{N}$ and $\mathfrak{M}$ satisfy the right Bezout identity $\widetilde{\mathfrak{X}}\mathfrak{M} - \widetilde{\mathfrak{Y}}\mathfrak{N} = 1$ in $TIC(U)$, hence also in $TIC_\omega(U)$ (see Lemma 2.6.8). Multiply this identity by $\mathfrak{M}^{-1}\mathfrak{M}_1$ to the right to get

$$\mathfrak{M}^{-1}\mathfrak{M}_1 = \widetilde{\mathfrak{X}}\mathfrak{M}_1 - \widetilde{\mathfrak{Y}}\mathfrak{N}_1$$

in $TIC_\omega(U)$. We get the conclusion of (a) by taking $\mathfrak{U} = \widetilde{\mathfrak{X}}\mathfrak{M}_1 - \widetilde{\mathfrak{Y}}\mathfrak{N}_1$.

(i)(b) If $\mathfrak{U} \in TIC(U)$ is invertible in $TIC(U)$, then $(\mathfrak{N}\mathfrak{U})(\mathfrak{M}\mathfrak{U})^{-1}$ is another right coprime fraction (multiply (8.3.1) by $\mathfrak{U}^{-1}$ to the left and by $\mathfrak{U}$ to the right and observe that, by Lemma 2.6.8, invertibility in *TIC* implies invertibility in TIC_ω). Conversely, suppose that we have two right coprime fractions $\mathfrak{N}\mathfrak{M}^{-1}$ and $\mathfrak{N}_1\mathfrak{M}_1^{-1}$. Let $\mathfrak{U} \in TIC(U)$ be the operator given by (a). Then $\mathfrak{N}_1 = \mathfrak{N}\mathfrak{U}$, $\mathfrak{M}_1 = \mathfrak{M}\mathfrak{U}$, and it follows from (a) with $\mathfrak{N}\mathfrak{M}^{-1}$ interchanged with $\mathfrak{N}_1\mathfrak{M}_1^{-1}$ that $\mathfrak{U}^{-1} = \mathfrak{M}_1^{-1}\mathfrak{M}$ can be extended to an operator in $TIC(U)$. Thus, $\mathfrak{U}$ is invertible in $TIC(U)$.

(ii) The proof of (ii) is completely analogous to the proof of (i). □

As the following lemma shows, every left and right coprime fraction can be completed to a double coprime factorization:

Lemma 8.3.10 *Let $\mathfrak{D} \in TIC_\omega(U;Y)$ for some $\omega \geq 0$. If $\mathfrak{D}$ has both a right coprime fraction $\mathfrak{N}\mathfrak{M}^{-1}$ and a left coprime fraction $\widetilde{\mathfrak{M}}^{-1}\widetilde{\mathfrak{N}}$, then these two fractions can be completed to a doubly coprime factorization* (8.3.3) *that contains the given operators $\mathfrak{N}$, $\mathfrak{M}$, $\widetilde{\mathfrak{M}}$, and $\widetilde{\mathfrak{N}}$. Moreover, in this factorization we can either choose $\widetilde{\mathfrak{X}}$ and $\widetilde{\mathfrak{Y}}$ to be an arbitrary solution of the right Bezout* *identity* (8.3.1)*, or alternatively, we can choose* $\mathfrak{Y}$ *and* $\mathfrak{X}$ *to be an arbitrary solution of the right Bezout identity* (8.3.1).

Proof Choose some operators $\widetilde{\mathfrak{Y}}$, $\widetilde{\mathfrak{X}}$, $\mathfrak{X}$, and $\mathfrak{Y}$ in *TIC* that together with the given operators satisfy the Bezout identities $\widetilde{\mathfrak{X}}\mathfrak{M} - \widetilde{\mathfrak{Y}}\mathfrak{N} = 1$ and $\widetilde{\mathfrak{M}}\mathfrak{X} - \widetilde{\mathfrak{N}}\mathfrak{Y} = 1$. Then a direct computation shows that

$$\begin{bmatrix} \widetilde{\mathfrak{M}} & -\widetilde{\mathfrak{N}} \\ -\widetilde{\mathfrak{Y}} & \widetilde{\mathfrak{X}} \end{bmatrix} \begin{bmatrix} \mathfrak{X} + \mathfrak{N}(\widetilde{\mathfrak{Y}}\mathfrak{X} - \widetilde{\mathfrak{X}}\mathfrak{Y}) & \mathfrak{N} \\ \mathfrak{Y} + \mathfrak{M}(\widetilde{\mathfrak{Y}}\mathfrak{X} - \widetilde{\mathfrak{X}}\mathfrak{Y}) & \mathfrak{M} \end{bmatrix} = 1.$$

By using the invertibility of $\mathfrak{M}$ and $\widetilde{\mathfrak{M}}$ in TIC_ω, we get

$$\begin{aligned} \mathfrak{X}\widetilde{\mathfrak{M}} - \mathfrak{D}\mathfrak{Y}\widetilde{\mathfrak{M}} &= 1, & \mathfrak{X}\widetilde{\mathfrak{N}} - \mathfrak{D}\mathfrak{Y}\widetilde{\mathfrak{N}} &= \mathfrak{D}, \\ \mathfrak{M}\widetilde{\mathfrak{X}} - \mathfrak{M}\widetilde{\mathfrak{Y}}\mathfrak{D} &= 1, & \mathfrak{N}\widetilde{\mathfrak{X}} - \mathfrak{N}\widetilde{\mathfrak{Y}}\mathfrak{D} &= \mathfrak{D}, \end{aligned}$$

and by using these identities we find that

$$\begin{bmatrix} \mathfrak{X} + \mathfrak{N}(\widetilde{\mathfrak{Y}}\mathfrak{X} - \widetilde{\mathfrak{X}}\mathfrak{Y}) & \mathfrak{N} \\ \mathfrak{Y} + \mathfrak{M}(\widetilde{\mathfrak{Y}}\mathfrak{X} - \widetilde{\mathfrak{X}}\mathfrak{Y}) & \mathfrak{M} \end{bmatrix} \begin{bmatrix} \widetilde{\mathfrak{M}} & -\widetilde{\mathfrak{N}} \\ -\widetilde{\mathfrak{Y}} & \widetilde{\mathfrak{X}} \end{bmatrix} = 1$$

in TIC_ω (as opposed to *TIC*). However, since all the operators above belong to *TIC*, and since $L^p|Reg_{\omega,0} \cap L^p|Reg_0$ is dense in $L^p|Reg_0$, we find that the same identity must be true in *TIC*, too. Thus, we have a doubly coprime factorization containing $\mathfrak{N}$, $\mathfrak{M}$, $\widetilde{\mathfrak{M}}$, $\widetilde{\mathfrak{N}}$, $\widetilde{\mathfrak{X}}$, and $\widetilde{\mathfrak{Y}}$. We leave the corresponding construction where $\widetilde{\mathfrak{X}}$ and $\widetilde{\mathfrak{Y}}$ have been replaced by $\mathfrak{Y}$ and $\mathfrak{X}$ to the reader. □

Our next lemma parametrizes the set of all doubly coprime factorizations of a given input/output map $\mathfrak{D}$:

Lemma 8.3.11 *Let* $\mathfrak{D} \in TIC_\omega(U;Y)$ *for some* $\omega \ge 0$*, and suppose that* $\mathfrak{D}$ *has a doubly coprime factorization* (8.3.3).

(i) *All possible choices of* $\mathfrak{Y}$, $\mathfrak{X}$, $\widetilde{\mathfrak{X}}$*, and* $\mathfrak{Y}$ *in* (8.3.3) *for a fixed set of* $\mathfrak{N}$, $\mathfrak{M}$, $\widetilde{\mathfrak{M}}$*, and* $\widetilde{\mathfrak{N}}$ *are parametrized by*

$$\left(\begin{bmatrix} \mathfrak{X} & \mathfrak{N} \\ \mathfrak{Y} & \mathfrak{M} \end{bmatrix} \begin{bmatrix} 1 & 0 \\ \mathfrak{Q} & 1 \end{bmatrix}\right)^{-1} = \begin{bmatrix} 1 & 0 \\ -\mathfrak{Q} & 1 \end{bmatrix} \begin{bmatrix} \widetilde{\mathfrak{M}} & -\widetilde{\mathfrak{N}} \\ -\widetilde{\mathfrak{Y}} & \widetilde{\mathfrak{X}} \end{bmatrix},$$

or equivalently, by

$$\begin{bmatrix} \mathfrak{X} + \mathfrak{N}\mathfrak{Q} & \mathfrak{N} \\ \mathfrak{Y} + \mathfrak{M}\mathfrak{Q} & \mathfrak{M} \end{bmatrix}^{-1} = \begin{bmatrix} \widetilde{\mathfrak{M}} & -\widetilde{\mathfrak{N}} \\ -(\widetilde{\mathfrak{Y}} + \mathfrak{Q}\widetilde{\mathfrak{M}}) & \widetilde{\mathfrak{X}} + \mathfrak{Q}\widetilde{\mathfrak{N}} \end{bmatrix},$$

where $\mathfrak{Q} \in TIC(Y;U)$.

(ii) *All possible completions of* $\left[\begin{smallmatrix} \mathfrak{N} \\ \mathfrak{M} \end{smallmatrix}\right]$ *to an invertible operator in* $TIC(\left[\begin{smallmatrix} Y \\ U \end{smallmatrix}\right])$ *are parametrized by*

$$\begin{bmatrix} \mathfrak{X} & \mathfrak{N} \\ \mathfrak{Y} & \mathfrak{M} \end{bmatrix} \begin{bmatrix} \mathfrak{V} & 0 \\ \mathfrak{Q} & 1 \end{bmatrix} = \begin{bmatrix} \mathfrak{X}\mathfrak{V} + \mathfrak{N}\mathfrak{Q} & \mathfrak{N} \\ \mathfrak{Y}\mathfrak{V} + \mathfrak{M}\mathfrak{Q} & \mathfrak{M} \end{bmatrix},$$

where $\mathfrak{V} \in TIC(Y)$ *is invertible in* $TIC(Y)$ *and* $\mathfrak{Q} \in TIC(Y;U)$.

(iii) *All doubly coprime factorizations of* $\mathfrak{D}$ *are parametrized by*

$$\left(\begin{bmatrix}\mathfrak{X} & \mathfrak{N}\\ \mathfrak{Y} & \mathfrak{M}\end{bmatrix}\begin{bmatrix}\mathfrak{V} & 0\\ \mathfrak{Q} & \mathfrak{U}\end{bmatrix}\right)^{-1} = \left(\begin{bmatrix}\mathfrak{V} & 0\\ \mathfrak{Q} & \mathfrak{U}\end{bmatrix}^{-1}\begin{bmatrix}\widetilde{\mathfrak{M}} & -\widetilde{\mathfrak{N}}\\ -\widetilde{\mathfrak{Y}} & \widetilde{\mathfrak{X}}\end{bmatrix}\right),$$

where $\mathfrak{U} \in TIC(U)$ *is invertible in* $TIC(U)$, $\mathfrak{V} \in TIC(Y)$ *is invertible in* $TIC(Y)$, *and* $\mathfrak{Q} \in TIC(Y;U)$.

(iv) *If* (8.3.3) *is a joint doubly coprime factorization of* $\mathfrak{D}$ *and* $\mathfrak{Q} \in TIC_\omega(Y;U)$, *then all joint doubly coprime factorizations of* $\mathfrak{D}$ *and* $\mathfrak{Q}$ *are parametrized by*

$$\left(\begin{bmatrix}\mathfrak{X} & \mathfrak{N}\\ \mathfrak{Y} & \mathfrak{M}\end{bmatrix}\begin{bmatrix}\mathfrak{V} & 0\\ 0 & \mathfrak{U}\end{bmatrix}\right)^{-1} = \left(\begin{bmatrix}\mathfrak{V} & 0\\ 0 & \mathfrak{U}\end{bmatrix}^{-1}\begin{bmatrix}\widetilde{\mathfrak{M}} & -\widetilde{\mathfrak{N}}\\ -\widetilde{\mathfrak{Y}} & \widetilde{\mathfrak{X}}\end{bmatrix}\right),$$

where $\mathfrak{U} \in TIC(U)$ *is invertible in* $TIC(U)$ *and* $\mathfrak{V} \in TIC(Y)$ *is invertible in* $TIC(Y)$.

Proof (i) Clearly, the given formula gives a valid doubly coprime factorization for all choices of $\mathfrak{Q} \in TIC(Y;U)$ (and it contains the fixed operators $\mathfrak{N}$, $\mathfrak{M}$, $\widetilde{\mathfrak{M}}$, and $\widetilde{\mathfrak{N}}$). Conversely, suppose that (8.3.3) holds also with $\mathfrak{Y}$, $\mathfrak{X}$, $\widetilde{\mathfrak{X}}$, and $\mathfrak{Y}$ replaced by $\mathfrak{Y}_1$, $\mathfrak{X}_1$, $\widetilde{\mathfrak{X}}_1$, and $\mathfrak{Y}_1$. Then

$$\begin{bmatrix}\widetilde{\mathfrak{M}} & -\widetilde{\mathfrak{N}}\\ -\widetilde{\mathfrak{Y}} & \widetilde{\mathfrak{X}}\end{bmatrix}\begin{bmatrix}\mathfrak{X}_1 & \mathfrak{N}\\ \mathfrak{Y}_1 & \mathfrak{M}\end{bmatrix} = \begin{bmatrix}1 & 0\\ \mathfrak{Q} & 1\end{bmatrix},$$

where $\mathfrak{Q} = \widetilde{\mathfrak{X}}\mathfrak{Y}_1 - \widetilde{\mathfrak{Y}}\mathfrak{X}_1$. Multiply this by $\left[\begin{smallmatrix}\mathfrak{X} & \mathfrak{N}\\ \mathfrak{Y} & \mathfrak{M}\end{smallmatrix}\right]$ to the left to get $\left[\begin{smallmatrix}\mathfrak{X}_1 & \mathfrak{N}\\ \mathfrak{Y}_1 & \mathfrak{M}\end{smallmatrix}\right] = \left[\begin{smallmatrix}\mathfrak{X} & \mathfrak{N}\\ \mathfrak{Y} & \mathfrak{M}\end{smallmatrix}\right]\left[\begin{smallmatrix}1 & 0\\ \mathfrak{Q} & 1\end{smallmatrix}\right]$.

(ii) Again, for all choices of $\mathfrak{V}$ and $\mathfrak{Q}$ we have a valid doubly coprime factorization. To prove the converse we argue as in the proof of (i), but this time we get only

$$\begin{bmatrix}\widetilde{\mathfrak{M}} & -\widetilde{\mathfrak{N}}\\ -\widetilde{\mathfrak{Y}} & \widetilde{\mathfrak{X}}\end{bmatrix}\begin{bmatrix}\mathfrak{X}_1 & \mathfrak{N}\\ \mathfrak{Y}_1 & \mathfrak{M}\end{bmatrix} = \begin{bmatrix}\mathfrak{V} & 0\\ \mathfrak{Q} & 1\end{bmatrix},$$

where $\mathfrak{V} = \widetilde{\mathfrak{M}}\mathfrak{X}_1 - \widetilde{\mathfrak{N}}\mathfrak{Y}_1$ and $\mathfrak{Q} = \widetilde{\mathfrak{X}}\mathfrak{Y}_1 - \widetilde{\mathfrak{Y}}\mathfrak{X}_1$. The invertibility of $\left[\begin{smallmatrix}\mathfrak{V} & 0\\ \mathfrak{Q} & 1\end{smallmatrix}\right]$ forces $\mathfrak{V}$ to be invertible in $TIC(Y)$; see Lemma A.4.2(ii).

(iii) We get this (with $\mathfrak{Q}$ replaced by $\mathfrak{U}\mathfrak{Q}$) by combining the parametrization in (ii) with the one in Lemma 8.3.9(i).

(iv) Combine (iii) with Lemma 8.3.9. □

The property of having a right or left coprime fraction is not preserved under addition of two arbitrary elements in TIC_ω with $\omega > 0$. However, it is preserved if one of the two terms is bounded.

Lemma 8.3.12 *Let $\mathfrak{D} \in TIC_\omega(U;Y)$ for some $\omega \geq 0$, and let $\mathfrak{U} \in TIC(U;Y)$.*

(i) *If $\mathfrak{D}$ has a right coprime fraction $\mathfrak{D} = \mathfrak{N}\mathfrak{M}^{-1}$, then $(\mathfrak{N} + \mathfrak{U}\mathfrak{M})\mathfrak{M}^{-1}$ is a right coprime fraction of $\mathfrak{D} + \mathfrak{U}$.*

(ii) *If $\mathfrak{D}$ has a left coprime fraction $\mathfrak{D} = \widetilde{\mathfrak{M}}^{-1}\widetilde{\mathfrak{N}}$, then $\widetilde{\mathfrak{M}}^{-1}(\widetilde{\mathfrak{N}} + \widetilde{\mathfrak{M}}\mathfrak{U})$ is a left coprime fraction of $\mathfrak{D} + \mathfrak{U}$.*

(iii) *If $\mathfrak{D}$ has a doubly coprime factorization* (8.3.3), *then*

$$\left(\begin{bmatrix} 1 & \mathfrak{U} \\ 0 & 1 \end{bmatrix}\begin{bmatrix} \mathfrak{X} & \mathfrak{N} \\ \mathfrak{Y} & \mathfrak{M} \end{bmatrix}\right)^{-1} = \begin{bmatrix} \widetilde{\mathfrak{M}} & -\widetilde{\mathfrak{N}} \\ -\widetilde{\mathfrak{Y}} & \widetilde{\mathfrak{X}} \end{bmatrix}\begin{bmatrix} 1 & -\mathfrak{U} \\ 0 & 1 \end{bmatrix},$$

or equivalently,

$$\begin{bmatrix} \mathfrak{X} + \mathfrak{U}\mathfrak{Y} & \mathfrak{N} + \mathfrak{U}\mathfrak{M} \\ \mathfrak{Y} & \mathfrak{M} \end{bmatrix}^{-1} = \begin{bmatrix} \widetilde{\mathfrak{M}} & -(\widetilde{\mathfrak{N}} + \widetilde{\mathfrak{M}}\mathfrak{U}) \\ -\widetilde{\mathfrak{Y}} & \widetilde{\mathfrak{X}} + \widetilde{\mathfrak{Y}}\mathfrak{U} \end{bmatrix},$$

is a doubly coprime factorization of $\mathfrak{D} + \mathfrak{U}$.

Proof (i) The numerator and denominator in this fraction are right coprime since $(\widetilde{\mathfrak{X}} + \widetilde{\mathfrak{Y}}\mathfrak{U})\mathfrak{M} - \widetilde{\mathfrak{Y}}(\mathfrak{N} + \mathfrak{U}\mathfrak{M}) = 1$ whenever $\widetilde{\mathfrak{X}}\mathfrak{M} - \widetilde{\mathfrak{Y}}\mathfrak{N} = 1$.

(ii) The numerator and denominator in this fraction are left coprime since $\widetilde{\mathfrak{M}}(\mathfrak{X} + \mathfrak{U}\mathfrak{Y}) - (\widetilde{\mathfrak{N}} + \widetilde{\mathfrak{M}}\mathfrak{U})\widetilde{\mathfrak{Y}} = 1$ whenever $\widetilde{\mathfrak{M}}\mathfrak{X} - \widetilde{\mathfrak{N}}\mathfrak{Y} = 1$.

(iii) This can be verified by a direct computation. □

8.4 Coprime stabilization and detection

As the following theorem shows, if a well-posed linear system is jointly input/output stabilizable and detectable, then its input/output map has a doubly coprime factorization. A converse to this statement is true as well.

Theorem 8.4.1

(i) *Let $\Sigma = \left[\begin{smallmatrix} \mathfrak{A} & \mathfrak{B} \\ \mathfrak{C} & \mathfrak{D} \end{smallmatrix}\right]$ be a jointly input/output stabilizable and detectable well-posed linear system (in the sense of Definition 8.2.4). Then, with the notation of Lemma 8.2.2 and Definition 8.3.2,*

$$\begin{bmatrix} \widetilde{\mathfrak{M}} & -\widetilde{\mathfrak{N}} \\ -\widetilde{\mathfrak{Y}} & \widetilde{\mathfrak{X}} \end{bmatrix}\begin{bmatrix} \mathfrak{X} & \mathfrak{N} \\ \mathfrak{Y} & \mathfrak{M} \end{bmatrix} = \begin{bmatrix} 1 - \mathfrak{G}^\times & -\mathfrak{D}^\times \\ -\mathfrak{E}^\times & 1 - \mathfrak{F}^\times \end{bmatrix}\begin{bmatrix} 1 + \mathfrak{G}_\times & \mathfrak{D}_\times \\ \mathfrak{E}_\times & 1 + \mathfrak{F}_\times \end{bmatrix}$$

is a doubly coprime factorization of $\mathfrak{D}$. If Σ is jointly exponentially input/output stabilizable and detectable, then this is an exponentially doubly coprime factorization of $\mathfrak{D}$.

(ii) *Conversely, every $\mathfrak{D} \in TIC_\omega(U;Y)$ (for some $\omega \geq 0$) which has a doubly coprime factorization (and, in the Reg-well-posed case satisfies the equivalent necessary conditions listed in Theorem 2.6.7) can be realized as the input/output map of a jointly strongly stabilizable and detectable* *well-posed linear system* $\Sigma = \left[\begin{smallmatrix} \mathfrak{A} & \mathfrak{B} \\ \mathfrak{C} & \mathfrak{D} \end{smallmatrix}\right]$. *It has a jointly exponentially stabilizable and detectable realization if (and only if)* $\mathfrak{D}$ *has an exponentially doubly coprime factorization.*

Proof (i) Let $\Sigma = \left[\begin{smallmatrix} \mathfrak{A} & \mathfrak{B} \\ \mathfrak{C} & \mathfrak{D} \end{smallmatrix}\right]$ be jointly stabilizable and detectable. Then both the systems drawn in Figures 8.2 and 8.3 are bounded. In particular, both the input/output map from $\left[\begin{smallmatrix} w \\ u_\times \end{smallmatrix}\right]$ to $\left[\begin{smallmatrix} w^\times \\ u \end{smallmatrix}\right]$ in Figure 8.2, and the input/output map from $\left[\begin{smallmatrix} w^\times \\ u \end{smallmatrix}\right]$ to $\left[\begin{smallmatrix} w \\ u_\times \end{smallmatrix}\right]$ in Figure 8.3 are bounded. The former one is given by $\left[\begin{smallmatrix} 1+\mathfrak{G}_\times & \mathfrak{D}_\times \\ \mathfrak{E}_\times & 1+\mathfrak{F}_\times \end{smallmatrix}\right]$ (cf. Remark 7.3.4), and the latter one by $\left[\begin{smallmatrix} 1-\mathfrak{G}^\times & -\mathfrak{D}^\times \\ -\mathfrak{E}^\times & 1-\mathfrak{F}^\times \end{smallmatrix}\right]$. By comparing the two figures to each other we immediately realize that they are equivalent in the sense that the relationships between the different signals with the same names are identical in the two diagrams. This means that the input/output maps given above are inverses of each other, i.e.,

$$\begin{bmatrix} 1+\mathfrak{G}_\times & \mathfrak{D}_\times \\ \mathfrak{E}_\times & 1+\mathfrak{F}_\times \end{bmatrix} = \begin{bmatrix} 1-\mathfrak{G}^\times & -\mathfrak{D}^\times \\ -\mathfrak{E}^\times & 1-\mathfrak{F}^\times \end{bmatrix}^{-1}.$$

Moreover, as is easily seen, $\mathfrak{D}_\times(1+\mathfrak{F}_\times)^{-1}$ is a right coprime fraction of $\mathfrak{D}$, and $(1-\mathfrak{G}^\times)^{-1}\mathfrak{D}^\times)$ is a left coprime fraction of $\mathfrak{D}$. If both the closed-loop systems are exponentially stable, then all the factors in this factorization are exponentially stable. This proves part (i) of the theorem.

(ii) Conversely, suppose that there exists a doubly coprime factorization of $\mathfrak{D}$. Our construction below starts with a realization of the closed-loop system $\Sigma_\times$; another equally good choice would be to start with a realization of $\Sigma^\times$. Motivated by the formula that we found above, we pick the input/output map of $\Sigma_\times$ to be given by

$$\begin{bmatrix} \mathfrak{G}_\times & \mathfrak{D}_\times \\ \mathfrak{E}_\times & \mathfrak{F}_\times \end{bmatrix} = \begin{bmatrix} \mathfrak{X}-1 & \mathfrak{N} \\ \mathfrak{Y} & \mathfrak{M}-1 \end{bmatrix}$$

and choose an arbitrary strongly stable realization of this input/output map, for example, the exactly observable shift realization presented in Example 2.6.5. Since $\mathfrak{M}$ is supposed to have an inverse in TIC_ω, the operator $\left[\begin{smallmatrix} 0 & 0 \\ 0 & -1 \end{smallmatrix}\right]$ is an admissible feedback operator for $\Sigma_\times$. Denote the resulting ω-bounded closed-loop system by Σ_{ext} and the system that we get by dropping the state feedback row and the output injection column from Σ_{ext} by Σ. By Lemma 7.2.6(iii) and (8.3.3), the input/output map of Σ_{ext} is

$$\begin{aligned} \begin{bmatrix} \mathfrak{G} & \mathfrak{D} \\ \mathfrak{E} & \mathfrak{F} \end{bmatrix} &= \begin{bmatrix} \mathfrak{X}-\mathfrak{N}\mathfrak{M}^{-1}\mathfrak{Y}-1 & \mathfrak{N}\mathfrak{M}^{-1} \\ \mathfrak{M}^{-1}\mathfrak{Y} & 1-\mathfrak{M}^{-1} \end{bmatrix} \\ &= \begin{bmatrix} \widetilde{\mathfrak{M}}^{-1}-1 & \widetilde{\mathfrak{M}}^{-1}\widetilde{\mathfrak{N}} \\ \widetilde{\mathfrak{Y}}\widetilde{\mathfrak{M}}^{-1} & 1-\widetilde{\mathfrak{X}}+\widetilde{\mathfrak{Y}}\widetilde{\mathfrak{M}}^{-1}\widetilde{\mathfrak{N}} \end{bmatrix}. \end{aligned} \tag{8.4.1}$$

Observe, in particular, that the input/output map of Σ is the desired $\mathfrak{D} = \mathfrak{N}\mathfrak{M}^{-1} = \widetilde{\mathfrak{M}}^{-1}\widetilde{\mathfrak{N}}$. It follows from Lemma 7.1.7 that the system Σ that we get in this way is strongly stabilizable (and that the closed-loop state feedback system is $\Sigma_\times$). Moreover, by Lemma 7.1.7 and Theorem 8.1.7, Σ is strongly detectable if the operator $-\left[\begin{smallmatrix}1 & 0\\ 0 & 1\end{smallmatrix}\right]$ is a stabilizing output feedback operator for $\Sigma_\times$, or equivalently, if

$$\begin{bmatrix}1 & 0\\ 0 & 1\end{bmatrix} + \begin{bmatrix}\mathfrak{X}-1 & \mathfrak{N}\\ \mathfrak{Y} & \mathfrak{M}-1\end{bmatrix} = \begin{bmatrix}\mathfrak{X} & \mathfrak{N}\\ \mathfrak{Y} & \mathfrak{M}\end{bmatrix}$$

has an inverse in $TIC\left(\left[\begin{smallmatrix}Y\\ U\end{smallmatrix}\right]\right)$. But this is true because of the doubly coprimeness assumption. Thus, Σ is jointly strongly stabilizable and detectable. In the case where $\mathfrak{D}$ has an exponentially coprime factorization we argue in the same way, but start with an exponentially stable realization of $\Sigma_\times$; cf. Example 2.6.5. □

Let us immediately apply this to the case where $\mathfrak{D}$ can be stabilized by static output feedback:

Corollary 8.4.2 *Let $\Sigma = \left[\begin{smallmatrix}\mathfrak{A} & \mathfrak{B}\\ \mathfrak{C} & \mathfrak{D}\end{smallmatrix}\right]$ be a well-posed linear system on (Y, X, U). Then $K \in \mathcal{B}(U; Y)$ is an input/output stabilizing output feedback operator if and only if, with the notation used in* (8.3.3),

$$\begin{bmatrix}\mathfrak{X} & \mathfrak{N}\\ \mathfrak{Y} & \mathfrak{M}\end{bmatrix} = \begin{bmatrix}1 & \mathfrak{D}(1-K\mathfrak{D})^{-1}\\ K & (1-K\mathfrak{D})^{-1}\end{bmatrix},$$

$$\begin{bmatrix}\widetilde{\mathfrak{M}} & -\widetilde{\mathfrak{N}}\\ -\widetilde{\mathfrak{Y}} & \widetilde{\mathfrak{X}}\end{bmatrix} = \begin{bmatrix}(1-\mathfrak{D}K)^{-1} & -(1-\mathfrak{D}K)^{-1}\mathfrak{D}\\ -K & 1\end{bmatrix},$$

is a doubly coprime factorization of $\mathfrak{D}$. This factorization is exponentially doubly coprime if and only if K is exponentially input/output stabilizing.

Proof Clearly, if this is a doubly coprime factorization, then $\mathfrak{D}(1-K\mathfrak{D})^{-1} \in TIC(U; Y)$ and K is input/output stabilizing. To prove the converse we can either simply check that all the operators above are in TIC and that (8.3.3) holds (see Lemma A.4.1), or use Lemma 8.2.6(iii) and Theorem 8.4.1 with $\left[\mathfrak{K}\,\middle|\,\mathfrak{F}\right] = K\left[\mathfrak{C}\,\middle|\,\mathfrak{D}\right]$, $\left[\begin{smallmatrix}\mathfrak{H}\\ \mathfrak{G}\end{smallmatrix}\right] = -\left[\begin{smallmatrix}\mathfrak{B}\\ \mathfrak{D}\end{smallmatrix}\right]K$, and $\mathfrak{E} = K - K\mathfrak{D}K$. The exponentially stable case is treated in the same way. □

In particular, an input/output map which can be stabilized by static output feedback always has both a right and a left coprime fraction.

Lemma 8.4.3 *Let $\Sigma = \left[\begin{smallmatrix}\mathfrak{A} & \mathfrak{B}\\ \mathfrak{C} & \mathfrak{D}\end{smallmatrix}\right]$ be a well-posed linear system on (Y, X, U), and let $K \in \mathcal{B}(Y; U)$.*

(i) *If $\mathfrak{D}$ has a right coprime fraction $\mathfrak{D} = \mathfrak{N}\mathfrak{M}^{-1}$, then K is an input/output stabilizing output feedback operator for Σ if and only if $\mathfrak{M} - K\mathfrak{N}$ has an inverse in $TIC(U)$, in which case $(1-K\mathfrak{D})^{-1} = \mathfrak{M}(\mathfrak{M}-K\mathfrak{N})^{-1}$ and*

$\mathfrak{D}(1-K\mathfrak{D})^{-1}=\mathfrak{N}(\mathfrak{M}-K\mathfrak{N})^{-1}$ *(these are the input/output maps from v to u and y in Figure 7.6).*

(ii) *If $\mathfrak{D}$ has a left coprime fraction $\mathfrak{D}=\widetilde{\mathfrak{M}}^{-1}\widetilde{\mathfrak{N}}$, then K is an input/output stabilizing output feedback operator for Σ if and only if $\widetilde{\mathfrak{M}}-\widetilde{\mathfrak{N}}K$ has an inverse in $TIC(Y)$, in which case $(1-\mathfrak{D}K)^{-1}=(\widetilde{\mathfrak{M}}-\widetilde{\mathfrak{N}}K)^{-1}\widetilde{\mathfrak{M}}$ and $(1-\mathfrak{D}K)^{-1}\mathfrak{D}=(\widetilde{\mathfrak{M}}-\widetilde{\mathfrak{N}}K)^{-1}\widetilde{\mathfrak{N}}$ (these are the input/output maps from w and v to y_1 in Figure 7.6).*

Proof (i) Clearly $1-K\mathfrak{D}=(\mathfrak{M}-K\mathfrak{N})\mathfrak{M}^{-1}$, hence $(1-K\mathfrak{D})^{-1}=\mathfrak{M}(\mathfrak{M}-K\mathfrak{N})^{-1}$ and $\mathfrak{D}(1-K\mathfrak{D})^{-1}=\mathfrak{N}(\mathfrak{M}-K\mathfrak{N})^{-1}$. If $(\mathfrak{M}-K\mathfrak{N})^{-1}\in TIC(U)$, then $\mathfrak{N}(\mathfrak{M}-K\mathfrak{N})^{-1}=\mathfrak{D}(1-K\mathfrak{D})^{-1}\in TIC(U;Y)$ and the closed-loop system is input/output stable (for this we did not need the assumption that $\mathfrak{N}$ and $\mathfrak{M}$ are right coprime). Conversely, suppose that the closed-loop system is input/output stable. Then, by Corollary 8.4.2, $\left[\mathfrak{D}(1-K\mathfrak{D})^{-1}\right]\left[(1-K\mathfrak{D})^{-1}\right]^{-1}$ is a right coprime fraction of $\mathfrak{D}$. By Lemma 8.3.9(i)(b), $(1-K\mathfrak{D})\mathfrak{M}=\mathfrak{M}-K\mathfrak{N}$ is invertible in $TIC(U)$.

(ii) We leave this proof to the reader. □

Let us illustrate Theorem 8.4.1 and Corollary 8.4.2 with a simple finite-dimensional example.

Example 8.4.4 *We consider the regular system*

$$\begin{aligned}\dot{x}_1(t)&=x_1(t)+u(t),\\ \dot{x}_2(t)&=2x_2(t)+u(t),\\ y(t)&=x_1(t),\qquad t\ge 0,\end{aligned}$$

with input space $U=\mathbb{C}$, state space $X=\mathbb{C}^2$, output space $Y=\mathbb{C}$, generators

$$\left[\begin{array}{c|c} A & B\\ \hline C & D\end{array}\right]=\left[\begin{array}{cc|c} 1 & 0 & 1\\ 0 & 2 & 1\\ \hline 1 & 0 & 0\end{array}\right],$$

and transfer function $\widehat{\mathfrak{D}}(\lambda)=1/(\lambda-1)$.

It follows from Lemma 8.2.8(ii) that this system is not detectable, since $\left[\begin{smallmatrix}\lambda-A\\ C\end{smallmatrix}\right]=\left[\begin{smallmatrix}\lambda-1 & 0\\ 0 & \lambda-2\\ 1 & 0\end{smallmatrix}\right]$ does not have full column rank for $\lambda=2$. However, it is exponentially input/output stabilizable by static output feedback. For example, if we use the output feedback $u=Ky+v=-2y+v$, then the closed-loop system

$$\begin{aligned}\dot{x}_1(t)&=-x_1(t)+v(t),\\ \dot{x}_2(t)&=-x_1(t)+2x_2(t)+v(t),\\ y(t)&=x_1(t),\qquad t\ge 0,\end{aligned}$$

is exponentially output stable (i.e., exponentially input/output and state/output stable). From Remark 8.3.7 and Corollary 8.4.2 we get the doubly coprime factorization

$$\begin{bmatrix} (\lambda-1)/(\lambda+1) & -1/(\lambda+1) \\ 2 & 1 \end{bmatrix} \begin{bmatrix} 1 & 1/(\lambda+1) \\ -2 & (\lambda-1)/(\lambda+1) \end{bmatrix} = \begin{bmatrix} 1 & 0 \\ 0 & 1 \end{bmatrix}$$

of $\widehat{\mathfrak{D}}(\lambda) = 1/(\lambda-1)$. In particular, we get the identical right and left coprime fractions

$$\frac{1}{(\lambda-1)} = \frac{1/(\lambda+1)}{(\lambda-1)/(\lambda+1)}.$$

This example is stabilizable (even controllable). One particular exponentially stabilizing state feedback is $u = 4x_1 - 9x_2 + u_\times$, which leads to a closed-loop system with generators (the last row is the feedback output)

$$\left[\begin{array}{c|c} A_\times & B_\times \\ \hline C_\times & D_\times \\ K_\times & F_\times \end{array}\right] = \left[\begin{array}{cc|c} 5 & -9 & 1 \\ 4 & -7 & 1 \\ \hline 1 & 0 & 0 \\ 4 & -9 & 0 \end{array}\right]$$

(we get this state feedback if we decide to place both the poles at the point $\lambda = -1$). We have

$$(\lambda - A_\times)^{-1} = (\lambda+1)^{-2} \begin{bmatrix} \lambda+7 & -9 \\ 4 & \lambda-5 \end{bmatrix}$$

and the closed-loop transfer function is

$$\begin{aligned} \begin{bmatrix} \widehat{\mathfrak{D}}_\times(\lambda) \\ \widehat{\mathfrak{F}}_\times(\lambda) \end{bmatrix} &= \begin{bmatrix} C_\times \\ K_\times \end{bmatrix} (\lambda - A_\times)^{-1} B_\times \\ &= (\lambda+1)^{-2} \begin{bmatrix} 1 & 0 \\ 4 & -9 \end{bmatrix} \begin{bmatrix} \lambda+7 & -9 \\ 4 & \lambda-5 \end{bmatrix} \begin{bmatrix} 1 \\ 1 \end{bmatrix} \\ &= (\lambda+1)^{-2} \begin{bmatrix} \lambda-2 \\ -5\lambda+1 \end{bmatrix}. \end{aligned}$$

The standard right coprime fraction formula formula for $\widehat{\mathfrak{D}}$ gives

$$\widehat{\mathfrak{D}}(\lambda) = \frac{\widehat{\mathfrak{D}}_\times(\lambda)}{1 + \widehat{\mathfrak{F}}(\lambda)_\times} = \frac{(\lambda-2)/(\lambda+1)^2}{(\lambda-2)(\lambda-1)/(\lambda+1)^2}.$$

However, this is *not a coprime fraction* because the numerator and denominator are not coprime in H^∞: they have a common zero at $\lambda = 2$ (which is the location of the original undetectable pole of the system), hence $\left[\begin{smallmatrix} \widehat{\mathfrak{D}}_\times \\ 1+\widehat{\mathfrak{F}}_\times \end{smallmatrix}\right]$ does not have a left inverse in H^∞.

It follows from Theorem 8.4.1 that it is impossible to find an output injection that would be jointly stabilizing and detecting together with the state feedback $u = 4x_1 - 9x_2 + u_\times$. This particular state feedback is exponentially stabilizing, so it will automatically stabilize any added output injection column. However, no output injection will stabilize the added state feedback row (because otherwise we would get an input/output stabilizing and detecting state feedback and output injection pair). This can be easily verified by direct computation: no output injection will stabilize the input/output map $\mathfrak{F}$ whose control operator is $B = \left[\begin{smallmatrix}1\\1\end{smallmatrix}\right]$ and observation operator $K = \begin{bmatrix}4 & -9\end{bmatrix}$.

As we have seen in the preceding example, some state feedbacks produce right coprime fractions of the input/output map, whereas others do not. Let us therefore introduce some additional terminology:

Definition 8.4.5 Let Σ be a well-posed linear system on (Y, X, U), and introduce the same notation as in Lemma 7.3.3.

(i) A stabilizing state feedback pair $\left[\mathfrak{K} \,\middle|\, \mathfrak{F}\right]$ is *right coprime stabilizing* if $\mathfrak{D}_\times$ and $1 + \mathfrak{F}_\times$ are right coprime. The system Σ is *right coprime stabilizable* if there exists a right coprime stabilizing state feedback pair for Σ.

(ii) A detecting output injection pair $\left[\begin{smallmatrix}\mathfrak{H}\\ \mathfrak{G}\end{smallmatrix}\right]$ is *left coprime detecting* if $1 - \mathfrak{G}^\times$ and $\mathfrak{D}^\times$ and are left coprime. The system Σ is *left coprime detectable* if there exists a left coprime detecting output injection pair for Σ.

(iii) To these definitions we add one of the words '*weakly*', '*strongly*', or '*exponentially*' whenever the closed-loop system and all the factors in the coprime fractions are stable in the corresponding sense (see Definition 8.1.1).

(iv) If to these definitions we add one or several of the qualifiers '*state/state*', '*input/state*', '*state/output*', '*input/output*', '*input*', or '*output*', then we mean that only the corresponding part of the closed-loop system has to be bounded or stable in the appropriate sense (we shall always assume at least input/output boundedness of the closed loop system since the coprimeness conditions require $\mathfrak{D}_\times, \mathfrak{F}_\times, \mathfrak{G}^\times, \mathfrak{D}^\times \in TIC$).

We have the following simple relationships between the right coprime stabilizability of a system and the existence of a right coprime fraction of its input/output map (and analogous results for left coprime detectability).

Theorem 8.4.6 *In this theorem we use the notation of Lemma 7.3.3.*

(i) *If* $\Sigma = \left[\begin{smallmatrix}\mathfrak{A} & \mathfrak{B}\\ \mathfrak{C} & \mathfrak{D}\end{smallmatrix}\right]$ *is a right coprime input/output stabilizable system then* $\mathfrak{D}_\times(1 + \mathfrak{F}_\times)^{-1}$ *is a right coprime fraction of* $\mathfrak{D}$. *If* Σ *is exponentially input/output right coprime stabilizable, then this is an exponentially right coprime fraction of* $\mathfrak{D}$.

(ii) *If* $\Sigma = \left[\begin{smallmatrix} \mathfrak{A} & \mathfrak{B} \\ \mathfrak{C} & \mathfrak{D} \end{smallmatrix}\right]$ *is a left coprime input/output detectable system then* $(1 - \mathfrak{G}^{\times})^{-1}\mathfrak{D}^{\times}$ *is a left coprime fraction of* $\mathfrak{D}$. *If* Σ *is left coprime exponentially input/output detectable, then this is an exponentially left coprime fraction of* $\mathfrak{D}$.

(iii) *Every* $\mathfrak{D} \in TIC_{\omega}(U; Y)$ *(for some* $\omega \geq 0$*) which has a right coprime fraction (and, in the Reg-well-posed case satisfies the equivalent necessary conditions listed in Theorem 2.6.7) can be realized as the input/output map of a right coprime strongly stabilizable well-posed linear system* $\Sigma = \left[\begin{smallmatrix} \mathfrak{A} & \mathfrak{B} \\ \mathfrak{C} & \mathfrak{D} \end{smallmatrix}\right]$. *It has a right coprime exponentially stabilizable realization if (and only if)* $\mathfrak{D}$ *has an exponentially right coprime fraction.*

(iv) *Every* $\mathfrak{D} \in TIC_{\omega}(U; Y)$ *(for some* $\omega \geq 0$*) which has a left coprime fraction (and, in the Reg-well-posed case satisfies the equivalent necessary conditions listed in Theorem 2.6.7) can be realized as the input/output map of a left coprime strongly detectable well-posed linear system* $\Sigma = \left[\begin{smallmatrix} \mathfrak{A} & \mathfrak{B} \\ \mathfrak{C} & \mathfrak{D} \end{smallmatrix}\right]$. *It has a left coprime exponentially detectable realization if (and only if)* $\mathfrak{D}$ *has an exponentially left coprime fraction.*

Proof (i)–(ii) See Lemma 7.3.3 and Definition 8.4.5.

(iii)–(iv) The proof of (iii) is a simplified version of the proof of part (ii) of Theorem 8.4.1. The proof of (iv) is similar, with $\Sigma_{\times}$ replaced by $\Sigma^{\times}$. □

Lemma 8.4.7 *Let* Σ *be a well-posed linear system on* (Y, X, U), *and introduce the same notation as in Lemma 7.3.3.*

(i) *If* Σ *is jointly input/output stabilizable and detectable, then* Σ *is both right coprime input/output stabilizable and left coprime input/output detectable. In particular, this is true if* Σ *is stabilizable by static output feedback. The same statements are true if throughout we add one or several of the qualifiers 'weakly', 'strongly', 'exponentially', 'input', 'output', 'state/state', 'input/state', 'state/output' (by adding the last three qualifiers we can remove 'input/output').*

(ii) *If* Σ *is both right coprime stabilizable and left coprime detectable then* $\mathfrak{D}$ *has both a right coprime fraction* $\mathfrak{D} = \mathfrak{D}_{\times}(1 + \mathfrak{F}_{\times})^{-1}$ *and a left coprime fraction* $\mathfrak{D} = (1 - \mathfrak{G}^{\times})^{-1}\mathfrak{D}^{\times}$. *Moreover, in this case* $\mathfrak{D}$ *has a jointly strongly stabilizable and detectable realization.*

Proof (i) See Lemma 8.2.6, Theorem 8.4.1 and Definition 8.4.5.

(ii) By Theorem 8.4.6, $\mathfrak{D}_{\times}(1 + \mathfrak{F}_{\times})^{-1}$ is a right coprime fraction and $(1 - \mathfrak{G}^{\times})^{-1}\mathfrak{D}^{\times}$ is a left coprime fraction of $\mathfrak{D}$. By Lemma 8.3.10, $\mathfrak{D}$ then has a doubly coprime factorization, hence, by Theorem 8.4.1(ii), it has a jointly stabilizable and detectable realization. □

By applying Definition 8.4.5 we also get the following modified version of Theorem 8.2.11.

Theorem 8.4.8 *Let Σ be a well-posed linear system.*

(i) *The following conditions are equivalent:*
  (a) *Σ is bounded;*
  (b) *Σ is input/output bounded and right coprime stabilizable;*
  (c) *Σ is input/output bounded and left coprime detectable.*

(ii) *In the L^p-well-posed case with $p < \infty$ the following conditions are equivalent:*
  (a) *Σ is strongly stable;*
  (b) *Σ is input/output bounded and strongly right coprime stabilizable;*
  (c) *Σ is input/output bounded and strongly left coprime detectable.*

(iii) *In the Reg-well-posed case the following conditions are equivalent:*
  (a) *Σ is strongly stable;*
  (b) *Σ is input/output bounded, strongly input/state stable, and strongly right coprime stabilizable;*
  (c) *Σ is input/output bounded, strongly state/output stable, and strongly left coprime detectable;*

(iv) *The following conditions are equivalent:*
  (a) *Σ is exponentially stable;*
  (b) *Σ is input/output bounded and exponentially right coprime stabilizable;*
  (c) *Σ is input/output bounded and exponentially left coprime detectable.*

Proof (i) Clearly (a) $\Rightarrow$ (b) and (a) $\Rightarrow$ (c). The proof that (c) $\Rightarrow$ (a) is analogous to the proof that (b) $\Rightarrow$ (a), so let us only prove the latter implication. Assume that (b) holds. Then the closed-loop system $\Sigma_\times$ in Lemma 7.3.3(i) is bounded and $\mathfrak{D}_\times(1+\mathfrak{F}_\times)^{-1}$ is a right coprime fraction of $\mathfrak{D}$. Assume further that $\mathfrak{D}$ is bounded. Then $\mathfrak{D}I^{-1}$ is another right coprime fraction of $\mathfrak{D}$. By Lemma 8.3.9, $1+\mathfrak{F}_\times = (1-\mathfrak{F})^{-1}$ is invertible in *TIC*(U), i.e, $1-\mathfrak{F} \in$ *TIC*(U). By Lemma 7.3.3(i)

$$\begin{bmatrix} \mathfrak{A} & \mathfrak{B}\tau \\ \hline \mathfrak{C} & \mathfrak{D} \\ \mathfrak{K} & \mathfrak{F} \end{bmatrix} = \begin{bmatrix} \mathfrak{A}_\times & \mathfrak{B}_\times\tau \\ \hline \mathfrak{C}_\times & \mathfrak{D}_\times \\ \mathfrak{K}_\times & \mathfrak{F}_\times \end{bmatrix} - \begin{bmatrix} \mathfrak{B}_\times\tau \\ \hline \mathfrak{D}_\times \\ \mathfrak{F}_\times \end{bmatrix} (1-\mathfrak{F}) \left[\mathfrak{K}_\times \,\middle|\, \mathfrak{F}_\times \right].$$

This shows that the original system Σ (and even the extended system) is bounded.

(ii) Again the implications (a) $\Rightarrow$ (b) and (a) $\Rightarrow$ (c) are trivial. To prove the converse implications we observe as in the proof of (i) that the extended system (where to Σ we have added either $\left[\, \mathfrak{K} \,\middle|\, \mathfrak{F} \,\right]$ or $\left[\frac{\mathfrak{H}}{\mathfrak{G}}\right]$) is bounded, and that it can be strongly stabilized by output feedback. In particular, it is strongly detectable. Thus, the strong stability of the extended system follows from Theorem 8.2.11(ii)(c).

(iii) This follows from part (i) and Theorem 8.2.11(ii).

(iv) As in (i), the implications (a) $\Rightarrow$ (b) and (a) $\Rightarrow$ (c) are trivial, and the proof of the implication (c) $\Rightarrow$ (a) is analogous to the proof of (b) $\Rightarrow$ (a). Assume (b). With the same notation as in the proof of (i), the system $\Sigma_\times$ is exponentially stable. The exponential stability of Σ then follows from (i) and Theorem 8.1.7(v)(j). □

Also right coprime stabilizability and left coprime detectability are preserved under static output feedback.

Lemma 8.4.9 *All the different notions of right coprime stabilizability and left coprime detectability listed in Definition 8.4.5 are preserved under (admissible) static output feedback, i.e., the closed-loop system is right coprime stabilizable or left coprime detectable in exactly the same sense as the original system.*

Proof The proof is the same as the proof of Lemma 8.2.7. □

Theorem 8.4.10 *Let Σ be a well-posed linear system.*

(i) *If Σ is right coprime stabilizable or left coprime detectable, then an arbitrary static output feedback operator is stabilizing if and only if it is input/output stabilizing.*
(ii) *In the L^p-well-posed case with $p < \infty$, if Σ is strongly right coprime stabilizable or strongly left coprime detectable, then an arbitrary static output feedback operator is strongly stabilizing if and only if it is input/output stabilizing.*
(iii) *In the Reg-well-posed case,*
    (a) *if Σ is strongly right coprime stabilizable, then an arbitrary static output feedback operator is strongly stabilizing if and only if it is input/output stabilizing and strongly input/state stabilizing, and*
    (b) *if Σ is strongly left coprime delectable, then an arbitrary static output feedback operator is strongly stabilizing if and only if it is input/output stabilizing and strongly state/output stabilizing.*
(iv) *If Σ is exponentially right coprime stabilizable or exponentially left coprime detectable, then an arbitrary static output feedback operator is exponentially stabilizing if and only if it is input/output stabilizing.*

Proof Apply first Lemma 8.4.9 and then Theorem 8.4.8 to the closed-loop system. □

Observe that, because of Corollary 8.4.2, the stabilization in Theorem 8.4.10 is always both right coprime stabilizing and left coprime detecting.

Slightly weaker versions of Theorem 8.4.10 are valid for state feedbacks and output injections:

Theorem 8.4.11 *Let Σ be a well-posed linear system.*

(i) *If Σ is right coprime stabilizable, then an arbitrary state feedback pair is right coprime stabilizing if and only if it is right coprime output stabilizing.*

(ii) *If Σ is left coprime detectable, then an arbitrary output injection pair is left coprime detecting if and only if it is left coprime input detecting.*

(iii) *If Σ is strongly right coprime stabilizable, then*

  (a) *in the L^p-well-posed case with $p < \infty$, an arbitrary state feedback pair is strongly right coprime stabilizing if and only if it is output right coprime stabilizing, and*

  (b) *in the Reg-well-posed case, an arbitrary state feedback pair is strongly right coprime stabilizing if and only if it is strongly input/state stabilizing and output right coprime stabilizing.*

(iv) *If Σ is strongly left coprime detectable, then,*

  (a) *in the L^p-well-posed case with $p < \infty$, an arbitrary output injection pair is strongly left coprime detecting if and only if it is left coprime input detecting, and*

  (b) *in the Reg-well-posed case, an arbitrary output injection pair is strongly left coprime detecting if and only if it is left coprime input detecting and strongly state/output detecting.*

(v) *If Σ is exponentially right coprime stabilizable, then an arbitrary state feedback pair is exponentially right coprime stabilizing if and only if it is input/output stabilizing.*

(vi) *If Σ is exponentially left coprime detectable, then an arbitrary output injection pair is exponentially left coprime detecting if and only if it is input/output detecting.*

Proof In all cases the 'only if' parts of the claims are trivial. Therefore, we only prove the 'if' claims below.

(i) Let $\left[\,\mathfrak{K} \,\middle|\, \mathfrak{F}\,\right]$ be a right coprime output stabilizing state feedback pair for Σ, let $\left[\,\mathfrak{K}^1 \,\middle|\, \mathfrak{F}^1\,\right]$ be a right coprime stabilizing state feedback pair for Σ, and consider the system

$$\Sigma_{\text{ext}} = \left[\begin{array}{c|c} \mathfrak{A} & \mathfrak{B} \\ \hline \mathfrak{C} & \mathfrak{D} \\ \mathfrak{K} & \mathfrak{F} \\ \mathfrak{K}^1 & \mathfrak{F}^1 \end{array}\right].$$

We denote the system that we get by using $\begin{bmatrix}0 & 1 & 0\end{bmatrix}$ as a state feedback operator by $\Sigma_\times$, and the system that we get by using $\begin{bmatrix}0 & 0 & 1\end{bmatrix}$ as a state feedback operator by $\Sigma_\natural$. We also use the same subindices to denote components of the corresponding closed-loop systems (cf. Lemmas 7.2.6(i) and 7.3.3(i)).

Clearly, the first, second, and fourth rows of $\Sigma_\natural$ are bounded since $\left[\mathfrak{K}^1 \middle| \mathfrak{F}^1\right]$ is stabilizing. We claim that also the third row

$$\left[\mathfrak{K}_\natural \,\middle|\, \mathfrak{F}_\natural\right] = \left[\mathfrak{K} + \mathfrak{F}(1-\mathfrak{F}^1)^{-1}\mathfrak{K}^1 \,\middle|\, \mathfrak{F}(1-\mathfrak{F}^1)^{-1}\right]$$

of $\Sigma_\natural$ is bounded, or in other words, we claim that $\Sigma_\natural$ is bounded. To prove this we use the assumption that $\left[\mathfrak{K} \middle| \mathfrak{F}\right]$ is right coprime output stabilizing. By the coprimeness assumption, $\mathfrak{D}_\times(1+\mathfrak{F}_\times)^{-1}$ is a right coprime fraction of $\mathfrak{D}$. On the other hand, we can also write $\mathfrak{D}$ in the form $\mathfrak{D} = \mathfrak{D}_\natural(1+\mathfrak{F}^1_\natural)^{-1}$. Hence, by Lemma 8.3.9(i)(a),

$$\mathfrak{U} = (1+\mathfrak{F}_\times)^{-1}(1+\mathfrak{F}^1_\natural) = (1-\mathfrak{F})(1-\mathfrak{F}^1)^{-1} \in TIC(U).$$

But

$$\mathfrak{F}_\natural = \mathfrak{F}(1-\mathfrak{F}^1)^{-1} = (1-\mathfrak{F}^1)^{-1} - (1-\mathfrak{F})(1-\mathfrak{F}^1)^{-1} = \mathfrak{F}^1_\natural + 1 - \mathfrak{U};$$

hence $\mathfrak{F}_\natural$ is bounded. To show that $\mathfrak{K}_\natural$ is bounded we use the fact that (by the assumption) all of $\mathfrak{C}_\times$, $\mathfrak{C}_\natural$, $\mathfrak{K}_\times$, and $\mathfrak{K}^1_\natural$ are bounded, and that (by some simple algebra),

$$\begin{aligned}
\mathfrak{K}_\times - \mathfrak{K}^1_\natural &= (1-\mathfrak{F})^{-1}[\mathfrak{K}_\natural - \mathfrak{K}^1_\natural],\\
\mathfrak{C}_\times - \mathfrak{C}_\natural &= \mathfrak{D}(1-\mathfrak{F})^{-1}[\mathfrak{K}_\natural - \mathfrak{K}^1_\natural].
\end{aligned}$$

By the coprimeness of $\mathfrak{D}_\times = \mathfrak{D}(1-\mathfrak{F})^{-1}$ and $(1+\mathfrak{F}_\times) = (1-\mathfrak{F})^{-1}$, we can choose some operators $\widetilde{\mathfrak{X}}$ and $\widetilde{\mathfrak{Y}}$ such that $\widetilde{\mathfrak{X}}(1-\mathfrak{F})^{-1} - \widetilde{\mathfrak{Y}}\mathfrak{D}(1-\mathfrak{F})^{-1} = 1$. Multiply the two equations above by $\widetilde{\mathfrak{X}}$ and $-\widetilde{\mathfrak{Y}}$, respectively, and add the result to get

$$\mathfrak{K}_\natural = \mathfrak{K}^1_\natural + \widetilde{\mathfrak{X}}(\mathfrak{C}_\times - \mathfrak{C}_\natural) - \widetilde{\mathfrak{Y}}(\mathfrak{K}_\times - \mathfrak{K}^1_\natural).$$

Thus, $\mathfrak{K}_\natural$ is bounded. This proves that $\Sigma_\natural$ is bounded. (So far we have not used the assumption that the pair $\left[\mathfrak{K}^1 \middle| \mathfrak{F}^1\right]$ is *right coprime* stabilizing.)

By interchanging the roles of the two feedback pairs $\left[\mathfrak{K} \middle| \mathfrak{F}\right]$ and $\left[\mathfrak{K}^1 \middle| \mathfrak{F}^1\right]$ and arguing as above we find that the system $\Sigma_\times$ is output stable. The system $\Sigma_\times$ can be interpreted as a feedback connection of $\Sigma_\natural$ with output feedback operator $\begin{bmatrix}0 & 1 & -1\end{bmatrix}$ (cf. the proof of Lemma 8.2.7). By Theorem 8.1.7(v), $\Sigma_\times$ is bounded.

(ii) The proof of (ii) is analogous to the proof of (i), and we leave it to the reader.

(iii) With the notation used in the proof of (i), both $\Sigma_\times$ and $\Sigma_\natural$ are bounded, and by Lemma 8.1.2(vii), $\Sigma_\natural$ is strongly stable. Since $\Sigma_\times$ can be obtained from $\Sigma_\natural$ by output feedback, the strong stability of $\Sigma_\times$ follows from Theorem 8.4.10(ii)–(iii).

(iv) This proof is similar to the proof of (iii).

(v) We introduce the same notation as in the proof of (i). The proof of the exponential stability of $\Sigma_\natural$ (including the third row) is now trivial, since the exponential growth bound of a system is determined by the growth bound of the semigroup (cf. Theorem 2.5.4(ii). The input/output stability of $\Sigma_\times$ is proved in the same way as in the proof of (i) (with $\Sigma_\times$ and $\Sigma_\natural$ interchanged). By Theorem 8.1.7(v)(j), $\Sigma_\times$ is exponentially stable. That $\mathfrak{D}_\times(1+\mathfrak{F}_\times)^{-1}$ is an exponentially right coprime fraction of $\mathfrak{D}$ follows from the exponential invertibility of $\mathfrak{U} = (1+\mathfrak{F}_\times)^{-1}(1+\mathfrak{F}_\natural^1)$ (both $\mathfrak{U} = \mathfrak{F}_\natural - \mathfrak{F}_\natural^1 - 1$ and $\mathfrak{U}^{-1} = \mathfrak{F}_\times^1 - \mathfrak{F}_\times - 1$ are exponentially stable) and the fact that $\mathfrak{D}_\natural(1+\mathfrak{F}_\natural)^{-1}$ is an exponentially right coprime fraction of $\mathfrak{D}$ (see Lemma 8.3.9(i)(b)).

(vi) This proof is similar to the proof of (v). □

Also Proposition 8.2.10 can be extended to the right coprime stabilizable and left coprime detectable case:

Proposition 8.4.12 *The list of stabilizability and detectability properties which are inherited by the systems listed in Proposition 8.2.10 can be extended to include all the versions of right coprime stabilizability and left coprime detectability listed in Definition 8.4.5(i)–(iv).*

Proof The proof is the same as the proof of Proposition 8.2.10, but we have to check, in addition, that the coprimeness assumption for the closed-loop system is preserved when we pass from the original system to the modified system. This is straightforward in cases (i)(a) and (i)(b). Cases (i)(c) and (i)(e) are output feedback connections covered by Lemma 8.4.9. Case (i)(d) contains two parts, i.e., the systems in parts (i) and (ii) of Theorem 7.1.2. The former is a special case of the cross-product in (ii)(a) to which we return later, and the latter is obtained from the former by addition of an identity feedthrough operator; by Lemma 8.3.12 this does not destroy the coprimeness condition. The right coprimeness in the case (ii)(a) follows from the fact that if $\mathfrak{N}_1\mathfrak{M}_1^{-1}$ is a right coprime fraction of $\mathfrak{D}_1$ and $\mathfrak{N}_2\mathfrak{M}_2^{-1}$ is a right coprime fraction of $\mathfrak{D}_2$, then $\left[\begin{smallmatrix} \mathfrak{N}_1 & 0 \\ 0 & \mathfrak{N}_2 \end{smallmatrix}\right]\left[\begin{smallmatrix} \mathfrak{M}_1 & 0 \\ 0 & \mathfrak{M}_2 \end{smallmatrix}\right]^{-1}$ is a right coprime fraction of $\left[\begin{smallmatrix} \mathfrak{D}_1 & 0 \\ 0 & \mathfrak{D}_2 \end{smallmatrix}\right]$. Preservation of left coprimeness is proved in the same way. Finally, (ii)(b) and (ii)(c) are static output feedback connections of (ii)(a) covered by Lemma 8.4.9. □

8.5 Dynamic stabilization

It is now time to take a closer look at the the notion of *dynamic feedback stabilization* which plays an important role in optimal control theory. For example, H^∞ control theory deals extensively with measurement feedback stabilization, which is a special case of dynamic feedback stabilization.

Definition 8.5.1 Let $\Sigma = \left[\begin{array}{c|c} \mathfrak{A} & \mathfrak{B} \\ \hline \mathfrak{C} & \mathfrak{D} \end{array}\right]$ be a well-posed linear system on (U, X, Y).

(i) The well-posed linear system $\Sigma_1 = \left[\begin{array}{c|c} \mathfrak{A}_1 & \mathfrak{B}_1 \\ \hline \mathfrak{C}_1 & \mathfrak{D}_1 \end{array}\right]$ on (Y, X_1, U) is called a *stabilizing dynamic feedback* system for Σ if the dynamic feedback connection in Figure 8.5 (cf. Example 7.2.5) is (admissible and) stable. The system Σ is *stabilizable by dynamic feedback* if there exists a stabilizing dynamic feedback system Σ_1 for Σ.

(ii) To this definition we add one of the words '*weakly*', '*strongly*', or '*exponentially*' whenever the closed-loop system is stable in the corresponding sense (see Definition 8.1.1).

(iii) If to these definitions we add one or several of the qualifiers '*state/state*', '*input/state*', '*state/output*', '*input/output*', '*input*', or '*output*', then we mean that only the corresponding part of the closed-loop system has to be bounded or stable in the appropriate sense.

Observe that Σ and Σ_1 can be interchanged with each other: Σ_1 stabilizes Σ if and only if Σ stabilizes Σ_1. Therefore we shall also say that the two systems *(dynamically) stabilize each other*.

In the sequel we shall primarily be concerned with *dynamic input/output stabilization*. This is partially motivated by the following facts:

Theorem 8.5.2 *Let* $\Sigma = \left[\begin{array}{c|c} \mathfrak{A} & \mathfrak{B} \\ \hline \mathfrak{C} & \mathfrak{D} \end{array}\right]$ *be a well-posed linear system on* (U, X, Y), *and let* $\Sigma_1 = \left[\begin{array}{c|c} \mathfrak{A}_1 & \mathfrak{B}_1 \\ \hline \mathfrak{C}_1 & \mathfrak{D}_1 \end{array}\right]$ *be a well-posed linear system on* (Y, X_1, U).

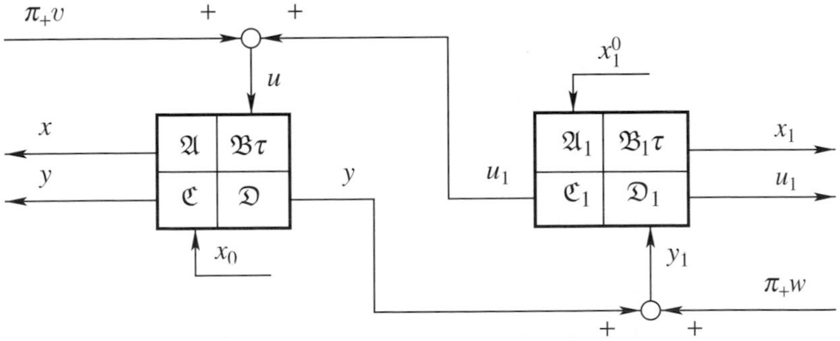

Figure 8.5 Dynamic stabilization

(i) *Suppose that at least one of the following conditions hold:*
   (a) *both* Σ *and* Σ_1 *are stabilizable and detectable (not necessarily jointly);*
   (b) *both* Σ *and* Σ_1 *are right coprime stabilizable;*
   (c) *both* Σ *and* Σ_1 *are left coprime detectable.*
   Then Σ *and* Σ_1 *stabilize each other if and only if they input/output stabilize each other.*

(ii) *Suppose both* Σ *and* Σ_1 *are* L^p*-well-posed with* $p < \infty$*, and that at least one of the following conditions hold:*
   (a) *both* Σ *and* Σ_1 *are strongly right coprime stabilizable;*
   (b) *both* Σ *and* Σ_1 *are stabilizable and strongly detectable;*
   (c) *both* Σ *and* Σ_1 *are strongly left coprime detectable.*
   Then, Σ *and* Σ_1 *stabilize each other strongly if and only if they input/output stabilize each other.*

(iii) *Suppose both* Σ *and* Σ_1 *are Reg-well-posed, detectable and strongly stabilizable. Then they stabilize each other strongly if and only if they input/output stabilize and strongly input/state stabilize each other.*

(iv) *Suppose both* Σ *and* Σ_1 *are Reg-well-posed, and that at least one of the following conditions hold:*
   (a) *both* Σ *and* Σ_1 *are stabilizable and strongly detectable;*
   (b) *both* Σ *and* Σ_1 *are strongly left coprime detectable.*
   Then, Σ *and* Σ_1 *stabilize each other strongly if and only if they input/output stabilize and strongly state/output stabilize each other.*

(v) *Suppose that at least one of the following conditions hold:*
   (a) *both* Σ *and* Σ_1 *are Reg-well-posed or* L^p*-well-posed with* $p > 1$*, detectable and exponentially stabilizable;*
   (b) *both* Σ *and* Σ_1 *are exponentially right coprime stabilizable;*
   (c) *both* Σ *and* Σ_1 *are* L^p*-well-posed with* $p < \infty$*, stabilizable and exponentially detectable;*
   (d) *both* Σ *and* Σ_1 *are exponentially left coprime detectable;*
   Then Σ *and* Σ_1 *stabilize each other exponentially if and only if they input/output stabilize each other.*

Proof First use Propositions 8.2.10(ii)(c) and 8.4.12 to show that the dynamic feedback connection of Σ and Σ_1 inherits the stabilizability and detectability properties of the two subsystems, and then use Theorems 8.2.11 and 8.4.8 to conclude that the dynamic feedback connection is stable in the appropriate sense. □

As we mentioned above, we shall in the sequel concentrate on the problem of dynamic input/output stabilization. This is motivated by Theorem 8.5.2, which says that, under appropriate assumptions, input/output stabilization implies stabilization of the full dynamic feedback connection.

As the following lemma shows, dynamic input/output stabilization implies that the input/output maps of the two systems have right and left fractions.

Lemma 8.5.3 *Suppose that* $\Sigma = \left[\begin{smallmatrix} \mathfrak{A} & \mathfrak{B} \\ \mathfrak{C} & \mathfrak{D} \end{smallmatrix}\right]$ *and* $\Sigma_1 = \left[\begin{smallmatrix} \mathfrak{A}_1 & \mathfrak{B}_1 \\ \mathfrak{C}_1 & \mathfrak{D}_1 \end{smallmatrix}\right]$ *input/output stabilize each other. Then* $\mathfrak{D}$ *and* $\mathfrak{D}_1$ *have right and left fractions* $\mathfrak{D} = \mathfrak{N}\mathfrak{M}^{-1} = \widetilde{\mathfrak{M}}^{-1}\widetilde{\mathfrak{N}}$ *and* $\mathfrak{D}_1 = \mathfrak{N}_1\mathfrak{M}_1^{-1} = \widetilde{\mathfrak{M}}_1^{-1}\widetilde{\mathfrak{N}}_1$, *where* $\mathfrak{N}$, $\mathfrak{M}$, $\widetilde{\mathfrak{M}}$, $\widetilde{\mathfrak{N}}$, $\mathfrak{N}_1$, $\mathfrak{M}_1$, $\widetilde{\mathfrak{M}}_1$, $\widetilde{\mathfrak{N}}_1 \in TIC$ *and the inverses belong to* TIC_ω *for some* $\omega \ge 0$. *In particular, we can choose these operators to be*

$$\begin{bmatrix} \mathfrak{M}_1 & \mathfrak{N} \\ \mathfrak{N}_1 & \mathfrak{M} \end{bmatrix} = \begin{bmatrix} \widetilde{\mathfrak{M}} & \widetilde{\mathfrak{N}} \\ \widetilde{\mathfrak{N}}_1 & \widetilde{\mathfrak{M}}_1 \end{bmatrix} = \begin{bmatrix} 1 & -\mathfrak{D} \\ -\mathfrak{D}_1 & 1 \end{bmatrix}^{-1}.$$

Proof See the formula for $\left[\begin{smallmatrix} 1 & -\mathfrak{D} \\ -\mathfrak{D}_1 & 1 \end{smallmatrix}\right]^{-1}$ given in Example 7.2.5 (and also Lemma A.4.1). □

Remark 8.5.4 Observe that we *do not claim the fractions in Lemma 8.5.3 to be right or left coprime*. We do not even claim that $\mathfrak{D}$ or $\mathfrak{D}_1$ have right or left coprime fractions. However, at least in one important case $\mathfrak{D}$ and $\mathfrak{D}_1$ in Lemma 8.5.3 do have left and right coprime fractions, namely in the L^2-well-posed case with finite-dimensional input and output spaces. This follows from the fact that a coprime fraction in TIC^2 of an input/output map is equivalent to a coprime fraction in H^∞ of the corresponding transfer function, and according to Smith (1989, Theorem 1), the existence of the frequency domain versions of the fractions in Lemma 8.5.3 implies that $\widehat{\mathfrak{D}}$ and $\widehat{\mathfrak{D}}_1$ have right and left coprime fractions in H^∞.

Lemma 8.5.5 *Let* $\Sigma = \left[\begin{smallmatrix} \mathfrak{A} & \mathfrak{B} \\ \mathfrak{C} & \mathfrak{D} \end{smallmatrix}\right]$ *be a well-posed linear system on* (Y, X, U), *let* $\Sigma_1 = \left[\begin{smallmatrix} \mathfrak{A}_1 & \mathfrak{B}_1 \\ \mathfrak{C}_1 & \mathfrak{D}_1 \end{smallmatrix}\right]$ *be a well-posed linear system on* (U, X_1, Y).

(i) *Suppose that* $\mathfrak{D}$ *and* $\mathfrak{D}_1$ *have right coprime fractions* $\mathfrak{D} = \mathfrak{N}\mathfrak{M}^{-1}$ *and* $\mathfrak{D}_1 = \mathfrak{Y}\mathfrak{X}^{-1}$. *Then* Σ *and* Σ_1 *input/output stabilize each other if and only if* $\left[\begin{smallmatrix} \mathfrak{X} & \mathfrak{N} \\ \mathfrak{Y} & \mathfrak{M} \end{smallmatrix}\right]$ *has an inverse in* $TIC(\left[\begin{smallmatrix} Y \\ U \end{smallmatrix}\right])$. *If this is the case, then by defining*

$$\begin{bmatrix} \widetilde{\mathfrak{M}} & -\widetilde{\mathfrak{N}} \\ -\widetilde{\mathfrak{Y}} & \widetilde{\mathfrak{X}} \end{bmatrix} = \begin{bmatrix} \mathfrak{X} & \mathfrak{N} \\ \mathfrak{Y} & \mathfrak{M} \end{bmatrix}^{-1}$$

we get a joint doubly coprime factorization of $\mathfrak{D}$ *and* $\mathfrak{D}_1$ *(in particular,* $\mathfrak{D} = \widetilde{\mathfrak{M}}^{-1}\widetilde{\mathfrak{N}}$ *and* $\mathfrak{D}_1 = \widetilde{\mathfrak{X}}^{-1}\widetilde{\mathfrak{Y}}$*). Moreover, the input/output map of the* *closed-loop system is given by*

$$\begin{bmatrix} 1 & -\mathfrak{D} \\ -\mathfrak{D}_1 & 1 \end{bmatrix}^{-1} - 1 = \begin{bmatrix} 0 & \mathfrak{N} \\ \mathfrak{Y} & 0 \end{bmatrix} \begin{bmatrix} \widetilde{\mathfrak{M}} & \widetilde{\mathfrak{N}} \\ \widetilde{\mathfrak{Y}} & \widetilde{\mathfrak{X}} \end{bmatrix}$$
$$= \begin{bmatrix} \mathfrak{X} & \mathfrak{N} \\ \mathfrak{Y} & \mathfrak{M} \end{bmatrix} \begin{bmatrix} 0 & \widetilde{\mathfrak{N}} \\ \widetilde{\mathfrak{Y}} & 0 \end{bmatrix}.$$

(ii) *Suppose that* $\mathfrak{D}$ *and* $\mathfrak{D}_1$ *have left coprime fractions* $\mathfrak{D} = \widetilde{\mathfrak{M}}^{-1}\widetilde{\mathfrak{N}}$ *and* $\mathfrak{D}_1 = \widetilde{\mathfrak{X}}^{-1}\widetilde{\mathfrak{Y}}$. *Then* Σ *and* Σ_1 *input/output stabilize each other if and only if* $\left[\begin{smallmatrix} \widetilde{\mathfrak{M}} & -\widetilde{\mathfrak{N}} \\ -\widetilde{\mathfrak{Y}} & \widetilde{\mathfrak{X}} \end{smallmatrix}\right]$ *has an inverse in* $TIC(\left[\begin{smallmatrix} Y \\ U \end{smallmatrix}\right])$. *If this is the case, then by defining*

$$\begin{bmatrix} \mathfrak{X} & \mathfrak{N} \\ \mathfrak{Y} & \mathfrak{M} \end{bmatrix} = \begin{bmatrix} \widetilde{\mathfrak{M}} & -\widetilde{\mathfrak{N}} \\ -\widetilde{\mathfrak{Y}} & \widetilde{\mathfrak{X}} \end{bmatrix}^{-1}$$

we get a joint doubly coprime factorization of $\mathfrak{D}$ *and* $\mathfrak{D}_1$, *and all the additional conclusions listed in (i) hold.*

Thus, if the two systems input/output stabilize each other, and both the input/output maps have coprime fractions from the same side, then they both have doubly coprime factorizations (and we have a formula for how to compute these factorizations).

Proof (i) By Example 7.2.5 and Lemma 8.4.3 with $\mathfrak{D}$, $\mathfrak{N}$, $\mathfrak{M}$, and K replaced by $\left[\begin{smallmatrix} 0 & \mathfrak{D} \\ \mathfrak{D}_1 & 0 \end{smallmatrix}\right]$, $\left[\begin{smallmatrix} 0 & \mathfrak{N} \\ \mathfrak{Y} & 0 \end{smallmatrix}\right]$, $\left[\begin{smallmatrix} \mathfrak{X} & 0 \\ 0 & \mathfrak{M} \end{smallmatrix}\right]$, and 1, respectively, the two systems input/output stabilize each other if and only if

$$\begin{bmatrix} \mathfrak{X} & 0 \\ 0 & \mathfrak{M} \end{bmatrix} - \begin{bmatrix} 0 & \mathfrak{N} \\ \mathfrak{Y} & 0 \end{bmatrix} = \begin{bmatrix} \mathfrak{X} & -\mathfrak{N} \\ -\mathfrak{Y} & \mathfrak{M} \end{bmatrix} = \begin{bmatrix} 1 & 0 \\ 0 & -1 \end{bmatrix} \begin{bmatrix} \mathfrak{X} & \mathfrak{N} \\ \mathfrak{Y} & \mathfrak{M} \end{bmatrix} \begin{bmatrix} 1 & 0 \\ 0 & -1 \end{bmatrix}$$

is invertible in $TIC(\left[\begin{smallmatrix} Y \\ U \end{smallmatrix}\right])$. This is equivalent to the invertibility of $\left[\begin{smallmatrix} \mathfrak{X} & \mathfrak{N} \\ \mathfrak{Y} & \mathfrak{M} \end{smallmatrix}\right]$. Also the two formulas for the closed-loop input/output map are found in Lemma 8.4.3. We leave the easy verification of the fact that the given factorization is a joint doubly coprime factorization of $\mathfrak{D}$ and $\mathfrak{D}_1$ to the reader.

(ii) This proof is analogous to the proof of (i). □

Proposition 8.5.6 *Let* $\Sigma = \left[\begin{smallmatrix} \mathfrak{A} & \mathfrak{B} \\ \mathfrak{C} & \mathfrak{D} \end{smallmatrix}\right]$ *be a well-posed linear system on* (Y, X, U), *and let* $\Sigma_1 = \left[\begin{smallmatrix} \mathfrak{A}_1 & \mathfrak{B}_1 \\ \mathfrak{C}_1 & \mathfrak{D}_1 \end{smallmatrix}\right]$ *be a well-posed linear system on* (U, X_1, Y).

(i) *Suppose that* $\mathfrak{D}$ *has a right coprime fraction* $\mathfrak{D} = \mathfrak{N}\mathfrak{M}^{-1}$. *Then the following conditions are equivalent.*
   (a) Σ *and* Σ_1 *input/output stabilize each other.*
   (b) $\mathfrak{D}_1$ *has a left coprime fraction* $\mathfrak{D}_1 = \widetilde{\mathfrak{X}}^{-1}\widetilde{\mathfrak{Y}}$ *satisfying* $\widetilde{\mathfrak{X}}\mathfrak{M} - \widetilde{\mathfrak{Y}}\mathfrak{N} = 1$.

(c) $\mathfrak{D}_1$ *has a left coprime fraction* $\mathfrak{D}_1 = \widetilde{\mathfrak{X}}^{-1}\widetilde{\mathfrak{Y}}$ *for which* $\widetilde{\mathfrak{X}}\mathfrak{M} - \widetilde{\mathfrak{Y}}\mathfrak{N}$ *is invertible in TIC*(U).

If $\widetilde{\mathfrak{X}}$ *and* $\widetilde{\mathfrak{Y}}$ *are chosen as in (c), then the closed-loop input/output map is given by*

$$\begin{bmatrix} 1 & -\mathfrak{D} \\ -\mathfrak{D}_1 & 1 \end{bmatrix}^{-1} - 1 = \begin{bmatrix} \mathfrak{N} \\ \mathfrak{M} \end{bmatrix} (\widetilde{\mathfrak{X}}\mathfrak{M} - \widetilde{\mathfrak{Y}}\mathfrak{N})^{-1} \begin{bmatrix} \widetilde{\mathfrak{Y}} & \widetilde{\mathfrak{X}} \end{bmatrix} - \begin{bmatrix} 0 & 0 \\ 0 & 1 \end{bmatrix}.$$

The same formula is valid with $(\widetilde{\mathfrak{X}}\mathfrak{M} - \widetilde{\mathfrak{Y}}\mathfrak{N})^{-1} = 1$ *if* $\widetilde{\mathfrak{X}}$ *and* $\widetilde{\mathfrak{Y}}$ *are chosen as in (b).*

(ii) *Suppose that* $\mathfrak{D}$ *has a left coprime fraction* $\mathfrak{D} = \widetilde{\mathfrak{M}}^{-1}\widetilde{\mathfrak{N}}$. *Then the following conditions are equivalent.*

(a) Σ *and* Σ_1 *input/output stabilize each other.*

(b) $\mathfrak{D}_1$ *has a right coprime fraction* $\mathfrak{D}_1 = \mathfrak{Y}\mathfrak{X}^{-1}$ *satisfying* $\widetilde{\mathfrak{M}}\mathfrak{X} - \widetilde{\mathfrak{N}}\mathfrak{Y} = 1$.

(c) $\mathfrak{D}_1$ *has a right coprime fraction* $\mathfrak{D}_1 = \mathfrak{Y}\mathfrak{X}^{-1}$ *for which* $\widetilde{\mathfrak{M}}\mathfrak{X} - \widetilde{\mathfrak{N}}\mathfrak{Y}$ *is invertible in TIC*(Y).

If $\mathfrak{Y}$ *and* $\mathfrak{X}$ *are chosen as in (c), then the closed-loop input/output map is given by*

$$\begin{bmatrix} 1 & -\mathfrak{D} \\ -\mathfrak{D}_1 & 1 \end{bmatrix}^{-1} - 1 = \begin{bmatrix} \mathfrak{X} \\ \mathfrak{Y} \end{bmatrix} (\widetilde{\mathfrak{M}}\mathfrak{X} - \widetilde{\mathfrak{N}}\mathfrak{Y})^{-1} \begin{bmatrix} \widetilde{\mathfrak{M}} & \widetilde{\mathfrak{N}} \end{bmatrix} - \begin{bmatrix} 1 & 0 \\ 0 & 0 \end{bmatrix}.$$

The same formula is valid with $(\widetilde{\mathfrak{M}}\mathfrak{X} - \widetilde{\mathfrak{N}}\mathfrak{Y})^{-1} = 1$ *if* $\mathfrak{Y}$ *and* $\mathfrak{X}$ *are chosen as in (b).*

(iii) *Suppose that* Σ *and* Σ_1 *input/output stabilize each other. Then*

(a) Σ *has a right coprime fraction if and only if* Σ_1 *has a left coprime fraction,*

(b) Σ *has a left coprime fraction if and only if* Σ_1 *has a right coprime fraction,*

and the following conditions are equivalent:

(c) Σ *has a doubly coprime factorization;*

(d) Σ_1 *has a doubly coprime factorization;*

(e) $\mathfrak{D}$ *and* $\mathfrak{D}_1$ *have a joint doubly coprime factorization;*

(f) *both* Σ *and* Σ_1 *have right coprime fractions;*

(g) *both* Σ *and* Σ_1 *have left coprime fractions.*

Proof (i) (a) $\Rightarrow$ (b): Suppose that the two systems input/output stabilize each other. Then the closed-loop input/output map belongs to $TIC(\left[\begin{smallmatrix} Y \\ U \end{smallmatrix}\right])$. By

Example 7.2.5 and Lemma A.4.1, this input/output map is given by

$$\begin{bmatrix} \mathfrak{D}(1-\mathfrak{D}_1\mathfrak{D})^{-1}\mathfrak{D}_1 & \mathfrak{D}(1-\mathfrak{D}_1\mathfrak{D})^{-1} \\ (1-\mathfrak{D}_1\mathfrak{D})^{-1}\mathfrak{D}_1 & (1-\mathfrak{D}_1\mathfrak{D})^{-1}-1 \end{bmatrix}$$
$$= \begin{bmatrix} \mathfrak{N}(\mathfrak{M}-\mathfrak{D}_1\mathfrak{N})^{-1}\mathfrak{D}_1 & \mathfrak{N}(\mathfrak{M}-\mathfrak{D}_1\mathfrak{N})^{-1} \\ \mathfrak{M}(\mathfrak{M}-\mathfrak{D}_1\mathfrak{N})^{-1}\mathfrak{D}_1 & \mathfrak{M}(\mathfrak{M}-\mathfrak{D}_1\mathfrak{N})^{-1}-1 \end{bmatrix}.$$

Thus, all the input/output maps listed above belong to *TIC* (over the appropriate spaces). By the coprimeness assumption, $\left[\begin{smallmatrix} \mathfrak{N} \\ \mathfrak{M} \end{smallmatrix}\right]$ has a left inverse in $TIC(\left[\begin{smallmatrix} Y \\ U \end{smallmatrix}\right]; U)$, hence $(\mathfrak{M}-\mathfrak{D}_1\mathfrak{N})^{-1}\mathfrak{D}_1 \in TIC(Y;U)$ and $(\mathfrak{M}-\mathfrak{D}_1\mathfrak{N})^{-1} \in TIC(U)$. Define $\widetilde{\mathfrak{X}} = (\mathfrak{M}-\mathfrak{D}_1\mathfrak{N})^{-1}$ and $\widetilde{\mathfrak{Y}} = (\mathfrak{M}-\mathfrak{D}_1\mathfrak{N})^{-1}\mathfrak{D}_1$. Then $\mathfrak{D}_1 = \widetilde{\mathfrak{X}}^{-1}\widetilde{\mathfrak{Y}}$, the closed-loop input/output map is given by $\left[\begin{smallmatrix} \mathfrak{N}\widetilde{\mathfrak{Y}} & \mathfrak{N}\widetilde{\mathfrak{X}} \\ \mathfrak{M}\widetilde{\mathfrak{Y}} & \mathfrak{M}\widetilde{\mathfrak{X}}-1 \end{smallmatrix}\right]$, and

$$\widetilde{\mathfrak{X}}\mathfrak{M} - \widetilde{\mathfrak{Y}}\mathfrak{N} = (\mathfrak{M}-\mathfrak{D}_1\mathfrak{N})^{-1}(\mathfrak{M}-\mathfrak{D}_1\mathfrak{N}) = 1;$$

in particular, $\widetilde{\mathfrak{X}}$ and $\widetilde{\mathfrak{Y}}$ are left coprime.

(b) ⇒ (a): Suppose that $\mathfrak{D}_1$ has a left coprime fraction $\mathfrak{D}_1 = \widetilde{\mathfrak{X}}^{-1}\widetilde{\mathfrak{Y}}$ such that $\widetilde{\mathfrak{X}}\mathfrak{M} - \widetilde{\mathfrak{Y}}\mathfrak{N} = 1$. Then $1 - \mathfrak{D}_1\mathfrak{D} = \widetilde{\mathfrak{X}}^{-1}(\widetilde{\mathfrak{X}}\mathfrak{M} - \widetilde{\mathfrak{Y}}\mathfrak{N})\mathfrak{M}^{-1} = \widetilde{\mathfrak{X}}^{-1}\mathfrak{M}^{-1}$, hence the closed-loop input/output map is given by

$$\begin{bmatrix} \mathfrak{D}(1-\mathfrak{D}_1\mathfrak{D})^{-1}\mathfrak{D}_1 & \mathfrak{D}(1-\mathfrak{D}_1\mathfrak{D})^{-1} \\ (1-\mathfrak{D}_1\mathfrak{D})^{-1}\mathfrak{D}_1 & (1-\mathfrak{D}_1\mathfrak{D})^{-1}-1 \end{bmatrix} = \begin{bmatrix} \mathfrak{N}\widetilde{\mathfrak{Y}} & \mathfrak{N}\widetilde{\mathfrak{X}} \\ \mathfrak{M}\widetilde{\mathfrak{Y}} & \mathfrak{M}\widetilde{\mathfrak{X}}-1 \end{bmatrix}.$$

Thus, the closed-loop system is input/output stable and all the additional claims are true.

(b) ⇔ (c): Trivially (b) ⇒ (c). If $\mathfrak{D}_1 = \widetilde{\mathfrak{X}}^{-1}\widetilde{\mathfrak{Y}}$ and $\mathfrak{U} := \widetilde{\mathfrak{X}}\mathfrak{M} - \widetilde{\mathfrak{Y}}\mathfrak{N}$ is invertible in $TIC(U)$, then, by Lemma 8.3.9(ii)(b), $\widetilde{\mathfrak{X}}^{-1}\widetilde{\mathfrak{Y}} = (\mathfrak{U}^{-1}\widetilde{\mathfrak{X}})^{-1}(\mathfrak{U}^{-1}\widetilde{\mathfrak{Y}})$ is another left coprime fraction of $\mathfrak{D}_1$ satisfying $\widetilde{\mathfrak{X}}\mathfrak{M} - \widetilde{\mathfrak{Y}}\mathfrak{N} = 1$.

The formula for the input/output map with $\widetilde{\mathfrak{X}}\mathfrak{M} - \widetilde{\mathfrak{Y}}\mathfrak{N} = 1$ was part of the proof of the equivalence (a) ⇔ (b), and this together with the proof of the implication (c) ⇒ (b) gives the general formula.

(ii) We leave this proof to the reader (it is analogous to the proof of (i).

(iii) See (i), (ii), and Lemmas 8.3.10 and 8.5.5. □

In the case where $\mathfrak{D}$ has a doubly coprime factorization we can develop Proposition 8.5.6 further.

Theorem 8.5.7 *Let $\Sigma = \left[\begin{smallmatrix} \mathfrak{A} & \mathfrak{B} \\ \mathfrak{C} & \mathfrak{D} \end{smallmatrix}\right]$ be a well-posed linear system on (Y, X, U), let $\Sigma_1 = \left[\begin{smallmatrix} \mathfrak{A}_1 & \mathfrak{B}_1 \\ \mathfrak{C}_1 & \mathfrak{D}_1 \end{smallmatrix}\right]$ be a well-posed linear system on (U, X_1, Y), and suppose that $\mathfrak{D}$ has a doubly coprime factorization* (8.3.3)*. Then the following conditions are equivalent:*

(i) *Σ and Σ_1 input/output stabilize each other.*
(ii) *$\mathfrak{D}_1$ has a right coprime fraction $\mathfrak{D}_1 = \mathfrak{T}\mathfrak{S}^{-1}$ satisfying $\widetilde{\mathfrak{M}}\mathfrak{S} - \widetilde{\mathfrak{N}}\mathfrak{T} = 1$.*

(iii) $\mathfrak{D}_1$ *has a left coprime fraction* $\mathfrak{D}_1 = \widetilde{\mathfrak{S}}^{-1}\widetilde{\mathfrak{T}}$ *satisfying* $\widetilde{\mathfrak{S}}\mathfrak{M} - \widetilde{\mathfrak{T}}\mathfrak{N} = 1$.

(iv) $\mathfrak{D}$ *and* $\mathfrak{D}_1$ *have a joint doubly coprime factorization* $\begin{bmatrix} \mathfrak{S} & \mathfrak{N} \\ \mathfrak{T} & \mathfrak{M} \end{bmatrix}^{-1} = \begin{bmatrix} \widetilde{\mathfrak{M}} & -\widetilde{\mathfrak{N}} \\ -\widetilde{\mathfrak{T}} & \widetilde{\mathfrak{S}} \end{bmatrix}$.

(v) $\mathfrak{D}_1$ *has a right coprime fraction* $\mathfrak{T}\mathfrak{S}^{-1}$ *of the type*

$$\begin{bmatrix} \mathfrak{S} \\ \mathfrak{T} \end{bmatrix} = \begin{bmatrix} \mathfrak{X} & \mathfrak{N} \\ \mathfrak{Y} & \mathfrak{M} \end{bmatrix} \begin{bmatrix} 1 \\ \mathfrak{Q} \end{bmatrix} = \begin{bmatrix} \mathfrak{X} + \mathfrak{N}\mathfrak{Q} \\ \mathfrak{Y} + \mathfrak{M}\mathfrak{Q} \end{bmatrix},$$

for some $\mathfrak{Q} \in TIC(Y; U)$.

(vi) $\mathfrak{D}_1$ *has a left coprime fraction* $\widetilde{\mathfrak{S}}^{-1}\widetilde{\mathfrak{T}}$ *of the type*

$$\begin{bmatrix} \widetilde{\mathfrak{T}} & \widetilde{\mathfrak{S}} \end{bmatrix} = \begin{bmatrix} \mathfrak{Q} & 1 \end{bmatrix} \begin{bmatrix} \widetilde{\mathfrak{M}} & \widetilde{\mathfrak{N}} \\ \widetilde{\mathfrak{Y}} & \widetilde{\mathfrak{X}} \end{bmatrix} = \begin{bmatrix} \widetilde{\mathfrak{Y}} + \mathfrak{Q}\widetilde{\mathfrak{M}} & \widetilde{\mathfrak{X}} + \mathfrak{Q}\widetilde{\mathfrak{N}} \end{bmatrix},$$

for some $\mathfrak{Q} \in TIC(Y; U)$.

All the operators appearing in (ii)–(vi) are unique (i.e., they are determined uniquely by the factorization (8.3.3) and $\mathfrak{D}_1$*), the operators* $\mathfrak{S}$ *and* $\mathfrak{T}$ *in (ii), (iv), and (v) are the same, the operators* $\widetilde{\mathfrak{S}}$ *and* $\widetilde{\mathfrak{T}}$ *in (iii), (iv), and (vi) are the same, and the operators* $\mathfrak{Q}$ *in (v) and (vi) are the same. In particular,*

$$\begin{aligned} \mathfrak{D}_1 &= (\mathfrak{Y} + \mathfrak{M}\mathfrak{Q})(\mathfrak{X} + \mathfrak{N}\mathfrak{Q})^{-1} \\ &= (\widetilde{\mathfrak{X}} + \mathfrak{Q}\widetilde{\mathfrak{N}})^{-1}(\widetilde{\mathfrak{Y}} + \mathfrak{Q}\widetilde{\mathfrak{M}}). \end{aligned} \tag{8.5.1}$$

The closed-loop input/output map is given by

$$\begin{aligned} \begin{bmatrix} 1 & -\mathfrak{D} \\ -\mathfrak{D}_1 & 1 \end{bmatrix}^{-1} - 1 &= \begin{bmatrix} \mathfrak{N} \\ \mathfrak{M} \end{bmatrix} \begin{bmatrix} \widetilde{\mathfrak{T}} & \widetilde{\mathfrak{S}} \end{bmatrix} - \begin{bmatrix} 0 & 0 \\ 0 & 1 \end{bmatrix} \\ &= \begin{bmatrix} \mathfrak{S} \\ \mathfrak{T} \end{bmatrix} \begin{bmatrix} \widetilde{\mathfrak{M}} & \widetilde{\mathfrak{N}} \end{bmatrix} - \begin{bmatrix} 1 & 0 \\ 0 & 0 \end{bmatrix} \\ &= \begin{bmatrix} \mathfrak{N} \\ \mathfrak{M} \end{bmatrix} \begin{bmatrix} \widetilde{\mathfrak{Y}} & \widetilde{\mathfrak{X}} \end{bmatrix} + \begin{bmatrix} \mathfrak{N} \\ \mathfrak{M} \end{bmatrix} \mathfrak{Q} \begin{bmatrix} \widetilde{\mathfrak{M}} & \widetilde{\mathfrak{N}} \end{bmatrix} - \begin{bmatrix} 0 & 0 \\ 0 & 1 \end{bmatrix} \\ &= \begin{bmatrix} \mathfrak{X} \\ \mathfrak{Y} \end{bmatrix} \begin{bmatrix} \widetilde{\mathfrak{M}} & \widetilde{\mathfrak{N}} \end{bmatrix} + \begin{bmatrix} \mathfrak{N} \\ \mathfrak{M} \end{bmatrix} \mathfrak{Q} \begin{bmatrix} \widetilde{\mathfrak{M}} & \widetilde{\mathfrak{N}} \end{bmatrix} - \begin{bmatrix} 1 & 0 \\ 0 & 0 \end{bmatrix}. \end{aligned} \tag{8.5.2}$$

Recall that the conditions in (ii)–(vi) contain the implicit assumptions that the denominators can be inverted in $TIC_\omega(U)$ or $TIC_\omega(Y)$ for some $\omega \geq 0$ (how large ω must be depends on the growth bounds of $\mathfrak{D}$ and $\mathfrak{D}_1$; cf. Lemma 8.3.3).

Proof (i) ⇔ (ii) ⇔ (iii): See Proposition 8.5.6(i)–(ii).

(i) ⇔ (iv): See Lemma 8.5.5 and Proposition 8.5.6(iii).

(iv) ⇒ (v) and (iv) ⇒ (vi): By Lemma 8.3.11(i), all possible choices of $\left[\begin{smallmatrix}\mathfrak{S}\\ \mathfrak{T}\end{smallmatrix}\right]$ and $\left[\widetilde{\mathfrak{T}}\ \widetilde{\mathfrak{S}}\right]$ in (iv) are of the form

$$\begin{aligned}\begin{bmatrix}\mathfrak{S}\\ \mathfrak{T}\end{bmatrix} &= \begin{bmatrix}\mathfrak{X} & \mathfrak{N}\\ \mathfrak{Y} & \mathfrak{M}\end{bmatrix}\begin{bmatrix}1\\ \mathfrak{Q}\end{bmatrix} = \begin{bmatrix}\mathfrak{X}+\mathfrak{N}\mathfrak{Q}\\ \mathfrak{Y}+\mathfrak{M}\mathfrak{Q}\end{bmatrix},\\ \left[\widetilde{\mathfrak{T}}\ \widetilde{\mathfrak{S}}\right] &= \left[\mathfrak{Q}\ \ 1\right]\begin{bmatrix}\widetilde{\mathfrak{M}} & \widetilde{\mathfrak{N}}\\ \widetilde{\mathfrak{Y}} & \widetilde{\mathfrak{X}}\end{bmatrix} = \left[\widetilde{\mathfrak{Y}}+\mathfrak{Q}\widetilde{\mathfrak{M}}\ \ \widetilde{\mathfrak{X}}+\mathfrak{Q}\widetilde{\mathfrak{N}}\right],\end{aligned}$$

where $\mathfrak{Q}\in TIC(Y;U)$. In addition (iv) contains the requirement that $\mathfrak{T}\mathfrak{S}^{-1}$ is a right and $\widetilde{\mathfrak{S}}^{-1}\widetilde{\mathfrak{T}}$ a left coprime fraction of $\mathfrak{D}_1$. Thus, (iv) implies both (v) and (vi) (with the same value of the parameter $\mathfrak{Q}$ in (v) and (vi)).

(v) ⇒ (iv): Choose $\left[\begin{smallmatrix}\mathfrak{S}\\ \mathfrak{T}\end{smallmatrix}\right]$ as in (v). Then, by Lemma 8.3.11(i), $\left[\begin{smallmatrix}\mathfrak{S} & \mathfrak{N}\\ \mathfrak{T} & \mathfrak{M}\end{smallmatrix}\right]$ is invertible in $TIC(\left[\begin{smallmatrix}Y\\ U\end{smallmatrix}\right])$, and $\left[\begin{smallmatrix}\mathfrak{S} & \mathfrak{N}\\ \mathfrak{T} & \mathfrak{M}\end{smallmatrix}\right]^{-1} = \left[\begin{smallmatrix}\widetilde{\mathfrak{M}} & -\widetilde{\mathfrak{N}}\\ -\widetilde{\mathfrak{T}} & \widetilde{\mathfrak{S}}\end{smallmatrix}\right]$, where $\left[\widetilde{\mathfrak{T}}\ \widetilde{\mathfrak{S}}\right]$ is given by the formula in (x) with the same parameter $\mathfrak{Q}$. According to Lemma 8.3.3(vi), this is a doubly coprime factorization of $\mathfrak{D}$ and $\mathfrak{D}_1$.

(vi) ⇒ (iv): This proof is similar to the one above.

The uniqueness of the fractions and factorizations follows from Lemma 8.3.3(vii), as does the claim that we get the same operators $\mathfrak{S}$, $\mathfrak{T}$, $\widetilde{\mathfrak{S}}$, and $\widetilde{\mathfrak{T}}$ in all cases. The uniqueness of $\mathfrak{Q}$ follows from the uniqueness of the other operators and the fact that $\mathfrak{Q}=\mathfrak{M}^{-1}(\mathfrak{T}-\mathfrak{N}) = (\widetilde{\mathfrak{T}}-\widetilde{\mathfrak{N}})\widetilde{\mathfrak{M}}^{-1}$, and we saw in the proof of the implications (iv) ⇒ (v) and (iv) ⇒ (vi) that we get the same parameter $\mathfrak{Q}$ in (v) as in (vi). Finally, we get (8.5.2) from Proposition 8.5.6. □

Theorem 8.5.7 immediately implies the following result:

Corollary 8.5.8 *Let* $\Sigma = \left[\begin{smallmatrix}\mathfrak{A} & \mathfrak{B}\\ \mathfrak{C} & \mathfrak{D}\end{smallmatrix}\right]$ *be a well-posed linear system on* (Y, X, U), *and suppose that* $\mathfrak{D}$ *has a doubly coprime factorization* (8.3.3). *Then the following conditions are equivalent:*[7]

(i) Σ *is input/output stabilizable by dynamic feedback.*
(ii) *The left Bezout identity* $\widetilde{\mathfrak{M}}\mathfrak{S}-\widetilde{\mathfrak{N}}\mathfrak{T}=1$ *has a solution pair* $\mathfrak{S}$ *and* $\mathfrak{T}$ *where* $\mathfrak{S}$ *has an inverse in* $TIC_\omega(Y)$ *for some* $\omega\ge 0$.
(iii) *The right Bezout identity* $\widetilde{\mathfrak{S}}\mathfrak{M}-\widetilde{\mathfrak{T}}\mathfrak{N}=I$ *has a solution pair* $\widetilde{\mathfrak{S}}$ *and* $\widetilde{\mathfrak{T}}$ *where* $\widetilde{\mathfrak{S}}$ *has an inverse in* $TIC_\omega(U)$ *for some* $\omega\ge 0$.
(iv) $\mathfrak{D}$ *has a doubly coprime factorization* $\left[\begin{smallmatrix}\mathfrak{S} & \mathfrak{N}\\ \mathfrak{T} & \mathfrak{M}\end{smallmatrix}\right]^{-1} = \left[\begin{smallmatrix}\widetilde{\mathfrak{M}} & -\widetilde{\mathfrak{N}}\\ -\widetilde{\mathfrak{T}} & \widetilde{\mathfrak{S}}\end{smallmatrix}\right]$, *where* $\mathfrak{S}$ *has an inverse in* $TIC_\omega(Y)$ *and* $\widetilde{\mathfrak{S}}$ *has an inverse in* $TIC_\omega(U)$ *for some* $\omega\ge 0$.

[7] In the *Reg*-well-posed case we assume throughout that all the operators appearing in this theorem, including those in (8.3.3), have *Reg*-well-posed realizations. See Theorem 2.6.7.

(v) *For some* $\mathfrak{Q} \in TIC(Y;U)$ *and some* $\omega \geq 0$, $\mathfrak{X} + \mathfrak{N}\mathfrak{Q}$ *is invertible in* $TIC_\omega(Y)$.

(vi) *For some* $\mathfrak{Q} \in TIC(Y;U)$ *and some* $\omega \geq 0$, $\widetilde{\mathfrak{X}} + \mathfrak{Q}\widetilde{\mathfrak{N}}$ *is invertible in* $TIC_\omega(U)$.

When these conditions hold, then the set of all input/output maps $\mathfrak{D}_1$ *of the dynamically input/output stabilizing systems are parametrized by* (8.5.1), *where* $\mathfrak{Q}$ *ranges over those* $\mathfrak{Q} \in TIC(Y;U)$ *for which* $\mathfrak{X} + \mathfrak{N}\mathfrak{Q}$ *is invertible in* $TIC_\omega(Y)$, *or equivalently,* $\widetilde{\mathfrak{X}} + \mathfrak{Q}\widetilde{\mathfrak{N}}$ *is invertible in* $TIC_\omega(Y)$, *for some* $\omega \geq 0$. *The corresponding closed-loop input/output maps are given by* (8.5.2).

The parametrization of all possible stabilizing input/output maps $\mathfrak{D}_1$ in (8.5.1) in terms of the parameter $\mathfrak{Q} \in TIC(Y;U)$ is usually referred to as the *Youla parametrization*. Observe, in particular, that the *closed-loop input/output map is affine in the Youla parameter* $\mathfrak{Q}$.

Proof By Theorem 8.5.7, condition (i) implies all the others. To go in the opposite directions we build a realization of the appropriate input/output map $\mathfrak{D}_1$, given by one of the expressions $\mathfrak{T}\mathfrak{S}^{-1}$, $\mathfrak{S}^{-1}\widetilde{\mathfrak{T}}$, $(\mathfrak{Y} + \mathfrak{M}\mathfrak{Q})(\mathfrak{X} + \mathfrak{N}\mathfrak{Q})^{-1}$, or $(\widetilde{\mathfrak{X}} + \mathfrak{Q}\widetilde{\mathfrak{N}})^{-1}(\widetilde{\mathfrak{Y}} + \mathfrak{Q}\widetilde{\mathfrak{M}})$, and apply Theorem 8.5.7 in the opposite direction. □

In principle it does not matter which doubly coprime factorization of $\mathfrak{D}$ we use in Theorem 8.5.7 and Corollary 8.5.8. However, for computational reasons, it makes sense to use a factorization which is a simple as possible. If the original system is jointly stabilizable and detectable, then Theorem 8.4.1 provides us with a realization of a particular doubly coprime factorization whose state space is the same as the state space of the original system (in particular, it has the same dimension). Below we shall describe this situation in more detail. We begin with the special case where the Youla parameter $\mathfrak{Q}$ in Theorem 8.5.7 and Corollary 8.5.8 is zero, a case which is interesting in its own right. The corresponding stabilizing system Σ_1 is often referred to as a *central controller*.

Lemma 8.5.9 *Let* $\Sigma = \left[\begin{smallmatrix} \mathfrak{A} & \mathfrak{B} \\ \mathfrak{C} & \mathfrak{D} \end{smallmatrix}\right]$ *be a jointly stabilizable and detectable well-posed linear system, and let* Σ_{ext}, $\Sigma_\times$, *and* $\Sigma^\times$ *denote the systems in Definition 8.2.1 and Lemma 8.2.2. Then the following conditions are equivalent:*

(i) *the operator* $\left[\begin{smallmatrix} -1 & 0 \\ 0 & 1 \end{smallmatrix}\right]$ *is an admissible output feedback operator for* Σ_{ext};

(ii) $1 + \mathfrak{G}_\times$ *has an inverse in* $TIC_\omega(Y)$ *for some* $\omega \geq 0$;

(iii) $1 - \mathfrak{F}^\times$ *has an inverse in* $TIC_\omega(U)$ *for some* $\omega \geq 0$.

In these cases the system $\Sigma^\times_\times$ *drawn in the upper half of Figure 8.6 with inputs* $\widetilde{w}^\times$ *and* $\tilde{u}_\times$ *and outputs* $\tilde{y}$, $\tilde{z}$, $\tilde{w}$ *and* $\tilde{u}$ *(i.e., the system that we get by using*

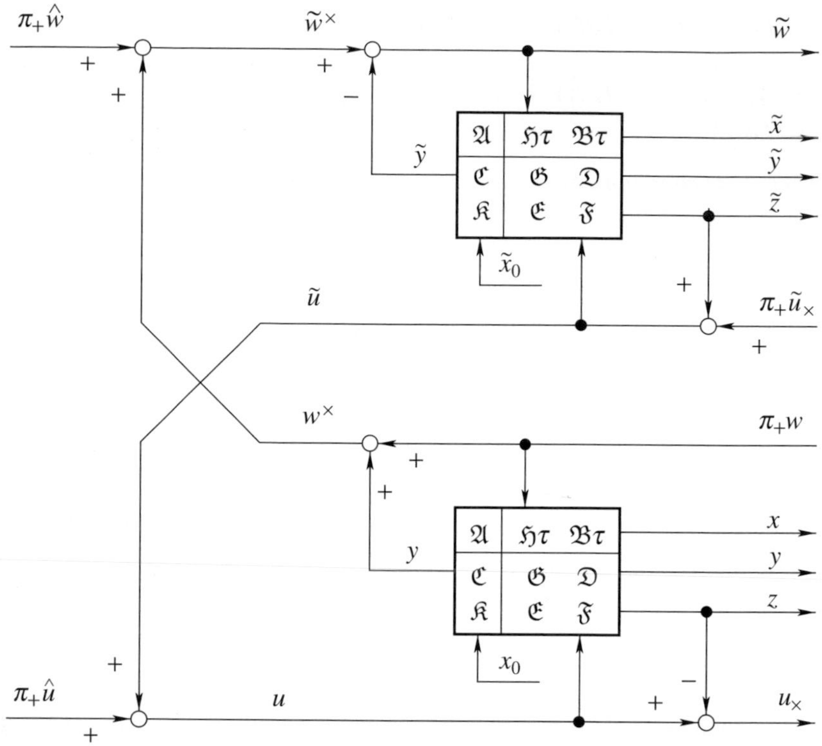

Figure 8.6 Dynamic stabilization

$\left[\begin{smallmatrix}-1 & 0\\ 0 & 1\end{smallmatrix}\right]$ *as an output feedback operator for* Σ_{ext}*) is a well-posed linear system, and the coprime fraction presented in Theorem 8.4.1 is a joint doubly coprime fraction of* $\mathfrak{D}$ *and* $\mathfrak{D}_1$*, where* $\mathfrak{D}_1 = \mathfrak{E}_\times(1 + \mathfrak{G}_\times)^{-1} = (1 - \mathfrak{F}^\times)^{-1}\mathfrak{E}^\times$.

This follows from Lemma 7.1.7.

In the situation described above the input/output map of the closed-loop system $\Sigma_\times^\times$ is equal to the stabilizing compensator $\mathfrak{D}_1$, and we can use the observer connection drawn in Figure 8.6 to stabilize the system:

Theorem 8.5.10 *Let* $\Sigma = \left[\begin{smallmatrix}\mathfrak{A} & \mathfrak{B}\\ \mathfrak{C} & \mathfrak{D}\end{smallmatrix}\right]$ *be a jointly stabilizable and detectable well-posed linear system, and let* $\Sigma_{\text{ext}} = \left[\begin{smallmatrix}\mathfrak{A} & \mathfrak{H} & \mathfrak{B}\\ \mathfrak{C} & \mathfrak{G} & \mathfrak{D}\\ \mathfrak{K} & \mathfrak{E} & \mathfrak{F}\end{smallmatrix}\right]$ *be the extended system in Definition 8.2.1. Then the connection drawn in Figure 8.6 defines a bounded well-posed linear system. This system is exponentially stable whenever one of the systems* $\Sigma_\times$ *or* $\Sigma^\times$ *in Lemma 8.2.2 is exponentially stable. In the* L^p*-case with* $p < \infty$*, this system is strongly stable whenever one of the systems* $\Sigma_\times$ *or* $\Sigma^\times$

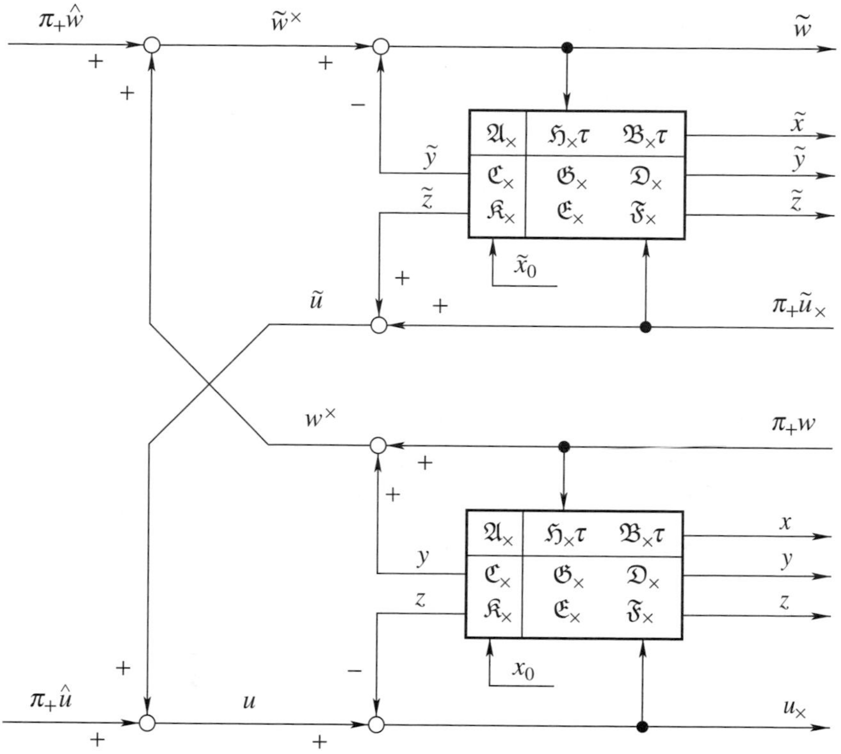

Figure 8.7 Equivalent version of dynamic stabilization

is strongly stable, and in the Reg-well-posed case it is strongly stable whenever both $\Sigma_\times$ *and* $\Sigma^\times$ *are strongly stable. Moreover, the two outputs* $\tilde{w}$ *and* $u_\times$ *are given by*

$$\begin{bmatrix} \tilde{w} \\ u_\times \end{bmatrix} = \begin{bmatrix} \pi_+ w \\ \pi_+ \tilde{u}_\times \end{bmatrix} + \begin{bmatrix} 1 - \mathfrak{G}^\times & \mathfrak{D}^\times \\ \mathfrak{E}^\times & 1 - \mathfrak{F}^\times \end{bmatrix} \begin{bmatrix} \pi_+ \widehat{w} \\ \pi_+ \hat{u} \end{bmatrix} + \begin{bmatrix} -\mathfrak{C}^\times \\ \mathfrak{K}^\times \end{bmatrix} (x_0 - \tilde{x}_0). \tag{8.5.3}$$

Proof By Remark 8.2.3, we can regard Σ_{ext} as a state feedback perturbed version of the closed-loop system $\Sigma_\times$; see Remark 8.2.3. By substituting this system for Σ_{ext} in Figure 8.6 we get the equivalent Figure 8.7, which can be interpreted as a feedback connection for a bounded system consisting of two copies of $\Sigma_\times$. By part (i) of Theorem 8.1.7, it suffices to show that the two internal inputs $u_\times$ and $\tilde{w}$ depend continuously in L^2_ω on the four inputs. The six summation junctions in Figure 8.7 are described by the equations

$$
\begin{aligned}
\widetilde{w}^\times &= \pi_+\widehat{w} + w^\times,\\
\tilde{w} &= \widetilde{w}^\times - \mathfrak{C}_\times \tilde{x}_0 - \mathfrak{G}_\times \tilde{w} - \mathfrak{D}_\times \pi_+ \tilde{u}_\times,\\
\tilde{u} &= \pi_+\tilde{u}_\times + \mathfrak{K}_\times \tilde{x}_0 + \mathfrak{E}_\times \tilde{w} + \mathfrak{F}_\times \pi_+ \tilde{u}_\times,\\
w^\times &= \pi_+ w + \mathfrak{C}_\times x_0 + \mathfrak{G}_\times \pi_+ w + \mathfrak{D}_\times u_\times,\\
u &= \tilde{u} + \pi_+\hat{u},\\
u_\times &= u - \mathfrak{K}_\times x_0 - \mathfrak{E}_\times \pi_+ w - \mathfrak{F}_\times u_\times.
\end{aligned}
$$

From here we eliminate the variables u, $\tilde{u}$, $w^\times$, and $\widetilde{w}^\times$ to get

$$
\begin{bmatrix} 1+\mathfrak{G}_\times & \mathfrak{D}_\times \\ \mathfrak{E}_\times & 1+\mathfrak{F}_\times \end{bmatrix}
\begin{bmatrix} \pi_+ w - \tilde{w} \\ u_\times - \pi_+\tilde{u}_\times \end{bmatrix}
= \begin{bmatrix} -\pi_+\widehat{w} \\ \pi_+\hat{u} \end{bmatrix}
- \begin{bmatrix} \mathfrak{C}_\times \\ \mathfrak{K}_\times \end{bmatrix} (x_0 - \tilde{x}_0).
$$

By Theorem 8.4.1, the operator on the left hand side has an inverse in TIC_ω. Inverting this operator we get the formula given in Theorem 8.5.10.

The extra claim about the exponential stability of the feedback system follows from Theorem 8.1.7(v)(j): since both $\Sigma_\times$ and $\Sigma^\times$ are bounded, if one of them is exponentially stable, then so is the other, hence so is the cross-product of $\Sigma_\times$ with itself, hence so is the system in Figure 8.7. The same argument applies to prove the strong stability in the L^p-well-posed case with $p < \infty$. In the *Reg*-well-posed case we instead appeal to Figure 8.9 (the argument given below shows that it is equivalent to Figure 8.6): the cascade connection of two strongly stable systems is always strongly stable (this is explained in more detail below). □

It is possible to derive (8.5.3) by using another argument which is conceptually more complicated, but which gives us a better understanding of the system drawn in Figure 8.6. Instead of interpreting that system as a feedback connection of two copies of $\Sigma_\times$ as we did in Figure 8.7 we may interpret it as a feedback connection of two copies of $\Sigma^\times$; see Figure 8.8. We then make a change of state variable by subtracting the bottom half of Figure 8.8 from the top half, replacing $\tilde{x}_0$ by $\tilde{x}_0 - x_0$, $\tilde{x}$ by $\tilde{x} - x$, $\tilde{y}$ by $\tilde{y} - y$, $\tilde{z}$ by $\tilde{z} - z$, etc. In particular, this means that we replace $\widetilde{w}^\times$ by $\widetilde{w}^\times - w^\times = \pi_+\widehat{w}$ and $\tilde{u}$ by $\tilde{u} - u = -\pi_+\hat{u}$. This has the effect of decoupling the upper part of Figure 8.8 from the lower part; cf. the upper part of Figure 8.9. There we have, in addition, changed the direction of the line below $\Sigma^\times$ (making $-\pi_+\hat{u}$ the input and $\pi_+\tilde{u}_\times - u_\times$ the output), changed the sign of this output, and added $\pi_+ w$ to $\tilde{w} - \pi_+ w$ and $\pi_+\tilde{u}_\times$ to $u_\times - \pi_+\tilde{u}_\times$ to recover $\tilde{w}$ and $u_\times$. Once we know $u_\times$ we can replace the lower half of Figure 8.8 by the equivalent lower half of Figure 8.7 and reverse the direction of the bottom line (making $u_\times$ the input and $\tilde{u}$ the output) to get Figure 8.9. The connection in Figure 8.9 is equivalent to the connections in

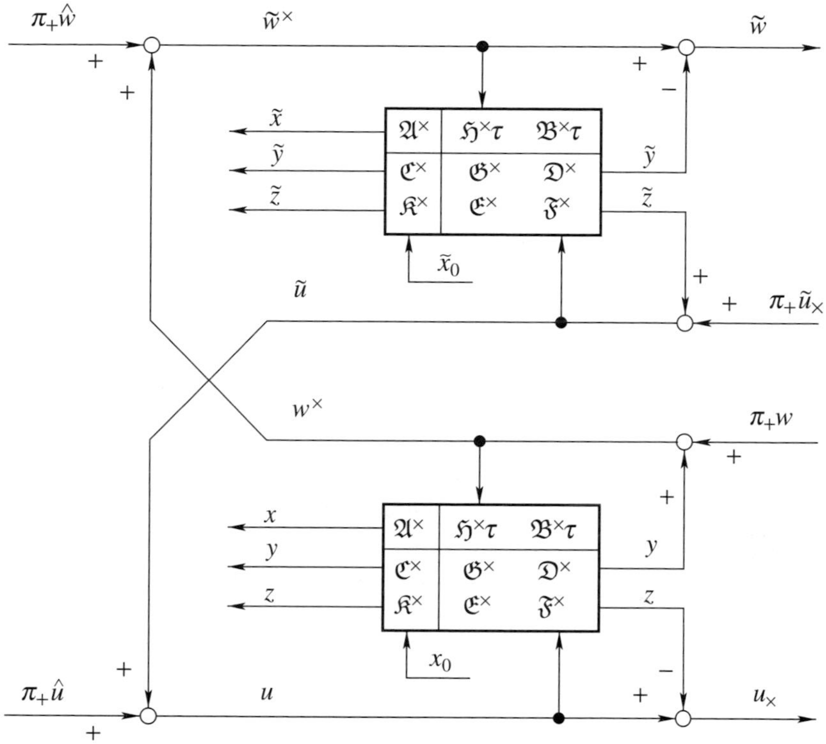

Figure 8.8 Second equivalent version of dynamic stabilization

Figures 8.6, 8.7, and 8.8, but it contains no feedback loops (it is a particular cascade connection of $\Sigma^\times$ and $\Sigma_\times$), hence it is automatically well posed and stable (and strongly stable if both $\Sigma^\times$ and $\Sigma_\times$ are strongly stable). From Figure 8.9 we can directly read off explicit formulas for the state and all the signals generated by the system, among others, the formulas for $\tilde{w}$ and $u_\times$ given in (8.5.3).

Theorem 8.5.11 *Let $\Sigma = \left[\begin{smallmatrix}\mathfrak{A} & \mathfrak{B}\\ \mathfrak{C} & \mathfrak{D}\end{smallmatrix}\right]$ be a jointly stabilizable and detectable well-posed linear system on (Y, X, U), let $\Sigma_{\text{ext}} = \left[\begin{smallmatrix}\mathfrak{A} & \mathfrak{H} & \mathfrak{B}\\ \mathfrak{C} & \mathfrak{G} & \mathfrak{D}\\ \mathfrak{K} & \mathfrak{E} & \mathfrak{F}\end{smallmatrix}\right]$ be the extended system in Definition 8.2.1, and let $\Sigma_\Omega = \left[\begin{smallmatrix}\mathfrak{A}_\Omega & \mathfrak{B}_\Omega\\ \mathfrak{C}_\Omega & \Omega\end{smallmatrix}\right]$ be a stable well-posed linear system on (U, X, Y). Then the connection drawn in Figure 8.10 defines a bounded well-posed linear system. This system is strongly stable or exponentially stable whenever Σ_Ω, $\Sigma_\times$, and $\Sigma^\times$ are strongly or exponentially stable (here $\Sigma_\times$ and $\Sigma^\times$ are the systems in Lemma 8.2.2). Moreover, the two outputs $\tilde{w}$ and $u_\times$ are*

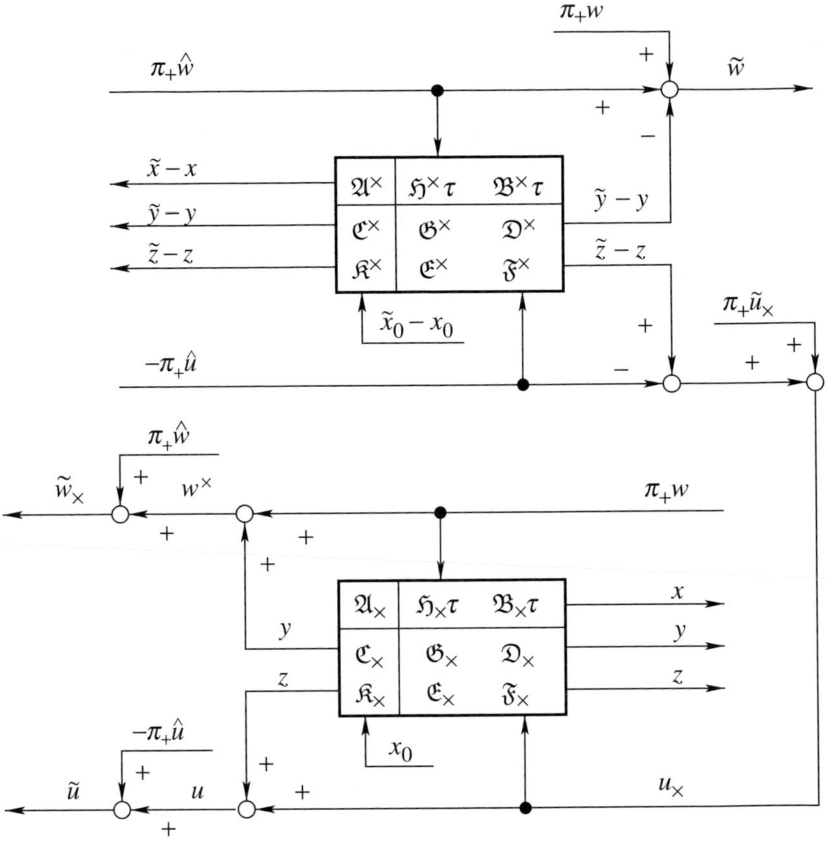

Figure 8.9 Third equivalent version of dynamic stabilization

given by

$$\begin{bmatrix} \tilde{w} \\ u_\times \end{bmatrix} = \begin{bmatrix} 1 & 0 \\ \mathfrak{Q} & 1 \end{bmatrix} \left(\begin{bmatrix} \pi_+ w \\ \pi_+ v \end{bmatrix} + \begin{bmatrix} 1 - \mathfrak{G}^\times & \mathfrak{D}^\times \\ \mathfrak{E}^\times & 1 - \mathfrak{F}^\times \end{bmatrix} \begin{bmatrix} \pi_+ \hat{w} \\ \pi_+ \hat{u} \end{bmatrix} + \begin{bmatrix} -\mathfrak{C}^\times \\ \mathfrak{K}^\times \end{bmatrix} (x_0 - \tilde{x}_0) \right). \tag{8.5.4}$$

Proof We can interpret the system in Figure 8.10 as a cascade connection of three stable systems, namely $\Sigma^\times$, $\Sigma_{\mathfrak{Q}}$, and $\Sigma_\times$: we simply replace the system in Figure 8.6 which is part of Figure 8.10 by the equivalent system in Figure 8.9. From there we can also read off (8.5.4) (as well as formulas for all the other signals in the system). □

The relationship between Theorems 8.5.7 and 8.5.11 is the following.

Theorem 8.5.12 *Let* $\Sigma = \left[\begin{smallmatrix} \mathfrak{A} & \mathfrak{B} \\ \mathfrak{C} & \mathfrak{D} \end{smallmatrix}\right]$ *be a jointly stabilizable and detectable well-posed linear system on* (Y, X, U).

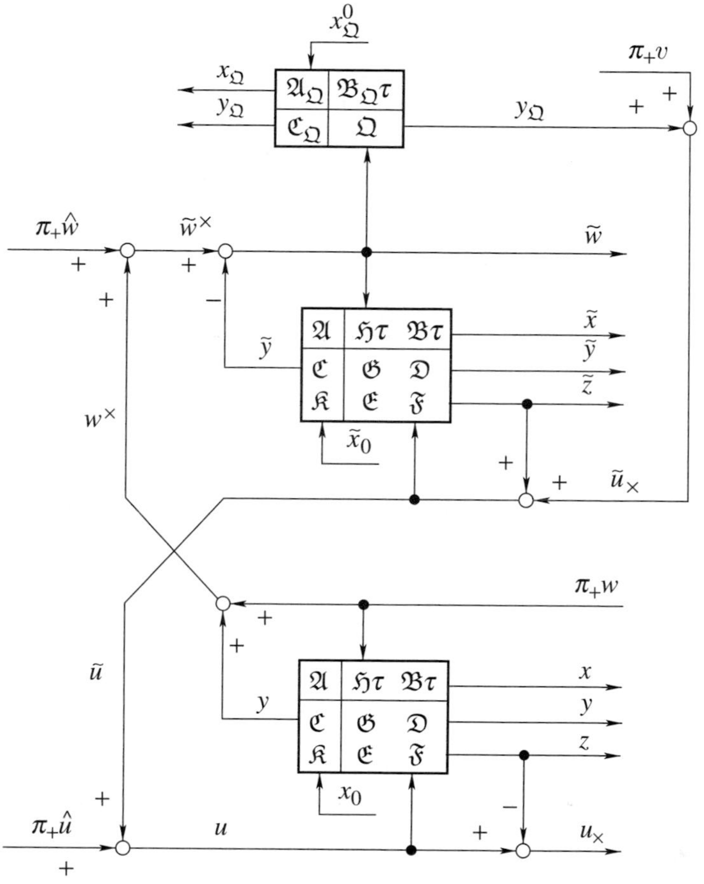

Figure 8.10 Youla parametrization

(i) *The system drawn in Figure 8.11 (which is a part of the system drawn in Figure 8.10) is a well-posed linear system if and only if* $1 + \mathfrak{G}_\times + \mathfrak{D}_\times \mathfrak{Q}$ *has an inverse in* $TIC_\omega(Y)$ *for some* $\omega \geq 0$*, or equivalently, if and only if* $1 - \mathfrak{F}^\times + \mathfrak{Q}\mathfrak{D}^\times$ *has an inverse in* $TIC_\omega(U)$ *for some* $\omega \geq 0$*, and when this is the case, then this system dynamically stabilizes* Σ *(cf. Figure 8.10). The input/output map* $\mathfrak{D}_1$ *of this system (from* $\widetilde{w}^\times$ *to* $\tilde{u}$*) has the right and left coprime fractions*

$$\begin{aligned} \mathfrak{D}_1 &= (\mathfrak{E}_\times + (1 + \mathfrak{F}_\times)\mathfrak{Q})(1 + \mathfrak{G}_\times + \mathfrak{D}_\times \mathfrak{Q})^{-1} \\ &= (1 - \mathfrak{F}^\times + \mathfrak{Q}\mathfrak{D}^\times)^{-1}(\mathfrak{E}^\times + \mathfrak{Q}(1 - \mathfrak{G}^\times)). \end{aligned} \tag{8.5.5}$$

(ii) *If* $\Sigma_1 = \left[\begin{array}{c|c} \mathfrak{A}_1 & \mathfrak{B}_1 \\ \hline \mathfrak{C}_1 & \mathfrak{D}_1 \end{array}\right]$ *is a well-posed linear system which dynamically stabilizes* Σ*, then it is possible to find some system* $\Sigma_{\mathfrak{Q}}$ *such that (i)*

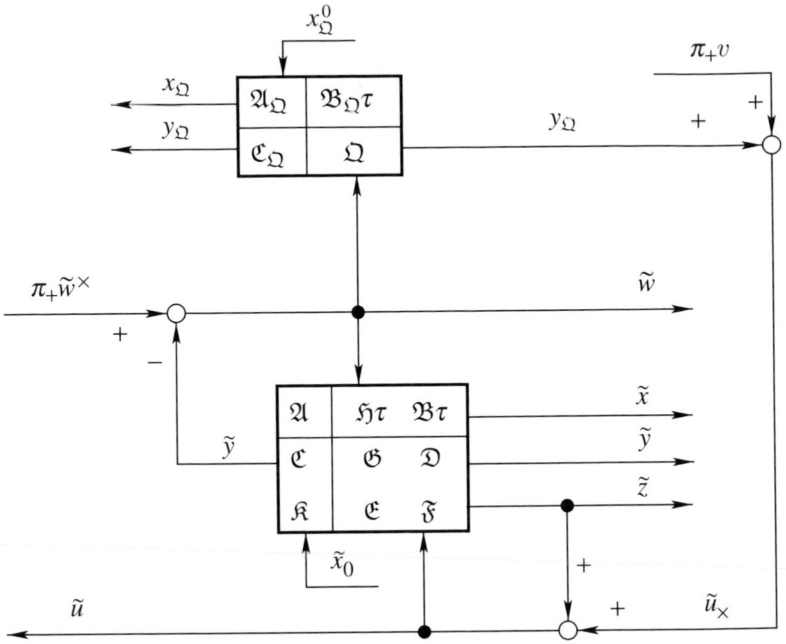

Figure 8.11 Youla parametrized stabilizing compensator

applies, and such that $\mathfrak{D}_1$ has the right and left coprime fractions given in (8.5.5). In particular, the system in Figure 8.11 (with input $\widetilde{w}^\times$ and output $\tilde{u}$) gives us another realization of $\mathfrak{D}_1$.

Proof (i) By using Remark 8.2.3, we can rewrite the system in Figure 8.11 in two equivalent ways as feedback connections of Σ_Ω and either $\Sigma_\times$ or $\Sigma^\times$; see Figures 8.12 and 8.13. We can get rid of the feedback loop in Figure 8.12 by reversing the line from $\widetilde{w}^\times$ to $\widetilde{w}$, making the latter signal the input and the former the output. The system that we get in this way is (well-posed and) stable. Thus, the system drawn in Figure 8.12 can be interpreted as a partial flow-inversion of a stable system, and it is well-posed if and only if the input/output map $1 + \mathfrak{G}_\times + \mathfrak{D}_\times \mathfrak{Q}$ from $\widetilde{w}$ to $\widetilde{w}^\times$ of that system has a well-posed inverse. The first half of (8.5.5) can be read off from Figure 8.12 (where we eliminate $\tilde{w}$). To get the second half of (8.5.5) we argue in essentially the same way, but replace Figure 8.12 by Figure 8.13, which we interpret as a partial inversion of the stable system that we get by reversing the line from v to $\tilde{u}$. That the system in Figures 8.11–8.13 dynamically stabilizes Σ (independently of whether it is well-posed or not) follows from Theorem 8.5.11.

(ii) See Theorem 8.5.7. □

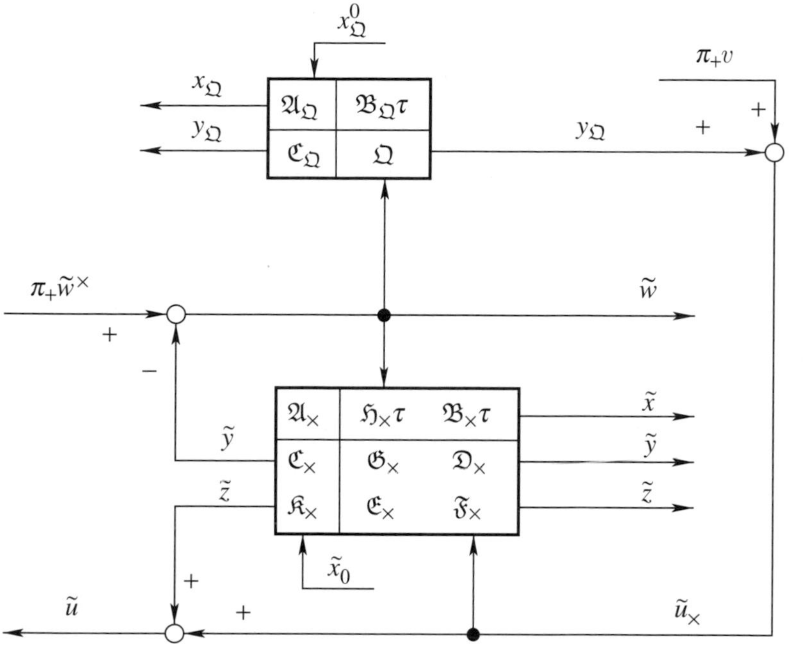

Figure 8.12 Youla parametrized stabilizing compensator

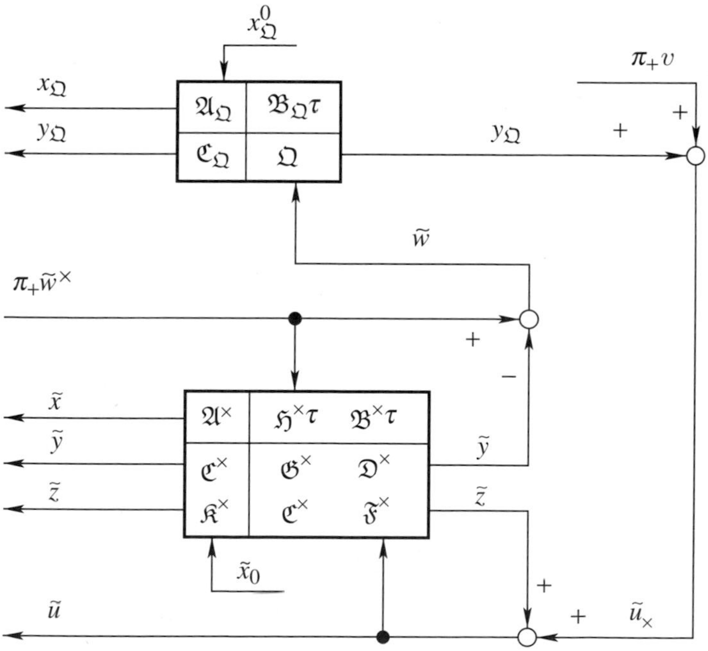

Figure 8.13 Youla parametrized stabilizing compensator

8.6 Comments

It is usually easier to work with exponential stability than with strong stability, and most of the existing literature uses exponential stability throughout. This is, in particular, true for all results on finite-dimensional systems, and also for, e.g., Curtain (1988) Curtain, Weiss, and Weiss (1996) Curtain and Zwart (1995), Lasiecka and Triggiani (2000a, b), Bensoussan *et al.* (1992), Helton (1976), Morris (1999), Rebarber (1993), van Keulen (1993), and Weiss and Rebarber (2001). Boundedness or strong stability is used in Arov and Nudelman (1996), Curtain and Oostveen (1998), Fuhrmann (1981), Jacob and Zwart (2001a, b, 2002), Staffans (1997, 1998a, b), and Weiss and Weiss (1997). Both types of stability are discussed in Mikkola (2002).

Section 8.1 Theorem 8.1.5 (or Theorem 3.11.8) is sometimes used to guarantee the exponential stability of the solution of some optimal control problems; see, for example, Lasiecka and Triggiani (2000a, b) or Weiss and Rebarber (2001).

Corollary 8.1.8 is trivially true in the L^p-well-posed case if we replace exponential stability by strong stability. Apparently it is not known if the inverse $\mathfrak{D}^{-1}$ inherits the strong stability of $\mathfrak{D}$ in the *Reg*-well-posed case.

Section 8.2 This section has been modeled after Staffans (1998a) and Weiss and Curtain (1997). Joint stabilizability and detectability were introduced independently by Staffans (1998a) and (for regular systems) by Weiss and Curtain (1997). Lemmas 8.2.6(iii), 8.2.7, and 8.2.12(ii) are generalizations of Proposition 3.3, Corollary 3.5, and Proposition 3.7, respectively, in Weiss and Curtain (1997). A special case of Theorem 8.2.11 was proved by Rebarber (1993) (he considers only the regular Hilbert space case with $p = 2$ and exponential stability). The same result is also found in Weiss and Curtain (1997, Proposition 3.2 and Theorem 3.8).

A different approach to Theorem 8.2.11 is taken by Weiss and Rebarber (2001, Theorem 6.3). There only the exponentially stable case is considered, but instead the stabilizability and detectability assumptions are replaced by optimizability and estimatability assumptions (these assumptions are weaker than exponential stabilizability and detectability).

An expanded version of Theorem 8.2.11 (which includes results related to optimizability and estimatability) is given in Mikkola (2002, Theorem 6.7.10).

Section 8.3 This section has been modeled after Francis (1987), Mikkola (2002), Staffans (1998a), and Vidyasagar (1985). We follow the sign convention of Francis (1987), which appears to be the most common one today.

Our definition of coprimeness is slightly nonstandard. It is possible to study coprime fractions in the quotient field of *TIC* without our additional assumption that $\mathfrak{D}$ belongs to $TIC_\omega(U;Y)$ and that $\mathfrak{M}$ and $\widetilde{\mathfrak{M}}$ are invertible in $TIC_\omega(U)$, respectively $TIC_\omega(Y)$, for some $\omega \geq 0$; see, e.g., Georgiou and Smith (1993), Logemann (1993), or Smith (1989). Much of the theory remains valid if we replace the invertibility assumptions on $\mathfrak{M}$ and $\widetilde{\mathfrak{M}}$ by the assumptions that the transfer functions of $\mathfrak{M}$ and $\widetilde{\mathfrak{M}}$ are invertible in at least one point in the half-plane $\Re z > 0$. Observe that if $\mathfrak{M}$ is invertible in any reasonable sense, then $\mathfrak{D} \in TIC_\omega(U;Y)$ iff $\mathfrak{M}^{-1} \in TIC_\omega(U)$ because $\mathfrak{D} = \mathfrak{N}\mathfrak{M}^{-1}$ and $\mathfrak{M}^{-1} = \widetilde{\mathfrak{X}} + \widetilde{\mathfrak{Y}}\mathfrak{D}$. Likewise, if $\widetilde{\mathfrak{M}}$ is invertible in any reasonable sense, then $\mathfrak{D} \in TIC_\omega(U;Y)$ iff $\widetilde{\mathfrak{M}}^{-1} \in TIC_\omega(Y)$ because $\mathfrak{D} = \widetilde{\mathfrak{M}}^{-1}\mathfrak{N}$ and $\widetilde{\mathfrak{M}}^{-1} = \mathfrak{X} + \mathfrak{D}\mathfrak{Y}$. According to Theorems 2.6.6 and 2.6.7, if $\mathfrak{D}$ does not belong to $TIC_\omega(U;Y)$ for any $\omega \geq 0$, then $\mathfrak{D}$ cannot be realized as the input/output map of a well-posed linear system on a triple of Hilbert spaces.

As is well-known, input/output maps with a finite-dimensional realization always have doubly coprime factorizations, and these factorizations can be computed by the method presented in Theorem 8.4.1; see, e.g., Francis (1987). A transfer function with finite-dimensional input and output spaces which is not meromorphic in the right half-plane cannot have right or left coprime fractions in H^∞ (see Definition 8.3.6 and Lemma 8.3.8), and it is not even true that all single-input single-output H^∞/H^∞ transfer functions have coprime fractions (see Logemann 1993, p. 108). However, many transfer functions that can be stabilized by a dynamic output feedback do; see Remark 8.5.4.

The converses of parts (i) and (ii) in Lemma 8.3.8 are closely related to the famous *corona theorem*, which holds in the L^2-well-posed Hilbert space case with finite-dimensional U and Y, but not for infinite-dimensional U and Y. See Mikkola (2002, Chapter 4) for an up-to-date discussion of these implications.

Notions of coprimeness which are weaker than ours (which is based on the Bezout identity) have been used in various places. For example, Fuhrmann (1981) uses 'weak' and 'strong' coprimeness, whereas Mikkola (2002) introduces the notions of 'quasi-coprime' and 'pseudo-coprime.' In the following, let us only discuss the relationships between these notions in the case of 'right' coprimeness; the situation is analogous in the 'left' case. The strongest out of the above mentioned weaker notions is the strong coprimeness (which is the same as Mikkola's pseudo-coprimeness), and it means that $\begin{bmatrix} \widehat{\mathfrak{N}}(z) \\ \widehat{\mathfrak{M}}(z) \end{bmatrix}$ has a left-inverse everywhere in $\mathbb{C}^+$ which is uniformly bounded (over all $z \in \mathbb{C}^+$). It follows from the corona theorem that if U is finite-dimensional and $p = 2$, then strong coprimeness is equivalent to the notion of coprimeness that we use in this work. The next weaker notion is weak coprimeness, which means that $\widehat{\mathfrak{N}}$ and $\widehat{\mathfrak{M}}$ do not have any nontrivial common right divisor in H^∞. Quasi-coprimeness means that if a function $u \in L^2_{\text{loc}}(\mathbb{R}^+;U)$ but $u \notin L^2(\mathbb{R}^+;U)$, then $\left[\begin{smallmatrix} \mathfrak{N} \\ \mathfrak{M} \end{smallmatrix}\right] u \notin L^2(\mathbb{R}^+, \left[\begin{smallmatrix} Y \\ U \end{smallmatrix}\right])$.

Section 8.4 Theorem 8.4.1 is essentially due to Staffans (1998a, Theorem 4.4) (there it was assumed that the system is jointly stabilizable and detectable, but the proof is the same). A special case of Theorem 8.4.1(i) was proved independently by Curtain, Weiss, and Weiss (1996, Theorem 3.2).

We have adopted the notions of right coprime stabilizability and left coprime detectability from Mikkola (2002), where the basic properties of these notions are explored. In particular, Theorem 8.4.8 is essentially due to Mikkola (2002, Theorem 6.7.10). An extension of Theorem 8.4.11 is given in Mikkola (2002, Theorem 6.7.15). (This chapter was written concurrently with Mikkola (2002), and there was a significant interaction. See the comments in Mikkola (2002, Section 6.7) for details.)

It appears to be an open problem under what conditions a right coprime stabilizing state feedback pair and left coprime detecting output injection pair can be completed to a jointly input/output stabilizing and detecting state feedback and output injection pair (cf. Lemma 8.4.7). In other words, with the notation of Lemma 7.3.3, under what conditions is it true that $\mathfrak{K}\mathfrak{H}$ is the Hankel operator of some $\mathfrak{E} \in TIC(Y;U)$ and that both the closed-loops systems $\Sigma_\times$ and $\Sigma^\times$ are input/output stable? (The operator $\mathfrak{E}$ is determined uniquely by $\mathfrak{K}$ and $\mathfrak{H}$ modulo a static term.)

Section 8.5 All the main results of this section are taken from Mikkola (2002), except Lemma 8.5.9 and Theorem 8.5.10 which are due (independently) to Staffans (1998a) and Weiss and Curtain (1997) (in the regular case). Our proof uses ingredients from both Staffans (1998a) and Weiss and Curtain (1997) (the proof in Staffans (1998a) is based on Figure 8.7, and the proof in Weiss and Curtain (1997) is based on Figure 8.9).

As we saw in Theorem 8.5.11, the system in Figure 8.10 is a (well-posed) bounded linear system even if the 'compensator' drawn in Figure 8.11 is not well-posed. Non-well-posed compensators of this type have been studied by Weiss and Curtain (1997), Curtain *et al.* (2001), and Mikkola (2002) where they are called 'compensators with internal loop' (because of the fact that they can be interpreted as non-well-posed feedback connections of well-posed linear systems). The use of such compensators makes it possible to stabilize some systems (for example, by short-circuiting an output) which cannot be stabilized by traditional methods. See the cited references for more details.

9

Realizations

By a *realization* of a given time-invariant causal map $\mathfrak{D}$ we mean a (often well-posed) linear system whose input/output map is $\mathfrak{D}$. In this chapter we study the basic properties of these realizations, such as minimality and similarity or pseudo-similarity of two different realizations of the same input/output map. We show that any two minimal realizations of the same input/output map are pseudo-similar to each other. Furthermore, in the stable case all minimal realizations of a given input/output map are (strongly) similar to each other if and only if the range of the Hankel operator is closed. The proof of this result is based on the fact that every realization induces a factorization of the Hankel operator of the input/output map. The converse is also true to some extent.

9.1 Minimal realizations

Let us begin by repeating the following definition (cf. Definition 2.6.3):

Definition 9.1.1 Let $1 \le p \le \infty$, let U and Y be Banach spaces, and let $\mathfrak{D} \in TIC^p_{\mathrm{loc}}(U;Y)$. By a L^p*-realization* of $\mathfrak{D}$ we mean an L^p-well-posed linear system on (Y, X, U) (for some Banach space X) with input/output map $\mathfrak{D}$. A *Reg-realization* of an operator $\mathfrak{D} \in TIC^{Reg}_{\mathrm{loc}}(U;Y)$ is defined in the same way, with L^p replaced by *Reg*.

In Section 2.6 we constructed some shift realizations of a given input/output map $\mathfrak{D}$, namely the *exactly controllable shift* realization $\left[\begin{array}{c|c} \tau_- & 1 \\ \hline \pi_+\mathfrak{D}\pi_- & \mathfrak{D} \end{array}\right]$ on $(Y, L^p_\omega(\mathbb{R}^-;U), U)$, and the *exactly observable shift* realization $\left[\begin{array}{c|c} \tau_+ & \pi_+\mathfrak{D}\pi_- \\ \hline 1 & \mathfrak{D} \end{array}\right]$ on $(Y, L^2_\omega(\mathbb{R}^+;Y), U)$. Neither of these realizations is, in general, *minimal*, i.e., it is often possible to replace the state space X by a 'smaller' state space. In order to see how this can be done we introduce the notions of *controllability* and *observability*. The former concept, also known under the name of *reachability*, is related to the properties of the input map, and the latter to the properties of the output map.

In the classical case with finite-dimensional state space it is possible to define controllability and observability in several different ways, but they are all equivalent and lead to the same concepts. In the infinite-dimensional case this is no longer true. We start by introducing the weakest possible versions of these concepts, and postpone a more detailed discussion to Section 9.4.

Definition 9.1.2 Let $\Sigma = \left[\begin{smallmatrix} \mathfrak{A} & \mathfrak{B} \\ \mathfrak{C} & \mathfrak{D} \end{smallmatrix}\right]$ be an $L^p|Reg$-well-posed linear system on (Y, X, U).

(i) The (approximately) *reachable subspace* $\overline{\mathcal{R}(\mathfrak{B})}$ is the closure in X of the range of $\mathfrak{B}$ as a mapping from $L^p|Reg_c(\mathbb{R}^-; U)$ into X.
(ii) The *unobservable subspace* $\mathcal{N}(\mathfrak{C})$ is the null space of $\mathfrak{C}\colon X \to L^p|Reg_{\mathrm{loc}}(\overline{\mathbb{R}}^+; Y)$.
(iii) Σ is (approximately) *controllable* if the reachable subspace is equal to X, and Σ is (approximately) *observable* if the unobservable subspace is 0.
(iv) A system is *minimal* if it is both controllable and observable.

In the preceding definition we regarded $\mathfrak{B}$ as an operator from $L^p|Reg_c(\mathbb{R}^-; U)$ into X and $\mathfrak{C}$ as an operator from X into $L^p|Reg_{\mathrm{loc}}(\overline{\mathbb{R}}^+; Y)$. As the following lemma shows, the notions of controllability and observability defined above do not change if we replace $L^p|Reg_c$ and $L^p|Reg_{\mathrm{loc}}$ by $L^p|Reg_\omega$ where $\omega > \omega_{\mathfrak{A}}$.

Lemma 9.1.3 *Let* $\Sigma = \left[\begin{smallmatrix} \mathfrak{A} & \mathfrak{B} \\ \mathfrak{C} & \mathfrak{D} \end{smallmatrix}\right]$ *be an* $L^p|Reg$*-well-posed linear system on* (Y, X, U) *with growth bound* $\omega_{\mathfrak{A}}$*, and let* $\omega > \omega_{\mathfrak{A}}$.

(i) *The reachable subspace* $\overline{\mathcal{R}(\mathfrak{B})}$ *is the closure in* X *of the range of* $\mathfrak{B}$ *as a mapping from* $L^p|Reg_\omega(\mathbb{R}^-; U)$ *into* X.
(ii) *The unobservable subspace* $\mathcal{N}(\mathfrak{C})$ *is the null space of* $\mathfrak{C}\colon X \to L^p|Reg_\omega(\overline{\mathbb{R}}^+; Y)$.
(iii) *The unobservable subspace is closed in* X.

Proof (i) The density of $L^p|Reg_c(\mathbb{R}^-; U)$ in $L^p|Reg_\omega(\mathbb{R}^-; U)$ and the continuity of $\mathfrak{B}\colon L^p|Reg_\omega(\mathbb{R}^-; U) \to X$ (see Theorem 2.5.4(ii)) imply that the image of $L^p|Reg_c(\mathbb{R}^-; U)$ under $\mathfrak{B}$ is dense in the image of $L^p|Reg_\omega(\mathbb{R}^-; U)$, thus these images have the same closure.

(ii) This is obvious, since $\mathfrak{C}$ maps X into $L^p|Reg_\omega(\overline{\mathbb{R}}^+; Y)$ (see Theorem 2.5.4(ii)).

(iii) The inverse image under a continuous map of the closed set $\{0\}$ is always closed. □

The advantage with the characterization of $\overline{\mathcal{R}(\mathfrak{B})}$ and $\mathcal{N}(\mathfrak{C})$ given in Lemma 9.1.3 is that we can regard $\mathfrak{B}$ and $\mathfrak{C}$ as bounded operators mapping one Banach space into another.

There is still another way of characterizing the reachable subspace in the well-posed case (i.e., we exclude the L^∞-case).

Lemma 9.1.4 *Let $\Sigma = \left[\begin{smallmatrix} \mathfrak{A} & \mathfrak{B} \\ \mathfrak{C} & \mathfrak{D} \end{smallmatrix}\right]$ be a well-posed linear system. Then the reachable subspace is the closure in X of the range of $\mathfrak{B}$ as a mapping from $C_c^\infty(\overline{\mathbb{R}}^-; U)$ into X.*

Proof In the L^p-well-posed case this follows from the fact that $C_c^\infty(\overline{\mathbb{R}}^-; U)$ is dense in $L_c^p(\mathbb{R}^-; U)$ for $p < \infty$. It is also dense in $BC_c(\overline{\mathbb{R}}^-; U)$, and the *Reg*-well-posed case then follows from Lemma 4.7.17. □

The purpose of this section is to show that every $L^p|Reg$-well-posed linear system can be reduced to a minimal system (Theorem 9.1.9). In this process the state space is replaced by another 'smaller' state space. The semigroup, the input map, and the output map change, but the input/output map stays the same. Thus, this means that every realization of a given input/output map can be reduced to a minimal realization. One part of the reduction consists of the replacement of the state space by the quotient of the state space over the unobservable subspace. Therefore, let us first make some preliminary comments about quotient spaces, and about the operator or semigroup induced on a quotient space by an operator or semigroups acting on the original space.

We recall from, e.g., Kato (1980, Section III.8) or Rudin (1973, pp. 29–30) that the quotient space X/Z consists of all the cosets $\pi x = x + Z = \{x + z \mid z \in Z\} \subset X$ where x varies over X. The norm in this space is given by

$$|\pi x|_{X/Z} = \inf_{z\in Z} |x - z|_X.$$

The quotient map $\pi : X \to X/Z$ is both continuous and open (both the image and the inverse image of an open set is open), and it maps the open unit ball of X onto the open unit ball of X/Z. In particular, $\|\pi\| = 1$. If X is a Hilbert space, then X/Z can be identified with the orthogonal complement to Z in X, and π can be identified with the orthogonal projection onto $Z^\perp$.

Let $K \in \mathcal{B}(X, Y)$, and suppose that K vanishes on Z, i.e., $Z \subset \mathcal{N}(K)$. Then we can define an operator $\widetilde{K} : X/Z \to Y$ by setting

$$\widetilde{K}\pi x = Kx.$$

This definition of $\widetilde{K}$ is possible since $K\pi(x_1) = K\pi(x_2)$ whenever $\pi(x_1) = \pi(x_2)$ (i.e., $x_1 - x_2 \in Z$, hence $Kx_1 = Kx_2$). This leads to a factorization of K as $K = \widetilde{K}\pi$. The new operator $\widetilde{K}$ satisfies $\|\widetilde{K}\| = \|K\|$ since π maps the unit ball of X onto the unit ball of X/Z. Obviously $\mathcal{R}(\widetilde{K}) = \mathcal{R}(K)$. The null space of $\widetilde{K}$ can be identified with $\pi\mathcal{N}(K)$; hence $\widetilde{K}$ is injective iff $\mathcal{N}(K) = Z$. In this case $\widetilde{K}$ has an inverse $Y \supset \mathcal{R}(K) \to X/Z$, which we can identify with the operator which maps $y \in Y$ into the inverse image $K^{-1}y \subset X$. If X is a Hilbert space, then we can identify $\widetilde{K}$ with the restriction of K to $Z^\perp$.

Definition 9.1.5 Let $K \in \mathcal{B}(X, Y)$ vanish on the closed subspace Z of X. By the *operator induced by* K *on* X/Z we mean the operator $\widetilde{K} \in \mathcal{B}(X/Z; Y)$ described above. We denote this operator, too, by K, whenever it is clear from the context which of the two operators we refer to.

We remark that it is also possible to factor out (a closed subspace of) the null space of a *closed* linear operator, and that the resulting operator is closed; see Kato (1980, p. 231).

Lemma 9.1.6 *Let $\mathfrak{A}$ be a C_0 semigroup on X, and suppose that $Z \subset X$ is closed and invariant under $\mathfrak{A}$ (i.e., $\mathfrak{A}^t Z \subset Z$ for $t \geq 0$). Let π be the quotient map $X \to X/Z$. For each $\pi x \in X/Z$, define*

$$\widetilde{\mathfrak{A}}^t \pi x = \pi \mathfrak{A}^t x.$$

Then $\widetilde{\mathfrak{A}}^t$ is a C_0 semigroup on X/Z. If $\mathfrak{A}$ is a contraction semigroup, then so is $\widetilde{\mathfrak{A}}$.

Proof First we must check that $\widetilde{\mathfrak{A}}$ is well-defined, i.e., $\pi\mathfrak{A}^t x = \pi\mathfrak{A}^t y$ whenever $\pi x = \pi y$. This follows from the invariance of Z under $\mathfrak{A}^t$: if $x - y \in Z$, then $\mathfrak{A}^t(x - y) \subset Z$, and $\pi\mathfrak{A}^t(x - y) = 0$. The same calculation shows that $Z \subset \mathcal{N}(\pi\mathfrak{A}^t)$, and that the operator $\widetilde{\mathfrak{A}}^t$ is nothing but the operator induced by $\pi\mathfrak{A}^t$ on X/Z.

Clearly, $\widetilde{\mathfrak{A}}^0$ is the identity on X/Z (it maps πx into πx). For $s, t > 0$, and $x \in X$, we have

$$\widetilde{\mathfrak{A}}^{s+t}\pi x = \pi\mathfrak{A}^{s+t}x = \pi\mathfrak{A}^s\mathfrak{A}^t x = \widetilde{\mathfrak{A}}^s\pi\mathfrak{A}^t x = \widetilde{\mathfrak{A}}^s\widetilde{\mathfrak{A}}^t\pi x,$$

and this shows that $\widetilde{\mathfrak{A}}$ is a semigroup on X/Z. The strong continuity of $\widetilde{\mathfrak{A}}$ is a consequence of the fact that

$$|\widetilde{\mathfrak{A}}^t\pi x - \pi x|_{X/Z} = |\pi(\mathfrak{A}^t x - x)|_{X/Z} \leq |\mathfrak{A}^t x - x|_X,$$

which tends to zero as $t \downarrow 0$. Thus, $\widetilde{\mathfrak{A}}$ is a C_0 semigroup on X/Z. If $\mathfrak{A}$ is a contraction semigroup, then for all $t \geq 0$ and $x \in X$,

$$\begin{aligned}|\widetilde{\mathfrak{A}}^t\pi x|_{X/Z} &= |\pi\mathfrak{A}^t x|_{X/Z} = \inf_{z\in Z}|\mathfrak{A}^t x - z|_X \leq \inf_{z\in Z}|\mathfrak{A}^t x - \mathfrak{A}^t z|_X \\ &= \inf_{z\in Z}|\mathfrak{A}^t(x - z)|_X \leq \inf_{z\in Z}|x - z|_X = |\pi x|_{X/Z},\end{aligned}$$

so $\widetilde{\mathfrak{A}}$ is a contraction semigroup in this case. □

Definition 9.1.7 Let $\mathfrak{A}$ be a C_0 semigroup on X, and suppose that $Z \subset X$ is closed and invariant under $\mathfrak{A}$. By the *quotient semigroup induced by* $\mathfrak{A}$ *on* X/Z we mean the C_0-semigroup in Lemma 9.1.6. We denote this semigroup by $\pi\mathfrak{A}$ (where π is the quotient map $X \to X/Z$).

Corollary 9.1.8 *If X is a Hilbert space, then the quotient semigroup in Lemma 9.1.6 can be written in the form $\pi_{Z^\perp}\mathfrak{A}^t_{|Z^\perp}$, where $\pi_{Z^\perp}$ is the orthogonal projection onto $Z^\perp$.*

This follows from our earlier comments about the interpretation of a quotient of a Hilbert space over one of its subspaces.

We now arrive at the main theorem of this section.

Theorem 9.1.9 *Let $\Sigma = \left[\begin{smallmatrix}\mathfrak{A} & \mathfrak{B} \\ \mathfrak{C} & \mathfrak{D}\end{smallmatrix}\right]$ be an $L^p|Reg$-well-posed linear system on (Y, X, U).*

(i) *Let $X_{\mathfrak{B}} = \overline{\mathcal{R}(\mathfrak{B})}$ be the reachable subspace, and define*

$$\mathfrak{A}_{\mathfrak{B}} = \mathfrak{A}_{|X_{\mathfrak{B}}}, \qquad \mathfrak{C}_{\mathfrak{B}} = \mathfrak{C}_{|X_{\mathfrak{B}}}.$$

Then $\left[\begin{smallmatrix}\mathfrak{A}_{\mathfrak{B}} & \mathfrak{B} \\ \mathfrak{C}_{\mathfrak{B}} & \mathfrak{D}\end{smallmatrix}\right]$ is a controllable $L^p|Reg$-well-posed linear system on $(Y, X_{\mathfrak{B}}, U)$. It is observable whenever Σ is observable.

(ii) *Let $X_{\mathfrak{C}} = X/\mathcal{N}(\mathfrak{C})$ (where $\mathcal{N}(\mathfrak{C})$ is the unobservable subspace), and let π be the quotient map $X \to X/\mathcal{N}(\mathfrak{C})$. Let $\mathfrak{A}_{\mathfrak{C}} = \pi\mathfrak{A}$ be the quotient semigroup induced by $\mathfrak{A}$ on $X/\mathcal{N}(\mathfrak{C})$, let $\mathfrak{B}_{\mathfrak{C}} = \pi\mathfrak{B}$, and let $\widetilde{\mathfrak{C}}$ be the operator induced on $X/\mathcal{N}(\mathfrak{C})$ by $\mathfrak{C}$ (see Definitions 9.1.5 and 9.1.7). Then $\left[\begin{smallmatrix}\mathfrak{A}_{\mathfrak{C}} & \mathfrak{B}_{\mathfrak{C}} \\ \widetilde{\mathfrak{C}} & \mathfrak{D}\end{smallmatrix}\right]$ is an observable $L^p|Reg$-well-posed linear system on $(Y, X_{\mathfrak{C}}, U)$. It is controllable whenever Σ is controllable.*

(iii) *Let $X_{\mathfrak{B}} = \overline{\mathcal{R}(\mathfrak{B})}$, let $X_{\mathfrak{B},\mathfrak{C}} = X_{\mathfrak{B}}/\mathcal{N}\left(\mathfrak{C}_{|X_{\mathfrak{B}}}\right)$, and let π be the quotient map $X \to X_{\mathfrak{B}}/\mathcal{N}\left(\mathfrak{C}_{|X_{\mathfrak{B}}}\right)$. Let $\mathfrak{A}_{\mathfrak{B},\mathfrak{C}}$ be the quotient semigroup induced by $\mathfrak{A}_{|X_{\mathfrak{B}}}$ on $X_{\mathfrak{B},\mathfrak{C}}$, let $\mathfrak{B}_{\mathfrak{C}} = \pi\mathfrak{B}$, and let $\widetilde{\mathfrak{C}}_{\mathfrak{B}}$ be the operator induced on $X_{\mathfrak{B},\mathfrak{C}}$ by $\mathfrak{C}_{|X_{\mathfrak{B}}}$. Then $\left[\begin{smallmatrix}\mathfrak{A}_{\mathfrak{B},\mathfrak{C}} & \mathfrak{B}_{\mathfrak{C}} \\ \widetilde{\mathfrak{C}}_{\mathfrak{B}} & \mathfrak{D}\end{smallmatrix}\right]$ is a minimal (i.e., controllable and observable) $L^p|Reg$-well-posed linear system on $(Y, X_{\mathfrak{B},\mathfrak{C}}, U)$.*

(iv) *Let $X_{\mathfrak{C}} = X/\mathcal{N}(\mathfrak{C})$, let π be the quotient map $X \to X/\mathcal{N}(\mathfrak{C})$, let $\mathfrak{B}_{\mathfrak{C}} = \pi\mathfrak{B}$, let $\widetilde{X}_{\mathfrak{B},\mathfrak{C}} = \overline{\mathcal{R}(\mathfrak{B}_{\mathfrak{C}})}$, let $\widetilde{\mathfrak{A}}_{\mathfrak{B},\mathfrak{C}} = \pi\mathfrak{A}_{|\widetilde{X}_{\mathfrak{B},\mathfrak{C}}}$, and let $\widetilde{\mathfrak{C}}$ be the restriction to $\widetilde{X}_{\mathfrak{B},\mathfrak{C}}$ of the operator induced by $\mathfrak{C}$ on $X_{\mathfrak{C}}$. Then $\left[\begin{smallmatrix}\widetilde{\mathfrak{A}}_{\mathfrak{B},\mathfrak{C}} & \mathfrak{B}_{\mathfrak{C}} \\ \widetilde{\mathfrak{C}}_{\mathfrak{B}} & \mathfrak{D}\end{smallmatrix}\right]$ is a minimal (i.e., controllable and observable) $L^p|Reg$-well-posed linear system on $(Y, \widetilde{X}_{\mathfrak{B},\mathfrak{C}}, U)$.*

Proof of Theorem 9.1.9 (i) It follows from the intertwining condition $\mathfrak{A}^t\mathfrak{B} = \mathfrak{B}\tau^t_-$ that $\mathcal{R}(\mathfrak{B})$ is invariant under $\mathfrak{A}$, hence so is its closure. Thus $\mathfrak{A}$ is a C_0 semigroup on $X_{\mathfrak{B}}$. The remaining conditions in Definition 2.2.1 are obviously satisfied on $(Y, X_{\mathfrak{B}}, U)$ whenever they are satisfied on (Y, X, U). The observability of the resulting system is obvious.

(ii) It follows from the intertwining condition $\mathfrak{C}\mathfrak{A}^t = \tau_+^t\mathfrak{C}$ that $\mathcal{N}(\mathfrak{C})$ is invariant under $\mathfrak{A}$. By Lemma 9.1.6, the quotient semigroup $\mathfrak{A}_{\mathfrak{C}}^t = \pi\mathfrak{A}$ is a C_0 semigroup on $X/\mathcal{N}(\mathfrak{C})$. The input intertwining condition $\mathfrak{A}_{\mathfrak{C}}^t\mathfrak{B}_{\mathfrak{C}} = \mathfrak{B}_{\mathfrak{C}}\tau_-^t$ is satisfied since

$$\mathfrak{A}_{\mathfrak{C}}^t\mathfrak{B}_{\mathfrak{C}} = \pi\mathfrak{A}^t\mathfrak{B} = \pi\mathfrak{B}\tau_-^t = \mathfrak{B}_{\mathfrak{C}}\tau_-^t,$$

and so is the output intertwining condition $\widetilde{\mathfrak{C}}\mathfrak{A}_{\mathfrak{C}} = \tau_+^t\mathfrak{A}_{\mathfrak{C}}$ since

$$\widetilde{\mathfrak{C}}\mathfrak{A}_{\mathfrak{C}}^t\pi x = \widetilde{\mathfrak{C}}\pi\mathfrak{A}^t x = \mathfrak{C}\mathfrak{A}^t x = \tau_+^t\mathfrak{C}^t x = \tau_+^t\widetilde{\mathfrak{C}}\pi x.$$

Finally,

$$\widetilde{\mathfrak{C}}\mathfrak{B}_{\mathfrak{C}} = \mathfrak{C}\mathfrak{B} = \pi_+\mathfrak{D}\pi_-,$$

so the Hankel operator of $\mathfrak{D}$ is $\widetilde{\mathfrak{C}}\mathfrak{B}_{\mathfrak{C}}$. This proves that the system in (ii) is $L^p|Reg$-well-posed. By construction, this system is observable ($\mathcal{N}(\mathfrak{C})$ is the zero element in $X/\mathcal{N}(\mathfrak{C})$). If Σ is controllable then the new system is also controllable, since the range of $\pi\mathfrak{B}$ is dense in $X/\mathcal{N}(\mathfrak{C})$ whenever the range of $\mathfrak{B}$ is dense in X.

(iii) We get (iii) by applying (ii) to the system obtained from (i).

(iv) We get (iv) by applying (i) to the system obtained from (ii). □

Corollary 9.1.10 *Let* $\Sigma = \left[\begin{smallmatrix}\mathfrak{A} & \mathfrak{B}\\ \mathfrak{C} & \mathfrak{D}\end{smallmatrix}\right]$ *be a* $L^p|Reg$*-well-posed linear system on* (Y, X, U)*, where* X *is a Hilbert space. Then* X *can be split into the orthogonal sum of three subspaces* $X = Z \oplus \widetilde{X} \oplus Z_*$ *which induces a block matrix decomposition of* Σ *of the following type (here* π_Z*,* $\pi_{\widetilde{X}}$*, and* π_{Z_*} *are the orthogonal projections onto* Z*,* $\widetilde{X}$*, respectively* Z_**)*

$$\left[\begin{array}{c|c}\mathfrak{A} & \mathfrak{B}\\ \hline \mathfrak{C} & \mathfrak{D}\end{array}\right] = \left[\begin{array}{ccc|c}\mathfrak{A}_{|Z} & \pi_Z\mathfrak{A}_{|\widetilde{X}} & \pi_Z\mathfrak{A}_{|Z_*} & \pi_Z\mathfrak{B}\\ 0 & \pi_{\widetilde{X}}\mathfrak{A}_{|\widetilde{X}} & \pi_{\widetilde{X}}\mathfrak{A}_{|Z_*} & \pi_{\widetilde{X}}\mathfrak{B}\\ 0 & 0 & \pi_{Z_*}\mathfrak{A}_{|Z_*} & 0\\ \hline 0 & \mathfrak{C}_{|\widetilde{X}} & \mathfrak{C}_{|Z_*} & \mathfrak{D}\end{array}\right] \tag{9.1.1}$$

in such a way that $\left[\begin{array}{c|c}\pi_{\widetilde{X}}\mathfrak{A}_{|\widetilde{X}} & \pi_{\widetilde{X}}\mathfrak{B}\\ \hline \mathfrak{C}_{|\widetilde{X}} & \mathfrak{D}\end{array}\right]$ *is a minimal realization of* $\mathfrak{D}$*. In particular,* Z *is an invariant subspace of* $\mathfrak{A}$ *which is contained in* $\mathcal{N}(\mathfrak{C})$ *and* $Z \oplus \widetilde{X}$ *is an invariant subspace of* $\mathfrak{A}$ *which contains* $\mathcal{R}(\mathfrak{B})$*. The decomposition* (9.1.1) *also gives two additional (possibly nonminimal) realizations of* $\mathfrak{D}$*, namely*

$$\left[\begin{array}{cc|c}\mathfrak{A}_{|Z} & \pi_Z\mathfrak{A}_{|\widetilde{X}} & \pi_Z\mathfrak{B}\\ 0 & \pi_{\widetilde{X}}\mathfrak{A}_{|\widetilde{X}} & \pi_{\widetilde{X}}\mathfrak{B}\\ \hline 0 & \mathfrak{C}_{|\widetilde{X}} & \mathfrak{D}\end{array}\right] \quad \textit{and} \quad \left[\begin{array}{cc|c}\pi_{\widetilde{X}}\mathfrak{A}_{|\widetilde{X}} & \pi_{\widetilde{X}}\mathfrak{A}_{|Z_*} & \pi_{\widetilde{X}}\mathfrak{B}\\ 0 & \pi_{Z_*}\mathfrak{A}_{|Z_*} & 0\\ \hline \mathfrak{C}_{|\widetilde{X}} & \mathfrak{C}_{|Z_*} & \mathfrak{D}\end{array}\right].$$

Proof We get one decomposition of the type described above from part (iv) of Theorem 9.1.9: take $Z = \mathcal{N}(\mathfrak{C})$, let $\widetilde{X}$ be the space $\widetilde{X}_{\mathfrak{B},\mathfrak{C}}$ in part (iv) of

Theorem 9.1.9, and let $Z_* = (Z \oplus \widetilde{X})^\perp$. Then $\mathfrak{A}$, $\mathfrak{B}$, and $\mathfrak{C}$ have the decompositions shown in (9.1.1), and by part (iv) of Theorem 9.1.9, $\left[\begin{array}{c|c} \pi_{\widetilde{X}}\mathfrak{A}_{|\widetilde{X}} & \pi_{\widetilde{X}}\mathfrak{B} \\ \hline \mathfrak{C}_{|\widetilde{X}} & \mathfrak{D} \end{array}\right]$ is a minimal realization of $\mathfrak{D}$.

That the first of the two given additional systems is an $L^p|Reg$-well-posed linear system is proved in the same way as in the proof of part (i) of Theorem 9.1.9 (using the fact that $Z \oplus \widetilde{X}$ is an invariant subspace of $\mathfrak{A}$ which contains $\mathcal{R}(\mathfrak{B})$). That the second of the two given additional systems is an $L^p|Reg$-well-posed linear system is proved in the same way as in the proof of part (ii) of Theorem 9.1.9 (using the fact that Z is an invariant subspace of $\mathfrak{A}$ which is contained in $\mathcal{N}(\mathfrak{C})$). □

Remark 9.1.11 In general, the two systems in parts (iii) and (iv) of Theorem 9.1.9 are *not (strongly) similar* to each other, only pseudo-similar (see the next section). Both of these systems give rise to decompositions of the type described in Corollary 9.1.10, so the decomposition in that corollary is far from unique: the subspace $\widetilde{X}$ is unique only up to pseudo-similarity, and even the dimensions of Z and Z_* need not be unique (the decomposition induced by part (iv) of Theorem 9.1.9 gives the maximal subspace $Z = \mathcal{N}(\mathfrak{C})$,[1] and the decomposition induced by part (iii) of Theorem 9.1.9 gives the maximal subspace $Z_* = \mathcal{R}(\mathfrak{B})^\perp$. Note that any reducing subspace of $\mathfrak{A}$ in $\mathcal{N}(\mathfrak{C}) \cap \mathcal{R}(\mathfrak{B})^\perp$ can be moved freely between Z and Z_*. See also Lemma 11.4.2 and Theorem 11.8.8.

9.2 Pseudo-similarity of minimal realizations

The purpose of this section is to show that any two minimal realizations of a given input/output map are similar to each other in the following weak sense.

Definition 9.2.1 Two $L^p|Reg$-well-posed linear systems $\Sigma_1 = \left[\begin{array}{c|c} \mathfrak{A}_1 & \mathfrak{B}_1 \\ \hline \mathfrak{C}_1 & \mathfrak{D} \end{array}\right]$ and $\Sigma_2 = \left[\begin{array}{c|c} \mathfrak{A}_2 & \mathfrak{B}_2 \\ \hline \mathfrak{C}_2 & \mathfrak{D} \end{array}\right]$ on (Y, X_1, U), respectively (Y, X_2, U) (with the same input/output map $\mathfrak{D}$) are *pseudo-similar* if there exists a closed and densely defined injective linear operator $E \colon X_1 \supset \mathcal{D}(E) \to \mathcal{R}(E) \subset X_2$ with the following properties: $\mathfrak{B}_1$ maps $L^p|Reg_c(\mathbb{R}^-; U)$ into $\mathcal{D}(E)$, $\mathfrak{B}_2$ maps $L^p|Reg_c(\mathbb{R}^-; U)$ into $\mathcal{R}(E)$, $\mathcal{D}(E)$ is invariant under $\mathfrak{A}_1$, $\mathcal{R}(E)$ is invariant under $\mathfrak{A}_2$, and

$$\begin{aligned} \mathfrak{A}_2^t E x_1 &= E \mathfrak{A}_1^t x_1, && x_1 \in \mathcal{D}(E), \\ \mathfrak{B}_2 u &= E \mathfrak{B}_1 u, && u \in L^p|Reg_c(\mathbb{R}^-; U), \\ \mathfrak{C}_2 E x_1 &= \mathfrak{C}_1 x_1, && x_1 \in \mathcal{D}(E). \end{aligned} \tag{9.2.1}$$

[1] By this we mean a subspace which contains the corresponding subspace Z of any other splitting.

Without any further assumptions the operator E in Definition 9.2.1 will not be unique (see Theorem 9.2.4 and the comments in Section 9.11). Note that this definition is symmetric with respect to Σ_1 and Σ_2: it stays the same if we interchange Σ_1 and Σ_2 and at the same time replace E by E^{-1}.

As we shall see in a moment, any two *minimal* realizations of the same input/output map $\mathfrak{D}$ are pseudo-similar to each other. The proof of this fact is based on the following lemma.

Lemma 9.2.2 *Let* $\Sigma = \left[\begin{smallmatrix}\mathfrak{A} & \mathfrak{B}\\ \mathfrak{C} & \mathfrak{D}\end{smallmatrix}\right]$ *be an* $L^p|Reg$*-well-posed linear system on* (Y, X, U) *with growth bound* $\omega_{\mathfrak{A}}$*. Let* $\omega > \omega_{\mathfrak{A}}$*. Then*

(i) *If* Σ *is observable, then* $\mathcal{N}(\mathfrak{B}) = \mathcal{N}(\pi_+\mathfrak{D}\pi_-)$ *(where the common domain of* $\mathfrak{B}$ *and* $\mathfrak{D}$ *is either* $L^p|Reg_c(\mathbb{R}^-; U)$ *or* $L^p|Reg_\omega(\mathbb{R}^-; U)$*).*
(ii) *If* Σ *is controllable, then* $\mathcal{R}(\mathfrak{C})$ *and* $\mathcal{R}(\pi_+\mathfrak{D}\pi_-)$ *have the same closure (both in* $L^p|Reg_{\text{loc}}(\overline{\mathbb{R}}^+; Y)$ *and* $L^p|Reg_\omega(\overline{\mathbb{R}}^+; Y)$*).*
(iii) *If* Σ *is minimal, then both (i) and (ii) apply.*

Proof (i) By assumption, $\pi_+\mathfrak{D}\pi_- = \mathfrak{C}\mathfrak{B}$. The controllability assumption says that $\mathfrak{C}$ is injective, hence $\mathcal{N}(\mathfrak{B}) = \mathcal{N}(\pi_+\mathfrak{D}\pi_-)$.

(ii) Let $T > 0$. By assumption, $\pi_{[0,T)}\mathfrak{C}\mathfrak{B} = \pi_{[0,T)}\mathfrak{D}\pi_-$. As $\mathcal{R}(\mathfrak{B})$ is dense in X,

$$\overline{\mathcal{R}\left(\pi_{[0,T)}\mathfrak{C}\right)} = \overline{\mathcal{R}\left(\pi_{[0,T)}\mathfrak{C}\mathfrak{B}\right)} = \overline{\mathcal{R}\left(\pi_{[0,T)}\mathfrak{D}\pi_-\right)}.$$

The same argument applies if we replace $L^p|Reg([0, T), Y)$ by $L^p|Reg_\omega(\overline{\mathbb{R}}^+, Y)$. □

Corollary 9.2.3 *Let* $\Sigma_1 = \left[\begin{smallmatrix}\mathfrak{A}_1 & \mathfrak{B}_1\\ \mathfrak{C}_1 & \mathfrak{D}\end{smallmatrix}\right]$ *and* $\Sigma_2 = \left[\begin{smallmatrix}\mathfrak{A}_2 & \mathfrak{B}_2\\ \mathfrak{C}_2 & \mathfrak{D}\end{smallmatrix}\right]$ *be two systems which are well-posed in the same sense on* (Y, X_1, U) *and* (Y, X_2, U)*, respectively, with the same input/output map* $\mathfrak{D}$ *(thus, they both realize the same input/output map).*

(i) *If both* Σ_1 *and* Σ_2 *are observable, then* $\mathcal{N}(\mathfrak{B}_1) = \mathcal{N}(\mathfrak{B}_2)$ *(where the common domain of* $\mathfrak{B}_1$ *and* $\mathfrak{B}_2$ *is either* $L^p|Reg_c(\mathbb{R}^-; U)$ *or* $L^p|Reg_\omega(\mathbb{R}^-; U)$*).*
(ii) *If both* Σ_1 *and* Σ_2 *are controllable, then* $\mathcal{R}(\mathfrak{C}_1)$ *and* $\mathcal{R}(\mathfrak{C}_2)$ *have the same closure (both in* $L^p|Reg_{\text{loc}}(\overline{\mathbb{R}}^+; Y)$ *and* $L^p|Reg_\omega(\overline{\mathbb{R}}^+; Y)$*).*
(iii) *If* Σ_1 *and* Σ_2 *are minimal, then both (i) and (ii) apply.*

This follows from Lemma 9.2.2.

Theorem 9.2.4 *Let* $\Sigma_1 = \left[\begin{smallmatrix}\mathfrak{A}_1 & \mathfrak{B}_1\\ \mathfrak{C}_1 & \mathfrak{D}\end{smallmatrix}\right]$ *and* $\Sigma_2 = \left[\begin{smallmatrix}\mathfrak{A}_2 & \mathfrak{B}_2\\ \mathfrak{C}_2 & \mathfrak{D}\end{smallmatrix}\right]$ *be two minimal realizations of the same input/output map* $\mathfrak{D}$ *(i.e., they are minimal* $L^p|Reg$*-well-posed linear systems with the same input/output map* $\mathfrak{D}$*). Then these systems are pseudo-similar. Moreover, all pseudo-similarities* E *in Definition 9.2.1 are* *restrictions of a* maximally defined *pseudo-similarity* $\overline{E}$*, which is characterized by the fact that*

$$\begin{aligned}\mathcal{D}\left(\overline{E}\right) &= \{x_1 \in X_1 \mid \mathfrak{C}_1 x_1 \in \mathcal{R}(\mathfrak{C}_2)\},\\ \mathcal{R}\left(\overline{E}\right) &= \{x_2 \in X_2 \mid \mathfrak{C}_2 x_2 \in \mathcal{R}(\mathfrak{C}_1)\},\end{aligned} \tag{9.2.2}$$

and they are all extensions of a unique minimally defined *pseudo-similarity* $\underline{E}$ *which we get by taking the closure of the restriction of* $\overline{E}$ *to the image of* $L^p|Reg_c(\mathbb{R}^-;U)$ *under* $\mathfrak{B}_1$.

Proof The output map $\mathfrak{C}_2$ is injective since Σ_2 is observable, so we may define $\overline{E}$ on its domain given in (9.2.2) by

$$\overline{E}x_1 = \mathfrak{C}_2^{-1}\mathfrak{C}_1 x_1, \quad x_1 \in \mathcal{D}\left(\overline{E}\right), \tag{9.2.3}$$

where $\mathfrak{C}_2^{-1}$ is the inverse of $\mathfrak{C}_2$ defined on $\mathcal{R}(\mathfrak{C}_2)$. If E is an arbitrary pseudo-similarity of the type described in Definition 9.2.1, then by the last equation in (9.2.1), $\mathfrak{C}_1 x \in \mathcal{R}(\mathfrak{C}_2)$ for all $x \in \mathcal{D}(E)$, i.e., $\mathcal{D}(E) \subset \mathcal{D}\left(\overline{E}\right)$. Moreover, it follows from the same equation that Ex_1 is given by (9.2.3) with $\overline{E}$ replaced by E for all $x \in \mathcal{D}(E)$. Thus, E is a restriction of $\overline{E}$. In particular, $\overline{E}$ is unique.

We next prove that $\overline{E}$ is closed. Fix $\omega > \max\{\omega_{\mathfrak{A}_1}, \omega_{\mathfrak{A}_2}\}$. By Theorem 2.5.4(ii),(iv), $\mathfrak{C}_i \in \mathcal{B}(X_i; L^p|Reg_\omega(\overline{\mathbb{R}}^+;Y))$ for $i = 1, 2$. Let $x_1^n \in \mathcal{D}\left(\overline{E}\right)$, $x_1^n \to x_1 \in X_1$, and $\overline{E}x_1^n \to x_2$ in X_2. We denote $\overline{E}x_1^n$ by x_2^n and $\mathfrak{C}_1 x_1^n$ by y^n. Then $x_2^n \in \mathcal{R}\left(\overline{E}\right)$, $y^n = \mathfrak{C}_1 x_1^n = \mathfrak{C}_2 x_2^n$, and by the continuity of $\mathfrak{C}_1$ and $\mathfrak{C}_2$,

$$\mathfrak{C}_1 x_1 = \lim_{n\to\infty} \mathfrak{C}_1 x_1^n = \lim_{n\to\infty} y_n = \lim_{n\to\infty} \mathfrak{C}_2 x_2^n = \mathfrak{C}_2 x_2.$$

This implies that $x_1 \in \mathcal{D}\left(\overline{E}\right)$ and that $\overline{E}x_1 = x_2$. Thus $\overline{E}$ is closed.

That $\mathfrak{B}_2 = \overline{E}\mathfrak{B}_1$ follows from the injectivity of $\mathfrak{C}_1$ and the fact that, for all $u \in L^p|Reg_c(\mathbb{R}^-;U)$,

$$\mathfrak{C}_1\mathfrak{B}_1 u = \pi_+\mathfrak{D}\pi_- u = \mathfrak{C}_2\mathfrak{B}_2 u,$$

hence $\mathfrak{B}_1 u \in \mathcal{D}\left(\overline{E}\right)$, $\mathfrak{B}_2 u \in \mathcal{R}\left(\overline{E}\right)$, and $\mathfrak{B}_2 = \overline{E}\mathfrak{B}_1$. In particular, $\mathcal{D}\left(\overline{E}\right)$ is dense in X_1 and $\mathcal{R}\left(\overline{E}\right)$ is dense in X_2 (the images of $L^p|Reg_c(\mathbb{R}^-;U)$ under $\mathfrak{B}_1$ and $\mathfrak{B}_2$ are dense in X_1, respectively X_2). The invariance of $\mathcal{D}\left(\overline{E}\right)$ under $\mathfrak{A}_1$, the invariance of $\mathcal{R}\left(\overline{E}\right)$ under $\mathfrak{A}_2$, and the identity $\overline{E}\mathfrak{A}_1^t = \mathfrak{A}_2^t\overline{E}$ follow from the intertwining properties of $\mathfrak{C}_1$ and $\mathfrak{C}_2$: if $x_1 \in \mathcal{D}\left(\overline{E}\right)$ and $x_2 = \overline{E}x_1 \in \mathcal{R}\left(\overline{E}\right)$, then for all $t \geq 0$,

$$\mathfrak{C}_1\mathfrak{A}_1^t x_1 = \tau_+^t\mathfrak{C}_1 x_1 = \tau_+^t\mathfrak{C}_2 x_2 = \mathfrak{C}_2\mathfrak{A}_2^t x_2,$$

hence $\mathfrak{A}_1^t x_1 \in \mathcal{D}\left(\overline{E}\right)$, $\mathfrak{A}_2^t x_2 \in \mathcal{R}\left(\overline{E}\right)$, and $\mathfrak{A}_2^t\overline{E}x_1 = \mathfrak{A}_2^t x_2 = \overline{E}\mathfrak{A}_1^t x_1$. By definition, $\mathfrak{C}_1 x_1 = \overline{E}\mathfrak{C}_2 x_2$ for all $x_1 \in \mathcal{D}\left(\overline{E}\right)$. Thus, $\overline{E}$ satisfies the conditions listed in Definition 9.2.1, and so the two systems are pseudo-similar.

Let us define $\underline{E}$ to be the closure of $\overline{E}_{|\underline{X}_1}$, where $\underline{X}_1$ is the restriction of $\overline{E}$ to the image of $L^p|Reg_c(\mathbb{R}^-;U)$ under $\mathfrak{B}_1$ (this operator is closable since $\overline{E}$ is closed). One of the properties required from the operator E in Definition 9.2.1 is that $\underline{X}_1 \subset \mathcal{D}(E)$. As E furthermore is closed, it must be an extension of $\underline{E}$. This proves the existence of a (unique) minimally defined pseudo-similarity $\underline{E}$. □

Unfortunately, the similarity in Theorem 9.2.4 is so weak that it is of limited value unless we have some additional information, such as the compactness of the resolvents of the semigroup generators, or exact controllability, or exact observability (these two notions are defined in Section 9.4); see Theorems 9.4.10 and 9.10.7. However, if X_1 or X_2 is finite-dimensional, then we recover the well-known fact that any two minimal realizations of a given rational transfer function are similar to each other (in particular, the dimensions of their state spaces are the same).

There are two other sets of results which resemble Theorem 9.2.4. They do not require the systems to be *minimal*, just *controllable or observable*, but instead they require either the pseudo-similarity E itself or its inverse to be bounded. Observe that if E or E^{-1} is bounded in Definition 9.2.1, then all the different similarities in Theorem 9.2.4 coincide, with $\mathcal{D}(E) = X_1$ (if E is bounded) or $\mathcal{R}(E) = X_2$ (if E^{-1} is bounded).

We begin with the observable case.

Theorem 9.2.5 *Let* $\Sigma_1 = \left[\begin{array}{c|c} \mathfrak{A}_1 & \mathfrak{B}_1 \\ \hline \mathfrak{C}_1 & \mathfrak{D} \end{array}\right]$ *and* $\Sigma_2 = \left[\begin{array}{c|c} \mathfrak{A}_2 & \mathfrak{B}_2 \\ \hline \mathfrak{C}_2 & \mathfrak{D} \end{array}\right]$ *on* (Y, X_1, U) *be two observable* $L^p|Reg$*-well-posed linear systems on* (Y, X_1, U)*, respectively* (Y, X_2, U) *(with the same input/output map* $\mathfrak{D}$*). Then* Σ_1 *and* Σ_2 *are pseudo-similar with a unique pseudo-similarity operator* E *if any one of the following conditions hold:*

(i) $\mathcal{R}(\mathfrak{C}_1) \subset \mathcal{R}(\mathfrak{C}_2)$ *and the range of* $\mathfrak{C}_2^{-1}\mathfrak{C}_1$ *is dense in* X_2*. In this case (and only in this case)* E *is bounded.*

(ii) $\mathcal{R}(\mathfrak{C}_2) \subset \mathcal{R}(\mathfrak{C}_1)$ *and the range of* $\mathfrak{C}_1^{-1}\mathfrak{C}_2$ *is dense in* X_1*. In this case (and only in this case)* E^{-1} *is bounded.*

(iii) $\mathcal{R}(\mathfrak{C}_1) = \mathcal{R}(\mathfrak{C}_2)$*. In this case (and only in this case) both* E *and* E^{-1} *are bounded.*

(iv) $\mathcal{R}(\mathfrak{C}_1) = \mathcal{R}(\mathfrak{C}_2)$*, and* $|x_1|_{X_1} = |x_2|_{X_2}$ *whenever* $x_1 \in X_1$*,* $x_2 \in X_2$*, and* $\mathfrak{C}_1 x_1 = \mathfrak{C}_2 x_2$*. In this case (and only in this case)* E *is an isometric isomorphism of* X_1 *onto* X_2*.*

In particular, if X_1*,* X_2*, and* Y *are Hilbert spaces, and if* Σ_1 *is state/output bounded in the* L^2*-sense (i.e.,* $\mathfrak{C}_1 \in \mathcal{B}(X_1; L^2(\mathbb{R}^+; Y))$*), then (iv) holds if and only if* Σ_2 *is state/output bounded in the* L^2*-sense and* $\mathfrak{C}_1\mathfrak{C}_1^* = \mathfrak{C}_2\mathfrak{C}_2^*$*.*

Proof (i) If (i) holds, then we can define E on X_1 by $E = \mathfrak{C}_2^{-1}\mathfrak{C}_1$ (since $\mathfrak{C}_2$ is injective and $\mathcal{R}(\mathfrak{C}_1) \subset \mathcal{R}(\mathfrak{C}_2)$). This operator is injective since $\mathfrak{C}_1$ is injective. Arguing as in the proof of Theorem 9.2.4 we find that E is closed. By the closed graph theorem, E is bounded, and by assumption, $\mathcal{R}(E)$ is dense in X_2. That (9.2.1) holds is shown as in the proof of Theorem 9.2.4. Conversely, if E is a bounded pseudo-similarity, then $\mathcal{D}(E) = X_1$, and it follows from Definition 9.2.1 that $\mathcal{R}(\mathfrak{C}_1) \subset \mathcal{R}(\mathfrak{C}_2)$ and that $E = \mathfrak{C}_2^{-1}\mathfrak{C}_1$. In particular, $\mathcal{R}\left(\mathfrak{C}_2^{-1}\mathfrak{C}_1\right) = \mathcal{R}(E)$ is dense in X_2.

(ii) This follows from (i) if we interchange Σ_1 and Σ_2.

(iii) Define $E = \mathfrak{C}_2^{-1}\mathfrak{C}_1$. Arguing as in the proof of part (i) we find that E is a bounded injection of X_1 into X_2. By interchanging the roles of Σ_1 and Σ_2 we find that $\mathcal{R}(E) = X_2$ and that E^{-1} is bounded. The rest follows from (i) and (iii).

(iv) This follows from (iii) (note that $\mathfrak{C}_2 x_2 = \mathfrak{C}_1 x_1$ if and only if $x_2 = E x_1$).

For the final claim, let us first observe that $\mathfrak{C}_1 \in \mathcal{B}(X_1; L^2(\mathbb{R}^+; Y))$ if and only if $\mathfrak{C}_2 \in \mathcal{B}(X_1; L^2(\mathbb{R}^+; Y))$ whenever E and E^{-1} are bounded, and that both $\mathfrak{C}_1\mathfrak{C}_1^*$ and $\mathfrak{C}_2\mathfrak{C}_2^*$ belong to $\mathcal{B}(L^2(\mathbb{R}^+; Y))$ in this (and only in this) case. If E is unitary and $\mathfrak{C}_1\mathfrak{C}_1^*$ and $\mathfrak{C}_2\mathfrak{C}_2^*$ are bounded, then $\mathfrak{C}_1\mathfrak{C}_1^* = \mathfrak{C}_2 E E^* \mathfrak{C}_2^* = \mathfrak{C}_2\mathfrak{C}_2^*$. Conversely, suppose that these two operators are bounded, and that $\mathfrak{C}_1\mathfrak{C}_1^* = \mathfrak{C}_2\mathfrak{C}_2^*$. Then, by Lemma A.2.5 (iv), $\mathcal{R}(\mathfrak{C}_1) = \mathcal{R}\left(\sqrt{\mathfrak{C}_1\mathfrak{C}_1^*}\right) = \mathcal{R}\left(\sqrt{\mathfrak{C}_2\mathfrak{C}_2^*}\right) = \mathcal{R}(\mathfrak{C}_2)$, so (iii) applies. Define $E = \mathfrak{C}_2^{-1}\mathfrak{C}_1$. Then $\mathfrak{C}_2\mathfrak{C}_2^* = \mathfrak{C}_1\mathfrak{C}_1^* = \mathfrak{C}_2 E E^* \mathfrak{C}_2^*$. Since $\mathfrak{C}_2$ is injective and $\mathfrak{C}_2^*$ has dense range, this implies that $EE^* = 1$. As E is invertible, this implies that E is unitary. $\square$

The controllable case is similar.

Theorem 9.2.6 *Let* $\Sigma_1 = \left[\begin{array}{c|c} \mathfrak{A}_1 & \mathfrak{B}_1 \\ \hline \mathfrak{C}_1 & \mathfrak{D} \end{array}\right]$ *and* $\Sigma_2 = \left[\begin{array}{c|c} \mathfrak{A}_2 & \mathfrak{B}_2 \\ \hline \mathfrak{C}_2 & \mathfrak{D} \end{array}\right]$ *on* (Y, X_1, U) *be two controllable* $L^p|Reg$*-well-posed linear systems on* (Y, X_1, U)*, respectively* (Y, X_2, U) *(with the same input/output map* $\mathfrak{D}$*). Then* Σ_1 *and* Σ_2 *are pseudo-similar with a unique pseudo-similarity operator E if any one of the following conditions hold:*

(i) *There is a finite constant M such that* $|\mathfrak{B}_2 u|_{X_2} \le M|\mathfrak{B}_1 u|_{X_1}$ *for all* $u \in L^p|Reg_c(\mathbb{R}^-; U)$*. In this case (and only in this case) E is bounded and* $\|E\| \le M$*.*

(ii) *There is a constant* $m > 0$ *such that* $|\mathfrak{B}_2 u|_{X_2} \ge m|\mathfrak{B}_1 u|_{X_1}$ *for all* $u \in L^p|Reg_c(\mathbb{R}^-; U)$*. In this case (and only in this case)* E^{-1} *is bounded and* $\|E^{-1}\| \le 1/m$*.*

(iii) *There exist constants* $0 < m \le M < \infty$ *such that* $m|\mathfrak{B}_1 u|_{X_1} \le |\mathfrak{B}_2 u|_{X_2} \le M|\mathfrak{B}_1 u|_{X_1}$ *for all* $u \in L^p|Reg_c(\mathbb{R}^-; U)$*. In this case (and only in this case) both E and* E^{-1} *are bounded,* $\|E\| \le M$*, and* $\|E^{-1}\| \le 1/m$*.*

(iv) $|\mathfrak{B}_2 u|_{X_2} = |\mathfrak{B}_1 u|_{X_1}$ *for all* $u \in L^p|Reg_c(\mathbb{R}^-; U)$. *In this case (and only in this case)* E *is an isometric isomorphism of* X_1 *onto* X_2.

In particular, if X_1, X_2, *and* Y *are Hilbert spaces, and if* Σ_1 *is input/state bounded in the* L^2*-sense (i.e.,* $\mathfrak{B}_1 \in \mathcal{B}(L^2(\mathbb{R}^-; X_1)))$*, then (iv) holds if and only if* Σ_2 *is input/state bounded in the* L^2*-sense and* $\mathfrak{B}_1^*\mathfrak{B}_1 = \mathfrak{B}_2^*\mathfrak{B}_2$.

Proof Choose $\omega > \max\{\omega_{\mathfrak{A}_1}, \omega_{\mathfrak{A}_2}\}$, where $\omega_{\mathfrak{A}_1}$ and $\omega_{\mathfrak{A}_2}$ are the growth bounds of Σ_1, respectively Σ_2. Then $\mathfrak{B}_i \in \mathcal{B}(L^p|Reg_\omega(\mathbb{R}^-; U); X_i)$ for $i = 1, 2$, and the conditions listed in (i)–(iv) are equivalent to the same conditions with $L^p|Reg_c(\mathbb{R}^-; U)$ replaced by $L^p|Reg_\omega(\mathbb{R}^-; U)$. Therefore, we may throughout replace $L^p|Reg_c(\mathbb{R}^-; U)$ by the Banach space $L^p|Reg_\omega(\mathbb{R}^-; U)$ in the proof below (see also Lemma 9.1.3(i)).

(i) The necessity of the given inequality for the boundedness of E is obvious. Conversely, suppose that the inequality in (i) holds. Then $\mathcal{N}(\mathfrak{B}_1) \subset \mathcal{N}(\mathfrak{B}_2)$. For each x_1 of the form $x_1 = \mathfrak{B}_1 u$ for some $u \in L^p|Reg_\omega(\mathbb{R}^-; U)$, define $Ex_1 = \mathfrak{B}_2\mathfrak{B}_1^{-1}x_1$, where $\mathfrak{B}_1$ and $\mathfrak{B}_2$ stand for the inductive operators induced by $\mathfrak{B}_1$ and $\mathfrak{B}_2$ on $L^p|Reg_\omega(\mathbb{R}^-; U)/\mathcal{N}(\mathfrak{B}_2)$ (see Definition 9.1.5). Then $\mathfrak{B}_2 u = E\mathfrak{B}_1 u$ for all $u \in L^p|Reg_\omega(\mathbb{R}^-; U)$, and by the controllability of Σ_1 and Σ_2, $\mathcal{D}(E) = \mathcal{R}(\mathfrak{B}_1)$ is dense in X_1 and $\mathcal{R}(E) = \mathcal{R}(\mathfrak{B}_1)$ is dense in X_1. The inequality in (i) implies that E is bounded with $\|E\| \le M$ on $\mathcal{D}(E)$, so E extends to a unique (closed) operator in $\mathcal{B}(X_1; X_2)$ (with the same norm), which we still denote by E. (After this $\mathcal{D}(E) = X_1$, and $\mathcal{R}(E)$ is still dense in X_2.)

It follows from our definition of E that $\mathfrak{B}_2 u = E\mathfrak{B}_1 u$ for all $u \in L^p|Reg_\omega(\mathbb{R}^-; U)$. Therefore

$$\mathfrak{C}_1\mathfrak{B}_1 u = \pi_+\mathfrak{D}\pi_- u = \mathfrak{C}_2\mathfrak{B}_2 u = \mathfrak{C}_2 E\mathfrak{B}_1 u.$$

As the range of $\mathfrak{B}_1$ is dense, this gives $\mathfrak{C}_1 = \mathfrak{C}_2 E$. By the intertwining conditions for $\mathfrak{B}_1$ and $\mathfrak{B}_2$, for all $u \in L^p|Reg_\omega(\mathbb{R}^-; U)$ and for all $t \ge 0$,

$$E\mathfrak{A}_1^t\mathfrak{B}_1 u = E\mathfrak{B}_1\tau_-^t u = \mathfrak{B}_2\tau_-^t u = \mathfrak{A}_2^t\mathfrak{B}_2 u = \mathfrak{A}_2^t E\mathfrak{B}_1 u.$$

As the range of $\mathfrak{B}_1$ is dense, this gives $E\mathfrak{A}_1^t = \mathfrak{A}_2^t E$. Thus, the two systems are pseudo-similar with the bounded similarity operator E.

(ii) This follows from (i) if we interchange Σ_1 and Σ_2.

(iii) This follows from (i) and (ii).

(iv) This is a special case of (iii).

For the final claim, it is obvious that Σ_1 is input/state bounded in the L^2-sense if and only if Σ_2 is so whenever the two systems are pseudo-similar bounded E and E^{-1}, and that this is equivalent to the boundedness of $\mathfrak{B}_1^*\mathfrak{B}_1$ and $\mathfrak{B}_2^*\mathfrak{B}_2$ in $L^2(\mathbb{R}^-; U)$. In this case the condition that $|\mathfrak{B}_2 u|_{X_2} = |\mathfrak{B}_1 u|_{X_1}$ can be written as

$$\begin{aligned}\langle \mathfrak{B}_1^*\mathfrak{B}_1 u, u\rangle_{L^2(\mathbb{R}^-;U)} &= \langle \mathfrak{B}_1 u, \mathfrak{B}_1 u\rangle_{X_1} = \langle \mathfrak{B}_2 u, \mathfrak{B}_2 u\rangle_{X_2} \\ &= \langle \mathfrak{B}_2^*\mathfrak{B}_2 u, u\rangle_{L^2(\mathbb{R}^-;U)},\end{aligned}$$

which is equivalent to $\mathfrak{B}_1^*\mathfrak{B}_1 = \mathfrak{B}_2^*\mathfrak{B}_2$. □

9.3 Realizations based on factorizations of the Hankel operator

The algebraic condition $\pi_+\mathfrak{D}\pi_- = \mathfrak{C}\mathfrak{B}$ in the definition of an $L^p|Reg$-well-posed linear system says that the Hankel operator $\pi_+\mathfrak{D}\pi_-$ of the input/output map $\mathfrak{D}$ of the system $\Sigma = \left[\begin{smallmatrix}\mathfrak{A} & \mathfrak{B} \\ \mathfrak{C} & \mathfrak{D}\end{smallmatrix}\right]$ factors into the product of the input map $\mathfrak{B}$ and the output map $\mathfrak{C}$. The purpose of this section is to prove the converse: if $\mathfrak{D}$ is time-invariant and causal (as all input/output maps are), then every factorization of the Hankel operator of $\mathfrak{D}$ which satisfies some additional necessary admissibility conditions induces an $L^p|Reg$-well-posed linear system. Actually, there are two such sets of implicit admissibility conditions (which are related to each other), one associated with the input map $\mathfrak{B}$, and another associated with the output map $\mathfrak{C}$. We begin by discussing the implicit admissibility conditions satisfied by every input map $\mathfrak{B}$.

Let $\mathfrak{B}$ be an $L^p|Reg$-admissible input map for the C_0 semigroup $\mathfrak{A}$. Then $\mathfrak{B}$ and $\mathfrak{A}$ satisfy the intertwining condition $\mathfrak{A}^t\mathfrak{B} = \mathfrak{B}\tau_-^t$. Since $\mathfrak{A}$ is locally bounded, this implies that for every $t > 0$ there exists a finite constant $K(t) > 0$ such that

$$\sup_{0<s<t} |\mathfrak{B}\tau_-^s u|_X \le K(t)|\mathfrak{B}u|_X, \quad u \in L^p|Reg_c(\mathbb{R}^-;U), \quad t > 0, \tag{9.3.1}$$

and the strong continuity of $\mathfrak{A}$ implies that

$$\lim_{t\downarrow 0}|\mathfrak{B}\tau_-^t u - \mathfrak{B}u|_X = 0, \quad u \in L^p|Reg_c(\mathbb{R}^-;U). \tag{9.3.2}$$

In particular, it follows from (9.3.1) that

$$\mathcal{N}\left(\mathfrak{B}\tau_-^t\right) \supset \mathcal{N}(\mathfrak{B}), \quad t \ge 0. \tag{9.3.3}$$

Out of these conditions the most important one is (9.3.1), since (9.3.2) is redundant in the L^p-well-posed case with $p < \infty$ (the left shift on $\mathbb{R}^-$ is strongly continuous in L^p with $p < \infty$; see Example 2.3.2(iii)), and since (9.3.3) is implied by (9.3.1).

An analogous calculation can be carried out for an $L^p|Reg$-admissible output map $\mathfrak{C}$ of a C_0 semigroup $\mathfrak{A}$. The intertwining condition $\mathfrak{C}\mathfrak{A}^t = \tau_+^t\mathfrak{C}$ implies that

$$\mathcal{R}\left(\tau_+^t \mathfrak{C}\right) \subset \mathcal{R}(\mathfrak{C}), \quad t \geq 0. \tag{9.3.4}$$

The operator $\mathfrak{C}$ induces an injective operator on $X/\mathcal{N}(\mathfrak{C})$ (see Definition 9.1.5). We denote the (closed but possibly unbounded) inverse of this operator, defined on $\mathcal{R}(\mathfrak{C})$, by $\mathfrak{C}^{-1}$ (see Definition 9.1.5). Because of (9.3.4), the operator $\mathfrak{C}^{-1}\tau_+^t\mathfrak{C}$ is well-defined (single-valued) as an operator from $X/\mathcal{N}(\mathfrak{C})$ into itself. This operator is closed (both $\mathfrak{C}$ and $\tau_+^t\mathfrak{C}$ are bounded), hence bounded, and it follows from the intertwining condition that the operator which $\mathfrak{C}^{-1}\tau_+^t\mathfrak{C}$ induces on $X/\mathcal{N}(\mathfrak{C})$ (see Definition 9.1.5) is nothing but the quotient semigroup $\pi\mathfrak{A}^t$ induced by $\mathfrak{A}^t$ in $X/\mathcal{N}(\mathfrak{C})$ (see Definition 9.1.7); here π is the quotient map $X \to X/\mathcal{N}(\mathfrak{C})$. Thus, by the strong continuity of the quotient semigroup,

$$\lim_{t\downarrow 0} |\mathfrak{C}^{-1}\tau_+^t\mathfrak{C}x - \pi x|_{X/\mathcal{N}(\mathfrak{C})} = 0, \quad x \in X. \tag{9.3.5}$$

Furthermore, the local boundedness of the quotient semigroup implies that

$$\sup_{0<s<t} \|\mathfrak{C}^{-1}\tau_+^s\mathfrak{C}\| < \infty, \quad t > 0, \tag{9.3.6}$$

but this property is less interesting than (9.3.4) and (9.3.5).

Theorem 9.3.1 *Let $\mathfrak{D} \in TIC_{\text{loc}}(U;Y)$ (i.e, $\mathfrak{D}$ is time-invariant and causal; see Definition 2.6.2), and suppose that the Hankel operator $\pi_+\mathfrak{D}\pi_-$ of $\mathfrak{D}$ factors into $\pi_+\mathfrak{D}\pi_- = \mathfrak{C}\mathfrak{B}$, where $\mathfrak{B}\colon L^p|Reg_c(\mathbb{R}^-;U) \to X$ and $\mathfrak{C}\colon X \to L^p|Reg_{\text{loc}}(\overline{\mathbb{R}}^+;Y)$ are bounded linear operators.*

(i) *If $\mathfrak{B}$ has dense range then* (9.3.1) *and* (9.3.2) *imply* (9.3.4)–(9.3.6) *(condition* (9.3.2) *is redundant in the L^p-case with $p < \infty$).*
(ii) *If $\mathfrak{C}$ is injective, then* (9.3.4) *and* (9.3.5) *imply* (9.3.1)–(9.3.3).
(iii) *Let $X_{\mathfrak{B}} = \overline{\mathcal{R}(\mathfrak{B})}$. Then the following conditions are equivalent:*
  (a) *conditions* (9.3.1) *and* (9.3.2) *hold (condition* (9.3.2) *is redundant in the L^p-case with $p < \infty$);*
  (b) *there is a (unique) semigroup $\mathfrak{A}_{\mathfrak{B}}$ on $X_{\mathfrak{B}}$ such that $\left[\begin{array}{c|c} \mathfrak{A}_{\mathfrak{B}} & \mathfrak{B} \\ \hline \mathfrak{C}_{\mathfrak{B}} & \mathfrak{D} \end{array}\right]$ is a controllable $L^p|Reg$-well-posed linear system on $(Y, X_{\mathfrak{B}}, U)$, where $\mathfrak{C}_{\mathfrak{B}}$ is the restriction of $\mathfrak{C}$ to $X_{\mathfrak{B}}$.*
(iv) *Let $X_{\mathfrak{C}} = X/\mathcal{N}(\mathfrak{C})$ and let π be the quotient map $X \to X/\mathcal{N}(\mathfrak{C})$. Then the following conditions are equivalent:*
  (a) *conditions* (9.3.4) *and* (9.3.5) *hold;*
  (b) *there is a (unique) semigroup $\mathfrak{A}_{\mathfrak{C}}$ on $X_{\mathfrak{C}}$ such that $\left[\begin{array}{c|c} \mathfrak{A}_{\mathfrak{C}} & \mathfrak{B}_{\mathfrak{C}} \\ \hline \widetilde{\mathfrak{C}} & \mathfrak{D} \end{array}\right]$ is an observable well-posed linear system on $(Y, X_{\mathfrak{C}}, U)$, where $\mathfrak{B}_{\mathfrak{C}} = \pi\mathfrak{B}$ and $\widetilde{\mathfrak{C}}$ is the (injective) operator induced on $X/\mathcal{N}(\mathfrak{C})$ by $\mathfrak{C}$.*
(v) *Suppose that the factorization is minimal, i.e., $\mathfrak{B}$ has dense range and $\mathfrak{C}$ is injective. Then the following conditions are equivalent:*

(a) *conditions* (9.3.1) *and* (9.3.2) *hold (condition* (9.3.2) *is redundant in the L^p-case with $p < \infty$);*
(b) *conditions* (9.3.4) *and* (9.3.5) *hold;*
(c) *there is a (unique) semigroup $\mathfrak{A}$ on X such that $\left[\begin{smallmatrix} \mathfrak{A} & \mathfrak{B} \\ \mathfrak{C} & \mathfrak{D} \end{smallmatrix}\right]$ is a minimal $L^p|Reg$-well-posed linear system.*

The proof of Theorem 9.3.1 is based on the following identity:

Lemma 9.3.2 *The Hankel operator $\pi_+\mathfrak{D}\pi_-$ of $\mathfrak{D}$ satisfies*

$$\tau_+^t \pi_+ \mathfrak{D}\pi_- = \pi_+\mathfrak{D}\pi_-\tau_-^t, \quad t \ge 0.$$

In particular, if $\pi_+\mathfrak{D}\pi_-$ factors into $\pi_+\mathfrak{D}\pi_- = \mathfrak{C}\mathfrak{B}$, then

$$\tau_+^t \mathfrak{C}\mathfrak{B} = \mathfrak{C}\mathfrak{B}\tau_-^t, \quad t \ge 0.$$

Proof Use the time-invariance of $\mathfrak{D}$ to get

$$\tau_+^t \pi_+\mathfrak{D}\pi_- = \pi_+\tau^t\mathfrak{D}\pi_- = \pi_+\mathfrak{D}\tau^t\pi_- = \pi_+\mathfrak{D}\pi_-\tau_-^t.$$

□

Proof of Theorem 9.3.1 (i) This follows from (iii) and the necessity of the conditions (9.3.4)–(9.3.6); cf. the derivation of these conditions given above.

(ii) This follows from (iv) and the necessity of the conditions (9.3.1); cf. the derivation of these conditions given above.

(iii) The argument that we used above to derive (9.3.1) shows that (b) ⇒ (a). (The condition $X_{\mathfrak{B}} = \overline{\mathcal{R}(\mathfrak{B})}$ is not needed for this part.)

Conversely, suppose that (a) holds. Without loss of generality, we may assume that $X_{\mathfrak{B}} = X$ (replace X by $X_{\mathfrak{B}}$). The idea is to use the intertwining condition $\mathfrak{A}^t\mathfrak{B} = \mathfrak{B}\tau_-^t$ as a definition of $\mathfrak{A}^t$. Clearly, for this to be possible, the range of $\mathfrak{B}$ must be dense in X.

We let $\mathfrak{B}^{-1}$ represent the (closed but possibly unbounded) inverse of the injective operator that $\mathfrak{B}$ induces on $L^p|Reg_c(\mathbb{R}^-;U)/\mathcal{N}(\mathfrak{B})$ (see Definition 9.1.5). By (9.3.3) (which is implied by (9.3.1)), the operator $\mathfrak{B}\tau_-^t$ induces an operator on $L^p|Reg_c(\mathbb{R}^-;U)/\mathcal{N}(\mathfrak{B})$. For simplicity, we use the same notation $\mathfrak{B}\tau_-^t$ for this induced operator. Then it follows from (9.3.1) that the operator $\mathfrak{B}\tau^t\pi_-\mathfrak{B}^{-1}$ is (well-defined and) bounded on $\mathcal{R}(\mathfrak{B})$. Thus, for each $x = \mathfrak{B}u \in \mathcal{R}(\mathfrak{B})$ and $t \in \overline{\mathbb{R}}^+$, we can define $\mathfrak{A}^t : \mathcal{R}(\mathfrak{B}) \to \mathcal{R}(\mathfrak{B}) \subset X$ by

$$\mathfrak{A}^t x = \mathfrak{B}\tau_-^t\mathfrak{B}^{-1}x.$$

Obviously $\mathfrak{A}^0 x = x$ for each $x \in \mathcal{R}(\mathfrak{B})$, and $\mathfrak{A}$ inherits the semigroup property $\mathfrak{A}^{s+t} = \mathfrak{A}^s\mathfrak{A}^t$ from τ_-^t. Condition (9.3.2) implies that $t \mapsto \mathfrak{A}^t x$ is continuous in X for every $x \in \mathcal{R}(\mathfrak{B})$ (this condition is redundant in the L^p-case with $p < \infty$ since τ_- is strongly continuous on $L^p_c(\mathbb{R}^-;U)$; see Example 2.3.2(iii)). Thus,

$\mathfrak{A}$ is a strongly continuous semigroup on $\mathcal{R}(\mathfrak{B})$ (where the continuity is in the norm of X). Because of (9.3.1), this semigroup is also locally bounded (in the norm of X).

Next we extend $\mathfrak{A}$ to a strongly continuous semigroup on X. For each t, $\mathfrak{A}^t$ is bounded and densely defined, so it can be extended to an operator in $\mathcal{B}(X)$ (which we still denote by $\mathfrak{A}^t$). Because of (9.3.1), we have a bound on $\|\mathfrak{A}^s\|$ which is uniform in s on any bounded interval. If we let $\mathcal{R}(\mathfrak{B}) \supset x_n \to x \in X$, then the sequence $\mathfrak{A}^s x_n$ is uniformly continuous in s and uniformly convergent on any bounded interval, so the limit $\mathfrak{A}^s x$ is (locally bounded and) continuous in s. The extended operator $\mathfrak{A}^t$ inherits not only the strong continuity, but also the semigroup property and the intertwining property $\mathfrak{A}^t\mathfrak{B} = \mathfrak{B}\tau_-^t$ from the original densely defined $\mathfrak{A}$.

It remains to show that $\mathfrak{A}$ also satisfies the second intertwining condition $\mathfrak{C}\mathfrak{A}^t = \tau_+^t\mathfrak{C}$ for all $t \geq 0$. By the density of $\mathcal{R}(\mathfrak{B})$ in X, it suffices to show that $\mathfrak{C}\mathfrak{A}^t\mathfrak{B} = \tau_+^t\mathfrak{C}\mathfrak{B}$, and this is an immediate consequence of Lemma 9.3.2: $\mathfrak{C}\mathfrak{A}^t\mathfrak{B} = \mathfrak{C}\mathfrak{B}\tau_-^t = \tau_+^t\mathfrak{C}\mathfrak{B}$.

(iv) We may assume, without loss of generality, that $\mathfrak{C}$ is injective, because otherwise we may replace X by $X/\mathcal{N}(\mathfrak{C})$, $\mathfrak{B}$ by $\pi\mathfrak{B}$, and $\mathfrak{C}$ by the operator induced on $X/\mathcal{N}(\mathfrak{C})$ by $\mathfrak{C}$.

For each $x \in X$ and $t \geq 0$, we define $\mathfrak{A}^t X \to X$ by

$$\mathfrak{A}^t x = \mathfrak{C}^{-1}\tau_+^t\mathfrak{C}x.$$

As we observed in the discussion leading up to (9.3.5), $\mathfrak{A}^t \in \mathcal{B}(X)$. Clearly $\mathfrak{A}^0 = 1$, and $\mathfrak{A}$ inherits the semigroup property $\mathfrak{A}^{s+t} = \mathfrak{A}^s\mathfrak{A}^t$ from τ_+. By (9.3.5), $\mathfrak{A}$ is strongly continuous. Thus, $\mathfrak{A}$ is a C_0 semigroup on X. Moreover, by construction, the intertwining condition $\mathfrak{C}\mathfrak{A}^t = \tau_+^t\mathfrak{C}$ holds for all $t \geq 0$.

It remains to show that $\mathfrak{A}$ also satisfies the second intertwining condition $\mathfrak{A}^t\mathfrak{B} = \mathfrak{B}\tau_-^t$ for all $t \geq 0$. Since $\mathfrak{C}$ is injective, it suffices to show that $\mathfrak{C}\mathfrak{A}^t\mathfrak{B} = \mathfrak{C}\mathfrak{B}\tau_-^t$, and this is an immediate consequence of Lemma 9.3.2: $\mathfrak{C}\mathfrak{A}^t\mathfrak{B} = \tau_+^t\mathfrak{C}\mathfrak{B} = \mathfrak{C}\mathfrak{B}\tau_-^t$.

(v) This follows from (i)–(iv). □

Corollary 9.3.3 *Let $\mathfrak{D} \in TIC_{\mathrm{loc}}(U;Y)$, and suppose that the Hankel operator $\pi_+\mathfrak{D}\pi_-$ of $\mathfrak{D}$ factors into $\pi_+\mathfrak{D}\pi_- = \mathfrak{C}\mathfrak{B}$, where $\mathfrak{B}\colon L^p|Reg_c(\mathbb{R}^-;U) \to X$ and $\mathfrak{C}\colon X \to L^p|Reg_{\mathrm{loc}}(\overline{\mathbb{R}}^+;Y)$ are bounded linear operators. If (9.3.1) or (9.3.4)–(9.3.5) hold, then $\mathfrak{D} \in TIC_\omega(U;Y)$ for some $\omega \in \mathbb{R}$ (i.e., $\mathfrak{D}$ has a finite growth bound).*

Proof This follows from Theorem 9.3.1 and its proof. In particular, we observe that the strong continuity of $\mathfrak{A}^t$ in part (iii) is not needed for the finite growth bound of $\mathfrak{A}$; see Remark 2.5.5. □

We remark that two of the shift realizations in Theorems 2.6.6 and 2.6.7 are special cases of Theorem 9.3.1. To get the exactly controllable shift realization we take $X = L^p_\omega(\mathbb{R}^-;U)$, $\mathfrak{B} = 1$, and $\mathfrak{C} = \pi_+\mathfrak{D}\pi_-$, and to get the exactly observable shift realization we take $X = L^p_\omega(\mathbb{R}^+;Y)$ or $X = BC_{0,\omega}(\overline{\mathbb{R}}^+;Y)$, $\mathfrak{B} = \pi_+\mathfrak{D}\pi_-$, and $\mathfrak{C} = 1$.

9.4 Exact controllability and observability

As we mentioned at the end of Section 9.2, we need stronger notions of controllability and/or observability if we want to get a stronger version of Theorem 9.2.4. The following is a list of commonly used controllability notions (the corresponding list of possible observability notions is given in Definition 9.4.2). For completeness we repeat the definition of controllability given in Definition 9.1.2(iii).

Definition 9.4.1 A $L^p|Reg$-well-posed linear system $\Sigma = \left[\begin{smallmatrix}\mathfrak{A} & \mathfrak{B}\\ \mathfrak{C} & \mathfrak{D}\end{smallmatrix}\right]$ on (Y, X, U) is

(i) *exactly controllable in time $t > 0$* if $\mathcal{R}\left(\mathfrak{B}\pi_{[-t,0)}\right) = X$;
(ii) *exactly controllable in finite time* if it is exactly controllable in time t for some $t > 0$;
(iii) (approximately) *controllable in time $t > 0$* if $\mathcal{R}\left(\mathfrak{B}\pi_{[-t,0)}\right)$ is dense in X;
(iv) (approximately) *controllable in finite time* if it is controllable in time t for some $t > 0$;
(v) (exactly) *null controllable in time $t > 0$* if $\mathcal{R}\left(\mathfrak{B}\pi_{[-t,0)}\right) \supset \mathcal{R}\left(\mathfrak{A}^t\right)$;
(vi) (exactly) *null controllable in finite time* if it is null controllable in time t for some $t > 0$;
(vii) (approximately) *controllable* (in infinite time) if $\mathfrak{B}$ maps $L^p|Reg_c(\mathbb{R}^-;U)$ onto a dense subset of X;
(viii) *exactly controllable in infinite time* if $\mathfrak{B}$ is stable (i.e., $\mathfrak{B} \in \mathcal{B}(L^p|Reg(\mathbb{R}^-;U);X))$ and $\mathfrak{B}$ maps $L^p|Reg(\mathbb{R}^-;U)$ onto X;
(ix) *exactly controllable in infinite time* with bound $\omega \in \mathbb{R}$ if $\mathfrak{B}$ is ω-bounded (i.e., $\mathfrak{B} \in \mathcal{B}(L^p|Reg_\omega(\mathbb{R}^-;U);X))$ and $\mathfrak{B}$ maps $L^p|Reg_\omega(\mathbb{R}^-;U)$ onto X.

The first seven of these notions have simple interpretations: according to Definition 2.2.7, the system Σ is

(i) exactly controllable in time t iff it is possible to find, for each $x_0 \in X$ and $x_1 \in X$, an input $u \in L^p|Reg([0,t);U)$ such that $x(t) = x_1$, where x is the state of Σ with initial time zero, initial state x_0, and input u,

(ii) exactly controllable in finite time iff there exists a $t > 0$ for which the conclusion of (i) is valid,
(iii) controllable in time t iff it is possible to find, for each $x_0 \in X$, $x_1 \in X$, and $\epsilon > 0$, an input $u \in L^p|Reg([0,t);U)$ such that $|x(t) - x_1| \le \epsilon$, where x is the state of Σ with initial time zero, initial state x_0, and input u,
(iv) controllable in finite time iff there exists a $t > 0$ for which the conclusion of (iii) is valid,
(v) null controllable in time t iff it is possible to find, for each $x_0 \in X$, an input $u \in L^p|Reg([0,t);U)$ such that $x(t) = 0$, where x is the state of Σ with initial time zero, initial state x_0, and input u,
(vi) null controllable in finite time iff there exists a $t > 0$ for which the conclusion of (v) is valid.
(vii) controllable iff it is possible to find, for each $x_1 \in X$ and $\epsilon > 0$, a time $t > 0$ and an input $u \in L^p|Reg([0,t);U)$ such that $|x(t) - x_1| \le \epsilon$, where x is the state of Σ with initial time zero, initial state zero, and input u.

The last two conditions (viii) and (ix) are those which are important in the strengthening of Theorem 9.2.4 (see Theorem 9.4.10). Condition (i) in strongest since it implies all the others. We shall not say much about (v) and (vi), but they can be used to insure that the 'finite cost condition' in optimal control is satisfied. If we exclude these two conditions, then (vii) is the weakest remaining one, i.e., it is implied by all the others.

The corresponding observability notions are defined as follows:

Definition 9.4.2 A $L^p|Reg$-well-posed linear system $\Sigma = \left[\begin{smallmatrix} \mathfrak{A} & \mathfrak{B} \\ \mathfrak{C} & \mathfrak{D} \end{smallmatrix}\right]$ on (Y, X, U) is

(i) *exactly observable in time* $t > 0$ if $\pi_{[0,t)}\mathfrak{C}$ has a bounded left inverse, defined on $\mathcal{R}\left(\pi_{[0,t)}\mathfrak{C}\right)$ (where we regard $\pi_{[0,t)}\mathfrak{C}$ as a mapping $X \to L^p|Reg((0,t);Y)$);
(ii) *exactly observable in finite time* if it is exactly observable in time t for some $t > 0$;
(iii) (approximately) *observable in time* $t > 0$ if $\pi_{[0,t)}\mathfrak{C}$ is injective on X;
(iv) (approximately) *observable in finite time* if it is observable in time t for some $t > 0$;
(v) (exactly) *final state observable in time* $t > 0$ if $\mathcal{N}\left(\mathfrak{A}^t\right) \supset \mathcal{N}\left(\pi_{[0,t)}\mathfrak{C}\right)$, and the operator[2] $\mathfrak{A}^t(\pi_{[0,t)}\mathfrak{C})^{-1}$ is bounded $X \to X$;

[2] Here $(\pi_{[0,t)}\mathfrak{C})^{-1}$ stands for the inverse of the injective operator that $\pi_{[0,t)}\mathfrak{C}$ induces on $X/\mathcal{N}\left(\pi_{[0,t)}\mathfrak{C}\right)$ and $\mathfrak{A}^t$ stands for the operator induced by $\mathfrak{A}^t$ on $X/\mathcal{N}\left(\pi_{[0,t)}\mathfrak{C}\right)$. See Definition 9.1.5.

(vi) (exactly) *final state observable in infinite time* if it is final state observable in time t for some $t > 0$;

(vii) (approximately) *observable* (in infinite time) if $\mathfrak{C}$ is injective as a mapping $X \to L^p|Reg_{loc}(\overline{\mathbb{R}}^+; U)$;

(viii) *exactly observable in infinite time* if $\mathfrak{C}$ is stable (i.e., $\mathfrak{C} \in \mathcal{B}(X; L^p|Reg(\overline{\mathbb{R}}^+; Y))$ and $\mathfrak{C}$ has a bounded left inverse defined on $\mathcal{R}(\mathfrak{C}) \subset L^p|Reg(\overline{\mathbb{R}}^+; Y)$;

(ix) *exactly observable in infinite time with bound* $\omega \in \mathbb{R}$ if $\mathfrak{C}$ is ω-bounded (i.e., $\mathfrak{C} \in \mathcal{B}(X; L^p|Reg_\omega(\overline{\mathbb{R}}^+; Y))$ and $\mathfrak{C}$ has a bounded left inverse defined on $\mathcal{R}(\mathfrak{C}) \subset L^p|Reg_\omega(\overline{\mathbb{R}}^+; Y)$.

As in the case of Definition 9.4.1, the last two conditions are of a technical nature, but the first seven have simple explanations: According to Definition 2.2.7, Σ is

(i) exactly observable in time t iff it is possible to reconstruct the initial state x_0 from the restriction of the observation $\mathfrak{C}x_0$ to the interval $[0, t)$, and the reconstruction operator (i.e., the left inverse of $\pi_{[0,t)}\mathfrak{C}$ defined on the range of $\pi_{[0,t)}\mathfrak{C}$) is bounded;

(ii) exactly observable in finite time iff there exists a $t > 0$ for which the conclusion of (i) is valid;

(iii) observable in time t iff it is possible to reconstruct the initial state x_0 from the restriction of the observation $\mathfrak{C}x_0$ to the interval $[0, t)$, but the reconstruction operator need not be bounded;

(iv) observable in finite time iff there exists a $t > 0$ for which the conclusion of (iii) is valid;

(v) final state observable in time t iff it is possible to reconstruct the final state $x(t) = \mathfrak{A}^t x_0$ from the restriction of the observation $\mathfrak{C}x_0$ to the interval $[0, t)$, and the reconstruction operator is bounded;

(vi) final state observable in finite time iff there exists a $t > 0$ for which the conclusion of (v) is valid;

(vii) observable iff it is possible to reconstruct the initial state x_0 from the observation $\mathfrak{C}x_0 \in L^p|Reg(\overline{\mathbb{R}}^+; Y)$, but the reconstruction operator (i.e., the left inverse of $\mathfrak{C}$ defined on the range of $\mathfrak{C}$) need not be bounded.

We leave the easy verifications of the following claims to the reader.

Example 9.4.3 *The delay line in Example 2.3.4 is exactly controllable and observable in time T, and it is not controllable or observable in time t for any $t < T$.*

Example 9.4.4 *Let $\Sigma = \left[\begin{smallmatrix} \mathfrak{A} & \mathfrak{B} \\ \mathfrak{C} & \mathfrak{D} \end{smallmatrix}\right]$ be an $L^p|Reg$-well-posed linear system on (Y, X, U). Then the following systems obtained from Σ have the same* *controllability/observability properties as Σ (i.e., any one of the conditions (i)–(ix) in Definition 9.4.1 or Definition 9.4.2 holds for the modified system iff the same condition holds for the original system):*

(i) *the exponentially shifted system Σ_α in Example 2.3.5 (provided the bound ω in parts (ix) is replaced by $\omega + \alpha$);*
(ii) *the time compressed system Σ_λ in Example 2.3.6 (provided the bound ω in parts (ix) is replaced by $\lambda\omega$);*
(iii) *the similarity transformed system Σ_E in Example 2.3.7.*

Example 9.4.5 *Let Σ_1 and Σ_2 be two systems which are $L^p|Reg$-well-posed in the same sense.*

(i) *The following systems obtained from Σ_1 and Σ_2 are controllable in the sense of any one of the conditions (i)–(ix) in Definition 9.4.1 if and only if both Σ_1 and Σ_2 are controllable in the same sense:*
   (a) *the cross-product of Σ_1 and Σ_2 in Example 2.3.10;*
   (b) *the sum junction of Σ_1 and Σ_2 in Example 2.3.11.*
(ii) *The following systems obtained from Σ_1 and Σ_2 are observable in the sense of any one of the conditions (i)–(ix) in Definition 9.4.2 if and only if both Σ_1 and Σ_2 are observable in the same sense:*
   (a) *the cross-product of Σ_1 and Σ_2 in Example 2.3.10;*
   (b) *the T-junction of Σ_1 and Σ_2 in Example 2.3.12.*

Using Definitions 9.4.1 and 9.4.2 we can add the following conclusions to Theorem 9.1.9.

Corollary 9.4.6 *Let $\Sigma = \left[\begin{array}{c|c} \mathfrak{A} & \mathfrak{B} \\ \hline \mathfrak{C} & \mathfrak{D} \end{array}\right]$ be an $L^p|Reg$-well-posed linear system on (Y, X, U).*

(i) *If Σ is controllable in any of the senses described in Definition 9.4.1, then the system $\left[\begin{array}{c|c} \mathfrak{A}_\mathfrak{C} & \mathfrak{B}_\mathfrak{C} \\ \hline \mathfrak{C} & \mathfrak{D} \end{array}\right]$ in Theorem 9.1.9(ii) is controllable in the same sense.*
(ii) *If Σ is observable in any of the senses described in Definition 9.4.2, then the system $\left[\begin{array}{c|c} \mathfrak{A}_\mathfrak{B} & \mathfrak{B} \\ \hline \mathfrak{C}_\mathfrak{B} & \mathfrak{D} \end{array}\right]$ in Theorem 9.1.9(i) is observable in the same sense.*

This follows immediately from Theorem 9.1.9 and Definitions 9.4.1 and 9.4.2.

The following two theorems are the main reason for the introduction of the notions of exact controllability and observability in infinite time with some bound $\omega \in \mathbb{R}$.

Theorem 9.4.7 *Let $1 \le p < \infty$, and let $\Sigma = \left[\begin{array}{c|c} \mathfrak{A} & \mathfrak{B} \\ \hline \mathfrak{C} & \mathfrak{D} \end{array}\right]$ be an L^p-well-posed linear system on (Y, X, U).*

(i) *If Σ is controllable, then, for each $\omega \in \mathbb{R}$ for which $\mathfrak{B}$ is ω-bounded, it is possible to construct a Banach space $\underline{X} \subset X$ (which may depend on ω) with the following properties. The embedding $\underline{X} \subset X$ is continuous and dense, $\underline{X}$ is invariant under $\mathfrak{A}$, $\mathcal{R}(\mathfrak{B}) \subset \underline{X}$, and if we define*

$$\underline{\mathfrak{A}} = \mathfrak{A}_{|\underline{X}}, \qquad \underline{\mathfrak{C}} = \mathfrak{C}_{|\underline{X}},$$

then $\underline{\Sigma} = \left[\begin{array}{c|c} \underline{\mathfrak{A}} & \mathfrak{B} \\ \hline \underline{\mathfrak{C}} & \mathfrak{D} \end{array}\right]$ is an L^p-well-posed linear system on $(Y, \underline{X}, U)$ which is exactly controllable in infinite time with bound ω. It is observable whenever Σ is observable.

(ii) *If Σ is observable, then, for each $\omega \in \mathbb{R}$ for which $\mathfrak{C}$ is ω-bounded, it is possible to construct a Banach space $\overline{X} \supset X$ (which may depend on ω) with the following properties. The embedding $X \subset \overline{X}$ is continuous and dense, $\mathfrak{A}$ can be extended to a C_0 semigroup $\overline{\mathfrak{A}}$ on $\overline{X}$, $\mathfrak{C}$ can be extended to an operator $\overline{\mathfrak{C}} \in \mathcal{B}(\overline{X}; L^p_\omega(\mathbb{R}^+; Y))$, and $\overline{\Sigma} = \left[\begin{array}{c|c} \overline{\mathfrak{A}} & \mathfrak{B} \\ \hline \overline{\mathfrak{C}} & \mathfrak{D} \end{array}\right]$ is an L^p-well-posed linear system on $(Y, \overline{X}, U)$ which is exactly observable in infinite time with bound ω. It is controllable whenever Σ is controllable.*

(iii) *If Σ is minimal, then both (i) and (ii) apply. Thus, it is possible to make Σ exactly controllable in infinite time with some bound $\omega \in \mathbb{R}$ by replacing the state space X by a smaller state space $\underline{X}$ (and strengthening the norm), and it is possible to make Σ exactly observable in infinite time with some bound $\omega \in \mathbb{R}$ by replacing the state space X by a larger state space $\overline{X}$ (and weakening the norm).*

Proof (i) Let $\underline{X}$ be the image in X of $L^p_\omega(\mathbb{R}^-; U)$ under $\mathfrak{B}$. Then, because of the intertwining condition $\mathfrak{A}^t\mathfrak{B} = \mathfrak{B}\pi_-\tau^t$, $\underline{X}$ is invariant under $\mathfrak{A}$. We equip this space with the norm induced by $\mathfrak{B}$. In other words, if we denote $\mathcal{U} = L^p_\omega(\mathbb{R}^-; U)/\mathcal{N}(\mathfrak{B})$, then $\mathfrak{B}$ induces an injective operator in $\mathcal{B}(\mathcal{U}; X)$, and we define the new (stronger) norm $|x|_{\underline{X}} = \|\mathfrak{B}^{-1}x\|_{\mathcal{U}}$ in X (cf. Lemma 4.3.13). This makes $\underline{X}$ a Banach space (even a Hilbert space if all the involved spaces are Hilbert spaces), and $\mathfrak{B}$ becomes an isometric isomorphism between $\mathcal{U}$ and $\underline{X}$. The left shift semigroup $\tau_-^t = \pi_-\tau^t$ on $\mathbb{R}^-$ induces a C_0 semigroup on this quotient, and it follows from the intertwining condition $\mathfrak{A}^t\mathfrak{B} = \mathfrak{B}\pi_-\tau^t$ that the restriction of $\mathfrak{A}$ to $\underline{X}$ is the image under $\mathfrak{B}$ of this quotient semigroup. Thus, $\mathfrak{A}$ is strongly continuous in the norm of $\underline{X}$. Obviously the restriction of $\mathfrak{C}$ to $\underline{X}$ is continuous (since the norm of $\underline{X}$ is stronger than the norm of X). This means that $\left[\begin{array}{c|c} \underline{\mathfrak{A}} & \mathfrak{B} \\ \hline \underline{\mathfrak{C}} & \mathfrak{D} \end{array}\right]$ is an L^p-well-posed linear system on $(Y, \underline{X}, U)$. By construction, it is exactly controllable in infinite time with bound ω.

(ii) By assumption, the operator $\mathfrak{C}$ is injective and bounded $X \to L^p_\omega(\mathbb{R}^+; Y)$. We can define a new norm on X, which is weaker than the original norm, by setting $|x|_{\overline{X}} = \|\mathfrak{C}x\|_{L^p_\omega(\mathbb{R}^+;Y)}$. The space X is not complete in this norm (unless

$\mathcal{R}(\mathfrak{C})$ is closed in $L^p_\omega(\mathbb{R}^+;Y)$), so we let $\overline{X}$ be the completion of X under this weaker norm. Then $\mathfrak{C}$ becomes an isometric isomorphism between $\overline{X}$ and $\mathcal{Y} = \overline{\mathcal{R}(\mathfrak{C})}$. Since $\mathcal{R}(\mathfrak{C})$ is invariant under the left shift $\tau^t_+ = \tau^t_+$ on $\mathbb{R}^+$ (this follows from the intertwining condition $\mathfrak{C}\mathfrak{A}^t = \tau^t_+\mathfrak{C}$), also $\mathcal{Y}$ is invariant under τ^t_+. Moreover, if we use the norm of $\overline{X}$ in X, then $\mathfrak{A}$ is (isometrically) similar to the restriction of τ^t_+ to $\mathcal{R}(\mathfrak{C})$ with the similarity transform $\mathfrak{A}^t = \mathfrak{C}^{-1}\tau^t_+\mathfrak{C}$. By continuity, $\mathfrak{A}$ can be extended to a C_0 semigroup on $\overline{X}$. The operator $\mathfrak{B}$ is obviously continuous with values in $\overline{X}$ (since the norm of $\overline{X}$ is weaker than the norm of X). Thus, $\left[\begin{smallmatrix} \mathfrak{A} & \mathfrak{B} \\ \mathfrak{C} & \mathfrak{D} \end{smallmatrix}\right]$ is an L^p-well-posed linear system on $(Y, \overline{X}, U)$. By construction, it is exactly observable in infinite time with bound ω.

(iii) This is obvious. □

Remark 9.4.8 Part (ii) of Theorem 9.4.7 is valid also for L^∞-well-posed and *Reg*-well-posed systems, for all $\omega \in \mathbb{R}$ such that $\mathcal{R}(\mathfrak{C}) \subset BC_{0,\omega}(\overline{\mathbb{R}}^+;Y)$. The proof remains the same (replace L^p_ω by $BC_{0,\omega}$).

Proposition 9.4.9 *Let $\Sigma = \left[\begin{smallmatrix} \mathfrak{A} & \mathfrak{B} \\ \mathfrak{C} & \mathfrak{D} \end{smallmatrix}\right]$ be a minimal L^2-well-posed linear system on the Hilbert spaces (Y, X, U) which is both input/state bounded and state/output bounded. Let $\sqrt{\mathfrak{B}\mathfrak{B}^*} \in \mathcal{B}(X)$ be the positive square root of the controllability gramian of Σ, and let $\sqrt{\mathfrak{C}^*\mathfrak{C}} \in \mathcal{B}(X)$ be the positive square root of the observability gramian of Σ (cf. Definition 10.4.1 and Lemma A.2.2).*

(i) *The spaces $\underline{X}$ and $\overline{X}$ constructed in the proof of Theorem 9.4.7 are Hilbert spaces: $\underline{X}$ is the range of $\sqrt{\mathfrak{B}\mathfrak{B}^*}$ equipped with the norm $|x|_{\underline{X}} = |\sqrt{\mathfrak{B}\mathfrak{B}^*}x|_X$, and $\overline{X}$ is the completion of X with respect to the norm $|x|_{\overline{X}} = |\sqrt{\mathfrak{C}^*\mathfrak{C}}x|_X$. Thus, $\sqrt{\mathfrak{B}\mathfrak{B}^*}$ is a unitary map of X onto $\underline{X}$, the extension $\sqrt{\mathfrak{C}^*\overline{\mathfrak{C}}} \in \mathcal{B}(\overline{X})$ of the operator $\sqrt{\mathfrak{C}^*\mathfrak{C}}$ is a unitary map of $\overline{X}$ onto X, and*

$$\langle x, y\rangle_X = \langle \sqrt{\mathfrak{B}\mathfrak{B}^*}x, \sqrt{\mathfrak{B}\mathfrak{B}^*}y\rangle_{\underline{X}}, \qquad x, y \in X,$$
$$\langle x, y\rangle_{\overline{X}} = \langle \sqrt{\mathfrak{C}^*\overline{\mathfrak{C}}}x, \sqrt{\mathfrak{C}^*\overline{\mathfrak{C}}}x\rangle_X, \qquad x, y \in \overline{X}.$$

(ii) *If we denote the adjoints of $\mathfrak{B}$ and $\underline{\mathfrak{C}}$ with respect to the inner product in $\underline{X}$ by $\mathfrak{B}'$, respectively $\underline{\mathfrak{C}}'$, then $\mathfrak{B}\mathfrak{B}' = 1$ and $\underline{\mathfrak{C}}'\underline{\mathfrak{C}} = \mathfrak{B}\mathfrak{B}^*\mathfrak{C}^*\mathfrak{C}_{|\underline{X}}$ (these are the controllability and observability gramians of the system $\underline{\Sigma}$).*

(iii) *If we denote the adjoints of $\mathfrak{B}$ and $\overline{\mathfrak{C}}$ with respect to the inner product in $\overline{X}$ by $\mathfrak{B}''$, respectively $\overline{\mathfrak{C}}''$, then $\mathfrak{B}\mathfrak{B}'' = \mathfrak{B}\mathfrak{B}^*\mathfrak{C}^*\overline{\mathfrak{C}}$ and $\overline{\mathfrak{C}}''\overline{\mathfrak{C}} = 1$; in particular, X is invariant under $\mathfrak{B}\mathfrak{B}''$ and $\mathfrak{B}\mathfrak{B}'' = \underline{\mathfrak{C}}'\overline{\mathfrak{C}}$ ($\mathfrak{B}\mathfrak{B}''$ and $\overline{\mathfrak{C}}''\overline{\mathfrak{C}}$ are the controllability and observability gramians of the system $\overline{\Sigma}$).*

(iv) *The operator $\sqrt{\mathfrak{B}\mathfrak{B}''} = \sqrt{\mathfrak{B}\mathfrak{B}^*\mathfrak{C}^*\overline{\mathfrak{C}}}$ is a unitary map of $\overline{X}$ onto $\underline{X}$, and*

$$\langle x, y\rangle_{\overline{X}} = \langle \sqrt{\mathfrak{B}\mathfrak{B}''}x, \sqrt{\mathfrak{B}\mathfrak{B}''}y\rangle_{\underline{X}}, \qquad x, y \in \overline{X}.$$

Proof (i) Let $x \in X$. Then $\mathfrak{B}\mathfrak{B}^*x \in \mathcal{R}(\mathfrak{B}) = \underline{X}$, and by the construction in the proof of Theorem 9.4.7(i),

$$\begin{aligned}|\mathfrak{B}\mathfrak{B}^*x|^2_{\underline{X}} &= |\mathfrak{B}^*x|^2_{L^2(\mathbb{R}^-;U)} = \langle \mathfrak{B}^*x, \mathfrak{B}^*x\rangle_X = \langle x, \mathfrak{B}\mathfrak{B}^*x\rangle_X \\ &= |\sqrt{\mathfrak{B}\mathfrak{B}^*}x|^2_X.\end{aligned}$$

The operator $\sqrt{\mathfrak{B}\mathfrak{B}^*}$ is injective and it has the same range as $\mathfrak{B}$ (see Lemma A.2.5(iii)), so $\sqrt{\mathfrak{B}\mathfrak{B}^*}$ maps X one-to-one onto $\underline{X}$. This space is dense in X, and therefore, it follows from the computation above that $|\sqrt{\mathfrak{B}\mathfrak{B}^*}x|_{\underline{X}} = |x|_X$ for all $x \in X$. This proves that $\sqrt{\mathfrak{B}\mathfrak{B}^*}$ is an isometric map of X onto $\underline{X}$, hence unitary.

Let $x \in X$. Then, by the construction in the proof of Theorem 9.4.7(ii),

$$\begin{aligned}|x|^2_{\overline{X}} &= |\mathfrak{C}x|^2_{L^2(\mathbb{R}^+;Y)} = \langle \mathfrak{C}x, \mathfrak{C}x\rangle_{L^2(\mathbb{R}^+;Y)} = \langle x, \mathfrak{C}^*\mathfrak{C}x\rangle_{L^2(\mathbb{R}^+;Y)} \\ &= |\sqrt{\mathfrak{C}^*\mathfrak{C}}x|^2_X.\end{aligned}$$

Thus, $\overline{X}$ is the completion of X with respect to the norm $|x|_{\overline{X}} = |\sqrt{\mathfrak{C}^*\mathfrak{C}}x|_X$, and $\sqrt{\mathfrak{C}^*\mathfrak{C}}$ can be extended to a unitary map of $\overline{X}$ onto X. The operator $E = \mathfrak{C}^*\mathfrak{C}$ has an obvious extension to an operator in $\mathcal{B}(\overline{X})$, namely $\mathfrak{C}^*\overline{\mathfrak{C}}$, which is self-adjoint in $\underline{X}$ (see the discussion leading up to Proposition 3.6.2). Let $\sqrt{\mathfrak{C}^*\overline{\overline{\mathfrak{C}}}}$ be the positive square root of this operator. Then the restriction of $\sqrt{\mathfrak{C}^*\overline{\overline{\mathfrak{C}}}}$ to X is $\sqrt{\mathfrak{C}^*\mathfrak{C}}$, so the (unique) extension of $\sqrt{\mathfrak{C}^*\mathfrak{C}}$ to an operator in $\mathcal{B}(\overline{X})$ is given by $\sqrt{\mathfrak{C}^*\overline{\overline{\mathfrak{C}}}}$. Thus, $\sqrt{\mathfrak{C}^*\overline{\overline{\mathfrak{C}}}}$ is an isometric, hence unitary map of $\overline{X}$ onto X.

(ii) We constructed the norm in $\underline{X}$ in such a way that $\mathfrak{B}_{|\mathcal{N}(\mathfrak{B})^\perp}$ is an isometric map of $\mathcal{N}(\mathfrak{B})^\perp$ onto $\underline{X}$. Therefore $\mathfrak{B}\colon L^2(\mathbb{R}^-;U) \to \underline{X}$ is co-isometric, and $\mathfrak{B}\mathfrak{B}' = 1$. By Proposition 3.6.2(v), $\underline{\mathfrak{C}}' = \mathfrak{B}\mathfrak{B}^*\mathfrak{C}^*$, so $\underline{\mathfrak{C}}'\overline{\mathfrak{C}} = \mathfrak{B}\mathfrak{B}^*\mathfrak{C}^*\mathfrak{C}_{|\underline{X}}$.

(iii) We constructed the norm in $\overline{X}$ in such a way that $\overline{\mathfrak{C}}\colon \underline{X} \to L^2(\mathbb{R}^+ : Y)$ is an isometry, so $\overline{\mathfrak{C}}''\overline{\mathfrak{C}} = 1$. By Proposition 3.6.2(ii), $\mathfrak{B}'' = \mathfrak{B}^*\mathfrak{C}^*\overline{\mathfrak{C}}$, so $\mathfrak{B}\mathfrak{B}'' = \mathfrak{B}\mathfrak{B}^*\mathfrak{C}^*\overline{\mathfrak{C}} = \underline{\mathfrak{C}}'\overline{\mathfrak{C}}$.

(iv) This follows from (i) with X replaced by $\underline{X}$ (the alternative formula for $\sqrt{\mathfrak{B}\mathfrak{B}''}$ is taken from (iii)). □

Another variation on the theme of Theorem 9.4.7 is the following modification of Theorem 9.2.4.

Theorem 9.4.10 *Let* $\Sigma_1 = \left[\begin{array}{c|c}\mathfrak{A}_1 & \mathfrak{B}_1 \\ \hline \mathfrak{C}_1 & \mathfrak{D}\end{array}\right]$ *and* $\Sigma_2 = \left[\begin{array}{c|c}\mathfrak{A}_2 & \mathfrak{B}_2 \\ \hline \mathfrak{C}_2 & \mathfrak{D}\end{array}\right]$ *be two minimal realizations of the same input/output map* $\mathfrak{D}$ *(i.e., they are minimal* $L^p|Reg$*-well-posed linear systems with the same input/output map* $\mathfrak{D}$*).*

(i) *If* Σ_1 *is exactly controllable in infinite time with bound* ω *and* $\mathfrak{B}_2$ *is* ω*-bounded, then the pseudo-similarity transform* E *in Theorem 9.2.4 has* $\mathcal{D}(E) = X_1$ *and* $E \in \mathcal{B}(X_1; X_2)$*. In this case it is possible to define a norm on* $\underline{X}_2 = \mathcal{R}(E) \subset X_2$ *which makes* $\left[\begin{array}{c|c}\underline{\mathfrak{A}}_2 & \mathfrak{B}_2 \\ \hline \underline{\mathfrak{C}}_2 & \mathfrak{D}\end{array}\right]$ *an* $L^p|Reg$*-well-posed* *linear system on* $(Y, \underline{X}_2, U)$*, where* $\underline{\mathfrak{A}}_2$ *and* $\underline{\mathfrak{C}}_2$ *are the restrictions of* $\mathfrak{A}_2$ *and* $\mathfrak{C}_2$ *to* $\underline{X}_2$*. This system is exactly controllable in infinite time with bound* ω*, and it is similar to* Σ_1 *(i.e.,* $E \in \mathcal{B}(X_1; \underline{X}_2)$ *and* $E^{-1} \in \mathcal{B}(\underline{X}_2; X_1)$*). It coincides with* Σ_2 *(in the sense that* $\underline{X}_2 = X_2$*, but the norms in* $\underline{X}_2$ *and* X_2 *may be different though equivalent) if and only if* Σ_2 *is exactly controllable in infinite time with bound* ω*. In particular, any two minimal realizations of the same input/output map* $\mathfrak{D}$ *which are exactly controllable in infinite time with the same bound* ω *are similar to each other.*

(ii) *If* Σ_2 *is exactly observable in infinite time with bound* ω *and* $\mathfrak{C}_1$ *is* ω*-bounded, then the pseudo-similarity transform in Theorem 9.2.4 has* $\mathcal{D}(E) = X_1$ *and* $E \in \mathcal{B}(X_1; X_2)$*. In this case it is possible to extend the state space* X_1 *of* Σ_1 *to a larger Banach space* $\overline{X}_1$ *with the following properties. The embedding* $X_1 \subset \overline{X}_1$ *is continuous and dense,* E *can be extended to an operator* $\overline{E} \in \mathcal{B}(\overline{X}_1; X_2)$ *with* $\overline{E}^{-1} \in \mathcal{B}(X_2; \overline{X}_1)$*,* $\mathfrak{A}_1$ *can be extended to a* C_0 *semigroup* $\overline{\mathfrak{A}}_1$ *on* $\overline{X}_1$*,* $\mathfrak{C}$ *can be extended to an operator* $\overline{\mathfrak{C}} \in \mathcal{B}(\overline{X}; L^p|Reg_\omega(\overline{\mathbb{R}}^+; Y)$*, and* $\left[\begin{smallmatrix} \overline{\mathfrak{A}} & \mathfrak{B} \\ \overline{\mathfrak{C}} & \mathfrak{D} \end{smallmatrix}\right]$ *is an* L^p*-well-posed linear system on* $(Y, X_{\mathfrak{B}}, U)$ *which is exactly observable in infinite time with bound* ω*. It is similar to* Σ_2 *with similarity operator* $\overline{E}$*, and it coincides with* Σ_1 *(in the sense that* $\overline{X}_1 = X_1$*, but the norms in* $\overline{X}_1$ *and* X_1*may be different though equivalent) if and only if* Σ_1 *is exactly observable in infinite time with bound* ω*. In particular, any two minimal realizations of the same input/output map* $\mathfrak{D}$ *which are exactly observable in infinite time with the same bound* ω *are similar to each other.*

Proof (i) Since $\pi_+\mathfrak{D}\pi_- = \mathfrak{C}_1\mathfrak{B}_1 = \mathfrak{C}_2\mathfrak{B}_2$, and both $\mathfrak{B}_1$ and $\mathfrak{B}_2$ are ω-bounded, it is possible to extend $\pi_+\mathfrak{D}\pi_-$ to an operator Γ mapping $L^p|Reg_\omega(\mathbb{R}^-; U)$ into $L^p|Reg_{\mathrm{loc}}(\mathbb{R}^+; Y)$, still satisfying $\Gamma = \mathfrak{C}_1\mathfrak{B}_1 = \mathfrak{C}_2\mathfrak{B}_2$. The exact controllability assumption on Σ_1 says that $\mathfrak{B}_1$ maps $L^p|Reg_\omega(\mathbb{R}^-; U)$ onto X_1, and hence

$$\mathcal{R}(\mathfrak{C}_1) = \mathcal{R}(\mathfrak{C}_1\mathfrak{B}_1) = \mathcal{R}(\Gamma) = \mathcal{R}(\mathfrak{C}_2\mathfrak{B}_2) \subset \mathcal{R}(\mathfrak{C}_2).$$

By Theorem 9.2.5(i), $\mathcal{D}(E) = X_1$ and $E \in \mathcal{B}(X_1; X_2)$.

We define $\underline{X}_2 = \mathcal{R}(E)$, and let $\underline{X}_2$ inherit the norm of X_1, i.e., let $|Ex_1|_{\underline{X}_2} = |x_1|_{X_1}$ for all $x \in X_1$. Then E is an isometric isomorphism of X_1 onto $\underline{X}_2$. Using E as the similarity operator in the standard similarity transform described in Example 2.3.7 we get another realization of $\mathfrak{D}$ with state space $\underline{X}_2$ which is exactly controllable in infinite time with bound ω. Comparing Example 2.3.7 with (9.2.1), we find that this system is the restriction of Σ_2 to $\underline{X}_2$.

We have $\underline{X}_2 = X_2$ if and only if E maps X_1 onto X_2. Since $\mathfrak{B}_2 = E\mathfrak{B}_1$ and $\mathfrak{B}_1$ maps $L^p|Reg_\omega(\mathbb{R}^-; U)$ onto X_1, this is true if and only if $\mathfrak{B}_2$ maps $L^p|Reg_\omega(\mathbb{R}^-; U)$ onto X_2, i.e., Σ_2 is exactly controllable in infinite time with bound ω.

(ii) Recall from Lemma 9.2.2 that $\mathcal{R}(\mathfrak{C}_1)$ and $\mathcal{R}(\mathfrak{C}_2)$ have the same closure in $L^p|Reg_\omega(\mathbb{R}^+;Y)$ (equal to the closure of $\mathcal{R}(\pi_+\mathfrak{D}\pi_-)$). The exact observability assumption on Σ_2 implies that $\mathcal{R}(\mathfrak{C}_2)$ is closed (see Lemma 9.10.2(iii)). Thus, $\mathcal{R}(\mathfrak{C}_1) \subset \mathcal{R}(\mathfrak{C}_2)$, and by Theorem 9.2.5(i), $\mathcal{D}(E) = X_1$ and $E \in \mathcal{B}(X_1; X_2)$.

We define $\overline{X}_1$ to be the completion of X_1 with respect to the norm induced by E, i.e., $|x_1|_{\overline{X}_1} = |Ex_1|_{X_2}$. Then E can be extended to an isometric isomorphism $\overline{E}$ of $\overline{X}_1$ onto X_2. Using $\overline{E}$ as the similarity operator in the standard similarity transform described in Example 2.3.7 we get another realization of $\mathfrak{D}$ with state space $\overline{X}_1$ which is exactly observable in infinite time with bound ω. Comparing Example 2.3.7 with (9.2.1), we find that the restriction of this system to X_1 is Σ_1.

We have $\overline{X}_1 = X_1$ if and only if E has a bounded inverse. By Theorem 9.2.5(iii), this is true if and only if $\mathcal{R}(\mathfrak{C}_1) = \mathcal{R}(\mathfrak{C}_2)$. Since $\mathcal{R}(\mathfrak{C}_2)$ is closed and $\mathcal{R}(\mathfrak{C}_1)$ and $\mathcal{R}(\mathfrak{C}_2)$ have the same closure in $L^p|Reg_\omega(\mathbb{R}^+;Y)$, this is true if and only if $\mathcal{R}(\mathfrak{C}_1)$ is closed in $L^p|Reg_\omega(\mathbb{R}^+;Y)$, or equivalently (see Lemma 9.10.2(iii)), if and only if Σ_1 is exactly observable in infinite time with bound ω. □

Corollary 9.4.11 *Let* $\Sigma = \left[\begin{smallmatrix} \mathfrak{A} & \mathfrak{B} \\ \mathfrak{C} & \mathfrak{D} \end{smallmatrix}\right]$ *be a minimal* L^p*-well-posed linear system with* $1 \le p < \infty$.

(i) *If* Σ *is state/output bounded and exactly observable in infinite time with bound* $\omega = 0$, *then* $\mathfrak{A}$ *is strongly stable.*
(ii) *If* Σ *is input/state bounded and exactly controllable in infinite time with bound* $\omega = 0$, *then* $\mathfrak{A}$ *is bounded.*

Proof (i) The exactly observable shift realization of $\mathfrak{D}$ described in Example 2.6.5 is strongly stable, but not necessarily minimal. We can make it minimal by restricting it to the reachable subspace, as described in Theorem 9.1.9(i). The restricted (shift) semigroup is still strongly stable. By Theorem 9.4.10(i), the system that we get in this way is similar to the given system Σ, so $\mathfrak{A}$ must be strongly stable.

(ii) The proof of part (i) is analogous to the proof of (i). □

The spaces $\overline{X}$ and $\underline{X}$ that we constructed in Theorem 9.4.7 (and also in Theorem 9.4.10) have the drawback that they depend on ω. It would be much better if they did not depend on ω, but this is not always possible, as the following counter-example shows.

Example 9.4.12

(i) *The exactly controllable shift realization in Example 2.6.5(i) is exactly controllable in infinite time with bound* ω, *but is not exactly controllable in infinite time for any other* $\omega' \neq \omega$. *Theorem 9.4.7(i) applies with* ω

replaced by $\omega' > \omega$*, and the space* $\underline{X}$ *that we get from this theorem is* $\underline{X} = L^p_{\omega'}(\mathbb{R}^-; U)$*. This realization is not (even approximately) controllable in finite time.*

(ii) *The exactly observable shift realization in Example 2.6.5(ii) is exactly observable in infinite time with bound* ω*, but is not exactly observable in infinite time for any other* $\omega' \neq \omega$*. Theorem 9.4.7(ii) applies with* ω *replaced by* $\omega' > \omega$*, and the space* $\overline{X}$ *that we get from this theorem is* $\overline{X} = L^p_{\omega'}(\mathbb{R}^+; Y)$*. This realization is not (even approximately) observable in finite time.*

Fortunately, there is a simple characterization of systems which are exactly controllable in infinite time with two different bounds ω; see Theorem 9.10.6.

9.5 Normalized and balanced realizations

The special realizations defined on $\overline{X}$ and $\underline{X}$ that we constructed in the proof of Theorem 9.4.7 are special examples of realizations that are either input normalized or output normalized.

Definition 9.5.1 Let Σ be an L^p-well-posed linear system on (Y, X, U) with $1 \le p < \infty$.

(i) Σ is *input normalized* if Σ is minimal and input/state bounded and the operator induced by $\mathfrak{B}$ on $L^p(\mathbb{R}^-; U)/\mathcal{N}(\mathfrak{B})$ is an isometry, i.e., if we denote the quotient map of $L^p(\mathbb{R}^-; U)$ onto $L^p(\mathbb{R}^-; U)/\mathcal{N}(\mathfrak{B})$ by π, then $\|\mathfrak{B}u\|_X = \|\pi u\|_{L^p(\mathbb{R}^-;U)/\mathcal{N}(\mathfrak{B})}$ for all $u \in L^p(\mathbb{R}^-; U)$.

(ii) Σ is *output normalized* if Σ is minimal and state/output bounded and $\mathfrak{C}$ is an isometry, i.e., $\|\mathfrak{C}x\|_{L^p(\mathbb{R}^+;Y)} = \|x\|_X$ for all $x \in X$.

It is also possible to consider input or output normalized system nodes rather than input or output normalized L^p-well-posed linear systems, but we shall not do so here (cf. Definition 10.1.1).

We have already encountered some input and output normalized systems:

Proposition 9.5.2 *Let* $1 \le p < \infty$*, and let* $\Sigma = \left[\begin{smallmatrix} \mathfrak{A} & \mathfrak{B} \\ \mathfrak{C} & \mathfrak{D} \end{smallmatrix}\right]$ *be a stable minimal* L^p*-well-posed linear system on* (Y, X, U)*.*

(i) *The system* $\underline{\Sigma}$ *constructed in the proof of Theorem 9.4.7(i) with* $\omega = 0$ *is input normalized, and it is unique in the sense that it is the only input normalized system* $\underline{\Sigma}$ *which has all the properties listed in Theorem 9.4.7(i). We call this the* input normalized system induced by Σ.

(ii) *The system* $\overline{\Sigma}$ *constructed in the proof of Theorem 9.4.7(ii) with* $\omega = 0$ *is output normalized if* $\omega = 0$*, and it is unique in the sense that it is the only* *output normalized system* $\overline{\Sigma}$ *which has all the properties listed in Theorem 9.4.7(ii) (if we identify any two completions of* X *with respect to the same norm). We call this the* output normalized system induced by Σ.

(iii) *The minimal realization of* $\mathfrak{D}$ *that we get by factoring out the unobservable subspace (cf. Theorem 9.1.9(ii)) from the exactly controllable shift realization of* $\mathfrak{D}$ *given in Example 2.6.5(i) with* $\omega = 0$ *is input normalized. We call this the* restricted exactly controllable shift realization *of* $\mathfrak{D}$.

(iv) *The minimal realization of* $\mathfrak{D}$ *that we get by restricting the exactly observable shift realization of* $\mathfrak{D}$ *given in Example 2.6.5(ii) with* $\omega = 0$ *to the reachable subspace (cf. Theorem 9.1.9(i)) is output normalized. We call this the* restricted exactly observable shift realization *of* $\mathfrak{D}$.

Proof (i)–(ii) That the systems constructed in the proof of Theorem 9.4.7(i),(ii) are input, respectively, output normalized follows immediately from Definition 9.5.1. The added normalization constraint determine the norms in $\underline{X}$ and $\overline{X}$ uniquely. This makes $\underline{X}$ and $\overline{X}$ unique, hence it makes the normalized versions of $\underline{\Sigma}$ and $\overline{\Sigma}$ unique.

(iii) In this case $X = L^p(\mathbb{R}^-; U)/\mathcal{N}(\pi_+\mathfrak{D}\pi_-)$ and $\mathfrak{B} = \pi$ is the quotient map of $L^p(\mathbb{R}^-; U)$ onto $X = L^p(\mathbb{R}^-; U)/\mathcal{N}(\mathfrak{B})$.

(iv) Here X is the closure of $\mathcal{R}(\pi_+\mathfrak{D}\pi_-)$ in $L^p(\mathbb{R}^+; Y)$ and $\mathfrak{C}$ is the restriction to X of the identity map in $L^p(\mathbb{R}^+; Y)$. $\square$

Theorem 9.5.3 *Let* $1 \le p < \infty$, *and let* $\mathfrak{D} \in TIC^p(U; Y)$, *and let* $\Gamma = \pi_+\mathfrak{D}\pi_- \in \mathcal{B}(L^p(\mathbb{R}^-; U); L^p(\mathbb{R}^+; Y))$ *be the Hankel operator induced by* $\mathfrak{D}$.

(i) *Any two input normalized realizations* $\Sigma_1 = \left[\begin{array}{c|c} \mathfrak{A}_1 & \mathfrak{B}_1 \\ \hline \mathfrak{C}_1 & \mathfrak{D} \end{array}\right]$ *and* $\Sigma_2 = \left[\begin{array}{c|c} \mathfrak{A}_2 & \mathfrak{B}_2 \\ \hline \mathfrak{C}_2 & \mathfrak{D} \end{array}\right]$ *of* $\mathfrak{D}$ *on* (Y, X_1, U), *respectively* (Y, X_2, U), *are isometrically isomorphic, i.e, the similarity operator* E *in Theorem 9.2.4 is a norm-preserving mapping of* X_1 *onto* X_2. *In particular, they are isometrically isometric to the restricted exactly controllable shift realization in Proposition 9.5.2(iii). Moreover,* $\mathcal{R}(\mathfrak{C}_1) = \mathcal{R}(\mathfrak{C}_2) = \mathcal{R}(\Gamma)$, *and* $|x_1|_{X_1} = |x_2|_{X_2}$ *whenever* $x_1 \in X_1$, $x_2 \in X_2$, *and* $\mathfrak{C}_1 x_1 = \mathfrak{C}_2 x_2$. *If* U *and* Y *are Hilbert spaces and* $p = 2$, *then* X_1 *and* X_2 *are Hilbert spaces,* $\mathfrak{B}_1\mathfrak{B}_1^* = \mathfrak{B}_2\mathfrak{B}_2^* = 1$, $\mathfrak{B}_1^*\mathfrak{B}_1 = \mathfrak{B}_2^*\mathfrak{B}_2$ *is the orthogonal projection in* $L^2(\mathbb{R}^-; U)$ *onto* $\mathcal{N}(\Gamma)^\perp$, $\mathfrak{C}_1\mathfrak{C}_1^* = \mathfrak{C}_2\mathfrak{C}_2^* = \Gamma\Gamma^*$, *and both* $\mathfrak{C}_1^*\mathfrak{C}_1$ *and* $\mathfrak{C}_2^*\mathfrak{C}_2$ *are unitarily similar to* $\Gamma^*\Gamma_{|\mathcal{N}(\Gamma)^\perp}$ *and to* $\Gamma\Gamma^*_{|\mathcal{N}(\Gamma^*)^\perp}$.

(ii) *Any two output normalized realizations* $\Sigma_1 = \left[\begin{smallmatrix} \mathfrak{A}_1 & \mathfrak{B}_1 \\ \mathfrak{C}_1 & \mathfrak{D} \end{smallmatrix}\right]$ *and* $\Sigma_2 = \left[\begin{smallmatrix} \mathfrak{A}_2 & \mathfrak{B}_2 \\ \mathfrak{C}_2 & \mathfrak{D} \end{smallmatrix}\right]$ *of* $\mathfrak{D}$ *on* (Y, X_1, U)*, respectively* (Y, X_2, U)*, are isometrically isomorphic. In particular, they are isometrically isometric to the restricted exactly observable shift realization in Proposition 9.5.2(iv). Moreover,* $\mathfrak{C}_1$ *and* $\mathfrak{C}_2$ *are isometric isomorphisms of* X_1*, respectively* X_2*, onto the closure of* $\mathcal{R}(\Gamma)$ *in* $L^p(\mathbb{R}^+; Y)$*. If* U *and* Y *are Hilbert spaces and* $p = 2$*, then* X_1 *and* X_2 *are Hilbert spaces,* $\mathfrak{C}_1^*\mathfrak{C}_1 = \mathfrak{C}_2^*\mathfrak{C}_2 = 1$*,* $\mathfrak{C}_1\mathfrak{C}_1^* = \mathfrak{C}_2\mathfrak{C}_2^*$ *is the orthogonal projection in* $L^2(\mathbb{R}^+; Y)$ *onto* $\overline{\mathcal{R}(\Gamma)}$*,* $\mathfrak{B}_1^*\mathfrak{B}_1 = \mathfrak{B}_2^*\mathfrak{B}_2 = \Gamma^*\Gamma$*, and both* $\mathfrak{B}_1\mathfrak{B}_1^*$ *and* $\mathfrak{B}_2\mathfrak{B}_2^*$ *are unitarily similar to* $\Gamma^*\Gamma_{|\mathcal{N}(\Gamma)^\perp}$ *and to* $\Gamma\Gamma^*_{|\mathcal{N}(\Gamma^*)^\perp}$.

(iii) *Let* $\underline{\Sigma}_1 = \left[\begin{smallmatrix} \underline{\mathfrak{A}}_1 & \underline{\mathfrak{B}}_1 \\ \underline{\mathfrak{C}}_1 & \mathfrak{D} \end{smallmatrix}\right]$ *and* $\underline{\Sigma}_2 = \left[\begin{smallmatrix} \underline{\mathfrak{A}}_2 & \underline{\mathfrak{B}}_2 \\ \underline{\mathfrak{C}}_2 & \mathfrak{D} \end{smallmatrix}\right]$ *be two input normalized realizations of* $\mathfrak{D}$ *on* $(Y, \underline{X}_1, U)$*, respectively* $(Y, \underline{X}_2, U)$*, and let* $\underline{E}$ *be the unitary similarity operator in (i) mapping* $\underline{X}_1$ *onto* $\underline{X}_2$*. Let* $\overline{\Sigma}_1 = \left[\begin{smallmatrix} \overline{\mathfrak{A}}_1 & \overline{\mathfrak{B}}_1 \\ \overline{\mathfrak{C}}_1 & \mathfrak{D} \end{smallmatrix}\right]$ *and* $\overline{\Sigma}_2 = \left[\begin{smallmatrix} \overline{\mathfrak{A}}_2 & \overline{\mathfrak{B}}_2 \\ \overline{\mathfrak{C}}_2 & \mathfrak{D} \end{smallmatrix}\right]$ *be the corresponding output normalized realizations of* $\mathfrak{D}$ *on* $(Y, \overline{X}_1, U)$*, respectively* $(Y, \overline{X}_2, U)$*, induced by* $\underline{\Sigma}_1$ *and* $\underline{\Sigma}_2$ *(cf. Proposition 9.5.2(ii), and let* $\overline{E}$ *be the isometric similarity operator in (ii) mapping* $\overline{X}_1$ *onto* $\overline{X}_2$*. Then* $\underline{E} = \overline{E}_{|\underline{X}}$*. In particular,* $\overline{E}$ *maps* $\underline{X}_1$ *isometrically onto* $\underline{X}_2$*.*

(iv) *Every input normalized system is observable and exactly controllable in infinite time with bound* $\omega = 0$*, and its semigroup is a strongly co-stable contraction semigroup.*

(v) *Every output normalized system is controllable and exactly observable in infinite time with bound* $\omega = 0$*, and its semigroup is a strongly stable contraction semigroup.*

Proof (i) To prove this claim it suffices to show that every input normalized realization $\Sigma = \left[\begin{smallmatrix} \mathfrak{A} & \mathfrak{B} \\ \mathfrak{C} & \mathfrak{D} \end{smallmatrix}\right]$ on (Y, X, U) of $\mathfrak{D}$ is isometrically isomorphic to the restricted exactly controllable shift realization in Proposition 9.5.2(iii). Let us denote this realization by $\Sigma' = \left[\begin{smallmatrix} \mathfrak{A}' & \mathfrak{B}' \\ \mathfrak{C}' & \mathfrak{D} \end{smallmatrix}\right]$. Its state space is $X' = L^p(\mathbb{R}^-; U)/\mathcal{N}(\pi_+\mathfrak{D}\pi_-)$ and $\mathfrak{B}' = \pi$ is the quotient map of $L^p(\mathbb{R}^-; U)$ onto X. By Lemma 9.2.2(i), $\mathcal{N}(\pi_+\mathfrak{D}\pi_-) = \mathcal{N}(\mathfrak{B})$. Therefore, by the condition imposed on $\mathfrak{B}$ in Definition 9.5.1, $|\mathfrak{B}u|_X = |\mathfrak{B}'u|_{X'}$ for all $u \in L^p(\mathbb{R}^-; U)$. By Theorem 9.2.6(iv), Σ and Σ' are similar to each other, with an isometric similarity operator E. That $\mathfrak{C}$ has the same range as $\mathfrak{C}' = \pi_+\mathfrak{D}\pi_-$ and that $|x_1|_{X_1} = |x_2|_{X_2}$ whenever $\mathfrak{C}_1 x_1 = \mathfrak{C}_2 x_2$ follows from Theorem 9.2.5(iv).

All the claims about the Hilbert space case in (i) are either obvious consequences of what has been said earlier, or else easy to prove (that $\Gamma^*\Gamma_{|\mathcal{N}(\Gamma)^\perp}$ and $\Gamma\Gamma^*_{|\mathcal{N}(\Gamma^*)^\perp}$ are unitarily similar follows from Lemma A.2.5). We leave this to the reader.

(ii) To prove this claim it suffices to show that every output normalized realization $\Sigma = \left[\begin{smallmatrix} \mathfrak{A} & \mathfrak{B} \\ \mathfrak{C} & \mathfrak{D} \end{smallmatrix}\right]$ on (Y, X, U) of $\mathfrak{D}$ is isometrically isomorphic to the restricted exactly observable shift realization in Proposition 9.5.2(iv). Let us denote this realization by $\Sigma'' = \left[\begin{smallmatrix} \mathfrak{A}'' & \mathfrak{B}'' \\ \mathfrak{C}'' & \mathfrak{D} \end{smallmatrix}\right]$. Its state space is $X'' = \overline{\mathcal{R}(\pi_+ \mathfrak{D} \pi_-)}$, and $\mathfrak{C}''$ is the identity (hence isometric). By Lemma 9.2.2(ii), $\overline{\mathcal{R}(\pi_+ \mathfrak{D} \pi_-)} = \overline{\mathcal{R}(\mathfrak{C})}$. But $\mathcal{R}(\mathfrak{C})$ is closed since $\mathfrak{C}$ is isometric, and so $\mathcal{R}(\mathfrak{C}) = \mathcal{R}(\mathfrak{C})'' = X''$. By Theorem 9.2.5, Σ and Σ'' are similar to each other, with an isometric similarity operator E.

We again leave the proof of the additional claims valid in the Hilbert space case to the reader.

(iii) Recall from (i) that $\mathcal{R}(\mathfrak{C}_1) = \mathcal{R}(\mathfrak{C}_2)$. We have $\underline{\mathfrak{C}}_1 = (\overline{\mathfrak{C}}_1)_{|\underline{X}}$, $\underline{\mathfrak{C}}_2 = (\overline{\mathfrak{C}}_2)_{|\underline{X}}$. Thus $(\overline{\mathfrak{C}}_2^{-1})_{|\mathcal{R}(\mathfrak{C}_1)} = \underline{\mathfrak{C}}_2^{-1}$, and by (9.2.3),

$$\overline{E}_{|\underline{X}} = \overline{\mathfrak{C}}_2^{-1}(\overline{\mathfrak{C}}_1)_{|\underline{X}} = \overline{\mathfrak{C}}_2^{-1}\underline{\mathfrak{C}}_1 = \underline{\mathfrak{C}}_2^{-1}\underline{\mathfrak{C}}_1 = \underline{E}.$$

(iv)–(v) This is true because the restricted exactly observable and exactly controllable shift realizations have these properties, and every input or output normalized realization is isometrically similar to one of these. □

Corollary 9.5.4 *Let U and Y be Hilbert spaces, and let $\mathfrak{D} \in TIC^2(U;Y)$. Then the observability gramian of any input normalized realization of $\mathfrak{D}$ is unitarily similar to the controllability gramian of any output normalized realization of $\mathfrak{D}$. In particular, if we denote the Hankel operator $\pi_+\mathfrak{D}\pi_-$ by Γ, then they are unitarily similar to $\Gamma^*\Gamma_{|\mathcal{N}(\Gamma)^\perp}$ and also to $\Gamma\Gamma^*_{|\mathcal{N}(\Gamma^*)^\perp}$.*

This follows from Theorem 9.5.3(i)–(ii).

In the Hilbert space L^2-well-posed case it is possible to go one step further and to construct a *balanced realization* by interpolating between an input normalized and an output normalized realization.

Definition 9.5.5 Let $\Sigma = \left[\begin{smallmatrix} \mathfrak{A} & \mathfrak{B} \\ \mathfrak{C} & \mathfrak{D} \end{smallmatrix}\right]$ be an L^2-well-posed linear system on the Hilbert spaces (Y, X, U). This system is *(Hankel) balanced* if it is minimal, input/state bounded and state/output bounded, and the two gramians $\mathfrak{B}\mathfrak{B}^*$ and $\mathfrak{C}^*\mathfrak{C}$ are equal.[3]

Theorem 9.5.6 *Let $\mathfrak{D} \in TIC^2(U;Y)$, where U and Y are Hilbert spaces. Then $\mathfrak{D}$ has a (Hankel) balanced realization $\Sigma = \left[\begin{smallmatrix} \mathfrak{A} & \mathfrak{B} \\ \mathfrak{C} & \mathfrak{D} \end{smallmatrix}\right]$, and this realization is unique up to a unitary similarity transformation in the state space.*

[3] Sometimes the word *parbalanced* is used in this case, and the word "balanced" is reserved for the case when these gramians are diagonal with respect to some orthonormal basis. This is true if and only if the gramians have a full set of eigenvectors which span the state space. The standard way to guarantee this is to assume that the Hankel operator of the input/output map is compact.

Both $\mathfrak{A}$ and $\mathfrak{A}^*$ are strongly stable contraction semigroups. If we denote the Hankel operator induced by $\mathfrak{D}$ by $\Gamma := \pi_+\mathfrak{D}\pi_- \in \mathcal{B}(L^2(\mathbb{R}^-;U);L^2(\mathbb{R}^+;Y))$, then $\mathfrak{C}\mathfrak{C}^* = \sqrt{\Gamma\Gamma^*}$, $\mathfrak{B}^*\mathfrak{B} = \sqrt{\Gamma^*\Gamma}$, and $\mathfrak{C}^*\mathfrak{C} = \mathfrak{B}\mathfrak{B}^*$ are unitarily similar to $\sqrt{\Gamma^*\Gamma}_{|\mathcal{N}(\Gamma)^\perp}$ and also to $\sqrt{\Gamma\Gamma^*}_{|\mathcal{N}(\Gamma^*)^\perp}$. If $\underline{\Sigma}$ is the input normalized system on $(Y,\underline{X},U)$ and $\overline{\Sigma}$ is the output normalized system on $(Y,\overline{X},U)$ induced by Σ (see Proposition 9.5.2(i)–(ii)), then $\sqrt{\mathfrak{B}\mathfrak{B}^*}$ is a unitary map of X onto $\underline{X}$, and $\overline{X}$ is the completion of X with respect to the norm $|x|_{\overline{X}} = |\sqrt{\mathfrak{B}\mathfrak{B}^*}x|_X$.

In Theorem 11.8.14 we shall encounter another type of balancing, where we interpolate between the minimal and maximal passive realizations of a given input/output map $\mathfrak{D}$. The main part of the argument is the same in both cases, and we have distilled it into the following lemma.

Lemma 9.5.7 *Let $\underline{\Sigma} = \left[\begin{array}{c|c} \underline{\mathfrak{A}} & \underline{\mathfrak{B}} \\ \hline \underline{\mathfrak{C}} & \mathfrak{D} \end{array}\right]$ and $\overline{\Sigma} = \left[\begin{array}{c|c} \overline{\mathfrak{A}} & \mathfrak{B} \\ \hline \overline{\mathfrak{C}} & \mathfrak{D} \end{array}\right]$ be two L^2-well-posed linear systems on the Hilbert spaces $(Y,\underline{X},U)$, respectively $(Y,\overline{X},U)$, where $\underline{X}$ is continuously and densely embedded in $\overline{X}$ (in particular, the two systems have the same input map $\mathfrak{B}$, the same input/output map $\mathfrak{D}$, $\underline{\mathfrak{A}} = \overline{\mathfrak{A}}_{|\underline{X}}$, and $\underline{\mathfrak{C}} = \overline{\mathfrak{C}}_{|\underline{X}}$). Suppose that $\underline{\mathfrak{A}}$ is a contraction semigroup on $\underline{X}$, and that $\overline{\mathfrak{A}}$ is a contraction semigroup on $\overline{X}$. Let $F \in \mathcal{B}(\overline{X})$ be the Gram operator of the embedding $\underline{X} \subset \overline{X}$ (i.e., $F > 0$, and $\langle x,y\rangle_{\overline{X}} = \langle x,Fy\rangle_{\underline{X}}$ for all $x,y \in \underline{X}$).*

(i) *There is a unique Hilbert space X such that $\underline{X} \subset X \subset \overline{X}$ with continuous and dense embeddings, and such that the Gram operator of the embedding $\underline{X} \subset X$ is the restriction to $\underline{X}$ of the Gram operator of the embedding $X \subset \overline{X}$. That is, there is a unique positive operator $E \in \mathcal{B}(\overline{X})$ (namely $E = \sqrt{F}$) such that $\langle x,y\rangle_X = \langle x,Ey\rangle_{\underline{X}}$ for all $x,y \in \underline{X}$ and $\langle x,y\rangle_{\overline{X}} = \langle x,Fy\rangle_{\underline{X}}$ for all $x,y \in X$. If we identify the dual of X with X itself, then $\overline{X}$ and $\underline{X}$ are duals of each other.*
(ii) *The space X is invariant under $\overline{\mathfrak{A}}$, and if we define $\mathfrak{A} = \overline{\mathfrak{A}}_{|X}$, $\mathfrak{C} = \overline{\mathfrak{C}}_{|X}$, and $\mathfrak{B} = \underline{\mathfrak{B}}$, then $\mathfrak{A}$ is a C_0 contraction semigroup on X and $\Sigma = \left[\begin{array}{c|c} \mathfrak{A} & \mathfrak{B} \\ \hline \mathfrak{C} & \mathfrak{D} \end{array}\right]$ is an L^2-well-posed linear system on (Y,X,U). Furthermore, $\mathfrak{A}$ is strongly stable (in X) whenever $\overline{\mathfrak{A}}$ is strongly stable (in $\overline{X}$), and $\mathfrak{A}$ is strongly co-stable (i.e., $\mathfrak{A}^*$ is strongly stable in X) whenever $\underline{\mathfrak{A}}$ is strongly co-stable (in $\underline{X}$).*
(iii) *If we use the superscripts $''$, *, and $'$ for adjoints computed with respect to the norms in $\overline{X}$, X, and $\underline{X}$, respectively, then*

$$\overline{\mathfrak{A}}'' = E^{-1}\mathfrak{A}^*E, \qquad \mathfrak{B}'' = \mathfrak{B}^*E, \qquad \mathfrak{C}^* = E\overline{\mathfrak{C}}'',$$
$$\mathfrak{A}^* = E^{-1}\underline{\mathfrak{A}}'E, \qquad \mathfrak{B}^* = \mathfrak{B}'E, \qquad \underline{\mathfrak{C}}' = E\mathfrak{C}^*,$$
$$\overline{\mathfrak{A}}'' = F^{-1}\underline{\mathfrak{A}}'F, \qquad \mathfrak{B}'' = \mathfrak{B}'F, \qquad \underline{\mathfrak{C}}' = F\overline{\mathfrak{C}}''.$$

Note that the growth bounds of the three systems is at most zero (since the semigroups are contraction semigroups), so that $\mathfrak{B}'$, $\mathfrak{B}^*$, and $\mathfrak{B}''$ map into $L^2_\omega(\mathbb{R}^-;U)$ and $\underline{\mathfrak{C}}'$, $\mathfrak{C}^*$, and $\overline{\mathfrak{C}}''$ are defined on $L^2_\omega(\mathbb{R}^+;Y)$ for every $\omega > 0$.

Proof (i) Define $E = \sqrt{F}$, and let X be the range of $\sqrt{E}$, with the induced norm: $|\sqrt{E}x|_X = |x|_{\underline{X}}$. This space has all the properties listed in (i) (see the discussion preceding Proposition 3.6.2 for an explanation of the duality). It is unique since a Hilbert subspace of $\underline{X}$ is determined uniquely by its norm, and the requirements listed in (i) imply that we must have $E = \sqrt{F}$.

(ii) By Lemma 9.5.8, X is invariant under $\overline{\mathfrak{A}}^t$, and $\mathfrak{A}^t := \overline{\mathfrak{A}}^t_{|X}$ is a contraction on X. It is clearly a semigroup on X. For all $x \in \underline{X}$, $\mathfrak{A}^t x = \underline{\mathfrak{A}}^t x$ is continuous in $\underline{X}$, hence in X. If $\overline{\mathfrak{A}}$ is strongly stable and $x \in \underline{X}$, then $\mathfrak{A}^t x$ tends to zero at infinity:

$$\begin{aligned}\|\mathfrak{A}^t x\|_X^2 &= \langle \mathfrak{A}^t x, \mathfrak{A}^t x\rangle_X = \langle \overline{\mathfrak{A}}^t x, \underline{\mathfrak{A}}^t x\rangle_{(\overline{X},\underline{X})} \le \|\overline{\mathfrak{A}}^t x\|_{\overline{X}} \|\underline{\mathfrak{A}}^t x\|_{\underline{X}} \\ &\le \|\overline{\mathfrak{A}}^t x\|_{\overline{X}} \|x\|_{\underline{X}} \to 0 \text{ as } t \to \infty.\end{aligned}$$

Thus, $\mathfrak{A}x \in BC(\overline{\mathbb{R}}^+;X)$ for all $x \in \underline{X}$, and if $\overline{\mathfrak{A}}$ is strongly stable, then $\mathfrak{A}x \in BC_0(\overline{\mathbb{R}}^+;X)$. Given $x \in X$, we can choose a sequence $x_n \in \underline{X}$ converging to x in X. Then $\mathfrak{A}^t x_n \to \mathfrak{A}^t x$ in X, uniformly in $t \in \overline{\mathbb{R}}^+$ (since $\mathfrak{A}$ is a contraction semigroup), and therefore the limit $\mathfrak{A}x$ is continuous, and it tends to zero at infinity if $\overline{\mathfrak{A}}$ is strongly stable. Thus, $\mathfrak{A}$ is strongly continuous (in X), and it is strongly stable whenever $\overline{\mathfrak{A}}$ is strongly stable.

That $\mathfrak{A}$ is strongly co-stable whenever $\underline{\mathfrak{A}}$ is strongly co-stable follows from the same argument, applied to the dual system (since $\overline{X}$ is the dual of $\underline{X}$ with X as pivot space).

Define $\mathfrak{C} = \overline{\mathfrak{C}}_{|X}$. Then $\Sigma = \left[\begin{smallmatrix}\mathfrak{A} & \mathfrak{B} \\ \mathfrak{C} & \mathfrak{D}\end{smallmatrix}\right]$ satisfies all the algebraic conditions in Definition 2.2.1, and $\mathfrak{A}$ is a C_0 contraction semigroup on X. The input map $\mathfrak{B}$ is continuous with values in X since it is continuous with values in $\underline{X}$, and the output map $\mathfrak{C}$ is continuous on X since it is continuous on $\overline{X}$. Thus, Σ is an L^2-well-posed linear system on X.

(iii) This follows from Proposition 3.6.2. □

Proof of Theorem 9.5.6 Let $\underline{\Sigma} = \left[\begin{smallmatrix}\underline{\mathfrak{A}} & \mathfrak{B} \\ \underline{\mathfrak{C}} & \mathfrak{D}\end{smallmatrix}\right]$ be an arbitrary input normalized realization of $\mathfrak{D}$ on $(Y, \underline{X}, U)$ (for example, the restricted exactly controllable shift realization in Proposition 9.5.2(iii)), and let $\overline{\Sigma} = \left[\begin{smallmatrix}\overline{\mathfrak{A}} & \overline{\mathfrak{B}} \\ \overline{\mathfrak{C}} & \mathfrak{D}\end{smallmatrix}\right]$ be the corresponding output normalized realization on $(Y, \overline{X}, U)$ (cf. Proposition 9.5.3(iii)). Then $\underline{\mathfrak{A}}$ is a strongly co-stable contraction semigroup, and $\overline{\mathfrak{A}}$ is a strongly stable contraction semigroup (see Theorem 9.5.3(iv)–(v)). Let Σ be the system in Lemma 9.5.7 with state space X. Then $\mathfrak{A}$ is a contraction semigroup which is both strongly stable and strongly co-stable.

With the notation introduced in part (iii) of Lemma 9.5.7, we have (since $\underline{\Sigma}$ is input normalized and $\overline{\Sigma}$ is output normalized; hence $\overline{\mathfrak{C}}''\mathfrak{C} = 1$ and $\mathfrak{B}\mathfrak{B}' = 1$)

$$\mathfrak{B}\mathfrak{B}^* = \mathfrak{C}^*\mathfrak{C} = E_{|X}, \quad \mathfrak{B}\mathfrak{B}'' = F, \quad \underline{\mathfrak{C}}'\underline{\mathfrak{C}} = F_{|\underline{X}}.$$

Thus, Σ is (Hankel) balanced. By Theorem 9.5.3(i),

$$\Gamma\Gamma^* = \mathfrak{C}\underline{\mathfrak{C}}' = \mathfrak{C}E\mathfrak{C}^* = \mathfrak{C}\mathfrak{C}^*\mathfrak{C}\mathfrak{C}^*,$$

hence $\mathfrak{C}\mathfrak{C}^* = \sqrt{\Gamma\Gamma^*}$. A similar computation shows that $\mathfrak{B}^*\mathfrak{B} = \sqrt{\Gamma^*\Gamma}$.

That a balanced realization is unique up to a unitary similarity transformation in its state space follows from the final claim in Theorem 9.2.5 and the fact that $\mathfrak{C}\mathfrak{C}^* = \sqrt{\Gamma\Gamma^*}$ is independent of the realization (as long as it is balanced). That $\mathfrak{C}\mathfrak{C}^*_{|\mathcal{N}(\mathfrak{C}^*)^\perp} = \sqrt{\Gamma\Gamma^*}_{|\mathcal{N}(\Gamma^*)^\perp}$, $\mathfrak{B}^*\mathfrak{B}_{|\mathcal{N}(\mathfrak{B})^\perp} = \sqrt{\Gamma^*\Gamma}_{|\mathcal{N}(\Gamma)^\perp}$ and $\mathfrak{C}^*\mathfrak{C} = \mathfrak{B}\mathfrak{B}^*$ are unitarily similar to each other follows from Lemma A.2.5. □

In the proof of Lemma 9.5.7 we use the following fundamental lemma.

Lemma 9.5.8 *Let U, Y, and $W \subset X \subset V$ be Hilbert spaces, where the embeddings are continuous and dense, let $E \in \mathcal{B}(V)$ be injective, self-adjoint (with respect to the inner product in V), and suppose that $\sqrt{E}$ maps V isometrically onto X and that $\sqrt{E}_{|X}$ maps X isometrically onto W. Let $\overline{A}$ be a contraction on V, and suppose that W is invariant under $\overline{A}$, and that $\underline{A} := \overline{A}_{|W}$ is a contraction on W. Then $A := \overline{A}_{|X}$ is a contraction on X (in particular, A maps X into X).*

Proof Let $\underline{A}^* \in \mathcal{B}(V)$ and $\overline{A}^* \in \mathcal{B}(W)$ be the adjoints of $\underline{A}$, respectively $\overline{A}$, when we identify the dual of W with V (with X as pivot space), let $\underline{A}' \in \mathcal{B}(W)$ be the adjoint of A with respect to the inner product in W, and let $\overline{A}'' \in \mathcal{B}(V)$ be the adjoint of $\overline{A}$ with respect to the inner product in V. All of these operators are contractions in the indicated spaces, $\overline{A}^* = \underline{A}^*_{|W}$, and by Proposition 3.6.2,

$$\underline{A}^* = E^{-1}\underline{A}'E, \qquad \overline{A}^* = E\overline{A}''E^{-1}, \qquad \overline{A}'' = E^{-2}\underline{A}'E^2.$$

Define $B := \underline{A}^*A = \overline{A}^*A$. Then $B \in \mathcal{B}(W)$, and B is positive with respect to the inner product in X on W since, for all $w \in W$,

$$\begin{aligned}\langle w, Bw\rangle_X &= \langle w, EBw\rangle_W = \langle w, \underline{A}'E\underline{A}w\rangle_W = \langle \underline{A}w, E\underline{A}w\rangle_W \\ &= \langle \underline{A}w, \underline{A}w\rangle_X \ge 0.\end{aligned}$$

This implies that B can be extended to a (possibly unbounded) positive self-adjoint operator (with respect to the inner product in X) mapping $\mathcal{D}(B) \subset X$ into X (see, e.g., Kato 1980, Theorem 3.4, p. 268, Corollary 1.28, p. 318, and Theorem 2.6, p. 323). We still denote the extended operator by the same letter B. Note that $W \subset \mathcal{D}(B^n)$ for all $n = 1, 2, 3, \ldots$ since B was originally defined on W and B maps W into itself. The operator B has a spectral resolution $F(d\lambda)$, and for all $n = 1, 2, 3, \ldots$ and $x \in \mathcal{D}(B^n)$ (in particular, for all $x \in W$)

$$B^n x = \int_{\overline{\mathbb{R}}^+} \lambda^n F(d\lambda)x.$$

Let π be the orthogonal projection $\pi x = F((1,\infty)) = \int_{(1,\infty)} F(d\lambda)x$, let $X_1 = \mathcal{N}(\pi)$ and $X_2 = \mathcal{R}(\pi)$, and let $B_1 = B(1-\pi)$, $B_2 = B\pi$. Then π commutes with B, B_1 is a contraction mapping X into X_1, $\mathcal{D}\left(B_2^n\right) = \mathcal{D}(B^n) \supset W$ for all $n = 1, 2, 3, \ldots$, and for all $x \in W$ and $n = 1, 2, 3, \ldots$,

$$\langle x, B^n x\rangle_X = \langle x, B_1^n x\rangle_X + \langle x, B_2^n x\rangle_X.$$

If we here let $n \to \infty$, then one of two things will happen. If $\pi x = 0$, then $B_2^n x = 0$ and $\langle x, B_1^n x\rangle_X \le |x|_X^2$, so in this case $\limsup_{n\to\infty}\langle x, B^n x\rangle_X \le |x|_X^2$. If $\pi x \ne 0$, then we still have $\langle x, B_1^n x\rangle_X \le |x|_X^2$ but, by the Lebesgue monotone convergence theorem

$$\langle x, B_2^n x\rangle_X = \int_{(1,\infty)} \lambda^n \langle x, F(d\lambda)x\rangle \to \infty.$$

Thus, we can test whether $x \in W$ satisfies $\pi x = 0$ by testing if $\limsup_{n\to\infty}$ $\langle x, B^n x\rangle_X$ is finite or not.

We claim that $\limsup_{n\to\infty}\langle x, B^n x\rangle_X$ is actually *finite for all* $x \in W$. This can be seen as follows. Let $x \in W$. We write Bx in the form

$$\begin{aligned} Bx &= E\overline{A}'' E^{-1}\underline{A} = E^{1/2}E^{1/2}\overline{A}'' E^{-1/2}E^{-1/2}\underline{A}E^{1/2}E^{-1/2} \\ &= E^{1/2}\overline{C}\underline{C}E^{-1/2}, \end{aligned}$$

where both $\overline{C} := E^{1/2}\overline{A}'' E^{-1/2}$ and $\underline{C} = E^{-1/2}\underline{A}E^{1/2}$ are contractions on X ($\overline{C}$ is unitarily similar to the contraction $\overline{A}''$ on V, and $\underline{C}$ is unitarily similar to the contraction $\underline{A}$ on W). Thus,

$$\begin{aligned} \langle x, B^n x\rangle_X &= \langle x, E^{1/2}[\overline{C}\underline{C}]^n E^{-1/2}x\rangle_X = \langle E^{1/2}x, [\overline{C}\underline{C}]^n E^{-1/2}x\rangle_X \\ &\le \langle E^{1/2}x, E^{-1/2}x\rangle_X = |x|_V|x|_W. \end{aligned}$$

This proves our earlier claim that $\limsup_{n\to\infty}\langle x, B^n x\rangle_X < \infty$ for all $x \in W$. As we observed earlier, this implies that $\pi x = 0$ for all $x \in W$, and since W is dense in X, we get $B_2 = 0$. We conclude that $B = B_1$ is a contraction on X.

Since B is a contraction on X, we get for all $x \in W$,

$$|\underline{A}x|_X^2 = \langle \underline{A}x, \underline{A}x\rangle_X = \langle x, Bx\rangle_X \le |x|_X^2.$$

This implies that $\underline{A}$ has a unique extension to a contraction operator on X. Since $\underline{A} = \overline{A}_{|W}$, this extension must coincide with $\overline{A}_{|X}$. Thus, $\overline{A}_{|X}$ is a contraction on X. □

9.6 Resolvent tests for controllability and observability

Our original definition of controllability and observability was given in the time domain in terms of the input map $\mathfrak{B}$ and output map $\mathfrak{C}$. It is useful to have some alternative characterizations in terms of the semigroup generator A, the control operator B, and the observation operator C. Some conditions of this type will be given below. See Section 9.10 for additional conditions.

We begin by establishing a number of different equivalent characterizations of the unobservable subspace.

Lemma 9.6.1 *Let Σ be an $L^p|Reg$-well-posed linear system on (Y, X, U) with semigroup generator A and observation operator C. Denote the component of $\rho(A)$ which contains some right half-plane by $\rho_\infty(A)$. Then the unobservable subspace $\mathcal{U}$ of Σ can be characterized in the following equivalent ways:*

(i) $\mathcal{U} = \mathcal{N}(\mathfrak{C})$,
(ii) $\mathcal{U} = \bigcap_{\lambda \in \rho_\infty(A)} \mathcal{N}\big(C(\lambda - A)^{-1}\big)$,
(iii) $\mathcal{U} = \bigcap_{n=1}^{\infty} \mathcal{N}\big(C(\lambda_n - A)^{-1}\big)$, *where $\{\lambda_n\}_{n=1}^{\infty}$ is an arbitrary sequence contained in $\rho_\infty(A)$ which has a cluster point in $\rho_\infty(A)$,*
(iv) $\mathcal{U} = \bigcap_{n=1}^{\infty} \mathcal{N}\big(C(\lambda - A)^{-n}\big)$, *where λ is an arbitrary point in $\rho_\infty(A)$,*
(v) *$\mathcal{U}$ is the largest $\mathfrak{A}$-invariant subspace which is contained in $\mathcal{N}\big(C(\lambda - A)^{-1}\big)$, where λ is an arbitrary point in $\rho_\infty(A)$.*

If $A \in \mathcal{B}(X)$, then

(vi) $\mathcal{U} = \bigcap_{n=0}^{\infty} \mathcal{N}(CA^n)$,

and if C has an extension to an operator $C_{|X} \in \mathcal{B}(X; Y)$, then

(vii) *$\mathcal{U}$ is the largest $\mathfrak{A}$-invariant subspace which is contained in $\mathcal{N}\big(C_{|X}\big)$.*

Proof We begin by observing that (i) is the original definition of the unobservable subspace.

Let $x_0 \in X$. Then $x_0 \in \mathcal{U}$ if and only if $\mathfrak{C}(x_0) = 0$, or equivalently, if and only if the Laplace transform of $\mathfrak{C}(x_0)$ vanishes. By Theorem 4.4.2(iv), this Laplace transform is given by

$$\widehat{\mathfrak{C}(x_0)}(\lambda) = C(\lambda - A)^{-1} x_0, \qquad \lambda \in \mathbb{C}^+_{\omega_\mathfrak{A}},$$

where $\omega_\mathfrak{A}$ is the growth bound of $\mathfrak{A}$. The right-hand side of this equation has a (unique) analytic extension to $\rho_\infty(A)$, so $\mathfrak{C}(x_0) = 0$ if and only if $(\lambda - A)^{-1} C x_0 = 0$ for all $\lambda \in \rho_\infty(A)$. This proves (ii). But (ii) is equivalent to (iii) since the analytic function $(\lambda - A)^{-1} C x_0$ vanishes identically if and only if its zero set has a cluster point in the domain of analyticity. It is also equivalent to (iv), because the same analytic function vanishes identically if and only if all its derivatives vanish at one point, and according to (3.2.6),

$$\frac{d^n}{d\lambda^n} C(\lambda - A)^{-1} x_0 = (-1)^n n! C(\lambda - A)^{-(n+1)} x_0.$$

If $A \in \mathcal{B}(X)$, then $C(\lambda - A)^{-1} x_0$ is analytic at infinity, and $\rho_\infty(A)$ contains a neighborhood of ∞. The function $C(\lambda - A)^{-1} x_0$ vanishes identically in this neighborhood if and only if its Taylor series at infinity vanishes. By Lemma 4.10.6, this series is given by

$$C(\lambda - A)^{-1} x_0 = \sum_{n=0}^{\infty} C A^n x_0 \lambda^{-n-1}.$$

Thus $C(\lambda - A)^{-1} x_0 = 0$ in a neighborhood of infinity if and only if $C A^n x_0 = 0$ for all $n = 0, 1, 2, \ldots$

It remains to prove (v) and (vii). We begin with (v). By (iv), $\mathcal{U} \subset \mathcal{N}\left(C(\lambda - A)^{-1}\right)$, and $\mathcal{U}$ is invariant under $(\lambda - A)^{-1}$ (since $(\lambda - A)^{-1} x_0$ satisfies (iv) whenever x_0 does). By Theorem 3.14.4, $\mathcal{U}$ is an invariant subspace of $\mathfrak{A}$. Let $V \subset \mathcal{N}\left(C(\lambda - A)^{-1}\right)$ be an arbitrary subspace which is invariant under $\mathfrak{A}$, hence under $(\lambda - A)^{-1}$, and let $x_0 \in V$. Then, for all $n = 0, 1, 2, \ldots, (\lambda - A)^{-n} x_0 \in V \subset \mathcal{N}\left(C(\lambda - A)^{-1}\right)$, so $C(\lambda - A)^{-n-1} x_0 = 0$. By (iv), $x_0 \in \mathcal{U}$. Thus $V \subset \mathcal{U}$, so $\mathcal{U}$ is the *largest* $\mathfrak{A}$-invariant subspace contained in $\mathcal{N}\left(C(\lambda - A)^{-1}\right)$.

Suppose, finally, that C has an extension $C_{|X} \in \mathcal{B}(X; Y)$. By (ii), if $x_0 \in \mathcal{U}$ then $C(\lambda - A)^{-1} x_0 = 0$ for all $\lambda \in \rho_\infty(A)$. Therefore, by Theorem 3.7.1(iii),

$$C_{|X} x_0 = C_{|X} \lim_{\lambda \to +\infty} \lambda(\lambda - A)^{-1} x_0 = \lim_{\lambda \to +\infty} \lambda C(\lambda - A)^{-1} x_0 = 0.$$

Thus $\mathcal{U} \subset \mathcal{N}\left(C_{|X}\right)$. The proof of the claim that $\mathcal{U}$ is the *largest* $\mathfrak{A}$-invariant subspace contained in $\mathcal{N}\left(C_{|X}\right)$ is analogous to the one above. □

This lemma immediately gives us the following characterization of observability:

Corollary 9.6.2 *Let Σ be an $L^p|Reg$-well-posed linear system on $(Y, X; U)$ with semigroup generator A and observation operator C. Denote the component of $\rho(A)$ which contains some right half-plane by $\rho_\infty(A)$. Then the following conditions are equivalent:*

(i) *Σ is observable,*
(ii) $\bigcap_{n=1}^{\infty} \mathcal{N}\left(C(\lambda_n - A)^{-1}\right) = 0$ *for some sequence (hence, for all sequences) $\{\lambda_n\}_{n=1}^{\infty} \subset \rho_\infty(A)$ which has a cluster point in $\rho_\infty(A)$,*
(iii) $\bigcap_{n=0}^{\infty} \mathcal{N}\left(C(\lambda - A)^{-n}\right) = 0$ *for some (hence, for all) $\lambda \in \rho_\infty(A)$,*
(iv) *for some (hence, for all) $\lambda \in \rho_\infty(A)$, $\mathcal{N}\left(C(\lambda - A)^{-1}\right)$ does not contain any nontrivial $\mathfrak{A}$-invariant subspace.*

If $A \in \mathcal{B}(X)$, then these conditions are equivalent to

(v) $\bigcap_{n=0}^{\infty} \mathcal{N}(CA^n) = 0$,

and if C has an extension to an operator $C_{|X} \in \mathcal{B}(X;Y)$, then they are further equivalent to

(vi) $\mathcal{N}\left(C_{|X}\right)$ *does not contain any nontrivial* $\mathfrak{A}$*-invariant subspace.*

Proof See Lemma 9.6.1. □

A result similar result to Lemma 9.6.1 is true for the input map $\mathfrak{B}$.

Lemma 9.6.3 *Let Σ be an $L^p|Reg$-well-posed linear system on $(Y, X; U)$ with semigroup generator A and control operator B. Denote the component of $\rho(A)$ which contains some right half-plane by $\rho_\infty(A)$. Then the reachable subspace $\overline{\mathcal{R}}$ of Σ can be characterized in the following equivalent ways:*

(i) $\overline{\mathcal{R}} = \overline{\mathcal{R}(\mathfrak{B})}$,
(ii) $\overline{\mathcal{R}}$ *is the closed linear span of* $\bigcup_{\lambda\in\rho_\infty(A)} \mathcal{R}\left((\lambda - A_{|X})^{-1}B\right)$,
(iii) $\overline{\mathcal{R}}$ *is the closed linear span of* $\bigcup_{n=1}^{\infty} \mathcal{R}\left((\lambda_n - A_{|X})^{-1}B\right)$, *where* $\{\lambda_n\}_{n=1}^{\infty}$ *is an arbitrary sequence contained in* $\rho_\infty(A)$ *which has a cluster point in* $\rho_\infty(A)$,
(iv) $\overline{\mathcal{R}}$ *is the closed linear span of* $\bigcup_{n=1}^{\infty} \mathcal{R}\left((\lambda - A_{|X})^{-n}B\right)$, *where λ is an arbitrary point in* $\rho_\infty(A)$,
(v) $\overline{\mathcal{R}}$ *is the smallest closed* $\mathfrak{A}$*-invariant subspace which contains* $\mathcal{R}\left((\lambda - A_{|X})^{-1}B\right)$, *where λ is an arbitrary point in* $\rho_\infty(A)$.

If $A \in \mathcal{B}(X)$, then

(vi) $\overline{\mathcal{R}}$ *is the closed linear span of* $\bigcup_{n=0}^{\infty} \mathcal{R}(A^n B)$,

and if $\mathcal{R}(B) \subset X$, then

(vii) $\overline{\mathcal{R}}$ *is the smallest closed* $\mathfrak{A}$*-invariant subspace which contains* $\mathcal{R}(B)$.

The proof of this lemma uses the following well-known result:

Lemma 9.6.4 *Let X be a Banach space. For each $E \subset X$ we define*

$$E^\perp = \{x^* \in X^* \mid \langle x, x^* \rangle = 0 \text{ for all } x \in E\},$$

and for each $F \subset X^*$ we define

$$^\perp F = \{x \in X \mid \langle x, x^* \rangle = 0 \text{ for all } x^* \in F\}.$$

Then, for each $E \subset X$, $^\perp(E^\perp)$ is the closed linear span of E.

See, for example, Kato (1980, p. 136) for a proof. The set $E^\perp$ is usually called the *annihilator* of E, and $^\perp F$ could be called the *pre-annihilator* (or simply annihilator) of F. Note that ${}^{\perp}F = F^{\perp} \cap X$ in the nonreflexive case, and that ${}^{\perp}F = F^{\perp}$ if X is reflexive.

Proof of Lemma 9.6.3 Clearly, (i) is just a repetition of the definition of $\overline{\mathcal{R}}$. For the duration of this proof, let us denote $\overline{\mathcal{R}} = \overline{\mathcal{R}(\mathfrak{B})}$, and let us denote the closed linear subspaces given in the different parts (ii)–(vii) by $\overline{\mathcal{R}}_{(ii)}$, $\overline{\mathcal{R}}_{(iii)}$, etc. Thus, for example, $\overline{\mathcal{R}}_{(ii)}$ is the closed linear span of $\cup_{\lambda\in\rho_\infty(A)}\mathcal{R}\left((\lambda - A_{|X})^{-1}B\right)$, and $\overline{\mathcal{R}}_{(v)}$ is the smallest closed $\mathfrak{A}$-invariant subspace which contains $\mathcal{R}\left((\lambda - A_{|X})^{-1}B\right)$, where λ is an arbitrary point in $\rho_\infty(A)$. By Lemma 9.6.4, to prove that these subspaces are identical it suffices to show that they have the same annihilator, and to prove that one is included in the other it suffices to prove the opposite inclusion of their annihilators.

We first show that $\overline{\mathcal{R}}_{(ii)} \subset \overline{\mathcal{R}}$. Let $x_0^* \in \overline{\mathcal{R}}^{\perp}$. Let $\lambda \in \mathbb{C}^+_{\omega_{\mathfrak{A}}}$ where $\omega_{\mathfrak{A}}$ is the growth bound of $\mathfrak{A}$, and define the functions $\mathrm{e}_{n,\lambda}$ as in Theorem 4.2.1(iii). Then $\mathfrak{B}(\mathrm{e}_{\lambda,n}u) \in \overline{\mathcal{R}}$, hence $\langle\mathfrak{B}(\mathrm{e}_{\lambda,n}u), x_0^*\rangle\rangle = 0$ for all $n = 0, 1, 2, \ldots$ and all $u \in U$. By Theorem 4.2.1(iii), this is equivalent to saying that $\langle(\lambda - A)^{-n}Bu, x_0^*\rangle\rangle = 0$ for all $n = 0, 1, 2, \ldots$ and all $u \in U$. The function $\langle(\lambda - A)^{-1}Bu, x_0^*\rangle\rangle$ is an analytic function on $\rho_\infty(A)$, and the above condition says that all its derivatives vanish at the point λ (see (3.2.6)). Therefore it must vanish identically on $\rho_\infty(A)$. We conclude that x_0^* belongs to the annihilator of $\mathcal{R}\left((\lambda - A)^{-1}B\right)$ for all $\lambda \in \rho_\infty(A)$, so $x_0^* \in \overline{\mathcal{R}}_{(ii)}^{\perp}$. This proves that $\overline{\mathcal{R}}_{(ii)} \subset \overline{\mathcal{R}}$.

A simplified version of the above argument also shows that $\overline{\mathcal{R}}_{(ii)} = \overline{\mathcal{R}}_{(iii)} = \overline{\mathcal{R}}_{(iv)}$ (an analytic function vanishes identically if all its derivatives vanish at some point, or if its zero set has a cluster point).

We next show that $\overline{\mathcal{R}}_{(v)} \subset \overline{\mathcal{R}}_{(iv)}$. It is clear that $\overline{\mathcal{R}}_{(iv)}$ contains $\mathcal{R}\left((\lambda - A)^{-1}B\right)$, and that $\overline{\mathcal{R}}_{(iv)}$ is invariant under $(\lambda - A)^{-1}$ (since $\cup_{n=2}^{\infty}\mathcal{R}\left((\lambda - A_{|X})^{-n}B\right) \subset \cup_{n=1}^{\infty}\mathcal{R}\left((\lambda - A_{|X})^{-n}B\right)$). By Theorem 3.14.4, it is $\mathfrak{A}$-invariant. It is also closed. Thus $\overline{\mathcal{R}}_{(iv)}$ must contain $\overline{\mathcal{R}}_{(v)}$.

To complete the proof of (ii)–(v) we must still show that $\overline{\mathcal{R}} \subset \overline{\mathcal{R}}_{(v)}$. As we saw above, $\mathcal{R}\left((\lambda - A)^{-1}B\right) \subset \overline{\mathcal{R}}$ for all $\lambda \in \rho_\infty(A)$. It follows from the intertwining condition $\mathfrak{A}^t\mathfrak{B} = \mathfrak{B}\tau_-^t$ that $\mathcal{R}(\mathfrak{B})$ is invariant under $\mathfrak{A}$, hence so is its closure $\overline{\mathcal{R}}$. Thus, $\overline{\mathcal{R}}$ is a closed $\mathfrak{A}$-invariant subspace which contains $\mathcal{R}\left((\lambda - A)^{-1}B\right)$. It remains to show that it is minimal. Let V be a closed $\mathfrak{A}$-invariant subspace containing $\mathcal{R}\left((\lambda - A_{|X})^{-1}B\right)$. Then, for each $u \in C_c^\infty(\mathbb{R}^-; U)$, the integral

$$\int_{-\infty}^{0} \mathfrak{A}^{-s}(\lambda - A_{|X})^{-1}Bu(s)\,ds$$

belongs to V. But this is equal to $(\lambda - A_{|X})^{-1}\mathfrak{B}u = (\lambda - A)^{-1}\mathfrak{B}u$ since $\mathfrak{A}^{-s}(\lambda - A_{|X})^{-1} = (\lambda - A_{|X})^{-1}\mathfrak{A}^{-s}_{|X_{-1}}$. Thus, $(\lambda - A)^{-1}\mathfrak{B}u \in V$. Applying $\lambda - A$ to this identity and using the fact that V is invariant under A (see

Theorem 3.14.4(iv)) we get $\mathfrak{B}u \in V$. Thus, V must contain $\mathcal{R}(\mathfrak{B})$ (where we regard $\mathfrak{B}$ as an operator mapping $C_c^\infty(\mathbb{R}^-;U)$ into X; cf. Lemma 9.1.4), and since V is closed, V must contain the reachable subspace. This shows that $\overline{\mathcal{R}} \subset \overline{\mathcal{R}}_{(v)}$, and it completes the proofs of (ii)–(v).

If $A \in \mathcal{B}(X)$, then, for each $x_0^* \in X^*$ and each $u \in U$, the function $\langle(\lambda - A)^{-1}Bu, x_0^*\rangle)$ is analytic at infinity. As we saw in the above proof, $X_0^* \in \overline{\mathcal{R}}^\perp$ if and only if this function vanishes identically on $\rho_\infty(A)$ for all $u \in U$. But this is true if and only if each coefficient in its Taylor expansion at infinity vanishes, i.e, $\langle A^n Bu, x_0^*\rangle = 0$ for all $u \in U$ (see Lemma 4.10.6). This is equivalent to $x_0^* \in \overline{\mathcal{R}}_{(vi)}^\perp$. Thus $\overline{\mathcal{R}} = \overline{\mathcal{R}}_{(vi)}$.

If $\mathcal{R}(B) \subset X$, then (by the closed graph theorem) $B \in \mathcal{B}(U;X)$. As we observed above, for every $\lambda \in \rho_\infty(A)$ and $u \in U$ we have $(\lambda - A)^{-1}Bu \in \mathcal{R}(\mathfrak{B})$, hence $\lambda(\lambda - A)^{-1}Bu \in \mathcal{R}(\mathfrak{B})$. Let $\lambda \to +\infty$. Then $\lambda(\lambda - A)^{-1}Bu \to Bu$, so $Bu \in \overline{\mathcal{R}(\mathfrak{B})}$. Thus $\overline{\mathcal{R}(\mathfrak{B})}$ is a closed $\mathfrak{A}$-invariant subspace which contains $\mathcal{R}(B)$. That it is the *smallest* such subspace follows from the fact that the condition $\mathcal{R}(B) \subset \overline{\mathcal{R}(\mathfrak{B})}$ is stronger than the original condition $\mathcal{R}\left((\lambda - A)^{-1}B\right) \subset \overline{\mathcal{R}(\mathfrak{B})}$ since $\overline{\mathcal{R}(\mathfrak{B})}$ is $\mathfrak{A}$-invariant, hence $(\lambda - A)^{-1}$-invariant. □

Lemma 9.6.3 immediately gives us the following characterization of controllability:

Corollary 9.6.5 *Let Σ be an $L^p|Reg$-well-posed linear system on $(Y, X; U)$ with semigroup generator A and control operator B. Denote the component of $\rho(A)$ which contains some right half-plane by $\rho_\infty(A)$. Then the following conditions are equivalent:*

(i) *Σ is controllable,*
(ii) *the linear span of $\bigcup_{n=1}^\infty \mathcal{R}\left((\lambda_n - A_{|X})^{-1}B\right)$ is dense in X for some sequence (hence, for all sequences) $\{\lambda_n\}_{n=1}^\infty \subset \rho_\infty(A)$ which has a cluster point in $\rho_\infty(A)$,*
(iii) *the linear span of $\bigcup_{n=0}^\infty \mathcal{R}\left((\lambda - A_{|X})^{-n}B\right)$ is dense in X for some (hence, for all) $\lambda \in \rho_\infty(A)$,*
(iv) *for some (hence, for all) $\lambda \in \rho_\infty(A)$, $\mathcal{R}\left((\lambda - A)^{-1}B\right)$ is not contained in any proper closed $\mathfrak{A}$-invariant subspace.*

If $A \in \mathcal{B}(X)$, then these conditions are equivalent to

(v) *the linear span of $\cup_{n=0}^\infty \mathcal{R}(A^nB)$ is dense in X,*

and if $\mathcal{R}(B) \subset X$, then they are further equivalent to

(vi) *$\mathcal{R}(B)$ is not contained in any proper closed $\mathfrak{A}$-invariant subspace.*

Proof See Lemma 9.6.3. □

In the case of a bounded semigroup generator we get the following *necessary* conditions for controllability and observability, often referred to as the *Hautus rank conditions*.

Lemma 9.6.6 *Let* $\Sigma = \left[\begin{smallmatrix} \mathfrak{A} & \mathfrak{B} \\ \mathfrak{C} & \mathfrak{D} \end{smallmatrix}\right]$ *be a linear system on* $(Y, X; U)$ *with bounded semigroup generator* $A \in \mathcal{B}(X)$*, bounded control observation* $B \in \mathcal{B}(U; X)$*, and bounded observation operator* $C \in \mathcal{B}(X; Y)$*. Then the following claims are true.*

(i) *If* Σ *is observable, then* $\mathcal{N}\left(\left[\begin{smallmatrix} A-\lambda \\ C \end{smallmatrix}\right]\right) = 0$ *for all* $\lambda \in C$.
(ii) *If* Σ *is controllable, then* $\mathcal{R}\left(\left[A - \lambda \;\; B\right]\right)$ *is dense in* X *for all* $\lambda \in \mathbb{C}$.

Of course, this condition is interesting only in the case where $\lambda \in \sigma(A)$, because the conditions in (i) and (ii) are trivially satisfied when $\lambda \in \rho(A)$. It says roughly that C detects every eigenvalue of A, and that B fills in the missing part of the range of $A - \lambda$ when λ belongs to the residual spectrum of A. See also Lemma 9.6.9.

Proof If Σ is controllable or observable, then the exponentially shifted system Σ_α described in Example 2.3.5 is controllable, respectively, observable for all $\alpha \in \mathbb{C}$ (see Example 9.4.4), so we may without loss of generality assume that $\lambda = 0$ both in (i) and in (ii).

(i) Assume that Σ is controllable. Let $x_0 \in X$. Suppose that both $Cx_0 = 0$ and $Ax_0 = 0$. Then $CA^n x_0 = 0$ for all $n = 0, 1, 2, \ldots$ By Lemma 9.6.1, $x_0 = 0$. Thus $\mathcal{N}\left(\left[\begin{smallmatrix} A \\ C \end{smallmatrix}\right]\right) = 0$.

(ii) This proof is similar to the proof of (i), but we replace Lemma 9.6.1 by 9.6.3. □

Remark 9.6.7 The converse of Lemma 9.6.6 is not true. This can be seen as follows. Let $A \in \mathcal{B}(X)$ have a spectrum consisting of the single point $\{0\}$ (i.e., A is quasi-nilpotent), and suppose that A has no eigenvalues. One such operator is the weighted right-shift which maps $x = \{x_n\}_{n=0}^{\infty} \in l^2(\mathbb{Z}_+; \mathbb{C})$ into

$$(Ax)_0 = 0, \qquad (Ax)_n = a_n x_{n-1}, \qquad n = 1, 2, 3, \ldots,$$

and a_n is an arbitrary sequence tending to zero as $n \to \infty$. Then condition (i) in Lemma 9.6.6 holds for all $\lambda \in \mathbb{C}$ since $\mathcal{N}(A - \lambda) = 0$ for all $\lambda \in \mathbb{C}$, independently of how we choose C. Taking, e.g., $C = 0$ we get a system which is not observable, and which still satisfies the Hautus condition.

However, although the converse to Lemma 9.6.6 is not true in general, it is true in many cases of interest. One such case is the one where A has a meromorphic resolvent, and we are interested in the *modal controllability* of each eigenmode (this notion will be introduced shortly in Definition 9.7.1).

Definition 9.6.8 Let Ω be an open subset of $\mathbb{C}$, let U and Y be Banach spaces, and let $f\colon \Omega \to \mathcal{B}(U;Y)$.

(i) f has a *pole* (of finite order) at $\lambda_0 \in \Omega$ if f is analytic in a neighborhood of λ_0, apart from at the point λ_0 itself, and for some $m = 1, 2, 3, \ldots$, the function $(\lambda - \lambda_0)^m f(\lambda)$ has a removable singularity at λ_0 (i.e., this function has an extension which is analytic in a full neighborhood of λ_0). The smallest such number $m \geq 0$ is the *order* of λ_0. (Thus, f has a pole of order zero if f is analytic or f has a removable singularity at λ_0.)

(ii) f is *meromorphic* in Ω if, at each point $\lambda \in \Omega$, f is (analytic at λ or) has a pole of finite order at λ.

An equivalent way of formulating (i) is to say that f has a Laurent expansion

$$f(\lambda) = \sum_{k=-m}^{\infty} f_k(\lambda - \lambda_0)^k, \tag{9.6.1}$$

valid in a neighborhood of λ_0 (not including the point λ_0 itself). The (local) *McMillan degree* of this pole is the rank of the operator $S \in \mathcal{B}(U^m; Y^m)$ given by

$$S = \begin{bmatrix} f_{-1} & f_{-2} & \cdots & f_{-m+1} & f_{-m} \\ f_{-2} & f_{-3} & \cdots & f_{-m} & 0 \\ \vdots & \vdots & \ddots & \vdots & \vdots \\ f_{-m+1} & f_{-m} & \cdots & 0 & 0 \\ f_{-m} & 0 & \cdots & 0 & 0 \end{bmatrix}. \tag{9.6.2}$$

Because of the presence of m copies of the operator f_{-m} (whose rank is at least one) in the above block matrix, the McMillan degree of a pole is at least as large as the order of the pole. It can even be infinite. For example, if A is an invertible operator on X, then zero is a first order pole of the $\mathcal{B}(X)$-valued function $\lambda^{-1}A$, and its MacMillan degree is the same as the dimension of X (which may be infinite).

Note that the counter-example in Remark 9.6.7 has a resolvent which is *not* meromorphic at zero (this follows from Lemma 4.10.6).

The following partial converse to Lemma 9.6.6 is true (see also Theorems 9.7.4 and 9.7.5).

Lemma 9.6.9 (Modal Hautus conditions) *Let $\Sigma = \left[\begin{smallmatrix} \mathfrak{A} & \mathfrak{B} \\ \mathfrak{C} & \mathfrak{D} \end{smallmatrix}\right]$ be a linear system on $(Y, X; U)$ with bounded semigroup generator $A \in \mathcal{B}(X)$, bounded control observation $B \in \mathcal{B}(U; X)$, and bounded observation operator $C \in \mathcal{B}(X; Y)$. Suppose further that A has a one point spectrum $\{\lambda_0\}$, and that λ_0 is a pole (of finite order) of $(\lambda - A)^{-1}$. Then the following claims are true.*

(i) Σ *is observable if and only if* $\mathcal{N}\left(\left[\begin{smallmatrix} A-\lambda_0 \\ C \end{smallmatrix}\right]\right) = 0$.
(ii) Σ *is controllable if and only if* $\mathcal{R}\left(\left[A - \lambda_0 \;\; B\right]\right)$ *is dense in* X.

The assumption that λ_0 is a pole of $(\lambda - A)^{-1}$ is equivalent to the assumption that $A - \lambda_0$ is *nilpotent*: if we denote by m the order of λ_0 as a pole of $(\lambda - A)^{-1}$, then by Lemma 4.10.6, $(A - \lambda_0)^{m-1} \neq 0$ but $(A - \lambda_0)^m = 0$.

Proof Without loss of generality we may take $\lambda_0 = 0$ (see the comment at the beginning of the proof of Lemma 9.6.6). The necessity of the conditions given in (i) and (ii) for observability, respectively, controllability follow from Lemma 9.6.6, so it suffices to prove the converse part.

(i) Suppose that $CA^k x_0 = 0$ for all $k = 0, 1, 2, \ldots$. According to Corollary 9.6.2(v), to show that Σ is observable it suffices to show that this condition implies that $x_0 = 0$. If $x_0 \neq 0$, then there is a number $n \geq 0$ such that $A^{n-1}x_0 \neq 0$ but $A^n x_0 = 0$ (since A is nilpotent; see the comment preceding this proof). But then $\left[\begin{smallmatrix} A \\ C \end{smallmatrix}\right] A^{n-1}x_0 = 0$, and so, by the Hautus condition in (i), $A^{n-1}x_0 = 0$. This contradiction shows that we cannot have $x_0 \neq 0$, and it proves that Σ is observable.

(ii) This proof is analogous to the one above, and we leave it to the reader. □

Theorem 9.6.10 *Let* $\Sigma = \left[\begin{smallmatrix} \mathfrak{A} & \mathfrak{B} \\ \mathfrak{C} & \mathfrak{D} \end{smallmatrix}\right]$ *be a linear system on* $(Y, X; U)$ *with bounded semigroup generator* $A \in \mathcal{B}(X)$, *bounded control observation* $B \in \mathcal{B}(U; X)$, *and bounded observation operator* $C \in \mathcal{B}(X; Y)$. *Suppose further that* A *has a one point spectrum* $\{\lambda_0\}$. *Then the following claims are true.*

(i) *If* λ_0 *is a pole of (finite) order* m *of* $(\lambda - A)^{-1}$, *then* λ_0 *is a pole of order at most* m *of* $\widehat{\mathfrak{D}}$, *and the McMillan degree of this pole is at most equal to* $\dim(X)$ *(finite or infinite). If* Σ *is controllable and observable, then the order of* λ_0 *as a pole of* $\widehat{\mathfrak{D}}$ *is exactly* m, *and its McMillan degree is equal to* $\dim(X)$.
(ii) *If* λ_0 *is a pole of* $\widehat{\mathfrak{D}}$ *with finite McMillan degree, then* Σ *is controllable and observable if and only if* $\dim(X)$ *is equal to the McMillan degree of this pole (hence* $\dim(X)$ *is finite,* λ_0 *is a pole of* $(\lambda - A)^{-1}$, *and (i) applies).*

Proof Again we shall, without loss of generality, take $\lambda_0 = 0$ (cf. the proof of Lemma 9.6.6).

(i) Since zero is a pole of order m of $(\lambda - A)^{-1}$ we have $A^{m-1} \neq 0$ but $A^m = 0$ (see Lemma 4.10.6). By (4.10.14), $\widehat{\mathfrak{D}}$ has a pole of order at most m at zero. That the McMillan degree of this pole cannot exceed $\dim(X)$ follows from the fact that the operator S in (9.6.2) can be factored as (see (4.10.14))

$$S = \begin{bmatrix} C \\ CA \\ \vdots \\ CA^m \end{bmatrix} \begin{bmatrix} B & AB & \cdots & A^m B \end{bmatrix}, \tag{9.6.3}$$

and the rank of both the operators on the right-hand side are bounded by $\dim(X)$.

Suppose that Σ is controllable and observable, and suppose that zero is a pole of order n of $\widehat{\mathfrak{D}}$. (For the argument in this paragraph it is irrelevant that zero is a pole of $(\lambda - A)^{-1}$ of finite order.) By (4.10.14), $CA^{n-1}B \neq 0$ but $CA^k B = 0$ for all $k \geq n$. Since $CA^{(n+k)}B = 0$ for all $k = 0, 1, 2, \ldots$, it follows from Corollaries 9.6.2(v) and 9.6.5(v) that $CA^n = 0$ and $A^n B = 0$. By the same corollaries, the first operator on the right-hand side of (9.6.3) is injective on X, and the second has dense range in X. Therefore $\dim(X) = \operatorname{rank}(S)$ equals the McMillan degree of zero as a pole of $\widehat{\mathfrak{D}}$.

To complete the proof of (i) we must still show that zero is a pole of order n of $(\lambda - A)^{-1}$, or equivalently, that $A^n = 0$. We know that $CA^n = 0$ and $A^n B = 0$. We also know that $A^m = 0$ but $A^{m-1} \neq 0$ (where $m \geq n$ is the order of zero as a pole of $(\lambda - A)^{-1}$). If $m > n$, then both $A^m = 0$ and $CA^{m-1} = 0$, hence $\left[\begin{smallmatrix} A \\ C \end{smallmatrix}\right] A^{m-1} = 0$. By Lemma 9.6.6, this would imply $A^{m-1} = 0$, and we specifically chose m so that $A^{m-1} \neq 0$. This shows that we must have $m = n$, so the order of zero as a pole of $(\lambda - A)^{-1}$ is the same as the order as a pole of $\widehat{\mathfrak{D}}$.

(ii) Suppose that Σ is controllable and observable. By the argument that we gave above, the McMillan degree of the pole at zero of $\widehat{\mathfrak{D}}$ is equal to $\dim(X)$ (in that part of the proof we did not use the assumption that zero is a pole of $(\lambda - A)^{-1}$). In particular, $\dim(X)$ is finite, and (i) applies. If, on the other hand, $\dim(X)$ is equal to the McMillan degree of the pole at zero of $\widehat{\mathfrak{D}}$, then in this case, too, $\dim(X)$ is finite and (i) applies. □

9.7 Modal controllability and observability

In this section we shall look at the controllability and observability properties of an isolated point of the spectrum of the main operator A.

Definition 9.7.1 Let $\Sigma = \left[\begin{smallmatrix} \mathfrak{A} & \mathfrak{B} \\ \mathfrak{C} & \mathfrak{D} \end{smallmatrix}\right]$ be an $L^p|Reg$-well-posed linear system on (Y, X, U) with generator A, and let λ_0 be an isolated point of $\sigma(A)$. Let π be the Riesz projection constructed in Theorem 3.14.10 with $\sigma_+(A) = \{\lambda_0\}$ and $\sigma_-(A) = \sigma(A) \setminus \{\lambda_0\}$, and let Σ_+ be the corresponding system constructed in Theorem 4.10.4. We call λ_0 *controllable* in any of the senses listed in

Definition 9.4.1 if Σ_+ is controllable in the same sense, and we call λ_0 *observable* in any of the senses listed in Definition 9.4.2 if Σ_+ is observable in the same sense. The system Σ is *modally controllable* if every isolated point of $\sigma(A)$ is controllable in the above sense, it is *modally observable* if every isolated point of $\sigma(A)$ is observable in the above sense, and it is *modally minimal* if it is both modally controllable and modally observable.

Observe that the restriction of A to each eigenspace is bounded, so we may use parts (v)–(vi) of Corollaries 9.6.2 and 9.6.5 to test modal controllability or observability. If $(\lambda - A)^{-1}$ is meromorphic (see Definition 9.6.8), then we can alternatively use the modal Hautus conditions in Lemma 9.6.9. If the eigenspace related to the spectral point λ_0 (i.e, the range of π) is finite-dimensional, then all the different controllability versions are equivalent, and so are all the different observability versions (and in this case Theorem 9.6.10(ii) applies).

If a system is controllable or observable, then all the isolated points in its spectrum are controllable and observable in the same sense. This is a consequence of the following lemma.

Lemma 9.7.2 *Let Σ be the parallel connection of two subsystems Σ_1 and Σ_2 (see Example 2.3.13). If Σ is controllable in any of the senses listed in Definition 9.4.1 then both Σ_1 and Σ_2 are controllable in the same sense, and if Σ is observable in any of the senses listed in Definition 9.4.2 then both Σ_1 and Σ_2 are observable in the same sense.*

Proof See Definitions 9.4.1, 9.4.2, 9.7.1, and Theorem 4.10.4. □

A more interesting question is to what extent the converse claim is true. Suppose that $\sigma(A)$ has no cluster points, and that each point in $\sigma(A)$ is controllable or observable. What additional assumptions do we need in order to conclude that Σ is controllable or observable? It is clear that some additional conditions are needed, because $\sigma(A)$ may, for example, be empty, in which case the modal controllability or observability assumption tells us absolutely nothing about the full system.

As a first step in our answer to this question, let us prove the following partial converse to Lemma 9.7.2.

Theorem 9.7.3 *Let Σ be an $L^p|Reg$-well-posed linear system on $(Y, X; U)$ with semigroup generator A. Suppose that $\sigma(A)$ can be split into two disjoint parts $\sigma_+(A)$ and $\sigma_-(A)$ by a piecewise continuously differentiable Jordan curve Γ contained in $\rho(A)$ encircling $\sigma_+(A)$. Let π be the Riesz projection constructed in Theorem 3.14.10, and let Σ_+ and Σ_- be the corresponding systems constructed in Theorem 4.10.4 (so that Σ is the parallel connection of Σ_- and Σ_-). Then Σ*

is observable if and only if both Σ_+ and Σ_- are observable and Σ is controllable if and only if both Σ_+ and Σ_- are controllable.

Proof One direction is contained in Lemma 9.7.2, so it suffices to prove that Σ is observable or controllable whenever both Σ_+ and Σ_- are observable or controllable.

Let $x_0 \in X$, and suppose that $\mathfrak{C}x_0 = 0$. Then, by Lemma 9.6.1, $C(\lambda - A)^{-1}x_0 = 0$ for all $\lambda \in \rho_\infty(A)$, and hence (see also Corollary 3.14.9)

$$C(\lambda - A_+)^{-1}\pi x_0 = -C(\lambda - A_-)^{-1}(1-\pi)x_0, \qquad \lambda \in \rho_\infty(A),$$

where $A_+ = A_{|X_+}$ and $A_- = A_{|X_-}$. The left-hand side has an analytic extension to $\mathbb{C} \setminus \sigma_+(A)$, and the right-hand side has an analytic extension to $\mathbb{C} \setminus \sigma_-(A)$. The union of these two sets is all of $\mathbb{C}$, so both sides can be extended to entire functions. The left-hand side tends to zero at infinity (recall that A_+ is bounded), so by Liouville's theorem, it must vanish identically. Therefore also the right-hand side vanishes identically. By Lemma 9.6.1, this implies that πx_0 is an unobservable element of Σ_+ and $(1-\pi)x_0$ is an unobservable element of Σ_-. If both Σ_+ and Σ_- is observable, then $\pi x_0 = 0$ and $(1-\pi)x_0 = 0$, hence $x_0 = 0$. Thus Σ is observable whenever both Σ_+ and Σ_- are observable.

To prove the controllability claim we let $x_0^* \in \overline{\mathcal{R}}^\perp$, where $\overline{\mathcal{R}}$ is the reachable subspace of Σ. Then by Lemma 9.6.3 (and its proof), $\langle(\lambda - A)^{-1}Bu, x_0^*\rangle = 0$ for all $\lambda \in \rho_\infty(A)$ and all $u \in U$. Arguing in the same way as above we find that both $\langle(\lambda - A_+)^{-1}Bu, x_0^*\rangle = 0$ and $\langle(\lambda - A_-)^{-1}Bu, x_0^*\rangle = 0$. Therefore, by Lemma 9.6.3, if we denote the reachable subspaces of Σ_+ and Σ_- by $\overline{\mathcal{R}}_+$ and $\overline{\mathcal{R}}_-$, then $x_0^* \in \overline{\mathcal{R}}_+^\perp \cap \overline{\mathcal{R}}_-^\perp = (\overline{\mathcal{R}}_+ \cup \overline{\mathcal{R}}_-)^\perp$. This means that $\overline{\mathcal{R}}$ contains the closed linear span of $\overline{\mathcal{R}}_+$ and $\overline{\mathcal{R}}_-$, which is all of X if both Σ_- and Σ_- are controllable. □

From Theorem 9.7.3 we easily get the following spectral characterization of observability.

Theorem 9.7.4 *Let Σ be an $L^p|Reg$-well-posed linear system on $(Y, X; U)$ with semigroup generator A. Suppose that $\sigma(A)$ contains a countable sequence of isolated points $\{\lambda_n\}_{n=1}^\infty$ such that the corresponding eigenspaces X_n span X in the sense that $x_0 \in X$ is the zero vector if and only if $\pi_n x_0 = 0$ for all $n = 1, 2, 3, \ldots$, where π_n is the Riesz projection onto the eigenspace X_n (see Theorem 3.14.10). Then Σ is observable if and only it is modally observable, or equivalently, if and only if $\mathcal{N}\left(\left[\begin{smallmatrix} A-\lambda_n \\ C \end{smallmatrix}\right]\right) = 0$ for all n.*

Proof If Σ is observable, then by Lemma 9.7.2 every isolated point of $\sigma(A)$ is observable, i.e, Σ is modally observable. Conversely, suppose that Σ is modally observable, and let $\mathfrak{C}x_0 = 0$. Arguing as in the proof of Theorem 9.7.3 with $\sigma_+(A) = \{\lambda_n\}$ we find that $\pi_n x_0 = 0$ for all n. Thus, $x_0 = 0$, and this proves that Σ is observable. The last equivalence follows from Lemmas 9.6.6(i) and 9.6.9(i). □

The corresponding controllability result is also true.

Theorem 9.7.5 *Let Σ be an $L^p|Reg$-well-posed linear system on $(Y, X; U)$ with semigroup generator A. Suppose that $\sigma(A)$ contains a countable sequence of isolated points $\{\lambda_n\}_{n=1}^{\infty}$ such that the linear span of the corresponding eigenspaces X_n is dense in X (see Theorem 3.14.10). Then Σ is controllable if and only it is modally controllable, or equivalently, if and only if $\mathcal{R}\left(\begin{bmatrix} A - \lambda_n & B \end{bmatrix}\right)$ is dense in X for all n.*

Proof If Σ is controllable, then by Lemma 9.7.2 every isolated point of $\sigma(A)$ is controllable, i.e, Σ is modally controllable. Conversely, suppose that Σ is modally controllable. Arguing as in the proof of Theorem 9.7.3 (first with $\sigma_+(A) = \{\lambda_1\}$, then with $\sigma_+(A) = \{\lambda_1, \lambda_2\}$, etc.) we find that the reachable subspace must contain the closed linear span of any finite union of the subspaces X_n. By the extra density assumption, the smallest closed subspace which contains all these subspaces is X itself. The last equivalence follows from Lemmas 9.6.6(ii) and 9.6.9(ii). □

9.8 Spectral minimality

In this section we say a few words about the *spectral minimality* of a particular realization of a given input/output map $\mathfrak{D}$.

Definition 9.8.1 Let $\Sigma = \left[\begin{smallmatrix} \mathfrak{A} & \mathfrak{B} \\ \mathfrak{C} & \mathfrak{D} \end{smallmatrix}\right]$ be an $L^p|Reg$-well-posed linear system on (Y, X, U) with main operator A and transfer function $\mathfrak{D}$. Denote the component of $\rho(A)$ which contains some right-half plane by $\rho_\infty(A)$. We call Σ *spectrally minimal* if the restriction of $\widehat{\mathfrak{D}}$ to $\rho_\infty(A)$ does not have an analytic continuation to any (boundary) point in $\sigma(A) \cap \overline{\rho_\infty(A)}$.

In general the spectrum of the main operator of a spectrally minimal realization of a given transfer function $\widehat{\mathfrak{D}}$ is not unique (this will typically be true of transfer functions which have branch points). A transfer function can also have different spectrally minimal realizations with very different spectra. If, for example, $\sigma(A) = \overline{\mathbb{C}^-}$, then Σ is spectrally minimal if and only if $\widehat{\mathfrak{D}}$ does not have an analytic continuation to any point on the imaginary axis. In this case it may sometimes be possible to find another realization with a smaller spectrum; see, for example, Fuhrmann (1981, p. 267).

Most systems occurring in practice are of the following type.

Theorem 9.8.2 *Let Σ be a minimal $L^p|Reg$-well-posed linear system with main operator A. If $\rho(A)$ is connected and $\sigma(A)$ is the closure of a totally disconnected set, then Σ is spectrally minimal.*

That a set is *totally disconnected* means that every point of this set is isolated.

Proof The assumption that $\rho(A)$ is connected means that $\rho_\infty(A) = \rho(A)$. Denote the totally disconnected subset of $\sigma(A)$ by Λ. Then every $\lambda_0 \in \Lambda$ is an isolated point of $\sigma(A)$. Let X_0 be the corresponding eigenspace of A (i.e., the range of the Riesz projection π; cf. Theorem 3.14.10), let $A_0 = A_{|X_0}$, and let $\widehat{\mathfrak{D}}_0$ be the transfer function of the corresponding subsystem (denoted by Σ_+ in Theorem 4.10.5). Then $\widehat{\mathfrak{D}}$ has an analytic continuation to λ_0 if and only if $\widehat{\mathfrak{D}}_0$ has an analytic continuation. However, by Theorem 9.6.10(ii), this is true if and only if $\dim(X_0) = 0$, in which case $\lambda_0 \notin \sigma(A)$. Thus, $\widehat{\mathfrak{D}}$ cannot have an analytic extension to any point in Λ. Since Λ is dense in $\sigma(A)$, $\widehat{\mathfrak{D}}$ cannot be analytically continued to any point in $\sigma(A)$ (a function which is analytic at a point is analytic in a full neighborhood of that point). □

Another quite common class of spectrally minimal systems is the following.

Theorem 9.8.3 *Let Σ be a minimal $L^p|Reg$-well-posed linear system whose state space X is a Hilbert space. If the main operator A is normal and $\rho(A)$ is connected, then Σ is spectrally minimal.*

Here the assumption that $\rho(A)$ is connected is important. For example, the two bilateral shift realizations in Example 2.6.5(iii),(iv) are normal if U and Y are Hilbert spaces, $p = 2$, and $\omega = 0$, but they will not, in general, be spectrally minimal in the sense of Definition 9.8.1. In both cases $\sigma(A) = j\mathbb{R}$.

Proof of Theorem 9.8.3 The assumption that $\rho(A)$ is connected means that $\rho_\infty(A) = \rho(A)$. Suppose, to get a contradiction, that $\widehat{\mathfrak{D}}$ can be analytically continued to some open disk D with center $\lambda_0 \in \sigma(A) \cap \overline{\rho(A)}$ and radius $\epsilon > 0$. Split A into two parts $A = A_+ + A_-$, where $\sigma(A_+) = \overline{\sigma(A) \cap D}$, and $\sigma(A_-) = \sigma(A) \setminus D$ as in Theorem 3.14.11 (note that $\sigma(A) \setminus D$ is closed). Note, furthermore, that (cf. Theorem 4.9.2(ii))

$$(z - A_+)^{-1}x = \int_{\sigma(A)\cap D} (z-\lambda)^{-1}E(d\lambda)x, \quad x \in \rho(A_+), \quad x \in X$$

(i.e., the boundary of D is excluded from the integral; this part goes into $(z - A_-)^{-1}x$). Write Σ as the parallel connection of two minimal systems Σ_+ and Σ_- as in Theorem 4.10.4 (the minimality follows from Lemma 9.7.2). This leads to a decomposition of $\widehat{\mathfrak{D}}$ into $\widehat{\mathfrak{D}}_+ + \widehat{\mathfrak{D}}_-$. Here $\widehat{\mathfrak{D}}_-$ is analytic in D since $D \subset \rho(A_-)$, and $\widehat{\mathfrak{D}}_+$ is analytic in $\mathbb{C} \setminus \overline{D}$ and has an analytic extension to D (since $\widehat{\mathfrak{D}}$ and $\widehat{\mathfrak{D}}_-$ are analytic there). We may further assume that at least one point on the boundary of D belongs to $\rho(A)$, hence to $\rho(A_+)$ (adjust ϵ, if necessary). Fix any $x, y \in X$. For all $z \in \rho(A_+)$,

$$\langle C_+(z - A_+)^{-1} B_+ x, y\rangle = \int_{\sigma(A)\cap D} (z - \lambda)^{-1} \langle C_+ E(d\lambda) B_+ x, y\rangle. \quad (9.8.1)$$

This is the Cauchy transform of the finite complex measure $\mu(d\lambda) = \langle C_+ E(d\lambda) B_+ x, y\rangle_{|\sigma(A)\cap D}$, and by, e.g., Gamelin (1984, Theorem 8.2, p. 46), μ vanishes on D (since $\widehat{\mathfrak{D}}_+ = C_+(z - A_+)^{-1} B_+$ is analytic there). By (9.8.1), $\langle \widehat{\mathfrak{D}}_+ x, y\rangle = 0$. This being true for all $x, y \in X$, we must have $\widehat{\mathfrak{D}}_+ = 0$. The zero transfer function has a trivial realization whose state space has dimension zero, and it follows from Theorem 9.2.4 that $\dim X_+ = 0$, too. Thus $A = A_-$, and therefore $D \subset \rho(A)$, contrary to our assumption that $\lambda_0 \in \sigma(A)$. This contradiction shows that $\widehat{\mathfrak{D}}$ cannot be analytically continued to any point $\lambda_0 \in \sigma(A) \cap \overline{\rho(A)}$, so Σ is spectrally minimal. □

9.9 Controllability and observability of transformed systems

In Chapters 6 and 7 we have studied a number of transformations of a given system. Here we investigate to what extent the different inversions and the feedback transformations preserve controllability and observability. (The duality transform will be discussed in Section 9.10.)

Lemma 9.9.1 *Let* $\Sigma = \left[\begin{smallmatrix} \mathfrak{A} & \mathfrak{B} \\ \mathfrak{C} & \mathfrak{D} \end{smallmatrix}\right]$ *be a flow-invertible well-posed linear system on* (Y, X, U), *with flow-inverse* $\Sigma_\times$. *Then the following claims are true.*

(i) Σ *and* $\Sigma_\times$ *have the same reachable subspace and the same unobservable subspace.*
(ii) $\Sigma_\times$ *is controllable in any of the senses listed in Definition 9.4.1(i)–(vii) if and only if* Σ *has the same property (i.e., we only exclude exact controllability in infinite time).*
(iii) $\Sigma_\times$ *is observable in any of the senses listed in Definition 9.4.2(i)–(vii) if and only if* Σ *has the same property (i.e., we only exclude exact observability in infinite time).*

Proof The flow-inverted input map is given by $\mathfrak{B}_\times = \mathfrak{B}\mathfrak{D}^{-1}$, and therefore $\Sigma_\times$ and Σ have the same reachable subspace. Because of the causality of $\mathfrak{D}$ and $\mathfrak{D}^{-1}$, we have for all $s < t$,

$$(\mathfrak{B}_\times)_s^t = \mathfrak{B}_\times \tau^t \pi_{[s,t)} = \mathfrak{B}\mathfrak{D}^{-1}\tau^t \pi_{[s,t)} = \mathfrak{B}\tau^t \pi_{[s,t)}\mathfrak{D}^{-1}\pi_{[s,t)} = \mathfrak{B}_s^t (\mathfrak{D}_s^t)^{-1},$$

where $(\mathfrak{D}_s^t)^{-1}$ maps $L^p|Reg([s,t);Y)$ one-to-one onto $L^p|Reg([s,t);U)$. This implies that all the different versions of controllability of Σ listed in (ii) are inherited by $\Sigma_\times$, except for null controllability. To get the null controllability we need, in addition, the fact that for all $s < t$,

$$(\mathfrak{A}_\times)_s^t = \mathfrak{A}_s^t - \mathfrak{B}_s^t(\mathfrak{D}_s^t)^{-1}\mathfrak{C}_s^t,$$

hence $\mathcal{R}\left((\mathfrak{B}_\times)_s^t\right) \supset \mathcal{R}\left((\mathfrak{A}_\times)_s^t\right)$ if and only if $\mathcal{R}\left(\mathfrak{B}_s^t\right) \supset \mathcal{R}\left(\mathfrak{A}_s^t\right)$.

The proof of the fact that Σ and $\Sigma_\times$ have the same unobservable subspace, and that all the different versions of observability of Σ listed in (ii) are inherited by $\Sigma_\times$ is similar (since $\mathfrak{C}_\times = -\mathfrak{D}^{-1}\mathfrak{C}$ and $(\mathfrak{C}_\times)_s^t = -(\mathfrak{D}_s^t)^{-1}\mathfrak{C}_s^t$ for all $s < t$). □

Above we only mentioned flow-inversion. The same proof works in many other cases, too.

Lemma 9.9.2 *Let* $\Sigma = \left[\begin{smallmatrix}\mathfrak{A} & \mathfrak{B}\\ \mathfrak{C} & \mathfrak{D}\end{smallmatrix}\right]$ *be a flow-invertible well-posed linear system on* (Y, X, U).

(i) *The conclusions of Lemma 9.9.1 remain true if we replace flow-inversion by partial flow-inversion (as in Theorem 6.6.1), or by static output feedback (as in Theorem 7.1.2).*
(ii) *Those conclusions of Lemma 9.9.1 which are related to controllability remain true if we replace flow-inversion by state feedback (as in Definition 7.3.1).*
(iii) *Those conclusions of Lemma 9.9.1 which are related to observability remain true if we replace flow-inversion by output injection (as in Definition 7.3.2).*

We leave the proof to the reader (it is the same as the proof of Lemma 9.9.1).

Let us next look at *time-inversion*, which has a more complicated behavior.

Lemma 9.9.3 *Let* $\Sigma = \left[\begin{smallmatrix}\mathfrak{A} & \mathfrak{B}\\ \mathfrak{C} & \mathfrak{D}\end{smallmatrix}\right]$ *be a time-invertible* $L^p|Reg$*-well-posed linear system on* (Y, X, U) *with main operator* A. *Denote the time-inverse by* $\Sigma^{\mathfrak{R}}$. *Then the following claims are true.*

(i) $\Sigma^{\mathfrak{R}}$ *is approximately or exactly controllable or observable in time* $t > 0$ *if and only if* Σ *has the same property.*
(ii) *Suppose that* $\rho_\infty(A) = \rho_{-\infty}(A)$, *where* $\rho_\infty(A)$ *is the component of* $\rho(A)$ *which contains some right half-plane, and* $\rho_{-\infty}(A)$ *is the component of* $\rho(A)$ *which contains some left half-plane. Then* Σ *and* $\Sigma^{\mathfrak{R}}$ *have the same reachable subspace and the same unobservable subspace. In particular,* $\Sigma^{\mathfrak{R}}$ *is controllable or observable if and only if* Σ *has the same property.*

Proof (i) Let Σ^a be the backward system in Theorem 6.4.1. Then, for all $t > 0$, $(\mathfrak{B}^a)_0^t = -\mathfrak{A}^{-t}\mathfrak{B}_0^t$ and $(\mathfrak{C}^a)_s^t = \mathfrak{C}_0^t\mathfrak{A}^{-t}$. Here $\mathfrak{A}^{-t}$ is boundedly invertible, so $\mathcal{R}\left((\mathfrak{B}^a)_0^t\right) = X$ if and only if $\mathcal{R}\left(\mathfrak{B}_0^t\right) = X$ and $\overline{\mathcal{R}\left((\mathfrak{B}^a)_0^t\right)} = X$ if and only if $\overline{\mathcal{R}\left(\mathfrak{B}_0^t\right)} = X$. From this the controllability part of (i) follows since $(\mathfrak{B}^{\mathbf{R}})^0_{-t} = (\mathfrak{B}^a)^t_0\, \boldsymbol{Я}$ and $\mathfrak{B}^0_{-t} = \mathfrak{B}^t_0\tau^{-t}$ where $\boldsymbol{Я}$ and τ^{-t} map $L^p|Reg$ $([-t, 0); U)$ one-to-one onto $L^p|Reg([0, t); U)$. The proof of the observability part of (i) is similar.

(ii) This follows from Lemmas 9.6.1(iv) and 9.6.3(iv) and the fact that $A^{\boldsymbol{Я}} = -A$, $B^{\boldsymbol{Я}} = -B$, and $C^{\boldsymbol{Я}} = C$. □

As the following example shows, part(ii) of this lemma does not hold without the assumption that $\rho_\infty(A) = \rho_{-\infty}(A)$.

Example 9.9.4 *Let $X = L^2(\mathbb{R}; \mathbb{C})$, $U = L^2(\mathbb{R}^+; \mathbb{C})$, and let Σ be the system whose semigroup is $\mathfrak{A} = \tau$ and whose (bounded) control operator is $B = \pi_+$. Then Σ is time-invertible and controllable, but the time-inverse $\Sigma^{\boldsymbol{Я}}$ is not (the reachable subspace of $\Sigma^{\boldsymbol{Я}}$ is $L^2(\mathbb{R}^+; \mathbb{C})$).*

Note that here $\sigma(A) = j\mathbb{R}$ separates $\rho_\infty(A)$ from $\rho_{-\infty}(A)$.

Proof We have $\mathcal{R}(B) = L^2(\mathbb{R}^+; \mathbb{C})$, and it follows from Lemma 9.6.3 that Σ is controllable. On the other hand, $(\mathfrak{A}^{\boldsymbol{Я}})^t = \tau^{-t}$ and $B^{\boldsymbol{Я}} = -B = -\pi_+$, so by the same lemma, the reachable subspace of $\Sigma^{\boldsymbol{Я}}$ is $L^2(\mathbb{R}^+; \mathbb{C})$. □

Out of the three inversions which we studied in Chapter 6, time-flow-inversion is the one which seems to be least well-behaved with respect to controllability and observability. It is easy to see that controllability and observability need not be preserved (see, e.g., Example 6.5.10). Of course, if the system is time-invertible, and the time-inverted system is flow-invertible (or the other way around), then we can say something about the controllability and observability of the time-flow-inverted system by applying Lemmas 9.9.1 and 9.9.3 in cascade. For example, we find that in this case exact or approximate controllability or invertibility in time t is preserved. However, we can even say something about controllability and observability in the general case.

Lemma 9.9.5 *Let $\Sigma = \left[\begin{smallmatrix} \mathfrak{A} & \mathfrak{B} \\ \mathfrak{C} & \mathfrak{D} \end{smallmatrix}\right]$ be a time-flow-invertible $L^p|Reg$-well-posed linear system on (Y, X, U) with main operator A. Denote its time-flow-inverse by $\Sigma^{Я}_{\times}$, and the time-flow-inverted main operator by $A^{Я}_{\times}$. Furthermore, suppose that $-\sigma_\infty(A) \cap \sigma_\infty(A^{Я}_{\times}) \neq \emptyset$. Then Σ and $\Sigma^{Я}_{\times}$ have the same reachable subspace and the same unobservable subspace. In particular, $\Sigma^{Я}_{\times}$ is controllable or observable if and only if Σ has the same property.*

Proof Let us for a change prove the observability claim, and leave the analogous proof of the controllability claim to the reader. Take some $\alpha \in -\sigma_\infty(A) \cap \sigma_\infty(A^{Я}_{\times})$. By Lemma 9.6.1(iv), the unobservable subspace $\mathcal{U}$ of Σ is given by $\mathcal{U} = \bigcap_{n=1}^\infty \mathcal{N}\left(C(\alpha + A)^{-n}\right)$, whereas the unobservable subspace $\mathcal{U}^{Я}_{\times}$ of $\Sigma^{Я}_{\times}$ is given by $\mathcal{U}^{Я}_{\times} = \bigcap_{n=1}^\infty \mathcal{N}\left(C^{Я}_{\times}(\alpha - A^{Я}_{\times})^{-n}\right)$. By Lemma 6.5.7(iii),

$$C^{Я}_{\times}(\alpha - A^{Я}_{\times})^{-1} = -[\widehat{\mathfrak{D}}(-\alpha)]^{-1}C(\alpha + A)^{-1},$$

so $\mathcal{N}\left(C^{\text{я}}_{\times}(\alpha - A^{\text{я}}_{\times})^{-1}\right) = \mathcal{N}\left(C(\alpha + A)^{-1}\right)$ (since $\widehat{\mathfrak{D}}(-\alpha)$ is boundedly invertible). To get higher powers of the resolvents we observe that

$$(\alpha - A^{\text{я}}_{\times})^{-1} = (\alpha + A)^{-1} + EC(\alpha + A)^{-1},$$

where $E = (\alpha + A_{|X})^{-1}B[\widehat{\mathfrak{D}}(-\alpha)]^{-1}$, and hence, by the binomial formula, for all $n \geq 1$,

$$(\alpha - A^{\text{я}}_{\times})^{-n} = (\alpha + A)^{-n} + \sum_{k=0}^{n-1}\binom{n}{k}(\alpha + A)^{-k}[EC(\alpha + A)^{-1}]^{n-k}.$$

This implies that for all $n \geq 1$,

$$(\alpha - A^{\text{я}}_{\times})^{-n}_{|\mathcal{N}(C(\alpha+A)^{-1})} = (\alpha + A)^{-n}_{|\mathcal{N}(C(\alpha+A)^{-1})},$$

and hence, for all $m \geq 1$,

$$\bigcap_{n=1}^{m}\mathcal{N}\left(C^{\text{я}}_{\times}(\alpha - A^{\text{я}}_{\times})^{-n}\right) = \bigcap_{n=1}^{m}\mathcal{N}\left(C(\alpha + A)^{-n}\right).$$

Letting $m \to \infty$ we find that $\mathcal{U} = \mathcal{U}^{\text{я}}_{\times}$. □

9.10 Time domain tests and duality

In this section we shall present several time domain characterizations which complement those given in Section 9.6 of the different controllability and observability notions that we introduced in Section 9.4. We shall also study the duality between observability and controllability.

The class of operators K which have a bounded left inverse, defined on $\mathcal{R}(K)$, will play an important role in the sequel. These operators can be characterized in several different ways:

Definition 9.10.1 Let X and Y be Banach spaces. The closed linear operator $K : X \supset \mathcal{D}(K) \to Y$ is *coercive* if there exists a constant $\epsilon > 0$ such that

$$|Kx|_Y \geq \epsilon |x|_X, \qquad x \in \mathcal{D}(K).$$

Lemma 9.10.2 *Let X and Y be Banach spaces, and let $K : X \supset \mathcal{D}(K) \to Y$ be a densely defined closed linear operator.*

(i) *The following conditions are equivalent:*
   (a) *K is injective;*
   (b) *The range of K^* is weak*-dense in X^*.*
   If X and Y are Hilbert spaces and $K \in \mathcal{B}(X; Y)$, then these conditions are equivalent to
   (c) $K^*K > 0$.

(ii) *The following conditions are equivalent:*
   (a) *the range of K is dense in Y;*
   (b) *K^* is injective.*
   If Y and Y are Hilbert spaces and $K \in \mathcal{B}(X; Y)$, then these conditions are equivalent to
   (c) $KK^* > 0$.

(iii) *The following conditions are equivalent:*
   (a) *K is coercive, i.e., there exists an $\epsilon > 0$ such that $|Kx|_Y \geq \epsilon |x|_X$ for all $x \in \mathcal{D}(K)$;*
   (b) *K is injective and $\mathcal{R}(K)$ is closed;*
   (c) *K has a bounded left inverse $K^{-1} \in \mathcal{B}(\mathcal{R}(K); X)$;*
   (d) *$\mathcal{R}(K^*) = X^*$;*
   (e) *the injective operator induced by K^* on $Y^*/\mathcal{N}(K^*)$ (see Definition 9.1.5) has a bounded right inverse $K^{-*} \in \mathcal{B}(X^*; Y^*/\mathcal{N}(K^*))$.*
   If X and Y are Hilbert spaces and $K \in \mathcal{B}(X; Y)$, then these conditions are equivalent to
   (f) $K^*K \gg 0$.

(iv) *The following conditions are equivalent:*
   (a) *$\mathcal{R}(K) = Y$;*
   (b) *the injective operator induced by K on $X/\mathcal{N}(K)$ (see Definition 9.1.5) has a bounded right inverse $K^{-1} \in \mathcal{B}(Y; X/\mathcal{N}(K))$;*
   (c) *K^* is coercive, i.e., there exists an $\epsilon > 0$ such that $|K^*y^*|_{X^*} \geq \epsilon |y^*|_{Y^*}$ for all $y^* \in \mathcal{D}(K^*)$;*
   (d) *K^* is injective and $\mathcal{R}(K^*)$ is closed;*
   (e) *K^* has a bounded left inverse $K^{-*} \in \mathcal{B}(\mathcal{R}(K^*); X^*)$.*
   If X and Y are Hilbert spaces and $K \in \mathcal{B}(X; Y)$, then these conditions are equivalent to
   (f) $KK^* \gg 0$.

The proof of Lemma 9.10.2 is based on the following well-known facts.

Lemma 9.10.3 *Let X and Y be Banach spaces, and let $K : X \supset \mathcal{D}(K) \to Y$ be a densely defined closed linear operator.*

(i) *The null spaces of K and K^* are given by*

$$\begin{aligned}\mathcal{N}(K) &= \mathcal{D}(K) \cap {}^{\perp}\mathcal{R}(K^*) \\ &= \{x \in \mathcal{D}(K) \mid \langle x, x^* \rangle = 0 \text{ for all } x^* \in \mathcal{R}(K^*)\}, \\ \mathcal{N}(K^*) &= \mathcal{R}(K)^{\perp} \\ &= \{y^* \in Y^* \mid \langle y, y^* \rangle = 0 \text{ for all } y \in \mathcal{R}(K)\}.\end{aligned}$$

(ii) *$\mathcal{N}(K)$ is closed in X and $\mathcal{N}(K^*)$ is weak*-closed in Y^* (hence closed).*

(iii) $\mathcal{N}(K)^{\perp}$ *is the weak*-closure of* $\mathcal{R}(K^*)$ *in* X^* *and* ${}^{\perp}\mathcal{N}(K^*)$ *is the norm-closure of* $\mathcal{R}(K)$ *in* Y.

(iv) *The following conditions are equivalent:*

(a) $\mathcal{R}(K)$ *is closed in* Y;

(b) $\mathcal{R}(K) = {}^{\perp}\mathcal{N}(K^*)$;

(c) $\mathcal{R}(K^*)$ *is weak*-closed in* X^*;

(d) $\mathcal{R}(K^*)$ *is norm-closed in* X^*;

(e) $\mathcal{R}(K^*) = \mathcal{N}(K)^{\perp}$;

Proof (i) By Lemma 3.5.1(iii), $\mathcal{D}(K^*)$ is weak*-dense in Y^*. This means that each of the following statements are equivalent to the one that immediately precedes or follows it:

$$\begin{aligned}
&x \in \mathcal{N}(K),\\
&x \in \mathcal{D}(K) \text{ and } Kx = 0,\\
&x \in \mathcal{D}(K) \text{ and } \langle Kx, y^*\rangle = 0 \text{ for all } y^* \in \mathcal{D}\left(K^*\right),\\
&x \in \mathcal{D}(K) \text{ and } \langle x, K^*y^*\rangle = 0 \text{ for all } y^* \in \mathcal{D}\left(K^*\right),\\
&x \in \mathcal{D}(K) \text{ and } \langle x, x^*\rangle = 0 \text{ for all } x^* \in \mathcal{R}\left(K^*\right),\\
&x \in \mathcal{D}(K) \cap {}^{\perp}\mathcal{R}\left(K^*\right).
\end{aligned}$$

Likewise, since $\mathcal{D}(K)$ is dense in X and (obviously) $\mathcal{R}(K)^{\perp} \subset \mathcal{D}(K^*)$, each of the following statements are equivalent to the one that immediately precedes or follows it:

$$\begin{aligned}
&y^* \in \mathcal{N}\left(K^*\right),\\
&y^* \in \mathcal{D}\left(K^*\right) \text{ and } K^*y^* = 0,\\
&y^* \in \mathcal{D}\left(K^*\right) \text{ and } \langle x, K^*y^*\rangle = 0 \text{ for all } x \in \mathcal{D}(K),\\
&y^* \in Y^* \text{ and } \langle x, K^*y^*\rangle = 0 \text{ for all } x \in \mathcal{D}(K),\\
&y^* \in Y^* \text{ and } \langle Kx, y^*\rangle = 0 \text{ for all } x \in \mathcal{D}(K),\\
&y^* \in Y^* \text{ and } \langle y, x^*\rangle = 0 \text{ for all } y \in \mathcal{R}(K),\\
&y^* \in \mathcal{R}(K)^{\perp}.
\end{aligned}$$

(ii) That $\mathcal{N}(K)$ is closed in X is an immediate consequence of the definition of a closed operator. That $\mathcal{N}(K^*)$ is weak*-closed in Y^* follows from (i) (the orthogonal complement of any subset of Y is always weak*-closed in Y^* (see, e.g., Rudin 1973, pp. 90–91).

(iii) This is true because (recall that $\mathcal{D}(K)$ is dense in X and see, e.g., Rudin 1973, p. 91)

$$\mathcal{N}(K)^{\perp} = (\mathcal{D}(K) \cap {}^{\perp}\mathcal{R}\left(K^*\right))^{\perp} = ({}^{\perp}\mathcal{R}\left(K^*\right))^{\perp}$$

and

$$^{\perp}\mathcal{N}\left(K^*\right) = {}^{\perp}(\mathcal{R}(K)^{\perp}).$$

(iv) This version of the closed range theorem can be found in, for example, Kato (1980, Theorem 5.13, p. 234) or Yosida (1974, p. 205). □

Proof of Lemma 9.10.2 (i) The equivalence of (a) and (b) follows from Lemma 9.10.3(i). Under the extra Hilbert space assumption

$$\langle x, K^*Kx\rangle_X = \langle Kx, Kx\rangle_Y = |Kx|_Y^2,$$

and this is strictly positive for all nonzero $x \in X$ iff K is injective.

(ii) This proof is almost identical to the proof of (i).

(iii) (a) $\Rightarrow$ (b): Clearly (a) implies that K is injective. Let $x_n \in X$ and $Kx_n \to y$. Then the coercivity of K implies that x_n is a Cauchy sequence in X, hence convergent to a limit $x \in X$. Since K is closed, this implies that $Kx = y$. Thus, $\mathcal{R}(K)$ is closed in Y.

(b) $\Rightarrow$ (c): This follows from the open mapping theorem since $\mathcal{R}(K)$ is a closed subspace of a Banach space, hence a Banach space in its own right.

(c) $\Rightarrow$ (a): This follows from the fact that for all $x \in \mathcal{D}(K)$, $x = LKx$, hence $|x|_X \le \|L\||Kx|_Y$. Take $\epsilon = 1/\|L\|$.

(b) $\Leftrightarrow$ (d): This follows from part (i) and Lemma 9.10.3(iv).

(d) $\Leftrightarrow$ (e): That (d) $\Rightarrow$ (e) follows from the open mapping theorem, and the opposite implication is trivial.

(a) $\Leftrightarrow$ (f): We leave this to the reader (see the proof of (i)).

(iv) The proof of the equivalence of (a) and (b) is the same as the proof of the equivalencc of (d) and (e) in (iii). The equivalence of (a) and (d) follows from part (ii) and Lemma 9.10.3(iv). The equivalence of (c), (d), (e), and (f) follows from part (i), applied to K^*. □

Lemma 9.10.2 applied to Definitions 9.4.1 and 9.4.2 gives the following result:

Theorem 9.10.4 *Let* $\Sigma = \left[\begin{smallmatrix} \mathfrak{A} & \mathfrak{B} \\ \mathfrak{C} & \mathfrak{D} \end{smallmatrix}\right]$ *be an* $L^p|Reg$*-well-posed linear system on* (Y, X, U)*, and let* $t > 0$.

(i) *The following conditions are equivalent:*
  (a) Σ *is controllable in time* t*, i.e.,* $\mathcal{R}\left(\mathfrak{B}\pi_{[-t,0)}\right)$ *is dense in* X*;*
  (b) $\pi_{[0,t)}\mathfrak{B}^*$ *is injective on* X^*.
  If X *and* U *are Hilbert spaces and* Σ *is* L^2*-well-posed, then these conditions are equivalent to*
  (c) $\mathfrak{B}\pi_{[0,t)}\mathfrak{B}^* > 0$.

(ii) *The following conditions are equivalent:*
  (a) Σ *is observable in time* t*, i.e.,* $\pi_{[0,t)}\mathfrak{C}$ *is one-to-one on* X*;*

(b) $\mathcal{R}\left((\pi_{[0,t)}\mathfrak{C})^*\right)$ *is weak*-dense in* X^*.

If X *and* Y *are Hilbert spaces and* Σ *is* L^2*-well-posed, then these conditions are equivalent to*

(c) $\mathfrak{C}^*\pi_{[0,t)}\mathfrak{C} > 0$.

(iii) *The following conditions are equivalent:*

(a) Σ *is exactly controllable in time* t*, i.e.,* $\mathcal{R}\left(\mathfrak{B}\pi_{[-t,0)}\right) = X$*;*

(b) *the injective operator that we get by factoring out the null space of* $\mathfrak{B}\pi_{[-t,0)}$ *(see Definition 9.1.5) has a bounded right inverse* $(\mathfrak{B}\pi_{[-t,0)})^{-1} \in \mathcal{B}(X;\mathcal{U})$*, where* $\mathcal{U} = L^p|Reg([-t,0);U)/\mathcal{N}\left(\mathfrak{B}\pi_{[-t,0)}\right)$*;*

(c) $(\mathfrak{B}\pi_{[-t,0)})^*$ *is coercive on* X^**;*

(d) $(\mathfrak{B}\pi_{[-t,0)})^*$ *is injective on* X^* *and has closed range;*

(e) $(\mathfrak{B}\pi_{[-t,0)})^*$ *has a bounded left inverse, defined on* $\mathcal{R}\left((\mathfrak{B}\pi_{[-t,0)})^*\right)$.

If X *and* U *are Hilbert spaces and* Σ *is* L^2*-well-posed, then these conditions are equivalent to*

(f) $\mathfrak{B}\pi_{[-t,0)}\mathfrak{B}^* \gg 0$.

(iv) *The following conditions are equivalent:*

(a) Σ *is exactly observable in time* t*, i.e.,* $\pi_{[0,t)}\mathfrak{C}$ *has a bounded left inverse, defined on* $\mathcal{R}\left(\pi_{[0,t)}\mathfrak{C}\right)$*;*

(b) $\pi_{[0,t)}\mathfrak{C}$ *is coercive, i.e., there exists a constant* $\epsilon > 0$ *such that* $\|\mathfrak{C}x\|_{L^p|Reg([0,t);Y)} \geq \epsilon|x|_X$ *for all* $x \in X$*;*

(c) $\pi_{[0,t)}\mathfrak{C}$ *is injective on* X *and has closed range in* $L^p|Reg([0,t);Y)$*;*

(d) $(\pi_{[0,t)}\mathfrak{C})^*$ *maps* $(L^p|Reg([0,t);Y))^*$ *onto* X^**;*

(e) *the injective operator that we get by factoring out the null space of* $(\pi_{[0,t)}\mathfrak{C})^*$ *has a bounded right inverse* $(\pi_{[0,t)}\mathfrak{C})^{-*} \in \mathcal{B}(X^*;\mathcal{Y}^*)$*, where* $\mathcal{Y}^* = (L^p|Reg([0,t);U))^*/\mathcal{N}\left((\pi_{[0,t)}\mathfrak{C})^*\right)$.

If X *and* Y *are Hilbert spaces and* Σ *is* L^2*-well-posed, then these conditions are equivalent to*

(f) $\mathfrak{C}^*\pi_{[0,t)}\mathfrak{C} \gg 0$.

Proof See Definitions 9.4.1 and 9.4.2, and Lemma 9.10.2. □

Theorem 9.10.5 *Let* $\Sigma = \left[\begin{smallmatrix}\mathfrak{A} & \mathfrak{B}\\ \mathfrak{C} & \mathfrak{D}\end{smallmatrix}\right]$ *be a* $L^p|Reg$*-well-posed linear system on* (Y, X, U) *with growth bound* $\omega_{\mathfrak{A}}$*, and let* $\omega > \omega_{\mathfrak{A}}$.

(i) *The following conditions are equivalent:*

(a) Σ *is controllable, i.e.,* $\mathfrak{B}\colon L^p|Reg_c(\mathbb{R}^-;U) \to X$ *has dense range;*

(b) $\mathfrak{B}\colon L^p|Reg_\omega(\mathbb{R}^-;U) \to X$ *has dense range;*

(c) $\mathfrak{B}^*\colon X^* \to (L^p|Reg_\omega(\mathbb{R}^-;U))^*$ *is injective.*

If X *and* U *are Hilbert spaces and* Σ *is* L^2*-well-posed, then these conditions are equivalent to*

(d) $\mathfrak{B}\mathfrak{B}^* > 0$.

(ii) *The following conditions are equivalent:*
   (a) *Σ is observable, i.e., $\mathfrak{C}\colon X \to L^p|Reg_{\mathrm{loc}}(\overline{\mathbb{R}}^+;Y)$ is injective;*
   (b) *$\mathfrak{C}\colon X \to L^p|Reg_\omega(\overline{\mathbb{R}}^+;Y)$ is injective;*
   (c) *$\mathfrak{C}^*\colon (L^p|Reg_\omega(\overline{\mathbb{R}}^+;Y))^* \to X^*$ has weak*-dense range.*
   If X and Y are Hilbert spaces and Σ is L^2-well-posed, then these conditions are equivalent to
   (d) *$\mathfrak{C}^*\mathfrak{C} > 0$.*

(iii) *The following conditions are equivalent:*
   (a) *Σ is exactly controllable in infinite time with bound ω, i.e., $\mathfrak{B}$ maps $L^p|Reg_\omega(\mathbb{R}^-;U)$ onto X;*
   (b) *the injective operator that we get by factoring out the null space of $\mathfrak{B}$ (see Definition 9.1.5) has a bounded right inverse $\mathfrak{B}^{-1} \in \mathcal{B}(X;\mathcal{U})$, where $\mathcal{U} = L^p|Reg_\omega(\mathbb{R}^-;U)/\mathcal{N}(\mathfrak{B})$;*
   (c) *$\mathfrak{B}^*\colon X^* \to (L^p|Reg_\omega(\mathbb{R}^-;U))^*$ is coercive on X^*;*
   (d) *$\mathfrak{B}^*\colon X^* \to (L^p|Reg_\omega(\mathbb{R}^-;U))^*$ is injective on X^* and has closed range.*
   (e) *$\mathfrak{B}^*\colon X^* \to (L^p|Reg_\omega(\mathbb{R}^-;U))^*$ has a bounded left inverse, defined on $\mathcal{R}(\mathfrak{B}^*)$.*
   If X and U are Hilbert spaces and Σ is L^2-well-posed, then these conditions are equivalent to
   (f) *$\mathfrak{B}\mathfrak{B}^* \gg 0$.*

(iv) *The following conditions are equivalent:*
   (a) *Σ is exactly observable in infinite time with bound ω, i.e., $\mathfrak{C}$ has a bounded left inverse, defined on $\mathcal{R}(\mathfrak{C}) \subset L^p|Reg_\omega(\overline{\mathbb{R}}^+;Y)$;*
   (b) *the injective operator that we get by factoring out the null space of $\mathfrak{C}^*$ has a bounded right inverse $\mathfrak{C}^{-*} \in \mathcal{B}(X^*;\mathcal{Y}^*)$, where $\mathcal{Y}^* = (L^p|Reg_\omega(\mathbb{R}^-;U))^*/\mathcal{N}(\mathfrak{C}^*)$;*
   (c) *$\mathfrak{C}\colon X \to L^p|Reg_\omega(\overline{\mathbb{R}}^+;Y)$ is coercive, i.e., there exists a constant $\epsilon > 0$ such that $\|\mathfrak{C}x\|_{L^p|Reg_\omega(\overline{\mathbb{R}}^+;Y)} \geq \epsilon|x|_X$ for all $x \in X$;*
   (d) *$\mathfrak{C}$ is injective on X and has closed range in $L^p|Reg_\omega(\overline{\mathbb{R}}^+;Y)$;*
   (e) *$\mathfrak{C}^*$ maps $(L^p|Reg_\omega(\overline{\mathbb{R}}^+;Y))^*$ onto X^*.*
   If X and Y are Hilbert spaces and Σ is L^2-well-posed, then these conditions are equivalent to
   (f) *$\mathfrak{C}^*\mathfrak{C} \gg 0$.*

Proof See Definitions 9.4.1 and 9.4.2, and Lemmas 9.1.3 and 9.10.2. □

As a first application of Theorems 9.10.4 and 9.10.5, let us present the following characterization of a system which is exactly controllable in infinite time with two (sufficiently large) different bounds ω:

Theorem 9.10.6 *Let* $\Sigma = \left[\begin{smallmatrix}\mathfrak{A} & \mathfrak{B}\\ \mathfrak{C} & \mathfrak{D}\end{smallmatrix}\right]$ *be an* $L^p|Reg$*-well-posed linear system on* (Y, X, U) *with growth bound* $\omega_{\mathfrak{A}}$.

(i) *The following conditions are equivalent:*
  (a) Σ *is exactly controllable in finite time;*
  (b) Σ *is exactly controllable in infinite time with bound* ω *for all those* ω *for which* $\mathfrak{B}$ *is* ω*-bounded;*
  (c) Σ *is exactly controllable in infinite time with bound* ω *for some* $\omega > \omega_{\mathfrak{A}}$.

(ii) *Also the following conditions are equivalent:*
  (a) Σ *is exactly observable in finite time;*
  (b) Σ *is exactly observable in infinite time with bound* ω *for all those* ω *for which* $\mathfrak{C}$ *is* ω*-bounded;*
  (c) Σ *is exactly observable in infinite time with bound* ω *for some* $\omega > \omega_{\mathfrak{A}}$;

Proof (i) Obviously (a) $\Rightarrow$ (b) $\Rightarrow$ (c), so it suffices to show that (c) $\Rightarrow$ (a). For each $t > 0$ we can write $\mathfrak{B}$ in the form

$$\mathfrak{B} = \mathfrak{B}\pi_{(-\infty,-t)} + \mathfrak{B}\pi_{[-t,0)} = \mathfrak{B}\tau_-^t\tau^{-t} + \mathfrak{B}\pi_{[-t,0)},$$

which by the intertwining condition $\mathfrak{A}^t\mathfrak{B} = \mathfrak{B}\tau_-^t$ can be written in the form

$$\mathfrak{B}\pi_{[-t,0)} = \mathfrak{B} - \mathfrak{A}^t\mathfrak{B}\tau^{-t}.$$

If (c) holds, then, by Theorem 2.5.4(i), $\|\mathfrak{A}^t\mathfrak{B}\tau^{-t}\| \to 0$ as $t \to \infty$. By Theorem 9.10.5(iii), there is a constant $\epsilon > 0$ such that $\|\mathfrak{B}^*x^*\| \ge \epsilon|x^*|$ for all $x^* \in X^*$. Thus, for $x^* \in X^*$,

$$\|\mathfrak{B}^*x^*\| \ge \epsilon|x^*| - \|\mathfrak{A}^t\mathfrak{B}\tau^{-t}\||x^*| \ge \epsilon/2|x^*|$$

for t large enough. By Theorem 9.10.4(iii), this implies that $\mathcal{R}\left(\mathfrak{B}\pi_{[-t,0)}\right) = X$; hence Σ is controllable in time t.

(ii) Obviously (a) $\Rightarrow$ (b) $\Rightarrow$ (c), so it suffices to show that (c) $\Rightarrow$ (a). For each $t > 0$ we can write $\mathfrak{C}$ in the form

$$\mathfrak{C} = \pi_{[0,t)}\mathfrak{C} + \pi_{(t,\infty)}\mathfrak{C} = \pi_{[0,t)}\mathfrak{C} + \tau^{-t}\tau_+^t\mathfrak{C},$$

which by the intertwining condition $\mathfrak{C}\mathfrak{A}^t = \tau_+^t\mathfrak{B}\pi_-$ can be written in the form

$$\pi_{[0,t)}\mathfrak{C} = \mathfrak{C} - \tau^{-t}\mathfrak{C}\mathfrak{A}^t.$$

By Theorem 2.5.4(i), $\|\mathfrak{A}^t\mathfrak{B}\tau^{-t}\| \to 0$ as $t \to \infty$. The proof is completed in the same way as the proof of (i) (do not pass to adjoints this time). □

As a second application of Theorem 9.10.5 we obtain the following simple description of systems that are both exactly controllable and exactly observable in infinite time with the same bound ω:

Theorem 9.10.7 *Let* $\Sigma = \left[\begin{smallmatrix} \mathfrak{A} & \mathfrak{B} \\ \mathfrak{C} & \mathfrak{D} \end{smallmatrix}\right]$ *be a minimal* ω*-bounded* $L^p|Reg$*-well-posed linear system on* (Y, X, U)*, where* $\omega \in \mathbb{R}$*. Then the following conditions are equivalent:*

(i) Σ *is both exactly controllable and exactly observable in infinite time with (the same) bound* ω*.*
(ii) *the range of the Hankel operator* $\pi_+\mathfrak{D}\pi_-\colon L^p|Reg_\omega(\mathbb{R}^-;U) \to L^p|Reg_\omega(\overline{\mathbb{R}}^+;Y)$ *is closed.*

In particular, (ii) is true (for all those $\omega \in \mathbb{R}$ *for which* Σ *is* ω*-bounded) if* Σ *is both exactly controllable and exactly observable in finite time.*

Proof (i) $\Rightarrow$ (ii): The Hankel operator $\pi_+\mathfrak{D}\pi_-$ factors into $\pi_+\mathfrak{D}\pi_- = \mathfrak{C}\mathfrak{B}$. If $\mathfrak{B}$ is onto, then $\mathcal{R}(\pi_+\mathfrak{D}\pi_-) = \mathcal{R}(\mathfrak{C})$, and the latter is closed (see Theorem 9.10.5(iv)).

(ii) $\Rightarrow$ (i): It is always true that $\mathcal{R}(\pi_+\mathfrak{D}\pi_-) \subset \mathcal{R}(\mathfrak{C})$ (since $\pi_+\mathfrak{D}\pi_- = \mathfrak{C}\mathfrak{B}$). By Lemma 9.2.2, controllability implies that $\mathcal{R}(\mathfrak{C}) \subset \overline{\mathcal{R}(\pi_+\mathfrak{D}\pi_-)}$. Thus, if the range of the Hankel operator is closed, then it must coincide with the range of $\mathfrak{C}$. By Theorem 9.10.5(iv), Σ is exactly observable in infinite time with bound ω. If we let $\mathfrak{C}^{-1}$ be the inverse of $\mathfrak{C}$, defined on $\mathcal{R}(\mathfrak{C}) = \mathcal{R}(\pi_+\mathfrak{D}\pi_-)$, then $\mathfrak{C}^{-1}$ maps $\mathcal{R}(\pi_+\mathfrak{D}\pi_-)$ onto X. We can write $\mathfrak{B}$ in the form

$$\mathfrak{B} = \mathfrak{C}^{-1}\pi_+\mathfrak{D}\pi_-.$$

The range of the operator on the right hand side is X, hence so is the range of the operator on the left hand side. □

The condition in Theorem 9.10.7 that the Hankel operator has a closed range is a quite strong one. A class of 'restricted shift systems' is studied in depth (in the discrete time Hilbert space case with $\omega = 0$) in Fuhrmann (1981, Chapter III), where it is shown that $\mathfrak{D}$ is *strictly noncyclic* whenever the range of the Hankel operator $\pi_+\mathfrak{D}\pi_-$ is closed. See, in particular, Fuhrmann (1981, Theorem 3–10, pp. 258–259).

The major part of the proof of the following theorem on the duality between all the different versions of observability and controllability is still another application of Theorems 9.10.4 and 9.10.5.

Theorem 9.10.8 *Let* $1 \le p < \infty$*, and let* Σ *be an* L^p*-well-posed linear system on the reflexive Banach spaces* (Y, X, U)*. Then* Σ *is controllable [observable] in the sense of any one of conditions (i)–(ix) in Definition 9.4.1 [Definition 9.4.2] iff the adjoint system* Σ^d *is observable [controllable] in the sense of the* *corresponding condition (with the same number) in Definition 9.4.2 [Definition 9.4.1], with one possible exception: it is not clear if null controllability of an L^1-well-posed system is a necessary condition for the final state observability of the adjoint L^∞-well-posed system.*

Proof With the exception of the duality of null controllability and final observability this follows from Definition 6.2.4 and Theorems 9.10.4 and 9.10.5. To get the duality of null controllability and final observability we use Theorem 9.10.9 below. □

Theorem 9.10.9 *Let $\Sigma = \left[\begin{smallmatrix} \mathfrak{A} & \mathfrak{B} \\ \mathfrak{C} & \mathfrak{D} \end{smallmatrix}\right]$ be an $L^p|Reg$-well-posed linear system on (Y, X, U), and let $t > 0$.*

(i) *If Σ is null controllable in time t, i.e., if $\mathcal{R}\left(\mathfrak{B}\pi_{[-t,0)}\right) \supset \mathcal{R}\left(\mathfrak{A}^t\right)$, then the following equivalent conditions hold:*
   (a) *$(\mathfrak{B}\pi_{[-t,0)})^*$ is coercive with respect to $\mathfrak{A}^{*t}$, i.e., there exists a constant $\epsilon > 0$ such that $|(\mathfrak{B}\pi_{[-t,0)})^* x^*| \geq \epsilon |\mathfrak{A}^{*t} x^*|$ for all $x^* \in X^*$;*
   (b) *$\mathcal{N}\left(\mathfrak{A}^{*t}\right) \subset \mathcal{N}\left((\mathfrak{B}\pi_{[-t,0)})^*\right)$ and the operator[4] $\mathfrak{A}^{*t}(\mathfrak{B}\pi_{[-t,0)})^{-*}$ is bounded $X^* \to X^*$;*
   (c) *there is an operator $M^* \in \mathcal{B}(\mathcal{R}\left((\mathfrak{B}\pi_{[-t,0)})^*\right); X^*)$ such that $\mathfrak{A}^{*t} = M^*(\mathfrak{B}\pi_{[-t,0)})^*$.*

   If X and U are Hilbert spaces and Σ is L^2-well-posed, then these conditions are also sufficient for the null controllability of Σ, and they are equivalent to
   (d) *$\mathfrak{B}\pi_{[-t,0)}\mathfrak{B}^* \geq \epsilon \mathfrak{A}^t \mathfrak{A}^{*t}$ for some $\epsilon > 0$.*

(ii) *The following conditions are equivalent:*
   (a) *Σ is finally observable in time t, i.e., $\mathcal{N}\left(\mathfrak{A}^t\right) \supset \mathcal{N}\left(\pi_{[0,t)}\mathfrak{C}\right)$, and the operator[5] $\mathfrak{A}^t(\pi_{[0,t)}\mathfrak{C})^{-1}$ is bounded $X \to X$;*
   (b) *$\pi_{[0,t)}\mathfrak{C}$ is coercive with respect to $\mathfrak{A}^t$, i.e., there exists a constant $\epsilon > 0$ such that $\|\mathfrak{C}x\|_{L^p|Reg([0,t);Y)} \geq \epsilon |\mathfrak{A}^t x|_X$ for all $x \in X$;*
   (c) *there is an operator $M \in \mathcal{B}(\mathcal{R}\left(\pi_{[0,t)}\mathfrak{C}\right); X)$ such that $\mathfrak{A}^t = M\pi_{[0,t)}\mathfrak{C}$;*
   (d) *$\mathcal{R}\left((\pi_{[0,t)}\mathfrak{C})^*\right) \supset \mathcal{R}\left(\mathfrak{A}^{*t}\right)$.*

   If X and Y are Hilbert spaces and Σ is L^2-well-posed, then these conditions are equivalent to
   (e) *$\mathfrak{C}\pi_{[0,t)}\mathfrak{C}^* \geq \epsilon \mathfrak{A}^{*t}\mathfrak{A}^t$ for some $\epsilon > 0$.*

This theorem follows from directly from the following lemma:

Lemma 9.10.10 *Let X, Y, and Z be Banach spaces.*

[4] Here $(\mathfrak{B}\pi_{[-t,0)})^{-*}$ stands for the inverse of the injective operator that $(\mathfrak{B}\pi_{[-t,0)})^*$ induces on $X^*/\mathcal{N}\left((\mathfrak{B}\pi_{[-t,0)})^*\right)$ and $\mathfrak{A}^{*t}$ stands for the operator induced by $\mathfrak{A}^{*t}$ on $X^*/\mathcal{N}\left((\mathfrak{B}\pi_{[-t,0)})^*\right)$. See Definition 9.1.5.

[5] See the footnote to Definition 9.4.2(v).

(i) *Let* $K \in \mathcal{B}(X;Y)$ *and* $L \in \mathcal{B}(X;Z)$. *Then the following conditions are equivalent*

(a) K *is coercive with respect to* L, *i.e., there exists an* $\epsilon > 0$ *such that* $|Kx|_Y \ge \epsilon |Lx|_Z$ *for all* $x \in X$;

(b) $\mathcal{N}(K) \subset \mathcal{N}(L)$, *and the operator*[6] LK^{-1} *is bounded* $\mathcal{R}(K) \to Z$;

(c) *there exists an operator* $M \in \mathcal{B}(\mathcal{R}(K); Z)$ *such that* L *factors into* $L = MK$;

(d) $\mathcal{R}(L^*) \subset \mathcal{R}(K^*)$;

(e) $\mathcal{R}(L^*) \subset \mathcal{R}(K^*)$, *and if we denote the inverse of the injective operator that* K^* *induces on* $Y^*/\mathcal{N}(K^*)$ *by* K^{-*}, *then* $K^{-*}L^* \in \mathcal{B}(Z^*; Y^*/\mathcal{N}(K^*))$;

(f) *there exists an operator* $M^* \in \mathcal{B}(Z^*; Y^*/\mathcal{N}(K^*))$ *such that* L^* *factors into* $L^* = K^*M^*$; *here* K^* *represents the operator induced by* K^* *on* $Y^*/\mathcal{N}(K^*)$.

If X, Y, *and* Z *are Hilbert spaces, then these conditions are equivalent to*

(g) $K^*K \ge \epsilon L^*L$ *for some* $\epsilon > 0$.

(ii) *Let* $K \in \mathcal{B}(Y;X)$ *and* $L \in \mathcal{B}(Z;X)$. *Then the following conditions are equivalent:*

(a) $\mathcal{R}(L) \subset \mathcal{R}(K)$;

(b) $\mathcal{R}(L) \subset \mathcal{R}(K)$, *and if we denote the inverse of the injective operator that* K *induces on* $Y/\mathcal{N}(K)$ *by* K^{-1}, *then* $K^{-1}L \in \mathcal{B}(Z; Y/\mathcal{N}(K))$;

(c) *there exists an operator* $M \in \mathcal{B}(Z; Y/\mathcal{N}(K))$ *such that* L *factors into* $L = KM$; *here* K *represents the operator induced by* K *on* $Y/\mathcal{N}(K)$.

These conditions imply the following set of equivalent conditions:

(d) K^* *is coercive with respect to* L^*, *i.e., there exists an* $\epsilon > 0$ *such that* $|K^*x|_{Y^*} \ge \epsilon |L^*x^*|_{Z^*}$ *for all* $x^* \in X^*$;

(e) $\mathcal{N}(K^*) \subset \mathcal{N}(L^*)$, *and the operator*[7] L^*K^{-*} *is bounded* $\mathcal{R}(K^*) \to Z^*$;

(f) *there exists an operator* $M^* \in \mathcal{B}(\mathcal{R}(K^*); Z^*)$ *such that* L^* *factors into* $L^* = M^*K^*$,

If Y *and* Z *are reflexive, then all the conditions (a)–(f) are equivalent, and if* X, Y, *and* Z *are Hilbert spaces, then they are equivalent to*

(g) $KK^* \ge \epsilon LL^*$ *for some* $\epsilon > 0$.

Proof (i) (a) $\Rightarrow$ (b): Clearly (a) implies that $\mathcal{N}(K) \subset \mathcal{N}(L)$. By the remarks leading up to Definition 9.1.5, we may without loss of generality assume that K is injective, because we may otherwise replace K by the operator induced

[6] Here K^{-1} stands for the inverse of the injective operator that K induces on $X/\mathcal{N}(K)$ and L stands for the operator induced by L on $X/\mathcal{N}(K)$. See Definition 9.1.5.

[7] Here K^{-*} stands for the inverse of the injective operator that K^* induces on $X^*/\mathcal{N}(K^*)$ and L^* stands for the operator induced by L^* on $X^*/\mathcal{N}(K^*)$. See Definition 9.1.5.

by K on $X/\mathcal{N}(K)$ and L by the operator induced by L on $X/\mathcal{N}(K)$. Then (a) implies that LK^{-1} is bounded $\mathcal{R}(K) \to Z$.

(b) $\Rightarrow$ (c): By (b), LK^{-1} can be extended to an operator $M \in \mathcal{B}(\mathcal{R}(K); Z)$. We claim that $L = MK$. To see that this is true we observe that $K^{-1}K$ is the quotient map $\pi : X \to X/\mathcal{N}(K)$, and that $LK^{-1}Kx$ is the operator induced by L on $X/\mathcal{N}(K)$ applied to πx. But, by definition, this is equal to Lx.

(c) $\Rightarrow$ (d): By continuity, we can extend M to an operator in $\mathcal{B}(\overline{\mathcal{R}(K)}; Z)$, where $\overline{\mathcal{R}(K)}$ is a Banach space with the norm induced by Y. Then $L^* = K^*M^*$, hence $\mathcal{R}(L^*) \subset \mathcal{R}(M^*)$.

(d) $\Rightarrow$ (e): The operator $K^{-*}L^*$ is closed, so by the closed graph theorem, it is bounded.

(e) $\Rightarrow$ (f): Take $M^* = K^{-*}$.

(f) $\Rightarrow$ (a): In this part of the proof we may, without loss of generality, assume that $\mathcal{R}(K)$ is dense in Y. If not, then we replace Y by $\overline{\mathcal{R}(K)}$. By, e.g., Rudin (1973, pp. 91) and Lemma 9.10.3(i), the dual of this space is

$$Y^*/\overline{\mathcal{R}(K)}^{\perp} = Y^*/\mathcal{R}(K)^{\perp} = Y^*/\mathcal{N}\left(K^*\right),$$

and the adjoint of the operator $K : X \to \overline{\mathcal{R}(K)}$ is the inductive operator induced by K^* on $Y^*/\mathcal{N}(K^*)$, i.e., exactly the operator which appears in (f).

Let (f) hold, and let $x \in X$ and $z^* \in Z^*$. Then

$$\begin{aligned}\langle Lx, z^*\rangle_{(Z,Z^*)} &= \langle x, L^*z^*\rangle_{(X,X^*)} = \langle x, K^*M^*z^*\rangle_{(X,X^*)} \\ &= \langle Kx, M^*z^*\rangle_{(Y,Y^*)},\end{aligned}$$

hence

$$\begin{aligned}|\langle Lx, z^*\rangle_{(Z,Z^*)}| &= \left|\langle Kx, M^*z^*\rangle_{(Y,Y^*)}\right| \\ &\le |Kx|_Y |M^*z^*|_{Y^*} \\ &\le |Kx|_Y \|M^*\| |z^*|_{Z^*}.\end{aligned}$$

Taking the supremum over all $z^* \in Z^*$ with $|z^*| = 1$ we find that $|Lx| \le \|M^*\| |Kx|_Y$. We thus get (a) with $\epsilon = 1/\|M^*\|$.

(a) $\Leftrightarrow$ (g): In the Hilbert space case condition (a) is equivalent to

$$\begin{aligned}0 &\le |Kx|_Y^2 - \epsilon^2 |Lx|_Z^2 \\ &= \langle Kx, Kx\rangle_Y - \epsilon^2 \langle Lx, Lx\rangle_Z \\ &= \langle x, K^*Kx\rangle_Y - \epsilon^2 \langle x, L^*Lx\rangle_Z \\ &= \langle x, (K^*K - \epsilon^2 L^*L)x\rangle_X,\end{aligned}$$

and this is equivalent to (g) (with ϵ replaced by ϵ^2).

(ii) The same proof that we used in (i) to show that (d) $\Rightarrow$ (e) $\Rightarrow$ (g) remains valid if we replace K^* by K and L^* by L, and this shows that

(a) $\Rightarrow$ (b) $\Rightarrow$ (c). Trivially, (c) $\Rightarrow$ (a). The equivalence of (d), (e), (f), and in the Hilbert space case also (g), follows from (i) applied to K^* and L^*. Thus, to complete the proof it suffices to show that, for example (c) $\Rightarrow$ (f), and that (d) $\Rightarrow$ (a) in the reflexive case.

(c) $\Rightarrow$ (f): By, e.g., Rudin (1973, pp. 92) and Lemma 9.10.3(iii), the dual of $Y/\mathcal{N}(K)$ can be identified with $\mathcal{N}(K)^\perp = ({}^\perp\mathcal{R}(K^*))^\perp$, which is equal to the weak*-closure of $\mathcal{R}(K^*)$ in Y^*. The adjoint of the operator induced by K on $Y/\mathcal{N}(K)$ is the operator $K^*\colon X^* \to \overline{\mathcal{R}(K^*)}$ (where the bar represents the weak*-closure). This means that we can choose M^* in (c) to be the restriction of the adjoint of the operator M in (f) to $\mathcal{R}(K^*)$.

(d) $\Rightarrow$ (a): By (i), applied to K^* and L^*, (d) is equivalent to $\mathcal{R}(L^{**}) \subset \mathcal{R}(K^{**})$. By a general rule for the second adjoint of an operator, $K = K^{**}_{|Y}$ and $L = L^{**}_{|Z}$. But $Y^{**} = Y$ and $Z^{**} = Z$ if these spaces are reflexive, so $K^{**} = K$ and $L^{**} = L$ in this case, hence $\mathcal{R}(L) \subset \mathcal{R}(K)$. □

9.11 Comments

Section 9.1 Theorem 9.1.9 appears in many different (usually somewhat less general) forms in the literature. See, for example, Arov and Nudelman (1996, Theorem 7.1) and Helton (1976, Theorem 4.3) (in continuous time) and Arov (1979b, Proposition 4) and Helton (1974, Theorem 3a.1) (in discrete time). A discrete time analogue of Corollary 9.1.10 is found in Helton (1974, Theorem 3a.1) (and most books on system theory contain a finite-dimensional version of this result).

Section 9.2 Early versions of Theorem 9.2.4 are found in Helton (1976, Theorem 4.4) and Arov and Nudelman (1996, Proposition 7.10) (in continuous time) and Arov (1979b, Proposition 6) and Helton (1974, Theorem 3b.1) (in discrete time). The pseudo-similarity given in these references is the minimal one, denoted by $\underline{E}$ in Theorem 9.2.4. We discovered the maximal pseudo-similarity $\overline{E}$ in Theorem 9.2.4 (without realizing that it was maximal) while writing the first draft of this section in 1998. The present formulation of Theorem 9.2.4 was obtained in discussions with Arov and Kaashoek in the spring of 2003. See Arov *et al.* (2004, Proposition 3.3) for the corresponding discrete time version. Theorems 9.2.5 and 9.2.6 may (formally) be new.

Section 9.3 Theorem 9.3.1 is the main result of Staffans (1999a). It is in the spirit Kalman *et al.* (1969, Part 4) (although the setting is different). The importance of the Hankel operator of the input/output map in realization theory has long been recognized. There is some formal resemblance between Theorem 9.3.1 and the factorizations results presented in Kalman (1963, Theorem 1), Kalman *et al.* (1969, Theorem (13.19)), and Brockett (1970, Theorem 1, p. 93), but there is a very significant non-technical difference: the realization presented there is intrinsically time-dependent (and time-reversible), and its state space dynamics is trivial. A much more closely related result is found in Kalman *et al.* (1969, Section 10.6) and Fuhrmann (1981, pp. 31–32): there we find the same algebraic construction (in discrete time), but without any continuity considerations of the type (9.3.1)–(9.3.6). Even closer to Theorem 9.3.1 is Baras and Brockett (1975, Theorem 6), Baras and Dewilde (1976, Theorem II.2.2) and Fuhrmann (1981, Theorem 6-3, p. 293), which give sufficient conditions for the existence of a realization with bounded control and observation operators in the case of finite-dimensional U and Y. As a special case of a stable factorization we can take either $\mathfrak{B}$ or $\mathfrak{C}$ to be the identity operator; this leads to the exactly controllable shift and exactly observable shift realizations in Proposition 2.6.5(i)–(ii). See also the comments to Section 9.5.

Jacob and Zwart (2002) have studied minimal realizations of a scalar inner transfer function with an invertible or exponentially stable semigroup.

Section 9.4 Theorem 9.4.7 is outlined in Salamon (1989, pp. 158–159), where it is called ‘well known’, and Theorem 9.4.10 is in the same spirit. A discrete time Hilbert space version with $p = 2$ of Theorem 9.4.10(i) is found in Fuhrmann (1981, Theorem 1–9(b), p. 247).

Section 9.5 Different versions of the restricted exactly controllable shift realization in Proposition 9.5.2(iii) (often referred to as ‘the $*$-restricted shift’) and the restricted exactly observable shift realization in Proposition 9.5.2(iv) (often referred to as ‘the restricted shift’) have appeared in, e.g., Baras and Dewilde (1976), Fuhrmann (1974, Theorem 2.6), Fuhrmann (1981, Section 3.2), Helton (1974, p. 31), Jacob and Zwart (2002, Theorem A.1), Ober and Wu (1996, Sections 5.2–5.3), and Salamon (1989, Theorem 4.3).

The existence and uniqueness of a (par)balanced realization in the infinite-dimensional (irrational) discrete time case was first proved by Young (1986, Theorems 1 and 2). That result was converted into a continuous time setting by Ober and Montgomery-Smith (1990, Theorem 8.5), using the Cayley transform (see Section 12.3). The setting used there is slightly different from ours: the systems need not be well-posed, but instead they are required to have a uniformly regular transfer function in the sense of Definition 5.6.1. That the semigroup and its adjoint are always strongly stable was first proved in Ober and Wu (1993, Theorem 4.9) (in discrete time) and in Ober and Wu (1996, Theorem 6.6) (in continuous time).

Balanced realizations have also been studied (in the infinite-dimensional case) in Curtain and Glover (1986a) and Glover *et al.* (1988). They are used extensively, e.g., in model reduction.

Lemma 9.5.8 is apparently due to Kreĭn and Petunin (1966). See Agler and McCarthy (2002, Appendix C) for further results in this direction and some historical comments. Our proof of Lemma 9.5.8 is based on some ideas found in the proof of Young (1986, Theorem 1).

Section 9.6 Corollaries 9.6.2 and 9.6.5 are, of course, straightforward and more or less known. We have not seen them in this exact form in the literature. Our definition of the (local) McMillan degree of a pole is based on one of the known finite-dimensional characterizations of McMillan degree; see, e.g., Bart *et al.* (1979, p. 77) or Rosenbrock (1970, Algorithm 5.3, p. 120). Lemmas 9.6.6 and 9.6.9 are taken from Uetake (2003, Section 4.1), and Theorem 9.6.10 is an extended version of Uetake (2003, Theorems 4.6 and 4.7).

Section 9.7 We do not know to what extent Theorem 9.7.3 has been known. Theorems 9.7.4 and 9.7.5 are extensions of Curtain and Zwart (1995, Theorems 4.2.1 and 4.2.3).

Section 9.8 Spectral minimality has been studied for a long time, but much still remains to be done. Sufficient conditions for spectral minimality have been given in, for example, Baras and Brockett (1975) (an example with two branch points), Baras *et al.* (1974) and Brockett and Fuhrmann (1976) (completely monotonic impulse response, or more generally, the main operator is normal and has a connected resolvent set, and $B = C^*$ is bounded), Feintuch (1976) (where the main operator is compact or spectral), and Ober (1996) (completely monotonic impulse response). It is also known that the input normalized, output normalized, and balanced realizations of a strictly noncyclic transfer function is spectrally minimal, see, e.g., Lax and Phillips (1967, Section 3.3), Helton (1976, Theorem 4.8), Sz.-Nagy and Foiaş (1970, Theorem 4.1, p. 259), Fuhrmann (1981, Section 3.4), Ober and Wu (1993, Corollary 4.12) and Ober and Wu (1996, Theorem 7.5). Theorem 9.8.2 is a reformulated and sharpened version of Helton (1976, Theorem 4.7) (which is based on Lax and Phillips (1973, Theorem 5.5)). Our proof of Theorem 9.8.3 has been modeled after the proofs of Brockett and Fuhrmann (1976, Theorem 2.2) and Feintuch (1976, Theorem 17). (Theorem 9.8.3 is actually true under the weaker assumption that A is a spectral operator; this can be shown by combining the argument given in the proof of Feintuch (1976, Theorem 17) with our proof of Theorem 9.8.3.)

Section 9.9 The feedback version of Lemma 9.9.1 has been known for a long time in the classical case where A, B, C, and D are bounded. An explicit infinite-dimensional well-posed version is found in G. Weiss (1994b, Remark 6.5). That exact controllability or observability is preserved under time-inversion has also been know for quite some time; see, e.g., Flandoli *et al.* (1988, Section 2.4). Lemmas 9.9.3(ii) and 9.9.5 may be new.

Section 9.10 Lemma 9.10.2–Theorem 9.10.5 and Theorems 9.10.8–9.10.9 are of a technical nature, and related (usually less general) results are found in many different places in the literature. See, in particular, Dolecki and Russell (1977). Theorem 9.10.6(ii) is a slight extension of Russell and Weiss (1994, Proposition 2.8). Early versions of Theorem 9.10.7 are found in Fuhrmann (1981, p. 249) and Helton (1974, Theorem 3c.1(2)) (in discrete time) and Helton (1976, Theorem 4.5(i)) (in continuous time).

10
Admissibility

In this chapter we give conditions under which the operators B and C can be interpreted as the control operator, respectively, observation operator of a $L^p|Reg$-well-posed linear system with a given semigroup generator A. In this case we call B and C *admissible* for A. If B and C can be interpreted as the control and observation operators of the *same* system, then they are *jointly admissible*. We are also interested in whether or not the system is *stable* (this is often refered to as *infinite time admissibility*). We furthermore discuss admissibility questions specifically related to the L^2-well-posed Hilbert space case (i.e, Y, X, and U are Hilbert spaces). This leads us to a study of H^2-spaces.

10.1 Introduction to admissibility

As we have shown in Chapter 4, every well-posed linear system $\Sigma = \left[\begin{smallmatrix} \mathfrak{A} & \mathfrak{B} \\ \mathfrak{C} & \mathfrak{D} \end{smallmatrix}\right]$ has a set of generators (which determine the system node). In the general case these consist of the semigroup generator A, the control operator B, and the combined observation/feedthrough operator $C\&D$. In the compatible and regular cases the operator $C\&D$ can be replaced by the extended observation operator $C_{|W}$ and the corresponding feedthrough operator D. We have given some sufficient conditions on $A, B, C\&D, C, C_{|W}$, and D for these operators to be the generators of a (possibly compatible) $L^p|Reg$-well-posed linear system. In particular, the Hille–Yosida Theorem 3.4.1 gives necessary and sufficient conditions on an operator A to generate a C_0 semigroup, Corollary 3.4.2 gives necessary and sufficient conditions on A to be the generator of a contraction semigroup, and the case of a diagonal semigroup in a Hilbert space is analyzed in Examples 3.3.3 and 3.3.5. We shall not add anything significant to these two semigroup generation theorems here, but refer the reader to, e.g., Pazy (1983) for additional results in this direction.

We have also obtained some necessary and some sufficient conditions for B to be a control operator, for C to be an observation operator, for $C\&D$ to be an observation/feedthrough operator, etc., of a well-posed linear system (necessary conditions are given in Theorems 4.2.1, 4.2.7, 4.4.2, 4.7.14, 5.4.3, 5.5.5, 5.6.5, and sufficient conditions are given in Corollary 4.2.8 and Theorem 4.4.7 and Theorems 4.3.4, 4.4.8, 4.7.14, and 5.7.3). However, most of the sufficient results are either rather restrictive (requiring B or C to be bounded), or very implicit, making them difficult to use. The main exception is Theorem 5.7.3, which is both simple to use and fairly general (as long as we restrict ourselves to the case where the semigroup is analytic).

The purpose of this chapter is to present a number of additional admissibility results. Some of these are valid only in the case where the state space, the input space, or the output space is a Hilbert space. Most of the results discussed here concern L^p-well-posedness with $1 \le p < \infty$, and some of them require $p = 2$.

In the sequel it will be convenient to call an operator *admissible* if it is one of the generators of a $L^p|Reg$-well-posed linear system.

Definition 10.1.1 Let A be the generator of a C_0 semigroup on X.

(i) The operator $B \in \mathcal{B}(U; X_{-1})$ is an *$L^p|Reg$-admissible control operator* for A (or for $\mathfrak{A}$) if the operator

$$\mathfrak{B}u = \int_{-\infty}^{0} \mathfrak{A}_{|X}^{-s} Bu(s)\, ds$$

maps $L^p|Reg_c(\mathbb{R}^-; U)$ into X (i.e., $\mathfrak{B}$ is an $L^p|Reg$-well-posed input map for $\mathfrak{A}$). We call B *stable or ω-bounded* if $\mathfrak{B}$ is stable or ω-bounded.

(ii) The operator $C \in \mathcal{B}(X_1; Y)$ is an *$L^p|Reg$-admissible observation operator* for A (or for $\mathfrak{A}$) if the map

$$(\mathfrak{C}x)(t) = C\mathfrak{A}^t x, \qquad x \in X_1, \qquad t \ge 0,$$

can be extended to a bounded operator $X \to L^p|Reg_{loc}(\overline{\mathbb{R}}^+; Y)$ (i.e., $\mathfrak{C}$ is an $L^p|Reg$-well-posed output map for $\mathfrak{A}$). We call C *stable or ω-bounded* if $\mathfrak{C}$ is stable or ω-bounded.

(iii) Let B be an admissible control operator for A, and let $A\&B$ be the restriction of $\begin{bmatrix} A_{|X} & B \end{bmatrix}$ to

$$\mathcal{D}(A\&B) = \left\{ \left[\begin{smallmatrix} x \\ u \end{smallmatrix}\right] \in \left[\begin{smallmatrix} X \\ U \end{smallmatrix}\right] \;\middle|\; A_{|X}x + Bu \in X \right\}.$$

The operator $C\&D \in \mathcal{B}(\mathcal{D}(A\&B); Y)$ is an *$L^p|Reg$-admissible* *observation/feedthrough* operator for the pair (A, B) if the operator

$$Cx = C\&D \begin{bmatrix} x \\ 0 \end{bmatrix}, \qquad x \in X_1,$$

is an admissible observation operator for A and the operator $\mathfrak{D}\colon C^2_{c,\mathrm{loc}}(\mathbb{R}; U) \to C_c(\mathbb{R}; Y)$ defined by

$$(\mathfrak{D}u)(t) = C\&D \begin{bmatrix} \mathfrak{B}\tau^t u \\ u(t) \end{bmatrix}, \qquad t \in \mathbb{R},$$

can be extended to a continuous operator $L^p|Reg_{c,\mathrm{loc}}(\mathbb{R}; U) \to L^p|Reg_{c,\mathrm{loc}}(\mathbb{R}; Y)$ (cf. Theorem 4.7.14). We call $C\&D$ *stable or ω-bounded* if both $\mathfrak{C}$ in (ii) and $\mathfrak{D}$ are stable or ω-bounded.

(iv) The operators $B \in \mathcal{B}(U; X_{-1})$ and $C \in \mathcal{B}(X_1; Y)$ are *jointly $L^p|Reg$-admissible* for A if B is an $L^p|Reg$-admissible control operator for A, C is an $L^p|Reg$-admissible observation operator for A, and there is an $L^p|Reg$-admissible observation/feedthrough operator $C\&D \in \mathcal{B}(\mathcal{D}(A\&B); Y)$ for the pair (A, B) such that

$$Cx = C\&D \begin{bmatrix} x \\ 0 \end{bmatrix}, \qquad x \in X_1.$$

We call B and C *jointly stable or ω-bounded* if the resulting $L^p|Reg$-well-posed linear system is stable or ω-bounded.

In this definition we say nothing about the admissibility of a feedthrough operator D. This is not an interesting issue, since every $D \in \mathcal{B}(U; Y)$ can be the feedthrough operator of an $L^p|Reg$-linear system (take, for example, $A = 0$, $B = 0$, and $C = 0$), and conversely, every feedthrough operator D belongs to $\mathcal{B}(U; Y)$. See Section 4.6 for a more detailed description of the relationship between parts (iii) and (iv) of Definition 10.1.1.

The conditions in Definition 10.1.1 on B, C, and $C\&D$ are not the weakest possible, and sometimes it is more convenient to use the following characterization of admissibility.

Lemma 10.1.2 *Let A be the generator of a C_0 semigroup on X.*

(i) *The operator $B \in \mathcal{B}(U; X_{-1})$ is an $L^p|Reg$-admissible control operator for A if and only if the operator*

$$\mathfrak{B}^0_{-T} u = \int_{-T}^{0} \mathfrak{A}^{-s}_{|X} B u(s)\, ds$$

maps $L^p|Reg([-T, 0); U)$ into X for some (hence all) $T > 0$.

(ii) *The operator $C \in \mathcal{B}(X_1;Y)$ is an $L^p|Reg$-admissible observation operator for A if the map*

$$(\mathfrak{C}_0^T x)(t) = C\mathfrak{A}^t x, \qquad x \in X_1, \qquad t \in [0,T),$$

can be extended to an bounded operator $X \to L^p|Reg_{loc}([0,T);Y)$ for some (hence all) $T > 0$.

(iii) *Let B be an admissible control operator for A. Define*

$$\mathcal{D}(A\&B) = \left\{\left[\begin{smallmatrix} x \\ u \end{smallmatrix}\right] \in \left[\begin{smallmatrix} X \\ U \end{smallmatrix}\right] \mid A_{|X}x + Bu \in X\right\}.$$

The operator $C\&D \in \mathcal{B}(\mathcal{D}(A\&B);Y)$ is an $L^p|Reg$-admissible observation/feedthrough operator for the pair (A, B) if the operator

$$Cx = C\&D\begin{bmatrix} x \\ 0 \end{bmatrix}, \qquad x \in X_1.$$

is an admissible observation operator for A and the operator $\mathfrak{D}_0^T : C^2([0,T];U) \to C([0,T];Y)$ defined by

$$(\mathfrak{D}_0^T u)(t) = C\&D\begin{bmatrix} \mathfrak{B}\tau^t u \\ u(t) \end{bmatrix}, \qquad t \in [0,T],$$

can be extended to a continuous operator $L^p|Reg([0,T);U) \to L^p|Reg([0,T);Y)$ for some (hence all) $T > 0$.

Proof In all cases, the conditions listed in Lemma 10.1.2 are weaker than those listed in Definition 10.1.1, so they are necessary. That they also are sufficient follows from the fact that we can recreate $\mathfrak{B}$, $\mathfrak{C}$ and $\mathfrak{D}$ from $\mathfrak{B}^0{}_{-T}$, $\mathfrak{C}_0^T$, and $\mathfrak{D}_0^T$ by using the formulas in Lemma 2.4.3. □

10.2 Admissibility and duality

In the reflexive case there is a simple connection between the admissibility of a semigroup generator A and the admissibility of its adjoint A^*, and there is also a simple connection between the admissibility of a control operator B for A and the admissibility of B^* as an observation operator of A^*. We begin with the L^p-well-posed case with $1 < p < \infty$.

Theorem 10.2.1 *Let U, X, and Y be reflexive, and let $1 < p < \infty$ and $1/p + 1/q = 1$.*

(i) *$A\colon X \supset \mathcal{D}(A) \to X$ is the generator of a C_0 semigroup on X iff A^* is the generator of a C_0 semigroup on X^*.*

(ii) *Let A be the generator of a C_0 semigroup on X. Then $B \in \mathcal{B}(U;X_{-1})$ is an L^p-admissible control operator for A if and only if $B^* \in \mathcal{B}(X_1^*;U^*)$ is* *an L^q-admissible observation operator for A^* and $C \in \mathcal{B}(X_1;Y)$ is an L^p-admissible observation operator for A iff $C^* \in \mathcal{B}(Y^*;X^*_{-1})$ is an L^q-admissible control operator for A^*.*

(iii) *Let A be the generator of a C_0 semigroup on X. Then $B \in \mathcal{B}(U;X_{-1})$ is an L^p-admissible control operator for A and $C\&D\colon \mathcal{D}(A\&B) \to Y$, with $\mathcal{D}(A\&B) = \{(x,u) \in \left[\begin{smallmatrix} X \\ U \end{smallmatrix}\right] \mid A_{|X}x + Bu \in X\}$, is an L^p-well-posed observation/feedthrough operator for (A, B) if and only if B^d is an L^q-admissible control operator for $A^d = A^*$ and $[C\&D]^d$ is an L^q-admissible observation/feedthrough operator for (A^d, B^d), where B^d and $[C\&D]^d$ are defined as follows: if we denote $C = C\&D\left[\begin{smallmatrix} 1 \\ 0 \end{smallmatrix}\right] \in \mathcal{B}(X_1;Y)$, then $B^d = C^* \in \mathcal{B}(Y^*;X^*_{-1})$, and $[C\&D]^d = P_{U^*}\left[\begin{smallmatrix} A\&B \\ C\&D \end{smallmatrix}\right]^*$, where P_{U^*} is the projection of $\left[\begin{smallmatrix} X^* \\ U^* \end{smallmatrix}\right]$ onto U^* (with $\mathcal{D}\left([C\&D]^d\right) = \mathcal{D}\left(\left[\begin{smallmatrix} A\&B \\ C\&D \end{smallmatrix}\right]^*\right)$).*

Proof Part (i) is contained in Theorem 3.5.6(v), part (ii) in Theorem 6.2.13, and part (iii) is contained in Lemma 6.2.12. □

A very similar result is true in the two remaining cases $p = 1$ and $p = \infty$. To prove that result we need the following characterization of a *Reg*-admissible control operator.

Theorem 10.2.2 *Let A be the generator of a C_0 semigroup $\mathfrak{A}$ on the Banach space X. Let U be another Banach space, and let $B \in \mathcal{B}(U;X_{-1})$. Then the following conditions are equivalent:*

(i) *B is a Reg-admissible control operator for A.*
(ii) *For some (hence for all) $\alpha \in \rho(A)$, there is a constant K such that, for each $x^* \in X^*$ with $|x^*| \le 1$, the total variation on the interval $[0, 1]$ of the function $t \mapsto B^*(\overline{\alpha} - A^*)^{-1}\mathfrak{A}^{*t}x^*$ is bounded by K.*

If X is reflexive, then (i) and (ii) are further equivalent to

(iii) *$B^* \in \mathcal{B}(X_1^*;U^*)$ is an L^1-admissible observation operator for the C_0 semigroup generated by A^*.*

See Section 3.6 and Remark 3.6.1 for the definitions of X_{-1} and X_1^*.

Proof Suppose that (i) holds, i.e., suppose that $B \in \mathcal{B}(U;X_{-1})$ is a *Reg*-admissible control operator for A. Denote the corresponding input map by $\mathfrak{B}$. Take $x^* \in X^*$ with $|x^*| \le 1$, and let $0 = t_0 < t_1 < \cdots < t_n = 1$. Choose a corresponding sequence of vectors $u_k \in U$ with $|u_k| \le 1$ so that

$$\begin{aligned}&\langle B^*(\overline{\alpha} - A^*)^{-1}(\mathfrak{A}^{*t_{k+1}} - \mathfrak{A}^{*t_k})x^*, u_k\rangle_{(U^*,U)}\\ &\quad \ge \tfrac{1}{2}|B^*(\overline{\alpha} - A^*)^{-1}(\mathfrak{A}^{*t_{k+1}} - \mathfrak{A}^{*t_k})x^*|_{U^*}\end{aligned}$$

(in particular, we require the left-hand side to be real). Define $u(s) = u_k$ for $-t_{k+1} < s \le -t_k$. Then $\sup_{-1\le s<0}|u(s)| \le 1$ and

$$\begin{aligned}(\alpha(\alpha - A)^{-1} - 1)\mathfrak{B}\pi_{[-1,0)}u &= A(\alpha - A)^{-1}\mathfrak{B}\pi_{[-1,0)}u\\ &= \int_{-1}^{0} A\mathfrak{A}^{-s}(\alpha - A_{|X})^{-1}Bu(s)\,ds\\ &= \sum_{k=0}^{n-1}\int_{-t_{k+1}}^{-t_k} A\mathfrak{A}^{-s}(\alpha - A_{|X})^{-1}Bu_k\,ds\\ &= \sum_{k=0}^{n-1}(\mathfrak{A}^{t_{k+1}} - \mathfrak{A}^{t_k})(\alpha - A_{|X})^{-1}Bu_k.\end{aligned}$$

Therefore

$$\begin{aligned}&\sum_{k=0}^{n-1}|B^*(\overline{\alpha} - A^*)^{-1}(\mathfrak{A}^{*t_{k+1}} - \mathfrak{A}^{*t_k})x^*|_{U^*}\\ &\quad\le 2\sum_{k=0}^{n-1}\langle B^*(\overline{\alpha} - A^*)^{-1}(\mathfrak{A}^{*t_{k+1}} - \mathfrak{A}^{*t_k})x^*, u_k\rangle_{(U^*,U)}\\ &\quad= 2\Big\langle x^*, \sum_{k=0}^{n-1}(\mathfrak{A}^{t_{k+1}} - \mathfrak{A}^{t_k})(\alpha - A_{|X})^{-1}Bu_k\Big\rangle_{(U^*,U)}\\ &\quad= 2\langle x^*, (\alpha(\alpha - A)^{-1} - 1)\mathfrak{B}\pi_{[-1,0)}u\rangle_{(U^*,U)}\\ &\quad\le 2\|(\alpha(\alpha - A)^{-1} - 1)\mathfrak{B}\pi_{[-1,0)}\|.\end{aligned}$$

Taking the supremum over all subdivisions of the interval [0, 1] we find that (ii) holds, with $K = 2\|(\alpha(\alpha - A)^{-1} - 1)\mathfrak{B}\pi_{[-1,0)}\|$ (and this is true for all $\alpha \in \rho(A)$).

Conversely, suppose that (ii) holds for some $\alpha \in \rho(A)$. Define $\mathfrak{B}u = \int_{-\infty}^{0} \mathfrak{A}_{|X}^{-s}Bu(s)\,ds$ for all $u \in Reg_c(\mathbb{R}^-; U)$. Then $\mathfrak{B}\pi_{[-1,0)}$ is a bounded linear operator from $Reg_c(\mathbb{R}^-; U)$ into X_{-1} (since $B \in \mathcal{B}(U; X_{-1})$), hence $(\alpha - A)^{-1}\mathfrak{B}\pi_{[-1,0)}$ is a bounded linear operator from $Reg_c(\mathbb{R}^-; U)$ into X. Let u be a function which is piecewise constant on $[-1, 0)$, i.e., suppose that there exist $0 = t_0 < t_1 < \cdots < t_n = 1$ and $u_k \in U$ so that $u(s) = u_k$ for $-t_{k+1} < s \le -t_k$. The same computation which we made above shows that $\mathfrak{B}\pi_{[-1,0)}u \in X$, and that

$$\mathfrak{B}\pi_{[-1,0)}u = \alpha(\alpha - A)^{-1}\mathfrak{B}\pi_{[-1,0)}u - \sum_{k=0}^{n-1}(\mathfrak{A}^{t_{k+1}} - \mathfrak{A}^{t_k})(\alpha - A_{|X})^{-1}Bu_k.$$

Therefore, for each $x^* \in X^*$ with $|x^*| \le 1$,

$$\begin{aligned}
&\big|\langle \mathfrak{B}\pi_{[-1,0)}u, x^*\rangle_{(X,X^*)}\big| \\
&\quad \le \big|\langle \alpha(\alpha - A)^{-1}\mathfrak{B}\pi_{[-1,0)}u, x^*\rangle_{(X,X^*)}\big| \\
&\qquad + \left|\left\langle \sum_{k=0}^{n-1} (\mathfrak{A}^{t_{k+1}} - \mathfrak{A}^{t_k})(\alpha - A_{|X})^{-1} B u_k, x^* \right\rangle_{(X,X^*)}\right| \\
&\quad \le \big\|\alpha(\alpha - A)^{-1}\mathfrak{B}\pi_{[-1,0)}\big\| \, \|u\|_{Reg([0,1]} \\
&\qquad + \sum_{k=0}^{n-1} \big|\langle u_k, B^*(\overline{\alpha} - A^*)^{-1}(\mathfrak{A}^{*t_{k+1}} - \mathfrak{A}^{*t_k})x^*\rangle_{(X,X^*)}\big| \\
&\quad \le \big\|\alpha(\alpha - A)^{-1}\mathfrak{B}\pi_{[-1,0)}\big\| \, \|u\|_{Reg([0,1]} \\
&\qquad + \Big(\sup_{0\le k\le n-1} |u_k|\Big) \sum_{k=0}^{n-1} |B^*(\overline{\alpha} - A^*)^{-1}(\mathfrak{A}^{*t_{k+1}} - \mathfrak{A}^{*t_k})x^*| \\
&\quad \le M\|u\|_{Reg([0,1]},
\end{aligned}$$

where $M = \big(K + \|\alpha(\alpha - A)^{-1}\mathfrak{B}\pi_{[-1,0)}\|\big)$. Taking the supremum over all x^* we find that $\mathfrak{B}\pi_{[-1,0)}$ is a bounded linear operator (with norm at most M) from a dense subset of $Reg([-1,0];U)$ into X, hence it must map all of $Reg([-1,0];U)$ into X (instead of just X_{-1}). By Lemma 10.1.2, B is a Reg-well-posed control operator for A.

Suppose that X^* is reflexive. Then A^* is the generator of the C_0 semigroup $t \mapsto \mathfrak{A}^{*t}$. If B^* is an L^1-well-posed observation operator for A^*, then by Theorem 6.2.13 (with $Y = 0$), B is L^∞-admissible for A, hence Reg-admissible for A and (i) holds. Conversely, suppose that (ii) holds (which is equivalent to (i)). Take any $x^* \in X_1^*$. Then $t \mapsto B^*\mathfrak{A}^{*t}x^*$ is continuous. Define $h(t) = \int_0^t |B^*\mathfrak{A}^{*s}x^*|\,ds$. Then

$$\begin{aligned}
h(t) &= \int_0^t |B^*\mathfrak{A}^{*s}x^*|\,ds \\
&= \int_0^t |B^*\mathfrak{A}^{*s}(\overline{\alpha} - A^*)^{-1}(\alpha - A)^*x^*|\,ds \\
&\le |\alpha|\|B^*(\overline{\alpha} - A^*)^{-1}\|\Big(\int_0^t \|\mathfrak{A}^{*s}\|\,ds\Big)|x^*|_{X^*} \\
&\quad + \int_0^t |B^*A^*\mathfrak{A}^{*s}(\overline{\alpha} - A^*)^{-1}x^*|\,ds.
\end{aligned}$$

The function $s \mapsto B^*A^*\mathfrak{A}^{*s}(\overline{\alpha} - A^*)^{-1}x^*$ is the continuous derivative of the function $s \mapsto B^*\mathfrak{A}^{*s}(\overline{\alpha} - A^*)^{-1}x^*$, so the last integral above is the total variation of $s \mapsto B^*\mathfrak{A}^{*s}(\overline{\alpha} - A^*)^{-1}x^*$ on $[0,t]$. By (ii), for $t \le 1$ this term is dominated by $K|x^*|_{X^*}$. Thus, there is a constant M such that $\int_0^1 |B^*\mathfrak{A}^{*s}x^*|\,ds \le M|x^*|_{X^*}$ for all $x^* \in X_1^*$. By Lemma 10.1.2, B^* is an L^1-admissible observation operator for A^*. □

Theorem 10.2.3 *Let U, X, and Y be reflexive, and let A be the generator of a C_0 semigroup on X.*

(i) *$B \in \mathcal{B}(U; X_{-1})$ is a Reg-admissible control operator for A if and only if $B^* \in \mathcal{B}(X_1^*; U^*)$ is an L^1-admissible observation operator for A^*.*
(ii) *$C \in \mathcal{B}(X_1; Y)$ is a Reg-admissible observation operator for A iff $C^* \in \mathcal{B}(Y^*; X_{-1}^*)$ is an L^1-admissible control operator for A^*, and this is true if and only if C is bounded, i.e., C has an extension to an operator in $\mathcal{B}(X; Y)$.*
(iii) *The operators $\left[\begin{smallmatrix} A & B \\ C & D \end{smallmatrix}\right]$, with $B \in \mathcal{B}(U; X_{-1})$, $C \in \mathcal{B}(X_1; Y)$ and $D \in \mathcal{B}(U; Y)$, generate a (regular) Reg-well-posed linear system on (Y, X, U) if and only if $\left[\begin{array}{c|c} A^* & C^* \\ \hline B^* & D^* \end{array}\right]$ generate a (regular) L^1-well-posed linear system on (U^*, X^*, Y^*).*

Proof Part (i) is contained in Theorem 10.2.2, part (ii) is contained in Theorems 4.2.7 and 4.4.2(ii), and part (iii) follows from (i)–(ii) and Theorems 4.3.4 and 4.4.8. □

Remark 10.2.4 Theorem 10.2.3 enables us to *define the dual Σ^d of a Reg-well-posed theorem* Σ with generators $\left[\begin{smallmatrix} A & B \\ C & D \end{smallmatrix}\right]$ on three reflexive spaces (Y, X, U) to be the L^1-well-posed system generated by $\left[\begin{array}{c|c} A^* & C^* \\ \hline B^* & D^* \end{array}\right]$. At the same time it enables us to *extend the original Reg-well-posed system to an L^∞-well-posed system* with the same generators: by Theorem 6.2.3, the dual of Σ^d is L^∞-well-posed, and by Theorem 6.2.13, the generators of this second dual are the same as the generators of the original system.

10.3 The Paley–Wiener theorem and H^∞

So far we have paid very little attention to the *Hilbert space case* where we have an L^2-well-posed linear system on three Hilbert spaces (Y, X, U). This case has some special properties. Most of these seem to be related in one way or another to some old results that go back to Paley and Wiener (1934). To present these results we have to first define the Hardy spaces H^p.

Definition 10.3.1 Let U and Y be Banach spaces, and let $1 \le p \le \infty$ and $\omega \in \mathbb{R}$.

(i) The space $H^p_\omega(U)$ consists of all analytic functions φ on $\Re z > \omega$ satisfying $\|\varphi\|_{H^p_\omega(U)} < \infty$, where

$$\|\varphi\|_{H^p_\omega(U)} = \begin{cases} \sup_{\alpha>\omega}\big(\int_{-\infty}^{\infty}|\varphi(\alpha+j\beta)|_U^p\,d\beta\big)^{1/p}, & 1 \le p < \infty, \\ \sup_{\Re z>\omega}|\varphi(z)|_U, & p=\infty. \end{cases}$$

In the case $\omega = 0$ we abbreviate $H^p_\omega(U)$ to $H^p(U)$.

(ii) The space $H^p_\omega(U;Y)$ consists of all analytic $\mathcal{B}(U;Y)$-valued functions ψ on $\Re z > \omega$ satisfying $\|\psi\|_{H^p_\omega(U;Y)} < \infty$, where

$$\|\psi\|_{H^p_\omega(U;Y)} = \begin{cases} \sup_{|u|_U\le 1,\alpha>\omega}\big(\int_{-\infty}^{\infty}|\psi(\alpha+j\beta)u|_Y^p\,d\beta\big)^{1/p}, & 1 \le p < \infty, \\ \sup_{\Re z>\omega}\|\psi(z)\|_{\mathcal{B}(U;Y)}, & p=\infty. \end{cases}$$

In the case $\omega = 0$ we abbreviate $H^p_\omega(U;Y)$ to $H^p(U;Y)$.

We remark that the norm in $H^\infty_\omega(U;Y)$ could have been written in a form that is analogous to the definition of the norm in $H^p_\omega(U;Y)$ with $p<\infty$, namely

$$\|\psi\|_{H^\infty_\omega(U;Y)} = \sup_{|u|_U\le 1,\alpha>\omega,\beta\in\mathbb{R}} |\psi(\alpha+j\beta)u|_Y.$$

Also note that we have defined $H^p_\omega(U;Y)$ so that it can be identified with the set of all bounded linear operators from U into $H^p_\omega(Y)$; thus it is a 'strong' version of $H^p_\omega(U;Y)$ rather than a 'uniform' version. (These coincide when $p=\infty$.)

The H^p-spaces are typical frequency domain spaces, and we have earlier encountered H^∞_ω-spaces more or less explicitly in various connections. The following two lemmas summarize some of the results that we have obtained so far in this direction:

Lemma 10.3.2 *The Laplace transform of a function $u \in L^p|Reg_\omega(\overline{\mathbb{R}}^+;U)$ $1\le p\le\infty$, belongs to $H^\infty_\alpha(U)$ for every $\alpha>\omega$.*

Proof By Definition 3.12.1,

$$\hat{u}(z) = \int_0^\infty \mathrm{e}^{-zt}u(t)\,dt = \int_0^\infty \mathrm{e}^{-(z-\omega)t}\mathrm{e}^{-\omega t}u(t)\,dt \qquad \Re z > \omega.$$

Without loss of generality we take $\omega = 0$ (denote $\mathrm{e}^{-\omega t}u(t)$ by $u(t)$).

The boundedness of $|\hat{u}(z)|_U$ on $\Re z > \alpha > 0$ follows from Hölder's inequality and the estimate (let $1/p + 1/q = 1$ and define $\mathrm{e}_{-\alpha}(t) = \mathrm{e}^{-\alpha t}$, $t \ge 0$)

$$|\hat{u}(z)|_U \le \int_0^\infty \mathrm{e}^{-\alpha t}|u(t)|_U\,dt \le \|\mathrm{e}_{-\alpha}\|_{L^q(\mathbb{R}^+)}\|u\|_{L^p(\mathbb{R}^+;U)}, \quad \Re z \ge \alpha.$$

To prove the analyticity of $\hat{u}$ it suffices to differentiate with respect to z under the integral sign; this is permitted because of Lemma 3.2.10 (and an estimate similar to the preceding one). □

Lemma 10.3.3 *Let $\Sigma = \left[\begin{smallmatrix} \mathfrak{A} & \mathfrak{B} \\ \mathfrak{C} & \mathfrak{D} \end{smallmatrix}\right]$ be an $L^p|Reg$-well-posed linear system on (Y, X, U) with semigroup generator A, control operator B, observation operator C, and transfer function $\widehat{\mathfrak{D}}$. Let $\omega > \omega_{\mathfrak{A}}$ where $\omega_{\mathfrak{A}}$ is the growth bound of Σ. Then*

(i) $\lambda \mapsto (\lambda - A)^{-1} \in H^\infty_\omega(X; X_1)$;
(ii) $\lambda \mapsto (\lambda - A_{|X})^{-1} B \in H^\infty_\omega(U; X)$;
(iii) $\lambda \mapsto C(\lambda - A)^{-1} \in H^\infty_\omega(X; Y)$;
(iv) $\widehat{\mathfrak{D}} \in H^\infty_\omega(U; Y)$, *and* $\|\widehat{\mathfrak{D}}\|_{H^\infty_\omega(U;Y)} \le \|\mathfrak{D}\|_{TIC_\omega(U;Y)}$.

The claim (iv) is also true for the transfer function of an arbitrary causal time-invariant operator $\mathfrak{D}$ which either maps $L^p_\omega(\mathbb{R}; U)$ continuously into $L^p_\omega(\mathbb{R}; Y)$ for some $\omega \in \mathbb{R}$ and p, $1 \le p < \infty$, or satisfies the assumptions of Theorem 2.6.7.

Proof (i) See Theorem 3.2.9(ii) and (3.2.6).

(ii) See Proposition 4.2.9 for the boundedness. The analyticity follows from the analyticity of the resolvent.

(iii) See Proposition 4.4.9 for the boundedness. The analyticity follows from the analyticity of the resolvent.

(iv) By Lemma 4.6.2 and Corollary 4.6.10, $\widehat{\mathfrak{D}} \in H^\infty_\omega(U; Y)$. The proof of this inclusion is ultimately based on Lemma 4.5.3 and Definition 4.6.1, and by inspecting the proof of Lemma 4.5.3 we get an explicit bound on $\|\widehat{\mathfrak{D}}\|_{H^\infty_\omega(U;Y)}$ as follows. For each α with $\Re\alpha > \omega$, let $\mathrm{e}_\alpha(t) = \mathrm{e}^{\alpha t}$, $t \in \mathbb{R}$. Then, for each $u \in U$ (cf. the proof of Lemma 4.5.3),

$$\begin{aligned}
\|\mathrm{e}_\alpha\|_{L^p|Reg_\omega(\mathbb{R}^-)} |\widehat{\mathfrak{D}}(\alpha)u|_Y &= \|\mathrm{e}_\alpha \widehat{\mathfrak{D}}(\alpha)u\|_{L^p|Reg_\omega(\mathbb{R}^-;Y)} \\
&= \|\mathfrak{D}(\mathrm{e}_\alpha u)\|_{L^p|Reg_\omega(\mathbb{R}^-;Y)} \\
&\le \|\mathfrak{D}\|_{TIC^2_\omega(U;Y)} \|\mathrm{e}_\alpha u\|_{L^p|Reg_\omega(\mathbb{R}^-;U)} \\
&= \|\mathfrak{D}\|_{TIC^2_\omega(U;Y)} |u|_U \|\mathrm{e}_\alpha\|_{L^p|Reg_\omega(\mathbb{R}^-)}.
\end{aligned}$$

Divide by $\|\mathrm{e}_\alpha\|_{L^p|Reg_\omega(\mathbb{R}^-)}$ and take the supremum over all $u \in U$ with $|u|_U = 1$ and all α with $\Re\alpha > \omega$ to get

$$\|\widehat{\mathfrak{D}}\|_{H^\infty_\omega(U;Y)} \le \|\mathfrak{D}\|_{TIC_\omega(U;Y)}.$$

□

In the Hilbert space case with $p = 2$ the two preceding lemmas can be made more precise. The key fact is the following representation theorem for the Laplace transform of a function in $L^2_\omega(\mathbb{R}^+; U)$:

Theorem 10.3.4 (Paley–Wiener) *Let U be a Hilbert space, and let $\omega \in \mathbb{R}$. Then the Laplace transform $\hat{u}$ of a function $u \in L^2_\omega(\mathbb{R}^+; U)$ belongs to $H^2_\omega(U)$,* *and conversely, every function φ in $H^2_\omega(U)$ is the Laplace transform of a function $u \in L^2_\omega(\mathbb{R}^+; U)$. Moreover,*

$$\|\hat{u}\|_{H^2_\omega(U)} = \sqrt{2\pi}\,\|u\|_{L^2_\omega(\mathbb{R}^+;U)}.$$

Proof Without loss of generality we take $\omega = 0$ (i.e., we denote $e^{-\omega t}u(t)$ by $u(t)$ and $\hat{u}(z-\omega)$ by $\hat{u}(z)$).

For each $\alpha > 0$, the function $\beta \mapsto \hat{u}(\alpha + j\beta)$ is the Fourier transform of the function $u_\alpha = (t \mapsto e^{-\alpha t}u(t)) \in L^1 \cap L^2(\mathbb{R}^+; U)$, and by Parseval's identity,

$$\frac{1}{2\pi}\int_{-\infty}^{\infty} |\hat{u}(\alpha + j\beta|_U^2\, d\beta = \|u_\alpha\|^2_{L^2(\mathbb{R}^+;U)} \le \|u\|^2_{L^2(\mathbb{R}^+;U)}.$$

Thus, $\hat{u} \in H^2(U)$, and $\|\hat{u}\|_{H^2(U)} \le \sqrt{2\pi}\,\|u\|_{L^2(\mathbb{R}^+;U)}$.

Conversely, let $\varphi \in H^2_\omega(U)$. The case $\omega \ne 0$ can be reduced to the case $\omega = 0$: we simply apply the case $\omega = 0$ to the function $z \mapsto \varphi(z - \omega)$. Thus, we may assume below that $\omega = 0$.

By the continuity of φ and the fact that the set of rational numbers is dense in $\Re z > 0$, the range of φ is separable-valued. Therefore we may assume that U is separable (i.e., we replace U by the closure of the range of φ). Choose an arbitrary orthonormal basis $\{e_i\}_{i=1}^\infty$ for U, and expand φ into

$$\varphi(z) = \sum_{i=1}^\infty \langle \varphi(z), e_i\rangle e_i = \sum_{i=1}^\infty \varphi_i(z) e_i.$$

Each function φ_i then belongs to the scalar H^2, and

$$\begin{aligned}
\sum_{i=1}^\infty \|\varphi_i\|^2_{H^2} &= \sum_{i=1}^\infty \sup_{\alpha>0} \frac{1}{2\pi}\int_{-\infty}^{\infty} |\varphi_i(\alpha + j\beta)|^2\, d\beta \\
&= \sup_{\alpha>0} \frac{1}{2\pi}\int_{-\infty}^{\infty} \sum_{i=1}^\infty |\varphi_i(\alpha + j\beta)|^2\, d\beta \\
&= \sup_{\alpha>0} \frac{1}{2\pi}\int_{-\infty}^{\infty} |\varphi(\alpha + j\beta)|_U^2\, d\beta \\
&= (1/2\pi)\|\varphi\|^2_{H^2(U)}.
\end{aligned}$$

By the standard scalar H^2-theorem (see, e.g., (Duren, 1970, Theorem 11.9) or (Hoffman, 1988, p. 131)), φ_i is the Laplace transform of a function $u_i \in L^2(\mathbb{R}^+)$ with $\|\varphi_i\|^2_{H^2} = \sqrt{2\pi}\,\|u_i\|_{L^2(\mathbb{R}^+)}$. Define

$$u = \sum_{i=m}^n u_i e_i.$$

This sum converges in $L^2(\mathbb{R}^+;U)$ since, for all $n \geq m$,

$$\Big\|\sum_{i=m}^{n} u_i e_i\Big\|^2_{L^2(\mathbb{R}^+;U)} = \sum_{i=m}^{n}\|u_i\|^2_{L^2(\mathbb{R}^+)} \leq \sum_{i=m}^{\infty}\|u_i\|^2_{L^2(\mathbb{R}^+)},$$

which tends to zero as $m \to \infty$. Moreover, since the terms in this sum are orthogonal,

$$\begin{aligned}\|u\|^2_{L^2(\mathbb{R}^+;U)} &= \Big\|\sum_{i=1}^{\infty} u_i e_i\Big\|^2_{L^2(\mathbb{R}^+;U)} = \sum_{i=1}^{\infty}\|u_i\|^2_{L^2(\mathbb{R}^+)} = (1/2\pi)\sum_{i=1}^{\infty}\|\varphi_i\|^2_{H^2}\\ &= (1/2\pi)\|\varphi\|^2_{H^2(U)}.\end{aligned}$$

By construction, for each $i \geq 1$, $\varphi_i = \hat{u}_i$, and

$$\begin{aligned}\langle \hat{u}(z), e_i\rangle &= \Big\langle \int_0^\infty \mathrm{e}^{-zt} u(t)\,dt, e_i\Big\rangle = \int_0^\infty \mathrm{e}^{-zt}\langle u(t), e_i\rangle\,dt\\ &= \hat{u}_i(z) = \varphi_i(z) = \langle \varphi(z), e_i\rangle.\end{aligned}$$

Thus, $\varphi = \hat{u}$. □

As a first application of Theorem 10.3.4 we get a simple characterization of $TIC^2_\omega(U;Y)$ whenever both U and Y are Hilbert spaces.

Theorem 10.3.5 *Let U and Y be Hilbert spaces, and let $\omega \in \mathbb{R}$. Then the transfer function $\widehat{\mathfrak{D}}$ of every operator $\mathfrak{D} \in TIC^2_\omega(U;Y)$ belongs to $H^\infty_\omega(U;Y)$ and conversely, every $\varphi \in H^\infty_\omega(U;Y)$ is the transfer function of some $\mathfrak{D} \in TIC^2_\omega(U;Y)$. Moreover, $\|\widehat{\mathfrak{D}}\|_{H^\infty_\omega(U;Y)} = \|\mathfrak{D}\|_{TIC^2_\omega(U;Y)}$. In particular, $TIC^2_\omega(U;Y)$ and $H^\infty_\omega(U;Y)$ are isometrically isomorphic.*

Proof That $\widehat{\mathfrak{D}} \in H^\infty_\omega(U;Y)$ whenever $\mathfrak{D} \in TIC^2_\omega(U;Y)$ and that $\|\widehat{\mathfrak{D}}\|_{H^\infty_\omega(U;Y)} \leq \|\mathfrak{D}\|_{TIC^2_\omega(U;Y)}$ is part of Lemma 10.3.3.

Conversely, suppose that $\varphi \in H^\infty_\omega(U;Y)$. We can then define an operator $\mathfrak{D}\colon L^2_\omega(\mathbb{R}^+;U) \to L^2_\omega(\mathbb{R}^+;U)$ as follows: we let $\mathfrak{D}u$ be the function whose Laplace transform is $\varphi\hat{u}$ (see Theorem 10.3.4). Then, for all $u \in L^2_\omega(\mathbb{R}^+;U)$,

$$\begin{aligned}\|\mathfrak{D}\pi_+ u\|_{L^2_\omega(\mathbb{R}^+;Y)} &= \sqrt{1/2\pi}\,\|\widehat{(\mathfrak{D}\pi_+ u)}\|_{H^2_\omega(Y)} = \sqrt{1/2\pi}\,\|\varphi\hat{u}\|_{H^2_\omega(Y)}\\ &\leq \sqrt{1/2\pi}\,\|\varphi\|_{H^\infty_\omega(U;Y)}\|\hat{u}\|_{H^2_\omega(U)} = \|\varphi\|_{H^\infty_\omega(U;Y)}\|u\|_{L^2_\omega(\mathbb{R}^+;U)}.\end{aligned}$$

This shows that $\|\mathfrak{D}\|_{\mathcal{B}(L^2_\omega(\mathbb{R}^+;U);L^2_\omega(\mathbb{R}^+;Y))} \leq \|\varphi\|_{H^\infty_\omega(U;Y)}$. For every $t \geq 0$ and $z > \omega$ (define $u(s) = 0$ for $s < 0$)

$$\widehat{\tau^{-t}u}(z) = \int_0^\infty \mathrm{e}^{-zs}u(s-t)\,ds = \int_0^\infty \mathrm{e}^{-z(s+t)}u(s)\,ds = \mathrm{e}^{-tz}\hat{u}(z),$$

and analogously,

$$\widehat{\tau^{-t}\mathfrak{D}u}(z) = \mathrm{e}^{-tz}\varphi(z)\hat{u}(z) = \varphi(z)\widehat{\tau^{-t}u}(z).$$

This implies that $\mathfrak{D}\tau^{-t}u = \tau^{-t}\mathfrak{D}u$ for all $t \geq 0$. We can then extend $\mathfrak{D}$ to a continuous operator mapping $L^2_{c,\omega}(\mathbb{R}; U) \to L^2_{c,\omega}(\mathbb{R}; U)$ by defining

$$\mathfrak{D}u = \tau^t \mathfrak{D} u \tau^{-t};$$

the result does not depend on t as long as t is large enough so that $\tau^{-t}u$ vanishes on $\mathbb{R}^-$. The extended operator satisfies

$$\|\mathfrak{D}\|_{\mathcal{B}(L^2_\omega(\mathbb{R};U);L^2_\omega(\mathbb{R};Y))} \leq \|\varphi\|_{H^\infty_\omega(U;Y)},$$

and it is time-invariant and causal. By continuity, it can be extended to an operator in $TIC^2_\omega(U; Y)$ satisfying $\|\mathfrak{D}\|_{TIC^2_\omega(U;Y)} \leq \|\varphi\|_{H^\infty_\omega(U;Y)}$. That $\widehat{\mathfrak{D}} = \varphi$ follows from Corollary 4.6.10. □

Our next theorem is an easy consequence of Theorems 10.3.4 and 10.3.5.

Theorem 10.3.6 *Let $\omega \in \mathbb{R}$, and let A be the generator of an ω-bounded C_0 semigroup on X.*

(i) *If U is a Hilbert space and X is reflexive, then $B \in \mathcal{B}(U; X_{-1})$ is an L^2-admissible ω-bounded control operator for A if and only*

$$\big(\lambda \mapsto B^*(\lambda - A^*)^{-1}\big) \in H^2_\omega(X; U).$$

(ii) *If Y is a Hilbert space, then $C \in \mathcal{B}(X_1; Y)$ is an L^2-admissible ω-bounded observation operator for A if and only if*

$$\big(\lambda \mapsto C(\lambda - A)^{-1}\big) \in H^2_\omega(X; Y).$$

(iii) *Suppose that both U and Y are Hilbert spaces. Let $B \in \mathcal{B}(U; X_{-1})$ be an L^2-admissible ω-bounded control operator for A (cf. (i)), and define $\mathcal{D}(A\&B)$ as in Definition 10.1.1(iii). Then $C\&D \in \mathcal{B}(\mathcal{D}(A\&B); Y)$ is an L^2-admissible ω-bounded observation/feedthrough operator for A if and only if the operator*

$$Cx = C\&D \begin{bmatrix} x \\ 0 \end{bmatrix}, \qquad x \in X_1.$$

is an L^2-admissible ω-bounded observation operator for A (cf. (ii)) and

$$\widehat{\mathfrak{D}} \in H^\infty_\omega(U; Y),$$

where

$$\widehat{\mathfrak{D}}(z) = C\&D \begin{bmatrix} (z - A_{|X})^{-1}B \\ 1 \end{bmatrix}, \qquad \Re z > \omega.$$

(iv) *Suppose that both U and Y are Hilbert spaces. The operators $B \in \mathcal{B}(U; X_{-1})$ and $C \in \mathcal{B}(X_1; U)$ are jointly L^2 admissible and ω-bounded iff B is an L^2-admissible ω-bounded control operator for A*

(cf. (i)), C is an L^2-admissible ω-bounded observation operator for A (cf. (ii)) and

$$\widehat{\mathfrak{D}} \in H^\infty_\omega(U;Y);$$

where

$$\widehat{\mathfrak{D}}(\lambda) = (\alpha - \lambda)C(\lambda - A)^{-1}(\alpha - A_{|X})^{-1}B + D_\alpha;$$

in the definition of $\widehat{\mathfrak{D}}$ the constant α with $\Re\alpha > \omega$ and the operator $D_\alpha \in \mathcal{B}(U;Y)$ can be chosen in an arbitrary manner.

Proof (i) This follows from (ii) and Theorems 6.2.3 and 6.2.13.

(ii) See Definition 10.3.1 and Theorems 4.4.2(iv), 4.7.14(iii), and 10.3.4.

(iii) The necessity of the given conditions follows from Definition 10.1.1 and Lemma 10.3.3(iv), and the sufficiency from Theorems 4.7.14, and 10.3.5.

(iv) See Definition 10.1.1, Theorem 4.6.7, and (iii). □

The following theorem is a restatement of Theorem 3.11.6.

Theorem 10.3.7 *Let A be the generator of a C_0 semigroup $\mathfrak{A}$ on a Hilbert space X. Then*

$$\omega_{\mathfrak{A}} = \inf\Big\{\omega \in \mathbb{R} \;\Big|\; \lambda \mapsto (\lambda - A)^{-1} \in H^\infty_\omega(X;X)\Big\}.$$

Proof By Lemma 10.3.3(i), to prove this theorem it suffices to show that $\omega_{\mathfrak{A}} < \omega$ whenever $\lambda \mapsto (\lambda - A)^{-1} \in H^\infty_\omega(X;X)$. Assume this, and consider the well-posed linear system with generators $\left[\begin{smallmatrix} A & 1 \\ 1 & 0 \end{smallmatrix}\right]$ on (X, X, X). The input/output map of this system is

$$(\mathfrak{D}u)(t) = \int_{-\infty}^{t} \mathfrak{A}^{t-s}u(s)\,ds,$$

the transfer function is

$$\widehat{\mathfrak{D}}(z) = (z - A)^{-1},$$

and, by Theorem 10.3.5, $\mathfrak{D}\pi_+$ maps $L^2_\omega(\mathbb{R}^+;U)$ into $L^2_\omega(\mathbb{R}^+;Y)$. By Theorem 3.11.8, $\omega_{\mathfrak{A}} < \omega$. □

Remark 10.3.8 Motivated by Theorem 10.3.4 we shall *extend the domain of definition* of the bilateral Laplace transform as follows. Let U be a Hilbert space, and let $u \in L^2_{\mathrm{loc}}(\mathbb{R};U)$. We recall from the discussion after Definition 3.12.1 that the usual domain of definition (if nonempty) of the bilateral Laplace transform $\hat{u}$ of u is a vertical strip $\{z \in \mathbb{C} \mid \Re z \in J\}$, where J is an interval in $\mathbb{R}$. For each α in the interior of J, the function $u_\alpha = (t \mapsto \mathrm{e}^{-\alpha t}u(t))$ belongs to $L^1 \cap L^2(\mathbb{R};U)$. Its Fourier transform is the function $\beta \mapsto 1/\sqrt{2\pi}\,\hat{u}(\alpha + j\beta)$, and it belongs to $BC_0 \cap L^2(\mathbb{R};U)$. If α is one of the end-points of J and $u_\alpha \notin L^1_\alpha(\mathbb{R};U)$, then our original bilateral Laplace transform is not defined on the vertical line $\Re z = \alpha$. However, it may still be possible that $u_\alpha \in L^2_\alpha(\mathbb{R}; U)$, and in this case we *define the bilateral Laplace transform of u on the vertical line* $\Re z = \alpha$ to be $\sqrt{2\pi}$ times the (almost everywhere defined) Fourier transform of u_α (in the L^2 sense). We still denote this transform by $\hat{u}(\alpha + j\beta)$, $\beta \in \mathbb{R}$, even if $\alpha \notin J$. In this case it is true for almost all $\beta \in \mathbb{R}$ that $\hat{u}(\alpha + j\beta)$ is the nontangential limit of $\hat{u}(z)$ with $\Re z \in J$ (this follows from well-known facts about Hardy spaces).

In particular, in the situation described in Theorem 10.3.4 we shall regard $\hat{u} \in H^2_\omega(U)$ as a function which is defined and analytic on $\mathbb{C}^+$, and, in addition, it is defined almost everywhere on the imaginary axis (and the restriction to the imaginary axis belongs to $L^2(j\mathbb{R}; U)$).

10.4 Controllability and observability gramians

In addition to the tests given in Theorem 10.3.6, there are some other admissibility and stability tests for control and observation operators in the Hilbert space case with $p = 2$. These tests are based on the existence of the *controllability and observability gramians.*

Definition 10.4.1 Let $\mathfrak{A}$ be a C_0 semigroup on the Hilbert space X, and let U and Y be Hilbert spaces.

(i) If $\mathfrak{B} \in \mathcal{B}(L^2(\mathbb{R}^-; U); X)$ is an L^2-well-posed stable input map for $\mathfrak{A}$, then $\mathfrak{B}\mathfrak{B}^* \in \mathcal{B}(X)$ is the *controllability gramian* of $\mathfrak{B}$.
(ii) If $\mathfrak{C} \in \mathcal{B}(X; L^2(\mathbb{R}^+; Y))$ is an L^2-well-posed stable output map for $\mathfrak{A}$, then $\mathfrak{C}^*\mathfrak{C} \in \mathcal{B}(X)$ is the *observability gramian* of $\mathfrak{C}$.

As our next two theorems show, the controllability and observability gramians have several different characterizations. We begin with the observability gramian.

Theorem 10.4.2 *Let $\mathfrak{A}$ be a C_0 semigroup with generator A on the Hilbert space X, let Y be a Hilbert space, let $X_1 = \mathcal{D}(A)$ and $X_1^* = \mathcal{D}(A^*)$ (cf. Remark 3.6.1), and let $C \in \mathcal{B}(X_1; Y)$. Then the following conditions are equivalent:*

(i) *C is an L^2-admissible stable observation operator for $\mathfrak{A}$.*
(ii) *There exists an operator $P \in \mathcal{B}(X)$ such that for all $x \in X_1$ and all $T \geq 0$,*

$$\int_0^T \langle x, \mathfrak{A}^{*t} C^* C \mathfrak{A}^t x \rangle_{(X_1, X^*_{-1})}\, dt \leq \langle x, Px \rangle_X. \tag{10.4.1}$$

(iii) *There exists an operator* $\Gamma_{\mathfrak{C}} \in \mathcal{B}(X)$ *such that for all* $x, y \in X_1$,

$$\langle y, \Gamma_{\mathfrak{C}} x \rangle_X = \lim_{T\to\infty} \int_0^T \langle y, \mathfrak{A}^{*t} C^* C \mathfrak{A}^t x \rangle_{(X_1, X^*_{-1})} \, dt. \tag{10.4.2}$$

(iv) *The* observability Lyapunov equation *(with terms in* $\mathcal{B}(X_1; X^*_{-1})$*)*

$$A^*\Pi + \Pi A = -C^*C \tag{10.4.3}$$

has at least one nonnegative solution $\Pi \in \mathcal{B}(X)$.

When these conditions hold, then the following additional claims are true for the operator $\Gamma_{\mathfrak{C}}$ *defined in* (10.4.2)*:*

(v) $\Gamma_{\mathfrak{C}} = \mathfrak{C}^*\mathfrak{C}$ *is the observability gramian of the output map* $\mathfrak{C}$ *generated by* $\mathfrak{A}$ *and* C.

(vi) $\Gamma_{\mathfrak{C}}$ *is the smallest operator for which* (10.4.1) *holds, i.e.,* (10.4.1) *holds with* P *replaced by* $\Gamma_{\mathfrak{C}}$*, and every other* P *for which* (10.4.1) *holds satisfies* $P \geq \Gamma_{\mathfrak{C}}$.

(vii) $\Gamma_{\mathfrak{C}}$ *is the smallest nonnegative solution of* (10.4.3)*, i.e.,* $\Gamma_{\mathfrak{C}}$ *is a solution of* (10.4.3)*, and every other nonnegative solution* Π *of* (10.4.3) *satisfies* $\Pi \geq \Gamma_{\mathfrak{C}}$.

(viii) *The set of all nonnegative solutions* $\Pi \in \mathcal{B}(X)$ *of* (10.4.3) *can be parametrized as* $\Pi = \Gamma_{\mathfrak{C}} + \Pi_\infty$*, where* Π_∞ *varies over all nonnegative solutions* $\mathcal{B}(X)$ *of the homogeneous Lyapunov equation (with terms in* $\mathcal{B}(X_1; X^*_1)$*)*

$$A^*\Pi_\infty + \Pi_\infty A = 0. \tag{10.4.4}$$

Here the residual cost operator Π_∞ *satisfies* $\mathfrak{A}^{*t}\Pi_\infty\mathfrak{A}^t = \Pi_\infty$ *for all* $t \geq 0$*, and it is determined by the fact that, for all* $x \in X$,

$$\lim_{t\to\infty} \mathfrak{A}^{*t}\Pi\mathfrak{A}^t x = \Pi_\infty x, \tag{10.4.5}$$

where the convergence is monotone decreasing in the sense that $\mathfrak{A}^{*t}\Pi\mathfrak{A}^t \leq \mathfrak{A}^{*s}\Pi\mathfrak{A}^s$ *for all* $t \geq s \geq 0$.

(ix) $\Gamma_{\mathfrak{C}}$ *is the unique nonnegative solution of* (10.4.3) *satisfying*

$$\lim_{t\to\infty} \langle \mathfrak{A}^t x, \Pi\mathfrak{A}^t x \rangle = 0, \qquad x \in X. \tag{10.4.6}$$

In particular, if $\mathfrak{A}$ *is strongly stable, then* $\Gamma_{\mathfrak{C}}$ *is the unique nonnegative solution of* (10.4.3).

(x) *If* $\Gamma_{\mathfrak{C}} \gg 0$ *then* $\mathfrak{A}$ *is strongly stable, and if* $\Gamma_{\mathfrak{C}} > 0$ *and* $\mathfrak{A}$ *is bounded then* $\mathfrak{A}$ *is weakly stable.*

Before proving Theorem 10.4.2, let us formulate the analogous result for the controllability gramian:

Theorem 10.4.3 *Let $\mathfrak{A}$ be a C_0 semigroup with generator A on the Hilbert space X, let U be a Hilbert spaces, let $X_1^* = \mathcal{D}(A^*)$ and $X_{-1} = \mathcal{D}(A^*)^*$ (cf. Remark 3.6.1), and let $B \in \mathcal{B}(U; X_{-1})$. Then the following conditions are equivalent:*

(i) *B is an L^2-admissible stable control operator for $\mathfrak{A}$.*
(ii) *There exists an operator $P \in \mathcal{B}(X)$ such that for all $x \in X_1^*$ and all $T \geq 0$,*

$$\int_0^T \langle x, \mathfrak{A}^t B B^* \mathfrak{A}^{t*} x \rangle_{(X_1^*, X_{-1})} \, dt \leq \langle x, Px \rangle_X. \tag{10.4.7}$$

(iii) *There exists an operator $\Gamma_{\mathfrak{B}} \in \mathcal{B}(X)$ such that for all $x, y \in X_1^*$,*

$$\langle y, \Gamma_{\mathfrak{B}} x \rangle_X = \lim_{T \to \infty} \int_0^T \langle y, \mathfrak{A}^t B B^* \mathfrak{A}^{t*} x \rangle_{(X_1^*, X_{-1})} \, dt. \tag{10.4.8}$$

(iv) *The* controllability Lyapunov equation *(with terms in $\mathcal{B}(X_1^*; X_{-1})$)*

$$A\Pi + \Pi A^* = -BB^* \tag{10.4.9}$$

has at least one nonnegative solution $\Pi \in \mathcal{B}(X)$.

When these conditions hold, then the following additional claims are true for the operator $\Gamma_{\mathfrak{B}}$ defined in (10.4.8)*:*

(v) *$\Gamma_{\mathfrak{B}} = \mathfrak{B}\mathfrak{B}^*$ is the controllability gramian of the input map $\mathfrak{B}$ generated by $\mathfrak{A}$ and B.*
(vi) *$\Gamma_{\mathfrak{B}}$ is the smallest operator for which* (10.4.7) *holds, i.e.,* (10.4.7) *holds with P replaced by $\Gamma_{\mathfrak{B}}$, and every other P for which* (10.4.7) *holds satisfies $P \geq \Gamma_{\mathfrak{B}}$.*
(vii) *$\Gamma_{\mathfrak{B}}$ is the smallest nonnegative solution of* (10.4.9)*, i.e., $\Gamma_{\mathfrak{B}}$ is a solution of* (10.4.9)*, and every other nonnegative solution Π of* (10.4.9) *satisfies $\Pi \geq \Gamma_{\mathfrak{B}}$.*
(viii) *The set of all nonnegative solutions $\Pi \in \mathcal{B}(X)$ of* (10.4.9) *can be parametrized as $\Pi = \Gamma_{\mathfrak{B}} + \Pi_\infty$, where Π_∞ varies over all nonnegative solutions $\mathcal{B}(X)$ of the homogeneous Lyapunov equation (with terms in $\mathcal{B}(X_1^*; X_{-1})$)*

$$A\Pi_\infty + \Pi_\infty A^* = 0. \tag{10.4.10}$$

Here the residual cost operator *Π_∞ satisfies $\mathfrak{A}^t \Pi_\infty \mathfrak{A}^{*t} = \Pi_\infty$ for all $t \geq 0$, and it is determined by the fact that, for all $x \in X$,*

$$\lim_{t \to \infty} \mathfrak{A}^t \Pi \mathfrak{A}^{*t} x = \Pi_\infty x, \tag{10.4.11}$$

where the convergence is monotone decreasing in the sense that $\mathfrak{A}^t \Pi \mathfrak{A}^{*t} \leq \mathfrak{A}^s \Pi \mathfrak{A}^{*s}$ for all $t \geq s \geq 0$.

(ix) $\Gamma_{\mathfrak{B}}$ *is the unique nonnegative solution of* (10.4.9) *satisfying*

$$\lim_{t\to\infty} \langle \mathfrak{A}^{*t}x, \Gamma_{\mathfrak{B}}\mathfrak{A}^{*t}x\rangle = 0, \qquad x \in X.$$

In particular, if $\mathfrak{A}$ *is strongly stable, then* $\Gamma_{\mathfrak{B}}$ *is the unique nonnegative solution of* (10.4.9)*;*

(x) *If* $\Gamma_{\mathfrak{B}} \gg 0$ *then* $\mathfrak{A}^*$ *is strongly stable, and if* $\Gamma_{\mathfrak{B}} > 0$ *and* $\mathfrak{A}$ *is bounded then* $\mathfrak{A}$ *is weakly stable.*

Clearly Theorem 10.4.3 follows from Theorem 10.4.2 by duality: $\mathfrak{B}$ is an L^2-well-posed stable input map for $\mathfrak{A}$ iff $\mathbf{Я}\mathfrak{B}^*$ is an L^2-well-posed stable output map for $\mathfrak{A}^*$; cf. Theorem 6.2.3. Thus, it suffices to prove Theorem 10.4.2.

Proof of Theorem 10.4.2. (ii) $\Rightarrow$ (i): Suppose that (ii) holds. For all $x \in X_1$ we define $(\mathfrak{C}x)(t) = C\mathfrak{A}^t x$, $t \ge 0$. Then (10.4.1) implies that, for $x \in X_1$,

$$\begin{aligned}\int_0^T \langle(\mathfrak{C}x)(t), (\mathfrak{C}x)(t)\rangle_Y\,dt &= \int_0^T \langle x, \mathfrak{A}^{*t}C^*C\mathfrak{A}^t x\rangle_{(X_1,X_1^*)}\,dt\\ &\le \langle x, Px\rangle_X \le \|P\|_{\mathcal{B}(X)}|x|_X^2.\end{aligned}$$

This means that $\mathfrak{C}$ can be extended to an operator in $\mathcal{B}(X; L^2(\mathbb{R}^+; Y))$. By Theorem 4.4.7(i), C is an L^2-admissible stable observation operator for $\mathfrak{A}$.

(i) $\Rightarrow$ (iii): Assume (i), and let $\Gamma_{\mathfrak{C}} = \mathfrak{C}^*\mathfrak{C}$ be the controllability gramian. We claim that (iii) then holds for this $\Gamma_{\mathfrak{C}}$, i.e., we claim that both (iii) and (v) hold. We prove this as follows. For each $x \in X$, $\pi_{[0,T)}\mathfrak{C}x \to \mathfrak{C}x$ in L^2 as $T \to \infty$, hence $\Gamma_{\mathfrak{C}}x = \lim_{T\to\infty} \mathfrak{C}^*\pi_{[0,T)}\mathfrak{C}x$ (a strong limit in X). In particular, for all x, $y \in X$, $\langle y, \Gamma_{\mathfrak{C}}x\rangle_X = \lim_{T\to\infty}\langle y, \mathfrak{C}^*\pi_{[0,T)}\mathfrak{C}x\rangle_X$. Thus, in order to prove (iii) it suffices to show that, for all x, $y \in X_1$, this is the same limit which appears in (10.4.2). Let x, $y \in X_1$. Then, by Theorem 4.4.2(i),

$$\begin{aligned}\langle y, \mathfrak{C}^*\pi_{[0,T)}\mathfrak{C}x\rangle_X &= \int_0^T \langle(\mathfrak{C}y)(t), (\mathfrak{C}x)(t)\rangle_Y\,dt\\ &= \int_0^T \langle C\mathfrak{A}^t y, C\mathfrak{A}^t x\rangle_Y\,dt\\ &= \int_0^T \langle y, \mathfrak{A}^{*t}C^*C\mathfrak{A}^t x\rangle_{(X_1,X_1^*)}\,dt.\end{aligned}$$

Thus, (iii) holds with $\Gamma_{\mathfrak{B}} = \mathfrak{C}^*\mathfrak{C}$.

(iii) $\Rightarrow$ (iv): Let $\Gamma_{\mathfrak{C}} \in \mathcal{B}(X)$ satisfy (10.4.2). Clearly $\Gamma_{\mathfrak{C}} \ge 0$, because $\langle x, \Gamma_{\mathfrak{C}}x\rangle \ge 0$ for all $x \in X_1$, hence (by continuity) for all $x \in X$. We claim that (10.4.3) holds with $\Pi = \Gamma_{\mathfrak{C}}$. Take x, $y \in X_2$, and define

$$f(t) = \langle C\mathfrak{A}^t y, C\mathfrak{A}^t x\rangle_X = \langle y, \mathfrak{A}^{*t}C^*C\mathfrak{A}^t x\rangle_{(X_1,X_1^*)}.$$

Then f is continuously differentiable and

$$\dot{f}(t) = \langle C\mathfrak{A}^t Ay, C\mathfrak{A}^t x\rangle_X + \langle C\mathfrak{A}^t y, C\mathfrak{A}^t Ax\rangle_X.$$

Integrating this equation over $[0, T]$ we get

$$\begin{aligned} f(T) - f(0) = &\int_0^T \langle Ay, \mathfrak{A}^{*t} C^* C\mathfrak{A}^t x\rangle_{(X_1, X_1^*)}\, dt \\ &+ \int_0^T \langle y, \mathfrak{A}^{*t} C^* C\mathfrak{A}^t Ax\rangle_{(X_1, X_1^*)}\, dt. \end{aligned}$$

Letting $T \to \infty$ and using (10.4.3) and the fact that $f(0) = \langle Cy, Cx\rangle_X$ we find that $f(\infty) = \lim f(T)_{T\to\infty}$ exists and satisfies

$$\begin{aligned} f(\infty) - \langle Cy, Cx\rangle_X &= \langle Ay, \Gamma_C x\rangle_{(X_1, X_1^*)} + \langle y, \Gamma_C Ax\rangle_{(X_1, X_1^*)} \\ &= \langle Ay, \Gamma_C x\rangle_X + \langle y, \Gamma_C Ax\rangle_X, \end{aligned}$$

where the re-interpretation of the right-hand side is possible since $\Gamma_C x \in X$ and $\Gamma_C Ax \in X$. By (10.4.2), the limit $\lim_{T\to\infty} \int_0^T f(t)\, dt$ $(= \langle y, \Gamma_{\mathfrak{C}} x\rangle)$ exists, hence $f(\infty) = 0$. We conclude that, for all $x, y \in X_2$,

$$\langle Ay, \Gamma_C x\rangle_X + \langle y, \Gamma_C Ax\rangle_X = -\langle Cy, Cx\rangle_X.$$

By the density of X_2 in X_1 and the continuity of the involved operators, the same equation must then be true for all $x, y \in X_1$. This implies that (10.4.3) is valid as an operator equation in $\mathcal{B}(X_1; X_1^*)$ with Π replaced by Γ_C.

(iv) $\Rightarrow$ (ii): Let $\Pi \in \mathcal{B}(X)$ be a nonnegative solution of (10.4.3). For every $t \geq 0$ we define $P(t) = \mathfrak{A}^{*t}\Pi\mathfrak{A}^t$. Then $P(0) = \Pi$ and $P(t) \geq 0$. Let $x \in X_1$, and define $g(t) = \langle x, P(t)x\rangle_X = \langle \mathfrak{A}^t x, \Pi\mathfrak{A}^t x\rangle_X$. Then g is continuously differentiable, and by (10.4.3),

$$\begin{aligned} \dot{g}(t) &= \langle A\mathfrak{A}^t x, \Pi\mathfrak{A}^t x\rangle_X + \langle \mathfrak{A}^t x, \Pi A\mathfrak{A}^t x\rangle_X \\ &= \langle \mathfrak{A}^t x, (A^*\Pi + \Pi A)\mathfrak{A}^t x\rangle_{(X_1, X_1^*)} \\ &= -\langle \mathfrak{A}^t x, C^* C\mathfrak{A}^t x\rangle_{(X_1, X_1^*)} \\ &= -\langle x, \mathfrak{A}^{*t} C^* C\mathfrak{A}^t x\rangle_{(X_1, X_1^*)}. \end{aligned}$$

Integrate over $[0, T]$ to get

$$\begin{aligned} \int_0^T \langle x, \mathfrak{A}^{*t} C^* C\mathfrak{A}^t x\rangle_{(X_1, X_1^*)}\, dt &= g(0) - g(T) \\ &= \langle x, \Pi x\rangle_X - \langle x, P(T)x\rangle_X \\ &\leq \langle x, \Pi x\rangle_X. \end{aligned}$$

Thus, (ii) holds with $P = \Pi$.

We have now proved the equivalence of (i)–(iv). We proceed to establish (v)–(x).

(v) The proof of the implication (i) $\Rightarrow$ (iii) given above shows (10.4.2) is true with $\Gamma_{\mathfrak{C}}$ equal to the controllability gramian $\mathfrak{C}^*\mathfrak{C}$, and the limit on the right-hand side of (10.4.2) determines $\Gamma_{\mathfrak{C}}$ uniquely.

(vi) The proof of the implications (iii) $\Rightarrow$ (iv) $\Rightarrow$ (ii) given above shows that (10.4.1) is true with $P = \Gamma_{\mathfrak{C}}$. That all other solutions P of (10.4.1) must satisfy $P \geq \Gamma_{\mathfrak{C}}$ follows from (10.4.2).

(vii) By the proof of the implication (iii) $\Rightarrow$ (iv), $\Gamma_{\mathfrak{C}}$ is a nonnegative solution of (10.4.3), and we saw in the proof of the implication (iv) $\Rightarrow$ (ii) that every nonnegative solution Π of (10.4.3) satisfies (10.4.1) with $P = \Pi$. By (vi), every nonnegative solution Π of (10.4.3) must therefore satisfy $\Pi \geq \Gamma_{\mathfrak{C}}$.

(viii) By (vii), if Π is a nonnegative solution of (10.4.3), then $\Pi_\infty = \Pi - \Gamma_{\mathfrak{C}} \geq 0$. Subtracting (10.4.3) with Π replaced by $\Gamma_{\mathfrak{C}}$ from (10.4.3) we find that $\Pi_\infty = \Pi - \Gamma_{\mathfrak{C}}$ satisfies (10.4.4). Conversely, if Π_∞ is a nonnegative solution of (10.4.4), then $\Pi = \Gamma_{\mathfrak{C}} + \Pi_\infty$ is a nonnegative solution of (10.4.3). To show that $\mathfrak{A}^{*t}\Pi_\infty\mathfrak{A}^t = \Pi_\infty$ for all $t \geq 0$ we repeat the argument in the proof of the implication (iv) $\Rightarrow$ (ii) with Π replaced by Π_∞ and (10.4.3) replaced by (10.4.4). Once we know this we find that

$$\mathfrak{A}^{*t}\Pi\mathfrak{A}^t - \Pi_\infty = \mathfrak{A}^{*t}(\Pi - \Pi_\infty)\mathfrak{A}^t = \mathfrak{A}^{*t}\Gamma_{\mathfrak{C}}\mathfrak{A}^t.$$

Thus, in order to prove (10.4.5) it suffices to show that, for all $x \in X$,

$$\lim_{t\to\infty} \mathfrak{A}^{*t}\Gamma_{\mathfrak{C}}\mathfrak{A}^t x = \lim_{t\to\infty} \mathfrak{A}^{*t}\mathfrak{C}^*\mathfrak{C}\mathfrak{A}^t x = 0.$$

By the intertwining condition $\mathfrak{C}\mathfrak{A} = \pi_+\tau\mathfrak{C}$, the limit above can be written in the form

$$\lim_{t\to\infty} \mathfrak{A}^{*t}\mathfrak{C}^*\mathfrak{C}\mathfrak{A}^t x = \lim_{t\to\infty} \mathfrak{C}^*\tau^{-t}\pi_+\tau^t\mathfrak{C}x = \lim_{t\to\infty} \mathfrak{C}^*\pi_{[t,\infty)}\mathfrak{C}x,$$

and we know this limit to be zero since $\pi_{[t,\infty)}\mathfrak{C}x \to 0$ in $L^2(\mathbb{R}^+; Y)$ as $t \to \infty$. Thus (10.4.5) holds. The same argument proves that the convergence is monotone since

$$\begin{aligned}\langle x, \mathfrak{A}^{*t}\Pi\mathfrak{A}^t - \Pi_\infty x\rangle_X &= \langle x, \mathfrak{A}^{*t}\mathfrak{C}^*\mathfrak{C}\mathfrak{A}^t x\rangle_X = \langle x, \mathfrak{C}^*\pi_{[t,\infty)}\mathfrak{C}x\rangle_X \\ &= \int_t^\infty \langle(\mathfrak{C}x)(s), (\mathfrak{C}x)(s)\rangle_Y\, ds\end{aligned}$$

is decreasing in t.

(ix) If Π satisfies (10.4.6), then by (viii), the corresponding operator Π_∞ satisfies $\langle x, \Pi_\infty\rangle_X = 0$ for all $x \in X$, i.e., $\Pi_\infty = 0$ and $\Pi = \Gamma_{\mathfrak{C}}$. The second claim is obvious since (10.4.6) is redundant whenever $\mathfrak{A}$ is strongly stable.

(x) The condition $\Gamma_{\mathfrak{C}} \gg 0$ in connection with (10.4.6) implies that $|\mathfrak{A}^t x| \to 0$ as $t \to \infty$, i.e., $\mathfrak{A}$ is strongly stable.

Suppose next that $\mathfrak{A}$ is stable and that $\Gamma_{\mathfrak{B}} > 0$. Let V be the positive square root of $\Gamma_{\mathfrak{B}}$. Then $\mathcal{R}(V)$ is dense in X since V is injective and $0 = \mathcal{N}(V) = \mathcal{R}(V^*)^{\perp} = \mathcal{R}(V)^{\perp}$. By (10.4.6), $V\mathfrak{A}^t x \to 0$ in X as $t \to \infty$. This implies that, for all $y \in X$,

$$\lim_{t\to\infty} \langle y, V\mathfrak{A}^t x\rangle = \lim_{t\to\infty} \langle Vy, \mathfrak{A}^t x\rangle = 0,$$

or equivalently, that for all $z \in \mathcal{R}(V)$,

$$\lim_{t\to\infty} \langle z, \mathfrak{A}^t x\rangle = 0.$$

Fix $x, y \in X$, and let $\epsilon > 0$. By the stability of $\mathfrak{A}$ and the density of $\mathcal{R}(V)$ in X, we can find some $z \in \mathcal{R}(V)$ such that $|z - y||\mathfrak{A}^t x| \le \epsilon/2$ for all $t \ge 0$. If we take t large enough so that $\left|\langle z, \mathfrak{A}^t x\rangle\right| \le \epsilon/2$, then

$$\left|\langle y, \mathfrak{A}^t x\rangle\right| \le \left|\langle z, \mathfrak{A}^t x\rangle\right| + |z - y||\mathfrak{A}^t x| \le \epsilon.$$

Thus $\langle y, \mathfrak{A}^t x\rangle \to 0$ as $t \to \infty$ for all $x, y \in X$, i.e., $\mathfrak{A}$ is weakly stable. □

In the case of a diagonal system it is possible to develop Theorems 10.4.2 and 10.4.3 a little further.

Definition 10.4.4 Let U and Y be separable Hilbert spaces spanned by the orthonormal bases $\{\phi_n\}_{n=1}^{N_U}$ and $\{\psi_n\}_{n=1}^{N_Y}$ (where N_U and N_Y are the finite or infinite dimensions of U and Y), respectively, and let $D \in \mathcal{B}(U;Y)$. By the *matrix* of D induced by the bases $\{\phi_n\}_{n=1}^{N_U}$ and $\{\psi_n\}_{n=1}^{N_Y}$ we mean the (possibly infinite) array $\{d_{ij}\}$, $1 \le i \le N_U$, $1 \le j \le N_Y$, where $d_{ij} = \psi_i^* D\phi_j = \langle D\phi_j, \psi_i\rangle$.

Lemma 10.4.5 *Let U and Y be separable Hilbert spaces spanned by the orthonormal bases $\{\phi_n\}_{n=1}^{N_U}$ and $\{\psi_n\}_{n=1}^{N_Y}$, respectively, let $D \in \mathcal{B}(U;Y)$, and let $\{d_{ij}\}$ be the matrix induced by D. Then for every $u = \sum_{j=1}^{N_U} u_j\phi_j \in U$, we have $Du = \sum_{i=1}^{N_Y} y_i\psi_i$, where*

$$y_i = \sum_{j=1}^{N_U} d_{ij}u_j, \tag{10.4.12}$$

and (10.4.12) *defines a bounded linear operator from $l^2(1, \dots, N_U)$ to $l^2(1, \dots, N_Y)$ with the same norm as D. Conversely, if* (10.4.12) *defines a bounded linear operator from $l^2(1, \dots, N_U)$ to $l^2(1, \dots, N_Y)$, then we get a bounded linear operator from U to Y by defining*

$$Du = \sum_{i=1}^{N_Y} \psi_i \sum_{j=1}^{N_U} d_{ij}\phi_j^* u, \tag{10.4.13}$$

and the matrix of this operator is $\{d_{ij}\}$.

We leave the easy proof of this lemma to the reader. (Observe that the mappings from $u \in U$ to its Fourier coefficients $u_j = \langle u, \phi_j \rangle \in l^2(1, \ldots, N_U)$ and from $y \in Y$ to its Fourier coefficients $y_i = \langle y, \psi_i \rangle \in l^2(1, \ldots, N_Y)$ are unitary.)

Theorem 10.4.6 *Let*

$$Ax = \sum_{n=1}^{N_X} \lambda_n \phi_n \phi_n^* x$$

be the generator of a diagonal semigroup $\mathfrak{A}$ with eigenvalues λ_n and eigenvectors ϕ_n on the separable N_X-dimensional Hilbert space X (see Examples 3.3.3 and 3.3.5 and Section 4.9), let $X_1 = \mathcal{D}(A)$, and let $C \in \mathcal{B}(X_1; Y)$ where Y is a Hilbert space. Then the following conditions are equivalent:

(i) *C is an L^2-admissible stable observation operator for $\mathfrak{A}$;*
(ii) *$C\phi_i = 0$ whenever $\Re\lambda_i \geq 0$, and the matrix*

$$\gamma_{ij} = \frac{-\langle C\phi_j, C\phi_i \rangle}{\overline{\lambda_i} + \lambda_j}$$

defines a bounded linear operator on $l^2(1, \ldots, N_X)$ (we interpret $0/0$ as zero).

When these conditions hold, then $\{\gamma_{ij}\}$ is the matrix induced by the observability gramian with respect to the basis $\{\phi_n\}_{n=1}^{N_X}$.

The requirement that $C\phi_i = 0$ whenever $\Re\lambda_i \geq 0$ means that all the eigenvectors which correspond to eigenvalues with nonnegative real part are unobservable.

Proof (i) $\Rightarrow$ (ii): Suppose that C is an L^2-admissible stable observation operator for $\mathfrak{A}$, and let $\Gamma_{\mathfrak{C}} = \mathfrak{C}^*\mathfrak{C}$ be the controllability gramian. For each basis vector ϕ_i, we have $(\mathfrak{C}\phi_i)(t) = e^{\lambda_i t} C\phi_i$. This function does not belong to $L^2(\mathbb{R}^+; Y)$ if $\Re\lambda_i \geq 0$ unless $C\phi_i = 0$, so we must have $C\phi_i = 0$ whenever $\Re\lambda_i \geq 0$. Moreover, for $\Re\lambda_i < 0$ and $\Re\lambda_j < 0$

$$\begin{aligned}\langle \Gamma_{\mathfrak{C}}\phi_j, \phi_i \rangle = \langle \mathfrak{C}\phi_j, \mathfrak{C}\phi_i \rangle &= \int_0^\infty \langle C\phi_j, C\phi_i \rangle e^{(\overline{\lambda_i}+\lambda_j)t}\, dt \\ &= \frac{-\langle C\phi_j, C\phi_i \rangle}{\overline{\lambda_i} + \lambda_j}.\end{aligned}$$

Thus, the matrix of $\Gamma_{\mathfrak{C}}$ is $\{\gamma_{ij}\}$, and this matrix defines a (positive) bounded linear operator on $l^2(1, \ldots, N_X)$ (see Lemma 10.4.5).

(ii) $\Rightarrow$ (i): Suppose that (ii) holds. Let $x \in X_1$ be a finite sum of the form $x = \sum_{j=1}^n x_j \phi_j$. Define $(\mathfrak{C}x)(t) = C\mathfrak{A}^t x$ for $t \geq 0$. Then $\mathfrak{C}x \in L^2(\mathbb{R}^+; Y)$ and a computation similar to the one above shows that

$$\int_0^\infty \|(C\mathfrak{A}^t x)(t)\|^2 = \sum_{i,j=1}^n \gamma_{ij} x_j \overline{x_i}.$$

Thus, if we denote the norm of the operator induced by the matrix $\{\gamma_{ij}\}$ on $L^2(1, \ldots, N_X)$ by M, then we get

$$\int_0^\infty \|(\mathfrak{C}x)(t)\|^2 \leq M\|x\|_X^2.$$

This being true for all x in a dense subset of X, we can extend the operator $\mathfrak{C}$ to a bounded linear operator from X to $L^2(\mathbb{R}^+; Y)$. □

There is an obvious dual counterpart to Theorem 10.4.6:

Theorem 10.4.7 *Let*

$$Ax = \sum_{n=1}^{N_X} \lambda_n \phi_n \phi_n^* x$$

be the generator of a diagonal semigroup $\mathfrak{A}$ *with eigenvalues* λ_n *and eigenvectors* ϕ_n *on the separable* N_X*-dimensional Hilbert space* X*, and let* $B \in \mathcal{B}(U; X_{-1})$ *where* U *is a Hilbert space. Then the following conditions are equivalent:*

(i) B *is an* L^2*-admissible stable observation operator for* $\mathfrak{A}$*;*
(ii) $B^*\phi_i = 0$ *whenever* $\Re\lambda_i \geq 0$*, and the matrix*

$$\gamma_{ij} = \frac{-\langle B^*\phi_j, B^*\phi_i\rangle}{\lambda_i + \overline{\lambda_j}}$$

defines a bounded linear operator on $l^2(1, \ldots, N_X)$ *(we interpret* $0/0$ *as zero).*

When these conditions hold, then $\{\gamma_{ij}\}$ *is the matrix induced by the controllability gramian with respect to the basis* $\{\phi_n\}_{n=1}^{N_X}$*.*

10.5 Carleson measures

The conditions for the L^2-admissibility of a control operator B with a Hilbert input space U or an observation operator C with a Hilbert output space in Theorem 10.3.6 can be developed further in the case where also the state space is a Hilbert space, and the semigroup is *normal*. The key result that we need for this is the Carleson measure theorem in the right half-plane. Several different formulations of this theorem are available, out of which we have chosen the one which has the sharpest constants known to us.

Theorem 10.5.1 *Let ν be a positive Borel measure on the closed right half-plane $\overline{\mathbb{C}}^+$, and let $\mathcal{H}$ be the harmonic extension operator*

$$(\mathcal{H}\varphi)(z) = \begin{cases} \varphi(z), & \Re z = 0, \\ \frac{1}{\pi}\int_{-\infty}^{\infty} \Re(z - j\omega)^{-1}\varphi(j\omega)\, d\omega, & \Re z > 0. \end{cases}$$

Then the following conditions are equivalent:

(i) *$\mathcal{H}$ maps $L^2(j\mathbb{R})$ continuously into $L^2(\nu)$;*
(ii) *$H^2 \subset L^2(\nu)$;*
(iii) *$a := \sup_{\Re\lambda>0} \pi^{-1} \int_{\Re z\geq 0} \Re\lambda |z+\overline{\lambda}|^{-2}\nu(dz) < \infty$;*
(iv) *$c := \sup_{\Re\lambda>0} \pi^{-1} \int_{0\leq\Re z\leq\Re\lambda} \Re\lambda |z+\overline{\lambda}|^{-2}\nu(dz) < \infty$;*
(v) *$M := \sup_{\omega\in\mathbb{R};r>0} \pi^{-1}\nu(D(j\omega, r)\cap\overline{\mathbb{C}}^+)/r < \infty$, where $D(j\omega, r) = \{z \in \mathbb{C} \mid |z - j\omega| \leq r\}$.*

Moreover,

$$c \leq a \leq \|\mathcal{H}\|^2_{\mathcal{B}(H^2;L^2(\nu))} \leq \|\mathcal{H}\|^2_{\mathcal{B}(L^2(j\mathbb{R});L^2(\nu))} \leq 32c$$

and

$$a \leq 2M \leq 8c.$$

Let us comment on how the statement should be interpreted. The space $L^2(\nu)$ consists of all ν-measurable functions φ on $\overline{\mathbb{C}}^+$ with a finite norm

$$\|\varphi\|_{L^2(\nu)} = \left(\int_{\overline{\mathbb{C}}^+} |\varphi(z)|^2\nu(dz)\right)^{1/2}$$

(where we identify functions with a zero norm with the zero function). We define $L^2(j\mathbb{R})$ to be the set of all L^2-functions on the imaginary axis, with the standard norm

$$\|\varphi\|_{L^2(j\mathbb{R})} = \left(\int_{-\infty}^{\infty} |\varphi(j\omega)|^2\, d\omega\right)^{1/2}.$$

In particular, if $\omega \mapsto \varphi(j\omega)$ is the boundary function of a function $\psi \in H^2$, then $\mathcal{H}\varphi = \psi$ (cf. the proof of Theorem 10.3.4). Statement (ii) means that every function $\varphi \in H^2$ has a finite norm in $L^2(\nu)$, but we do not claim that H^2 is *embedded* in $L^2(\nu)$, i.e., we do not claim that every nonzero $\varphi \in H^2$ has a nonzero $L^2(\nu)$-norm. Clearly the integral in (iii) is the norm in $L^2(\nu)$ of the H^2-function

$$\varphi_\lambda(z) = \frac{\sqrt{\Re\lambda}}{\sqrt{\pi}(z+\overline{\lambda})}, \quad \Re z \geq 0, \quad \Re\lambda > 0.$$

Observe that these functions have been normalized so that

$$\|\varphi_\lambda\|_{H^2} = \|\varphi_\lambda\|_{L^2(j\mathbb{R})} = \left(\frac{1}{\pi}\int_{-\infty}^{\infty} \frac{\Re\lambda}{|\lambda - j\omega|^2}\,d\omega\right)^{1/2} = 1.$$

The norm $\|\mathcal{H}\|_{\mathcal{B}(H^2;L^2(\nu))}$ in the last statement refers to the norm of the operator in (ii) mapping H^2 into $L^2(\nu)$, and the norm $\|\mathcal{H}\|_{\mathcal{B}(L^2(j\mathbb{R});L^2(\nu))}$ refers to the norm of the operator $\mathcal{H}$ in (i).

In the proof of Theorem 10.5.1 we need the following two lemmas:

Lemma 10.5.2 *Let $y, z \in \mathbb{C}$ with $\Re y > 0$, $\Re z \neq 0$. Then*

$$\frac{1}{\pi^2}\int_{-\infty}^{\infty} \Re(y - j\omega)^{-1}\Re(z - j\omega)^{-1}\,d\omega = \begin{cases} \pi^{-1}\Re(y + \overline{z})^{-1}, & \Re z > 0, \\ -\pi^{-1}\Re(y - z)^{-1}, & \Re z < 0. \end{cases}$$

Observe that we always, without loss of generality, take $\Re y > 0$ (otherwise we can replace the integration variable ω by $-\omega$).

Proof The case where $\Re z < 0$ can be reduced to the case where $\Re z > 0$ since

$$\Re(y - j\omega)^{-1}\Re(z - j\omega)^{-1} = \frac{\Re y\,\Re z}{|y - j\omega|^2|z - j\omega|^2}$$

and

$$\begin{aligned} |z - j\omega|^2 &= |\overline{z} + j\omega|^2 = |-\overline{z} - j\omega|^2, \\ \Re z &= -\Re(-\overline{z}). \end{aligned}$$

Thus, we may suppose that $\Re z > 0$. Then

$$\begin{aligned} &\frac{1}{\pi^2}\int_{-\infty}^{\infty} \Re(y - j\omega)^{-1}\Re(z - j\omega)^{-1}\,d\omega \\ &= \frac{1}{\pi^2}\int_{-\infty}^{\infty} \frac{\Re y\,\Re z}{|y - j\omega|^2|z - j\omega|^2}\,d\omega \\ &= \frac{1}{\pi^2}\int_{-\infty}^{\infty} \frac{\Re y\,\Re z}{|\omega + jy|^2|\omega + jz|^2}\,d\omega \\ &= \frac{1}{\pi^2}\int_{-\infty}^{\infty} \frac{\Re y \Re z}{(\omega + jy)(\omega - j\overline{y})(\omega + jz)(\omega - j\overline{z})}\,d\omega. \end{aligned}$$

The last integral can be evaluated (e.g., by residue calculus) to

$$\pi^{-1}\Re(y + \overline{z})^{-1} = \pi^{-1}\Re(z + \overline{y})^{-1}.$$

□

Lemma 10.5.3 *Let ν be a nonnegative measure on a measure space Π, and let k be nonnegative and measurable on $\Pi \times \Pi$. If it is true for almost all*

$(y, z) \in \Pi \times \Pi$ and some finite constant c that

$$\int_{\Pi} k(s, y)k(s, z)\nu(ds) \le c(k(y, z) + k(z, y)),$$

then

$$Q := \iint_{\Pi\times\Pi} k(y, z)\psi(y)\psi(z)\nu(dy)\nu(dz) \le 2c$$

for all nonnegative $\psi \in L^2(\nu)$ with $\|\psi\|_{L^2(\nu)} \le 1$ for which $Q < \infty$.

Proof By Hölder's inequality,

$$\begin{aligned}
Q^2 &= \left(\int_{\Pi} \psi(s)\left(\int_{\Pi} k(s, y)\psi(y)\nu(dy)\right)\nu(ds)\right)^2 \\
&\le \|\psi\|_{L^2(\nu)}^2 \int_{\Pi}\left(\int_{\Pi} k(s, y)\psi(y)\nu(dy)\right)^2 \nu(ds) \\
&\le \iiint_{\Pi\times\Pi\times\Pi} k(s, y)k(s, z)\psi(y)\psi(z)\nu(dy)\nu(dz)\nu(ds) \\
&= \iint_{\Pi\times\Pi}\left(\int_{\Pi} k(s, y)k(s, z)\nu(ds)\right)\psi(y)\psi(z)\nu(dy)\nu(dz) \\
&\le c\iint_{\Pi\times\Pi} \big(k(y, z) + k(z, y)\big)\psi(y)\psi(z)\nu(dy)\nu(dz) \\
&= 2c\iint_{\Pi\times\Pi} \psi(y)\psi(z)k(y, z)\nu(dy)\nu(dz) \\
&= 2cQ.
\end{aligned}$$

Divide by Q to get the conclusion of Lemma 10.5.3. □

Proof of Theorem 10.5.1. It is clear from the remarks made above that (i) ⇒ (ii) ⇒ (iii) ⇒ (iv) and that $c \le a \le \|\mathcal{H}\|^2_{\mathcal{B}(H^2;L^2(\nu))} \le \|\mathcal{H}\|^2_{\mathcal{B}(L^2(j\mathbb{R});L^2(\nu))}$.

(iv) ⇒ (v): Let $\lambda = r + j\omega$ with $r > 0$. Then $|z + \overline{\lambda}| = |z + r - j\omega|$. If $z \in D(j\omega, r)$, then $|z - j\omega| \le r$, and

$$|z + \overline{\lambda}| \le |z - j\omega| + r \le 2r.$$

Squaring this inequality we get

$$r^{-1} \le 4r|z + \overline{\lambda}|^{-2}, \quad z \in D(j\omega, r).$$

Thus,

$$\begin{aligned} r^{-1}\nu(D(j\omega,r)\cap\overline{\mathbb{C}}^+) &= r^{-1}\int_{\Re z\ge 0;|z-r|\le r}\nu(dz) \\ &\le 4\int_{\Re z\ge 0;|z-j\omega|\le r} r|z+\overline{\lambda}|^{-2}\nu(dz) \\ &\le 4\int_{0\le\Re z\le\Re\lambda}\Re\lambda|z+\overline{\lambda}|^{-2}\nu(dz)\le 4c. \end{aligned}$$

This proves that (iv) $\Rightarrow$ (v) and that $M \le 4c$.

(v) $\Rightarrow$ (iii): Let $\lambda = r + j\omega$ with $r > 0$, and suppose that $\nu(D(j\omega, s)) \le Ms$ for all $s > 0$. For each $t > 0$ and z with $\Re z \ge 0$, define

$$g(t,z) = \begin{cases} 1, & \text{if } t < |z+\overline{\lambda}|^{-2}, \\ 0, & \text{if } t \ge |z+\overline{\lambda}|^{-2}. \end{cases}$$

Then

$$\int_0^\infty g(t,z)\,dt = |z+\overline{\lambda}|^{-2},$$

and we can estimate the integral in (iii) as follows (note that $\Re z \ge 0$ and $|z + \overline{\lambda}|^2 = |z + r - j\omega|^2 < t^{-1}$ imply that $|z - j\omega|^2 < t^{-1}$)

$$\begin{aligned} \int_{\Re z\ge 0}\frac{\Re\lambda}{|z+\overline{\lambda}|^2}\nu(dz) &= r\int_{\Re z\ge 0}\int_0^\infty g(t,z)\,dt\,\nu(dz) \\ &= r\int_0^\infty\int_{\Re z\ge 0} g(t,z)\nu(dz)\,dt \\ &= r\int_0^\infty \nu\big(\{z \mid |z+\overline{\lambda}|^{-2} > t\}\big)\,dt \\ &= r\int_0^{1/r^2} \nu\big(\{z \mid |z+\overline{\lambda}|^{2} < t^{-1}\}\big)\,dt \\ &\le r\int_0^{1/r^2} \nu\big(\{z \mid |z-j\omega|^{2} < t^{-1}\}\big)\,dt \\ &= r\int_0^{1/r^2} \nu\big(D(j\omega,t^{-1/2})\big)\,dt \\ &\le r\int_0^{1/r^2} Mt^{-1/2}\,dt \\ &= 2M. \end{aligned}$$

Thus (v) $\Rightarrow$ (iii) and $a \le 2M$.

(iv) $\Rightarrow$ (i): We begin with the case where $\nu(j\mathbb{R}) = 0$, and study the (formal) adjoint $\mathcal{H}^*$ of $\mathcal{H}$, given by

$$(\mathcal{H}^*\psi)(j\omega) = \frac{r}{\pi}\int_{\Re z>0} \Re(z - j\omega)^{-1}\psi(z)\nu(dz). \tag{10.5.1}$$

We claim that the integral in (10.5.1) converges (absolutely) for almost all $\omega \in \mathbb{R}$, that $\mathcal{H}^* \in \mathcal{B}(L^2(\nu); L^2(j\mathbb{R}))$ and that $\|\mathcal{H}^*\|^2_{\mathcal{B}(L^2(\nu);L^2(j\mathbb{R})} \le 32c$. By passing to adjoints we then get $\mathcal{H} \in \mathcal{B}(L^2(j\mathbb{R}; L^2(\nu)))$ and $\|\mathcal{H}\|^2_{\mathcal{B}(L^2(j\mathbb{R});L^2(\nu))} \le 32c$ (once we know that $\mathcal{H}^*$ is bounded it follows from Fubini's theorem that the adjoint of $\mathcal{H}^*$ is $\mathcal{H}$).

Suppose first that ψ is continuous and supported on a compact subset of the *open* right half-plane Π, with $\|\psi\|_{L^2(\nu)} \le 1$. The ν-measure of the support of ψ is finite (this follows from (v)), hence we know that the integral in (10.5.1) converges (absolutely) in this case, and the resulting function (is continuous and) belongs to $L^2(j\mathbb{R})$. The L^2-norm of this function is given by (cf. Lemma 10.5.2)

$$\begin{aligned}
\|\mathcal{H}^*\psi\|^2_{L^2(j\mathbb{R})} &= \frac{1}{\pi^2}\int_{-\infty}^{\infty}\iint_{\Pi\times\Pi} \Re(y - j\omega)^{-1}\Re(z - j\omega)^{-1} \\
&\qquad\qquad \times \psi(y)\overline{\psi}(z)\nu(dy)\nu(dz)\,d\omega \\
&= \iint_{\Pi\times\Pi}\frac{1}{\pi^2}\int_{-\infty}^{\infty} \Re(y - j\omega)^{-1}\Re(z - j\omega)^{-1}\,d\omega \\
&\quad \times \psi(y)\overline{\psi}(z)\nu(dy)\nu(dz) \\
&= \frac{1}{\pi}\iint_{\Pi\times\Pi} \Re(y + \overline{z})^{-1}\psi(y)\overline{\psi}(z)\nu(dy)\nu(dz) \\
&\le \frac{1}{\pi}\iint_{\Pi\times\Pi} \Re(y + \overline{z})^{-1}|\psi(y)||\psi(z)|\nu(dy)\nu(dz) \\
&\le \frac{1}{\pi}\Big(\iint_{\Re y\le\Re z} + \iint_{\Re y>\Re z}\Big) \\
&\quad \times \Re(y + \overline{z})^{-1}|\psi(y)||\psi(z)|\nu(dy)\nu(dz) \\
&\le 2Q
\end{aligned}$$

where

$$\begin{aligned}
Q &= \iint_{\Pi\times\Pi} k(y,z)|\psi(y)||\psi(z)|\nu(dy)\nu(dz) \\
k(y,z) &= \begin{cases} \frac{1}{\pi}\Re(y + \overline{z})^{-1}, & \Re y \le \Re z, \\ 0, & \Re y > \Re z. \end{cases}
\end{aligned} \tag{10.5.2}$$

We have for all $s, y, z \in \Pi$,

$$|y + \overline{z}| \le |y + \overline{s} - \overline{s} + \overline{z}| \le |s + \overline{y}| + |s - z| \le |s + \overline{y}| + |s + \overline{z}|,$$

hence

$$\frac{1}{|y+\overline{z}|} \geq \frac{1}{|s+\overline{y}|+|s+\overline{z}|} = \frac{|s+\overline{y}|^{-1}|s+\overline{z}|^{-1}}{|s+\overline{y}|^{-1}+|s+\overline{z}|^{-1}}$$

and

$$\frac{1}{|s+\overline{y}|^2|s+\overline{z}|^2} \leq \frac{2}{|y+\overline{z}|^2}\Big(|s+\overline{y}|^{-2}+|s+\overline{z}|^{-2}\Big).$$

This implies that the kernel k defined in (10.5.2) satisfies, for $\Re y \leq \Re z$,

$$\begin{aligned}\int_{\Pi} k(s,y)k(s,z)\nu(ds) &\leq \frac{1}{\pi^2}\int_{\Re s\leq \Re y} \frac{\Re(s+\overline{y})\Re(s+\overline{z})}{|s+\overline{y}|^2|s+\overline{z}|^2}\nu(ds)\\ &\leq \frac{4}{\pi^2}\frac{\Re y\,\Re(y+\overline{z})}{|y+\overline{z}|^2}\int_{\Re s\leq\Re y}\Big(\frac{1}{|s+\overline{y}|^2}+\frac{1}{|s+\overline{z}|^2}\Big)\nu(ds)\\ &\leq \frac{4}{\pi^2}\frac{\Re(y+\overline{z})}{|y+\overline{z}|^2}\int_{\Re s\leq\Re y}\Big(\frac{\Re\overline{y}}{|s+\overline{y}|^2}+\frac{\Re\overline{z}}{|s+\overline{z}|^2}\Big)\nu(ds)\\ &\leq 8ck(y,z) \leq 8c(k(y,z)+k(z,y)),\end{aligned}$$

where c is the constant in (iv). By Lemma 10.5.3,

$$Q \leq 16c,$$

hence

$$\|\mathcal{H}^*\psi\|^2_{L^2(j\mathbb{R})} \leq 32c. \tag{10.5.3}$$

Now let us still suppose that $\nu(j\mathbb{R})=0$, but that ψ is an arbitrary function in $L^2(\nu)$ with $\|\psi\|_{L^2(\nu)} \leq 1$. Then by (10.5.3) and the density of functions with compact support in $L^2(\nu)$, we find that (10.5.3) still holds. This proves that $\mathcal{H}^* \in \mathcal{B}(L^2(\nu);L^2(j\mathbb{R}))$ and that $\|\mathcal{H}^*\|^2_{\mathcal{B}(L^2(\nu);L^2(j\mathbb{R})} \leq 32c$, hence $\mathcal{H} \in \mathcal{B}(L^2(j\mathbb{R};L^2(\nu))) \gg$ and $\|\mathcal{H}\|^2_{\mathcal{B}(L^2(j\mathbb{R});L^2(\nu))} \leq 32c$.

We still have to remove the assumption that $\nu(j\mathbb{R})=0$. Fix $\epsilon>0$, and study the shifted operator

$$(\mathcal{H}_\epsilon\phi)(z) = (\mathcal{H}\phi)(z+\epsilon), \quad \Re z \geq 0.$$

Clearly

$$\begin{aligned}\|\mathcal{H}_\epsilon\phi\|^2_{L^2(\nu)} &= \int_{\overline{\mathbb{C}}^+} |(\mathcal{H}_\epsilon\phi)(z)|^2\nu(dz)\\ &= \int_{\overline{\mathbb{C}}^+} |(\mathcal{H}\phi)(z+\epsilon)|^2\nu(dz)\\ &= \int_{\Re z\geq\epsilon} |(\mathcal{H}\phi)(z)|^2\nu_\epsilon(dz),\end{aligned}$$

where ν_ϵ is the shifted measure $\nu_\epsilon(E) = \nu((E-\epsilon)\cap\overline{\mathbb{C}}^+)$ for each Borel subset E of $\overline{\mathbb{C}}^+$. Observe that ν_ϵ is supported on $\Re z \geq \epsilon$, hence the version of

Theorem 10.5.1 that we have proved above can be applied to this measure. For each λ with $\Re\lambda > 0$, the function $\Re(z+\overline{\lambda})^{-1}$ is decreasing in $\Re z$, hence it follows from (iv) that (iv) holds with the original constant c if we replace ν by ν_ϵ, for all $\epsilon > 0$. Thus, $\mathcal{H}_\epsilon \in \mathcal{B}(L^2(j\mathbb{R}); L^2(\nu))$ and

$$\|\mathcal{H}_\epsilon\phi\|_{L^2(\nu)} \le 32c\|\phi\|_{L^2(j\mathbb{R})}.$$

Let $\epsilon \downarrow 0$. Then $(\mathcal{H}_\epsilon\phi)(z) \to (\mathcal{H}\phi)(z)$ for all z with $\Re z > 0$ and almost all z with $\Re z = 0$; hence, by Fatou's lemma, $\mathcal{H} \in \mathcal{B}(L^2(j\mathbb{R}); L^2(\nu))$ and

$$\|\mathcal{H}\phi\|_{L^2(\nu)} \le 32c\|\phi\|_{L^2(j\mathbb{R})}.$$

□

Definition 10.5.4 A positive Borel measure ν on $\mathbb{C}$ is a *Carleson measure* if ν is supported on the closed right half-plane $\overline{\mathbb{C}}^+ = \Re z \ge 0$ and the equivalent conditions in Theorem 10.5.1 are satisfied.

10.6 Admissible control and observation operators for diagonal and normal semigroups

We now continue our study of diagonal and normal systems begun in Section 4.9. This time we take all the spaces (Y, X, U) to be Hilbert spaces, and require the system to be L^2-well-posed. In this case it is possible to employ Theorem 10.5.1 to derive necessary and sufficient conditions for the admissibility of a control operator B with a scalar input space, and for the admissibility of an observation operator C with a scalar output space. Some necessary conditions for the admissibility of B and C in the multi-input multi-output cases will also be given. We begin with the (notationally) simpler case where the semigroup is diagonal.

Theorem 10.6.1 *Let $\mathfrak{A}$ be a diagonal semigroup on the separable Hilbert space X generated by the operator A with eigenvalues λ_n and eigenvectors ϕ_n, $n = 1, 2, 3, \ldots$. Define X_{-1} and X^*_{-1} as in Section 4.9.*

(i) *A vector $b = \sum_{n=1}^{\infty} b_n\phi_n \in X_{-1}$ defines an L^2-admissible and stable control operator $B\colon u \mapsto bu$ for $\mathfrak{A}$ (with the scalar input space $U = \mathbb{C}$) if and only if $b_n = 0$ whenever $\Re\lambda_n \ge 0$ and the measure μ which consists of point masses of size $|b_n|^2$ at the points $-\lambda_n$ is a Carleson measure on Π. In this case the norm of $\mathfrak{B}$ as an operator in $\mathcal{B}(L^2(\mathbb{R}^-); X)$ is equal to the norm of the embedding operator $H^2 \to L^2(\mu)$ in Theorem 10.5.1(ii).*

(ii) *A vector* $c^* = \sum_{n=1}^{\infty} c_n\phi_n \in X_{-1}^*$ *defines an* L^2*-admissible and stable observation operator* $C\colon x \mapsto cx = \langle x, c^*\rangle_{(X_1, X_{-1}^*)}$ *(with the scalar output space* $Y = \mathbb{C}$*) if and only if* $c_n = 0$ *whenever* $\Re\lambda_n \geq 0$ *and the measure* ν *which consists of point masses of size* $|c_n|^2$ *at the points* $-\overline{\lambda}_n$ *is a Carleson measure on* Π*. In this case the norm of* $\mathfrak{C}$ *as an operator in* $\mathcal{B}(X; L^2(\mathbb{R}^+))$ *is equal to the norm of the embedding operator* $H^2 \to L^2(\mu)$ *in Theorem 10.5.1(ii).*

Before proving this theorem, let take at closer look at the coefficients b_n and c_n. Recall from Section 4.9 that the conditions $b \in X_{-1}$ and $c^* \in X_{-1}^*$ are equivalent to

$$\|b\|_{X_{-1}}^2 = \sum_{n=1}^{\infty}(1 + |\lambda_n|^2)^{-1}|b_n|^2 < \infty, \quad b_n = \phi_n^* b,$$

$$\|c^*\|_{X_{-1}}^2 = \sum_{n=1}^{\infty}(1 + |\lambda_n|^2)^{-1}|c_n|^2 < \infty, \quad c_n = \phi_n^* c^*,$$

where the notation is the same as in Section 4.9. We recall from Theorems 4.2.1 and 4.4.2 that the conditions $b \in X_{-1}$ and $c^* \in X_{-1}^*$ are *necessary* for the admissibility of the control operator $B\colon u \mapsto bu$ and the observation operator $C\colon x \mapsto \langle x, c^*\rangle$.

Proof of Theorem 10.6.1. It suffices to prove, e.g., (i), because (ii) then follows by duality (see Theorems 6.2.3 and 6.2.13).

By Theorem 4.2.1(i) (and Definition 2.5.6(ii)) the operator B in (i) is an L^2-admissible and stable observation operator for $\mathfrak{A}$ if and only if it is true that $\int_{-\infty}^0 \mathfrak{A}^{-s} bu(s)\,ds \in X$ for all $u \in L_c^2(\mathbb{R}^-)$, and, in addition, there is a constant $M < \infty$ such that

$$\left|\int_{-\infty}^0 \mathfrak{A}^{-s} bu(s)\,ds\right|_X \leq M\|u\|_{L^2(\mathbb{R}^-)}, \quad u \in L_c^2(\mathbb{R}^-)$$

(it suffices to test this condition for functions u with bounded support since $L_c^2(\mathbb{R}^+)$ is dense in $L^2(\mathbb{R}^+)$). The assumption $b \in X_{-1}$ enables us to repeat the proof of Theorem 4.9.1(viii) to show that

$$\int_{-\infty}^0 \mathfrak{A}^{-s} bu(s)\,ds = \sum_{n=1}^{\infty} \phi_n b_n \int_{-\infty}^0 \mathrm{e}^{-\lambda_n s} u(s)\,ds,$$

where the sum converges in X_{-1} and the terms are orthogonal to each other. The result belongs to X if and only if it has a finite X-norm

$$\left(\sum_{n=1}^{\infty}\left|b_n \int_{-\infty}^0 \mathrm{e}^{-\lambda_n s} u(s)\,ds\right|^2\right)^{1/2} = \left(\sum_{n=1}^{\infty}\left|b_n \int_0^{\infty} \mathrm{e}^{\lambda_n s} u(-s)\,ds\right|^2\right)^{1/2}.$$

We replace u by the reflected $(\mathbf{Я}u)(s) = u(-s)$ to conclude that B is an L^2-admissible and stable control operator for $\mathfrak{A}$ if and only if there exists a finite constant M such that

$$\left(\sum_{n=1}^{\infty} |b_n \hat{u}(-\lambda_n)|^2\right)^{1/2} \leq M \|u\|_{L^2(\mathbb{R}^+)}, \quad u \in L^2_c(\mathbb{R}^+),$$

where $\hat{u}(-\lambda_n) = \int_0^\infty e^{\lambda_n s} u(s)\, ds$ is the Laplace transform of u, evaluated at $-\lambda_n$ ($\hat{u}$ of u is entire since u has bounded support). Moreover, the infimum over all such M gives us $\|\mathfrak{B}\|_{\mathcal{B}(L^2(\mathbb{R}^-);X)}$. Observe that, by the Paley–Wiener Theorem 10.3.4, this can alternatively be written as

$$\left(\sum_{n=1}^{\infty} |b_n \hat{u}(-\lambda_n)|^2\right)^{1/2} \leq M \|\hat{u}\|_{H^2}, \quad u \in L^2_c(\mathbb{R}^+). \tag{10.6.1}$$

Suppose that $b_n = 0$ whenever $\Re\lambda_n > 0$, and that the measure μ defined in (i) is a Carleson measure. The left hand side of (10.6.1) is simply $\|\hat{u}\|^2_{L^2(\mu)}$, hence by Theorem 10.5.1 and Definition 10.5.4, condition (10.6.1) holds with $M = \|\mathcal{H}\|^2_{\mathcal{B}(H^2;L^2(\mu))}$. In particular, we observe that the norm of $\mathfrak{B}$ as an operator in $\mathcal{B}(L^2(\mathbb{R}^-); X)$ is equal to the norm of the embedding operator $H^2 \to L^2(\mu)$.

Conversely, suppose that B is L^2-admissible and stable, i.e., suppose that (10.6.1) holds for some finite M. Then, for each fixed n,

$$|b_n \hat{u}(-\lambda_n)| \leq M \|\hat{u}\|_{H^2}, \quad u \in L^2_c(\mathbb{R}^+),$$

or equivalently (by the Paley–Wiener theorem)

$$\left| b_n \int_0^\infty e^{\lambda_n s} u(s)\, ds \right| \leq M \|u\|_{L^2(\mathbb{R}^+)}, \quad u \in L^2_c(\mathbb{R}^+).$$

By the density of $L^2_c(\mathbb{R}^+)$ in $L^2(\mathbb{R}^+)$, this inequality is true for some fixed finite M for all $u \in L^2_c(\mathbb{R}^+)$ iff it is true for all $u \in L^2(\mathbb{R}^+)$. It is easy to see that this inequality cannot hold if $b_n \neq 0$ and $\Re\lambda_n \geq 0$ (take $u(t) = e^{-\lambda_n t}$ if $\Re\lambda_n > 0$, and $u(t) = (1+t)^{-3/4} e^{-\lambda_n t}$ if $\Re\lambda_n = 0$). Thus, we conclude that $b_n = 0$ whenever $\Re\lambda_n \geq 0$. This means that the terms in the sum on the left-hand side of (10.6.1) where $\Re\lambda_n > 0$ drop out. Since $L^2_c(\mathbb{R}^+)$ is dense in $L^2(\mathbb{R}^+)$, the set of Laplace transforms of functions in $L^2_c(\mathbb{R}^+)$ is dense in H^2, and (10.6.1) then implies that $H^2 \subset L^2(\mu)$, where μ is the measure defined in (ii). Thus, by Theorem 10.5.1(ii) and Definition 10.5.4, μ is a Carleson measure. □

It is possible to reformulate Theorem 10.6.1 in a way which makes no reference to Carleson measures:

Theorem 10.6.2 *Let $\mathfrak{A}$ be a diagonal semigroup on the separable Hilbert space X generated by the operator A. Define X_{-1} and X^*_{-1} as in Section 4.9.*

(i) *A vector $b \in X_{-1}$ defines an L^2-admissible stable control operator $B: u \mapsto bu$ for $\mathfrak{A}$ (with the scalar input space $U = \mathbb{C}$) if and only if*

$$a_b := \sup_{\substack{\lambda \in \rho(A) \\ \Re\lambda > 0}} \pi^{-1} \Re\lambda |(\lambda - A_{|X})^{-1} b|_X^2 < \infty. \tag{10.6.2}$$

Moreover, $a_b \le \|\mathfrak{B}\|^2_{\mathcal{B}(L^2(\mathbb{R}^-);X)} \le 32 a_b$.

(ii) *A vector $c^* \in X^*_{-1}$ defines an L^2-admissible and stable observation operator $C: x \mapsto cx = \langle x, c^* \rangle_{(X_1, X^*_{-1})}$ (with the scalar output space $Y = \mathbb{C}$) if and only if*

$$a_c := \sup_{\substack{\lambda \in \rho(A^*) \\ \Re\lambda > 0}} \pi^{-1} \Re\lambda |(\lambda - A^*_{|X})^{-1} c^*|_X^2 < \infty. \tag{10.6.3}$$

Moreover, $a_c \le \|\mathfrak{C}\|^2_{\mathcal{B}(X;L^2(\mathbb{R}^+))} \le 32 a_c$.

Proof Again it suffices to prove (i), since (ii) then follows by duality. For this we introduce the same notation as in Theorem 10.6.1.

Let $\lambda \in \rho(A)$. The assumption $b \in X_{-1}$ enables us to repeat the proof of Theorem 4.9.1(ii)–(iii) (with X replaced by X_{-1}) to show that

$$(\lambda - A_{|X})^{-1} b = \sum_{n=1}^{\infty} (\lambda - \lambda_n)^{-1} b_n \phi_n.$$

The X-norm of this vector is given by

$$|(\lambda - A_{|X})^{-1} b|_X^2 = \sum_{n=1}^{\infty} |\lambda - \lambda_n|^{-2} |b_n|^2, \quad \lambda \in \rho(A). \tag{10.6.4}$$

Suppose that B is L^2-admissible and stable. Then, by Theorem 10.6.1(i), $b_n = 0$ whenever $\Re\lambda_n \ge 0$, and the measure μ defined there is a Carleson measure. Condition (10.6.2) then follows from (10.6.4) and the fact that μ satisfies condition (iii) in Theorem 10.5.1.

Conversely, suppose that (10.6.2) holds. Then it follows from Lemma 3.2.8(iii) that $b_n = 0$ whenever $\Re\lambda_n \ge 0$. Moreover, if we define the measure μ as in Theorem 10.6.1(i), then, because of (10.6.4), (10.6.2) is equivalent to condition (iii) in Theorem 10.5.1. Thus, μ is a Carleson measure, and by Theorem 10.6.1(i), B is L^2-admissible and stable.

The given bound on $\|\mathfrak{B}\|$ follows from Theorems 10.5.1 and 10.6.1 (since a_b is equal to the constant a in Theorem 10.5.1(iii)). □

Theorem 10.6.3 *Let $\mathfrak{A}$ be a normal semigroup on the Hilbert space X with generator A and spectral resolution E (cf. Example 3.3.6). Define X_{-1} and X^*_{-1} as in Section 4.9.*

(i) *A vector* $b \in X_{-1}$ *defines an* L^2*-admissible and stable control operator* $B\colon u \mapsto bu$ *for* $\mathfrak{A}$ *(with the scalar input space* $U = \mathbb{C}$*) if and only if the measure* $\langle Eb, b\rangle$ *is a Carleson measure (supported on* $\overline{\mathbb{C}}^+$*).*

(ii) *A vector* $c^* \in X^*_{-1}$ *defines an* L^2*-admissible and stable observation operator* $C\colon x \mapsto cx = \langle x, c^*\rangle_{(X_1, X^*_{-1})}$ *(with the scalar output space* $Y = \mathbb{C}$*) if and only if the measure* $\langle \widetilde{E}c^*, c^*\rangle$ *is a Carleson measure (supported on* $\overline{\mathbb{C}}^+$*), where* $\widetilde{E}$ *is the spectral resolution of* A^**.*

The proof of this theorem is the same as the proof of Theorem 10.6.1 except for the fact that all sums are replaced by integrals. We leave the necessary modifications to the reader.

Theorem 10.6.4 *Let* $\mathfrak{A}$ *be a normal semigroup on the Hilbert space* X *with generator* A *and spectral resolution* E *(cf. Example 3.3.6). Define* X_{-1} *and* X^*_{-1} *as in Section 4.9.*

(i) *A vector* $b \in X_{-1}$ *defines an* L^2*-admissible stable control operator* $B\colon u \mapsto bu$ *for* $\mathfrak{A}$ *(with the scalar input space* $U = \mathbb{C}$*) if and only if*

$$a_b := \sup_{\substack{\lambda\in\rho(A)\\ \Re\lambda>0}} \pi^{-1}\Re\lambda|(\lambda - A_{|X})^{-1}b|^2_X < \infty. \tag{10.6.5}$$

Moreover, $a_b \le \|\mathfrak{B}\|^2_{\mathcal{B}(L^2(\mathbb{R}^-);X)} \le 32a_b$.

(ii) *A vector* $c^* \in X^*_{-1}$ *defines an* L^2*-admissible and stable observation operator* $C\colon x \mapsto cx = \langle x, c^*\rangle_{(X_1, X^*_{-1})}$ *(with the scalar output space* $Y = \mathbb{C}$*) if and only if*

$$a_c := \sup_{\substack{\lambda\in\rho(A^*)\\ \Re\lambda>0}} \pi^{-1}\Re\lambda|(\lambda - A^*_{|X})^{-1}c^*|^2_X < \infty. \tag{10.6.6}$$

Moreover, $a_c \le \|\mathfrak{C}\|^2_{\mathcal{B}(X;L^2(\mathbb{R}^+))} \le 32a_c$.

We leave this proof, too, to the reader. (It is again a question of replacing sums by integrals.)

10.7 Admissible control and observation operators for contraction semigroups

Theorem 10.6.4 can be extended to the case where the semigroup is a contraction semigroup.

Theorem 10.7.1 *Let* $\mathfrak{A}$ *be a contraction semigroup on the Hilbert space* X *with generator* A*. Define* X_{-1} *and* X^*_{-1} *as in Section 4.9.*

(i) *A vector $b \in X_{-1}$ defines an L^2-admissible stable control operator $B\colon u \mapsto bu$ for $\mathfrak{A}$ (with the scalar input space $U = \mathbb{C}$) if and only if*

$$\sup_{\Re\lambda>0} \Re\lambda |(\lambda - A_{|X})^{-1}b|_X^2 < \infty. \tag{10.7.1}$$

(ii) *A vector $c^* \in X_{-1}^*$ defines an L^2-admissible and stable observation operator $C\colon x \mapsto cx = \langle x, c^*\rangle_{(X_1, X_{-1}^*)}$ (with the scalar output space $Y = \mathbb{C}$) if and only if*

$$\sup_{\Re\lambda>0} \Re\lambda |(\lambda - A_{|X}^*)^{-1}c^*|_X^2 < \infty. \tag{10.7.2}$$

The proof of part (i) of this theorem can be reduced to the proof of part (ii) by duality, and part (ii) can, in turn, be reduced to two special cases: the normal case which we already encountered in Theorem 10.6.4, and the case where $\mathfrak{A}$ is the (outgoing) left-shift τ_- on $L^2(\mathbb{R}^-; U)$.

Lemma 10.7.2 *Let U be a Hilbert space, let τ_- be the left-shift semigroup on $\mathcal{U} = L^2(\mathbb{R}^-; U)$ with generator $\frac{\mathrm{d}}{\mathrm{d}s}_-$ (cf. Example 3.2.3), and denote*

$$\mathcal{U}_1 = \mathcal{D}\left(\frac{\mathrm{d}}{\mathrm{d}s}_-\right) = W_0^{1,2}(\overline{\mathbb{R}}^-; U) = \{x \in W^{1,2}(\overline{\mathbb{R}}^-; U) \mid x(0) = 0\}.$$

Then $c \in (\mathcal{U}_1)^*$ is an L^2-admissible stable observation operator (with the scalar output space $Y = \mathbb{C}$) if and only if (10.7.2) holds with A replaced by $\frac{\mathrm{d}}{\mathrm{d}s}_-$.

Before proving this lemma, let us formulate and prove some auxiliary results. In these auxiliary lemmas it will be convenient to distinguish between the Hilbert space U and its dual $U^* = \mathcal{B}(U; \mathbb{C})$. (Recall that there is a conjugate-linear correspondence between these two spaces.)

Lemma 10.7.3 *Every bounded linear functional $c\colon W_0^{1,2}(\overline{\mathbb{R}}^-; U) \to \mathbb{C}$ (where U is a Hilbert space) has a unique representation of the form*

$$cu = \int_{-\infty}^0 c_1(-s)(u(s) - \dot{u}(s))\,ds, \qquad u \in W_0^{1,2}(\overline{\mathbb{R}}^-; U), \tag{10.7.3}$$

for some $c_1 \in L^2(\mathbb{R}^+; U^*)$ (where $U^* = \mathcal{B}(U; \mathbb{C})$). Conversely, every $c_1 \in L^2(\mathbb{R}^+; U^*)$ defines via (10.7.3) a (unique) bounded linear functional on $W_0^{1,2}(\overline{\mathbb{R}}^-; U)$.

Proof Denote $\mathcal{U} = L^2(\mathbb{R}^-; U)$ and $\mathcal{U}_1 = W_0^{1,2}(\overline{\mathbb{R}}^-; U)$, and let $\frac{\mathrm{d}}{\mathrm{d}s}_-$ be the generator of the left-shift semigroup τ_- on $L^2(\mathbb{R}^-; U)$. Then $1 - \frac{\mathrm{d}}{\mathrm{d}s}_-$ maps $\mathcal{U}$ one-to-one onto $\mathcal{U}_1$ with a bounded inverse (cf. Example 3.3.1). Thus, $c\big(1 - \frac{\mathrm{d}}{\mathrm{d}s}_-\big)^{-1}$ defines a bounded linear functional on $\mathcal{U}$, which can be written in the form $c(1 - \frac{\mathrm{d}}{\mathrm{d}s}_-)^{-1}u = \langle u, c_0\rangle$ for a unique $c_0 \in \mathcal{U}$ and all $u \in \mathcal{U}$. Thus, for every

$u \in \mathcal{U}_1$, we have

$$cu = c\Big(1 - \frac{\mathrm{d}}{\mathrm{d}s_-}\Big)^{-1}\Big(1 - \frac{\mathrm{d}}{\mathrm{d}s_-}\Big)u = \Big\langle\Big(1 - \frac{\mathrm{d}}{\mathrm{d}s_-}\Big)u, c_0\Big\rangle = \langle u - \dot{u}, c_0\rangle.$$

From here we get (10.7.3) by defining $c_1(s)u = \langle u, c_0(-s)\rangle$ for $u \in U$ and $s \geq 0$.

The proof of the converse claim is trivial. □

Lemma 10.7.4 *Let c be a bounded linear functional on $W_0^{1,2}(\overline{\mathbb{R}}^-; U)$, and define c_1 as in Lemma 10.7.3. For each $u \in W_{0,\mathrm{loc}}^{1,2}(\mathbb{R}; U)$, define $\mathfrak{D}u \in BC_{0,\mathrm{loc}}(\mathbb{R}; U)$ by*

$$(\mathfrak{D}u)(t) = \int_{-\infty}^{t} c_1(t-s)(u(s) - \dot{u}(s))\,ds, \qquad t \in \mathbb{R}.$$

Then $\mathfrak{D}$ is time-invariant and causal, and for all $u \in W_0^{1,2}(\overline{\mathbb{R}}^-; U)$,

$$(\mathfrak{D}u)(t) = c\tau_-^t u, \qquad t \geq 0.$$

Moreover, c is an L^2-admissible stable output operator for τ_- if and only if there is a finite constant K such that

$$\int_0^\infty |(\mathfrak{D}u)(t)|^2\,dt \leq K^2 \|u\|^2_{L^2(\mathbb{R}^-)}, \qquad u \in W_0^{1,2}(\overline{\mathbb{R}}^-; U).$$

In this case, if we denote the output operator induced by c by $\mathfrak{C}$, then $\mathfrak{C}u = \pi_+\mathfrak{D}\pi_- u$ for all $u \in W_0^{1,2}(\overline{\mathbb{R}}^-; U)$, and

$$\|\mathfrak{C}\| = \sup_{\substack{u \in W_0^{1,2}(\overline{\mathbb{R}}^-; U)\\ \|u\|_{L^2(\mathbb{R}^-)} \leq 1}} \|\mathfrak{D}u\|_{L^2(\mathbb{R}^+)}.$$

Proof The causality of $\mathfrak{D}$ is obvious. To show that $\mathfrak{D}$ is time-invariant, and that $(\mathfrak{D}u)(t) = c\tau_-^t u$ for all $u \in W_0^{1,2}(\overline{\mathbb{R}}^-; U)$ and all $t \geq 0$ it suffices to make a change of variable in the integral defining $\mathfrak{D}\tau^t u$. The remaining claims are obvious (see Definition 10.1.1(ii)). □

To proceed further we have to move over to the frequency domain, and to introduce the transfer function of the operator $\mathfrak{D}$ in Lemma 10.7.4.

Lemma 10.7.5 *Let c be a bounded linear functional on $W_0^{1,2}(\overline{\mathbb{R}}^-; U)$, and define c_1 and $\mathfrak{D}$ as in Lemmas 10.7.3 and 10.7.4. Then, for each $u \in W^{1,2}(\mathbb{R}; U)$ whose support is bounded to the left, the bilateral Laplace transform of $\mathfrak{D}u$ is given by*

$$\widehat{\mathfrak{D}u}(z) = \widehat{\mathfrak{D}}(z)\hat{u}(z), \qquad \Re z > 0,$$

where $\hat{u}$ is the bilateral Laplace transform of u, $\widehat{\mathfrak{D}}(z) = (1 - z)\hat{c}_1(z)$ for $\Re z > 0$, and $\hat{c}_1 = \int_0^\infty \mathrm{e}^{-zs} c_1(s)\,ds$ is the Laplace transform of c_1 (cf. Definition 3.12.1).

Proof To prove this it suffices to use Fubini's theorem, change the integration variable, and do an integration by parts: for all $\Re z > 0$,

$$\begin{aligned}\widehat{\mathfrak{D}}u(z) &= \int_{-\infty}^{\infty} \mathrm{e}^{-zt} \int_{-\infty}^{t} c_1(t-s)(u(s) - \dot{u}(s))\,ds\,dt \\ &= \int_{-\infty}^{\infty} \int_{s}^{-\infty} \mathrm{e}^{-zt} c_1(t-s)\,dt\,(u(s) - \dot{u}(s))\,ds \\ &= \int_{0}^{-\infty} \mathrm{e}^{-zv} c_1(v)\,dv \int_{-\infty}^{\infty} \mathrm{e}^{-zs}(u(s) - \dot{u}(s))\,ds \\ &= \hat{c}_1(z)(1-z)\hat{u}(z).\end{aligned}$$

□

Definition 10.7.6 The function $\widehat{\mathfrak{D}}$ in Lemma 10.7.5 is called the *transfer function* of the operator $\mathfrak{D}$.

Lemma 10.7.7 *For all $\Re\lambda > 0$ and $u \in L^2(\mathbb{R}^-; U)$, we have*

$$c\Big(\lambda - \frac{\mathrm{d}}{\mathrm{d}s}_-\Big)^{-1} u = \int_{-\infty}^{0} d_\lambda(-s)u(s)\,ds,$$

where $d_\lambda \in L^2(\mathbb{R}^+; U^*)$ has the Laplace transform

$$\hat{d}_\lambda(z) = (\lambda - z)^{-1}(\widehat{\mathfrak{D}}(z) - \widehat{\mathfrak{D}}(\lambda)), \qquad \Re z > 0.$$

In particular, the norm of $c\big(\lambda - \frac{\mathrm{d}}{\mathrm{d}s}_-\big)^{-1}$ is given by

$$\Big\| c\Big(\lambda - \frac{\mathrm{d}}{\mathrm{d}s}_-\Big)^{-1} \Big\| = \|d_\lambda\|_{L^2(\mathbb{R}^+;U^*)}.$$

Proof By Example 3.3.2 and Lemma 10.7.3,

$$\begin{aligned}c\Big(\lambda - \frac{\mathrm{d}}{\mathrm{d}s}_-\Big)^{-1} u &= \int_{-\infty}^{0} c_1(-t)(u(t) + \dot{u}(t))\,dt \\ &= \int_{-\infty}^{0} c_1(-t)\Big(u(t) + (1-\lambda)\int_{t}^{0} \mathrm{e}^{\lambda(t-s)}u(s)\,ds\Big)\,dt \\ &= \int_{-\infty}^{0} c_1(-t)u(t)\,dt \\ &\quad + (1-\lambda)\int_{-\infty}^{0}\int_{-\infty}^{s} c_1(-t)\mathrm{e}^{\lambda(t-s)}u(s)\,dt\,ds \\ &= \int_{-\infty}^{0} d_\lambda(-s)u(s)\,ds,\end{aligned}$$

where

$$d_\lambda(s) = c_1(s) + (1-\lambda)\int_{s}^{\infty} \mathrm{e}^{\lambda(s-t)}c_1(t)\,dt.$$

The Laplace transform of d_λ is, for $\Re z > 0$,

$$\begin{aligned}
\hat{d}_\lambda(z) &= \hat{c}_1(z) + (1-\lambda)\int_0^\infty \mathrm{e}^{-zs}\int_s^\infty \mathrm{e}^{\lambda(s-t)}c_1(t)\,dt\,ds \\
&= \hat{c}_1(z) + (1-\lambda)\int_0^\infty \mathrm{e}^{-\lambda t}\int_0^t \mathrm{e}^{(\lambda-z)s}\,ds\,c_1(t)\,dt \\
&= \hat{c}_1(z) + \frac{1-\lambda}{\lambda-z}\int_0^\infty (\mathrm{e}^{-zt}-\mathrm{e}^{-\lambda t})c_1(t)\,dt \\
&= \hat{c}_1(z) + \frac{1-\lambda}{\lambda-z}(\hat{c}_1(z)-\hat{c}_1(\lambda)) \\
&= \frac{(1-z)\hat{c}_1(z)+(1-\lambda)\hat{c}_1(\lambda)}{\lambda-z} \\
&= (\lambda-z)^{-1}(\widehat{\mathfrak{D}}(z)-\widehat{\mathfrak{D}}(\lambda)).
\end{aligned}$$

□

In our proof of Lemma 10.7.2 we shall also need the spaces *BMO* and *BMOA* which are defined as follows.

Definition 10.7.8 Let X be a Hilbert space.

(i) The space $BMO(j\mathbb{R};X)$ consists of all $f \in L^2_{\mathrm{loc}}(j\mathbb{R};X)$ satisfying

$$\|f\|_{BMO} = \sup_I \frac{1}{|I|}\int_I |f(j\omega)-f_I|\,d\omega < \infty,$$

where the supremum is taken over all intervals I of finite length $|I| > 0$, and

$$f_I = \frac{1}{|I|}\int_I f(j\omega)\,d\omega$$

is the mean value of f over I.

(ii) The space $BMOA(j\mathbb{R};X)$ consists of all functions f such that the function $z \mapsto f(z)/(1+z)$ belongs to $H^2(\mathbb{C}^+;X)$ and the corresponding boundary function belongs to $BMO(j\mathbb{R};X)$.

As is well-known, the space $BMO(j\mathbb{R};X)$ is a Banach space (after we identify any two functions which differ by a constant), with the norm given above. To prove that a function belongs to $BMO(j\mathbb{R};X)$ the following simple argument is often useful. Assume that, for each interval I with $0 < |I| < \infty$ there is a constant c_I such that

$$\sup_I \frac{1}{|I|}\int_I |f(j\omega)-c_I|\,d\omega = M < \infty.$$

Then trivially $|f_I - c_I| \le M$, so that, by the triangle inequality,

$$\|f\|_{BMO} \le 2 \sup_I \frac{1}{|I|} \int_I |f(j\omega) - c_I| \, d\omega. \tag{10.7.4}$$

Proof of Lemma 10.7.2 It is easy to show that condition (10.7.2) is always necessary for the L^2-admissibility and stability of c.[1] If c is L^2-admissible and stable, and if we denote the corresponding output map by $\mathfrak{C}$, then, according to Theorem 4.4.2(iv), we have for all $u \in \mathcal{U}$,

$$c(\lambda - A)^{-1}u = \int_0^\infty e^{-\lambda s}(\mathfrak{C}u)(s)\, ds, \qquad \Re\lambda > 0.$$

Thus, by Hölder's inequality,

$$\begin{aligned} \left|c(\lambda - A)^{-1}u\right|^2 &\le \int_0^\infty e^{-\Re\lambda s}\, ds \int_0^\infty |(\mathfrak{C}u)(s)|^2\, ds \\ &\le (\Re\lambda)^{-1} \|\mathfrak{C}\|_{\mathcal{U}^*} \|u\|_{\mathcal{U}}. \end{aligned}$$

Taking the supremum over all $u \in \mathcal{U}$ with $|u| \le 1$ we get

$$\left|c(\lambda - A)^{-1}\right|^2 \le (\Re\lambda)^{-1} \|\mathfrak{C}\|_{\mathcal{U}^*}, \qquad \Re\lambda > 0.$$

Conversely, suppose that (10.7.2) holds. Let c_1 be the function in Lemma 10.7.3, let $\widehat{\mathfrak{D}}(z) = (1 - z)\hat{c}_1(z)$ (as in Lemma 10.7.5), and define $\hat{d}_\lambda$ as in Lemma 10.7.7. Then both $z \mapsto (1 + z)^{-1}\widehat{\mathfrak{D}}(z)$ and $\hat{d}_\lambda$ belong to $H^2(\mathbb{C}^+; U^*)$, so they have boundary functions in $L^2(j\mathbb{R}; U^*)$. We denote these boundary functions simply by $j\omega \mapsto (1 + j\omega)^{-1}\widehat{\mathfrak{D}}(j\omega)$, respectively $j\omega \mapsto \hat{d}_\lambda(j\omega)$. With this notation, it follows from Lemma 10.7.7 that

$$M := \sup_{\Re\lambda > 0} \Re\lambda \|\hat{d}_\lambda\|^2_{L^2(j\mathbb{R}; U^*)} < \infty. \tag{10.7.5}$$

We claim that $\widehat{\mathfrak{D}} \in BMOA(\mathbb{C}^+; B^*)$. To prove this we use (10.7.4) with $f = \widehat{\mathfrak{D}}$ and with a specific choice of c_I. For each $\alpha > 0$ and $\beta \in \mathbb{R}$ we define

$$E_{\alpha,\beta} = \frac{1}{2\alpha} \int_{\beta-\alpha}^{\beta+\alpha} |\widehat{\mathfrak{D}}(j\omega) - \widehat{\mathfrak{D}}(\alpha + j\beta)|\, d\omega,$$

and observe from (10.7.4) that

$$\|\widehat{\mathfrak{D}}\|_{BMO} \le 2 \sup_{\alpha>0,\, \beta\in\mathbb{R}} E_{\alpha,\beta}.$$

We estimate $E_{\alpha,\beta}$ as follows (using the Cauchy–Schwarz inequality

[1] The following computation is identical to the one that we used to prove Proposition 4.4.9.

and (10.7.5)):

$$\begin{aligned} E_{\alpha,\beta} &= \frac{1}{2\alpha}\int_{\beta-\alpha}^{\beta+\alpha} \frac{|\widehat{\mathfrak{D}}(j\omega) - \widehat{\mathfrak{D}}(\alpha + j\beta)|}{|j\omega - (\alpha + j\beta)|}|j\omega - (\alpha + j\beta)|\, d\omega \\ &= \frac{1}{2\alpha}\int_{\beta-\alpha}^{\beta+\alpha} |\hat{d}_{\alpha+j\beta}(j\omega)||j\omega - (\alpha + j\beta)|\, d\omega \\ &\le \frac{1}{2\alpha}\|\hat{d}_{\alpha+j\beta}\|_{L^2(j\mathbb{R};U^*)}\left(\int_{\beta-\alpha}^{\beta+\alpha} |j\omega - (\alpha + j\beta)|^2\, d\omega\right)^{1/2} \\ &= \sqrt{2\alpha/3}\|\hat{d}_{\alpha+j\beta}\|_{L^2(j\mathbb{R};U^*)} \le \sqrt{2M/3}. \end{aligned}$$

This proves our claim that $\widehat{\mathfrak{D}} \in BMOA(\mathbb{C}^+; B^*)$ (and that $\|\widehat{\mathfrak{D}}\|_{BMO} \le 2\sqrt{2M/3}$).

We now invoke Fefferman's theorem, which says that the Hankel operator induced by $\mathfrak{D}$ (i.e., the operator $\mathfrak{C}$ in Lemma 10.7.4) is bounded if (and only if) its transfer function $\widehat{\mathfrak{D}}$ belongs to $BMOA(\mathbb{C}^+; U^*)$.[2] By Lemma 10.7.4, this is equivalent to the L^2-admissibility and stability of c. □

Remark 10.7.9 It follows from the above proof that the *BMO*-norm of $\widehat{\mathfrak{D}}$ is equivalent to the supremum on the left-hand side of (10.7.2) with A replaced by $\frac{\mathrm{d}}{\mathrm{d}s}_-$. This is the same thing as

$$\sup_{\Re\lambda>0} \Re\lambda \|\hat{d}_\lambda\|_{H^2(\mathbb{C}^+;U^*)},$$

and it is also equal to $\sqrt{M}$ with M as in (10.7.5). In the case where $U = \mathbb{C}$ a slightly more general result (which does not require analyticity in $\mathbb{C}^+$) is found in Garnett (1981, Corollary 2.4, p. 234), and the proof given there applies also when U is a general Hilbert space.

We have now arrived at a point where we are able to prove the following limited version of Theorem 10.7.1.

Lemma 10.7.10 *Let $\mathfrak{A}$ be an isometric semigroup on the Hilbert space X with generator A. Define X_1 as in Section 4.9. Then $c \in X_1^*$ is an L^2-admissible stable observation operator (with the scalar output space $Y = \mathbb{C}$) if and only if* (10.7.2) *holds.*

Proof The necessity of (10.7.2) was proved above (see the beginning of the proof of Lemma 10.7.2).

The proof of the sufficiency of (10.7.2) is based on Lemma 10.7.2 and the universal model for isometric semigroups presented in Corollary 11.6.9. This

[2] It is shown in Jacob and Partington (2001, p. 240) how to derive this version of Fefferman's theorem from the version which says that $BMOA(\mathbb{C}^+; U^*)$ can be interpreted as the dual of $H^1(\mathbb{C}^-; U)$. The latter version is found in, e.g., Blasco (1997).

model employs a unitary similarity transformation in the state space of the type discussed in Example 2.3.7. The statement of Lemma 10.7.10 is invariant under such transformations in the following sense. If E is a unitary map of X onto $\widetilde{X}$, then for all $x \in X$, $\|Ex\|_{\widetilde{X}} = \|x\|_X$, and $\widetilde{\mathfrak{A}}^t = E^{-1}\mathfrak{A}^t E$ is a C_0-semigroup (still isometric) on $\widetilde{X}$ with generator $\widetilde{A} = E^{-1}AE$ and resolvent $(\lambda - \widetilde{A})^{-1} = E^{-1}(\lambda - A)^{-1}E$. Thus, if we define $\tilde{c} = cE$, then the conclusion of Lemma 10.7.10 is true for a particular choice of X, $\mathfrak{A}$, and c if and only if it is true for $\widetilde{X}$, $\widetilde{\mathfrak{A}}$, and $\tilde{c}$.

The Wold decomposition of $\mathfrak{A}$ (see Corollary 11.6.9) tells us that there is a subspace $Z \subset X$, a Hilbert space U, and unitary map E of $L^2(\mathbb{R}^-; U)$ onto $Z^\perp$ such that, if we decompose X into $X = \left[\begin{smallmatrix} Z \\ Z^\perp \end{smallmatrix}\right]$, then

$$\mathfrak{A} = \begin{bmatrix} \mathfrak{A}_1^t & 0 \\ 0 & E\tau_- E^{-1} \end{bmatrix},$$

where $\mathfrak{A}_1$ is a unitary semigroup on Z. Let c_1 be the restriction of c to Z, and let c_2 be the restriction of $c\left[\begin{smallmatrix} 1 & 0 \\ 0 & E \end{smallmatrix}\right]$ to $L^2(\mathbb{R}^-; U)$. Then c is an L^2-admissible stable observation operator for $\mathfrak{A}$ if and only if c_1 and c_2 are L^2-admissible and stable observation operators for $\mathfrak{A}_1$ and τ_-, respectively, and (10.7.2) holds if and only if it holds both when c and $\mathfrak{A}$ are replaced by c_1 and $\mathfrak{A}_1$, and when c and $\mathfrak{A}$ are replaced by c_2 and τ_-.

Suppose that (10.7.2) holds. Then, by Theorem 10.6.4(ii), c_1 is an L^2-admissible stable observation operator for $\mathfrak{A}_1$, and by Lemma 10.7.2, c_2 is an L^2-admissible stable observation operator for τ_-. Thus, c is an L^2-admissible stable observation operator for $\mathfrak{A}$. □

Proof of Theorem 10.7.1 It suffices to prove part (ii), since (i) then follows by duality.

The necessity of (10.7.2) was proved above (see the beginning of the proof of Lemma 10.7.2).

The proof of the sufficiency is based on a reduction to the isometric case treated in Lemma 10.7.10. By Theorem 11.4.5(iii), the contraction semigroup $\mathfrak{A}$ on X can be dilated to an isometric semigroup $\widetilde{\mathfrak{A}} = \left[\begin{smallmatrix} \mathfrak{A}_{11} & \mathfrak{A}_{12} \\ 0 & \mathfrak{A} \end{smallmatrix}\right]$ on $\left[\begin{smallmatrix} Z \\ X \end{smallmatrix}\right]$, where Z is some other Hilbert space. We denote the generator of $\widetilde{\mathfrak{A}}$ by $\widetilde{A}$. Trivially, if $\left[\begin{smallmatrix} z \\ x \end{smallmatrix}\right] \in \mathcal{D}(\widetilde{A})$, then $x \in \mathcal{D}(A)$, so we can define an operator $\tilde{c}$ on $\mathcal{D}(\widetilde{A})$ by

$$\tilde{c} = \begin{bmatrix} 0 & c \end{bmatrix}.$$

We claim that $\tilde{c}$ is an L^2-well-posed stable observation operator for $\widetilde{\mathfrak{A}}$ whenever (10.7.2) holds, and prove this as follows.

By Theorem 3.2.9(i), for all $\Re\lambda > 0$,

$$(\lambda - \widetilde{A})^{-1} = \int_0^\infty e^{-\lambda s}\widetilde{\mathfrak{A}}^s\, ds = \begin{bmatrix} R_{11}(\lambda) & R_{12}(\lambda) \\ 0 & R_{22}(\lambda) \end{bmatrix},$$

where $R_{22}(\lambda) = (\lambda - A)^{-1}$ is the resolvent of A. Therefore, for all $\left[\begin{smallmatrix} z \\ x \end{smallmatrix}\right] \in \left[\begin{smallmatrix} Z \\ X \end{smallmatrix}\right]$, $\tilde{c}(\lambda - \widetilde{A})^{-1}\left[\begin{smallmatrix} z \\ x \end{smallmatrix}\right] = c(\lambda - A)^{-1}x$. Thus, if (10.7.2) holds, then it also holds with c and $\mathfrak{A}$ replaced by $\tilde{c}$ and $\widetilde{\mathfrak{A}}$. By Lemma 10.7.10, this implies that $\tilde{c}$ is an L^2-admissible and stable observation operator for $\widetilde{\mathfrak{A}}$, as claimed above.

Let $\mathfrak{C} = \left[\mathfrak{C}_1 \ \mathfrak{C}_2\right]$ be the output map induced by $\tilde{c}$. Then, by Theorem 4.4.2(iv), for all $\left[\begin{smallmatrix} z \\ x \end{smallmatrix}\right] \in \left[\begin{smallmatrix} Z \\ X \end{smallmatrix}\right]$ and all $\Re\lambda > 0$,

$$\begin{aligned} \int_0^\infty e^{-\lambda s}\big((\mathfrak{C}_1 z)(s) + (\mathfrak{C}_1 x)(s)\big)\, ds &= [0 \ c]\begin{bmatrix} R_{11}(\lambda) & R_{12}(\lambda) \\ 0 & R_{22}(\lambda) \end{bmatrix}\begin{bmatrix} z \\ x \end{bmatrix} \\ &= cR_{22}(\lambda)x. \end{aligned}$$

Thus, the Laplace transform of $\mathfrak{C}_1 z$ is identically zero for all $z \in Z$, and this means that $\mathfrak{C}_1 = 0$, i.e., $\mathfrak{C} = \left[0 \ \mathfrak{C}_2\right]$.

We claim that $\mathfrak{C}_2$ is an L^2-well-posed stable output map for $\mathfrak{A}$. Clearly $\mathfrak{C}_2$ maps X into $L^2(\mathbb{R}^+)$, so it suffices to show that $\mathfrak{C}_2\mathfrak{A}^t = \tau_+^t\mathfrak{C}_2$ for all $t \geq 0$. However, this is true since, for all $\left[\begin{smallmatrix} z \\ x \end{smallmatrix}\right] \in \left[\begin{smallmatrix} Z \\ X \end{smallmatrix}\right]$ and all $t \geq 0$,

$$\mathfrak{C}_2\mathfrak{A}^t x = \begin{bmatrix} 0 & \mathfrak{C}_2 \end{bmatrix}\begin{bmatrix} \mathfrak{A}_{11}^t & \mathfrak{A}_{12}^t \\ 0 & \mathfrak{A}^t \end{bmatrix}\begin{bmatrix} z \\ x \end{bmatrix} = \tau_+^t \begin{bmatrix} 0 \ \mathfrak{C}_2 \end{bmatrix}\begin{bmatrix} z \\ x \end{bmatrix} = \tau_+^t\mathfrak{C}_2.$$

This confirms that $\mathfrak{C}_2$ is an L^2-well-posed stable output map for $\mathfrak{A}$.

There is only one more thing left to prove, namely that the observation operator of $\mathfrak{C}_2$ is c. Equivalently, we must show that for all $x \in X_1$, $(\mathfrak{C}_2 x)(t) = c\mathfrak{A}^t x$ for all $t \geq 0$. We know that $\left(\mathfrak{C}\left[\begin{smallmatrix} z \\ x \end{smallmatrix}\right]\right)(t) = c\mathfrak{A}^t x$ for all $\left[\begin{smallmatrix} z \\ x \end{smallmatrix}\right] \in \mathcal{D}(\widetilde{A})$, so it suffices show that for every $x \in X_1$, it is possible to find some $z \in Z$ so that $\left[\begin{smallmatrix} z \\ x \end{smallmatrix}\right] \in \mathcal{D}(\widetilde{A})$. Take $x \in X_1$, $\Re\lambda > 0$, and $x_1 = (\lambda - A)x$, and define

$$\begin{bmatrix} z \\ x_2 \end{bmatrix} = (\lambda - \widetilde{A})^{-1}\begin{bmatrix} 0 \\ x_1 \end{bmatrix} = \begin{bmatrix} R_{11}(\lambda) & R_{12}(\lambda) \\ 0 & R_{22}(\lambda) \end{bmatrix}\begin{bmatrix} 0 \\ x_1 \end{bmatrix}.$$

Then $\left[\begin{smallmatrix} z \\ x \end{smallmatrix}\right] \in \mathcal{D}(\widetilde{A})$, $x_2 = R_{22}(\lambda)x_1 = (\lambda - A)^{-1}x_1 = x$, and $\big(\mathfrak{C}_2 x\big)(t) = \left(\mathfrak{C}\left[\begin{smallmatrix} z \\ x \end{smallmatrix}\right]\right)(t) = c\mathfrak{A}^t x$ for all $t \geq 0$. □

10.8 Admissibility results based on the Lax–Phillips model

According to Corollary 2.7.7, there is a one-to-one correspondence between the class of all L^p-well-posed linear systems and all Lax–Phillips scattering models of type L^p. This means that we can reduce the study of the generators of a well-posed linear system to the study of the generators of the Lax–Phillips semigroup. This way we can obtain, e.g., necessary and sufficient conditions for the admissibility of a control operator B or an observation operator C. For simplicity we here only study the L^p-admissible case with $1 \leq p < \infty$.

Theorem 10.8.1 *Let $\omega \in \mathbb{R}$, $1 \le p < \infty$, and let A be the generator of an ω-bounded C_0 semigroup on X.*

(i) *$B \in \mathcal{B}(U; X_{-1})$ is an L^p-admissible ω-bounded control operator for A if and only if there is a constant $M > 0$ such that, for all $u \in L^p_\omega(\mathbb{R}^+; U)$, $\lambda > \omega$, and $n = 0, 1, 2, \ldots$,*

$$\left| \frac{\partial^n}{\partial \lambda^n} (\lambda - A_{|X})^{-1} B \hat{u}(\lambda) \right|_X \le \frac{M n!}{(\lambda - \omega)^{n+1}} \|u\|_{L^p_\omega(\mathbb{R}^+;U)}. \tag{10.8.1}$$

(ii) *$C \in \mathcal{B}(X_1; Y)$ is an L^p-admissible ω-bounded observation operator for A if and only if there is a constant $M > 0$ such that, for all $x_0 \in X$, $\lambda > \omega$, and $n = 0, 1, 2, \ldots$,*

$$\left(\int_0^\infty \left| \frac{\partial^n}{\partial \lambda^n} \mathrm{e}^{-(\lambda-\omega)t} C(\lambda - A)^{-1} x_0 \right|_Y^p dt \right)^{1/p} \le \frac{M n!}{(\lambda - \omega)^{n+1}} |x_0|_X. \tag{10.8.2}$$

(iii) *Let $B \in \mathcal{B}(U; X_{-1})$ be an L^p-admissible ω-bounded control operator for A (cf. (i)), and define $\mathcal{D}(A\&B)$ as in Definition 10.1.1(iii). Then $C\&D \in \mathcal{B}(\mathcal{D}(A\&B); Y)$ is an L^p-admissible ω-bounded observation/feedthrough operator for A if and only if the operator*

$$Cx = C\&D \begin{bmatrix} x \\ 0 \end{bmatrix}, \qquad x \in X_1.$$

is an admissible ω-bounded observation operator for A (cf. (ii)) and there is a constant $M > 0$ such that, for all $u \in L^p_\omega(\mathbb{R}^+; U)$, $\lambda > \omega$, and $n = 0, 1, 2, \ldots$,

$$\left(\int_0^\infty \left| \frac{\partial^n}{\partial \lambda^n} \mathrm{e}^{-(\lambda-\omega)t} \widehat{\mathfrak{D}}(\lambda) \hat{u}(\lambda) \right|_Y^p dt \right)^{1/p} \le \frac{M n!}{(\lambda - \omega)^{n+1}} \|u\|_{L^p_\omega(\mathbb{R}^+;U)}, \tag{10.8.3}$$

where

$$\widehat{\mathfrak{D}}(\lambda) = C\&D \begin{bmatrix} (\lambda - A_{|X})^{-1} B \\ 1 \end{bmatrix}, \qquad \Re\lambda > \omega.$$

(iv) *The operators $B \in \mathcal{B}(U; X_{-1})$ and $C \in \mathcal{B}(X_1; Y)$ are jointly L^p admissible and ω-bounded iff B is an L^p-admissible ω-bounded control operator for A (cf. (i)), C is an L^p-admissible ω-bounded observation operator for A (cf. (ii)) and there is a constant $M > 0$ such that* (10.8.3) *holds for all $u \in L^p_\omega(\mathbb{R}^+; U)$, $\lambda > \omega$, and $n = 0, 1, 2, \ldots$; this time the function $\widehat{\mathfrak{D}}$ in* (10.8.3) *is given by*

$$\widehat{\mathfrak{D}}(\lambda) = (\alpha - \lambda) C (\lambda - A)^{-1} (\alpha - A_{|X})^{-1} B + D_\alpha,$$

where α with $\Re\alpha > \omega$ and $D_\alpha \in \mathcal{B}(U; Y)$ can be chosen in an arbitrary manner.

Proof Let us first prove the necessity of (10.8.1), (10.8.2), and (10.8.3) for admissibility. If B is an L^p-admissible ω-bounded control operator for A, then $\left[\begin{smallmatrix}\mathfrak{A} & \mathfrak{B}\\ 0 & 0\end{smallmatrix}\right]$ is an ω-bounded L^p-well-posed linear system on (Y, X, U) (where the output space Y is irrelevant). If C is an L^p-admissible ω-bounded observation operator for A, then $\left[\begin{smallmatrix}\mathfrak{A} & 0\\ \mathfrak{C} & 0\end{smallmatrix}\right]$ is an ω-bounded L^p-well-posed linear system on (Y, X, U) (where the input space U is irrelevant). Finally, if $C\&D$ is an L^p-admissible ω-bounded observation operator for A, then $\left[\begin{smallmatrix}\mathfrak{A} & \mathfrak{B}\\ \mathfrak{C} & \mathfrak{D}\end{smallmatrix}\right]$ is an ω-bounded L^p-well-posed linear system on (Y, X, U). Thus, in all cases we get an ω-bounded L^p-well-posed linear system on (Y, X, U). The corresponding Lax–Phillips model $\mathfrak{T}$ of type L^p_ω is an ω-bounded C_0 semigroup. We can therefore apply the Hille–Yosida Theorem 3.4.1 to this semigroup. We denote the generator of the Lax–Phillips model $\mathfrak{T}$ by T, and split the resolvent $(\lambda - T)^{-1}$ into its components (corresponding to the splitting of the state space into $\left[\begin{smallmatrix}\mathcal{Y}\\ X\\ \mathcal{U}\end{smallmatrix}\right] = \left[\begin{smallmatrix}L^p_\omega(\mathbb{R}^-;Y)\\ X\\ L^p(\mathbb{R}^+;U)\end{smallmatrix}\right]$)

$$(\lambda - T)^{-1} = \begin{bmatrix} R_{11}(\lambda) & R_{12}(\lambda) & R_{13}(\lambda)\\ 0 & R_{22}(\lambda) & R_{23}(\lambda)\\ 0 & 0 & R_{33}(\lambda)\end{bmatrix}.$$

The derivatives $\frac{\partial^n}{\partial \lambda^n}$ of this operator satisfy the Hille–Yosida condition in Theorem 3.4.1(ii′) if and only if each of the components $R_{ij}(\lambda)$ satisfies the same condition (as an operator between the appropriate spaces); this follows from the fact that the norm of a 3×3 matrix is equivalent to the maximum of the norms of its elements. The diagonal elements $R_{ii}(\lambda)$, $i = 1, 2, 3$, do satisfy this Hille–Yosida condition, since they are the resolvents of the generators τ_-, A, and τ_+ of ω-bounded C_0 semigroups on $\mathcal{Y}$, X, and $\mathcal{U}$. Conditions (10.8.1) (10.8.2), and (10.8.3) are simply the corresponding Hille–Yosida conditions for the elements $R_{23}(\lambda)$, $R_{12}(\lambda)$, and $R_{13}(\lambda)$, respectively. This proves the necessity of (10.8.1), (10.8.2), and (10.8.3) for the admissibility of B, C, and/or $C\&D$.

The proof of the sufficiency of (10.8.1), (10.8.2), and (10.8.3) for the admissibility of B, C, and $C\&D$ is essentially the same. We define the operator T as in Theorem 4.8.3(i), i.e., the domain of T consists of all the vectors $\left[\begin{smallmatrix}y_0\\ x_0\\ u_0\end{smallmatrix}\right] \in \left[\begin{smallmatrix}W^{1,p}_\omega(\overline{\mathbb{R}}^-;Y)\\ X\\ W^{1,p}_\omega(\overline{\mathbb{R}}^+;U)\end{smallmatrix}\right]$ which satisfy $A x_0 + B u_0(0) \in X$ and $y_0(0) = C\&D \left[\begin{smallmatrix}x_0\\ u_0(0)\end{smallmatrix}\right]$, and on its domain T is given by

$$T \begin{bmatrix} y_0\\ x_0\\ u_0\end{bmatrix} = \begin{bmatrix} \dot{y}_0\\ A_{|X} x_0 + B u_0(0)\\ \dot{u}_0\end{bmatrix}.$$

The domain of T is dense in $\left[\begin{smallmatrix} \mathcal{Y} \\ X \\ \mathcal{U} \end{smallmatrix}\right]$, and arguing as in the proof of Theorem 4.8.3(iii) we find that the resolvent of S is given by the formulas in Theorem 4.8.3(iii)(b). It follows from the Hille–Yosida Theorem 3.4.1 that T generates an ω-bounded C_0 semigroup $\mathfrak{T}$. By Theorem 2.7.6, from this semigroup we get an ω-bounded well-posed linear system if we can show that the $\mathfrak{T}$ satisfies the causality conditions (2.7.2). Let us denote the state at time $t \geq 0$ by $\left[\begin{smallmatrix} y_t \\ x_t \\ u_t \end{smallmatrix}\right] = \mathfrak{T}\left[\begin{smallmatrix} y_0 \\ x_0 \\ u_0 \end{smallmatrix}\right]$. For $\left[\begin{smallmatrix} y_0 \\ x_0 \\ u_0 \end{smallmatrix}\right] \in \mathcal{D}(T)$ we have $\left[\begin{smallmatrix} y_t \\ x_t \\ u_t \end{smallmatrix}\right] \in \mathcal{D}(T)$ for all $t \geq 0$, and

$$\begin{bmatrix} \frac{d}{dt} y_t(s) \\ \dot{x}_t \\ \frac{d}{dt} u_t(s) \end{bmatrix} = \begin{bmatrix} \frac{d}{ds} y_t(s) \\ A_{|X} x_t + B u_t(0) \\ \frac{d}{ds} u_t(s) \end{bmatrix}.$$

The equation for u_t does not depend on x_0 and y_0, and $u_t(s) = (\tau_+^t u)(s) = u_0(s+t)$ for all $s,\ t \geq 0$. In particular for data in $\mathcal{D}(T)$ we have $u_0 \in W^{1,p}_\omega(\overline{\mathbb{R}}^+; U)$, and $u_t(0) = u_0(t)$. The equation for $\dot{x}_t$ and the condition $\left[\begin{smallmatrix} y_t \\ x_t \\ u_t \end{smallmatrix}\right] \in \mathcal{D}(T)$ give

$$\begin{aligned} \dot{x}_t &= A x_t + B u_0(t), \\ y_t(0) &= C\&D \left[\begin{smallmatrix} x_t \\ u_0(t) \end{smallmatrix}\right], \qquad t \geq 0. \end{aligned}$$

The latter condition combined with the equation for $\dot{y}_t$ gives

$$y_t(s) = \begin{cases} y_0(s+t), & s \leq -t, \\ C\&D \left[\begin{smallmatrix} x_{s+t} \\ u_0(s+t) \end{smallmatrix}\right], & -t < s \leq 0. \end{cases}$$

In particular, this shows that y_t and x_t do not depend on $u_0(s)$ for $s > t$, and that $\pi_{(-\infty,-t]}$ does not depend on x_0 and u_0. This implies the causality conditions (2.7.2) for data in $\mathcal{D}(T)$. Since $\mathcal{D}(T)$ is dense in $\left[\begin{smallmatrix} \mathcal{Y} \\ X \\ \mathcal{U} \end{smallmatrix}\right]$, we get (2.7.2) for arbitrary initial data.

(iv) See Definition 10.1.1, Theorem 4.6.7, and (iii). □

10.9 Comments

Some of the results in this chapter are rather old, whereas others are very recent.

Section 10.2 The part of Theorem 10.2.1 which describes the duality between B as an L^p-admissible control operator and B^* as an admissible observation operator has been known for a long time; see, e.g., G. Weiss (1989b, Theorem 6.9). Part (iii) of Theorem 10.2.1 appears to be new in this form. The proof of Theorem 10.2.3 has been adapted from the proof of Desch *et al.* (1985, Theorem 2.2), and it is essentially contained in Desch *et al.* (1985, Corollary 2). (There it is required, in addition, that the total variation over $[0, t]$ tends to zero as $t \downarrow 0$. That condition is used in Desch *et al.* (1985, Theorem 2.1) in the same way as in Corollary 7.1.9 to prove the *Reg*-well-posedness of a state feedback connection.)

Section 10.3 As we already mentioned, the basic results presented in this section essentially date back to Paley and Wiener (1934) (where the scalar version of Theorem 10.3.4 is proved). The version of Theorem 10.3.5 which we give is due to Fourès and Segal (1955). See G. Weiss (1991a, pp. 195–196) for additional comments on the history of this theorem. Theorem 10.3.7 (which is a restatement of Theorem 3.11.6) was proved independently by Herbst (1983), Huang (1985), and Prüss (1984). Our proof follows the one in G. Weiss (1988b, Theorem 4.2).

Section 10.4 The equivalence of (i) and (iv) in Theorem 10.4.2 was proved by Grabowski (1991, Theorem 3), and the uniqueness of the nonnegative solution of (10.4.3) in the strongly continuous case is proved in Theorem 4 of the same article. Theorem 10.4.3 is a slight extension of Hansen and Weiss (1997, Theorem 3.1 and Proposition 3.2). Theorems 10.4.6 and 10.4.7 are contained in Grabowski and Callier (1999, Theorem 2.1) and Hansen and Weiss (1997, Proposition 5.1).

Section 10.5 Theorem 10.5.1 is due to Carleson (1958, 1962). Our proof of this theorem is essentially the same as the one given in Nikol'skiĭ (1986, pp. 151, 258, 259) (and contributed there to Vinogradov), but it has been translated to the right half-plane from the unit disk. See Nikol'skiĭ (1986, pp. 170–171) for further comments on this proof.

Section 10.6 The sufficiency part of Theorem 10.6.1 (i.e., the fact that b and c are L^2-admissible and stable whenever the corresponding measure is a Carleson measure) was discovered by Ho and Russell (1983), and the (easier) necessity part was added by G. Weiss (1988a).

Multi-dimensional analogues of Theorems 10.6.3 and 10.6.4 (where U and Y are allowed to be infinite-dimensional Hilbert spaces) have been studied by, e.g., Hansen and Weiss. In Hansen and Weiss (1991) an operator-valued version of the Carleson measure condition is introduced (in the diagonal case), and it is conjectured that this condition is equivalent to the L^2-admissibility and stability of an observation or control operator. There it was also proved that the conjecture holds in the case where the semigroup $\mathfrak{A}$ generated by the main operator A is (normal and) exponentially stable and, in addition, analytic or a group. In Hansen and Weiss (1997) the same conjecture was proved to be true under some additional condition on the spectrum of A. It is not known if it is true in its full generality for normal semigroups (see the comments below for the case when $\mathfrak{A}$ is not normal).

Section 10.7 Lemma 10.7.2 was proved by Partington and Weiss (2000) in the case where $U = \mathbb{C}$, and our proof is a slight modification of their proof (they use the right-shift on $L^2(\mathbb{R}^+)$ instead of the left-shift on $L^2(\mathbb{R}^-)$, and their proof applies equally well to the case of a general Hilbert space U). The discrete time version of this lemma is found in Peller (2003, Theorem 6.1). Theorem 10.7.1 was discovered by Jacob and Partington (2001). They give a more complicated proof based on the universal model of a general contraction semigroup (instead of the simpler model of an isometric semigroup combined with a isometric dilation that we use).

George Weiss conjectured in G. Weiss (1991b) and G. Weiss (1999) that the growth estimates (10.7.1) and (10.7.2) (with b replaced by B, c replaced by C, and the norms in X replaced by operator-norms) would be equivalent to the L^2-stability and boundedness of the control operator B and the observation operator C (even in the case where U and Y are infinite-dimensional Hilbert spaces). This conjecture is now known to be partially true, but false in general. See the comments that we made above on Section 10.6 for the case when the semigroup $\mathfrak{A}$ is normal. Theorem 10.7.1 shows that the conjecture holds in the case where $Y = U = \mathbb{C}$ and $\mathfrak{A}$ is a contraction semigroup. The first counter-example with $Y = \mathbb{C}$ was produced by Jacob and Zwart (2000) (by Theorem 10.7.1, their semigroup cannot be similar to a contraction semigroup). Another counter-example with infinite-dimensional Y is given in Jacob *et al.* (2002), and in that example $\mathfrak{A}$ is even a contraction semigroup (it is an outgoing shift of infinite multiplicity). Thus, Theorem 10.7.1 is not valid if we allow U or Y to be infinite-dimensional. Two more counter-examples are given in Zwart *et al.* (2003). Additional conditions on the system under which the conjecture is valid are investigated in Jacob *et al.* (2003).

Section 10.8 Part (ii) of Theorem 10.8.1 (in the Hilbert space case with $p = 2$) was proved by Grabowski and Callier (1996, Theorem 2.3), and part (i) can be obtained from part (ii) by duality (see also Engel (1998)). Part (iii) of this theorem first appeared in Staffans (2001b).

11

Passive and conservative scattering systems

In this chapter we study passive and conservative systems in a scattering setting. These systems are L^2-well-posed, and the input, state, and output spaces are Hilbert spaces. Passivity means that the system has no internal energy sources. Conservativity means that the system is passive, and neither the system itself nor the dual system has any energy sinks. Our study of conservative systems leads us to universal models of energy preserving, co-energy preserving, and conservative systems whose semigroup is a compression of the bilateral shift.

11.1 Passive systems

Passive systems (in a scattering setting) are characterized by the fact that at any time, the sum of the 'final energy' and the 'output energy' can be no larger than the sum of the 'initial energy' and the 'input energy.'

Definition 11.1.1 By a (scattering) *passive system* Σ on three Hilbert spaces (Y, X, U) we mean an L^2-well-posed linear system with the following property: for all initial states $x_0 \in X$ and all input functions $u \in L^2_{\mathrm{loc}}(\mathbb{R}^+; U)$, the state trajectory x and the output function y of Σ with initial time zero satisfy

$$|x(t)|_X^2 + \int_0^t |y(s)|_Y^2\, ds \le |x_0|_X^2 + \int_0^t |u(s)|_U^2\, ds, \qquad t \ge 0. \quad (11.1.1)$$

Instead of starting with a system Σ we may as well start with a system node S, in which case the corresponding definition reads as follows.

Definition 11.1.2 A *system node* S on three Hilbert spaces (Y, X, U) is (scattering) *passive* if for all $x_0 \in X$ and $u \in W^{2,1}_{\mathrm{loc}}(\mathbb{R}^+; U)$ with $\left[\begin{smallmatrix} x_0 \\ u(0) \end{smallmatrix}\right] \in \mathcal{D}(S)$ the state trajectory x and the output function y in Lemma 4.7.8 with $s = 0$ satisfy (11.1.1).

As the following lemma shows, these two definitions are equivalent:

Lemma 11.1.3 *There is a one-to-one correspondence between passive system nodes and passive systems on the Hilbert spaces (Y, X, U): the system node of a passive system on (Y, X, U) is passive, and conversely, every passive system node S is L^2-well-posed, and the system generated by S is passive.*

Proof That the system node of a passive system is passive follows from Theorem 4.6.11(i). Conversely, suppose that S is a passive system node. By Theorem 4.7.13, S is L^2-well-posed and, since the set of data in Lemma 4.7.8 is dense in $\begin{bmatrix} X \\ L^2([0,t);U) \end{bmatrix}$, formula (11.1.1) extends by continuity to arbitrary initial data $x_0 \in X$ and $u \in L^2_{\mathrm{loc}}(\mathbb{R}^+; U)$. □

Passivity can be characterized in different ways. One way is to work on the 'system level' (as opposed to the system node level).

Lemma 11.1.4 *Let Σ be an L^2-well-posed system on three Hilbert spaces (Y, X, U). Then the following conditions are equivalent:*

(i) *Σ is (scattering) passive.*
(ii) *The causal dual system Σ^d is (scattering) passive.*
(iii) *In the notation of Definition 2.2.6, for all $t > 0$, the operator $\Sigma_0^t = \begin{bmatrix} \mathfrak{A}^t & \mathfrak{B}_0^t \\ \mathfrak{C}_0^t & \mathfrak{D}_0^t \end{bmatrix}$ is a contraction from $\begin{bmatrix} X \\ L^2([0,t);U) \end{bmatrix}$ to $\begin{bmatrix} X \\ L^2([0,t);Y) \end{bmatrix}$. (Here we use the norm $\left\| \begin{bmatrix} x \\ u \end{bmatrix} \right\| = \left(|x|_X^2 + \|u\|_{L^2([0,t);U)}^2 \right)^{1/2}$ in $\begin{bmatrix} X \\ L^2([0,t);U) \end{bmatrix}$ and the analogous norm in $\begin{bmatrix} X \\ L^2([0,t);Y) \end{bmatrix}$.)*
(iv) *In the notation of Definition 2.2.6, for all $-\infty < s < t < \infty$, the operator $\Sigma_s^t = \begin{bmatrix} \mathfrak{A}_s^t & \mathfrak{B}_s^t \\ \mathfrak{C}_s^t & \mathfrak{D}_s^t \end{bmatrix}$ is a contraction from $\begin{bmatrix} X \\ L^2([s,t);U) \end{bmatrix}$ to $\begin{bmatrix} X \\ L^2([s,t);Y) \end{bmatrix}$.*
(v) *Let $s \in \mathbb{R}$, $x_s \in X$ and $u \in L^2_{\mathrm{loc}}([s,\infty); U)$, and let x be the state and let y be the output of Σ with initial time s, initial state x_s, and input function u. Then, for all $t \ge r \ge s$,*

$$|x(t)|_X^2 + \int_r^t |y(v)|_Y^2 \, dv \le |x(r)|_X^2 + \int_r^t |u(v)|_U^2 \, dv. \qquad (11.1.2)$$

(vi) *The corresponding Lax–Phillips model with parameter $\omega = 0$ is a contraction semigroup.*

Moreover, every passive system is stable, its semigroup is a contraction semigroup, and the input map $\mathfrak{B} \in \mathcal{B}(L^2(\mathbb{R}^-; U); X)$, the output map $\mathfrak{C} \in \mathcal{B}(X; L^2(\mathbb{R}^+; Y))$ and the input/output map $\mathfrak{D} \in \mathcal{B}(L^2(\mathbb{R}; U); L^2(\mathbb{R}; Y))$ are contractions.

Proof (i) ⇔ (iii): See Definitions 2.2.6, 2.2.7 and 11.1.1. (Observe that it suffices to take $t > 0$ in (iii) since Σ_0^0 is the identity operator on $\begin{bmatrix} X \\ 0 \end{bmatrix}$.)

(i) ⇔ (ii): This follows from the equivalence of (i) and (iii) and the fact that

$$(\Sigma^d)_0^t = \begin{bmatrix} 1 & 0 \\ 0 & \mathbf{Я}_{t/2} \end{bmatrix} (\Sigma_0^t)^* \begin{bmatrix} 1 & 0 \\ 0 & \mathbf{Я}_{t/2} \end{bmatrix},$$

where the two operators denoted by $\left[\begin{smallmatrix} 1 & 0 \\ 0 & \mathbf{Я}_{t/2} \end{smallmatrix}\right]$ are unitary $\Big($one on $\left[\begin{smallmatrix} X \\ L^2([0,t);U) \end{smallmatrix}\right]$ and the other on $\left[\begin{smallmatrix} X \\ L^2([0,t);Y) \end{smallmatrix}\right]\Big)$ and $(\Sigma_0^t)^*$ is a contraction iff Σ_0^t is a contraction.

(iii) ⇔ (iv): See Lemma 2.2.8(iii).

(iv) ⇔ (v): See Definitions 2.2.6 and 2.2.7.

(i) ⇔ (vi): We use the same notation as in Section 2.7. Let $y_0 \in L^2(\mathbb{R}^-; Y)$, $x_0 \in X$, $u_0 \in L^2(\mathbb{R}^+; U)$, and let x be the state and y the output of Σ with initial time zero, initial state x_0, and input function u_0. Then, for all $t \geq 0$,

$$\begin{aligned}
\left\| \mathfrak{T}^t \begin{bmatrix} y_0 \\ x_0 \\ u_0 \end{bmatrix} \right\|^2 - \left\| \begin{bmatrix} y_0 \\ x_0 \\ u_0 \end{bmatrix} \right\|^2 &= \int_{-\infty}^{-t} |y_0(s+t)|_Y^2 \, ds + \int_{-t}^{0} |y(s+t)|_Y^2 \, ds \\
&\quad + |x(t)|_X^2 + \int_0^\infty |u_0(s+t)|_U^2 \, ds \\
&\quad - |x_0|_X^2 - \int_{-\infty}^0 |y_0(s)|_Y^2 \, ds - \int_0^\infty |u_0(s)|_U^2 \, ds \\
&= \int_0^t |y(s)|_Y^2 \, ds + |x(t)|_X^2 - |x_0|_X^2 - \int_0^t |u_0(s)|_U^2 \, ds.
\end{aligned}$$

The left-hand side is nonpositive for all y_0, x_0, and u_0 iff $\mathfrak{T}$ is a contraction semigroup, and the right-hand side is nonpositive for all x_0, and u_0 iff Σ is passive.

The final claim about the stability of a passive system and the contractivity of its different parts follows from Definitions 8.1.1 and 11.1.1. □

Alternatively, we may study the passivity of a system node S without any reference to the system that it generates.

Theorem 11.1.5 *Let $S = \left[\begin{smallmatrix} A\&B \\ C\&D \end{smallmatrix}\right]$ be an operator node on the three Hilbert spaces (Y, X, U) with main operator A, control operator B, observation operator C, and transfer function $\widehat{\mathfrak{D}}$. Then the following conditions are equivalent.*

(i) *S is (scattering) passive (in particular, A generates a C_0 contraction semigroup and S is L^2-well-posed).*
(ii) *A generates a C_0 semigroup, and for all $x_0 \in X$ and $u \in W^{2,1}_{\mathrm{loc}}(\mathbb{R}^+; U)$ with $\left[\begin{smallmatrix} x_0 \\ u(0) \end{smallmatrix}\right] \in \mathcal{D}(S)$ the state trajectory x and the output function y in Lemma 4.7.8 with $s = 0$ satisfy*

$$\frac{\mathrm{d}}{\mathrm{d}t} |x(t)|_X^2 + |y(t)|_Y^2 \leq |u(t)|_U^2, \qquad t \geq 0. \tag{11.1.3}$$

(iii) $\rho(A) \cap \mathbb{C}_+ \neq \emptyset$, *and for all* $\left[\begin{smallmatrix} x_0 \\ u_0 \end{smallmatrix}\right] \in \mathcal{D}(S)$,

$$2\Re\left\langle A\&B \left[\begin{smallmatrix} x_0 \\ u_0 \end{smallmatrix}\right], x_0\right\rangle_X + \left|C\&D \left[\begin{smallmatrix} x_0 \\ u_0 \end{smallmatrix}\right]\right|_Y^2 \leq |u_0|_U^2. \tag{11.1.4}$$

(iv) $\rho(A) \cap \mathbb{C}_+ \neq \emptyset$, *and for all* $\alpha \in \rho(A)$ *we have, with* $E_\alpha = \left[\begin{smallmatrix} 1 & (\alpha - A_{|X})^{-1}B \\ 0 & 1 \end{smallmatrix}\right]$,

$$E_\alpha^* \left[\begin{smallmatrix} 1 & 0 \\ 0 & 0 \end{smallmatrix}\right] SE_\alpha + (SE_\alpha)^* \left[\begin{smallmatrix} 1 & 0 \\ 0 & 0 \end{smallmatrix}\right] E_\alpha + (SE_\alpha)^* \left[\begin{smallmatrix} 0 & 0 \\ 0 & 1 \end{smallmatrix}\right] SE_\alpha \leq \left[\begin{smallmatrix} 0 & 0 \\ 0 & 1 \end{smallmatrix}\right], \tag{11.1.5}$$

which should be interpreted as an operator inequality in $\mathcal{B}\left(\left[\begin{smallmatrix} X_1 \\ U \end{smallmatrix}\right]; \left[\begin{smallmatrix} X_{-1}^* \\ U \end{smallmatrix}\right]\right)$, *and we identify the dual of* $\left[\begin{smallmatrix} X_1 \\ U \end{smallmatrix}\right]$ *with* $\left[\begin{smallmatrix} X_{-1}^* \\ U \end{smallmatrix}\right]$. *Explicitly, this means that for all* $x \in X_1$ *and all* $u \in U$,

$$\begin{aligned}
&\left\langle E_\alpha \left[\begin{smallmatrix} x \\ u \end{smallmatrix}\right]\left[\begin{smallmatrix} 1 & 0 \\ 0 & 0 \end{smallmatrix}\right] SE_\alpha \left[\begin{smallmatrix} x \\ u \end{smallmatrix}\right]\right\rangle_{\left[\begin{smallmatrix} X \\ Y \end{smallmatrix}\right]} + \left\langle\left[\begin{smallmatrix} 1 & 0 \\ 0 & 0 \end{smallmatrix}\right] SE_\alpha \left[\begin{smallmatrix} x \\ u \end{smallmatrix}\right] E_\alpha \left[\begin{smallmatrix} x \\ u \end{smallmatrix}\right]\right\rangle_{\left[\begin{smallmatrix} X \\ Y \end{smallmatrix}\right]} \\
&\quad + \left\langle SE_\alpha \left[\begin{smallmatrix} x \\ u \end{smallmatrix}\right]\left[\begin{smallmatrix} 0 & 0 \\ 0 & 1 \end{smallmatrix}\right] SE_\alpha \left[\begin{smallmatrix} x \\ u \end{smallmatrix}\right]\right\rangle_{\left[\begin{smallmatrix} X \\ Y \end{smallmatrix}\right]} \leq \left\langle\left[\begin{smallmatrix} x \\ u \end{smallmatrix}\right]\left[\begin{smallmatrix} 0 & 0 \\ 0 & 1 \end{smallmatrix}\right]\left[\begin{smallmatrix} x \\ u \end{smallmatrix}\right]\right\rangle_{\left[\begin{smallmatrix} X \\ U \end{smallmatrix}\right]}.
\end{aligned} \tag{11.1.6}$$

(v) $\rho(A) \cap \mathbb{C}_+ \neq \emptyset$, *and the inequality* (11.1.5) *holds for some* $\alpha \in \rho(A)$.

(vi) $\rho(A) \cap \mathbb{C}_+ \neq \emptyset$, *and for all* $\alpha \in \rho(A)$, *we have*

$$\begin{aligned}
&\begin{bmatrix} A + A_{|X}^* & (\alpha + A_{|X}^*)(\alpha - A_{|X})^{-1}B \\ B^*(\overline{\alpha} - A^*)^{-1}(\overline{\alpha} + A) & B^*(\overline{\alpha} - A^*)^{-1}(2\Re\alpha)(\alpha - A_{|X})^{-1}B \end{bmatrix} \\
&\qquad + \begin{bmatrix} C^*C & C^*\widehat{\mathfrak{D}}(\alpha) \\ \widehat{\mathfrak{D}}(\alpha)^*C & \widehat{\mathfrak{D}}(\alpha)^*\widehat{\mathfrak{D}}(\alpha) \end{bmatrix} \leq \begin{bmatrix} 0 & 0 \\ 0 & 1 \end{bmatrix},
\end{aligned} \tag{11.1.7}$$

which is an operator inequality in $\mathcal{B}\left(\left[\begin{smallmatrix} X_1 \\ U \end{smallmatrix}\right]; \left[\begin{smallmatrix} X_{-1}^d \\ U \end{smallmatrix}\right]\right)$.

(vii) $\rho(A) \cap \mathbb{C}_+ \neq \emptyset$, *and the inequality* (11.1.5) *holds for some* $\alpha \in \rho(A)$.

(viii) $\mathbb{C}_+ \subset \rho(A)$, *and for all* $\alpha \in \mathbb{C}_+$, *the operator*

$$\begin{bmatrix} \mathbf{A}(\alpha) & \mathbf{B}(\alpha) \\ \mathbf{C}(\alpha) & \mathbf{D}(\alpha) \end{bmatrix} = \begin{bmatrix} (\overline{\alpha} + A)(\alpha - A)^{-1} & \sqrt{2\Re\alpha}(\alpha - A_{|X})^{-1}B \\ \sqrt{2\Re\alpha}C(\alpha - A)^{-1} & \widehat{\mathfrak{D}}(\alpha) \end{bmatrix}$$

is a contraction.

(ix) *For some* $\alpha \in \rho(A) \cap \mathbb{C}^+$, *the operator* $\left[\begin{smallmatrix} \mathbf{A}(\alpha) & \mathbf{B}(\alpha) \\ \mathbf{C}(\alpha) & \mathbf{D}(\alpha) \end{smallmatrix}\right]$ *defined in (viii) is a contraction.*

If both $\rho(A) \cap \mathbb{C}^+ \neq \emptyset$ *and* $\rho(A) \cap \mathbb{C}^- \neq \emptyset$, *then these conditions are further equivalent to the following two conditions:*

(x) *For all* $\alpha \in \rho(A) \cap \mathbb{C}^-$, *the operator* $\left[\begin{smallmatrix} \mathbf{A}(\alpha) & \mathbf{B}(\alpha) \\ \mathbf{C}(\alpha) & \mathbf{D}(\alpha) \end{smallmatrix}\right]$ *defined in (viii) (with* $\sqrt{2\Re\alpha} = j\sqrt{2|\Re\alpha|}$*) satisfies*

$$\begin{bmatrix} \mathbf{A}(\alpha) & \mathbf{B}(\alpha) \\ \mathbf{C}(\alpha) & \mathbf{D}(\alpha) \end{bmatrix}^* \begin{bmatrix} -1 & 0 \\ 0 & 1 \end{bmatrix} \begin{bmatrix} \mathbf{A}(\alpha) & \mathbf{B}(\alpha) \\ \mathbf{C}(\alpha) & \mathbf{D}(\alpha) \end{bmatrix} \le \begin{bmatrix} -1 & 0 \\ 0 & 1 \end{bmatrix}. \tag{11.1.8}$$

(xi) *The inequality* (11.1.8) *holds for some* $\alpha \in \rho(A) \cap \mathbb{C}^-$.
If $\rho(A) \cap j\mathbb{R} \neq \emptyset$, *then (i)–(xi) are further equivalent to the following two conditions:*

(xii) *For all* $\alpha \in \rho(A) \cap j\mathbb{R}$,

$$\begin{aligned} &\begin{bmatrix} A + A^* & (\alpha + A^*)(\alpha - A_{|X})^{-1}B \\ B^*(\alpha + A^*)^{-1}(\alpha - A) & 0 \end{bmatrix} \\ &\qquad + \begin{bmatrix} C^*C & C^*\widehat{\mathfrak{D}}(\alpha) \\ \widehat{\mathfrak{D}}(\alpha)^*C & \widehat{\mathfrak{D}}(\alpha)^*\widehat{\mathfrak{D}}(\alpha) \end{bmatrix} \le \begin{bmatrix} 0 & 0 \\ 0 & 1 \end{bmatrix}. \end{aligned} \tag{11.1.9}$$

(xiii) *The inequality* (11.1.9) *holds for some* $\alpha \in \rho(A) \cap j\mathbb{R}$.

Proof (i) ⇒ (ii): Assume (i). By Lemmas 11.1.3 and 11.1.4, the data in (ii) satisfy the identity (11.1.2). In this inequality we move the term $|x(r)|_X^2$ over to the left-hand side, divide by $t - r$, and let $t \downarrow r$ to get (11.1.3) with t replaced by r.

(ii) ⇒ (i): We get (11.1.1) by integrating (11.1.3)).

(ii) ⇒ (iii): By Theorem 4.6.11, it follows from (ii) that (11.1.4) holds with $\left[\begin{smallmatrix} x_0 \\ u_0 \end{smallmatrix}\right]$ replaced by $\left[\begin{smallmatrix} x(t) \\ u(t) \end{smallmatrix}\right]$ for all $t \ge 0$ (note that $\frac{\mathrm{d}}{\mathrm{d}t}|x(t)|^2 = 2\Re\langle \dot{x}(t), x(t)\rangle$). In particular, taking $t = 0$ we get (11.1.4) for all $\left[\begin{smallmatrix} x_0 \\ u(0) \end{smallmatrix}\right] \in \mathcal{D}(S)$.

(iii) ⇒ (ii): If (iii) holds, then it follows from Theorem 3.4.8 that A generates a C_0 semigroup (take $u_0 = 0$ in (11.1.4)). If x_0 and u satisfy the conditions in (ii), then by Theorem 4.6.11, $\left[\begin{smallmatrix} x(t) \\ u(t) \end{smallmatrix}\right] \in \mathcal{D}(S)$ and $\left[\begin{smallmatrix} \dot{x}(t) \\ y(t) \end{smallmatrix}\right] = \left[\begin{smallmatrix} A\&B \\ C\&D \end{smallmatrix}\right]\left[\begin{smallmatrix} x(t) \\ u(t) \end{smallmatrix}\right]$ for all $t \ge 0$, hence (11.1.3) holds.

(iii) ⇒ (iv) ⇒ (v) ⇒ (iii): As $\left[\begin{smallmatrix} A\&B \\ 0 \;\; 0 \end{smallmatrix}\right] = \left[\begin{smallmatrix} 1 & 0 \\ 0 & 0 \end{smallmatrix}\right] S$ and $\left[\begin{smallmatrix} 0 \;\; 0 \\ C\&D \end{smallmatrix}\right] = \left[\begin{smallmatrix} 0 & 0 \\ 0 & 1 \end{smallmatrix}\right] S$, we can rewrite (11.1.4) in the following form: for all $\left[\begin{smallmatrix} x \\ u \end{smallmatrix}\right] \in \mathcal{D}(S)$,

$$\begin{aligned} &\left\langle \left[\begin{smallmatrix} x \\ u \end{smallmatrix}\right], \left[\begin{smallmatrix} 1 & 0 \\ 0 & 0 \end{smallmatrix}\right] S \left[\begin{smallmatrix} x \\ u \end{smallmatrix}\right] \right\rangle + \left\langle S\left[\begin{smallmatrix} x \\ u \end{smallmatrix}\right], \left[\begin{smallmatrix} 1 & 0 \\ 0 & 0 \end{smallmatrix}\right] \left[\begin{smallmatrix} x \\ u \end{smallmatrix}\right] \right\rangle \\ &\qquad + \left\langle S\left[\begin{smallmatrix} x \\ u \end{smallmatrix}\right], \left[\begin{smallmatrix} 0 & 0 \\ 0 & 1 \end{smallmatrix}\right] S \left[\begin{smallmatrix} x \\ u \end{smallmatrix}\right] \right\rangle \le \left\langle \left[\begin{smallmatrix} x \\ u \end{smallmatrix}\right], \left[\begin{smallmatrix} 0 & 0 \\ 0 & 1 \end{smallmatrix}\right] \left[\begin{smallmatrix} x \\ u \end{smallmatrix}\right] \right\rangle. \end{aligned}$$

From here we get (11.1.6) by replacing $\left[\begin{smallmatrix} x \\ u \end{smallmatrix}\right]$ by $E_\alpha \left[\begin{smallmatrix} z \\ u \end{smallmatrix}\right]$ where $z \in X_1$ and $u \in U$ are arbitrary (recall that, by Lemma 4.7.18(ii), E_α maps $\left[\begin{smallmatrix} X_1 \\ U \end{smallmatrix}\right]$ one-to-one onto $\mathcal{D}(S)$, and observe that $E_\alpha^* \left[\begin{smallmatrix} 0 & 0 \\ 0 & 1 \end{smallmatrix}\right] E_\alpha = \left[\begin{smallmatrix} 0 & 0 \\ 0 & 1 \end{smallmatrix}\right]$).

(iv) ⇔ (vi) and (v) ⇔ (vii): The inequality in (vi) is just an equivalent way of writing (11.1.5) (see Lemma 4.7.18), so (iv) ⇔ (vi) and (v) ⇔ (vii).

(vi) ⇒ (viii) ⇒ (ix) ⇒ (vii): For any particular $\alpha \in \rho(A) \cap \mathbb{C}^+$, the operator $\left[\begin{smallmatrix} \mathbf{A}(\alpha) & \mathbf{B}(\alpha) \\ \mathbf{C}(\alpha) & \mathbf{D}(\alpha) \end{smallmatrix}\right]$ is a contraction if and only if

$$\begin{bmatrix} \mathbf{A}(\alpha) & \mathbf{B}(\alpha) \\ \mathbf{C}(\alpha) & \mathbf{D}(\alpha) \end{bmatrix}^* \begin{bmatrix} \mathbf{A}(\alpha) & \mathbf{B}(\alpha) \\ \mathbf{C}(\alpha) & \mathbf{D}(\alpha) \end{bmatrix} \le \begin{bmatrix} 1 & 0 \\ 0 & 1 \end{bmatrix}. \tag{11.1.10}$$

By substituting the definition of $\left[\begin{smallmatrix} \mathbf{A}(\alpha) & \mathbf{B}(\alpha) \\ \mathbf{C}(\alpha) & \mathbf{D}(\alpha) \end{smallmatrix}\right]$ given in (viii) into this identity and simplifying the resulting expression, we observe that (11.1.10) is equivalent to the inequality that we get by multiplying (11.1.7) by $\left[\begin{smallmatrix} \sqrt{2\Re s}\,(s-A)^{-1} & 0 \\ 0 & 1 \end{smallmatrix}\right]$ to the right and by the adjoint of this operator to the left. Thus $\left[\begin{smallmatrix} \mathbf{A}(\alpha) & \mathbf{B}(\alpha) \\ \mathbf{C}(\alpha) & \mathbf{D}(\alpha) \end{smallmatrix}\right]$ is a contraction if and only if (11.1.7) holds. If (vi) holds, then Σ is passive, and by Lemma 11.1.4, $\mathbb{C}^+ \in \rho(A)$. We conclude that (vi) ⇒ (viii) and that (ix) ⇒ (vii). The implication (viii) ⇒ (ix) is obvious.

(vi) ⇒ (x) ⇒ (xi) ⇒ (vii): This is proved in almost exactly the same way as the chain of implications (vi) ⇒ (viii) ⇒ (ix) ⇒ (vii). The only difference is that $\sqrt{2\Re\alpha} = j\sqrt{2|\Re\alpha|}$ is pure imaginary, so that the complex conjugate of $\sqrt{2\Re\alpha}$ is $-\sqrt{2\Re\alpha}$ instead of $\sqrt{2\Re\alpha}$.

(xii) and (xiii): These are specialized version of (iv)–(vii) where we take $\Re\alpha = 0$. □

Every passive system has certain characteristic operators which describe the amount of energy which is 'trapped into the state space forever.'

Theorem 11.1.6 *Let* $\Sigma = \left[\begin{smallmatrix} \mathfrak{A} & \mathfrak{B} \\ \mathfrak{C} & \mathfrak{D} \end{smallmatrix}\right]$ *be passive on* (Y, X, U). *With the notation of Definition 2.2.6, the following strong self-adjoint limits exist in* $\mathcal{B}\left(\left[\begin{smallmatrix} X \\ L^2(\mathbb{R}^+;U) \end{smallmatrix}\right]\right)$ *and* $\mathcal{B}\left(\left[\begin{smallmatrix} X \\ L^2(\mathbb{R}^-;Y) \end{smallmatrix}\right]\right)$, *respectively:*

$$\begin{aligned} \begin{bmatrix} Q_{\mathfrak{A}^*,\mathfrak{A}} & Q_{\mathfrak{A}^*,\mathfrak{B}} \\ Q_{\mathfrak{B}^*,\mathfrak{A}} & Q_{\mathfrak{B}^*,\mathfrak{B}} \end{bmatrix} &= \lim_{t\to\infty} \begin{bmatrix} \mathfrak{A}^{*t} \\ (\mathfrak{B}_0^t)^* \end{bmatrix} \begin{bmatrix} \mathfrak{A}^t & \mathfrak{B}_0^t \end{bmatrix} \ge 0, \\ \begin{bmatrix} Q_{\mathfrak{C},\mathfrak{C}^*} & Q_{\mathfrak{C},\mathfrak{A}^*} \\ Q_{\mathfrak{A},\mathfrak{C}^*} & Q_{\mathfrak{A},\mathfrak{A}^*} \end{bmatrix} &= \lim_{t\to\infty} \begin{bmatrix} \mathfrak{C}_{-t}^0 \\ \mathfrak{A}^t \end{bmatrix} \begin{bmatrix} (\mathfrak{C}_{-t}^0)^* & \mathfrak{A}^{*t} \end{bmatrix} \ge 0. \end{aligned} \tag{11.1.11}$$

Moreover, the following inequalities hold:

$$\begin{aligned} \begin{bmatrix} Q_{\mathfrak{A}^*,\mathfrak{A}} & Q_{\mathfrak{A}^*,\mathfrak{B}} \\ Q_{\mathfrak{B}^*,\mathfrak{A}} & Q_{\mathfrak{B}^*,\mathfrak{B}} \end{bmatrix} + \begin{bmatrix} \mathfrak{C}^* \\ \pi_+\mathfrak{D}^* \end{bmatrix} \begin{bmatrix} \mathfrak{C} & \mathfrak{D}\pi_+ \end{bmatrix} &\le \begin{bmatrix} 1 & 0 \\ 0 & \pi_+ \end{bmatrix}, \\ \mathfrak{D}^*\pi_-\mathfrak{D} + \mathfrak{B}^*\mathfrak{B} &\le \pi_-, \\ \begin{bmatrix} Q_{\mathfrak{C},\mathfrak{C}^*} & Q_{\mathfrak{C},\mathfrak{A}^*} \\ Q_{\mathfrak{A},\mathfrak{C}^*} & Q_{\mathfrak{A},\mathfrak{A}^*} \end{bmatrix} + \begin{bmatrix} \pi_-\mathfrak{D} \\ \mathfrak{B} \end{bmatrix} \begin{bmatrix} \mathfrak{D}^*\pi_- & \mathfrak{B}^* \end{bmatrix} &\le \begin{bmatrix} \pi_- & 0 \\ 0 & 1 \end{bmatrix}, \\ \mathfrak{C}\mathfrak{C}^* + \mathfrak{D}\pi_+\mathfrak{D}^* &\le \pi_+. \end{aligned} \tag{11.1.12}$$

In particular, if Σ is observable then $Q_{\mathfrak{A}^*,\mathfrak{A}} < 1$, if Σ is exactly observable in infinite time then $Q_{\mathfrak{A}^*,\mathfrak{A}} \ll 1$, if Σ is controllable then $Q_{\mathfrak{A},\mathfrak{A}^*} < 1$, and if Σ is exactly controllable in infinite time then $Q_{\mathfrak{A},\mathfrak{A}^*} \ll 1$.

The above inequalities imply that both the backward wave operator W_- and the forward wave operator W_+ are contractions (these operators are defined in Section 2.7).

The proof of Theorem 11.1.6 is based on the following lemma.

Lemma 11.1.7 *Let A_t, $t \geq 0$, be a nondecreasing [nonincreasing] family of self-adjoint operators on a Hilbert space X which is bounded from above [from below]. Then the strong limit $A = \lim_{t\to\infty} A_t$ exists, and $A_0 \leq A_t \leq A$ [$A_0 \geq A_t \geq A$] for all $t \geq 0$.*

The monotonicity and boundedness assumptions of this lemma mean the following: there exists a self-adjoint operator $B \in \mathcal{B}(X)$ such that $A_{t_1} \leq A_{t_2} \leq B$ [$A_{t_1} \geq A_{t_2} \geq B$] for all $t_2 \geq t_1 \geq 0$.

Proof of Lemma 11.1.7 It suffices to prove the second case where the family is nonincreasing. Then, for all $x \in X$, the function $t \mapsto \langle x, A_t x\rangle$ is real, nonincreasing, and bounded from below, so it has a limit. By using polarization we can strengthen this result as follows: for all $x \in X$ and $y \in Y$,

$$\begin{aligned}\langle x, A_t y\rangle = \frac{1}{4}[&\langle x+y, A_t(x+y)\rangle - \langle x-y, A_t(x-y)\rangle \\ &+ j\langle x+jy, A_t(x+jy)\rangle - j\langle x-jy, A_t(x-jy)\rangle] = 0,\end{aligned} \tag{11.1.13}$$

hence $\lim_{t\to\infty}\langle y, A_t x\rangle$ exists, i.e., A_t tends weakly to a self-adjoint limit $A \in \mathcal{B}(X)$ as $t \to \infty$. Clearly $A_0 \geq A_t \geq A$ for all $t \geq 0$.

We claim that the convergence of A_t to A is, in fact, strong. Let $x \in X$, and let $B_t = (A_t - A)^{1/2}$ be the positive square root of $A_t - A$ in $\mathcal{B}(X)$ (see Lemma A.2.2). Then B_t is uniformly bounded for $t \geq 0$, and

$$|B_t x|^2 = \langle B_t x, B_t x\rangle = \langle x, (B_t)^2 x\rangle = \langle x, (A_t - A)x\rangle \to 0 \text{ as } t \to \infty;$$

hence also

$$|(A_t - A)x|^2 = |(B_t)^2 x|^2 \leq \sup_{t\geq 0}\|B_t\|^2 |B_t x|^2 \to 0 \text{ as } t \to \infty.$$

□

Proof of Theorem 11.1.6 Let $x_0 \in X$ and $u \in L^2(\mathbb{R}^+; U)$, and let x be the state and y the output of Σ with initial time zero, initial state x_0, and input function

u. It follows from (11.1.2) that the following function is nonincreasing:

$$
\begin{aligned}
t \mapsto\ & |x(t)|_X^2 + \int_0^t |y(v)|_Y^2\, dv - \int_0^t |u(v)|_U^2\, dv \\
&= \left\langle \begin{bmatrix} x \\ u \end{bmatrix}, \begin{bmatrix} \mathfrak{A}^t \\ (\mathfrak{B}_0^t)^* \end{bmatrix} \begin{bmatrix} \mathfrak{A}^t & \mathfrak{B}_0^t \end{bmatrix} \begin{bmatrix} x \\ u \end{bmatrix} \right\rangle \\
&+ \left\langle \begin{bmatrix} x \\ u \end{bmatrix}, \begin{bmatrix} (\mathfrak{C}_0^t)^* \\ (\mathfrak{D}_0^t)^* \end{bmatrix} \begin{bmatrix} \mathfrak{C}_0^t & \mathfrak{D}_0^t \end{bmatrix} \begin{bmatrix} x \\ u \end{bmatrix} \right\rangle \\
&- \left\langle \begin{bmatrix} x \\ u \end{bmatrix}, \begin{bmatrix} 0 & 0 \\ 0 & \pi_{[0,t)} \end{bmatrix} \begin{bmatrix} x \\ u \end{bmatrix} \right\rangle .
\end{aligned}
\tag{11.1.14}
$$

This function is bounded from above by $|x_0|_X^2$ and from below by $-\|u\|_{L^2(\mathbb{R}^+)}$. By Lemma 11.1.7 (the nonincreasing case), the sum of the operator families on the right-hand side of this identity has a strong limit as $t \to \infty$. The middle family of operators on the right-hand side is nondecreasing, and it tends strongly to $\left[\begin{smallmatrix} \mathfrak{C}^* \\ \mathfrak{D}^* \end{smallmatrix}\right] \left[\begin{smallmatrix} \mathfrak{C} & \mathfrak{D} \end{smallmatrix}\right]$ as $t \to \infty$ (once more, see Lemma 11.1.7). Clearly, $\left[\begin{smallmatrix} 0 & 0 \\ 0 & \pi_{[0,t)} \end{smallmatrix}\right] \to \left[\begin{smallmatrix} 0 & 0 \\ 0 & \pi_+ \end{smallmatrix}\right]$ as $t \to \infty$. Thus, the first of the limits in (11.1.11) exist in the strong sense, and the first inequality in (11.1.12) is true. To get the second inequality in (11.1.12) we let $u \in L^2(\mathbb{R}^-; U)$, replace the initial time zero in (11.1.14) by s, replace x_0 by zero, replace the final time t by zero, and let $s \to -\infty$. The existence of the second limit in (11.1.11) and the last two inequalities in (11.1.12) are proved in the same way, with the original system Σ replaced by the causal dual Σ^d.

The final claims follow from (11.1.12) and Theorem 9.10.5. □

Let us end this section by establishing a decomposition of a passive system into a 'trivial' unitary part and a *purely passive part*. Here we use the following definition.

Definition 11.1.8 Let X, U, and Y be Hilbert spaces.

(i) A C_0-semigroup $\mathfrak{A}$ on X is *isometric*, or *co-isometric*, or *unitary* if $\mathfrak{A}^t$ is isometric, or co-isometric, or unitary, respectively, for all $t \geq 0$ (see Definition A.2.3). It is *completely nonunitary* if there is no subspace $Y \subset X$, $Y \neq 0$, such that Y is invariant under $\mathfrak{A}$ (i.e., $\mathfrak{A}^t Y \subset Y$ for all $t \geq 0$) and $\mathfrak{A}_{|Y}$ is unitary.

(ii) A function $\widehat{\mathfrak{D}} \in H_0^\infty(U; Y)$ is *purely contractive* if $|\widehat{\mathfrak{D}}(z)u|_Y < |u|_U$ for all $z \in \mathbb{C}^+$ and all nonzero $u \in U$.

(iii) An operator $\mathfrak{D} \in TIC^2(U; Y)$ is (scattering) *purely passive* if its transfer function is purely contractive.

We warn the reader that *a purely passive input/output map may very well be unitary*, i.e., it may satisfy $\mathfrak{D}^*\mathfrak{D} = 1$ and $\mathfrak{D}\mathfrak{D}^* = 1$. (The transfer functions of these input/output maps are usually called *bi-inner*.)

Theorem 11.1.9 *Let $\Sigma = \left[\begin{smallmatrix} \mathfrak{A} & \mathfrak{B} \\ \mathfrak{C} & \mathfrak{D} \end{smallmatrix}\right]$ be a passive system on the Hilbert spaces (Y, X, U). Then there exist unique orthogonal splittings of Y, X, and U into $Y = \left[\begin{smallmatrix} Y_1 \\ Y_0 \end{smallmatrix}\right]$, $X = \left[\begin{smallmatrix} X_0 \\ X_1 \end{smallmatrix}\right]$, $U = \left[\begin{smallmatrix} U_1 \\ U_0 \end{smallmatrix}\right]$ with the following properties: if we split $\left[\begin{smallmatrix} \mathfrak{A} & \mathfrak{B} \\ \mathfrak{C} & \mathfrak{D} \end{smallmatrix}\right]$ according to the splittings of Y, X, and U, then*

$$\left[\begin{array}{c|c} \mathfrak{A} & \mathfrak{B} \\ \hline \mathfrak{C} & \mathfrak{D} \end{array}\right] = \left[\begin{array}{cc|cc} \mathfrak{A}_0 & 0 & 0 & 0 \\ 0 & \mathfrak{A}_1 & \mathfrak{B}_1 & 0 \\ \hline 0 & \mathfrak{C}_1 & \mathfrak{D}_1 & 0 \\ 0 & 0 & 0 & \mathfrak{D}_0 \end{array}\right], \tag{11.1.15}$$

where $\mathfrak{A}_0 = \mathfrak{A}_{|X_0}$ is a unitary C_0-semigroup on X_0, $\mathfrak{A}_1 = \mathfrak{A}_{|X_1}$ is a completely nonunitary C_0-semigroup on X_1, $\mathfrak{C}_1 = \mathfrak{C}_{|X_1} \in \mathcal{B}(X_1; L^2(\mathbb{R}^+; Y_1))$, $\mathfrak{B}_1 = \mathfrak{B}_{|L^2(\mathbb{R}^-;U_1)} \in \mathcal{B}(L^2(\mathbb{R}^-; U_1); X_1)$, $\mathfrak{D}_1 = \mathfrak{D}_{|L^2(\mathbb{R};U_1)} \in TIC(U; Y)$ is purely passive, and $\mathfrak{D}_0$ is a static unitary operator from $L^2(\mathbb{R}; U_0)$ to $L^2(\mathbb{R}; Y_0)$ (i.e., it acts as multiplication with a unitary operator $D \in \mathcal{B}(U_0; Y_0)$). In particular, both X_0 and X_1 are invariant under $\mathfrak{A}$, $L^2(\mathbb{R}^-; U_0) \subset \mathcal{N}(\mathfrak{B})$, $\mathcal{R}(\mathfrak{B}) \subset X_1$, $X_0 \subset \mathcal{N}(\mathfrak{C})$, $\mathcal{R}(\mathfrak{C}) \subset L^2(\mathbb{R}^+; Y_1)$, and U_0 and Y_0 have the same dimension.

The proof of this theorem uses the following lemma.

Lemma 11.1.10 *If $T \in \mathcal{B}(U; Y)$ (where U and Y are Hilbert spaces) is a contraction, then $|Tu|_Y = |u|_U$ if and only if $u = T^*Tu$.*

Proof Clearly, if $u = T^*Tu$, then

$$|u|^2 = \langle u, u\rangle = \langle u, T^*Tu\rangle = \langle Tu, Tu\rangle = |Tu|^2.$$

Conversely, suppose that $|Tu| = |u|$. Denoting the positive square root of the positive operator $1 - T^*T$ by $(1 - T^*T)^{1/2}$ (see Lemma A.2.2), we get

$$0 = |u|^2 - |Tu|^2 = \langle u, (1 - T^*T)u\rangle = |(1 - T^*T)^{1/2}u|^2,$$

so $(1 - T^*T)^{1/2}u = 0$. This implies that also

$$(1 - T^*T)u = (1 - T^*T)^{1/2}(1 - T^*T)^{1/2}u = 0,$$

i.e., $u = T^*Tu$. □

Proof of Theorem 11.1.9 Let X_0 denote the set of all $x \in X$ for which

$$|\mathfrak{A}^t x| = |x| = |\mathfrak{A}^{*t} x|, \qquad t \geq 0. \tag{11.1.16}$$

This condition can alternatively be written as

$$\lim_{t\to\infty} |\mathfrak{A}^t x| = |x| = \lim_{t\to\infty} |\mathfrak{A}^{*t} x|, \tag{11.1.17}$$

since both $|\mathfrak{A}^{*t}x|$ and $|\mathfrak{A}^t x|$ are nonincreasing (because of the fact that $\mathfrak{A}$ is a contraction semigroup). Thus, with the notation of Theorem 11.1.6, $X_0 = \mathcal{N}(1 - Q_{\mathfrak{A}^*,\mathfrak{A}}) \cap \mathcal{N}(1 - Q_{\mathfrak{A},\mathfrak{A}^*})$, hence X_0 is a closed linear subspace of X_0. We claim that X_0 is invariant under both $\mathfrak{A}$ and $\mathfrak{A}^*$. Let $x \in X_0$, and let $s \geq 0$. Then, for all $t \geq 0$, $|\mathfrak{A}^t\mathfrak{A}^s x| = |\mathfrak{A}^{t+s}x| = |x| = |\mathfrak{A}^s x|$, so the first part of (11.1.16) holds with x replaced by $\mathfrak{A}^s x$. Moreover, for $t \geq s$, we have, by Lemma 11.1.10,

$$|\mathfrak{A}^{*t}\mathfrak{A}^s x| = |\mathfrak{A}^{*(t-s)}\mathfrak{A}^{*s}\mathfrak{A}^s x| = |\mathfrak{A}^{*(t-s)}x| = |x| = |\mathfrak{A}^s x|,$$

so the second half of (11.1.17) holds. This shows that $\mathfrak{A}^s x \in X_0$, i.e., X_0 is invariant under $\mathfrak{A}$. An analogous argument shows that X_0 is invariant also under $\mathfrak{A}^*$. Moreover, the same argument shows that the restrictions of $\mathfrak{A}$ and $\mathfrak{A}^*$ to X_0 are inverses of each other, so $\mathfrak{A}_0$ is unitary.

Let $X_1 = X_0^\perp$. Then X_1 is invariant under both $\mathfrak{A}$ and $\mathfrak{A}^*$ (the invariance of X_0 under $\mathfrak{A}$ implies that X_1 is invariant under $\mathfrak{A}^*$, and the invariance of X_0 under $\mathfrak{A}^*$ implies that X_1 is invariant under $\mathfrak{A}$). Clearly $\mathfrak{A}_{|X_1}$ must be completely nonunitary, since no nonzero $x \in X_1$ satisfies (11.1.16) (by the definition of X_0). It is also clear that the decomposition of X into $\left[\begin{smallmatrix} X_0 \\ X_1 \end{smallmatrix}\right]$ is unique.

By construction, both $Q_{\mathfrak{A}^*,\mathfrak{A}}$ and $Q_{\mathfrak{A},\mathfrak{A}^*}$ act like identity operators on X_0, so (11.1.12) implies that for all $x \in X_0$

$$\langle x, \mathfrak{C}^*\mathfrak{C}x\rangle \leq 0, \qquad \langle x, \mathfrak{B}\mathfrak{B}^*x\rangle \leq 0,$$

i.e., $X_0 \subset \mathcal{N}(\mathfrak{C})$ and $X_0 \subset \mathcal{N}(\mathfrak{B}^*)$, or equivalently, $\mathcal{R}(\mathfrak{B}) \subset X_0^\perp = X_1$. Therefore, if we split the system Σ according to the splitting of the state space $X = \left[\begin{smallmatrix} X_0 \\ X_1 \end{smallmatrix}\right]$, then we get

$$\left[\begin{array}{c|c} \mathfrak{A} & \mathfrak{B} \\ \hline \mathfrak{C} & \mathfrak{D} \end{array}\right] = \left[\begin{array}{cc|c} \mathfrak{A}_0 & 0 & 0 \\ 0 & \mathfrak{A}_1 & \mathfrak{B} \\ \hline 0 & \mathfrak{C}_1 & \mathfrak{D} \end{array}\right],$$

where $\mathfrak{A}_0 = \mathfrak{A}_{|X_0}$, $\mathfrak{A}_1 = \mathfrak{A}_{|X_1}$, and $\mathfrak{C}_1 = \mathfrak{C}_{|X_1}$. We have now completed the first part of the proof, related to the splitting of the state space X into $\left[\begin{smallmatrix} X_0 \\ X_1 \end{smallmatrix}\right]$. It remains to split the system $\left[\begin{array}{c|c} \mathfrak{A}_1 & \mathfrak{B} \\ \hline \mathfrak{C}_1 & \mathfrak{D} \end{array}\right]$ by further splitting Y and U.

Our splitting of Y and U into $\left[\begin{smallmatrix} Y_1 \\ Y_0 \end{smallmatrix}\right]$ and $\left[\begin{smallmatrix} U_1 \\ U_0 \end{smallmatrix}\right]$ is based on properties of the transfer function $\widehat{\mathfrak{D}}$. By Lemma 11.1.4 and Theorem 10.3.5, we have $\widehat{\mathfrak{D}} \in H^\infty_0(U;Y)$ and $\|\widehat{\mathfrak{D}}(z)\| \leq 1$ for all $z \in \mathbb{C}^+$. Let U_0 consist of all those $u \in U$ which satisfy

$$|\widehat{\mathfrak{D}}(1)u| = |u|, \tag{11.1.18}$$

and let Y_0 consist of all those $y \in Y$ which satisfy

$$|\widehat{\mathfrak{D}}(1)^* y| = |y|. \tag{11.1.19}$$

By Lemma 11.1.10, $U_0 = \mathcal{N}\left(1 - \widehat{\mathfrak{D}}(1)^*\widehat{\mathfrak{D}}(1)\right)$ and $Y_0 = \mathcal{N}\left(1 - \widehat{\mathfrak{D}}(1)\widehat{\mathfrak{D}}(1)^*\right)$, so U_0 and Y_0 are closed linear subspaces of U. For each $u \in U_0$ we have $|\widehat{\mathfrak{D}}(1)^*\widehat{\mathfrak{D}}(1)u| = |u| = |\widehat{\mathfrak{D}}(1)u|$, so $\widehat{\mathfrak{D}}(1)u \in Y_0$. Conversely, for each $y \in Y_0$ we have $|\widehat{\mathfrak{D}}(1)\widehat{\mathfrak{D}}(1)^* y| = |y| = |\widehat{\mathfrak{D}}(1)^* y|$, so $\widehat{\mathfrak{D}}(1)^* y \in U_0$. Thus, $\widehat{\mathfrak{D}}(1)$ is a unitary operator from U_0 to Y_0.

We claim that the restriction of $\mathfrak{D}$ to $L^2(\mathbb{R}; U_0)$ is a static unitary operator with values in $L^2(\mathbb{R}; Y_0)$. More precisely, we claim that for all $u \in L^2(\mathbb{R}; U_0)$, we have $(\mathfrak{D}u)(t) = Du(t)$ for almost all $t \in \mathbb{R}$, where $D = \widehat{\mathfrak{D}}(1)_{|U_0}$. To prove this claim we first show that $\widehat{\mathfrak{D}}(z)_{|U_0}$ is independent of z for all $z \in \mathbb{C}^+$.

Take some nonzero $u \in U_0$. Then $\widehat{\mathfrak{D}}(z)u$ is a bounded analytic function in the disk $|z - 1| \le 1/2$, so by Cauchy's theorem,

$$\widehat{\mathfrak{D}}(1)u = \frac{1}{2\pi}\int_0^{2\pi} \widehat{\mathfrak{D}}(1 + \tfrac{1}{2}\mathrm{e}^{j\varphi})u \, d\varphi.$$

In particular, by the Schwarz inequality,

$$\begin{aligned}
|u|^2 = |\widehat{\mathfrak{D}}(1)u|^2 &= \langle \widehat{\mathfrak{D}}(1)u, \widehat{\mathfrak{D}}(1)u\rangle \\
&= \frac{1}{2\pi}\int_0^{2\pi} \langle \widehat{\mathfrak{D}}(1 + \tfrac{1}{2}\mathrm{e}^{j\varphi})u, \widehat{\mathfrak{D}}(1)u\rangle \, d\varphi \\
&\le \frac{1}{2\pi}\left\{\int_0^{2\pi} |\widehat{\mathfrak{D}}(1 + \tfrac{1}{2}\mathrm{e}^{j\varphi})u|^2 \, d\varphi\right\}^{1/2} |\widehat{\mathfrak{D}}(1)u| \\
&\le \left\{\int_0^{2\pi} \frac{1}{2\pi}|u|^2 \, d\varphi\right\}^{1/2} |\widehat{\mathfrak{D}}(1)u| = |u|^2.
\end{aligned}$$

Thus, the inequalities above are, in fact, equalities. Since the Schwarz inequality becomes an equality, we must therefore have $\widehat{\mathfrak{D}}(1 + \frac{1}{2}\mathrm{e}^{j\varphi})u = \alpha\widehat{\mathfrak{D}}(1)u$ for some $\alpha \in \mathbb{C}$ and for (almost) all φ, and this together with the Cauchy representation formula implies that $\widehat{\mathfrak{D}}(z)u = \alpha\widehat{\mathfrak{D}}(1)u$ for all z in the disk $|z - 1| \le 1/2$; in particular, $\alpha = 1$. By analytic continuation, it must then be true that $\widehat{\mathfrak{D}}(z)u = \widehat{\mathfrak{D}}(1)u$ for all $u \in \mathbb{C}^+$.

Define $U_1 = U_0^\perp$ and $Y_1 = Y_0^\perp$, and let $\mathfrak{B}_1 = \mathfrak{B}_{|L^2(\mathbb{R};U_1)}$, $\mathfrak{D}_1 = \mathfrak{D}_{|L^2(\mathbb{R};U_1)}$, and $\mathfrak{D}_0 = \mathfrak{D}_{|L^2(\mathbb{R};U_0)}$. Then, by the preceding argument, $\widehat{\mathfrak{D}}_0 = \widehat{\mathfrak{D}}_{|U_0} = \widehat{\mathfrak{D}}(1)_{|U_0} = D$, where $D = \widehat{\mathfrak{D}}(1)_{|U_0}$ is a unitary operator from U_0 onto Y_0. By Theorem 10.3.5, for all $u \in L^2(\mathbb{R}^+; U_0)$ we therefore have $\mathfrak{D}u \in L^2(\mathbb{R}^+; Y_0)$. This combined with the shift-invariance of $\mathfrak{D}$ and the fact that functions whose support is bounded to the left implies that $\mathcal{R}(\mathfrak{D}_0) \subset L^2(\mathbb{R}; Y_0)$. This gives us one of the two zeros in the decomposition of $\mathfrak{D}$ given in (11.1.15). To get the other zero we apply the same argument to the dual system.

There are still two more zeros in (11.1.15) to be accounted for, namely the one to the right of $\mathfrak{B}_1$ and the one below $\mathfrak{C}_1$. These are obtained as follows. We know that $\mathfrak{D}_0$ is a unitary static operator mapping $L^2(\mathbb{R}; U_0)$ onto $L^2(\mathbb{R}; Y_0)$, and this together with (11.1.12) implies that $L^2(\mathbb{R}^-; U_0) \subset \mathcal{N}(\mathfrak{B})$ and that $L^2(\mathbb{R}^+; Y_0) \subset \mathcal{N}(\mathfrak{C}_1^*)$, or equivalently, that $\mathcal{R}(\mathfrak{C}) = \mathcal{R}(\mathfrak{C}_1) \subset L^2(\mathbb{R}^+; Y_1)$. □

Definition 11.1.11 Let $\mathfrak{A}$ be a contraction semigroup on a Hilbert space X, and let $\mathfrak{D}$ be a contraction in $TIC^2(U; Y)$ where U and Y are Hilbert spaces.

(i) By the *unitary part* of $\mathfrak{A}$ we mean the semigroup $\mathfrak{A}_0$ in Theorem 11.1.9, and by the *completely nonunitary part* of $\mathfrak{A}$ we mean the semigroup $\mathfrak{A}_1$ in Theorem 11.1.9.
(ii) By the *purely passive part* of $\mathfrak{D}$ we mean the operator $\mathfrak{D}_1$ in Theorem 11.1.9 (the splitting of $\mathfrak{D}$ into $\mathfrak{D} = \mathfrak{D}_0 + \mathfrak{D}_1$ in that theorem only depends on $\widehat{\mathfrak{D}}$, and not on $\mathfrak{A}$, $\mathfrak{B}$, or $\mathfrak{C}$).
(iii) A passive system $\Sigma = \left[\begin{smallmatrix} \mathfrak{A} & \mathfrak{B} \\ \mathfrak{C} & \mathfrak{D} \end{smallmatrix}\right]$ is *completely nonunitary* if $\mathfrak{A}$ is completely nonunitary, it is *purely passive* of $\mathfrak{D}$ is purely passive, and it is *simple* if it is both completely nonunitary and purely passive.

In other words, in the notation of Theorem 11.1.9, in part (iii) we have $X_0 = U_0 = Y_0 = 0$. This definition is motivated by the fact that both $\mathfrak{A}_0$ and $\mathfrak{D}_0$ represent 'trivial' behavior where there is no interaction between the semigroup and the input/output map. We warn the reader our use of the word 'simple' is not quite the same as in, e.g., Arov and Nudelman (1996), where 'simple' stands for our 'completely nonunitary' without any requirement of pure passivity. (Although we have defined 'simple' for an arbitrary passive system, we shall primarily apply it only to conservative systems.)

We shall occasionally need to refer to the maximal isometric or co-isometric part of a contraction semigroup.

Lemma 11.1.12 *Let $\mathfrak{A}$ be a C_0 contraction semigroup on a Hilbert space X. Then there is a unique maximal invariant (closed) subspace of X on which $\mathfrak{A}$ is isometric, and there is also a unique maximal co-invariant (closed) subspace on which $\mathfrak{A}^*$ is isometric. The intersection of these two subspaces coincides with the maximal subspace X_0 in Theorem 11.1.9 on which $\mathfrak{A}$ is unitary.*

Proof Let X_i be the set of all $x \in X$ for which $|\mathfrak{A}^t x| = |x|$ for all $t \geq 0$. As in the proof of Theorem 11.1.9 we find that X_i is an invariant subspace of $\mathfrak{A}$, and clearly there is no larger subspace of X on which $\mathfrak{A}$ can be an isometry. The claim about the existence of a maximal co-invariant subspace X_{ci} on which $\mathfrak{A}^*$ is isometric is proved in a similar manner. Finally, that the intersection of these two subspaces is equal to the subspace X_0 in Theorem 11.1.9 follows from the proof of that theorem. □

11.2 Energy preserving and conservative systems

In the *conservative* case we require (11.1.1) to hold with the inequality replaced by equality, both for the original system and for its adjoint. This can be alternatively characterized as follows.

Definition 11.2.1 By a (scattering) *energy preserving system* Σ on three Hilbert spaces (Y, X, U) we mean an L^2-well-posed linear system with the following property: for all initial states $x_0 \in X$ and all input functions $u \in L^2_{\mathrm{loc}}(\mathbb{R}^+; U)$, the state trajectory x and the output function y of Σ with initial time zero satisfy

$$|x(t)|_X^2 + \int_0^t |y(s)|_Y^2\, ds = |x_0|_X^2 + \int_0^t |u(s)|_U^2\, ds, \quad t \geq 0. \quad (11.2.1)$$

By a (scattering) *co-energy preserving* system we mean a system whose causal dual system Σ^d is energy preserving, and by a *conservative* system we mean a system which is both energy preserving and co-energy preserving. In these cases we shall also call the system node S of Σ (scattering) *energy preserving*, or *co-energy preserving*, or *conservative*.

The name 'conservative' in Definition 11.2.1 is motivated by the fact that the energy is 'conserved' both in the original system and in the adjoint system.

The following two lemmas are the energy preserving and conservative versions of Lemma 11.1.4.

Lemma 11.2.2 *Let Σ be an L^2-well-posed system on three Hilbert spaces (Y, X, U). Then the following conditions are equivalent:*

(i) *Σ is (scattering) energy preserving.*
(ii) *In the notation of Definition 2.2.6, for all $-\infty < s < t < \infty$, the operator $\Sigma_s^t = \begin{bmatrix} \mathfrak{A}_s^t & \mathfrak{B}_s^t \\ \mathfrak{C}_s^t & \mathfrak{D}_s^t \end{bmatrix}$ is an isometry from $\begin{bmatrix} X \\ L^2([s,t);U) \end{bmatrix}$ to $\begin{bmatrix} X \\ L^2([s,t);Y) \end{bmatrix}$.*
(iii) *Let $s \in \mathbb{R}$, $x_s \in X$, and $u \in L^2_{\mathrm{loc}}([s, \infty); U)$, and let x be the state and let y be the output of Σ with initial time s, initial state x_s, and input function u. Then, for all $t \geq r \geq s$,*

$$|x(t)|_X^2 + \int_r^t |y(s)|_Y^2\, ds = |x(r)|_X^2 + \int_r^t |u(s)|_U^2\, ds. \quad (11.2.2)$$

(iv) *The corresponding Lax–Phillips model with parameter $\omega = 0$ is an isometric semigroup.*

The proof of this lemma is a simplified version of the proof of Lemma 11.1.4, and we leave it to the reader.

Lemma 11.2.3 *Let Σ be an L^2-well-posed system on three Hilbert spaces (Y, X, U). Then the following conditions are equivalent:*

(i) Σ *is (scattering) conservative.*

(ii) *In the notation of Definition 2.2.6, for all* $-\infty < s < t < \infty$, *the operator* $\Sigma_s^t = \begin{bmatrix} \mathfrak{A}_s^t & \mathfrak{B}_s^t \\ \mathfrak{C}_s^t & \mathfrak{D}_s^t \end{bmatrix}$ *is a unitary operator from* $\begin{bmatrix} X \\ L^2([s,t);U) \end{bmatrix}$ *to* $\begin{bmatrix} X \\ L^2([s,t);Y) \end{bmatrix}$.

(iii) *Let* $s \in \mathbb{R}$, $x_s \in X$, *and* $u \in L^2_{\mathrm{loc}}([s,\infty);U)$, *and let* x *be the state and let* y *be the output of* Σ *with initial time* s, *initial state* x_s, *and input function* u. *Then, for all* $t \geq r \geq s$,

$$|x(t)|_X^2 + \int_r^t |y(s)|_Y^2\, ds = |x(r)|_X^2 + \int_r^t |u(s)|_U^2\, ds, \qquad (11.2.3)$$

and the same statement is true for the dual system, too.

(iv) *The corresponding Lax–Phillips model with parameter* $\omega = 0$ *is a unitary semigroup.*

This follows Lemma 11.2.2, applied both to the original system and its causal dual.

In particular, we observe the following fact:

Lemma 11.2.4 *Every conservative system* Σ *is time-flow-invertible, and its time-flow-inverse* $\Sigma_\times^{\mathsf{R}}$ *coincides with its causal dual* Σ^d.

This follows from Lemma 11.2.2, applied both to the original system Σ and its causal dual Σ^d, and the definition of a time-flow-invertible system.

A result similar to the one in Theorem 11.1.5 is true also for energy preserving, co-energy preserving, and conservative systems. For simplicity, let us only treat the energy preserving case (since the others can be reduced to this one).

Theorem 11.2.5 *Let* $S = \begin{bmatrix} A\&B \\ C\&D \end{bmatrix}$ *be an operator node on the three Hilbert spaces* (Y, X, U) *with main operator* A, *control operator* B, *observation operator* C, *and transfer function* $\widehat{\mathfrak{D}}$. *Then the following conditions are equivalent.*

(i) S *is (scattering) energy preserving (in particular,* A *generates a* C_0 *continuous semigroup and* S *is* L^2*-well-posed).*

(ii) A *generates a* C_0 *semigroup, and for all* $x_0 \in X$ *and* $u \in W^{2,1}_{\mathrm{loc}}(\mathbb{R}^+;U)$ *with* $\begin{bmatrix} x_0 \\ u(0) \end{bmatrix} \in \mathcal{D}(S)$ *the state trajectory* x *and the output function* y *in Lemma 4.7.8 with* $s = 0$ *satisfy*

$$\frac{\mathrm{d}}{\mathrm{d}t}|x(t)|_X^2 + |y(t)|_Y^2 = |u(t)|_U^2, \qquad t \geq 0. \qquad (11.2.4)$$

(iii) $\rho(A) \cap \mathbb{C}_+ \neq \emptyset$, *and for all* $\begin{bmatrix} x_0 \\ u_0 \end{bmatrix} \in \mathcal{D}(S)$,

$$2\Re\left\langle A\&B \begin{bmatrix} x_0 \\ u_0 \end{bmatrix}, x_0 \right\rangle_X + \left| C\&D \begin{bmatrix} x_0 \\ u_0 \end{bmatrix} \right|_Y^2 = |u_0|_U^2. \qquad (11.2.5)$$

(iv) $\rho(A) \cap \mathbb{C}_+ \neq \emptyset$, *and for all* $\alpha, \beta \in \rho(A)$ *we have, with* $E_\alpha = \left[\begin{smallmatrix} 1 & (\alpha - A_{|X})^{-1}B \\ 0 & 1 \end{smallmatrix}\right]$ *and* $E_\beta = \left[\begin{smallmatrix} 1 & (\beta - A_{|X})^{-1}B \\ 0 & 1 \end{smallmatrix}\right]$,

$$E_\alpha^* \left[\begin{smallmatrix} 1 & 0 \\ 0 & 0 \end{smallmatrix}\right] SE_\beta + (SE_\alpha)^* \left[\begin{smallmatrix} 1 & 0 \\ 0 & 0 \end{smallmatrix}\right] E_\beta + (SE_\alpha)^* \left[\begin{smallmatrix} 0 & 0 \\ 0 & 1 \end{smallmatrix}\right] SE_\beta = \left[\begin{smallmatrix} 0 & 0 \\ 0 & 1 \end{smallmatrix}\right], \tag{11.2.6}$$

which should be interpreted as an operator equation in $\mathcal{B}\left(\left[\begin{smallmatrix} X_1 \\ U \end{smallmatrix}\right]; \left[\begin{smallmatrix} X_{-1}^* \\ U \end{smallmatrix}\right]\right)$, *and we identify the dual of* $\left[\begin{smallmatrix} X_1 \\ U \end{smallmatrix}\right]$ *with* $\left[\begin{smallmatrix} X_{-1}^* \\ U \end{smallmatrix}\right]$.

(v) $\rho(A) \cap \mathbb{C}_+ \neq \emptyset$, *and the identity* (11.2.6) *holds for some* $\alpha, \beta \in \rho(A)$.

(vi) $\rho(A) \cap \mathbb{C}_+ \neq \emptyset$, *and for all* $\alpha, \beta, \gamma \in \rho(A)$, *we have*

$$\begin{aligned} A + A_{|X}^* + C^*C &= 0, \\ (\alpha + A_{|X}^*)(\alpha - A_{|X})^{-1}B + C^*\widehat{\mathfrak{D}}(\alpha) &= 0, \\ B^*(\overline{\beta} - A^*)^{-1}(\overline{\beta} + \gamma)(\gamma - A_{|X})^{-1}B + \widehat{\mathfrak{D}}(\beta)^*\widehat{\mathfrak{D}}(\gamma) &= 1, \end{aligned} \tag{11.2.7}$$

which should be interpreted as operator equations in $\mathcal{B}(X_1; X_{-1}^*)$, $\mathcal{B}(U; X_{-1}^*)$, *and* $\mathcal{B}(U; Y)$, *respectively.*

(vii) $\rho(A) \cap \mathbb{C}_+ \neq \emptyset$, *and the identities* (11.2.7) *hold for some* $\alpha, \beta, \gamma \in \rho(A)$.

(viii) $\mathbb{C}^+ \subset \rho(A)$, *and for all* $\alpha \in \mathbb{C}^+$, *the operator*

$$\begin{bmatrix} \mathbf{A}(\alpha) & \mathbf{B}(\alpha) \\ \mathbf{C}(\alpha) & \mathbf{D}(\alpha) \end{bmatrix} = \begin{bmatrix} (\overline{\alpha} + A)(\alpha - A)^{-1} & \sqrt{2\Re\alpha}\,(\alpha - A_{|X})^{-1}B \\ \sqrt{2\Re\alpha}\,C(\alpha - A)^{-1} & \widehat{\mathfrak{D}}(\alpha) \end{bmatrix}$$

is an isometry.

(ix) *For some* $\alpha \in \rho(A) \cap \mathbb{C}^+$, *the operator* $\left[\begin{smallmatrix} \mathbf{A}(\alpha) & \mathbf{B}(\alpha) \\ \mathbf{C}(\alpha) & \mathbf{D}(\alpha) \end{smallmatrix}\right]$ *defined in (viii) is an isometry.*

If both $\rho(A) \cap \mathbb{C}^+ \neq \emptyset$ *and* $\rho(A) \cap \mathbb{C}^- \neq \emptyset$, *then these conditions are further equivalent to the following two conditions:*

(x) *For all* $\alpha \in \rho(A) \cap \mathbb{C}^-$, *the operator* $\left[\begin{smallmatrix} \mathbf{A}(\alpha) & \mathbf{B}(\alpha) \\ \mathbf{C}(\alpha) & \mathbf{D}(\alpha) \end{smallmatrix}\right]$ *defined in (viii) (with* $\sqrt{2\Re\alpha} = j\sqrt{2|\Re\alpha|}$*) satisfies*

$$\begin{bmatrix} \mathbf{A}(\alpha) & \mathbf{B}(\alpha) \\ \mathbf{C}(\alpha) & \mathbf{D}(\alpha) \end{bmatrix}^* \begin{bmatrix} -1 & 0 \\ 0 & 1 \end{bmatrix} \begin{bmatrix} \mathbf{A}(\alpha) & \mathbf{B}(\alpha) \\ \mathbf{C}(\alpha) & \mathbf{D}(\alpha) \end{bmatrix} = \begin{bmatrix} -1 & 0 \\ 0 & 1 \end{bmatrix}. \tag{11.2.8}$$

(xi) *The identity* (11.2.8) *holds for some* $\alpha \in \rho(A) \cap \mathbb{C}^-$.

If $\rho(A) \cap j\mathbb{R} \neq \emptyset$ *then (i)–(xi) are further equivalent to the following two conditions:*

(xii) *For all* $\alpha \in \rho(A)$ *and all* $\beta \in \rho(A) \cap j\mathbb{R}$, *we have*

$$\begin{aligned} A + A_{|X}^* + C^*C &= 0, \\ (\alpha + A_{|X}^*)(\alpha - A_{|X})^{-1}B + C^*\widehat{\mathfrak{D}}(\alpha) &= 0, \\ \widehat{\mathfrak{D}}(\beta)^*\widehat{\mathfrak{D}}(\beta) &= 1. \end{aligned} \tag{11.2.9}$$

(xiii) *The identities in* (11.2.9) *hold for some* $\alpha \in \rho(A)$ *and some* $\beta \in \rho(A) \cap j\mathbb{R}$.

Proof Most of the proof is identical to the proof of Theorem 11.1.5, except that we replace $\le$ by $=$. In particular the equivalence of the conditions (i), (ii), (iii), (viii), (ix), (x), and (xi) is proved in this way, and so are the equivalence of these conditions to any one of (iv)–(vii) and (xii)–(xiii) as long as we take $\alpha = \beta = \gamma$ in (iv)–(vii) and (xii)–(xiii). In the case of (vi), (vii), (xii), and (xiii) this follows from Lemma 11.2.6 below. Thus, it remains to show that we may allow different α and β in (iv) and (v).

(iii) $\Leftrightarrow$ (iv): As we observed above, (iii) is equivalent to (11.2.6) with the extra restriction that $\beta = \alpha$. To get rid of this restriction it suffices to observe that $E_\alpha^{-1}E_\beta$ is an isomorphism of $\left[\begin{smallmatrix} X_1 \\ U \end{smallmatrix}\right]$ onto itself for all $\alpha, \beta \in \rho(A)$, that $\left[\begin{smallmatrix} 0 & 0 \\ 0 & 1 \end{smallmatrix}\right] E_\alpha^{-1}E_\beta = \left[\begin{smallmatrix} 0 & 0 \\ 0 & 1 \end{smallmatrix}\right]$, and to multiply the restricted version of (11.2.6) (with $\beta = \alpha$) to the right by $E_\alpha^{-1}E_\beta$ to get the general version of (11.2.6).

(iii) $\Leftrightarrow$ (v): This proof is analogous to the proof of the equivalence of (iii) and (iv) given above. □

Lemma 11.2.6 *Make the same assumptions as in Theorem 11.2.5.*

(i) *If* $A + A_{|X}^* + C^*C = 0$ *(as an operator identity in* $\mathcal{B}(X_1; X_{-1}^*)$*), then the operator* $Q(\alpha) \in \mathcal{B}(X_1; Y)$ *defined by*

$$Q(\alpha) := B^*(\overline{\alpha} - A^*)^{-1}(\overline{\alpha} + A) + \widehat{\mathfrak{D}}(\alpha)^*C$$

does not depend on $\alpha \in \rho(A)$.

(ii) *If (i) holds and* $Q(\alpha) = 0$ *for some (and hence, for all)* $\alpha \in \rho(A)$*, then the operator* $R(\beta, \gamma) \in \mathcal{B}(U)$ *defined by*

$$R(\beta, \gamma) := B^*(\overline{\beta} - A^*)^{-1}(\overline{\beta} + \gamma)(\gamma - A_{|X})^{-1}B + \widehat{\mathfrak{D}}(\beta)^*\widehat{\mathfrak{D}}(\gamma)$$

is self-adjoint and independent of $\beta, \gamma \in \rho(A)$.

Proof (i) By using the Lyapunov equation $A + A_{|X}^* + C^*C = 0$ we can rewrite $Q(\alpha)$ in the form

$$\begin{aligned} Q(\alpha) &= B^*(\overline{\alpha} - A_{|X}^*)^{-1}\big[\overline{\alpha} - A_{|X}^* - C^*C\big] + \widehat{\mathfrak{D}}(\alpha)^*C \\ &= B^*\big[1 - (\overline{\alpha} - A_{|X}^*)^{-1}C^*C\big] + \widehat{\mathfrak{D}}(\alpha)^*C. \end{aligned}$$

If also $\beta \in \rho(A)$, then, by the dual version of (4.7.2)

$$\begin{aligned} Q(\alpha) - Q(\beta) = &-B^*\big[(\overline{\alpha} - A_{|X}^*)^{-1} - (\overline{\beta} - A_{|X}^*)^{-1}\big]C^*C \\ &+ \big[\widehat{\mathfrak{D}}(\alpha)^* - \widehat{\mathfrak{D}}(\beta)^*\big]C = 0. \end{aligned}$$

(ii) We first show that $R(\beta,\gamma)$ is independent of β. Let $\alpha,\beta,\gamma\in\rho(A)$. Since $Q(\gamma)=0$, also its adjoint is zero, i.e.,

$$Q(\gamma)^* = (\gamma + A^*_{|X})(\gamma - A_{|X})^{-1}B + C^*\widehat{\mathfrak{D}}(\gamma) = 0.$$

By using this equation we can rewrite $R(\beta,\gamma)$ in the form

$$\begin{aligned} R(\beta,\gamma) &= B^*(\overline{\beta} - A^*)^{-1}\big[(\overline{\beta} - A^*_{|X})(\gamma - A_{|X})^{-1}B - C^*\widehat{\mathfrak{D}}(\gamma)\big] \\ &\quad + \widehat{\mathfrak{D}}(\beta)^*\widehat{\mathfrak{D}}(\gamma) \\ &= B^*\big[(\gamma - A_{|X})^{-1}B - (\overline{\beta} - A^*_{|X})^{-1}C^*\widehat{\mathfrak{D}}(\gamma)\big] + \widehat{\mathfrak{D}}(\beta)^*\widehat{\mathfrak{D}}(\gamma). \end{aligned}$$

Therefore, by the dual version of (4.7.2),

$$\begin{aligned} R(\alpha,\gamma) - R(\beta,\gamma) &= -B^*\big[(\overline{\beta} - A^*_{|X})^{-1} - (\overline{\alpha} - A^*_{|X})^{-1}\big]C^*\widehat{\mathfrak{D}}(\gamma) \\ &\quad + \big[\widehat{\mathfrak{D}}(\beta)^* - \widehat{\mathfrak{D}}(\alpha)^*\big]\widehat{\mathfrak{D}}(\gamma) = 0. \end{aligned}$$

This shows that $R(\beta,\gamma)$ is independent of β. It must also be independent of γ because of the fact that obviously, $R(\beta,\gamma) = R(\gamma,\beta)^*$, and $R(\gamma,\beta)$ does not depend on γ. To see that $R(\beta,\gamma)$ is self-adjoint it suffices to take $\beta=\gamma$. □

From Theorem 11.2.5 we immediately derive the following result:

Corollary 11.2.7 *Let A be a Hilbert space, let $A\colon X\supset\mathcal{D}(A)\to X$ be a linear operator with dense domain, and suppose that $\rho(A)\cap\mathbb{C}^+\neq\emptyset$. Then*

(i) *A is the generator of a C_0 isometric semigroup if and only if $\mathcal{D}(A)\subset\mathcal{D}(A^*)$ and $A^*x=-Ax$ for all $x\in\mathcal{D}(A)$;*
(ii) *A is the generator of a C_0 co-isometric semigroup if and only if $\mathcal{D}(A^*)\subset\mathcal{D}(A)$ and $A^*x=-Ax$ for all $x\in\mathcal{D}(A^*)$; and*
(iii) *A is the generator of a C_0 unitary semigroup if and only if $\mathcal{D}(A^*)=\mathcal{D}(A)$ and $A^*=-A$.*

Proof It suffices to prove (i), since (ii) follows from (i) by duality, and (iii) is a combination of (i) and (ii).

Suppose that A is the generator of a C_0 contraction semigroup $\mathfrak{A}$. Then, for all $x\in\mathcal{D}(A)$, we have $Ax=\lim_{h\downarrow 0}\frac{1}{h}(\mathfrak{A}^h x - x)$. Since $\mathfrak{A}^{*h}$ tends strongly to zero as $h\downarrow 0$ and $\mathfrak{A}^{*h}\mathfrak{A}^h = 1$ for all $h>0$, this implies that also the following limit exists:

$$Ax = \lim_{h\downarrow 0}\frac{1}{h}\mathfrak{A}^{*h}(\mathfrak{A}^h x - x) = \lim_{h\downarrow 0}\frac{1}{h}(x - \mathfrak{A}^{*h}x).$$

Thus, $x\in\mathcal{D}(A^*)$ and $A^*x=-Ax$.

Conversely, suppose that $\mathcal{D}(A)\subset\mathcal{D}(A^*)$ and that $A^*x=-Ax$ for all $x\in\mathcal{D}(A)$. Then, for all $x\in\mathcal{D}(A)$,

$$2\Re\langle x, Ax\rangle = \langle x, Ax\rangle + \langle Ax, x\rangle = \langle x, (A+A^*)x\rangle = 0.$$

Thus, A is passive, and it generates a C_0 contraction semigroup. That this semigroup is isometric follows from Theorem 11.2.5(iii) with $U = Y = 0$ (i.e., there is neither an input nor an output). □

Whereas the descriptions of an energy preserving system given in Theorem 11.2.5 are more or less directly inherited from Theorem 11.1.5, the following alternative characterization is of a slightly different nature.

Theorem 11.2.8 *A system node $S = \left[\begin{smallmatrix} A\&B \\ C\&D \end{smallmatrix}\right]$ with main operator A on three Hilbert spaces (Y, X, U) is energy preserving if and only if $\rho(A) \cap \mathbb{C}_+ \neq \emptyset$, $\left[\begin{smallmatrix} 1 & 0 \\ C\&D \end{smallmatrix}\right]$, maps $\mathcal{D}(S)$ into $\mathcal{D}(S^*)$ and*

$$S^* \begin{bmatrix} 1 & 0 \\ C\&D \end{bmatrix} = \begin{bmatrix} -A\&B \\ 0 \quad 1 \end{bmatrix} \text{ (on } \mathcal{D}(S)). \tag{11.2.10}$$

Proof The proof is based on the equivalence of (i) and (iii) in Theorem 11.2.5.

Suppose that S is energy preserving. By regrouping the terms in (11.2.5) we get the equivalent identity

$$\left\langle \begin{bmatrix} A\&B \\ C\&D \end{bmatrix} \begin{bmatrix} x_0 \\ u_0 \end{bmatrix}, \begin{bmatrix} 1 & 0 \\ C\&D \end{bmatrix} \begin{bmatrix} x_0 \\ u_0 \end{bmatrix} \right\rangle_{\left[\begin{smallmatrix} X \\ Y \end{smallmatrix}\right]} = \left\langle \begin{bmatrix} x_0 \\ u_0 \end{bmatrix}, \begin{bmatrix} -A\&B \\ 0 \quad 1 \end{bmatrix} \begin{bmatrix} x_0 \\ u_0 \end{bmatrix} \right\rangle_{\left[\begin{smallmatrix} X \\ Y \end{smallmatrix}\right]}, \tag{11.2.11}$$

valid for all $\left[\begin{smallmatrix} x_0 \\ u_0 \end{smallmatrix}\right] \in \mathcal{D}(S)$. This together with the polarization identity gives

$$\left\langle \begin{bmatrix} A\&B \\ C\&D \end{bmatrix} \begin{bmatrix} x_1 \\ u_1 \end{bmatrix}, \begin{bmatrix} 1 & 0 \\ C\&D \end{bmatrix} \begin{bmatrix} x_0 \\ u_0 \end{bmatrix} \right\rangle_{\left[\begin{smallmatrix} X \\ Y \end{smallmatrix}\right]} = \left\langle \begin{bmatrix} x_1 \\ u_1 \end{bmatrix}, \begin{bmatrix} -A\&B \\ 0 \quad 1 \end{bmatrix} \begin{bmatrix} x_0 \\ u_0 \end{bmatrix} \right\rangle_{\left[\begin{smallmatrix} X \\ Y \end{smallmatrix}\right]}, \tag{11.2.12}$$

for all $\left[\begin{smallmatrix} x_0 \\ u_0 \end{smallmatrix}\right] \in \mathcal{D}(S)$ and all $\left[\begin{smallmatrix} x_1 \\ u_1 \end{smallmatrix}\right] \in \mathcal{D}(S)$. Here the right-hand side has an obvious extension to a bounded linear functional on $\left[\begin{smallmatrix} X \\ Y \end{smallmatrix}\right]$ (with respect to the argument $\left[\begin{smallmatrix} x_1 \\ u_1 \end{smallmatrix}\right]$) for all $\left[\begin{smallmatrix} x_0 \\ u_0 \end{smallmatrix}\right] \in \mathcal{D}(S)$, so $\left[\begin{smallmatrix} 1 & 0 \\ C\&D \end{smallmatrix}\right]\left[\begin{smallmatrix} x_0 \\ u_0 \end{smallmatrix}\right] \in \mathcal{D}(S^*)$ and (11.2.10) holds.

The same computation done backwards proves the converse claim. □

Our following theorem is a specialized and, at the same time, expanded version of Theorem 11.1.6.

Theorem 11.2.9 *Let $\Sigma = \left[\begin{smallmatrix} \mathfrak{A} & \mathfrak{B} \\ \mathfrak{C} & \mathfrak{D} \end{smallmatrix}\right]$ be a passive system on (Y, X, U) with system node $S = \left[\begin{smallmatrix} A\&B \\ C\&D \end{smallmatrix}\right]$, observation operator C, and transfer function $\widehat{\mathfrak{D}}$. Introduce the same notation as in Theorem 11.1.6.*

(i) *If Σ is energy preserving, then*

$$\begin{bmatrix} Q_{\mathfrak{A}^*,\mathfrak{A}} & Q_{\mathfrak{A}^*,\mathfrak{B}} \\ Q_{\mathfrak{B}^*,\mathfrak{A}} & Q_{\mathfrak{B}^*,\mathfrak{B}} \end{bmatrix} + \begin{bmatrix} \mathfrak{C}^* \\ \pi_+\mathfrak{D}^* \end{bmatrix} \begin{bmatrix} \mathfrak{C} & \mathfrak{D}\pi_+ \end{bmatrix} = \begin{bmatrix} 1 & 0 \\ 0 & \pi_+ \end{bmatrix}, \qquad (11.2.13)$$
$$\mathfrak{D}^*\pi_-\mathfrak{D} + \mathfrak{B}^*\mathfrak{B} = \pi_-,$$

the unobservable subspace $\mathcal{N}(\mathfrak{C})$ is the largest invariant subspace of $\mathfrak{A}$ on which $\mathfrak{A}$ is an isometry, and the space U_0 in the canonical decomposition presented in Theorem 11.1.9 is given by $U_0 = \mathcal{N}(B)$. In particular, $\left[\begin{smallmatrix} \pi_-\mathfrak{D} \\ \mathfrak{B} \end{smallmatrix}\right]$ maps $L^2(\mathbb{R}^-;U)$ isometrically into $\left[\begin{smallmatrix} L^2(\mathbb{R}^-;Y) \\ X \end{smallmatrix}\right]$, and $\left[\begin{smallmatrix} \pi_-\mathfrak{D} \\ \mathfrak{B} \end{smallmatrix}\right]\left[\mathfrak{D}^*\pi_- \;\; \mathfrak{B}^*\right]$ is a self-adjoint projection in $\left[\begin{smallmatrix} L^2(\mathbb{R}^-;Y) \\ X \end{smallmatrix}\right]$.

(ii) *If Σ is co-energy preserving then*

$$\begin{bmatrix} Q_{\mathfrak{C},\mathfrak{C}^*} & Q_{\mathfrak{C},\mathfrak{A}^*} \\ Q_{\mathfrak{A},\mathfrak{C}^*} & Q_{\mathfrak{A},\mathfrak{A}^*} \end{bmatrix} + \begin{bmatrix} \pi_-\mathfrak{D} \\ \mathfrak{B} \end{bmatrix} \begin{bmatrix} \mathfrak{D}^*\pi_- & \mathfrak{B}^* \end{bmatrix} = \begin{bmatrix} \pi_- & 0 \\ 0 & 1 \end{bmatrix}, \qquad (11.2.14)$$
$$\mathfrak{C}\mathfrak{C}^* + \mathfrak{D}\pi_+\mathfrak{D}^* = \pi_+,$$

$\mathcal{N}(\mathfrak{B}^*)$ is the largest invariant subspace of $\mathfrak{A}^*$ on which $\mathfrak{A}^*$ is an isometry, and the space Y_0 in the canonical decomposition presented in Theorem 11.1.9 is given by $Y_0 = (\mathcal{R}(C))^\perp$. In particular, $\left[\begin{smallmatrix} \mathfrak{C}^* \\ \pi_+\mathfrak{D}^* \end{smallmatrix}\right]$ maps $L^2(\mathbb{R}^+;Y)$ isometrically into $\left[\begin{smallmatrix} X \\ L^2(\mathbb{R}^+;U) \end{smallmatrix}\right]$, and $\left[\begin{smallmatrix} \mathfrak{C}^* \\ \pi_+\mathfrak{D}^* \end{smallmatrix}\right]\left[\mathfrak{C} \;\; \mathfrak{D}\pi_+\right]$ is a self-adjoint projection in $\left[\begin{smallmatrix} X \\ L^2(\mathbb{R}^+;U) \end{smallmatrix}\right]$.

Proof (i) In the case of an energy preserving system the function defined in (11.1.14) is a constant, hence the first identity in (11.2.13) holds. To get the second identity in (11.2.13) we take $x(r) = 0$ and $t = 0$ in (11.2.2) and let $r \to -\infty$. We know that $\mathcal{N}(\mathfrak{C})$ is invariant under $\mathfrak{A}$, and it follows from (11.2.2) with $r = 0$ and $u = 0$ that $\|\mathfrak{A}^t x_0\|_X = \|x_0\|_X$ for all $t \geq 0$ if and only if $\mathfrak{C}x_0 = 0$.

We defined U_0 to be the set of $u \in U$ satisfying $|\widehat{\mathfrak{D}}(1)u| = |u|$. By the last identity in (11.2.7) with $\beta = \gamma = 1$,

$$2|(1 - A_{|X})^{-1}Bu|^2 + |\widehat{\mathfrak{D}}(1)u|^2 = |u|^2,$$

so $|\widehat{\mathfrak{D}}(1)u| = |u|$ if and only if $Bu = 0$. This shows that $U_0 = \mathcal{N}(B)$.

The second identity in (11.2.13) says that $\left[\begin{smallmatrix} \pi_-\mathfrak{D} \\ \mathfrak{B} \end{smallmatrix}\right]$ is an isometry, i.e., if we denote this operator by R, then $R^*R = 1$. This implies that RR^* is a projection, since $(RR^*)^2 = R(R^*R)R^* = RR^*$.

(ii) To prove the claims concerning co-energy preserving systems we apply the energy preserving results to the dual system. □

Remark 11.2.10 The two identities (11.2.13) and (11.2.14) have the following interesting interpretation in terms of the backward and forward wave operators W_- and W_+ of the L^2-well-posed Lax–Phillips model induced by the system.

The identity (11.2.13) says that if the system is energy preserving, then the backward wave operator W_- is an isometry, and $\left[\begin{smallmatrix} Q_{\mathfrak{A}^*,\mathfrak{A}} & Q_{\mathfrak{A}^*,\mathfrak{B}} \\ Q_{\mathfrak{B}^*,\mathfrak{A}} & Q_{\mathfrak{B}^*,\mathfrak{B}} \end{smallmatrix}\right]$ measures how much the forward wave operator W_+ differs from an isometry. The identity (11.2.14) says that if the system is co-energy preserving, then the forward wave operator W_+ is a co-isometry, and $\left[\begin{smallmatrix} Q_{\mathfrak{C},\mathfrak{C}^*} & Q_{\mathfrak{C},\mathfrak{A}^*} \\ Q_{\mathfrak{A},\mathfrak{C}^*} & Q_{\mathfrak{A},\mathfrak{A}^*} \end{smallmatrix}\right]$ measures how much the backward wave operator W_- differs from a co-isometry.

In the conservative case we can say a little bit more.

Theorem 11.2.11 *Let $\Sigma = \left[\begin{smallmatrix} \mathfrak{A} & \mathfrak{B} \\ \mathfrak{C} & \mathfrak{D} \end{smallmatrix}\right]$ be a conservative system on (Y, X, U) with system operator $S = \left[\begin{smallmatrix} A\&B \\ C\&D \end{smallmatrix}\right]$, observation operator C, and transfer function $\widehat{\mathfrak{D}}$. Introduce the same notation as in Theorem 11.1.6. Then all the conclusions listed in parts (i) and (ii) of Theorem 11.2.9 hold, and the following additional claims hold.*

(i) *The two operators on the left-hand side of the identity*

$$\begin{bmatrix} Q_{\mathfrak{A}^*,\mathfrak{A}} & Q_{\mathfrak{A}^*,\mathfrak{B}} \\ Q_{\mathfrak{B}^*,\mathfrak{A}} & Q_{\mathfrak{B}^*,\mathfrak{B}} \end{bmatrix} + \begin{bmatrix} \mathfrak{C}^* \\ \pi_+\mathfrak{D}^* \end{bmatrix} \begin{bmatrix} \mathfrak{C} & \mathfrak{D}\pi_+ \end{bmatrix} = \begin{bmatrix} 1 & 0 \\ 0 & \pi_+ \end{bmatrix}$$

are self-adjoint (and complementary) projections on $\left[\begin{smallmatrix} X \\ L^2(\mathbb{R}^+;U) \end{smallmatrix}\right]$.

(ii) *The two operators on the left-hand side of the identity*

$$\begin{bmatrix} Q_{\mathfrak{C},\mathfrak{C}^*} & Q_{\mathfrak{C},\mathfrak{A}^*} \\ Q_{\mathfrak{A},\mathfrak{C}^*} & Q_{\mathfrak{A},\mathfrak{A}^*} \end{bmatrix} + \begin{bmatrix} \pi_-\mathfrak{D} \\ \mathfrak{B} \end{bmatrix} \begin{bmatrix} \mathfrak{D}^*\pi_- & \mathfrak{B}^* \end{bmatrix} = \begin{bmatrix} \pi_- & 0 \\ 0 & 1 \end{bmatrix}$$

are self-adjoint (and complementary) projections on $\left[\begin{smallmatrix} L^2(\mathbb{R}^-;Y) \\ X \end{smallmatrix}\right]$.

(iii) *The maximal subspace X_0 in Theorem 11.1.9 on which $\mathfrak{A}$ is unitary is given by $X_0 = \mathcal{N}(\mathfrak{B}^*) \cap \mathcal{N}(\mathfrak{C})$. In particular, $\mathfrak{A}$ is completely nonunitary if and only if the intersection of the unreachable and unobservable subspaces is trivial.*

Proof Clearly, all the conclusions of Theorem 11.1.6 hold, since a unitary system is both energy preserving and co-energy preserving.

(i) Let us denote $R = \begin{bmatrix} \mathfrak{C} & \mathfrak{D}\pi_+ \end{bmatrix}^*$. Then, by Theorem 11.2.9(i), R is isometric and RR^* is a projection. Therefore also $\left[\begin{smallmatrix} Q_{\mathfrak{A}^*,\mathfrak{A}} & Q_{\mathfrak{A}^*,\mathfrak{B}} \\ Q_{\mathfrak{B}^*,\mathfrak{A}} & Q_{\mathfrak{B}^*,\mathfrak{B}} \end{smallmatrix}\right] = 1 - RR^*$ is a projection, and these two projections are obviously self-adjoint.

(ii) The proof of (ii) is analogous to the proof of (i).

(iii) This follows from Lemma 11.1.12 and Theorem 11.2.9. □

11.3 Semi-lossless and lossless systems

In an energy preserving system no energy is lost, but it may be first transferred from the input to the state, and then 'trapped' in the state space forever, so that it can no longer be retrieved from the outside. Thus, from the point of view of an external observer, a conservative system may be 'lossy'. To specifically exclude this case we need another notion, that we shall refer to as *losslessness*.

Definition 11.3.1

(i) An operator $\mathfrak{D} \in TIC^2(U;Y)$ (where U and Y are Hilbert spaces) is (scattering) semi-lossless if it is isometric, i.e.,

$$\int_0^\infty |(\mathfrak{D}\pi_+ u)(s)|_Y^2\, ds = \int_0^\infty |u(s)|_U^2\, ds$$

for all $u \in L^2(\mathbb{R}^+;U)$. It is (scattering) *co-lossless* if the causal dual operator $\mathfrak{D}^d = \mathbf{Я}\mathfrak{D}^*\mathbf{Я}$ is semi-lossless, and it is (scattering) *lossless* if it is both semi-lossless and co-lossless.

(ii) By a *semi-lossless*, *co-lossless*, or *lossless system* we mean an L^2-well-posed linear system on three Hilbert spaces (Y, X, U) whose input/output map is semi-lossless or lossless.

Thus, semi-losslessness, co-losslessness, and losslessness can be interpreted as the input/output versions of energy preservation, co-energy preservation, or conservativity, respectively.

Semi-losslessness of an operator $\mathfrak{D} \in TIC^2(U;Y)$ can alternatively be interpreted as a property of the transfer function $\widehat{\mathfrak{D}}$ of $\mathfrak{D}$:

Proposition 11.3.2 *An operator $\mathfrak{D} \in TIC^2(U;Y)$ (where U and Y are Hilbert spaces) is semi-lossless if and only if its transfer function $\widehat{\mathfrak{D}}$ is left-inner in the following sense: $\widehat{\mathfrak{D}}$ is a contractive analytic function on $\mathbb{C}^+$, the restriction of $\widehat{\mathfrak{D}}$ to every separable subspace of U has a strong limit from the right a.e. at the imaginary axis, and this limit is isometric a.e.*[1]

Proof By Proposition 10.3.5, $\mathfrak{D} \in TIC^2(U;Y)$ if and only if $\widehat{\mathfrak{D}} \in H^\infty(U;Y)$, and $\widehat{\mathfrak{D}}$ is a contractive analytic function on $\mathbb{C}^+$ whenever $\mathfrak{D}$ is a contraction (in particular, if it is isometric). Without loss of generality, we may assume that U and Y are separable, because any function $u \in L^2(\mathbb{R}^+;U)$ to which we may apply $\mathfrak{D}$ is almost separable-valued, and so is $\mathfrak{D}\pi_+ u$. The separable case is

[1] The limit from the right at the imaginary axis of $\widehat{\mathfrak{D}}u$ exists almost everywhere for all $u \in U$ even if U is nonseparable. By restricting $\widehat{\mathfrak{D}}$ to a separable subspace of U we can ensure that the limiting operator is defined almost everywhere. The continuity of $\widehat{\mathfrak{D}}$ in $\mathbb{C}^+$ implies that the values of $\widehat{\mathfrak{D}}$ lie in a separable subspace of Y in this case.

well-known and found in many places (in slightly different settings); see, e.g., Duren (1970, pp. 187–192), Hoffman (1988), or Sz.-Nagy and Foiaş (1970, Section V.2, pp. 186–192). □

Note that, if $\mathfrak{D}$ is the input/output map of an L^2-well-posed system with main operator A, then the transfer function $\widehat{\mathfrak{D}}$ has an analytic extension to all of $\rho(A)$. In particular, if $\rho(A) \cap j\mathbb{R} \neq \emptyset$, then at every point $\alpha \in \rho(A) \cap j\mathbb{R}$, $\widehat{\mathfrak{D}}(\alpha)$ will be isometric if $\mathfrak{D}$ is semi-lossless and unitary if $\mathfrak{D}$ is lossless (since $\widehat{\mathfrak{D}}$ is continuous at every $\alpha \in \rho(A)$).

As the following theorem says, if a system is controllable and passive and semi-lossless, then it must be energy preserving and output normalized.

Theorem 11.3.3 *A controllable semi-lossless passive system* $\Sigma = \left[\begin{smallmatrix} \mathfrak{A} & \mathfrak{B} \\ \mathfrak{C} & \mathfrak{D} \end{smallmatrix}\right]$ *on* (Y, X, U) *is necessarily energy preserving and strongly stable, and* $\mathfrak{C}^*\mathfrak{C} = 1$. *In particular,* Σ *is minimal and output normalized, and it is unitarily similar to the restricted exactly observable shift realization of* $\mathfrak{D}$ *given in Proposition 9.5.2(iv).*

Proof We begin by showing that the state trajectory x and the output function y of Σ with initial time zero, initial state zero, and input function u satisfy

$$|x(t)|_X^2 + \int_0^t |y(s)|_Y^2\, ds = \int_0^t |u(s)|_U^2\, ds, \qquad t \geq 0. \tag{11.3.1}$$

Because of the passivity, we know that

$$|x(t)|_X^2 + \textstyle\int_0^t |y(s)|_Y^2\, ds \leq \int_0^t |u(s)|_U^2\, ds, \qquad t \geq 0. \tag{11.3.2}$$

Fix $v > 0$, and assume that $u(s) = 0$ for $s \geq v$. Then

$$|x(t)|_X^2 + \int_0^t |y(s)|_Y^2\, ds \leq \int_0^v |u(s)|_U^2\, ds, \qquad t \geq v.$$

By the lossless property, $\int_0^\infty |y(s)|_Y^2\, ds = \int_0^v |u(s)|_U^2\, ds$, so by letting $t \to \infty$ we conclude that $x(t) \to 0$ as $t \to \infty$. Furthermore, by the passivity and the fact that (x, y) is the state and output of S on the time interval $[v, \infty)$ with initial state $x(v)$ and input function 0, and by (11.3.2),

$$\begin{aligned} 0 &\leq |x(v)|_X^2 - \int_v^\infty |y(s)|_Y^2\, ds \\ &\leq \int_0^v |u(s)|_U^2\, ds - \int_0^v |y(s)|_Y^2\, ds - \int_v^\infty |y(s)|_Y^2\, ds = 0. \end{aligned}$$

Thus, both the inequalities in this chain must be equalities, and this implies (11.3.1).

Subtracting two copies of (11.3.1) from each other, with t replaced by v in one of them, we get

$$|x(t)|_X^2 + \int_v^t |y(s)|_Y^2\,ds = |x(v)|_X^2 + \int_v^t |u(s)|_U^2\,ds, \qquad t \ge v.$$

Denote $u_1(t) = u(t-v)$, $x_1(t) = x(t-v)$, and $y_1(t) = y(t-v)$ for $t \ge 0$, and replace $x(v)$ by x_0. Then we get (11.2.1) with u, x, and y replaced by u_1, x_1, and y_1. By varying v and u we obtain all possible $u_1 \in L^2(\mathbb{R}^+;U)$ and all those x_0 that lie in the range of $\mathfrak{B}_0^v$ for some $v \ge 0$. The latter set is dense in X (since Σ is controllable), and by continuity, (11.2.1) must therefore hold for all $u \in L^2(\mathbb{R}^+;U)$ and $x_0 \in X$. This means that Σ is energy preserving.

In the first part of the proof (where we took $x_0 = 0$) we observed that $x(t) \to 0$ as $t \to \infty$. Using this fact in the next part of the proof we find that $x(t) \to 0$ as $t \to \infty$ for all x_0 in a dense set of X and all u with bounded support. Furthermore, by (11.1.1), $|x(t)|^2 \le |x_0|_X^2 + \int_0^\infty |u(s)|_U^2\,ds$. By density and continuity, the same claim remains true for all $x_0 \in X$ and all $u \in L^2(\mathbb{R}^+;Y)$. Thus Σ is strongly stable (see Definition 8.1.1). That $\mathfrak{C}^*\mathfrak{C} = 1$ (i.e., the controllability gramian is the identity operator) follows from (11.2.13) ($Q_{\mathfrak{A}^*,\mathfrak{A}} = 0$ since $\mathfrak{A}$ is strongly stable).

The final claim about the similarity to the restricted exactly observable shift realization in Proposition 9.5.2(iv) follows from Theorem 9.5.3. □

In Theorem 11.3.3 we assumed Σ to be controllable, passive, and lossless, and we concluded that Σ must be energy preserving and strongly stable. If we instead start by assuming that Σ is energy preserving, then the strong stability of Σ can be characterized in several equivalent ways.

Theorem 11.3.4 *Let* $\Sigma = \left[\begin{smallmatrix} \mathfrak{A} & \mathfrak{B} \\ \mathfrak{C} & \mathfrak{D} \end{smallmatrix}\right]$ *be an (scattering) energy preserving linear system on* (Y, X, U). *Then, with the notation of Theorem 11.1.6, the following conditions are equivalent:*

(i) $\mathfrak{A}$ *is strongly stable;*
(ii) $Q_{\mathfrak{A}^*,\mathfrak{A}} = 0$;
(iii) $\begin{bmatrix} Q_{\mathfrak{A}^*,\mathfrak{A}} & Q_{\mathfrak{A}^*,\mathfrak{B}} \\ Q_{\mathfrak{B}^*,\mathfrak{A}} & Q_{\mathfrak{B}^*,\mathfrak{B}} \end{bmatrix} = 0$;
(iv) $Q_{\mathfrak{A}^*,\mathfrak{A}} \ll 1$;
(v) $\mathfrak{C}^*\mathfrak{C} = 1$;
(vi) $\left[\begin{smallmatrix} \mathfrak{C}^* \\ \pi_+\mathfrak{D}^* \end{smallmatrix}\right]\left[\begin{smallmatrix} \mathfrak{C} & \mathfrak{D}\pi_+ \end{smallmatrix}\right] = 1$;
(vii) Σ *is exactly observable in infinite time.*

If these conditions hold, then

(viii) Σ *is semi-lossless.*

If Σ *is controllable, then (viii) is equivalent to (i)–(vi).*

The equivalent conditions listed above have simple interpretations in terms of the forward wave operator W_+ and the backward wave operator W_- of the corresponding Lax–Phillips model. We know from (11.2.13) that W_- is isometric whenever the system is energy preserving. Condition (iii) holds (hence so do (i)–(viii)) if and only if also W_+ is an isometry.

Proof of Theorem 11.3.4 (i) $\Rightarrow$ (viii): Assume (i). Let x be the state trajectory and y the output function of Σ with initial time zero, initial state zero, and input function $u \in L^2(\mathbb{R}^+; U)$. If u is supported on $[0, T]$, then $x(t) \to 0$ as $t \to \infty$ (because $x(t - T) = \mathfrak{A}^{t-T} x(T)$ for $t \geq T$). Arguing as in the last part of the proof of Theorem 11.3.3 (or alternatively, as in the proof of Lemma 8.1.2 (iii)) we find that $x(t) \to 0$ also in the case where u does not have bounded support. This combined with (11.2.1) (with $x_0 = 0$) implies that Σ is semi-lossless.

(i) $\Leftrightarrow$ (ii): This follows from the definition of $Q_{\mathfrak{A}^*,\mathfrak{A}}$.

(i) $\Rightarrow$ (iii): Assume (i). Then the definitions of $Q_{\mathfrak{A}^*,\mathfrak{A}}$ and $Q_{\mathfrak{B}^*,\mathfrak{A}}$ imply that these operators are zero, and also $Q_{\mathfrak{A}^*,\mathfrak{B}} = Q^*_{\mathfrak{B}^*,\mathfrak{A}} = 0$. As we have shown above, Σ is semi-lossless, and this together with (11.2.13) implies that $Q_{\mathfrak{B}^*,\mathfrak{B}} = 0$.

(iii) $\Rightarrow$ (ii): This is obvious.

(ii) $\Rightarrow$ (iv): This is obvious, too.

(iv) $\Rightarrow$ (i): Assume (iv). Then $\|Q_{\mathfrak{A}^*,\mathfrak{A}}\| < 1$. Pick some η satisfying $\|Q_{\mathfrak{A}^*,\mathfrak{A}}\| < \eta < 1$, and let $x_0 \in X$. By the definition of $Q_{\mathfrak{A}^*,\mathfrak{A}}$, we can find some $t_1 > 0$ such that $x_1 = \mathfrak{A}^{t_1} x_0$ satisfies $|x_1|_X \leq \eta |x_0|_X$. Repeating the same argument with x_0 replaced by x_1 we can find some $t_2 > t_1$ so that $x_2 = \mathfrak{A}^{t_2 - t_1} \mathfrak{A}^{t_1} x_0 = \mathfrak{A}^{t_2} x_0$ satisfies $|x_2|_X \leq \eta^2 |x_0|_X$. Continuing in the same way, for each integer $k > 0$, we can find some t_k such that $|\mathfrak{A}^{t_k} x_0|_X \leq \eta^k |x_0|_X$. In particular, $\mathfrak{A}^{t_k} x_0 \to 0$ as $k \to \infty$. This combined with the fact that $\mathfrak{A}$ is a contraction semigroup implies that $\lim_{t\to\infty} \mathfrak{A}^t x_0 = 0$ for all $x_0 \in X$.

(ii) $\Leftrightarrow$ (v) and (iii) $\Leftrightarrow$ (vi): See (11.2.13).

(iv) $\Leftrightarrow$ (vii): This follows from (11.2.13), Definition 9.4.2, and Lemma 9.10.2(iii)).

(vii) $\Rightarrow$ (i): That (vii) $\Rightarrow$ (i) in the controllable case is a part of Theorem 11.3.3. □

With the help of Theorem 11.3.4 it is easy to prove the following alternative characterizations of lossless energy preserving systems.

Theorem 11.3.5 *Let* $\Sigma = \left[\begin{smallmatrix} \mathfrak{A} & \mathfrak{B} \\ \mathfrak{C} & \mathfrak{D} \end{smallmatrix}\right]$ *be an energy preserving linear system on* (Y, X, U). *Then, with the notation of Theorem 11.1.6, the following conditions are equivalent (and they are implied by the conditions listed in Theorem 11.3.4):*

(i) Σ *is (scattering) semi-lossless;*
(ii) $\mathfrak{A}$ *is strongly input/state stable;*

(iii) $Q_{\mathfrak{B}^*,\mathfrak{B}} = 0$;
(iv) $\mathfrak{B}^* Q_{\mathfrak{A}^*,\mathfrak{A}} \mathfrak{B} = 0$;
(v) $\mathfrak{B}^*(1 - \mathfrak{C}^*\mathfrak{C})\mathfrak{B} = 0$;
(vi) $\left[\begin{smallmatrix} Q_{\mathfrak{A}^*,\mathfrak{A}} & 0 \\ 0 & 0 \end{smallmatrix}\right] + \left[\begin{smallmatrix} \mathfrak{C}^* \\ \pi_+\mathfrak{D}^* \end{smallmatrix}\right]\left[\begin{smallmatrix} \mathfrak{C} & \mathfrak{D}\pi_+ \end{smallmatrix}\right] = \left[\begin{smallmatrix} 1 & 0 \\ 0 & \pi_+ \end{smallmatrix}\right]$;
(vii) $\mathfrak{D}^*\mathfrak{D} = 1$;
(viii) *the restriction of* $\mathfrak{A}$ *to the reachable subspace is strongly stable;*
(ix) *the restriction of* Σ *to the reachable subspace is exactly observable in infinite time.*

If U and Y are separable, then these conditions are further equivalent to

(x) $\widehat{\mathfrak{D}}(j\omega)^*\widehat{\mathfrak{D}}(j\omega) = 1$ *for almost all* $\omega \in \mathbb{R}$.

If Σ *is controllable, then the conditions listed above are equivalent to those listed in Theorem 11.3.4.*

Proof Recall that the conditions (i)–(vii) in Theorem 11.3.4 imply that Σ is semi-lossless (which is condition (i) in the present theorem), and that they are equivalent to semi-losslessness if Σ is controllable. Also recall from Proposition 11.3.2 that (i) $\Leftrightarrow$ (x) in the separable case.

(i) $\Leftrightarrow$ (iii): This follows from (11.2.13) (the bottom right corner of the first identity; note that Σ is semi-lossless iff $\pi_+\mathfrak{D}^*\mathfrak{D}\pi_+ = \pi_+$).

(ii) $\Rightarrow$ (iii): Compare Definition 8.1.1(ii)(b) to the definition of $Q_{\mathfrak{B}^*,\mathfrak{B}}$, and take u to be supported on $\mathbb{R}^+$.

(iii) $\Rightarrow$ (ii): Suppose that $Q_{\mathfrak{B}^*,\mathfrak{B}} = 0$, i.e, suppose that for all $u \in L^2(\mathbb{R}^+; U)$, $\mathfrak{B}_0^t u = \mathfrak{B}\tau^t\pi_+ u \to 0$ as $t \to \infty$. If $u \in L^2(\mathbb{R}; U)$ is supported on $[-T, \infty)$, then $\tau^{-T}u$ is supported on $\mathbb{R}^+$, so $\mathfrak{B}\tau^{t-T}u = \tau^t(\tau^{-T}u) \to 0$ as $t \to \infty$. This means that $\mathfrak{B}\tau^t u \to 0$ as $t \to \infty$ for all u whose support is bounded to the left. This set of functions is dense in $L^2(\mathbb{R}; U)$ and $\mathfrak{B}\tau^t$ is uniformly bounded, so $\mathfrak{B}\tau^t u \to 0$ for all $u \in L^2(\mathbb{R}; U)$, and this proves that Σ is strongly input/state stable.

(iii) $\Leftrightarrow$ (vi): By (11.2.13), (vi) holds iff $Q_{\mathfrak{B}^*,\mathfrak{B}} = 0$ and $Q_{\mathfrak{A}^*,\mathfrak{B}} = 0$. In particular, (vi) $\Rightarrow$ (iii). Conversely, if $Q_{\mathfrak{B}^*,\mathfrak{B}} = 0$, then it follows from the definition of $Q_{\mathfrak{B}^*,\mathfrak{B}}$ that $\mathfrak{B}_0^t$ tends strongly to zero as $t \to \infty$, and this combined with the definition of $Q_{\mathfrak{A}^*,\mathfrak{B}}$ implies that also $Q_{\mathfrak{A}^*,\mathfrak{B}} = 0$. Thus (iii) $\Rightarrow$ (vi).

(i) $\Leftrightarrow$ (vii): We get the identity $\pi_+\mathfrak{D}^*\mathfrak{D}\pi_+ = \pi_+$ (which is equivalent to (i)) from (vii) by multiplying by π_+ to the left and to the right. To go in the opposite direction we observe that if (i) holds and u has bounded support, then for some sufficiently large T, $\tau^{-T}u$ is supported on $\mathbb{R}^+$, hence $\langle u, \mathfrak{D}^*\mathfrak{D}u\rangle = \langle u, \tau^T\mathfrak{D}^*\mathfrak{D}\tau^{-T}u\rangle = \langle \tau^{-T}u, \mathfrak{D}^*\mathfrak{D}\tau^{-T}u\rangle = \langle \tau^{-T}u, \tau^{-T}u\rangle = \langle u, u\rangle$. This set of functions is dense in $L^2(\mathbb{R}; U)$, so we must then have $\langle u, \mathfrak{D}^*\mathfrak{D}u\rangle = \langle u, u\rangle$ for all $u \in L^2(\mathbb{R}; U)$.

(ii) $\Leftrightarrow$ (viii): See Lemma 8.1.2(iii).

(viii) ⇔ (ix): Apply the equivalence of (i) and (vii) in Theorem 11.3.4 to the system which we get by replacing the state space by the reachable subspace $\overline{\mathcal{R}(\mathfrak{B})}$ (see Theorem 9.1.9(i)).

(viii) ⇔ (iv): We apply the equivalence of (i) and (ii) in Theorem 11.3.4 to the same reduced system as above. If (viii) holds, then $Q_{\mathfrak{A}^*,\mathfrak{A}}$ vanishes on $\overline{\mathcal{R}(\mathfrak{B})}$, so (iv) holds. Conversely, if (iv) holds, then $Q_{\mathfrak{A}^*,\mathfrak{A}}$ vanishes on $\mathcal{R}(\mathfrak{B})$, hence on $\overline{\mathcal{R}(\mathfrak{B})}$.

(iv) ⇔ (v): By (11.2.13), $Q_{\mathfrak{A}^*,\mathfrak{A}} = 1 - \mathfrak{C}^*\mathfrak{C}$, so $\mathfrak{B}^* Q_{\mathfrak{A}^*,\mathfrak{A}}\mathfrak{B} = \mathfrak{B}^*(1 - \mathfrak{C}^*\mathfrak{C})\mathfrak{B}$ and (iv) and (v) are equivalent. □

In the case of a conservative system we can strengthen the conclusion of Theorem 11.3.4 slightly.

Theorem 11.3.6 *In the case where $\Sigma = \left[\begin{smallmatrix} \mathfrak{A} & \mathfrak{B} \\ \mathfrak{C} & \mathfrak{D} \end{smallmatrix}\right]$ is (scattering) conservative the conditions (i)–(vii) in Theorem 11.3.4 are equivalent to the following conditions:*

- (xi) *$Q_{\mathfrak{A}^*,\mathfrak{A}}$ is the orthogonal projection onto the maximal subspace X_0 in Theorem 11.1.9 on which $\mathfrak{A}$ in unitary.*
- (xii) *$\mathfrak{C}^*\mathfrak{C}$ is the orthogonal projection onto the subspace X_1 in Theorem 11.1.9 on which $\mathfrak{A}$ in completely nonunitary.*
- (xiii) *$\mathfrak{C}^*\mathfrak{C}$ is a projection and $\mathcal{R}(\mathfrak{B}) \subset \mathcal{R}(\mathfrak{C}^*)$.*
- (xiv) *The restriction of $\mathfrak{A}$ to the subspace X_1 in Theorem 11.1.9 is strongly stable.*

In particular, if Σ is semi-lossless, conservative and completely nonunitary, then Σ is observable, and the conditions (i)–(vii) in Theorem 11.3.4 are equivalent to the conditions (i)–(ix) in Theorem 11.3.5 and to the conditions (xi)–(xiv) above.

Proof (xi) ⇔ (xii): By (11.2.13), $Q_{\mathfrak{A}^*,\mathfrak{A}} + \mathfrak{C}^*\mathfrak{C} = 1$, so $Q_{\mathfrak{A}^*,\mathfrak{A}}$ is a projection if and only if $\mathfrak{C}^*\mathfrak{C}$ is a projection, and the ranges of these two projections are the orthogonal complements of each other. Thus (xi) and (xii) are equivalent.

(xii) ⇔ (xiii): Both (xii) and (xiii) require $\mathfrak{C}^*\mathfrak{C}$ to be a projection. The range of this projection is equal to $\mathcal{R}(\mathfrak{C}^*)$ (which is closed since $\mathfrak{C}^*\mathfrak{C}$ is a projection), so to prove that (xii) and (xiii) are equivalent we must show that $\mathcal{R}(\mathfrak{C}^*) = X_1 = X_0^\perp$ if and only if $\mathcal{R}(\mathfrak{B}) \subset \mathcal{R}(\mathfrak{C}^*)$, or equivalently, that $\mathcal{N}(\mathfrak{C}) = X_0$ if and only if $\mathcal{N}(\mathfrak{C}) \subset \mathcal{N}(\mathfrak{B}^*)$. However, this is true because by Theorem 11.2.11(iii), $X_0 = \mathcal{N}(\mathfrak{C}) \cap \mathcal{N}(\mathfrak{B}^*)$, and this is equal to $\mathcal{N}(\mathfrak{C})$ if and only if $\mathcal{N}(\mathfrak{C}) \subset \mathcal{N}(\mathfrak{B}^*)$.

(xi) ⇔ (xiv): It follows from the definition of $Q_{\mathfrak{A}^*,\mathfrak{A}}$ that $Q_{\mathfrak{A}^*,\mathfrak{A}}x = 0$ if and only if $\mathfrak{A}^t x \to 0$ as $t \to \infty$. Thus (xi) implies (xiv). Conversely, if (xiv) holds, then $Q_{\mathfrak{A}^*,\mathfrak{A}}x = 0$ for every $x \in X_1$. From the proof of Theorem 11.1.9 we can infer that $Q_{\mathfrak{A}^*,\mathfrak{A}}x = x$ for all $x \in X_0$. Thus $Q_{\mathfrak{A}^*,\mathfrak{A}}$ is the projection operator onto X_0.

(i)–(vii) ⇒ (xiii): Assume that the equivalent conditions listed in Theorem 11.3.4 hold. By Theorem 11.2.11, the operator $\begin{bmatrix} Q_{\mathfrak{A}^*,\mathfrak{A}} & Q_{\mathfrak{A}^*,\mathfrak{B}} \\ Q_{\mathfrak{B}^*,\mathfrak{A}} & Q_{\mathfrak{B}^*,\mathfrak{B}} \end{bmatrix}$ is a self-adjoint projection, so if furthermore $Q_{\mathfrak{B}^*,\mathfrak{B}} = 0$ (i.e., (iii) in Theorem 11.3.4 holds), then $Q_{\mathfrak{A}^*,\mathfrak{B}} = 0$ and $Q_{\mathfrak{A}^*,\mathfrak{A}}$ is a self-adjoint projection (this can be verified with a short computation). As we saw above, this implies that $\mathfrak{C}^*\mathfrak{C}$ is a projection. By condition (iv) in Theorem 11.3.4, if $x \in \mathcal{R}(\mathfrak{B})$, then $\langle x, Q_{\mathfrak{A}^*,\mathfrak{A}}x\rangle = 0$, or equivalently, $Q_{\mathfrak{A}^*,\mathfrak{A}}x = 0$ and $\mathfrak{C}^*\mathfrak{C}x = x$. Thus $\mathcal{R}(\mathfrak{B}) \subset \mathcal{R}(\mathfrak{C}^*)$ and (xiii) holds.

(xiii) ⇒ (v): Assume (xiii). Then $\mathfrak{C}^*$ acts as an identity on $\mathcal{R}(\mathfrak{C}^*) \supset \mathcal{R}(\mathfrak{B})$, so $\mathfrak{C}^*\mathfrak{B} = \mathfrak{B}$ and $\mathfrak{B}^*\mathfrak{B} = \mathfrak{B}^*\mathfrak{C}\mathfrak{C}^*\mathfrak{B}$. Thus condition (v) in Theorem 11.3.4 holds in this case.

The final statement of Theorem 11.3.6 follows from the fact that if $\mathfrak{A}$ is completely nonunitary, then $X_0 = 0$, hence $Q_{\mathfrak{A}^*,\mathfrak{A}} = 0$ and $\mathfrak{C}^*\mathfrak{C} = 1$. □

By applying Theorems 11.3.5 and 11.3.6 both to the original system and its dual we can derive the following characterizations of lossless conservative systems.

Corollary 11.3.7 *Let $\Sigma = \left[\begin{smallmatrix} \mathfrak{A} & \mathfrak{B} \\ \mathfrak{C} & \mathfrak{D} \end{smallmatrix}\right]$ be a (scattering) conservative completely non-unitary system on (Y, X, U). Then the following conditions are equivalent:*

(i) *Σ is lossless;*
(ii) *both $\mathfrak{A}$ and $\mathfrak{A}^*$ are strongly stable;*
(iii) *$\mathfrak{B}^*\mathfrak{B} = 1$ and $\mathfrak{C}\mathfrak{C}^* = 1$;*
(iv) *S is exactly observable and exactly controllable in infinite time.*

This follows immediately from Theorems 11.3.5 and 11.3.6. The systems studied in Lax and Phillips (1967) are of this type.

There is also an exponentially stable version of Theorem 11.3.4.

Theorem 11.3.8 *Let $\Sigma = \left[\begin{smallmatrix} \mathfrak{A} & \mathfrak{B} \\ \mathfrak{C} & \mathfrak{D} \end{smallmatrix}\right]$ be an (scattering) energy preserving linear system on the three Hilbert spaces (Y, X, U) with main operator A. Then the following conditions are equivalent:*

(i) *$j\mathbb{R} \in \rho(A)$ and $\sup_{\omega\in\mathbb{R}}\|(j\omega - A)^{-1}\|$ is finite;*
(ii) *Σ is exponentially stable;*
(iii) *Σ is exactly observable in some finite time T.*

If these conditions hold, then Σ is semi-lossless and $\widehat{\mathfrak{D}}(\alpha)$ is isometric for all $\alpha \in j\mathbb{R}$.

Proof (i) ⇒ (ii): This is trivial.

(ii) ⇒ (i): If (i) holds, then by Lemma 3.2.8(ii), $\|(z - A)^{-1}\|$ is uniformly bounded on some strip $\{z \in \mathbb{C} \mid |\Re z| \le \epsilon\}$ around $j\mathbb{R}$. Since $\mathfrak{A}$ is a contraction semigroup, by Corollary 3.4.5, $\|(z-A)^{-1}\|$ is also uniformly bounded on $\mathbb{C}^+_\epsilon$. By Theorem 3.11.6, $\mathfrak{A}$ is exponentially stable.

(ii) $\Leftrightarrow$ (iii): By Theorem 2.5.4(i), $\mathfrak{A}$ is exponentially stable if and only if $\|\mathfrak{A}^T\| < 1$ for some $T > 0$. Equivalently, there is some $\epsilon > 0$ such that $|\mathfrak{A}^T x_0|^2_X \le (1-\epsilon)|x_0|^2_X$. Since Σ is energy preserving, this identity is equivalent to the identity (take $u = 0$ in (11.2.1)) $\int_0^T |(\mathfrak{C}x_0)(t)|^2_Y \ge \epsilon |x_0|^2_X$, i.e., S is exactly observable in time T.

The final claim follows from Theorem 11.3.4 and (11.2.9). □

Theorem 11.3.9 *Let $\Sigma = \left[\begin{smallmatrix}\mathfrak{A} & \mathfrak{B}\\ \mathfrak{C} & \mathfrak{D}\end{smallmatrix}\right]$ be a (scattering) conservative linear system on the three Hilbert spaces (Y, X, U) with main operator A. Then the following conditions are equivalent:*

(i) *$j\mathbb{R} \in \rho(A)$ and $\sup_{\omega\in\mathbb{R}} \|(j\omega - A)^{-1}\|$ is finite;*
(ii) *Σ is exponentially stable;*
(iii) *Σ is exactly observable in some finite time T;*
(iv) *Σ is exactly controllable in some finite time T.*

If these conditions hold, then Σ is lossless and $\widehat{\mathfrak{D}}(\alpha)$ is unitary for all $\alpha \in j\mathbb{R}$.

This follows directly from Theorem 11.3.8.

11.4 Isometric and unitary dilations of contraction semigroups

We begin by defining what we mean by an orthogonal compression or dilation of a C_0 semigroup on a Hilbert space.[2]

Definition 11.4.1

(i) Let $\widetilde{\mathfrak{A}}$ be a C_0 semigroup on a Hilbert space $\widetilde{X}$, and let X be a closed subspace of $\widetilde{X}$. Then the (orthogonal) *compression* $\mathfrak{A}$ of $\widetilde{\mathfrak{A}}$ to X is given by $\mathfrak{A}^t = \pi_X \widetilde{\mathfrak{A}}^t_{|X}$, $t \ge 0$, where π_X is the orthogonal projection of $\widetilde{X}$ onto X.
(ii) Let $\mathfrak{A}$ be a C_0 semigroup on a Hilbert space X. By an (orthogonal) *dilation* of $\mathfrak{A}$ we mean a C_0 semigroup on a Hilbert space $\widetilde{X}$ such that X is a closed subspace of $\widetilde{X}$ and $\mathfrak{A}$ is the compression of $\widetilde{\mathfrak{A}}$ to X.
(iii) If the semigroup $\widetilde{\mathfrak{A}}$ in (ii) is isometric, co-isometric, or unitary on $\widetilde{X}$, then we refer to $\widetilde{\mathfrak{A}}$ as an *isometric*, *co-isometric*, or *unitary dilation* of $\mathfrak{A}$.

[2] Nonorthogonal dilations and compressions are defined in a similar way, with the orthogonal projection replaced by an oblique projection. For the moment we shall only need orthogonal projections.

(iv) Let $\widetilde{\mathfrak{A}}$ be a dilation of $\mathfrak{A}$ on the space $\widetilde{X} \supset X$.
  (a) $\widetilde{\mathfrak{A}}$ is a *minimal isometric dilation* of $\mathfrak{A}$ if the span of the set $\{\widetilde{\mathfrak{A}}^t x \mid x \in X,\ t \geq 0\}$ is dense in $\widetilde{X}$.
  (b) $\widetilde{\mathfrak{A}}$ is a *minimal co-isometric dilation* of $\mathfrak{A}$ if the span of the set $\{\widetilde{\mathfrak{A}}^{*t} x \mid x \in X,\ t \geq 0\}$ is dense in $\widetilde{X}$.
  (c) $\widetilde{\mathfrak{A}}$ is a *minimal unitary dilation* of $\mathfrak{A}$ if the span of the set $\{\widetilde{\mathfrak{A}}^t x \mid x \in X,\ t \in \mathbb{R}\}$ is dense in $\widetilde{X}$ (where $\mathfrak{A}^{-t} = \mathfrak{A}^{*t}$ for all $-t < 0$).

Note that the compression $\mathfrak{A}$ in (i) *need not be a semigroup*, although it is always strongly continuous (this is discussed in more detail in Lemma 11.4.2 below). Obviously $\mathfrak{A}^t$ is a contraction for all $t \geq 0$ if $\widetilde{\mathfrak{A}}$ is isometric or co-isometric. Also note that $\widetilde{\mathfrak{A}}$ is a co-isometric dilation of $\mathfrak{A}$ if and only if $\widetilde{\mathfrak{A}}^*$ is an isometric dilation of $\mathfrak{A}^*$. In particular, a necessary condition for the existence of an isometric, co-isometric, or unitary dilation of $\mathfrak{A}$ is that $\mathfrak{A}$ is a contraction semigroup. As we shall see later, this condition is also sufficient.

Lemma 11.4.2 *The operator family $\mathfrak{A}^t = \pi_X \widetilde{\mathfrak{A}}^t_{|X}$ in part (i) of Definition 11.4.1 is a semigroup if and only if $\widetilde{X}$ is a direct sum $X = Z \oplus X \oplus Z_*$ such that the block matrix decomposition of $\widetilde{\mathfrak{A}}$ with respect to this decomposition of $\widetilde{X}$ has the form*

$$\widetilde{\mathfrak{A}} = \begin{bmatrix} \widetilde{\mathfrak{A}}_{|Z} & \pi_Z \widetilde{\mathfrak{A}}_{|X} & \pi_Z \widetilde{\mathfrak{A}}_{|Z_*} \\ 0 & \pi_X \widetilde{\mathfrak{A}}_{|X} & \pi_X \widetilde{\mathfrak{A}}_{|Z_*} \\ 0 & 0 & \pi_{Z_*} \widetilde{\mathfrak{A}}_{|Z_*} \end{bmatrix}. \tag{11.4.1}$$

In particular, Z and $Z \oplus X$ are invariant subspaces of $\widetilde{\mathfrak{A}}$, $X \oplus Z_*$ and Z_* are invariant subspaces of $\widetilde{\mathfrak{A}}^*$, and $\widetilde{\mathfrak{A}}_{|Z}$, $\pi_X \widetilde{\mathfrak{A}}_{|X}$ and $\pi_{Z_*} \widetilde{\mathfrak{A}}_{|Z_*}$ are C_0 semigroups on Z, $\widetilde{X}$, respectively Z_*.

The subspaces Z and Z_* are not unique in general (see Remark 9.1.11).

Proof This is a special case of Lemma 11.5.2. □

In the sequel we shall often need to discuss *unitary similarity*. By this we mean the following.

Definition 11.4.3 Let X, U, Y, X_1, U_1, and Y_1 be Hilbert spaces.

(i) The spaces X and X_1 are *unitarily similar* if there is a unitary map E mapping X onto X_1.
(ii) Two operators $A \in \mathcal{B}(U; Y)$ and $A_1 \in \mathcal{B}(U_1; Y_1)$ are *unitarily similar* if there exist unitary operators F mapping U onto U_1 and G mapping Y onto Y_1 such that $A_1 = GAF^{-1}$ (if $U = Y$ and $U_1 = Y_1$ then we require, in addition, that $F = G$).

(iii) Two semigroups $\mathfrak{A}$ on X and $\mathfrak{A}_1$ on X_1 are *unitarily similar* if there is a unitary operator E mapping X onto X_1 such that $\mathfrak{A}_1^t = E\mathfrak{A}^t E^{-1}$ for all $t \geq 0$.

(iv) Two maps $\mathfrak{D} \in TIC^2(U;Y)$ and $\mathfrak{D}_1 \in TIC^2(U_1;Y_1)$ are *unitarily similar* if there exist unitary operators F mapping U onto U_1 and G mapping Y onto Y_1 such that $\mathfrak{D}_1 = G\mathfrak{D}F^{-1}$ (if $U = Y$ and $U_1 = Y_1$ then we require, in addition, that $F = G$).

(v) Two L^2-well-posed linear systems $\left[\begin{array}{c|c} \mathfrak{A} & \mathfrak{B} \\ \hline \mathfrak{C} & \mathfrak{D} \end{array}\right]$ an (Y, X, U) and $\left[\begin{array}{c|c} \mathfrak{A}_1 & \mathfrak{B}_1 \\ \hline \mathfrak{C}_1 & \mathfrak{D}_1 \end{array}\right]$ on (Y_1, X_1, U_1) are *unitarily similar* if there exist unitary operators E, F, and G, mapping X onto X_1, U onto U_1, and Y onto Y_1, respectively, such that

$$\left[\begin{array}{c|c} \mathfrak{A}_1 & \mathfrak{B}_1 \\ \hline \mathfrak{C}_1 & \mathfrak{D}_1 \end{array}\right] = \begin{bmatrix} E & 0 \\ 0 & G \end{bmatrix} \left[\begin{array}{c|c} \mathfrak{A} & \mathfrak{B} \\ \hline \mathfrak{C} & \mathfrak{D} \end{array}\right] \begin{bmatrix} E^{-1} & 0 \\ 0 & F^{-1} \end{bmatrix}.$$

As a first application of unitary similarity, let us prove that isometric, co-isometric, and unitary dilations are unique, up to a unitary similarity.

Lemma 11.4.4 *Let $\widetilde{\mathfrak{A}}_1$ and $\widetilde{\mathfrak{A}}_2$ be two minimal isometric, co-isometric, or unitary dilations of $\mathfrak{A}$ on $\widetilde{X}_1 = \left[\begin{smallmatrix} X \\ X_1 \end{smallmatrix}\right]$, respectively $\widetilde{X}_2 = \left[\begin{smallmatrix} X \\ X_2 \end{smallmatrix}\right]$. Then there is unitary map E of X_1 onto X_2 such that*

$$\widetilde{\mathfrak{A}}_2^t = \begin{bmatrix} 1 & 0 \\ 0 & E \end{bmatrix} \widetilde{\mathfrak{A}}_1^t \begin{bmatrix} 1 & 0 \\ 0 & E^{-1} \end{bmatrix};$$

this is true for all $t \geq 0$ in the isometric and co-isometric cases, and for all $t \in \mathbb{R}$ in the unitary case.

Proof The co-isometric case is reduced to the isometric case if we replace $\mathfrak{A}$ by $\mathfrak{A},^*$, so it suffices to prove the isometric and unitary cases. Let us begin with the isometric case.

Let $x, y \in X$, and let $t \geq s \geq 0$. If $\widetilde{\mathfrak{A}}$ is an isometric dilation of $\mathfrak{A}$ to $\widetilde{X}$, then $\widetilde{\mathfrak{A}}^{*s}\widetilde{\mathfrak{A}}^s = 1$ on $\widetilde{X}$, so

$$\langle \widetilde{\mathfrak{A}}^s x, \widetilde{\mathfrak{A}}^t y\rangle_{\widetilde{X}} = \langle x, \widetilde{\mathfrak{A}}_1^{*s}\widetilde{\mathfrak{A}}_1^s\widetilde{\mathfrak{A}}^{t-s} y\rangle_{\widetilde{X}} = \langle x, \widetilde{\mathfrak{A}}^{t-s} y\rangle_{\widetilde{X}} = \langle x, \mathfrak{A}^{t-s} y\rangle_X.$$

Thus, this expression does not depend on the particular dilation, as long as it is isometric. In the same way we get $\langle \widetilde{\mathfrak{A}}^t y, \widetilde{\mathfrak{A}}^s x\rangle_{\widetilde{X}} = \langle \mathfrak{A}^{t-s} y, x\rangle_X$, so we can remove the restriction $s \leq t$. Applying this argument with $\widetilde{\mathfrak{A}}$ replaced by $\widetilde{\mathfrak{A}}_1$ and $\widetilde{\mathfrak{A}}_2$ we find that

$$\langle \widetilde{\mathfrak{A}}_1^s x, \widetilde{\mathfrak{A}}_1^t y\rangle_{\widetilde{X}_1} = \langle \widetilde{\mathfrak{A}}_2^s x, \widetilde{\mathfrak{A}}_2^t y\rangle_{\widetilde{X}_2}$$

for all $x, y \in X$ and all $s, t \geq 0$. Therefore, if we take arbitrary sequences of vectors $x_i \in X$ and numbers $t_i \geq 0$, $1 \leq i \leq n$, and define $\tilde{x}_1 \in \widetilde{X}_1$ and $\tilde{x}_2 \in \widetilde{X}_2$

by

$$\widetilde{x}_1 = \sum_{i=1}^{n} \widetilde{\mathfrak{A}}_1^{t_i} x_i, \qquad \widetilde{x}_2 = \sum_{i=1}^{n} \widetilde{\mathfrak{A}}_2^{t_i} x_i,$$

then $\|\tilde{x}_1\|_{\widetilde{X}_1} = \|\tilde{x}_2\|_{\widetilde{X}_2}$. In particular, $\widetilde{x}_1 = 0$ iff $\widetilde{x}_2 = 0$, so we can define a norm-preserving operator F from the span of $\{\widetilde{\mathfrak{A}}_1^t x \mid x \in X,\ t \geq 0\}$ to the span of $\{\widetilde{\mathfrak{A}}_2^t x \mid x \in X,\ t \geq 0\}$ by setting $F\tilde{x}_1 = \tilde{x}_2$. For all $t \geq 0$, the preceding construction maps the point $\widetilde{\mathfrak{A}}_1^t \tilde{x}_1 = \sum_{i=1}^n \widetilde{\mathfrak{A}}_1^{t+t_i} x_i$ onto the point $F\widetilde{\mathfrak{A}}_1^t \tilde{x}_1 = \sum_{i=1}^n \widetilde{\mathfrak{A}}_2^{t+t_i} x_i = \widetilde{\mathfrak{A}}_2^t \tilde{x}_2 = \mathfrak{A}_2^t F\tilde{x}_1$. By density, we can extend this operator to a norm-preserving, hence unitary, operator (which we still denote by F) from $\widetilde{X}_1$ onto $\widetilde{X}_2$, which satisfies $F\widetilde{\mathfrak{A}}_1^t = \widetilde{\mathfrak{A}}_2^t F$.

Let us decompose F into $F = \left[\begin{smallmatrix} F_{11} & F_{12} \\ F_{21} & F_{22} \end{smallmatrix}\right]$ according to the decompositions $\widetilde{X}_1 = \left[\begin{smallmatrix} X \\ X_1 \end{smallmatrix}\right]$ and $\widetilde{X}_2 = \left[\begin{smallmatrix} X \\ X_2 \end{smallmatrix}\right]$. By construction, $Fx = x$ for every $x \in X$, so $F_{11} = 1$. As F is unitary, we have $F^*F = 1$, i.e.,

$$\begin{aligned} 1 + F_{21}^* F_{21} &= 1, & F_{12}^* + F_{22}^* F_{12} &= 0, \\ F_{12} + F_{21}^* F_{22} &= 0, & F_{12}^* F_{12} + F_{22}^* F_{22} &= 1. \end{aligned}$$

The first equation implies that $F_{21} = 0$, after which we see from the third equation that $F_{12} = 0$ and from the fourth equation that F_{22} is an isometry. From the corresponding adjoint identity we find that F_{22} is unitary. Taking $E = F_{22}$ we get the desired unitary map of $\widetilde{X}_1$ onto $\widetilde{X}_2$.

The proof of the *unitary* case is identical to the one above, except that we drop the restrictions $s \geq 0$ and $t \geq 0$. □

Theorem 11.4.5 *Let $\mathfrak{A}$ be a C_0 contraction semigroup on a Hilbert space X.*

(i) *There is a Hilbert space Y and an admissible observation operator $C\colon X_1 \to Y$ with dense range such that, if $\mathfrak{C}$ is the output map induced by C and $\mathfrak{A}$, then $\left[\begin{smallmatrix} \mathfrak{A} \\ \mathfrak{C} \end{smallmatrix}\right]$ is an (scattering) energy preserving system on $(Y, X, 0)$ (i.e., there is no input, only a state and an output).*

(ii) *The space Y and the operator C are unique up to unitary similarity: if Y_1 is another Hilbert space, if $C_1\colon X_1 \to Y_1$ is another admissible observation operator with dense range, if $\mathfrak{C}_1$ is the corresponding output map, and if $\left[\begin{smallmatrix} \mathfrak{A} \\ \mathfrak{C}_1 \end{smallmatrix}\right]$ is an energy preserving system on $(Y, X, 0)$, then there is a unitary map $G\colon Y \to Y_1$ so that $C_1 = GC$ and $\mathfrak{C}_1 = G\mathfrak{C}$.*

(iii) *The corresponding Lax–Phillips model with parameter $\omega = 0$ (and with $U = 0$, i.e., there is no input) is a minimal isometric dilation of $\mathfrak{A}$.*

Proof (i) By Theorem 3.4.8, the contractivity of $\mathfrak{A}$ implies that the generator A of $\mathfrak{A}$ is dissipative (this can also be seen from Theorem 11.1.5 with $U = Y = 0$). Fix an arbitrary $\alpha \in \mathbb{C}^+$. Then the operator $-(\overline{\alpha} - A_{|X}^*)^{-1}(A + A_{|X}^*)(\alpha - A)^{-1}$ is positive, and it has a positive square root (see Lemma A.2.2). For all $x \in$

$X_1 = \mathcal{D}(A)$, we define

$$Cx = \left[-(\overline{\alpha} - A^*_{|X})^{-1}(A + A^*_{|X})(\alpha - A)^{-1}\right]^{1/2}(\alpha - A)x$$

(we ignore the fact that this operator may depend on α). Let Y be the closure of $\mathcal{R}(C)$ in X. Then $C \in \mathcal{B}(X_1; Y)$, $C^* \in \mathcal{B}(Y; X^*_{-1})$, and

$$C^*C = -(A + A^*_{|X}).$$

By Theorem 11.2.5 with $U = 0$, C is an admissible observation operator for $\mathfrak{A}$, and $\left[\begin{smallmatrix}\mathfrak{A}\\ \mathfrak{C}\end{smallmatrix}\right]$ is an energy preserving system on $(Y, X, 0)$. This proves (i).

(ii) Let C_1 and Y_1 be as in (ii), and let $C = E|C|$ and $C_1 = E_1|C_1|$ be the polar decompositions of C and C_1 (see Lemma A.2.5). Then both E and E_1 are onto since both C and C_1 have dense ranges, and $|C| = E^*C$. By Theorem 11.2.5, $C^*C = -(A + A^*_{|X}) = C_1^*C_1$. Thus,

$$C_1 = E_1|C_1| = E_1|C| = E_1E^*C,$$

where E_1E^* is a unitary operator mapping Y onto Y_1. Clearly also $\mathfrak{C}_1 = E_1E^*\mathfrak{C}$.

(iii) By Lemma 11.2.2, the Lax–Phillips model $\mathfrak{T}$ with parameter $\omega = 0$ is isometric, and it is obvious that it is a dilation of $\mathfrak{A}$. It remains to show that it is minimal, i.e., that the span of elements of the type $\mathfrak{T}^t\left[\begin{smallmatrix}0\\ x\end{smallmatrix}\right]$ with $x \in X$ and $t \geq 0$ is dense in the state space $\left[\begin{smallmatrix}\mathcal{Y}\\ X\end{smallmatrix}\right]$, where $\mathcal{Y} = L^2(\mathbb{R}^-; Y)$. Obviously $\left[\begin{smallmatrix}0\\ X\end{smallmatrix}\right]$ is contained in this span, so it suffices to prove that no nonzero element of the type $\left[\begin{smallmatrix}y\\ 0\end{smallmatrix}\right]$ is orthogonal to this set. Suppose that $y \in \mathcal{Y}$, and that $\langle y, \tau^t\mathfrak{C}_0^t x\rangle_{\mathcal{Y}}$ for all $x \in X$ and $t \geq 0$. In particular, this is then true for all $x \in X_1$, and we get from Theorem 4.4.2 for all $x \in X_1$ and all $t \geq 0$,

$$\begin{aligned} 0 &= \int_{-t}^0 \langle C\mathfrak{A}^{s-t}x, y(s)\rangle_Y\, ds = \int_0^t \langle x, \mathfrak{A}^{*(t-s)}_{|X^*_{-1}}C^*y(-s)\rangle_{(X_1,X^*_{-1})}\, ds \\ &= \left\langle x, \int_0^t \mathfrak{A}^{*(t-s)}_{|X^*_{-1}}C^*y(-s)\, ds\right\rangle_{(X_1,X^*_{-1})}. \end{aligned}$$

This being true for all $x \in X_1$, we must have $\int_0^t \mathfrak{A}^{*(t-s)}_{|X^*_{-1}}C^*y(-s)_X\, ds = 0$ for all $t \geq 0$. We now refer to Theorems 3.5.6(v), 4.3.4, 6.2.3, and 6.2.13, and let $\mathfrak{D}^*$ be the input/output map of the L^2-well-posed system on (X, X, Y) with growth bound ≤ 0 whose semigroup is $\mathfrak{A}^*$, control operator is C^*, observation operator is 1, and feedthrough operator is zero. If we denote $z(s) = y(-s)$ for $s \geq 0$, then $z \in L^2(\mathbb{R}^+; Y)$, and the condition above says that $\mathfrak{D}^*z$ vanishes on $\mathbb{R}^+$. Therefore we must also have, for all $\lambda \in \mathbb{C}^+$,

$$0 = \widehat{\mathfrak{D}^*z}(\lambda) = \widehat{\mathfrak{D}^*}(\lambda)\hat{z}(\lambda) = (\lambda - A^*)^{-1}C^*\hat{z}(\lambda).$$

The operator $(\lambda - A^*)^{-1}$ is injective, and so is C^* since we assume that C has dense range. Therefore $\hat{z}(\lambda) = 0$ for all $\lambda \in C^+$, so $z = 0$ and $y = 0$. This proves the minimality of the extension of $\mathfrak{A}$. □

By applying this result to the dual system we immediately get the following corollary:

Corollary 11.4.6 *Let $\mathfrak{A}$ be a C_0 contraction semigroup on a Hilbert space X.*

(i) *There is a Hilbert space U and an admissible control operator $B\colon U \to X_{-1}$ with $\mathcal{N}(B) = 0$ such that, if $\mathfrak{B}$ is the input map induced by B and $\mathfrak{A}$, then $\left[\mathfrak{A} \,\middle|\, \mathfrak{B}\right]$ is a (scattering) co-energy preserving system on $(0, X, U)$ (i.e., there is no output, only a state and an input).*

(ii) *The space U and the operator B are unique up to unitary similarity: if U_1 is another Hilbert space, if $B_1\colon U_1 \to X_{-1}$ is another admissible control operator with $\mathcal{N}(B_1) = 0$, if $\mathfrak{B}_1$ is the corresponding output map, and if $\left[\mathfrak{A} \,\middle|\, \mathfrak{B}_1\right]$ is a co-energy preserving system on $(0, X, U_1)$, then there is a unitary map $F\colon U \to U_1$ so that $B_1 = BF^{-1}$ and $\mathfrak{B}_1 = \mathfrak{B}F^{-1}$.*

(iii) *The corresponding Lax–Phillips model with parameter $\omega = 0$ (and with $Y = 0$, i.e., there is no input) is a minimal co-isometric dilation of $\mathfrak{A}$.*

Theorem 11.4.5 and Corollary 11.4.6 imply the following result:

Theorem 11.4.7 *Let $\mathfrak{A}$ be a contraction semigroup on a Hilbert space X.*

(i) *Let $\mathfrak{A}_-$ be a minimal isometric dilation of $\mathfrak{A}$ on $\left[\begin{smallmatrix} X_- \\ X \end{smallmatrix}\right]$ (where X_- is the orthogonal complement of X in the extended state space). Then X_- is unitarily similar to $L^2(\mathbb{R}^-; Y)$ for some Hilbert space Y, and $\mathfrak{A}_-$ decomposes into*

$$\mathfrak{A}_- = \begin{bmatrix} \mathfrak{A}_{--} & \mathfrak{A}_{-0} \\ 0 & \mathfrak{A} \end{bmatrix},$$

where $\mathfrak{A}_{--}$ is unitarily similar to the left shift τ_-^t on $L^2(\mathbb{R}^-; Y)$. In particular, X_- is invariant under $\mathfrak{A}_-$, X is invariant under $\mathfrak{A}_-^*$, and $\mathcal{R}\left(\mathfrak{A}_{--}^t\right) \perp \mathcal{R}\left(\mathfrak{A}_{-0}^t\right)$ for all $t \geq 0$. The space Y is unique up to a unitary similarity transformation.

(ii) *Let $\mathfrak{A}_+$ be a minimal co-isometric dilation of $\mathfrak{A}$ on $\left[\begin{smallmatrix} X \\ X_+ \end{smallmatrix}\right]$ (where X_- is the orthogonal complement of X in the extended state space). Then X_+ is unitarily similar to $L^2(\mathbb{R}^+; U)$ for some Hilbert space U, and $\mathfrak{A}_+$ decomposes into*

$$\mathfrak{A}_+ = \begin{bmatrix} \mathfrak{A} & \mathfrak{A}_{0+} \\ 0 & \mathfrak{A}_{++} \end{bmatrix},$$

where $\mathfrak{A}_{++}$ is unitarily similar to the left shift τ_+^t on $L^2(\mathbb{R}^+; U)$. In particular, X is invariant under $\mathfrak{A}_+$, X_+ is invariant under $\mathfrak{A}_+^*$, and $\mathcal{N}\left(\mathfrak{A}_{++}^t\right) \perp \mathcal{N}\left(\mathfrak{A}_{+0}^t\right)$ for all $t \geq 0$. The space U is unique up to a unitary similarity transformation.

Proof Properties (i) and (ii) obviously hold for the minimal isometric and co-isometric dilations presented in Theorem 11.4.5, and Corollary 11.4.6. That all minimal isometric or co-isometric dilations must have the same properties follows from Lemma 11.4.4. □

We next proceed to construct a *unitary* dilation of a given contraction semigroup. This construction is based on the following result:

Theorem 11.4.8 *Let $\mathfrak{A}$ be a contraction semigroup on a Hilbert space X. Then $\mathfrak{A}$ is isometric if and only if every minimal co-isometric dilation of $\mathfrak{A}$ is a minimal unitary dilation, and $\mathfrak{A}$ is co-isometric if and only if every minimal isometric dilation of $\mathfrak{A}$ is a minimal unitary dilation.*

Proof It suffices to prove the case where, e.g., $\mathfrak{A}$ is isometric, and the dilation is co-isometric (we get the other case by repeating the same argument with $\mathfrak{A}$ replaced by $\mathfrak{A}^*$).

We begin with the necessity of $\mathfrak{A}$ to be isometric. By Theorem 11.4.7, X is invariant under every minimal co-isometric dilation, so it coincides with the original semigroup $\mathfrak{A}$ on X. This implies that $\mathfrak{A}$ must be isometric in order for the dilated semigroup to be unitary.

We next turn to the converse. Obviously, if a dilation is minimal in the co-isometric sense, and if it happens to be unitary, it is also minimal in the unitary sense. Without loss of generality, we may assume that the dilation is the Lax–Phillips model described in Corollary 11.4.6 (every dilation is unitarily similar to this one). According to Lemma 11.2.2, to prove that the dilation is unitary we must show that the system in Corollary 11.4.6 is conservative whenever $\mathfrak{A}$ is isometric. By construction, we know that it is co-energy preserving, so it suffices to show that it is energy preserving.

Since $\mathfrak{A}$ is isometric, by Corollary 11.2.7, $\mathcal{D}(A) \subset \mathcal{D}(A^*)$ and $A^*x = -Ax$ for all $x \in \mathcal{D}(A)$. Thus, the first condition in (11.2.7) holds with $C = 0$. The co-isometric dilation is the dual of the isometric dilation of $\mathfrak{A}^*$, and by the construction in Theorem 11.4.5, the added control operator B satisfies $BB^*x = -(A_{|X} + A^*)x$ for all $x \in \mathcal{D}(A^*)$. This implies that $BB^*x = 0$ for all $x \in \mathcal{D}(A)$, hence $B^*x = 0$ for all $x \in \mathcal{D}(A)$. Since $A^*x = -Ax$ for all $x \in \mathcal{D}(A)$, we have, for all $\alpha \in \mathbb{C}^+$ and all $x \in \mathcal{D}(A)$,

$$B^*(\overline{\alpha} - A^*)^{-1}(\overline{\alpha} + A)x = B^*(\overline{\alpha} - A^*)^{-1}(\overline{\alpha} - A^*)x = B^*x = 0.$$

This is the adjoint of the second condition in (11.2.7) with $C = 0$. Thus, the first two conditions in (11.2.7) hold.

The verification of the third condition in (11.2.7) requires a closer study of the operator B^*. Fix $\alpha \in \mathbb{C}^+$, and define $\mathbf{A} = (\overline{\alpha} + A)(\alpha - A)^{-1}$. Then, by Theorem 11.2.5(viii) with $U = Y = 0$ (no input and no output), $\mathbf{A}$ is an isometry, so $\mathbf{A}^*\mathbf{A} = 1$. This implies that

$$(\mathbf{A}\mathbf{A}^*)^2 = \mathbf{A}\mathbf{A}^*\mathbf{A}\mathbf{A}^* = \mathbf{A}\mathbf{A}^*,$$

so $\mathbf{A}\mathbf{A}^*$ is a self-adjoint projection on X, hence so is $1 - \mathbf{A}\mathbf{A}^*$. Let us denote this projection by π. Then

$$\begin{aligned}\pi &= 1 - (\alpha - A_{|X})^{-1}(\overline{\alpha} + A_{|X})(\alpha + A^*)(\overline{\alpha} - A^*)^{-1} \\ &= (\alpha - A_{|X})^{-1}[(\alpha - A_{|X})(\overline{\alpha} - A^*) + (\overline{\alpha} + A_{|X})(\alpha + A^*)](\overline{\alpha} - A^*)^{-1} \\ &= 2\Re\alpha\,(\alpha - A_{|X})^{-1}(A_{|X} + A^*)(\overline{\alpha} - A^*)^{-1}.\end{aligned}$$

Comparing this to the definition of C in the proof of Theorem 11.4.5, and replacing C by B^* and A by A^* we realize that, if we define B^* as in the proof of that theorem, then $B^* = 1/\sqrt{2\Re\alpha}\,\pi^{1/2}(\overline{\alpha} - A^*)$. As π (this time) is a self-adjoint projection, this is equivalent to $B^* = 1/\sqrt{2\Re\alpha}\,\pi(\overline{\alpha} - A^*)$, or equivalently,

$$\pi = \sqrt{2\Re\alpha}\,B^*(\overline{\alpha} - A^*)^{-1}.$$

In particular, by our earlier definition of U we have $U = \mathcal{R}(B^*) = \mathcal{R}(\pi)$. Thus,

$$2\Re\alpha\,B^*(\alpha - A^*)^{-1}(\alpha - A_{|X})^{-1}B = \pi,$$

and the restriction of this operator to $U = \mathcal{R}(\pi)$ is the identity, as required by the third condition in (11.2.7) with $\beta = \alpha$ and $\widehat{\mathfrak{D}} = 0$. By Theorem 11.2.5, the dilated system is conservative. □

Theorem 11.4.9 *Let $\mathfrak{A}$ be a C_0 contraction semigroup on a Hilbert space X.*

(i) *There exist two Hilbert space U and Y and a (scattering) conservative linear system $\left[\begin{smallmatrix}\mathfrak{A} & \mathfrak{B} \\ \mathfrak{C} & \mathfrak{D}\end{smallmatrix}\right]$ on (Y, X, U) (where the semigroup is the given semigroup $\mathfrak{A}$) whose control operator is injective on U and whose observation operator has dense range in Y. The input/output map of this system is purely passive.*

(ii) *The system in (i) is unique in the following sense: if we have two different conservative systems $\left[\begin{smallmatrix}\mathfrak{A} & \mathfrak{B} \\ \mathfrak{C} & \mathfrak{D}\end{smallmatrix}\right]$ on (Y, X, U) and $\left[\begin{smallmatrix}\mathfrak{A} & \mathfrak{B}_1 \\ \mathfrak{C}_1 & \mathfrak{D}_1\end{smallmatrix}\right]$ on (Y_1, X, U_1) with purely passive input/output maps, then there exist unitary operators $F: U \to U_1$ and $G: Y \to Y_1$ so that*

$$\begin{bmatrix}\mathfrak{A} & \mathfrak{B}_1 \\ \mathfrak{C}_1 & \mathfrak{D}_1\end{bmatrix} = \begin{bmatrix}1 & 0 \\ 0 & G\end{bmatrix}\begin{bmatrix}\mathfrak{A} & \mathfrak{B} \\ \mathfrak{C} & \mathfrak{D}\end{bmatrix}\begin{bmatrix}1 & 0 \\ 0 & F^{-1}\end{bmatrix}.$$

(iii) *The corresponding Lax–Phillips model with parameter $\omega = 0$ is a minimal unitary dilation of $\mathfrak{A}$.*

Proof (i) The idea of this proof is very simple: first we use Corollary 11.4.6 to create an input space U and an injective control operator B such that the corresponding system (with no output) is co-energy preserving. Then we apply Theorem 11.4.5 to this system to create an output space Y and a combined observation/feedthrough operator $C\&D$ with dense range so that the resulting system is energy preserving. The details are as follows.

Let $\Sigma_+ = \left[\mathfrak{A} \middle| \mathfrak{B}\right]$ be the co-energy preserving system with input space U and control operator B that we get from Corollary 11.4.6. The domain $\mathcal{D}(S)$ of the corresponding system node $S = A\&B$ is given by $\mathcal{D}(S) = \left\{\left[\begin{smallmatrix} x \\ u \end{smallmatrix}\right] \in \left[\begin{smallmatrix} X \\ U \end{smallmatrix}\right] \,\middle|\, A_{|X}x + Bu \in X\right\}$. We denote the corresponding Lax–Phillips model on $\mathbf{X}_+ = \left[\begin{smallmatrix} X \\ L^2(\mathbb{R}^+;U) \end{smallmatrix}\right]$ by

$$\mathfrak{T}_+^t = \begin{bmatrix} \mathfrak{A}^t & \mathfrak{B}^t \\ 0 & \tau_+^t \end{bmatrix}.$$

This semigroup is co-isometric. We now use Theorem 11.4.5 to create an observation operator $\mathbf{C}_+$ for $\mathfrak{T}_+$ such that the corresponding system is energy preserving. Let $\mathfrak{T}$ be the corresponding Lax–Phillips semigroup on $\mathbf{X} = \left[\begin{smallmatrix} L^2(\mathbb{R}^-;Y) \\ \mathbf{X}_+ \end{smallmatrix}\right]$ (with zero input space). By Theorems 11.4.5(iii) and 11.4.8, $\mathfrak{T}$ is a minimal unitary dilation of $\mathfrak{T}_+$. By decomposing the state space of $\mathfrak{T}$ into $\left[\begin{smallmatrix} L^2(\mathbb{R}^-;Y) \\ X \\ L^2(\mathbb{R}^+;U) \end{smallmatrix}\right]$ we can write $\mathfrak{T}$ in block matrix form as

$$\mathfrak{T}^t = \begin{bmatrix} \tau_-^t & \mathfrak{C}^t & \mathfrak{D}^t \\ 0 & \mathfrak{A}^t & \mathfrak{B}^t \\ 0 & 0 & \tau_+^t \end{bmatrix}.$$

By construction, $\mathfrak{T}$ is a Lax–Phillips model on $\left[\begin{smallmatrix} L^2(\mathbb{R}^-;Y) \\ \mathbf{X}_+ \end{smallmatrix}\right]$ (with zero input space), but we do not yet know if it can also be interpreted as a Lax–Phillips model with input space U and output space Y. This will be true if and only if $\mathfrak{T}$ satisfies the causality conditions (2.7.2). Some of these causality conditions we get for free. Since $\mathfrak{T}_+$ is a Lax–Phillips model, the condition $\mathfrak{B}^t = \mathfrak{B}^t\pi_{[0,t)}$ holds for all $t \geq 0$, and since $\mathfrak{T}$ is a Lax–Phillips model on $\left[\begin{smallmatrix} L^2(\mathbb{R}^-;Y) \\ \mathbf{X}_+ \end{smallmatrix}\right]$ (with zero input space), the condition $\left[\mathfrak{C}^t \;\; \mathfrak{D}^t\right] = \pi_{[-t,0)}\left[\mathfrak{C}^t \;\; \mathfrak{D}^t\right]$ holds for all $t \geq 0$. Thus, the only part of (2.7.2) which remains to be verified is that $\pi_{[-t,0)}\mathfrak{D}^t\pi_{[t,\infty)} = 0$. To do this we investigate the generator of $\mathfrak{T}_+$.

Denote the generator of $\mathfrak{T}_+$ by $\mathbf{A}_+$. By Theorem 4.8.3,

$$\mathcal{D}(\mathbf{A}_+) = \left\{\left[\begin{smallmatrix} x \\ u \end{smallmatrix}\right] \in \left[\begin{smallmatrix} X \\ W^{1,2}(\mathbb{R}^+;U) \end{smallmatrix}\right] \,\middle|\, \left[\begin{smallmatrix} x \\ u(0) \end{smallmatrix}\right] \in \mathcal{D}(S)\right\},$$

and for all $\left[\begin{smallmatrix} x \\ u \end{smallmatrix}\right] \in \mathcal{D}(\mathbf{A}_+)$,

$$\mathbf{A}_+ \begin{bmatrix} x \\ u \end{bmatrix} = \begin{bmatrix} A\&B\left[\begin{smallmatrix} x \\ u(0) \end{smallmatrix}\right] \\ \dot{u} \end{bmatrix}.$$

In particular, for all $\left[\begin{smallmatrix} x \\ u \end{smallmatrix}\right] \in \mathcal{D}(\mathbf{A}_+)$,

$$\begin{aligned} 2\Re\left\langle \left[\begin{smallmatrix} x \\ u \end{smallmatrix}\right], \mathbf{A}_+ \left[\begin{smallmatrix} x \\ u \end{smallmatrix}\right] \right\rangle &= 2\Re\left\langle A\&B \left[\begin{smallmatrix} x \\ u(0) \end{smallmatrix}\right], x \right\rangle + 2\Re \int_{\mathbb{R}^+} \langle u(s), \dot{u}(s) \rangle \, ds \\ &= 2\Re\left\langle A\&B \left[\begin{smallmatrix} x \\ u(0) \end{smallmatrix}\right], x \right\rangle - |u(0)|_U^2. \end{aligned}$$

This vanishes (at least) if $x = 0$ and $u(0) = 0$, so the operator $\mathbf{A}_+ + \mathbf{A}^*_{+|\mathbf{X}_+}$ (which maps $\mathcal{D}(\mathbf{A}_+)$ into the dual of $\mathcal{D}(\mathbf{A}_+)$ with $\mathbf{X}_+$ as a pivot space) vanishes on the subspace $\left[\begin{smallmatrix} 0 \\ W_0^{1,2}(\mathbb{R}^+;U) \end{smallmatrix}\right]$. The observation operator $\mathbf{C}_+ \in \mathcal{B}(\mathcal{D}(\mathbf{A}_+); Y)$ constructed in the proof of Theorem 11.4.5 satisfies $\mathbf{A}_+ + \mathbf{A}^*_{+|\mathbf{X}_+} = -\mathbf{C}^*_+\mathbf{C}_+$. By the above argument, $\mathbf{C}^*_+\mathbf{C}_+$ vanishes on $\left[\begin{smallmatrix} 0 \\ W_0^{1,2}(\mathbb{R}^+;U) \end{smallmatrix}\right]$, hence so does $\mathbf{C}_+$. Therefore $\mathbf{C}_+ \left[\begin{smallmatrix} x \\ u \end{smallmatrix}\right]$ can depend only on $\left[\begin{smallmatrix} x \\ u(0) \end{smallmatrix}\right] \in \mathcal{D}(S)$, i.e., $\mathbf{C}_+ \left[\begin{smallmatrix} x \\ u \end{smallmatrix}\right] = C\&D \left[\begin{smallmatrix} x \\ u(0) \end{smallmatrix}\right]$ for some operator $C\&D \in \mathcal{B}(\mathcal{D}(S); Y)$.

We now apply $\mathfrak{T}^t$ to data of the type $\left[\begin{smallmatrix} 0 \\ 0 \\ u \end{smallmatrix}\right]$, where $u \in W^{1,2}(\mathbb{R}^+; U)$ vanishes on $[0, t)$. Then the restriction of the first component of $\mathfrak{T}^t$ to $[-t, 0)$ is $\pi_{[-t,0)}\mathfrak{D}^t u$. On the other hand, by the representation of $\mathfrak{T}^t$ found in Theorem 2.7.1 (with zero input space) and by Theorem 4.4.2(i) (applied to $\mathfrak{T}_+$ and $\mathbf{C}_+$), it is given by τ^t applied to the function $s \mapsto \mathbf{C}_+\mathfrak{T}^s_+ \left[\begin{smallmatrix} 0 \\ u \end{smallmatrix}\right] = \mathbf{C}_+ \left[\begin{smallmatrix} x(s) \\ \tau^s u \end{smallmatrix}\right] = C\&D \left[\begin{smallmatrix} x(s) \\ u(s) \end{smallmatrix}\right]$, where x is the unique strong solution of the equation $\dot{x}(s) = A_{|X}x(s) + u(s)$, $s \geq 0$, with initial state zero. But $x(s) = 0$ for $s \in [0, t)$ (since we took u to vanish on $[0, t)$). Thus, $\mathbf{C}_+\mathfrak{T}^s_+ \left[\begin{smallmatrix} 0 \\ u \end{smallmatrix}\right] = C\&D \left[\begin{smallmatrix} x(s) \\ u(s) \end{smallmatrix}\right] = 0$ for $s \in [0, t)$, and this shows that $\pi_{[-t,0)}\mathfrak{D}^t u = 0$ whenever $u \in W^{1,2}(\mathbb{R}^+; U)$ vanishes on $[0, t)$. By continuity, the same statement is true with $W^{1,2}(\mathbb{R}^+; U)$ replaced by $L^2(\mathbb{R}^+; U)$. Thus, $\pi_{[-t,0)}\mathfrak{D}^t\pi_{[t,\infty)} = 0$, and we have proved that $\mathfrak{T}$ is a Lax–Phillips model with input space U and output space Y.

Let Σ be the L^2-well-posed system corresponding to the Lax–Phillips model (given by Theorem 2.7.6). Then, by Lemma 11.2.3, Σ is conservative. The control operator B of this system is injective, hence by Theorem 11.2.9(i), the space U_0 in the decomposition presented in Theorem 11.1.9 is zero. Thus also $Y_0 = 0$ (since Y_0 is a unitary similar to U_0), and by Theorem 11.2.9(ii), the observation operator C has dense range. By Theorem 11.1.9, $\mathfrak{D}$ is purely passive. This proves (i).

(ii) By Theorem 11.2.9, if $\mathfrak{D}$ and $\mathfrak{D}_1$ are purely passive, then C and C_1 have dense ranges, and B and B_1 are injective. The existence of a unitary operator F such that $U_1 = FU$, $B_1 = BF^{-1}$, and $\mathfrak{B}_1 = \mathfrak{B}F^{-1}$ follows from Corollary 11.4.6. Recall that the system denoted by $\mathfrak{T}_+$ in the proof above is a co-isometric semigroup. We can extend this semigroup to a conservative system whose output operator has a dense range in two ways, namely by choosing the output space to be Y and the output map to be $\left[\mathfrak{C} \;\; \mathfrak{D}\right]$, or by choosing the output space to be Y_1 and the output map to be $\begin{bmatrix}\mathfrak{C}_1 & \mathfrak{D}_1 F\end{bmatrix}$ (the ranges of the two observation operators contain the ranges of C, respectively C_1, which are dense). By Theorem 11.4.5 with $\mathfrak{A}$ replaced by $\mathfrak{T}_+$, there is a unitary operator G such that $Y_1 = GY$ and $\begin{bmatrix}\mathfrak{C}_1 & \mathfrak{D}_1 F\end{bmatrix} = G\begin{bmatrix}\mathfrak{C}_1 & \mathfrak{D}_1\end{bmatrix}$.

(iii) It is clear by now that $\mathfrak{T}$ is a unitary dilation of $\mathfrak{A}$. We still need to show that it is minimal.

On $\begin{bmatrix}0\\ X\\ L^2(\mathbb{R}^+;U)\end{bmatrix}$, $\mathfrak{T}^*$ coincides with $\mathfrak{T}_+^*$, and by the construction of $\mathfrak{T}_+$, the span of $\mathfrak{T}_+^{*t}$, $t \ge 0$, acting on $\begin{bmatrix}0\\ X\\ 0\end{bmatrix}$ is dense in this space. Thus, the span of $\mathfrak{T}_+^{*t}$ acting on $\begin{bmatrix}0\\ X\\ 0\end{bmatrix}$ is dense in $\begin{bmatrix}0\\ X\\ L^2(\mathbb{R}^+;U)\end{bmatrix}$. Let us denote the system $\begin{bmatrix}\mathfrak{A}\\ \hline \mathfrak{C}\end{bmatrix}$ that we get from Σ by dropping the input by Σ_- and the corresponding Lax–Phillips model by $\mathfrak{T}_-$. Then $\mathfrak{T}$ coincides with $\mathfrak{T}_-$ on $\begin{bmatrix}L^2(\mathbb{R}^+;Y)\\ X\\ 0\end{bmatrix}$, so we have to show that the span of $\mathfrak{T}_-$ acting on $\begin{bmatrix}0\\ X\\ 0\end{bmatrix}$ is dense in $\begin{bmatrix}L^2(\mathbb{R}^+;Y)\\ X\\ 0\end{bmatrix}$. To do this it suffices to show that $\mathfrak{T}_-$ is the minimal isometric dilation of $\mathfrak{A}$ constructed in Theorem 11.4.5. A direct comparison of the observation operator in the proof of Theorem 11.4.5 and the observation operator of Σ shows that they are, indeed, the same (take $u = 0$ in all the formulas). Thus the span of $\mathfrak{T}$ acting on $\begin{bmatrix}0\\ X\\ 0\end{bmatrix}$ is dense in $\begin{bmatrix}L^2(\mathbb{R}^+;Y)\\ X\\ 0\end{bmatrix}$. We conclude that $\mathfrak{T}$ is a minimal unitary dilation of $\mathfrak{A}$. □

Corollary 11.4.10 *Two (scattering) conservative systems* $\left[\begin{array}{c|c}\mathfrak{A} & \mathfrak{B}\\ \hline \mathfrak{C} & \mathfrak{D}\end{array}\right]$ *on* (Y, X, U) *and* $\left[\begin{array}{c|c}\mathfrak{A}_1 & \mathfrak{B}_1\\ \hline \mathfrak{C}_1 & \mathfrak{D}_1\end{array}\right]$ *on* (Y_1, X_1, U_1) *with purely passive input/output maps* $\mathfrak{D}$ *and* $\mathfrak{D}_1$ *are unitarily similar if and only if the two semigroups* $\mathfrak{A}$ *and* $\mathfrak{A}_1$ *are unitarily similar. In particular, the semigroup* $\mathfrak{A}$ *of a conservative system* $\left[\begin{array}{c|c}\mathfrak{A} & \mathfrak{B}\\ \hline \mathfrak{C} & \mathfrak{D}\end{array}\right]$ *determines the input map* $\mathfrak{B}$, *the output map* $\mathfrak{C}$, *and the purely passive part of the input/output map* $\mathfrak{D}$ *uniquely up to unitary similarity transformations in the input and output spaces.*

Recall that, by the purely passive part of $\mathfrak{D}$ we mean the operator $\mathfrak{D}_1$ in Theorem 11.1.9.

Proof Obviously, if the two systems are unitarily similar, then so are their semigroups (see Definition 11.4.3). Conversely, if there is a unitary map E of X onto X_1 such that $\mathfrak{A}_1 = E\mathfrak{A}E^{-1}$, then we find that the two systems are unitarily similar by applying Theorem 11.4.9(ii) to the systems $\left[\begin{array}{c|c}\mathfrak{A} & \mathfrak{B}\\ \hline \mathfrak{C} & \mathfrak{D}\end{array}\right]$ and $\begin{bmatrix}E & 0\\ 0 & 1\end{bmatrix}\left[\begin{array}{c|c}\mathfrak{A}_1 & \mathfrak{B}_1\\ \hline \mathfrak{C}_1 & \mathfrak{D}_1\end{array}\right]\begin{bmatrix}E^{-1} & 0\\ 0 & 1\end{bmatrix}$. □

Corollary 11.4.11 *A (scattering) conservative system* $\Sigma = \left[\begin{array}{c|c}\mathfrak{A} & \mathfrak{B}\\ \hline \mathfrak{C} & \mathfrak{D}\end{array}\right]$ *on three Hilbert spaces* (Y, X, U) *is unitarily similar to its causal dual system* Σ^d *if and only if the semigroup* $\mathfrak{A}$ *is unitarily similar to its adjoint* $\mathfrak{A}^*$.

Proof If $\mathfrak{D}$ is purely passive, then it follows from Corollary 11.4.10 that Σ is unitarily similar to Σ^d if and only if $\mathfrak{A}$ is unitarily similar to $\mathfrak{A}^*$. The general case follows from this special case and the decomposition established in Theorem 11.1.9. □

Definition 11.4.12 By the *characteristic function* of a C_0 contraction semigroup $\mathfrak{A}$ on a Hilbert space X we mean the transfer function $\widehat{\mathfrak{D}}$ of a (scattering) conservative system $\left[\begin{smallmatrix} \mathfrak{A} & \mathfrak{B} \\ \mathfrak{C} & \mathfrak{D} \end{smallmatrix}\right]$ with this semigroup and with a purely passive input/output map $\mathfrak{D}$. (By Theorem 11.4.9, such a conservative system always exists, and $\widehat{\mathfrak{D}}$ is unique up to unitary similarity transformations in the input and output spaces.)

Theorem 11.4.13 *Two C_0 contraction semigroups $\mathfrak{A}$ and $\mathfrak{A}_1$ have the same characteristic functions (unique up to unitary similarity transformations in the input and output spaces) if and only if the completely nonunitary parts of $\mathfrak{A}$ and $\mathfrak{A}_1$ are unitarily similar. In particular, the characteristic function of $\mathfrak{A}$ does not depend on the unitary part of $\mathfrak{A}$.*

Proof With the notation of Theorem 11.1.9, if $\mathfrak{D}$ is purely passive, then $\left[\begin{smallmatrix} \mathfrak{A} & \mathfrak{B} \\ \mathfrak{C} & \mathfrak{D} \end{smallmatrix}\right]$ is conservative if and only if the system $\left[\begin{smallmatrix} \mathfrak{A}_1 & \mathfrak{B}_1 \\ \mathfrak{C}_1 & \mathfrak{D} \end{smallmatrix}\right]$ is conservative (here the latter system is the compression of the former to the space where $\mathfrak{A}$ is completely nonunitary). Thus, the full semigroup $\mathfrak{A}$ has the same characteristic function as the completely nonunitary part of $\mathfrak{A}$. Hence, if the completely nonunitary parts of the semigroups are unitarily similar, then by Corollary 11.4.10, their characteristic functions are the same. The converse statement, i.e., that the input/output map determines the semigroups uniquely up to a unitary similarity transformation in the state space follows from Corollary 11.6.4. □

Theorem 11.4.8 also provides us with the following alternative characterization of a conservative system.

Theorem 11.4.14 *Let $\Sigma = \left[\begin{smallmatrix} \mathfrak{A} & \mathfrak{B} \\ \mathfrak{C} & \mathfrak{D} \end{smallmatrix}\right]$ be a (scattering) energy preserving system on (Y, X, U) with system node $S = \left[\begin{smallmatrix} A\&B \\ C\&D \end{smallmatrix}\right]$, main operator A, and control operator B. Then Σ is conservative if and only if the range of $C\&D$ is dense in Y and $A_{|X}x + A^*x + BB^*x = 0$ for all $x \in \mathcal{D}(A^*)$.*

Proof We begin with the necessity. Suppose that Σ is conservative. Then, by the dual version of Theorem 11.2.5, $A_{|X}x + A^*x + BB^*x = 0$ for all $x \in \mathcal{D}(A^*)$. Moreover, Σ is time-flow-invertible with a stable time-flow inverse Σ^d, so by Corollary 6.5.8(iii), the operator $\left[\begin{smallmatrix} \alpha & 0 \\ 0 & 0 \end{smallmatrix}\right] + S$ maps $\mathcal{D}(S)$ onto $\left[\begin{smallmatrix} X \\ Y \end{smallmatrix}\right]$ for all $\alpha \in \mathbb{C}_+$. Thus, the range of $C\&D$ is not just dense in Y; it is even equal to Y.

Conversely, suppose that $A_{|X}x + A^*x + BB^*x = 0$ for all $x \in \mathcal{D}(A^*)$ and that the range of $C\&D$ is dense in Y. We first apply the dual version of

Theorem 11.2.5 to conclude that the system $\Sigma_+ = \left[\mathfrak{A}\middle|\,\mathfrak{B}\right]$ is co-energy preserving, hence the corresponding Lax–Phillips model $\mathfrak{T}_+$ is co-isometric. We assumed Σ to be energy preserving, and this means that the Lax–Phillips model $\mathfrak{T}$ induced by Σ is an isometric dilation of $\mathfrak{T}_+$. Moreover, this dilation is minimal; this can be seen in the same way as in the last part of the proof of Theorem 11.4.9. By Theorem 11.4.8, $\mathfrak{T}$ is unitary, or equivalently, Σ is conservative. □

11.5 Energy preserving and conservative extensions of passive systems

In the preceding section we studied dilations and compressions of a *semigroup*. Here we study dilations and compressions of *systems*. Since the proofs are very similar to those in Section 11.4, we leave most of them to the reader.

Definition 11.5.1 Let $\Sigma = \left[\begin{smallmatrix}\mathfrak{A} & \mathfrak{B}\\ \mathfrak{C} & \mathfrak{D}\end{smallmatrix}\right]$ and $\widetilde{\Sigma} = \left[\begin{smallmatrix}\widetilde{\mathfrak{A}} & \widetilde{\mathfrak{B}}\\ \widetilde{\mathfrak{C}} & \widetilde{\mathfrak{D}}\end{smallmatrix}\right]$ be two $L^p|Reg$-well-posed linear systems on (Y, X, U), respectively $(Y, \widetilde{X}, U)$. We say that Σ is an (orthogonal) *compression* of $\widetilde{\Sigma}$ and that $\widetilde{\Sigma}$ is an (orthogonal) *dilation* of Σ if

$$\left[\begin{array}{c|c}\mathfrak{A} & \mathfrak{B}\\ \hline \mathfrak{C} & \mathfrak{D}\end{array}\right] = \left[\begin{array}{c|c}\pi_X\widetilde{\mathfrak{A}}_{|X} & \pi_X\widetilde{\mathfrak{B}}\\ \hline \widetilde{\mathfrak{C}}_{|X} & \widetilde{\mathfrak{D}}\end{array}\right], \tag{11.5.1}$$

where π_X is the orthogonal projection of $\widetilde{X}$ onto X.

The fact that the right-hand side of (11.5.1) is required to be an $L^p|Reg$-well-posed linear system puts a rather strong restriction on the system $\widetilde{\Sigma}$.

Lemma 11.5.2 *Let* $\widetilde{\Sigma} = \left[\begin{smallmatrix}\widetilde{\mathfrak{A}} & \widetilde{\mathfrak{B}}\\ \widetilde{\mathfrak{C}} & \widetilde{\mathfrak{D}}\end{smallmatrix}\right]$ *be an* $L^p|Reg$*-well-posed linear system on* $(Y, \widetilde{X}, Y)$*, let* X *be a closed subspace of* $\widetilde{X}$*, and define* $\Sigma = \left[\begin{smallmatrix}\mathfrak{A} & \mathfrak{B}\\ \mathfrak{C} & \mathfrak{D}\end{smallmatrix}\right]$ *by* (11.5.1)*. Then* Σ *is an* $L^p|Reg$*-well-posed linear system if and only if* $\widetilde{X}$ *is a direct sum* $X = Z \oplus X \oplus Z_*$ *such that the block matrix decomposition of* $\widetilde{\Sigma}$ *with respect to this decomposition of* $\widetilde{X}$ *has the form*

$$\widetilde{\Sigma} = \left[\begin{array}{ccc|c}\widetilde{\mathfrak{A}}_{|Z} & \pi_Z\widetilde{\mathfrak{A}}_{|X} & \pi_Z\widetilde{\mathfrak{A}}_{|Z_*} & \pi_Z\widetilde{\mathfrak{B}}\\ 0 & \pi_X\widetilde{\mathfrak{A}}_{|X} & \pi_X\widetilde{\mathfrak{A}}_{|Z_*} & \pi_X\widetilde{\mathfrak{B}}\\ 0 & 0 & \pi_{Z_*}\widetilde{\mathfrak{A}}_{|Z_*} & 0\\ \hline 0 & \widetilde{\mathfrak{C}}_{|X} & \widetilde{\mathfrak{C}}_{|Z_*} & \widetilde{\mathfrak{D}}\end{array}\right]. \tag{11.5.2}$$

Thus, Z is invariant and unobservable, whereas Z_* is co-invariant and unreachable.

Proof First suppose that $\widetilde{\Sigma}$ is of the form (11.5.2). Trivially, $\mathfrak{A}$ is strongly continuous, and $\mathfrak{A}^0 = 1$. For all $s, t \geq 0$,

$$\mathfrak{A}^s\mathfrak{A}^t = \pi_X\widetilde{\mathfrak{A}}^s\pi_X\widetilde{\mathfrak{A}}^t_{|X} = \pi_X\widetilde{\mathfrak{A}}^s\widetilde{\mathfrak{A}}^t_{|X} = \pi_X\widetilde{\mathfrak{A}}^{s+t}_{|X} = \mathfrak{A}^{s+t}.$$

Thus, $\mathfrak{A}$ is a semigroup. For all $t \geq 0$,

$$\mathfrak{A}^t\mathfrak{B} = \pi_X\widetilde{\mathfrak{A}}^t\pi_X\widetilde{\mathfrak{B}} = \pi_X\widetilde{\mathfrak{A}}^t\widetilde{\mathfrak{B}} = \pi_X\widetilde{\mathfrak{B}}\tau^t_- = \mathfrak{B}\tau^t_-,$$

i.e., $\mathfrak{B}$ intertwines $\mathfrak{A}^t$ and τ^t_-. For all $t \geq 0$,

$$\mathfrak{C}\mathfrak{A}^t = \widetilde{\mathfrak{C}}\pi_X\widetilde{\mathfrak{A}}^t_{|X} = \widetilde{\mathfrak{C}}\widetilde{\mathfrak{A}}^t_{|X} = \tau^t_+\widetilde{\mathfrak{C}}_{|X} = \tau^t_+\mathfrak{C},$$

i.e., $\mathfrak{C}$ intertwines $\mathfrak{A}^t$ and τ^t_+. Finally,

$$\pi_+\mathfrak{D}\pi_- = \pi_+\widetilde{\mathfrak{D}}\pi_- = \widetilde{\mathfrak{C}}\widetilde{\mathfrak{B}} = \widetilde{\mathfrak{C}}\pi_X\widetilde{\mathfrak{B}} = \mathfrak{C}\mathfrak{B},$$

i.e., the Hankel operator of $\mathfrak{D}$ is $\mathfrak{C}\mathfrak{B}$. This shows that $\mathfrak{A}$ is an $L^p|Reg$-well-posed linear system.

We now turn to the converse claim, and suppose that $\mathfrak{A}$ is an $L^p|Reg$-well-posed linear system. Let us add another output to this system, with output space X and output operator π_X. Thus, the added output map $\mathfrak{C}_1$ is given by $(\mathfrak{C}_1x)(t) = \pi_X\widetilde{\mathfrak{A}}^tx, t \geq 0$. Let Z be the unobservable subspace for the pair $\left[\begin{smallmatrix}\mathfrak{C}_1\\ \widetilde{\mathfrak{C}}\end{smallmatrix}\right]$, i.e., $x \in Z$ if and only if $\pi_X\widetilde{\mathfrak{A}}^t = 0$ for all $t \geq 0$ and $\widetilde{\mathfrak{C}}x = 0$. By Lemma 9.6.1, Z is an invariant subspace of $\widetilde{\mathfrak{A}}$ contained in $\mathcal{N}(\pi_X) = X^\perp$. Let $Z_* = (Z \oplus X)^\perp$. Then $\widetilde{X} = Z \oplus X \oplus Z_*$, $\pi_X\widetilde{\mathfrak{A}}_{|Z} = 0$ and $\pi_{Z_*}\widetilde{\mathfrak{A}}_{|Z} = 0$ (since Z is invariant and both X and Z_* are orthogonal to Z), and $\widetilde{\mathfrak{C}}_{|Z} = 0$ (since Z is unobservable). This gives us the three zeros in the first column of the decomposition of $\widetilde{\Sigma}$ given in (11.5.2).

It remains to account for the zeros in the second and fourth columns of the third row of (11.5.2). To get these zeros we have to use the assumption that Σ is a system. The fact that $\widetilde{\mathfrak{A}}$ and $\mathfrak{A}$ are semigroups gives, for all $s, t \geq 0$,

$$\pi_X\widetilde{\mathfrak{A}}^s\widetilde{\mathfrak{A}}^t_{|X} = \pi_X\widetilde{\mathfrak{A}}^{s+t}_{|X} = \mathfrak{A}^{s+t} = \mathfrak{A}^s\mathfrak{A}^t = \pi_X\widetilde{\mathfrak{A}}^s\pi_X\widetilde{\mathfrak{A}}^t_{|X},$$

hence

$$\pi_X\widetilde{\mathfrak{A}}^s(1 - \pi_X)\widetilde{\mathfrak{A}}^t_{|X} = \pi_X\widetilde{\mathfrak{A}}^s(\pi_Z + \pi_{Z_*})\widetilde{\mathfrak{A}}^t_{|X} = 0$$

for all $s, t \geq 0$. The output intertwining conditions for $\widetilde{\Sigma}$ and Σ give, for all $t \geq 0$,

$$\widetilde{\mathfrak{C}}\widetilde{\mathfrak{A}}^t_{|X} = \tau^t_+\widetilde{\mathfrak{C}}_{|X} = \tau^t_+\mathfrak{C}_{|X} = \mathfrak{C}\mathfrak{A}^t_{|X} = \widetilde{\mathfrak{C}}\pi_X\widetilde{\mathfrak{A}}^t_{|X},$$

hence

$$\widetilde{\mathfrak{C}}(1 - \pi_X)\widetilde{\mathfrak{A}}^t_{|X} = \widetilde{\mathfrak{C}}(\pi_Z + \pi_{Z_*})\widetilde{\mathfrak{A}}^t_{|X} = 0.$$

This says that, for each $t \ge 0$ and each $x \in X$, both $\mathfrak{C}_1(\pi_Z + \pi_{Z_*})\widetilde{\mathfrak{A}}^t x = 0$ and $\widetilde{\mathfrak{C}}(\pi_Z + \pi_{Z_*})\widetilde{\mathfrak{A}}^t x = 0$. By the definition of Z as the unobservable subspace (with respect to these two output maps), $(\pi_Z + \pi_{Z_*})\widetilde{\mathfrak{A}}^t x \in Z$, i.e., $\pi_{Z_*}\widetilde{\mathfrak{A}}^t_{|X} = 0$. This gives us the zero in the second column of the third row of (11.5.2).

The zero in the last column of the third row of (11.5.2) still remains. The input intertwining conditions for $\widetilde{\Sigma}$ and Σ give, for all $s \ge 0$,

$$\pi_X \widetilde{\mathfrak{A}}^s \widetilde{\mathfrak{B}} = \pi_X \widetilde{\mathfrak{B}} \tau_-^s = \pi_X \mathfrak{B} \tau_-^s = \pi_X \mathfrak{A}^s \mathfrak{B} = \pi_X \widetilde{\mathfrak{A}}^s \pi_X \widetilde{\mathfrak{B}},$$

hence

$$\pi_X \widetilde{\mathfrak{A}}^s (1 - \pi_X)\widetilde{\mathfrak{B}} = \pi_X \widetilde{\mathfrak{A}}^s (\pi_Z + \pi_{Z_*})\widetilde{\mathfrak{B}} = 0.$$

The Hankel conditions for the two systems give

$$\widetilde{\mathfrak{C}}\widetilde{\mathfrak{B}} = \pi_+ \widetilde{\mathfrak{D}} \pi_- = \pi_+ \mathfrak{D} \pi_- = \mathfrak{C}\mathfrak{B} = \widetilde{\mathfrak{C}} \pi_X \widetilde{\mathfrak{B}},$$

hence

$$\widetilde{\mathfrak{C}}(1 - \pi_X)\widetilde{\mathfrak{B}} = \widetilde{\mathfrak{C}}(\pi_Z + \pi_{Z_*})\widetilde{\mathfrak{B}} = 0.$$

This says that both $\mathfrak{C}_1(\pi_Z + \pi_{Z_*})\widetilde{\mathfrak{B}} = 0$ and $\widetilde{\mathfrak{C}}(\pi_Z + \pi_{Z_*})\widetilde{\mathfrak{B}} = 0$. By the definition of Z as the unobservable subspace, the range of $(\pi_Z + \pi_{Z_*})\widetilde{\mathfrak{B}}$ is contained in Z, i.e., $\pi_{Z_*}\widetilde{\mathfrak{B}} = 0$. This gives us the final zero in the second column of the third row of (11.5.2). □

Theorem 11.4.5 has the following analogue for systems which have no input. (This is a preliminary version of the more general Theorem 11.5.4.)

Lemma 11.5.3 *Let $\left[\frac{\mathfrak{A}}{\mathfrak{C}}\right]$ be a (scattering) passive L^2-well-posed linear system (with no input) on the Hilbert spaces $(Y, X, 0)$.*

(i) *There is a Hilbert space Y_1 and an admissible observation operator $C_1 \colon X_1 \to Y_1$ with dense range such that, if $\mathfrak{C}_1$ is the output map induced by C_1 and $\mathfrak{A}$, then $\begin{bmatrix} \mathfrak{A} \\ \hline \mathfrak{C}_1 \\ \mathfrak{C} \end{bmatrix}$ is an (scattering) energy preserving system on $\left(\left[\begin{smallmatrix} Y_1 \\ Y \end{smallmatrix}\right], X, 0\right)$ (with no input).*

(ii) *The space Y_1 and the operator C_1 are unique up to unitary similarity: if Y_2 is another Hilbert space, if $C_2 \colon X_1 \to Y_2$ is another admissible observation operator with dense range, if $\mathfrak{C}_2$ is the corresponding output map, and if $\begin{bmatrix} \mathfrak{A} \\ \hline \mathfrak{C}_2 \\ \mathfrak{C} \end{bmatrix}$ is an energy preserving system on $\left(\left[\begin{smallmatrix} Y_2 \\ Y \end{smallmatrix}\right], X, 0\right)$, then there is a unitary map $G \colon Y_1 \to Y_2$ so that $C_2 = GC_1$ and $\mathfrak{C}_2 = G\mathfrak{C}_1$.*

(iii) *The Lax–Phillips model with parameter $\omega = 0$ induced by $\begin{bmatrix} \mathfrak{A} \\ \hline \mathfrak{C}_1 \\ \mathfrak{C} \end{bmatrix}$ (with no input) is a minimal isometric dilation of the corresponding Lax–Phillips model induced by $\left[\frac{\mathfrak{A}}{\mathfrak{C}}\right]$.*

Proof The proof is very similar to the proof of Theorem 11.4.5, and for this reason we leave most of the details to the reader. In the proof of (i) we define $C_1 x$ for all $x \in X_1 = \mathcal{D}(A)$ by

$$C_1 x = \big[-(\overline{\alpha} - A^*_{|X})^{-1}(A + A^*_{|X} - C^*C)(\alpha - A)^{-1}\big]^{1/2}(\alpha - A)x,$$

and let Y_1 be the closure of $\mathcal{R}(C_1)$. The proofs of (ii) and (iii) are essentially the same as the proofs of (ii) and (iii) in Theorem 11.4.5. □

The restriction that the system Σ in Lemma 11.5.3 has no input can be removed as follows.

Theorem 11.5.4 *Let $\Sigma = \left[\begin{smallmatrix} \mathfrak{A} & \mathfrak{B} \\ \mathfrak{C} & \mathfrak{D} \end{smallmatrix}\right]$ be a (scattering) passive L^2-well-posed linear system on the Hilbert spaces (Y, X, U).*

(i) *There is a Hilbert space Y_1 such that Σ can be extended to a (scattering) energy preserving system $\left[\begin{smallmatrix} \mathfrak{A} & \mathfrak{B} \\ \mathfrak{C}_1 & \mathfrak{D}_1 \\ \mathfrak{C} & \mathfrak{D} \end{smallmatrix}\right]$ on $\big(\left[\begin{smallmatrix} Y_1 \\ Y \end{smallmatrix}\right], X, U\big)$ with the additional property that the range of the combined observation/feedthrough operator corresponding to the first output has dense range.*

(ii) *The extended system is unique up to unitary similarity: if Y_2 is another Hilbert space, if $\left[\begin{smallmatrix} \mathfrak{A} & \mathfrak{B} \\ \mathfrak{C}_2 & \mathfrak{D}_2 \\ \mathfrak{C} & \mathfrak{D} \end{smallmatrix}\right]$ is another system on $\big(\left[\begin{smallmatrix} Y_2 \\ Y \end{smallmatrix}\right], X, U\big)$ with the properties described in (i), then there is a unitary map $G \colon Y_1 \to Y_2$ so that $\mathfrak{C}_2 = G\mathfrak{C}_1$, and $\mathfrak{D}_2 = G\mathfrak{D}_1$, and the two system nodes S_1 and S_2 satisfy $S_2 = \left[\begin{smallmatrix} 1 & 0 & 0 \\ 0 & G & 0 \\ 0 & 0 & 1 \end{smallmatrix}\right] S_1$.*

(iii) *The Lax–Phillips model with parameter $\omega = 0$ induced by $\left[\begin{smallmatrix} \mathfrak{A} & \mathfrak{B} \\ \mathfrak{C}_1 & \mathfrak{D}_1 \\ \mathfrak{C} & \mathfrak{D} \end{smallmatrix}\right]$ is a minimal isometric dilation of the corresponding Lax–Phillips model induced by $\left[\begin{smallmatrix} \mathfrak{A} & \mathfrak{B} \\ \mathfrak{C} & \mathfrak{D} \end{smallmatrix}\right]$.*

Proof The proof of this theorem is essentially the same as the proof of Theorem 11.4.9, with Theorem 11.4.5 replaced by Lemma 11.5.3. We leave it to the reader. □

Corollary 11.5.5 *Let $\Sigma = \left[\begin{smallmatrix} \mathfrak{A} & \mathfrak{B} \\ \mathfrak{C} & \mathfrak{D} \end{smallmatrix}\right]$ be a (scattering) passive L^2-well-posed linear system on the Hilbert spaces (Y, X, U).*

(i) *There is a Hilbert space U_1 such that Σ can be extended to a (scattering) co-energy preserving system $\left[\begin{smallmatrix} \mathfrak{A} & \mathfrak{B} & \mathfrak{B}_1 \\ \mathfrak{C} & \mathfrak{D} & \mathfrak{D}_1 \end{smallmatrix}\right]$ on $\big(Y, X, \left[\begin{smallmatrix} U \\ U_1 \end{smallmatrix}\right]\big)$ with the additional property that*

$$\mathcal{N}(S_1) \cap \left[\begin{smallmatrix} 0 \\ 0 \\ U_1 \end{smallmatrix}\right] = 0,$$

where S_1 is the system node of the extended system.

(ii) *The extended system is unique up to unitary similarity. If* U_2 *is another Hilbert space, if* $\left[\begin{array}{c|cc} \mathfrak{A} & \mathfrak{B} & \mathfrak{B}_2 \\ \hline \mathfrak{C} & \mathfrak{D} & \mathfrak{D}_2 \end{array}\right]$ *is another system on* $\left(Y, X, \left[\begin{smallmatrix} U \\ U_2 \end{smallmatrix}\right]\right)$ *with the properties described in (i), then there is a unitary map* $F\colon Y_1 \to Y_2$ *so that* $\mathfrak{B}_1 = \mathfrak{B}_2 F$, $\mathfrak{D}_1 = \mathfrak{D}_2 F$, *and the two system nodes* S_1 *and* S_2 *satisfy* $S_1 = S_2 \left[\begin{smallmatrix} 1 & 0 & 0 \\ 0 & 1 & 0 \\ 0 & 0 & F \end{smallmatrix}\right]$.

(iii) *The Lax–Phillips model with parameter* $\omega = 0$ *induced by* $\left[\begin{array}{c|cc} \mathfrak{A} & \mathfrak{B} & \mathfrak{B}_1 \\ \hline \mathfrak{C} & \mathfrak{D} & \mathfrak{D}_1 \end{array}\right]$ *is a minimal co-isometric dilation of the corresponding Lax–Phillips model induced by* $\left[\begin{array}{c|c} \mathfrak{A} & \mathfrak{B} \\ \hline \mathfrak{C} & \mathfrak{D} \end{array}\right]$.

Proof This follows from Theorem 11.5.4 by duality. □

By combining Theorem 11.5.4 and Corollary 11.5.5 we get the following theorem on the existence of a *conservative extension* of a given system.

Theorem 11.5.6 *Let* $\Sigma = \left[\begin{array}{c|c} \mathfrak{A} & \mathfrak{B} \\ \hline \mathfrak{C} & \mathfrak{D} \end{array}\right]$ *be a (scattering) passive* L^2*-well-posed linear system on the Hilbert spaces* (Y, X, U).

(i) *There exist two Hilbert space* U_1 *and* Y_1 *and a (scattering) conservative linear system* $\left[\begin{array}{c|cc} \mathfrak{A} & \mathfrak{B} & \mathfrak{B}_1 \\ \hline \mathfrak{C}_1 & \mathfrak{D}_{10} & \mathfrak{D}_{11} \\ \mathfrak{C} & \mathfrak{D} & \mathfrak{D}_{01} \end{array}\right]$ *on* $\left(\left[\begin{smallmatrix} Y_1 \\ Y \end{smallmatrix}\right], X, \left[\begin{smallmatrix} U \\ U_1 \end{smallmatrix}\right]\right)$ *with the additional minimality properties mentioned in part (i) of Theorem 11.5.4 and Corollary 11.5.5.*

(ii) *The extended system is unique up to unitary similarity: if* U_2 *and* Y_2 *are two other Hilbert spaces, and if* $\left[\begin{array}{c|cc} \mathfrak{A} & \mathfrak{B} & \mathfrak{B}_2 \\ \hline \mathfrak{C}_2 & \mathfrak{D}_{20} & \mathfrak{D}_{22} \\ \mathfrak{C} & \mathfrak{D} & \mathfrak{D}_{02} \end{array}\right]$ *is another system on* $\left(\left[\begin{smallmatrix} Y_2 \\ Y \end{smallmatrix}\right], X, \left[\begin{smallmatrix} U \\ U_2 \end{smallmatrix}\right]\right)$ *with the properties described in (i), then there exist unitary maps* $F\colon U_1 \to U_2$ *and* $G\colon Y_1 \to Y_2$ *so that*

$$\begin{bmatrix} 1 & 0 & 0 \\ 0 & G & 0 \\ 0 & 0 & 1 \end{bmatrix} \left[\begin{array}{c|cc} \mathfrak{A} & \mathfrak{B} & \mathfrak{B}_1 \\ \hline \mathfrak{C}_1 & \mathfrak{D}_{10} & \mathfrak{D}_{11} \\ \mathfrak{C} & \mathfrak{D} & \mathfrak{D}_{01} \end{array}\right] = \left[\begin{array}{c|cc} \mathfrak{A} & \mathfrak{B} & \mathfrak{B}_2 \\ \hline \mathfrak{C}_2 & \mathfrak{D}_{20} & \mathfrak{D}_{22} \\ \mathfrak{C} & \mathfrak{D} & \mathfrak{D}_{02} \end{array}\right] \begin{bmatrix} 1 & 0 & 0 \\ 0 & 1 & 0 \\ 0 & 0 & F \end{bmatrix},$$

$$\begin{bmatrix} 1 & 0 & 0 \\ 0 & G & 0 \\ 0 & 0 & 1 \end{bmatrix} S_1 = S_2 \begin{bmatrix} 1 & 0 & 0 \\ 0 & 1 & 0 \\ 0 & 0 & F \end{bmatrix},$$

where S_1 *and* S_1 *are the system nodes of the two extended systems.*

(iii) *The Lax–Phillips model with parameter* $\omega = 0$ *induced by* $\left[\begin{array}{c|cc} \mathfrak{A} & \mathfrak{B} & \mathfrak{B}_1 \\ \hline \mathfrak{C}_1 & \mathfrak{D}_{10} & \mathfrak{D}_{11} \\ \mathfrak{C} & \mathfrak{D} & \mathfrak{D}_{01} \end{array}\right]$ *is a minimal isometric dilation of the corresponding Lax–Phillips model induced by* $\left[\begin{array}{c|c} \mathfrak{A} & \mathfrak{B} \\ \hline \mathfrak{C} & \mathfrak{D} \end{array}\right]$.

Proof Once more we leave the details of the proof to the reader. The idea is the same as in the proof of Theorem 11.4.9: first Σ is extended to a co-energy preserving system as in Corollary 11.5.5, and then this system is extended to an energy preserving system as in Theorem 11.5.4. □

Remark 11.5.7 Theorem 11.5.6 can be used in a weak form of *Darlington synthesis*, i.e., in the construction of a lossless extension $\left[\begin{smallmatrix} \mathfrak{D}_{10} & \mathfrak{D}_{11} \\ \mathfrak{D} & \mathfrak{D}_{01} \end{smallmatrix}\right]$ of a given contractive time-invariant causal map $\mathfrak{D}$. Let $\left[\begin{smallmatrix} \mathfrak{A} & \mathfrak{B} \\ \mathfrak{C} & \mathfrak{D} \end{smallmatrix}\right]$ be a passive realization of $\mathfrak{D}$. The extension $\left[\begin{smallmatrix} \mathfrak{D}_{10} & \mathfrak{D}_{11} \\ \mathfrak{D} & \mathfrak{D}_{01} \end{smallmatrix}\right]$ of $\mathfrak{D}$ obtained from Theorem 11.5.6 is lossless if and only if the resulting conservative system is lossless. A sufficient condition for this is that both the original system semigroup $\mathfrak{A}$ and its adjoint are strongly stable (see Theorem 11.3.4). Conversely, if $\mathfrak{D}$ has a lossless extension $\left[\begin{smallmatrix} \mathfrak{D}_{10} & \mathfrak{D}_{11} \\ \mathfrak{D} & \mathfrak{D}_{01} \end{smallmatrix}\right]$, then this extension has a conservative realization with the property that both the system semigroup $\mathfrak{A}$ and its adjoint are strongly stable (see Theorem 11.8.1). By dropping the first output and second input we get a realization of $\mathfrak{D}$ with the same semigroup. Thus, a contractive time-invariant causal map $\mathfrak{D}$ admits weak Darlington synthesis if and only if it has a passive realization with the property that both the system semigroup $\mathfrak{A}$ and its adjoint are strongly stable.[3]

11.6 The universal model of a contraction semigroup

We shall use the results presented in the preceding section to develop a representation of an arbitrary completely nonunitary contraction semigroup on a Hilbert space as the compression of a bilateral left-shift to a suitable subspace of an L^2-space.

We begin by decomposing the state space of a conservative Lax–Phillips group into a number of invariant subspaces.

Definition 11.6.1 Let $\Sigma = \left[\begin{smallmatrix} \mathfrak{A} & \mathfrak{B} \\ \mathfrak{C} & \mathfrak{D} \end{smallmatrix}\right]$ be a passive linear system on the Hilbert spaces (Y, X, U), and let $\mathfrak{T}$ be the corresponding passive Lax–Phillips semigroup on the state space $\mathcal{X} = \begin{bmatrix} L^2(\mathbb{R}^-;Y) \\ X \\ L^2(\mathbb{R}^+;U) \end{bmatrix}$, with backward wave operator $W_- = \begin{bmatrix} \pi_-\mathfrak{D} \\ \mathfrak{B} \\ \pi_+ \end{bmatrix}$ and forward wave operator $W_+ = \begin{bmatrix} \pi_- & \mathfrak{C} & \mathfrak{D}\pi_+ \end{bmatrix}$.

(i) By the *reachable part* $\mathcal{X}_R$ *of* $\mathcal{X}$ we mean the closure of the range of W_-. The orthogonal complement $\mathcal{X}_R^\perp = \mathcal{N}\left(W_-^*\right)$ of $\mathcal{X}_R$ in $\mathcal{X}$ is called the *unreachable part* $\mathcal{X}_R$ *of* $\mathcal{X}$.

(ii) By the *observable part* $\mathcal{X}_O$ *of* $\mathcal{X}$ we mean the closure of the range of W_+^*. The orthogonal complement $\mathcal{X}_O^\perp = \mathcal{N}\left(W_+\right)$ of $\mathcal{X}_O$ in $\mathcal{X}$ is called the *unobservable part* $\mathcal{X}_R$ *of* $\mathcal{X}$.

[3] In the strong version of Darlington synthesis the extended system is required to have minimal losses. See Arov (2002) for details.

Next we develop some shift representation of the various parts of a conservative system.

Theorem 11.6.2 *Let* $\Sigma = \left[\begin{smallmatrix} \mathfrak{A} & \mathfrak{B} \\ \mathfrak{C} & \mathfrak{D} \end{smallmatrix}\right]$ *be a (scattering) passive linear system on* (Y, X, U)*, and let* $\mathfrak{T}$ *be the corresponding Lax–Phillips semigroup on the state space* $\mathcal{X} = \left[\begin{smallmatrix} L^2(\mathbb{R}^-;Y) \\ X \\ L^2(\mathbb{R}^+;U) \end{smallmatrix}\right]$*. (If* Σ *is conservative, then we denote the corresponding Lax–Phillips* group *by* $\mathfrak{T}$*.)*

(i) *If* Σ *is energy preserving, then the reachable part* $\mathcal{X}_R$ *of* $\mathcal{X}$ *is invariant under* $\mathfrak{T}^t$ *for all* $t \ge 0$*, and* $\mathfrak{T}_{|\mathcal{X}_R}$ *is unitarily similar to the bilateral left-shift* τ *on* $L^2(\mathbb{R}; U)$*. In particular,* $\mathfrak{T}_{|\mathcal{X}_R}$ *can be extended to a group. More precisely, the backward wave operator* W_- *is a unitary map of* $L^2(\mathbb{R}; U)$ *onto* $\mathcal{X}_R$*, and for all* $t \ge 0$*,*

$$\mathfrak{T}^t W_- u = W_- \tau^t u, \qquad u \in L^2(\mathbb{R}; U).$$

This identity holds also for $t < 0$ *if* Σ *is conservative (so that* $\mathfrak{T}$ *is a group) or if we replace* $\mathfrak{T}^t$ *by the (group extension of)* $\mathfrak{T}^t_{|\mathcal{X}_R}$*.*

(ii) *If* Σ *is co-energy preserving, then the observable part* $\mathcal{X}_O$ *of* $\mathcal{X}$ *is co-invariant under* $\mathfrak{T}^t$ *for all* $t \ge 0$*, and* $\pi_{\mathcal{X}_O}\mathfrak{T}_{|\mathcal{X}_O}$ *is unitarily similar to the bilateral left-shift* τ *on* $L^2(\mathbb{R}; Y)$ *(where* $\pi_{\mathcal{X}_O}$ *is the orthogonal projection onto* $\mathcal{X}_O$*). In particular,* $\pi_{\mathcal{X}_O}\mathfrak{T}_{|\mathcal{X}_O}$ *can be extended to a group. More precisely, the restriction of the forward wave operator* W_+ *to* $\mathcal{X}_O$ *is a unitary map of* $\mathcal{X}_O$ *onto* $L^2(\mathbb{R}; Y)$*, and for all* $t \ge 0$*,*

$$W_+ \mathfrak{T}^t \left[\begin{smallmatrix} y \\ x \\ u \end{smallmatrix}\right] = \tau^t W_+ \left[\begin{smallmatrix} y \\ x \\ u \end{smallmatrix}\right], \qquad \left[\begin{smallmatrix} y \\ x \\ u \end{smallmatrix}\right] \in \mathcal{X}.$$

This identity holds also for $t < 0$ *if* Σ *is conservative (so that* $\mathfrak{T}$ *is a group) or if we replace* $\mathcal{X}$ *by* $\mathcal{X}_O$ *and* $\mathfrak{T}^t$ *by the (group extension of)* $\pi_{\mathcal{X}_O}\mathfrak{T}^t_{|\mathcal{X}_O}$*.*

(iii) *If* Σ *is conservative, then the unreachable part* $\mathcal{X}_R^\perp$ *of* $\mathcal{X}$ *is invariant under* $\mathfrak{T}^t$ *for all* $t \in \mathbb{R}$*. If, moreover,* $\mathfrak{A}$ *is completely nonunitary, then the restriction of* $\mathfrak{T}$ *to* $\mathcal{X}_R^\perp$ *is unitarily similar to the bilateral left-shift* τ *on the closure of the range of* $1 - \mathfrak{D}\mathfrak{D}^*$ *in* $L^2(\mathbb{R}; Y)$ *(which is invariant under* τ^t *for all* $t \in \mathbb{R}$*). More precisely, let* $(1 - \mathfrak{D}\mathfrak{D}^*)^{1/2}$ *be the positive square root of* $1 - \mathfrak{D}\mathfrak{D}^*$*, and let* $(1 - \mathfrak{D}\mathfrak{D}^*)^{-1/2}$ *be the inverse of the restriction of* $(1 - \mathfrak{D}\mathfrak{D}^*)^{1/2}$ *to*

$$\mathcal{N}\left((1 - \mathfrak{D}\mathfrak{D}^*)^{1/2}\right)^\perp = \overline{\mathcal{R}\left((1 - \mathfrak{D}\mathfrak{D}^*)^{1/2}\right)} = \overline{\mathcal{R}(1 - \mathfrak{D}\mathfrak{D}^*)}$$

(defined on the range of $(1 - \mathfrak{D}\mathfrak{D}^*)^{1/2}$*). Then* $(1 - \mathfrak{D}\mathfrak{D}^*)^{1/2}$ *has the same range as* $W_+(1 - W_- W_-^*)$*, and the operator*

$$E_R = (1 - \mathfrak{D}\mathfrak{D}^*)^{-1/2} W_+ (1 - W_- W_-^*)$$

is a partial isometry with initial space $\mathcal{X}_R^\perp$ *and final space* $\overline{\mathcal{R}(1-\mathfrak{D}\mathfrak{D}^*)}$*, the restriction of* E_R *to* $\mathcal{X}_R^\perp$ *is a unitary operator of* $\mathcal{X}_R^\perp$ *onto* $\overline{\mathcal{R}(1-\mathfrak{D}\mathfrak{D}^*)}$*, and for all* $t \in \mathbb{R}$*,*

$$E_R \mathfrak{T}^t \begin{bmatrix} y \\ x \\ u \end{bmatrix} = \tau^t E_R \begin{bmatrix} y \\ x \\ u \end{bmatrix}, \qquad \begin{bmatrix} y \\ x \\ u \end{bmatrix} \in \mathcal{X}.$$

(iv) *If* Σ *is conservative, then the unobservable part* $\mathcal{X}_O^\perp$ *of* $\mathcal{X}$ *is invariant under* $\mathfrak{T}^t$ *for all* $t \in \mathbb{R}$*. If, moreover,* $\mathfrak{A}$ *is completely nonunitary, then the restriction of* $\mathfrak{T}$ *to* $\mathcal{X}_O^\perp$ *is unitarily similar to the bilateral left-shift* τ *on the closure of the range of* $1-\mathfrak{D}^*\mathfrak{D}$ *in* $L^2(\mathbb{R};U)$ *(which is invariant under* τ^t *for all* $t \in \mathbb{R}$*). More precisely, define* $(1-\mathfrak{D}^*\mathfrak{D})^{1/2}$ *and* $(1-\mathfrak{D}^*\mathfrak{D})^{-1/2}$ *as in (iii) with* $\mathfrak{D}$ *replaced by* $\mathfrak{D}^*$ *and* U *interchanged with* Y*. Then* $W_-^*(1-W_+^*W_+)$ *has the same range as* $(1-\mathfrak{D}^*\mathfrak{D})^{1/2}$*, and the operator*

$$E_O = (1-\mathfrak{D}^*\mathfrak{D})^{-1/2} W_-^*(1-W_+^*W_+)$$

is a partial isometry with initial space $\mathcal{X}_O^\perp$ *and final space* $\overline{\mathcal{R}(1-\mathfrak{D}^*\mathfrak{D})}$*, the restriction of* E_O *to* $\mathcal{X}_O^\perp$ *is a unitary operator of* $\mathcal{X}_O^\perp$ *onto* $\overline{\mathcal{R}(1-\mathfrak{D}^*\mathfrak{D})}$*, and for all* $t \in \mathbb{R}$*,*

$$E_O \mathfrak{T}^t \begin{bmatrix} y \\ x \\ u \end{bmatrix} = \tau^t E_O \begin{bmatrix} y \\ x \\ u \end{bmatrix}, \qquad \begin{bmatrix} y \\ x \\ u \end{bmatrix} \in \mathcal{X}.$$

Proof (i) It follows from Theorem 11.2.9(i) that $W_-^*W_- = 1$ and that $W_-W_-^*$ is a projection operator. This implies that W_- is injective and has closed range, and that it is a unitary map of $L^2(\mathbb{R};U)$ onto $\mathcal{R}(W_-)$. Since $\overline{\mathcal{R}(W_-)} = \mathcal{X}_R$, this means that W_- is a unitary map of $L^2(\mathbb{R};U)$ onto $\mathcal{R}(W_-)$. That $\mathcal{X}_R$ is invariant under $\mathfrak{T}^t$ and that W_- intertwines $\mathfrak{T}^t$ with τ^t for all $t \ge 0$ follows from Lemma 2.7.8. The intertwining condition $\mathfrak{T}^t W_- = W_-\tau^t$ for $t \ge 0$ implies that $\mathfrak{T}^t_{|\mathcal{X}_R}$ is onto (since W_- and τ^t are onto), so $\mathfrak{T}^t_{|\mathcal{X}_R}$ can be extended to a group by defining $\mathfrak{T}^t_{\mathcal{X}_R} = (\mathfrak{T}^{-t}_{|\mathcal{X}_R})^{-1}$ for $t < 0$. If Σ is conservative, then $\mathfrak{T}$ is a group, and $(\mathfrak{T}^t)^{-1} = \mathfrak{T}^{-t} = \mathfrak{T}^{*t}$ maps $\mathcal{X}_R$ onto $\mathcal{X}_R$. Thus, $\mathcal{X}_R$ is invariant under $\mathfrak{T}^t$ for all $t \in \mathbb{R}$ in this case, and $(\mathfrak{T}^t)_{|\mathcal{X}_R} = (\mathfrak{T}_{|\mathcal{X}_R})^t = \mathfrak{T}^t_{\mathcal{X}_R}$ for all $t \in \mathbb{R}$. Multiplying the condition $\mathfrak{T}^t_{\mathcal{X}_R} W_- = \mathfrak{T}^t W_- = W_-\tau^t$ (with $t \ge 0$) by $\mathfrak{T}^{-t}_{|\mathcal{X}_R}$ to the left and by τ^{-t} to the right we get $W_-\tau^{-t} = \mathfrak{T}^{-t}_{|\mathcal{X}_R} W_-$, so the intertwining condition $\mathfrak{T}^t_{\mathcal{X}_R} W_- = W_-\tau^t$ holds also for $t < 0$. If Σ is conservative, then $\mathfrak{T}^t W_- = W_-\tau^t$ for all $t \in \mathbb{R}$.

(ii) We leave the proof of (ii) to the reader (either apply (i) to the dual system, or imitate the proof of (i)).

(iii) We know from (i) that $\mathfrak{T}^t$ maps $\mathcal{X}_R$ onto $\mathcal{X}_R$ for all $t \in \mathbb{R}$. As $\mathfrak{T}^t$ is isometric, it preserves orthogonality, so $\mathfrak{T}^t$ must map $\mathcal{X}_R^\perp$ into $\mathcal{X}_R^\perp$ (and even onto since $\mathfrak{T}$ is invertible). Thus $\mathcal{X}_R^\perp$ is invariant under $\mathfrak{T}^t$ for all $t \in \mathbb{R}$.

Next we construct an operator which maps $L^2(\mathbb{R};Y)$ onto a dense subset of $\mathcal{X}_R^\perp$. We know from (ii) that W_+^* maps $L^2(R;Y)$ one-to-one onto $\mathcal{X}_O$. As we saw in the proof of (i), the orthogonal projection onto $\mathcal{X}_R$ is $W_-W_-^*$, so the orthogonal projection onto $\mathcal{X}_R^\perp$ is $1-W_-W_-^*$. Thus, $(1-W_-W_-^*)W_+^*$ maps $L^2(\mathbb{R};Y)$ into $X_R^\perp$. A direct computation shows that (cf. (2.7.8) and Theorem 11.2.11)

$$\begin{aligned}[(1-W_-W_-^*)W_+^*]^*(1-W_-W_-^*)W_+^* &= W_+W_+^* - W_+W_-W_-^*W_+^* \\ &= 1-\mathfrak{D}\mathfrak{D}^*.\end{aligned}$$

Thus, with reference to Lemma A.2.5, $W_+(1-W_-W_-^*)$ has the same range as $|(1-W_-W_-^*)W_+^*| = (1-\mathfrak{D}\mathfrak{D}^*)^{1/2}$, and the operator E_R defined in (iii) is the partial isometry in the polar decomposition

$$W_+(1-W_-W_-^*) = |(1-W_-W_-^*)W_+^*|E_R = (1-\mathfrak{D}\mathfrak{D}^*)^{1/2}E_R$$

of $W_+(1-W_-W_-^*)$. By Lemmas A.2.2 and A.2.5, the common final space of $W_+(1-W_-W_-^*)$ and E_R is $\mathcal{N}\left((1-\mathfrak{D}\mathfrak{D}^*)^{1/2}\right)^\perp = \overline{\mathcal{R}\left((1-\mathfrak{D}\mathfrak{D}^*)^{1/2}\right)} = \overline{\mathcal{R}(1-\mathfrak{D}\mathfrak{D}^*)}$. We know that $W_+(1-W_-W_-^*)$ vanishes on $\mathcal{X}_R$, so to show that the common initial space of $W_+(1-W_-W_-^*)$ and E_R is $\mathcal{X}_R^\perp$ we must show that $W_+(1-W_-W_-^*)\left[\begin{smallmatrix} y \\ x \\ u\end{smallmatrix}\right] \neq 0$ for every nonzero $\left[\begin{smallmatrix} y \\ x \\ u\end{smallmatrix}\right] \in \mathcal{X}_R^\perp$, or equivalently, that $W_+\left[\begin{smallmatrix} y \\ x \\ u\end{smallmatrix}\right] \neq 0$ for every nonzero $\left[\begin{smallmatrix} y \\ x \\ u\end{smallmatrix}\right] \in \mathcal{X}_R^\perp$. Suppose that $\left[\begin{smallmatrix} y \\ x \\ u\end{smallmatrix}\right] \in \mathcal{X}_R^\perp$ and that $W_+\left[\begin{smallmatrix} y \\ x \\ u\end{smallmatrix}\right] = 0$. By the definitions of W_-, W_+, and $\mathcal{X}_R$, the condition $\left[\begin{smallmatrix} y \\ x \\ u\end{smallmatrix}\right] \in \mathcal{X}_R^\perp$ implies that $u=0$, and the condition $W_+\left[\begin{smallmatrix} y \\ x \\ u\end{smallmatrix}\right] = 0$ implies that $y=0$, and by taking $u=0$ and $y=0$ in the same conditions we get $\mathfrak{B}^*x = 0$ and $\mathfrak{C}x=0$. Thus $x \in \mathcal{N}(\mathfrak{B}^*)\cap\mathcal{N}(\mathfrak{C})$ (i.e., x is both unobservable and unreachable). By Theorem 11.2.11, this implies that $Q_{\mathfrak{A}^*,\mathfrak{A}}x = Q_{\mathfrak{A},\mathfrak{A}^*}x = x$, i.e., x belongs to the subspace X_0 in Theorem 11.1.9 on which $\mathfrak{A}$ reduces to a unitary operator. Thus, since we assume $\mathfrak{A}$ to be completely nonunitary, we must have $x=0$. We have now proved that the common initial space of $W_+(1-W_-W_-^*)$ and E_R is $\mathcal{X}_R^\perp$. Moreover, E_R is a unitary operator of $\mathcal{X}_R^\perp$ onto $\overline{\mathcal{R}(1-\mathfrak{D}\mathfrak{D}^*)}$.

It remains to prove that E_R intertwines $\mathfrak{T}^t$ with τ^t. The operator $1-W_-W_-^*$ is the orthogonal projection onto $\mathcal{X}_R^\perp$, and both $\mathcal{X}_R$ and $\mathcal{X}_R^\perp$ are invariant under $\mathfrak{T}^t$, so $\mathfrak{T}^t(1-W_-W_-^*) = (1-W_-W_-^*)\mathfrak{T}^t$ for all $t\in\mathbb{R}$. By (ii), $W_+\mathfrak{T}^t = \tau^t W_+$ for all $t\in\mathbb{R}$ (on the whole space $\mathcal{X}$). Thus, $W_-(1-W_-W_-^*)\mathfrak{T}^t = \tau^t W_-(1-W_-W_-^*)$ for all $t\in\mathbb{R}$. By the shift-invariance of $\mathfrak{D}$ (and $\mathfrak{D}^*$), $\overline{\mathcal{R}(1-\mathfrak{D}\mathfrak{D}^*)}$ is invariant under τ^t, and $\tau^t(1-\mathfrak{D}\mathfrak{D}^*) = (1-\mathfrak{D}\mathfrak{D}^*)\tau^t$ for all $t\in\mathbb{R}$. By Lemma A.2.6 (with $X=\mathcal{X}$, $Y=\overline{\mathcal{R}(1-\mathfrak{D}\mathfrak{D}^*)}$, $A=W_-(1-W_-W_-^*)$, $B=\mathfrak{T}^t$,

$C = \tau^t|_Y$, $U = E_R$, and $A^*A = 1 - \mathfrak{D}\mathfrak{D}^*$), we get $E_R \mathfrak{T}^t = \tau^t E_R$, as claimed in (iii).

(iv) We leave the proof of (iv) to the reader (either apply (iii) to the dual system, or imitate the proof of (iii)). □

By developing this result further we get the *universal model* of an arbitrary completely nonunitary conservative system.

Theorem 11.6.3 *Let $\Sigma = \left[\begin{smallmatrix} \mathfrak{A} & \mathfrak{B} \\ \mathfrak{C} & \mathfrak{D} \end{smallmatrix}\right]$ be a (scattering) conservative completely nonunitary linear system on (Y, X, U). Then the following claims are true.*

(i) *The range of $\mathfrak{C} - \mathfrak{D}\mathfrak{B}^*$ is contained in the range of $(1 - \mathfrak{D}\mathfrak{D}^*)^{1/2}$, and the operator*

$$F_R = \begin{bmatrix} (1 - \mathfrak{D}\mathfrak{D}^*)^{-1/2}(\mathfrak{C} - \mathfrak{D}\mathfrak{B}^*) \\ \mathfrak{B}^* \end{bmatrix}$$

is an isometry from X to $\left[\begin{smallmatrix} L^2(\mathbb{R};Y) \\ L^2(\mathbb{R}^-;U) \end{smallmatrix}\right]$ with final space

$$\begin{aligned} X_R &= \mathcal{R}\left(\begin{bmatrix} (1 - \mathfrak{D}\mathfrak{D}^*)^{-1/2}(\mathfrak{C} - \mathfrak{D}\mathfrak{B}^*) \\ \mathfrak{B}^* \end{bmatrix}\right) \\ &= \begin{bmatrix} \overline{\mathcal{R}(1 - \mathfrak{D}\mathfrak{D}^*)} \\ L^2(\mathbb{R}^-;U) \end{bmatrix} \cap \mathcal{R}\left(\begin{bmatrix} (1 - \mathfrak{D}\mathfrak{D}^*)^{1/2}\pi_- \\ \mathfrak{D}^*\pi_- \end{bmatrix}\right)^{\perp} \\ &= \begin{bmatrix} \overline{\mathcal{R}(1 - \mathfrak{D}\mathfrak{D}^*)} \\ L^2(\mathbb{R}^-;U) \end{bmatrix} \cap \mathcal{N}\left(\begin{bmatrix} \pi_-(1 - \mathfrak{D}\mathfrak{D}^*)^{1/2} & \pi_-\mathfrak{D} \end{bmatrix}\right). \end{aligned}$$

Let π_R be the orthogonal projection in $L^2(\mathbb{R};Y)$ onto $\overline{\mathcal{R}(1 - \mathfrak{D}\mathfrak{D}^*)}$. Then the orthogonal projection π_{X_R} in $\left[\begin{smallmatrix} L^2(\mathbb{R};Y) \\ L^2(\mathbb{R}^-;U) \end{smallmatrix}\right]$ onto X_R is given by

$$\begin{aligned} \pi_{X_R} &= \begin{bmatrix} (1 - \mathfrak{D}\mathfrak{D}^*)^{-1/2}(\mathfrak{C} - \mathfrak{D}\mathfrak{B}^*) \\ \mathfrak{B}^* \end{bmatrix} \begin{bmatrix} (1 - \mathfrak{D}\mathfrak{D}^*)^{-1/2}(\mathfrak{C} - \mathfrak{D}\mathfrak{B}^*) \\ \mathfrak{B}^* \end{bmatrix}^* \\ &= \begin{bmatrix} \pi_R & 0 \\ 0 & \pi_- \end{bmatrix} - \begin{bmatrix} (1 - \mathfrak{D}\mathfrak{D}^*)^{1/2} \\ \mathfrak{D}^* \end{bmatrix} \pi_- \begin{bmatrix} (1 - \mathfrak{D}\mathfrak{D}^*)^{1/2} & \mathfrak{D} \end{bmatrix} \\ &= \begin{bmatrix} \pi_R - (1 - \mathfrak{D}\mathfrak{D}^*)^{1/2}\pi_-(1 - \mathfrak{D}\mathfrak{D}^*)^{1/2} & -(1 - \mathfrak{D}\mathfrak{D}^*)^{1/2}\pi_-\mathfrak{D} \\ -\mathfrak{D}^*\pi_-(1 - \mathfrak{D}\mathfrak{D}^*)^{1/2} & \pi_- - \mathfrak{D}^*\pi_-\mathfrak{D} \end{bmatrix}. \end{aligned}$$

Define

$$\begin{aligned} \mathfrak{A}_R &= \pi_{X_R} \begin{bmatrix} \tau & 0 \\ 0 & \tau_- \end{bmatrix}_{|X_R}, \quad \mathfrak{B}_R = \begin{bmatrix} -(1 - \mathfrak{D}\mathfrak{D}^*)^{1/2}\pi_-\mathfrak{D} \\ \pi_- - \mathfrak{D}^*\pi_-\mathfrak{D} \end{bmatrix}, \\ \mathfrak{C}_R &= \begin{bmatrix} \pi_+(1 - \mathfrak{D}\mathfrak{D}^*)^{1/2} & \pi_+\mathfrak{D}\pi_- \end{bmatrix}_{|X_R}. \end{aligned}$$

Then $\left[\begin{smallmatrix} \mathfrak{A}_R & \mathfrak{B}_R \\ \mathfrak{C}_R & \mathfrak{D} \end{smallmatrix}\right]$ *is a conservative realization of* $\mathfrak{D}$ *which is unitarily similar to* Σ *(in the sense of Example 2.3.7):*

$$\mathfrak{A}_R = F_R \mathfrak{A} F_R^*, \qquad \mathfrak{B}_R = F_R \mathfrak{B}, \qquad \mathfrak{C}_R = \mathfrak{C} F_R^*.$$

In particular, $\mathfrak{A}$ *is unitarily similar to* $\mathfrak{A}_R$*. Moreover, the space* X_R *is co-invariant under the semigroup* $\left[\begin{smallmatrix} \tau^t & 0 \\ 0 & \tau_-^t \end{smallmatrix}\right]$ *on* $\left[\begin{smallmatrix} L^2(\mathbb{R};Y) \\ L^2(\mathbb{R}^-;U) \end{smallmatrix}\right]$ *(i.e., it is invariant under the adjoint of this semigroup).*

(ii) *The range of* $\mathfrak{B}^* - \mathfrak{D}^*\mathfrak{C}$ *is contained in the range of* $(1 - \mathfrak{D}^*\mathfrak{D})^{1/2}$*, and the operator*

$$F_O = \begin{bmatrix} \mathfrak{C} \\ (1 - \mathfrak{D}^*\mathfrak{D})^{-1/2}(\mathfrak{B}^* - \mathfrak{D}^*\mathfrak{C}) \end{bmatrix}$$

is an isometry from X *to* $\left[\begin{smallmatrix} L^2(\mathbb{R}^+;Y) \\ L^2(\mathbb{R};U) \end{smallmatrix}\right]$ *with final space*

$$\begin{aligned} X_O &= \mathcal{R}\left(\begin{bmatrix} \mathfrak{C} \\ (1 - \mathfrak{D}^*\mathfrak{D})^{-1/2}(\mathfrak{B}^* - \mathfrak{D}^*\mathfrak{C}) \end{bmatrix}\right) \\ &= \begin{bmatrix} L^2(\mathbb{R}^+;Y) \\ \overline{\mathcal{R}(1 - \mathfrak{D}^*\mathfrak{D})} \end{bmatrix} \cap \mathcal{R}\left(\begin{bmatrix} \mathfrak{D}\pi_+ \\ (1 - \mathfrak{D}^*\mathfrak{D})^{1/2}\pi_+ \end{bmatrix}\right)^{\perp} \\ &= \begin{bmatrix} L^2(\mathbb{R}^+;Y) \\ \overline{\mathcal{R}(1 - \mathfrak{D}^*\mathfrak{D})} \end{bmatrix} \cap \mathcal{N}\left(\begin{bmatrix} \pi_+\mathfrak{D}^* & \pi_+(1 - \mathfrak{D}^*\mathfrak{D})^{1/2} \end{bmatrix}\right). \end{aligned}$$

Let π_O *be the orthogonal projection in* $L^2(\mathbb{R};U)$ *onto* $\overline{\mathcal{R}(1 - \mathfrak{D}^*\mathfrak{D})}$*. Then the orthogonal projection* π_{X_O} *in* $\left[\begin{smallmatrix} L^2(\mathbb{R};Y) \\ L^2(\mathbb{R};U) \end{smallmatrix}\right]$ *onto* X_O *is given by*

$$\begin{aligned} \pi_{X_O} &= \begin{bmatrix} \mathfrak{C} \\ (1 - \mathfrak{D}^*\mathfrak{D})^{-1/2}(\mathfrak{B}^* - \mathfrak{D}^*\mathfrak{C}) \end{bmatrix} \begin{bmatrix} \mathfrak{C} \\ (1 - \mathfrak{D}^*\mathfrak{D})^{-1/2}(\mathfrak{B}^* - \mathfrak{D}^*\mathfrak{C}) \end{bmatrix}^* \\ &= \begin{bmatrix} \pi_+ & 0 \\ 0 & \pi_O \end{bmatrix} - \begin{bmatrix} \mathfrak{D} \\ (1 - \mathfrak{D}^*\mathfrak{D})^{1/2} \end{bmatrix} \pi_+ \begin{bmatrix} \mathfrak{D}^* & (1 - \mathfrak{D}^*\mathfrak{D})^{1/2} \end{bmatrix}. \end{aligned}$$

Define

$$\mathfrak{A}_O = \begin{bmatrix} \tau_+ & 0 \\ 0 & \tau \end{bmatrix}_{|X_O}, \quad \mathfrak{B}_O = \begin{bmatrix} \pi_+\mathfrak{D}\pi_- \\ (1 - \mathfrak{D}^*\mathfrak{D})^{1/2}\pi_- \end{bmatrix},$$

$$\mathfrak{C}_O = \begin{bmatrix} 1 - \mathfrak{D}\pi_+\mathfrak{D}^* & -\mathfrak{D}\pi_+(1 - \mathfrak{D}^*\mathfrak{D})^{1/2} \end{bmatrix}_{|X_O}.$$

Then $\left[\begin{smallmatrix} \mathfrak{A}_O & \mathfrak{B}_O \\ \mathfrak{C}_O & \mathfrak{D} \end{smallmatrix}\right]$ *is a conservative realization of* $\mathfrak{D}$ *which is unitarily similar to* Σ *(in the sense of Example 2.3.7):*

$$\mathfrak{A}_O = F_O \mathfrak{A} F_O^*, \qquad \mathfrak{B}_O = F_O \mathfrak{B}, \qquad \mathfrak{C}_O = \mathfrak{C} F_O^*.$$

In particular, $\mathfrak{A}$ *is unitarily similar to* $\mathfrak{A}_O$*, and the space* X_O *is invariant under the semigroup* $\left[\begin{smallmatrix} \tau_+ & 0 \\ 0 & \tau \end{smallmatrix}\right]$ *on* $\left[\begin{smallmatrix} L^2(\mathbb{R}^+;Y) \\ L^2(\mathbb{R};U) \end{smallmatrix}\right]$.

Proof We shall prove only part (i), and leave the proof of (ii) to the reader (for example, one can derive (ii) by applying (i) to the dual system).

By Theorem 11.6.2, the operator $E = \left[\begin{smallmatrix} E_R \\ W_-^* \end{smallmatrix}\right]$ (where the matrix form of the right-hand side represents a splitting of $\mathcal{X}$ into $\left[\begin{smallmatrix} \mathcal{X}_R^\perp \\ \mathcal{X}_R \end{smallmatrix}\right]$) is a unitary map of $\mathcal{X}$ onto $\left[\begin{smallmatrix} \overline{\mathcal{R}(1-\mathfrak{D}\mathfrak{D}^*)} \\ L^2(\mathbb{R};U) \end{smallmatrix}\right]$, and, for all $t \in \mathbb{R}$, $E\mathfrak{T}^t = \tau^t E \left[\begin{smallmatrix} E_R \\ W_-^* \end{smallmatrix}\right]\mathfrak{T}^t = \tau^t \left[\begin{smallmatrix} E_R \\ W_-^* \end{smallmatrix}\right]$. The subspaces $\mathcal{Y} = L^2(\mathbb{R}^-; Y)$, X, and $\mathcal{U} = L^2(\mathbb{R}^+; U)$ are orthogonal complementary subspaces of $\mathcal{X}$, so their images $E\mathcal{Y}$, EX, and $E\mathcal{U}$ are orthogonal complementary subspaces of $\left[\begin{smallmatrix} \overline{\mathcal{R}(1-\mathfrak{D}\mathfrak{D}^*)} \\ L^2(\mathbb{R};U) \end{smallmatrix}\right]$. If we denote the projections in $\mathcal{X}$ onto $\mathcal{Y}$, X, and $\mathcal{U}$ by π_-, π_0, and π_+, respectively, then $\pi_- + \pi_0 + \pi_+ = 1$, and the corresponding orthogonal projections in $\left[\begin{smallmatrix} \overline{\mathcal{R}(1-\mathfrak{D}\mathfrak{D}^*)} \\ L^2(\mathbb{R};U) \end{smallmatrix}\right]$ are $E\pi_- E^*$, $E\pi_0 E^*$, and $E\pi_+ E^*$. For all $x \in X$ and $t \geq 0$ we have

$$\mathfrak{A}^t x = \pi_0 \mathfrak{T}^t x = \pi_0 E^* \tau^t E x,$$

or equivalently,

$$E\mathfrak{A}^t x = E\pi_0 E^* \tau^t E x = \pi_{X_R} \tau^t E x,$$

where $\pi_{X_R} = E\pi_0 E^*$ is the orthogonal projection onto $X_R = \mathcal{R}(E\pi_0)$. This is of the form

$$F_R \mathfrak{A}^t x = \pi_{X_R} \tau^t F_R x,$$

where F_R is the restriction of E to X. Thus, if we define

$$\mathfrak{A}_R = \pi_{X_R} \tau^t, \qquad \mathfrak{B}_R = F_R \mathfrak{B}, \qquad \mathfrak{C}_R = \mathfrak{C} F_R^*,$$

then $\left[\begin{smallmatrix} \mathfrak{A}_R & \mathfrak{B}_R \\ \mathfrak{C}_R & \mathfrak{D} \end{smallmatrix}\right]$ is a unitary similarity transformation of the system Σ; hence it is a conservative system on (Y, X_R, U).

To complete the proof of (i) we must still derive the specific formulas for F_R, X_R, π_{X_R}, $\mathfrak{A}_R$, $\mathfrak{B}_R$, and $\mathfrak{C}_R$ listed in (i). Recall that $E = \left[\begin{smallmatrix} E_R \\ W_-^* \end{smallmatrix}\right]$. The splitting of the bottom row of E is already built into the definition of W_-^*, which says that $W_-^* = \begin{bmatrix} \mathfrak{D}^*\pi_- & \mathfrak{B}^* & \pi_+ \end{bmatrix}$. The splitting of the top row can readily be computed from the formula for E_R given in Theorem 11.6.2 as follows. We have

$$\begin{aligned}
(1 - \mathfrak{D}\mathfrak{D}^*)^{1/2} E_R &= \big(W_+ - W_+(1 - W_- W_-^*)\big) \\
&= \begin{bmatrix} \pi_- & \mathfrak{C} & \mathfrak{D}\pi_+ \end{bmatrix} - W_+ W_- \begin{bmatrix} \mathfrak{D}^*\pi_- & \mathfrak{B}^* & \pi_+ \end{bmatrix} \\
&= \begin{bmatrix} \pi_- & \mathfrak{C} & \mathfrak{D}\pi_+ \end{bmatrix} - \mathfrak{D} \begin{bmatrix} \mathfrak{D}^*\pi_- & \mathfrak{B}^* & \pi_+ \end{bmatrix} \\
&= \begin{bmatrix} (1 - \mathfrak{D}\mathfrak{D}^*)\pi_- & \mathfrak{C} - \mathfrak{D}\mathfrak{B}^* & 0 \end{bmatrix}.
\end{aligned}$$

Obviously, the range of $(1 - \mathfrak{D}\mathfrak{D}^*)\pi_-$ is contained in $\mathcal{R}(1 - \mathfrak{D}\mathfrak{D}^*) \subset \mathcal{R}\big((1 - \mathfrak{D}\mathfrak{D}^*)^{1/2}\big)$, so the preceding identity shows that the range of $\mathfrak{C} - \mathfrak{D}\mathfrak{B}^*$

is contained in $\mathcal{R}\left((1-\mathfrak{D}\mathfrak{D}^*)^{1/2}\right)$, and that we therefore may write

$$E_R = (1-\mathfrak{D}\mathfrak{D}^*)^{-1/2}\begin{bmatrix}(1-\mathfrak{D}\mathfrak{D}^*)\pi_- & \mathfrak{C}-\mathfrak{D}\mathfrak{B}^* & 0\end{bmatrix}.$$

Thus,

$$E = \begin{bmatrix}(1-\mathfrak{D}\mathfrak{D}^*)^{1/2}\pi_- & (1-\mathfrak{D}\mathfrak{D}^*)^{-1/2}(\mathfrak{C}-\mathfrak{D}\mathfrak{B}^*) & 0\\ \mathfrak{D}^*\pi_- & \mathfrak{B}^* & \pi_+\end{bmatrix}.$$

From here we immediately see that if we remove the $\mathcal{U}$-component of $\mathcal{X}$, then this simply replaces the image of E by $\begin{bmatrix}\overline{\mathcal{R}(1-\mathfrak{D}\mathfrak{D}^*)}\\ L^2(\mathbb{R}^-;U)\end{bmatrix}$ and also replaces the bilateral shift τ^t in both components by the partially truncated shift $\begin{bmatrix}\tau^t & 0\\ 0 & \tau^t_-\end{bmatrix}$. Moreover, the alternative formulas for X_R and for the projection π_{X_R} onto X_R can be read off directly from this representation. That X_R is co-invariant under the truncated shift follows from the fact that both $\mathcal{R}\left(\begin{bmatrix}\pi_R & 0\\ 0 & \pi_-\end{bmatrix}\right)$ and $\mathcal{N}\left(\begin{bmatrix}\pi_-(1-\mathfrak{D}\mathfrak{D}^*)^{1/2} & \pi_-\mathfrak{D}\end{bmatrix}\right)$ are invariant under the (truncated) right-shift.

The only remaining parts of the proof of (i) are the explicit formulas for $\mathfrak{B}_R$ and $\mathfrak{C}_R$. By Theorem 11.2.9,

$$\begin{aligned}\mathfrak{B}_R = F_R\mathfrak{B} &= \begin{bmatrix}(1-\mathfrak{D}\mathfrak{D}^*)^{-1/2}(\mathfrak{C}\mathfrak{B}-\mathfrak{D}\mathfrak{B}^*\mathfrak{B})\\ \mathfrak{B}^*\mathfrak{B}\end{bmatrix}\\ &= \begin{bmatrix}(1-\mathfrak{D}\mathfrak{D}^*)^{-1/2}(\pi_+\mathfrak{D}\pi_- - \mathfrak{D}\pi_- + \mathfrak{D}\mathfrak{D}^*\pi_-\mathfrak{D})\\ \pi_- - \mathfrak{D}^*\pi_-\mathfrak{D}\end{bmatrix}\\ &= \begin{bmatrix}-(1-\mathfrak{D}\mathfrak{D}^*)^{-1/2}(1-\mathfrak{D}\mathfrak{D}^*)\mathfrak{D}\pi_-\\ \pi_- - \mathfrak{D}^*\pi_-\mathfrak{D}\end{bmatrix}\\ &= \begin{bmatrix}-(1-\mathfrak{D}\mathfrak{D}^*)^{1/2}\mathfrak{D}\pi_-\\ \pi_- - \mathfrak{D}^*\pi_-\mathfrak{D}\end{bmatrix},\end{aligned}$$

which is the formula that we give for $\mathfrak{B}_R$ in (i). A similar computation verifies the formula for $\mathfrak{C}_R$. □

Corollary 11.6.4 *Two (scattering) conservative completely non-unitary systems* $\left[\begin{smallmatrix}\mathfrak{A} & \mathfrak{B}\\ \mathfrak{C} & \mathfrak{D}\end{smallmatrix}\right]$ *on* (Y, X, U) *and* $\left[\begin{smallmatrix}\mathfrak{A}_1 & \mathfrak{B}_1\\ \mathfrak{C}_1 & \mathfrak{D}_1\end{smallmatrix}\right]$ *on* (Y_1, X_1, U_1) *are unitarily similar if and only if* $\mathfrak{D}$ *and* $\mathfrak{D}_1$ *are unitarily similar (see Definition 11.4.3). In particular, if two completely non-unitary semigroups* $\mathfrak{A}$ *and* $\mathfrak{A}_1$ *have the same characteristic function, then* $\mathfrak{A}$ *and* $\mathfrak{A}_1$ *are unitarily similar.*

Proof Obviously, if the two systems are unitarily similar, then so are their input/output maps. Conversely, suppose first that the two input/output maps are not just similar but identical. Then both systems are unitarily similar to the system described in part (i) of Theorem 11.6.3 (and to the system described in part (ii); both of the special systems described in Theorem 11.6.3 are completely determined by $\mathfrak{D}$, and do not depend on the rest of the original system). Therefore they are also unitarily similar to each other. The general case where $\mathfrak{D}$ and $\mathfrak{D}_1$ are just unitarily similar instead of identical follows from this special case applied to $\left[\begin{smallmatrix} 1 & 0 \\ 0 & G \end{smallmatrix}\right]\left[\begin{array}{c|c} \mathfrak{A} & \mathfrak{B} \\ \hline \mathfrak{C} & \mathfrak{D} \end{array}\right]\left[\begin{smallmatrix} 1 & 0 \\ 0 & F^{-1} \end{smallmatrix}\right]$ and $\left[\begin{array}{c|c} \mathfrak{A}_1 & \mathfrak{B}_1 \\ \hline \mathfrak{C}_1 & \mathfrak{D}_1 \end{array}\right]$, where F and G are chosen so that $\mathfrak{D}_1 = G\mathfrak{D}F^{-1}$. □

Corollary 11.6.5 *Let* $\Sigma = \left[\begin{array}{c|c} \mathfrak{A} & \mathfrak{B} \\ \hline \mathfrak{C} & \mathfrak{D} \end{array}\right]$ *and* $\Sigma_1 = \left[\begin{array}{c|c} \mathfrak{A}_1 & \mathfrak{B}_1 \\ \hline \mathfrak{C}_1 & \mathfrak{D}_1 \end{array}\right]$ *be two simple (i.e., completely nonunitary and purely passive) conservative systems on* (Y, X, U)*, respectively* (Y_1, X_1, U_1)*. Then the following conditions are equivalent:*

(i) Σ *is unitarily similar to* Σ_1.
(ii) $\mathfrak{A}$ *is unitarily similar to* $\mathfrak{A}_1$.
(iii) $\mathfrak{D}$ *is unitarily similar to* $\mathfrak{D}_1$.

This follows from Corollaries 11.4.10 and 11.6.4.

Corollary 11.6.6 *Let* $\Sigma = \left[\begin{array}{c|c} \mathfrak{A} & \mathfrak{B} \\ \hline \mathfrak{C} & \mathfrak{D} \end{array}\right]$ *be a simple (i.e., completely nonunitary and purely passive) conservative system, and let* $\Sigma^d = \left[\begin{array}{c|c} \mathfrak{A}^d & \mathfrak{B}^d \\ \hline \mathfrak{C}^d & \mathfrak{D}^d \end{array}\right]$ *be the causal dual of* Σ*. Then the following conditions are equivalent:*

(i) Σ *is unitarily similar to* Σ^d.
(ii) $\mathfrak{A}$ *is unitarily similar to* $\mathfrak{A}^*$.
(iii) $\mathfrak{D}$ *is unitarily similar to* $\mathfrak{D}^d$.

This follows from Corollary 11.6.5.

As our following corollary shows, if $\mathfrak{D}$ is semi-lossless or co-lossless, or equivalently, if $\mathfrak{A}$ is completely non-unitary and strongly stable or strongly co-stable (i.e., $\mathfrak{A}^*$ is strongly stable), then the model in Theorem 11.6.3 simplifies considerably.

Corollary 11.6.7 *Let* $\Sigma = \left[\begin{array}{c|c} \mathfrak{A} & \mathfrak{B} \\ \hline \mathfrak{C} & \mathfrak{D} \end{array}\right]$ *be a (scattering) conservative linear system on* (Y, X, U)*. Then the following claims are true.*

(i) *If* $\mathfrak{A}^*$ *is strongly stable, or equivalently, if* $\mathfrak{A}$ *is completely nonunitary and* Σ *is co-lossless, then* $\mathfrak{B}^*$ *is an isometry from* X *to* $L^2(\mathbb{R}^-; U)$ *with final space*

$$X_R = \mathcal{R}\left(\mathfrak{B}^*\right) = \mathcal{R}\left(\mathfrak{D}^*\pi_-\right)^\perp = \mathcal{N}\left(\pi_-\mathfrak{D}\right).$$

The orthogonal projection π_{X_R} *in* $L^2(\mathbb{R}^-; U)$ *onto* X_R *is given by*

$$\pi_{X_R} = \mathfrak{B}^*\mathfrak{B} = \pi_- - \mathfrak{D}^*\pi_-\mathfrak{D}.$$

Define

$$\mathfrak{A}_R = \pi_{X_R}\tau_{-|X_R}, \qquad \mathfrak{B}_R = \pi_{X_R}, \qquad \mathfrak{C}_R = \pi_+\mathfrak{D}\pi_{-|X_R}.$$

Then $\left[\begin{smallmatrix} \mathfrak{A}_R & \mathfrak{B}_R \\ \mathfrak{C}_R & \mathfrak{D} \end{smallmatrix}\right]$ *is a conservative realization of* $\mathfrak{D}$ *which is unitarily similar to* Σ *(in the sense of Example 2.3.7):*

$$\mathfrak{A}_R = F_R\mathfrak{A}F_R^*, \qquad \mathfrak{B}_R = F_R\mathfrak{B}, \qquad \mathfrak{C}_R = \mathfrak{C}F_R^*.$$

In particular, $\mathfrak{A}$ *is unitarily similar to* $\mathfrak{A}_R$*. Moreover, the space* X_R *is co-invariant under the semigroup* τ_- *on* $L^2(\mathbb{R}^-;U)$ *(i.e., it is invariant under the adjoint of this operator).*

(ii) *If* $\mathfrak{A}$ *is strongly stable, or equivalently, if* $\mathfrak{A}$ *is completely nonunitary and* Σ *is lossless, then* $\mathfrak{C}$ *is an isometry from* X *to* $L^2(\mathbb{R}^+;Y)$ *with final space*

$$X_O = \mathcal{R}(\mathfrak{C}) = \mathcal{R}(\mathfrak{D}\pi_+)^\perp = \mathcal{N}\left(\pi_+\mathfrak{D}^*\right).$$

The orthogonal projection π_{X_O} *in* $L^2(\mathbb{R}^+;Y)$ *onto* X_O *is given by*

$$\pi_{X_O} = \mathfrak{C}\mathfrak{C}^* = \pi_+ - \mathfrak{D}\pi_+\mathfrak{D}^*.$$

Define

$$\mathfrak{A}_O = \tau_{+|X_O}, \qquad \mathfrak{B}_O = \pi_+\mathfrak{D}\pi_-, \qquad \mathfrak{C}_O = 1_{|X_O}.$$

Then $\left[\begin{smallmatrix} \mathfrak{A}_O & \mathfrak{B}_O \\ \mathfrak{C}_O & \mathfrak{D} \end{smallmatrix}\right]$ *is a conservative realization of* $\mathfrak{D}$ *which is unitarily similar to* Σ*:*

$$\mathfrak{A}_O = F_O\mathfrak{A}F_O^*, \qquad \mathfrak{B}_O = F_O\mathfrak{B}, \qquad \mathfrak{C}_O = \mathfrak{C}F_O^*.$$

In particular, $\mathfrak{A}$ *is unitarily similar to* $\mathfrak{A}_O$*, and the space* X_O *is invariant under* τ_+ *on* $L^2(\mathbb{R}^+;Y)$*.*

This follows from Theorems 11.3.6 and 11.6.3. (It is also possible to give a direct proof which is significantly simpler than the proof of Theorem 11.6.3.) See Theorem 11.3.6 for an additional equivalent set of conditions under which (i) and (ii) apply.

Corollary 11.6.8 *Let* $\mathfrak{A}$ *be a* C_0 *semigroup on a Hilbert space* X*.*

(i) $\mathfrak{A}$ *is unitarily similar to the outgoing left-shift* τ_- *on* $L^2(\mathbb{R}^-;U)$ *for some Hilbert space* U *if and only if* $\mathfrak{A}$ *is isometric and* $\mathfrak{A}^*$ *is strongly stable, or equivalently, if and only if* $\mathfrak{A}$ *is isometric and completely nonunitary.*

(ii) $\mathfrak{A}$ *is unitarily similar to the incoming left-shift* τ_+ *on* $L^2(\mathbb{R}^+;Y)$ *for some Hilbert space* Y *if and only if* $\mathfrak{A}$ *is co-isometric and strongly stable, or equivalently, if and only if* $\mathfrak{A}$ *is co-isometric and completely nonunitary.*

Proof (i) Clearly, every semigroup which is unitarily similar to the outgoing shift τ_- is isometric and its adjoint is strongly stable. It is also clear that if $\mathfrak{A}^*$

is strongly stable, then $\mathfrak{A}$ is completely nonunitary. Thus, in order to prove (i) it suffices to show that $\mathfrak{A}$ is unitarily similar to an outgoing left-shift whenever $\Sigma = \mathfrak{A}$ is isometric and completely nonunitary. Let us assume the latter, and let $\left[\begin{smallmatrix} \mathfrak{A} & \mathfrak{B} \\ \mathfrak{C} & \mathfrak{D} \end{smallmatrix}\right]$ be the corresponding conservative system described in Theorem 11.4.9. The fact that $\mathfrak{A}$ is isometric implies that $Q_{\mathfrak{A}^*,\mathfrak{A}} = 1$ (see (11.1.11)), hence, by (11.2.13), $\mathfrak{C}^*\mathfrak{C} = 0$. Thus, $\mathfrak{C} = 0$, and since the range of the observation operator is dense in Y, this means that $Y = 0$. Thus, Σ collapses into $\Sigma = \left[\mathfrak{A} \,|\, \mathfrak{B}\right]$. This means that the $L^2(\mathbb{R}; Y)$-component of the model in Theorem 11.6.3 vanishes, and that the operator F_R, the projection π_{X_R}, and the semigroup $\mathfrak{A}_R$ in that theorem become $F_R = \mathfrak{B}^*$, $\pi_{X_R} = \pi_-$, and $\mathfrak{A}_R = \tau_-$. By Theorem 11.6.3, $\mathfrak{A} = \mathfrak{B}\mathfrak{A}\mathfrak{B}^*$, where $\mathfrak{B}^*$ is a unitary map of X onto $L^2(\mathbb{R}^-; U)$.

(ii) This is proved in the analogous way (or it follows from (i) by duality). □

Corollary 11.6.9 (Wold decomposition) *Let $\mathfrak{A}$ be an isometric C_0 semigroup on a Hilbert space X. Then there is a subspace $Z \subset X$, a Hilbert space U, and unitary map E of $Z^\perp$ onto $L^2(\mathbb{R}^-; U)$ such that, if we decompose X into $X = \left[\begin{smallmatrix} Z \\ Z^\perp \end{smallmatrix}\right]$, then*

$$\mathfrak{A} = \begin{bmatrix} \mathfrak{A}_0 & 0 \\ 0 & E^{-1}\tau_- E \end{bmatrix},$$

where $\mathfrak{A}_0$ is a unitary semigroup on Z and τ_- is the outgoing left-shift on $L^2(\mathbb{R}^-; U)$.

Proof An isometric semigroup can be interpreted as an energy preserving, hence passive, system with zero input and output spaces; cf. Definition 11.1.1. Let Z be the space denoted by X_0 in Theorem 11.1.9 (with $U = Y = 0$). Then $\mathfrak{A}$ is decomposed into $\mathfrak{A} = \left[\begin{smallmatrix} \mathfrak{A}_0 & 0 \\ 0 & \mathfrak{A}_1 \end{smallmatrix}\right]$, where $\mathfrak{A}_0$ is unitary and $\mathfrak{A}_1$ is isometric and completely nonunitary. By Corollary 11.6.8(i), $\mathfrak{A}_1$ is unitarily similar to the outgoing left-shift τ_- on $L^2(\mathbb{R}^-; U)$ for some Hilbert space U. □

11.7 Conservative realizations

In the preceding section we showed that every C_0 contraction semigroup can be 'dilated' into a conservative system, which is unique up to unitary similarity if we require the input/output map to be purely passive. There is a related question which starts with the input/output map: given a contractive map $\mathfrak{D} \in TIC^2(U; Y)$ (where U and Y are Hilbert spaces), is it always possible to find a (scattering) *conservative realization* of $\mathfrak{D}$, i.e., a conservative system $\left[\begin{smallmatrix} \mathfrak{A} & \mathfrak{B} \\ \mathfrak{C} & \mathfrak{D} \end{smallmatrix}\right]$ which has this input/output map? The purpose of this section is to give a positive answer to this question.

As we saw in Corollary 11.6.4, a conservative realization of $\mathfrak{D}$ (if it exists) is unique up to unitary similarity if we require $\mathfrak{A}$ to be completely nonunitary. Moreover, Theorem 11.6.3 gives us explicit descriptions in terms of $\mathfrak{D}$ of two possible realizations $\left[\begin{smallmatrix} \mathfrak{A}_R & \mathfrak{B}_R \\ \mathfrak{C}_R & \mathfrak{D} \end{smallmatrix}\right]$ and $\left[\begin{smallmatrix} \mathfrak{A}_O & \mathfrak{B}_O \\ \mathfrak{C}_O & \mathfrak{D} \end{smallmatrix}\right]$ of $\mathfrak{D}$. Thus, to prove that every contractive map $\mathfrak{D} \in TIC^2(U;Y)$ has a conservative realization it suffices to show that if we *define*, e.g., $\left[\begin{smallmatrix} \mathfrak{A}_R & \mathfrak{B}_R \\ \mathfrak{C}_R & \mathfrak{D} \end{smallmatrix}\right]$, using the formulas in Theorem 11.6.3, then this is indeed a conservative system with input/output map $\mathfrak{D}$ whenever $\mathfrak{D} \in TIC^2(U;Y)$ is contractive.

Theorem 11.7.1 *Let U and Y be Hilbert spaces, and let $\mathfrak{D} \in TIC^2(U;Y)$ be a contraction. Then the operator*

$$G = \begin{bmatrix} (1 - \mathfrak{D}\mathfrak{D}^*)^{1/2} \\ \mathfrak{D}^* \end{bmatrix}$$

is an isometry from $L^2(\mathbb{R};Y)$ into $\left[\begin{smallmatrix} L^2(\mathbb{R};Y) \\ L^2(\mathbb{R};U) \end{smallmatrix}\right]$. Let X be the closed subspace of $\left[\begin{smallmatrix} L^2(\mathbb{R};Y) \\ L^2(\mathbb{R};U) \end{smallmatrix}\right]$ defined by

$$X = \begin{bmatrix} \overline{\mathcal{R}(1 - \mathfrak{D}\mathfrak{D}^*)} \\ L^2(\mathbb{R}^-;U) \end{bmatrix} \cap \mathcal{R}(G\pi_-)^\perp = \begin{bmatrix} \overline{\mathcal{R}(1 - \mathfrak{D}\mathfrak{D}^*)} \\ L^2(\mathbb{R}^-;U) \end{bmatrix} \cap \mathcal{N}\left(\pi_- G^*\right).$$

Let π_1 be the orthogonal projection in $L^2(\mathbb{R};Y)$ onto $\overline{\mathcal{R}(1 - \mathfrak{D}\mathfrak{D}^*)}$. Then the orthogonal projection π_X in $\left[\begin{smallmatrix} L^2(\mathbb{R};Y) \\ L^2(\mathbb{R};U) \end{smallmatrix}\right]$ onto X is given by

$$\pi_X = \begin{bmatrix} \pi_1 & 0 \\ 0 & \pi_- \end{bmatrix} - G\pi_- G^*.$$

Define

$$\mathfrak{A} = \pi_X \begin{bmatrix} \tau & 0 \\ 0 & \tau_- \end{bmatrix}_{|X}, \quad \mathfrak{B} = \pi_X \begin{bmatrix} 0 \\ \pi_- \end{bmatrix} = \begin{bmatrix} 0 \\ \pi_- \end{bmatrix} - G\pi_- \mathfrak{D},$$

$$\mathfrak{C} = \pi_+ G^* \begin{bmatrix} \pi_1 & 0 \\ 0 & \pi_- \end{bmatrix}_{|X}.$$

Then $\left[\begin{smallmatrix} \mathfrak{A} & \mathfrak{B} \\ \mathfrak{C} & \mathfrak{D} \end{smallmatrix}\right]$ is a (scattering) conservative realization of $\mathfrak{D}$, $\mathfrak{A}$ is completely nonunitary, and the space X is co-invariant under the semigroup $\left[\begin{smallmatrix} \tau & 0 \\ 0 & \tau_- \end{smallmatrix}\right]$ on $\left[\begin{smallmatrix} L^2(\mathbb{R};Y) \\ L^2(\mathbb{R}^-;U) \end{smallmatrix}\right]$ (i.e., it is invariant under the adjoint of this operator).

Proof Let us begin with some comments on the state space X. Clearly G is an isometry on $L^2(\mathbb{R};U)$ since

$$G^*G = (1 - \mathfrak{D}\mathfrak{D}^*) + \mathfrak{D}\mathfrak{D}^* = 1.$$

Likewise, $G\pi_-$ is an isometry on $L^2(\mathbb{R}^-;U)$. Therefore $G\pi_- G^*$ is a projection, whose range is a subset of $\left[\begin{smallmatrix} \overline{\mathcal{R}(1-\mathfrak{D}\mathfrak{D}^*)} \\ L^2(\mathbb{R}^-;U) \end{smallmatrix}\right]$ (see Lemma A.2.2). The space X is the orthogonal complement in the subspace $\begin{bmatrix} \overline{\mathcal{R}(1-\mathfrak{D}\mathfrak{D}^*)} \\ L^2(\mathbb{R}^-;U) \end{bmatrix}$ to $\mathcal{R}(G)\pi_- = \mathcal{R}(G\pi_- G^*)$, and therefore the projection π_X is given by

$$\begin{aligned} \pi_X &= \begin{bmatrix} \pi_1 & 0 \\ 0 & \pi_- \end{bmatrix} - G\pi_- G^* \\ &= \begin{bmatrix} \pi_1 - (1-\mathfrak{D}\mathfrak{D}^*)^{1/2}\pi_-(1-\mathfrak{D}\mathfrak{D}^*)^{1/2} & -(1-\mathfrak{D}\mathfrak{D}^*)^{1/2}\pi_-\mathfrak{D} \\ -\mathfrak{D}^*\pi_-(1-\mathfrak{D}\mathfrak{D}^*)^{1/2} & \pi_- - \mathfrak{D}^*\pi_-\mathfrak{D} \end{bmatrix}. \end{aligned}$$

Observe, in particular, that the two different formulas given for $\mathfrak{B}$ are equivalent.

We next take a closer look at $\mathfrak{A}$. The space X is invariant under $\begin{bmatrix} \tau^t & 0 \\ 0 & \tau_-^t \end{bmatrix}^*$ since both $\mathcal{R}\left(\begin{bmatrix} \pi_1 & 0 \\ 0 & \pi_- \end{bmatrix}\right)$ and $\mathcal{N}(\pi_- G^*)$ are invariant under $\begin{bmatrix} \tau^t & 0 \\ 0 & \tau_-^t \end{bmatrix}^*$. Thus, the restriction of $\begin{bmatrix} \tau^t & 0 \\ 0 & \tau_-^t \end{bmatrix}^*$ to X is a C_0 contraction semigroup, hence so is its adjoint, which is exactly the semigroup $\mathfrak{A}$ given in the theorem. Moreover, the invariance means that, for all $t \ge 0$, $\pi_X \begin{bmatrix} \tau^t & 0 \\ 0 & \tau_-^t \end{bmatrix}^* \pi_X = \begin{bmatrix} \tau^t & 0 \\ 0 & \tau_-^t \end{bmatrix}^* \pi_X$, hence

$$\pi_X \begin{bmatrix} \tau^t & 0 \\ 0 & \tau_-^t \end{bmatrix} = \pi_X \begin{bmatrix} \tau^t & 0 \\ 0 & \tau_-^t \end{bmatrix} \pi_X,$$

and this is true on the space $\begin{bmatrix} L^2(\mathbb{R};Y) \\ L^2(\mathbb{R}^-;U) \end{bmatrix}$.

Let us next show that $\mathfrak{B}$ is an input map for $\mathfrak{A}$. We have, for all $t \ge 0$,

$$\begin{aligned} \mathfrak{B}\tau_-^t &= \pi_X \begin{bmatrix} 0 \\ 1 \end{bmatrix} \tau_-^t = \pi_X \begin{bmatrix} \tau^t & 0 \\ 0 & \tau_-^t \end{bmatrix} \begin{bmatrix} 0 \\ 1 \end{bmatrix} \\ &= \pi_X \begin{bmatrix} \tau^t & 0 \\ 0 & \tau_-^t \end{bmatrix} \pi_X \begin{bmatrix} 0 \\ 1 \end{bmatrix} = \mathfrak{A}^t\mathfrak{B}. \end{aligned}$$

This shows that $\mathfrak{B}$ is an input map for $\mathfrak{A}$.

We turn to the output map $\mathfrak{C}$. Formally $\mathfrak{C}$ is defined on X, but it has an obvious extension to $\begin{bmatrix} L^2(\mathbb{R};Y) \\ L^2(\mathbb{R}^-;U) \end{bmatrix}$ that we denote by the same symbol. We claim that this extended map satisfies $\mathfrak{C} = \mathfrak{C}\pi_X$. The proof of this claim is not difficult. Applying $\mathfrak{C}$ to the first column of π_X we get (after removing some redundant projections)

$$\begin{aligned} \mathfrak{C}\pi_X \begin{bmatrix} 1 \\ 0 \end{bmatrix} &= \pi_+(1-\mathfrak{D}\mathfrak{D}^*)^{1/2} - \pi_+(1-\mathfrak{D}\mathfrak{D}^*)\pi_-(1-\mathfrak{D}\mathfrak{D}^*)^{1/2} \\ &\quad - \pi_+\mathfrak{D}\mathfrak{D}^*\pi_-(1-\mathfrak{D}\mathfrak{D}^*)^{1/2} \\ &= \pi_+(1-\mathfrak{D}\mathfrak{D}^*)^{1/2}, \end{aligned}$$

which is the first column of $\mathfrak{C}$. Applying $\mathfrak{C}$ to the second column of π_X we get (after again removing some redundant projections)

$$\begin{aligned}\mathfrak{C}\pi_X \begin{bmatrix}0\\1\end{bmatrix} &= -\pi_+(1-\mathfrak{D}\mathfrak{D}^*)\pi_-\mathfrak{D} + \pi_+\mathfrak{D}\pi_- - \pi_+\mathfrak{D}\mathfrak{D}^*\pi_-\mathfrak{D}\\ &= \pi_+\mathfrak{D}\pi_-.\end{aligned}$$

This proves our claim that the extended operator $\mathfrak{C}$ satisfies $\mathfrak{C} = \mathfrak{C}\pi_X$. Once we know this we get, for all $t \geq 0$,

$$\begin{aligned}\mathfrak{C}\mathfrak{A}^t &= \mathfrak{C}\pi_X \begin{bmatrix}\tau^t & 0\\ 0 & \tau_-^t\end{bmatrix} = \mathfrak{C}\begin{bmatrix}\tau^t & 0\\ 0 & \tau_-^t\end{bmatrix}\\ &= \pi_+(1-\mathfrak{D}\mathfrak{D}^*)^{1/2}\tau^t + \pi_+\mathfrak{D}\tau^t\pi_-\\ &= \pi_+\tau^t(1-\mathfrak{D}\mathfrak{D}^*)^{1/2} + \pi_+\tau^t\mathfrak{D}\pi_- = \tau_+^t\mathfrak{C}.\end{aligned}$$

Thus, $\mathfrak{C}$ is an output map for $\mathfrak{A}$.

To complete the proof of the fact that $\left[\begin{smallmatrix}\mathfrak{A} & \mathfrak{B}\\ \mathfrak{C} & \mathfrak{D}\end{smallmatrix}\right]$ is an L^2-well-posed linear system we must still show that $\mathfrak{C}\mathfrak{B} = \pi_+\mathfrak{D}\pi_-$. However, this is immediate:

$$\mathfrak{C}\mathfrak{B} = -\pi_-(1-\mathfrak{D}\mathfrak{D}^*)\pi_-\mathfrak{D} + \pi_+\mathfrak{D}\pi_- - \pi_+\mathfrak{D}\mathfrak{D}^*\pi_-\mathfrak{D} = \pi_+\mathfrak{D}\pi_-.$$

Let us next show that $\mathfrak{A}$ is completely nonunitary. Take $\left[\begin{smallmatrix}y\\u\end{smallmatrix}\right]$ in X_0, where X_0 is the maximal subspace of X on which $\mathfrak{A}$ is unitary (see Theorem 11.1.9). Then $\left|\mathfrak{A}^{*t}\left[\begin{smallmatrix}y\\u\end{smallmatrix}\right]\right| = \left|\left[\begin{smallmatrix}y\\u\end{smallmatrix}\right]\right|$ for all $t \geq 0$. Since $|\tau^{-t}y| = |y|$ for all $t \geq 0$ and $|\tau_-^{*t}u| \to 0$ as $t \to \infty$, we must have $u = 0$. On the other hand, we also have $\left|\mathfrak{A}^t\left[\begin{smallmatrix}y\\u\end{smallmatrix}\right]\right| = \left|\mathfrak{A}^t\left[\begin{smallmatrix}y\\0\end{smallmatrix}\right]\right| = |y|$ for all $t \geq 0$, where

$$\mathfrak{A}^t\begin{bmatrix}y\\0\end{bmatrix} = \begin{bmatrix}\pi_1 - (1-\mathfrak{D}\mathfrak{D}^*)^{1/2}\pi_-(1-\mathfrak{D}\mathfrak{D}^*)^{1/2}\\ -\mathfrak{D}^*\pi_-(1-\mathfrak{D}\mathfrak{D}^*)^{1/2}\end{bmatrix}\tau^t y.$$

Since X_0 is invariant under $\mathfrak{A}$, the second component of $\mathfrak{A}^t\left[\begin{smallmatrix}y\\0\end{smallmatrix}\right]$ must vanish for all $t \geq 0$, i.e., $\mathfrak{D}^*\pi_-(1-\mathfrak{D}\mathfrak{D}^*)^{1/2}\tau^t y = 0$ for all $t \geq 0$, and hence

$$\begin{aligned}|y| &= \left|\tau^t y - (1-\mathfrak{D}\mathfrak{D}^*)^{1/2}\pi_-(1-\mathfrak{D}\mathfrak{D}^*)^{1/2}\tau^t y\right|\\ &= \left|y - (1-\mathfrak{D}\mathfrak{D}^*)^{1/2}\tau^{-t}\pi_-\tau^t(1-\mathfrak{D}\mathfrak{D}^*)^{1/2}y\right|.\end{aligned}$$

Letting $t \to \infty$ we get $|y| = |\mathfrak{D}\mathfrak{D}^*y|$. By Lemma 11.1.10, this implies

$$(1-\mathfrak{D}\mathfrak{D}^*\mathfrak{D}\mathfrak{D}^*)y = (1+\mathfrak{D}\mathfrak{D}^*)(1-\mathfrak{D}\mathfrak{D}^*)y = 0.$$

The operator $(1+\mathfrak{D}\mathfrak{D}^*)$ is strictly positive, hence injective, so $(1-\mathfrak{D}\mathfrak{D}^*)y = 0$. On the other hand, since $\left[\begin{smallmatrix}y\\0\end{smallmatrix}\right] \in X$, we have $y \in \overline{\mathcal{R}(1-\mathfrak{D}\mathfrak{D}^*)} = \mathcal{N}(1-\mathfrak{D}\mathfrak{D}^*)^\perp$. Therefore $y = 0$. This proves that $\mathfrak{A}$ is completely non-unitary.

It remains to show that $\left[\begin{smallmatrix}\mathfrak{A} & \mathfrak{B}\\ \mathfrak{C} & \mathfrak{D}\end{smallmatrix}\right]$ is conservative. We begin by showing that it is isometric. Let x be the state and let y be the output of $\left[\begin{smallmatrix}\mathfrak{A} & \mathfrak{B}\\ \mathfrak{C} & \mathfrak{D}\end{smallmatrix}\right]$ with initial state $\left[\begin{smallmatrix} y_0 \\ u_0 \end{smallmatrix}\right] \in X$, input $u \in L^2(\mathbb{R}^+;U)$, and initial time zero. Then, for all $t \geq 0$, the state at time t is given by

$$\begin{aligned} x(t) &= \mathfrak{A}^t \begin{bmatrix} y_0 \\ u_0 \end{bmatrix} + \mathfrak{B}\tau^t \pi_- u = \pi_X \left(\begin{bmatrix} \tau^t & 0 \\ 0 & \tau_-^t \end{bmatrix} \begin{bmatrix} y_0 \\ u_0 \end{bmatrix} + \begin{bmatrix} 0 \\ \tau^t \pi_{[0,t)} u \end{bmatrix} \right) \\ &= (1 - G\pi_- G^*)\tau^t \begin{bmatrix} y_0 \\ \pi_- u_0 + \pi_{[0,t)} u \end{bmatrix}. \end{aligned}$$

As $G\pi_- G^*$ is an orthogonal projection, this implies that

$$\begin{aligned} |x|^2 &= \left| \begin{bmatrix} y_0 \\ \pi_- u_0 + \pi_{[0,t)} u \end{bmatrix} \right|^2 - \left| G\pi_- G^* \begin{bmatrix} y_0 \\ \pi_- u_0 + \pi_{[0,t)} u \end{bmatrix} \right|^2 \\ &= \left| \begin{bmatrix} y_0 \\ \pi_- u_0 + \pi_{[0,t)} u \end{bmatrix} \right|^2 \\ &\quad - \left\langle \begin{bmatrix} y_0 \\ \pi_- u_0 + \pi_{[0,t)} u \end{bmatrix}, G\pi_- G^* \begin{bmatrix} y_0 \\ \pi_- u_0 + \pi_{[0,t)} u \end{bmatrix} \right\rangle. \end{aligned}$$

The restriction of the output to $[0,t)$ is given by

$$\begin{aligned} \pi_{[0,t)} y &= \pi_{[0,t)} \left(\mathfrak{C} \begin{bmatrix} y_0 \\ u_0 \end{bmatrix} + \mathfrak{D}\pi_{[0,t)} u \right) = \pi_{[0,t)} \left(G^* \begin{bmatrix} y_0 \\ \pi_- u_0 \end{bmatrix} + \mathfrak{D}\pi_{[0,t)} u \right) \\ &= \pi_{[0,t)} G^* \begin{bmatrix} y_0 \\ \pi_- u_0 + \pi_{[0,t)} u \end{bmatrix}. \end{aligned}$$

Since $\left[\begin{smallmatrix} y_0 \\ u_0 \end{smallmatrix}\right] \in X \subset \mathcal{N}(\pi_- G^*)$, and since $\mathfrak{D}$ is causal, the function $G^* \left[\begin{smallmatrix} y_0 \\ \pi_- u_0 + \pi_{[0,t)} u \end{smallmatrix}\right]$ vanishes on $\mathbb{R}^-$, so we may alternatively write this as $\pi_{[0,t)} y = \pi_{(-\infty,t)} G^* \left[\begin{smallmatrix} y_0 \\ \pi_- u_0 + \pi_{[0,t)} u \end{smallmatrix}\right]$, or equivalently,

$$\tau^t \pi_{[0,t)} y = \pi_- G^* \tau^t \begin{bmatrix} y_0 \\ (\pi_- u_0 + \pi_{[0,t)} u) \end{bmatrix}.$$

In particular,

$$|\pi_{[0,t)} y|^2 = \left\langle \begin{bmatrix} y_0 \\ \pi_- u_0 + \pi_{[0,t)} u \end{bmatrix}, G\pi_- G^* \begin{bmatrix} y_0 \\ \pi_- u_0 + \pi_{[0,t)} u \end{bmatrix} \right\rangle.$$

Thus, we conclude that

$$x(t) = \tau^t \begin{bmatrix} y_0 \\ \pi_- u_0 + \pi_{[0,t)} u \end{bmatrix} - G\tau^t \pi_{[0,t)} y,$$

and that

$$|x(t)|^2 = \left\| \begin{bmatrix} y_0 \\ \pi_- u_0 + \pi_{[0,t)} u \end{bmatrix} \right\|^2 - |\pi_{[0,t)} y|^2$$
$$= \left\| \begin{bmatrix} y_0 \\ \pi_- u_0 \end{bmatrix} \right\|^2 + |\pi_{[0,t)} u|^2 - |\pi_{[0,t)} y|^2.$$

In particular, $\left[\begin{smallmatrix} \mathfrak{A} & \mathfrak{B} \\ \mathfrak{C} & \mathfrak{D} \end{smallmatrix}\right]$ is energy preserving.

To complete the proof of the fact that $\left[\begin{smallmatrix} \mathfrak{A} & \mathfrak{B} \\ \mathfrak{C} & \mathfrak{D} \end{smallmatrix}\right]$ is conservative we must still show that the mapping from the pair $\left(\left[\begin{smallmatrix} y_0 \\ u_0 \end{smallmatrix}\right], \pi_{[0,t)} u\right)$ to the pair $(x(t), \pi_{[0,t)} y)$ is invertible. From the computation above we get an obvious candidate for the inverse: let $x(t) = \left[\begin{smallmatrix} y_t \\ u_t \end{smallmatrix}\right] \in X$ and $\pi_{[0,t)} y \in L^2([0,t); Y)$ be arbitrary, and let us take a closer look at the pair of functions

$$\begin{bmatrix} y_0 \\ u_0 \end{bmatrix} = \begin{bmatrix} 1 & 0 \\ 0 & \pi_- \end{bmatrix} (\tau^{-t} x(t) + G\pi_{[0,t)} y),$$
$$u = \begin{bmatrix} 0 & \pi_+ \end{bmatrix} (\tau^{-t} x(t) + G\pi_{[0,t)} y).$$

Then

$$\begin{bmatrix} y_0 \\ u_0 \end{bmatrix} + \begin{bmatrix} 0 \\ u \end{bmatrix} = (\tau^{-t} x(t) + G\pi_{[0,t)} y).$$

The top component of $(\tau^{-t} x(t) + G\pi_{[0,t)} y)$ belongs to $\mathcal{R}(\pi_1)$ since both $y_t \subset \mathcal{R}(\pi_1)$ and $\mathcal{R}\left((1 - \mathfrak{D}\mathfrak{D}^*)^{1/2}\right) \subset \mathcal{R}(\pi_1)$, and since $\mathcal{R}(\pi_1)$ is invariant under τ^{-t}. Thus, $\left[\begin{smallmatrix} y_0 \\ u_0 \end{smallmatrix}\right] \subset \mathcal{R}\left(\left[\begin{smallmatrix} \pi_1 & 0 \\ 0 & \pi_- \end{smallmatrix}\right]\right)$. As G is energy preserving and $x(t) \in X \subset \mathcal{N}(\pi_- G^*)$, we find that

$$G^*(\tau^{-t} x(t) + G\pi_{[0,t)} y) = \tau^{-t} G^* x(t) + \pi_{[0,t)} y$$

vanishes on $\mathbb{R}^-$. The second component of G^* is $\mathfrak{D}$ which is causal, so $\pi_- G^* \left[\begin{smallmatrix} y_0 \\ u_0 \end{smallmatrix}\right] = 0$. This proves that $\left[\begin{smallmatrix} y_0 \\ u_0 \end{smallmatrix}\right] \in X$.

The remaining component u of $(\tau^{-t} x(t) + G\pi_{[0,t)} y)$ is explicitly given by

$$u = \pi_+(\tau^{-t} u_t + \mathfrak{D}^* \pi_{[0,t)} y).$$

Since u_t vanishes on $\mathbb{R}^+$ and $\mathfrak{D}^*$ is anti-causal, this means that u vanishes outside of the interval $[0, t)$, and we are done. □

The construction in Theorem 11.7.1 was based on the model given in part (i) of Theorem 11.6.3. There is a similar result which uses the model in part (ii) of that theorem instead.

Corollary 11.7.2 *Let U and Y be Hilbert spaces, and let $\mathfrak{D} \in TIC^2(U;Y)$ be a contraction. Then the operator*

$$H = \begin{bmatrix} \mathfrak{D} \\ (1 - \mathfrak{D}^*\mathfrak{D})^{1/2} \end{bmatrix}$$

is an isometry from $L^2(\mathbb{R};U)$ into $\left[\begin{smallmatrix} L^2(\mathbb{R};Y) \\ L^2(\mathbb{R};U) \end{smallmatrix}\right]$. Let X be the closed subspace of $\left[\begin{smallmatrix} L^2(\mathbb{R};Y) \\ L^2(\mathbb{R};U) \end{smallmatrix}\right]$ defined by

$$X = \begin{bmatrix} L^2(\mathbb{R}^+;Y) \\ \overline{\mathcal{R}(1 - \mathfrak{D}^*\mathfrak{D})} \end{bmatrix} \cap \mathcal{R}(H\pi_+)^\perp = \begin{bmatrix} L^2(\mathbb{R}^+;Y) \\ \overline{\mathcal{R}(1 - \mathfrak{D}^*\mathfrak{D})} \end{bmatrix} \cap \mathcal{N}\left(\pi_+ H^*\right).$$

Let π_2 be the orthogonal projection in $L^2(\mathbb{R};U)$ onto $\overline{\mathcal{R}(1 - \mathfrak{D}^*\mathfrak{D})}$. Then the orthogonal projection π_X in $\left[\begin{smallmatrix} L^2(\mathbb{R};Y) \\ L^2(\mathbb{R};U) \end{smallmatrix}\right]$ onto X is given by

$$\pi_X = \begin{bmatrix} \pi_+ & 0 \\ 0 & \pi_2 \end{bmatrix} - H\pi_+ H^*.$$

Define

$$\mathfrak{A} = \begin{bmatrix} \tau_+ & 0 \\ 0 & \tau \end{bmatrix}_{|X}, \qquad \mathfrak{B} = \begin{bmatrix} \pi_+ & 0 \\ 0 & \pi_2 \end{bmatrix} H\pi_-,$$

$$\mathfrak{C} = \begin{bmatrix} \pi_+ \\ 0 \end{bmatrix}_{|X} = \left(\begin{bmatrix} \pi_+ \\ 0 \end{bmatrix} - \mathfrak{D}^*\pi_+ H^*\right)_{|X}.$$

Then $\left[\begin{smallmatrix} \mathfrak{A} & \mathfrak{B} \\ \mathfrak{C} & \mathfrak{D} \end{smallmatrix}\right]$ is a (scattering) conservative completely non-unitary realization of $\mathfrak{D}$. (In particular, the space X is invariant under the semigroup $\left[\begin{smallmatrix} \tau_+ & 0 \\ 0 & \tau \end{smallmatrix}\right]$ on $\left[\begin{smallmatrix} L^2(\mathbb{R}^+;Y) \\ L^2(\mathbb{R};U) \end{smallmatrix}\right]$.)

Proof This follows from Theorem 11.7.1 by duality (i.e., we apply Theorem 11.7.1 to the causal dual $\mathbf{Я}\mathfrak{D}^*\mathbf{Я}$ of $\mathfrak{D}$.) □

In particular, every contractive $\mathfrak{D} \in TIC^2(U;Y)$ has a conservative realization:

Corollary 11.7.3 *Every contractive $\mathfrak{D} \in TIC^2(U;Y)$ (where U and Y are Hilbert spaces) has a (scattering) conservative completely nonunitary realization. This realization is unique up to a unitary similarity transformation in the state space.*

Proof The two realizations in Theorem 11.7.1 and Corollary 11.7.2 are conservative completely nonunitary realizations of $\mathfrak{D}$. That any two conservative completely nonunitary realizations are unitarily similar to each other was proved in Corollary 11.6.4. □

In particular the realization in Corollary 11.7.2 is unitarily similar to the one in Theorem 11.7.1.

Corollary 11.7.4 *Let U and Y be Hilbert spaces, and let $\mathfrak{D} \in TIC^2(U;Y)$ be (scattering) semi-lossless (cf. Definition 11.3.1). Let $X := \mathcal{N}(\pi_+ D^*) \subset L^2(\mathbb{R}^+;Y)$, let $\pi_X = 1 - \mathfrak{D}\pi_+\mathfrak{D}^*$ be the orthogonal projection in $L^2(\mathbb{R}^+;Y)$ onto X, and define*

$$\mathfrak{A} = \tau_{+|X}, \quad \mathfrak{B} = \pi_+\mathfrak{D}\pi_-, \quad \mathfrak{C} = 1_{|X}.$$

Then $\left[\begin{smallmatrix} \mathfrak{A} & \mathfrak{B} \\ \mathfrak{C} & \mathfrak{D} \end{smallmatrix}\right]$ is a (semi-lossless) conservative realization of $\mathfrak{D}$ which is strongly stable and exactly observable in infinite time.

Proof This follows from Corollary 11.7.2 since $1 - \mathfrak{D}^*\mathfrak{D} = 0$ in this case. The strong stability of $\mathfrak{A}$ is obvious (τ^+ is strongly stable on $L^2(\mathbb{R}^+;Y)$), and the exact observability follows from Theorem 11.3.4. □

Corollary 11.7.5 *Let U and Y be Hilbert spaces, and let $\mathfrak{D} \in TIC^2(U;Y)$ be (scattering) co-lossless. Let $X := \mathcal{N}(\pi_- D) \subset L^2(\mathbb{R}^-;U)$, let $\pi_X = 1 - \mathfrak{D}^*\pi_-\mathfrak{D}$ be the orthogonal projection in $L^2(\mathbb{R}^-;U)$ onto X, and define*

$$\mathfrak{A} = \pi_X\tau_{-|X}, \quad \mathfrak{B} = \pi_X, \quad \mathfrak{C} = \pi_+\mathfrak{D}\pi_{-|X}.$$

Then $\left[\begin{smallmatrix} \mathfrak{A} & \mathfrak{B} \\ \mathfrak{C} & \mathfrak{D} \end{smallmatrix}\right]$ is a (co-lossless) conservative realization of $\mathfrak{D}$ which is strongly co-stable (i.e., the causal dual Σ^d is strongly stable) and exactly controllable in infinite time.

This follows from Corollary 11.7.4 by duality.

11.8 Energy preserving and passive realizations

In the last section we proved the existence of *conservative completely non-unitary* realizations of a given contractive $\mathfrak{D} \in TIC^2(U;Y)$. In general, this realization will not be *minimal*, unless $\mathfrak{D}$ happens to be *lossless* (see part (iii) of Theorem 11.8.1 below). In many cases it is natural to relax the requirement that the realization should be *conservative*, and allow it to be just *passive*. Then we can instead require it to be minimal.

The situation is especially simple if $\mathfrak{D}$ is semi-lossless or co-lossless (or even lossless).

Theorem 11.8.1 *Let U and Y be Hilbert spaces.*

(i) *Every semi-lossless $\mathfrak{D} \in TIC^2(U;Y)$ has a minimal (scattering) passive realization. This realization is unique up to a unitary similarity* *transformation in the state space, and it is strongly stable, output normalized, and energy preserving.*

(ii) *Every co-lossless $\mathfrak{D} \in TIC^2(U;Y)$ has a minimal (scattering) passive realization. This realization is unique up to a unitary similarity transformation in the state space, and it is strongly co-stable (i.e., the dual system is strongly stable), input normalized, and co-energy preserving.*

(iii) *Every lossless $\mathfrak{D} \in TIC^2(U;Y)$ has a minimal (scattering) passive realization. This realization is unique up to a unitary similarity transformation in the state space, and it is strongly stable, strongly co-stable, input and output normalized, and conservative.*

Proof (i) Let $\mathfrak{D}$ be semi-lossless. Then the conservative realization in Corollary 11.7.4 is observable, but it need not be controllable. If we restrict the system to the reachable subspace as described in Theorem 9.1.9(i), then the resulting system need no longer be conservative, but it is still passive (even energy preserving). This proves the existence of a minimal passive realization of $\mathfrak{D}$. That it is strongly stable, energy preserving, output normalized, and unique up to a unitary similarity transformation in the state space follows from Theorem 11.3.3.

(ii) This follows from (i) by duality.

(iii) This follows from (i) and (ii). □

The idea used in the preceding proof can also be used in a slightly more general context to produce controllable energy preserving or observable co-energy preserving realizations of an arbitrary contractive input/output map $\mathfrak{D}$.

Theorem 11.8.2 *Let U and Y by Hilbert spaces, and let $\mathfrak{D} \in TIC^2(U;U)$.*

(i) *$\mathfrak{D}$ has a controllable energy preserving realization. This realization is unique up to a unitary similarity transformation in the state space.*

(ii) *$\mathfrak{D}$ has an observable co-energy preserving realization. This realization is also unique up to a unitary similarity transformation.*

Proof (i) We proceed in the same way as we did in the proof of part (i) of Theorem 11.8.1, but start with an arbitrary conservative completely nonunitary realization of $\mathfrak{D}$ (see Theorem 11.7.1). This realization need not be controllable or observable. However, if we restrict this system to the reachable subspace as described in Theorem 9.1.9(i), then the resulting system is controllable and energy preserving. Thus, $\mathfrak{D}$ has at least one controllable energy preserving realization.

To prove the uniqueness of this realization, suppose that we have two controllable energy preserving realizations $\left[\begin{array}{c|c} \mathfrak{A}_1 & \mathfrak{B}_1 \\ \hline \mathfrak{C}_1 & \mathfrak{D} \end{array}\right]$ and $\left[\begin{array}{c|c} \mathfrak{A}_2 & \mathfrak{B}_2 \\ \hline \mathfrak{C}_2 & \mathfrak{D} \end{array}\right]$ of $\mathfrak{D}$ on (Y, X_1, U), respectively (Y, X_2, U). Then, by Theorem 11.2.9, both $\left[\begin{smallmatrix}\pi_-\mathfrak{D}\\ \mathfrak{B}_1\end{smallmatrix}\right]$ and $\left[\begin{smallmatrix}\pi_-\mathfrak{D}\\ \mathfrak{B}_2\end{smallmatrix}\right]$ are isometric. Thus, for all $u \in L^2(\mathbb{R}^-;U)$,

$$\|\mathfrak{D}u\|^2_{L^2(\mathbb{R}^-;U)} + |\mathfrak{B}_1u|^2_{X_1} = \|\mathfrak{D}u\|^2_{L^2(\mathbb{R}^-;U)} + |\mathfrak{B}_2u|^2_{X_2},$$

or equivalently, $|\mathfrak{B}_1u|^2_{X_1} = |\mathfrak{B}_2u|^2_{X_2}$. By Theorem 9.2.6(iv), Σ_1 and Σ_2 are unitarily similar.

(ii) This follows from (i) by duality. □

We can easily get *shift models* for the two realizations in Theorem 11.8.2 from the universal models in Theorem 11.6.3.

Corollary 11.8.3 *Let U, X, and Y be Hilbert spaces, and let Σ be a controllable (scattering) energy preserving linear system on (Y, X, U).*

(i) *Define $\mathfrak{B}_1 \colon L^2(\mathbb{R}^-;U) \to \left[\begin{smallmatrix}L^2(\mathbb{R};Y)\\ L^2(\mathbb{R}^-;U)\end{smallmatrix}\right]$ by*

$$\mathfrak{B}_1 = \begin{bmatrix}-(1-\mathfrak{D}\mathfrak{D}^*)^{1/2}\pi_-\mathfrak{D}\\ \pi_- - \mathfrak{D}^*\pi_-\mathfrak{D}\end{bmatrix},$$

and denote $X_1 := \overline{\mathcal{R}(\mathfrak{B}_1)} = \mathcal{N}(\mathfrak{B}_1)^\perp$. Let π_{X_1} be the orthogonal projection in $\left[\begin{smallmatrix}L^2(\mathbb{R};Y)\\ L^2(\mathbb{R}^-;U)\end{smallmatrix}\right]$ onto X_1, and define

$$\mathfrak{A}_1 = \pi_{X_1}\begin{bmatrix}\tau & 0\\ 0 & \tau_-\end{bmatrix}_{|X_1},$$

$$\mathfrak{C}_1 = \begin{bmatrix}\pi_+(1-\mathfrak{D}\mathfrak{D}^*)^{1/2} & \pi_+\mathfrak{D}\pi_-\end{bmatrix}_{|X_1}.$$

Then $\Sigma_1 = \left[\begin{smallmatrix}\mathfrak{A}_1 & \mathfrak{B}_1\\ \mathfrak{C}_1 & \mathfrak{D}\end{smallmatrix}\right]$ is a controllable (scattering) energy preserving linear system on (Y, X_1, U) which is unitarily similar to Σ:

$$\mathfrak{A}_1 = E_1\mathfrak{A}E_1^*, \qquad \mathfrak{B}_1 = E_1\mathfrak{B}, \qquad \mathfrak{C}_1 = \mathfrak{C}E_1^*,$$

where $E_1 = (\mathfrak{B}_1^*)^{-1}\mathfrak{B}^*$ is a unitary map of X onto X_1.

(ii) *Define $\mathfrak{B}_2 \colon L^2(\mathbb{R}^-;U) \to \left[\begin{smallmatrix}L^2(\mathbb{R}^+;Y)\\ L^2(\mathbb{R};U)\end{smallmatrix}\right]$ by*

$$\mathfrak{B}_2 = \begin{bmatrix}\pi_+\mathfrak{D}\pi_-\\ (1-\mathfrak{D}^*\mathfrak{D})^{1/2}\pi_-\end{bmatrix},$$

and denote $X_2 := \overline{\mathcal{R}(\mathfrak{B}_2)} = \mathcal{N}(\mathfrak{B}_2)^\perp$. Let π_{X_2} be the orthogonal projection in $\left[\begin{smallmatrix}L^2(\mathbb{R}^+;Y)\\ L^2(\mathbb{R};U)\end{smallmatrix}\right]$ onto X_2, and define

$$\mathfrak{A}_2 = \begin{bmatrix}\tau_+ & 0\\ 0 & \tau\end{bmatrix}_{|X_2},$$

$$\mathfrak{C}_2 = \begin{bmatrix}1-\mathfrak{D}\pi_+\mathfrak{D}^* & -\mathfrak{D}\pi_+(1-\mathfrak{D}^*\mathfrak{D})^{1/2}\end{bmatrix}_{|X_2}.$$

Then $\Sigma_2 = \left[\begin{smallmatrix} \mathfrak{A}_2 & \mathfrak{B}_2 \\ \mathfrak{C}_2 & \mathfrak{D} \end{smallmatrix}\right]$ *is another controllable (scattering) energy preserving linear system on* (Y, X_2, U) *which is unitarily similar to* Σ*:*

$$\mathfrak{A}_2 = E_2 \mathfrak{A} E_2^*, \qquad \mathfrak{B}_2 = E_2 \mathfrak{B}, \qquad \mathfrak{C}_2 = \mathfrak{C} E_2^*,$$

where $E_2 = (\mathfrak{B}_2^*)^{-1}\mathfrak{B}^*$ *is a unitary map of* X *onto* X_2*.*

Proof We get the model in (i) from Theorem 11.6.3(i) by restricting that system to the reachable subspace. The model in (ii) is obtained in the same way from Theorem 11.6.3(ii). As we saw in the proof of Theorem 11.8.2, this leads to controllable energy preserving systems. That these two systems are unitarily similar to the original one follows from Theorem 11.8.2. Let us denote the unitary similarity operator mapping X into X_1 by E_1. Then $\mathfrak{B}_1 = E_1\mathfrak{B}$, and $\mathfrak{B}^* = \mathfrak{B}_1^*(E_1^{-1})^* = \mathfrak{B}_1^* E$. In particular, $\mathcal{R}\left(\mathfrak{B}_1^*\right) = \mathcal{R}\left(\mathfrak{B}^*\right)$. The controllability of Σ_1 implies that $\mathfrak{B}_1^*$ is injective, and so $E_1 = (\mathfrak{B}_1^*)^{-1}\mathfrak{B}^*$. For the same reason the unitary similarity operator mapping X into X_2 is given by $E_2 = (\mathfrak{B}_2^*)^{-1}\mathfrak{B}^*$. □

Corollary 11.8.4 *Let* U*,* X*, and* Y *be Hilbert spaces, and let* Σ *be an observable (scattering) co-energy preserving linear system on* (Y, X, U)*.*

(i) *Let* X_1 *be the orthogonal complement to the null space of the operator* $\left[\pi_+(1-\mathfrak{D}\mathfrak{D}^*)^{1/2} \;\; \pi_+\mathfrak{D}\pi_-\right] : \left[\begin{smallmatrix} L^2(\mathbb{R};Y) \\ L^2(\mathbb{R}^-;U) \end{smallmatrix}\right] \to L^2(\mathbb{R}^+;Y)$*. Let* π_{X_1} *be the orthogonal projection in* $\left[\begin{smallmatrix} L^2(\mathbb{R};Y) \\ L^2(\mathbb{R}^-;U) \end{smallmatrix}\right]$ *onto* X_1*, and define*

$$\mathfrak{A}_1 = \pi_{X_1}\begin{bmatrix} \tau & 0 \\ 0 & \tau_- \end{bmatrix}_{|X_1}, \quad \mathfrak{B}_1 = \pi_{X_1}\begin{bmatrix} -(1-\mathfrak{D}\mathfrak{D}^*)^{1/2}\pi_-\mathfrak{D} \\ \pi_- - \mathfrak{D}^*\pi_-\mathfrak{D} \end{bmatrix},$$
$$\mathfrak{C}_1 = \left[\pi_+(1-\mathfrak{D}\mathfrak{D}^*)^{1/2} \quad \pi_+\mathfrak{D}\pi_-\right]_{|X_1}.$$

Then $\Sigma_1 = \left[\begin{smallmatrix} \mathfrak{A}_1 & \mathfrak{B}_1 \\ \mathfrak{C}_1 & \mathfrak{D} \end{smallmatrix}\right]$ *is an observable (scattering) co-energy preserving linear system on* (Y, X_1, U) *which is unitarily similar to* Σ*:*

$$\mathfrak{A}_1 = E_1 \mathfrak{A} E_1^*, \qquad \mathfrak{B}_1 = E_1 \mathfrak{B}, \qquad \mathfrak{C}_1 = \mathfrak{C} E_1^*,$$

where $E_1 = \mathfrak{C}_1^{-1}\mathfrak{C}$ *is a unitary map of* X *onto* X_1*.*

(ii) *Let* X_2 *be the orthogonal complement to the null space of the operator* $\left[1-\mathfrak{D}\pi_+\mathfrak{D}^* \;\; -\mathfrak{D}\pi_+(1-\mathfrak{D}^*\mathfrak{D})^{1/2}\right] : \left[\begin{smallmatrix} L^2(\mathbb{R};Y) \\ L^2(\mathbb{R}^-;U) \end{smallmatrix}\right] \to L^2(\mathbb{R}^+;Y)$*. Let* π_{X_2}

be the orthogonal projection in $\left[\begin{smallmatrix} L^2(\mathbb{R};Y) \\ L^2(\mathbb{R}^-;U) \end{smallmatrix}\right]$ *onto* X_2*, and define*

$$\mathfrak{A}_2 = \pi_{X_2} \begin{bmatrix} \tau_+ & 0 \\ 0 & \tau \end{bmatrix}_{|X_2}, \quad \mathfrak{B}_2 = \pi_{X_2} \begin{bmatrix} \pi_+ \mathfrak{D} \pi_- \\ (1 - \mathfrak{D}^* \mathfrak{D})^{1/2} \pi_- \end{bmatrix},$$
$$\mathfrak{C}_2 = \begin{bmatrix} 1 - \mathfrak{D}\pi_+\mathfrak{D}^* & -\mathfrak{D}\pi_+(1 - \mathfrak{D}^*\mathfrak{D})^{1/2} \end{bmatrix}_{|X_2}.$$

Then $\Sigma_2 = \left[\begin{array}{c|c} \mathfrak{A}_2 & \mathfrak{B}_2 \\ \hline \mathfrak{C}_2 & \mathfrak{D} \end{array}\right]$ *is another observable (scattering) co-energy preserving linear system on* (Y, X_2, U) *which is unitarily similar to* Σ*:*

$$\mathfrak{A}_2 = E_2 \mathfrak{A} E_2^*, \qquad \mathfrak{B}_2 = E_2 \mathfrak{B}, \qquad \mathfrak{C}_2 = \mathfrak{C} E_2^*,$$

where $E_2 = \mathfrak{C}_2^{-1}\mathfrak{C}$ *is a unitary map of* X *onto* X_2*.*

Proof This follows from Corollary 11.8.3 by duality. □

From the realization in Theorem 11.8.2(i) we can obtain a passive realization by projecting the state space onto the orthogonal complement of the unobservable subspace (as in Theorem 9.1.9(ii)). Another way to get a passive realization is to restrict the realization in Theorem 11.8.2(ii) to the reachable subspace (as in Theorem 9.1.9(i)). Unfortunately, the two realizations obtained this way are *not* unitarily similar to each other (only pseudo-similar; see Theorem 9.2.4).

To get a more explicit description of the two special passive realizations described above we first observe a few facts.

Lemma 11.8.5 *Let* $\Sigma_1 = \left[\begin{array}{c|c} \mathfrak{A}_1 & \mathfrak{B}_1 \\ \hline \mathfrak{C}_1 & \mathfrak{D} \end{array}\right]$ *and* $\Sigma_2 = \left[\begin{array}{c|c} \mathfrak{A}_2 & \mathfrak{B}_2 \\ \hline \mathfrak{C}_2 & \mathfrak{D} \end{array}\right]$ *be two* L^2*-well-posed linear systems on the Hilbert spaces* (Y, X_1, U)*, respectively* (Y, X_2, U) *(with the same transfer function* $\mathfrak{D}$*) which are both input/state and state/output bounded.*

(i) *If* Σ_1 *is controllable and* $\mathfrak{B}_1^*\mathfrak{B}_1 \geq \mathfrak{B}_2^*\mathfrak{B}_2$*, then* $\mathfrak{C}_1\mathfrak{C}_1^* \leq \mathfrak{C}_2\mathfrak{C}_2^*$*.*
(ii) *If* Σ_2 *is observable and* $\mathfrak{C}_1\mathfrak{C}_1^* \leq \mathfrak{C}_2\mathfrak{C}_2^*$*, then* $\mathfrak{B}_1^*\mathfrak{B}_1 \geq \mathfrak{B}_2^*\mathfrak{B}_2$*.*

Note that the condition $\mathfrak{B}_1^*\mathfrak{B}_1 \geq \mathfrak{B}_2^*\mathfrak{B}_2$ is equivalent to the requirement that $|\mathfrak{B}_1 u|_{X_1} \geq |\mathfrak{B}_2 u|_{X_2}$ for every $u \in L^2(\mathbb{R}^-; U)$. To see this, observe that $|\mathfrak{B}_i u|^2 = \langle \mathfrak{B}_i u, \mathfrak{B}_i u\rangle_{X_i} = \langle u, \mathfrak{B}_i^*\mathfrak{B}_i u\rangle_{L^2(\mathbb{R}^-;U)}$ for $i = 1, 2$. Analogously, $\mathfrak{C}_1\mathfrak{C}_1^* \leq \mathfrak{C}_2\mathfrak{C}_2^*$ if and only if $|\mathfrak{C}_1^* y|_{X_1} \leq |\mathfrak{C}_2^* y|_{X_2}$ for every $y \in L^2(\mathbb{R}^+; Y)$.

Proof of Lemma 11.8.5 (i) Recall that $\mathfrak{C}_1\mathfrak{B}_1 = \pi_+\mathfrak{D}\pi_- = \mathfrak{C}_2\mathfrak{B}_2$. This means that for all $u \in L^2(\mathbb{R}^-; U)$ and $y \in L^2(\mathbb{R}^+; Y)$,

$$\begin{aligned} \langle \mathfrak{B}_1 u, \mathfrak{C}_1^* y\rangle_{X_1} &= \langle \mathfrak{C}_1\mathfrak{B}_1 u, y\rangle_{L^2(\mathbb{R}^+;Y)} \\ &= \langle \mathfrak{C}_2\mathfrak{B}_2 u, y\rangle_{L^2(\mathbb{R}^+;Y)} = \langle \mathfrak{B}_2 u, \mathfrak{C}_2^* y\rangle_{X_2}. \end{aligned}$$

By the controllability assumption on Σ_1, $\mathcal{R}(\mathfrak{B}_1)$ is dense in X_1, and so

$$\begin{aligned}\|\mathfrak{C}_1^* y\|_{X_1} &= \sup_{x\in X_1,|x|\le 1} |\langle x, \mathfrak{C}_1^* y\rangle_{X_1}| &&= \sup_{\substack{u\in L^2(\mathbb{R}^-;U)\\ \|\mathfrak{B}_1 u\|\le 1}} |\langle \mathfrak{B}_1 u, \mathfrak{C}_1^* y\rangle_{X_1}| \\ &= \sup_{\substack{u\in L^2(\mathbb{R}^-;U)\\ \|\mathfrak{B}_1 u\|\le 1}} |\langle \mathfrak{B}_2 u, \mathfrak{C}_2^* y\rangle_{X_2}| &&\le \sup_{\substack{u\in L^2(\mathbb{R}^-;U)\\ \|\mathfrak{B}_2 u\|\le 1}} |\langle \mathfrak{B}_2 u, \mathfrak{C}_2^* y\rangle_{X_2}| \\ &\le \sup_{x\in X_2,|x|\le 1} |\langle x, \mathfrak{C}_2^* y\rangle_{X_2}| &&= \|\mathfrak{C}_2^* y\|_{X_2}.\end{aligned}$$

(ii) This follows from (i) by duality. □

Lemma 11.8.6 *Let* $\Sigma_1 = \left[\begin{smallmatrix}\mathfrak{A}_1 & \mathfrak{B}_1\\ \mathfrak{C}_1 & \mathfrak{D}\end{smallmatrix}\right]$ *and* $\Sigma_2 = \left[\begin{smallmatrix}\mathfrak{A}_2 & \mathfrak{B}_2\\ \mathfrak{C}_2 & \mathfrak{D}\end{smallmatrix}\right]$ *be two (scattering) passive systems on the Hilbert spaces* (Y, X_1, U)*, respectively* (Y, X_2, U) *(with the same transfer function* $\mathfrak{D}$*).*

(i) *If* Σ_1 *is energy preserving, then* $\mathfrak{B}_1^*\mathfrak{B}_1 \ge \mathfrak{B}_2^*\mathfrak{B}_2$*. If, furthermore,* Σ_1 *is controllable, then* $\mathfrak{C}_1\mathfrak{C}_1^* \le \mathfrak{C}_2\mathfrak{C}_2^*$*.*
(ii) *If* Σ_1 *is co-energy preserving, then* $\mathfrak{C}_1\mathfrak{C}_1^* \ge \mathfrak{C}_2\mathfrak{C}_2^*$*. If, furthermore,* Σ_1 *is observable, then* $\mathfrak{B}_1^*\mathfrak{B}_1 \le \mathfrak{B}_2^*\mathfrak{B}_2$*.*

Proof (i) Since Σ_1 is energy preserving and Σ_2 is passive, we have for every $u \in L^2(\mathbb{R}^-; U)$ (see Theorems 11.1.6 and 11.2.9),

$$\|\mathfrak{D}u\|^2_{L^2(\mathbb{R}^-;Y)} + |\mathfrak{B}_1 u|^2_{X_1} = \|u\|^2_{L^2(\mathbb{R}^-;U)} \ge \|\mathfrak{D}u\|^2_{L^2(\mathbb{R}^-;U)} + |\mathfrak{B}_2 u|^2_{X_2},$$

or equivalently, $|\mathfrak{B}_1 u|^2_X \ge |\mathfrak{B}_2 u|^2_X$. The second half of (i) follows from Lemma 11.8.5(i).

(ii) This follows from (i) by duality. □

From this result we can derive the following characterization of controllable energy preserving and observable co-energy preserving systems.

Corollary 11.8.7 *Let* U*, and* Y *be Hilbert spaces, and let* $\mathfrak{D} \in TIC^2(U; Y)$*.*

(i) *A controllable (scattering) passive realization* $\Sigma = \left[\begin{smallmatrix}\mathfrak{A} & \mathfrak{B}\\ \mathfrak{C} & \mathfrak{D}\end{smallmatrix}\right]$ *of* $\mathfrak{D}$ *is energy preserving if and only if* $\mathfrak{B}^*\mathfrak{B} \ge \mathfrak{B}_1^*\mathfrak{B}_1$ *for every passive realization* $\Sigma_1 = \left[\begin{smallmatrix}\mathfrak{A}_1 & \mathfrak{B}_1\\ \mathfrak{C}_1 & \mathfrak{D}\end{smallmatrix}\right]$ *of* $\mathfrak{D}$*, or equivalently, for every controllable passive realization* $\Sigma_1 = \left[\begin{smallmatrix}\mathfrak{A}_1 & \mathfrak{B}_1\\ \mathfrak{C}_1 & \mathfrak{D}\end{smallmatrix}\right]$ *of* $\mathfrak{D}$*.*
(ii) *An observable (scattering) passive realization* $\Sigma = \left[\begin{smallmatrix}\mathfrak{A} & \mathfrak{B}\\ \mathfrak{C} & \mathfrak{D}\end{smallmatrix}\right]$ *of* $\mathfrak{D}$ *is co-energy preserving if and only if* $\mathfrak{C}\mathfrak{C}^* \ge \mathfrak{C}_1\mathfrak{C}_1^*$ *for every passive realization* $\Sigma_1 = \left[\begin{smallmatrix}\mathfrak{A}_1 & \mathfrak{B}_1\\ \mathfrak{C}_1 & \mathfrak{D}\end{smallmatrix}\right]$ *of* $\mathfrak{D}$*, or equivalently, for every observable passive realization* $\Sigma_1 = \left[\begin{smallmatrix}\mathfrak{A}_1 & \mathfrak{B}_1\\ \mathfrak{C}_1 & \mathfrak{D}\end{smallmatrix}\right]$ *of* $\mathfrak{D}$*.*

Proof The necessity (of both the given conditions) for Σ to be energy preserving follows from Lemma 11.8.6. Conversely, suppose that the weaker of the two conditions holds, i.e., suppose that $\mathfrak{B}^*\mathfrak{B} \ge \mathfrak{B}_1^*\mathfrak{B}_1$ for every controllable passive realization $\Sigma_1 = \left[\begin{smallmatrix} \mathfrak{A}_1 & \mathfrak{B}_1 \\ \mathfrak{C}_1 & \mathfrak{D} \end{smallmatrix}\right]$. If we take Σ_1 to be the controllable energy preserving realization in Theorem 11.8.2(i), then by Lemma 11.8.6(i), the opposite inequality is also true, i.e., $\mathfrak{B}^*\mathfrak{B} = \mathfrak{B}_1^*\mathfrak{B}_1$. By Theorem 9.2.6, Σ is unitarily similar to Σ_1, hence energy preserving.

(ii) This follows from (i) by duality. □

Another consequence of Lemma 11.8.6 is the following.

Theorem 11.8.8 *Let* $\Sigma = \left[\begin{smallmatrix} \mathfrak{A} & \mathfrak{B} \\ \mathfrak{C} & \mathfrak{D} \end{smallmatrix}\right]$ *be a (scattering) passive system on* (Y, X, U).

(i) *The following conditions are equivalent:*
   (a) Σ *is controllable, and* $\mathfrak{B}^*\mathfrak{B} \le \mathfrak{B}_1^*\mathfrak{B}_1$ *for every passive realization* $\Sigma_1 = \left[\begin{smallmatrix} \mathfrak{A}_1 & \mathfrak{B}_1 \\ \mathfrak{C}_1 & \mathfrak{D} \end{smallmatrix}\right]$ *of* $\mathfrak{D}$, *or equivalently, for every minimal passive realization* $\Sigma_1 = \left[\begin{smallmatrix} \mathfrak{A}_1 & \mathfrak{B}_1 \\ \mathfrak{C}_1 & \mathfrak{D} \end{smallmatrix}\right]$ *of* $\mathfrak{D}$.
   (b) Σ *is observable, and* $\mathfrak{C}\mathfrak{C}^* \ge \mathfrak{C}_1\mathfrak{C}_1^*$ *for every controllable passive realization* $\Sigma_1 = \left[\begin{smallmatrix} \mathfrak{A}_1 & \mathfrak{B}_1 \\ \mathfrak{C}_1 & \mathfrak{D} \end{smallmatrix}\right]$ *of* $\mathfrak{D}$, *or equivalently, for every minimal passive realization* $\Sigma_1 = \left[\begin{smallmatrix} \mathfrak{A}_1 & \mathfrak{B}_1 \\ \mathfrak{C}_1 & \mathfrak{D} \end{smallmatrix}\right]$ *of* $\mathfrak{D}$.
   (c) Σ *is unitarily similar to the realization of* $\mathfrak{D}$ *that we get from an observable co-energy preserving realization by restricting it to the reachable subspace (as in Theorem 9.1.9(i)).*

   In particular, a realization with these properties always exists, it is minimal, and it is unique up to a unitary similarity transformation in the state space.

(ii) *The following conditions are equivalent:*
   (a) Σ *is observable, and* $\mathfrak{C}\mathfrak{C}^* \le \mathfrak{C}_1\mathfrak{C}_1^*$ *for every passive realization* $\Sigma_1 = \left[\begin{smallmatrix} \mathfrak{A}_1 & \mathfrak{B}_1 \\ \mathfrak{C}_1 & \mathfrak{D} \end{smallmatrix}\right]$ *of* $\mathfrak{D}$, *or equivalently, for every minimal passive realization* $\Sigma_1 = \left[\begin{smallmatrix} \mathfrak{A}_1 & \mathfrak{B}_1 \\ \mathfrak{C}_1 & \mathfrak{D} \end{smallmatrix}\right]$ *of* $\mathfrak{D}$.
   (b) Σ *is controllable, and* $\mathfrak{B}^*\mathfrak{B} \ge \mathfrak{B}_1^*\mathfrak{B}_1$ *for every observable passive realization* $\Sigma_1 = \left[\begin{smallmatrix} \mathfrak{A}_1 & \mathfrak{B}_1 \\ \mathfrak{C}_1 & \mathfrak{D} \end{smallmatrix}\right]$ *of* $\mathfrak{D}$, *or equivalently, for every minimal passive realization* $\Sigma_1 = \left[\begin{smallmatrix} \mathfrak{A}_1 & \mathfrak{B}_1 \\ \mathfrak{C}_1 & \mathfrak{D} \end{smallmatrix}\right]$ *of* $\mathfrak{D}$.
   (c) Σ *is unitarily similar to the realization of* $\mathfrak{D}$ *that we get from a controllable energy preserving realization by projecting the system onto the orthogonal complement of the unobservable subspace (as in Theorem 9.1.9(ii)).*

In particular, a realization with these properties always exists, it is minimal, and it is unique up to a unitary similarity transformation in the state space.

Proof Parts (i) and (ii) are dual to each other, so it suffices to prove part (i).

(c) $\Rightarrow$ (a): By Lemma 11.8.6(i), every system Σ which is observable and co-energy preserving (but not necessarily controllable) has the property that $\mathfrak{B}^*\mathfrak{B} \le \mathfrak{B}_1^*\mathfrak{B}_1$ for every passive realization $\Sigma_1 = \left[\begin{smallmatrix} \mathfrak{A}_1 & \mathfrak{B}_1 \\ \mathfrak{C}_1 & \mathfrak{D} \end{smallmatrix}\right]$ of $\mathfrak{D}$. The input map $\mathfrak{B}$ does not change when we restrict this system to the reachable subspace, so the realization described in (c) still has the same property. This property is invariant under unitary similarity transforms in the state space, so every system which is unitarily similar to the system described in (c) satisfies (a).

(a) $\Rightarrow$ (c): Assume that the weaker of the two conditions listed in (a) holds (the one where Σ_1 is minimal). Choose Σ_1 to be the (minimal) system described in (c). We know by now that (c) implies (a), so we have at the same time both the original $\mathfrak{B}^*\mathfrak{B} \le \mathfrak{B}_1^*\mathfrak{B}_1$ and the reverse inequality $\mathfrak{B}_1^*\mathfrak{B}_1 \le \mathfrak{B}^*\mathfrak{B}$. Thus, $\mathfrak{B}^*\mathfrak{B} = \mathfrak{B}_1^*\mathfrak{B}_1$. By Theorem 9.2.6, Σ is unitarily similar to Σ_1.

(a) $\Rightarrow$ (b): Assume (a). As we have seen, (a) implies (c), so Σ must be observable. That $\mathfrak{C}\mathfrak{C}^* \ge \mathfrak{C}_1\mathfrak{C}_1^*$ for every controllable passive realization $\Sigma_1 = \left[\begin{smallmatrix} \mathfrak{A}_1 & \mathfrak{B}_1 \\ \mathfrak{C}_1 & \mathfrak{D} \end{smallmatrix}\right]$ of $\mathfrak{D}$ follows from Lemma 11.8.5(i). Thus (a) implies (b).

(b) $\Rightarrow$ (c): Assume that the weaker of the two conditions listed in (b) holds (the one where Σ_1 is minimal). Choose Σ_1 to be the (minimal) system described in (c). We know by now that (c) implies (a) which implies (b), so we have at the same time both the original $\mathfrak{C}\mathfrak{C}^* \ge \mathfrak{C}_1\mathfrak{C}_1^*$ and the reverse inequality $\mathfrak{C}_1\mathfrak{C}_1^* \ge \mathfrak{C}\mathfrak{C}^*$. Thus, $\mathfrak{C}\mathfrak{C}^* = \mathfrak{C}_1\mathfrak{C}_1^*$. By Theorem 9.2.5, Σ is unitarily similar to Σ_1. □

Notice that it is not possible to remove the word 'controllable' in condition (b) in part (i), since that would force Σ to be co-energy preserving (see Corollary 11.8.7(ii)). Likewise, it is not possible to remove the word "observable" in condition (b) in part (ii), since that would force Σ to be energy preserving.

Definition 11.8.9 A system satisfying the equivalent conditions (a)–(c) in part (i) of Theorem 11.8.8 is called *optimal*. A system satisfying the equivalent conditions (a)–(c) in part (ii) of Theorem 11.8.8 is called $*$-*optimal*.

It is obvious from Theorem 11.8.8 that Σ is optimal if and only if its causal dual Σ^d is $*$-optimal.

Theorem 11.8.8 implies that the optimal and $*$-optimal realizations of a given input/output map are the two minimal passive realizations whose state space norms are the smallest, respectively, largest possible, i.e., $\mathfrak{B}\mathfrak{B}^*$ is either as small as possible or as large as possible (among all passive realizations). The following alternative characterization has an interesting physical interpretation (see Remark 11.8.11 below):

Theorem 11.8.10 *Let* $\Sigma = \left[\begin{smallmatrix} \mathfrak{A} & \mathfrak{B} \\ \mathfrak{C} & \mathfrak{D} \end{smallmatrix}\right]$ *be a minimal (scattering) passive system on* (Y, X, U).

(i) *The following two conditions are equivalent:*
  (a) Σ *is optimal (i.e., conditions (a)–(c) in part (i) of Theorem 11.8.8 hold);*
  (b) *for all* $x \in X$,

$$|x|_X^2 = \sup_{u \in L^2(\mathbb{R}^+;U)} \left(\|\mathfrak{C}x + \mathfrak{D}\pi_+ u\|^2_{L^2(\mathbb{R}^+;U)} - \|u\|^2_{L^2(\mathbb{R}^+;U)} \right). \qquad (11.8.1)$$

Thus, if Σ *is optimal, then* X *is the completion of the state space* $\underline{X}$ *of the input normalized realization* $\underline{\Sigma}$ *induced by* Σ *(see Proposition 9.5.2) with respect to the norm* (11.8.1).

(ii) *The following two conditions are equivalent:*
  (a) Σ *is* $*$*-optimal (i.e., conditions (a)–(c) in part (ii) of Theorem 11.8.8 hold).*
  (b) *the norm of every* $x \in X$ *which is of the form* $x = \mathfrak{B}u$ *for some* $u \in L^2(\mathbb{R}^-; U)$ *is given by*

$$|x|_X^2 = \inf_{v \in L^2(\mathbb{R}^-;U),\, \mathfrak{B}v = x} \left(\|v\|^2_{L^2(\mathbb{R}^-;U)} - \|\mathfrak{D}v\|^2_{L^2(\mathbb{R}^-;U)} \right). \qquad (11.8.2)$$

Thus, if Σ *is* $*$*-optimal, then* X *is the completion of the state space* $\underline{X}$ *of the input normalized realization* $\underline{\Sigma}$ *induced by* Σ *(see Proposition 9.5.2) with respect to the norm* (11.8.2).

Proof (i) Assume that Σ is optimal. By Theorem 9.1.9 and part (i)(c) of Theorem 11.8.8, we may assume that X is the reachable subspace of an observable co-energy preserving realization $\widetilde{\Sigma} = \left[\begin{smallmatrix} \widetilde{\mathfrak{A}} & \mathfrak{B} \\ \widetilde{\mathfrak{C}} & \mathfrak{D} \end{smallmatrix}\right]$ of $\mathfrak{D}$, with $\mathfrak{A} = \widetilde{\mathfrak{A}}_{|X}$ and $\mathfrak{C} = \widetilde{\mathfrak{C}}_{|X}$. Clearly, if we can show that (11.8.1) holds for the system $\widetilde{\Sigma}$, then it will hold for Σ, too. To simplify the notation we write Σ, $\mathfrak{A}$, and $\mathfrak{C}$ instead of $\widetilde{\Sigma}$, $\widetilde{\mathfrak{A}}$, and $\widetilde{\mathfrak{C}}$, and suppose that Σ is observable and co-energy preserving (instead of being optimal).

We first show that

$$|x|_X^2 \le \sup_{u \in L^2(\mathbb{R}^+;U)} \left(\|\mathfrak{C}x + \mathfrak{D}\pi_+ u\|^2_{L^2(\mathbb{R}^+;U)} - \|u\|^2_{L^2(\mathbb{R}^+;U)} \right) \qquad (11.8.3)$$

whenever x is of the form $x = \mathfrak{C}^* y^*$ for some $y^* \in L^2(\mathbb{R}^+; Y)$. By Theorem 11.2.9(ii), $\pi := \left[\begin{smallmatrix} \mathfrak{C}^* \\ \pi_+ \mathfrak{D}^* \end{smallmatrix}\right] \left[\begin{smallmatrix} \mathfrak{C} & \mathfrak{D}\pi_+ \end{smallmatrix}\right]$ is a self-adjoint projection in $\left[\begin{smallmatrix} X \\ L^2(\mathbb{R}^+;U) \end{smallmatrix}\right]$. Let $y^* \in L^2(\mathbb{R}^+; Y)$, and define $\left[\begin{smallmatrix} x \\ u \end{smallmatrix}\right] = \left[\begin{smallmatrix} \mathfrak{C}^* \\ \pi_+ \mathfrak{D}^* \end{smallmatrix}\right] y^*$. Then $\left[\begin{smallmatrix} x \\ u \end{smallmatrix}\right]$ belongs to the range

π, and hence $\pi\left[\begin{smallmatrix} x \\ u \end{smallmatrix}\right] = \left[\begin{smallmatrix} x \\ u \end{smallmatrix}\right]$. Explicitly, this identity says that

$$\mathfrak{C}^*(\mathfrak{C}x + \mathfrak{D}\pi_+ u) = x, \qquad \pi_+\mathfrak{D}^*(\mathfrak{C}x + \mathfrak{D}\pi_+ u) = u.$$

From these two identities we conclude that

$$\begin{aligned} &\|\mathfrak{C}x + \mathfrak{D}\pi_+ u\|^2_{L^2(\mathbb{R}^+;U)} \\ &= \langle \mathfrak{C}x + \mathfrak{D}\pi_+ u, \mathfrak{C}x + \mathfrak{D}\pi_+ u\rangle_{L^2(\mathbb{R}^+;U)} \\ &= \langle x, \mathfrak{C}^*(\mathfrak{C}x + \mathfrak{D}\pi_+ u)\rangle_X + \langle u, \pi_+\mathfrak{D}^*(\mathfrak{C}x + \mathfrak{D}\pi_+ u)\rangle_{L^2(\mathbb{R}^+;U)} \\ &= |x|^2_X + |u|_{L^2(\mathbb{R}^+;U)}. \end{aligned}$$

Thus, (11.8.3) holds for all $x \in \mathcal{R}(\mathfrak{C}^*)$. The range of $\mathfrak{C}^*$ is dense in X (since Σ is observable), and therefore (11.8.3) must hold for all $x \in X$.

The opposite inequality is a trivial consequence of the first inequality in (11.1.12) (and it is true for all *passive* systems Σ).

Conversely, suppose that (11.8.1) holds. Let $\Sigma_1 = \left[\begin{array}{c|c} \mathfrak{A}_1 & \mathfrak{B}_1 \\ \hline \mathfrak{C}_1 & \mathfrak{D} \end{array}\right]$ of $\mathfrak{D}$ be a minimal passive realization of $\mathfrak{D}$ with state space X_1. As we noticed in the preceding paragraph, for all $x_1 \in X_1$,

$$|x_1|^2_{X_1} \geq \sup_{u \in L^2(\mathbb{R}^+;U)} \left(\|\mathfrak{C}_1 x_1 + \mathfrak{D}\pi_+ u\|^2_{L^2(\mathbb{R}^+;U)} - \|u\|^2_{L^2(\mathbb{R}^+;U)}\right),$$

whereas for all $x \in X$,

$$|x|^2_X = \sup_{u \in L^2(\mathbb{R}^+;U)} \left(\|\mathfrak{C}x + \mathfrak{D}\pi_+ u\|^2_{L^2(\mathbb{R}^+;U)} - \|u\|^2_{L^2(\mathbb{R}^+;U)}\right).$$

Let E be the maximally defined pseudo-similarity between Σ_1 and Σ in Theorem 9.2.4. Then, with the notation introduced above, $x = Ex_1$ if and only if $\mathfrak{C}x = \mathfrak{C}_1 x_1$. By the preceding argument, if $\mathfrak{C}x = \mathfrak{C}_1 x_1$, then $|x|^2_X \leq |x_1|^2_{X_1}$. Thus, E is a contraction, defined on all of X, and $\mathfrak{C}_1\mathfrak{C}_1^* = \mathfrak{C}EE^*\mathfrak{C}^* \leq \mathfrak{C}\mathfrak{C}^*$. By part (i)(b) of Theorem 11.8.8, Σ is optimal.

(ii) Assume that Σ is $*$-optimal. By Theorem 9.1.9 and part (ii)(c) of Theorem 11.8.8, we may assume that X is the orthogonal complement to the unobservable subspace $\mathcal{N}(\widetilde{\mathfrak{C}})$ of a controllable energy preserving realization $\widetilde{\Sigma} = \left[\begin{array}{c|c} \widetilde{\mathfrak{A}} & \widetilde{\mathfrak{B}} \\ \hline \widetilde{\mathfrak{C}} & \mathfrak{D} \end{array}\right]$ of $\mathfrak{D}$, and that $\mathfrak{A} = \pi_X \widetilde{\mathfrak{A}}_{|X}$, $\mathfrak{B} = \pi_X \widetilde{\mathfrak{B}}$, and $\mathfrak{C} = \widetilde{\mathfrak{C}}_{|X}$, where π_X is the orthogonal projection of the state space $\widetilde{X}$ onto X. If $u \in L^2(\mathbb{R}^-;U)$ and if we denote $\tilde{x} = \widetilde{\mathfrak{B}}u$, then (see the second identity in (11.2.13))

$$|\tilde{x}|^2_{\widetilde{X}} = |\widetilde{\mathfrak{B}}u|^2_{\widetilde{X}} = \|u\|^2_{L^2(\mathbb{R}^-;U)} - \|\mathfrak{D}u\|^2_{L^2(\mathbb{R}^-;U)}.$$

Let $x = \pi_X \tilde{x}$ be the corresponding state of Σ. Then $|x|^2_X = \inf_{z \in \widetilde{X}, \widetilde{\mathfrak{C}}z = \widetilde{\mathfrak{C}}\tilde{x}} |z|^2_{\widetilde{X}}$. Replace z by $\widetilde{\mathfrak{B}}v$, and let v vary over $L^2(\mathbb{R}^-;U)$ to get

$$|x|^2_X = \inf_{v \in L^2(\mathbb{R}^-;U), \widetilde{\mathfrak{C}}\widetilde{\mathfrak{B}}v = \widetilde{\mathfrak{C}}\tilde{x}} \left(\|v\|^2_{L^2(\mathbb{R}^-;U)} - \|\mathfrak{D}v\|^2_{L^2(\mathbb{R}^-;U)}\right).$$

Here $\widetilde{\mathfrak{C}}\widetilde{\mathfrak{B}}v = \pi_+\mathfrak{D}\pi_-v = \mathfrak{C}\mathfrak{B}v$, and $\widetilde{\mathfrak{C}}\tilde{x} = \widetilde{\mathfrak{C}}\pi_X\tilde{x} = \mathfrak{C}x$, so the side condition $\widetilde{\mathfrak{C}}\widetilde{\mathfrak{B}}v = \widetilde{\mathfrak{C}}\tilde{x}$ is equivalent to $\mathfrak{C}\mathfrak{B}v = \mathfrak{C}x$. But $\mathfrak{C}$ is injective (since Σ is observable), so this is equivalent to $x = \mathfrak{B}u$, and we arrive at (11.8.2).

Conversely, suppose that (11.8.2) holds. For every passive minimal realization $\Sigma_1 = \left[\begin{smallmatrix} \mathfrak{A}_1 & \mathfrak{B}_1 \\ \mathfrak{C}_1 & \mathfrak{D} \end{smallmatrix}\right]$ of $\mathfrak{D}$ with state space X_1 and for every $u \in L^2(\mathbb{R}^-; U)$ we have $|x_1|^2_{X_1} \le \|u\|^2_{L^2(\mathbb{R}^-;U)} - \|\mathfrak{D}u\|^2_{L^2(\mathbb{R}^-;U)}$ where $x_1 = \mathfrak{B}_1 u$ (see the second inequality in (11.1.12)). Take the infimum of the right-hand side over all $v \in L^2(\mathbb{R}^-; U)$ with $\mathfrak{B}_1 v = x_1$ to get

$$|x_1|^2_{X_1} \le \inf_{v \in L^2(\mathbb{R}^-;U),\, \mathfrak{B}_1 v = x_1} \left(\|v\|^2_{L^2(\mathbb{R}^-;U)} - \|\mathfrak{D}v\|^2_{L^2(\mathbb{R}^-;U)}\right).$$

On the other hand, by (11.8.2),

$$|x|^2_X = \inf_{v \in L^2(\mathbb{R}^-;U),\, \mathfrak{B}v = x} \left(\|v\|^2_{L^2(\mathbb{R}^-;U)} - \|\mathfrak{D}v\|^2_{L^2(\mathbb{R}^-;U)}\right).$$

where $x = \mathfrak{B}u$. Let E be a pseudo-similarity between Σ and Σ_1 (as defined in Definition 9.2.1). Then, with the notation introduced above, $Ex = E\mathfrak{B}u = \mathfrak{B}_1 u = x_1$ and $\mathfrak{B}v = x$ if and only if $\mathfrak{B}_1 v = E\mathfrak{B}v = Ex = x_1$. It follows from the above computation that $|x_1|^2_{X_1} = |Ex|^2_{X_1} \le |x|_X$. Thus, E is a contraction, defined on all of X, and $\mathfrak{B}_1^*\mathfrak{B}_1 = \mathfrak{B}^*E^*E\mathfrak{B} \le \mathfrak{B}^*\mathfrak{B}$. By part (ii)(b) of Theorem 11.8.8, Σ is $*$-optimal. □

Remark 11.8.11 Both (11.8.1) and (11.8.2) have obvious physical interpretations. Formula (11.8.1) says that the square of the norm of the state of an optimal system coincides with the maximal amount of energy that can be withdrawn from the system. For this reason the square of this norm is called the *available storage*. Formula (11.8.2) says that the square of the norm of the state of a $*$-optimal system coincides with the minimal amount of external energy which is needed to reach this particular state. For this reason the square of this norm is called the *required supply*. Note that in order to compute these two norms it is not necessary to start with a system that is optimal or $*$-optimal in the first place. As we saw in the proof of Theorem 11.8.10, given an arbitrary minimal passive system $\Sigma = \left[\begin{smallmatrix} \mathfrak{A} & \mathfrak{B} \\ \mathfrak{C} & \mathfrak{D} \end{smallmatrix}\right]$, the right-hand side of (11.8.1) is finite for every $x \in X$, and the right-hand side of (11.8.2) is finite for every $x \in \mathcal{R}(\mathfrak{B})$. Thus, the square root of the right-hand side of (11.8.1) is a norm on X, and we get an optimal realization of $\mathfrak{D}$ by completing X with respect to this norm. Analogously, the square root of the right-hand side of (11.8.1) is a norm on $\underline{X}$, and we get a $*$-optimal realization of $\mathfrak{D}$ by completing $\underline{X}$ with respect to this norm. (See also Theorem 11.8.12 below.)

As our following result shows, *optimal and $*$-optimal* systems play the same role for minimal *passive* realizations as the *input and output normalized* do for *input and output bounded* realizations (compare this to Theorem 9.4.7(i)–(ii) and Proposition 9.5.2(i)–(ii)).

Theorem 11.8.12 *Let* $\Sigma = \left[\begin{smallmatrix} \mathfrak{A} & \mathfrak{B} \\ \mathfrak{C} & \mathfrak{D} \end{smallmatrix}\right]$ *be a minimal (scattering) passive system on* (Y, X, U)*. Then it is possible to make* Σ $*$*-optimal by restricting it to a subspace* $\underline{X}$ *of* X *(and strengthening the norm), and it is possible to make it optimal by completing* X *into a larger space* $\overline{X}$ *(and weakening the norm). More precisely, the following claims are true.*

(i) *X has a unique Hilbert subspace $\underline{X}$ with the following properties. The embedding $\underline{X} \subset X$ is continuous and dense, $\underline{X}$ is invariant under $\mathfrak{A}$, $\mathcal{R}(\mathfrak{B}) \subset \underline{X}$, and if we define*

$$\underline{\mathfrak{A}} = \mathfrak{A}_{|\underline{X}}, \qquad \underline{\mathfrak{C}} = \mathfrak{C}_{|\underline{X}},$$

then $\underline{\Sigma} = \left[\begin{smallmatrix} \underline{\mathfrak{A}} & \mathfrak{B} \\ \underline{\mathfrak{C}} & \mathfrak{D} \end{smallmatrix}\right]$ is a $*$-optimal realization of $\mathfrak{D}$ (with state space $\underline{X}$). The embedding of $\underline{X} \subset X$ is contractive, i.e., $|x|_X \le |x|_{\underline{X}}$ for every $x \in \underline{X}$.

(ii) *$\underline{X}$ has a unique Hilbert subspace $\underline{\underline{X}}$, with a contractive and dense embedding, such that the system $\underline{\underline{\Sigma}}$ which we get by restricting Σ to $\underline{\underline{X}}$ (i.e., we repeat the construction in (i) with $\underline{X}$ replaced by $\underline{\underline{X}}$) is input normalized.*

(iii) *X can be completed into a unique Hilbert space $\overline{X}$ with the following properties. The embedding $X \subset \overline{X}$ is dense and continuous, $\mathfrak{A}$ can be extended to a C_0 semigroup $\overline{\mathfrak{A}}$ on $\overline{X}$, $\mathfrak{C}$ can be extended to an operator $\overline{\mathfrak{C}} \in \mathcal{B}(\overline{X}; L^2(\mathbb{R}^+; Y))$, and $\overline{\Sigma} = \left[\begin{smallmatrix} \overline{\mathfrak{A}} & \mathfrak{B} \\ \overline{\mathfrak{C}} & \mathfrak{D} \end{smallmatrix}\right]$ is an optimal realization of $\mathfrak{D}$ (with state space $\overline{X}$). The embedding of $X \subset \overline{X}$ is contractive, i.e., $|x|_{\overline{X}} \le |x|_X$ for every $x \in X$.*

(iv) *$\overline{X}$ can be further completed into a unique Hilbert space $\overline{\overline{X}}$, such that the embedding of $\overline{X}$ in $\overline{\overline{X}}$ is (dense and) contractive, and such that the corresponding system $\overline{\overline{\Sigma}}$ which we get by repeating the construction in (iii) with $\overline{X}$ replaced by $\overline{\overline{X}}$ is output normalized.*

Proof (i) Let $\Sigma_1 = \left[\begin{smallmatrix} \mathfrak{A}_1 & \mathfrak{B}_1 \\ \mathfrak{C}_1 & \mathfrak{D} \end{smallmatrix}\right]$ of $\mathfrak{D}$ be an arbitrary $*$-optimal realization of $\mathfrak{D}$ with state space X_1. Then both Σ and Σ_1 are minimal realizations of $\mathfrak{D}$, so they are pseudo-similar in the sense of Theorem 9.2.4. It follows from property (a) in Theorem 11.8.8(i) and Theorem 9.2.6(i) that the pseudo-similarity operator $E \colon X \to X_1$ is a contraction (hence bounded with $\mathcal{D}(E) = X$). We let $\underline{X} = \mathcal{R}(E)$, and equip it with the range norm: $|Ex_1|_{\underline{X}} = |x_1|_{X_1}$. Then E maps X_1 unitarily onto $\underline{X}$. The intertwining condition $\mathfrak{A}^t E = E\mathfrak{A}_1^t$ implies that $\underline{X}$ is invariant under $\mathfrak{A}$, and $\mathcal{R}(\mathfrak{B}) \subset \underline{X}$ since $\mathfrak{B} = E\mathfrak{B}_1$. Clearly, the system $\underline{\Sigma}$

described in (i) is unitarily similar to Σ_1, hence $*$-optimal. The embedding is contractive since E is contractive.

(ii) We get the space $\underline{\underline{X}}$ by using the construction in part (i) of Theorem 9.4.7. By Proposition 9.5.2, the system $\underline{\underline{\Sigma}}$ is input normalized. To see that the embedding is contractive it suffices to observe that for all $u \in L^2(\mathbb{R}^-; U)$ (since $\underline{\Sigma}$ is passive and $\underline{\underline{\Sigma}}$ is input normalized),

$$|\mathfrak{B}u|^2_{\underline{X}} \le |\mathfrak{B}u|^2_{\underline{X}} + \|\mathfrak{D}u\|^2_{L^2(\mathbb{R}^-;Y)} \le \|u\|^2_{L^2(\mathbb{R}^-;U)} = |\mathfrak{B}u|^2_{\underline{\underline{X}}},$$

and to use Theorem 9.2.6(i).

(iii) This proof is similar to the proof of (i), and we leave it to the reader (either use duality, or imitate the proof given above).

(iv) Use the construction in Theorem 9.4.7(ii) (or use (ii) and duality). □

Once we have the two realizations $\underline{\Sigma}$ and $\overline{\Sigma}$ described in Theorems 11.8.12 and 11.8.10 we can interpolate between them to get a *passive balanced* minimal realization of $\mathfrak{D}$. The method is exactly the same as that which we used to interpolate between the input normalized realization and the output normalized realization in Theorem 9.5.6 (these are the realizations that we have denoted by $\underline{\underline{\Sigma}}$, respectively $\overline{\overline{\Sigma}}$, in Theorem 11.8.12).

Definition 11.8.13 Let $\Sigma = \left[\begin{smallmatrix}\mathfrak{A} & \mathfrak{B} \\ \mathfrak{C} & \mathfrak{D}\end{smallmatrix}\right]$ be an L^2-well-posed linear system on the Hilbert spaces (Y, X, U). This system is *(scattering) passive balanced* if it is minimal, passive, and it has the following property: if $\underline{X}$ and $\overline{X}$ are the state spaces of the corresponding $*$-optimal and optimal systems described in parts (i) and (iii) of Theorem 11.8.12, then the Gram operator of the embedding $\underline{X} \subset X$ is the restriction to $\underline{X}$ of the Gram operator of the embedding $X \subset \overline{X}$ (this property is discussed in more detail in Lemma 9.5.7).

Theorem 11.8.14 *Every contractive $\mathfrak{D} \in TIC^2(U; Y)$ (where U and Y are Hilbert spaces) has a (scattering) passive balanced realization. This realization is unique up to a unitary similarity transformation in the state space.*

Proof We know from Theorem 11.8.8(i) (see also Definition 11.8.9) that $\mathfrak{D}$ has a $*$-optimal realization $\underline{\Sigma}$. Let $\overline{\Sigma}$ be the corresponding optimal system described in Theorem 11.8.12(iii). Let X be the subspace of $\overline{X}$ in Lemma 9.5.7(i), and let $\Sigma = \left[\begin{smallmatrix}\mathfrak{A} & \mathfrak{B} \\ \mathfrak{C} & \mathfrak{D}\end{smallmatrix}\right]$ be the system described in Lemma 9.5.7(ii). Then the Gram operator of the embedding $\underline{X} \subset X$ is the restriction to $\underline{X}$ of the Gram operator of the embedding $X \subset \overline{X}$, and Σ is a minimal L^2-well-posed linear system on (Y, X, U).

To show that Σ is passive we argue as follows. Denote the Lax–Phillips semigroups induced by $\underline{\Sigma}$, Σ, and $\overline{\Sigma}$ by $\underline{\mathfrak{T}}$, $\mathfrak{T}$, and $\overline{\mathfrak{T}}$, respectively. By Lemma 11.1.4, $\underline{\mathfrak{T}}$ and $\overline{\mathfrak{T}}$ are contraction semigroups on their respective state spaces on $\begin{bmatrix} L^2(\mathbb{R}^-;Y) \\ \underline{X} \\ L^2(\mathbb{R}^+;U) \end{bmatrix}$ and $\begin{bmatrix} L^2(\mathbb{R}^-;Y) \\ \overline{X} \\ L^2(\mathbb{R}^+;U) \end{bmatrix}$. It is easy to see that the state space $\begin{bmatrix} L^2(\mathbb{R}^-;Y) \\ X \\ L^2(\mathbb{R}^+;U) \end{bmatrix}$ of $\mathfrak{T}$ interpolates half-way between the two other state spaces (as described in Lemma 9.5.8), and it follows from Lemma 9.5.8 that also $\mathfrak{T}$ is a contraction semigroup. Therefore, by Lemma 11.1.4, Σ is passive.

It remains to prove uniqueness. To do this we take a closer look at the construction in Lemma 9.5.7. Suppose that we have two different balanced passive realizations $\Sigma_1 = \left[\begin{smallmatrix} \mathfrak{A}_1 & \mathfrak{B}_1 \\ \mathfrak{C}_1 & \mathfrak{D} \end{smallmatrix}\right]$ and $\Sigma_2 = \left[\begin{smallmatrix} \mathfrak{A}_2 & \mathfrak{B}_2 \\ \mathfrak{C}_2 & \mathfrak{D} \end{smallmatrix}\right]$ of $\mathfrak{D}$. Denote the corresponding optimal and $*$-optimal systems in Theorem 11.8.12(i)–(ii) by $\underline{\Sigma}_1 = \left[\begin{smallmatrix} \underline{\mathfrak{A}}_1 & \underline{\mathfrak{B}}_1 \\ \underline{\mathfrak{C}}_1 & \mathfrak{D} \end{smallmatrix}\right]$, $\overline{\Sigma}_1 = \left[\begin{smallmatrix} \overline{\mathfrak{A}}_1 & \overline{\mathfrak{B}}_1 \\ \overline{\mathfrak{C}}_1 & \mathfrak{D} \end{smallmatrix}\right]$, $\underline{\Sigma}_2 = \left[\begin{smallmatrix} \underline{\mathfrak{A}}_2 & \underline{\mathfrak{B}}_2 \\ \underline{\mathfrak{C}}_2 & \mathfrak{D} \end{smallmatrix}\right]$, and $\overline{\Sigma}_2 = \left[\begin{smallmatrix} \overline{\mathfrak{A}}_2 & \overline{\mathfrak{B}}_2 \\ \overline{\mathfrak{C}}_2 & \mathfrak{D} \end{smallmatrix}\right]$, with state spaces $\underline{X}_1$, $\overline{X}_1$, $\underline{X}_2$, and $\overline{X}_2$, respectively. By Theorem 11.8.8, $\underline{\Sigma}_1$ and $\underline{\Sigma}_2$ are unitarily similar, and so are $\overline{\Sigma}_1$ and $\overline{\Sigma}_2$. Denote the unitary similarity transformation of $\overline{X}_1$ onto $\overline{X}_2$ by $\overline{W}$. Then $\overline{W} = \overline{\mathfrak{C}}_2^{-1}\overline{\mathfrak{C}}_1$ (see (9.2.3)). Analogously, the unitary similarity transformation of $\underline{X}_1$ onto $\underline{X}_2$ is given by $\underline{W} = \underline{\mathfrak{C}}_2^{-1}\underline{\mathfrak{C}}_1$. In particular, $\underline{W} = \overline{W}_{|\underline{X}}$. Let F_1 and F_2 be the Gram operators of the embeddings $\underline{X}_1 \subset \overline{X}_1$ respectively $\underline{X}_2 \subset \overline{X}_2$. Then, by Lemma 9.5.7(iii), $F_1 = \underline{\mathfrak{C}}_1'(\overline{\mathfrak{C}}_1'')^{-1}$ and $F_2 = \underline{\mathfrak{C}}_2'(\overline{\mathfrak{C}}_2'')^{-1}$ (where we have used the notation in that lemma). Here $\overline{\mathfrak{C}}_1 = \overline{\mathfrak{C}}_2\overline{W}$ and $\underline{\mathfrak{C}}_1 = \underline{\mathfrak{C}}_2\underline{W}$, so $\overline{\mathfrak{C}}_1'' = \overline{W}''\overline{\mathfrak{C}}_2'' = \overline{W}^{-1}\overline{\mathfrak{C}}_2''$ and $\underline{\mathfrak{C}}_1' = \underline{W}'\underline{\mathfrak{C}}_2' = \underline{W}^{-1}\underline{\mathfrak{C}}_2' = \overline{W}^{-1}\underline{\mathfrak{C}}_2'$ (since $\underline{W}^{-1} = \overline{W}^{-1}_{|\underline{X}_2}$). Thus,

$$F_1 = \underline{\mathfrak{C}}_1'(\overline{\mathfrak{C}}_1'')^{-1} = \overline{W}^{-1}\underline{\mathfrak{C}}_2'(\overline{\mathfrak{C}}_2'')^{-1}\overline{W} = \overline{W}^{-1}F_2\overline{W}.$$

In particular, F_1 and F_2 are unitarily similar, hence so are their square roots: $\sqrt{F_1} = \overline{W}^{-1}\sqrt{F_2}\overline{W}$. By Lemma 9.5.7(iii),

$$\mathfrak{C}_1\mathfrak{C}_1^* = \mathfrak{C}_1\sqrt{F_1}\overline{\mathfrak{C}}_1'' = \mathfrak{C}_1\overline{W}^{-1}\sqrt{F_2}\overline{W}\mathfrak{C}_1'' = \mathfrak{C}_2\sqrt{F_2}\overline{\mathfrak{C}}_2'' = \mathfrak{C}_2\mathfrak{C}_2^*.$$

This together with Theorem 9.2.5 implies that Σ_1 and Σ_2 are unitarily similar. (Arguing as above one finds that the unitary similarity transform is the restriction of $\overline{W}$ to X_1.) □

As we saw in the preceding proof, the operator $\mathfrak{C}\mathfrak{C}^*$ does not depend on the particular realization, as long as it is balanced passive. It is natural to suggest that this operator can be computed in terms of the two operators $\mathfrak{C}\mathfrak{C}''$ (which is the corresponding unique operator for an optimal realization) and $\mathfrak{C}\mathfrak{C}'$ (which is the corresponding unique operator for a $*$-optimal realization). Indeed, this can be done. One way to do this is to choose the realization $\overline{\overline{\Sigma}}$ in Theorem 11.8.12(iv) to be the restricted exactly observable shift realization, whose state space is a closed subspace of $L^2(\mathbb{R}^+;Y)$, namely $\overline{\mathcal{R}(\pi_+\mathfrak{D}\pi_-)}$. Denote this space by $\overline{\overline{X}}$. Then the semigroups of the different systems $\overline{\Sigma}$, Σ, and $\underline{\Sigma}$ are the left-shifts on the respective state spaces $\overline{X}$, X, and $\underline{X}$, the observation operators are the inclusion maps of these subspaces into $\overline{\mathcal{R}(\pi_+\mathfrak{D}\pi_-)}$, and the input maps are the Hankel operator $\pi_+\mathfrak{D}\pi_-$. The Gram operators of the embeddings of $\overline{X}$, X, and $\underline{X}$ are given by $\overline{Q} := \mathfrak{C}\mathfrak{C}''$, $Q := \mathfrak{C}\mathfrak{C}^*$, and $\underline{Q} := \mathfrak{C}\mathfrak{C}'$. By continuing the computation given at the end of the proof of Theorem 11.8.14 one finds that $Q = \mathfrak{C}\mathfrak{C}^*$ is given by

$$\begin{aligned} Q &= \overline{Q}^{1/2}\Big[\overline{Q}^{-1/2}\underline{Q}\,\overline{Q}^{-1/2}\Big]^{1/2}\overline{Q}^{1/2} \\ &= \underline{Q}^{1/2}\Big[\underline{Q}^{1/2}\overline{Q}^{-1}\underline{Q}^{1/2}\Big]^{-1/2}\underline{Q}^{1/2}. \end{aligned}$$

In these formulas $\underline{Q}^{1/2}\overline{Q}^{-1/2}$ can be interpreted as the extension to $L^2(\mathbb{R}^+; Y)$ of a densely defined contraction which vanishes on $\mathcal{N}(\pi_-\mathfrak{D}\pi_+)$, whose adjoint is the (everywhere defined) contraction $\overline{Q}^{-1/2}\underline{Q}^{1/2}$. A similar formula can be obtained for the uniquely determined operator $\mathfrak{B}^*\mathfrak{B}$ in terms of $\mathfrak{B}'\mathfrak{B}$ and $\mathfrak{B}''\mathfrak{B}$ (for example by duality, or by choosing the system $\underline{\underline{\Sigma}}$ in Theorem 11.8.12(iv) to be the restricted exactly controllable shift realization).

11.9 The spectrum of a conservative system

The spectrum of the main operator of a conservative system has the following basic properties.

Theorem 11.9.1 *Let $S = \left[\begin{smallmatrix} A\&B \\ C\&D \end{smallmatrix}\right]$ be a conservative system node on (Y, X, U), with main operator A, control operator B, observation operator C, and transfer function $\widehat{\mathfrak{D}}$.*

(i) *$\alpha \in \rho(A)$ if and only if $\left[\begin{smallmatrix} \alpha & 0 \\ 0 & 0 \end{smallmatrix}\right] + S^*$ is invertible, in which case*

$$\left(\begin{bmatrix} \alpha & 0 \\ 0 & 0 \end{bmatrix} + S^*\right)^{-1} = \begin{bmatrix} (\alpha - A)^{-1} & (\alpha - A_{|X})^{-1}B \\ C(\alpha - A)^{-1} & \widehat{\mathfrak{D}}(\alpha) \end{bmatrix}. \tag{11.9.1}$$

In particular, this formula holds for all $\alpha \in \mathbb{C}^+$.

(ii) *$\alpha \in \rho(A) \cap \mathbb{C}^-$ if and only if $\widehat{\mathfrak{D}}^*(-\overline{\alpha})$ is invertible, in which case*

$$\begin{aligned} \begin{bmatrix} (\alpha - A)^{-1} & (\alpha - A_{|X})^{-1}B \\ C(\alpha - A)^{-1} & \widehat{\mathfrak{D}}(\alpha) \end{bmatrix} &= \begin{bmatrix} (\alpha + A^*)^{-1} & 0 \\ 0 & 0 \end{bmatrix} \\ &+ \begin{bmatrix} -(\alpha + A^*_{|X})^{-1}C^* \\ 1 \end{bmatrix} [\widehat{\mathfrak{D}}^*(-\overline{\alpha})]^{-1} \begin{bmatrix} -B^*(\alpha + A^*)^{-1} & 1 \end{bmatrix}. \end{aligned} \tag{11.9.2}$$

In particular, U and Y are isomorphic (hence they have the same dimension) whenever $\rho(A) \cap \mathbb{C}^- \neq \emptyset$.

Proof By Lemma 11.2.4, S is time-flow-invertible, and the time-flow-inverted system node $S^{\mathfrak{R}}_{\times}$ coincides with S^*. Theorem 11.9.1 now follows from Lemma

6.5.7(ii),(iv) with the $S^{\mathfrak{A}}_{\times}$ replaced by S^*. (We have simplified some of the statements of that lemma by using the fact that A and A^* are maximal dissipative, hence $\mathbb{C}^+ \subset \rho(A)$ and $\mathbb{C}^+ \subset \rho(A^*)$.) □

According to Corollary 11.7.3, every contractive $\mathfrak{D} \in TIC^2(U;Y)$ has a conservative realization (which is unique up to a unitary similarity transformation in the state space). This result is of limited value due to the following fact: without any further conditions on $\mathfrak{D}$, the spectrum of the main operator A is likely to contain the whole left half-plane, or at least the whole imaginary axis, even if we do not count the trivial case where U and Y have different dimension (cf. Theorem 11.9.1(ii)). This can be seen from the universal model in Theorem 11.6.2, but it can be seen even more directly from the following simple facts:

Corollary 11.9.2 *Let $S = \left[\begin{smallmatrix} A\&B \\ C\&D \end{smallmatrix}\right]$ be a conservative system node on (Y, X, U) with main operator A and transfer function $\widehat{\mathfrak{D}}$.*

(i) *$\widehat{\mathfrak{D}}(\alpha)$ is unitary for every $\alpha \in \rho(A) \cap j\mathbb{R}$.*
(ii) *If the measure of $\sigma(A) \cap j\mathbb{R}$ is zero, then S is lossless. (This is, in particular, true if $\sigma(A)$ is countable.)*

Proof (i) Apply part (xii) of Theorem 11.2.5 both to S and S^*.

(ii) By (i), $\widehat{\mathfrak{D}}(\beta)$ is unitary for almost all $\beta \in j\mathbb{R}$. This implies that $\mathfrak{D}$ is lossless (apply Proposition 11.3.2 both to $\mathfrak{D}$ and its causal dual $\widehat{\mathfrak{D}}^d = \boldsymbol{\mathfrak{R}}\mathfrak{D}^*\boldsymbol{\mathfrak{R}}$). □

11.10 Comments

Most of the results of this chapter are in one way or another based on the theory presented in Sz.-Nagy and Foiaş (1970) (with some minor additions and modifications) and on Arov and Nudelman (1996). The main difference (apart from notation) is that we work in continuous time throughout, whereas Sz.-Nagy and Foiaş (1970) mainly work in discrete time, and derive most of the continuous time results from the corresponding discrete time results by means of the Cayley transform (which is discussed in Chapter 12). The same method is used in Arov and Nudelman (1996).

Section 11.1 The main part of this section is found in Arov and Nudelman (1996) (in a less detailed form). Our presentation has been modeled after Malinen *et al.* (2003, Section 2), Staffans (2002a, b), and Staffans and Weiss (2002, Section 7). In particular, Lemma 11.1.4 is essentially Staffans and Weiss (2002, Proposition 7.2) and Theorem 11.1.5 is a slightly expanded version of Malinen *et al.* (2003, Proposition 5.1) (see also Arov and Nudelman (1996, Proposition 4.1 and Theorem 5.2); a slightly less general version of this result is found in Staffans and Weiss (2002, Theorem 7.4)). A discrete time version of Theorem 11.1.9 can be obtained from Sz.-Nagy and Foiaş (1970, Theorem 3.2, p. 9 and Proposition 2.1, p. 188) (and our proof of Theorem 11.1.9 resembles the proofs given in Sz.-Nagy and Foiaş (1970)). In the case where $\mathfrak{A}$ is isometric and $U = Y = 0$ this becomes the well-known Wold decomposition of $\mathfrak{A}$ given in Corollary 11.6.9.

Section 11.2 The main part of this section is also found in Arov and Nudelman (1996) (see, in particular, Arov and Nudelman (1996, Proposition 4.5 and Theorem 5.2)). Our presentation has been modeled after Malinen *et al.* (2003, Section 2), Staffans (2002a, b), and Staffans and Weiss (2003). Lemma 11.2.3 is essentially contained in Staffans and Weiss (2004, Theorem 7.2), Theorems 11.2.5 and 11.2.8 extend Malinen *et al.* (2003, Proposition 3.2), and Lemma 11.2.6 is Malinen *et al.* (2003, Lemma 3.3).

Section 11.3 This section has been modeled after Staffans (2002b, Section 10). Proposition 11.3.2 is originally due to Fourès and Segal (1955). Theorem 11.3.3 is Staffans (2002b, Theorem 10.1), and Theorem 11.3.4 is an expanded version of Staffans (2002b, Theorem 10.2). A discrete time version of Theorem 11.3.4 (without conditions (iv) and (vii)) is essentially contained in Sz.-Nagy and Foiaş (1970, Theorem 2.3, p. 248). Lax and Phillips (1967) assume throughout that the equivalent conditions in Corollary 11.3.7 hold. There also exist other scattering conservative versions of Corollary 11.3.7, for example, those found in Malinen *et al.* (2003), Tucsnak and Weiss (2003, Theorem 1.5 and Proposition 3.4) Weiss *et al.* (2001, Propositions 6 and 9), and Tucsnak and Weiss (2003, Proposition 3.4). Theorem 11.3.9 is due to Tucsnak and Weiss (2003, Proposition 3.3) (which further contains some additional material), and some related results are found in Tucsnak and Weiss (2003, Theorem 1.3, Proposition 3.2) and Weiss *et al.* (2001, Proposition 8). See also Jacob and Zwart (2002).

Section 11.4 The discrete time version of the theory presented in this section is found in, e.g., Sz.-Nagy and Foiaş (1970). For example, Theorem 11.4.5 corresponds to Sz.-Nagy and Foiaş (1970, Theorem 4.1, p. 11), and Theorem 11.4.9 corresponds to Sz.-Nagy and Foiaş (1970, Theorem 4.2, p. 13 and Section I.5, pp. 16–19). Our proofs are also quite similar to those in Sz.-Nagy and Foiaş (1970). We carry out the proofs in continuous time, imitating the discrete time proofs, and with the machinery that we have developed by now they are no more difficult in the continuous time setting than in the discrete time setting. Another option would have been to translate the discrete time results into continuous time by means of the Cayley transform (as is done in Sz.-Nagy and Foiaş (1970)). We have preferred the more direct approach since it gives a clearer picture of the final structure. A historical discussion is given in the notes on Sz.-Nagy and Foiaş (1970, pp. 51–56).

Lemma 11.4.2 is called 'Sarason's lemma on dilation of operators' in Arov and Nudelman (1996, p. 28). Theorem 11.4.14 is due to Malinen *et al.* (2003).

Section 11.5 The results in this section could alternatively have been proved by first proving the corresponding discrete time results, and then using the Cayley transform (presented in Section 12.3). The 'only if' part of Lemma 11.5.2 may be formally new. Theorem 11.5.6 appears as Arov and Nudelman (1996, Proposition 7.7) (where it is proved with the aid of the Cayley transform).

We get a minimal conservative dilation of a given passive system Σ from Theorem 11.5.6 by absorbing the first output and the second input into the semigroup (as in Remark 2.7.5). This dilated conservative system need not always be completely non-unitary. Arov and Nudelman (1996) say Σ has *minimal losses* if it has a *completely nonunitary conservative dilation*. For example, optimal and $*$-optimal systems have minimal losses, whereas the balanced passive realization need not have minimal losses.

Section 11.6 This section can also be seen as a continuous time version of the corresponding discrete time results found in Sz.-Nagy and Foiaş (1970). Corollary 11.6.9 is the same as Sz.-Nagy and Foiaş (1970, Theorem 9.3, p. 151), Theorem 11.6.2 corresponds roughly to Sz.-Nagy and Foiaş (1970, Section II.1), and the rest of the section corresponds roughly to Sz.-Nagy and Foiaş (1970, Section VI.2) and parts of Sz.-Nagy and Foiaş (1970, Section VI.3). In Sz.-Nagy and Foiaş (1970, pp. 279–280) our continuous time model is derived from the discrete time model by means of the Cayley transform, but we have preferred to prove the continuous time case directly by imitating the discrete time proofs (it is no more difficult, and it gives a better intuitive feeling for the final result). Again, for a historical discussion we refer the reader to the notes on Sz.-Nagy and Foiaş (1970, pp. 276–280).

Section 11.7 This is a continuous time version of Sz.-Nagy and Foiaş (1970, Section VI.3). Theorem 11.7.1 corresponds to the main result (Sz.-Nagy and Foiaş, 1970, Theorem 3.1, p. 255) of that section. Corollary 11.7.3 was first published in Arov and Nudelman (1996, Theorem 6.4).

Section 11.8 Most of the results in this section are found in Arov and Nudelman (1996, Section 8) (but the proofs are different). A discrete time version of Theorem 11.8.1(iii) is found in Arov (2002, Theorem 1, p. 106). We do not know if Theorem 11.8.2 and Corollaries 11.8.3 and 11.8.4 have been explicitly published before (as of spring 2004). Arov and Nudelman (1996, Theorems 8.1 and 8.2) define 'optimal' in a slightly different way, i.e., they do not always require an optimal system to be controllable (hence, their optimal system is not unique up to a similarity transformation).

The same notion was studied (independently) in Willems (1972a, b) in a setting which resembles the one in Theorem 11.8.10 and Remark 11.8.11. In particular, this seems to be the source of the terms *available storage* and *required supply*.

If the optimal and $*$-optimal realization of a given input/output map coincide (i.e., they are unitarily equivalent to each other), then it follows from Theorem 11.8.12 that *all minimal passive realizations coincide*. Necessary and sufficient conditions for this to happen are given in Arov and Nudelman (2000). If these two realizations instead are similar to each other (not just pseudo-similar), then *all minimal passive realizations are similar to each other*. Necessary and sufficient conditions for this are given in Arov and Nudelman (2002). By Theorem 11.8.1, if $\mathfrak{D}$ is semi-lossless, then all passive minimal realizations of $\mathfrak{D}$ coincide with the output normalized realization, and if $\mathfrak{D}$ is co-lossless, then all passive minimal realizations of $\mathfrak{D}$ coincide with the input normalized realization.

The strong stability of the semigroup $\mathfrak{A}$ and its adjoint $\mathfrak{A}^*$ has been studied in Arov *et al.* (2004, Section 6) (in discrete time, but the same results carry over to continuous time by means of the Cayley transform).

Theorem 11.8.14 may be new in this form. The balanced passive realization was first introduced in a finite-dimensional setting in Opdenacker and Jonckheere (1988). It has been studied in some detail in Ober (1991, Section 5) under the name *bounded real balanced realization*. The standard approach in the finite-dimensional case is to balance between the minimal and the maximal solution of the Kalman–Yakubovich–Popov inequality. It is possible to use this approach in the infinite-dimensional case, too, but then a more general version of the KYP inequality is needed, where the Riccati operator is allowed to be unbounded and to have an unbounded inverse. See Arov *et al.* (2004) for a fairly complete treatment of the discrete-time version of this inequality.

Section 11.9 Lax and Phillips (1967, Section II.3) contains a more complete discussion of the spectral theory of a *lossless* conservative system than what we give here. The lossless version of Theorem 11.9.1(ii) is found in Lax and Phillips (1967, Theorem 3.2(i)).

12

Discrete time systems

In this chapter we give a short overview of discrete time systems. Especially in the L^2-well-posed case on a triple of Hilbert spaces there is a fairly close connection between the discrete-time and continuous-time theory. The crucial transformation between these two settings is the Cayley transform (which we have already encountered implicitly in Chapter 11) and its time-domain version, the Laguerre transform. In the last section we return to continuous time, using the discrete time theory to develop the continuous time reciprocal transformation.

12.1 Discrete time systems

Throughout this book we have been working with systems whose time variable is continuous (as opposed to discrete). Below we shall develop an analogous theory for systems in discrete time. We have already encountered such systems in passing in Section 2.4, where we discretized the time to be able to carry out some of the proofs in discrete time. In the next section we shall see another method to pass between continuous and discrete time.

As in the case of a continuous time system, it is possible to define a discrete time system by specifying either its 'generating operators,' or the input/state/output maps. In the continuous time case the latter approach is simpler because of the unboundedness of the generating operators, but in discrete time the situation is the opposite: it is common to start with the generating operators (which are bounded in this case), and to use these to define the input/state/output maps.

A discrete time system is usually written in difference form

$$\begin{aligned} x_{k+1} &= \mathbf{A}x_k + \mathbf{B}u_k, \\ y_k &= \mathbf{C}x_k + \mathbf{D}u_k, \qquad k \in \mathbb{Z}^+ = \{0, 1, 2, \ldots\}, \end{aligned} \tag{12.1.1}$$

where $\mathbf{A} \in \mathcal{B}(X)$, $\mathbf{B} \in \mathcal{B}(U; X)$, $\mathbf{C} \in \mathcal{B}(X; Y)$, $\mathbf{D} \in \mathcal{B}(U; Y)$ and U, X, and Y are Banach spaces. Here $\mathbf{A}$ is the *main operator*, $\mathbf{B}$ is the *control operator*, $\mathbf{C}$ is the *observation operator*, $\mathbf{D}$ is the *feedthrough operator*. The discrete time semigroup $\mathbb{A}$, input map $\mathbb{B}$, output map $\mathbb{C}$, and input/output map $\mathbb{D}$ are given by

$$
\begin{aligned}
(\mathbb{A}x)_k &= \mathbf{A}^k x, && k \in \mathbb{Z}^+,\\
\mathbb{B}\mathbf{u} &= \sum_{k=0}^{\infty} \mathbf{A}^k \mathbf{B} u_{-k-1}, && \\
(\mathbb{C}x)_k &= \mathbf{C}\mathbf{A}^k x, && k \in \mathbb{Z}^+,\\
(\mathbb{D}\mathbf{u})_k &= \sum_{i=0}^{\infty} \mathbf{C}\mathbf{A}^i \mathbf{B} u_{k-i-1} + \mathbf{D} u_k, && k \in \mathbb{Z} = \{0, \pm 1, \pm 2, \ldots\},
\end{aligned}
\tag{12.1.2}
$$

where $\mathbf{u} = \{u_k\}_{k\in\mathbb{Z}}$ represents a U-valued sequence with finite support and $x \in X$.

To get the *input/output representation* of the system in (12.1.1) we replace the operators $\left[\begin{smallmatrix} \mathbf{A} & \mathbf{B} \\ \mathbf{C} & \mathbf{D} \end{smallmatrix}\right]$ by the operators $\left[\begin{smallmatrix} \mathbb{A} & \mathbb{B} \\ \mathbb{C} & \mathbb{D} \end{smallmatrix}\right]$. This quadruple of operators satisfies a set of conditions similar to those listed in Definition 2.2.1. For each sequence $\mathbf{u} = \{u_k\}$ and each $j \in \mathbb{Z}$ we define

$$
(\pi_-\mathbf{u})_k = \begin{cases} u_k, & k \in \mathbb{Z}^-,\\ 0, & k \in \mathbb{Z}^+, \end{cases} \qquad (\pi_+\mathbf{u})_k = \begin{cases} 0, & k \in \mathbb{Z}^-,\\ u_k, & k \in \mathbb{Z}^+, \end{cases}
$$

$$
(\sigma\mathbf{u})_k = u_{k+1}, \quad k \in \mathbb{Z}, \qquad \mathbf{e}_k^j = \begin{cases} 1, & k = j,\\ 0, & k \neq j. \end{cases}
$$

Here $\mathbb{Z}^- = \{-1, -2, \ldots\}$. In the ℓ^2-case π_- and π_- are complementary orthogonal projections in $\ell^2(\mathbb{Z})$, σ is the (bilateral) left shift in $\ell^2(\mathbb{Z})$, $\sigma_+ = \pi_+\sigma$ is the (unilateral) left-shift on $\ell^2(\mathbb{Z}^+)$, and $\sigma_- = \sigma\pi_-$ is the (unilateral) left-shift semigroup on $\ell^2(\mathbb{Z}^-)$. The vectors $\{\mathbf{e}^j\}_{j\in\mathbb{Z}}$ form an orthonormal basis in $\ell^2(\mathbb{Z})$, and we get orthonormal bases of $\ell^2(\mathbb{Z}^+)$ and $\ell^2(\mathbb{Z}^-)$ by restricting the index set to $\mathbb{Z}^+$, respectively $\mathbb{Z}^-$. Clearly, the operators σ, π_+, and π_- have natural extensions to the case where the sequence $\mathbf{u}$ is U-valued instead of scalar-valued. In the Hilbert space case, if $\{u_\alpha\}_{\alpha\in A}$ is an orthonormal basis in U (for some index set A), then $\{u_\alpha \mathbf{e}^j\}_{\alpha\in A, j\in\mathbb{Z}}$ is an orthonormal basis for $\ell^2(\mathbb{Z}; U)$, $\{u_\alpha \mathbf{e}^j\}_{\alpha\in A, j\in\mathbb{Z}^+}$ is an orthonormal basis for $\ell^2(\mathbb{Z}^+; U)$, and $\{u_\alpha \mathbf{e}^j\}_{\alpha\in A, j\in\mathbb{Z}^-}$ is an orthonormal basis for $\ell^2(\mathbb{Z}^-; U)$.

The operators $\left[\begin{smallmatrix} \mathbb{A} & \mathbb{B} \\ \mathbb{C} & \mathbb{D} \end{smallmatrix}\right]$ arising from a discrete time system on (Y, X, U) are characterized by the fact that they satisfy the following four algebraic conditions:

(i) $\mathbb{A}$ is a discrete time semigroup on $\mathbb{Z}^+$ with $(\mathbb{A})_0 = 1$. We denote the generator of $\mathbb{A}$ by $\mathbf{A} = (\mathbb{A})_1$;
(ii) $\mathbb{B}$ satisfies $\mathbb{B}\boldsymbol{\sigma}_- = \mathbf{A}\mathbb{B}$;
(iii) $\mathbb{C}$ satisfies $\mathbb{C}\mathbf{A} = \boldsymbol{\sigma}_+\mathbb{C}$;
(iv) $\mathbb{D}$ satisfies $\boldsymbol{\sigma}\mathbb{D} = \mathbb{D}\boldsymbol{\sigma}$, $\boldsymbol{\pi}_-\mathbb{D}\boldsymbol{\pi}_+ = 0$, and $\boldsymbol{\pi}_+\mathbb{D}\boldsymbol{\pi}_- = \mathbb{C}\mathbb{B}$.

In particular, $\mathbb{D}$ is again time-invariant and causal, and its Hankel operator $\boldsymbol{\pi}_+\mathbb{D}\boldsymbol{\pi}_-$ factors into $\boldsymbol{\pi}_+\mathbb{D}\boldsymbol{\pi}_- = \mathbb{C}\mathbb{B}$. We call a quadruple of operators $\left[\begin{smallmatrix} \mathbb{A} & \mathbb{B} \\ \mathbb{C} & \mathbb{D} \end{smallmatrix}\right]$ which satisfies (i)–(iv) a *discrete time linear system in input/output form* on (Y, X, U). The corresponding operators $\mathbf{A}$, $\mathbf{B}$, $\mathbf{C}$, and $\mathbf{D}$ can be recovered from $\mathbb{A}$, $\mathbb{B}$, $\mathbb{C}$, and $\mathbb{D}$ through

$$\mathbf{A}x = (\mathbb{A}x)_1, \quad \mathbf{B}u = \mathbb{B}(u\mathbf{e}^{-1}), \quad \mathbf{C}x = (\mathbb{C}x)_0, \quad \mathbf{D}u = (\mathbb{D}(u\mathbf{e}^0))_0. \tag{12.1.3}$$

We shall use the alternative notation $\boldsymbol{\Sigma} = \left[\begin{smallmatrix} \mathbb{A} & \mathbb{B} \\ \mathbb{C} & \mathbb{D} \end{smallmatrix}\right]$ and $\boldsymbol{\Sigma} = \left[\begin{smallmatrix} \mathbf{A} & \mathbf{B} \\ \mathbf{C} & \mathbf{D} \end{smallmatrix}\right]$ for the same system, and let the context determine which set of operators we mean.

A discrete time system is always well-posed, so the discrete time version of most of the theory in Chapters 3–5 is much simpler than the continuous time theory (we shall here study only the case where the discrete time generating operators $\mathbf{A}$, $\mathbf{B}$, $\mathbf{C}$, and $\mathbf{D}$ are bounded). Observe, in particular, that the distinction that we have made between L^∞-well-posed and *Reg*-well-posed systems disappears in discrete time.

We introduce definitions of ℓ^p-boundedness and r-weighted ℓ^p-boundedness by imitating Definition 2.5.6, replacing L^p_ω by ℓ^p_r. Here ℓ^p_r, with $1 \le p \le \infty$ and $r > 0$, stands for the set of sequences $\mathbf{u} = \{u_k\}$ for which the sequence $r^{-k}u_k$ belongs to ℓ^p. We also use the space $\ell^p_{0,r}$, which coincides with ℓ^p_r if $p < \infty$, and which consists of those $\mathbf{u} \in \ell^\infty_r$ which satisfy $\lim_{k\to\infty} r^{-k}u_k = 0$ if $p = \infty$. A discrete time system $\boldsymbol{\Sigma} = \left[\begin{smallmatrix} \mathbb{A} & \mathbb{B} \\ \mathbb{C} & \mathbb{D} \end{smallmatrix}\right]$ is ℓ^p_r-*bounded* (where $1 \le p \le \infty$ and $r > 0$) if all its components are ℓ^p_r-bounded. Here boundedness of $\mathbb{A}$ means that $\sup_{k\in\mathbb{Z}^+} r^{-k}\|A^k\| < \infty$, boundedness of $\mathbb{C}$ means that $\mathbb{C} \in \mathcal{B}(X; l^p_r(\mathbb{Z}^+; Y))$, and boundedness of $\mathbb{B}$ or $\mathbb{D}$ means that these operators can be extended to (bounded) operators $\mathbb{B} \in \mathcal{B}(\ell^p_{0,r}(\mathbb{Z}^-; U); X)$ and

$$\mathbb{D}\colon \ell^p_{0,r}(\mathbb{Z}^-; U) \dotplus \ell^p_r(\mathbb{Z}^+; U) \to \ell^p_{0,r}(\mathbb{Z}^-; Y) \dotplus \ell^p_r(\mathbb{Z}^+; Y).$$

Most of the results in Chapter 2 have natural discrete time analogues. For example, every discrete time system has a well-defined growth bound $r_\mathbf{A}$ equal to the spectral radius of its semigroup generator $\mathbf{A}$, and the analogue of Theorem 2.5.4 holds, i.e., the system is ℓ^p_r-bounded for every $r > r_\mathbf{A}$ and every $p \in [1, \infty]$. The exactly controllable and the exactly observable shift realization presented in Example 2.6.5 have natural (and obvious) extensions to discrete time, and so does the Lax–Phillips model in Section 2.7 (including the wave operators).

There are two different commonly used methods to pass from the time domain to the frequency domain in discrete time. One of them uses the (bilateral) Z-transform, defined by

$$\hat{\mathbf{u}}(z) = \sum_{k\in\mathbb{Z}} z^{-k} u_k$$

if $\mathbf{u} = \{u_k\}$ is defined on $\mathbb{Z}$, and with $\mathbb{Z}$ replaced by $\mathbb{Z}^+$ or $\mathbb{Z}^-$ if $\mathbf{u}$ is defined on this subset of $\mathbb{Z}$. This choice is common in control theory. The other common choice, which is the dominant one in mathematics, is to use the same transform but replace z by $1/z$, i.e., to replace z^{-k} by z^k. In this work we shall use the former of the two alternatives, following the control theory tradition.

The (right-sided) Z-transform of a sequence in $\ell^p_r(\mathbb{Z}^+)$ is analytic *in the exterior* $\mathbb{D}^+_r$ of the disk around the origin with radius r, and the (left-sided) Z-transform of a sequence in $\ell^p_r(\mathbb{Z}^-)$ is analytic *in the interior* $\mathbb{D}^-_r$ of the disk around the origin with radius r. For $p = 1$ and $r = 1$ we get the usual Fourier series, i.e.,

$$\hat{\mathbf{u}}(\mathrm{e}^{j\varphi}) = \sum_{k\in\mathbb{Z}} \mathrm{e}^{-jk\varphi} u_k, \qquad 0 \le \varphi < 2\pi,$$

with uniform pointwise convergence. In the Hilbert space case with $p = 2$ and $r = 1$ we can also use this formula in the L^2-sense, i.e., we have norm-convergence in $L^2(\mathbb{T})$. In both cases the Fourier coefficients u_k of $\mathbf{u}$ can be recovered from the boundary function through the integral

$$u_k = \frac{1}{2\pi}\int_0^{2\pi} \mathrm{e}^{jk\varphi}\hat{\mathbf{u}}(\mathrm{e}^{j\varphi})\,d\varphi = \frac{1}{2\pi j}\oint_{|z|=1} z^{-k}\hat{\mathbf{u}}(z)\,dz/z. \tag{12.1.4}$$

In the Hilbert space case with $r = 1$ and $p = 2$ the Z-transform becomes a unitary operator from $\ell^2(\mathbb{Z}; U)$ to $L^2(\mathbb{T}; U)$ if we divide it by the constant factor $\sqrt{2\pi}$, and it maps $\ell^2(\mathbb{Z}^+; U)$ onto $H^2(\mathbb{D}^+; U)$ and $\ell^2(\mathbb{Z}^-; U)$ onto $H^2(\mathbb{D}^-; U)$. In particular, Parseval's identity is valid in the form

$$\begin{aligned}\sum_{k\in\mathbb{Z}}\langle u_k, v_k\rangle^2_U &= \frac{1}{2\pi}\int_0^{2\pi}\langle\hat{\mathbf{u}}(\mathrm{e}^{j\varphi}), \hat{\mathbf{v}}(\mathrm{e}^{j\varphi})\rangle_U\,d\varphi \\ &= \frac{1}{2\pi j}\oint_{|z|=1} z^{-1}\langle\hat{\mathbf{u}}(z), \hat{\mathbf{v}}(z)\rangle_U\,dz,\end{aligned} \tag{12.1.5}$$

for all $u, v \in \ell^2(\mathbb{Z}; U)$. For later use, let us remark that, by making the change of variable $z = (\overline{\alpha} + j\omega)/(\alpha - j\omega)$ where $\Re\alpha > 0$ we can write the integral above in the alternative forms

$$\begin{aligned}
&\frac{1}{2\pi j}\oint_{|z|=1} z^{-1}\langle \hat{\mathbf{u}}(z), \hat{\mathbf{v}}(z)\rangle\, dz \\
&= \frac{1}{2\pi}\int_{-\infty}^{\infty} \frac{2\Re\alpha}{(\Re\alpha)^2 + (\Im\alpha - \omega)^2}\Big\langle \hat{\mathbf{u}}\Big(\frac{\overline{\alpha} + j\omega}{\alpha - j\omega}\Big), \hat{\mathbf{v}}\Big(\frac{\overline{\alpha} + j\omega}{\alpha - j\omega}\Big)\Big\rangle\, d\omega \\
&= \frac{1}{2\pi}\int_{-\infty}^{\infty}\Big\langle \frac{\sqrt{2\Re\alpha}}{\overline{\alpha} + j\omega}\hat{\mathbf{u}}\Big(\frac{\overline{\alpha} + j\omega}{\alpha - j\omega}\Big), \frac{\sqrt{2\Re\alpha}}{\overline{\alpha} + j\omega}\hat{\mathbf{v}}\Big(\frac{\overline{\alpha} + j\omega}{\alpha - j\omega}\Big)\Big\rangle\, d\omega.
\end{aligned} \tag{12.1.6}$$

Here the kernel in the middle integral is the Poisson kernel for the right half-plane.

The fact that we use the Z-transform to derive our frequency domain results leads to the definition of the *transfer function* of the system $\Sigma = \left[\begin{smallmatrix} \mathbf{A} & \mathbf{B} \\ \mathbf{C} & \mathbf{D} \end{smallmatrix}\right]$ to be given by

$$\widehat{\mathfrak{D}}(z) = \mathbf{C}(z - \mathbf{A})^{-1}\mathbf{B} + \mathbf{D}, \qquad z \in \rho(\mathbf{A}),$$

and it is analytic (at least) in $\mathbb{D}^{+}_{r_{\mathbf{A}}}$, where $r_{\mathbf{A}}$ is the growth bound (i.e, the spectral radius of $\mathbf{A}$). In particular, it is analytic at infinity, and $\widehat{\mathfrak{D}}(\infty) = \mathbf{D}$. If $\mathbf{u} \in \ell^p_r(\mathbb{R}^+; U)$ for some p, $1 \le p \le \infty$ and $r \ge r_{\mathbf{A}}$, then the Z-transform of $\mathfrak{D}\mathbf{u}$ is given by

$$\widehat{\mathfrak{D}\mathbf{u}}(z) = \widehat{\mathfrak{D}}(z)\hat{\mathbf{u}}(z), \qquad |z| > r.$$

We leave the proof of this fact to the reader (it is possible to give a short proof for the ℓ^2-bounded Hilbert space case based on the theory in the next section, but the general case requires arguments similar to those that we have presented for the continuous time case).

There is a frequency domain description of the discrete time bilateral shift $\boldsymbol{\sigma}$, the incoming shift $\boldsymbol{\sigma}+$, and the outgoing shift $\boldsymbol{\sigma}_{-}$, with the Laplace transform replaced by the Z-transform. The discrete time analogue of Proposition 3.13.1 is the following. The transform of a bilateral shift is given by $\widehat{\boldsymbol{\sigma}\mathbf{u}}(z) = z\hat{\mathbf{u}}(z)$ (assuming that one of these transforms converge absolutely). The corresponding formula for the incoming shift $\boldsymbol{\sigma}_{+}$ is much simpler in discrete time than in continuous time, namely

$$\widehat{\boldsymbol{\sigma}_{+}\mathbf{u}}(z) = z[\hat{\mathbf{u}}(z) - u_0] = z[\hat{\mathbf{u}}(z) - \hat{\mathbf{u}}(\infty)].$$

The outgoing shift satisfies the same formula as the bilateral shift, i.e., $\widehat{\boldsymbol{\sigma}_{-}\mathbf{u}}(z) = z\hat{\mathbf{u}}(z)$.

We get the *dual* of a system by simply taking the adjoint of the system block matrix $\left[\begin{smallmatrix} \mathbf{A} & \mathbf{B} \\ \mathbf{C} & \mathbf{D} \end{smallmatrix}\right]$, so that the dual main operator is $\mathbf{A}^*$, the dual control operator is $\mathbf{C}^*$, the dual observation operator is $\mathbf{B}^*$, and the dual feedthrough operator is $\mathbf{D}^*$. As in continuous time, the (causal) dual transfer function $\widehat{\mathfrak{D}}^d$ is given by

$\widehat{\mathbb{D}}^d(z) = \widehat{\mathbb{D}}(\overline{z})^*$. The discrete time analogues of the following results are valid: Theorem 6.2.10 and Corollary 6.2.11 (which describe the connections between a primal and a dual trajectory), and Theorem 6.2.12 (which describes the dual of the Lax–Phillips semigroup). We leave the exact formulations and their proofs to the reader.

A discrete time system $\boldsymbol{\Sigma}$ with generators $\left[\begin{smallmatrix} \mathbf{A} & \mathbf{B} \\ \mathbf{C} & \mathbf{D} \end{smallmatrix}\right]$ is flow-invertible if and only if $\mathbf{D}$ is invertible. We get the formulas for the generators $\left[\begin{smallmatrix} \mathbf{A}_\times & \mathbf{B}_\times \\ \mathbf{C}_\times & \mathbf{D}_\times \end{smallmatrix}\right]$ of the flow-inverted system $\boldsymbol{\Sigma}_\times$ by solving for $\mathbf{x}$ and $\mathbf{u}$ in terms of $\mathbf{y}$ in (12.1.1):

$$\begin{aligned} x_{k+1} &= (\mathbf{A} - \mathbf{B}\mathbf{D}^{-1}\mathbf{C})x_k + \mathbf{B}\mathbf{D}^{-1}y_k, \\ u_k &= -\mathbf{D}^{-1}\mathbf{C}x_k + \mathbf{D}^{-1}y_k, \qquad k \in \mathbb{Z}^+. \end{aligned} \tag{12.1.7}$$

Thus,

$$\begin{aligned} \begin{bmatrix} \mathbf{A}_\times & \mathbf{B}_\times \\ \mathbf{C}_\times & \mathbf{D}_\times \end{bmatrix} &= \begin{bmatrix} \mathbf{A} & \mathbf{B} \\ 0 & 1 \end{bmatrix} \begin{bmatrix} 1 & 0 \\ \mathbf{C} & \mathbf{D} \end{bmatrix}^{-1} \\ &= \begin{bmatrix} \mathbf{A} - \mathbf{B}\mathbf{D}^{-1}\mathbf{C} & \mathbf{B}\mathbf{D}^{-1} \\ -\mathbf{D}^{-1}\mathbf{C} & \mathbf{D}^{-1} \end{bmatrix}. \end{aligned}$$

Note that this is *the same formula as in the classical continuous time case* (see (6.3.7)). As expected, the flow-inverted transfer function $\widehat{\mathbb{D}}_\times$ is given by $\widehat{\mathbb{D}}_\times(z) = \widehat{\mathbb{D}}^{-1}(z)$. The spectrum of the flow-inverted main operator $\mathbf{A}_\times$ can be obtained from the discrete time analogue of Lemma 6.3.8. As in continuous time, flow-inversion does not affect the reachable and the unobservable subspaces. In particular, it preserves controllability and observability.

A discrete time system $\boldsymbol{\Sigma}$ with generators $\left[\begin{smallmatrix} \mathbf{A} & \mathbf{B} \\ \mathbf{C} & \mathbf{D} \end{smallmatrix}\right]$ is time-invertible if and only if $\mathbf{A}$ is invertible. We get the formulas for the generators $\left[\begin{smallmatrix} \mathbf{A}^{я} & \mathbf{B}^{я} \\ \mathbf{C}^{я} & \mathbf{D}^{я} \end{smallmatrix}\right]$ of the time-inverted system $\boldsymbol{\Sigma}^{я}$ by solving for x_k and y_k in terms of x_{k+1} and u_k in (12.1.1):

$$\begin{aligned} x_k &= \mathbf{A}^{-1}x_{k+1} - \mathbf{A}^{-1}\mathbf{B}u_k, \\ y_k &= \mathbf{C}\mathbf{A}^{-1}x_{k+1} + (\mathbf{D} - \mathbf{C}\mathbf{A}^{-1}\mathbf{B})y_k, \qquad k \in \mathbb{Z}^+. \end{aligned} \tag{12.1.8}$$

Thus,

$$\begin{aligned} \begin{bmatrix} \mathbf{A}^{я} & \mathbf{B}^{я} \\ \mathbf{C}^{я} & \mathbf{D}^{я} \end{bmatrix} &= \begin{bmatrix} 1 & 0 \\ \mathbf{C} & \mathbf{D} \end{bmatrix} \begin{bmatrix} \mathbf{A} & \mathbf{B} \\ 0 & 1 \end{bmatrix}^{-1} \\ &= \begin{bmatrix} \mathbf{A}^{-1} & -\mathbf{A}^{-1}\mathbf{B} \\ \mathbf{C}\mathbf{A}^{-1} & \mathbf{D} - \mathbf{C}\mathbf{A}^{-1}\mathbf{B} \end{bmatrix}. \end{aligned}$$

Note the similarity to flow-inversion, with the role of $\mathbf{D}$ taken over by $\mathbf{A}$. (Time-inversion becomes formally flow-inversion if we permute the two rows and the two columns in $\left[\begin{smallmatrix} \mathbf{A} & \mathbf{B} \\ \mathbf{C} & \mathbf{D} \end{smallmatrix}\right]$.) The transfer function of the time-inverted system is $\widehat{\mathbb{D}}_{я}(z) = \widehat{\mathfrak{D}}(1/z)$. Time-inversion does not change the reachable and unobservable subspaces if $0 \in \sigma_\infty(A)$, but they may change if $0 \notin \sigma_\infty(A)$ (the counter-example is essentially the same as in continuous time: let $\mathbf{A} = \boldsymbol{\sigma}$ on $\ell^2(\mathbb{Z})$, let $U = \ell^2(\mathbb{Z}^+)$, and let $\mathbf{B} = \pi_+$; then the original system is controllable but the time-inverted system is not). However, exact or approximate controllability in finite time and exact or approximate observability in finite time is preserved.

A discrete time system $\boldsymbol{\Sigma}$ with generators $\left[\begin{smallmatrix} \mathbf{A} & \mathbf{B} \\ \mathbf{C} & \mathbf{D} \end{smallmatrix}\right]$ is time-flow-invertible if and only if the block matrix $\left[\begin{smallmatrix} \mathbf{A} & \mathbf{B} \\ \mathbf{C} & \mathbf{D} \end{smallmatrix}\right]$ is invertible, in which case the generators of the time-flow-inverted system are given by $\left[\begin{smallmatrix} \mathbf{A}^{я}_{\times} & \mathbf{B}^{я}_{\times} \\ \mathbf{C}^{я}_{\times} & \mathbf{D}^{я}_{\times} \end{smallmatrix}\right] = \left[\begin{smallmatrix} \mathbf{A} & \mathbf{B} \\ \mathbf{C} & \mathbf{D} \end{smallmatrix}\right]^{-1}$ (we solve (12.1.1) for x_k and u_k in terms of x_{k+1} and y_k). The transfer function of the time-flow-inverted system is $\widehat{\mathfrak{D}}^{я}_{\times}(z) = \widehat{\mathfrak{D}}^{-1}(1/z)$. If the original system is flow-invertible and the flow-inverted system is time-invertible, then the original system is time-flow-invertible and its time-flow-inverse is the time-inverse of the flow-inverse of the original system. A similar result is true if the original system is time-invertible and the time-inverted system is flow-invertible. Time-flow-inversion need not preserve the reachable and unobservable subspaces.

The results on partial flow-inversion in Section 6.6 and the results on feedback in Chapter 7 have natural (and obvious) discrete time analogues. We leave the formulation of these to the reader.

Also the notion of stabilization and detection discussed in Chapter 8 carry over to discrete time. We transfer the definition of stability in Definition 8.1.1, replacing L^p_ω by ℓ^p_r and $L^p_{0,\omega}$ by $\ell^p_{0,r}$. In particular, the notions of strong or weak stability are still relevant: a system is *strongly or weakly l^p_r-stable* if it is l^p_r-bounded and A is strongly or weakly stable, i.e., $A^k x \to 0$ as $k \to \infty$ strongly or weakly for all $x \in Z$. A system is *power stable* if it is ℓ^p_r stable for some $r < 1$ (the value of p is irrelevant here), i.e., if the spectral radius of $\mathbf{A}$ is less than one. The results presented in Sections 8.1–8.2 carry over with nominal changes. So do the results on coprimeness in Sections 8.3–8.4, once we replace the H^∞-spaces over the right half-planes $\mathbb{C}^+_\omega$ by the corresponding H^∞-spaces over the exterior $\mathbb{D}^+_r$ of the disk around the origin with radius r. The dynamic stabilization in Section 8.5 remains virtually the same.

All the controllability and observability notions in Sections 9.1 and 9.4 carry over to discrete time in an obvious way. So do the results on pseudo-similarity in Section 9.2 (it is interesting to observe that the boundedness of the generators $\left[\begin{smallmatrix} \mathbf{A} & \mathbf{B} \\ \mathbf{C} & \mathbf{D} \end{smallmatrix}\right]$ does not play any significant role here; we still have only pseudo-similarity and not ordinary similarity). Also the realizations result in Section 9.3 remain valid, as do the results on normalized and balanced realizations in Section 9.5. The controllability and observability tests in Section 9.6 apply without any changes (in particular, so do the tests which require the main operator to be bounded). The modal controllability and observability described in Section 9.7 stays the same, and so does the spectral minimality in Section 9.8. The results in Section 9.9 on how controllability and observability is preserved under various transformations have more or less obvious discrete time counterparts, as have the time domain tests in Section 9.10.

Chapter 10 deals with admissibility. The *admissibility* results of this chapter become trivial in the discrete time case, but the *stability* results do not: they essentially stay the same, as do their proofs. Of course, we have to replace the right half-spaces $\mathbb{C}^+_\omega$ by the exterior disks $\mathbb{D}^+_r$, and also replace the H^p spaces and Carleson measures over $\mathbb{C}^+_\omega$ by the corresponding spaces and measures over $\mathbb{D}^+_r$.

Chapter 11 has a discrete time counterpart, and many of the results proved there were first obtained in discrete time. The proofs essentially stay the same and they are neither significantly easier nor significantly more difficult in discrete time than in continuous time. However, the characterization of a passive, energy preserving, or conservative system given in Theorem 11.2.5 simplifies: a system $\boldsymbol{\Sigma}$ with generating operators $\mathbf{S} = \left[\begin{smallmatrix} \mathbf{A} & \mathbf{B} \\ \mathbf{C} & \mathbf{D} \end{smallmatrix}\right]$ is (scattering) passive if and only if $\mathbf{S}$ is a contraction, it is energy preserving if and only if $\mathbf{S}$ is isometric, and it is conservative if and only if $\mathbf{S}$ is unitary.

12.2 The internal linear fractional transform

In Section 2.4 we defined one transformation, namely time discretization, which maps a continuous time system into a discrete time system. This transformation has the drawback that the input and output spaces of the discrete time system differ from the input and output spaces of the continuous time system; in particular, they are always infinite-dimensional, even if the original input and output spaces are finite-dimensional. One advantage with this transformation is that the discrete time variable has a natural interpretation in terms of the continuous time variable; it is just a discretization.

Below we shall present a different transformation, *the internal linear fractional transform*, which also maps continuous systems into discrete time systems, but which preserves the original input, output, and state spaces.

Proposition 12.2.1 *Let $S = \left[\begin{smallmatrix} A\&B \\ C\&D \end{smallmatrix}\right]$ be an operator node on the Banach spaces (Y, X, U), with main operator A, control operator B, observation operator C, and transfer function $\widehat{\mathfrak{D}}$. Let $\alpha, \beta, \gamma, \delta \in \mathbb{C}$ with $\alpha \in \rho(A)$, $\beta - \alpha\gamma \neq 0$, and $\delta \neq 0$. Let $\boldsymbol{\Sigma}$ be the discrete time system on (Y, X, U) with generating operators*

$$\mathbf{A} = (\beta - \gamma A)(\alpha - A)^{-1}, \qquad \mathbf{B} = \delta(\alpha - A_{|X})^{-1} B,$$

$$\mathbf{C} = (\beta - \alpha\gamma)\delta^{-1} C(\alpha - A)^{-1}, \qquad \mathbf{D} = \widehat{\mathfrak{D}}(\alpha).$$

Then, $\alpha \neq \lambda \in \rho(A)$ *if and only if* $\gamma \neq z = (\beta - \gamma\lambda)/(\alpha - \lambda) \in \rho(\mathbf{A})$, *or equivalently,* $\gamma \neq z \in \rho(\mathbf{A})$ *if and only if* $\alpha \neq \lambda = (\alpha z - \beta)/(z - \gamma) \in \rho(A)$, *and*

$$\widehat{\mathbb{D}}(z) = \widehat{\mathfrak{D}}(\lambda), \quad z = \frac{\beta - \gamma\lambda}{\alpha - \lambda}, \quad \lambda = \frac{\alpha z - \beta}{z - \gamma}, \quad z \in \rho(\mathbf{A}), \ \lambda \in \rho(A).$$

Moreover, $\mathbf{A} - \gamma$ *is injective and has dense range.*

Thus, for every nontrivial linear fractional transform which maps infinity into $\rho(A)$, it is possible to find a discrete time system which has a transfer function which is obtained from the continuous time transfer function through the given linear fractional transform. Observe that the discrete time generator $\mathbf{A}$ of the discrete time system is obtained from the continuous time operator by applying the same linear fractional transform. In this result the parameter δ plays a trivial role, but we shall later want to take $\delta \neq 1$ (in order to balance the sizes of $\mathbf{B}$ and $\mathbf{C}$).

Proof A short computation shows that $\alpha \neq \lambda = (\alpha z - \beta)/(z - \gamma)$ if and only if $\gamma \neq z = (\beta - \gamma\lambda)/(\alpha - \lambda)$. Take $\lambda \in \rho(A), \lambda \neq \alpha$, and let $z = (\beta - \gamma\lambda)/(\alpha - \lambda)$. Then a short algebraic computation (where we factor out $(\alpha - A)^{-1}$ to the right) shows that

$$z - \mathbf{A} = \frac{\beta - \alpha\gamma}{\alpha - \lambda}(\lambda - A)(\alpha - A)^{-1}.$$

This implies that $z \neq \gamma$ belongs to $\rho(\mathbf{A})$ iff $\lambda \neq \alpha$ belongs to $\rho(A)$. By inverting the equation above and using Lemma 4.7.5(i) we get

$$\begin{aligned}\widehat{\mathbb{D}}(z) &= \mathbf{C}(z - \mathbf{A})^{-1}\mathbf{B} + \mathbf{D} \\ &= (\alpha - \lambda)C(\lambda - A)^{-1}(\alpha - A_{|X})^{-1}B + \widehat{\mathfrak{D}}(\alpha) \\ &= \widehat{\mathfrak{D}}(\lambda).\end{aligned}$$

That $\mathbf{A} - \gamma$ is injective and has dense range follows from the fact that

$$(\mathbf{A} - \gamma) = (\beta - \alpha\gamma)(\alpha - A)^{-1}.$$

□

Definition 12.2.2 We call the set of generating operators $\left[\begin{smallmatrix} \mathbf{A} & \mathbf{B} \\ \mathbf{C} & \mathbf{D} \end{smallmatrix}\right]$ of the system $\mathbf{\Sigma}$ in Proposition 12.2.1 the (internal) *linear fractional transform* with parameters α, β, γ, and δ of the operator node S.

In order to prove a converse to Proposition 12.2.1 we need a more exact formula for the correspondence between $\left[\begin{smallmatrix} A\&B \\ C\&D \end{smallmatrix}\right]$ and $\left[\begin{smallmatrix} \mathbf{A} & \mathbf{B} \\ \mathbf{C} & \mathbf{D} \end{smallmatrix}\right]$ in that proposition.

Lemma 12.2.3 *In the situation described in Proposition 12.2.1 the continuous time operator node* $S = \left[\begin{smallmatrix} A\&B \\ C\&D \end{smallmatrix}\right]$ *and the discrete time generators* $\left[\begin{smallmatrix} \mathbf{A} & \mathbf{B} \\ \mathbf{C} & \mathbf{D} \end{smallmatrix}\right]$ *correspond to each other in the following way.*

(i) *The operator* $\mathbf{A} - \gamma$ *maps* X *one-to-one onto* $\mathcal{D}(A)$*, the operator* $\left[\begin{smallmatrix} 1/\delta & 0 \\ 0 & 1 \end{smallmatrix}\right]\left[\begin{smallmatrix} \mathbf{A}-\gamma & \mathbf{B} \\ 0 & 1 \end{smallmatrix}\right]$ *maps* $\left[\begin{smallmatrix} X \\ U \end{smallmatrix}\right]$ *one-to-one onto* $\mathcal{D}(S)$*, and*

$$\mathbf{A} - \gamma = 2\Re\alpha(\alpha - A)^{-1},$$

$$\begin{bmatrix} 1/\delta & 0 \\ 0 & 1 \end{bmatrix}\begin{bmatrix} \mathbf{A}-\gamma & \mathbf{B} \\ 0 & 1 \end{bmatrix}\begin{bmatrix} \delta/(\beta-\alpha\gamma) & 0 \\ 0 & 1 \end{bmatrix} = \left(\begin{bmatrix} \alpha & 0 \\ 0 & 1 \end{bmatrix} - \begin{bmatrix} A\&B \\ 0\ 0 \end{bmatrix}\right)^{-1}.$$

(ii) *The discrete time generators* $\left[\begin{smallmatrix} \mathbf{A} & \mathbf{B} \\ \mathbf{C} & \mathbf{D} \end{smallmatrix}\right]$ *can be computed from the continuous time operator node* S *by means of either of the following two formulas:*

$$\begin{aligned}\begin{bmatrix} \mathbf{A} & \mathbf{B} \\ \mathbf{C} & \mathbf{D} \end{bmatrix} &= \begin{bmatrix} \delta/(\beta-\alpha\gamma) & 0 \\ 0 & 1 \end{bmatrix}\left(\begin{bmatrix} \beta\ 0 \\ C\&D \end{bmatrix} - \gamma\begin{bmatrix} A\&B \\ 0\ 0 \end{bmatrix}\right) \\ &\quad\times\left(\begin{bmatrix} \alpha & 0 \\ 0 & 1 \end{bmatrix} - \begin{bmatrix} A\&B \\ 0\ 0 \end{bmatrix}\right)^{-1}\begin{bmatrix} (\beta-\alpha\gamma)/\delta & 0 \\ 0 & 1 \end{bmatrix}, \\ \begin{bmatrix} \mathbf{A}-\gamma & \mathbf{B} \\ \mathbf{C} & \mathbf{D} \end{bmatrix} &= \begin{bmatrix} \delta & 0 \\ 0 & 1 \end{bmatrix}\begin{bmatrix} 1 & 0 \\ C\&D \end{bmatrix} \\ &\quad\times\left(\begin{bmatrix} \alpha & 0 \\ 0 & 1 \end{bmatrix} - \begin{bmatrix} A\&B \\ 0\ 0 \end{bmatrix}\right)^{-1}\begin{bmatrix} (\beta-\alpha\gamma)/\delta & 0 \\ 0 & 1 \end{bmatrix}.\end{aligned}$$

(iii) *The continuous time operator node* S *can be computed from the discrete time generators* $\left[\begin{smallmatrix} \mathbf{A} & \mathbf{B} \\ \mathbf{C} & \mathbf{D} \end{smallmatrix}\right]$ *by means of either of the following two formulas:*

$$S = \begin{bmatrix} A\&B \\ C\&D \end{bmatrix} = \begin{bmatrix} 1/\delta & 0 \\ 0 & 1 \end{bmatrix}\begin{bmatrix} \alpha\mathbf{A}-\beta & \alpha\mathbf{B} \\ \mathbf{C} & \mathbf{D} \end{bmatrix}\begin{bmatrix} \mathbf{A}-\gamma & \mathbf{B} \\ 0 & 1 \end{bmatrix}^{-1}\begin{bmatrix} \delta & 0 \\ 0 & 1 \end{bmatrix},$$

$$\begin{bmatrix} A\&B \\ C\&D \end{bmatrix} - \begin{bmatrix} \alpha & 0 \\ 0 & 0 \end{bmatrix} = \begin{bmatrix} (\beta-\alpha\gamma)/\delta & 0 \\ 0 & 1 \end{bmatrix}\begin{bmatrix} -1 & 0 \\ \mathbf{C} & \mathbf{D} \end{bmatrix}\begin{bmatrix} \mathbf{A}-\gamma & \mathbf{B} \\ 0 & 1 \end{bmatrix}^{-1}\begin{bmatrix} \delta & 0 \\ 0 & 1 \end{bmatrix}.$$

In particular, the main operator A*, the control operator* B*, and the observation operator* C *are given by*

$$\begin{aligned} A &= (\alpha\mathbf{A}-\beta)(\mathbf{A}-\gamma)^{-1} = \alpha - (\beta-\alpha\gamma)(\mathbf{A}-\gamma)^{-1}, \\ B &= \delta^{-1}(\alpha - A_{|X})\mathbf{B}, \quad C = \delta\mathbf{C}(\mathbf{A}-\gamma)^{-1}. \end{aligned}$$

Proof All of this follows from some straightforward algebraic computations, based on Lemma 4.7.18 and the definitions of **A**, **B**, **C**, and **D**. □

This lemma immediately implies the following result:

Proposition 12.2.4 *Let* $\left[\begin{smallmatrix} \mathbf{A} & \mathbf{B} \\ \mathbf{C} & \mathbf{D} \end{smallmatrix}\right] \in \mathcal{B}(\left[\begin{smallmatrix} X \\ U \end{smallmatrix}\right], \left[\begin{smallmatrix} X \\ Y \end{smallmatrix}\right])$*, where* U*,* X*, and* Y *are Banach spaces. Then* $\left[\begin{smallmatrix} \mathbf{A} & \mathbf{B} \\ \mathbf{C} & \mathbf{D} \end{smallmatrix}\right]$ *is the linear fractional transform of an operator node* $\left[\begin{smallmatrix} A\&B \\ C\&D \end{smallmatrix}\right]$ *with parameters* α*,* β*,* γ*, and* δ *(where* $\beta - \alpha\gamma \neq 0$ *and* $\delta \neq 0$*) if and only if* $\mathbf{A} - \gamma$ *is injective and has dense range.*

Proof One direction is contained in Proposition 12.2.1. For the other direction we use the formulas in part (iii) of Lemma 12.2.3 to define the operator node S in terms of A, B, C, and $\widehat{\mathfrak{D}}(\alpha) = \mathbf{D}$, as in Lemma 4.7.6. The linear fractional transform of S will then be $\left[\begin{smallmatrix} \mathbf{A} & \mathbf{B} \\ \mathbf{C} & \mathbf{D} \end{smallmatrix}\right]$. □

Observe that we do not claim that S in Proposition 12.2.4 is always a *system node*, i.e., we *do not claim that* A *generates a* C_0 *semigroup*. Neither do we claim that S is well-posed in any sense. This will, however, be the case under some extra conditions, such as those given in Theorems 12.3.10 and 12.3.11.

Above we have presented the internal linear fractional transform as a transformation mapping a continuous time system into a discrete time system. The same transform can be used to map a discrete time system into another discrete time system (and the formulas for the new generators remain the same). We have already encountered one particular instance of this transform, namely the *discrete time time-inversion*, which corresponds to the linear fractional transform with $z \mapsto 1/z$ (take $\alpha = 0$, $\beta = -1$, $\gamma = 0$, and $\delta = 1$). The same map is also used in continuous time, where it is known as *the reciprocal transform*.

Let us take a closer look at to what extent the linear fractional transform preserves the reachable and unobservable subspaces. However, before doing so, let us observe that for a system with *bounded* generating operators, as far as the reachable and unobservable subspaces are concerned, it does not matter if we interpret this as a continuous time system or as a discrete time system.

Lemma 12.2.5 *Let* U*,* X*, and* Y *be Banach spaces, and let* $\left[\begin{smallmatrix} A & B \\ C & D \end{smallmatrix}\right] \in \mathcal{B}(\left[\begin{smallmatrix} X \\ U \end{smallmatrix}\right]; \left[\begin{smallmatrix} X \\ Y \end{smallmatrix}\right])$*. Then the continuous time system* $\Sigma = \left[\begin{smallmatrix} \mathfrak{A} & \mathfrak{B} \\ \mathfrak{C} & \mathfrak{D} \end{smallmatrix}\right]$ *with the generating operators* $\left[\begin{smallmatrix} A & B \\ C & D \end{smallmatrix}\right]$ *and the discrete time system* $\mathbf{\Sigma} = \left[\begin{smallmatrix} \mathbb{A} & \mathbb{B} \\ \mathbb{C} & \mathbb{D} \end{smallmatrix}\right]$ *with the same generating operators* $\left[\begin{smallmatrix} A & B \\ C & D \end{smallmatrix}\right]$ *have the same reachable and unobservable subspaces.*

Proof By Definition 9.1.2, the unobservable subspace for Σ is $\mathcal{N}(\mathfrak{C})$, and analogously, the unobservable subspace for $\mathbf{\Sigma}$ is $\mathcal{N}(\mathbb{C})$. By the definition of $\mathbb{C}$, the latter is equal to $\bigcap_{n=0}^{\infty} \mathcal{N}(CA^n)$. That $\mathcal{N}(\mathfrak{C}) = \bigcap_{n=0}^{\infty} \mathcal{N}(CA^n)$ follows from Lemma 9.6.1(vi). The claim about the reachable subspaces is proved analogously (replace Lemma 9.6.1(vi) by Lemma 9.6.3(vi)). □

Lemma 12.2.6 *If* $\alpha \in \rho_\infty(A)$*, then the linear fractional transform presented in Proposition 12.2.1 preserves the reachable and unobservable subspaces.*

Proof Writing $\mathbf{A}$ in the form $\mathbf{A} = \gamma + (\beta + \alpha\gamma)(\alpha - A)^{-1}$ we immediately observe that, for all $N = 1, 2, 3, \ldots,$

$$\bigcap_{n=0}^{N-1} \mathcal{N}\left(\mathbf{C}\mathbf{A}^n\right) = \bigcap_{n=1}^{N} \mathcal{N}\left(C(\alpha - A)^{-n}\right),$$

$$\bigcup_{n=0}^{N-1} \mathcal{R}\left(\mathbf{A}^n\mathbf{B}\right) = \bigcup_{n=1}^{N} \mathcal{R}\left((\alpha - A)^{-n}B\right).$$

This, together with Lemmas 9.6.1(iv) and 9.6.3(iv) implies (as in the proof of Lemma 12.2.5) that the two systems have the same reachable and unobservable subspaces. □

The conclusion of Lemma 12.2.6 does *not* remain true without the assumption that $\alpha \in \rho_\infty(A)$. The counterexample is the same which we gave for discrete time time-inversion: let $\mathbf{A} = \sigma$ on $\ell^2(\mathbb{Z})$, let $U = \ell^2(\mathbb{Z}^+)$, and let $\mathbf{B} = \pi_+$; then the original system is controllable but the time-inverted system is not (and time-inversion is a special case of a linear fractional transform).

12.3 The Cayley and Laguerre transforms

There is one important special case of the linear fractional transform that we shall take a closer look at, namely the (internal) *Cayley transform*. It is used primarily in the L^2-well-posed case where the input and output spaces are Hilbert spaces. It is possible to apply this transform also to unstable systems, but for simplicity, we mainly discuss the stable case (the unstable case can be reduced to the stable case by an exponential shift, as explained in Example 2.3.5). We obtain the most complete correspondence between continuous and discrete time in the case where the semigroup of the system is a contraction semigroup in a Hilbert space, and for this reason we shall require the state space, too, to be a Hilbert space.

The idea behind the Cayley transform is to map the input and output signal spaces $L^2(\mathbb{R}; U)$ and $L^2(\mathbb{R}; Y)$ isometrically onto $\ell^2(\mathbb{R}; U)$ and $\ell^2(\mathbb{R}; Y)$, using (rescaled versions of) Laguerre series. This is done in a causal way, so that $L^2(\mathbb{R}^+)$ is mapped onto $\ell^2(\mathbb{Z}^+)$ and $L^2(\mathbb{R}^-)$ is mapped onto $\ell^2(\mathbb{Z}^-)$. We begin by describing these transformations.

Theorem 12.3.1 *Let U be a Hilbert space, and let $\alpha \in \mathbb{C}^+$.*

(i) *The sequence of functions $\{\varphi_k\}_{k\in\mathbb{Z}}$ defined via their bilateral Laplace transforms through the relation*

$$\widehat{\varphi}_k(\lambda) = \frac{\sqrt{2\Re\alpha}(\alpha - \lambda)^k}{(\overline{\alpha} + \lambda)^{k+1}}, \qquad \lambda \in j\mathbb{R},$$

is a complete orthonormal sequence in $L^2(\mathbb{R})$. Moreover, $\varphi_k \in L^2(\mathbb{R}^+)$ for $k \in \mathbb{Z}^+$ and $\varphi_k \in L^2(\mathbb{R}^-)$ for $k \in \mathbb{Z}^-$, hence $\{\varphi_k\}_{k\in\mathbb{Z}^+}$ is a complete *orthonormal sequence in* $L^2(\mathbb{R}^+)$ *and* $\{\varphi_k\}_{k\in\mathbb{Z}^-}$ *is a complete orthonormal sequence in* $L^2(\mathbb{R}^-)$*. Explicitly, for all* $k = 0, 1, 2, \ldots$*,*

$$\begin{aligned}\varphi_k(t) &= (-1)^k\sqrt{2\Re\alpha}\, e^{-\overline{\alpha}t}\sum_{i=0}^{k}\binom{k}{i}\frac{(-2\Re\alpha\, t)^i}{i!}, \qquad t\in\mathbb{R}^+,\\ \varphi_{-(k+1)}(t) &= (-1)^k\sqrt{2\Re\alpha}\, e^{\alpha t}\sum_{i=0}^{k}\binom{k}{i}\frac{(2\Re\alpha\, t)^i}{i!}, \qquad t\in\mathbb{R}^-.\end{aligned} \tag{12.3.1}$$

(ii) *The transformation which takes* $u \in L^2(\mathbb{R}; U)$ *into the sequence* $\mathbf{u} = \{u_k\}$ *where*

$$u_k = \int_{\mathbb{R}} u(t)\overline{\varphi}_k(t)\, dt$$

is a unitary transformation of $L^2(\mathbb{R}; U)$ *onto* $\ell^2(\mathbb{Z}; U)$*, which maps* $L^2(\mathbb{R}^+; U)$ *onto* $\ell^2(\mathbb{Z}^+; U)$ *and* $L^2(\mathbb{R}^-; U)$ *onto* $\ell^2(\mathbb{Z}^-; U)$*. The inverse of this transformation is given by*

$$u = \sum_{k\in\mathbb{Z}} u_k\varphi_k,$$

where the sum converges unconditionally (in norm) in $L^2(\mathbb{R}; U)$*.*

(iii) *Let* $u, v \in L^2(\mathbb{R}; U)$ *and define* $\mathbf{u}, \mathbf{v} \in \ell^2(\mathbb{Z}; U)$ *as in (ii). Then*

$$\sum_{k\in\mathbb{Z}}\langle u_k, v_k\rangle = \int_{\mathbb{R}}\langle u(t), v(t)\rangle\, dt.$$

If instead $v \in L^2(\mathbb{R})$ *(i.e., u is still vector-valued but v is scalar-valued), then*

$$\sum_{k\in\mathbb{Z}} u_k\overline{v}_k = \int_{\mathbb{R}} u(t)\overline{v}(t)\, dt.$$

(iv) *If we let* $\hat{u}$ *be the (bilateral) Laplace transform of u and let* $\hat{\mathbf{u}}$ *be the Z-transform of* $\mathbf{u}$*, where u and* $\mathbf{u}$ *are related as in (ii), then*

$$\hat{u}(\lambda) = \frac{\sqrt{2\Re\alpha}}{\overline{\alpha}+\lambda}\,\hat{\mathbf{u}}(z) = \frac{1+1/z}{\sqrt{2\Re\alpha}}\,\hat{\mathbf{u}}(z)$$

where $z = \dfrac{\overline{\alpha}+\lambda}{\alpha-\lambda}$ *and* $\lambda = \dfrac{\alpha z-\overline{\alpha}}{z+1}$*.*

This identity is always valid in the L^2*-sense for* $\lambda \in j\mathbb{R}$ *and* $z \in \mathbb{T}$*. If* $u \in L^2(\mathbb{R}^+; U)$*, or equivalently,* $\mathbf{u} \in \ell^2(\mathbb{Z}^+; U)$*, then this identity is valid, in addition, for all* $\lambda \in \mathbb{C}^+$ *and* $z \in \mathbb{D}^+$*, and if* $u \in L^2(\mathbb{R}^-; U)$*, or equivalently,* $\mathbf{u} \in \ell^2(\mathbb{Z}^-; U)$*, then this identity is valid, in addition, for all* $\lambda \in \mathbb{C}^-$ *and* $z \in \mathbb{D}^-$*.*

Proof As we observed in Section 12.1, the sequence $\{\mathbf{e}^k\}_{k\in\mathbb{Z}}$, where $\mathbf{e}_k^j = 1$ iff $j = k$, is a complete orthonormal sequence in $\ell^2(\mathbb{Z})$, and by restricting the index set to either $\mathbb{Z}^+$ or $\mathbb{Z}^-$ we get complete orthonormal sequences in $\ell^2(\mathbb{Z}^+)$ and $\ell^2(\mathbb{Z}^-)$. The Z-transform maps $\mathbf{e}^k$ onto the function $\hat{\mathbf{e}}^k(z) = z^{-k}$. The sequence $\{z^{-k}\}_{k\in\mathbb{Z}}$ is a complete orthogonal sequence in $L^2(\mathbb{T})$, and it becomes orthonormal if we divide each function by the constant $\sqrt{2\pi}$. By restricting the index set to $\mathbb{Z}^+$ or $\mathbb{Z}^-$ we get complete orthogonal sequences in $H^2(\mathbb{D}^+)$ and $H^2(\mathbb{D}^-)$, respectively (these are the set of functions in $L^2(\mathbb{T})$ which have analytic extensions to $\mathbb{D}^+$, respectively $\mathbb{D}^-$). By instead considering sequences in $\ell^2(\mathbb{Z};U)$ we get analogous results. The Z-transform is a bijection of $\ell^2(\mathbb{Z};U)$ onto $L^2(\mathbb{T};U)$, of $\ell^2(\mathbb{Z}^+;U)$ onto $L^2(\mathbb{D}^+;U)$, and of $\ell^2(\mathbb{Z}^-;U)$ onto $L^2(\mathbb{D}^-;U)$, and it becomes a unitary operator if we divide it by the scalar $\sqrt{2\pi}$.

The reversible change of variable from z to $\lambda = (\alpha z - \overline{\alpha})/(z+1)$ maps $z \in \mathbb{T}$ onto $\lambda \in j\mathbb{R}$, $z \in \mathbb{D}^+$ onto $\lambda \in \mathbb{C}^+$, and $z \in \mathbb{D}^-$ onto $\lambda \in \mathbb{C}^+$. Formula (12.1.6) shows that the transformation described in (iii) is a unitary map between $L^2(\mathbb{T};U)$ and $L^2(j\mathbb{R};U)$, which maps $H^2(\mathbb{D}^+;U)$ onto $H^2(\mathbb{C}^+;U)$ and $H^2(\mathbb{D}^+;U)$ onto $H^2(\mathbb{C}^+;U)$. The images of the functions $(\hat{\mathbf{e}}^k)(z) = z^{-k}$ (from discrete to continuous time) under this transformation are the functions $\widehat{\varphi}_k$ listed in (i).

The inverse Laplace transform becomes a unitary operator from $L^2(j\mathbb{R};U)$ onto $L^2(\mathbb{R};U)$ if we multiply it by the scalar constant $\sqrt{2\pi}$. Thus, starting with a sequence $\{u_k\}_{k\in\mathbb{Z}} \in \ell^2(\mathbb{Z};U)$, we can first use the Z-transform to map it onto a function $\hat{\mathbf{u}} \in L^2(\mathbb{T};U)$, then use the change of variable described in (iii) to map it onto a function $\hat{u} \in L^2(j\mathbb{R};U)$, and finally use the inverse Laplace transform to map it onto a function $u \in L^2(\mathbb{R};U)$. The resulting transformation is unitary from $\ell^2(\mathbb{Z};U)$ onto $L^2(\mathbb{R};U)$, and it maps $\ell^2(\mathbb{Z}^+;U)$ onto $L^2(\mathbb{R}^+;U)$ and $\ell^2(\mathbb{Z}^-;U)$ onto $L^2(\mathbb{R}^-;U)$. The image of $\mathbf{e}^k$ under the scalar-valued version of this transformation is the function φ_k.

The formulas in (iii) are two different versions of the standard Parseval identity, the first one for U-valued functions, and the second for the case when one of the two functions is scalar-valued.

We get the formulas for the coefficients u_k given in (ii) by using the second formula in (iii) with v replaced by φ_k. The formula expressing u in terms of u_k is simply the transform of the identity $\mathbf{u} = \sum_{k\in\mathbb{Z}} u_k \mathbf{e}^k$.

We leave the verification of (12.3.1) to the reader. □

Definition 12.3.2 Below we use the notation of Theorem 12.3.1.

(i) We call $\mathbf{u} \in \ell^2(\mathbb{Z};U)$ in part (ii) of Theorem 12.3.1 the *Laguerre transform* (with parameter α) of $u \in L^2(\mathbb{R};U)$ and denote it by $\mathbf{u} = \mathcal{L}u$.

We call u the *inverse Laguerre transform* of $\mathbf{u}$ and denote it by $u = \mathcal{L}^{-1}\mathbf{u}$.

(ii) We call $\hat{\mathbf{u}} \in L^2(\mathbb{T}; U)$ in part (iv) of Theorem 12.3.1 the *Cayley transform* (with parameter α) of $\hat{u} \in L^2(j\mathbb{R}; U)$ and denote it by $\hat{\mathbf{u}} = \mathcal{C}\hat{u}$. We call $\hat{u}$ the *inverse Cayley transform* of $\hat{\mathbf{u}}$ and denote it by $\hat{u} = \mathcal{C}^{-1}\hat{\mathbf{u}}$.

Thus, the Cayley transform is the frequency domain version of the Laguerre transform. The Laguerre transform is a unitary map from $L^2(\mathbb{R}; U)$ onto $\ell^2(\mathbb{Z}; U)$ which maps $L^2(\mathbb{R}^+; U)$ onto $\ell^2(\mathbb{Z}^+; U)$ and $L^2(\mathbb{R}^-; U)$ onto $\ell^2(\mathbb{Z}^-; U)$, and the Cayley transform is a unitary map from $L^2(j\mathbb{R}; U)$ onto $L^2(\mathbb{T}; U)$ which maps $H^2(\mathbb{C}^+; U)$ onto $H^2(\mathbb{D}^+; U)$ and $H^2(\mathbb{C}^-; U)$ onto $H^2(\mathbb{D}^-; U)$. This is true for all values of the free parameter $\alpha \in \mathbb{C}^+$.

The discrete and continuous time shifts and projection operators behave as follows under the Laguerre transform.

Theorem 12.3.3 *Let $\boldsymbol{\sigma}$ be the bilateral discrete time left shift, let $\boldsymbol{\pi}_-$ and $\boldsymbol{\pi}_+$ be the discrete time anti-causal and causal projections, let τ^t be the bilateral continuous time left shift, and let π_- and π_+ be the continuous time anti-causal and causal projections.*

(i) *The Laguerre transform preserves causality in the sense that*

$$\boldsymbol{\pi}_- = \mathcal{L}\pi_-\mathcal{L}^{-1}, \qquad \boldsymbol{\pi}_+ = \mathcal{L}\pi_+\mathcal{L}^{-1}.$$

(ii) *The operator $\mathcal{L}^{-1}\boldsymbol{\sigma}\mathcal{L}$ is time-invariant and anti-causal, and it is given by*

$$\mathcal{L}^{-1}\boldsymbol{\sigma}\mathcal{L} = 2\Re\alpha\left(\alpha - \frac{\mathrm{d}}{\mathrm{d}s}\right)^{-1} - 1,$$

where $\frac{\mathrm{d}}{\mathrm{d}s}$ is the generator of τ. Thus, for all $u \in L^2(\mathbb{R}; U)$ and almost all $t \in \mathbb{R}$,

$$(\mathcal{L}^{-1}\boldsymbol{\sigma}\mathcal{L}u)(t) = -u(t) + 2\Re\alpha \int_t^\infty \mathrm{e}^{\alpha(t-s)}u(s)\,ds.$$

In the frequency domain this operator acts as multiplication by the function $z(\lambda) = (\overline{\alpha} + \lambda)/(\alpha - \lambda)$, $\lambda \in \mathbb{C}$.

(iii) *The operator $\mathcal{L}\tau^t\mathcal{L}^{-1}$ is the discrete time time-invariant operator which in the frequency domain acts as multiplication by $z \mapsto \mathrm{e}^{\lambda(z)t}$ where $\lambda(z) = (\alpha z - \overline{\alpha})/(z + 1)$, $z \neq -1$. It is anti-causal if $t \geq 0$ and causal if $t \leq 0$.*

(iv) *The Laguerre transform preserves time-invariance in the sense that $\mathfrak{D} \in \mathcal{B}(L^2(\mathbb{R}; U); L^2(\mathbb{R}; Y))$ is time-invariant (in continuous time) if and only if $\mathbb{D} = \mathcal{L}\mathfrak{D}\mathcal{L}^{-1} \in \mathcal{B}(\ell^2(\mathbb{Z}; U); \ell^2(\mathbb{Z}; Y))$ is time-invariant (in discrete time). If, in addition, $\mathfrak{D}$ is causal, then the transfer functions $\widehat{\mathfrak{D}}$ of $\mathfrak{D}$ and*

$\widehat{\mathbb{D}}$ *of* $\mathbb{D}$ *are related as follows:*

$$\widehat{\mathfrak{D}}(\lambda) = \widehat{\mathbb{D}}(z) \textit{ for all } z = \frac{\overline{\alpha} + \lambda}{\alpha - \lambda} \in \mathbb{D}^+$$
$$\textit{and all } \lambda = \frac{\alpha z - \overline{\alpha}}{z + 1} \in \mathbb{C}^+.$$

Proof (i) This is obvious (the (causal) basis functions φ_k, $k \in \mathbb{Z}^+$, for $L^2(\mathbb{R}^+)$ are mapped onto the (causal) basis functions $\mathbf{e}^k$, $k \in \mathbb{Z}^+$ of $\ell(\mathbb{Z}^+)$, and the (anti-causal) basis functions $\varphi_k, k \in \mathbb{Z}^-$, for $L^2(\mathbb{R}^-)$ are mapped onto the (anti-causal) basis functions $\mathbf{e}^k$, $k \in \mathbb{Z}^-$ of $\ell(\mathbb{Z}^+)$).

(ii) The anti-causality of $\mathcal{L}^{-1}\boldsymbol{\sigma}\mathcal{L}$ follows from (i) and the anti-causality of $\boldsymbol{\sigma}$.

Let $\mathbf{u} \in \ell^2(\mathbb{Z}; U)$, and denote $u \in \mathcal{L}^{-1}\mathbf{u}$. Then (as is easily seen) $\widehat{\boldsymbol{\sigma}\mathbf{u}}(z) = z\hat{\mathbf{u}}(z)$ for almost all $z \in \mathbb{T}$, so by Theorem 12.3.1(iv), the bilateral Laplace transform of $\mathcal{L}^{-1}\boldsymbol{\sigma}\mathbf{u}$ is given by $\lambda \mapsto z(\lambda)\hat{\mathbf{u}}(z(\lambda)) = z(\lambda)\hat{u}(\lambda)$, $\Re\lambda = 0$, where $z(\lambda) = (\overline{\alpha} + \lambda)/(\alpha - \lambda) = 2\Re\alpha(\alpha - \lambda)^{-1} - 1$. If $u \in L^1 \cap L^2(\mathbb{R}; U)$, then, by Proposition 3.13.3(i), this is also the Laplace transform of $2\Re\alpha\big(\alpha - \frac{\mathrm{d}}{\mathrm{d}s}\big)^{-1}u - u$, so

$$\mathcal{L}^{-1}\boldsymbol{\sigma}\mathbf{u} = 2\Re\alpha\left(\alpha - \frac{\mathrm{d}}{\mathrm{d}s}\right)^{-1} u - u.$$

By the density of $L^1 \cap L^2(\mathbb{R}; U)$ in $L^2(\mathbb{R}; U)$, the same identity must be true for all $\mathbf{u} \in \ell^2(\mathbb{Z}; U)$. In particular, $\mathcal{L}^{-1}\boldsymbol{\sigma}\mathcal{L}$ is time-invariant (since $\frac{\mathrm{d}}{\mathrm{d}s}$ is the generator of the bilateral shift; cf. Theorem 3.2.9(iv)).

(iii) That $\mathcal{L}\tau^t\mathcal{L}^{-1}$ is time-invariant and anti-causal or causal depending on the value of t follows from (i) and (iv) and the corresponding properties of τ^t (the proof of (iv) does not use (iii)). If $u \in L^1(\mathbb{R}; U)$, then by Proposition 3.13.1(i), the Laplace transform of $\tau^t u$ is the function $\lambda \mapsto \mathrm{e}^{\lambda t}\hat{u}(\lambda)$, so by Theorem 12.3.1(iv), the Z-transform of $\mathcal{L}\tau^t u$ is the function $z \mapsto \mathrm{e}^{\lambda(z))t}\hat{u}(\lambda(z)) = \mathrm{e}^{\lambda(z)t}\hat{\mathbf{u}}(z)$, where $\mathbf{u} = \mathcal{L}u$. Since $L^1 \cap L^2(\mathbb{R}; U)$ is dense in $L^2(\mathbb{R}; U)$, the same identity must be true (in the L^2-sense) for all $u \in L^2(\mathbb{R}; U)$.

(iv) Clearly, $\mathcal{L}\mathfrak{D}\mathcal{L}^{-1}$ commutes with $\boldsymbol{\sigma}$ if and only if $\mathfrak{D}$ commutes with $\mathcal{L}^{-1}\boldsymbol{\sigma}\mathcal{L}$, so to prove (iv) we must show that $\mathfrak{D}$ is time-invariant if and only if $\mathfrak{D}$ commutes with $\mathcal{L}^{-1}\boldsymbol{\sigma}\mathcal{L}$.[1] By (ii) the latter statement is true if and only if $\mathfrak{D}$

[1] Strictly speaking, $\mathcal{L}\mathfrak{D}\mathcal{L}^{-1}$ does not 'commute with $\boldsymbol{\sigma}$' but it intertwines a left-shift $\boldsymbol{\sigma}$ in the input space U with a left shift $\boldsymbol{\sigma}$ in the output space Y, and these can be of different multiplicity if U and Y have different dimensions (we use the same notation for both of these shifts.) Likewise, in (iv) $\mathcal{L}$ and τ^t stand for two different Laguerre transforms and shifts, one of them defined on $L^2(\mathbb{R}; U)$ and the other on $L^2(\mathbb{R}; Y)$. This has, of course, to do with the general definition of time-invariance: a time-invariant operator intertwines a shift in the input space U with a shift in the output space Y.

commutes with $\left(\alpha - \frac{\mathrm{d}}{\mathrm{d}s}\right)^{-1}$, and by Theorem 3.14.15, this is true if and only if $\mathfrak{D}$ commutes with τ^t for all $t \in \mathbb{R}$.

To prove the last claim about the transfer functions we take an arbitrary $u \in L^2(\mathbb{R}^+; U)$, and let $\mathbf{u} = \mathcal{L}u$ be the corresponding discrete time signal. By Theorem 12.3.1(iv), for $\lambda \in \mathbb{C}^+$ and $z \in \mathbb{D}^+$ which are related as described in the theorem,

$$\widehat{\mathbb{D}\mathbf{u}}(z) = \widehat{\mathcal{L}\mathfrak{D}u}(z) = \frac{\overline{\alpha} + \lambda}{\sqrt{2\Re\alpha}}\,\widehat{\mathfrak{D}u}(\lambda) = \frac{\overline{\alpha} + \lambda}{\sqrt{2\Re\alpha}}\,\widehat{\mathfrak{D}}(\lambda)\hat{u}(\lambda) = \widehat{\mathfrak{D}}(\lambda)\hat{\mathbf{u}}(z).$$

Thus, $\widehat{\mathbb{D}}(z) = \widehat{\mathfrak{D}}(\lambda)$. □

Definition 12.3.4 The operator $\mathcal{L}^{-1}\sigma\mathcal{L}$ in Theorem 12.3.3 is called the *Laguerre shift*.

The (internal) Cayley transform can be interpreted as a special case of the internal linear fractional transform presented in Proposition 12.2.1, with parameters $\Re\alpha > 0$, $\beta = \overline{\alpha}$, $\gamma = -1$, and $\delta = \sqrt{2\Re\alpha}$. This particular set of parameters α, β and γ give a linear fractional transform which maps the right half-plane $\mathbb{C}^+$ is mapped onto the exterior $\mathbb{D}^+$ of the unit disk.[2] Let us establish some additional properties of the linear fractional transform in Proposition 12.2.1 with this special set of parameters.

Theorem 12.3.5 *Let $\Sigma = \left[\begin{smallmatrix} \mathfrak{A} & \mathfrak{B} \\ \mathfrak{C} & \mathfrak{D} \end{smallmatrix}\right]$ be an L^2-well-posed linear system on (Y, X, U), where Y and U are Hilbert spaces, with system node $S = \left[\begin{smallmatrix} A\&B \\ C\&D \end{smallmatrix}\right]$, control operator C, transfer function $\widehat{\mathfrak{D}}$, and growth rate $\omega_{\mathfrak{A}}$. Let $\alpha > \max\{0, \omega_{\mathfrak{A}}\}$, and let $\mathbf{\Sigma} = \left[\begin{smallmatrix} \mathbb{A} & \mathbb{B} \\ \mathbb{C} & \mathbb{D} \end{smallmatrix}\right]$ be the discrete time system on (Y, X, U) (in input/output form) whose generating operators are*

$$\mathbf{A} = (\overline{\alpha} + A)(\alpha - A)^{-1}, \qquad \mathbf{B} = \sqrt{2\Re\alpha}\,(\alpha - A_{|X})^{-1}B,$$
$$\mathbf{C} = \sqrt{2\Re\alpha}\,C(\alpha - A)^{-1}, \qquad \mathbf{D} = \widehat{\mathfrak{D}}(\alpha).$$

Then $\mathfrak{B}$, $\mathfrak{C}$, or $\mathfrak{D}$ is bounded if and only if $\mathbb{B}$, $\mathbb{C}$, or $\mathbb{D}$ is bounded, respectively, in which case

$$\mathbb{B} = \mathfrak{B}\mathcal{L}^{-1}, \quad \mathbb{C} = \mathcal{L}\mathfrak{C}, \quad \mathbb{D} = \mathcal{L}\mathfrak{D}\mathcal{L}^{-1},$$

where $\mathcal{L}^{-1}$ stands for the inverse Laguerre transform from $\ell^2(\mathbb{Z}; U)$ to $L^2(\mathbb{R}; U)$ and $\mathcal{L}$ stands for the Laguerre transform from $L^2(\mathbb{R}; Y)$ to $\ell^2(\mathbb{Z}; Y)$. Moreover,

$$\widehat{\mathbb{D}}(z) = \widehat{\mathfrak{D}}(\lambda), \quad z = \frac{\overline{\alpha} + \lambda}{\alpha - \lambda}, \quad \lambda = \frac{\alpha z - \overline{\alpha}}{z + 1}, \quad \lambda \in \rho(A), \quad z \in \rho(\mathbf{A}).$$

[2] We have normalized the parameters in the Cayley transform so that the point at infinity in continuous time is mapped to the point -1 in discrete time. Here the point -1 could be replaced by any other point on the unit circle $\mathbb{T}$ without any significant consequences.

Thus, $\mathfrak{B}$ is unitarily similar to $\mathbb{B}$, $\mathfrak{C}$ is unitarily similar to $\mathbb{C}$, and $\mathfrak{D}$ is unitarily similar to $\mathbb{D}$, whenever these operators are bounded.

Proof The formulas for $\left[\begin{smallmatrix} \mathbf{A} & \mathbf{B} \\ \mathbf{C} & \mathbf{D} \end{smallmatrix}\right]$ and $\widehat{\mathbb{D}}$ are found in Proposition 12.2.1 (take $\Re\alpha > 0$, $\beta = \overline{\alpha}$, $\gamma = -1$, and $\delta = \sqrt{2\Re\alpha}$).

Suppose that $\mathfrak{D}$ is bounded. By Theorem 12.3.3(i) and (iv) and Proposition 12.2.1, $\mathcal{L}\mathfrak{D}\mathcal{L}^{-1}$ is a bounded discrete time time-invariant causal operator which has the same transfer function as the input/output map $\mathbb{D}$ of the given discrete time system. Thus $\mathbb{D} = \mathcal{L}\mathfrak{D}\mathcal{L}^{-1}$. In particular, $\mathbb{D}$ is bounded.

Conversely, suppose that $\mathbb{D}$ is bounded. Then, by the discrete time analogue of Theorem 10.3.5, $\widehat{\mathbb{D}}$ has an analytic bounded extension to $\mathbb{D}^+$ (the outside of the unit disk), including the point at infinity. The mapping from z to λ in the formula $\widehat{\mathbb{D}}(z) = \widehat{\mathfrak{D}}(\lambda)$ maps $\mathbb{D}^+$ onto $\mathbb{C}^+$, and this means that $\widehat{\mathfrak{D}}$ can be extended to a function in $H^\infty(U; Y)$. By Theorem 10.3.5, $\mathfrak{D}$ is bounded.

Assume next that $\mathfrak{B}$ is bounded. To prove that $\mathbb{B} = \mathfrak{B}\mathcal{L}^{-1}$ it suffices to show that, with the notation of Theorem 12.3.1, for all $k \in \mathbb{Z}^-$ and $u \in U$, $\mathbb{B}\mathbf{e}^k u = \mathfrak{B}\mathcal{L}^{-1}\mathbf{e}^k u = \mathfrak{B}\varphi_k u$, since the set of linear combinations of functions of this type is dense in $\ell^2(\mathbb{Z}^-; U)$. By Theorem 3.12.6, for these k and u (see also Section 3.9 and recall that $k < 0$)

$$\begin{aligned}\mathfrak{B}\varphi_k u &= \int_{-\infty}^{0} \varphi_k(t)\mathfrak{A}_{|X_{-1}}^{-t} Bu\,dt = \widehat{\varphi}_k(A_{|X})Bu \\ &= \sqrt{2\Re\alpha}\,(\overline{\alpha} + A_{|X})^{-k-1}(\alpha - A_{|X})^k Bu = \mathbf{A}^{-k-1}\mathbf{B}u = \mathbb{B}\mathbf{e}^k u.\end{aligned}$$

This proves that $\mathbb{B} = \mathfrak{B}\mathcal{L}^{-1}$. In particular, $\mathbb{B}$ is bounded.

Actually, in the above computation we did not really use the boundedness of $\mathfrak{B}$. More precisely, $\mathfrak{B}$ can be applied to $\varphi_k u$ and the preceding computation is valid since $\mathfrak{B} \in \mathcal{B}(L^2_\omega(\mathbb{R}^-, U); X)$ for every $\omega > \omega_{\mathfrak{A}}$ (we assumed that $\alpha > \omega_{\mathfrak{A}}$, and $\varphi_k u \in L^2_\omega(\mathbb{R}^-, U)$ for every $\omega < \alpha$; take $\omega_{\mathfrak{A}} < \omega < \alpha$). The same computation, carried out in the reverse direction, shows that for all $k \in \mathbb{Z}^-$ and $u \in U$, $\mathbb{B}\mathbf{e}^k u = \mathfrak{B}\mathcal{L}^{-1}\mathbf{e}^k u = \mathfrak{B}\varphi_k u$. By linearity, for all finitely supported sequences $\mathbf{u}$ we have $\mathbb{B}\mathbf{u} = \mathfrak{B}\mathcal{L}^{-1}\mathbf{u}$. Since $\mathbb{B}$ is bounded, $\mathcal{L}$ is unitary, and the set of all finitely supported sequences is dense in $\ell^2(\mathbb{Z}^-; U)$, this implies that $\mathfrak{B}$ is bounded.

Finally, assume that $\mathfrak{C}$ is bounded. To prove that $\mathbb{C} = \mathcal{L}\mathfrak{C}$ it suffices to show that, with the same notation as above, for all $k \in \mathbb{Z}^+$ and $x \in X_1$, $(\mathbb{C}x)_k = (\mathcal{L}\mathfrak{C}x)_k$. By Definition 12.3.2 and Corollary 3.12.7, we have for these k and x,

$$\begin{aligned}(\mathcal{L}\mathfrak{C}x)_k &= \int_0^\infty \overline{\varphi}_k(t)\, C\mathfrak{A}^t x\,dt = C\widehat{\overline{\varphi}}(-A)x \\ &= \sqrt{2\Re\alpha}\, C(\overline{\alpha} + A)^k(\alpha - A)^{-k-1}x = \mathbf{C}\mathbf{A}^k x = (\mathbb{C}x)_k.\end{aligned}$$

In particular, $\mathbb{C}$ is bounded.

We leave the proof that boundedness of $\mathbb{C}$ implies boundedness of $\mathfrak{C}$ to the reader (it is analogous to the corresponding proof for the input maps). □

Definition 12.3.6 The discrete time system $\left[\begin{smallmatrix}\mathbb{A} & \mathbb{B}\\ \mathbb{C} & \mathbb{D}\end{smallmatrix}\right]$ in Theorem 12.3.5 is called the *(internal) Cayley transform* (with parameter α) of the continuous time system $\left[\begin{smallmatrix}\mathfrak{A} & \mathfrak{B}\\ \mathfrak{C} & \mathfrak{D}\end{smallmatrix}\right]$. The operator $\mathbf{A} = (\overline{\alpha} + A)(\alpha - A)^{-1}$ is called the *co-generator* of $\mathfrak{A}$ (with parameter α).

Theorem 12.3.5 has the following partial converse:

Theorem 12.3.7 *Let $\mathbf{\Sigma} = \left[\begin{smallmatrix}\mathbb{A} & \mathbb{B}\\ \mathbb{C} & \mathbb{D}\end{smallmatrix}\right]$ be a discrete time ℓ^2-well-posed linear system on (Y, X, U), where Y and U are Hilbert spaces, with generating operators $\left[\begin{smallmatrix}\mathbf{A} & \mathbf{B}\\ \mathbf{C} & \mathbf{D}\end{smallmatrix}\right]$, and let $\alpha \in \mathbb{C}^+$. Suppose that $\mathbb{B}$, $\mathbb{C}$ and $\mathbb{D}$ are stable, that -1 is not an eigenvalue of $\mathbf{A}$, and that*

$$A = (\alpha\mathbf{A} - \overline{\alpha})(\mathbf{A} + 1)^{-1}[= \alpha - 2\Re\alpha(\mathbf{A} + 1)^{-1}],$$

with $\mathcal{D}(A) = \mathcal{R}(\mathbf{A} + 1)$, is the generator of a semigroup $\mathfrak{A}$ with growth rate $\omega_{\mathfrak{A}} < \Re\alpha$. Then $\Sigma = \left[\begin{smallmatrix}\mathfrak{A} & \mathfrak{B}\\ \mathfrak{C} & \mathfrak{D}\end{smallmatrix}\right]$ is a continuous time L^2-well-posed linear system, where

$$\mathfrak{B} = \mathbb{B}\mathcal{L}, \quad \mathfrak{C} = \mathcal{L}^{-1}\mathbb{C}, \qquad \mathfrak{D} = \mathcal{L}^{-1}\mathbb{D}\mathcal{L};$$

here $\mathcal{L}$ stands for the Laguerre transform from $L^2(\mathbb{R}; U)$ to $\ell^2(\mathbb{Z}; U)$ and $\mathcal{L}^{-1}$ stands for the inverse Laguerre transform from $\ell^2(\mathbb{Z}; Y)$ to $L^2(\mathbb{R}; Y)$. In particular, $\mathfrak{B}$, $\mathfrak{C}$, and $\mathfrak{D}$ are stable. The Cayley transform of Σ is $\mathbf{\Sigma}$, and the system node of Σ is

$$\begin{bmatrix} A\&B \\ C\&D \end{bmatrix} = \begin{bmatrix} 1/\sqrt{2\Re\alpha} & 0 \\ 0 & 1 \end{bmatrix} \begin{bmatrix} \alpha\mathbf{A} - \overline{\alpha} & \alpha\mathbf{B} \\ \mathbf{C} & \mathbf{D} \end{bmatrix} \begin{bmatrix} \mathbf{A} + 1 & \mathbf{B} \\ 0 & 1 \end{bmatrix}^{-1} \begin{bmatrix} \sqrt{2\Re\alpha} & 0 \\ 0 & 1 \end{bmatrix}.$$

The control operator B, observation operator C, and the transfer function of Σ are given by

$$\begin{aligned} B &= 1/\sqrt{2\Re\alpha}\,(\alpha - A_{|X})\mathbf{B}, \\ C &= \sqrt{2\Re\alpha}\,\mathbf{C}(\mathbf{A} + 1)^{-1}, \\ \widehat{\mathfrak{D}}(\lambda) &= \widehat{\mathbb{D}}\Big(\frac{\overline{\alpha} + \lambda}{\alpha - \lambda}\Big), \qquad \lambda \in \rho(A). \end{aligned}$$

Proof The stability of $\mathfrak{B}$, $\mathfrak{C}$, and $\mathfrak{D}$ is obvious. Thus, to show that Σ is an L^2-well-posed linear system it suffices to verify the algebraic conditions in Definition 2.2.1. We assume condition (i), so only the remaining three conditions (ii)–(iv) need to be checked.

We begin with the conditions satisfied by $\mathfrak{D}$. By Theorem 12.3.3, $\mathfrak{D}$ is a causal time-invariant L^2-bounded operator. By the same theorem, the Hankel of $\mathfrak{D}$ is given by

$$\pi_+\mathfrak{D}\pi_- = \mathcal{L}^{-1}\pi_-\mathcal{L}\mathcal{L}^{-1}\mathbb{D}\mathcal{L}\mathcal{L}^{-1}\pi_+\mathcal{L} = \mathcal{L}^{-1}\pi_-\mathbb{D}\pi_+\mathcal{L} = \mathcal{L}^{-1}\mathbb{C}\mathbb{B}\mathcal{L} = \mathfrak{C}\mathfrak{B}.$$

Thus, part (iv) of Definition 2.2.1 is true.

Let us next check part (ii), i.e., the intertwining condition $\mathfrak{B}\tau_-^t = \mathfrak{A}^t\mathfrak{B}$ for all $t \ge 0$. By Theorem 3.14.15 and Example 3.2.3(iii), this condition is true if and only if $\mathfrak{B}\big(\alpha - \frac{\mathrm{d}}{\mathrm{d}x}{}_-\big)^{-1} = (\alpha - A)^{-1}\mathfrak{B}$. By Example 3.3.2(iii) and Theorems 3.12.9 and 12.3.3(ii),

$$\begin{aligned}2\Re\alpha\mathfrak{B}\Big(\alpha - \frac{\mathrm{d}}{\mathrm{d}x}{}_-\Big)^{-1} &= \mathfrak{B}(\mathcal{L}^{-1}\sigma\mathcal{L} + 1) = \mathbb{B}\sigma\mathcal{L}u + \mathfrak{B}\\ &= \mathbf{A}\mathbb{B}\mathcal{L}u + \mathfrak{B} = (\mathbf{A} + 1)\mathfrak{B} = 2\Re\alpha(\alpha - A)^{-1}\mathfrak{B}.\end{aligned}$$

Dividing this by $2\Re\alpha$ we get $\mathfrak{B}\big(\alpha - \frac{\mathrm{d}}{\mathrm{d}x}{}_-\big)^{-1} = (\alpha - A)^{-1}\mathfrak{B}$, and hence $\mathfrak{B}\tau_-^t = \mathfrak{A}^t\mathfrak{B}$ for all $t \ge 0$.

A completely analogous proof shows that $\mathfrak{C}\mathfrak{A}^t = \tau_+^t\mathfrak{C}$ for all $t \ge 0$.

Thus, according to Definitions 2.2.1 and 8.1.1, Σ is an L^2-well-posed linear system.

By Theorem 12.3.5 and Definition 12.3.6, $\mathbf{\Sigma}$ is the Cayley transform of Σ. We get the formulas for the system operator of Σ and the different parts of this operator from Lemma 12.2.3(iii). □

By combining Theorems 12.3.5 and 12.3.7 with an exponential shift of the continuous time system (see Example 2.3.5) we get the following two results.

Corollary 12.3.8 *Let $\Sigma = \left[\begin{smallmatrix}\mathfrak{A} & \mathfrak{B}\\ \mathfrak{C} & \mathfrak{D}\end{smallmatrix}\right]$ be an L^2-well-posed linear system on (Y, X, U), where Y and U are Hilbert spaces, with system node $S = \left[\begin{smallmatrix}A\&B\\ C\&D\end{smallmatrix}\right]$, control operator C, transfer function $\widehat{\mathfrak{D}}$, and growth rate $\omega_{\mathfrak{A}}$. Let $\omega \in \mathbb{R}$, $\Re\alpha > \max\{\omega, \omega_{\mathfrak{A}}\}$, and let $\mathbf{\Sigma} = \left[\begin{smallmatrix}\mathbb{A} & \mathbb{B}\\ \mathbb{C} & \mathbb{D}\end{smallmatrix}\right]$ be the discrete time system on (Y, X, U) (in input/output form) whose generating operators are*

$$\begin{aligned}\mathbf{A} &= (\alpha - 2\omega + A)(\alpha - A)^{-1}, & \mathbf{B} &= \sqrt{2\Re(\alpha - \omega)}\,(\alpha - A_{|X})^{-1}B,\\ \mathbf{C} &= \sqrt{2\Re(\alpha - \omega)}\,C(\alpha - A)^{-1}, & \mathbf{D} &= \widehat{\mathfrak{D}}(\alpha).\end{aligned}$$

If $\mathfrak{B}$, $\mathfrak{C}$, and $\mathfrak{D}$ are ω-bounded, then

$$\mathbb{B} = \mathfrak{B}\mathrm{e}_\omega\mathcal{L}^{-1}, \quad \mathbb{C} = \mathcal{L}\mathrm{e}_{-\omega}\mathfrak{C}, \qquad \mathbb{D} = \mathcal{L}\mathrm{e}_{-\omega}\mathfrak{D}\mathrm{e}_\omega\mathcal{L}^{-1},$$

where e_ω stands for multiplication by the function $t \mapsto \mathrm{e}^{\omega t}$ ($t \in \mathbb{R}$), $\mathcal{L}^{-1}$ stands for the inverse Laguerre transform from $\ell^2(\mathbb{Z}; U)$ to $L^2(\mathbb{R}; U)$ and $\mathcal{L}$ stands for the Laguerre transform from $L^2(\mathbb{R}; Y)$ to $\ell^2(\mathbb{Z}; Y)$. In particular, $\mathbb{B}$, $\mathbb{C}$, and $\mathbb{D}$ are stable. Moreover,

$$\widehat{\mathbb{D}}(z) = \widehat{\mathfrak{D}}(\lambda), \qquad z = \frac{\overline{\alpha} - 2\omega + \lambda}{\alpha - \lambda}, \qquad \lambda \in \mathbb{C}_\omega^+, \qquad z \in \mathbb{D}^+.$$

To prove this corollary we first use Example 2.3.5 to map Σ into a system $\Sigma_{-\omega}$ whose input, output, and input/output maps are stable, then apply Theorem 12.3.5 to map this system into discrete time, and finally replace $\alpha + \omega$ by α. Observe that the linear fractional transform $\lambda \mapsto z(\lambda) = (\overline{\alpha} - 2\omega + \lambda)/(\alpha - \lambda)$ maps the line $\Re\lambda = \omega$ onto the unit circle $|z| = 1$, $\mathbb{C}^+_\omega$ onto $\mathbb{D}^+$, and $\mathbb{C}^-_\omega$ onto $\mathbb{C}^-$.

The proof of the next corollary is analogous, except that we replace Theorem 12.3.5 by 12.3.7.

Corollary 12.3.9 *Let $\boldsymbol{\Sigma} = \left[\begin{smallmatrix} \mathbb{A} & \mathbb{B} \\ \mathbb{C} & \mathbb{D} \end{smallmatrix}\right]$ be a discrete time ℓ^2-well-posed linear system on (Y, X, U), where Y and U are Hilbert spaces, with generating operators $\left[\begin{smallmatrix} \mathbf{A} & \mathbf{B} \\ \mathbf{C} & \mathbf{D} \end{smallmatrix}\right]$, let $\omega \in \mathbb{R}$, and let $\alpha \in \mathbb{C}^+_\omega$. Suppose that $\mathbb{B}$, $\mathbb{C}$ and $\mathbb{D}$ are stable, that -1 is not an eigenvalue of $\mathbf{A}$, and that*

$$A = \alpha - 2\Re(\alpha - \omega)(\mathbf{A} + 1)^{-1},$$

with $\mathcal{D}(A) = \mathcal{R}(\mathbf{A} + 1)$ is the generator of a semigroup $\mathfrak{A}$ with growth rate $\omega_{\mathfrak{A}} < \Re\alpha$. Then $\Sigma = \left[\begin{smallmatrix} \mathfrak{A} & \mathfrak{B} \\ \mathfrak{C} & \mathfrak{D} \end{smallmatrix}\right]$ is a continuous time L^2-well-posed linear system, where

$$\mathfrak{B} = \mathbb{B}\mathcal{L}\mathrm{e}_{-\omega}, \qquad \mathfrak{C} = \mathrm{e}_\omega\mathcal{L}^{-1}\mathbb{C}, \qquad \mathfrak{D} = \mathrm{e}_\omega\mathcal{L}^{-1}\mathbb{D}\mathcal{L}\mathrm{e}_{-\omega};$$

here e_ω stands for the multiplication operator by the function $t \mapsto \mathrm{e}^{\alpha t}$, $t \in \mathbb{R}$, $\mathcal{L}$ stands for the Laguerre transform from $L^2(\mathbb{R}; U)$ to $\ell^2(\mathbb{Z}; U)$ and $\mathcal{L}^{-1}$ stands for the inverse Laguerre transform from $\ell^2(\mathbb{Z}; Y)$ to $L^2(\mathbb{R}; Y)$. In particular, $\mathfrak{B}$, $\mathfrak{C}$, and $\mathfrak{D}$ are ω-bounded. If we transform this system as described in Corollary 12.3.8 then we recover $\boldsymbol{\Sigma}$. The control operator B, observation operator C, and combined observation/feedthrough operator $C\&D$ of Σ are given by

$$\begin{aligned} B &= 1/\sqrt{2\Re(\alpha - \omega)}\,(\alpha - A_{|X})\mathbf{B}, \\ C &= \sqrt{2\Re(\alpha - \omega)}\,\mathbf{C}(\mathbf{A} + 1)^{-1}, \\ C\&D &= \begin{bmatrix} \mathbf{C} & \mathbf{D} \end{bmatrix} \begin{bmatrix} \mathbf{A} + 1 & \mathbf{B} \\ 0 & 1 \end{bmatrix}^{-1} \begin{bmatrix} \sqrt{2\Re(\alpha - \omega)} & 0 \\ 0 & 1 \end{bmatrix}. \end{aligned}$$

Theorems 12.3.5 and 12.3.7 and Corollaries 12.3.8 and 12.3.9 give useful information on how the stability or ω-boundedness of the continuous time input map, the output map, and the input/output map is reflected in the boundedness of the corresponding discrete time maps, but they do not say anything specific about how the stability or ω-boundedness of the continuous time semigroup is reflected in the boundedness of the powers of the co-generator. Actually, there seems to be no simple general connection.[3] However, there is one exception,

[3] Of course, we can map the resolvent conditions in the Hille–Yosida Theorem 3.4.1 into discrete time, but the resulting conditions do not seem to be very informative.

namely the case where $e^{-\omega t}\mathfrak{A}^t$ is a *contraction* semigroup on a *Hilbert space*. This is the content of the following theorem. For simplicity we treat only the case where $\omega = 0$, and leave the (easy) conversion to the case $\omega \neq 0$ to the reader.

Theorem 12.3.10 *Let X be a Hilbert space, and let $\alpha \in \mathbb{C}^+$.*

(i) *Let A be the generator of a contraction semigroup $\mathfrak{A}$ on X. Then the co-generator $\mathbf{A} = (\overline{\alpha} + A)(\alpha - A)^{-1}$ is a contraction, and -1 is not an eigenvalue of $\mathbf{A}$.*

(ii) *Conversely, let $\mathbf{A}$ be a contraction on X which does not have -1 as an eigenvalue. Then $A = (\alpha \mathbf{A} + \overline{\alpha})(\mathbf{A} + 1)^{-1}$ is the generator of a contraction semigroup, and $\mathbf{A}$ is its co-generator, i.e., $\mathbf{A} = (\overline{\alpha} + A)(\alpha - A)^{-1}$.*

(iii) *Let A and $\mathbf{A}$ be related as in (i)–(ii). Then*

  (a) *A closed subspace of X is invariant or co-invariant under $\mathfrak{A}$ if and only if it is invariant or co-invariant under $\mathbf{A}$.*

  (b) *$\mathfrak{A}$ is isometric, or co-isometric, or unitary if and only if $\mathbf{A}$ has the same property.*

  (c) *$\mathfrak{A}$ is completely non-unitary if and only if $\mathbf{A}$ is completely non-unitary.*

  (d) *$\mathfrak{A}$ is strongly stable if and only if $\mathbf{A}$ is strongly stable.*

  (e) *$\mathfrak{A}^*$ is strongly stable if and only if $\mathbf{A}^*$ is strongly stable.*

Proof Parts (i) and (ii) are contained in Theorem 3.4.9. Thus, at this point we only need to prove (iii).

(a) See Theorem 3.14.4 (and recall that $\alpha \in \rho_\infty(A)$) for the invariance. The co-invariance claim follows from this by duality.

(b) In the isometric case this is a special case of the equivalence of (i) and (viii) in Theorem 11.2.5 (take $U = Y = 0$, so that the input and output are absent). The co-isometric and unitary cases follow from this by duality.

(c) This follows from (a) and (b).

(d) Let $\Sigma = \left[\frac{\mathfrak{A}}{\mathfrak{C}}\right]$ be an (scattering) energy preserving system on $(Y, X, 0)$ constructed in Theorem 11.4.5. Then, by Theorem 11.3.4, $\mathfrak{A}$ is strongly stable if and only if $\mathfrak{C}^*\mathfrak{C} = 1$. Let $\boldsymbol{\Sigma} = \left[\frac{\mathbb{A}}{\mathbb{C}}\right]$ be the corresponding Cayley transformed system. Then, by Theorem 12.3.5, $\mathbb{C}^*\mathbb{C} = \mathfrak{C}^*\mathfrak{C}$ (since $\mathcal{L}$ is unitary), hence $\mathbb{C}^*\mathbb{C} = 1$ if and only if $\mathfrak{A}$ is strongly stable. By the discrete time version of Theorem 11.3.4 (the proof of this theorem is exactly the same in discrete time), this is equivalent to the strong stability of $\mathbb{A}$, or equivalently, of $\mathbf{A}$.

(e) This follows from (d) by duality. □

As we remarked at the end of Section 12.1, the theory presented in Chapter 11 on (scattering) passive and conservative systems has a discrete time analogue, which can be developed by simply repeating the proofs given there, but working in discrete time instead of continuous time. An alternative way to derive this theory would be to use the Cayley transform, which provides us with an almost one-to-one correspondence between a continuous time and a discrete time passive system. The only difference is that the discrete time systems that we get through the Cayley transform have the property that -1 is not an eigenvalue of the main operator. We have listed some of the properties which are conserved under the Cayley transform in the following theorem.

Theorem 12.3.11 *Let U, X, and Y be Hilbert spaces, let $\Sigma = \left[\begin{smallmatrix} \mathfrak{A} & \mathfrak{B} \\ \mathfrak{C} & \mathfrak{D} \end{smallmatrix}\right]$ be a bounded L^2-well-posed system on (Y, X, U), let $\alpha \in \mathbb{C}^+$, and let $\mathbf{\Sigma} = \left[\begin{smallmatrix} \mathbb{A} & \mathbb{B} \\ \mathbb{C} & \mathbb{D} \end{smallmatrix}\right]$ be the corresponding Cayley transformed system. Then the following claims are true.*

(i) *The reachable and unobservable subspaces of Σ and $\mathbf{\Sigma}$ are the same.*
(ii) *$\mathbf{\Sigma}$ is (scattering) passive if and only if Σ is (scattering) passive. In this case the deficiency operators of the two systems are the same (the continuous time deficiency operators are defined in Theorem 11.1.6, and the discrete time deficiency operators are defined in the analogous way). In particular, $\mathbf{\Sigma}$ is energy preserving, or co-energy preserving, or conservative, or simple, or semi-lossless, or co-lossless, or lossless, or strongly stable, or strongly co-stable if and only if Σ has the same property.*

Proof (i) This follows from Lemma 12.2.6 (the boundedness of Σ implies that $\mathbb{C}^+ \subset \rho_\infty(A)$, hence $\alpha \in \rho_\infty(A)$).

(ii) That passivity, energy preservation, co-energy preservation, and conservativity are preserved follows from Definition 11.2.1 and Theorems 11.1.5 and 11.2.5 (note that the quadruple of operators $\left[\begin{smallmatrix} \mathbf{A}(\alpha) & \mathbf{B}(\alpha) \\ \mathbf{C}(\alpha) & \mathbf{D}(\alpha) \end{smallmatrix}\right]$ appearing in these theorems are exactly the generators of the Cayley transformed system). That semi-losslessness, co-losslessness, and losslessness are preserved follows from Proposition 11.3.2, and that strong stability and strong co-stability are preserved follows from Theorem 12.3.10.

It remains to prove that the deficiency operators are the same. We start by showing that, in the notation of Theorem 11.1.6, $Q_{\mathfrak{A}^*,\mathfrak{A}} = Q_{\mathbb{A}^*,\mathbb{A}}$, i.e., that $\lim_{t\to\infty} \mathfrak{A}^{*t}\mathfrak{A}^t = \lim_{n\to\infty} \mathbf{A}^{*n}\mathbf{A}^n$. To do this we ignore the input part of the original system Σ, and replace $\mathfrak{C}$ temporarily by the output map $\mathfrak{C}_1$ given in Theorem 11.4.5 which makes $\left[\begin{smallmatrix} \mathfrak{A} \\ \mathfrak{C}_1 \end{smallmatrix}\right]$ energy preserving, and look at the corresponding Cayley transformed system $\left[\begin{smallmatrix} \mathbb{A} \\ \mathbb{C}_1 \end{smallmatrix}\right]$. By Theorem 12.3.5, $\mathbb{C}_1^*\mathbb{C}_1 = \mathfrak{C}_1^*\mathfrak{C}_1$ (since $\mathcal{L}$ is unitary), and by Theorem 11.2.9 and its discrete time counterpart,

$$Q_{\mathfrak{A}^*,\mathfrak{A}} + \mathfrak{C}_1^*\mathfrak{C}_1 = 1 = Q_{\mathbb{A}^*,\mathbb{A}} + \mathbb{C}_1^*\mathbb{C}_1.$$

Thus, $Q_{\mathfrak{A}^*,\mathfrak{A}} = Q_{\mathbb{A}^*,\mathbb{A}}$.

The same argument can be used to show that $\begin{bmatrix} Q_{\mathfrak{A}^*,\mathfrak{A}} & Q_{\mathfrak{A}^*,\mathfrak{B}} \\ Q_{\mathfrak{B}^*,\mathfrak{A}} & Q_{\mathfrak{B}^*,\mathfrak{B}} \end{bmatrix}$ coincides with its discrete time counterpart. To do this we ignore the output part of the system, and look at the Lax–Phillips semigroup

$$\mathfrak{T}^t = \begin{bmatrix} \mathfrak{A}^t & \mathfrak{B}_0^t \\ 0 & \tau_+^t \end{bmatrix}$$

induced by $\begin{bmatrix} \mathfrak{A} | \mathfrak{B} \end{bmatrix}$ on $\begin{bmatrix} X \\ L^2(\mathbb{R}^+;U) \end{bmatrix}$ (the output part of this semigroup is absent). By Lemma 11.1.4(vi), $\mathfrak{T}$ is a contraction semigroup. Clearly

$$\mathfrak{T}^{*t}\mathfrak{T}^t = \begin{bmatrix} \mathfrak{A}^{*t} \\ (\mathfrak{B}_0^t)^* \end{bmatrix} \begin{bmatrix} \mathfrak{A}^t & \mathfrak{B}_0^t \end{bmatrix} + \begin{bmatrix} 0 & 0 \\ 0 & \pi_{[t,\infty)} \end{bmatrix}.$$

Letting $t \to \infty$ we get

$$\begin{bmatrix} Q_{\mathfrak{A}^*,\mathfrak{A}} & Q_{\mathfrak{A}^*,\mathfrak{B}} \\ Q_{\mathfrak{B}^*,\mathfrak{A}} & Q_{\mathfrak{B}^*,\mathfrak{B}} \end{bmatrix} = \lim_{t\to\infty} \mathfrak{T}^{*t}\mathfrak{T}^t.$$

The corresponding result is also true in discrete time (we leave it to the reader to check that the Cayley transform of the Lax–Phillips system is the Lax–Phillips system induced by the Cayley transform). We have just seen that $Q_{\mathfrak{A}^*,\mathfrak{A}} = Q_{\mathbb{A}^*,\mathbb{A}}$ (whenever $\mathfrak{A}$ is a contraction semigroup), and by applying this result with $\mathfrak{A}$ replaced by $\mathfrak{T}$ we find that $\begin{bmatrix} Q_{\mathfrak{A}^*,\mathfrak{A}} & Q_{\mathfrak{A}^*,\mathfrak{B}} \\ Q_{\mathfrak{B}^*,\mathfrak{A}} & Q_{\mathfrak{B}^*,\mathfrak{B}} \end{bmatrix} = \begin{bmatrix} Q_{\mathfrak{A}^*,\mathfrak{A}} & Q_{\mathfrak{A}^*,\mathfrak{B}} \\ Q_{\mathfrak{B}^*,\mathfrak{A}} & Q_{\mathfrak{B}^*,\mathfrak{B}} \end{bmatrix}$.

That also the adjoint deficiency operators for the two systems coincide follows by duality. □

12.4 The reciprocal transform

Another important linear fractional transform is the *reciprocal transform*, which corresponds to the linear fractional transform $z \mapsto 1/z$. We have already encountered it in Section 12.1, where we interpreted it as the *time-inversion* of a discrete time system. This interpretation is related to the fact that the mapping $\lambda \mapsto 1/\lambda$ maps the outside $\mathbb{D}^+$ of the unit disk one-to-one onto the inside $\mathbb{D}^-$ of the unit disk. However, the same transform also maps the right half-plane $\mathbb{C}^+$ one-to-one onto itself, and this makes it possible to alternatively interpret this as a transform which maps *a continuous time system into another continuous time system*. It is in this context that the name 'reciprocal transform' has been used lately.[4]

Since much of the theory is very similar to the theory of the Cayley transform, we shall be rather brief at this point.

[4] Instead of using the transform $\lambda \mapsto 1/\lambda$ one may equally well use the transform $\lambda \mapsto 1/(\lambda - j\alpha)$ for some fixed $j\alpha \in j\mathbb{R}$, $\alpha \neq 0$. The inverse transform is then given by $z \mapsto 1/z + j\alpha$.

Lemma 12.4.1 *Let U and Y be Hilbert spaces.*

(i) *The operator $\mathcal{R}$ which takes $u \in L^2(\mathbb{R};U)$ into the inverse Laplace transform of the function $\lambda \mapsto \frac{1}{\lambda}\hat{u}(\frac{1}{\lambda})$ is a unitary map of $L^2(\mathbb{R};U)$ onto itself. (An equivalent way of writing this is that $\widehat{\mathcal{R}u}(\lambda) = \frac{1}{\lambda}\hat{u}(\frac{1}{\lambda})$.) It is causal in the sense that it maps $L^2(\mathbb{R}^+;U)$ and $L^2(\mathbb{R}^-;U)$ onto themselves. Moreover, $\mathcal{R}^2 = 1$.*

(ii) *If $\mathfrak{D} \in TI^2(U;Y)$, then the operator $\mathfrak{D}^{\downarrow} := \mathcal{R}\mathfrak{D}\mathcal{R}$ is also in $TI^2(U;Y)$ (i.e., it is time-invariant and bounded from $L^2(\mathbb{R};U)$ to $L^2(\mathbb{R};Y)$). If $\mathfrak{D} \in TIC^2(U;Y)$, then $\mathfrak{D}^{\downarrow} \in TIC^2(U;Y)$.*

Proof (i) The Laplace transform is a unitary map of $L^2(\mathbb{R}^+;U)$ onto $H^2(U)$, of $L^2(\mathbb{R}^+;U)$ onto H^2 over $\mathbb{C}^-$ with values in U, and of $L^2(\mathbb{R};U)$ onto $L^2(j\mathbb{R};U)$ (after we divide the norm in H^2 by $\sqrt{2\pi}$; see Theorem 10.3.4). Therefore, it suffices that the map which takes $\hat{u}$ into the function $\lambda \mapsto \frac{1}{\lambda}\hat{u}(\frac{1}{\lambda})$ is a unitary map of $H^2(U)$ onto itself, of H^2 over $\mathbb{C}^-$ with values in U onto itself, and of $L^2(j\mathbb{R};U)$ onto itself. But this is more or less obvious.

(ii) Clearly $\mathfrak{D}^{\downarrow}$ is bounded. To show that $\mathfrak{D}^{\downarrow}$ is time-invariant it suffices to show that it commutes with the Laguerre shift $\boldsymbol{\mu} := \mathcal{L}^{-1}\sigma\mathcal{L}$ (see the proof of part (iv) of Theorem 12.3.3). Let us choose the parameter α in the definition of the Laguerre shift to be real. Then it follows from Theorem 12.4.3 below that $\boldsymbol{\mu}\mathcal{R} = -\mathcal{R}\boldsymbol{\mu}$. We know that $\boldsymbol{\mu}$ commutes with $\mathfrak{D}$ since $\mathfrak{D}$ is time-invariant. Thus,

$$\mathfrak{D}^{\downarrow}\boldsymbol{\mu} = \mathcal{R}\mathfrak{D}\mathcal{R}\boldsymbol{\mu} = -\mathcal{R}\mathfrak{D}\boldsymbol{\mu}\mathcal{R} = -\mathcal{R}\boldsymbol{\mu}\mathfrak{D}\mathcal{R} = \boldsymbol{\mu}\mathcal{R}\mathfrak{D}\mathcal{R} = \boldsymbol{\mu}\mathfrak{D}^{\downarrow}.$$

This proves that $\mathfrak{D}^{\downarrow}$ is time-invariant. □

Definition 12.4.2 We call the operator $\mathcal{R}$ defined in Lemma 12.4.1 the *reciprocal transform* on $L^2(\mathbb{R};U)$.

The reciprocal transform on $L^2(\mathbb{R};U)$ can easily be expressed in terms of the Laguerre basis in Theorem 12.3.1 as follows.

Theorem 12.4.3 *Let $\alpha = |\alpha|e^{j\theta} \in \mathbb{C}^+$ and $u \in L^2(\mathbb{R}^+;U)$, and expand u into the (orthogonal) Laguerre series $u = \sum_{k=0}^{\infty} a_k\varphi_{\alpha,k}$ (where the subindex α stands for the parameter α used in the definition of the Laguerre functions in Theorem 12.3.1). Then the coefficients b_k in the Laguerre expansion $\mathcal{R}u = \sum_{k=0}^{\infty} b_k\varphi_{1/\alpha,k}$ of $\mathcal{R}u$ (note that we here use the Laguerre functions with parameter $1/\alpha$) is given by*

$$b_k = (-1)^k e^{(2k+1)j\theta} a_k.$$

In particular, $|b_k| = |a_k|$.

If we take take α to be real, then $\theta = 0$, and

$$b_k = (-1)^k a_k, \qquad k = 0, 1, 2, \ldots$$

for all k. If, in addition, $\alpha = 1$, then both expansions use the *same* Laguerre functions.

Proof To prove this result it suffices to prove that

$$\mathcal{R}\varphi_{\alpha,k} = (-1)^k e^{(2k+1)j\theta} \varphi_{1/\alpha,k}$$

(since both $\{\varphi_{\alpha,k}\}_{k=0}^{\infty}$ and $\{\varphi_{1/\alpha,k}\}_{k=0}^{\infty}$ are orthogonal bases in $L^2(\mathbb{R}^+; U)$). This is a simple algebraic manipulation based on the explicit formulas for the Laplace transforms of the Laguerre functions given in Theorem 12.3.1, and we leave this to the reader. □

In our presentation of the fractional linear transform in Section 12.2 we throughout assumed that $\alpha \in \rho(A)$, where A is the main operator of the system and α is the point which gets mapped into infinity. The reason for this was that we wanted the generators of the transformed system to be bounded, which is a natural requirement if we want to interpret the new system as a discrete time system. In the case of the reciprocal transform we shall interpret the transformed system as a continuous time system, and it is possible to allow the generators to be unbounded, as long as the formal transform of the original system node is a system node. However, let us begin with the simplest case, where the generators of the transformed system are bounded.

Definition 12.4.4 Let $\Sigma = \left[\begin{smallmatrix} \mathfrak{A} & \mathfrak{B} \\ \mathfrak{C} & \mathfrak{D} \end{smallmatrix}\right]$ be an L^2-well-posed linear system on (Y, X, U) with main operator A, control operator B, observation operator C, and transfer function $\widehat{\mathfrak{D}}$. Suppose that $0 \in \rho(A)$. Then the system $\Sigma^{\downarrow}$ generated by the bounded operators

$$S^{\downarrow} = \left[\begin{array}{c|c} A^{\downarrow} & B^{\downarrow} \\ \hline C^{\downarrow} & D^{\downarrow} \end{array}\right] := \left[\begin{array}{c|c} A^{-1} & -A_{|X}^{-1} B \\ \hline CA^{-1} & \widehat{\mathfrak{D}}(0) \end{array}\right]$$

is called the *reciprocal transform* of Σ.

Since this is a special case of the linear fractional transform (with $\alpha = 0$, $\beta = -1$, $\gamma = 0$, and $\delta = 1$), we can immediately say something about the reciprocal system.

Lemma 12.4.5 *Let $\Sigma = \left[\begin{smallmatrix} \mathfrak{A} & \mathfrak{B} \\ \mathfrak{C} & \mathfrak{D} \end{smallmatrix}\right]$ be an L^2-well-posed linear system on the Hilbert spaces (Y, X, U), with main operator A and transfer function $\widehat{\mathfrak{D}}$. Assume that $0 \in \rho(A)$, and denote the reciprocal system by $\Sigma^{\downarrow} = \left[\begin{smallmatrix} \mathfrak{A}^{\downarrow} & \mathfrak{B}^{\downarrow} \\ \mathfrak{C}^{\downarrow} & \mathfrak{D}^{\downarrow} \end{smallmatrix}\right]$, its generating operators by $\left[\begin{smallmatrix} A^{\downarrow} & B^{\downarrow} \\ C^{\downarrow} & D^{\downarrow} \end{smallmatrix}\right]$, and its transfer function by $\widehat{\mathfrak{D}}^{\downarrow}$.*

(i) $\rho(A^{\downarrow}) = 1/\rho(A) \cup \{0\}$ *if* A *is bounded,* $\rho(A^{\downarrow}) = 1/\rho(A)$ *if* A *is unbounded, and* $\widehat{\mathfrak{D}}^{\downarrow}(\lambda) = \widehat{\mathfrak{D}}(\frac{1}{\lambda})$ *for all* $\lambda \in \rho(A^{\downarrow})$ *(where we interpret* $1/0 = \infty$*).*

(ii) *If* $0 \in \rho_\infty(A)$*, then* $\Sigma^{\downarrow}$ *has the same reachable and unobservable subspaces as* Σ*.*

(iii) $\mathfrak{B}$*,* $\mathfrak{C}$*, or* $\mathfrak{D}$ *is bounded if and only if* $\mathfrak{B}^{\downarrow}$*,* $\mathfrak{C}^{\downarrow}$*, or* $\mathfrak{D}^{\downarrow}$ *is bounded, respectively, in which case*

$$\mathfrak{B}^{\downarrow} = \mathfrak{B}\mathcal{R}, \quad \mathfrak{C}^{\downarrow} = -\mathcal{R}\mathfrak{C}, \qquad \mathfrak{D}^{\downarrow} = \mathcal{R}\mathfrak{D}\mathcal{R}.$$

Proof (i)–(ii) This follows from Proposition 12.2.1.

(iii) Let us begin this time with the claim about the output maps. Take $\omega > \max\{\omega_{\mathfrak{A}}, \omega_{\mathfrak{A}^{\downarrow}}\}$. Then, for all $x \in X$ and all $\lambda \in \mathbb{C}^+_\omega$,

$$(\lambda - A)^{-1}Cx = -\frac{1}{\lambda}\Big(\frac{1}{\lambda} - A^{-1}\Big)^{-1} A^{-1}Cx = -\frac{1}{\lambda}\Big(\frac{1}{\lambda} - A^{\downarrow}\Big)^{-1} C^{\downarrow}x.$$

The left-hand side has an analytic extension to a function in $H^2(\mathbb{C}^+; U)$ if and only if $\mathfrak{C}$ is bounded. Let us denote this extension by $-\hat{y}$. Then also the the function $\lambda \mapsto \frac{1}{\lambda}\hat{y}(\frac{1}{\lambda})$ belongs to $H^2(\mathbb{C}^+; U)$, and

$$\lambda \mapsto \frac{1}{\lambda}\hat{y}\Big(\frac{1}{\lambda}\Big) = (\lambda - A^{\downarrow})^{-1}C^{\downarrow}x$$

whenever $\frac{1}{\lambda} \in \mathbb{C}^+_\omega$. Thus, the function $(\lambda - A^{\downarrow})^{-1}C^{\downarrow}x$ has an analytic extension to a function in $H^2(\mathbb{C}^+; U)$. This implies that $\mathfrak{C}^{\downarrow}$ is bounded. Furthermore, it follows from the above computation that $\mathfrak{C}^{\downarrow}x = -\mathcal{R}\mathfrak{C}x$. The proof of the fact that boundedness of $\mathfrak{C}^{\downarrow}$ implies boundedness of $\mathfrak{C}$ is essentially the same (interchange the roles of $\mathfrak{C}$ and $\mathfrak{C}^{\downarrow}$).

That $\mathfrak{B}$ is bounded if and only if $\mathfrak{B}^{\downarrow}$ is bounded, and that $\mathfrak{B}^{\downarrow} = \mathfrak{B}\mathcal{R}$ can be proved in a similar way (or it can be derived from the above result by duality).

This only leaves the two input/output maps $\mathfrak{D}$ and $\mathfrak{D}^{\downarrow}$. Again take $\omega > \max\{\omega_{\mathfrak{A}}, \omega_{\mathfrak{A}^{\downarrow}}\}$. Then by (i), for all $\lambda \in \mathbb{C}^+_\omega$, $\widehat{\mathfrak{D}}^{\downarrow}(\lambda) = \widehat{\mathfrak{D}}(\frac{1}{\lambda})$. By Theorem 10.3.5, $\mathfrak{D}$ is bounded if and only if $\widehat{\mathfrak{D}}$ can be extended to a function in $H^\infty(U; Y)$, whereas $\mathfrak{D}^{\downarrow}$ is bounded if and only if $\widehat{\mathfrak{D}}^{\downarrow}$ can be extended to a function in $H^\infty(U; Y)$. The proof of the equivalence of these two conditions is essentially the same one which we gave for the two output maps $\mathfrak{C}$ and $\mathfrak{C}^{\downarrow}$. Also the identity $\mathfrak{D}^{\downarrow} = \mathcal{R}\mathfrak{D}\mathcal{R}$ follows immediately. □

To treat the more general case where the generators of the transformed system are allowed to be unbounded we can use the same method as in Theorem 12.3.7 and assume that $\mathfrak{B}$, $\mathfrak{C}$, and $\mathfrak{D}$ are bounded.

Theorem 12.4.6 *Let* $\Sigma = \left[\begin{smallmatrix} \mathfrak{A} & \mathfrak{B} \\ \mathfrak{C} & \mathfrak{D} \end{smallmatrix}\right]$ *be an* L^2*-well-posed linear system on* (Y, X, U)*, where* Y *and* U *are Hilbert spaces, with main operator* A*. Suppose* *that A is injective, that A^{-1} is the generator of a C_0 semigroup $\mathfrak{A}^{\downarrow}$, and that $\mathfrak{B}$, $\mathfrak{C}$ and $\mathfrak{D}$ are stable. Then $\Sigma^{\downarrow} = \left[\begin{array}{c|c} \mathfrak{A}^{\downarrow} & \mathfrak{B}^{\downarrow} \\ \hline \mathfrak{C}^{\downarrow} & \mathfrak{D}^{\downarrow} \end{array}\right]$ is a continuous time L^2-well-posed linear system, where*

$$\mathfrak{B}^{\downarrow} = \mathfrak{B}\mathcal{R}, \quad \mathfrak{C}^{\downarrow} = -\mathcal{R}\mathfrak{C}, \qquad \mathfrak{D}^{\downarrow} = \mathcal{R}\mathfrak{D}\mathcal{R};$$

here $\mathcal{R}$ is the reciprocal transform on $L^2(\mathbb{R}; U)$ or $L^2(\mathbb{R}; Y)$. In particular, $\mathfrak{B}^{\downarrow}$, $\mathfrak{C}^{\downarrow}$, and $\mathfrak{D}^{\downarrow}$ are stable. The transfer function of Σ is

$$\widehat{\mathfrak{D}}^{\downarrow}(\lambda) = \widehat{\mathfrak{D}}\Big(\frac{1}{\lambda}\Big), \qquad \lambda \in \rho(A) \cap \rho(A^{\downarrow}).$$

If $A \in \mathcal{B}(X)$, then this system coincides with the one in Definition 12.4.4.

It is possible to prove this theorem directly by imitating the proof of Theorem 12.3.7, but to save some space we instead reduce it to a special case of that theorem.

Proof Choose some real $\alpha > \max\{0, \omega_{\mathfrak{A}}\}$, and let $\mathbf{\Sigma} = \left[\begin{array}{c|c} \mathbb{A} & \mathbb{B} \\ \hline \mathbb{C} & \mathbb{D} \end{array}\right]$ be the Cayley transform of Σ with generating operators $\left[\begin{smallmatrix} \mathbf{A} & \mathbf{B} \\ \mathbf{C} & \mathbf{D} \end{smallmatrix}\right]$. By Theorem 12.3.5, $\mathbb{B}$, $\mathbb{C}$, and $\mathbb{D}$ are bounded. Let us apply the linear fractional transform to this system which maps $z \mapsto -z$ (this particular transformation is not among the ones listed in Proposition 12.2.1, but it has the same type of properties as those listed there; in particular, it preserves controllability and observability since $\mathbf{A}$ is bounded). That is, we replace $\left[\begin{smallmatrix} \mathbf{A} & \mathbf{B} \\ \mathbf{C} & \mathbf{D} \end{smallmatrix}\right]$ by $\left[\begin{smallmatrix} -\mathbf{A} & \mathbf{B} \\ -\mathbf{C} & \mathbf{D} \end{smallmatrix}\right]$. Let us denote the corresponding system by $\widetilde{\mathbf{\Sigma}} = \left[\begin{array}{c|c} \widetilde{\mathbb{A}} & \widetilde{\mathbb{B}} \\ \hline \widetilde{\mathbb{C}} & \widetilde{\mathbb{D}} \end{array}\right]$. We obtain $\widetilde{\mathbb{B}}$ and $\widetilde{\mathbb{C}}$ from $\mathbb{B}$ and $\mathbb{C}$ by simply changing the sign of every second term, so $\widetilde{\mathbb{B}}$ and $\widetilde{\mathbb{C}}$ are bounded. The transfer functions satisfy $\widehat{\widetilde{\mathbb{D}}}(z) = \widehat{\mathbb{D}}(-z)$, and it follows from the discrete time analogue of Theorem 10.3.5 (with $\mathbb{C}^+$ replaced by $\mathbb{D}^+$) that $\widetilde{\mathbb{D}}$ is bounded.

The next (and final) step is to apply the inverse Cayley transform with parameter $1/\alpha$ to $\widetilde{\mathbf{\Sigma}}$. For this to be possible we need to know that $1 - \mathbf{A}$ is injective, and that $1/\alpha - 2/\alpha(1 - \mathbf{A})^{-1}$ generates a C_0 semigroup. That

$$1 - \mathbf{A} = 1 - (\alpha + A)(\alpha - A)^{-1} = -2A(\alpha - A)^{-1}$$

follows from our assumptions that $\alpha \in \rho(A)$ and A is injective, and

$$\frac{1}{\alpha} - \frac{2}{\alpha}(1 - \mathbf{A})^{-1} = \frac{1}{\alpha}\Big(1 + (\alpha - A)A^{-1}\Big) = A^{-1}$$

was explicitly assumed to generate a C_0 semigroup. Thus, by Theorem 12.3.7, the inverse Cayley transform is an L^2-well-posed linear system. We denote this system by $\Sigma^{\downarrow} = \left[\begin{array}{c|c} \mathfrak{A}^{\downarrow} & \mathfrak{B}^{\downarrow} \\ \hline \mathfrak{C}^{\downarrow} & \mathfrak{D}^{\downarrow} \end{array}\right]$.

We claim that the system which we have obtained has all the properties listed in Theorem 12.4.6. We already know that the main operator is A^{-1}, and we also know from Theorem 12.3.7 that $\mathfrak{B}^{\downarrow}$, $\mathfrak{C}^{\downarrow}$, and $\mathfrak{D}^{\downarrow}$ are stable. That $\mathfrak{B}^{\downarrow} = \mathfrak{B}\mathcal{R}$,

$\mathfrak{C}^{\downarrow} = -\mathcal{R}\mathfrak{C}$, and $\mathfrak{D}^{\downarrow} = \mathcal{R}\mathfrak{D}\mathcal{R}$ follows from Theorems 12.3.7 and 12.4.3. In particular, if $A \in \mathcal{B}(X)$, then by Lemma 12.4.5(iii), this is the same system as in Definition 12.4.4. The formula for the transform function can be proved by tracing the three separate transforms:

$$\widehat{\mathfrak{D}}^{\downarrow}(\lambda) = \widehat{\mathbb{D}}\Big(\frac{1/\alpha + \lambda}{1/\alpha - \lambda}\Big) = \widehat{\mathbb{D}}\Big(\frac{\lambda + 1/\alpha}{\lambda - 1/\alpha}\Big) = \widehat{\mathbb{D}}\Big(\frac{\alpha + 1/\lambda}{\alpha - 1/\lambda}\Big) = \widehat{\mathfrak{D}}\Big(\frac{1}{\lambda}\Big).$$

□

Definition 12.4.7 In the sequel we shall also refer to the system $\Sigma^{\downarrow}$ in Theorem 12.4.6 as the *reciprocal transform* of Σ (cf. Definition 12.4.4).

Theorem 12.4.6 does not yet contain a statement about the connection between the system node $S = \left[\begin{smallmatrix} A\&B \\ C\&D \end{smallmatrix}\right]$ of Σ and the system node $S^{\downarrow} = \left[\begin{smallmatrix} [A\&B]^{\downarrow} \\ [C\&D]^{\downarrow} \end{smallmatrix}\right]$ of $\Sigma^{\downarrow}$. A formal computation (based on Definition 12.4.4) indicates that we ought to have

$$\begin{bmatrix} A\&B \\ C\&D \end{bmatrix} = \begin{bmatrix} 1 & 0 \\ [C\&D]^{\downarrow} \end{bmatrix} \begin{bmatrix} [A\&B]^{\downarrow} \\ 0 \quad 1 \end{bmatrix}^{-1} \quad (\text{on } \mathcal{D}(S)), \tag{12.4.1}$$

$$\begin{bmatrix} [A\&B]^{\downarrow} \\ [C\&D]^{\downarrow} \end{bmatrix} = \begin{bmatrix} 1 & 0 \\ C\&D \end{bmatrix} \begin{bmatrix} A\&B \\ 0 \quad 1 \end{bmatrix}^{-1} \quad (\text{on } \mathcal{D}(S^{\downarrow})). \tag{12.4.2}$$

Indeed, the latter of these does define an operator node $S^{\downarrow} = \left[\begin{smallmatrix} [A\&B]^{\downarrow} \\ [C\&D]^{\downarrow} \end{smallmatrix}\right]$ under very natural assumptions.

Lemma 12.4.8 *Let $S = \left[\begin{smallmatrix} A\&B \\ C\&D \end{smallmatrix}\right]$ be an operator node on the Banach spaces (Y, X, U), with main operator A. If A is injective and has dense range, then formula* (12.4.2) *defines an operator node $S^{\downarrow} = \left[\begin{smallmatrix} [A\&B]^{\downarrow} \\ [C\&D]^{\downarrow} \end{smallmatrix}\right]$. The operator $\left[\begin{smallmatrix} A\&B \\ 0 \;\; 1 \end{smallmatrix}\right]$ maps $\mathcal{D}(S)$ one-to-one onto $\mathcal{D}(S^{\downarrow})$. The main operator $A^{\downarrow}$ of $S^{\downarrow}$ is injective and has dense range, and S can be recovered from $S^{\downarrow}$ by formula* (12.4.1). *The operator node $S^{\downarrow}$ is bounded if and only if $0 \in \rho(A)$, in which case it coincides with the system node in Definition 12.4.4*

Proof Our proof is based on Lemma 4.7.7, so we have to verify the assumptions (i)–(iv) of that lemma.

We begin by showing that $S^{\downarrow}$ is closed with $\mathcal{D}(S^{\downarrow}) = \mathcal{R}\left(\left[\begin{smallmatrix} A\&B \\ 0 \;\; 1 \end{smallmatrix}\right]\right)$. By Lemma 4.7.3, the operator $A\&B$ is closed, and this implies that also $\left[\begin{smallmatrix} A\&B \\ 0 \;\; 1 \end{smallmatrix}\right]$ is closed. It is injective since we assume A to be injective. Its inverse is closed (since $\left[\begin{smallmatrix} A\&B \\ 0 \;\; 1 \end{smallmatrix}\right]$ itself is closed). Let $\mathcal{R}\left(\left[\begin{smallmatrix} A\&B \\ 0 \;\; 1 \end{smallmatrix}\right]\right) \supset \left[\begin{smallmatrix} x_n \\ u_n \end{smallmatrix}\right] \to \left[\begin{smallmatrix} x \\ u \end{smallmatrix}\right]$ in $\left[\begin{smallmatrix} X \\ U \end{smallmatrix}\right]$, and suppose that $\left[\begin{smallmatrix} z_n \\ y_n \end{smallmatrix}\right] := S^{\downarrow}\left[\begin{smallmatrix} x \\ u \end{smallmatrix}\right] \to \left[\begin{smallmatrix} z \\ y \end{smallmatrix}\right]$ in $\left[\begin{smallmatrix} X \\ Y \end{smallmatrix}\right]$. Then $\left[\begin{smallmatrix} z_n \\ u_n \end{smallmatrix}\right] = \left[\begin{smallmatrix} A\&B \\ 0 \;\; 1 \end{smallmatrix}\right]^{-1}\left[\begin{smallmatrix} x_n \\ u_n \end{smallmatrix}\right]$, and $\left[\begin{smallmatrix} z_n \\ y_n \end{smallmatrix}\right] = \left[\begin{smallmatrix} 1 \;\; 0 \\ C\&D \end{smallmatrix}\right]\left[\begin{smallmatrix} z_n \\ u_n \end{smallmatrix}\right]$. Therefore $\left[\begin{smallmatrix} z_n \\ y_n \end{smallmatrix}\right] \to \left[\begin{smallmatrix} z \\ u \end{smallmatrix}\right]$ (since $\left[\begin{smallmatrix} A\&B \\ 0 \;\; 1 \end{smallmatrix}\right]^{-1}$ is closed), and $y_n \to C\&D\left[\begin{smallmatrix} z \\ u \end{smallmatrix}\right]$ since $C\&D$ is continuous with respect to the graph norm of

$A\&B$ (see Lemma 4.7.3(v) and (ix)). Thus $\left[\begin{smallmatrix} x \\ u \end{smallmatrix}\right] = \left[\begin{smallmatrix} A\&B \\ 0 \ 1 \end{smallmatrix}\right]\left[\begin{smallmatrix} z \\ u \end{smallmatrix}\right] \in \mathcal{R}\left(\left[\begin{smallmatrix} A\&B \\ 0 \ 1 \end{smallmatrix}\right]\right) = \mathcal{D}\left(S^{\downarrow}\right)$ and $\left[\begin{smallmatrix} z \\ y \end{smallmatrix}\right] = \left[\begin{smallmatrix} 1 \ 0 \\ C\&D \end{smallmatrix}\right]\left[\begin{smallmatrix} z \\ u \end{smallmatrix}\right]$, $y = C\&D\left[\begin{smallmatrix} z \\ u \end{smallmatrix}\right]$, $\left[\begin{smallmatrix} z \\ y \end{smallmatrix}\right] = S^{\downarrow}\left[\begin{smallmatrix} x \\ u \end{smallmatrix}\right]$. This proves that $S^{\downarrow}$ is closed.

It follows from (12.4.2) that $\left[\begin{smallmatrix} [A\&B]^{\downarrow} \\ 0 \ 1 \end{smallmatrix}\right] = \left[\begin{smallmatrix} A\&B \\ 0 \ 1 \end{smallmatrix}\right]^{-1}$. In particular, this implies that $\left[\begin{smallmatrix} [A\&B]^{\downarrow} \\ 0 \ 1 \end{smallmatrix}\right]$ is closed, hence so is $[A\&B]^{\downarrow}$. It also implies that $A^{\downarrow} = A^{-1}$, with $\mathcal{D}\left(A^{\downarrow}\right) = \mathcal{R}(A)$. Thus, $A^{\downarrow}$ is injective, and by assumption, $\mathcal{D}\left(A^{\downarrow}\right) = \mathcal{R}(A)$ is dense in X.

The last assumption of Lemma 4.7.7 requires that for every $u \in U$ there should exist an $x \in X$ with $\left[\begin{smallmatrix} x \\ u \end{smallmatrix}\right] \in \mathcal{D}\left(S^{\downarrow}\right)$. But this follows from the same part of Lemma 4.7.7 applied to S itself: there is a $z \in X$ such that $\left[\begin{smallmatrix} x \\ u \end{smallmatrix}\right] \in \mathcal{D}(S)$. Let $x = A\&B\left[\begin{smallmatrix} z \\ u \end{smallmatrix}\right]$. Then $\left[\begin{smallmatrix} x \\ u \end{smallmatrix}\right] \in \mathcal{R}\left(\left[\begin{smallmatrix} A\&B \\ 0 \ 1 \end{smallmatrix}\right]\right) = \mathcal{D}\left(S^{\downarrow}\right)$.

By Lemma 4.7.7, $S^{\downarrow}$ is an operator node on (Y, X, U).

Let us next check that S can be recovered from $S^{\downarrow}$ by formula (12.4.1). By (12.4.1),

$$\begin{bmatrix} 1 & 0 \\ \multicolumn{2}{c}{[C\&D]^{\downarrow}} \end{bmatrix}\begin{bmatrix} \multicolumn{2}{c}{A\&B} \\ 0 & 1 \end{bmatrix} = \begin{bmatrix} A\&B \\ C\&D \end{bmatrix}$$

(more precisely, the top row is trivial, and the second row follows from (12.4.1)). Replacing $\left[\begin{smallmatrix} A\&B \\ 0 \ 1 \end{smallmatrix}\right]$ by $\left[\begin{smallmatrix} [A\&B]^{\downarrow} \\ 0 \ 1 \end{smallmatrix}\right]^{-1}$ we get (12.4.1).

Clearly, $0 \in \rho(A)$ if and only if $A^{\downarrow}$ is bounded, and this is true if and only if $S^{\downarrow}$ is bounded (because then $X_{-1}^{\downarrow} = X$). In this case (use (12.4.2) and Lemma 4.7.18 with $\alpha = 0$)

$$\begin{aligned} S^{\downarrow} &= \begin{bmatrix} A^{\downarrow} & B^{\downarrow} \\ C^{\downarrow} & D^{\downarrow} \end{bmatrix} = \begin{bmatrix} 1 & 0 \\ \multicolumn{2}{c}{C\&D} \end{bmatrix}\begin{bmatrix} \multicolumn{2}{c}{A\&B} \\ 0 & 1 \end{bmatrix}^{-1} = \begin{bmatrix} 1 & 0 \\ \multicolumn{2}{c}{C\&D} \end{bmatrix}\begin{bmatrix} A^{-1} & -A_{|X}^{-1}B \\ 0 & 1 \end{bmatrix} \\ &= \begin{bmatrix} A^{-1} & -A_{|X}^{-1}B \\ CA^{-1} & \widehat{\mathfrak{D}}(0) \end{bmatrix}, \end{aligned}$$

which is the same system node as in Definition 12.4.4. □

Definition 12.4.9 We call the operator node $S^{\downarrow}$ the *reciprocal* of the operator node S. (It is defined whenever the main operator of S is injective and has dense range.)

Lemma 12.4.10 *The operator nodes* $S = \left[\begin{smallmatrix} A\&B \\ C\&D \end{smallmatrix}\right]$ *and* $S^{\downarrow} = \left[\begin{smallmatrix} [A\&B]^{\downarrow} \\ [C\&D]^{\downarrow} \end{smallmatrix}\right]$ *in Lemma 12.4.8 have the following additional properties. (We denote the main operator, the control operator, the observation operator, and the transfer function of* S *by* A, B, C, *and* $\widehat{\mathfrak{D}}$, *and the corresponding entities for* $S^{\downarrow}$ *by* $A^{\downarrow}$, $B^{\downarrow}$, $C^{\downarrow}$, *and* $\widehat{\mathfrak{D}}^{\downarrow}$.)

(i) $A^{\downarrow} = A^{-1}$ *with* $\mathcal{D}(A^{\downarrow}) = \mathcal{R}(A)$, *and* $C^{\downarrow} = CA^{-1}$.
(ii) $0 \in \rho(A)$ *if and only if* $A^{\downarrow}$ *is bounded,* $0 \in \rho(A^{\downarrow})$ *if and only if* A *is bounded, and a nonzero* $\alpha \in \mathbb{C}$ *belongs to* $\rho(A)$ *if and only if* $1/\alpha \in \rho(A^{\downarrow})$.
(iii) *For all nonzero* $\alpha \in \rho(A)$, *we have*

$$(1 - \alpha A^{\downarrow}_{|X})^{-1} B^{\downarrow} = (\alpha - A_{|X})^{-1} B,$$
$$C^{\downarrow}(1 - \alpha A^{\downarrow})^{-1} = -C(\alpha - A)^{-1},$$
$$\widehat{\mathfrak{D}}^{\downarrow}(1/\alpha) = \widehat{\mathfrak{D}}(\alpha).$$

These formulas are also valid for $\alpha = 0$ *if* $0 \in \rho(A)$ *(i.e.,* $A^{\downarrow}$ *is bounded).*

Proof (i) We already observed in the proof of Lemma 12.4.8 that $A^{\downarrow} = A^{-1}$. Applying (12.4.2) to a vector $\left[\begin{smallmatrix} x \\ 0 \end{smallmatrix}\right]$ with $x \in \mathcal{D}(A^{\downarrow}) = \mathcal{R}(A)$ we get $C^{\downarrow} = CA^{-1}$.

(ii) This follows from (i).

(iii) The formulas in (iii) with $\alpha = 0$ reduces to the formulas in Definition 12.4.4, and we know from Lemma 12.4.8 that they are valid if $0 \in \rho(A)$. Thus, in the sequel we may take $\alpha \neq 0$.

By (i),

$$C^{\downarrow}(1 - \alpha A^{\downarrow})^{-1} = CA^{-1}(1 - \alpha A^{-1})^{-1} = -C(\alpha - A)^{-1}.$$

Multiplying (12.4.2) by $\left[\begin{smallmatrix} A\&B \\ 0 \;\; 1 \end{smallmatrix}\right]\left[\begin{smallmatrix} (\alpha - A_{|X})^{-1}B \\ 1 \end{smallmatrix}\right]$ to the right we get (cf. Lemma 4.7.18)

$$\begin{bmatrix} [A\&B]^{\downarrow} \\ [C\&D]^{\downarrow} \end{bmatrix} \begin{bmatrix} \alpha(\alpha - A_{|X})^{-1}B \\ 1 \end{bmatrix} = \begin{bmatrix} (\alpha - A_{|X})^{-1}B \\ \widehat{\mathfrak{D}}(\alpha) \end{bmatrix}.$$

The top entry gives (after splitting $[A\&B]^{\downarrow}$ into $[A\&B]^{\downarrow} = \begin{bmatrix} A^{\downarrow}_{|X} & B^{\downarrow} \end{bmatrix}$)

$$\alpha A^{\downarrow}_{|X}(\alpha - A_{|X})^{-1}B + B^{\downarrow} = (\alpha - A_{|X})^{-1}B,$$

or equivalently, $(1 - \alpha A^{\downarrow}_{|X})^{-1}B^{\downarrow} = (\alpha - A_{|X})^{-1}B$. Thus

$$\begin{aligned} \widehat{\mathfrak{D}}(\alpha) &= [C\&D]^{\downarrow} \begin{bmatrix} \alpha(\alpha - A_{|X})^{-1}B \\ 1 \end{bmatrix} = [C\&D]^{\downarrow} \begin{bmatrix} (1/\alpha - A^{\downarrow}_{|X})^{-1}B^{\downarrow} \\ 1 \end{bmatrix} \\ &= \widehat{\mathfrak{D}}^{\downarrow}(1/\alpha). \end{aligned}$$

□

Lemma 12.4.11 *Let* $\Sigma = \left[\begin{smallmatrix} \mathfrak{A} & \mathfrak{B} \\ \mathfrak{C} & \mathfrak{D} \end{smallmatrix}\right]$ *be an* L^2*-well-posed linear system on the Hilbert spaces* (Y, X, U), *with main operator* A *and transfer function* $\widehat{\mathfrak{D}}$. *Let* $\Sigma^{\downarrow} = \left[\begin{smallmatrix} \mathfrak{A}^{\downarrow} & \mathfrak{D}^{\downarrow} \\ \mathfrak{C}^{\downarrow} & \mathfrak{D}^{\downarrow} \end{smallmatrix}\right] = \left[\begin{smallmatrix} \mathfrak{A}^{\downarrow} & \mathfrak{D}^{\downarrow} \\ \mathfrak{C}^{\downarrow} & \mathfrak{D}^{\downarrow} \end{smallmatrix}\right] = \left[\begin{smallmatrix} \mathfrak{A}^{\downarrow} & \mathfrak{D}^{\downarrow} \\ \mathfrak{C}^{\downarrow} & \mathfrak{D}^{\downarrow} \end{smallmatrix}\right]$ *be the reciprocal system, as defined* *in either Definition 12.4.4 or Definition 12.4.7. Then the system node $S^{\downarrow}$ of $\Sigma^{\downarrow}$ is the reciprocal of the system node S of Σ.*

Proof This follows from Lemmas 12.4.8 and 12.4.10. □

Theorem 12.4.12 *Let X be a Hilbert spaces, and let A be the generator of a contraction semigroup $\mathfrak{A}$ on X. If A is injective, then A has dense range, and $A^{\downarrow} = A^{-1}$ is the generator of a contraction semigroup $\mathfrak{A}^{\downarrow}$. The following additional claims also hold.*

(i) *A closed subspace of X is invariant or co-invariant under $\mathfrak{A}$ if and only if it is invariant or co-invariant under $\mathfrak{A}^{\downarrow}$.*
(ii) *$\mathfrak{A}$ is isometric, or co-isometric, or unitary if and only if $\mathfrak{A}^{\downarrow}$ has the same property.*
(iii) *$\mathfrak{A}$ is completely non-unitary if and only if $\mathfrak{A}^{\downarrow}$ is completely non-unitary.*
(iv) *$\mathfrak{A}$ is strongly stable if and only if $\mathfrak{A}^{\downarrow}$ is strongly stable.*
(v) *$\mathfrak{A}^*$ is strongly stable if and only if $(\mathfrak{A}^{\downarrow})^*$ is strongly stable.*

Proof We shall prove this theorem by reducing it to Theorem 12.3.10, arguing as in the proof of Theorem 12.4.6. Let $\mathbf{A}$ be the Cayley transform of A with parameter $\alpha = 1$. By Theorem 12.3.10, $\mathbf{A}$ is a contraction, and it has one of the properties listed in (i)–(v) if and only if $\mathfrak{A}$ has the same property. Equivalently, $-\mathbf{A}$ is a contraction, and it has one of the properties listed in (i)–(v) if and only if $\mathfrak{A}$ has the same property. The operator $1 - \mathbf{A}$ is injective, since A is injective and $1 - \mathbf{A} = -2A(1 - A)^{-1}$. We can therefore take the inverse Cayley transform of $-A$, still with parameter $\alpha = 1$. The resulting operator is A^{-1} (cf. the proof of Theorem 12.4.6). By Theorem 12.3.10, A^{-1} generates a contraction semigroup $\mathfrak{A}^{\downarrow}$, which has one of the properties listed in (i)–(v) if and only if $-\mathbf{A}$ has the same property, or equivalently, $\mathfrak{A}$ has the same property. □

As our final theorem shows, the reciprocal transform preserves (scattering) passivity and many other properties.

Theorem 12.4.13 *Let U, X, and Y be Hilbert spaces, and let $\Sigma = \left[\begin{smallmatrix}\mathfrak{A} & \mathfrak{B}\\ \mathfrak{C} & \mathfrak{D}\end{smallmatrix}\right]$ be a bounded L^2-well-posed system on (Y, X, U), with main operator A. Suppose that A is injective and that A^{-1} generates a C_0 semigroup. We denote the reciprocal system by $\Sigma^{\downarrow} = \left[\begin{smallmatrix}\mathfrak{A}^{\downarrow} & \mathfrak{D}^{\downarrow}\\ \mathfrak{C}^{\downarrow} & \mathfrak{D}^{\downarrow}\end{smallmatrix}\right]$ (see Definition 12.4.4). Then the following claims are true.*

(i) *The reachable and unobservable subspaces of Σ and $\Sigma^{\downarrow}$ are the same.*
(ii) *$\Sigma^{\downarrow}$ is (scattering) passive if and only if Σ is (scattering) passive. In this case the deficiency operators of the two systems are the same. In particular, $\Sigma^{\downarrow}$ is energy preserving, or co-energy preserving, or conservative,*

or simple, or semi-lossless, or co-lossless, or lossless, or strongly stable, or strongly co-stable if and only if Σ has the same property.

Proof This follows from Theorem 12.3.11 by the same method that we used to prove Theorem 12.4.6 (take, e.g., $\alpha = 1$). We leave the details to the reader. □

12.5 Comments

Section 12.1 The results presented in this section are classic. Many of the available classical results on infinite-dimensional system theory are given in a discrete time setting (especially those with an operator theory background). One reason for this is that the well-posedness problems of Chapters 3–5 can be avoided. See, e.g., the monographs by Fuhrmann (1981) and Sz.-Nagy and Foiaş (1970).

Sections 12.2–12.3 The Cayley transform is well-known and much used in different settings, but there does not seem to be a good source describing the properties of the Cayley transform of the full system. In particular, it is not easy to find explicit references in the literature to the relationships between the discrete and continuous time input maps, input/output maps, and output maps described in Theorem 12.3.5. This transform has been used in, e.g, Arov and Nudelman (1996), Sz.-Nagy and Foiaş (1970), and Staffans (2002b) (and in many other places in different settings) to convert results originally proved for discrete time systems to continuous time. The mapping which takes a continuous time semigroup generator to its co-generator is found in most books on operator theory. For example, Sz.-Nagy and Foiaş (1970) contains a discussion of this transform which is more detailed than the one that we present here. The frequency domain version mentioned in Definition 12.3.2 is also well-known, and it is found in many books on complex function theory and harmonic analysis, such as Hoffman (1988). Laguerre series appear primarily in books on special functions and mathematical physics (and also in, e.g., Rosenblum and Rovnyak (1985)). The most common choices of the parameter α is to take either $\alpha = 1$ or $\alpha = 1/2$ (the latter choice makes the factor $\sqrt{2\Re\alpha} = 1$ disappear from the formulas). In many operator theory books a 'rotated' version of the Cayley transform is used (mapping the upper half-plane onto the unit disk). As noticed in Theorems 12.3.10 and 12.3.11, the Cayley transforms have a natural place in the theory of scattering passive and conservative systems (and in the theory of impedance scattering systems as well; see Staffans (2002b)).

Section 12.4 As we commented earlier, the reciprocal transform corresponds to time-inversion in a discrete time setting, but it has also been applied in continuous time for a long time. Livšic (1973, Section 6.3) uses the reciprocal transform to convert continuous time equations with unbounded operators into systems whose generating operators are bounded. The same idea has recently been used extensively in Curtain (2003a, b) and Opmeer and Curtain (2004) (see Curtain (2002) for an overview). Apparently this transform has been used implicitly by a number of researchers from time to time (as described in Curtain (2002)).

Appendix

In this appendix we present a number of auxiliary results about regulated functions, the positive square root and polar decomposition of a closed unbounded operator convolutions, and inversion of block matrices.

A.1 Regulated functions

Definition A.1.1 Let X be a Banach space, and let $I = [a, b] \subset \mathbb{R}$ be a closed bounded interval.

(i) A function $f : I \to X$ is a *step function* if f is constant on successive intervals $I_k = [a_k, a_{k+1})$, $0 \le k < n$, where $a_0 = a$ and $a_n = b$.
(ii) A function $f : I \to X$ is *regulated* if f is right-continuous and has a left-hand limit at every point of $[a, b]$. We denote the class of all regulated X-valued functions on $I = [a, b]$ by $Reg(I; X)$.
(iii) By an X-valued step function on a closed unbounded subinterval J of $\mathbb{R}$ we mean a function whose restriction to any closed bounded interval is a step function in the sense of (i).
(iv) By an X-valued regulated function on a closed unbounded subinterval J of $\mathbb{R}$ we mean a function whose restriction to any closed bounded interval is regulated in the sense of (ii). We denote this class of functions by $Reg_{\text{loc}}(J; X)$, and let $Reg(J; X)$ be the space of all functions which are both regulated and bounded.

Proposition A.1.2 *Let X be a Banach space, and let $I = [a, b]$ be a closed bounded interval. Then every function in $Reg(I; X)$ is bounded, and it is the uniform limit of a sequence of step functions. The converse is also true: the uniform limit of a sequence of X-valued step functions on I is regulated.*

Proof Let $f \in Reg(I; X)$, and let $\epsilon > 0$. By the right-continuity of f and the existence of left-hand limits of f, for each $t \in I$ there is an interval $(t - \delta_t, t + \delta_t)$ such that $|f(x) - f(t)| \le \epsilon$ if $x \in [t, t + \delta_t) \cap I$ and such that $|f(x) - f(y)| \le \epsilon$ if both $x \in (t - \delta_t, t] \cap I$ and $y \in (t - \delta_t, t] \cap I$. Since I is compact, we can cover I by a finite number of these intervals, and by removing the overlapping parts of this finite number of intervals we get a sequence of points t_i such that $t_0 = a$, $t_n = b$, and such that $|f(x) - f(t_i)| \le \epsilon$ if $x \in I_i = [a_i, a_{i+a})$. Thus, for all x, $|f(x) - g(x)| \le \epsilon$, where g is the step function whose value in the interval I_i is $f(t_i)$. This implies that f is bounded. Repeating the same construction with ϵ replaced by $1/n$, $n = 1, 2, 3 \ldots$, we get a sequence of step functions which converges uniformly to f.

Conversely, let g_i be a uniformly converging sequence of X-valued step functions on I. All of these are right-continuous and have left-hand limits at each point, and these two properties are preserved under uniform convergence. Thus, the limit function must still have the same properties, i.e., it is regulated. □

Corollary A.1.3 *Let X be a Banach space, and let J be a closed (bounded or unbounded) interval. Then $Reg(J; X)$ is a Banach space with the norm $\|f\|_{Reg(J;X)} = \sup_{x \in J} |f(x)|$. If the interval J is bounded, then the set of X-valued step functions is dense in $Reg(J; X)$.*

Proof As is well known, the set of all bounded X-valued functions on J is a Banach space with the sup-norm. The same argument that we used in the last part of the proof of Proposition A.1.2 shows that $Reg(J; X)$ is closed in this space. The density of the set of all step functions in $Reg(J; X)$ when J is bounded follows from Proposition A.1.2. □

Proposition A.1.4 *Let X be a Banach space, and let J be a closed (bounded or unbounded) interval. Then set space $BC(J; X)$ is a closed subspace of $Reg(J; X)$.*

Proof The uniform limit of a sequence of continuous functions is continuous. □

Proposition A.1.5 *Let X be a Banach space, and let $I = [a, b]$ be a closed bounded interval, and let $f \in Reg(I; X)$. Then*

(i) *the range of f is totally bounded, and*
(ii) *f has only countably many discontinuities in I.*

Proof (i) As we saw in the preceding proof, for each $\epsilon > 0$ the range of f is contained in the finite union of balls with radius ϵ and centers $f(t_i)$.

(ii) Since f is regulated it can only have jump discontinuities where $f(x+) \neq f(x-)$. It suffices to show that for each $n = 1, 2, 3, \dots$, there are only finitely many points x_i where $|f(x_i+) - f(x_i-)| \geq 1/n$. To do this we approximate f by a step function g so that $|f(x) - g(x)| \leq 1/(3n)$ for all $x \in I$. Then g must be discontinuous at a point x_i where $|f(x_i+) - f(x_i-)| \geq 1/n$, so there can only be a finite number of such points. □

Proposition A.1.6 *Let X be a Banach space.*

(i) *Let $I = (-\infty, b]$ or $I = [a, b]$ be a closed interval which is bounded to the right. Then every $f \in Reg(I; X)$ belongs to $L^\infty(I; X)$ and*

$$\operatorname*{ess\,sup}_{x\in I} |f(x)| = \sup_{x\in I, x<b} |f(x)|.$$

In particular, the space $\{f \in Reg(I; X) \mid f(b) = f(b-)\}$ is a closed subspace of $L^\infty(I; X)$.

(ii) *Let I be a closed interval which is unbounded to the right. Then every $Reg(I; X)$ is a closed subspace of $L^\infty(I; X)$, and, for every $f \in Reg(I; X)$,*

$$\operatorname*{ess\,sup}_{x\in I} |f(x)| = \sup_{x\in I} |f(x)|.$$

Proof (i) Trivially $\operatorname{ess\,sup}_{x\in I}|f(x)| \leq \sup_{x\in I,x<b}|f(x)|$, so only the opposite inequality is nontrivial. To get this inequality we fix $\epsilon > 0$ and approximate f by a step function g so that $|f(x) - g(x)| \leq \epsilon$ for all $x \in I$. Since the step function g has the property that $\operatorname{ess\,sup}_{x\in I}|g(x)| = \sup_{x\in I,x<b}|g(x)|$, we find that $\operatorname{ess\,sup}_{x\in I}|f(x)| \geq \sup_{x\in I,x<b}|f(x)| - \epsilon$. This being true for all $\epsilon > 0$ we must have $\operatorname{ess\,sup}_{x\in I}|f(x)| \geq \sup_{x\in I,x<b}|f(x)|$.

(ii) The proof of (ii) is similar to the proof of (i). □

Corollary A.1.7 *Let X be a Banach space, and let $I = [a, b]$ be a closed bounded interval. Then the closure in $L^\infty(I; X)$ of the set of all X-valued step functions on X can be identified with the set $\{f \in Reg(I; X) \mid f(b) = f(b-)\}$.*

Proof This follows from Propositions A.1.2 and A.1.6. □

Corollary A.1.8 *Let X be a Banach space, and let $I = [a, b]$ be a closed bounded interval. Then the closure in $L^\infty(I; X)$ of the set of all X-valued step functions on X can be identified with the set $\{f \in Reg(I; X) \mid f(b) = f(b-)\}$.*

Proof This follows from Propositions A.1.2 and A.1.6. □

A.2 The positive square root and the polar decomposition

Definition A.2.1 By a *positive operator* in a Hilbert space X we mean a (densely defined, closed and) self-adjoint (possibly unbounded) operator $A\colon X \supset \mathcal{D}(A) \to X$ satisfying $\langle x, Ax\rangle \geq 0$ for all $x \in \mathcal{D}(A)$. The notation $A \geq 0$ means that A is positive in the above sense, and the notation $A \geq B$ means that both A and B are self-adjoint operators on X and that $A - B \geq 0$ (here we require, in addition, that $B \in \mathcal{B}(X)$). If $A \geq \epsilon I$ for some $\epsilon > 0$, then we write $A \gg 0$ and call A *uniformly positive*

Lemma A.2.2 *Let A be a positive operator in a Hilbert space X.*

(i) *A has a unique positive square root $A^{1/2}$, i.e., there is a unique positive operator $A^{1/2}$ such that $A = (A^{1/2})^2$. In particular, $\mathcal{D}(A) = \{x \in \mathcal{D}(A^{1/2}) \mid A^{1/2}x \in \mathcal{D}(A^{1/2})\}$.*
(ii) *$A^{1/2} \in \mathcal{B}(X)$ if and only if $A \in \mathcal{B}(X)$.*
(iii) *$\mathcal{N}(A^{1/2}) = \mathcal{N}(A)$ and $\mathcal{R}(A) \subset \mathcal{R}(A^{1/2}) \subset \overline{\mathcal{R}(A)}$. In particular, $\overline{\mathcal{R}(A^{1/2})} = \overline{\mathcal{R}(A)}$, and $A^{1/2}$ is injective if and only if A is injective.*
(iv) *$\mathcal{R}(A^{1/2})$ is closed if and only if $\mathcal{R}(A)$ is closed.*
(v) *$A \gg 0$ if and only if $A^{1/2} \gg 0$, or equivalently, $A^{1/2}$ has a bounded inverse if and only if A has a bounded inverse.*
(vi) *$A^{1/2} = A$ if and only if A is a projection operator, or equivalently, if and only if $A^{1/2}$ is a projection operator (i.e., $A^{1/2} = A = A^2$).*
(vii) *An operator $B \in \mathcal{B}(X)$ commutes with A if and only if B commutes with $A^{1/2}$ (i.e., $AB = BA$ if and only if $A^{1/2}B = BA^{1/2}$).*

Proof (i), (ii), (v), and (vii): See, for example, Kato (1980, Theorem 3.35, p. 281) or Rudin (1973, Theorems 13.24 and 13.31).

(iii): Obviously, $\mathcal{N}(A^{1/2}) \subset \mathcal{N}(A)$ (if $A^{1/2}x = 0$, then $Ax = (A^{1/2})^2x = 0$). Conversely, if $Ax = 0$, then

$$|A^{1/2}x|^2 = \langle A^{1/2}x, A^{1/2}x\rangle = \langle x, Ax\rangle = 0,$$

so $A^{1/2}x = 0$. The fact that $\mathcal{N}(A^{1/2}) = \mathcal{N}(A)$ implies that also their orthogonal complements are equal (see Lemma 9.10.3(iii)): $\overline{\mathcal{R}(A^{1/2})} = \overline{\mathcal{R}(A)}$. The inclusion $\mathcal{R}(A) \subset \mathcal{R}(A^{1/2})$ is trivial. Thus, $\mathcal{R}(A) \subset \mathcal{R}(A^{1/2}) \subset \overline{\mathcal{R}(A^{1/2})} = \overline{\mathcal{R}(A)}$.

(iv): If $\mathcal{R}(A)$ is closed, then it follows from (iii) that also $\mathcal{R}(A^{1/2})$ is closed. Conversely, if $\mathcal{R}(A^{1/2})$ is closed, then it follows from the closed graph theorem that $A^{1/2}$ is a boundedly invertible operator from $\mathcal{D}(A^{1/2})$ onto $\mathcal{N}(A^{1/2})^\perp = \overline{\mathcal{R}(A)}$ and from $\mathcal{D}(A)$ onto $\mathcal{D}(A^{1/2})$, hence A is a boundedly invertible operator from $\mathcal{D}(A)$ onto $\overline{\mathcal{R}(A)}$. Thus, $\mathcal{R}(A)$ is closed.

(vi) Trivially, $A^{1/2} = A$ iff $A^{1/2} = (A^{1/2})^2$, i.e., iff $A^{1/2}$ is a projection operator; hence also $A(= A^{1/2})$ is a projection operator. Conversely, if A is a projection, then A is a positive solution of the equation $B^2 = A$, so by the uniqueness of the square root, $A = A^{1/2}$. □

Definition A.2.3 Let $A\colon X \supset \mathcal{D}(A) \to Y$ be a closed linear operator, where X and Y are Hilbert spaces.

(i) We call $\mathcal{N}(A)^{\perp}$ *the initial space* of A and $\overline{\mathcal{R}(A)}$ (the closure of the range of A) the *final space* of A.
(ii) A is *isometric* if $A \in \mathcal{B}(X; Y)$ and $|Ax| = |x|$ for all $x \in X$.
(iii) A is *a partial isometry* if the restriction of A to its initial space is isometric.
(iv) A is *co-isometric* if A is a partial isometry whose final space is equal to Y.
(v) A is *unitary* if it is both isometric and co-isometric.

Lemma A.2.4 *Let $A\colon X \supset \mathcal{D}(A) \to Y$ be a closed linear operator, where X and Y are Hilbert spaces.*

(i) *The initial space of A is the final space of A^*, and the final space of A is the initial space of A^*.*
(ii) *A^* is a partial isometry if and only if A is a partial isometry.*
(iii) *A is isometric if and only if A^* is co-isometric.*
(iv) *A is unitary if and only if A^* is unitary. In this case both A and A^* are invertible, and $A^{-1} = A^*$.*
(v) *If A is a partial isometry, then the restriction of A to its initial space is a unitary operator from this space onto the final space of A, whose inverse is the restriction of A^* to the final space of A.*

See, for example, Kato (1980, pp. 267 and 257–259) for the proof.

Lemma A.2.5 *Let $A\colon X \supset \mathcal{D}(A) \to Y$ be a closed linear operator, where X and Y are Hilbert spaces.*

(i) *A has a unique decomposition of the form $A = U|A|$, where $U \in \mathcal{B}(X; Y)$ is a partial isometry with the same initial and final spaces as A and $|A|\colon X \supset \mathcal{D}(|A|) \to X$ is positive. Moreover, $|A| = (A^*A)^{1/2}$ (the unique positive square root of A^*A), $\mathcal{D}(A^*A) \subset \mathcal{D}(|A|) = \mathcal{D}(A)$, and $|A| = U^*A$. In particular, $|A| \in \mathcal{B}(X)$ if and only if $A \in \mathcal{B}(X; Y)$.*
(ii) *A also has a unique decomposition of the form $A = |A^*|U$, where $U \in \mathcal{B}(X; Y)$ is a partial isometry with the same initial and final spaces as A and $|A^*|\colon Y \supset \mathcal{D}(|A^*|) \to Y$ is positive. Here the operator U is the same operator as in (i), and $|A^*| = (AA^*)^{1/2}$ is the unique operator which we get by applying (i) with A replaced by A^*. In particular,*

$\mathcal{D}(AA^*) \subset \mathcal{D}(|A^*|) = \mathcal{D}(A^*)$, $|A^*| = UA^* = AU^* = U|A|U^*$, *and* $|A^*| \in \mathcal{B}(X)$ *if and only if* $A \in \mathcal{B}(X;Y)$.

(iii) *The unique decompositions of* A^* *that we get by applying (i) and (ii) with* A *replaced by* A^* *are given by* $A^* = |A|U^* = U^*|A^*|$, *where* U *is the same operator as in (i) and (ii).*

(iv) $\mathcal{N}(A) = \mathcal{N}(|A|)$, $\mathcal{R}(A) = \mathcal{R}(|A^*|)$, $\mathcal{N}(A^*) = \mathcal{N}(|A^*|)$, *and* $\mathcal{R}(A^*) = \mathcal{R}(|A|)$.

See Kato (1980, pp. 334–336) for the proof of this lemma.

Lemma A.2.6 *Let* $A \in \mathcal{B}(X;Y)$, *where* X *and* Y *are Hilbert spaces. Let* $X_I = \mathcal{N}(A)^\perp$ *and* $Y_F = \overline{\mathcal{R}(A)}$ *be the initial and final spaces of* A, *respectively, let* $A = |A^*|U = U|A|$ *be the polar decomposition of* A *(cf. Lemma A.2.5), and let* π_{X_I} *and* π_{X_F} *be the orthogonal projections onto* X_I *(in* X*) and onto* Y_F *(in* Y*), respectively. Suppose further that* $B \in \mathcal{B}(X)$ *and* $C \in \mathcal{B}(Y)$, *and that* $AB = CA$. *Then the following conditions are equivalent.*

(i) $\pi_{X_I} BA^*A = A^*AB$.
(ii) $\pi_{X_I} B|A| = |A|B$.
(iii) $\pi_{X_I} BA^* = A^*C\pi_{X_F}$.
(iv) $CAA^* = AA^*C\pi_{Y_F}$.
(v) $C|A^*| = |A^*|C\pi_{Y_F}$.
(vi) $UB = CU$.

Proof We begin by observing that we may, without loss of generality, assume that $\mathcal{R}(B) \subset X_I$ (otherwise we throughout replace B by $\pi_{X_I} B$; the conclusion of the original theorem is valid if and only if it is valid with this replacement). With this additional assumption, $\mathcal{N}(B) = \mathcal{N}(AB)$, so the identity $AB = CA$ shows that $\mathcal{N}(A) \subset \mathcal{N}(B)$, i.e., B vanishes on $X_I^\perp = \mathcal{N}(A) = \mathcal{N}(|A|) = \mathcal{N}(U) = \mathcal{R}(A^*)^\perp = \mathcal{R}(|A|)^\perp$. This means that we may, without loss of generality, replace X_I by X, i.e., we may assume that A is injective. The same argument applied to the adjoint identity $A^*C^* = B^*A^*$ shows that we may, without loss of generality, replace Y_F by Y, i.e., we may assume that A has dense range. Thus, it suffices to prove the lemma in the special case where A is injective and has dense range, and the projections listed in (i)–(vi) become identity operators. In this special case (i)–(vi) are proved as follows.

(i) $\Leftrightarrow$ (ii) and (iv) $\Leftrightarrow$ (v): This follows from Lemmas A.2.2(vii) and A.2.5(i).

(i) $\Leftrightarrow$ (iii): If $BA^* = A^*C$ then (using the intertwining property $AB = CA$) we get $A^*AB = A^*CA = BA^*A$. Conversely, if $A^*AB = BA^*A$, then $A^*CA = A^*AB = BA^*A$. Since A has dense range, this implies that $A^*C = BA^*$.

(iii) ⇔ (iv): This proof is similar to the proof of the equivalence (i) ⇔ (iii) given above.

(ii) ⇔ (vi): Assume that $B|A| = |A|B$. Then the intertwining property $AB = CA$ gives $CU|A| = CA = AB = U|A|B = UB|A|$. As $|A|$ has dense range, this implies that $CU = UB$. Conversely, if $UB = CU$, then $UB|A| = CU|A| = CA = AB = U|A|B$. Thus $B|A| = U^*U|A|B = |A|B$. □

A.3 Convolutions

Definition A.3.1 The *convolution* $A * u$ of two (strongly measurable and a.e. defined) functions $A\colon \mathbb{R} \to \mathcal{B}(X;Y)$ and $u\colon \mathbb{R} \to X$ is the function

$$(A * u)(t) = \int_{\mathbb{R}} A(t-s)u(s)\,ds,$$

defined for all $t \in \mathbb{R}$ for which the integral exists (as a Bochner integral). If A or u is defined only on $\mathbb{R}^+$, then we define $A * u$ in the same way, except that we interpret $A(t-s)$ and $u(s)$ as zero for $s > t$ or $s < 0$, respectively.

Convolutions interact with shifts and with the multiplication by an exponential function in the following way:

Lemma A.3.2 *Let* $A\colon \mathbb{R} \to \mathcal{B}(X;Y)$ *and* $u\colon \mathbb{R} \to X$ *be (strongly) measurable. Let* $\omega \in \mathbb{R}$, *and define* $\mathrm{e}_\omega(t) = \mathrm{e}^{\omega t}$ *for* $t \in \mathbb{R}$.

(i) *For each* $h \in \mathbb{R}$, *the following conditions are equivalent:*
   (a) $(A * u)(t+h)$ *exists;*
   (b) $((\tau^h A) * u)(t)$ *exists;*
   (c) $(A * (\tau^h u))(t)$ *exists.*
   When these conditions hold, then

$$(A * u)(t+h) = (A * (\tau^h u))(t) = ((\tau^h A) * u)(t).$$

(ii) $\big((\mathrm{e}_\omega A) * (\mathrm{e}_\omega u)\big)(t)$ *exists if and only if* $(A * u)(t)$ *exists, and*

$$\big((\mathrm{e}_\omega A) * (\mathrm{e}_\omega u)\big)(t) = \mathrm{e}^{\omega t}(A * u)(t).$$

Proof (i) This is obvious (by a change of integration variable).

(ii) This is true because, for all $s \in \mathbb{R}$,

$$(\mathrm{e}_\omega A)(t-s)(\mathrm{e}_\omega u)(s) = \mathrm{e}^{t-s} A(t-s)\mathrm{e}^s u(s) = \mathrm{e}^{\omega t} A(t-s)u(s).$$

□

The following lemma can be used in many cases to show that $A * u$ is (almost) everywhere defined and measurable.

Lemma A.3.3 *Let $A\colon \mathbb{R} \to \mathcal{B}(X;Y)$ and $u\colon \mathbb{R} \to X$ be (strongly) measurable, and suppose that the function $t \mapsto \|A(t)\|_{\mathcal{B}(X;Y)}$ is measurable (this assumption is redundant if X is separable).*

(i) *The convolution $A * u$ exists for all those t for which $(\|A\|_{\mathcal{B}(X;Y)} * |u|_X)(t)$ exists, and $(|A * u|_Y)(t) \le (\|A\|_{\mathcal{B}(X;Y)} * |u|_X)(t)$.*

(ii) *If $\|A\|_{\mathcal{B}(X;Y)} \in L^1_{\mathrm{loc}}(\mathbb{R})$ and if $u\colon \mathbb{R} \to X$ has bounded support and a totally bounded range, then $A * u$ is defined everywhere and $A * u \in C(\mathbb{R}; Y)$.*

(iii) *If $\|A\|_{\mathcal{B}(X;Y)} \in L^1(\mathbb{R})$ and if $u\colon \mathbb{R} \to X$ has a totally bounded range, then $A * u$ is defined everywhere and $A * u \in BUC(\mathbb{R}; Y)$.*

(iv) *If $\|A\|_{\mathcal{B}(X;Y)} \in L^1(\mathbb{R})$ and if $u \in BUC(\mathbb{R}; X)$, then $A * u$ is defined everywhere and $A * u \in BUC(\mathbb{R}; Y)$.*

(v) *If there is a sequence of functions $u_n\colon \mathbb{R} \to X$, where each u_n has bounded support and a totally bounded range (the latter condition is true, for example, if u_n is continuous, or u_n is regulated, or u_n is simple), such that*
   (a) *$A * u_n \in C(\mathbb{R}; Y)$ for each n, and*
   (b) *$(\|A\|_{\mathcal{B}(X;Y)} * |u - u_n|_X)(t) \to 0$ as $n \to \infty$ for almost all $t \in \mathbb{R}$,*

   then $A * u$ is measurable. In particular, (a) is true whenever $\|A\|_{\mathcal{B}(X;Y)} \in L^1_{\mathrm{loc}}(\mathbb{R})$.

That $\|A\|_{\mathcal{B}(X;Y)}$ is measurable whenever X is separable follows from, e.g., Dinculeanu (1974, Corollary 15.17).

Proof (i) For each $t \in \mathbb{R}$, the function $s \mapsto A(t-s)$ is strongly measurable. Since the product of a strongly measurable function with a measurable function is measurable (Dinculeanu, 1974, Proposition 15.15), we find that $s \mapsto A(t-s)u(s)$ is measurable. It is furthermore integrable at those points where $(\|A\|_{\mathcal{B}(X;Y)} * |u|_X)(t) < \infty$ since

$$\begin{aligned}\int_{\mathbb{R}} |A(t-s)u(s)|_Y\, ds &\le \int_{\mathbb{R}} \|A(t-s)\|_{\mathcal{B}(X;Y)} |u(s)|_X\, ds \\ &= (\|A\|_{\mathcal{B}(X;Y)} * |u|_X)(t).\end{aligned}$$

Thus $A * u$ exists almost everywhere. Moreover,

$$\begin{aligned}\left| \int_{\mathbb{R}} A(t-s)u(s)\, ds \right|_Y &\le \int_{\mathbb{R}} |A(t-s)u(s)|_Y\, ds \\ &\le (\|A\|_{\mathcal{B}(X;Y)} * |u|_X)(t).\end{aligned}$$

(ii) Suppose that $\|A\|_{\mathcal{B}(X;Y)} \in L^1_{\mathrm{loc}}(\mathbb{R})$. We begin with the case where $u = xv$, with $x \in X$ and $v \in L^\infty(\mathbb{R}; \mathbb{C})$ with bounded support. Choose T so large that the support of v is contained in $(-T, T)$. Then, for all $t \in \mathbb{R}$

$$\begin{aligned}
&\int_{\mathbb{R}} |A(t-s)|_{\mathcal{B}(X;Y)} |xv(s)|_X \, ds \\
&\quad \le \Big(\int_{-T}^{T} |A(t-s)|_{\mathcal{B}(X;Y)} \, ds \Big) \|x\|_X \sup_{-T<s<T} |u(s)| < \infty, \qquad \text{(A.1)}
\end{aligned}$$

and we conclude from (i) that $(A * xv)(t)$ is defined for all $t \in \mathbb{R}$. Moreover, for all $t \in \mathbb{R}$ and $h \in (-1, 1)$,

$$\begin{aligned}
&|(A * xv)(t+h) - (A * (xv))(t)|_Y \\
&\quad = \Big| \int_{-T}^{T} (A(t+h-s)x - A(t-s)x) v(s) \, ds \Big|_Y \\
&\quad \le \Big(\int_{-T}^{T} |A(t+h-s)x - A(t-s)x|_Y \, ds \Big) \sup_{-T<s<T} |v(s)|.
\end{aligned}$$

The right hand side tends to zero as $h \to 0$ for the following reason: we observe that $Ax \in L^1_{\mathrm{loc}}(\mathbb{R}; Y)$ since Ax is measurable and $\int_{-T}^{T} |A(s)x|_Y \, ds < \infty$ for every finite T, and by Lemma 2.3.3(i), translation is a continuous operation in $L^1_{\mathrm{loc}}(\mathbb{R}; Y)$. Thus, $A * xv \in C(\mathbb{R}; Y)$.

Each simple function u with finite support can be written as a finite sum of functions of the type $x_n v_n$, where $X_n \in X$ and v_n is the characteristic function of a bounded subset of $\mathbb{R}$. This, combined with the fact that each $A * x_n v_n \in C(\mathbb{R}; Y)$ shows that $A * u \in C(\mathbb{R}; Y)$ whenever u is a simple function.

If u has a totally bounded range, then we claim that it is possible to find a sequence of simple functions u_n converging uniformly to u, and such that the support of each u_n is contained in the support of u. To prove this it suffices to show that, for each $\epsilon > 0$, it is possible to find a simple function v whose support is contained in the support of u such that $|u(s) - v(s)|_X \le \epsilon$ for all $s \in \mathbb{R}$. Such a function can be obtained as follows. Since the range of u is totally bounded, we can find a finite collection of vectors $x_k \in X$ such that the union of balls with center x_k and radius ϵ covers $\mathcal{R}(u)$. Define

$$\begin{aligned}
E_1 &= \{s \in \mathbb{R} \mid |u(s) - x_1| < \epsilon\}, \\
E_2 &= \{s \in \mathbb{R} \mid |u(s) - x_2| < \epsilon\} \setminus E_1, \\
E_3 &= \{s \in \mathbb{R} \mid |u(s) - x_2| < \epsilon\} \setminus E_1 \setminus E_2,
\end{aligned}$$

etc., and let $v = \chi_E \sum_k x_k \chi_{E_k}$, where E is the support of u. Clearly this function v is of the desired form.

Let $u\colon \mathbb{R} \to Y$ be an arbitrary measurable function with totally bounded range and bounded support. Let u_n be a sequence of simple functions with support contained in some interval $(-T, T)$ converging uniformly to u. Then, for all $t \in \mathbb{R}$,

$$\begin{aligned}
&|(A * u)(t) - (A * u_n)(t)|_Y \\
&= \left| \int_{\mathbb{R}} A(t-s)(u(s) - u_n(s))\, ds \right|_Y \\
&\le \int_{-T}^{T} \|A(t-s)\|_{\mathcal{B}(X;Y)} |u(s) - u_n(s)|_X\, ds \\
&\le \left(\int_{t-T}^{t+T} \|A(s)\|_{\mathcal{B}(X;Y)}\, ds \right) \left(\sup_{s\in\mathbb{R}} |u(s) - u_n(s)|_X \right).
\end{aligned}$$

This implies that $(A * u_n)(t) \to (A * u)(t)$, uniformly for t in bounded intervals, and hence $A * u \in C(\mathbb{R}; Y)$ (the limit of a locally uniformly convergent sequence of continuous functions is continuous).

(iii) This proof is the same as the proof of (ii), with $(-T, T)$ replaced by $(-\infty, \infty)$.

(iv) That $(A * u)(t)$ is defined for all t and uniformly bounded is proved as in (ii) and (iii). We get the uniform continuity of $A * u$ from the uniform continuity of u and the fact that, for all $h \in \mathbb{R}$,

$$\begin{aligned}
&|(A * u)(t+h) - (A * u)(t)|_Y \\
&= \left| \int_{\mathbb{R}} \big(A(t+h-s)u(s) - A(t-s)u(s) \big)\, ds \right|_Y \\
&= \left| \int_{\mathbb{R}} A(s)(u(t+h-s) - u(t-s))\, ds \right|_Y \\
&\le \int_{\mathbb{R}} \|A(s)\|_{\mathcal{B}(X;Y)} |u(t+h-s) - u(t-s)|_X\, ds \\
&\le \left(\int_{\mathbb{R}} \|A(s)\|_{\mathcal{B}(X;Y)}\, ds \right) \left(\sup_{s\in\mathbb{R}} |u(s+h) - u(s)|_X \right).
\end{aligned}$$

(v) For each n, $A * u_n$ is continuous, hence measurable. The computation in the proof of (ii), with u replaced by $u - u_n$, shows that $(A * u_n)(t) \to (A * u)(t)$ at every point where $(\|A\|_{\mathcal{B}(X;Y)} * |u - u_n|_X)(t) \to 0$. Thus, $A * u_n \to A * u$ a.e., and $A * u$ is measurable. □

Lemma A.3.3 enables us to prove several results for the convolution operator $u \mapsto A * u$ by reducing the claims to the corresponding scalar claims. For example, the following result is true:

Theorem A.3.4 *Let $A\colon \mathbb{R} \to \mathcal{B}(X;Y)$ be strongly measurable, and let $\omega \in \mathbb{R}$.*

(i) *If $\|A\|_{\mathcal{B}(X;Y)} \in L^1_\omega(\mathbb{R})$, then the convolution operator $u \mapsto A * u$ is a bounded linear operator between the following spaces:*
   (a) $u \mapsto A * u\colon L^p_\omega(\mathbb{R}; U) \to L^p_\omega(\mathbb{R}; Y)$, $1 \le p < \infty$;
   (b) $u \mapsto A * u\colon BUC_\omega(\mathbb{R}; U) \to BUC_\omega(\mathbb{R}; Y)$;

(c) $u \mapsto A * u \colon BC_\omega(\mathbb{R}; U) \to BC_\omega(\mathbb{R}; Y)$;
(d) $u \mapsto A * u \colon BC_{0,\omega}(\mathbb{R}; U) \to BC_{0,\omega}(\mathbb{R}; Y)$;
(e) $u \mapsto A * u \colon Reg_\omega(\mathbb{R}; U) \to BUC_\omega(\mathbb{R}; Y)$;
(f) $u \mapsto A * u \colon Reg_{0,\omega}(\mathbb{R}; U) \to BC_{0,\omega}(\mathbb{R}; Y)$.

The norms of all these operators are no bigger than the L^1_ω-norm of $\|A\|_{\mathcal{B}(X;Y)}$.

(ii) *If $\|A\|_{\mathcal{B}(X;Y)} \in L^\infty_\omega(\mathbb{R})$, then the convolution operator $u \mapsto A * u$ maps $L^1_\omega(\mathbb{R}; U)$ continuously into $BUC_\omega(\mathbb{R}; Y)$. The norm of this operator is no bigger than the L^∞_ω-norm of $\|A\|_{\mathcal{B}(X;Y)}$.*

(iii) *If $\|A\|_{\mathcal{B}(X;Y)} \in L^q_\omega(\mathbb{R})$ for some q, $1 < q < \infty$, and if $1/p + 1/q = 1$, then the convolution operator $u \mapsto A * u$ maps $L^p_\omega(\mathbb{R}; U)$ continuously into $BC_{0,\omega}(\mathbb{R}; Y)$. The norm of this operator is no bigger than the L^q_ω-norm of $\|A\|_{\mathcal{B}(X;Y)}$.*

(iv) *If $\|A\|_{\mathcal{B}(X;Y)} \in L^q_\omega(\mathbb{R})$ for some q, $1 < q < \infty$, and if p, q, and r satisfy $1 \le p < r < \infty$ and $1/r = 1/p + 1/q - 1$, then the convolution operator $u \mapsto A * u$ maps $L^p_\omega(\mathbb{R}; U)$ continuously into $L^r_\omega(\mathbb{R}; Y)$. The norm of this operator is no bigger than the L^q_ω-norm of $\|A\|_{\mathcal{B}(X;Y)}$.*

In all the cases listed above, the operator $u \mapsto A * u$ is time-invariant. It is causal if $A(s) =$ for $s < 0$. In this case the transfer function of $u \mapsto A * u$ (see Corollary 4.6.10) is given by

$$\hat{A}(z)x = \int_{\mathbb{R}^+} e^{-zs} A(s)x\, ds, \qquad x \in X, \qquad \Re z > \omega \tag{A.2}$$

(this is the Laplace transform of $Ax \in L^1_\omega(\mathbb{R}^+; X)$).

The norm inequality resulting from part (iv) is often referred to as *Young's inequality*.

Proof Thanks to Lemma A.3.2(ii), we can throughout this proof, without loss of generality, take $\omega = 0$.

The time-invariance of $u \mapsto A * u$ follows from Lemma A.3.2, and the causality from Definition A.3.1. That the transfer function is given by (A.2) follows from Corollary 4.6.10 and Definition A.3.1 (and a change of integration variable $s \to -s$).

In all cases, the fact that $(A * u)(t)$ is defined everywhere or almost everywhere and the needed norm estimates on $|(A * u)(t)|_Y$ follow from Lemma A.3.3(i) and the corresponding scalar result. For the needed scalar results we refer the reader to, e.g., Gripenberg *et al.* (1990, Theorem 2.2(i) and (vii), p. 39) (cases (i)–(iii)) and Stein and Weiss (1971, p. 178) (case (iv)). To prove the measurability of $A * u$ in cases (i)(a) and (iv) we use Lemma A.3.3(v) and the fact that $C_c(\mathbb{R}; U)$ is dense in $L^p(\mathbb{R}; U)$. If $u_n \in C_c(\mathbb{R}; U)$ and $u_n \to u$

in $L^p(\mathbb{R};U)$, then $\|A\|_{\mathcal{B}(X;Y)} * |u - u_n|_U \to 0$ in $L^p(\mathbb{R})$ or $L^r(\mathbb{R})$, hence there is a subsequence u_{n_k} for which we have convergence almost everywhere. By Lemma A.3.3, this implies that $(A * u_n)(t) \to (A * u)(t)$ almost everywhere. Thus $A * u$ is measurable in these cases.

It remains to prove the claims about the continuity or uniform continuity of $A * u$. Here we give a separate argument for each case.

(i) The uniform continuity of $A * u$ follows from Lemma A.3.3(iv) in case (b), and from Lemma A.3.3(iii) in cases (d)–(f). In case (c) we observe that $(\|A\|_{\mathcal{B}(X;Y)} * |u - \tau^h u|_U)(t) \to 0$ as $h \to 0$ (use Lebesgue's dominated convergence theorem and the fact that $|u(t+h) - a(t)|_U \to 0$ as $h \to 0$). The continuity of $A * u$ then follows from Lemma A.3.3(i).

(ii)–(iii) The uniform continuity of $A * u$ follows from the same norm inequality in (ii) and (iii) which gives (put $p = 1$ and $q = \infty$ in case (ii) and denote the L^q-norm of $\|A\|_{\mathcal{B}(X;Y)}$ by M)

$$\begin{aligned}
&\sup_{t\in\mathbb{R}} |(A * u)(t+h) - (A * u)(t)|_Y \\
&\quad = \sup_{t\in\mathbb{R}} |(A * (\tau^h u - u)(t)|_Y \\
&\quad \le M \|\tau^h u - u\|_{L^p(\mathbb{R};U)},
\end{aligned}$$

and the fact that translation is continuous in $L^p(\mathbb{R};U)$ (see Example 2.3.2(i)). □

Theorem A.3.4 is not optimal in the sense that the conditions $\|A\|_{\mathcal{B}(X;Y)} \in L^1_\omega(\mathbb{R})$ or $\|A\|_{\mathcal{B}(X;Y)} \in L^q_\omega(\mathbb{R})$ that we used there are far from necessary (unless X or Y is finite-dimensional). For example, in the case where $p = 1$ the following sharper result holds (see Remark A.3.6):

Theorem A.3.5 *Let $\mathfrak{C} \in \mathcal{B}(X; L^p_\omega(\mathbb{R};Y))$ for some p, $1 \le p \le \infty$ and $\omega \in \mathbb{R}$. Then there is a unique time-invariant operator $\mathfrak{F}$ which maps $L^1_\omega(\mathbb{R};X)$ continuously into $L^p_\omega(\mathbb{R};Y)$ and satisfies*

$$(\mathfrak{F}(xv))(t) = \int_{-\infty}^{\infty} (\mathfrak{C}x)(t-s)v(s)\,ds \tag{A.3}$$

for all $x \in X$ and $v \in L^1_\omega(\mathbb{R};\mathbb{C})$ and almost all $t \in \mathbb{R}$. Moreover,

$$\|\mathfrak{F}\|_{\mathcal{B}(L^1_\omega(\mathbb{R};X);L^p_\omega(\mathbb{R};Y))} \le \|\mathfrak{C}\|_{\mathcal{B}(X;L^p_\omega(\mathbb{R}^+;Y))}. \tag{A.4}$$

If $p = \infty$, then $\mathfrak{F}$ maps $L^1_\omega(\mathbb{R};X)$ continuously into $BUC_\omega(\mathbb{R};Y)$, and if the range of $\mathfrak{C}$ belongs to $L^p_\omega(\mathbb{R}^+;Y)$ (i.e., if $\pi_-\mathfrak{C} = 0$), then $\mathfrak{F}$ is causal. In this *case the transfer function of $\mathfrak{F}$ (see Corollary 4.6.10) is given by*

$$\widehat{\mathfrak{F}}(z)x = \int_{\mathbb{R}^+} e^{-zs}(\mathfrak{C}x)(s)\,ds, \qquad x \in X, \qquad \Re z > \omega \tag{A.5}$$

(this is the Laplace transform of $\mathfrak{C}x \in L^1_\omega(\mathbb{R}^+; Y)$).

Proof Without loss of generality we may take $\omega = 0$ (see Lemma A.3.2(ii)).

By Theorem A.3.4(iv) with $p = 1$ and $r = q$, for each $x \in X$, the formula $\mathfrak{F}_x v = v \mapsto \left(t \mapsto \int_{-\infty}^{\infty}(\mathfrak{C}x)(t-s)v(s)\,ds\right)$ defines a continuous linear operator $L^1(\mathbb{R};\mathbb{C}) \to L^p(\mathbb{R};Y)$. Clearly $\mathfrak{F}_x$ is time-invariant for each fixed $x \in X$, and it is causal if $\pi_-\mathfrak{C} = 0$. Moreover, if we denote $K = \|\mathfrak{C}\|_{\mathcal{B}(U;L^p(\mathbb{R}^+;Y))}$, then

$$\|\mathfrak{F}_x v\|_{L^p(\mathbb{R};Y)} \le K\|x\|_X\|v\|_{L^1(\mathbb{R};\mathbb{C})}.$$

If x is a simple integrable X-valued function $x = \sum_{k=1}^n x_k\chi_{E_k}$ where $x_k \in X$ and $E_k \cap E_j = \emptyset$ for $k = j$, then we define

$$(\mathfrak{F}x)(t) = \sum_{k=1}^n \mathfrak{F}_{x_k}\chi_{E_k}.$$

By the preceding argument and the triangle inequality,

$$\begin{aligned}\|\mathfrak{F}x\|_{L^p(\mathbb{R};Y)} &\le \sum_{k=1}^n \|\mathfrak{F}_{x_k}\chi_{E_k}\|_{L^1(\mathbb{R};Y)}\\ &\le \sum_{k=1}^n K\|x_k\|_X\|\chi_{E_k}\|_{L^1(\mathbb{R};\mathbb{C})} = K\|x\|_{L^1(\mathbb{R};X)}.\end{aligned}$$

The set of simple functions is dense in $L^1(\mathbb{R};X)$, so we can extend the operator $\mathfrak{F}$ by continuity to an operator in $\mathcal{B}(L^1(\mathbb{R};X);L^p(\mathbb{R};Y))$. This extended operator is time-invariant and satisfies (A.4), and it is causal if $\pi_-\mathfrak{C} = 0$. That $\mathfrak{F}x \in BUC(\mathbb{R};Y)$ if $p = \infty$ follows from the time-invariance of $\mathfrak{F}$ and the fact that $t \mapsto \tau^t x$ is uniformly continuous $\mathbb{R} \to L^1(\mathbb{R};U)$ for each $x \in L^1(\mathbb{R};X)$. The uniqueness of $\mathfrak{F}$ follows from the fact that (A.3) determines $\mathfrak{F}$ uniquely on the set of all simple functions, which is dense in $L^1(\mathbb{R};X)$. □

Remark A.3.6 If $A\colon \mathbb{R} \to \mathcal{B}(X;Y)$ is strongly measurable and $\|A\|_{\mathcal{B}(X;Y)} \in L^p_\omega(\mathbb{R})$, then we can define $(\mathfrak{C}x)(t) = A(t)x$ to get an operator which satisfies the hypothesis of Theorem A.3.5, and $\|\mathfrak{C}\|_{\mathcal{B}(X;L^p_\omega(\mathbb{R}^+;Y))}$ is no bigger than the L^p_ω-norm of $\|A\|_{\mathcal{B}(X;Y)}$. The converse is not true: not every $\mathfrak{C}$ in Theorem A.3.5 arises in this way, as the following counter-example shows. If $\mathfrak{C}$ is defined in this way, then k defined by $k(t) := \sup_{|x|_X\le 1}|(\mathfrak{C}x)(t)| = k(t)$ satisfies $x \in L^p_\omega(\mathbb{R})$ (since $k = \|A\|_{\mathcal{B}(X;Y)}$). Thus, it suffices to construct an example where k defined above does not belong to $L^p_\omega(\mathbb{R})$. We take $\omega = 0$, $p < \infty$, $Y = \mathbb{C}$, and $X = l^1$, i.e., the set of sequences $x_k \in \mathbb{C}$, $k = 0, 1, 2, \ldots$, which satisfy $|x|_{l^1} = \sum_k |x_k| < \infty$.

This space has the basis ϕ_n, $n = 0, 1, 2, \dots$, where $(\phi_n)_k = 1$ if $k = n$ and $(\phi_n)_k = 0$ otherwise, i.e., every $x \in l^1$ can be written in the form $x = \sum_n x_n \phi_n$, with $|x|_{l^1} = \sum_k |x_k| < \infty$. Choose some sequence t_n which is dense in $\mathbb{R}^+$, and let

$$(\mathfrak{C}\phi_n)(t) = \begin{cases} \log|1/(t - t_n)|, & |t - t_n| < 1,\ t \geq 0 \\ 0 & \text{otherwise.} \end{cases}$$

For an arbitrary $x = \sum_n x_n \phi_n \in l^1$ we define $\mathfrak{C}x$ by $\mathfrak{C}x = \sum_n x_n \mathfrak{C}\phi_n$. This sum converges (absolutely) in $L^p(\mathbb{C})$ since

$$\|\phi_n\|_{L^p} \leq K := \left(\int_{-1}^{1} (\log(1/|s|))^p \, ds \right)^{1/p}$$

for all n and $\sum_n |x_n| < \infty$. Moreover, $\|\mathfrak{C}x\|_{L^p(\mathbb{R})} \leq K|x|_{l_1}$. Thus, $\mathfrak{C}$ satisfies the hypothesis of Theorem A.3.5, and it induces a causal time-invariant operator from $L^1(\mathbb{R}; l^1)$ to $L^p(\mathbb{R}; \mathbb{C})$. On the other hand,

$$k(t) := \sup_{x \in l^1} |(\mathfrak{C}x)(t)| = \infty, \qquad t \geq 0,$$

so $k \notin L^p(\mathbb{R})$.

In the causal cases it is also possible to formulate local versions of Theorems A.3.4 and A.3.5:

Theorem A.3.7 *Let $A\colon \mathbb{R}^+ \to \mathcal{B}(X; Y)$ be strongly measurable.*

(i) *If $\|A\|_{\mathcal{B}(X;Y)} \in L^1_{\text{loc}}(\mathbb{R}^+)$, then the convolution operator $u \mapsto A * u$ is a bounded linear operator between the following spaces:*
  (i) *$u \mapsto A * u\colon L^p_{c,\text{loc}}(\mathbb{R}; U) \to L^p_{c,\text{loc}}(\mathbb{R}; Y)$, $1 \leq p < \infty$;*
  (ii) *$u \mapsto A * u\colon Reg_{c,\text{loc}}(\mathbb{R}; U) \to BC_{c,\text{loc}}(\mathbb{R}; Y)$.*

(ii) *If $\|A\|_{\mathcal{B}(X;Y)} \in L^q_{\text{loc}}(\mathbb{R}^+)$ for some q, $1 < q \leq \infty$, and if $1/p + 1/q = 1$, then the convolution operator $u \mapsto A * u$ maps $L^p_{c,\text{loc}}(\mathbb{R}; U)$ continuously into $BC_{c,\text{loc}}(\mathbb{R}; Y)$.*

(iii) *If $\|A\|_{\mathcal{B}(X;Y)} \in L^q_{\text{loc}}(\mathbb{R}^+)$ for some q, $1 < q < \infty$, and if p, q, and r satisfy $1 \leq p < r < \infty$ and $1/r = 1/p + 1/q - 1$, then the convolution operator $u \mapsto A * u$ maps $L^p_{c,\text{loc}}(\mathbb{R}; U)$ continuously into $L^r_{c,\text{loc}}(\mathbb{R}; Y)$.*

In all the cases listed above, the operator $u \mapsto A * u$ is time-invariant and causal.

Theorem A.3.8 *Let $\mathfrak{C} \in \mathcal{B}(X; L^p_{\text{loc}}(\mathbb{R}^+; Y))$ for some p, $1 \leq p \leq \infty$. Then there is a unique time-invariant operator $\mathfrak{F}$ which maps $L^1_{c,\text{loc}}(\mathbb{R}; X)$*

continuously into $L^p_{c,\mathrm{loc}}(\mathbb{R};Y)$ *and satisfies*

$$(\mathfrak{F}(xv))(t) = \int_{-\infty}^{t} (\mathfrak{C}x)(t-s)v(s)\,ds$$

for all $x \in X$ *and* $v \in L^1_{c,\mathrm{loc}}(\mathbb{R};\mathbb{C})$ *and almost all* $t \in \mathbb{R}$. *If* $p = \infty$, *then* $\mathfrak{F}$ *maps* $L^1_{c,\mathrm{loc}}(\mathbb{R};X)$ *continuously into* $C_c(\mathbb{R};Y)$.

Proofs of Theorems A.3.7 and A.3.8. The proof is the same in both cases (except for the notation), so let us prove only Theorem A.3.8.

For each $T > 0$ we can use Theorem A.3.5 to construct the time-invariant causal operator $\mathfrak{F}_T\colon L^1(\mathbb{R};X) \to L^p(\mathbb{R};Y)$ which corresponds to the operator $\pi_{[0,T)}\mathfrak{C}$. It is clear from (A.3) that $\pi_{[s,t)}\mathfrak{F}_T\pi_{[s,t)}$ is independent of T for all $T \geq s+t$. Therefore $\mathfrak{F} = \lim_{T\to\infty}\mathfrak{F}_T$ defines a continuous time-invariant causal operator $\mathfrak{F}\colon L^1_{c,\mathrm{loc}}(\mathbb{R};X) \to L^p_{c,\mathrm{loc}}(\mathbb{R};Y)$. □

A.4 Inversion of block matrices

Lemma A.4.1 *Let* X *and* Y *be Banach spaces, and let* $K \in \mathcal{B}(X;Y)$ *and* $L \in \mathcal{B}(Y;X)$. *Then the following claims hold.*

(i) *If* $\left[\begin{smallmatrix} A & B \\ C & D \end{smallmatrix}\right]$ *is a left [right] inverse of* $\left[\begin{smallmatrix} 1 & -K \\ -L & 1 \end{smallmatrix}\right]$, *then* A *is a left [right] inverse of* $1 - KL$ *and* D *is a left [right] inverse of* $1 - LK$. *In particular, if* $\left[\begin{smallmatrix} A & B \\ C & D \end{smallmatrix}\right]$ *is invertible, then both* $1 - KL$ *and* $1 - LK$ *are invertible.*

(ii) *If* $(1 - LK)^{-1}$ *is an arbitrary left inverse of* $(1 - LK)$, *then*

$$(1 - KL)^{-1} = \big(1 + K(1 - LK)^{-1}L\big), \tag{A.1}$$

is a left inverse of $(1 - KL)$,

$$\begin{bmatrix} 1 & -K \\ -L & 1 \end{bmatrix}^{-1} = \begin{bmatrix} 1 + K(1 - LK)^{-1}L & K(1 - LK)^{-1} \\ (1 - LK)^{-1}L & (1 - LK)^{-1} \end{bmatrix} \tag{A.2}$$

is a left inverse of $\left[\begin{smallmatrix} 1 & -K \\ -L & 1 \end{smallmatrix}\right]$, *and*

$$(1 - KL)^{-1}K = K(1 - LK)^{-1}. \tag{A.3}$$

If, in addition,

$$L(1 - KL)^{-1} = (1 - LK)^{-1}L, \tag{A.4}$$

then these left inverses are also right inverses. Conversely, if any one of $1 - LK$, $1 - KL$, *or* $\left[\begin{smallmatrix} 1 & -K \\ -L & 1 \end{smallmatrix}\right]$ *is invertible, then so are the others, and* (A.1)–(A.4) *hold.*

(iii) *If* $(1-KL)^{-1}$ *is an arbitrary left inverse of* $(1-KL)$, *then*

$$(1-LK)^{-1} = \big(1 + L(1-KL)^{-1}K\big), \tag{A.5}$$

is a left inverse of $(1-LK)$,

$$\begin{bmatrix} 1 & -K \\ -L & 1 \end{bmatrix}^{-1} = \begin{bmatrix} (1-KL)^{-1} & (1-KL)^{-1}K \\ L(1-KL)^{-1} & 1 + L(1-KL)^{-1}K \end{bmatrix} \tag{A.6}$$

is a left inverse of $\left[\begin{smallmatrix} 1 & -K \\ -L & 1 \end{smallmatrix}\right]$, *and* (A.4) *holds. These left inverses are also right inverses if and only if* (A.3) *holds.*

(iv) *Let* $(1-LK)^{-1}$ *be a left inverse of* $(1-LK)$ *and let* $(1-KL)^{-1}$ *be a left inverse of* $(1-KL)$. *Then the following conditions are equivalent:*

(a) $1-LK$ *is invertible (with inverse* $(1-LK)^{-1}$*);*

(b) $1-KL$ *is invertible (with inverse* $(1-KL)^{-1}$*);*

(c) $\left[\begin{smallmatrix} 1 & -K \\ -L & 1 \end{smallmatrix}\right]$ *is invertible;*

(d) *The two left inverses of* $\left[\begin{smallmatrix} 1 & -K \\ -L & 1 \end{smallmatrix}\right]$ *given in* (A.2) *and* (A.6) *coincide (and they are equal to the inverse of* $\left[\begin{smallmatrix} 1 & -K \\ -L & 1 \end{smallmatrix}\right]$*);*

(e) *Both* (A.3) *and* (A.4) *hold;*

(f) *Both* (A.1) *and* (A.5) *hold.*

(v) $(1-KL)$ *is left [right] invertible in* $\mathcal{B}(Y)$ *iff* $(1-LK)$ *is left [right] invertible in* $\mathcal{B}(X)$, *and this is true iff* $\left[\begin{smallmatrix} 1 & -K \\ -L & 1 \end{smallmatrix}\right]$ *is left [right] invertible in* $\mathcal{B}(\left[\begin{smallmatrix} X \\ Y \end{smallmatrix}\right])$.

(vi) $(1-KL)$ *is invertible in* $\mathcal{B}(Y)$ *iff* $(1-LK)$ *is invertible in* $\mathcal{B}(X)$ *iff* $\left[\begin{smallmatrix} 1 & -K \\ -L & 1 \end{smallmatrix}\right]$ *is invertible in* $\mathcal{B}(X;Y)$, *and the inverses satisfy* (A.1)–(A.6).

(vii) *With the possible exception of the zero point,* KL *and* LK *have the same spectrum.*

Proof (i) The operator $\left[\begin{smallmatrix} A & B \\ C & D \end{smallmatrix}\right]$ is a left inverse of $\left[\begin{smallmatrix} 1 & -K \\ -L & 1 \end{smallmatrix}\right]$ iff $\left[\begin{smallmatrix} A & B \\ C & D \end{smallmatrix}\right]\left[\begin{smallmatrix} 1 & -K \\ -L & 1 \end{smallmatrix}\right] = \left[\begin{smallmatrix} 1 & 0 \\ 0 & 1 \end{smallmatrix}\right]$, or equivalently, iff

$$\begin{aligned} B &= AK, & A(1-KL) &= 1, \\ C &= DL, & D(1-LK) &= 1. \end{aligned}$$

Thus, A is a left inverse of $(1-KL)$ and D is a left inverse of $(1-LK)$. The proof of the right invertibility is similar.

(ii) Suppose that $(1-LK)$ is left invertible, and let $(1-LK)^{-1}$ stand for one of its left inverses. Then it is easy to show (we leave these computations to the reader) that the right hand side of (A.1) is a left inverse of $(1-KL)$, that the right hand side of (A.2) is a left inverse of $\left[\begin{smallmatrix} 1 & -K \\ -L & 1 \end{smallmatrix}\right]$, and that (A.3) holds.

If, in addition, (A.4) holds, then

$$\begin{aligned}(1-LK)(1-LK)^{-1} &= (1-LK)^{-1} - LK(1-LK)^{-1}\\ &= (1-LK)^{-1} - L(1-KL)^{-1}K\\ &= (1-LK)^{-1} - (1-LK)^{-1}LK\\ &= (1-LK)^{-1}(1-LK)\\ &= 1,\end{aligned}$$

hence $(1-LK)^{-1}$ is also a right inverse of $1-LK$. Similar computations show that the two other left inverses are right inverses, too, in this case.

Conversely, suppose that $1-LK$ is invertible. Then

$$\begin{aligned}L(1-KL)^{-1} &= L(1+K(1-LK)^{-1}L)\\ &= L+LK(1-LK)^{-1}L\\ &= (1+LK(1-LK)^{-1})L\\ &= (1-LK+LK)(1-LK)^{-1})L\\ &= (1-LK)^{-1})L,\end{aligned}$$

hence (A.4) holds. By the preceding argument, this implies that $1-KL$ and $\left[\begin{smallmatrix}1 & -K\\ -L & 1\end{smallmatrix}\right]$ are invertible, too.

That the invertibility of $\left[\begin{smallmatrix}1 & -K\\ -L & 1\end{smallmatrix}\right]$ implies the invertibility of both $1-LK$ and $1-KL$ follows from (i). To show that the invertibility of $1-KL$ implies the invertibility of both $1-LK$ and $\left[\begin{smallmatrix}1 & -K\\ -L & 1\end{smallmatrix}\right]$ one argues in a similar way; cf. (iii).

(iii) The proof of (iii) is completely analogous to the proof of (ii).

(iv) It follows from (i)–(iii) that (a) $\Leftrightarrow$ (b) $\Leftrightarrow$ (c) $\Leftrightarrow$ (d) $\Leftrightarrow$ (e) $\Rightarrow$ (f). Thus, it only remains to show that (f) $\Rightarrow$ (e).

Suppose that (A.1) holds. We multiply this equation by K to the right to get

$$\begin{aligned}(1-KL)^{-1}K &= \big(1+K(1-LK)^{-1}L\big)K\\ &= K\big(1+(1-LK)^{-1}LK\big)\\ &= K(1-LK)^{-1}\big(1-LK+LK\big)\\ &= K(1-LK)^{-1},\end{aligned}$$

i.e., (A.3) holds. A similar computation shows that (A.5) implies (A.4).

(v) The left-invertible case follows from (i)–(iii), and the right invertible case is proved in a similar way.

(vi) This is a shortened version of (iv).

(vii) Let $\alpha \in \mathbb{C}$ be nonzero, and apply (vi) with K replaced by K/α to show that $(\alpha - KL)$ is invertible iff $(\alpha - LK)$ is invertible. $\square$

Lemma A.4.2 *Let* $A = \left[\begin{smallmatrix} A_{11} & A_{12} \\ A_{21} & A_{22} \end{smallmatrix}\right]$, $B = \left[\begin{smallmatrix} B_{11} & B_{12} \\ B_{21} & B_{22} \end{smallmatrix}\right] \in \mathcal{B}\left(\left[\begin{smallmatrix} Y \\ U \end{smallmatrix}\right]\right)$, *where* Y *and* U *are Banach spaces.*

(i) *If* A *is of the form* $A = \left[\begin{smallmatrix} A_{11} & A_{12} \\ 0 & A_{22} \end{smallmatrix}\right]$ *(i.e.,* $A_{21} = 0$*), and if* A_{11} *is invertible, then* A *is invertible if and only if* A_{22} *is invertible, in which case*

$$A^{-1} = \begin{bmatrix} A_{11} & A_{12} \\ 0 & A_{22} \end{bmatrix}^{-1} = \begin{bmatrix} A_{11}^{-1} & -A_{11}^{-1}A_{12}A_{22}^{-1} \\ 0 & A_{22}^{-1} \end{bmatrix}. \tag{A.7}$$

(ii) *Suppose that* A *is of the form* $A = \left[\begin{smallmatrix} A_{11} & A_{12} \\ 0 & A_{22} \end{smallmatrix}\right]$ *(i.e.,* $A_{21} = 0$*). If* A *is invertible with* $A^{-1} = B$*, then the following conditions are equivalent.*
   (a) A_{11} *is right invertible;*
   (b) A_{11} *is invertible;*
   (c) B *is of the type* $B = \left[\begin{smallmatrix} B_{11} & B_{12} \\ 0 & B_{22} \end{smallmatrix}\right]$ *(i.e.,* $B_{21} = 0$*);*
   (d) A_{22} *is invertible;*
   (e) A_{22} *is left invertible.*
   In this case the inverse of A *is given by* (A.7).

(iii) *Suppose that* A_{11} *is invertible. Then*

$$\begin{aligned} \begin{bmatrix} A_{11} & A_{12} \\ A_{21} & A_{22} \end{bmatrix} &= \begin{bmatrix} A_{11} & 0 \\ A_{21} & 1 \end{bmatrix} \begin{bmatrix} 1 & A_{11}^{-1}A_{12} \\ 0 & A_{22} - A_{21}A_{11}^{-1}A_{12} \end{bmatrix} \\ &= \begin{bmatrix} 1 & 0 \\ A_{21}A_{11}^{-1} & A_{22} - A_{21}A_{11}^{-1}A_{12} \end{bmatrix} \begin{bmatrix} A_{11} & A_{12} \\ 0 & 1 \end{bmatrix} \\ &= \begin{bmatrix} 1 & 0 \\ A_{21}A_{11}^{-1} & 1 \end{bmatrix} \begin{bmatrix} A_{11} & 0 \\ 0 & A_{22} - A_{21}A_{11}^{-1}A_{12} \end{bmatrix} \begin{bmatrix} 1 & A_{11}^{-1}A_{12} \\ 0 & 1 \end{bmatrix}. \end{aligned} \tag{A.8}$$

In particular, in this case A *is invertible if and only if* $A_{22} - A_{21}A_{11}^{-1}A_{12}$ *is invertible and*

$$\begin{aligned} A^{-1} &= \begin{bmatrix} 1 & A_{11}^{-1}A_{12} \\ 0 & A_{22} - A_{21}A_{11}^{-1}A_{12} \end{bmatrix}^{-1} \begin{bmatrix} A_{11} & 0 \\ A_{21} & 1 \end{bmatrix}^{-1} \\ &= \begin{bmatrix} A_{11} & A_{12} \\ 0 & 1 \end{bmatrix}^{-1} \begin{bmatrix} 1 & 0 \\ A_{21}A_{11}^{-1} & A_{22} - A_{21}A_{11}^{-1}A_{12} \end{bmatrix}^{-1} \\ &= \begin{bmatrix} 1 & -A_{11}^{-1}A_{12} \\ 0 & 1 \end{bmatrix} \begin{bmatrix} A_{11}^{-1} & 0 \\ 0 & (A_{22} - A_{21}A_{11}^{-1}A_{12})^{-1} \end{bmatrix} \begin{bmatrix} 1 & 0 \\ -A_{21}A_{11}^{-1} & 1 \end{bmatrix} \\ &= \begin{bmatrix} A_{11}^{-1} + A_{11}^{-1}A_{12}(A_{22} - A_{21}A_{11}^{-1}A_{12})^{-1}A_{21}A_{11}^{-1} & -A_{11}^{-1}A_{12}(A_{22} - A_{21}A_{11}^{-1}A_{12})^{-1} \\ -(A_{22} - A_{21}A_{11}^{-1}A_{12})^{-1}A_{21}A_{11}^{-1} & (A_{22} - A_{21}A_{11}^{-1}A_{12})^{-1} \end{bmatrix}, \end{aligned} \tag{A.9}$$

where $\begin{bmatrix} A_{11} & 0 \\ A_{21} & 1 \end{bmatrix}^{-1} = \begin{bmatrix} A_{11}^{-1} & 0 \\ -A_{21}A_{11}^{-1} & 1 \end{bmatrix}$ *and* $\begin{bmatrix} A_{11} & A_{12} \\ 0 & 1 \end{bmatrix}^{-1} = \begin{bmatrix} A_{11}^{-1} & -A_{11}^{-1}A_{12} \\ 0 & 1 \end{bmatrix}$.

(iv) *Suppose that A is invertible with $A^{-1} = B$. Then A_{11} is invertible iff B_{22} is invertible. In this case the identities in* (A.9) *are valid,* $B_{22} = (A_{22} - A_{21}A_{11}^{-1}A_{12})^{-1}$, *and*

$$\begin{aligned} A^{-1} &= \begin{bmatrix} 1 & B_{12} \\ 0 & B_{22} \end{bmatrix} \begin{bmatrix} A_{11} & 0 \\ A_{21} & 1 \end{bmatrix}^{-1} \\ &= \begin{bmatrix} A_{11} & A_{12} \\ 0 & 1 \end{bmatrix}^{-1} \begin{bmatrix} 1 & 0 \\ B_{21} & B_{22} \end{bmatrix}, \end{aligned} \tag{A.10}$$

where

$$\begin{bmatrix} 1 & B_{12} \\ 0 & B_{22} \end{bmatrix} = \begin{bmatrix} 1 & A_{11}^{-1}A_{12} \\ 0 & A_{22} - A_{21}A_{11}^{-1}A_{12} \end{bmatrix}^{-1},$$

$$\begin{bmatrix} 1 & 0 \\ B_{21} & B_{22} \end{bmatrix} = \begin{bmatrix} 1 & 0 \\ A_{21}A_{11}^{-1} & A_{22} - A_{21}A_{11}^{-1}A_{12} \end{bmatrix}^{-1}.$$

The decompositions in parts (iii) and (iv) are often referred to as *Schur decompositions.*

Proof (i) If A_{22} is invertible, then we get the identity operator if we multiply A with $\begin{bmatrix} A_{11}^{-1} & -A_{11}^{-1}A_{12}A_{22}^{-1} \\ 0 & A_{22}^{-1} \end{bmatrix}$ to the left or right. Thus, A is invertible in this case, and (A.7) holds. That A_{22} is invertible whenever A is invertible follows from the equivalence of (b) and (d) in part (ii) of this lemma.

(ii) We begin by writing out the the identities $AB = BA = 1$ componentwise as

$$\begin{aligned} A_{11}B_{11} + A_{12}B_{21} &= 1, & B_{11}A_{11} &= 1, \\ A_{11}B_{12} + A_{12}B_{22} &= 0, & B_{11}A_{12} + B_{21}A_{22} &= 0, \\ A_{22}B_{21} &= 0, & B_{21}A_{11} &= 0, \\ A_{22}B_{22} &= 1, & B_{21}A_{12} + B_{22}A_{22} &= 1. \end{aligned}$$

In particular, B_{11} is a left inverse of A_{11} and B_{22} is a right inverse of A_{22}, so (a) $\Leftrightarrow$ (b) and (d) $\Leftrightarrow$ (e). Moreover, from the third pair of identities we conclude that (b) $\Rightarrow$ (c) and (d) $\Rightarrow$ (c). If $B_{21} = 0$, then the first and last identities show that B_{11} is the inverse of A_{11} and that B_{22} is the inverse of A_{22}, hence (c) $\Rightarrow$ (b) and (c) $\Rightarrow$ (d). That (A.7) holds follows from (i).

(iii) To verify (A.8) it suffices to multiply the given block matrices and check that the result is A in each case. That A is invertible iff

$A_{22} - A_{21}A_{11}^{-1}A_{12}$ is invertible follows from (i) and (ii). We get (A.9) by inverting (A.8).

(iv) If A_{11} is invertible, then we conclude from (iii) that $B_{22} = (A_{22} - A_{21}A_{11}^{-1}A_{12})^{-1}$, hence B_{22} is invertible. If B_{22} is invertible, then we interchange A and B and argue in the same way to prove that A_{11} is invertible. To verify (A.10) it suffices to expand the products and check the result. □

Bibliography

Adamajan, A. and D. Z. Arov (1970) On unitary couplings of semiunitary operators. In *Eleven Papers in Analysis*, volume **95** of *American Mathematical Society Translations*, pp. 75–129, Providence, RI: American Mathematical Society.

Agler, J. and J. E. McCarthy (2002) *Pick Interpolation and Hilbert Function Spaces*, volume **44** of *Graduate Studies in Mathematics*. Providence, RI: American Mathematical Society.

Arov, D. Z. (1974a) Scattering theory with dissipation of energy. *Dokl. Akad. Nauk SSSR.*, **216**:713–716. Translated in *Soviet Math. Dokl.*, **15** (1974), 848–854.

Arov, D. Z. (1974b) Unitary couplings with losses (a theory of scattering with losses). *Funkcional. Anal. i Priložen.*, **8**:5–22. Translated in *Functional Analysis and its Applications*, **8**, 280–294.

Arov, D. Z. (1979a) Optimal and stable passive systems. *Dokl. Akad. Nauk SSSR.*, **247**: 265–268. Translated in *Soviet Math. Dokl.*, **20** (1979), 676–680.

Arov, D. Z. (1979b) Passive linear stationary dynamic systems. *Sibir. Mat. Zh.*, **20**: 211–228. Translated in *Sib. Math. J.*, **20** (1979), 149–162.

Arov, D. Z. (1979c) Stable dissipative linear stationary dynamical scattering systems. *J. Operator Theory*, **1**:95–126. Translated in Arov (2002).

Arov, D. Z. (1995) A survey on passive networks and scattering sytstems which are lossless or have minimal losses. *Arch. Electron. Übertrag. tech.*, **49**:252–265.

Arov, D. Z. (1999) Passive linear systems and scattering theory. In *Dynamical Systems, Control Coding, Computer Vision*, volume **25** of *Progress in Systems and Control Theory*, pp. 27–44, Basel, Boston, and Berlin: Birkhäuser-Verlag.

Arov, D. Z. (2002) Stable dissipative linear stationary dynamical scattering systems. In *Interpolation Theory, Systems Theory, and Related Topics. The Harry Dym Anniversary Volume*, volume **134** of *Operator Theory: Advances and Applications*, pp. 99–136, Basel, Boston, and Berlin: Birkhäuser-Verlag. English translation of the article in *J. Operator Theory*, **1** (1979), 95–126.

Arov, D. Z., M. A. Kaashoek, and D. R. Pik (2004) The Kalman–Yakubovich–Popov inequality and infinite dimensional discrete time dissipative systems. Manuscript.

Arov, D. Z. and M. A. Nudelman (1996) Passive linear stationary dynamical scattering systems with continuous time. *Integral Equations Operator Theory*, **24**:1–45.

Arov, D. Z. and M. A. Nudelman (2000) Criterion of unitary similarity of minimal passive scattering systems with a given transfer function. *Ukr. Math. J.*, **52**:161–172.

Arov, D. Z. and M. A. Nudelman (2002) Tests for the similarity of all minimal passive realizations of a fixed transfer function (scattering or resistance matrix). *Sbornik: Math.*, **193**:791–810.

Arov, D. Z. and O. J. Staffans (2004) State/signal linear time-invariant systems theory, Part I: Discrete time systems. Manuscript.

Avalos, G. and I. Lasiecka (1996) Differential Riccati equation for the active control of a problem in structural acoustics. *J. Optim. Theory Appl.*, **91**:695–728.

Avalos, G., I. Lasiecka, and R. Rebarber (1999) Lack of time-delay robustness for stabilization of a structural acoustics model. *SIAM J. Control Optim.*, **37**:1394–1418.

Balakrishnan, A. V. (1966) On the state space theory of linear systems. *J. Math. Anal. Appl.*, **14**:371–391.

Ball, J. A. and O. J. Staffans (2003) Conservative state-space realizations of dissipative system behaviors. Technical Report 37, 2002/2003, spring: Institute Mittag-Leffler.

Baras, J. S. and R. W. Brockett (1975) H^2-functions and infinite-dimensional realization theory. *SIAM J. Control*, **13**:221–241.

Baras, J. S., R. W. Brockett, and P. Fuhrmann (1974) State-space models for infinite dimensional systems. *IEEE Trans. Autom. Control*, **19**:693–700.

Baras, J. S. and P. Dewilde (1976) Invariant subspace methods in linear multivariable-distributed systems and lumped-distributed network synthesis. *Proc. IEEE*, **64**: 160–178.

Barbu, V. (1998) *Partial Differential Equations and Boundary Value Problems*, volume 441 of *Mathematics and its Applications*. Dordrecht: Kluwer Academic Publishers.

Bart, H., I. Gohberg, and M. A. Kaashoek (1979) *Minimal Factorization of Matrix and Operator Functions*, volume **1** of *Operator Theory: Advances and Applications*. Basel, Boston, and Berlin: Birkhäuser-Verlag.

Bensoussan, A., G. Da Prato, M. C. Delfour, and S. K. Mitter (1992) *Representation and Control of Infinite Dimensional Systems*, vols. **1** and **2**. Boston, Basel, and Berlin: Birkhäuser.

Blasco, O. (1997) Vector-valued analytic functions of bounded mean oscillation and geometry of Banach spaces. *Illinois J. Math.*, **41**:532–558.

Bradley, M. E. and M. A. Horn (1995) Global stabilization of the von Karman plate with boundary feedback acting via bending moments only. *Nonlinear Anal. Theory Methods Appl.*, **25**:417–435.

Brockett, R. W. (1970) *Finite Dimensional Linear Systems*. New York and London: John Wiley.

Brockett, R. W. and P. A. Fuhrmann (1976) Normal symmetrical dynamical systems. *SIAM J. Control Optim.*, **14**:107–119.

Brodskiĭ, M. S. (1971) *Triangular and Jordan Representations of Linear Operators*, volume **32** of *Translations of Mathematical Monographs*. Providence, RI: American Mathematical Society.

Bucci, F. and L. Pandolfi (1998) The value function of the singular quadratic regulator problem with distributed control action. *SIAM J. Control Optim.*, **36**:115–136.

Callier, F. M. and L. Dumortier (1998) Partially stabilizing LQ-optimal control for stabilizable semigroup systems. *Integral Equations Operator Theory*, **32**:119–151.

Callier, F. M., L. Dumortier, and J. Winkin (1995) On the nonnegative self-adjoint solutions of the operator Riccati equation for infinite-dimensional systems. *Integral Equations Operator Theory*, **22**:162–195.

Callier, F. M. and J. Winkin (1990) Spectral factorization and LQ-optimal regulation for multivariable distributed systems. *Internat. J. Control*, **52**:55–75.

Callier, F. M. and J. Winkin (1992) LQ-optimal control of infinite-dimensional systems by spectral factorization. *Automatica*, **28**:757–770.

Carleson, L. (1958) An interpolation theorem for bounded analytic functions. *Amer. J. Math.*, **80**:921–930.

Carleson, L. (1962) Interpolation by bounded analytic functions and the corona theorem. *Ann. Math.*, **76**:547–559.

Chang, S. and I. Lasiecka (1986) Riccati equations for nonsymmetric and nondissipative hyperbolic systems with L^2-boundary controls. *J. Math. Anal. Appl.*, **116**:378–414.

Curtain, R. F. (1984) Finite dimensional compensators for parabolic distributed systems with unbounded control and observation. *SIAM J. Control Optim.*, **22**:255–276.

Curtain, R. F. (1985) Sufficient conditions for infinite-rank Hankel operators to be nuclear. *IMA J. Math. Control Inform.*, **2**:171–182.

Curtain, R. F. (1988) Equivalence of input-output stability and exponential stability for infinite-dimensional systems. *Math. Syst. Theory*, **21**:19–48.

Curtain, R. F. (1989a) Equivalence of input-output stability and exponential stability. *Syst. Control Lett.*, **12**:235–239.

Curtain, R. F. (1989b) Representations of infinite-dimensional systems. In H. Nijmeijer and J. M. Schumacher (eds.) *Three Decades of Mathematical Systems Theory*, volume **135** of *Lecture Notes in Control and Information Sciences*, pp. 101–128, Berlin: Springer-Verlag.

Curtain, R. F. (1990) Robust stabilizability of normalized coprime factors: the infinite-dimensional case. *Internat. J. Control*, **51**:1173–1190.

Curtain, R. F. (1992) A synthesis of time and frequency domain methods for the control of infinite-dimensional systems: a system theoretic approach. In H. T. Banks (ed.) *Control and Estimation in Distributed Parameter Systems*, Frontiers in Applied Mathematics, pp. 171–224, Philadelphia: SIAM.

Curtain, R. F. (1993) The strict bounded real lemma in infinite dimensions. *Syst. Control Lett.*, **20**:113–116.

Curtain, R. F. (1996) The Kalman–Yakubovich–Popov lemma for Pritchard–Salamon systems. *Syst. Control Lett.*, **27**:67–72.

Curtain, R. F. (1997) The Salamon–Weiss class of well-posed infinite-dimensional linear systems: A survey. *IMA J. Math. Control Inform.*, **14**:207–223.

Curtain, R. F. (2002) Reciprocals of regular linear systems: a survey. In *Electronic Proceedings of the 15th International Symposium on the Mathematical Theory of Networks and Systems*, Indiana: University of Notre Dame, South Bend.

Curtain, R. F. (2003a) Regular linear systems and their reciprocals: applications to Riccati equations. *Syst. Control Lett.*, **49**:81–89.

Curtain, R. F. (2003b) Riccati equations for stable well-posed linear systems; the generic case. *SIAM J. Control Optim.*, **42**:1681–1702.

Curtain, R. F. and K. Glover (1986a) Balanced realisations for infinite dimensional systems. In *Operator Theory and Systems*, volume **19** of *Operator Theory: Advances and Applications*, pp. 87–104, Basel, Boston, and Berlin: Birkhäuser-Verlag.

Curtain, R. F. and K. Glover (1986b) Controller design for distributed systems based on Hankel-norm approximations. *IEEE Trans. Autom. Control*, **31**:173–176.

Curtain, R. F. and A. Ichikawa (1996) The Nehari problem for infinite-dimensional linear systems of parabolic type. *Integral Equations Operator Theory*, **26**:29–45.

Curtain, R. F., H. Logemann, S. Townley, and H. Zwart (1994) Well-posedness, stabilizability and admissibility for Pritchard–Salamon systems. *J. Math. Syst. Estim. Control*, **4**:493–496.

Curtain, R. F. and J. C. Oostveen (1998) The Nehari problem for nonexponentially stable systems. *Integral Equations Operator Theory*, **31**:307–320.

Curtain, R. F. and A. J. Pritchard (1978) *Infinite Dimensional Linear Systems Theory*, volume **8** of *Lecture Notes in Control and Information Sciences*. New York and Berlin: Springer-Verlag.

Curtain, R. F. and A. J. Pritchard (1994) Robust stabilization of infinite-dimensional systems with respect to coprime factor perturbations. In Elworthy (1994), pp. 437–456.

Curtain, R. F. and A. C. M. Ran (1989) Explicit formulas for Hankel norm approximations of infinite-dimensional systems. *Integral Equations Operator Theory*, **12**: 455–469.

Curtain, R. F. and L. Rodman (1990) Comparison theorems for infinite-dimensional Riccati equations. *Syst. Control Lett.*, **15**:153–159.

Curtain, R. F. and D. Salamon (1986) Finite dimensional compensators for infinite dimensional systems with unbounded input operators. *SIAM J. Control Optim.*, **24**: 797–816.

Curtain, R. F. and G. Weiss (1989) Well posedness of triples of operators (in the sense of linear systems theory). In *Control and Optimization of Distributed Parameter Systems*, volume **91** of *International Series of Numerical Mathematics*, pp. 41–59, Basel, Boston, and Berlin: Birkhäuser-Verlag.

Curtain, R. F., G. Weiss, and M. Weiss (1996) Coprime factorization for regular linear systems. *Automatica*, **32**:1519–1531.

Curtain, R. F., G. Weiss, and M. Weiss (2001) Stabilization of irrational transfer functions by controllers with internal loop. In *Systems, Approximation, Singular Integral Operators, and Related Topics*, volume **129** of *Operator Theory: Advances and Applications*, pp. 179–208, Basel, Boston, and Berlin: Birkhäuser-Verlag.

Curtain, R. F., M. Weiss, and Y. Zhou (1996) Closed formulae for a parametric-mixed-sensitivity problem for Pritchard–Salamon systems. *Syst. Control Lett.*, **27**:157–167.

Curtain, R. F. and H. Zwart (1994) The Nehari problem for the Pritchard–Salamon class of infinite-dimensional linear systems: a direct approach. *Integral Equations Operator Theory*, **18**:130–153.

Curtain, R. F. and H. Zwart (1995) *An Introduction to Infinite-Dimensional Linear Systems Theory*. New York: Springer-Verlag.

Da Prato, G. and M. Ianelli (1985) Existence and regularity for a class of integrodifferential equations of parabolic type. *J. Math. Anal. Appl.*, **112**:36–55.

Da Prato, G. and A. Ichikawa (1985) Riccati equations with unbounded coefficients. *Ann. Mat. Pura Appl. (4)*, **140**:209–221.

Da Prato, G. and A. Ichikawa (1993) Optimal control for integrodifferential equations of parabolic type. *SIAM J. Control Optim.*, **31**:1167–1182.

Da Prato, G., I. Lasiecka, and R. Triggiani (1986) A direct study of the Riccati equation arising in hyperbolic boundary control problems. *J. Differential Equations*, **64**: 26–47.

Da Prato, G. and A. Lunardi (1988) Solvability on the real line of a class of linear Volterra integrodifferential equations of parabolic type. *Ann. Mat. Pura Appl. (4)*, **150**:67–117.

Da Prato, G. and A. Lunardi (1990) Stabilizability of integrodifferential parabolic equations. *J. Integral Equations Appl.*, **2**:281–304.

Datko, R. (1970) Extending a theorem of A. M. Liapunov to Hilbert space. *J. Math. Anal. Appl.*, **32**:610–616.

Davies, E. B. (1980) *One-Parameter Semigroups*. London and New York. Academic Press.

Delfour, M. C. and J. Karrakchou (1987) State space theory of linear time-invariant systems with delays in state, control, and observation variables, I and II. *J. Math. Anal. Appl.*, **125**:361–399 and 400–450.

Delfour, M. C., J. E. Lagnese, and M. P. Polis (1986) Stabilization of hyperbolic systems using concentrated sensors and actuators. *IEEE Trans. Autom. Control*, **31**:1091–1096.

Desch, W., I. Lasiecka, and W. Schappacher (1985) Feedback boundary control problems for linear semigroups. *Israel J. Math.*, **51**:177–207.

Dewilde, P. (1971) Input-output description of roomy systems. *SIAM J. Control Optim.*, **14**:712–736.

Diestel, J. and J. J. Uhl (1977) *Vector Measures*. Providence, RI: American Mathematical Society.

Dinculeanu, N. (1974) *Integration on Locally Compact Spaces*. Leyden: Noordhoff International Publishing.

Dolecki, S. and D. L. Russell (1977) A general theory of observation and control. *SIAM J. Control Optim.*, **15**:185–220.

Dunford, N. and J. T. Schwartz (1958) *Linear Operators. I. General Theory*. London: Interscience Publishers.

Dunford, N. and J. T. Schwartz (1963) *Linear Operators. II. Spectral Theory. Self Adjoint Operators in Hilbert Space*. New York and London: Interscience Publishers, John Wiley & Sons.

Dunford, N. and J. T. Schwartz (1971) *Linear Operators. Part III. Spectral Operators*. New York and London: Interscience Publishers, John Wiley & Sons.

Duren, P. L. (1970) *Theory of H^p Spaces*. Boston: Academic Press.

Elworthy, K. D. (ed.) (1994) *Differential Equations, Dynamical Systems, and Control Science. A Festschrift in Honor of Lawrence Markus*, New York: Marcel Dekker.

Engel, K.-J. (1998) On the characterization of admissible control- and observation operators. *Syst. Control Lett.*, **34**:225–227.

Fattorini, H. O. (1968) Boundary control systems. *SIAM J. Control*, **6**:349–385.

Fattorini, H. O. (1999) *Infinite Dimensional Optimization and Control Theory*, volume **62** of *Encyclopedia of Mathematics and its Applications*. Cambridge and New York: Cambridge University Press.

Feintuch, A. (1976) Spectral minimality for infinite-dimensional linear systems. *SIAM J. Control*, **14**:945–950.

Feintuch, A. (1998) *Robust Control Theory in Hilbert Space*. Applied Mathematical Sciences. New York, Berlin, and Heidelberg: Springer-Verlag.

Feintuch, A. and R. Saeks (1982) *System Theory: A Hilbert Space Approach*. New York and London: Pure and Applied Mathematics. Academic Press.

Flandoli, F. (1984) Riccati equation arising in a boundary control problem with distributed parameters. *SIAM J. Control Optim.*, **22**:76–86.

Flandoli, F. (1986) Direct solution of a Riccati equation arising in a stochastic control problem with control and observation on the boundary. *Appl. Math. Optim.*, **14**: 107–129.

Flandoli, F. (1987) Algebraic Riccati equation arising in boundary control problems. *SIAM J. Control Optim.*, **25**:612–636.

Flandoli, F. (1993) On the direct solution of Riccati equations arising in boundary control theory. *Ann. Mat. Pura Appl. (4)*, **163**:93–131.

Flandoli, F., I. Lasiecka, and R. Triggiani (1988) Algebraic Riccati equations with non-smoothing observation arising in hyperbolic and Euler–Bernoulli boundary control problems. *Ann. Mat. Pura Appl. (4)*, **153**:307–382.

Foiaş, C. and A. E. Frazho (1990) *The Commutant Lifting Approach to Interpolation Problems*, volume **44** of *Operator Theory: Advances and Applications*. Basel, Boston, and Berlin: Birkhäuser-Verlag.

Fourès, Y. and I. E. Segal (1955) Causality and analyticity. *Trans. Amer. Math. Soc.*, **78**: 385–405.

Francis, B. A. (1987) *A Course in H_∞ Control Theory*, volume **88** of *Lecture Notes in Control and Information Sciences*. Berlin: Springer-Verlag.

Fuhrmann, P. A. (1972) On weak and strong reachability and controllability of infinite-dimensional linear systems. *J. Optim. Theory Appl.*, **9**:77–89.

Fuhrmann, P. A. (1974) On realization of linear systems and applications to some questions of stability. *Math. Syst. Theory*, **8**:132–140.

Fuhrmann, P. A. (1981) *Linear Systems and Operators in Hilbert Space*. New York: McGraw-Hill.

Gamelin, T. W. (1984) *Uniform Algebras*, second edition. New York: Chelsea Publishing Company.

Garnett, J. B. (1981) *Bounded Analytic Functions*. New York: Academic Press.

Georgiou, T. T. and M. C. Smith (1993) Graphs, causality, and stabilizability: Linear, shift-invariant systems on $L^2[0, \infty)$. *Math. Control Signals Syst.*, **6**:195–223.

Glover, K., R. F. Curtain, and J. R. Partington (1988) Realisation and approximation of linear infinite-dimensional systems with error bounds. *SIAM J. Control Optim.*, **26**: 863–898.

Goldstein, J. A. (1985) *Semigroups of Linear Operators and Applications*. New York and Oxford: Clarendon Press.

Grabowski, P. (1991) On the spectral-Lyapunov approach to parametric optimization of distributed-parameter systems. *IMA J. Math. Control Inform.*, **7**:317–338.

Grabowski, P. and F. M. Callier (1996) Admissible observation operators. Semigroup criteria of admissibility. *Integral Equations Operator Theory*, **25**:182–198.

Grabowski, P. and F. M. Callier (1999) Admissible observation operators. Duality of observation and control using factorizations. *Dyn. Contin. Discrete Impulsive Syst.*, **6**:87–119.

Greiner, G., J. Voigt, and M. Wolff (1981) On the spectral bound of the generator of semigroups of positive operators. *J. Operator Theory*, **5**:245–256.

Gripenberg, G., S.-O. Londen, and O. Staffans (1990) *Volterra Integral and Functional Equations*, volume **34** of *Encyclopedia of Mathematics and its Applications*. Cambridge and New York: Cambridge University Press.

Hansen, S. and G. Weiss (1991) The operator Carleson measure criterion for admissibility of control operators for diagonal semigroups on l^2. *Syst. Control Lett.*, **16**: 219–227.

Hansen, S. and G. Weiss (1997) New results on the operator Carleson measure criterion. *IMA J. Math. Control Inform.*, **14**:3–32.

Hansen, S. W. and B.-Y. Zhang (1997) Boundary control of a linear thermoelastic beam. *J. Math. Anal. Appl.*, **210**:182–205.

Helton, J. W. (1974) Discrete time systems, operator models, and scattering theory. *J. Funct. Anal.*, **16**:15–38.

Helton, J. W. (1976) Systems with infinite-dimensional state space: the Hilbert space approach. *Proc. IEEE*, **64**:145–160.

Hendrickson, E. and I. Lasiecka (1993) Numerical approximations and regularizations of Riccati equations arising in hyperbolic dynamics with unbounded control operators. *Comput. Optim. Appl.*, **2**:343–390.

Hendrickson, E. and I. Lasiecka (1995) Finite dimensional approximations of boundary control problems arising in partially observed hyperbolic systems. *Dyn. Contin. Discrete Impulsive Syst.*, **1**:101–142.

Herbst, I. (1983) The spectrum of Hilbert space semigroups. *J. Operator Theory*, **10**: 87–94.

Hewitt, E. and K. Stromberg (1965) *Real and Abstract Analysis*. Berlin and New York: Springer-Verlag.

Hille, E. and R. S. Phillips (1957) *Functional Analysis and Semi-Groups*, revised edition. Providence, RI: American Mathematical Society.

Hinrichsen, D. and A. J. Pritchard (1994) Robust stability of bilinear evolution operators on Banach spaces. *SIAM J. Control Optim.*, **32**:1503–1541.

Ho, L. F. and D. L. Russell (1983) Admissible input elements for systems in Hilbert space and a Carleson measure criterion. *SIAM J. Control Optim.*, **21**:614–640.

Hoffman, K. (1988) *Banach Spaces of Analytic Functions*. New York: Dover Publications.

Horn, M. A. (1992a) Exact controllability of the Euler–Bernoulli plate via bending moments only on the space of optimal regularity. *J. Math. Anal. Appl.*, **167**:557–581.

Horn, M. A. (1992b) Uniform decay rates for the solutions to the Euler–Bernoulli plate equation with boundary feedback acting via bending moments. *Differential Integral Equations*, **5**:1121–1150.

Horn, M. A. (1994) Uniform stabilization of the Kirchhoff plate equation with boundary conditions containing moments of inertia. *J. Math. Syst. Estim. Control*, **4**:39–65.

Horn, M. A. (1996) Boundary control and stabilization of Kirchhoff and von Karman plates. *Z. Angew. Math. Mech.*, **76**:437–440.

Horn, M. A. (1998a) Implications of sharp trace regularity results on boundary stabilization of the system of linear elasticity. *J. Math. Anal. Appl.*, **223**:126–150.

Horn, M. A. (1998b) Sharp trace regularity for the solutions of the equations of dynamic elasticity. *J. Math. Syst. Estim. Control*, **8**:1–11.

Horn, M. A. and I. Lasiecka (1994) Asymptotic behavior with respect to thickness of boundary stabilizing feedback for the Kirchoff plate. *J. Differential Equations*, **114**: 396–433.

Huang, F. (1985) Characteristic conditions for exponential stability of linear dynamical systems in Hilbert space. *Ann. Differential Equations*, **1**:43–56.

Jacob, B. and J. R. Partington (2001) The Weiss conjecture on admissibility of observation operators for contraction semigroups. *Integral Equations Operator Theory*, **40**:231–243.

Jacob, B., J. R. Partington, and S. Pott (2002) Admissible and weakly admissible observation operators for the right shift semigroup. *Proc. Edinburgh Math. Soc.*, **45**: 353–362.

Jacob, B., J. R. Partington, and S. Pott (2003) Conditions for admissibility of observation operators and boundedness of Hankel operators. *Integral Equations Operator Theory*, **47**:315–338.

Jacob, B. and H. Zwart (2000) Disproof of two conjectures of George Weiss. Memorandum 1546, Twente, the Netherlands: Faculty of Mathematical Sciences, University of Twente.

Jacob, B. and H. Zwart (2001a) Exact observability of diagonal systems with a finite-dimensional output operator. *Syst. Control Lett.*, **43**:101–109.

Jacob, B. and H. Zwart (2001b) Exact observability of diagonal systems with a one-dimensional output operator. *Internat. J. Appl. Math. Comput. Sci.*, **11**:1277–1283.

Jacob, B. and H. Zwart (2002) Properties of the realization of inner functions. *Math. Control Signals Syst.*, **15**:356–379.

Kaashoek, M. A., C. van der Mee, and A. C. M. Ran (1997) Weighting operator patterns of Pritchard–Salamon realizations. *Integral Equations Operator Theory*, **27**:48–70.

Kalman, R. E. (1963) Mathematical description of linear dynamical systems. *SIAM J. Control*, **1**:152–192.

Kalman, R. E., P. L. Falb, and M. A. Arbib (1969) *Topics in Mathematical System Theory*. New York: McGraw-Hill.

Kamen, E. W. (1975) On an algebraic theory of systems defined by convolution operators. *Math. Syst. Theory*, **9**:57–74.

Kato, T. (1980) *Perturbation Theory for Linear Operators*. Berlin, Heidelberg, and New York: Springer-Verlag.

Komornik, V. (1997) Rapid boundary stabilization of linear distributed systems. *SIAM J. Control Optim.*, **35**:1591–1613.

Köthe, G. (1969) *Topological Vector Spaces I*. New York: Springer-Verlag.

Kreĭn, S. G. and J. I. Petunin (1966) Scales of Banach spaces. *Uspehi Mat. Nauk SSSR*, **21**:89–168.

Lagnese, J. E. (1977) Boundary value control of a class of hyperbolic equations in a general region. *SIAM J. Control Optim.*, **15**:973–983.

Lagnese, J. E. (1978) Exact boundary value controllability of a class of hyperbolic equations. *SIAM J. Control Optim.*, **16**:1000–1017.

Lagnese, J. E. (1980) Boundary patch control of the wave equation in some nonstar complemented regions. *J. Math. Anal. Appl.*, **77**:364–380.

Lagnese, J. E. (1983a) Boundary stabilization of linear elastodynamic systems. *SIAM J. Control Optim.*, **21**:968–984.

Lagnese, J. E. (1983b) Decay of solutions of wave equations in a bounded region with boundary dissipation. *J. Differential Equations*, **50**:163–182.

Lagnese, J. E. (1989) *Boundary Stabilization of Thin Plates*, volume **10** of *SIAM Studies in Applied Mathematics*. Philadelphia: Society for Industrial and Applied Mathematics.

Lagnese, J. E. (1995) Modelling and controllability of plate-beam systems. *J. Math. Syst. Estim. Control*, **5**:141–187.

Lagnese, J. E. and J. Lions (1988) *Modelling Analysis and Control of Thin Plates*, volume **6** of *Recherches en Mathématiques Appliquées*. Paris: Masson.

Lasiecka, I. (1980) Unified theory for abstract parabolic boundary problems – a semigroup approach. *Appl. Math. Optim.*, **6**:287–333.

Lasiecka, I. (1992) Exponential decay rates for the solutions of Euler–Bernoulli equations with boundary dissipation occurring in the moments only. *J. Differential Equations*, **95**:169–182.

Lasiecka, I., D. Lukes, and L. Pandolfi (1995) Input dynamics and non-standard Riccati equations with applications to boundary control of damped wave and plate equations. *J. Optim. Theory Appl.*, **84**:549–574.

Lasiecka, I., L. Pandolfi, and R. Triggiani (1997) A singular control approach to highly damped second-order equations and applications. *Appl. Math. Optim.*, **36**:67–107.

Lasiecka, I. and R. Triggiani (1981) A cosine operator approach to modeling $L^2(0, T; L^2(\Gamma))$-boundary input hyperbolic equations. *Appl. Math. Optim.*, **7**:35–93.

Lasiecka, I. and R. Triggiani (1983a) Dirichlet boundary control problem for parabolic equations with quadratic cost: analyticity and Riccati's feedback synthesis. *SIAM J. Control Optim.*, **21**:41–67.

Lasiecka, I. and R. Triggiani (1983b) Dirichlet boundary control problem for parabolic equations with quadratic cost: analyticity and Riccati's feedback synthesis. *SIAM J. Control Optim.*, **21**:41–67.

Lasiecka, I. and R. Triggiani (1983c) Regularity of hyperbolic equations under $L^2(0, T; L^2(\Gamma))$-Dirichlet boundary terms. *Appl. Math. Optim.*, **10**:275–286.

Lasiecka, I. and R. Triggiani (1986) Riccati equations for hyperbolic partial differential equations with $L_2(0, T; L_2(\Gamma))$-Dirichlet boundary terms. *SIAM J. Control Optim.*, **24**:884–925.

Lasiecka, I. and R. Triggiani (1987a) The regulator problem for parabolic equations with Dirichlet boundary control. Part I: Riccati's feedback synthesis and regularity of optimal solution. *Appl. Math. Optim.*, **16**:147–168.

Lasiecka, I. and R. Triggiani (1987b) The regulator problem for parabolic equations with Dirichlet boundary control. II: Galerkin approximation. *Appl. Math. Optim.*, **16**:187–216.

Lasiecka, I. and R. Triggiani (1987c) Uniform exponential energy decay of wave equations in a bounded region with $L_2(0, \infty; L_2(\Gamma))$-feedback control in the Dirichlet boundary conditions. *J. Differential Equations*, **66**:340–390.

Lasiecka, I. and R. Triggiani (1988) A lifting theorem for the time regularity of solutions to abstract equations with unbounded operators and applications to hyperbolic equations. *Proc. Amer. Math. Soc.*, **104**:745–755.

Lasiecka, I. and R. Triggiani (1989a) Exact controllability of the Euler–Bernoulli equation with controls in the Dirichlet and Neumann boundary conditions: a nonconservative case. *SIAM J. Control Optim.*, **27**:330–373.

Lasiecka, I. and R. Triggiani (1989b) Exact controllability of the wave equation with Neumann boundary control. *Appl. Math. Optim.*, **19**:243–290.

Lasiecka, I. and R. Triggiani (1990a) Exact controllability of the Euler–Bernoulli equation with boundary controls for displacement and moment. *J. Math. Anal. Appl.*, **146**:1–33.

Lasiecka, I. and R. Triggiani (1990b) Sharp regularity theory for second order hyperbolic equations of Neumann type. I: L_2 nonhomogeneous data. *Ann. Mat. Pura Appl. (4)*, **157**:285–367.

Lasiecka, I. and R. Triggiani (1991a) *Differential and Algebraic Riccati Equations with Applications to Boundary/Point Control Problems: Continuous Theory and Approximation Theory*, volume **164** of *Lecture Notes in Control and Information Sciences*. Berlin: Springer-Verlag.

Lasiecka, I. and R. Triggiani (1991b) Differential Riccati equations with unbounded coefficients: applications to boundary control/boundary observation hyperbolic problems. *Nonlinear Anal. Theory Methods Appl.*, **17**:655–682

Lasiecka, I. and R. Triggiani (1991c) Exact controllability and uniform stabilization of Kirchoff plates with boundary control only on $\Delta w|_\Sigma$ and homogeneous boundary displacement. *J. Differential Equations*, **93**:62–101.

Lasiecka, I. and R. Triggiani (1991d) Regularity theory of hyperbolic equations with non-homogeneous Neumann boundary conditions. II: general boundary data. *J. Differential Equations*, **94**:112–164.

Lasiecka, I. and R. Triggiani (1992a) Optimal regularity, exact controllability and uniform stabilization of Schroedinger equations with Dirichlet control. *Differential Integral Equations*, **5**:521–535.

Lasiecka, I. and R. Triggiani (1992b) Riccati differential equations with unbounded coefficients and nonsmooth terminal condition – the case of analytic semigroups. *SIAM J. Math. Anal.*, **23**:449–481.

Lasiecka, I. and R. Triggiani (1992c) Uniform stabilization of the wave equation with Dirichlet or Neumann feedback control without geometrical conditions. *Appl. Math. Optim.*, **25**:189–224.

Lasiecka, I. and R. Triggiani (1993a) Algebraic Riccati equations arising from systems with unbounded input-output solution operator: applications to boundary control problems for wave and plate equations. *Nonlinear Anal. Theory Methods Appl.*, **20**: 659–693.

Lasiecka, I. and R. Triggiani (1993b) Sharp trace estimates of solutions to Kirchoff and Euler–Bernoulli equations. *Appl. Math. Optim.*, **28**:277–306.

Lasiecka, I. and R. Triggiani (2000a) *Control Theory for Partial Differential Equations: Continuous and Approximation Theories. I Abstract Parabolic Systems*, volume **74** of *Encyclopedia of Mathematics and its Applications*. Cambridge and New York: Cambridge University Press.

Lasiecka, I. and R. Triggiani (2000b) *Control Theory for Partial Differential Equations: Continuous and Approximation Theories. II Abstract Hyperbolic-Like Systems over a Finite Time Horizon*, volume **75** of *Encyclopedia of Mathematics and its Applications*. Cambridge and New York: Cambridge University Press.

Lax, P. D. and R. S. Phillips (1967) *Scattering Theory*. New York: Academic Press.

Lax, P. D. and R. S. Phillips (1973) Scattering theory for dissipative hyperbolic systems. *J. Funct. Anal.*, **14**:172–235.

Lions, J. L. (1971) *Optimal Control of Systems Governed by Partial Differential Equations*, volume **170** of *Die Grundlehren der mathematishen Wissenschaften in Einzeldarstellungen*. Berlin: Springer-Verlag.

Lions, J. L. (1988) Exact controllability, stabilization and perturbations for distributed systems. *SIAM Rev.*, **30**:1–68.

Lions, J. L. and E. Magenes (1972) *Non-Homogeneous Boundary Value Problems and Applications, **I***, volume **181** of *Die Grundlehren der mathematishen Wissenschaften in Einzeldarstellungen*. Berlin: Springer-Verlag.

Livšic, M. S. (1973) *Operators, Oscillations, Waves (Open Systems)*, volume **34** of *Translations of Mathematical Monographs*. Providence, RI: American Mathematical Society.

Livšic, M. S. and A. A. Yantsevich (1977) *Operator Colligations in Hilbert Spaces*. New York: John Wiley & Sons.

Logemann, H. (1993) Stabilization and regulation of infinite-dimensional systems using coprime factorizations. In R. F. Curtain (ed.) *Analysis and Optimization of Systems: State and Frequency Domain Approaches for Infinite-Dimensional Systems*, volume **185** of *Lecture Notes in Control and Information Sciences*, pp. 102–139, Berlin and New York: Springer-Verlag.

Logemann, H., R. Rebarber, and G. Weiss (1996) Conditions for robustness and non-robustness of the stability of feedback systems with respect to small delays in the feedback loop. *SIAM J. Control Optim.*, **34**:572–600.

Logemann, H. and E. P. Ryan (2000) Time-varying and adaptive integral control of infinite-dimensional regular linear systems with input nonlinearities. *SIAM J. Control Optim.*, **38**:1120–1144.

Logemann, H., E. P. Ryan, and S. Townley (1998) Integral control of infinite-dimensional linear systems subject to input saturation. *SIAM J. Control Optim.*, **36**:1940–1961.

Logemann, H., E. P. Ryan, and S. Townley (2000) Integral control of linear systems with actuator nonlinearities: lower bounds for the maximal regulating gain. *IEEE Trans. Autom. Control*, **44**:1315–1319.

Logemann, H. and S. Townley (1997a) Discrete-time low-gain control of uncertain infinite-dimensional systems. *IEEE Trans. Autom. Control*, **42**:22–37.

Logemann, H. and S. Townley (1997b) Low gain control of uncertain regular linear systems. *SIAM J. Control Optim.*, **35**:78–116.

Lukes, D. L. and D. L. Russell (1969) The quadratic criterion for distributed systems. *SIAM J. Control*, **7**:101–121.

Lunardi, A. (1995) *Analytic Semigroups and Optimal Regularity in Parabolic Equations*. Basel, Boston, and Berlin: Birkhäuser-Verlag.

Malinen, J. (2000) Discrete time H^∞ algebraic Riccati equations. Doctoral dissertation, Helsinki University of Technology.

Malinen, J., O. J. Staffans, and G. Weiss (2003) When is a linear system conservative? Technical Report **46**, 2002/2003, spring: Institute Mittag-Leffler.

McMillan, C. A. and R. Triggiani (1994a) Min-max game theory and algebraic Riccati equations for boundary control problems with analytic semigroups. II: the general case. *Nonlinear Anal. Theory Methods Appl.*, **22**:431–465.

McMillan, C. A. and R. Triggiani (1994b) Min-max game theory and algebraic Riccati equations for boundary control problems with analytic semigroups: the stable case. In Elworthy (1994) pp. 757–780.

McMillan, C. A. and R. Triggiani (1994c) Min-max game theory and algebraic Riccati equations for boundary control problems with continuous input-solution map. II: the general case. *Appl. Math. Optim.*, **29**:1–65.

Mikkola, K. (2002) Infinite-dimensional linear systems, optimal control and algebraic Riccati equations. Doctoral dissertation, Helsinki University of Technology.

Morgül, Ö. (1990) Control and stabilization of a flexible beam attached to a rigid body. *Internat. J. Control*, **51**:11–31.

Morgül, Ö. (1994) A dynamic control law for the wave equation. *Automatica*, **30**:1785–1792.

Morris, K. A. (1994) State feedback and estimation of well-posed systems. *Math. Control Signals Syst.*, **7**:351–388.

Morris, K. A. (1999) Justification of input-output methods for systems with unbounded control and observation. *IEEE Trans. Autom. Control*, **44**:81–84.

Nagel, R. (ed.) (1986) *One-Parameter Semigroups of Positive Operators*, volume **1184** of *Lecture Notes in Mathematics*, Berlin: Springer-Verlag.

Nikol'skiĭ, N. K. (1986) *Treatise on the Shift Operator*. Berlin: Springer-Verlag.

Ober, R. (1991) Balanced parameterization of classes of linear operators. *SIAM J. Control Optim.*, **29**:1251–1287.

Ober, R. (1996) System theoretic aspects of completely symmetric systems. In *Recent Developments in Operator Theory and its Applications (Winnipeg, MB, 1994)*, volume **87** of *Operator Theory: Advances and Applications*, pp. 233–262, Basel, Boston, and Berlin: Birkhäuser-Verlag.

Ober, R. and S. Montgomery-Smith (1990) Bilinear transformation of infinite-dimensional state-space systems and balanced realizations of nonrational transfer functions. *SIAM J. Control Optim.*, **28**:438–465.

Ober, R. and Y. Wu (1993) Asymptotic stability of infinite-dimensional discrete time balanced realizations. *SIAM J. Control Optim.*, **31**:1321–1339.

Ober, R. and Y. Wu (1996) Infinite-dimensional continuous-time linear systems: stability and structure analysis. *SIAM J. Control Optim.*, **34**:757–812.

Oostveen, J. (1999) Strongly stabilizable infinite-dimensional systems. Doctoral dissertation, Rijksuniversiteit Groningen.

Opdenacker, P. C. and E. A. Jonckheere (1988) A contraction mapping preserving balanced reduction scheme and its infinity norm error bounds. *IEEE Trans. Circuits Syst.*, **35**:184–189.

Opmeer, M. R. and R. F. Curtain (2004) New Riccati equations for well-posed linear systems. *Syst. Control Lett.*, **52**:339–347.

Paley, R. E. A. C. and N. Wiener (1934) *Fourier Transforms in the Complex Domain*. Providence, RI: American Mathematical Society.

Pandolfi, L. (1989) Some properties of the frequency domain description of boundary control systems. *J. Math. Anal. Appl.*, **142**:219–241.

Pandolfi, L. (1990) Generalized control systems, boundary control systems, and delayed control systems. *Math. Control Signals Syst.*, **3**:165–181.

Pandolfi, L. (1992) Some properties of distributed control systems with finite-dimensional input space. *SIAM J. Control Optim.*, **30**:926–941.

Pandolfi, L. (1995) The standard regulator problem for systems with input delays. An approach through singular control theory. *Appl. Math. Optim.*, **31**:19–136.

Pandolfi, L. (1998) Dissipativity and Lur'e problem for parabolic boundary control systems. *SIAM J. Control Optim.*, **36**:2061–2081.

Partington, J. R. and G. Weiss (2000) Admissible observation operators for the right shift semigroup. *Math. Control Signals Syst.*, **13**:179–192.

Pazy, A. (1972) On the applicability of Lyapunov's theorem in Hilbert space. *SIAM J. Math. Anal.*, **3**:291–294.

Pazy, A. (1983) *Semi-Groups of Linear Operators and Applications to Partial Differential Equations*. Berlin: Springer-Verlag.

Peller, V. V. (2003) *Hankel Operators and their Applications*. New York: Springer-Verlag.

Pritchard, A. J. and D. Salamon (1985) The linear-quadratic control problem for retarded systems with delays in control and observation. *IMA J. Math. Control Inform.*, **2**: 335–362.

Pritchard, A. J. and D. Salamon (1987) The linear quadratic control problem for infinite dimensional systems with unbounded input and output operators. *SIAM J. Control Optim.*, **25**:121–144.

Pritchard, A. J. and A. Wirth (1978) Unbounded control and observation systems and their duality. *SIAM J. Control Optim.*, **16**:535–545.

Pritchard, A. J. and J. Zabczyk (1981) Stability and stabilizability of infinite dimensional systems. *SIAM Rev.*, **23**:25–52.

Prüss, J. (1984) On the spectrum of C_0-semigroups. *Trans. Amer. Math. Soc.*, **284**: 847–857.

Rebarber, R. (1993) Conditions for the equivalence of internal and external stability for distributed parameter systems. *IEEE Trans. Autom. Control*, **38**:994–998.

Rebarber, R. (1995) Exponential stability of coupled beams with dissipative joints: a frequency domain approach. *SIAM J. Control Optim.*, **33**:1–28.

Rosenblum, M. and J. Rovnyak (1985) *Hardy Classes and Operator Theory*. New York and Oxford: Oxford University Press.

Rosenbrock, H. H. (1970) *State-Space and Multivariable Theory*. New York: John Wiley (Wiley Interscience Division).

Rudin, W. (1973) *Functional Analysis*. New York: McGraw-Hill.

Rudin, W. (1987) *Real and Complex Analysis*, 3rd edition. New York: McGraw-Hill.

Russell, D. L. (1971) Boundary value control theory of the higher-dimensional wave equation. *SIAM J. Control*, **9**:401–419.

Russell, D. L. (1973a) Quadratic performance criteria in boundary control of linear symmetric hyperbolic systems. *SIAM J. Control*, **11**:475–509.

Russell, D. L. (1973b) A unified boundary controllability theory for hyperbolic and parabolic partial differential equations. *Studies Appl. Math.*, **52**:189–211.

Russell, D. L. (1975) Decay rates for weakly damped systems in Hilbert space obtained with control-theoretic methods. *J. Differential Equations*, **19**:344–370.

Russell, D. L. (1978) Controllability and stabilizability theory for linear partial differential equations: Recent progress and open questions. *SIAM Rev.*, **20**:639–739.

Russell, D. L. and G. Weiss (1994) A general necessary condition for exact observability. *SIAM J. Control Optim.*, **32**:1–23.

Salamon, D. (1984) *Control and Observation of Neutral Systems*. London: Pitman.

Salamon, D. (1987) Infinite dimensional linear systems with unbounded control and observation: a functional analytic approach. *Trans. Amer. Math. Soc.*, **300**:383–431.

Salamon, D. (1989) Realization theory in Hilbert space. *Math. Syst. Theory*, **21**:147–164.

Slemrod, M. (1976) Asymptotic behavior of C_0 semi-groups as determined by the spectrum of the generator. *SIAM J. Control Optim.*, **25**:783–792.

Smith, M. C. (1989) On stabilization and the existence of coprime factorizations. *IEEE Trans. Autom. Control*, **34**:1005–1007.

Šmuljan, Y. L. (1986) Invariant subspaces of semigroups and the Lax–Phillips scheme. Deposited in VINITI, N. 8009-B86, Odessa, 49p.

Staffans, O. J. (1985) Feedback stabilization of a scalar functional differential equation. *J. Integral Equations*, **10**:319–342.

Staffans, O. J. (1991) Stabilization of a distributed system with a stable compensator. *Math. Control Signals Syst.*, **5**:1–22.

Staffans, O. J. (1994) Well-posedness and stabilizability of a viscoelastic equation in energy space. *Trans. Amer. Math. Soc.*, **345**:527–575.

Staffans, O. J. (1995a) A neutral FDE is similar to the product of an ODE and a shift. *J. Math. Anal. Appl.*, **192**:627–654.

Staffans, O. J. (1995b) Quadratic optimal control of stable systems through spectral factorization. *Math. Control Signals Syst.*, **8**:167–197.

Staffans, O. J. (1996) On the discrete and continuous time infinite-dimensional algebraic Riccati equations. *Syst. Control Lett.*, **29**:131–138.

Staffans, O. J. (1997) Quadratic optimal control of stable well-posed linear systems. *Trans. Amer. Math. Soc.*, **349**:3679–3715.

Staffans, O. J. (1998a) Coprime factorizations and well-posed linear systems. *SIAM J. Control Optim.*, **36**:1268–1292.

Staffans, O. J. (1998b) Quadratic optimal control of well-posed linear systems. *SIAM J. Control Optim.*, **37**:131–164.

Staffans, O. J. (1998c) Feedback representations of critical controls for well-posed linear systems. *Internat. J. Robust Nonlinear Control*, **8**:1189–1217.

Staffans, O. J. (1998d) On the distributed stable full information H^∞ minimax problem. *Internat. J. Robust Nonlinear Control*, **8**:1255–1305.

Staffans, O. J. (1998e) Quadratic optimal control of a parabolic equation. In A. Berghi, L. Finesso, G. Picci (eds.) *Mathematical Theory of Networks and Systems, Proceedings of the MTNS98 symposium in Padova, Italy*, pp. 535–538, Padova, Italy: Il Poligrafo.

Staffans, O. J. (1999a) Admissible factorizations of Hankel operators induce well-posed linear systems. *Syst. Control Lett.*, **37**:301–307.

Staffans, O. J. (1999b) Quadratic optimal control through coprime and spectral factorisation. *Eur. J. Control*, **5**:167–179.

Staffans, O. J. (2001a) J-energy preserving well-posed linear systems. *Internat. J. Appl. Math. Comput. Sci.*, **11**:1361–1378.

Staffans, O. J. (2001b) Well-posed linear systems, Lax–Phillips scattering and L^p-multipliers. In *Systems, Approximation, Singular Integral Operators, and Related Topics*, volume **129** of *Operator Theory: Advances and Applications*, pp. 445–464, Basel, Boston, and Berlin: Birkhäuser-Verlag.

Staffans, O. J. (2002a) Passive and conservative continuous-time impedance and scattering systems. Part I: Well-posed systems. *Math. Control Signals Syst.*, **15**:291–315.

Staffans, O. J. (2002b) Passive and conservative infinite-dimensional impedance and scattering systems (from a personal point of view). In *Mathematical Systems Theory in Biology, Communication, Computation, and Finance*, volume **134** of *IMA Volumes in Mathematics and its Applications*, pp. 375–414, New York: Springer-Verlag.

Staffans, O. J. (2002c) Stabilization by collocated feedback. In *Directions in Mathematical Systems Theory and Optimization*, volume **286** of *Lecture Notes in Control and Information Sciences*, pp. 261–278, New York: Springer-Verlag.

Staffans, O. J. and G. Weiss (1998) Well-posed linear systems in L^1 and L^∞ are regular. In S. Domek, R. Kaszynski, and L. Tarasiejski (eds.) *Methods and Models in Automation and Robotics, Proceedings of the Fifth International Symposium on*

Methods and Models in Automation and Robotics, Poland, pp. 75–80, Szczećin, Poland: Technical University of Szczećin.

Staffans, O. J. and G. Weiss (2002) Transfer functions of regular linear systems. Part II: the system operator and the Lax–Phillips semigroup. *Trans. Amer. Math. Soc.*, **354**:3229–3262.

Staffans, O. J. and G. Weiss (2004) Transfer functions of regular linear systems. Part III: inversions and duality. *Integral Equations Operator Theory*. To appear.

Stein, E. M. and G. Weiss (1971) *Introduction to Fourier Analysis on Euclidean Spaces*. Princeton, NJ: Princeton University Press.

Sz.-Nagy, B. and C. Foiaş (1970) *Harmonic Analysis of Operators on Hilbert Space*. Amsterdam: North-Holland.

Triggiani, R. (1975) On the stabilizability problem in Banach space. *J. Math. Anal. Appl.*, **52**:383–403.

Triggiani, R. (1979) On Nambu's boundary stabilizability problem for diffusion processes. *J. Differential Equations*, **33**:189–200.

Triggiani, R. (1980a) Boundary feedback stabilizability of parabolic equations. *Appl. Math. Optim.*, **6**:201–220.

Triggiani, R. (1980b) Well-posedness and regularity of boundary feedback parabolic systems. *J. Differential Equations*, **36**:347–362.

Triggiani, R. (1988) Exact boundary controllability on $L_2(\Omega) \times H^{-1}(\Omega)$ of the wave equation with Dirichlet boundary control acting on a portion of the boundary $\partial\Omega$, and related problems. *Appl. Math. Optim.*, **18**:241–277.

Triggiani, R. (1989) Wave equation on a bounded domain with boundary dissipation: an operator approach. *J. Math. Anal. Appl.*, **137**:438–461.

Triggiani, R. (1991) Regularity of some structurally damped problems with point control and with boundary control. *J. Math. Anal. Appl.*, **161**:299–331.

Triggiani, R. (1992) Regularity with interior point control of Schrödinger equations. *J. Math. Anal. Appl.*, **171**:567–577.

Triggiani, R. (1993a) Interior and boundary regularity of the wave equation with interior point control. *Differential Integral Equations*, **6**:111–129.

Triggiani, R. (1993b) Regularity with interior point control. II. Kirchhoff equations. *J. Differential Equations*, **103**:394–420.

Triggiani, R. (1994a) Optimal boundary control and new Riccati equations for highly damped second-order equations. *Differential Integral Equations*, **7**:1109–1144.

Triggiani, R. (1994b) Optimal quadratic boundary control problem for wave- and plate-like equations with high internal damping: an abstract approach. In *Control of Partial Differential Equations. IFIP WG 7.2 Conference in Trento, Italy, January 4–9, 1993*, volume **165** of *Lecture Notes in Pure and Applied Mathematics*, pp. 215–270, New York: Marcel Dekker.

Triggiani, R. (1997) An optimal control problem with unbounded control operator and unbounded observation operator where the algebraic Riccati equation is satisfied as a Lyapunov equation. *Appl. Math. Lett.*, **10**:95–102.

Tucsnak, M. and G. Weiss (2003) How to get a conservative well-posed linear system out of thin air. Part II. Controllability and stability. *SIAM J. Control Optim.*, **42**: 907–935.

Uetake, Y. (2004) Some local properties of spectrum of linear dynamical systems on Hilbert space. *Integral Equations Operator Theory*. To appear.

van Keulen, B. (1993) *H_∞-Control for Distributed Parameter Systems: A State Space Approach*. Basel, Boston, and Berlin: Birkhäuser-Verlag.

van Neerven, J. (1996) *The Asymptotic Behavior of Semigroups of Linear Operators*, volume **88** of *Operator Theory, Advances and Applications*. Basel, Boston, and Berlin: Birkhäuser-Verlag.

Vidyasagar, M. (1985) *Control System Synthesis: A Factorization Approach*. Cambridge, MA: MIT Press.

Washburn, D. (1979) A bound on the boundary input map for parabolic equations with application to time optimal control. *SIAM J. Control Optim.*, **17**:652–671.

Weiss, G. (1988a) Admissibility of input elements for diagonal semigroups on l^2. *Syst. Control Lett.*, **10**:79–82.

Weiss, G. (1988b) Weak L^p-stability of a linear semigroup on a Hilbert space implies exponential stability. *J. Differential Equations*, **76**:269–285.

Weiss, G. (1989a) Admissibility of unbounded control operators. *SIAM J. Control Optim.*, **27**:527–545.

Weiss, G. (1989b) Admissible observation operators for linear semigroups. *Israel J. Math.*, **65**:17–43.

Weiss, G. (1989c) The representation of regular linear systems on Hilbert spaces. In *Control and Optimization of Distributed Parameter Systems*, volume **91** of *International Series of Numerical Mathematics*, pp. 401–416, Basel, Boston, and Berlin: Birkhäuser-Verlag.

Weiss, G. (1991a) Representations of shift-invariant operators on L^2 by H^∞ transfer functions: an elementary proof, a generalization to L^p, and a counterexample for L^∞. *Math. Control Signals Syst.*, **4**:193–203.

Weiss, G. (1991b) Two conjectures on the admissibility of control operators. In *Control and Optimization of Distributed Parameter Systems*, pp. 367–378, Basel, Boston, and Berlin: Birkhäuser-Verlag.

Weiss, G. (1994a) Transfer functions of regular linear systems. Part I: characterizations of regularity. *Trans. Amer. Math. Soc.*, **342**:827–854.

Weiss, G. (1994b) Regular linear systems with feedback. *Math. Control Signals Syst.*, **7**:23–57.

Weiss, G. (1999) A powerful generalization of the Carleson measure theorem? In *Open Problems in Mathematical Systems and Control Theory*, pp. 267–272, London: Springer-Verlag.

Weiss, G. (2003) Optimal control of systems with a unitary semigroup and with colocated control and observation. *Syst. Control Lett.*, **48**:329–340.

Weiss, G. and R. F. Curtain (1997) Dynamic stabilization of regular linear systems. *IEEE Trans. Autom. Control*, **42**:4–21.

Weiss, G. and R. F. Curtain (1999) Exponential stabilization of vibrating systems by collocated feedback. In *Proceedings of the 7th IEEE Mediterranean Conference on Control and Systems*, pp. 1705–1722, Haifa, Israel: CD-ROM.

Weiss, G. and M. Häfele (1999) Repetitive control of MIMO systems using H^∞ design. *Automatica*, **35**:1185–1199.

Weiss, G. and R. Rebarber (1998) Dynamic stabilization of well-posed linear systems. In S. Domek, R. Kaszynski, L. Tarasiejski (eds.) *Methods and Models in Automation and Robolics, Proceedings of MMAR98*, Miedzyzdroje, Poland, pp. 69–74. Szczećin Poland: Technical University of Szczećin.

Weiss, G. and R. Rebarber (2001) Optimizability and estimatability for infinite-dimensional linear systems. *SIAM J. Control Optim.*, **39**:1204–1232.

Weiss, G., O. J. Staffans, and M. Tucsnak (2001) Well-posed linear systems – a survey with emphasis on conservative systems. *Internat. J. Appl. Math. Comput. Sci.*, **11**: 7–34.

Weiss, G. and H. Zwart (1998) An example in linear quadratic optimal control. *Syst. Control Lett.*, **33**:339–349.

Weiss, M. (1994) Riccati equations in Hilbert space: a Popov function approach. Doctoral dissertation, Rijksuniversiteit Gronigen.

Weiss, M. (1997) Riccati equation theory for Pritchard–Salamon systems: a Popov function approach. *IMA J. Math. Control Inform.*, **14**:1–37.

Weiss, M. and G. Weiss (1997) Optimal control of stable weakly regular linear systems. *Math. Control Signals Syst.*, **10**:287–330.

Willems, J. C. (1972a) Dissipative dynamical systems Part I: General theory. *Arch. Rational Mech. Anal.*, **45**:321–351.

Willems, J. C. (1972b) Dissipative dynamical systems Part II: Linear systems with quadratic supply rates. *Arch. Rational Mech. Anal.*, **45**:352–393.

Willems, J. C. (1991) Paradigms and puzzles in the theory of dynamical systems. *IEEE Trans. Autom. Control*, **36**:259–294.

Yamamoto, Y. (1981) Realization theory of infinite-dimensional linear systems, parts I and II. *Math. Syst. Theory*, **15**:55–77,169–190.

Yosida, K. (1974) *Functional Analysys*. Berlin, Heidelberg, and New York: Springer-Verlag.

Young, N. J. (1986) Balanced realizations in infinite dimensions. In *Operator Theory and Systems*, volume **19** of *Operator Theory: Advances and Applications*, pp. 449–471. Basel, Boston, and Berlin: Birkhäuser-Verlag.

Zabczyk, J. (1975) A note on C_0-semigroups. *Bull. Acad. Polon. Sci. Sér. Sci. Math. Astronom. Phys*, **23**:895–898.

Zwart, H., B. Jacob, and O. J. Staffans (2003) Weak admissibility does not imply admissibility for analytic semigroups. *Syst. Control Lett.*, **48**:341–350.

Index